Apollo 11

Lunar Science Conference

Papers presented at the Lunar Science Conference
held at Houston, Texas, from 5 to 8 January 1970

American Association for the Advancement of Science
Washington, D.C.

Pages 449 through 784 were printed in *Science*, 30 January 1970
AAAS Miscellaneous Publication Number 70-1
International Standard Book Number 087168-091-2
Library of Congress Catalogue Card Number 74-119890

Printed in the United States of America
McCall Printing Company, Washington, D.C.

Meteoritic body (4 millimeters in diameter), composed of Ni-Fe and FeS, mimics man's image of the moon. This fragment crashed on the moon. The event produced a liquid drop, which collapsed as it cooled in the lunar gravitational field. The upper side of the fragment was abraded by lunar dust and struck by splashed lunar fragments traveling at crater-forming speeds. See page 664. [V. E. Krantz, Smithsonian Institution]

Contents

30 January 1970, Volume 167, Number 3918

SCIENCE

The Moon Issue

The successful Apollo 11 mission placed experimental equipment on the moon and brought back 22 kilograms of lunar materials. The samples presented unique scientific opportunities, but they also presented potential problems. One was the possible contamination of the earth from the moon. Less talked about, but more real, was the possibility of a black market in lunar specimens. Even more likely was the hazard of a wild scramble for publication immediately after samples became available. It was necessary to ensure receipt of the materials by qualified investigators and to prevent flagrant duplication of effort. While many kinds of approaches were to be encouraged, excessive allocation or destruction of a limited supply was to be avoided.

In meeting its responsibilities in these matters, the National Aeronautics and Space Administration employed the power of government wisely. In the choice of investigators and the allocation of samples it followed the advice of groups of highly qualified scientists, including the Lunar Sample Analysis Planning Team (see page 449 of this issue). These groups did their work well. They selected a distinguished body of American scientists and responded generously to qualified applicants from abroad. To ensure proper custody of samples and orderly procedures with respect to publication, NASA released materials only after suitable contracts had been signed. One provision was that investigators were not to divulge information prematurely; instead, all were to participate in a Lunar Science Conference held at Houston, Texas, from 5 to 8 January. On arrival at the meeting, each principal investigator was to submit a written report.

In accord with its goal to publish these articles promptly in a journal of wide circulation, NASA negotiated with *Science*. Our decision to publish these reports was a close one. The material was to be four times the volume of a usual issue. There were worrisome problems of quality control, coupled with the difficulty of handling a large number of reports on a tight schedule. There was also the financial burden. On the positive side was the fact that the examination of lunar samples was a unique event and that *Science* with its broad international circulation (120 countries) could best serve as publisher. A small but nontrivial aspect was the challenge that the task presented to the staff of *Science*.

The financial question was resolved by a contract with NASA on essentially a no-loss, no-gain basis. Obtaining the best possible quality in the manuscripts while avoiding excessive length took some doing. For most investigators, the time between receipt of samples and submittal of manuscripts was only 3 months. The natural tendency was to make measurements until the last moment and then to throw together an article hastily. *Science* insisted that manuscripts be subjected to the reviewing process. Publication of defective material was to be delayed until satisfactory revisions had been made. Rigid limitations on length were established. Authors responded well to the guidelines by submitting manuscripts that were better than usual and that were within the length limitations. *Science* sent a team of editors to Houston and assembled a group of reviewers (page 781), they completed their work on the spot and interacted with authors to make improvements.

Some readers may feel that this issue of *Science* provides more than they care to know about the moon, but others will treasure it as an important part of the scientific record of a great accomplishment.

—Philip H. Abelson

30 January 1970, Volume 167, Number 3918 SCIENCE

Summary of Apollo 11 Lunar Science Conference

On 24 July 1969 the first samples from our sister planet, the moon, were returned to earth for direct scientific investigation. Prior to this, our understanding of the extraterrestrial universe derived from study of electromagnetic radiation from stars and planets, from study of cosmic rays, and from analysis of meteorites. Meteorites were, until the return of Apollo 11, the only extraterrestrial objects we could actually hold in our hands and scrutinize in the laboratory. Unlike meteorites, the lunar samples come from a good sized planetary object whose location is well known.

This volume is a compilation of the first systematic studies of these samples by more than 500 scientists from nine countries; it represents the initial scientific fruit of the courageous undertaking in which three men made the first voyage to the surface of another planet.

The variety and sophistication of the techniques used in the study of these samples draws heavily on the experience gained in the last 25 years in the study of meteorites and terrestrial petrology and mineralogy. The scope and quality of the analyses applied to the returned samples far exceed anything we could hope to achieve with remotely controlled devices even with the most advanced methods of modern technology.

The results of the studies of lunar materials are significant both for what they are and for the way in which they were achieved. They represent an effort which crossed many of the usual barriers between disciplines and scientific interest groups. The common objective, the study of the moon, has gone a long way toward establishing a new cosmopolitan science. We are confident that this close interaction of a large and diverse group of scientists will have the good effect of previous occasions in the history of science when circumstances caused scientists from a variety of backgrounds to work on a common effort. It is not surprising that at this early stage in a new era of study of a planet that there should be diversity in interpretation of observations and even some differences between the observations themselves. We do not propose to resolve these differences in this introductory summary. Our objective is to furnish a guide to the results and conclusions that are still in a state of ferment.

The samples from Tranquillity Base consist of basaltic igneous rocks, microbreccias, which are a mechanical mixture of soil and small rock fragments compacted into a coherent rock, and lunar soil. The soil is a diverse mixture of crystalline fragments and glassy fragments with a variety of most interesting shapes; it also contains small fragments of iron meteorites. Most of the rock fragments are similar to the larger igneous rocks and apparently were derived from them; the rocks in turn were probably once part of the underlying bedrock. A small number of the crystalline fragments are totally different from any of the igneous rocks of the Tranquillity site. There is a strong possibility that these fragments represent samples from the nearby Highlands.

Many of the rock surfaces and individual fragments in the soil show evidence of surface erosion by hypervelocity impacts. Examination of the surfaces of glassy objects which are themselves formed by impact processes shows that they contain beautifully preserved microscopic pits as small as 10 microns in diameter, which are the result of impacts by tiny high velocity particles. There is also evidence that the impact processes are accompanied by local melting, splashing, evaporation, and condensation.

The crystalline rocks, which have typical igneous textures, range from very fine-grained vesicular rocks to vuggy, medium-grained equigranular rocks. The most common minerals are pyroxene, often highly zoned with iron-rich rims, plagioclase, ilmenite, olivine, and cristobalite. Three new minerals occur in the igneous rocks. They are pyroxmanganite (a triclinic pyroxene-like mineral), ferropseudobrookite, and a chromium-titanium spinel. Free metallic iron and troilite, both of which are extremely rare on earth, are common accessory minerals in the igneous rocks. All of the silicate minerals are unusually transparent and clear, because of the complete absence of hydrothermal alteration or weakening. Laboratory experiments with silicate liquids similar in composition to the lunar liquids show that, at the time of crystallization, the observed phases can have coexisted only in a very dry, highly reducing system; the partial pressure of oxygen in this system is estimated to be 10^{-13} atmosphere. This is more than 5 orders of magnitude lower than that for typical terrestrial basaltic magmas. The very low abundance of ferric ions in pyroxenes, determined by Mössbauer spectroscopy and electron spin resonance, is further evidence for the low oxidation level of the magmas. The melting experiments also indicate that 98 percent of the primary igneous liquid crystallized in the temperature range 1210° to 1060°C, with minor interstitial liquids continuing to crystallize down to temperatures around 950°C. Microscopic and microprobe examination provides clear-cut evidence for the existence of an interstitial liquid rich in potassium and aluminum which probably was immiscible with the main liquid. It is further calculated that the viscosity of the lunar magmas was about an order of magnitude lower than that of terrestrial ba-

This summary was prepared by the Lunar Sample Analysis Planning Team (LSAPT). Members at the time of preparation of the summary were J. Arnold and G. Arrhenius, University of California at San Diego; G. Eglinton, University of Bristol, England; C. Frondel, Harvard University; P. Gast, Columbia University; I. MacGregor, University of California at Davis; R. Pepin, University of Minnesota; D. Strangway, University of Toronto; R. Walker, Washington University; G. Wasserburg, California Institute of Technology; P. Zill, Ames Research Center.

saltic magmas. This characteristic may play a significant role in the explanation of the textural features, the differentiation mechanisms that produced the observed chemical composition, and the morphological features of the lunar seas themselves.

The chemical compositions of all the igneous rocks are remarkably similar except for the concentration of potassium, rubidium, cesium, uranium, thorium, and barium. These elements distinguish two groups of igneous rocks; in general, the fine grained rocks contain more of these elements than the coarse grained rocks. All rocks have unusually high concentrations of titanium, scandium, zirconium, hafnium, yttrium, and trivalent rare earth elements and low concentrations of sodium. One of the most striking features of the igneous rocks is the low abundance of europium relative to the other rare earth elements. Some of the more volatile elements—for example, bismuth, mercury, zinc, cadmium, thallium, lead, germanium, chlorine, and bromine—are significantly depleted with respect to their presumed abundance in the primitive solar system. The composition of the soils and breccias is similar to but distinguishable from the igneous rocks and clearly shows that they contain at least one other "rock" component distinct from the lunar basalts we have sampled. The soil is also enriched in nickel and volatile elements (cadmium, zinc, silver, gold, copper, and thallium) that occur in high abundances in carbonaceous chondrites. This enrichment is consistent with the observed occurrence of meteorite material in the soil.

Many elements such as potassium, rubidium, cesium, chlorine, and thallium that occur in very low abundance in the lunar igneous rocks are strongly enriched in terrestrial crustal rocks. These terrestrial rocks are the product of igneous differentiation and are either residual liquids or the low melting fraction. On a larger scale such processes have resulted in the strong enrichment of these elements in the crust of the earth. It has been suggested that the low abundance of these elements in rocks derived from residual lunar liquids is an indication that the whole moon is depleted in a number of volatile elements. If this inference about the bulk composition is correct, one can infer that the lunar material separated from a high-temperature dispersed nebula at 1000°C or higher.

The chemistry of the igneous liquids, in particular the coincidence of the high abundance of titanium, the separation of europium from the other rare earth elements, and the separation of barium and strontium, suggests that these liquids are the end product of an extensive fractional crystallization process. An alternative but less likely possibility is that these liquids were produced by small amounts of partial melting and subsequent segregation of liquid in the lunar interior.

High pressure and temperature experiments show that materials with the chemical composition of Tranquillity Base basalts would, at the conditions inferred for the lunar interior, have densities which far exceed the average density of the moon. It follows from these experiments that these basalts cannot represent the bulk composition of the moon.

The ages of the basaltic crystalline rocks were determined by ^{87}Pb-^{87}Sr and ^{40}K-^{40}Ar methods and show that they formed at 3.7×10^9 years ago. This age is supported by U-Th-Pb data. These results show that igneous rocks of the Sea of Tranquillity were melted (possibly extruded) and crystallized about 10^9 years after the formation of the moon. A single exotic rock fragment yielded an age of 4.4×10^9 years and indicates variability in age between different areas on the lunar surface. The relatively young age of the basalts shows that the moon has not been a completely dead planet from its formation but has undergone significant differentiation, at least locally, in a thin lunar crust. The time period prior to the 3.7×10^9 year event almost certainly is recorded in older Highland areas; it represents an interval where the earth's record has been obliterated. Therefore, much of the surface of the moon is of great importance in understanding the early evolution and differentiation of planets.

Some samples of soil and breccia give concordant Pb-U-Th and Sr-Rb ages of 4.6×10^9 years. These samples, which are aggregates of highly varied rocks, minerals, and glasses of younger age, appear to provide an effective average of the lunar crust. This crust, originally of material 4.6×10^9 years old, underwent differentiation at later times but without the injection into the lunar crust of younger differentiated material which is enriched in rubidium, uranium, and thorium.

All of these observations show unequivocally that the materials on the surface of the moon are the product of considerable geochemical and petrological evolution. Thus, the ancient planetary surface we have sampled gives us a picture of the early evolutionary processes of a terrestrial planet which have hitherto been obscured from view.

Physical properties of surface materials are important to the interpretation of a number of telescopic observations. The optical properties of the soil measured in the laboratory are in good agreement with those inferred from telescopic studies, indicating that a thin layer of soil must cover most of the moon. Spectral features determined in the laboratory owing to iron can be identified in the soil. Correlation of these features with telescopic studies suggests that the iron content of the lunar surface may vary. The measured dielectric constant of the soil is in accord with the known radar reflectivity of the moon, and the penetration depth suggests that the upper layers are electrically transparent. The measured thermal conductivity suggests that diurnal variations in temperature should extend to less than 1 meter in the soil regions. Thermoluminescence studies on two core samples showed clear evidence that there is a substantial diurnal variation in temperature 12 centimeters below the surface.

The interpretation of seismic experiments also requires a knowledge of material properties of the moon. Laboratory measurements of velocities of both P and S waves show that velocities crease rapidly with pressure to values similar to those observed for terrestrial rocks at 5 kilobars. The attenuation of sound waves decreases with pressure. The high pressure velocities are quite consistent with those observed in the Apollo 12 seismic experiments. The sharp increase in velocity and decrease in attenuation with pressure further suggest the possibility of a seismic wave guide at shallow depths on the moon.

Natural remanent magnetization has been found in the crystalline rocks and in the breccias. This suggests that the ancient moon may have had a magnetic field with a strength of a few percent of the earth's field. The field existing 3.7×10^9 years ago may have been the result of fluid motions in the moon's interior, an effect of the earth's field when the moon and the earth were closer together, or the result of processes not yet understood.

The atmosphere-free surface of the moon is a monitor of radiation from both the galaxy and the sun. The lunar samples are particularly valuable for

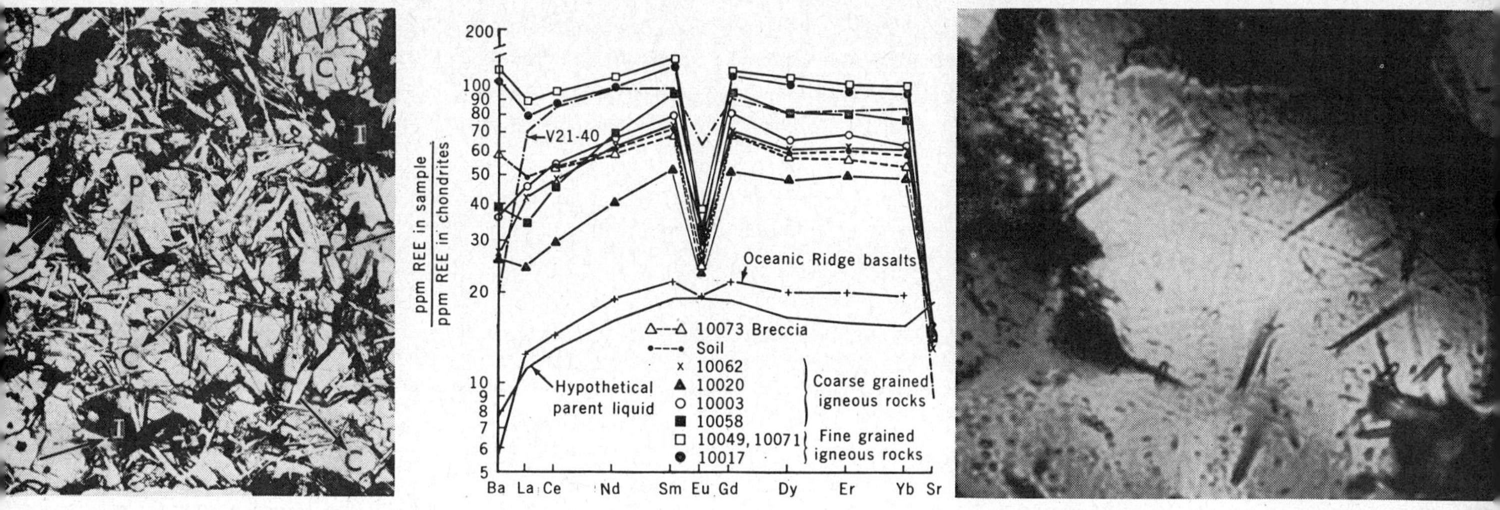

(Left) Thin section of lunar sample showing major minerals: (O) olivine, (P) plagioclase, (I) ilmenite, and (C) clinopyroxene. (Center) Chondrite normalized abundance patterns for nine rare earth elements and Sr and Ba. (Right) Etched cosmic ray tracks in a clinopyroxene crystal. Track densities indicate that this rock spent 6 million years exposed directly to space.

the study of low energy, weakly penetrating radiations. Furthermore, the lunar rocks have provided us with a sample of the gases blown off the sun in the solar wind. Thus the isotopic composition and relative abundance of certain elements in the sun can now be inferred from direct measurements. Studies of these gases (hydrogen, helium, neon, argon, krypton, and xenon) have given an upper limit on the deuterium content and an estimate of the magnitude of the $^{3}He/^{4}He$ ratio in the sun. Both of these results will lead to a better understanding of the evolution of the sun. Precise isotopic compositions determined for heavier rare gases permit us to better infer the evolution of the earth's atmosphere from the more primitive material of the sun.

Studies of stored nuclear tracks and induced radioactivities show clearly the effects of bombardment by solar flare particles (as well as galactic cosmic rays) and suggest that this activity has persisted without gross modification for long periods of time. Imbedded solar wind particles indicate large rare gas concentrations in both the breccias and fine particles and show that grains now well below the surface must at one time have been exposed at the very top. The studies of nuclear tracks confirm both the solar wind origin of the implanted rare gases and the inference regarding surface exposure of the samples.

By combining evidence from radioactive and stable nuclear reaction products and stored nuclear tracks, it is also possible to give new information on both lunar surface processes and the cosmic ray flux over long time periods. For example, it has been possible to determine the recent "top" surface of rock 10017 and at the same time demonstrate that the present "top" and "bottom" must have interchanged in the past; this rock was tumbled around on the lunar surface and partially buried. These studies have also shown that the rate of erosion of rocks on the lunar surface is about 10^{-7} centimeter per year. The abundance of cosmic ray produced nuclides shows that some rocks have been on or within several centimeters of the surface for at least 10 million years and within 1 or 2 meters for at least 500 million years.

The existence of complex carbon compounds in extraterrestrial materials would have great significance for the origin of life both on earth and other planets. The search for important protobiological compounds, such as purines, pyrimidines, amino acids, and porphyrins, was carried on with some of the most sophisticated and sensitive analytical techniques ever devised. Nevertheless, no unambiguous identification of indigenous compounds was made at extremely low levels of detection (usually less than 10 parts per billion). Contaminants from the exhaust from the lunar module and from terrestrial handling of the samples have been detected at these levels. The bulk carbon content of the soils, breccias, and igneous rocks ranges from 50 to 250 parts per million. The highest concentrations occur in the fine-grained portion of the soil; the lowest concentrations occur in the igneous rocks. Some methane and carbon monoxide are released by acid dissolution, but the majority of the carbon is not released until samples are heated above 800°C in a vacuum, when it comes off as carbon monoxide and carbon dioxide. The concentrations of carbon observed in the soils and breccias are approximately those expected from solar wind deposition. Graphite has been observed, but the precise nature of most of the lunar carbon remains to be defined.

Micropaleontological examination of the lunar sample by optical microscopy and by electron and scanning electron microscopy produced uniformly negative results. An intensive search for viable organisms employing a multitude of environmental and media combinations produced negative results, as did the one quarantine study.

The Apollo 11 samples were collected from a very tiny fraction of the moon's surface. Nevertheless, they have given a vast new insight into the processes that have shaped this surface and have established some significant limits on the rates and mechanisms by which it evolved. The results reported do not resolve the problem of the origin of the moon. However, the number of constraints that must be met by any theory have been greatly increased. For example, if the moon formed from the earth, it can now be stated with some confidence that this separation took place prior to 4.3×10^{9} years ago. Furthermore, such a hypothesis must now take account of certain definite differences in chemical composition. There is clear evidence in the chemical and petrologic observations that the surface of the moon is variable in both composition and age. It is therefore of great scientific importance to obtain materials from a variety of terrains and sites. Samples from the Highlands and from deeper in the lunar crust will probably be the next important milestones in the advance of lunar science.

Lunar Regolith at Tranquillity Base

Abstract. *The regolith at Tranquillity Base is a layer of fragmental debris that ranges in thickness from about 3 to 6 meters. The thickness of the regolith and the exposure histories of its constituent fragments can be related, by means of a relatively simple model, to the observed crater distribution.*

Tranquillity Base, the landing site of Apollo 11, is a relatively smooth and level part of the lunar surface in the southern part of the Sea of Tranquillity. At this site the lunar surface is underlain by weakly coherent fragmental material, the lunar regolith, which ranges in size from particles too fine to be seen with the naked eye to blocks more than a meter across.

Near the lunar module the surface of the regolith is pockmarked by abundant craters ranging in diameter from less than 2 cm to several tens of meters. A number of craters in the area, in the size range from 30 to 55 m, have sharply formed raised rims and anomalously shallow flat floors or floors with central humps. The floors of these craters are believed to have been formed on a more cohesive rocky substratum that underlies the regolith (*1*). We refer to this substratum as the bedrock. On the basis of the observed depths of the anomalous craters, the thickness of the regolith is estimated to range from 3 to 6 m. One of the anomalous craters, 33 m in diameter and 4 m deep, was visited by Armstrong, and a series of photographs were obtained of a low pile of coarse broken rock, which forms a central hump.

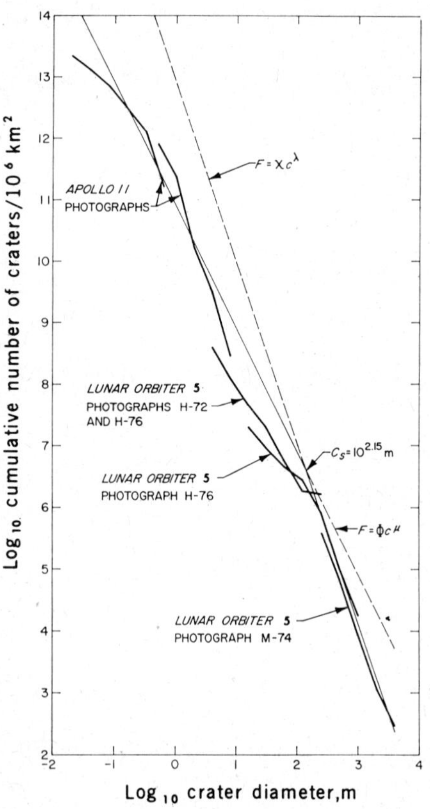

Fig. 1. Cumulative size-frequency distribution of craters on the lunar surface in the vicinity of Tranquillity Base, determined from Apollo 11 photographs and Lunar Orbiter 5 photographs. Curve shown for Lunar Orbiter 5 photograph M-74 is based upon unpublished data by E. C. Morris.

About 400 m east of the module is a prominent blocky-rimmed crater, informally named West Crater, which is about 180 m in diameter and about 30 m deep. West Crater is surrounded by a rim deposit of blocky ejecta which extends about 250 m out from the rim crest on all sides of the crater. Rays of coarse blocks extend up to several hundred meters beyond the continuous deposit of ejecta; one of these rays, with associated secondary craters, passes about 50 m north of the module. The surface of the regolith in the immediate vicinity of the module is relatively free of extremely coarse blocks, although some rock fragments nearby may have been derived from West Crater.

The size-frequency distribution of craters at and near Tranquillity Base has been estimated from Apollo 11 and Lunar Orbiter 5 photographs (Fig 1). The distribution of craters less than 100 m in diameter roughly follows a function, $F = \Phi c^\mu$, where F is the cumulative number of craters with diameter equal to or higher than c per 10^6 km^2, c is crater diameter, Φ is a constant ($10^{10.9}$/m^{-2} 10^6 km^2), and μ is a constant (−2.00). This function, determined initially from Ranger photographs (*2*), has been found to fit the distribution of small craters at all Surveyor landing sites very closely (*3*). It is interpreted as the steady-state distribution of craters produced by repetitive bombardment of level surfaces on the moon.

The distribution of craters larger than a few hundred meters diameter in the vicinity of Tranquillity Base does not follow the steady-state distribution but can be fitted closely by another power function, $F = \chi c^\lambda$, where χ is $10^{12.9}$/m^{-2} 10^6 km^2 and λ is −2.93. The intersection of $F = \chi c^\lambda$ with $F = \Phi c^\mu$, at 141 m, is taken to be the upper limiting crater diameter of the steady-state distribution, here designated c_s. Soderbloom has shown (*4*) that the exponent μ of the steady-state distribution can be derived exactly and the constant Φ can be derived approximately, if a function close to $F = \chi c^\lambda$ is assumed to represent the distribution of all craters produced that are smaller than the steady-state limit.

The thickness distribution of the regolith can also be derived from the assumption that the function $F = \chi c^\lambda$ represents distribution of all craters that have been produced which are smaller than the steady-state limit. Under this assumption the difference between $F = \chi c^\lambda$ and $F = \Phi c^\mu$, at crater diameters less than c_s, represents the number of craters lost by erosion or burial as a result of repetitive cratering and other processes of material transport, such as mass wasting. The diameter of the largest crater that has just disappeared, as a consequence of these processes, is c_s. A crater of this size is lost partly by erosion of the rim but primarily by being filled in by debris derived from the rim and from more distant sources. The maximum thickness of the regolith in an area of 10^6 km is given by the original

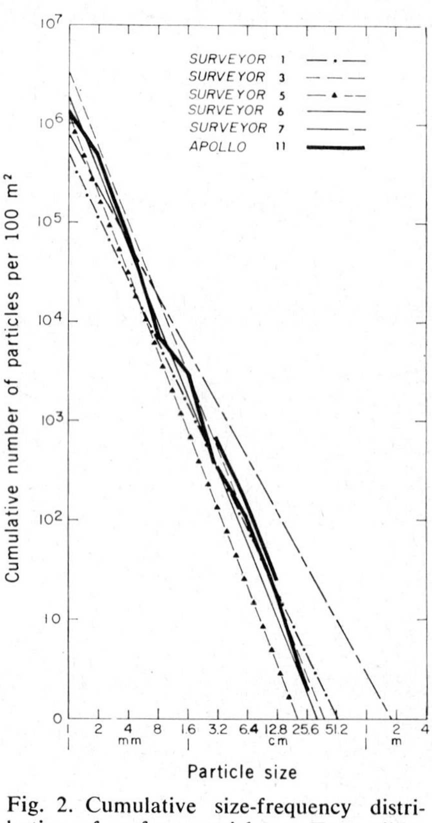

Fig. 2. Cumulative size-frequency distribution of surface particles at Tranquillity Base, compared with power functions fitted to observed size-frequency distributions of surface particles at the Surveyor landing sites.

depth of a crater with diameter c_s minus the original height of the rim. Elsewhere the base of the regolith is formed by segments of floors of smaller craters that have been filled in.

An approximate solution for the frequency distribution of regolith thickness has been given by Shoemaker *et al.* (3) as

$$H = \frac{100\,\lambda\,\pi\,\chi\,(1/q)^{\lambda+2}}{4(\lambda+2)A_m} \times [h^{\lambda+2} - h_s^{\lambda+2}], \quad h_{min} \leq h \leq h_s \qquad (1)$$

where H is cumulative percentage of the surface underlain by a regolith of thickness equal to or greater than h, h is the thickness of the regolith, q is the ratio of the original depth (minus rim height) to diameter of the craters ($q \approx 1/5$), A_m is the total area of craters whose connected floors constitute the base of the regolith ($A_m \approx 2 \times 10^6$ km^2), h_s is the maximum thickness of the regolith ($h_s = qc_s$), and h_{min} is an artificially defined "minimum" thickness of the regolith

$$h_{min} = q\left[c_s^{\lambda+2} + \frac{4(\lambda+2)A_m}{\lambda\,\pi\,\chi}\right]^{1/(\lambda+2)} \qquad (2)$$

Substituting the values for χ and λ at Tranquillity Base, we find the median thickness of the regolith to be 4.1 m.

The regolith has been turned over to the median depth of 4.1 m by cratering just once since the mare surface was formed, presumably as a lava flow, at Tranquillity Base. At shallower depths, the regolith has been turned over repeatedly by the formation and filling of small craters. The time for turnover to any median depth less than median thickness of the regolith is given, to good approximation, by

$$t = \frac{4A_r q^{\lambda+2}(\lambda+2)}{\pi\,\lambda\,\chi(d^{\lambda+2} - h_s^{\lambda+2})} \qquad (3)$$

where t is time of turnover expressed as a fraction of the age of the mare surface at Tranquillity Base, $A_r = 10^6$ km^2, and d is the median depth of turnover. Solutions to Eq. 3 for median depths of turnover: for 1 mm, time of turnover is 0.00037; for 1 cm, 0.0031; for 10 cm, 0.027; for 1 m, 0.24. In these calculations the crater production function, $F = \chi c^\lambda$, has been extrapolated to crater diameters of the order of a few millimeters.

Craters or pits smaller than a millimeter have been observed on the rocks returned to earth by the Apollo 11 mission. Craters smaller than 4 mm in diameter commonly are glass-lined, whereas the craters larger than 4 mm are only rarely glass-lined and are commonly irregular in form. The small craters are generally abundant on rounded sides of the rocks and are less abundant or rare on angular sides. A light halo, probably produced by minute fractures, surrounds most of the small craters, and the rounded surfaces of most rocks are underlain by a light rind, about a millimeter thick, that probably has been produced by repetitive cratering. These observations suggest that repetitive bombardment by small particles has been an important mechanism of erosion of the rocks.

If the crater production function, $F = \chi c^\lambda$, is extrapolated to very small craters, the following expression for the expected lifetime of a rock fragment can be obtained by integration of the volumes removed by repeated small cratering events,

$$l_1 = \frac{50A_r(\lambda+3)D}{\pi\,\lambda\,\chi\,(c_{min}^{\lambda+3} - D^{\lambda+3})} \qquad (4)$$

where l_1 is the mean expected lifetime, expressed in units of the age of the mare surface, of a rock fragment eroded by bombardment of particles that produce craters smaller in diameter than D; D is the initial diameter of the rock fragment; and c_{min} is the smallest crater considered in the integration. Relatively few particles smaller than 1 μm may be expected to form craters on the moon's surface, and a value of 10 μm has been somewhat arbitrarily adopted for c_{min}. With the observed value of λ, an order of magnitude error in c_{min} has a relatively small effect on the calculated lifetimes. Solutions to Eq. 4 are listed in Table 1. A 10-cm rock fragment, for example, has a mean expected lifetime, if exposed continuously to small particle bombardment, of about 1 percent of the age of the mare surface.

Table 1. Mean expected lifetime of rock fragments expressed as fractions of the age of the mare surface at Tranquillity Base.

	Initial size of rock fragment			
	1 mm	1 cm	10 cm	1 m
Mean lifetime of rock eroded by small* particle bombardment	0.00034	0.0017	0.012	0.087
Mean lifetime of rock destroyed by large† particle bombardment	0.00026	0.0022	0.019	0.16
Mean expected lifetime of rock bombarded by both small and large particles	0.00015	0.00097	0.0073	0.056

* A small particle is defined as a particle that produces a crater of diameter smaller than the rock fragment. † A large particle is defined as a particle that produces a crater of diameter equal to or larger than the rock fragment.

In addition to being worn down gradually by small particle bombardment, a rock fragment may be catastrophically broken up or destroyed if it is struck by a sufficiently large or energetic particle. Experiments suggest that almost all rocks are destroyed if they are hit by a particle that would have formed a crater equal in diameter to the rock. Impacts by projectiles somewhat smaller or less energetic than those required for catastrophic disruption tend to produce extensive spalling from the sides of rocks (5). The recovery of some rocks in the Apollo 11 mission with one or two rounded, pitted sides and several angular sides suggests that splitting or fragmentation by large particle impact is a significant process of rock destruction. The outer parts of many of the rocks are broken by a series of concentric fractures into a series of exfoliation shells. These fractures may have been produced in part by impact spalling.

The mean expected lifetime of a rock fragment destroyed by large particle bombardment is given by

$$l_2 = \frac{4A_r}{\pi\,\chi\,D^{\lambda+2}} \qquad (5)$$

where l_2 is the mean expected lifetime, expressed in units of the age of the mare surface, of a rock fragment destroyed by a particle that would produce a crater of diameter equal to or larger than D, the diameter of the rock fragment. As shown in Table 1, the mean expected lifetime of a rock destroyed by large particle bombardment is of the same order as the mean expected lifetime calculated for small particle bombardment. The calculations suggest that, for rocks 10 cm in diameter and larger, small particle bombardment is the dominant process of rock destruction. This result is sensitive to the assumed crater diameter at which catastrophic disruption occurs but is consistent with the observation that most large rocks have well-rounded, pitted surfaces on most sides.

The true mean expected lifetime of a

rock must be calculated by combining the rate of erosion by small particles and the hazard of catastrophic disruption,

$$L = \frac{1}{1/l_1 + 1/l_2} \qquad (6)$$

where L is the mean expected lifetime, expressed in units of the age of the mare surface, of a rock fragment bombarded by both small and large particles. Solutions of Eq. 6 are listed in Table 1.

A typical exposure history of a rock fragment may be deduced by comparing Table 1 with solutions to Eq. 3. Most rocks, once they have been ejected onto the lunar surface, may be expected to be destroyed before they become buried in the regolith. A rock will tend to be buried if the local time of turnover to a depth equal to the diameter of the rock is less than the lifetime of the rock. According to the calculations presented above, this happens in only a minority of cases because the time of turnover to a given median depth is about three times as great as the lifetime of a rock of that dimension. A correction should be made, however, for the fact that a crater produced in the regolith by a given release of energy will be about six times as deep as a crater produced in bedrock. With this correction, a rock fragment may be expected to be buried, on the average, before it is destroyed. Most rocks that are buried are likely to be buried only at shallow depth and are likely to be reexhumed in a time given by the time of turnover to a median depth which is equal to the depth of burial.

The mean expected exposure age for a rock fragment of a given size is approximately the same as the mean expected lifetime of a rock of the same initial size. On the average, an observed rock will have started out about twice as large but still will have half its lifetime left. Exposure ages ranging from 10 to 160 m.y. were reported for rocks ranging in average dimension from 3 to 11 cm, on the basis of the preliminary examination of the Apollo 11 samples (6). If the age of the mare surface at Tranquillity Base is of the order of 4 b.y. (6), the mean expected exposure ages for rocks in this size range are 10 to 30 m.y. A few exposure ages substantially greater than the mean should be observed for rock fragments that have been buried at shallow depth or that have been derived by breakup of much larger rocks. Some rock fragments should also be observed with exposure ages substantially less than the mean.

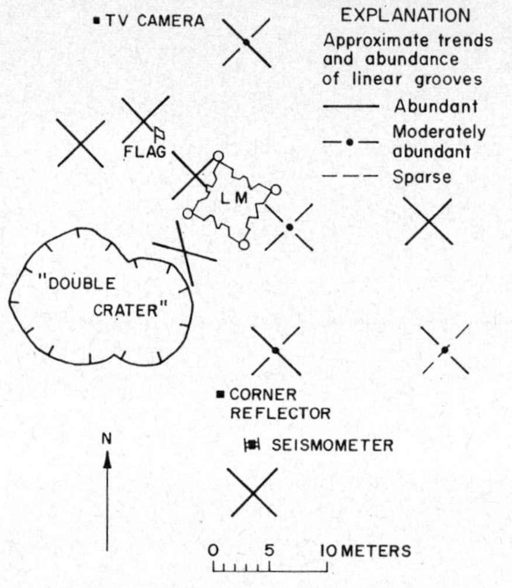

Fig. 3. Map of trend of linear grooves observed from extravehicular activity and module window photographs at Tranquillity Base.

The rough agreement between predicted and observed exposure ages suggests the extrapolation of the crater production function, $F = \chi c^{\lambda}$, to craters less than a millimeter in diameter is approximately correct.

Most rocks have been tumbled about on the lunar surface at least one or more times during their exposure to small particle bombardment. Small craters are generally found on all sides of a rock, showing that the bottom of the rock has been exposed at some time as a free surface. In the cases of three larger rocks whose position at time of collection during the Apollo 11 mission has been determined, one rock was oriented with the most rounded, pitted surface down, one rock was oriented on edge, with part of a heavily pitted, rounded side buried, and the third rock was oriented with the most rounded, pitted surface on top.

Rocks on the surface of the regolith may be tumbled by two somewhat different mechanisms: (i) A rock may be hit by one or more particles, which will impart sufficient angular momentum to turn the rock over. The efficiency of this mechanism depends on the degree to which the rock is embedded in the regolith. In many cases the impacting particles will destroy the rock, rather than turn it over. (ii) A rock may be thrown out of a crater. Rocks are tumbled most frequently by ejection from craters with diameters about twice as great as the tumbled rocks. This may be expected to occur about once in the lifetime of a rock, on the basis of the crater production function, $F = \chi c^{\lambda}$. However, $F = \chi c^{\lambda}$ has been determined for craters produced almost entirely in bedrock. For an impacting particle of a given energy, a crater produced in the regolith may be about twice the diameter of a crater in bedrock. If a correction is made for this difference, rocks may be expected to be tumbled about 10 times in their lifetime. The available evidence that most rocks collected in the Apollo 11 mission have been tumbled indicates that the mass flux of impacting projectiles is not dominated by very small particles.

The size-frequency distribution of rock fragments in the regolith at Tranquillity Base has been estimated from counts of fragments on the surface shown in Apollo 11 photographs (Fig. 2). The observed surface particle distribution is similar to that observed at other mare localities from Surveyor television pictures. This observed distribution follows fairly closely a power function with an exponent $\gamma = -2.4$. The size-frequency distribution of fragments in the regolith should be expected to follow a power function with about the same exponent as a power function which describes the size-frequency distribution of the impacting particles (7), or

$$\gamma \approx \lambda + 1 \qquad (7)$$

It may be seen that the agreement is rough, $\gamma = -2.4$ whereas $\lambda + 1 = -1.93$.

About half of the rock specimens returned in the Apollo 11 mission are fragments of microbreccia. The size-frequency distribution of the constituent particles of the microbreccia specimens is similar to the observed size distribution of particles in the regolith. Furthermore, the lithology and composition of the particles and the abundance of solar wind gases is about the same in the microbreccias as in the regolith (6). The inference seems secure that the microbreccias have been formed by induration of regolith material. This induration probably has been produced by shock compaction during formation of small and large craters in the regolith.

Some specimens of microbreccia exhibit a faint but distinct foliation that strongly resembles stratification. This foliation may be partly formed or enhanced by shock-produced shear, but in some cases there is a very slight sorting of fine and coarse particles in alternate layers. We believe the foliation is sedimentary stratification. Such stratification may be produced by deposition of successive layers of ejecta from small craters formed in the regolith. Stratification formed in this way will be preserved from destruction by repetitive cratering only where the rate of deposition exceeds the rate of turnover. This occurs primarily in the floors of craters that are gradually being filled and on fillets on the sides of rocks. Elsewhere the regolith is not stratified but has a rubbly structure (*8*) produced by repetitive cratering and turnover. This rubbly structure is also observed in some specimens of microbreccia.

The surface of the regolith at the Apollo 11 landing site is marked by intersecting sets of small shallow grooves. Most of the grooves are a fraction of a centimeter to a centimeter deep, about ½ to 3 cm wide, and 3 to 50 cm long a few were noted which are 2 to 3 m long. Where most strongly developed, the grooves are spaced 3 to 5 cm apart. One set of grooves trends northwest, another trends northeast. A few sets were observed that trend in other directions. The average trends and abundance of the grooves were estimated from ten surface photographs (Fig. 3). There appears to be no spatial or geometric relationship of the grooves to the location of the module or to local relief. Many of the grooves trend across small depressions and hummocks, with no apparent change in size, shape, or orientation.

The grooves may be produced by drainage of fine-grained surface material into fractures in the underlying bedrock or by vibration or jostling of joint blocks in the bedrock. If this interpretation is correct, it suggests that fine-grained regolith material is in direct contact with the bedrock. The drainage may be triggered or accelerated by tidal deformation or seismic disturbances. The lack of patterns that are radial or concentric to the module, or within craters, indicates that the grooves are not caused primarily by the descent exhaust or by downslope creep of material.

As the upper surface of the regolith is relatively rapidly changed by small cratering events, the grooves must be fairly young features or they must be frequently or continuously renewed. If they were not renewed, the lifetime of the grooves would be approximately equal to the time of turnover of the regolith to a depth equal to the depth of the grooves. The life expectancy of grooves 1 mm to 1 cm deep would be 0.0004 to 0.003 times the age of the mare surface.

E. M. SHOEMAKER
M. H. HAIT
G. A. SWANN
D. L. SCHLEICHER
D. H. DAHLEM, G. G. SCHABER
R. L. SUTTON

U.S. Geological Survey,
Flagstaff, Arizona 86001

References and Notes

1. V. R. Oberbeck and W. L. Quaide, *J. Geophys. Res.* **72**, 4697 (1968).
2. N. J. Trask, *Ranger VIII and IX. Part II: Experimenters' Analyses and Interpretation, Tech. Rept. 32-800* (Jet Propulsion Laboratory, 1966), p. 249.
3. E. M. Shoemaker *et al.*, *J. Geophys. Res.* **74**, 6081 (1969).
4. L. Soderbloom, *ibid.*, in press.
5. D. E. Gault, *Trans. Amer. Geophys. Union* **50**, (abstract) 219 (1969).
6. Lunar Sample Preliminary Examination Team, *Science* **165**, 1211 (1969).
7. E. M. Shoemaker and E. C. Morris, *Surveyor Project Final Report. Part II: Science Results, Tech. Rept. 32-1265* (Jet Propulsion Laboratory, 1968), p. 103.
8. E. M. Shoemaker, R. M. Batson, E. E. Holt, E. C. Morris, J. J. Rennilson, E. A. Whitaker, *Science* **158**, 642 (1967).
9. This report was prepared under NASA contract T-65253G.

4 January 1970

Passive Seismic Experiment

Abstract. *Seismometer operation for 21 days at Tranquillity Base revealed, among strong signals produced by the Apollo 11 lunar module descent stage, a small proportion of probable natural seismic signals. The latter are long-duration, emergent oscillations which lack the discrete phases and coherence of earthquake signals. From similarity with the impact signal of the Apollo 12 ascent stage, they are thought to be produced by meteoroid impacts or shallow moonquakes. This signal character may imply transmission with high Q and intense wave scattering, conditions which are mutually exclusive on earth. Natural background noise is very much smaller than on earth, and lunar tectonism may be very low.*

The Passive Seismic Experiment is the principal means for determining the moon's internal structure, physical state, and tectonic activity and for relating these deductions to the various surface observations. Detailed investigation of the lunar structure must await the establishment and operation of a network of seismic stations; however, a single, suitable, well-recorded seismic event can provide information that is of fundamental importance and that could not be gained in any other way.

The Passive Seismic Experiment Package (PSEP) operated for 21 earth days at Tranquillity Base. It demonstrated that the goals of lunar seismic exploration could indeed be achieved.

The PSEP, which has been described elsewhere (*1*), was placed 16.8 m from the nearest foot-pad of the LM. Termination of the experiment resulted from failure of the system to execute commands from earth.

It became evident in the early stages of the experiment that the LM is a source of a great variety of seismic signals of unexpectedly large amplitudes, apparently generated by venting gases or circulating fluids, or both, within the LM. Thermoelastic deformation of the LM structure may also be a significant source of seismic noise.

Prior to the ascent of the LM, many signals corresponding to various activities of the astronauts, both on the surface and within the LM, were recorded, primarily on the short-period vertical (SPZ) seismometer. The astronauts' footfalls were detected at all points along their traverse (maximum range approximately 40 m). Signals were particularly strong when the astronauts were in physical contact with the LM. The velocity (compressional) at which such signals travel through the lunar surface material (debris layer?), as measured during a thruster test at the Apollo 12 site, is approximately 108 m/sec.

During the post-ascent period a great variety of high-frequency (2 to 18 hz) signals were recorded from the short-period (SP) seismometer. These signals have been separated into a number of descriptive categories based primarily upon the shapes of signal envelopes, the time-history of occurrence, and the spectral characteristics of each type. The following classification is tentatively assigned: (i) signals with impulsive onsets and relatively short durations (types I and X); and (ii) signals with emergent onsets and relatively long durations (types B, M, T, and L). Examples of seismic signals corresponding to several of these categories are shown in Fig. 1. Only the L events, which are believed to be of natural origin, are described below. The other signals are described elsewhere (*2*).

Type L signals have emergent beginnings and relatively long durations (up to 1 hour). The wave trains build up slowly to a maximum and then decrease slowly into the background. Of the various types of events, L events show the greatest variability in signal character, location of spectral peaks, and time of occurrence. They are observed on both

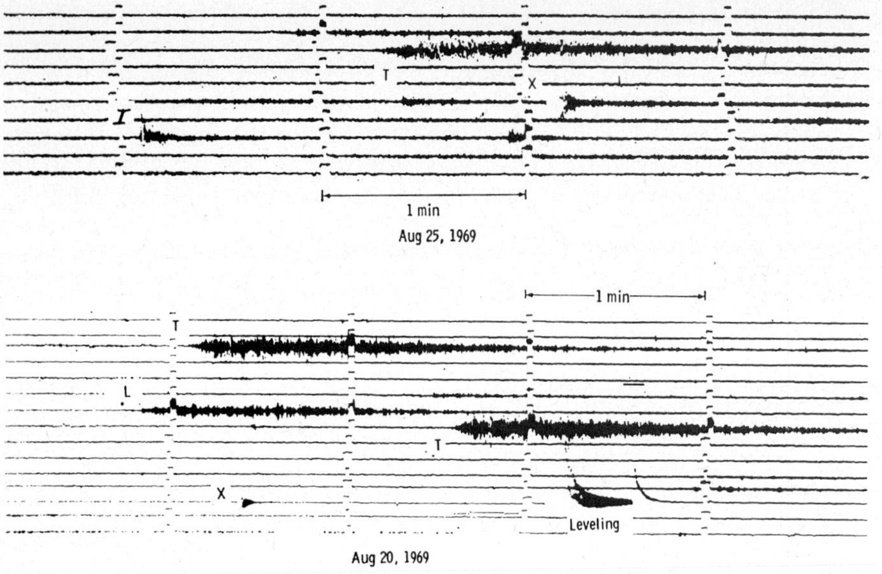

Fig. 1. Seismograms from the short-period (SP) seismometer showing signals of types I, T, X, and L.

the long-period (LP) and SP seismometers. For L events detected by the LP seismometers, little phase correlation between components is observed. The spectral peaks related to L events are broad, with frequencies of spectral maxima ranging between 1 and 6 hz. About 83 L events have been identified on the Apollo 11 records. Nearly all events thus far observed on the Apollo 12 records are of the L type, and the signal received from the impact of the LM ascent stage is also classified as L. The signal from the LM impact is shown in Fig. 2, along with the two largest L events recorded to date during the Apollo 12 mission.

An important result of the experiment is the discovery that the background noise level on the moon is extremely low. Before PSEP data were obtained, the level of lunar background noise was uncertain by several orders of magnitude. According to one popular hypothesis, the large diurnal thermal variations within the lunar surface material produced stresses that would lead to spallation and fracture and, consequently, to a high level of background noise. Another hypothesis proposed that meteoroid impacts were sufficiently numerous to create a significant continuous noise level. Still another hypothesis cited the absence of an atmosphere, oceans, and cultural activities to support the contention that the background noise level on the moon would be much lower than the level on the earth. Apollo 11 and 12 experience now suggests that the last of these is most accurate. The noise level is frequently below the threshold sensitivity of the instrument (0.3 nm at 1 hz) and far below the noise level at any known earth site.

The following possible source mechanisms for the observed signals are suggested: (i) venting gases from the LM and Portable Life Support Systems and circulating fluids within the LM (types B, M, and T); (ii) thermoelastic stress relief within the LM and PSEP structures (types I and X); (iii) instrumental effects (thermal effects, telemetry dropout, and the like); (iv) meteoroid impacts on the LM, the PSEP, and the lunar surface (types I, X, and L); (v) displacement of rock material along steep lunar slopes; (vi) moonquakes. Mechanisms i and ii are quantitatively important on the Apollo 11 seismograms because of the proximity of the LM to the seismometers and because of the single-package construction of the entire experiment. The differences in the Apollo 12 configuration make these sources relatively unimportant in the data being received from that instrument. Mechanism iv is thought to be a source of significant activity on both missions. The question of the existence of mechanisms v and vi is still open.

The occurrence of similar L events during both the Apollo 11 and 12 missions greatly strengthens our present belief that, of the various types of signals described above, L events are most likely to be of natural origin. Equally important, the inclusion of the impact signal of the Apollo 12 LM ascent stage in this family shows that signals of the L type can be generated by impulsive sources on the moon. This observation suggests that all L events were produced either by meteoroid impacts or by shallow moonquakes. On this basis, comparing the various L phases with the LM impact signal, we conclude that few, if any, L events originated significantly farther from the seismometer than the LM impact, which occurred at a distance of 75.9 km.

The average rate of L events detected during the Apollo 11 recording period is 4 per day. This rate is consistent with the number of detectable signals from meteoroid impacts predicted by Laster and Press (*3*) and by McGarr *et al.* (*4*) for a high-*Q* moon. Thus it is possible that all of the observed L signals were produced by meteoroid impacts. Calculation of meteoroid flux based upon this hypothesis has not yet been completed.

Few, if any, of the observed signals have patterns normally observed in recordings of seismic activity on earth (with the possible exceptions of volcanic tremors and landslide signals). Distinct phases corresponding to the various types of body and surface waves are not as apparent in any of the recorded seismic signals, and most wave trains are of long duration with less phase correlation between components than in the case of the earth.

The unexpected long duration of most of the observed signals must be explained either by a prolonged source mechanism or as a propagation effect. Prolongation

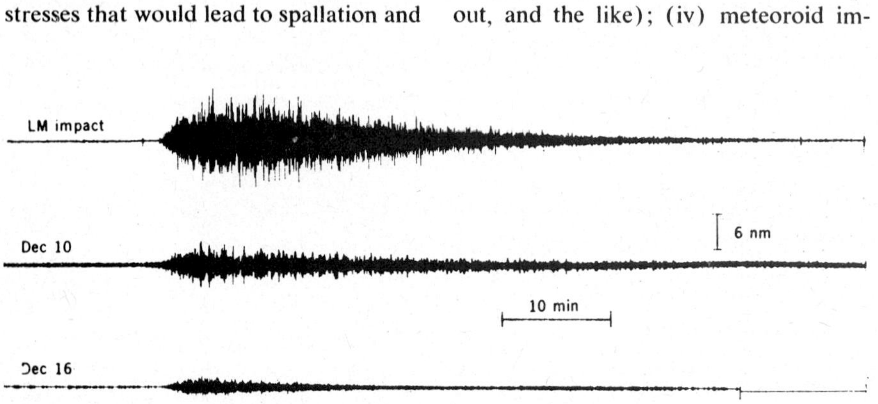

Fig. 2. Seismic signals received on the LP (long-period) vertical component seismometer from the LM impact and from natural sources on 16 December 1969.

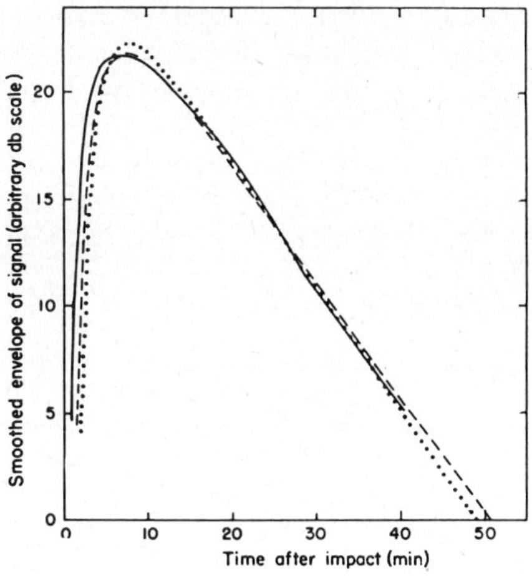

Fig. 3. Observed envelope of wave train received from the LM impact compared with the envelope predicted by diffusion theory. Solid line, observed; dashed line, two-dimensional ($vl = 5$ km^2/sec, $Q = 3600$); dotted line, three-dimensional ($vl = 7$ km^2/sec, $Q = 3600$ at 1 hz).

of an impact source might occur if the impact debris fell in the vicinity of the seismometers or if the debris triggered landslides. The small amount of debris thrown out by the LM impact weakens this hypothesis. A remote possibility is prolongation by collapse of "fairy castle" structures due to seismic waves excited by impacts.

The seismic signal detected from the Apollo 12 LM impact (Fig. 2) demonstrated that prolonged wave trains can be produced on the moon by relatively small impulsive sources. This result is important to our interpretation of the long reverberation as a propagation effect. It could be explained by postulating extremely low attenuation (high Q) for the moon. Although the evidence for this is hard to deny, this phenomenon raises some difficult questions as to its mechanism. Regardless of the explanation of signal duration, the similarity between the impact signal and other prolonged signals suggests that the latter were produced by meteoroid impacts or by near-surface moonquakes at ranges mostly within 100 km of the seismometer.

One hypothesis that could explain the signal character of L phases is that the moon not only has a high Q but also is very heterogeneous, at least in its outer regions. Heterogeneity is also implied by surface evidence. The scattering of seismic waves that would occur through a highly heterogeneous material would tend to increase the duration of the observed seismic wave and to suppress the appearance of distinct phases within the wave train. A medium which shows high Q and high scattering efficiency is unlike anything observed within the earth. Cold blocks of different composition in welded contact might show these properties. A welded aggregate might scatter seismic waves and still maintain high Q. Whatever the mechanism turns out to be, it will provide important evidence concerning the origin and evolution of the lunar interior.

In a highly heterogeneous medium, seismic waves may be scattered to the extent that propagation might properly be described by diffusion theory in which seismic energy "flow" is proportional to the gradient of energy density. If we define a characteristic length, l, as the distance over which ½ of the outward radiating seismic energy is reflected back toward the source, the flow of seismic energy is given by

$$\frac{\delta E}{\delta t} = \tfrac{1}{4} vl \Delta^2 E \qquad (1)$$

where E is seismic energy density and v is the average velocity of propagation. This is the diffusion equation with diffusion constant

$$D = \tfrac{1}{4} vl \qquad (2)$$

The solution of Eq. 1 leads to amplitude-time functions shown in Fig. 3. It can be seen that diffusion theory accurately predicts the observed signal amplitude for times long in comparison with the direct travel time (approximately 23.5 sec). Although this result does not verify the scattering hypothesis, it is sufficiently encouraging to warrant further consideration. More detailed analysis of this problem will be given in the Apollo 12 Mission Science Report.

Alternatively, a velocity-depth function might be found which could explain the character of these signals by ray theory without resort to a scattering mechanism. Some preliminary calculations show that this may be a possibility.

In interpreting present data, we are not concerned with the structure of the moon's deep interior, since most of the recorded events appear to have occurred at relatively short ranges. The relatively thick zone of self-compaction in which elastic wave velocity increases with depth would be perhaps 20 to 40 km thick. The large increase in velocity with depth results in a surface sound channel which may carry the seismic energy of these events. The impact of the S-IVB (third stage of the Apollo booster) planned for April 1970 at a range of 200 km should enable us to extend our interpretation to depths approaching 50 km into the moon. It will also be of great importance to see whether the same propagation effects are observed in nonmaria (highland) regions.

Because of the high sensitivity at which the instrument was operated, many distinctive seismic events would have been detected if the moon had seismic activity and transmission characteristics similar to those of the earth. The fact that a large number of seismic events of familiar earth types were not observed during 21 days of operation is a major result. If lunar seismicity is much lower than that of the earth, this would imply the absence of tectonic processes within the moon similar to those associated with major crustal movements on earth. This, in turn, implies lower thermal energy in the lunar interior than is present in the interior of the earth. When a seismic array is established on the moon more details on structure and seismicity will become evident.

GARY V. LATHAM
MAURICE EWING
Lamont-Doherty Geological Observatory, Columbia University, Palisades, New York 10964
FRANK PRESS
Department of Earth and Planetary Sciences, Massachusetts Institute of Technology, Cambridge 02139
GEORGE SUTTON
Hawaii Institute of Geophysics, University of Hawaii, Honolulu 96822
JAMES DORMAN
Lamont-Doherty Geological Observatory
YOSIO NAKAMURA
Fort Worth Division of General Dynamics, Fort Worth, Texas 76101
NAFI TOKSÖZ, RALPH WIGGINS, JOHN DERR
Department of Earth and Planetary Sciences, Massachusetts Institute of Technology
FREDERICK DUENNEBIER
Hawaii Institute of Geophysics

References and Notes

1. G. V. Latham, M. Ewing, F. Press, G. H. Sutton, *Science* **165**, 241 (1969).
2. G. V. Latham, M. Ewing, F. Press, G. Sutton, J. Dorman, N. Toksöz, R. Wiggins, Y. Nakamura, J. Derr, F. Duennebier, NASA SP-214, 143 (1969).
3. S. J. Laster and F. Press, *Phys. Earth Planet. Interiors* **1**, 151 (1968).
4. A. McGarr, G. V. Latham, D. Gault, *J. Geophys. Res.* **74**, 5981 (1969).
5. Lamont-Doherty Geological Observatory Contribution No. 1450. Hawaii Institute of Geophysics Contribution No. 305. Supported by NASA contracts NAS 9-5957 and NGR-22-009-123.

6 January 1970

Laser Ranging Retro-Reflector: Continuing Measurements and Expected Results

Abstract. *After successful acquisition in August of reflected ruby laser pulses from the Apollo 11 laser ranging retro-reflector (LRRR) with the telescopes at the Lick and McDonald observatories, repeated measurements of the round-trip travel time of light have been made from the McDonald Observatory in September with an equivalent range precision of ±2.5 meters. These acquisition period observations demonstrated the performance of the LRRR through lunar night and during sunlit conditions on the moon. Instrumentation activated at the McDonald Observatory in October has yielded a precision of ±0.3 meter, and improvement to ± 0.15 meter is expected shortly. Continued monitoring of the changes in the earth-moon distance as measured by the round-trip travel time of light from suitably distributed earth stations is expected to contribute to our knowledge of the earth-moon system.*

The performance (*1*) of the Apollo 11 laser ranging retro-reflector (LRRR) left on the moon, as well as that of the instrumentation at the ground observatories, has been in accord with the original expectations (*2–5*). The scientific objectives are such that lengthy analysis of a long-continuing series of frequent measurements is required before results are available. The experiment had its origin in discussions among members of the experimental gravitational research group at Princeton University (*6*). Because of the many areas of science and technology involved in the experiment, a group (*7*) was organized to carry it out.

The compact array of high-precision optical retro-reflectors (cube corners) deployed on the moon (*1, 8*) is intended to serve as a reference point in measuring precise ranges between the array and points on the earth by using the technique of short-pulse laser ranging. The atmospheric fluctuations in the index of refraction diverge a laser beam and prevent the spot on the moon from being smaller than approximately 1.6 km in diameter. The curvature of the lunar surface results in part of the short pulse being reflected before the rest, producing a reflected pulse measured in microseconds, even if the incident pulse is measured in nanoseconds. The retro-reflector array eliminates this spreading because of the small size of the array. (The maximum spreading of a pulse because of optical libration tipping of the array will be approximately ± 0.125 nsec.) In addition, the retro-reflective property causes a much larger amount of light to be directed back to the telescope from the array than is reflected from the entire surface area illuminated by the laser beam.

The basic uncertainty in measuring the approximately 2.5-sec round-trip travel time is associated with the performance of photomultipliers at the single photoelectron level. This uncertainty is estimated to be approximately 1 nsec. When the entire system is calibrated and the effects of the atmospheric delay are calculated from local temperature, pressure, and humidity measurements and subtracted from the travel time, where the uncertainty in this correction is estimated to be less than 0.5 nsec, an overall uncertainty of ± 15 cm in one-way range seems achievable.

The present uncertainty of three parts in 10^7 in the knowledge of the velocity of light will not affect the scientific aims of the experiment, since it is the practice to measure astronomical distances in light travel time. Primary scientific objectives include the study of gravitation and relativity (secular variation in the gravitational constant), the physics of the earth (fluctuation in rotation rate, motion of the pole, large-scale crustal motions), and the physics of the moon (physical librations, center-of-mass motion, size, and shape) (*2–5*). Estimates of improvements expected in some of these categories are shown in Tables 1 to 3. The estimated uncertainty for each quantity is intended to be an upper limit.

Reflected signals from the LRRR were acquired 1 August (and 3 August, with a different laser system) with the 120-inch telescope of the Lick Observatory (*9*) at Mt. Hamilton, California, and 20 August with the 107-inch tele-

Table 1. Lunar orbital data parameters.

Quantity	Present uncertainty (approximate)	0.15-m Range accuracy* Uncertainty	Time (yr)
Mean distance	500 m	25 m	1
Eccentricity	1×10^{-7}	4×10^{-9}	1
Angular position of Moon			
With respect to perigee	2×10^{-6} rad	4×10^{-8} rad	1
With respect to Sun	5×10^{-7} rad	4×10^{-8} rad	1
Time necessary to check predictions of Brans-Dicke scalar-tensor gravitational theory			8

* Three observing stations.

Table 2. Lunar libration and relation of LRRR to center of mass.

Quantity	Present uncertainty (approximate)	1.5-m Range accuracy* Uncertainty	Time (yr)
Libration parameters†			
$\beta = (C-A)/B$†	1×10^{-5}	3×10^{-8}	4
$\gamma = (B-A)/C$	5×10^{-5}	2×10^{-7}	1.5
Coordinates of LRRR with respect to center of mass‡			
X_1	500 m	25 m	1
X_2	200 m	7 m	1
X_3	200 m	5 m	3

* Three observing stations are assumed. † A is the moment of inertia about the principal axis toward the earth, B is the moment about the principal axis tangent to the orbit, and C is the moment about the moon's rotation axis. Knowledge of the parameters β and γ is important in determining the mass distribution within the moon. ‡ X_1, X_2, and X_3 are measured along the principal axes about which the moments A, B, and C are defined.

Table 3. Geophysical data determinable from LRRR.

Quantity	Present uncertainty (estimated)	0.15-m Range accuracy
Rotation period of earth (sec)	5×10^{-3}	1×10^{-3}
Distance of station from axis of rotation (m)	10	0.3
Distance of station from equatorial plane (m)*	20	0.6 to 2†
Motion of the pole (m)*	1 to 2	0.15
East-west continental drift rate observable in 5 years (cm/yr)*	30 to 60	3
Time for observing predicted drift of 10 cm/yr of Hawaii toward Japan (years)	15 to 30	1.5

* Three or more observing stations are required. † Depending upon the latitude of the station.

Table 4. Measurements at the McDonald Observatory of round-trip travel time during acquisition. The residuals represent the difference between the observed time of travel and that predicted on the basis of the JPL LE16 ephemeris.

Day	U.T.	Residual round-trip travel time (nsec)
20 August	03:00	96 ± 15
3 September	11:10	490 ± 15
4 September	10:10	795 ± 24
22 September	04:00	−1430 ± 15
17 October	01:44	−798 ± 15*
18 October	01:17	−978 ± 15*
1 November	11:40	−2034 ± 15*
16 December	01:45	1232 ± 15*

* The present accuracy is ±15 nsec in the knowledge of the electronic time delays. Upon completion of the current calibration, the accuracy will be determined by the present overall resolution, less than ±2 nsec, limited by the laser pulse length and photomultiplier jitter.

scope of the McDonald Observatory (*10*) at Mt. Locke, Texas. These observations showed that the LRRR did not suffer any major degradation, if any at all, from debris generated during lift-off of the lunar module. The signals are consistent with the return expected from the LRRR design, within the uncertainties of atmospheric "seeing," telescope transmission, and other optical losses. Continued acquisition period measurements at McDonald in September (*10*), taken with the initial observations, have demonstrated the successful performance of the LRRR at several sun illumination angles, as well as during and after a lunar night, confirming the prediction of thermal design analyses.

A first "geodetic result" from the acquisition observations at Lick (*9*) was the discovery, from the drift of the residual round-trip travel time with respect to the JPL lunar ephemeris 16 (LE16) predictions, that the coordinates for the 120-inch telescope are not those given for Mt. Hamilton (Lick Observatory) in the *American Ephemeris and Nautical Almanac* (*9a*). The Lick Observatory participated in the acquisition phase of the experiment to increase the probability of getting early returns. The weather and seeing are generally excellent there in the summer. Laser ranging activities ceased at Lick in August.

In October a custom-built four-stage ruby laser, made by Korad, was installed at the McDonald Observatory. This laser was built to specifications developed for long-term precision measurements in the lunar ranging experiment. The pulse length is produced by a time-varying reflectivity mode of operation and can be as short as 2.8 nsec, although a more typical value is 4 nsec. It transmits 5 joules with a beam divergence (full width) of 1.4 mradian at a repetition rate capability of one shot every 3 seconds. At the same time, vernier timing circuits (*11–13*) shown schematically in Fig. 1 were activated. This instrumentation allows a resolution uncertainty of ±2 nsec on each measured return. The accuracy depends on thorough calibration of all electronic delays. This will be completed soon to the 1-nsec level. The measurements made in October, November, and December, and which are shown in Table 4, still have a calibration uncertainty in accuracy of ±15 nsec. These later measurements are subject to reduction of the accuracy uncertainty upon completion of the final calibration.

As more experience is gained in the use of the new 107-inch McDonald telescope, the goal is three measurement periods daily. Each period would last about 15 minutes, enabling several hundred laser shots to be fired; the periods would be scheduled near the time of meridian crossing, several hours before, and several hours after.

From these measurements, one can obtain the minimum range and its epoch of occurrence. Harmonic analysis of this range time series will permit the determination of the qualities listed in Tables 1 through 3.

In order to satisfy all the scientific aims of the experiment, it is hoped that more U.S. and foreign ground stations can be established to carry out regular precision ranging to the LRRR. The deployment of several more LRRR's on the moon would allow a more detailed study of the lunar physical librations, independent of any model. One of these should be designed to give a larger return than the Apollo 11 LRRR, so as to allow participation in the ranging program by smaller telescopes.

C. O. ALLEY, R. F. CHANG
D. G. CURRIE, S. K. POULTNEY
Department of Physics and Astronomy, University of Maryland, College Park
P. L. BENDER
Joint Institute for Laboratory Astrophysics, National Bureau of Standards and *University of Colorado, Boulder 80302*
R. H. DICKE, D. T. WILKINSON
Palmer Physical Laboratory, Princeton University, Princeton, New Jersey 08540
J. E. FALLER
Department of Physics, Wesleyan University, Middletown, Connecticut
W. M. KAULA
Institute of Geophysics and Planetary Physics, University of California, Los Angeles 90024
G. J. F. MACDONALD
University of California, Santa Barbara
J. D. MULHOLLAND
Jet Propulsion Laboratory, California Institute of Technology, Pasadena 91103
H. H. PLOTKIN, W. CARRION
Goddard Space Flight Center, Greenbelt, Maryland 20771
E. J. WAMPLER
Lick Observatory, University of California, Santa Cruz 95060

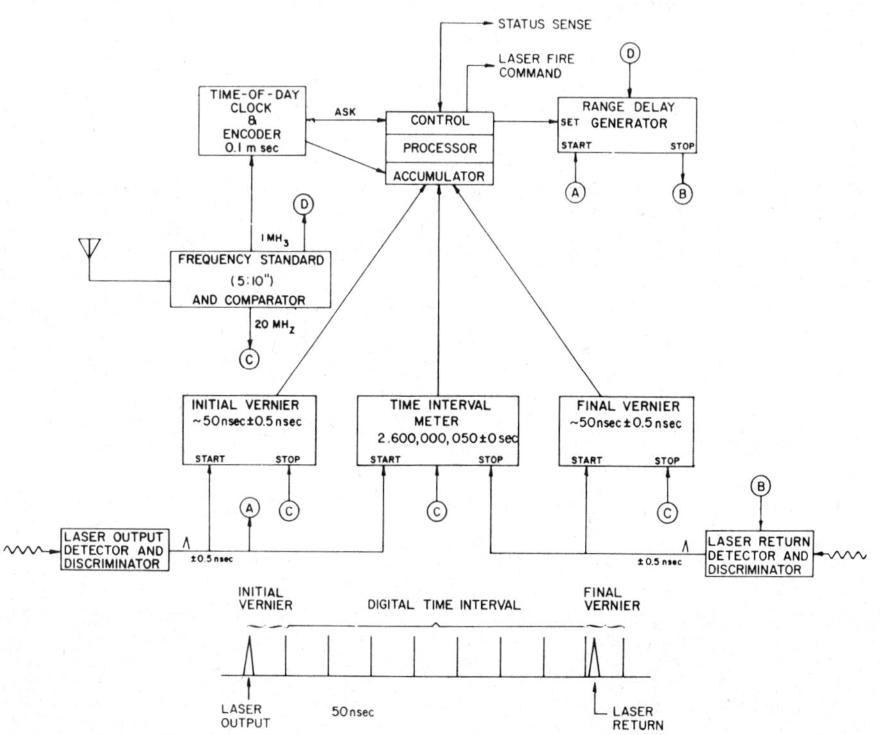

Fig. 1. Representation of the nanosecond-resolution time-interval measurement system now in use at the McDonald Observatory. Special circuits eliminate any ±1 count uncertainty in the 20-Mhz, digitally measured interval. The vernier components are time to pulse height converters.

References and Notes

1. C. O. Alley *et al.*, "Laser ranging retro-reflector" in *Apollo 11 Preliminary Science Report* (NASA Special Publication SP-214, 1969).
2. C. O. Alley, P. L. Bender, R. H. Dicke, J. E. Faller, P. A. Franken, H. H. Plotkin, D. T. Wilkinson, *J. Geophys. Res.* **70**, 2267 (1965).
3. C. O. Alley and P. L. Bender, in *Symposium No. 32 of the International Astronomical Union on Continental Drift, Secular Motion of the Pole, and Rotation of the Earth,* W. Markowitz and B. Guinot, Eds. (Reidel, Holland, 1968).
4. C. O. Alley, P. L. Bender, D. G. Currie, R. H. Dicke, J. E. Faller, *Proceedings of the N.A.T.O. Advanced Study Institute on the Application of Modern Physics to the Earth and Planetary Interiors,* S. K. Runcorn, Ed. (Wiley, London, 1969).
5. G. J. F. MacDonald, *Science* **157**, 204 (1967).
6. W. F. Hoffman, R. Krotkov, R. H. Dicke, *Inst. Radio Eng. IRE Trans. Military Electron.* **4**, 28 (1960). The authors served as a committee for the whole group: C. O. Alley, J. Brault, D. Brill, R. H. Dicke, J. Faller, W. F. Hoffman, L. Jordan, R. Krotkov, S. Liebes, R. Moore, J. Peebles, J. Stoner, and K. Turner.
7. Formal responsibility has rested with the following group: principal investigator, C. O. Alley (University of Maryland); co-investigators: P. L. Bender (National Bureau of Standards), R. H. Dicke (Princeton University), J. E. Faller (Wesleyan University), W. M. Kaula (University of California at Los Angeles), G. J. F. MacDonald (University of California at Santa Barbara), J. D. Mulholland (Jet Propulsion Laboratory), H. H. Plotkin (NASA Goddard Space Flight Center), and D. T. Wilkinson (Princeton University); participating scientists: W. Carrion (NASA Goddard Space Flight Center), R. F. Chang (University of Maryland), D. G. Currie (University of Maryland) and S. K. Poultney (University of Maryland).
8. Responsibility for the detailed design of the LRRR to perform continuously in the lunar environment has been carried primarily by J. E. Faller, D. G. Currie, R. F. Chang, and C. O. Alley, supported by the following engineering companies: Arthur D. Little, Inc. (P. Glaser, J. Burke, F. Gabron, and D. Comstock); Perkin-Elmer Corporation (J. Atwood, P. Forman, G. Watt, D. Corbett, and S. Lauffer); and the Bendix Aerospace Corporation (C. Weatherred, R. Hill, J. Brueger, R. Wolford, and K. Moore); and by project engineer H. Kriemelmeyer (University of Maryland).
9. J. E. Faller *et al.*, "Observations of the First Returns from a Laser Beam Directed at the Lunar Retro-Reflector Array," *Science* **166**, 99 (1969).

9a. U.S. Government Printing Office, Washington, D.C. (1969).

10. C. O. Alley, R. F. Chang, D. G. Currie, J. Mullendore, S. K. Poultney, J. D. Rayner, E. C. Silverberg, C. A. Steggerda, H. H. Plotkin, W. Williams, B. Warner, H. Richardson, B. Bopp, *Science* **167**, 368 (1970).
11. S. K. Poultney, *The Concept of the Time Interval Measurement and Control Circuitry for the Lunar Ranging Experiment: a History* (University of Maryland Department of Physics and Astronomy Technical Report No. 70–068, December 1969).
12. C. A. Steggerda, *A Description of the Time Interval Measurement and Laser Control Circuitry for the Lunar Ranging Experiment* (University of Maryland Department of Physics and Astronomy Technical Report No. 70–049, November 1969).
13. J. D. Rayner, *A Description of the Control Program for the Lunar Ranging Experiment* (University of Maryland Department of Physics and Astronomy Technical Report No. 70–064, November 1969).

6 January 1970

Age Measurements

Age of the Moon: An Isotopic Study of Uranium-Thorium-Lead Systematics of Lunar Samples

Abstract. *Concentrations of U, Th, and Pb in Apollo 11 samples studied are low (U, 0.16 to 0.87; Th, 0.53 to 3.4; Pb, 0.29 to 1.7, in ppm) but the extremely radiogenic lead in samples allows radiometric dating. The fine dust and the breccia have a concordant age of 4.66 billion years on the basis of* $^{207}Pb/^{206}Pb$, $^{206}Pb/^{238}U$, $^{207}Pb/^{235}U$, *and* $^{208}Pb/^{232}Th$ *ratios. This age is comparable with the age of meteorites and with the age generally accepted for the earth. Six crystalline and vesicular samples are distinctly younger than the dust and breccia. The* $^{238}U/^{235}U$ *ratio is the same as that in earth rocks, and* ^{234}U *is in radioactive equilibrium with parent* ^{238}U.

Patterson and others (*1*) first applied isotopic lead data to determine the age of meteorites and the earth as 4.55 ± 0.07 billion years, assuming that lead in stone meteorites and modern terrestrial lead have evolved from a primordial lead compositionally like the lead in iron meteorites. We have used this method to determine the age of lunar samples and to obtain insight into the chemical differentiation history of lunar surface material. The abundance ratios of lead isotopes and uranium isotopes (^{238}U, ^{235}U, and ^{234}U) were determined by mass spectrometry, and the radioactivity ratios of the uranium isotopes and the thorium isotopes (^{232}Th, ^{230}Th, and ^{228}Th) by α spectrometry.

Details of analytical procedures will be discussed in a later report. Lead was first separated by barium coprecipitation (*2*) and dithizone extraction after HF-$HClO_4$ sample decomposition. Uranium and thorium were separated by Dowex-1 resin from 6*N* HCl. Uranium was further purified by hexone extraction, and thorium was separated by Dowex-1 resin from 7*N* HNO_3 media (*2*) for isotopic ratio measurement.

The concentrations of lead, uranium, and thorium were determined by isotope dilution (*2*). Owing to the importance of blank corrections, each set of analyses for the determination of both the isotopic composition and the concentration was accompanied by a blank analysis. Blanks yielded 0.015 to 0.032 μg for Pb concentration and composition separations; 0.0001 and 0.00015 μg for U and Th concentrations, respectively. Blanks for U and Th composition are about 10 times the blanks of the concentration runs. Lead was run by mass spectrometer with phosphoric acid–silica gel (*3*) with some modification. The standard deviations of the lead isotope ratios are $^{206}Pb/^{204}Pb$ (±1 percent, except sample 10050 which has ±3 percent uncertainty), $^{206}Pb/^{207}Pb$ (±0.2 percent), $^{206}Pb/^{208}Pb$ (±0.3 percent). However, because of the low concentration of lead in samples, the procedure for making blank correction of the data introduces relatively large overall uncertainties. The blank correction has the most effect on ^{204}Pb abundance; and, because corrections essentially change the location of points along isochrons (Fig. 1), no serious difficulty seems to arise for the age measurements.

Results are summarized in Table 1. The uranium and thorium concentrations are in good agreement with those reported by the Preliminary Examination Team (*4*). The average values of uranium and thorium concentrations are higher than those in stone meteorites (*5*); however, the lowest value of uranium and thorium observed for sample 10050 is similar to values of the Nuevo Laredo achondrite. Lead contents of the lunar samples are in the range of stone meteorites (*5*). The uranium and thorium concentration is 2 to 4 times higher than that of terrestrial oceanic basalts and Hawaiian tholeiites, and it is similar to the concentration in oceanic alkali basalts (*6*). Th/U ratios are about 3.9 with some variation. The uranium isotopic ratio ($^{238}U/^{235}U$) in five different lunar rocks that were measured is the same as for terrestrial rocks, within the experimental error (137.8 ± 0.4). The ^{234}U daughter is in radioactive equilibrium with parent ^{238}U within experimental error of ±1.5 percent.

The lead isotopic composition in all Apollo 11 samples examined is very

Table 1. Concentrations of lead, uranium, and thorium and isotopic compositions of lead in Apollo 11 samples.

Sample No.	Type	Concentration* Pb (ppm)	U (ppm)	Th (ppm)	Atomic ratios $^{206}Pb/^{204}Pb$	$^{207}Pb/^{204}Pb$	$^{208}Pb/^{204}Pb$	$^{238}U/^{204}Pb$	$^{232}Th/^{238}U$
10003	Crystalline	0.51	0.268	1.029	423.9 (311.9)†	198.0 (147.6)	448.9 (333.3)	491.3	3.96
10017	Crystalline	1.56	0.854	3.363	410.0 (367.7)	191.9 (173.7)	435.0 (390.2)	492.8	4.07
10020	Vesicular	0.37	0.202	0.694	288.7 (238.9)	139.0 (116.3)	289.7 (243.5)	338.3	3.55
10050	Crystalline	0.29	0.156	0.531	295.8 (187.7)	147.0 (95.8)	287.3 (190.2)	397.2	3.53
10057	Vesicular	1.68	0.865	3.415	1241.5 (938.6)	590.1 (447.9)	1281.3 (973.6)	1392	4.08
10071	Vesicular	1.71	0.873	3.434	199.7 (191.2)	95.8 (92.0)	206.7 (198.7)	222.5	4.06
10061	Breccia	1.74	0.674	2.572	249.1 (238.8)	163.2 (152.8)	258.2 (242.7)	255.6	3.94
10084	Fine material	1.39	0.544	2.092	261.9 (237.4)	171.0 (155.3)	270.1 (246.7)	239.6	3.97

*Estimated errors for the concentrations are better than 2 percent for Pb and 1 percent for U and Th.
†Values in parentheses are measured ratios before blank correction.
Erratum: The lead isotopic composition (corrected for blank) of sample 10071 was recalculated to be $^{206}Pb/^{204}Pb = 735.4$, $^{207}Pb/^{204}Pb = 361.2$, $^{208}Pb/^{204}Pb = 779.5$, and $^{238}U/^{204}Pb = 839.2$.

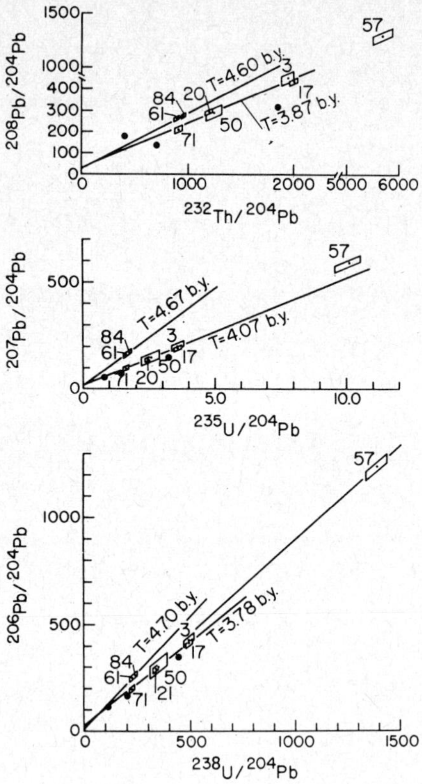

Fig. 1. The ^{206}Pb-^{238}U, ^{207}Pb-^{235}U, and ^{208}Pb-^{232}Th isochrons for Apollo 11 samples. Indicated errors (boxes) include blank corrections (see Table 1). Solid circles are for the mineral composite fraction of sample 10071. The errors in the data on the fractions are not indicated. Numbers are final digits of sample numbers.

radiogenic, which makes radiometric dating possible. The present $^{238}U/^{204}Pb$ is in the range of 200 to 1400. Patterson *(1)*, in his lead isotope study of meteorites, calculated the uranium and thorium concentrations required to produce the lead concentration and isotopic composition. However, these calculated values of uranium and thorium were an order of magnitude higher than those actually determined later in stony meteorites *(5)*. On the other hand, almost all lead that exists in the measured lunar samples is supported by its uranium and thorium parents, as is shown by the nearly concordant ages in Table 2.

The results are plotted in three ways in Fig. 1. By this plotting technique we may be able to obtain the initial $^{206}Pb/^{204}Pb$, $^{207}Pb/^{204}Pb$, and $^{208}Pb/^{204}Pb$, as well as isochron ages, on the assumption that some samples are coeval and were formed from sources of uniform lead isotopic composition. The dust (sample 10084) and breccia (sample 10061) have clearly older ages than crystalline rocks. The tie lines between these data points and primordial lead *(7)* correspond to 4.70, 4.67, and 4.60 billion years, respectively, in plots of ^{238}U-^{206}Pb, ^{235}U-^{207}Pb, and ^{232}Th-^{208}Pb. The data points of the crystalline rocks, except sample 10057, fall in a group and indicate, respectively, isochrons of 3.78, 4.07, and 3.87 billion years in the three-way plot. The initial lead ratios are obtained from the intercepts on the *Y* axes in Fig. 1. The initial values obtained by least-squares method *(8)* and based on these six crystalline rocks, if coeval, are 23.6 ± 5.4 for $^{206}Pb/^{204}Pb$, 13.3 ± 4.0 for $^{207}Pb/^{204}Pb$, and 11.9 ± 17.4 for $^{208}Pb/^{204}Pb$. These ratios are, however, highly uncertain. The $^{208}Pb/^{204}Pb$ and $^{207}Pb/^{204}Pb$ values appear to be too small, compared with those of the primordial lead of meteorites.

The $^{206}Pb/^{204}Pb$ intercept value is higher compared with that of primordial lead but seems reasonable if the $(^{238}U/^{204}Pb)_0$ value on the moon surface is high as seen in these Apollo 11 samples. The $(^{238}U/^{204}Pb)_0$ is calculated to be about 70 based on the observed intercept value. Of course the common lead

Table 2. Apparent ages of Apollo 11 samples (in million years). Constants used were
$\lambda_{^{238}U} = 1.5369 \times 10^{-10} y^{-1}$
$\lambda_{^{235}U} = 9.7216 \times 10^{-10} y^{-1}$
$\lambda_{^{232}Th} = 4.990 \times 10^{-11} y^{-1}$
$^{238}U/^{235}U = 137.8$

Sample No.	$^{206}Pb/^{238}U$	$^{207}Pb/^{235}U$	$^{207}Pb/^{206}Pb$	$^{208}Pb/^{232}Th$
10003*	3878	3976	4025	3812
10017*	3767	3935	4022	3598
10020*	3765	3903	3996	3773
10050*	3760	3952	4051	3680
10057*	4089	4146	4173	3959
10071*	3794	3815	3826	3374
10061†	4710	4678	4663	4594
10084†	4685	4668	4659	4525

*Assumed isotopic composition for common lead: $^{206}Pb/^{204}Pb$, 23.6; $^{207}Pb/^{204}Pb$, 31.5; $^{208}Pb/^{204}Pb$ = 40.9. †Assumed initial lead for common lead: $^{206}Pb/^{204}Pb$, 9.346; $^{207}Pb/^{204}Pb$, 10.218; $^{208}Pb/^{204}Pb$, 28.96.

correction does not introduce a large effect in the age, owing to the large sample age and to the very radiogenic nature of the lead.

Crystalline sample 10017 and dust sample 10084 were separated into "mineral composites," and the fractions were analyzed for lead isotopes and for lead, uranium, and thorium content. The isochrons from mineral fractions are slightly younger than isochrons of total rocks. The reason for the difference is not well understood, but it may be due to a large lead contamination, to leaching out of daughters during the mineral separation, or to the rocks not being coeval. However, it should be mentioned that the intercepts for the initial $^{206}Pb/^{204}Pb$ and $^{207}Pb/^{204}Pb$ obtained from sample 10017 fractions are the same, within the experimental error, as the intercept of the six crystalline rocks (see Fig. 1). Therefore, we chose the following values, as a maximum, to correct common lead for the crystalline

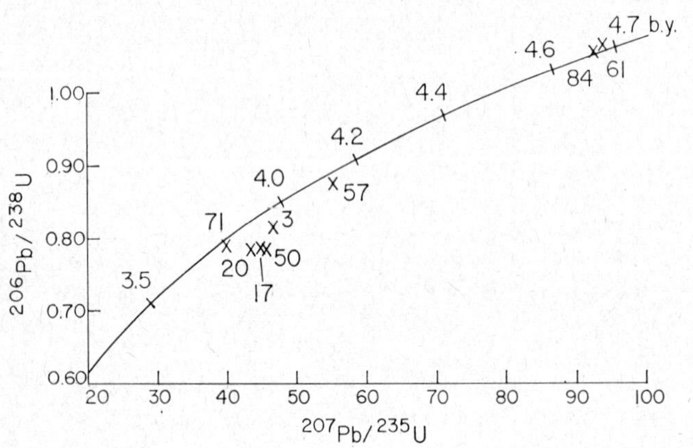

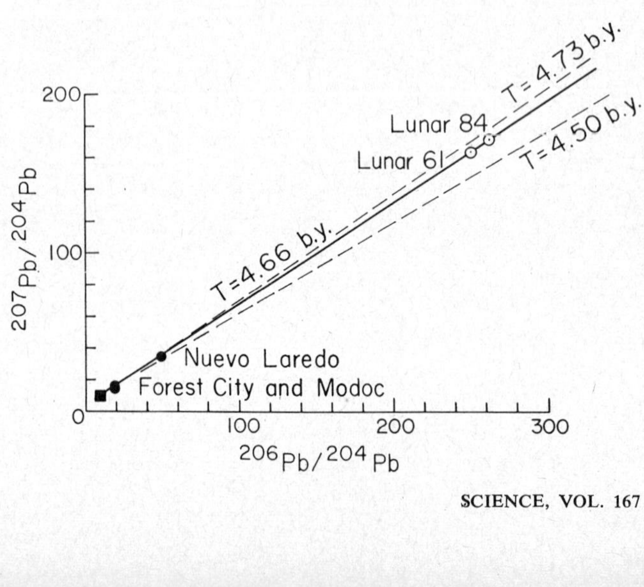

Fig. 2 (above). Concordia diagram showing lead-uranium ratios for Apollo 11 samples. Numbers are final digits of sample numbers. Fig. 3 (right). The ^{207}Pb-^{206}Pb evolution diagram for moon and meteorites.

rocks: $^{206}Pb/^{204}Pb$ = 23.6, $^{207}Pb/^{204}Pb$ =31.5 and $^{208}Pb/^{204}Pb$=40.9. These values were calculated from the $^{206}Pb/^{204}Pb$ intercept and on the assumption that lead had evolved from 4.7 to 3.7 billion years prior to crystalline rock formation. The intercepts obtained from 10084 fractions are the same within experimental error as the value obtained from meteorite lead.

The apparent ages are shown in Table 2. The initial leads used for correction are primeval lead of meteorite (*7*) for samples 10061 and 10084 and common lead (given above) for crystalline rocks. The ages indicate that sample 10071 may be younger and 10057 older than the other four crystalline rocks. Thus it might be conceivable that the common lead used for age calculation remains small because the isochrons in Fig. 1 are strongly controlled by the position of sample 10071.

The results corrected for initial lead are plotted in the Pb-U evolution diagram (Fig. 2). The points of the dust and breccia are above the concordia line, but those of crystalline rocks are below the concordia line. The apparent ages of 3.8 to 4.1 billion years for crystallized rocks indicate that they solidified several hundred million years after the formation of the moon. The dust apparently contains only minor amounts of fine fragments of the younger rocks that were analyzed. The most feasible interpretation is that dust and breccia probably were formed at an early stage in the history of the moon.

In Fig. 3, $^{206}Pb/^{204}Pb$ and $^{207}Pb/^{204}Pb$ for lunar dust and breccia are plotted with Patterson's data (*1*) on stone meteorite and Oversby's measurement (*7*) of primordial lead on the Canyon Diablo troilite. These data indicate that the system, of which earth, moon, and meteorites are parts, probably was formed about $4.66\ ^{+0.07}_{-0.16}$ billion years ago. We set the upper limit from our experimental error, which amounts to about 1.5 percent, and the lower limit to include Patterson's meteorite data. It might be argued that the obtained age for the moon is from breccia and dust—not the lunar crystalline rock—and there may be doubt that those two samples represent the oldest rock of the moon. However, inasmuch as (i) the observed data fit into the extension of the meteorite lead array, (ii) the data lie near the U-Pb evolution curve, and (iii) the obtained age is very similar to the "older" meteorite ages measured by the Rb-Sr method (*10*), the age may be correct.

Thorium-232, with a 4.0-Mev α particle emission, has been used as a natural yield tracer for the determination of the state of radioactive equilibrium between ^{230}Th and parent ^{238}U in crustal silicate rocks from the earth (*11*). This technique of using ^{232}Th, of known concentration in a sample, as the yield tracer for ^{230}Th was used on several samples from Apollo 11. Although, within experimental error, ^{234}U is in radioactive equilibrium with ^{238}U, preliminary data indicate a value for the $^{230}Th/^{238}U$ activity ratio that is greater than unity in three of the samples analyzed (10017, 10020, 10057). A $^{228}Th/^{232}Th$ activity ratio of unity would indicate that ^{228}Th is in radioactive equilibrium with parent ^{232}Th. In two of the rocks (10020, 10057) the ^{228}Th α activity was measured to be about 10 percent greater than the 4.0-Mev α activity of ^{232}Th. The reasons for these variations are not understood at this time.

MITSUNOBU TATSUMOTO
JOHN N. ROSHOLT

U.S. Geological Survey, Denver, Colorado 80225

References and Notes

1. C. C. Patterson, *Geochim. Cosmochim. Acta* **7**, 151 (1955); *ibid.*, **10**, 230 (1956); V. R. Murthy and C. C. Patterson, *J. Geophys. Res.* **67**, 1161 (1962).
2. M. Tatsumoto, *J. Geophys. Res.* **71**, 1721 (1966); J. N. Rosholt, B. R. Doe, M. Tatsumoto, *Bull. Geol. Soc. Amer.* **77**, 987 (1966).
3. A. E. Cameron, D. H. Smith, R. L. Walker, *Anal. Chem.* **41**, 525 (1969).
4. Lunar Sample Preliminary Examination Team, *Science* **165**, 1211 (1969).
5. H. Hamaguchi, G. W. Reed, A. Turkevich, *Geochim. Cosmochim. Acta* **12**, 337 (1957); G. L. Bate, J. R. Huizenga, H. A. Potratz, *ibid.* **16**, 88 (1959); G. W. Reed, K. Kigoshi, A. Turkevich, *ibid.* **20**, 122 (1960).
6. M. Tatsumoto, *Science* **153**, 1094 (1966);——— *J. Geophys. Res.* **71**, 1721 (1966);———and R. J. Knight (abstract), *Geol. Soc. Amer. Spec. Paper 101*, 218 (1968).
7. V. M. Oversby, *J. Geophys. Res.*, in press.
8. D. York, *Can. J. Phys.* **44**, 1079 (1966).
9. T. J. Chow and C. C. Patterson, *Geochim. Cosmochim. Acta* **26**, 263 (1962); C. C. Patterson and M. Tatsumoto, *ibid.* **28**, 1 (1964); R. D. Russell and R. M. Farquar, *ibid.* **19**, 41 (1960).
10. G. J. Wasserburg, D. S. Burnett, C. Frondel, *Science* **150**, 1814 (1965); D. S. Burnett and G. J. Wasserburg, *Earth Planet. Sci. Lett.* **2**, 137 (1967); *ibid.* **2**, 397 (1967).
11. J. N. Rosholt, Jr., Z. E. Peterman, A. J. Bartel, *U.S. Geol. Survey Prof. Paper 575–B*, 133 (1967).
12. We thank R. J. Knight and D. M. Unruh for laboratory assistance, M. Delevaux for mass spectrometer maintenance, and other branch personnel for help in many phases of this investigation. We acknowledge comments by Z. E. Peterman and R. E. Zartman in reviewing the manuscript. This study was financed by NASA grant T–75445.

4 January 1970

Ages, Irradiation History, and Chemical Composition of Lunar Rocks from the Sea of Tranquillity

Abstract. *The ^{87}Rb-^{87}Sr internal isochrons for five rocks yield an age of $3.65 \pm 0.05 \times 10^9$ years which presumably dates the formation of the Sea of Tranquillity. Potassium-argon ages are consistent with this result. The soil has a model age of 4.5×10^9 years, which is best regarded as the time of initial differentiation of the lunar crust. A peculiar rock fragment from the soil gave a model age of 4.44×10^9 years. Relative abundances of alkalis do not suggest differential volatilization. The irradiation history of lunar rocks is inferred from isotopic measurements of gadolinium, vanadium, and cosmogenic rare gases. Spallation xenon spectra exhibit a high and variable $^{131}Xe/^{126}Xe$ ratio. No evidence for ^{129}I was found. The isotopic composition of solar-wind xenon is distinct from that of the atmosphere and of the average for carbonaceous chondrites, but the krypton composition appears similar to average carbonaceous chondrite krypton.*

Samples investigated were 10017, 10044, 10050, 10057, 10059, 10069, 10071, soil (10084), coarse fines (10085), and core 2 (referred to by Nos. 17, 44, etc.). Contamination levels were negligible. Standards were regularly run to control precision *to the levels reported.*

Rubidium-strontium analyses on total rocks and minerals were obtained. The Sr is concentrated in plagioclase and the K in a fine-grained or glassy interstitial phase, containing up to 12 percent K. Maximum Rb/Sr was obtained in cristobalite and ilmenite separates due to the interstitial high-K phase. Precise internal isochrons on five rocks give the same age of 3.65×10^9 years (Fig. 1). All $(^{87}Sr/^{86}Sr)_I$ are distinctly above BABI (*1*). The $(^{87}Sr/^{86}Sr)_I$ for No. 44 is distinct from that of other samples, indicating that the samples represent at least two different rock bodies (Fig. 2). The possibility that the isochrons are two-phase mixing lines cannot be excluded, considering the distribution of alkalis, but the coincidence of ages for rocks of different $(^{87}Sr/Sr^{86})_I$ and texture argues that 3.65×10^9 years is the time of crystallization.

The possibility that the age of 3.65×10^9 years represents closed-system metamorphism of rocks initially formed 4.6 $\times$ 10^9 years ago can be ruled out, as this would require values of $^{87}Sr/^{86}Sr$ at 4.6 $\times$ 10^9 years which are considerably less than BABI for Nos. 17, 57, 69, and 71. Rock textures and chemical heterogeneity in minerals do not support post-crystallization metamorphism. From

$$\Delta^{44} = (^{87}Sr/^{86}Sr)_I^{44} - (^{87}Sr/^{86}Sr)_{BABI} = 0.00011$$

the time required to evolve this difference in an environment of chondritic Rb/Sr is 11 million years (4 million years for solar). Material which makes up these rocks was therefore physically separated from the protosun at the same time as other planetary objects. This separation was accompanied by a large reduction in Rb/Sr. Thus the moon as a whole cannot have chondritic Rb/Sr unless a rapid development of a lunar crust occurred with enrichment of Sr relative to Rb in contrast to the earth. The rocks studied cannot be the source of basaltic achondrites, as all lunar $(^{87}Sr/^{86}Sr)_I$ exceed the $(^{87}Sr/^{86}Sr)$ of several achondrites. The high $^{87}Sr/^{86}Sr$ in tektites cannot be produced from any of the materials studied except by an elaborate process.

On the basis of soil analyses, lunar Rb/Sr is similar to that of the earth's upper mantle. For modern terrestrial $^{87}Sr/^{86}Sr \geq 0.702$, the value of $(^{87}Sr/^{86}Sr)_I^{44}$ means that the moon cannot have formed from the earth unless this took place prior to 4.3 $\times$ 10^9 years ago.

Rubidium-strontium analyses of soil give a model age of 4.5 $\times$ 10^9 years, assuming $(^{87}Sr/^{86}Sr)_I$ = BABI. Magnetic separates form a linear array which yields an "age" of 4.07 $\times$ 10^9 years. This is a mixing line whose lower intercept is fixed by Sr-rich feldspar; the resulting "isochron" does not represent a meaningful age. The soil is an aggregate of particles with different ages and complex histories, and, as shown by the glass fragment (5 mg) which falls far off the isochron, individual particles are not necessarily compatible with the model age. The 4.5 $\times$ 10^9 year model age of the soil (which includes fragments of 3.6 $\times$ 10^9 year rocks) is striking and should not be considered accidental. A reasonable interpretation of the data is that the soil represents a nearly valid sampling of lunar crust which was formed about 4.6 $\times$ 10^9 years ago. This is not surprising, since the soil is produced by an impact process which is capable of sampling and averaging large segments of the lunar crust. The younger rocks on the lunar surface were produced by subsequent melting in this ancient crust. The whole crust is considered to behave as a closed system without the injection of large volumes of highly differentiated material from depth within the moon after 4.6 $\times$ 10^9 years ago. A fortuitous coincidence with the age of meteorites and the earth is demanded if the 4.5 $\times$ 10^9 year model age of the soil was produced by Rb loss through volatilization. The general pattern of K, Rb, and Cs abundances does not suggest differential losses of these elements by volatilization.

Luny rock 1 (No. 85) is a fine-grained rock fragment (shocked) consisting of low-Ca pyroxene $[(Ca_{0.04}Mg_{0.65}Fe_{0.31})SiO_3]$, isotropic "plagioclase" $(An_{90}Ab_{10})$ and "K-feldspar," ilmenite, troilite, chlorofluorapatite, and whitlockite. It contains 0.56 percent K and 16.7 ppm Rb, has $(^{87}Sr/^{86}Sr)$ of 0.71399 $\pm$ 0.00011, and yields a model age (T_{BABI}) of 4.44 $\times$ 10^9 years. An assumed age of 3.65 $\times$ 10^9 years would require $(^{87}Sr/^{86}Sr)_I$ = 0.70174. This peculiar rock may represent a non-mare sample.

The ^{40}K-^{40}Ar ages are for No. 17: whole rock, 2.45 $\times$ 10^9 years; the ^4He age, 2.5 $\times$ 10^9 years [U-Th from (*2*)]; plagioclase, 3.2 $\times$ 10^9 years. For No. 44: whole rock, 3.45 $\times$ 10^9 years; pyroxene, 3.6 $\times$ 10^9 years. For No. 69: whole rock, 2.9 $\times$ 10^9 years. For soil: feldspar glass, 4.9 $\pm$ 0.4 $\times$ 10^9 years; brown glass, 1.6 $\times$ 10^9 years.

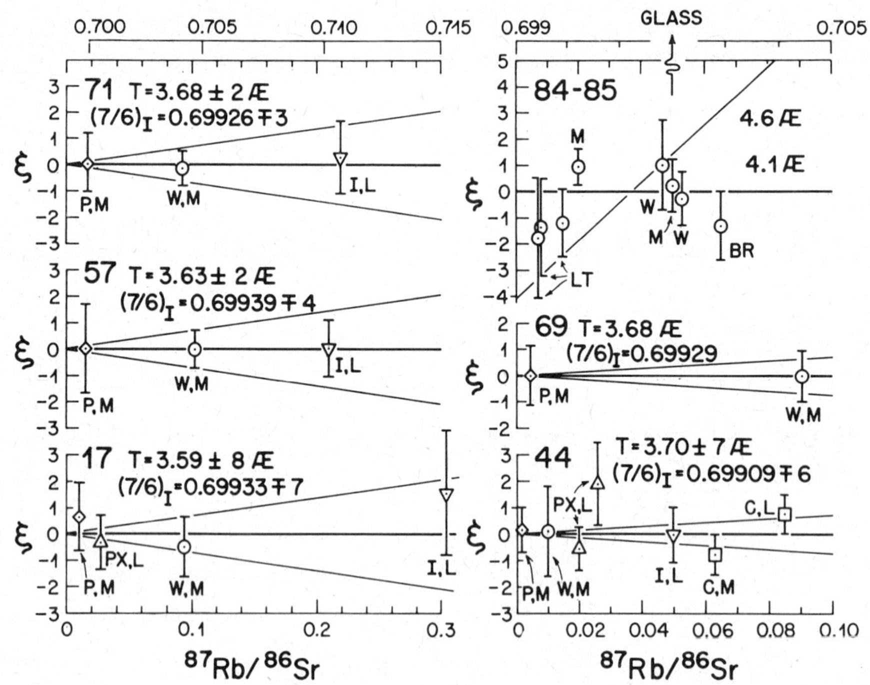

Fig. 1. Displacement of measured samples (m) from best fit (BF) (York) isochron

$$\left(\xi = \frac{(^{87}Sr/^{86}Sr)_m - (^{87}Sr/^{86}Sr)_{BF}}{(^{87}Sr/^{86}Sr)_{BF}} \times 10^4\right)$$

vs. $(^{87}Rb/^{86}Sr)_m$. Errors on data points are 2σ (mean). Ages T ($\lambda = 1.39 \times 10^{-11}$ yr^{-1}) and initial $(^{87}Sr/^{86}Sr)_I$ are given (errors correspond to last digit). Lines on either side of $\xi = 0$ correspond to $\pm$ 50 million years. Symbols are whole rock (W), plagioclase (P), pyroxene (PX), cristobalite (C), and ilmenite (I). For dust (Nos. 84, 85) $\xi = 0$ corresponds to $T = 4.1 \pm 0.1$ AE and $^{87}Sr/^{86}Sr)_I = 0.69927 \mp 11$ calculated from lithic fragments (LT), dust (W,M), and No. 59 breccia (BR). A glass fragment lies off scale. Mechanical separate (M), heavy liquids (L).

Comparison of mineral and rock data demonstrates gas loss. The plagioclase for No. 17 yields a much higher age than the total rock, indicating Ar loss from the fine-grained, K-rich, interstitial phases. The concordance of He and Ar ages must be fortuitous. The maximum age is equal to the Rb-Sr age, and the general pattern is compatible with the Sr results. Assuming no inheritance of Ar, the age of the brown glass fragment shows that the soil contains particles produced by events of intermediate age (~ 10^9 years).

A widespread 3.6 $\times$ 10^9-year event due either to impact or to internal melting would cause major crustal differentiation and melting of a significant part

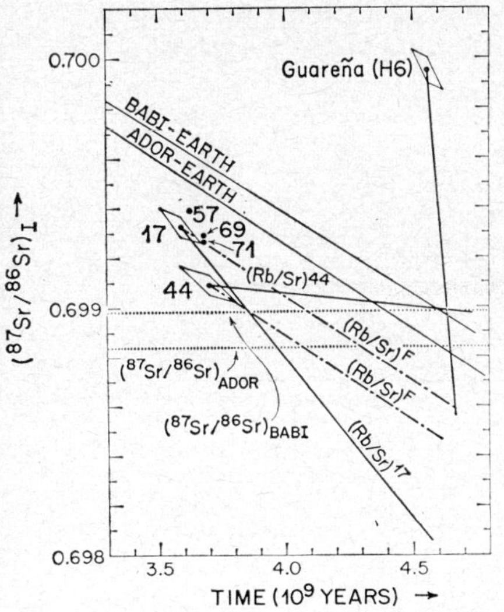

Fig. 2. Age-($^{87}Sr/^{86}Sr$)$_I$ anticorrelation plot. Distinct error envelopes of Nos. 17 and 44 correspond to 2σ errors. Similar error envelopes for Nos. 57, 69, and 71 are omitted for clarity. The Guareña chondrite is shown for comparison. BABI-EARTH and ADOR-EARTH lines represent evolution of terrestrial Sr for corresponding ($^{87}Sr/^{86}Sr$)$_I$. Extrapolation of Nos. 17 and 44 to times before 3.6 AE are shown assuming Rb/Sr is either (i) as measured in the whole rock or (ii) as for the fines (F).

of the moon. The high U and Th content (*2*) of surface rocks must have been derived from material of much lower U and Th if the moon is not extensively molten today (U < 10 ppb). This requires that mare materials be differentiated from a layer at least 100 km thick. If the 3.6 × 10^9-year event was due to internal melting, the moon should be differentiated to a depth greater than 100 km, it should be partially molten today, and volcanic events should occur at the present time. If the event was the result of widespread impacts, the earth was also subjected to severe bombardment at that time. This may explain the absence of a geologic record before 3.6 × 10^9 years ago.

Microprobe analyses of phosphates showed fluorapatite in No. 44 and in lithic fragment 85-1-11 (F ~ 3.4 percent; Y and rare earths (REE) ~ 2 percent). Sufficient F and Cl are present so that no OH is required. Whitlockite in No. 69 contains no REE. Luny rock 1 (No. 85) contains chlorofluorapatite (F ~ 2.4 percent; Cl ~ 1.2 percent) with no REE and co-existing whitlockite with ~ 9 percent REE.

Glasses (balls and crusts) commonly contain Fe-Ni spherules with 10 percent Ni, troilite, and schreibersite. By contrast Ni-free Fe blebs within troilite in crystalline rocks could not have been in equilibrium with a silicate melt containing significant Ni. The Ni-free metal must have crystallized from a melt already depleted in Ni or have been produced by secondary reduction of Fe in Ni-free phases already separated from the melt. The Fe-Ni spherules in glasses, as well as Fe-Ni fragments in soil, are evidence of meteorite impact, since the crystalline rocks contain almost no Ni.

Chemical abundance measurements (Fig. 3) show the alkaline earths, Gd, and Li strongly enhanced relative to chondrites. Soil, breccia, and fine-grained rocks are not highly depleted in K-Rb-Cs, but the coarse-grained rock has K-Rb-Cs abundances similar to those found in Ca-rich achondrites. Concentrations of alkalis (except Na) and Ba are highly correlated. Strontium has a definite anticorrelation with the alkalis, whereas Ca is almost constant. Gross abundance features such as the alkaline earth, Gd, and Li enhancement probably reflect the composition of the source material, whereas the "internal correlations" (for example, Ba vs. alkalis) are a result of magmatic processes. It is not possible to form the soil precisely from a mixture of the rocks analyzed. In general, a component with high Rb, Cs, and Ba relative to K is required. Except for Na and Ca, the chemical composition of the breccia (No. 59) can be described as a mixture of local soil and fine-grained rocks. No evidence for fractionation of volatile alkalis can be seen in comparing soil and rocks.

Integrated thermal neutron fluxes were determined by measuring the depletion of ^{157}Gd and ^{155}Gd and the corresponding enrichment of ^{158}Gd and ^{156}Gd (Fig. 4). Taking a cosmic ray exposure age of 5.5 × 10^8 years for No. 17 and 7 × 10^8 years for soil, we obtain fluxes of 1.1 and 0.4 n/cm^2 sec. The neutron exposure of the top and bottom of core 2 is the same. If the flux at the core bottom is twice that at the top, the soil has been mixed to 13 cm at least once in the last 10^8 years. Assuming the soil to be well mixed over 4.5 × 10^9 years, the total depth of this zone must be less than about 25 times the thickness of the neutron production zone (~ 2 to 4 m) using an average flux of 1.5 n/cm^2 sec.

Vanadium isotopic composition measurements on whole rock samples of Nos. 17 and 84 showed that ($^{50}V/^{51}V$) has the terrestrial value to within 3 percent and 2 percent. Vanadium appears to be concentrated in ilmenite

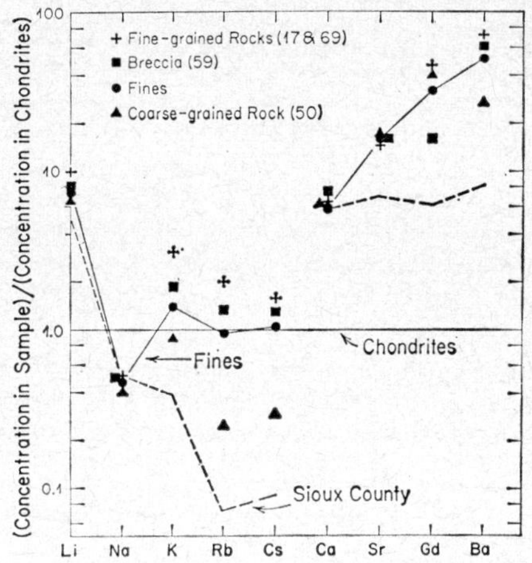

Fig. 3. Chemical abundances in lunar samples relative to chondrites. Assumed chondritic abundances in ppm: Li = 1.72, Na = 6600, K = 786, Rb = 2.81, Cs = 0.099, Ca = 12,200, Sr = 9.96, Ba = 3.37, Gd = 0.40. Typical Ca-rich achondrite (Sioux County) shown for reference.

(~ 100 ppm). A limit of ~ 4 × 10^{17} particles/cm^2 ($E \gtrsim 50$ Mev) for a model of uniform differential exposure of terrestrial, meteoritic, and lunar material can be set.

Cosmic ray exposure ages from ^{21}Ne, ^{36}Ar, and ^{126}Xe for each sample are essentially self-consistent, giving 550, 100, and 40 million years for Nos. 17, 44, and 69, respectively. Production rates are estimated from meteorite data, making allowance for chemical composition and geometry. Solar flare contributions are ignored. These ages may be in error because of erosion, variable shielding, and greater shielding than in meteorites.

Xe and Kr concentrations [ ] are

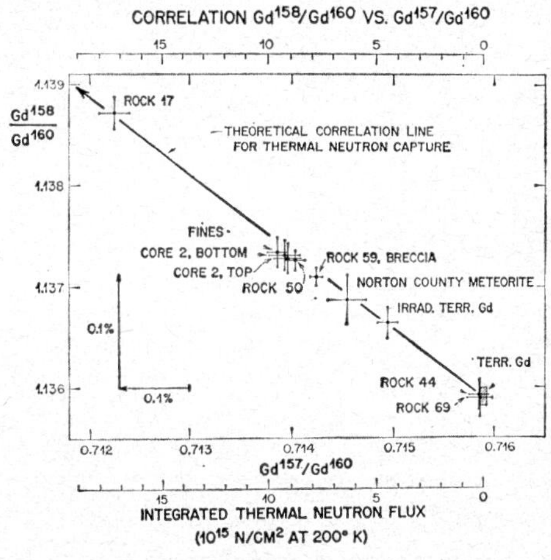

Fig. 4. Isotopic shifts of the Gd isotopic ratios due to thermal neutron capture. Mass fractionation is corrected by normalizing ratios to $^{156}Gd/^{160}Gd$ = 0.9361. Errors correspond to 2σ.

given in units of 10^{-12} cm^3 (STP)/g. Cosmogenic (c) Xe and Kr spectra were derived for No. 17, giving

$$\text{Xe } 124/126/128/129/130/131/132/134/136 = 0.55/\equiv 1/1.58/1.67/1.10/7.90/0.91/0.099 \equiv 0;$$
$$^{126}Xe_c = [170];$$
$$^{136}Xe_{tr} = [30] \text{ and}$$
$$\text{Kr } 80/82/83/84/86 = 0.484/0.743/\equiv 1/0.384/\equiv 0;$$
$$^{83}Kr_c = [1400]$$
$$^{86}Kr_{tr} = [70].$$

We assumed ^{136}Xe and ^{86}Kr represented trapped (tr) gas of atmospheric composition after correcting for ^{238}U spontaneous fission ($^{136}Xe_f/^{136}Xe_{tr} \approx 0.1$). For No. 69 ($^{126}Xe_c = [14]$; $^{136}Xe_{tr} = [13]$), cosmogenic spectra derived with these assumptions agree well with those for No. 17. Using soil Xe for Xe_{tr} leaves the Xe_c spectra for No. 17 relatively unaffected but destroys good agreement between Nos. 69 and 17, suggesting that either large amounts of atmospheric Xe contamination are present or that Xe_{tr} in lunar rocks has atmospheric composition.

Xe_c and Kr_c spectra for No. 17 are compatible with meteoritic spallation spectra except at masses 129 and 131. The 129 and 131 excesses appear cosmogenic in that, for chemically similar Nos. 17 and 69, $(^{129}Xe/^{126}Xe)_c$ and $(^{131}Xe/^{126}Xe)_c$ are the same to 10 percent, although the exposure ages differ by a factor of 14. The ratio $(^{131}Xe/^{126}Xe)_c = 3.51$ in No. 44 is reasonable for meteoritic spallation, but the $(^{129}Xe/^{126}Xe)_c = 1.47$ is still anomalous and similar to that in No. 17. Measured neutron exposures and estimated target concentrations are too low to account for excesses in ^{131}Xe and ^{129}Xe by capture on Ba or Te. Solar proton reactions (15 to 30 Mev) on Ba would predominantly produce ^{129}Xe and ^{131}Xe (from ^{130}Ba and ^{132}Ba); however, excess $^{83,84}Kr$ from similar reactions on ^{84}Sr and ^{85}Rb are not observed. Excess ^{129}Xe may indicate that meteoritic $(^{129}Xe/^{126}Xe)_c$ is too low. Excess lunar ^{129}Xe is not attributable to ^{129}I.

A fraction ($< 37\ \mu m$) of soil (No. 85) was used for rare gas analysis to enhance solar-wind Xe relative to spallation. Measured spectra are

$$\text{Xe } 124/126/128/129/130/131/132/134/136 = 0.590/0.658/8.68/105.0/16.68/83.08/\equiv 100/37.13/30.24$$

with

$$^{132}Xe = [54{,}000]$$

and

$$\text{Kr } 80/82/83/84/86 = 12.87/65.00/65.51/323.6/\equiv 100$$
$$^{86}Kr = [80{,}000].$$

Abundances of heavier Xe isotopes are distinct from both the average for carbonaceous chondrites (AVCC) and atmosphere. Enrichments in lighter Xe isotopes are presumably due to spallation. The $^{130-136}Xe$ data can be accounted for by a mass fractionation of 1.5 percent per mass unit relative to AVCC. However, subtraction of this fractionated AVCC component from $^{124-8}Xe$ yields residuals

$$\text{Xe } 124/126/128 = 0.37 \pm 0.04/1.0/0.5 \pm 0.2$$

which are not compatible with known Xe spallation spectra. The quoted errors arise from uncertainties in AVCC. A similar calculation relative to atmosphere gives a more unreasonable "spallation" spectrum. The Kr spectrum can be accounted for by a mixture of AVCC Kr plus spallation. It is not possible from our data to establish a genetic relationship between solar-wind Kr and Xe and AVCC (or atmosphere).

The average depth to which the soil is well mixed is about 6 m, assuming that the integrated ^{126}Xe production occurred over 3.6×10^9 years.

The Lunatic Asylum of the Charles Arms Laboratory of Geological Sciences, California Institute of Technology, Pasadena 91109*

* The inmates are A. L. ALBEE, D. S. BURNETT, A. A. CHODOS, O. J. EUGSTER, J. C. HUNEKE, D. A. PAPANASTASSIOU, F. A. PODOSEK, G. PRICE RUSS II, H. G. SANZ, F. TERA, G. J. WASSERBURG.

References and Notes

1. BABI is the initial $(^{87}Sr/^{86}Sr)_I$ in basaltic achondrites; ADOR is the more primitive value found in Angra dos Reis. D. A. Papanastassiou and G. J. Wasserburg, *Earth Planet. Sci. Lett.* **5**, 361 (1969); D. A. Papanastassiou, thesis, California Inst. of Technology (1970).
2. Preliminary Examination of Lunar Samples from Apollo 11, *Science* **165**, 1211 (1969).
3. We thank our fellow inmates Pai Young and Uwe Derksen for the sophisticated skill and knowledge they have brought to this study, without which this work would not have been accomplished. We also acknowledge the unstinting efforts of J. Brown, H. L. Derksen, L. Ray, and T. Wen. This work was supported by NASA contract 64049.

4 January 1970

Argon-40/Argon-39 Dating of Lunar Rock Samples

Abstract. *Seven crystalline rock samples returned by Apollo 11 have been analyzed in detail by means of the ^{40}Ar-^{39}Ar dating technique. The extent of radiogenic argon loss in these samples ranges from 7 percent to $\geqslant$ 48 percent. Potassium-argon ages, corrected for the effects of this loss, cluster relatively closely around the value of 3.7×10^9 years. Most of the vulcanism associated with the formation of the Mare Tranquillitatis presumably occurred around 3.7×10^9 years ago. A major cause of the escape of gas from lunar rock is probably the impact event which ejected the rock from its place of origin to its place of discovery. Upper limits for the times at which these impact events occurred have been estimated.*

The determination of accurate and meaningful ages of rocks and soil from many parts of the lunar surface will be of prime importance in unraveling the sequence of events which have occurred there since the moon formed as an independent object in space. In this paper I report on the measurement and interpretation of potassium-argon ages of crystalline rocks returned by Apollo 11.

In its simplest form the determination of a potassium-argon age involves the measurement of the total amount of potassium and the total amount of radiogenic ^{40}Ar in the sample. The assumptions are made that the rock was free of argon when formed and that it has quantitatively retained ^{40}Ar, from the decay of ^{40}K, since that time. The assumption of quantitative argon retention is particularly *inappropriate* for the lunar rocks. The rocks returned to earth have been picked up loose from the surface of the moon, presumably at some distance from their place of origin. The presence of shock effects in some, if not all, of the crystalline rocks indicates that high-energy events, possibly meteorite impacts, may have transported the rocks from their place of origin to their place of discovery and it is very probable that argon loss occurred at the time of transfer. In an attempt to estimate the extent of gas loss and to apply a suitable correction to the potassium-argon age, an activation technique, the ^{40}Ar-^{39}Ar method, has been applied to seven of the crystalline lunar rocks.

The ^{40}Ar-^{39}Ar method has been described more fully elsewhere (*1, 2*). The technique consists of converting a measured fraction of ^{39}K in the rock to ^{39}Ar by neutron activation, and then heating the sample in stages to release this ^{39}Ar, together with radiogenic ^{40}Ar, by thermal diffusion. The argon is subsequently analyzed in a mass spectrometer. In a sample which has quantita-

Table 1. Argon-40/argon-39 ages of lunar samples. Ages calculated on the basis of $\lambda\beta = 4.72 \times 10^{-10}$ year^{-1}; $\lambda\varepsilon = 0.584 \times 10^{-10}$ year^{-1}; $^{40}K/K = 0.0119$ atom percent; J values are 0.341 ± 0.007 for samples 10003,43 to 10022,46 and 0.319 ± 0.007 for the remainder. The two groups of samples were irradiated at the same time in separate Al canisters. An explanation of the significance of the "errors" in the ratios is given in the text. Quoted errors in ages include also the uncertainties in the J values which arise in part from nonuniformity of the measured neutron flux. Absolute uncertainties in the monitor age (~2 percent) have not been included as all samples are affected equally.

Sample No.	Fractional ^{40}Ar loss	Total argon		High-temperature argon		Low-temperature argon	
		$^{40}Ar/^{39}Ar$	"Age" × 10^9 years	$^{40}Ar/^{39}Ar$	"Age" × 10^9 years	$^{40}Ar/^{39}Ar$	"Age" × 10^9 years
10003,43	0.10	185.1 ± 1.4	3.76 ± 0.05	206.2 ± 6.8	3.93 ± 0.07	< 21	< 1.0
10017,49	⩾ 0.48	69.8 ± 0.5	2.30 ± 0.05	132.9 ± 3.0	⩾ 3.23 ± 0.06	< 4.2	< 0.25
10022,46	0.19	135.5 ± 0.7	3.26 ± 0.05	166.3 ± 3.8	3.59 ± 0.06	< 14	< 0.75
10024,26	⩾ 0.45	91.5 ± 0.5	2.58 ± 0.05	166.2 ± 1.4	⩾ 3.48 ± 0.05	< 3.7	< 0.2
10044,40	0.07	182.0 ± 0.9	3.62 ± 0.05	195.6 ± 2.5	3.74 ± 0.05	< 16	< 0.8
10062,30	0.22	160.3 ± 1.6	3.42 ± 0.05	206.5 ± 3.7	3.83 ± 0.06	< 22	< 1.0
10072,45	0.14	148.2 ± 0.8	3.30 ± 0.05	172.5 ± 2.6	3.53 ± 0.05	< 12	< 0.6

tively retained argon the two isotopes are expected to reside in fixed proportions in equivalent crystal sites since they have both been produced from potassium. Since the two isotopes have nearly identical diffusion coefficients, they will be released in proportionate amounts during a thermal diffusion experiment and the (constant) $^{40}Ar/^{39}Ar$ ratio will be related in a simple way to the potassium-argon age of the rock.

In contrast, a rock which has lost some of its argon at an intermediate stage of its history will contain easily outgassed crystal sites which are low in ^{40}Ar relative to ^{40}K. These sites may be, for example, minerals of high diffusion coefficient, small mineral grains, surface regions of large mineral grains, or regions surrounding crystal defects. When such a sample is analyzed by the ^{40}Ar-^{39}Ar method the gas released in the early stages of the diffusion experiment is expected to come from the more easily outgassed sites and therefore to be low in ^{40}Ar relative to ^{39}Ar. As the experiment proceeds, the more retentive sites will release their argon and the $^{40}Ar/^{39}Ar$ ratio will increase. The $^{40}Ar/^{39}Ar$ ratio of the gas released at high temperatures may be converted to an age which will be closer, though not necessarily equal, to the "true" age of the rock, whereas the initial $^{40}Ar/^{39}Ar$ ratio may, in suitable circumstances, indicate the most recent time at which argon loss occurred.

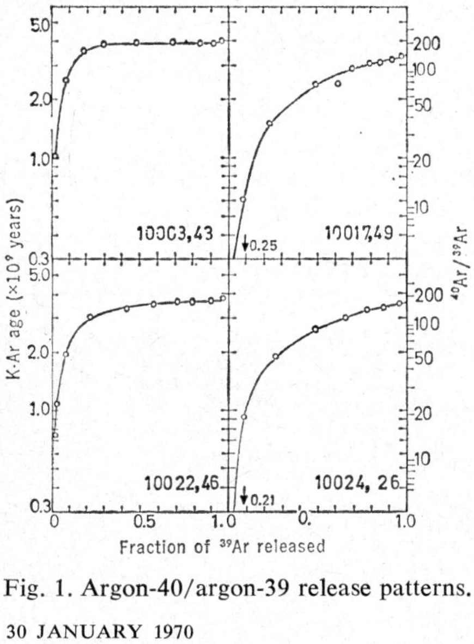

Fig. 1. Argon-40/argon-39 release patterns.

I have applied the ^{40}Ar-^{39}Ar technique to seven lunar rock samples; each sample weighed approximately 100 mg. The samples were broken into small chips 1 to 3 mm in diameter for analysis. They came into contact with the atmosphere during this operation, and also briefly when being loaded into the gas extraction system, but during the neutron irradiation they were vacuum-encapsulated in quartz ampules to minimize contamination with atmospheric argon. The samples were irradiated for 2 days in the core of the Herald reactor at the Atomic Weapons Research Establishment (AWRE), Aldermaston, and received an integrated fast neutron flux of 6.0×10^{18} neutron cm^{-2}. The irradiation was monitored by the inclusion of two terrestrial hornblende samples of known age. After the irradiation, samples were heated for 1-hour periods at successively higher temperatures from 400°C up to the melting point and the gas released at each stage was analyzed.

Corrections for atmospheric argon released from the gas extraction system were applied by measuring system blanks before and after each series of extractions. In general, these corrections were small (~ 1 to 2 percent) except for the first and the last extraction for each sample. A small correction was made for trapped ^{40}Ar on the assumption that the $^{40}Ar/^{36}Ar$ ratio of the trapped gas is unity (*3*). It was also necessary to apply a correction for ^{39}Ar produced by the $^{42}Ca(n,\alpha)^{39}Ar$ reaction. This was very conveniently done for each extraction by measuring the amount of ^{37}Ar produced by the reaction $^{40}Ca(n,\alpha)^{37}Ar$. The release patterns of ^{37}Ar and ^{39}Ar are quite distinct, the ^{37}Ar not being released until the very highest temperatures. From a consideration of the $^{40}Ar/^{39}Ar$ and $^{37}Ar/^{39}Ar$ ratios in two ^{37}Ar-rich fractions of sample 10003,43, it is possible to show that the $^{37}Ar/^{39}Ar$ ratio of the calcium-derived argon is 1250. All ^{39}Ar abundances have therefore been corrected by subtracting 1/1250 of the ^{37}Ar abundance. In most fractions where significant radiogenic argon is released, this correction is also small.

One may convert the corrected $^{40}Ar/^{39}Ar$ ratio to a potassium-argon age t by using the expression

$$t = T \ln [1 + J\,(^{40}Ar/^{39}Ar)]$$

where

$$J = (1 + \exp(t_s/T)/(^{40}Ar/^{39}Ar)_s$$

T is the mean life of ^{40}K, and the subscript s refers to measurements on the monitor.

All the samples analyzed have been crystalline igneous rocks of either type A or type B (*3*). Breccias and soil

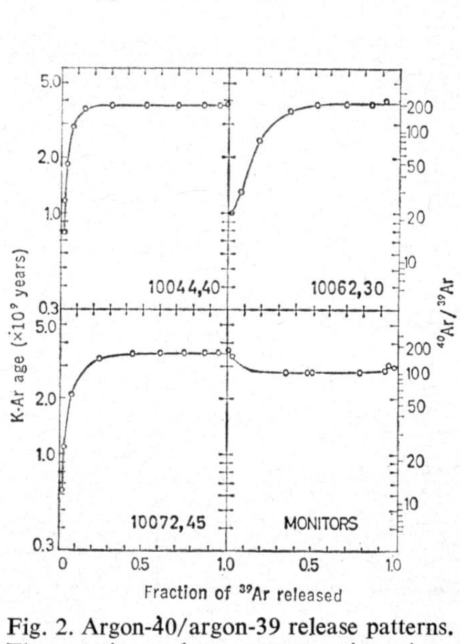

Fig. 2. Argon-40/argon-39 release patterns. The monitor release pattern given is a composite of three experiments and is characteristic of the monitor used (*2*).

samples have not so far been analyzed because of their complex history and their extremely high solar wind content. Figures 1 and 2 summarize the results obtained so far. Without exception, the release patterns are consistent with a recent episode of argon loss having occurred in relatively old rocks, after the fashion described above.

The evidence, in the low-temperature argon release, of regions within each of the samples that are low in radiogenic ^{40}Ar, relative to potassium, makes it possible in principle for one to estimate the time of the event which produced the loss of argon. In practice, with the present samples, it has been possible to place only upper limits on the time of outgassing because in most cases the gas loss was relatively small. These times are included in Table 1 and indicate that the events concerned, possibly the impact events which transported the rocks across the lunar surface, occurred relatively recently. It will be necessary to examine cosmic-ray exposure ages for details of this part of the rocks' history.

In all but two samples (10017,49 and 10024,26) the $^{40}Ar/^{39}Ar$ ratio attains a constant value in the last five or so high-temperature extractions. The average value of these high-temperature ratios is presented in Table 1 together with the root-mean-square deviation of the individual ratios. The ages calculated from these ratios are also included in the table and are a truer indication of the time of origin of the rocks than ages based on the total gas release. The ages based on total gas release are included for comparison and also included is an estimate of the extent of argon loss. This is simply the fractional difference between the high-temperature $^{40}Ar/^{39}Ar$ ratio and the total gas $^{40}Ar/^{39}Ar$ ratio.

Samples 10017,49 and 10024,26 do not show plateaus in the argon release pattern, and it is probable that the "high-temperature" age, calculated on the basis of the last significant argon extraction, represents only a lower limit on the true age of the rock. Likewise the fractional argon losses calculated (48 percent and 45 percent, respectively) can only be regarded as lower limits.

The effect of applying the correction for argon loss is impressive in that the spread of ages is considerably reduced. The ages based on total argon cover a range of 1.4×10^9 years, from 2.30×10^9 years to 3.76×10^9 years. If we take account of the fact that the "high-temperature" ages of samples 10017,49 and 10024,26 are now to be regarded only as lower limits, the significant spread in ages is reduced to 0.4×10^9 years (from 3.53×10^9 years to 3.93×10^9 years).

Once one has been able to largely eliminate the effects of argon loss, the problem of the interpretation of these ages is considerably eased, provided some estimate can be made of the area of the lunar surface represented by the samples. It is probable that most of the regolith is of comparatively local origin. This viewpoint is supported by the relative uniformity of composition of many of the Apollo 11 rocks (*3*), the existence of well-defined color boundaries associated with differences in underlying rock type (*4*), and the relationship of regolith thickness to local cratering density (*5*). Despite the fact that only a limited number of age determinations have been made, we can therefore now say with a fair degree of confidence that much of the vulcanism associated with this region of Mare Tranquillitatis occurred around 3.7×10^9 years ago.

It is impossible at the present time to reach any firm conclusions about the significance of the remaining spread in the measured ages. It is difficult to escape the conclusion that the spread is real and represents different rock-forming events which have occurred in the vicinity of the Apollo 11 site. However, before reading too much into these differences, it will be useful to seek corroboration by a comparison of the ^{40}Ar-^{39}Ar ages with either Rb-Sr or U-Pb ages.

GRENVILLE TURNER
Department of Physics, University of Sheffield, Sheffield, England

References and Notes

1. C. M. Merrihue and G. Turner, *J. Geophys. Res.* **71**, 2852 (1966); G. Turner, *Meteorite Research* (Reidel, Dordrecht, Holland, 1969), p. 407.
2. G. Turner, *Origin and Distribution of the Elements* (Pergamon, London, 1968), p. 387.
3. Lunar Sample Preliminary Examination Team, *Science* **165**, 1211 (1969).
4. E. A. Whitaker, *The Nature of the Lunar Surface* (Johns Hopkins Press, Baltimore, 1966), p. 79.
5. V. R. Oberbeck and W. L. Quaide, *Icarus* **9**, 446 (1968).
6. I thank D. Trotman of AWRE for performing the neutron irradiation and M. Ashworth for invaluable help in writing the computer programs for the reduction of mass spectrometer data. I thank the staff of the University of Sheffield Computing Laboratory for performing the calculations.

4 January 1970

Uranium-Thorium-Lead Isotope Relations in Lunar Materials

Abstract. *The lead isotopic compositions and uranium, thorium, and lead concentrations have been measured on six samples of material from the Sea of Tranquillity. The leads are moderately to very radiogenic; the initial lead concentrations are very low; the uranium and thorium levels are 0.26 to 0.88 and 0.87 to 3.35 parts per million, respectively. The Th/U ratios cluster about a 3.6 value. Apparent ages calculated for four rocks are 4.1 to 4.2 × 10^9 years. Dust and breccia yield apparent ages of 4.60 to 4.63 × 10^9 years. The uranium-lead ages are concordant, or nearly so, in all cases. The lunar surface is an ancient region with an extended record of events in the early history of the solar system. The discrepancy between the rock ages and dust ages poses a fundamental question about rock genesis on the moon.*

Elemental and isotopic abundances of uranium, thorium, and lead have been studied in six samples of lunar materials collected from the Sea of Tranquillity. The initial analytical results contain some fundamental information bearing upon the early history of the moon in matters of genesis, chemical differentiation, and timing. They raise some new questions concerning lunar evolution.

The samples include lunar dust (fines < 1 mm, sample 10084,35) a fine-grained dark breccia, apparently a sample of indurated lunar regolith (10060,-15), and chips of four rocks fairly representative of the textural and compositional ranges encountered in the Apollo 11 collections (10017,34; 10045,30; 10047,34; and 10072,39). The general character and lunar setting of these materials have been described previously (*1*).

The rocks are basaltic or gabbroic in composition and texture. Rocks 10072 and 10045 are fine-grained, vesicular to vuggy, and olivine-bearing. Rock 10017 has not been definitely identified as containing olivine and is a microgabbro in texture, with striking spherical vesicles. Rock 10047 is a medium-grained, cristobalite-bearing gabbro. The microbreccia, 10060, is very dark and fine-grained with a few scattered glass spherules suggesting its affinity to the dust sample.

The analytical techniques consisted of chemical decomposition with HF and $HClO_4$, extraction of Pb, U, and Th from the residue with concentrated HNO_3 and various procedures of coprecipitation and organic solvent extraction

Table 1. U-Th-Pb isotope data for some Apollo 11 materials.

Sample	Type	Pb Composition						Concentration (ppm)*				
		Observed			Corrected for blank†			U	Th	Th/U	Pb total	Initial lead limits‡
		206/204	207/204	208/204	206/204	207/204	208/204					
10084,35	D Fines	141.18	92.80	153.79	171.22	111.16	181.74	0.56_2	2.17_2	3.86	1.46_5	0.15_7
10060,15	C Breccia	77.77	51.97	82.31	83.18	55.29	86.53	0.60_6	1.46_6	2.42	1.78_5	0.40_0
10017,34	A Rock	206.9	99.8	225.6	257.0	122.1	290.0	0.73_5	2.72_4	3.71	1.65_2	0.13_2
10072,39	A Rock	242.2	115.8	266.7	563.3	259.0	594.1	0.88_4	3.3_5	3.79	1.58_6	0.06_1
10045,30	A Rock	83.20	45.46	102.0	117.81	61.24	135.90			Incomplete		
10047,34	B Rock	67.79	39.25	82.25	79.03	44.41	92.14	0.25_9	0.86_9	3.36	0.74_1	0.18_2
Averages (for 3 rocks)								0.62_6	2.31_3	3.62	1.32_6	

* General concentration error limits, ±3 percent; a few are larger. †Individual blanks determined in all cases where available; general blank for samples 10060,15 and 10045,30. ‡Maximum values, nonradiogenic lead.

for each element. The Ba $(NO_3)_2$ coprecipitation method for lead of Tatsumoto (*2*) was particularly useful.

Concentrations of the elements were determined by isotope dilution and mass spectrometry. Replicate sample analyses show reproducibility within 2 to 3 percent, with greatest uncertainty derived from application of blank corrections. Blank analyses have been run in parallel with all samples, but not all have been completed and specifically applied at this time. Lead blanks vary with the size of the sample analyzed and range from 0.04 to 0.12 μg per analysis. Lead in the sample ranged from 0.75 to 2.0 μg. The size and composition of the lead blanks make them significant corrections in treating the analytical results. Some general blank corrections have been applied which will be refined as the study continues. Uranium and thorium blanks are consistently at the level of 0.002 ± 0.001 μg and in no case have affected the observed data by as much as 1 percent.

Isotopic analyses were performed on a 12-inch (30.5-cm) solid-source mass spectrometer with digital data output. The precision of the analyses were generally better than 1 percent. Small corrections for observed mass-dependent fractionation (0.2 percent per mass unit) have been applied to the lead data. They do not influence the results significantly.

The analytical results are presented in Table 1. The observed lead isotopic compositions are shown along with composition ratios corrected for a common lead blank. The correction lead has values $^{206}Pb/^{204}Pb$, 18.2; $^{207}Pb/^{204}Pb$, 15.6; $^{208}Pb/^{204}Pb$, 38.0; values were measured directly. The uranium, thorium, and lead concentrations have all been corrected for blanks. The lead is reported as total lead and includes both radiogenic and nonradiogenic components. Limiting values for initial nonradiogenic lead, based upon model compositions compatible with the apparent ages of the samples and assuming no extralunar contamination, have been calculated. These must be considered as maximum values. The Th/U ratio has also been calculated on a weight ratio basis.

The observed lead isotopic compositions are strikingly radiogenic, from 75 to more than 95 percent. Comparable enrichments have not been reported for terrestrial basaltic materials. The enrichment reflects (i) very low concentrations of initial or original lead relative to uranium and thorium and (ii) very great ages for these materials.

Pure initial leads have not yet been extracted from the samples. To estimate the quantities, model lead compositions based upon earth and meteorite lead evolution data have been assumed. Because the blank corrections apply only to contamination in our laboratory procedures, previous contaminations are possible and uncorrected. The calculated original lead values are probably maximum limits. The values range from 0.06 to 0.40 ppm and are lower than concentrations found in primitive oceanic tholeites (*2*).

The uranium and thorium concentrations of the three rocks average about 0.63 ppm and 2.31 ppm, respectively; these values are similar to those found in oceanic alkali basalts and are enriched 5 to 10 times compared to oceanic tholeites (*3*). The Th/U ratios also resemble those found in alkali basalts, while contrasting with the oceanic tholeites.

The concentration averages for the limited number of rock samples closely approximate the concentration values for the mixed debris of the lunar regolith represented by the fines and the breccia. This suggests that most of the sources of these fine-grained fragmental aggregates may be comparatively local and related to the source of the rocks. Many other compositional similarities which support this possibility can also be noted in the preliminary data (*1*). Thus, this relatively limited sampling probably has significance for the U-Th-Pb characteristics of some of the Sea of Tranquillity region but may not represent the entire moon.

With this cautionary statement on the sampling, it is interesting to observe that the Th/U ratios found in the lunar ma-

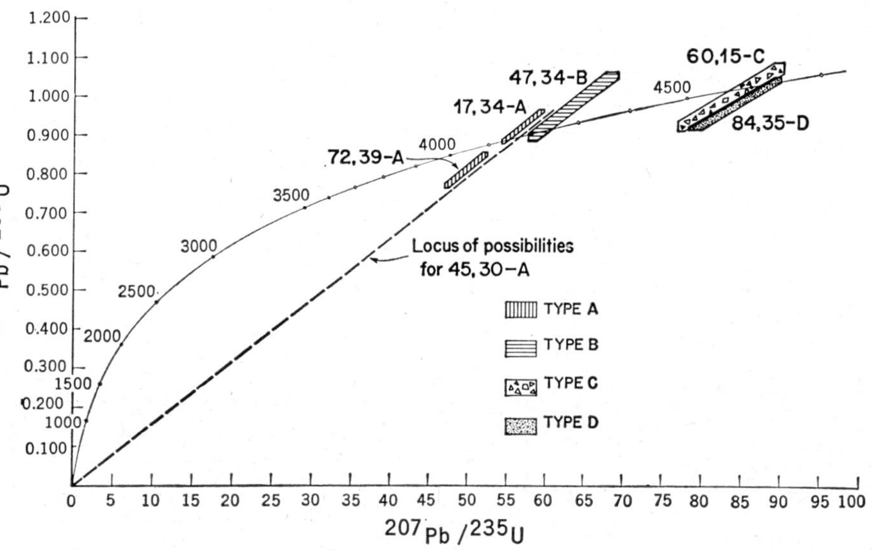

Fig. 1. Concordia diagram showing U-Pb isotope relationships in the analyzed lunar materials.

terials closely approximate those which are calculated for a general evolutionary model for the earth and meteorites (*4*). This is in marked contrast to the K/U ratios and Pb/U ratios observed. The Apollo 11 Preliminary Science Report (*1*) indicates K values ranging from 0.05 to 0.25 weight percent for the various samples of rocks and dust. This indicates K/U ratios of 2000 to 4000, values significantly lower than those found in meteorites or in the terrestrial crust.

An even more striking contrast with other solar system materials we know is found in the U/(initial lead) ratios. Expressed as observed μ-values ($^{238}U/^{204}Pb$) they are, for all samples studied, greater than 70 and probably greater than 200 to 700. Terrestrial basalts commonly are in the range of $\mu = 5$ to 10, and μ rarely exceeds 30 for terrestrial rocks of any kind.

One must conclude that K and Pb fractionation relative to U and Th has been extreme. Unidentified processes which discriminate the volatile K and Pb from the refractory U and Th to an extent unrecognized terrestrially appear to have been operative. These processes must be fundamental to lunar evolution.

The systematics of the U-Th-Pb isotope systems in the lunar materials are reported in Table 2. They are derived from the data in Table 1, utilizing a uranium isotopic composition of $^{238}U/^{235}U = 137.8$. Direct observations on the uranium isotopic composition have not been completed, but preliminary measurements support this or perhaps a very slightly lower value. Real differences would require some absolute but not relative changes in the values reported.

In order to calculate the isotope ratios in Table 2, it is necessary to select appropriate values for the nonradiogenic lead correction. These values can be either the initial (original) leads of the samples or a nonradiogenic contamination lead; it is probably some combination of both. Table 2 shows pairs of values based on both limits; the best values lie between the pairs. The following initial values have been inferred from earth and meteorite lead evolution models and from the apparent $^{207}Pb/^{206}Pb$ ages which are insensitive to these corrections:

	206/204	207/204	208/204
10084, 10060	9.6	10.4	29.7
10017, 10045 10047, 10072	10.5	12.2	31.0
Contamination lead	18.2	15.6	38.0

Apparent ages from the relevant isotope ratios are also shown in Table 2. The choice of nonradiogenic lead has a negligible effect on the $^{207}Pb/^{206}Pb$ apparent age and affects the other apparent ages less as the sample leads are more radiogenic.

Assignment of compounded errors to all of the values in Table 2 is complex. They are best shown in dimensions of uncertainty in Fig. 1, where the U-Pb isotope ratios are plotted on a "Concordia" diagram (*5*). The error figures incorporate analytical error assignments and the uncertainties due to the selection of nonradiogenic lead corrections. Although a complete set of values is not available for sample 10045,30, the dashed line represents the locus of possibilities for its position, based upon a good set of lead isotopic composition data.

Inspection of the U-Pb systematics in Table 2 and Fig. 1 reveals some important relations, which are supported by the $^{208}Pb/^{232}Th$ data.

1) The dust and breccia samples have 4630 and 4600 million year $^{207}Pb/^{206}Pb$ ages, have U-Pb ages in near agreement, and are apparently as old as any earth or meteorite materials measured.

2) The four individual rocks have $^{207}Pb/^{206}Pb$ ages in the range of 4130 to 4220 million years and appear to be at least 400 million years younger than the composite dust and breccia samples. They are older than any known terrestrial rocks. The three rocks for which complete data are available are nearly concordant, but it appears that the slight discordance of 10017,34 is real.

If the four rocks represent sources which have contributed to the dust also, and many data suggest this, then it may be inferred that the dust contains some materials with higher $^{207}Pb/^{206}Pb$ ratios than that measured from the composite sample. This would imply an even greater age than 4.63 billion years for some part of the lunar surface. It is difficult to reconcile the significant differences between the rocks and the regolith if material of the regolith is comprised principally of constituents isotopically similar to the rocks. Unfortunately, the patterns of discordance are too loosely defined by the present data to permit evaluation of several possibilities which might exist. It will be necessary to refine these data and add others before a complete understanding is possible.

The relative ages of the four rock samples appear to be partially resolvable. Rock 10047,34, a type B cristobalite-gabbro, appears distinctly older than the other three rocks, which appear to be close to the same age. It is possible, of course, that rock fragments 10017, 10045, and 10072 were torn from the same outcrop and are different parts of the same genetic mass. More petrologic and mineralogical information about these samples is necessary for evaluation of this possibility.

It is noteworthy that the apparent age of the regolith is similar to ages determined from many meteorite samples. The lunar surface has been a postulated source for some meteorites. The only stony meteorite which has yielded a significantly radiogenic lead is Nuevo Laredo, a basaltic achondrite, which also yields a 4.6×10^9 $^{207}Pb/^{206}Pb$ age (*6*, *7*).

The U-Th-Pb isotope relations among

Table 2. U-Th-Pb isotope data for some Apollo 11 materials.

Sample	Type	Isotope ratio*				Apparent age (10^6 years)			
		$^{207}Pb/^{206}Pb$	$^{206}Pb/^{238}U$	$^{207}Pb/^{235}U$	$^{208}Pb/^{232}Th$	$^{207}Pb/^{206}Pb$	$^{206}Pb/^{238}U$	$^{207}Pb/^{235}U$	$^{208}Pb/^{232}Th$
10084,35	D Fines	0.623_2	1.05_2	$90._4$	0.24_8	4630	4675	4645	4535
		0.627_6	$.99_6$	$85._8$	0.23_4	4635	4500	4600	4310
10060,15	C Breccia	0.609_5	1.05_1	$88._3$	0.32_5	4595	4675	4620	5760
		0.610_8	$.92_8$	$78._1$	0.17_8	4600	4270	4495	3365
10017,34	A Rock	0.445_9	0.95_9	$59._1$	0.26_3	4130	4375	4210	4790
		0.446	0.93	$57._3$	0.25_6	4130	4275	4180	4670
10072,39	A Rock	0.446_4	0.81_2	$49._9$	0.21_1	4130	3865	4045	3925
		0.446_4	0.80	$49._3$	0.20_8	4130	3825	4030	3880
10045,30	A Rock	0.457				4170			
		0.458_2				4170			
10047,34	B Rock	0.470	1.05_4	$68._3$	0.27_1	4210	4680	4360	4950
		0.473_6	0.93_6	$61._1$	0.24	4220	4300	4245	4410

*No initial leads have been determined; upper values corrected with model initial lead; lower values with contamination lead composition.

the rocks and the regolithic material indicate an extended history of rock-forming events for the moon which may be studied with profit by these techniques, particularly when coupled with systematic geologic documentation of sample sources. The question posed by the discrepancy between the apparent ages of the rock fragments and the associated lunar dust and breccia raises the possibility that fundamental and unexpected rock-forming processes exist.

LEON T. SILVER
California Institute of Technology, Pasadena 91109

References and Notes

1. "Apollo 11 Preliminary Science Report," *NASA SP 214* (Washington, D.C., 1969).
2. M. Tatsumoto, *Science* **153**, 1094 (1966).
3. M. Tatsumoto, C. E. Hedge, A. E. Engel, *Science* **150**, 886 (1965).
4. V. R. Murthy and C. C. Patterson, *J. Geophys. Res.* **67**, 116b (1962).
5. G. W. Wetherill, *Am. Geophys. Union Trans.* **37**, 320 (1956).
6. M. D. Duke, L. T. Silver, *Geochim. Cosmochim. Acta* **31**, 1661 (1967).
7. C. C. Patterson, *ibid.* **7**, 151 (1955).
8. Essential to this work was the dedicated laboratory participation and cheer of Geraldine Baenteli and Dolly Pearson. E. V. Nenow and C. Bauman provided important technical support. E. Brown wrestled untiringly with NASA, which supported the major part of this research under contract NAS 9-7963. California Institute of Technology Contribution No. 1707.

4 January 1970

Rubidium-Strontium, Uranium, and Thorium-Lead Dating of Lunar Material

Abstract. *Rubidium and strontium concentrations and strontium isotopic compositions have been measured on whole rock samples and density fractions of microgabbro. Density fractions on two rocks define isochrons of 3400 and 4500 million years with large uncertainties owing to low enrichment of radiogenic strontium. Lead from fine surface material is highly radiogenic. An age of 4750 million years has been calculated from the ratio of $^{207}Pb/^{206}Pb$. The concentrations of uranium, thorium, and lead isotopes are consistent with the evolution of lead in a 4700-million-year-old closed system characterized by the ratios of uranium to lead and of thorium to lead in this surface material.*

A time scale established by dating significant events in lunar history will permit one to relate lunar history to that of the earth and solar system and will place absolute values on relative chronologies established by photogeology. The first step in this program is to assess the usefulness of available lunar materials for dating purposes and to establish the approximate range of ages with which we will be dealing in the dating of lunar rocks.

Initial Rb-Sr dating has been carried out on samples of crystalline rock, in the belief that the interpretation of results on such rocks should be more straightforward than that on the microbreccia samples. The concentrations of K, Rb, and Sr found by mass spectrometric isotope dilution analysis of five such rocks, as well as in a sample of fine lunar surface material, are given at the top of Table 1. The data are plotted in Fig. 1. These rocks are characterized by low Rb/Sr ratios, resulting in small (~1 percent) enrichments in radiogenic strontium-87.

These values of $^{87}Sr/^{86}Sr$, not greatly enriched over the primordial values found from meteorite studies, indicate that, like the earth, the source of these lunar rocks acquired a Rb/Sr ratio much lower than the chondritic or solar value very early in the history of the solar system.

These small enrichments place constraints on the interpretation of the Rb-Sr data in terms of ages. Consequently, any such interpretations given herein must be regarded as preliminary. Improvements in our techniques for measuring Sr isotope ratios, such as conversion to digital recording, are currently being made, and more definitive results will soon be obtained by remeasurement of these same Sr samples.

Rb-Sr measurements define the time at which a Rb-Sr fractionation took place in the chemical system being studied. In the case of rock systems, this requires measurement of at least two samples with differing Rb-Sr ratios, which had the same Sr isotopic composition at the time of the event being dated. This is the case for cogenetic materials in which the Sr isotopic composition was initially homogeneous. For whole-rock dating, it is therefore preferable that independent evidence exists for the cogenesis of the rocks, in order that the entire burden of the argument for cogenesis does not rest upon the colinearity of the data when plotted on a Sr evolution diagram. This independent evidence is lacking in the case of these lunar rocks. Therefore, our initial effort has been directed to the study of the "internal ages" of single rock samples, obtaining varying Rb-Sr ratios by separation of density fractions of finely ground (< 88 μm) rock. These are not pure mineral separates, but have the advantage that an insignificant quantity of material is rejected in the process of separation.

The separation procedures were described in work on density fractions from enstatite and shocked hypersthene chondrites (*1*). The chemical procedures are identical to those employed in our published work on meteorites (*2*). As in this work, K, Rb, and Sr blanks are ~50, 0.1, and 1 ng, respectively, and they represent a negligible contribution to our results.

The most radiogenic fraction from two samples of microgabbro (Table 1) has a density of ~3.3 g/cm^3 and probably represents a pyroxene concentrate. The least radiogenic fraction has a density of less than 2.96 g/cm^3 and was found to consist almost entirely of plagioclase. The $^{87}Rb/^{86}Sr$ ratios range over a factor of about three, and with

Table 1. The Rb and Sr analytical results.

Sample	K (μg/g)	Rb (μg/g)	Sr (μg/g)	$^{87}Sr/^{86}Sr$ (atomic)
84,25 (fines)	1100	2.83	164.7	0.7017
72,38	2539	5.72	168.8	0.7043
24,24	2814	6.20	183.9	0.7044
17,41	2500	5.80	174.2	0.7044
50,30	665	0.788	188.9	0.7002
22,45		5.66	165.5	0.7047
17,41 fractions:				
$\rho<2.96$		9.52	548.3	0.7022
$2.96<\rho<3.32$ (coarse)	4090	9.19	257.3	0.7047
$3.32<\rho$ (coarse)	1545	4.01	77.9	0.7070
$3.15<\rho<3.25$ (fine)		7.79	134.0	0.7080
$3.25<\rho$ (fine)	843	2.30	45.8	0.7065
24,24 fractions:				
$\rho<2.96$		14.27	598.2	0.7026
$2.96<\rho<3.15$		19.42	520.0	0.7050
$3.15<\rho<3.32$		15.43	286.2	0.7081
$3.32<\rho$		2.82	47.7	0.7083

improvements in measurement of $^{87}Sr/^{86}Sr$ ratios should permit age measurements of reasonable quality.

The data from these density fractions seem to define linear arrays with slopes corresponding to ages of 3400 ± 600 and 4050 ± 700 million years, respectively ($\lambda = 1.39 \times 10^{-11}$ $year^{-1}$) (Fig. 2). The errors represent the extreme isochrons which can be drawn through the error brackets of the individual points. As stated previously, identification of these results in terms of ages is tentative, pending improvements in the measurement of the $^{87}Sr/^{86}Sr$ ratios. The error brackets represent mean deviations of sets of at least ten $^{87}Sr/^{86}Sr$ ratios, normalized to $^{86}Sr/^{88}Sr$ = 0.1194. A limited amount of work done with isotope ratio standards indicates that we can resolve differences in $^{87}Sr/^{86}Sr$ ratios of 0.0005. Absolute values could have somewhat larger systematic errors, pending more complete study of the linearity of our recording system in this range of 1 per mil (*3*).

Uranium, Th-Pb measurements have been carried out on fine lunar surface material (sample 84,25). Isotope dilution measurements give a concentration of uranium of 0.549 μg/g and a concentration of thorium of 2.08 μg/g. Blank corrections are less than 1 percent. Two isotope dilution lead measurements have been made, the first with ^{208}Pb tracer, and the second with ^{206}Pb tracer. This permits calculation of the abundance and concentration of all the Pb isotopes, after subtraction of the contribution from the tracers. In this way it was found that the lead in the lunar surface was highly radiogenic. Uncorrected for the blank, the ratio $^{206}Pb/^{204}Pb$ was 108 and the ratio $^{207}Pb/^{204}Pb$ was 76.5. After correction for a total lead blank of about 0.2 μg, the ratios become 231 and 161, respectively. These ratios may well be still higher, since the residual ^{204}Pb may represent terrestrial contamination beyond that introduced by the blank. Assuming the small amount of nonradiogenic lead remaining after subtraction of the blank to be of primordial composition (*4*), a $^{207}Pb/^{235}U$ age of 4720 ± 100 million years and a $^{206}Pb/^{238}U$ age of 4655 ± 100 million years can be calculated. The age calculated from the $^{207}Pb/^{206}Pb$ ratio is 4750 ± 50 million years. The data from the analysis with ^{206}Pb tracer were of poor quality. The $^{208}Pb/^{232}Th$ age calculated from these data is 4720 million years, but this value may be in error as much as 10 percent. Of course, these numbers cannot be interpreted as the age of any particular rock. The fine surface material is probably a sample of many different rocks, possibly from a large area of the lunar surface. However the $^{207}Pb/^{206}Pb$ age can be interpreted as the time of the major U-Pb fractionation which took place at least in the surficial regions of the moon, and also as an "age of the moon," analogous to Houterman's (*5*) age of the earth and subject to similar assumptions. The U-Pb ages indicate that Houterman's closed system model is characterized by a $^{238}U/^{204}Pb$ ratio essentially that of the present surface material.

These Pb results indicate that the lunar surface material has a very high ratio of $^{238}U/^{204}Pb$ (>210) in contrast with the terrestrial value of 9 for this ratio. Furthermore, the $^{207}Pb/^{206}Pb$ ratio indicates that this enrichment of uranium with respect to lead took place very early in the history of the solar system. If this very old and highly radiogenic lead proves to be generally characteristic of lunar material, this result argues against a terrestrial origin for the moon at any time significantly later than the time of

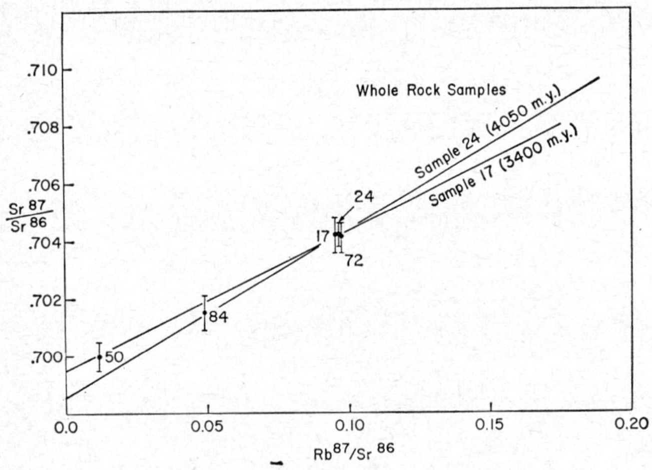

Fig. 1 (top left). Strontium evolution in whole rock samples. the lines drawn are the 3400- and 4050-million-year isochrons fitted to the data obtained from the density fractions. Fig. 2 (below). (A) Strontium evolution for density fractions from sample 17,41. (B) Strontium evolution for density fractions from sample 24,24.

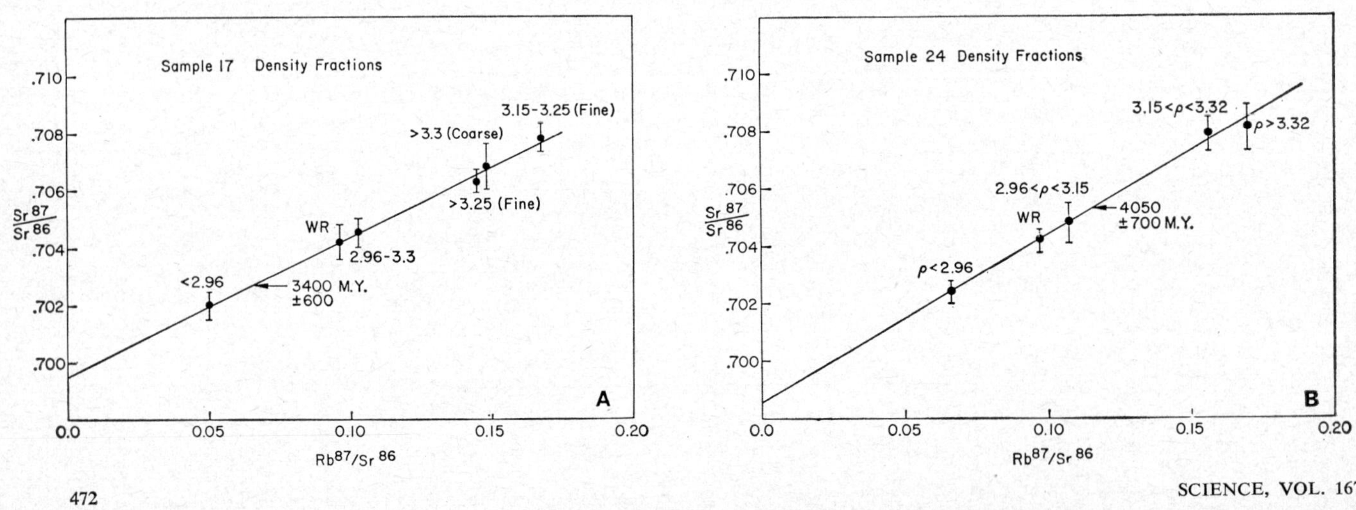

formation of the earth. Also, since Tilton has shown that tektites contain lead resembling closely the isotopic composition of modern terrestrial lead (6), a lunar origin of these objects can be ruled out unless some lead much more like that found on earth is found elsewhere on the moon.

K. GOPALAN, S. KAUSHAL*
C. LEE-HU, G. W. WETHERILL
Department of Planetary and Space Science, University of California, Los Angeles 90024

References and Notes

1. K. Gopalan and G. W. Wetherill, *J. Geophys. Res.*, in press.
2. ———, *ibid.* **73**, 7133 (1968); **74**, 4349 (1969); S. K. Kaushal and G. W. Wetherill, *ibid.*, p. 2717.
3. The ratio of ^{87}Sr to ^{86}Sr in the Pasamonte basaltic achondrite is found to be 0.6994.
4. V. Rama Murthy and C. C. Patterson, *J. Geophys. Res.* **67**, 1161 (1962).
5. F. G. Houtermans, *Nuovo Cimiento* **10**, 1623 (1953).
6. G. R. Tilton, *Geochim. Cosmochim. Acta* **14**, 323 (1958).
7. We thank A. D. Sharbaugh for technical assistance. Supported by NASA contract 9-8095.
* Deceased 2 December 1969.

4 January 1970

Rubidium-Strontium Relations in Tranquillity Base Samples

Abstract. *Preliminary total rock analyses disclosed a greatly different rubidium depletion between two groups of these igneous rocks, and ratios of strontium-87 to strontium-86 indicate that the rubidium depletion in these materials must have occurred during or shortly after the accretion of the terrestrial planets.*

Five rock samples weighing 0.82 g (No. 22.43), 1.27 g (No. 20.28), 1.14 g (No. 47.25), 2.36 g (No. 72.25), and 1.2 g (No. 58.36) and a 5.16-g sample of fines from the lunar soil (No. 84.39) were available for analysis. This progress report gives analyses on total rock samples only; work on separated minerals and the fined-grained soil material has not been completed. Rubidium, strontium, and the isotopic composition of Sr were determined by mass spectrometric techniques in separate whole pieces of rock ranging upward in size from a few tens of milligrams. These samples were not necessarily assumed to be representative of the rock chips selected by the Lunar Receiving Laboratory, but, by relating the isotopic data from each sample separately and using different whole fragments in repeated analyses, we hoped to obtain additional information on the homogeneity of the rocks, test the assumption of a lack of short range migration of components, and derive meaningful age evaluations and average compositions.

The sample fragments were gently broken in a boron carbide mortar, weighed, and digested in a Parr Industries Teflon-lined bomb at 120° to 140°C. Before acid digestion the samples were spiked with ^{87}Rb. Minimum amounts of reagents were used: 10 ml of HCl carrying the spike, 3 ml HF and 1 ml $HClO_4$ per 0.15 g of sample were the approximate amounts. The solution was washed into a Pt dish, covered with a Teflon lid, and heated with a hot plate below and a heat lamp above. After achieving total solution (no visible mineral or chemical residues) the solution was evaporated to get rid of excess HF and SiF_4 and diluted with 0.5*N* HCl to 50 ml in a volumetric flask, and aliquots were taken for Sr spiking and for Rb and Sr isotope dilution and Sr isotope ratio analysis. Element separations were achieved on cation exchange columns using Dowex 50, 8 percent cross-linked, 350 mesh.

The mass spectrometer used for this work consisted of a 12 inch radius, 60° sector analyzer section (Consolidated Electrodynamics Corporation) with detachable source, in-line valve, and cold finger in source housing using single Ta filament surface ionization, Faraday cup collector, Cary vibrating reed electrometer, and W. Shields design expanded scale system of recording residual voltages (after 70 to 90 percent precise nullification) on a modified Leeds-Northrop Speedomax G recorder. All scale settings on vibrating reed electrometer and expanded scale system were kept the same for all runs, including the calibration and standardization analyses, so that instrumental fixed bias was removed.

In analyses of Sr isotopic ratios the internal consistency of repeated ratio measurements in a single analysis gave a standard deviation of the mean of less than 0.01 percent. Allowing for possible contamination, we believe that the overall error will not exceed a standard deviation of 0.00015.

Mass spectrometer errors in analyses of isotope dilution are generally not more than 1 percent (standard deviation) for Sr and 2 percent for Rb, but in these cases of low Rb content the error due to contamination may outweigh the measurement error. Repeated analyses on separate small rock fragments show also the inhomogeneity of the rock, so that overall errors determined on this basis would reflect this added factor. Because the Rb, Sr, and Sr isotope analyses are performed on each piece separately, the inhomogeneity does not affect the age relationship. The uncertainty in the error at this stage of the investigation has resulted in an assignment of a larger error than is probably necessary.

Analyses of the California Institute of Technology (CIT) Rb and Sr normal solutions are given in Table 1, relative

Table 1. Preliminary contamination and spike analyses.

Reagent	Rubidium (10^{-9} g/ml)	Strontium (10^{-9} g/ml)
H_2O	0.003	
HCl	0.005	0.05
HNO_3	0.009	
HF	0.05	
$HClO_4$	0.15	
Total blank	1.5	
Comparison of normal solutions		
CIT Sr	14.61 $\times 10^{-9}$ mole $^{88}Sr/g$	
MIT Sr	14.70 $\times 10^{-9}$ mole $^{88}Sr/g$	
CIT Rb	0.440 $\times 10^{-9}$ mole $^{87}Rb/g$	
MIT Rb	0.449 $\times 10^{-9}$ mole $^{87}Rb/g$	
Fused silica	**Rb, 10^{-9} g/g**	**Sr, 10^{-9} g/g**
C-17-A	<9	<190
F-19-A	<3	<160

Table 2. Tranquillity Base samples : total rock fragments.

Sample	Rubidium (10^{-6} g/g)	Strontium (10^{-6} g/g)	$^{87}Rb/^{86}Sr$ Atomic ratio	$^{87}Sr/^{86}Sr$ Normalized*
22.43	6.0†	174†	0.099	0.7051‡
72.25	5.4	168	0.093	0.7051
20.28	0.77	152	0.0146	0.7000
58.36	0.62	173	0.0103	0.6998
47.24	1.18	189	0.0181	0.6996
Sargasso Sea water				0.7090

* Normalized to $^{86}Sr/^{88}Sr = 0.1194$. † Preliminary values only; errors not yet established by replicate analyses to ascertain inhomogeneity between fragments. ‡ Error in replicate ratio measurements in single analysis $\sigma_m = 0.0001$.

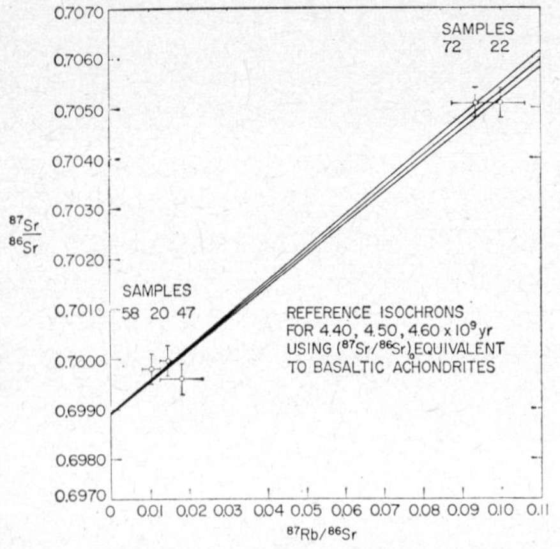

Fig. 1. Rubidium-strontium relations in the total rock fragment.

to the current Massachusetts Institute of Technology (MIT) normal solutions. All analyses of lunar materials are calculated relative to the CIT normal solutions for consistency.

The MIT value of 0.7090 for $^{87}Sr/^{86}Sr$ in Sargasso Sea water may be compared with 0.7091 reported by Papanastassiou and Wasserburg (*1*) and 0.7094 by Gopalan and Wetherill (*2*) for Pacific seawater. The MIT value for the MIT $SrCO_3$ standard is 0.7080.

Analyses are listed in Table 2 and plotted in Fig. 1. Reference isochrons for 4.4, 4.5, and 4.6 $\times$ 10^9 years are given in Fig. 1 for comparison, based on $(^{87}Sr/^{86}Sr)_0 = 0.6990$ found for basaltic achondrites (*1*) but adjusted by a correction of -0.0001 to allow for the difference in the value for standard seawater.

We believe that the most significant result of this study has been the disclosure that the Tranquillity Base sampling has yielded at least two distinctly different materials. In all the samples analyzed the Rb/Sr ratio is much smaller than that in chondritic meteorites and that estimated for the planetary system as a whole. Two of the five samples analyzed (Nos. 72 and 22) have ratios of Rb/Sr comparable to the estimated average earth value. The remaining three samples (20, 58, and 47) have ratios that are much lower, which causes values for the ratio of $^{87}Sr/^{86}Sr$ lower than any found in earth or meteoritic rock materials. The combination of $^{87}Sr/^{86}Sr$ ratios and Rb/Sr ratios indicates that the process which caused this relative loss of Rb must have occurred during or shortly after the accretion of the planetary system.

Thus the lunar material scattered on the surface at the landing site must have been derived from material that has been heterogeneous since close to the time of origin of the solar system. The material is volcanic in nature, and the heterogeneity has been preserved despite the volcanism. The enrichment of nonvolatile but hyperfusible components such as the rare and alkaline earths, and the relative depletion in volatile components such as the alkalies, would require a combination of both chemical enrichment by crystal fractionation and a physical loss of some elements in a hot gaseous environment by selective volatility.

Thus these analyses suggest that the moon is made up of an assortment of materials from different condensation regimes in the solar nebula, some of which materials were greatly depleted in alkalies, and that a subsequent volcanic or magmatic stage of crystal fractionation must have caused enrichments in some trace elements while still preserving the inherent discrepancy in alkalies. Whether this second stage occurred in the moon, or in earlier bodies which later made up the moon, is not indicated by these measurements.

P. M. HURLEY
W. H. PINSON, JR.
Department of Earth and Planetary Sciences, Massachusetts Institute of Technology, Cambridge 02139

References and Notes

1. D. A. Papanastassiou and G. J. Wasserburg, *Earth Planet. Sci. Lett.* **5**, 361 (1969).
2. K. Gopalan and G. W. Wetherill, personal communication.
3. Research supported by the U.S. Atomic Energy Commission as part of a project under contract AT(30-1)-1381.

5 January 1970

Rubidium-Strontium Chronology and Chemistry of Lunar Material

Abstract. *Igneous lunar rocks divide into two chemical types, probably representing two rock units. They form separate close groups on the isochron diagram; no total rock age is valid unless the rocks are cogenetic. Mineral isochrons prove that one type has an age of 3.80 ± 0.11 billion years, equal to the line joining total-rock groups, and the initial ratio of strontium-87 to strontium-86 of both types is close to 0.6994. Soil and breccias chemically resemble a mixture of the two igneous types, with a superimposed variation of mineral components, plus a small transferred component rich in nickel, copper, zinc, and possibly strontium-87.*

We obtained 1- to 5-g specimens of seven igneous rocks, two breccias, and the soil for a combined dating and chemical study of Apollo 11 material. Analyses for major and minor elements were made primarily to see which of these rocks might be cogenetic and, hence, which might be validly grouped together on a ^{87}Sr evolution diagram to define a total rock isochron. The Rb-Sr analyses on mineral separates ultimately proved essential for a conclusive result.

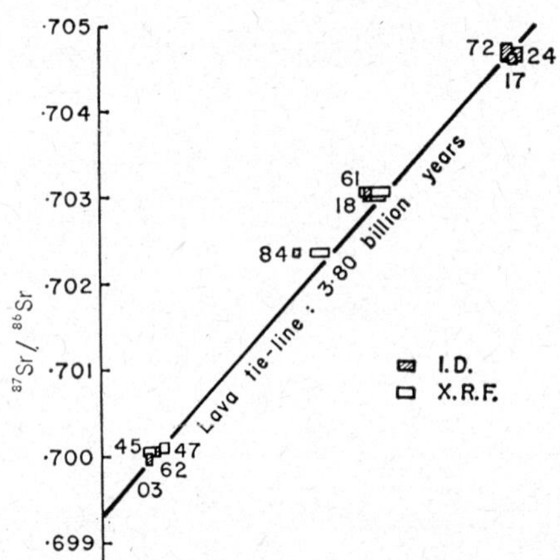

Fig. 1. Isochron diagram showing total-rock analyses of Apollo 11 samples. Sample numbers are identified in Table 1. I. D., isotope dilution analysis; X.R.F., x-ray flourescence analysis.

The trace elements (Table 1) were determined nondestructively by x-ray fluorescence spectrometry on 1 g of powdered sample using the method of direct measurement of absorption coefficient (*1*). The recovered powder was used for Na determinations by flame photometry, for the other major elements (Table 1) by x-ray measurements on a fused sample (*2*), and for the mass spectrometric measurements. Preliminary results for four of the samples have been reported (*3*). There is good agreement for the major and for many minor elements, although our data indicate lower Mn, Zr, Y, Cr, and Cu and higher Sr and Co.

The igneous rocks divide into two related chemical types, here termed group 1 and group 2. Group 1 (specimens 72, 17, and 24) is particularly uniform and is much higher than group 2 (specimens 03, 45, 47, and 62) in Rb, K, Ba, and Th; higher in Ti, Na, P, S, Zr, Y, and rare earth elements (REE); and lower in Al and Ca. Analyses by the preliminary examination team show that specimens 22 and 57 also belong to group 1, and specimens 20 and 58 belong to group 2. It seems likely that two separate rock units were sampled, including textural variants of both (types A and B as defined by the preliminary examination team). It is unlikely that the strong enrichment of K, Rb, Ba, and Th in group 1 is due to crystal fractionation of a single parent magma, as the two groups have only a restricted variation in major elements (*4*). Different parent magmas are required, generated either from chemically different source regions or from similar sources but with different degrees of partial melting. For both groups, our data show that the lighter REE resemble Ti, Zr, and Y in their extreme enrichment relative to terrestrial or cosmic abundances (*3*).

The breccias resemble the soil very closely in their chemistry (Table 1), which is consistent with their derivation from the soil (*3*). For most elements the breccia-soil composition is between groups 1 and 2 rocks, which indicates that both groups have contributed to their production and that the soil may be largely of local derivation. The exceptions are Fe and Ti, which are lower in the soil, and Al, which is higher. The differences for Fe and Ti may be explained by removing approximately 6 percent (by weight) of ilmenite from a more primitive soil, which would also account for the slightly higher abundances of Si, Mg, and Ca in the soil. Such removal might well occur by mechanical separation of the denser phase during the gardening processes associated with meteorite impact. However, the Al content remains anomalously high, so that a rock or mineral phase rich in this element must be added. The soil is also enriched in Ni, Cu, and Zn as minor transferred components.

Table 2 lists the mean values for Rb and Sr obtained by isotope-dilution, and mean $^{87}Sr/^{86}Sr$ ratios from both unspiked and spiked mass analyses. Our value of $^{87}Sr/^{86}Sr$ for the Eimer and Amend $SrCo_3$ standard is 0.70810. Our previous practice of determining Rb and Sr in separate aliquots of the dissolved sample was not followed for the lunar samples because representative aliquots in HCl solution could not be taken. The failure to obtain aliquots has been shown by current experiments to be associated with the high ilmenite content of the lunar samples; it is probably due to ionic adsorption onto a Ti colloid. Good reproducibility and agreement with the x-ray spectrometric results were obtained by adding a mixed ^{87}Rb and ^{84}Sr spike solution before dissolution.

The group 1 rocks plot at approximately one point in $^{87}Sr/^{86}Sr$, $^{87}Rb/^{86}Sr$ coordinates, whereas group 2 rocks plot in a narrow range averaging about 8 times lower in $^{87}Rb/^{86}Sr$ (Fig. 1). The tie line joining both groups need not have time significance, as each group had a different parent magma,

Table 1. Analyses of major and minor elements in ten lunar samples by x-ray fluoresence spectrometry. All values are expressed as ppm, T.E., trace elements as oxides.

	Igneous rocks									
	Group 1			Group 2				Breccias Type C		Soil Type D
	Type A	Type B		Type B						
	10072	10017	10024	10003	10045	10047	10062	10018	10061	10084
SiO_2	40.49	40.69	40.25	39.76	39.04	42.16	39.80	41.81	41.87	41.79
TiO_2	11.99	11.92	11.90	10.50	11.32	9.43	10.74	7.99	7.84	7.55
Al_2O_3	7.74	7.78	8.09	10.43	9.51	9.89	10.22	12.34	12.62	13.44
FeO	19.38	19.49	19.46	19.80	19.40	19.11	19.22	16.46	16.45	15.91
MnO	0.24	0.28	0.24	0.30	0.27	0.28	0.30	0.22	0.22	0.21
MgO	7.45	7.51	7.53	6.69	7.73	5.67	7.08	7.79	7.83	7.66
CaO	10.56	10.76	10.66	11.13	11.28	12.15	11.47	12.00	11.96	12.14
Na_2O	0.50	0.51	0.52	0.40	0.36	0.45	0.41	0.46	0.47	0.43
K_2O	0.29	0.30	0.30	0.06	0.05	0.11	0.08	0.17	0.18	0.14
P_2O_5	0.19	0.18	0.20	0.12	0.10	0.11	0.12	0.15	0.14	0.13
S	0.23	0.23	0.22	0.18	0.15	0.18	0.16	0.15	0.15	0.14
T.E.	0.52		0.56		0.39	0.30		0.45	0.43	0.43
Total	99.58	99.65	99.93	99.37	99.60	99.84	99.60	99.99	100.16	99.97
O = S	0.11	0.11	0.11	0.09	0.07	0.09	0.08	0.07	0.07	0.07
Total	99.47	99.54	99.82	99.28	99.53	99.75	99.52	99.92	100.09	99.90
					Trace elements					
Ba	300		310		10	88		175	128	134
Rb	5.61	5.72	5.96	0.62	0.62	1.11	0.89	3.60	3.68	2.96
Sr	168.4	163.7	178.3	160.9	137.7	209.2	187.8	158.5	161.6	164.8
Pb	3.4	7.1	3.0	6.6	3.3	3.2	6.3	4.2	3.1	2.9
Th	3.5	3.7	4.1	1.1	0.4	0.6	0.9	2.4	2.7	2.5
Zr	497	476	375	309	194	334	319	328	342	318
Nb	25		25		14	23		19	19	18
Y	162	159	168	112	73	134	103	106	108	99
La	43		39		16	20		24	23	21
Ce	94		108		32	48		67	37	58
Pr	16		12		6	13		11	15	10
Nd	49		55		17	36		29	20	33
V	22		37		98	13		51	34	36
Cr	2280		2610		2400	1220		1950	1940	1850
Co	34		32		23	16		35	23	34
Ni	< 20		< 20		< 20	< 20		200	170	230
Cu	22		16		20	16		32	25	33
Zn	34		14		14	13		54	37	37
Ga	4	4	5	3	3	4	3	4	5	4

Table 2. Isotope-dilution analyses for Rb, Sr, and $^{87}Sr/^{86}Sr$ of lunar rocks and mineral concentrates. t.r. denotes total-rock; plag., plagioclase; pyrox., pyroxene; and meso., mesostasis.

		Rb (ppm)	Sr (ppm)	$^{87}Rb/^{86}Sr$	$^{87}Sr/^{86}Sr$
		Group 1 lavas			
10024,20	t.r.	5.75	169.5	.0980	.70464
10072,30	t.r.	5.60	168.5	.0959	.70467
	plag.	3.87	531.0	.0210	.70063
	pyrox.	1.87	85.9	.0629	.70306
10017,68	t.r.	5.59	167.1	.0966	.70460
	meso. A	19.5	282.7	.1993	.71034
	meso. B	20.9	292.8	.2066	.71058
	plag.	0.92	499.9	.0053	.69970
		Group 2 lavas			
10003,20	t.r.	0.63	159.9	.0114	.69996
10062,36	t.r.	0.88	190.0	.0133	.70003
10045,20	t.r.	0.62*	137.7*	.0130*	.70005
	plag.	1.33	299.0	.0129	.69999
	pyrox.	0.65	86.7	.0220	.70036
10047,21	t.r.	1.11*	209.2*	.0153*	.70009
	plag.	0.53	574.6	.00268	.69949
		Breccias			
10061,25	t.r.	3.69	166.1	.0631	.70306
10018,20	t.r.	3.64	166.1	.0633	.70299
		Soil			
10084,44	t.r.	2.65	164.4	.0466	.70233

* X-ray fluorescence result.

which need not have the same age or the same primary $^{87}Sr/^{86}Sr$. Nor can precise total-rock isochrons be defined within each group, as the internal dispersion in Rb/Sr is too low. Analyses of separated minerals were thus essential. Nevertheless, assuming that both groups are precisely the same in age and initial $^{87}Sr/^{86}Sr$, we obtain a total-rock age of 3.80 ± 0.12 billion years and an initial $^{87}Sr/^{86}Sr$ of 0.69935 ± 0.00012. Using the x-ray spectrometric values alone for $^{87}Rb/^{86}Sr$, these parameters become 3.77 ± 0.20 billion years and 0.69934 ± 0.00017. In both cases, the scatter of points slightly exceeds their assigned experimental errors, and the lines were fitted by assigning greater *y* variance.

By heavy liquid and magnetic separation, mineral concentrates that ranged from 50 to 200 mg in weight were made from two samples of each group (Table 2); sample 47 of plagioclase was handpicked. Data for plagioclase, pyroxene, and total-rock from sample 72 are well correlated, which suggests an age of crystallization of 3.8 ± 0.3 billion years. The same age is given by the mineral data for sample 17: 3.80 ± 0.11 billion years with an initial $^{87}Sr/^{86}Sr$ of 0.6994 ± 0.0002. Most of the Rb in the rock is located in the glassy mesostasis (Table 2), and the big difference in $^{87}Rb/^{86}Sr$ between this phase and the plagioclase produces a precise result. The age of the group 2 rocks is comparatively poorly defined as there is much less spread in $^{87}Rb/^{86}Sr$ between the mineral concentrates. Both groups were included in the seven specimens dated by the preliminary examination team as 3.0 ± 0.7 billion years (*3*), which also implies a single value of age for both.

Equality in initial $^{87}Sr/^{86}Sr$ for both groups is consistent with their cogenesis. In general, it now seems likely that the lava tie line of Fig. 1 does have time significance and that both groups have the same age and the same source region. Combining all rock and mineral data on this model, we obtain 3.81 ± 0.07 billion years as our best present estimate for this age, with initial $^{87}Sr/^{86}Sr$ of 0.69936 ± 0.00008.

The data for the soil and breccias (Fig. 1) are displaced from the rock isochron toward lower $^{87}Rb/^{86}Sr$. This might be due to loss of a few percent of original Rb during the production of glass in the soil or to the presence of a transferred Sr component having a higher $^{87}Sr/^{86}Sr$.

WILLIAM COMPSTON
PIETER A. ARRIENS
MICHAEL J. VERNON

Department of Geophysics and Geochemistry, Australian National University, Canberra

BRUCE W. CHAPPELL

Department of Geology, Australian National University

References and Notes

1. K. Norrish and B. W. Chappell, *Physical Methods in Determinative Mineralogy* (Academic Press, New York, 1967), chap. 4.
2. K. Norrish and J. T. Hutton, *Geochim. Cosmochim. Acta* **33**, 431 (1969).
3. Lunar Sample Preliminary Examination Team, *Science* **165**, 1211 (1969).
4. E. J. Essene and A. E. Ringwood, personal communication.
5. We thank H. Berry and R. Rudowski for sample preparation and mineral separation; J. Pennington, Z. Wasik, D. Millar, Mrs. A. Brown, and Mrs. K. Phillips for assistance with the experimental work; and Professor A. E. Ringwood, Dr. E. J. Essene, Dr. A. J. R. White and N. G. Ware for helpful discussions.

4 January 1970

Rubidium-Strontium Age and Elemental and Isotopic Abundances of Some Trace Elements in Lunar Samples

Abstract. *Data on six lunar crystalline rocks give an apparent Rb-Sr isochron age of 4.42 ± 0.24 × 10^9 years (95 percent confidence limits) and initial $^{87}Sr/^{86}Sr$ ratio similar to that in a basaltic achondrite. Relationships between K, Rb, Sr, and Ba and depletion of Eu in these samples point to plagioclase separation from the melts that produced these rocks. The abundance of ^{157}Gd in the three lunar samples is similar to terrestrial abundance within <0.2 percent, thus setting a limit of $<6 \times 10^{15}$ neutrons per square centimeter for the integrated thermal neutron flux difference between lunar and terrestrial materials.*

This paper deals with the abundance and distribution of K, Rb, Sr, and Ba and the isotopic composition of strontium and the rare earth elements Sm, Eu, and Gd in Apollo 11 lunar samples. Six crystalline rocks of types A and B (*1*), a breccia, and the lunar soil sample were analyzed for K, Rb, Sr, and Ba abundances by stable isotope dilution methods. We measured the isotopic composition of Sr to a precision of 0.6 to 2.6 parts in 10^4 by the use of a digital data acquisition system in conjunction with programmed magnetic-field switching with a field-sensitive device in a 60° solid source mass spectrometer of 30.5-cm radius. Repeat analyses of two lunar samples and multiple analyses of the Eimer and Amend standard reagent (Table 1) show that a reproducibility of about 2 parts in 10^4 has been achieved. The abundances of the rare earth elements were determined by neutron activation analyses, and their isotopic composition was determined in a manner similar to that used in Sr-isotopic analyses.

The data for the abundances of K,

Table 1. Elemental abundances of K, Rb, Sr, and Ba and the $^{87}Sr/^{86}Sr$ ratios in Apollo 11 lunar samples.

Lunar sample	K (μg/g)	Rb (μg/g)	Sr (μg/g)	Ba (μg/g)	K/Ba	K/Sr	Ba/Sr	$^{87}Rb/^{86}Sr$ ($\times 10^2$)	$^{87}Sr/^{86}Sr$*	Δ† ($\times 10^4$)
10045,23(B)	424	0.823	109	69.4	6.1	3.9	0.63	2.17	0.69905	0.92
10044,21(A)	816	1.15	224	128	6.4	3.6	0.59	1.48	0.69876	0.56
10058,29(B)	853	1.23	201	124	6.9	4.1	0.59	1.72	0.69911	0.69
10084,38(D)	1020	2.88	162	162	6.3	6.3	1.00	5.13	0.70093	0.85
10065,19(C)	1406	3.69	168	200	7.0	8.4	1.25	6.39	0.70212	0.77
10017,11(B)	2207	5.33	169	290	7.6	13.1	1.67	9.18	0.70387	1.6
10022,30(A)	2289	5.57	163	277	8.3	14.0	1.67	9.90	0.70404	0.65
10069,22(A)	2295	5.60	165	277	8.3	13.9	1.67	9.84	0.70397	2.6
Nuevo Laredo										
	433	0.437	85.1			5.1		1.49	0.69882	1.8
Eimer and Amend reagent $SrCO_3$										
									0.70734	2.1

* Normalized to $^{86}Sr/^{88}Sr = 0.1194$.

$$\dagger\ \Delta = 2 \times \left[\frac{\sum_{i=1}^{n} (\bar{X}_i - m)^2}{n(n-1)} \right]^{1/2}$$

where n is the number of sets of ten scans each, and ($\bar{X}_i$-m) is the difference between the mean value of the ith set and the mean of all sets.

Rb, Sr, and Ba in eight lunar samples are shown in Table 1. There is excellent coherence between data for K, Ba, Sr, and, to a lesser extent, Rb. The ratios K/Ba, K/Sr, and Ba/Sr increase systematically with increase in K content. The K/Rb ratios range from 350 to 700; although no pronounced trends are seen, there is a suggestion of a decrease of the K/Rb ratio with increase in K content.

The trends shown by the various trace element ratios in the lunar samples could be due to mixing relationships between two end members. The two end members themselves (type A and type B rocks) can be products of fractionation processes involving liquid-solid equilibria from an original source material. Until data on the major mineral phases of lunar rocks are available, the choice between fractional crystallization and partial melting seems equivocal. However, we suggest that the pronounced increase in the K/Sr ratio with the increase in K content is suggestive of trends produced by fractional crystallization of plagioclase from the melts. Data on the distribution coefficients for K and Sr between liquid and basaltic plagioclase (*2*) show that plagioclase concentrates Sr relative to K and to a lesser extent relative to Ba, whereas the liquid tends to be enriched in K relative to plagioclase. Thus, a simple mechanism that can account for the K/Sr and Ba/Sr trends in lunar rocks is the separation of significant amounts of plagioclase from the melts which formed these rocks. Model calculations along these lines will be reported later. Plagioclase separation is also suggested by the pronounced Eu depletion observed in the lunar rocks as discussed below.

If plagioclase separation is the cause of the above trace element trends, the original chemical composition of the lunar melts would be richer in Ca and Al than was the case in the samples analyzed. The major elemental composition of such melts would approach that of gabbroic anorthosite as suggested by Olsen (*3*).

The Rb-Sr isotopic data for the lunar samples are shown in Table 1. The total range observed in the $^{87}Sr/^{86}Sr$ ratio is 0.8 percent. From these data, we have attempted to determine the solidification ages of the lunar rocks. The data have been grouped in different ways on the basis of assumptions regarding their genesis and internal relationships and checked for the requirements of isochronism, that is, a uniform initial $^{87}Sr/^{86}Sr$ ratio and chemical closure for the same time. If these assumptions are satisfied, we should obtain a linear relationship between the present values of $^{87}Sr/^{86}Sr$ and $^{87}Rb/^{86}Sr$ ratios. From the best-fit lines corresponding to various groupings, both the initial $^{87}Sr/^{86}Sr$ ratio and the age can be obtained.

The initial $^{87}Sr/^{86}Sr$ values, the slopes of the best-fit lines, and the ages thus obtained for various groups of lunar rocks are shown in Table 2. Although there is no a priori reason to group the data, this was deliberately done to search for any differences between the initial Sr-isotopic compositions and ages of the various types of rocks.

All the lunar samples studied here can

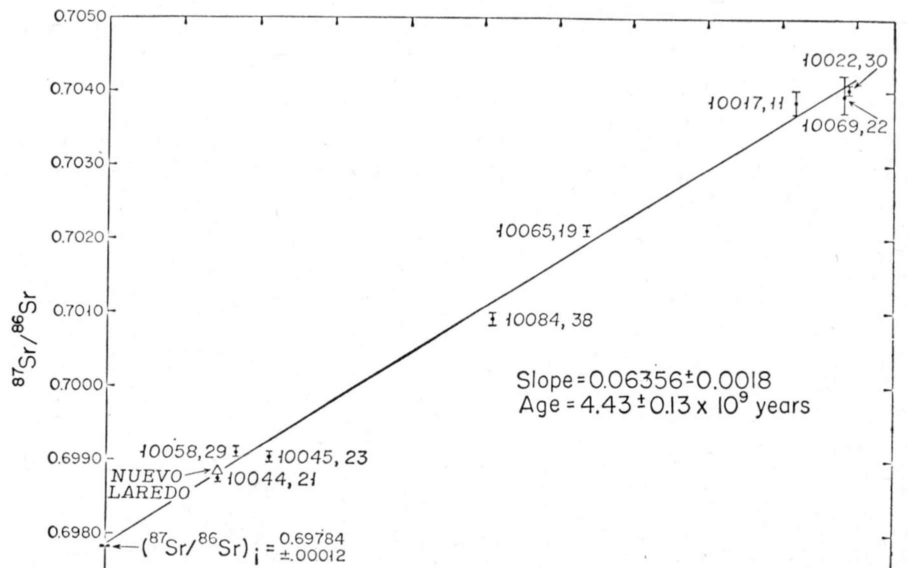

Fig. 1. Strontium evolution diagram for Apollo 11 lunar samples. The straight line represents the least-squares fit of all the lunar sample data points. The data point for the basaltic achondrite Nuevo Laredo falls on the lunar sample regression line (see Table 2).

Table 2. Values of initial $^{87}Sr/^{86}Sr$ ratios and ages for various groupings of lunar samples. The errors shown are 1 standard deviation.

Group of samples	$(^{87}Sr/^{86}Sr)_i$	Slope	Age ($\times 10^9$ years)
Eight samples (all samples received in this study)	0.69784 ± 0.00012	0.06356 ± 0.0018	4.43 ± 0.13
Eight lunar samples and Nuevo Laredo	0.69785 ± 0.00010	0.06346 ± 0.0016	4.43 ± 0.11
Six crystalline rocks of types A and B	0.69785 ± 0.00012	0.06330 ± 0.0018	4.42 ± 0.12
Seven rock samples of types A, B, and C	0.69788 ± 0.00014	0.06350 ± 0.0021	4.43 ± 0.14
Three rock samples of type A	0.69783 ± 0.00023	0.06252 ± 0.0003	4.36 ± 0.02
Three rock samples of type B	0.69780 ± 0.00023	0.06598 ± 0.0043	4.60 ± 0.29
Five samples of types B, C, and D	0.69776 ± 0.00019	0.06623 ± 0.0034	4.61 ± 0.23

fit a linear relationship on the Sr-evolution diagram (Fig. 1) with an age of 4.43 ± 0.13 × 10^9 years and an initial $^{87}Sr/^{86}Sr$ ratio [$(^{87}Sr/^{86}Sr)_i$] of 0.69784 ± 0.00012 (Table 2). If the breccia and lunar soil are excluded on the grounds that they may have had a more complex chemical history, the data for six crystalline rocks of types A and B yield a similar age and a value for $(^{87}Sr/^{86}Sr)_i$ that is virtually identical to that obtained for all samples. Higher age values of 4.61 ± 0.23 × 10^9 years are obtained for groups that include type B samples, the breccia, and the dust, but the errors, both in the age and the value of $(^{87}Sr/^{86}Sr)_i$, become much larger, probably implying that the conditions for isochron are not adequately met.

The initial $^{87}Sr/^{86}Sr$ values for the lunar samples reported here are lower by about 0.07 to 0.1 percent than values based upon initial $^{87}Sr/^{86}Sr$ values from basaltic achondrites (*4*). This appears to be due to differences in instrumental bias between different laboratories. Our measurements on the achondrite Nuevo Laredo and on the Eimer and Amend reagent $SrCO_3$ are shown in Table 1. The value of the initial $^{87}Sr/^{86}Sr$ ratio obtained for Nuevo Laredo, after correction for a growth of 4.5 × 10^9 years, is 0.69786 and is identical with the initial value obtained for all the lunar samples, or the six crystalline rocks. The point described by Nuevo Laredo in the Sr-evolution diagram is located on the lunar data best-fit line (Fig. 1).

The precision of the present Sr-isotopic analyses permits an evaluation of the *time interval* between the formation of the lunar rocks and that of the achondritic meteorites, on the basis of certain assumptions concerning the Rb/Sr ratio of the reservoir from which these objects were made. For example, for a reservoir Rb/Sr ratio similar to that of chondrites (~0.25), the present data suggest that the achondrites and the lunar rocks are coeval within about 1.5 × 10^7 years. On the other hand, if achondrites originate from the surface of the moon, with a Rb/Sr ratio of ~0.02, the time interval corresponds to no more than 1.8 × 10^8 years.

The data on the Rb-Sr age indicate that melting processes on the lunar surface date back to the early history of the solar system. By whatever mechanism the earth acquired the moon, by fission from the earth or by capture, this process must have taken place in the earliest history of the earth, probably contemporaneously with the terminal stages of accretion of the earth.

The elemental and isotopic abundance ratios for Sm, Eu, and Gd in two lunar crystalline rocks and in the soil sample are shown in Table 3. Because of the large thermal neutron capture cross sections of ^{151}Eu, ^{153}Eu, ^{149}Sm, and, in particular, ^{157}Gd (254,000 barns) (*5*), the abundances of these isotopes in lunar samples serve as sensitive indicators of the integrated thermal neutron flux to which these samples were exposed (*6*). Data (Table 3) show that the $^{157}Gd/^{158}Gd$ ratios in the lunar samples are similar to the values for terrestrial rocks to within 0.2 percent. These results set an upper limit of 6 × 10^{15} neutrons per square centimeter for the differential integrated thermal neutron flux between terrestrial and lunar materials. In models that postulate a primordial irradiation of only a fraction of the solar system material and subsequent mixing with unirradiated material before accretion of planetary bodies, the present data show that the lunar material and the earth contain the same proportions of irradiated and unirradiated material to within 1 to 2 percent, if originally 5 to 10 percent of the solar system material was irradiated (*7*).

The inferred differential flux for the lunar materials can be discussed in terms of secondary neutrons produced in the lunar material by cosmic ray interactions. The lunar neutron flux has been estimated to be about 18 neutron cm^{-2} sec^{-1} with a surface leakage of 30 percent for basaltic composition with a H/Si atom ratio of 0.22 (*8*). The lunar rocks are entirely devoid of hydrous minerals (*1*), and probably a lower H/Si ratio is appropriate, which would lead to higher leakage and a lower thermal-to-nonthermal flux ratio. The flux limit inferred from the Gd data on the lunar soil is compatible with an exposure age of ~ 10^8 years. We conclude that the flux limit estimated from the data is more in keeping with secondary neutron flux on the lunar surface; there seems to be no reason to call on differential primordial irradiation effects.

The Sm/Eu and Gd/Eu elemental ratios in the lunar samples analyzed (Table 3) are distinctly higher than those observed in chondrites and Ca-rich achondrites (*9*), thus indicating that the Eu is depleted in these samples. In terrestrial rocks, Eu depletion is due to fractionation effects produced by plagioclase which tends to concentrate Eu relative to Sm and Gd (*9*). The trace element trends discussed earlier and the Eu depletion in lunar samples indicate that a significant amount of plagioclase has separated from the melts that produced the analyzed lunar rocks. Therefore, we suggest that there will be a direct correlation between the K content and the magnitude of the Eu depletion anomaly, and that rocks rich in

Table 3. Elemental and isotopic abundance ratios of Sm, Eu, and Gd. Errors are 1 standard deviation of the mean.

Sample	Sm/Eu	Gd/Eu	$^{151}Eu/^{153}Eu$	$^{149}Sm/^{148}Sm$*	$^{157}Gd/^{158}Gd$†
Reagent			0.9147 ± 0.0013	1.232 ± 0.001	0.6307 ± 0.0002
10069,22(A)	7.4	13.7	0.9144 ± 0.0027	1.235 ± 0.002	0.6306 ± 0.0003
10045,23(B)	4.5	5.48	0.9147 ± 0.0007	1.235 ± 0.001	0.6304 ± 0.0004
10084,38(D)	5.2	10.9	0.9127 ± 0.0007	1.231 ± 0.001	0.6299 ± 0.0007

* $^{149}Sm/^{148}Sm$ normalized to $^{152}Sm/^{154}Sm$ = 1.1770. † $^{157}Gd/^{158}Gd$ normalized to $^{156}Gd/^{160}Gd$ = 0.9361.

plagioclase (gabbroic anorthosite and anorthosite) may be found on the moon in future Apollo missions.

V. RAMA MURTHY
Department of Geology and Geophysics, University of Minnesota, Minneapolis 55455

R. A. SCHMITT, P. REY
Department of Chemistry, Oregon State University, Corvallis

References and Notes

1. Lunar Sample Preliminary Examination Team, *Science* **165**, 1211 (1969).
2. C. C. Schnetzler and J. A. Philpotts, program abstracts of the annual meeting of the Geological Society of America, 1968, p. 265; W. L. Griffin and V. Rama Murthy, *Geochim. Cosmochim. Acta* **33**, 1389 (1969).
3. E. Olsen, *Science* **166**, 401 (1969).
4. D. A. Papanastassiou and G. J. Wasserburg, *Earth Planet. Sci. Lett.* **5**, 361 (1969); P. W. Gast, *Geochim. Cosmochim. Acta* **26**, 927 (1962).
5. H. B. Moller, F. J. Shore, V. L. Sailor, *Nucl. Sci. Eng.* **8**, 183 (1960).
6. V. Rama Murthy and R. A. Schmitt, *J. Geophys. Res.* **68**, 911 (1963).
7. W. A. Fowler, J. L. Greenstein, F. Hoyle, *Geophys. J.* **6**, 148 (1962); D. S. Burnett, W. A. Fowler, F. Hoyle, *Geochim. Cosmochim. Acta* **29**, 1209 (1965).
8. R. E. Lingenfelter, E. H. Canfield, W. N. Hess, *J. Geophys. Res.* **66**, 2665 (1966).
9. L. A. Haskin, F. A. Frey, R. A. Schmitt, R. H. Smith, *Phys. Chem. Earth* **7**, 167 (1966).
10. Supported by NASA Contract NAS9-7884. We are indebted to M. R. Coscio for the chemical procedures, and to N. M. Evensen and Bor-Ming Jahn for help in the mass spectrometric analyses. We thank Professor H. T. Hall for many valuable discussions.

4 January 1970

Age Determinations and Isotopic Abundance Measurements on Lunar Samples

Abstract. *A K-Ar age of 2300 × 10^6 years has been determined for a sample of type A crystalline rock (57,34). The presence of an anomalously large quantity of ^{40}Ar, in a sample of type C breccia (65,35) precluded the calculation of its K-Ar age. Both of the rock types are characterized by low Rb/Sr ratios and consequently low $^{87}Sr/^{86}Sr$ values. The U-Th-Pb results for a sample of type D fines (84,33) yield a $^{207}Pb/^{206}Pb$ age of 4760 × 10^6 years, but ages based on U-Pb and Th-Pb ratios are anomalously high. Isotopic compositions of Li, K, Rb, Sr, U, and Th are very close to the accepted values for terrestrial materials.*

Isotopic determinations were undertaken for several geochronologically important elements found in three types (A, C, and D) of lunar material acquired during the flight of Apollo 11. Two additional samples are currently being studied, and the results will be reported elsewhere. The object of the study was (i) to precisely determine the concentrations of certain parent-daughter elemental pairs; (ii) to measure isotopic abundances for comparison with values for terrestrial elements; and (iii) to determine the geological age of the specimens.

Our routine experimental procedures were slightly modified to facilitate the measurement of isotopic abundances in addition to the determination of elemental concentrations. Isotope dilution techniques were exclusively used in determining the latter. Analyses of Li, K, Rb, Sr, Pb, U, and Th were carried out on two (one 15-cm radius and one 25-cm radius) 90-degree deflection, solid-source mass spectrometers, both of which are equipped with triple-filament source assemblies and electron multiplier ion detection systems. A modified M.S.-10 mass spectrometer (Associated Electrical Industries), operated in the static mode, was used for all argon analyses. All chemical procedures and vacuum fusions were monitored for contamination through the analysis of "blanks." Contamination corrections applied for Rb, Sr, and Ar were <0.5 percent; corrections for U and Th amounted to about 1 percent, as did those for K, with the exception of one extraction of the latter which required a 2 percent correction; a 10 percent correction was required for terrestrial Pb contamination. Our determinations of the concentrations of Rb and Sr in the Cal. Tech.

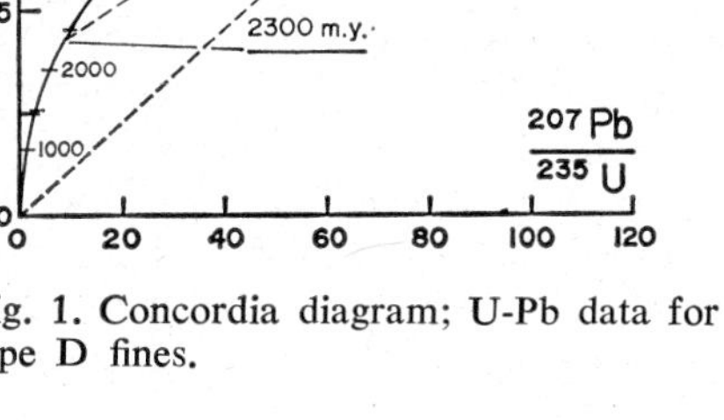

Fig. 1. Concordia diagram; U-Pb data for type D fines.

standard solutions supplied with the lunar samples agree with the stated values to within 0.25 percent.

Argon was extracted from two samples of type A crystalline rock (57,34) under slightly differing experimental conditions. In the first instance (extraction No. 2, Table 1) the purified gas was collected in two calibrated volumes having ratios of about 1:2. The smaller volume of gas was equilibrated with ^{38}Ar tracer on the mass spectrometer inlet line just before isotopic analysis, while the larger portion was used unspiked to determine the relative isotopic abundances. The $^{36}Ar/^{38}Ar$ ratio determined for the latter was used to correct the results of the spiked analysis. In calculating the age it was assumed that there was no primordial ^{40}Ar component. In the second instance (extraction No. 3, Table 1) the gas was equilibrated with the ^{38}Ar tracer in the usual manner during sample fusion. The $^{36}Ar/^{38}Ar$ ratio measured for the unspiked portion of extraction No. 2 was again used to correct the observed abundances for this sample.

The K measurements were made on chips of material broken from the larger samples fused for Ar extraction and should therefore provide a realistic estimate of the K content of the latter. An examination of the ^{40}Ar and K values (Table 1) reveals that the concentrations of both are slightly different in the two portions of this sample.

The greatest variation however was noted for the ^{36}Ar content. While both samples yielded essentially the same quantity of ^{40}Ar, extraction No. 2 contained nearly twice as much ^{36}Ar as extraction No. 3. In spite of this variation the calculated ages for the two samples are very close at 2300 × 10^6 and 2270 × 10^6 years, respectively.

Two extractions were carried out for a sample (65,35) of type C breccia; the first was equilibrated with the ^{38}Ar tracer in the routine fashion during fusion, and the second was not mixed with tracer. The $^{36}Ar/^{38}Ar$ ratio determined for the latter was to be used to correct the results obtained for extraction No. 1. However, this material contained very large quantities of both ^{36}Ar and ^{40}Ar, and consequently it has not been possible to calculate a realistic age for the sample. This is in agreement with preliminary data reported for breccia samples (*1*). The K concentration was again determined on numerous chips from the larger sample used for the Ar extraction in an attempt to ensure the

Table 1. Isotopic and elemental abundance data. For ^{40}K, $\lambda\beta$ (decay constant for β decay) $= 4.72 \times 10^{-10}$ year^{-1}, $\lambda\varepsilon$ (decay constant for electron capture) $= 0.585 \times 10^{-10}$ year^{-1}, and atomic abundance $= 1.19 \times 10^{-4}$.

Extraction No.	Rb (ppm)	Sr (ppm)	$^{87}Sr/^{86}Sr$ Spiked	$^{87}Sr/^{86}Sr$ Unspiked	$^{87}Sr/^{86}Sr$ Avg.	$^{87}Rb/^{86}Sr$	K (ppm)	K/Rb	^{40}Ar ($\times 10^{-8}$ cm^3/g)	$^{40}Ar/^{36}Ar$	$^{36}Ar/^{38}Ar$	K-Ar age ($\times 10^6$ years)	Remarks
						Type A – crystalline, No. 57,34							
1	5.853	171.0	0.7052	0.7043 0.7055	0.7050	0.0991	2520	430.5					
2							2588		4543	83.988 83.671	 2.768	2300	Spiked Unspiked
3							2647		4646	160.418		2270	Spiked
						Type C – breccia, No. 65,35							
1	3.741	165.5	0.7033	0.7036	0.7034	0.0654	1470	392.9	140,800	2.361			Spiked
2										2.087	5.383		Unspiked

Table 2. The U-Th-Pb analyses, type D fines No. 84,33. For ^{238}U, λ (total decay constant) $= 1.537 \times 10^{-10}$ year^{-1}; for ^{235}U, $\lambda = 9.722 \times 10^{-10}$ year^{-1}; for ^{232}Th, $\lambda = 4.881 \times 10^{-11}$ year^{-1}. Ratio $^{238}U/^{235}U = 137.7$.

U (ppm)	Th (ppm)	Pb Total (ppm)	Pb Radiogenic (ppm)	Pb Atomic ratios 206/204	Pb Atomic ratios 207/204	Pb Atomic ratios 208/204	Age ($\times 10^6$ years) $^{207}Pb/^{206}Pb$	Age $^{206}Pb/^{238}U$	Age $^{207}Pb/^{235}U$	Age $^{208}Pb/^{232}Th$
0.530	1.31	2.26	1.55	45.8* 51.9†	33.6* 37.6†	64.8* 70.6†	4670	5410	4885	8200

*Measured ratios. †Corrected for Pb blank.

determination of a K content representative of the total sample. The K concentration for this material is considerably lower than that determined for the type A crystalline rock (Table 1).

Rubidium and Sr results for one extraction from each of the two samples (57,34 and 65,35; types A and C) are given in Table 1. The Rb concentrations are very low, and consequently the ^{87}Sr enrichment is very small. However the experimental data points do fall reasonably close to a 3200×10^6 years reference isochron on a Rb-Sr evolution diagram, thus providing some evidence for higher ages than the K-Ar ages calculated for sample 57,34. Rubidium and Sr analyses for the remaining lunar samples may serve to confirm the greater age indicated.

A major portion of the sample of fines (84,33; type D) was used for U, Th, and Pb determinations. Considerable difficulty was experienced in working with the larger sample, but reasonable results were obtained. As mentioned above, about 10 percent of the Pb was attributed to contamination arising in the laboratory and the isotopic ratios were corrected by use of the relative abundances of modern day terrestrial lead. In addition, a correction for primordial lead amounting to about 32 percent of the sample was made, based on the relative abundances of meteoritic lead as determined by Patterson (*2*) (Table 2). The $^{207}Pb/^{206}Pb$ age has been calculated to be 4670×10^6 years which, although a bit high, is much more realistic than the other ages based on the Pb-U and Pb-Th ratios (Table 2). The data were plotted on a concordia diagram (Fig. 1), and dashed lines were drawn from this experimental point to zero age and to the 2300×10^6 years age point (the K-Ar age for our sample of type A crystalline rock). In the latter case the line cuts concordia at 4490×10^6 years. The position of the experimental point above the curve is considered to be indicative of either loss of U from the sample or gain of modern terrestrial Pb. The former is in good agreement with the preliminary results (*1*), and the latter has been corrected for the Pb of the "blank" extraction processed with the sample. While it is possible that the sample was contaminated before processing, exacting precautions were taken to avoid this possibility. In support of the proposed loss of U, it should be noted that the Th result also appears to be low. If indeed both parent elements have been lost, this loss, expressed as a percentage of the U and Th required to produce concordant ages at 4670×10^6 years, amounts to 19 percent for U and 48 percent for Th.

Isotopic abundances of Li, K, Rb, and Sr were measured for samples 57,34 and 65,35 and of U and Th for sample 83,33. Since no major deviations from accepted terrestrial values were found, the experimental data have been omitted but will be presented in our final report.

Evidence has been presented to support two contrasting ages for lunar materials. One based on the $^{207}Pb/^{206}Pb$ ratio determined for a bulk sample of type D fines is in essential agreement with the accepted age of the earth. The lower K-Ar age of 2300×10^6 years assigned to a single sample of type A crystalline rock is considered to be a valid determination and may indicate a stage in the evolutionary history of the moon.

R. K. WANLESS
W. D. LOVERIDGE
R. D. STEVENS

Geological Survey of Canada,
601 Booth Street,
Ottawa 4, Ontario

References and Notes

1. Lunar Sample Preliminary Examination Team, *Science* **165**, 1211 (1969).
2. C. C. Patterson, *Geochim. Cosmochim. Acta* **7**, 151 (1955).
3. We thank R. W. Sullivan for chemical extractions of Rb, Sr, and K; J. C. Bisson for chemical preparation of Pb, U, and Th; F. B. Quigg for Ar extractions and K analyses; Miss M. J. Swerdfager for Ar analyses; J. L. MacRae for preparation of the mass spectrometer filament assemblies; and Mrs. K. Neidy for assisting with the calculations and preparation of the tables.

4 January 1970; revised 7 January 1970

Lead and Thallium Isotopes in Mare Tranquillitatis Surface Material

Abstract: *Lead from Apollo 11 fines is more radiogenic than any meteoritic lead reported and older than any terrestrial radiogenic lead:* $^{204}Pb/^{206}Pb/^{207}Pb/^{208}Pb = 1/99_{.6}/69_{.0}/117_{.1}$. *Comparison with primordial lead from meteoritic troilite yields a* $^{207}Pb/^{206}Pb$ *age of* $4.7 \pm 0.1 \times 10^9$ *years. The* $^{238}U/^{204}Pb$ *ratio is* $\gtrsim 90$ *and the* $^{232}Th/^{238}U$ *ratio is* 3.9 ± 0.1. *The lead content is* $\gtrsim 1.7 \times 10^{-6}$. *Evidently Pb was strongly depleted and Th and U strongly enriched in the formation of this material. Thallium was too low* ($\lesssim 5 \times 10^{-9}$) *to yield mass spectra, but indications are favorable for eventual observation of extinct natural radioactivity of* ^{205}Pb.

We have made an isotopic analysis of lead from a sample of fines from the Apollo 11 site, from which an age of the lunar surface can be derived, and have initiated a search for isotopic anomalies due to certain extinct natural radionuclides, from which fine time-resolution of the earliest lunar history might be derived. In particular, we have undertaken to determine the abundance and isotopic composition of thallium, which might be altered by the electron-capture decay of ^{205}Pb (half-life ~ 24×10^6 years) to ^{205}Tl. Previous searches for variations in the $^{205}Tl/^{203}Tl$ ratio in meteoritic thallium have given negative results (*1*). Lead and thallium are isolated by volatilization under vacuum at temperatures close to the melting point of the bulk rock. Anion exchange and electrodeposition are used for further purification. A single-filament mass-spectrometric technique modified from Cameron *et al.* (*2*) is used. Sub-microgram quantities of silica gel, the purified Tl and Pb in 1*M* HNO_3, and H_3PO_4 are successively loaded onto the filament; drying is effected by a combination of resistance and radiant heating in air. The mass spectrometer is a Nuclide Corporation Model SU 2.4 (90° deflection, 30-cm radius). Filament temperatures are monitored by an optical pyrometer. The Pb and Tl data are collected by repeated magnet-current switching between peak pairs; approximately 12 cycles are used for each ratio. The Tl is measured at ~740°C and Pb at ~1170°C.

Elemental abundances and yields of various parts of the procedure are determined by isotopic dilution with the use of enriched ^{203}Tl and ^{206}Pb. The ion-exchange purification yields were 95 to 100 percent for both Tl and Pb. The electrodeposition yields were 60 to 70 percent for Tl and 90 to 95 percent for Pb. Because isotopic mixing of spikes and elements in natural solids cannot be achieved, in actual runs with such materials the spikes are added to the solutions of the vapor condensates. Volatilization yields were estimated by comparison of recovered Tl and Pb from standard rocks AGV-1 and BCR-1 with published determinations (*3*) as 70 to 100 percent for Tl and 60 to 90 percent for Pb in heatings of 24 to 48 hours. Yields seem to be higher at temperatures just below the melting point than just above. The highest temperatures tried have been 1050°C.

The first lunar material used was a 3-g sample (A) of less than 1-mm fines from the Apollo 11 site (sample 10084,45) ground in a motor-shaken tungsten-carbide capsule-pestle to pass through a 200-mesh nylon screen. It was heated successively at 800°, 850°, and 900°C for 20 hours each. The vapor condensate solution was made up to 10 ml. Two 4-ml aliquots (**1** and **3**) were processed for isotope-abundance measurements and a 2-ml aliquot (**2**) was spiked for abundance determinations. In neither of the unspiked samples could Tl be seen in the mass spectrometer. In run **1**, Pb isotope ratios were obtained only with a chart recorder with ~2 percent accuracy; the signal appeared at a relatively low temperature (820°C) and disappeared before digital-voltmeter measurements could be made. In run **3** moderately stable ion beams of ~ 10^{-11} amp were obtained at 1170°C, and data were collected with the digital voltmeter. Digital-voltmeter data were obtained for both Tl and Pb in run **2**.

The results are given in Table 1. The results of run **3** were corrected for a typical blank of 0.07 μg of modern terrestrial Pb (*4*). Additional contamination during collection, handling, and grinding of the sample would not have been corrected for.

The $^{206}Pb/^{204}Pb$ and $^{207}Pb/^{204}Pb$ ratios are plotted along with similar data on meteoritic and common terrestrial lead in Fig. 1a. The troilite point represents the mean of several measurements selected by Murthy and Patterson (*4*), who adopt $^{204}Pb/^{206}Pb/^{207}Pb/^{208}Pb = 1/9.56/10.42/29.71$ for this and for primordial lead of the solar system. Subtraction of these ratios from the corresponding **A-3** lunar values yields a radiogenic $^{207}Pb/^{206}Pb$ ratio of 0.65_1. With disintegration constants $\lambda(^{238}U) = 1.537 \times 10^{-10}$ year^{-1} and $\lambda(^{235}U) = 9.722 \times 10^{-10}$ year^{-1}, and $(^{238}U/^{235}U)_{now} = 137.8$ (*5*), the age of the lunar surface sample is calculated to be $T = 4.6_9 \times 10^9$ years. Isochrons corresponding to several ages are shown on the figure. The effects of possible corrections for error in the ^{204}Pb peak measurements, for contamination by modern terrestrial lead, and for isotope fractionation in the mass spectrometer are shown by dashed lines. Since the

Table 1. Mass-spectrometric data and calculations.

Quantity	Run A-1 (unspiked)	Run A-3 (unspiked) Data	Run A-3 (unspiked) Corrected	Run A-2 (spiked)
Effective sample mass	1.2 g	1.2 g		0.6 g
Mass Pb spike*				1.094 μg
Mass Tl spike†				221.0 ng
Measured $^{205}Tl/^{203}Tl$	Not seen	Not seen		0.093
Measured $^{206}Pb/^{204}Pb$	$71._2$‡	90.2 ± 0.3‡	$99._6$	
Measured $^{207}Pb/^{206}Pb$	0.70_6	$0.6965 \pm .0005$	0.69_4	
Measured $^{208}Pb/^{206}Pb$	1.25_6	$1.1978 \pm .0006$	1.17_8	0.333
Calculated $^{207}Pb/^{204}Pb$	$50._3$‡	62.8 ± 0.2‡	$69._0$	
Calculated $^{208}Pb/^{204}Pb$	$89._6$‡	108.0 ± 0.3‡	$117._1$	
Mass natural Tl				3.3 ng §
Mass natural Pb				1.01_7 μg
Tl abundance in sample				$<5._5 \times 10^{-9}$ ‖
Pb abundance in sample				$>1.7_0 \times 10^{-6}$ ¶

*$^{204}Pb/^{206}Pb/^{207}Pb/^{208}Pb = 1/2688/189.9/36.34$. †$^{205}Tl/^{203}Tl = 0.081_6$. ‡Probably lower limits because of possibility of lead contamination. Not corrected for possible isotope fractionation. §Calculated assuming for natural thallium $^{205}Tl/^{203} = 2.389$ as observed in this instrument (*1*). ‖ Probably an upper limit because of nonzero blank. ¶ Probably a lower limit because of incomplete volatilization.

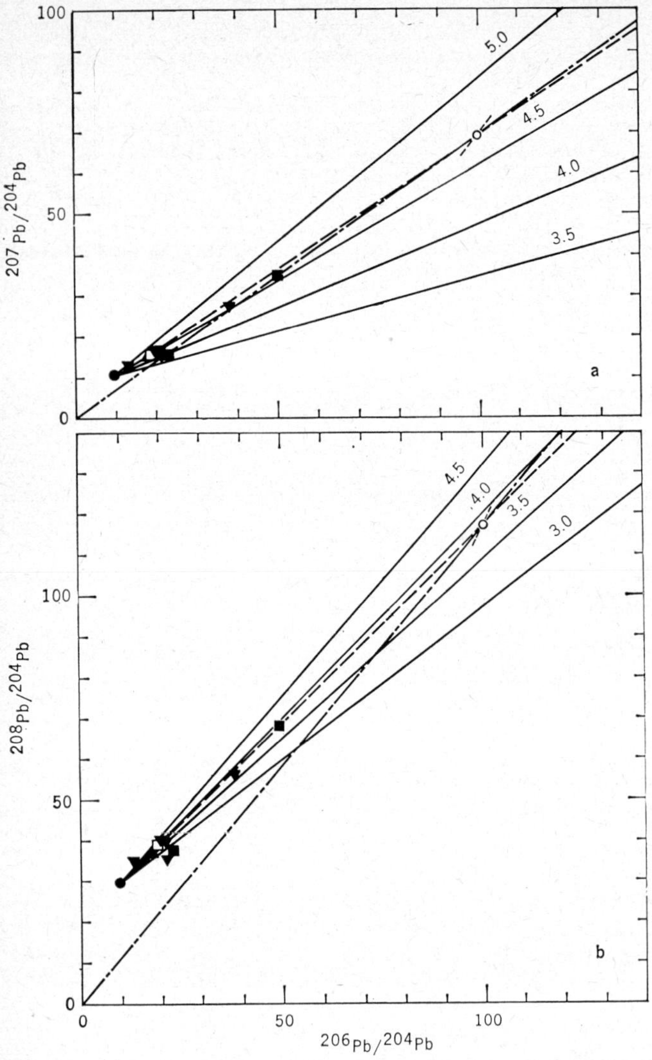

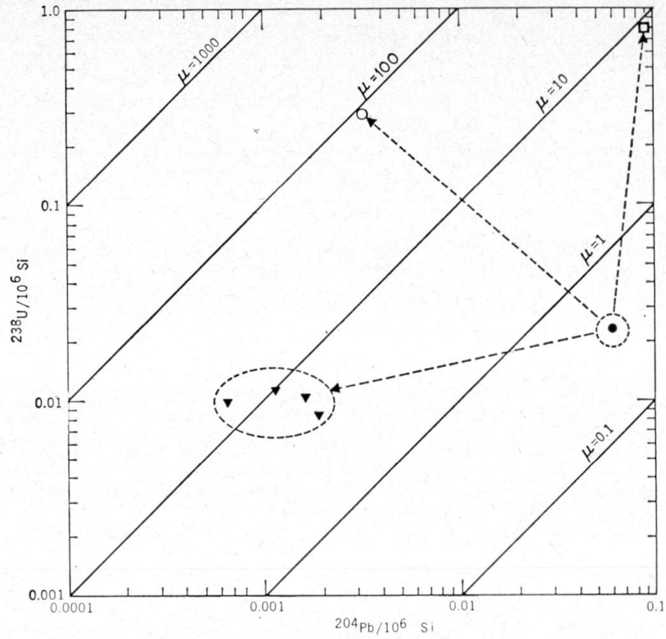

Fig. 1 (left). Comparison of isotopic composition of lead from Apollo 11 lunar surface fines with that of meteorites and the earth. (a) $^{207}Pb/^{204}Pb$ versus $^{206}Pb/^{204}Pb$. ——— Calculated isochrons for single-stage ages indicated in units of 10^9 years. (b) $^{208}Pb/^{204}Pb$ versus $^{206}Pb/^{204}Pb$. ——— Calculated locus lines for indicated values of $^{232}Th/^{238}U$ at the present time following 4.7×10^9 years of single-stage evolution. In both parts: (○) lunar fines sample 10084,45, A-3; (●) meteorite troilite; (▼) chondrite meteorite; (■) achondrite meteorite; (▲) iron meteorite; (□) typical modern terrestrial common lead; —·— common-lead contamination line; — — — ^{204}Pb error line; - - - isotope fractionation line. Fig. 2 (above). Comparison of ^{238}U and ^{204}Pb contents of Apollo 11 lunar surface fines (○) with those of type I carbonaceous chondrites (●), ordinary chondrites (▼), and the earth (□). μ = present atomic ratio $^{238}U/^{204}Pb$.

contamination line through the lunar point is almost identical with the isochron which it defines, and the other two corrections are not likely to be appreciable, the error in the age due to these effects should be small. For example, an uncertainty in the $^{207}Pb/^{206}Pb$ ratio of as much as 2 percent corresponds to an uncertainty of only $\sim 0.03 \times 10^9$ years in T. Considering all uncertainties we express our result as $4.7 \pm 0.1 \times 10^9$ years. This age is not substantially different from those derived from the meteorite Pb isochron, or for the earth by a similar calculation with mean oceanic lead, 4.55×10^9 years (*4*, *5*), or from meteoritic ^{87}Rb-^{87}Sr isochron ages. However, our radiogenic $^{207}Pb/^{206}Pb$ ratio, 0.65, seems significantly greater than that of chondrite and achondrite meteorites, 0.59 or 0.60, suggesting that the lunar surface may actually be the oldest sample of solar-system matter yet available. On the other hand, a growth model involving two or more stages could be consistent with a more recent formation of the Mare Tranquillitatis surface material.

The $^{208}Pb/^{204}Pb$ and $^{206}Pb/^{204}Pb$ ratios are similarly plotted (Fig. 1b) along with iso-compositional lines corresponding to various $^{232}Th/^{238}U$ ratios, calculated for $T = 4.69 \times 10^9$ years with $\lambda(^{232}Th) = 4.99 \times 10^{-11}$ year^{-1} (*5*). The above-mentioned subtraction yields a radiogenic $^{208}Pb/^{206}Pb$ ratio of 0.97_1, which corresponds to an atomic $^{232}Th/^{238}U$ ratio of 3.8_9 and an elemental Th/U ratio of 3.7_6. Any uncertainty in the $^{208}Pb/^{206}Pb$ ratio corresponds to the same fractional uncertainty in these parent ratios. The effects of ^{204}Pb error, contamination, and fractionation are likewise shown in the figure; again these effects are very small. Considering all uncertainties, including that in the age, we express our result as $^{232}Th/^{238}U = 3.9 \pm 0.1$ and $Th/U = 3.8 \pm 0.1$.

Our lead isotopic composition and the assumption of a closed system aged 4.69×10^9 years corresponds to a present-day ratio $^{238}U/^{204}Pb$ (μ) ≈ 86. This is undoubtedly a minimum value, because a small amount of common-lead contamination could substantially increase the ^{204}Pb in the isolated sample. We conclude that $\mu \gtrsim 90$. The preliminary γ-ray measurement of similar Apollo 11 fines (*6*) indicated a U content of $\sim 0.46 \times 10^{-6}$. Combining this with $\mu = 86$ and our Pb isotopic composition yields a total Pb abundance of $\sim 1.3_3 \times 10^{-6}$, relatively insensitive to common-lead contamination. This is close to the value calculated from the isotope-dilution analysis of the vapor-condensate, $1.7_0 \times 10^{-6}$, which we regard as a lower limit because of probably incomplete volatilization. The γ-ray measurements (*6*) gave $\sim 1.6 \times 10^{-6}$ for the Th content, or ~ 3.5 for the Th/U ratio, consistent with our result.

Comparison of the ^{238}U and ^{204}Pb contents of the Apollo 11 fines, which, because of the natural mixing processes by which they were produced, may be fairly representative of the lunar surface, with the same values in type I carbonaceous chondrites (*7*) ($\mu \sim$ 0.4), widely regarded as good samples of nonvolatile solar-system matter, is made in Fig. 2. Evidently U and Th were highly enriched and Pb highly depleted in the formation of the lunar surface. This process must have been distinctly different from that producing the characteristic elemental assemblages of the ordinary chondrites (*7*) ($\mu \sim 4$) and the earth ($\mu \approx 9$) (Fig. 2).

Unfortunately, too little Tl was present in the 1.2-g sample aliquots to yield mass spectra. The $3_{.3}$ ng of Tl calculated for the 0.6-g aliquote by isotope dilution is lower than that observed in any previous blank and probably still includes a residual blank. The Tl content of this composite lunar surface material, $< 5 \times 10^{-9}$, is at least tenfold lower than in type I carbonaceous chondrites, $\sim 74 \times 10^{-9}$ (*8*), and may be similar to that observed in some highly equilibrated ordinary chondrites, $\lesssim 1 \times 10^{-9}$ (*8*). The depletion of Tl seems to be greater than that of ^{204}Pb. This, combined with the extreme antiquity indicated for the lunar surface, indicates favorable conditions for the eventual detection of radiogenic ^{205}Tl from extinct natural radioactivity of ^{205}Pb (*1, 9*).

TRUMAN P. KOHMAN
LANCE P. BLACK
HARUHIKO IHOCHI, JAMES M. HUEY
Department of Chemistry, Carnegie-Mellon University, Pittsburgh, Pennsylvania 15213

References and Notes

1. E. Anders and C. M. Stevens, *J. Geophys. Res.* **65**, 3043 (1960); R. G. Ostic, H. M. El-Badry, T. P. Kohman, *Earth Planet. Sci. Lett.* **7**, 72 (1969).
2. A. E. Cameron, D. H. Smith, R. L. Walker, *Anal. Chem.* **41**, 525 (1969).
3. F. J. Flanagan, *Geochim. Cosmochim. Acta* **33**, 81 (1969); M. Tatsumoto, private communication to F. J. Flanagan; W. Wahler, *Neues Jahrb. Mineral. Abhandl.* **108**, 36 (1968).
4. V. R. Murthy and C. C. Patterson, *J. Geophys. Res.* **67**, 1161 (1962).
5. E. R. Kanasewich, in *Radiometric Dating for Geologists*, E. I. Hamilton and R. M. Farquhar, Eds. (Interscience-Wiley, New York, 1968), p. 147.
6. Lunar Sample Preliminary Examination Team, *Science* **165**, 1211 (1969).
7. W. D. Ehmann, in *Encyclopedia of the Chemical Elements*, C. A. Hampel, Ed. (Reinhold, New York, 1968), p. 567.
8. J. C. Laul, I. Pelly, M. E. Lipshutz, in preparation.
9. T. P. Kohman, *Ann. N.Y. Acad. Sci.* **62**, Art. 21, 503 (1956); T. P. Kohman, *J. Chem. Educ.* **38**, 73 (1961).
10. We thank R. G. Ostic, J. T. Tanner, M. J. Toia and M. W. Haramic of this University, and W. R. Shields and co-workers of the National Bureau of Standards, for assistance in various portions of this investigation. Supported by NASA through contract NAS-9-8073 and by the U.S. Atomic Energy Commission through contract AT(30-1)-844. Acquisition of the mass spectrometer was assisted by grant GA-349 from the National Science Foundation. J.M.H. is a NASA graduate trainee.

4 January 1970

Trace Elements

Abundance of Alkali Metals, Alkaline and Rare Earths, and Strontium-87/Strontium-86 Ratios in Lunar Samples

Abstract. *The variation in trace element abundance patterns indicates that lunar igneous rocks are the product of extensive igneous fractionation. Variations in the* Sr^{87}/Sr^{86} *ratio indicate that these rocks crystallized* $3.5 \pm 0.3 \times 10^9$ *years ago.*

The concentrations of K, Rb, Cs, Sr, Ba, La, Ce, Nd, Sm, Eu, Gd, Dy, Er, Yb, Lu and the isotopic composition of Sr have been determined mass spectrometrically. By using quantitative ion exchange separation procedures the concentrations of Na, Ca, Zn, and Ti were determined on the same samples using other techniques. Results of the analysis of individual 200- to 300-milligram samples of seven igneous rocks, one breccia, and one soil sample are given in Table 1.

The chemical data obtained in this study and in the preliminary examination (*1*) show that the nine samples analyzed are remarkably similar in terms of a wide range of chemical characteristics. These characteristics clearly distinguish the Tranquillity Base basalts from basaltic meteorites and terrestrial basalts. The primary distinguishing characteristics are (i) the high concentration of titanium, zirconium, and rare earth elements (REE); (ii) the two- to fourfold depletion of Eu relative to Sm and Gd; (iii) the low abundance of sodium relative to other major elements.

On a finer scale the data clearly distinguish three groups of samples from each other. The fine grained (type A) samples 10071, 10017, and 10049 are very similar. In fact, 10071 and 10049 are almost identical for all the elements reported here.

Coarser grained (type B) rocks 10020, 10003, 10062, and 10058 form a second distinguishable chemical group. In the rocks we have analyzed, the fine grained rocks have higher K and Rb contents, lower K/Rb ratios, higher REE abundances, higher Sm/Eu ratios, lower La/Sm ratios, and higher Ba/Sr ratios than the coarse grained rocks.

Our soil and breccia samples are

Table 1. Concentrations of 20 elements in nine samples from Tranquillity Base. All concentrations except CaO and TiO_2 are given in parts per million. Those shown in roman were determined by stable isotope dilution; others were determined by the methods indicated.

Element	Soil sample 10084	Analytical error (%)*	Breccia sample 10073	Fine grained igneous rocks			Coarse grained igneous rocks			
				10049	10017	10071	10062	10020	10003	10058
Na†	*3200*	±*10*	*3500*	*4000*	*3800*	*3900*	*3300*	*2800*	*2900*	*2900*
K	1200	±3	1200	2730	2610	2770	628	486	470	877
Rb	2.79	±3	2.89	6.24	5.63	5.93	0.90	0.63	0.49	0.98
Cs	0.104	±10	0.098	0.166	0.155	0.170	0.032		0.022	0.027
Sr	171.4	±2.5	167.5	160.8	174.8	160.9	141.6	149.8	158.6	218.3
Ba	176	±5	175	338	308	327	79.9	77.1	108	117
La	16.3	±3		29.2	26.2	27.8	13.8	7.95	15.1	11.7
Ce	47.6	±3	46.5	84.2	77.2	83.5	42.7	25.8	47.2	40.2
Nd	36.8	±3	35.4	64.2	59.7	64.8	37.0	24.1	40.0	41.2
Sm	13.1	±3	12.4	22.5	21.0	22.7	13.3	9.49	14.5	17.1
Eu	1.77	±3	1.70	2.31	2.15	2.29	2.02	1.61	1.82	2.64
Gd	17.2	±5	16.9	28.0	27.0	29.2	17.2	12.8	20.2	23.6
Dy	19.7	±5	19.2	36.1	34.1	34.5	20.4	16.2	22.4	27.3
Er	12.1	±5	11.5	19.9	19.2	21.1	12.0	9.99	13.7	16.3
Yb	11.5	±5	10.7	20.2	18.8	20.2	12.1	9.77	12.7	15.2
Lu	1.58	±7	1.56		2.66	2.87	1.73	1.43	1.81	2.14
Ni‡	*198*	±15	*161*	*12*	*17*	*24*	*13*	*10*	*18*	*13*
Zn†	*24*	±15	*24*	*9.3*	*30*	*11*	*9.9*	*10*	*7.4*	
CaO%§	*11.4*	±3	*11.0*	*10.4*	*10.1*	*10.9*	*10.8*	*10.8*	*10.3*	*12.3*
TiO_2%§	*7.7*	±5	*7.5*	*11.5*	*12.4*	*13.8*	*10.7*	*10.4*	*9.2*	*8.3*
La/Sm	1.24			1.29	1.25	1.22	1.04	.837	1.04	.679
Sm/Eu	7.40		7.30	9.74	9.76	9.90	7.58	5.90	7.92	6.47
Ba/Sr	1.025		1.045	2.10	1.77	2.03	0.564	.517	0.677	.542
K/Rb	430		415	437	464	467	698	772	959	895
$Sr^{87/86}$‖	0.7015 ±.0005		0.7025 ±0.0006	0.7043 ±0.0006	0.7046 ±.0003	0.7041 ±.0004	0.7002 ±.0004	0.7005 ±.0006	0.6993 ±.0004	0.6996 ±.0005

*90 percent confidence limits based on replicate analysis of other samples. †Atomic absorption analysis of separated fraction. ‡X-ray fluorescence analysis of powder. §Colorimetric or titrimetric analysis of separated fraction. ‖ Normalized to 8.375.

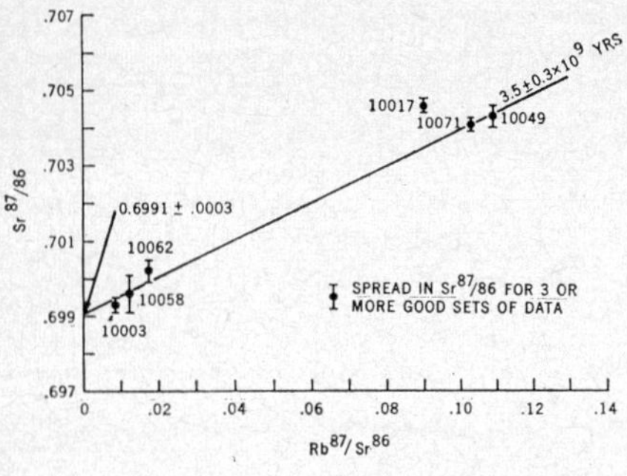

Fig. 1. Chondrite normalized abundance patterns for nine rare earth elements and Sr and Ba. Chondrite REE concentrations are from Frey *et al.* (*9*), Dy is interpolated to be 0.335 ppm, Ba is taken to be 3 ppm, and Sr 11 ppm. Data for sample V21-40 and average ORB are from Kay *et al* (*5*).

nearly identical and in most respects chemically intermediate between the coarse and fine grained igneous rocks. However, both samples of these materials are clearly distinguished from the igneous rocks by higher nickel and zinc contents and lower TiO_2 contents. (We conjecture that the high zinc content of our sample—a surface chip—from rock 10017 is an indication that the surface composition of this rock may be different from the bulk of the rock.) The soil and breccia samples may have been modified with respect to the underlying bedrock by (i) a mixture of meteorite impact material, (ii) addition of material by the solar wind and (iii) enrichment in materials that may have condensed out of the lunar atmosphere after vaporization out of superheated silicate liquids produced by impact melting. The high nickel content is probably due to the first of these causes—that is, the admixture of about 0.1 percent of nickel iron or approximately 1 percent of average chondrites. The low TiO_2 content of the soil and breccia indicates that rocks with TiO_2 contents much lower than those analyzed here make up a significant fraction of the soil.

The large variations in K, Rb, Cs, and Ba content within the otherwise similar igneous rocks are quite puzzling. If explained in terms of simple fractional crystallization processes they would require 80 to 90 percent crystallization of the "coarse grained" liquid to produce the "fine grained" liquid. Such extensive crystallization seems inconsistent with the near identity of the major element composition of these two groups of rocks. Another possible explanation of the nearly fivefold variation on K, Rb, Cs, and Ba content involves the existence of a third liquid containing 2 to 4 percent K, 50 to 100 ppm rubidium, and 2000 to 4000 ppm barium. Several percent of such a liquid added to the "coarse grained" liquid would explain the difference between these two groups of rocks. The low K, Rb, and Cs contents may also be explained if the coarse grained rocks are locally accumulated from a liquid having most of the observed characteristics—that is, large Eu anomalies and low Sr content.

Tentatively accepting this hypothesis, we now consider the origin of most of the chemical characteristics of both the coarse and fine grained igneous rocks. At the outset we observe that it is quite unlikely that the observed composition is entirely primary in the sense that it is material formed directly from dispersed matter in the solar nebula. The extreme enrichment of Ti, REE, U, and Th (*1*), the fractionation of the REE, and the depletion of Eu, Ba, Sr, Ti, and Al are all difficult to explain by the volatilization or fractional condensation processes. On the other hand, the low abundance of elements which may be volatile at high temperature, e.g., the alkalis, Zn, and Ni, can be ascribed to the primitive composition of the moon. Both the low K/U ratio (*1*) and Rb/Sr ratio (see below) indicate that some chemical fractionation has taken place during the formation of both the moon and the earth. Therefore, one need not call upon igneous fractionation processes to explain all of the nonchondritic relative abundances observed in the lunar samples. In the following discussion we examine the possible fractionation of large refractory ions with respect to each other in terms of igneous processes. The abundances of these ions relative to chondritic abundances are illustrated in Fig. 1. The dominant characteristics illustrated are (i) the general and extensive depletions of Sr and Eu with respect to the other ions and (ii) the increasing depletion of the elements Nd, Ce, La and sometimes Ba (opposite to the general case for terrestrial basalts). The REE and Ba abundance patterns in the Tranquillity Base basalts are similar to those observed for oceanic ridge basalts (ORB) but much higher in total REE plus Ba abundances.

It has been shown that most mafic minerals (e.g., pyroxenes, amphiboles, garnets, and spinels) that contain significant concentrations of REE and Ba tend to fractionate the REE with respect to each other (*2, 3*). Thus any igneous processes that involve a significant distribution of REE between such mafic phases and a basaltic liquid would tend to fractionate the REE in the liquid relative to the associated crystalline phases. In spe-

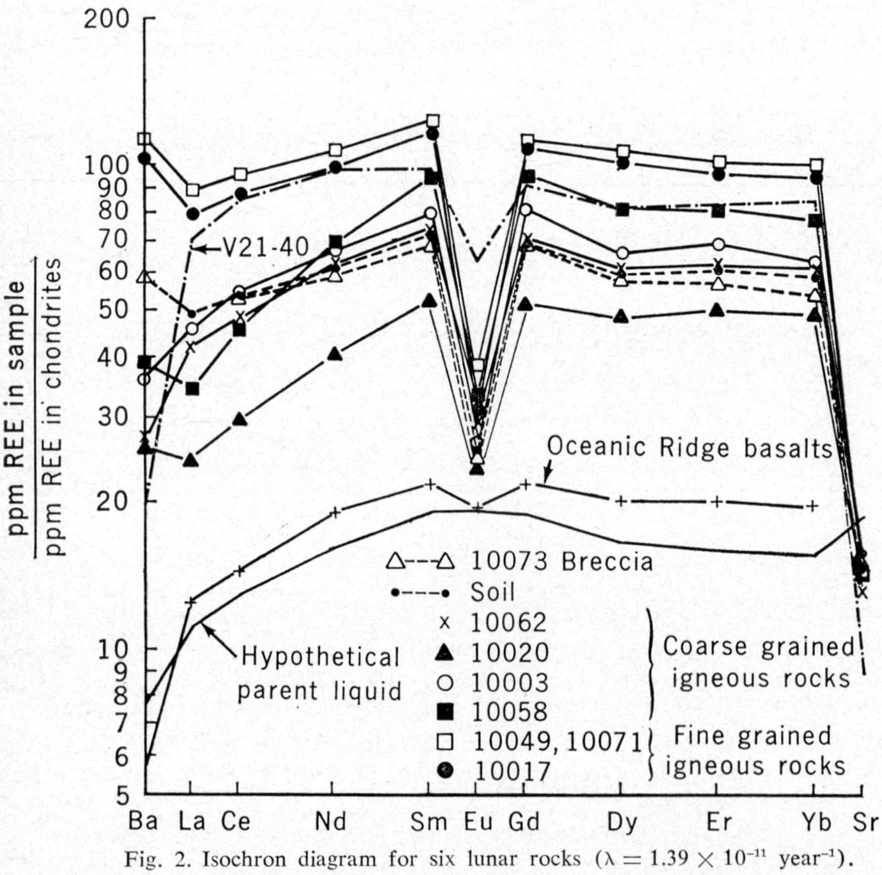

Fig. 2. Isochron diagram for six lunar rocks ($\lambda = 1.39 \times 10^{-11}$ year^{-1}).

cific cases it may be possible by arranging proper combination of phases, e.g., basaltic liquid, pyroxene, and spinel, to obtain a situation where the liquid REE abundances are unfractionated with respect to the bulk solid. This is, however, a special case that would probably not produce the other element abundances and ratios, e.g., Ti, Zr, Al, and Cr (*1*), observed in the Tranquillity Base basalts. It has been proposed that the relative lack of REE fractionation in the ORB is due to complete extraction of the REE from the earth's mantle (*4*). The low Al content and high Ti content found for the lunar materials is quite inconsistent with this hypothesis for the lunar case. A detailed study of the ORB shows that the total REE, FeO, and TiO_2 contents increase together and, furthermore, that those rocks high in total REE have small negative Eu anomalies (*5*). The explanation proposed for these variations in the ORB abundance patterns is the removal of basic plagioclase and olivine (*5*). A similar hypothesis is proposed here to explain the abundance patterns observed in the lunar basalts. The original composition and pattern of a liquid consisting of 6 parts of plagioclase (An 96), 1½ parts olivine (Fo 80) and 2½ parts of "coarse grained" liquid (10062 and 10003 averaged) are shown in Fig. 1. We recognize that absence of pyroxene from such a liquid may be unusual. If pyroxene were removed from the parent liquid in significant amounts, less than 10 percent, the Cr content of this initial liquid would have to be over 1 percent by weight (*4*). Assuming that the distribution coefficients between liquid and plagioclase for Eu, Sr, Ba, and REE are 1.3, 0.7, 10, and 20 (*2*), respectively, the CaO, Al_2O_3, TiO_2, FeO, and MgO concentrations of such an original liquid would be about 15, 25, 3, 7½, and 8 percent, respectively. If this hypothesis is correct, we infer that cumulates rich in plagioclase should occur on the lunar surface. It is also of some interest that the results of the Surveyor 7 analysis (*6*) indicate unusually high Al_2O_3 (21 percent) and CaO (15 percent) contents and relatively low Fe contents are found in highland regions.

The depletion of Ba and La relative to heavy REE and broad peak at Sm and Gd is not explained by this process. These features could be due to other minor phases that have crystallized and separated from the liquid or to fractionation that takes place during production of parent liquid in the lunar interior. If the proposed original liquid is produced by partial melting in a lunar mantle (< 20 percent) we infer that the U, Th, and K contents of the parent lunar mantle are less than 0.08, 0.3 and 100 ppm, respectively.

The Sr^{87}/Sr^{86} ratios and Rb^{87}/Sr^{86} ratios of six igneous rocks are shown in an isochron diagram in Fig. 2. With the exception of sample 10017, these data define an isochron indicating that the time of crystallization was 3.5 ± 0.3 billion years ago. We have no explanation for sample 10017 but note again that our sample was a surface chip. The limits placed on the difference between the initial ratio (0.6991) and the achondrite initial ratio, 0.6989 (*7*), indicate that the source of these igneous rocks has a rubidium-strontium ratio less than one-tenth that of chondrites. If the igneous liquids were derived from the interior of the moon 3.5 ± 0.3 billion years ago, we conclude that large portions of the moon must be even more depleted in rubidium relative to strontium than the earth (*8*). The possibility that the high abundance of refractory elements e.g., Ca, Al, TiO_2, REE, and so forth, is a primary characteristic of a large part of the moon should not be overlooked. Unless this possibility is excluded one cannot infer general differentiation of the moon from the existence of basaltic rocks on the lunar surface.

PAUL W. GAST
NORMAN J. HUBBARD
Lamont-Doherty Geological Observatory, Palisades, New York

References and Notes

1. Lunar Sample Preliminary Examination Team, *Science*, **165**, 1211 (1969).
2. C. C. Schnetzler and J. A. Philpotts, *Geochim. Cosmochim. Acta*, in press.
3. F. A. Fay, *ibid.* **33**, 1429 (1969).
4. P. W. Gast, *ibid.* **32**, 1057 (1968).
5. R. Kay, N. J. Hubbard, P. W. Gast, *J. Geophys. Res.*, in press.
6. J. H. Patterson, E. J. Franzgrote, A. L. Turkevich, W. A. Anderson, T. E. Economou, H. E. Griffin, S. L. Grotch, K. P. Sowinski, *ibid.* **74**, 6120 (1969).
7. D. A. Papanastassiou and G. J. Wasserburg, *Earth Planet. Sci. Lett.* **5**, 361 (1969).
8. P. W. Gast, in *History of the Earth's Crust*, R. A. Phinney, Ed. (Princeton Univ. Press, Princeton, N.J., 1968), pp. 15-27.
9. F. A. Frey, M. Haskin, J. Poetz, L. Haskin, *J. Geophys. Res.* **73**, 6085 (1968).
10. This research was supported by NASA grant NAS 9-7895. The chemical separations, data reductions, and Na and Zn determinations were done by H. Wiesmann. We thank R. Kay for operating the computer and Chi-Yu Shih for doing the CaO and TiO_2 determinations. Lamont-Doherty Geological Observatory contribution No. 1441.

4 January 1970

Rare Earth Elements in Returned Lunar Samples

Abstract. *A linear correlation between concentrations of Sm and ratios of Sm to Eu for nine lunar samples suggests that those samples could correspond to liquids from equilibrium partial melting of a common source. On the basis of partition coefficients in terrestrial systems, the fraction of melting would not have exceeded about 15 percent and the immediate source could have been composed of olivine, orthopyroxene, and opaque minerals plus at least 25 percent feldspar, with at most a few percent calcic clinopyroxene and less than 1 percent apatite. The large Eu depletions could also have been produced by fractional crystallization if the ratio of Eu^{2+} to Eu^{3+} in lunar magmas significantly exceeds the values for terrestrial magmas.*

We present here concentrations for 12 of the rare earth elements (REE) in nine samples of lunar material returned to earth by the Apollo 11 astronauts (*1*). Included are five fine-grained igneous rocks (type A), one medium-grained igneous rock (type B), two breccias (type C), and "fine" material less than 1 mm in diameter (type D) from the lunar soil. The igneous rocks were received as 1-g chips, and the breccias and soil as cuts of powders. The concentrations of the REE were measured by neutron activation analysis; the REE were separated as a group from the irradiated samples and then were assayed by Ge-Li gamma-ray spectrometry.

The REE concentrations for the lunar samples plus those for a terrestrial basalt which was analyzed in the same irradiation as the lunar breccias are given in Table 1. The accompanying uncertainties are based strictly on counting statistics and are applicable to the comparison of one REE with the others in the same rock. Values for all the REE in any particular sample may be systematically in error by 1 to 2 percent.

Relative and absolute REE abundances for these samples are compared in Fig. 1 with those for a composite of nine chondritic meteorites (*2*). The ratios of the concentrations in parts per million for each element in a sample to the concentration in parts per million for the same element in the chondrites are plotted as a function of REE atomic number (*3*). Included as a dashed line

Table 1. Rare earth concentrations in parts per million for nine lunar materials and an oceanic andesite*.

Element	Sample number and type									
	10022 A	10057 A	10072 A	10003 B	10020 B	10045 B	10048 C	10060 C	10084 D	EPR-D-3 basalt
La	26.4 ± 0.7	28.2 ± 0.4	22.7 ± 0.5	14.1 ± 0.2	8.4 ± 0.2	6.7 ± 0.3	20.2 ± 0.3	20.8 ± 0.4	16.6 ± 0.4	20.4 ± 0.4
Ce	68 ± 2	75 ± 1	69 ± 1	41.3 ± 0.7	25.6 ± 0.6	22.5 ± 0.4	56 ± 1	58 ± 1	47 ± 1	61 ± 1
Nd	66 ± 4	69 ± 4	51 ± 4	42.5 ± 3	31 ± 3	21 ± 2	41 ± 7	46 ± 15	43 ± 6	57 ± 10
Sm	21.2 ± 0.1	20.8 ± 0.1	17.9 ± 0.2	13.1 ± 0.1	9.94 ± 0.06	7.94 ± 0.05	14.8 ± 0.2	15.4 ± 0.1	13.7 ± 0.1	15.9 ± 0.1
Eu	2.04 ± 0.01	2.18 ± 0.02	2.07 ± 0.04	1.80 ± 0.01	1.75 ± 0.01	1.52 ± 0.03	1.95 ± 0.02	2.06 ± 0.02	1.77 ± 0.02	4.05 ± 0.03
Gd	25 ± 2	26 ± 3	26 ± 2	17 ± 2	16 ± 2	12.3 ± 0.9	20 ± 1	24 ± 2	15 ± 2	22 ± 2
Tb	4.7 ± 0.1	5.0 ± 0.1	4.3 ± 0.1	3.26 ± 0.07	2.59 ± 0.06	2.11 ± 0.06	3.5 ± 0.1	3.6 ± 0.1	3.2 ± 0.2	3.6 ± 0.1
Dy	31.2 ± 0.3	34.7 ± 0.2	31.2 ± 0.3	22.4 ± 0.3	17.8 ± 0.1	14.4 ± 0.1	25.0 ± 0.3	26.3 ± 0.2	20.1 ± 0.1	25.1 ± 0.2
Ho	5.5 – 8.7	5.8 – 7.0	5.2 – 8.4	3.7 – 4.4	3.0 – 3.6	2.2 – 3.5	4.2 – 5.2	4.4 – 5.4	4.2 – 4.7	4.5 – 5.5
Er	16 ± 3	19 ± 2	16 ± 3	12 ± 1	9 ± 1	8.7 ± 2	14 ± 4	16 ± 5	12 ± 2	19 ± 6
Yb	17.7 ± 0.5	18.8 ± 0.2	16.6 ± 0.5	11.9 ± 0.3	9.8 ± 0.2	8.6 ± 0.3	12.5 ± 0.3	13.2 ± 0.3	11.3 ± 0.3	14.8 ± 0.3
Lu	2.55 ± 0.02	2.66 ± 0.02	2.24 ± 0.01	1.69 ± 0.01	1.41 ± 0.01	1.17 ± 0.01	2.10 ± 0.02	2.00 ± 0.02	1.66 ± 0.01	2.29 ± 0.02

*This rock was classified as a basalt in reference *4* but is now known to be an andesite (S. Hart, private communication).

in the upper part of Fig. 1 are the ratios relative to chondrites of a composite of shales (*2*). The REE abundances of these shales are typical of surface matter from the terrestrial continents.

The curves for rocks 10057 and 10045 (Fig. 1) represent the extremes of REE concentrations among our nine samples. The REE concentrations for the soil, which presumably represents some kind of average for the area around Tranquillity Base, fall between the extremes and are about 10 percent lower for the lighter REE and 20 percent lower for the heavier REE than the average for the other eight samples. Except for Eu, the relative abundances of the REE in these lunar samples are rather similar to those in the chondrites, in sharp contrast to terrestrial continental matter. Relative to the other REE, and with respect to chondrites, Eu is much more depleted (0.25 to 0.5 times the interpolated values from Fig. 1) in these lunar materials than in the shales (0.67 times). The similarity in REE abundances (except for Eu) between the two lunar breccias and the basalt from the East Pacific Rise (lower part of Fig. 1) is remarkable.

There is no compelling reason to presume that the REE abundances in the samples we analyzed are representative of the entire Sea of Tranquillity or of lunar maria in general. The maria comprise about one-fifth of the lunar surface. If the average abundance of a REE, for example, Sm, in the whole moon is 0.18 ppm as in the chondrites, then at an average density of 3.3 g/cm^3, for every kilometer of depth, the mare could contain as much as 15 percent of all the lunar Sm.

The similarity between the REE abundances in these lunar samples and those in the oceanic andesite does not suggest that the moon came from the Pacific Ocean basin. The sample chosen for comparison, EPR-D-3, is three times richer in REE than most other subalkaline basalts from the ocean ridges and has a significant Eu anomaly which most of these basalts do not have (*4*). There are significant differences in bulk composition (*1, 4*). Calcium-rich achondrites also have relative REE abundances similar to those in chondrites but the absolute REE abundances are several times greater (*5*). We suggest that similar processes produced the similar REE abundances in the oceanic and lunar rocks. Both types of rocks are igneous, most are fine-grained, and they presumably crystallized fairly rapidly from silicate melts (*1*).

Processes which act on or generate silicate melts and which can reasonably be expected to alter REE abundances in them are fractional crystallization and partial melting. We consider here only the problem of generating by those processes the variations in abundances of Eu relative to Sm and Gd in these lunar rocks. We presume also that the *relative* REE abundances found in chondrites are those of the starting material for objects native to the solar system (*5*) and would like to account for the Eu anomalies relative to chondrites by those processes.

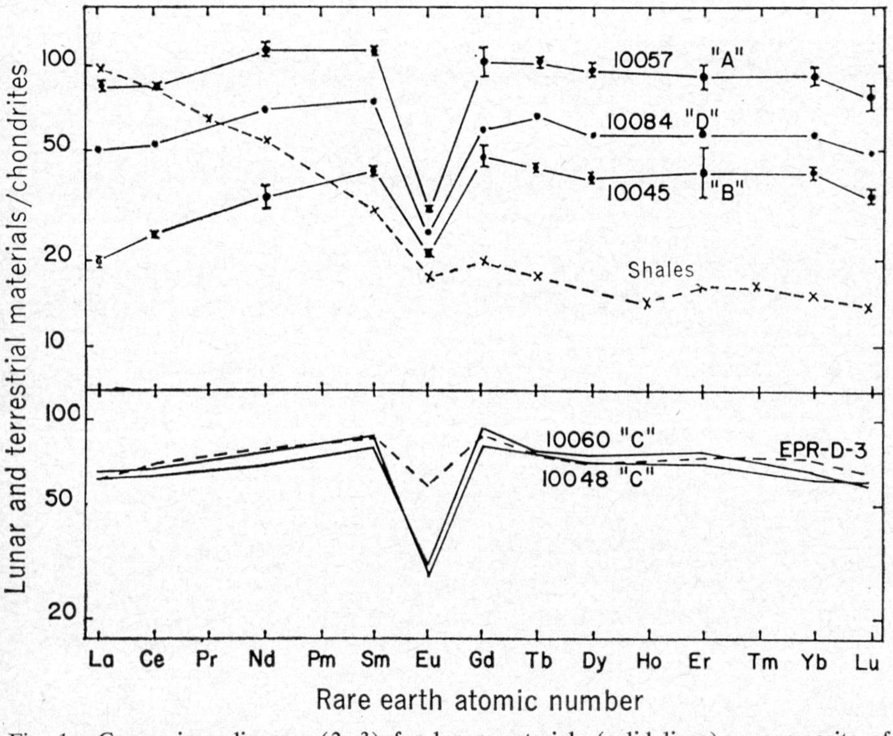

Fig. 1. Comparison diagram (*2, 3*) for lunar materials (solid lines), a composite of North American shales (*2*) (upper dashed line), and a submarine andesite (*4*) from the East Pacific Rise (lower dashed line).

It is convenient for this purpose to use the ratios of concentrations of Sm to Eu. From Figs. 1 and 2 it can be seen that the ratios of Sm to Eu increase as the absolute abundances of Sm and the other REE in the rocks increase. Europium is taken up in preference to the other REE by feldspars, presumably through its reduction to the 2+ oxidation state (*5, 6*). When minerals separated from a basic rock are analyzed, most of the Eu is usually found in the feldspar; the bulk of the other REE is found in clinopyroxenes and in apatite and other minor REE-concentrating minerals when they are present. Europium has not been found to be preferentially concentrated in any other mineral. Crystallization of feldspar from a silicate liquid should therefore give rise to a smooth variation of the ratio of Sm and Eu in the residual liquid. Alternatively, partial melting to different extents of a rock containing feldspar should yield liquids with smoothly varying ratios of Sm and Eu.

Mathematical models for the behavior of trace elements during these processes have been developed (*7*). Their use depends on knowledge of partition coefficients for those elements between the liquid and solid phases present. Partition coefficients, D, taken here to be the ratio of the concentration of the element in a solid phase to its concentration in the liquid, have been obtained by analysis of phenocrysts and host matrices, treatment of associated gabbros and ultramafics as melt-residue pairs, analyses of minerals in the Skaergaard and Kiglapait intrusions, and laboratory studies of synthetic systems (*8*). Some variations have been observed, but the general agreement among these several methods indicates that the values of interest can be used with confidence, at least for common terrestrial silicate liquids. We have used the following values in our calculations; D(Eu, plagioclase), 0.35; D(Sm, plagioclase), 0.03; D(Eu and Sm, clinopyroxene), 0.4; D(Eu and Sm, apatite), 10; D(Eu and Sm, olivine, orthopyroxene, and opaque minerals), <0.01.

During fractional crystallization, when the fraction of a system with average Sm content $C_{\mathrm{Sm},A}$ that has crystallized is x, the concentration of Sm in the residual liquid $C_{\mathrm{Sm},L}$ is

$$C_{\mathrm{Sm},L} = C_{\mathrm{Sm},A}\,(1-x)^{D(\mathrm{Sm})-1} \qquad (1)$$

The ratio of the concentration of Sm to that of Eu in the system is

$$C_{\mathrm{Sm},L}/C_{\mathrm{Eu},L} = (C_{\mathrm{Sm},A}/C_{\mathrm{Eu},A})\,(1-x)^{D(\mathrm{Sm})-D(\mathrm{Eu})} \qquad (2)$$

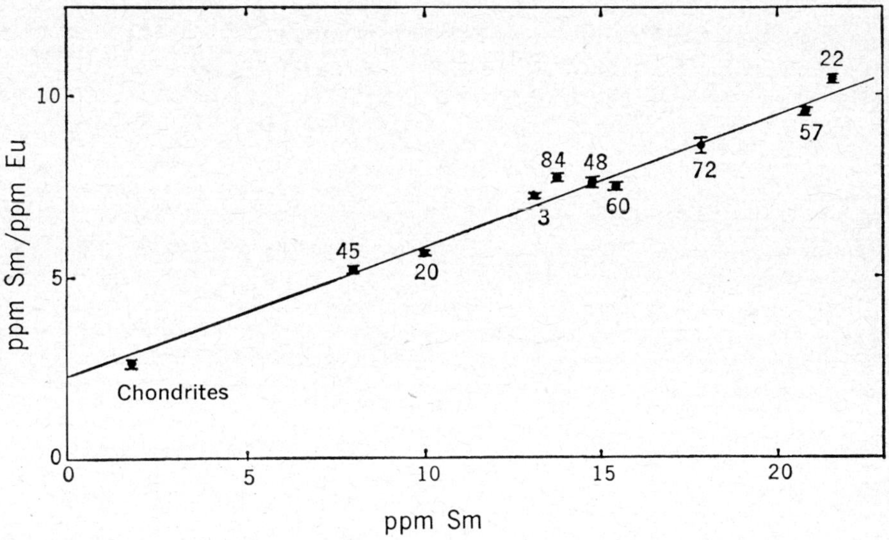

Fig. 2. Values for the ratios of Sm to Eu in the lunar samples vary linearly with concentrations of Sm. The line was obtained by a least-squares analysis of the nine points for the lunar samples.

The extent of separation between Eu and Sm depends on the *difference* between the partition coefficients. If we take a value for D(Eu, plagioclase) of 0.35, the maximum published value, and a value for D(Sm, plagioclase) of 0.03, the minimum conceivable value, and begin with a melt which has the same ratio of Sm to Eu as chondrites, 97 percent of the melt would have to crystallize *as feldspar* to produce the ratio found in the lunar soil, 88 percent of the melt would have to crystallize as feldspar to produce the ratio of rock 10045, and 99 percent of the melt would have to crystallize as feldspar to produce the ratio of rock 10022. This is clearly impossible. Simultaneous crystallization of other minerals would increase the necessary fraction crystallized, since the difference between the partition coefficients would be decreased.

Could the lunar rocks represent residual liquids from a system with an initial ratio of Sm to Eu lower than that of chondrites? This idea not only leaves unsolved the problem of generating such a ratio of Sm to Eu in the starting material, but also can be ruled out on the basis of the following argument. Elimination of x between Eq. 1 for Sm and the corresponding equation for Eu gives Eq. 3, which is linear in ln $C_{\mathrm{Sm},L}/C_{\mathrm{Eu},L}$ and ln $C_{\mathrm{Sm},L}$.

$$\ln\frac{C_{\mathrm{Sm},L}}{C_{\mathrm{Eu},L}} = \frac{D(\mathrm{Sm})-D(\mathrm{Eu})}{D(\mathrm{Sm})-1}\ln C_{\mathrm{Sm},L} + \ln\left[\frac{C_{\mathrm{Sm},A}}{C_{\mathrm{Eu},A}}\right]^{[D(\mathrm{Eu})-1]/[D(\mathrm{Sm})-1]} \qquad (3)$$

A plot of ln ($C_{\mathrm{Sm},L}/C_{\mathrm{Eu},L}$) as a function of ln ($C_{\mathrm{Sm},L}$) (not shown) is approximately linear. The slope, as determined by least-squares analysis, is 0.70; this requires that the value for D(Eu) exceed 0.71 for $D(\mathrm{Sm}) > 0$, which is twice as high as the value allowed from measurements of partition coefficients. On the basis of the above partition coefficients, the lunar rocks would not appear to be solidified residues from fractional crystallization. If lunar magmas are sufficiently reducing, however, that their ratio of free ion activity of Eu^{2+} to Eu^{3+} exceeds the ratio for terrestrial magmas, the variations in ratios of Sm to Eu might still be explained by fractional crystallization.

To test the possibility that the lunar samples might correspond to different extents of partial melting of a common starting material with a ratio of Sm to Eu equal to that in chondrites, we have made several calculations using the formulation of Gast (*7*). Partial melting of a rock consisting of 50 percent feldspar and 50 percent orthopyroxene, with equal amounts of the two minerals forming each increment of the melt, provides the ratio of Sm to Eu in rock 10022 in 4 to 5 percent partial melting, and that in rock 10045 in 15 percent partial melting. Reducing D(Eu, plagioclase) to 0.20 still makes it possible to produce the ratios in rocks 10022 and 10045 in less than 1 percent and 7 percent partial melting. Both models yield straight lines when plots like Fig. 2 are made.

The maximum ratio of Sm to Eu, relative to the ratio in the starting material, that can be obtained (by the minimum of partial melting) is equal to the ratio of the partition coefficients between the liquid and the bulk rock, which varies according to the constituent minerals and their proportions. As a consequence, models that contained less than 25 percent of plagioclase, or more than a few percent of clinopyroxene, or as much as 1 percent of apatite, failed.

The linear relationship of Fig. 2 does not correspond to simple mixing of two materials with different ratios of Sm to Eu; such mixtures generate parabolic curves. This may explain why the values for the soil and one of the breccias lie off the line. Such mixing also excludes any significant contribution from nonequilibrium partial melting of homogeneous minerals.

If the lunar rocks represent partial melts from a starting material composed of feldspar and mafic constituents such as olivine, orthopyroxene, and opaque oxides, the absolute contents of the REE and the magnitudes of the ratios of Sm to Eu should be inversely correlated with the relative amount of felsic components in the rocks. Comparison of our results with analyses for the major constituents (*1*) indicates that such is the case.

Thus we conclude that these lunar rocks may very well represent liquids from different stages of equilibrium partial melting from a common starting material which could have had a ratio of Sm to Eu equal to that in chondrites. The very limited deviation of the points from the least-squares line in Fig. 2 indicates that the compositions of these liquids were not appreciably modified by fractional crystallization before freezing.

Those models that were successful in producing the requisite ratios of Sm to Eu in the lunar rocks from that in chondrites failed by a factor of 2 to 3 to concentrate those elements as much as was required in order that the starting material have chondritic absolute abundances. That plus the requirement of a high content of plagioclase in the starting material suggest that the bulk composition of the moon more closely resembles that of basalts (or basaltic achondrites) than that of chondrites, or that, in analogy with the upper mantle of the earth (*9*), the components of plagioclase and the REE were strongly concentrated toward the exterior of the moon.

LARRY A. HASKIN
PHILIP A. HELMKE
RALPH O. ALLEN

Department of Chemistry, University of Wisconsin, Madison 53706

References and Notes

1. Lunar Sample Preliminary Examination Team, *Science* **165**, 1211 (1969).
2. L. A. Haskin, M. A. Haskin, F. A. Frey, T. R. Wildeman, in *Proceedings of the Symposium of the International Association of Geochemists and Cosmochemists, Paris, 1967* (Permagon Press, New York, 1968), pp. 889–912.
3. C. D. Coryell, J. W. Chase, J. W. Winchester, *J. Geophys. Res.* **68**, 559 (1963).
4. F. A. Frey, L. A. Haskin, *ibid.* **69**, 775 (1964); F. A. Frey, M. A. Haskin, J. A. Poetz, L. A. Haskin, *ibid.* **73**, 6085 (1968).
5. L. A. Haskin, F. A. Frey, R. A. Schmitt, R. H. Smith, *Phys. Chem. Earth* **7**, 167 (1966); L. A. Haskin and F. A. Frey, *Science* **152**, 299 (1966).
6. D. G. Towell, J. W. Winchester, R. V. Spirn, *J. Geophys. Res.* **70**, 3485 (1965); J. A. Philpotts and C. C. Schnetzler, *Chem. Geol.* **3**, 5 (1968); D. G. Towell, R. V. Spirn, J. W. Winchester, *ibid.* **4**, 461 (1969); J. A. Philpotts and C. C. Schnetzler, *ibid.*, p. 464.
7. W. L. McIntire, *Geochim. Cosmochim. Acta* **27**, 1209 (1963); P. W. Gast, *ibid.* **32**, 1057 (1968).
8. C. C. Schnetzler and J. A. Philpotts, in *Proceedings of the Symposium of the International Association of Geochemists and Cosmochemists, Paris, 1967* (Permagon Press, New York, 1968), pp. 929–939; N. Onuma, H. Higuchi, H. Wakita, H. Nagasawa, *Earth Planet. Sci. Lett.* **5**, 47 (1968); F. A. Frey, *J. Geophys. Res.*, in press; R. Cullers, G. Medaris, L. Haskin, in preparation; T. Paster and L. Haskin, in preparation.
9. P. W. Gast, *J. Geophys. Res.* **65**, 1287 (1960).
10. Supported in part by NASA contract NAS 9-7975. Samples were irradiated in the University of Wisconsin nuclear reactor. Dr. T. P. Paster and R. L. Korotev assisted with chemical separations and data processing.

4 January 1970; revised 5 January 1970

Trace Elements and Radioactivity in Lunar Rocks: Implications for Meteorite Infall, Solar-Wind Flux, and Formation Conditions of Moon

Abstract. *Lunar soil and type C breccias are enriched 3- to 100-fold in Ir, Au, Zn, Cd, Ag, Br, Bi, and Tl, relative to type A, B rocks. Smaller enrichments were found for Co, Cu, Ga, Pd, Rb, and Cs. The solar wind at present intensity can account for only 3 percent of this enrichment; an upper limit to the average proton flux during the last* 4.5×10^9 *years thus is* 8×10^9 cm^{-2} yr^{-1}. *The remaining enrichment seems to be due to a 1.5 to 2 percent admixture of carbonaceous-chondrite-like material, corresponding to an average influx rate of meteoritic and cometary matter of* 2.9×10^{-9} *g* cm^{-2} yr^{-1} *at Tranquility Base. This is about one-quarter the terrestrial rate. Type A, B rocks are depleted 10- to 100-fold in Ag, Au, Zn, Cd, In, Tl, and Bi, relative to terrestrial basalts. This suggests loss by high-temperature volatilization, before or after accretion of the moon. Positron activities due mainly to* ^{22}Na *and* ^{26}Al *range from 90 to 220* β^+ min^{-1} kg^{-1} *in five small rocks or fragments (9 to 29 g). The higher activities presumably indicate surface locations. Th and U contents generally agree with those found by the preliminary examination team.*

In a series of experiments, carried out by Keays, Ganapathy, Laul, and Anders, 12 lunar samples of types A to D and four samples of terrestrial rocks were analyzed for 16 trace elements by a radiochemical neutron activation procedure (*1*). Sample weights were typically ~0.1 g; the integrated neutron flux was 1×10^{20} cm^{-2} (Table 1).

The reproducibility of the method is indicated by triplicate analyses of lunar soil 84,49 (the third analysis, 84,49W, actually is a weighted mean of three individual analyses on sieve fractions, 84,49,1-3) and duplicates of Columbia River basalt BCR-1. Radiochemical purity was checked with γ-spectrum and half-life, in the case of a few Tl samples also by recycling. The absolute accuracy seems to be better than 10 percent for homogeneously distributed elements, judging from published analyses of basalt BRC-1.

Italicized values are considered meaningless or doubtful. They include all In and two Ag analyses on lunar samples, which were contaminated up to 10,000-fold the intrinsic level by In-Ag vacuum gaskets on sample return containers. One of our Ir values was surprisingly high, probably due to laboratory contamination. No data were obtained on Te, because of severe interference from fission-produced ^{131}Te-^{131}I. The Pd-numbers (determined via ^{111}Ag) are somewhat uncertain, because ^{111}Ag is also made from ^{109}Ag by double-neutron capture. Corrections for this effect were rather large, and not too well determined: ~10 percent for Ag/Pd = 0.5 and ~60 percent for Ag/Pd = 5. Sample 47,32 (a rock powder prepared at the LRL) gave higher values for 7 elements than did any other A or B sample. We suspect that this was due to contamination during crushing, and therefore did not use any of the data for this sample.

The first 9 elements in Table 1 are 2 to 100 times more abundant in type C and D material than in average A, B rock (last 3 lines of Table 1). Inasmuch as C (breccia) and D (soil) probably are largely derived from A, B rock, it seems likely that the enrichment is due to addition of extraneous material: meteoritic, cometary, and solar wind. (Other factors may also play a role: mechanical sorting, volatilization by shock melting, and so forth. It is difficult to correct for them, though; and the overall consistency of our results for elements greatly differing in geochemical character such as Ir, Br, and Bi suggests that these effects are not large.)

In order to characterize this extraneous component, we have plotted in Fig. 1 the "excess" of each element in type C and D material over that in average A, B rock. The data were normalized to type I carbonaceous chondrites (solid line). This approach had previously been used by Mazor and Anders (*2*) in a study of gas-rich meteorites.

It is striking that the data for most elements lie parallel to the carbonaceous chondrite line, at an abundance of 0.01 to 0.02. Insofar as these elements are concerned, types C and D could have been made from type A, B rocks by adding 1 to 2 percent of carbonaceous-chondrite-like material. Opik predicted a meteoritic component of 3 percent of unspecified composition (*2a*).

The match is not perfect: Zn, Cd, and one of the Ag points fall higher, at 3 to 8 percent. Perhaps our A, B "average" is unrepresentative, failing to include some Zn, Cd-rich rock. Indeed, data on major elements, lanthanides, and Hf (*3, 4*) suggest that a straight A, B average does not quite match C, D composition; some particular B rocks (for example 44 and 45) perhaps having contributed to a disproportionate degree (*4*). Alternatively, the "extraneous" component may have been richer in Zn, Cd than are type I carbonaceous chondrites. Figure 1 sets a few limits on the composition of the extraneous component. Type I and II carbonaceous chondrites and type I enstatite chondrites are acceptable source materials, as are other types of relatively undifferentiated cosmic matter: solar-wind and cometary material. (The high water content of some of these materials does not disqualify them, as such water would evaporate during impact and quickly be lost from the moon.) Ordinary chondrites and iron meteorites are excluded as major contributors: the former contain far too little Ag, Br, Bi, and Tl; and the latter, too much Ir and Au in relation to the remaining elements.

The solar wind, too, seems to have made only a minor contribution. The present-day proton flux is 2×10^8 cm^{-2} sec^{-1} (*5*); assuming a cosmic Ir/H ratio of 1.65×10^{-11} (*6*), we obtain an integrated flux of 4.9×10^{-7} g Ir/cm^2 in 4.5×10^9 years. The average thickness

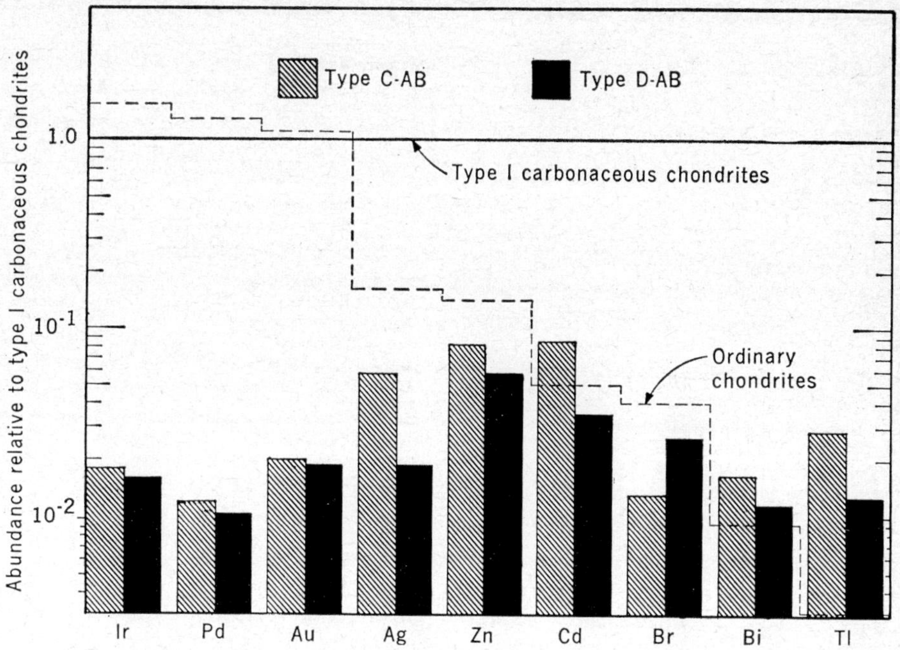

Fig. 1. Compared to A, B rocks, breccia (C) and soil (D) are enriched in nine trace elements. "Excess" component, obtained by subtracting average A, B abundances from C and D, shows abundance pattern resembling that of carbonaceous chondrites, not of ordinary chondrites. Apparently the enrichment was caused by the addition of 1 to 2 percent carbonaceous chondrite-like material.

Table 1. Abundances in Apollo 11 lunar samples and terrestrial rocks. Doubtful values are italicized. BCR-1 = Columbia River basalt; Knippa = Knippa basalt; APG = titanium-rich Adirondack pyroxene gabbro, 1 mile SSW of Brown Point, Willsboro Quadrangle. We are indebted to A. F. Buddington for this sample. Powdered material is available from G. G. Goles, Department of Geology, University of Oregon.

Sample	Type	Ir (ppb)	Au (ppb)	Zn (ppm)	Cd (ppb)	Ag (ppb)	Bi (ppb)	Tl (ppb)	Br (ppb)	Pd (ppb)	Co (ppm)	Cu (ppm)	Ga (ppm)	Rb (ppm)	Cs (ppb)	In (ppb)
20,25	A	0.027	0.075	1.29	6.37	2.27	0.15	0.33	19.2	1.5	5.65	6.57	1.90	0.74	30.6	*15*
57,31	A	0.023	0.017	1.71	3.15	0.69	0.27	1.09	21.7	7.3	27.2	3.52	>3.5	3.68	159	*3*
72,23	A	*4.02*	0.14	1.81	6.47	*17.3*	0.73	0.92	76	3	27.2	4.94	4.73	5.98	159	*179*
50,26	B	0.007	0.031	1.75	2.56	1.42	≦0.16	0.33	7.6	1.4	15.2	15.2	4.41	0.60	25.9	*4*
47,32*	B	*0.240*	*0.33*	*5.76*	*255*	*24.7*	*2.15*	0.57	*133*	2	14.4	13.3	5.35	1.25	44.6	*109*
48,31	C	6.88	2.66	28.6	78.3	23.6	1.62	2.83	119	13.0	34.8	9.19	5.85	4.15	128	*96*
61,31	C	9.18	3.42	29.2	106	*163*	2.79	2.70	69.2	7	34.2	22.0	5.79	3.99	146	*1430*
84,49	D	7.62	4.15	21.1	53.3	8.89	1.55	1.65	126	11.0	26.8	8.07	5.41	3.33	98	*524*
84,49	D	6.93	2.01	21.0	35.4	8.60	1.37	1.47	159	9.4	28.1	7.75	5.24	3.09	94	*768*
84,49W†	D	6.88	2.38	19.5	29.6	8.67	1.97	3.54	187	8.9	27.0	8.64	5.25	3.22	96	*1470*
84,49,1†	D	6.28	2.87	10.6	28.3	4.33	1.43	7.35	139	8.3	26.1	6.99	4.89	3.36	99	*12*
84,49,2†	D	6.67	1.99	21.0	28.3	8.53	1.96	1.91	189	9.2	26.7	8.94	5.34	3.17	94	*1950*
84,49,3†	D	10.77	4.41	34.2	46.0	24.1	3.77	6.74	317	8.6	32.5	11.2	5.72	3.27	104	*1690*
BCR-1	Bas.		0.43	127	99.0	26.4	40.2	274	97	13.0	35.4	17.1	20.1	44	930	96
BCR-1	Bas.		1.33	126		26.7	44.4	283	152	11.0	36.7	18.8	22.5	48	920	107
Knippa	Bas.		2.85	125			88.3	20.3	1300		76.8	15.0	20.2	26	745	115
APG	Gab.		3.59	125	160	34.8	5.84	24.5	73.8		67.6	17.6	1.41	3.31	104	112
A, B ave.		0.066	0.041	1.64	4.64	1.46	0.33	0.63	31.1	3.4	18.8	7.54	3.68	2.75	99	≦4
C ave.		8.03	3.04	28.9	92	23.6	2.20	2.76	94	9.9	34.5	15.6	5.82	4.07	137	≦100
D ave.		7.14	2.85	20.5	39.4	8.72	1.63	2.22	157	9.1	27.3	8.15	5.30	3.21	96	≦500

*Powder prepared at LRL; probably contaminated. †Sieve fractions: 1 = +100 mesh (24%); 2 = 100 to 325 mesh (69%); 3 = −325 mesh (7%); W = weighted average of individual analyses of fractions 1, 2, 3.

of the lunar regolith in the southern Sea of Tranquillity is 4.6 m (7). For a mean packed density of 1.8 g/cm^3, a 10 percent content of large rock fragments, and complete mixing, one would thus expect a mean Ir content of 2.0×10^{-10} grams of Ir/g, only about 2.7 percent of the observed value in our C and D samples. Perhaps the solar wind flux was higher in the past, but it is interesting that the observed Xe content in lunar soil, 3.6×10^{-7} cm^3 per gram at standard temperature and pressure (8) agrees with the amount expected for the present-day flux, 3.9×10^{-7} cm^3 per gram at standard temperature and pressure. The Ir content of C and D material can be used, though, to set an upper limit of $\leq 8 \times 10^9$ cm^{-2} sec^{-1} to the mean solar wind proton flux during the past 4.5 eons.

It appears that the bulk of the extraneous component is meteoritic or cometary. With the same assumptions as before, we find that an influx of 2.9×10^{-9} g cm^{-2} yr^{-4} of carbonaceous–chondrite-like material is needed to account for the observed average excess of Ag, Au, Bi, Br, Ir, Pd, and Tl. This figure agrees fairly well with an analogous estimate for the earth, $(1.2 \pm 0.6) \times 10^{-8}$ g cm^{-2} yr^{-1}, obtained by Barker and Anders (9) from the Ir and Os content of Pacific sediments. The agreement is even better if one allows for the disproportionately greater gravitational enhancement of the earth's capture cross section.

Not too much should be made of this agreement, though. The terrestrial value is an average for the last 10^5 years over some 10^7 to 10^8 km^2, while the lunar value is an average for $>10^9$ years at a single location. Both the mechanisms and efficiencies of lateral and vertical mixing are surely different for the two bodies, particularly for large projectiles that contribute most of the incoming mass. The compositional difference between C and D material, and the presence of a stratigraphic boundary in one of the Apollo 11 core samples (8) show that the Tranquillity Base regolith is not well mixed; thus our results for C and D samples may not be representative of the regolith as a whole. The apparent predominance of carbonaceous chondrite material is of interest, however. Several authors (2, 10) have previously suggested that carbonaceous chondrites comprise the major part of the meteoritic influx, but are underrepresented in collections because their friability leads to preferential destruction in the atmosphere.

Table 2. Thorium and uranium abundances in these lunar samples.

	^{232}Th (ppm)	^{238}U (ppm)	Sample
Chicago	2.0 ± 0.2	0.70 ± 0.8	72,41
Houston	2.9 ± 0.4	0.75 ± 0.11	
Chicago	2.5 ± 0.2	0.38 ± 0.15	18,24
Houston	2.3 ± 0.3	0.60 ± 0.09	
Chicago	2.2 ± 0.2	0.73 ± 0.06	61,21

The trace element enrichment in the − 325-mesh fraction of the fines (84, 49,3) is consistent with either an impact or a solar wind origin. Material closest to the impact center will be comminuted most severely and will receive the greatest admixture of projectile matter. Grain crushing by subsequent impacts would be an additional mechanism tending to enrich the finest fraction in meteorite material.

Some highly tentative conclusions about the moon may be drawn from a comparison of A, B rocks with other materials. A suitable terrestrial reference material is the Adirondack pyroxene gabbro (APG) (Table 1) whose mineralogy and titanium content match that of the Apollo 11 rocks (11). Of the elements analyzed by us, Ag, Au, Cd, Zn, In, Tl, and Bi are 1 or 2 orders of magnitude less abundant in A, B rocks than in the pyroxene gabbro, while most others, including some major elements measured by Schmitt (private communication), agree within a factor of 3. Essentially similar trends are found if average continental basalts (12) are used for comparison.

It is surprising that siderophile elements (Co, Ir, Au) are more depleted in A, B lunar rocks than in their terrestrial analogs. Apparently the removal of siderophiles from the crust was more efficient on the moon, although its metal-phase content is at most 1/10 that of the earth. Perhaps conditions were more reducing, or the scarcity of water and sulfur depressed the solubility of siderophiles in the silicate phase.

The remaining elements (Ag, Cd, Zn, In, T1, and Bi) are volatile and have large ionic radii; they are therefore expected to concentrate rather strongly in the crust (13, 14). The most straightforward inference to be drawn from their low abundance is that the moon accreted at higher temperatures than did the earth (that is, above 600° to 700°K), and acquired a lesser proportion of low-temperature volatile-rich material in the terminal stages of accretion: perhaps only 10^{-4} to 10^{-5} lunar masses. The scarcity of water and carbon certainly agrees with this view. But the alternative, that the volatiles were first accreted and then lost in a high-temperature stage, cannot be completely excluded. Loss by exospheric escape requires unreasonably high temperatures, around 3000°K, for elements such as Bi and Tl. But loss by vaporization into a low molecular weight atmosphere which is subsequently swept away requires only about 1500°K. Although this mechanism requires rather special circumstances, one cannot yet rule it out altogether. We must also bear in mind the possibility that the major part of the moon's Bi, Tl, and so on is located elsewhere in the lunar crust, or retained in the interior to a greater degree than on earth.

Our data show numerous interelement correlations in A, B rocks, only one of which will be mentioned here. The covariance of Tl with Rb and Cs suggests that Tl is mainly lithophile in the lunar crust, as in the earth but not in the chondrites.

In a separate set of experiments carried out by authors Herzog and Jeffery, the positron activities of five small lunar samples (9 to 29 g) were measured by gamma-gamma coincidence spectrometry. One of the spectrometers has been described previously (15); the other, with 7.6 cm × 3.8 cm and 7.6 cm × 10.2 cm NaI detectors, had circuitry for two-parameter analysis and was operated at the bottom of a 16-m well for maximum background reduction.

Counting efficiencies for ^{22}Na, ^{26}Al, ^{44}Ti, ^{46}Sc, ^{56}Co, ^{58}Co, ^{60}Co, ^{232}Th, and ^{238}U were determined by means of individual mock-ups (15, 16) containing known amounts of the radionuclides dispersed in a matrix of Fe, MgO, or dunite powder matching the lunar sample in bulk density.

In order to determine total positron activities, gross counting rates in the 511-511 kev coincidence channel were corrected for contributions from ^{232}Th and ^{238}U, typically about 5 percent. All but 5 percent of the positron activity was due to ^{22}Na and ^{26}Al. Rock 18,24 exhibited a positron activity of 107 ± 10 β^+ min^{-1} kg^{-1} in good agreement with the preliminary result (8) of 135 ± 25 β^+ min^{-1} kg^{-1} despite the fact that our sample was only a 25.6 g fragment of the 213 g rock counted by the preliminary examination team. Our result for sample 72,41, of 190 ± 20 β^+ min^{-1} kg^{-1}, on the other hand, lies nearly a factor of two above that found for the whole rock (8) (97 ± 18 β^+

min^{-1} kg^{-1}). Since variations in composition are presumably too small to account for such a large difference we infer that sample 72,41 came from near the top surface of the rock where it was exposed to the low energy component of solar and galactic cosmic radiation. On the same basis we deduce that 26,10 (an entire small rock with an activity of 217 ± 20 β^+ min^{-1} kg^{-1}) was also exposed while 44,36 (an 11.2 g fragment of a 248 g rock with an activity of 88 ± 7 β^+ min^{-1} kg^{-1}) was shielded. The type C rock fragment 61,21 had a positron activity of 103 ± 5 β^+ min^{-1} kg^{-1} (*17*).

While the positron activities of these small samples can be determined to within 5 percent in a week's counting, accurate resolution of ^{22}Na and ^{26}Al requires counting times of a month or more. We hope to present data for ^{22}Na and ^{26}Al at a later time. In contrast, Th and U can typically be determined to within better than 20 percent in a week's counting time; data for three samples are given in Table 2. The results are generally in accord with those of the preliminary examination team (*8*).

REID R. KEAYS, R. GANAPATHY
J. C. LAUL, EDWARD ANDERS
G. F. HERZOG, P. M. JEFFERY
Enrico Fermi Institute and
Department of Chemistry, University of Chicago, Chicago, Illinois 60637

References and Notes

1. R. R. Keays and R. Ganapathy, in preparation.
2. E. Mazor and E. Anders, *Geochim. Cosmochim. Acta* **31**, 1441 (1967).
2a. E. J. Öpik, *Annu. Rev. Astron. Astrophys.* **7**, 473 (1969).
3. G. G. Goles, R. A. Schmitt, W. D. Ehmann, H. Wakita, K. Randle, M. Osawa, J. W. Morgan, *Science*, this issue.
4. G. G. Goles, M. Osawa, D. Randle, R. L. Beyer, D. Y. Jérome, D. J. Lindstrom, M. R. Martin, S. M. McKay, T. L. Steinborn, *ibid.*
5. D. Tilles, *ibid.* **148**, 1085 (1965).
6. A. G. W. Cameron, *Origin and Distribution of the Elements*, L. H. Ahrens, Ed. (Pergamon Press, Oxford, 1968), p. 125.
7. V. R. Oberbeck and W. L. Quaide, *Icarus* **9**, 446 (1968).
8. Lunar Sample Preliminary Examination Team, *Science* **165**, 1211 (1969).
9. J. L. Barker, Jr., and E. Anders, *Geochim. Cosmochim. Acta* **32**, 627 (1968).
10. R. E. McCrosky, *Smithsonian Astrophysical Observatory, Special Report 280* (1968); E. M. Shoemaker, *Meteorites* **3**, 123 (1967).
11. E. Olsen, *Science* **166**, 401 (1969); A. F. Buddington, *Geol. Soc. Amer. Mem.* **7**, 36 (1939).
12. S. R. Taylor, *Geochim. Cosmochim. Acta* **28**, 1273 (1964).
13. J. W. Larimer, *ibid.* **31**, 1215 (1967); J. W. Larimer and E. Anders, *ibid.*, p. 1239 (1967); E. Anders, *Acc. Chem. Res.* **1**, 289 (1968).
14. A. E. Ringwood, *Geochim. Cosmochim. Acta* **30**, 41 (1966).
15. D. Heymann and E. Anders, *ibid.* **31**, 1793 (1967).
16. We are indebted to G. D. O'Kelley and J. Eldridge of the Oak Ridge National Laboratory for the ^{26}Al, ^{44}Ti, ^{46}Sc, ^{56}Co, and ^{58}Co standards, to J. A. S. Adams of Rice University for the ^{232}Th and ^{238}U standards, and to H. R. Heydegger of the University of Chicago for the ^{60}Co standard.
17. Positron activities quoted have been corrected to 12/1/69.
18. We are indebted to A. Pierce and F. Quinn for technical assistance, and to M. E. Lipshutz of Purdue University for the use of counting equipment. We gratefully acknowledge the assistance of G. F. Herman in counting the samples, and of J. R. Naples in programming and data reduction. This project was supported by AEC contract AT(11-1)-382 and NASA contract NAS 9-7887.

4 January 1970

Potassium, Rubidium, Strontium, Barium, and Rare-Earth Concentrations in Lunar Rocks and Separated Phases

Abstract. *Concentrations of potassium, rubidium, strontium, barium, and rare-earth elements have been determined by mass spectrometric isotope dilution for eight Apollo 11 lunar samples and for some separated phases. Potassium and rubidium are at chondritic levels, strontium at 15 times, and barium and rare earths at 30 to 100 times chondritic levels. There are trace element similarities between the lunar samples and basaltic achondrites, terrestrial dredge basalts and the bulk earth. The trace element data appear to be consistent with these lunar samples being the result of limited partial fusion of some material similar to the brecciated eucrite meteorites.*

Concentrations of K, Rb, Sr, Ba, and rare-earth elements in eight Apollo 11 lunar samples, in a fine fraction from the soil, and in mineral fractions separated from two of the igneous rocks have been determined. Preliminary values for K, Rb, Sr, Ba, and Yb contents in Apollo 11 samples, obtained by optical spectrography and gamma-ray spectrometry, have been presented (*1*). Our trace-element analyses were done by mass-spectrometric stable-isotope dilution (*2*). The precision of our whole-rock data is, with a few random exceptions, better than about ±2 percent; accuracy is believed to be quite similar. The quality of the mineral separate data is somewhat poorer because of significant blank corrections, which exceed 10 percent in three cases, and because of less than optimum spiking. Our analytical data are given in Table 1. Values for four of the whole rocks, normalized to average chondritic meteorite concentrations (given in Table 1), and covering the observed concentration range of our samples, are shown in Fig. 1. The order in which the elements are presented in the table and figures was selected to permit ready estimation of important elemental ratios.

It is apparent from Table 1 and from Fig. 1 that to a first approximation, all our whole-rock samples are quite similar in trace-element characteristics. It may be significant that two of our three igneous rocks are nearly identical in all of the trace elements studied, and that the four breccias are very similar to each other; a limited number of provenance sites may be indicated for the Apollo 11 samples.

The relationships between the trace element concentrations in the lunar samples and in the average chondritic meteorite are apparent in Fig. 1. The lunar samples have trace element characteristics similar to those of the basaltic achondrite meteorites (*3*, *4*). The similarity is brought out in Fig. 1 by our data for Juvinas, a typical brecciated eucrite. There is some difference in the absolute level of the trace element concentrations. The level in Stannern is the highest that we know of in the basaltic achondrites, being about double that of Juvinas; Stannern is also of interest in showing a definite negative Eu anomaly. The similarity in K/Ba for the lunar samples and the basaltic achondrites may be particularly significant, inasmuch as this ratio tends not to change much during igneous differentiation (*5*). Angra dos Reis, a unique titaniferous-augite achondrite, has rare-earth concentrations (*4*), including a negative Eu anomaly, that are distinctly similar to those in the lunar samples. However, Ba is lower, and K and Rb much lower, in Angra dos Reis. It might be significant that sample 62 displays a tendency in this direction relative to the other lunar samples; sample 62 may have contained considerable cumulative feldspar and pyroxene at some stage in its evolution.

Trace-element concentrations in the terrestrial continental crust (*6*) are quite different from those in the lunar samples. K and Rb are about an order of magnitude higher, Sr about three times higher, and the heavier rare earths a factor of 3 to 4 lower than in the lunar samples. Tektite trace element contents resemble those of the continental crust rather than those of the lunar samples. The lunar samples show a marked resemblance to later differentiates dredged from the ocean floor (*7*, *8*). The similarity might arise from their both possibly represent-

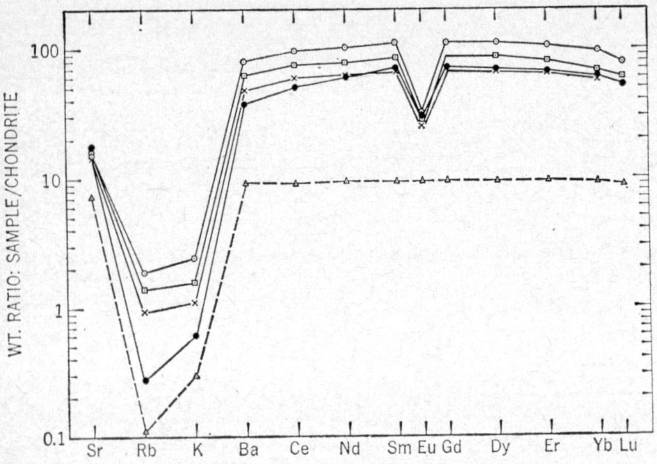

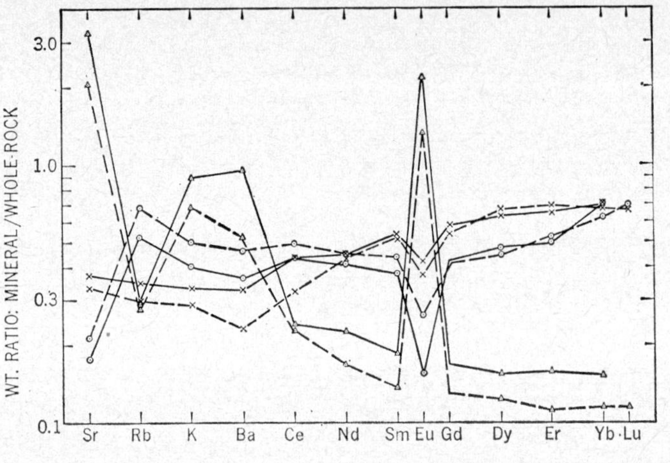

Fig. 1. (left) Chondrite-normalized Sr, Rb, K, Ba, and rare-earth concentrations in crystalline rocks 17,44 (circled dot) and 62,29 (solid circle); breccia 60,19 (square); soil 84,130 (times sign); and basaltic achondrite Juvinas (triangle). Fig. 2. (right) Whole-rock normalized Sr, Rb, K, Ba and rare-earth concentrations in plagioclases (dotted triangle), pyroxenes (times sign), and opaques (circled dot) separated from crystalline rocks 44,24 (connected by solid lines) and 62,29 (connected by dashed lines).

ing low pressure differentiation products from similar source materials. Differences, such as the higher K and lower Ba relative concentrations, and the smaller Eu anomaly of the terrestrial samples, might be due to the residual nature of their source rocks (*7*) or to the influence of hydrous or high-pressure phases and less reducing conditions. It is worth noting that, although there are small differences such as the slightly lower K/Ba, the lunar sample trace element concentrations are similar in a relative sense to those expected for the bulk earth (*6*). This might tend to favor a fission hypothesis, or a propinquitous aggregation hypothesis, rather than the capture hypothesis for the origin of the moon. Alternatively, the similar trace element characteristics of the moon, bulk earth, and the basaltic achondrites may reflect similar processes occurring in different parts of the solar system.

Trace element data for plagioclase, pyroxene, and opaque (presumably largely ilmenite) fractions separated from the igneous rocks 44 and 62, are presented in Fig. 2, normalized to the concentrations in the respective whole rocks. The mineral separates were obtained by using a combination of electromagnetic, heavy-liquid, and hand-picking techniques; they were estimated to be more than 97 percent free of contaminating grains, but some foreign material included within grains may have gone unnoticed, particularly in the case of the opaque separates. Because of the fine grain-size of the rocks and the stringent purification, the mineral yields were very small. Hence the separated phases were probably not typical of these phases in the whole rock—the procedures favor separation of coarse grains, presumably grains that started to crystallize earlier. This might explain the fact, apparent in Fig. 2, that the mineral separates do not contain the various trace elements in concentrations sufficient, in general, to yield the whole-rock values; that is, other pyroxene, feldspar, or ilmenite fractions in the whole rock, presumably forming later than the phases separated, may well have higher trace element concentra-

Table 1. Abundances of Sr, Rb, K, Ba, and rare earths, in parts per million by weight.

Sample Nos. (and rock type)	Element												
	Sr	Rb	K	Ba	Ce	Nd	Sm	Eu	Gd	Dy	Er	Yb	Lu
17,44 (A)	165	5.70	2390	287	75.5	66.1	23.4	2.26	28.4	32.8	19.1	18.1	2.63
44,22(B)	167	5.64	2430	285	76.6	66.1	23.4	2.21	28.6	33.6	19.3	16.6	
62,29 (B)	194	0.832	601	134	40.2	39.9	14.7	2.07	18.1	20.6	11.8	11.3	1.76
18,22 (C)	164	3.79	1490	200	52.8	45.4	16.3	1.84	20.5	21.8	12.8	11.8	1.87
21,24 (C)	165	4.03	1550	211	57.2	48.9	17.2	1.91		25.2	13.0	12.7	1.99
46,18 (C)	165	4.22	1620	219	58.0	50.2	17.7	1.94	21.5	24.9	14.8	13.0	
60,19 (C)	167	4.13	1570	224	59.0	50.0	17.5	1.98		26.2	14.5	12.7	1.92
84,130 (D)	162	2.78	1120	170	46.1	40.5	13.9	1.77		19.5	11.7	10.6	
84,130 (<400 mesh)	173	2.92	1100	180	46.8	40.9	14.0	1.84	16.8	18.9	11.1	9.91	1.54
44,24													
Pyroxene	62.6	1.96	811	93.7	33.3	29.9	12.6	0.93	16.6	21.3	12.7	11.6	1.85
Plagioclase	541	1.67	2160	271	18.5	15.0	4.34	4.79		4.43	3.04	2.54	0.401
Opaque	29.6	2.95	987	104	33.3	27.2	8.84	0.345		13.6	9.64	11.8	2.06
62,29													
Pyroxene	64.6	0.248	173	30.9	13.0	17.3	7.71	0.768	9.7	13.8	8.14	7.57	1.19
Plagioclase	396	0.230	413	70.1	9.19	6.75	2.0	2.75		2.56	1.32	1.29	0.20
Opaque	41.3	0.571	303	62.1	20.2	17.9	6.42	0.539		9.33	6.04	7.02	1.24
Normalizing values	11	3	1000	3.6	0.787	0.652	0.208	0.071	0.256	0.30[illegible]	0.182	0.188	0.034

tions, in general, than those measured. At the same time it is also probable that the differences in absolute trace-element concentration between the whole rocks and the separated phases are in part due to the presence of interstitial, accessory phases, such as phosphates, alkali feldspar, or glass, that are rich in trace elements. That our separated pyroxene and feldspar may represent early formed material is supported to some extent by the fact that the trace element ratios between these minerals and their respective whole rocks are close to those expected on the basis of measured trace-element partition coefficients (*5, 9*) between such minerals and the liquids from which they are crystallizing. The consistency of the mineral to whole-rock trace element ratios, in spite of the differences between the whole-rock concentrations themselves, tends to indicate that the separates were fairly pure. In the case of the ilmenite separates the consistency is problematic; we did not expect such high trace element concentrations in this phase, particularly the high concentration of the large Rb cation. Perhaps Rb is substituting for oxygen, or possibly the high trace element concentrations reflect late crystallization of the separated ilmenites. Perhaps the consistency is fortuitous and we are seeing contamination by late phases, a possibility favored to some extent by the complementary nature of the ilmenite and feldspar Sr, Rb, and K concentrations, for example. These questions cannot be answered at the present time. A matter on which the mineral data do throw some light is the redox state of the lunar samples. The partitioning of Eu (an element that differs from the other trivalent rare-earths in showing significant natural stability in the divalent state) between two phases, is a function of a number of parameters, including redox conditions (*10*). Measured partition coefficients (*9*) indicate that augite will contain more trivalent Eu than coexisting calcic plagioclase, but that the plagioclase will contain considerably more divalent Eu than the augite. Hence, the ratio of total Eu between feldspar and pyroxene will be a function of the divalent to trivalent Eu ratio of the system which in turn is a function of the redox conditions. The ratio of Eu between plagioclase and pyroxene in the lunar rocks is higher than between comparable terrestrial feldspar and pyroxene, and this indicates a higher divalent to trivalent Eu ratio than on earth, reflecting more reducing conditions. Finally, it is interesting that the relative trace-element concentrations of 44 plagioclase, and of 44 pyroxene, are very similar to those of pyroxene and plagioclase separated from the Juvinas achondrite (*4*).

The overall similarity in the trace-element characteristics of the breccias, the soil, and two of our igneous rocks, in particular, is apparent from Table 1 and Fig. 1. It appears that the process, probably meteorite impact, that produced the fragmental rocks, did not greatly alter their trace element concentrations from those in their presumably igneous parent rocks. The differences that exist between the two similar igneous rocks, the breccias and the soil, most noticeably the differences in absolute concentrations and the small differences in the ratios of Sr and of Eu to the other elements, are not those expected for a volatilization process (the similarity between the < 400 mesh fraction and the bulk soil is pertinent in this respect) but rather are just those observed for igneous differentiation involving feldspar (*8*). It appears that these lunar rocks might well represent liquids produced by fractional crystallization or varying degrees of partial fusion. If these rocks were derived from material with chondritic trace elements, then some process must be invoked to produce enrichment of the other trace elements relative to Rb and K. An igneous process does not seem likely; perhaps volatilization is required. In contrast, the trace element characteristics of these lunar rocks, including Rb/Sr, K/Rb, K/Ba, and the negative Eu anomaly, when considered in the light of appropriate partition coefficient data, appear to be consistent with these rocks being the result of limited partial fusion of source material similar to the brecciated eucrites. It has been suggested that these meteorites are of lunar origin (*11*); the lunar whole rock major element analyses (*1, 12*) and our trace element analyses seem not inconsistent with this hypothesis.

JOHN A. PHILPOTTS
C. C. SCHNETZLER

Planetology Branch,
Goddard Space Flight Center,
Greenbelt, Maryland

References and Notes

1. Lunar Sample Preliminary Examination Team, *Science* **165**, 1211 (1969).
2. C. C. Schnetzler, H. H. Thomas, J. A. Philpotts, *Geochim. Cosmochim. Acta* **31**, 95 (1966); *Anal. Chem.* **39**, 1888 (1967).
3. R. A. Schmitt, R. H. Smith, D. A. Olehy, *Geochim. Cosmochim. Acta* **28**, 67 (1964); P. W. Gast, *Science* **147**, 858 (1965).
4. C. C. Schnetzler and J. A. Philpotts, *Meteorite Research*, P. M. Millman, Ed. (Reidel, Dordrecht, Holland, 1969), p. 206.
5. J. A. Philpotts and C. C. Schnetzler, *Geochim. Cosmochim. Acta*, in press.
6. P. M. Hurley, *ibid.* **32**, 273 (1968); L. A. Haskin, F. A. Frey, R. A. Schmitt, R. H. Smith, *Physics and Chemistry of the Earth*, L. H. Ahrens, F. Press, K. Rankama, S. K. Runcorn, Eds. (Pergamon, London, 1966), vol. 7, p. 167.
7. P. W. Gast, *Geochim. Cosmochim. Acta* **32**, 1057 (1968); F. A. Frey, M. A. Haskin, J. A. Poetz, L. A. Haskin, *J. Geophys. Res.* **73**, 6085 (1968).
8. J. A. Philpotts, C. C. Schnetzler, S. R. Hart, *Earth Planetary Sci. Letters*, in press.
9. C. C. Schnetzler and J. A. Philpotts, *Geochim. Cosmochim. Acta*, in press.
10. D. G. Towell, J. W. Winchester, R. V. Spirn, *J. Geophys. Res.* **70**, 3485 (1965); J. A. Philpotts and C. C. Schnetzler, *Chem. Geol.* **3**, 5 (1968).
11. M. B. Duke and L. T. Silver, *Geochim. Cosmochim. Acta* **31**, 1637 (1967).
12. J. H. Patterson, E. J. Franzgrote, A. L. Turkevich, W. A. Anderson, T. E. Economon, H. E. Griffin, S. L. Grotch, K. P. Sowinski, *J. Geophys. Res.* **74**, 6120 (1969).
13. We thank S. Schuhmann, C. W. Kouns, and P. A. Shadid for their assistance.

4 January 1970

Total Carbon and Nitrogen Abundances in Lunar Samples

Abstract. *Total carbon and nitrogen abundances were determined by using combustion-gas chromatographic techniques. Weighted mean analyses for bulk fines were 225 and 140 ppm total carbon and 150 and 100 ppm total nitrogen. Total carbon in fine breccia was 230 ppm; in coarse breccia, 100 ppm; in fine-grained rock, 70 ppm; and in medium-grained rock, 64 ppm. The finest sieved fraction had the highest carbon content. Total nitrogen in fine breccia was 125 ppm; in coarse breccia, 100 ppm; in fine-grained rock, 115 ppm; and in medium-grained rock, 30 ppm. The total carbon and nitrogen fines appear to be mixtures of indigenous lunar material together with meteoritic and solar wind components.*

Total carbon and nitrogen contents were determined in Apollo 11 lunar samples with analytical techniques that were developed for these elements in meteorites (*1*). These elements are important because of their crucial role as organogenic elements. Their abundances give an indication of the oxidation-reduction conditions and, owing to their volatility, of thermal conditions near the lunar surface.

For the determination of total carbon, samples were burned in a flowing oxygen atmosphere at approximately

Table 1. Total carbon in Apollo 11 lunar rocks. The symbol △ denotes results obtained by heating for 24 hours at 300°C in a flowing helium atmosphere.

Sample weight (g)	Total carbon (μg/g)	Weighted mean (μg/g)
Fines, sample 86,A		
0.3433	130 ± 4	. . .
0.3094	162 ± 6	. . .
0.0782 △	160 ± 25 △	. . .
0.1555 △	122 ± 13 △	. . .
(0.8864)	. . .	142 ± 10
Fines, sample 86,B		
0.1080	190 ± 20	. . .
0.2586	181 ± 8	. . .
0.2565	109 ± 8	. . .
0.3482	353 ± 6	. . .
(0.9713)	. . .	226 ± 10
"C" breccia, sample 2,54		
0.2831	228 ± 7	. . .
0.2874	231 ± 7	. . .
(0.5705)	. . .	230 ± 7
"C-B" coarse breccia, sample 44,37		
0.2093	108 ± 9	. . .
0.3677	76 ± 6	. . .
0.1058	180 ± 18	. . .
(0.6828)	. . .	102 ± 10
"A" basalt, sample 49,23		
0.2268	68 ± 9	. . .
0.3464	71 ± 6	. . .
(0.5732)	. . .	70 ± 8
"B" basalt, sample 50,33		
0.2394	152 ± 8	. . .
0.3234	15 ± 7	. . .
0.3007	47 ± 6	. . .
(0.8635)	. . .	64 ± 8

1800°C to form CO_2. After necessary purification and trapping of the effluent gases, the CO_2 was detected by a gas chromatographic technique in a LECO 589-400 low carbon analyzer. Lunar samples were heated to about 2400°C in a graphite crucible in a helium atmosphere to reduce all nitrogen compounds to N_2, which was detected in a LECO Nitrox-6 gas chromatographic analyzer. Differential thermal conductivity was used as the detection method in both systems. National Bureau of Standards low carbon and nitrogen steel standards were used to construct standard analytical curves for both determinations. The combustion-gas chromatographic detection technique determines total carbon and nitrogen but does not discriminate their chemical state in the lunar samples.

The results of the analyses are listed in Table 1 for total carbon and in Table 2 for total nitrogen. No attempt was made to homogenize individual samples, and thus the possibility of contamination was minimized and sample inhomogeneity was tested. Rock samples were crushed by one stroke in a clean diamond mortar. All the rock specimens were crushed quite easily to a moderately fine material. To guard against contamination, they were not sieved or run through a mechanical splitter.

The precision (indicated as ± in Tables 1 and 2) is taken from the 90 percent confidence level on the line of regression for the reference standards. Analytical precision of the nitrogen analyses is superior to that for carbon because the difference between the thermal conductivities of nitrogen and helium is greater than that for carbon dioxide and helium.

The carbon and nitrogen abundances show some similarities. Both are highest in the fines samples 86,A and 86,B and in the fine-grained breccia 2,54. The coarse-grained basalt 50,33 contains the lowest concentrations of both carbon and nitrogen. With respect to carbon, the fine-grained basalt 49,23 is similar to the coarse-grained basalt, but its nitrogen content is relatively high. The analyzed sample of rock 44,37 had the macroscopic characteristics of a coarse-grained basalt but was catalogued as a coarse-grained breccia. In terms of both carbon and nitrogen, it is intermediate to the fine-grained breccia 2,54 and the coarse-grained basalt 50,33.

The carbon and nitrogen in Apollo 11 lunar samples may be from four sources: (i) indigenous, (ii) meteorite or comet impact, (iii) solar wind, and (iv) contamination. Contamination may have resulted from the landing of the lunar module, from astronaut activity, or from subsequent laboratory handling. The elements may be indigenous to the moon as elemental carbon, as interstitial atoms of carbon or nitrogen, as organic material, or as inorganic compounds (carbonate, carbides, ammoniacal nitrogen, molecular nitrogen). If the elements were accumulated on the lunar surface by both meteorite impact, solar wind, or both, it may be difficult to distinguish which process was dominant. Both processes are rather catastrophic with respect to the components involved. Meteoritic material arrives with hypervelocity impacts, causing volatilization and atomization of the projectile and part of the target. Solar wind particles arrive as single atoms or parts of atoms.

The total carbon trends seem to support the idea that the values of 70 and 64 μg/g (ppm) found in the basalts are primarily due to indigenous carbon. The close agreement for the two splits of the fine-grained basalt 49,23 indicates some degree of homogeneity with respect to the analyzed sample, but the variation between the splits of the coarse-grained basalt 50,33 indicates a more sporadic distribution of carbon among or between the large crystalline phases.

Table 2. Total nitrogen in Apollo 11 lunar rocks. The symbol △ denotes results obtained by heating for 24 hours at 300°C in a flowing helium atmosphere.

Sample weight (g)	Total nitrogen (μg/g)	Weighted mean (μg/g)
Fines, sample 86,A		
0.0971	142 ± 5	. . .
0.1822	87 ± 2	. . .
0.1239	100 ± 3	. . .
0.0749 △	107 ± 5 △	. . .
0.1192 △	92 ± 4 △	. . .
(0.5973)	. . .	102 ± 4
Fines, sample 86,B		
0.1969	150 ± 2	. . .
0.1299	164 ± 5	. . .
0.1024	144 ± 4	. . .
(0.4292)	. . .	153 ± 4
"C" breccia, sample 2,54		
0.0896	113 ± 5	. . .
0.1588	135 ± 4	. . .
0.1146	119 ± 4	. . .
(0.3630)	. . .	125 ± 4
"C-B" coarse breccia, sample 44,37		
0.1666	135 ± 4	. . .
0.1266	95 ± 4	. . .
0.1993	69 ± 3	. . .
(0.4925)	. . .	98 ± 4
"A" basalt, sample 49,23		
0.1510	85 ± 3	. . .
0.1293	35 ± 4	. . .
0.1159	247 ± 5	. . .
(0.3962)	. . .	116 ± 4
"B" basalt, sample 50,33		
0.1888	32 ± 2	. . .
0.2067	28 ± 2	. . .
(0.3955)	. . .	30 ± 2

The two fines samples 86,A and 86,B and the fine dark breccia 2,54 are significantly higher in total carbon than are the basalts. The fact that sample 86,A received less handling in the Lunar Receiving Laboratory may account for its lower carbon values. There is no evidence to support this assumption, and the overlap in the range of values in 86,A and 86,B, together with a constant carbon-nitrogen ratio, may indicate a normal sample variation. To investigate the distribution with respect to particle size, a split of sample 86,A was sieved and analyzed for carbon. Material from 60 to 140 mesh had 115 ± 20 μg/g total carbon; from 140 to 300 mesh, 210 ± 10 μg/g; and minus 300 mesh, 500 ± 20 μg/g. Evidently the carbon is concentrated in the finest-sized material. Microscopic examination of the material larger than 60 mesh, and the 60- to 140-mesh fraction indicated that it is made up primarily of fine-grained basalt, glassy impactite material, and some crystal fragments. Samples 86,A and 86,B may contain different proportions of coarse and fine fragments and, hence, have different total carbon abundances. The fine-grained breccia appears to consist primarily of compacted fines, and it apparently lost no carbon during lithification.

The rare gas analyses reported in the preliminary examination of lunar samples from Apollo 11 (*2*) also showed enrichment in the fines and breccia. Assuming that all the rare gas content is due to solar wind, some rough calculations may be made to indicate how much of the carbon may be attributed to solar wind. The major unknown in the calculation is a "sticking" or retention factor relating the relative retention on the lunar fines to each of the rare gases. A comparison of Apollo 11 rare gas abundances with Cameron's 1967 (*3*) table of abundances in the solar system indicates a fractionation in the lunar materials favoring the retention of higher atomic weight gases. If the solar carbon/neon is normalized to the neon in the lunar samples, the solar wind contribution of carbon to the lunar fines is 6 μg/g or about 4 percent of the total carbon. If the carbon abundance is calculated from the krypton abundance, the solar wind contribution is about 50 μg/g, and, if it is calculated from xenon, the solar wind contribution would be 300 μg/g. Assuming a similar "sticking" factor, the solar wind contribution would be about 5 times lower for nitrogen than for carbon.

A general calculation may also be made to estimate the possible meteoritic and cometary contribution to the lunar fines. The major unknown in this case is the composition of the meteoritic influx. A useful start might be to take material with the composition of carbonaceous chondrites. Such a choice is strengthened by comparing the increase in nickel from the lunar igneous rocks with the lunar breccia and fines as reported in the preliminary examination of samples from Apollo 11 (*2*). Nickel and carbon both show significant increases, and nickel, like carbon, has a much higher concentration in carbonaceous chondrites than in lunar rocks.

Mason's (*4*) review of carbonaceous chondrites gives typical concentrations of 2 percent carbon and 1 percent nickel in these meteorites. Using these values to calculate the approximate contribution of meteoritic extra-lunar material indicates that about 0.5 to 2 percent has been mixed with lunar material. The nitrogen contribution of meteorites is about 10 percent that of carbon and, again, this element does not give an apparent satisfactory material balance. Based on our simplifying assumptions, both solar wind and meteorite influx fail to account for the relatively high nitrogen abundances. They may be explained by the possibility that indigenous nitrogen is high in the lunar rocks, as indicated by the fine-grained basalt 49,23, or that solar wind or meteoritic nitrogen is retained to a greater degree than is carbon.

The possibilities for contamination from nitrogen in returned lunar samples appear to be greater than for carbon. Possible sources are the exhaust gases from the lunar module and the nitrogen atmosphere to which the fresh lunar rocks were exposed at the Lunar Receiving Laboratory.

To see if nitrogen or carbon could be easily desorbed from lunar fines, a split of sample 86,A was heated for 24 hours at 300°C in a flowing helium atmosphere. No carbon or nitrogen appeared to be removed by this treatment. If contamination by atmospheric nitrogen has taken place, it is strongly bonded to the fines material.

CARLETON B. MOORE
CHARLES F. LEWIS
Arizona State University, Tempe
EVERETT K. GIBSON
NASA Manned Spacecraft Center, Houston, Texas 77085
WALTER NICHIPORUK
Arizona State University

References and Notes

1. C. B. Moore and C. F. Lewis, *Science* **149**, 317 (1965); *Earth Planet. Sci. Lett.* **1**, 376 (1966); *J. Geophys. Res.* **72**, 6289 (1967); C. B. Moore and E. K. Gibson, *Science* **163**, 174 (1969); C. B. Moore, E. K. Gibson, K. Keil, *Earth Planet. Sci. Lett.* **6**, 457 (1969); E. K. Gibson and C. B. Moore, *Anal. Chem.*, in press.
2. Lunar Sample Preliminary Examination Team, *Science* **165**, 1211 (1969).
3. A. G. W. Cameron, *Origin and Distribution of the Elements*, L. H. Ahrens, Ed. (Pergamon, Oxford, England, 1968), pp. 125–143.
4. B. Mason, *Space Sci. Rev.* **1**, 621 (1962-1963).
5. We thank H. M. Schwartz and P. Juchnevics for their help in keeping the analytical instrumentation operating. Discussions with Dr. J. Larimer and Dr. C. F. Royse regarding the analytical results were helpful. Supported in part by NSF grant GA-909, in addition to the NASA grants and contracts.

4 January 1970

Instrumental Neutron Activation Analyses of Lunar Specimens

Abstract. *Ten Apollo 11 specimens were divided into 24 samples. Sodium contents of 8 diverse specimens cluster tightly about 0.3 percent. Plagioclase separated from sample 10044 contains about 1.09 percent Na; barium is not enriched in the plagioclase separate. Contents of the rare earths are strikingly high, and relative abundances resemble those of calcium-rich achondrites or abyssal basalts but are depleted in Eu by factors of 2 to 3 and in La by about 20 percent. The plagioclase separate is enriched in Eu and pyroxenes and opaque minerals are Eu-depleted. Fine fractions of 10044 are about 20 to 40 percent richer in most rare earths (50 percent for Eu) than coarse fractions, probably because of the presence of small grains in which rare earths are markedly concentrated. "Microgabbro" 10045 is impoverished, relative to the soil, in rare earths and Hf. Ratios by mass of Zr to Hf are comparatively low. Abundances of Mn, Co, Fe, Sc and Cr suggest systematic differences between igneous rocks on one hand and breccias and "soil" on the other. From the Co abundances, no more than about 3 percent of the present "soil" can consist of chondritic meteorite contamination.*

We have analyzed ten Apollo 11 specimens by techniques similar to those of Gordon *et al.*(*1*), with such modifications as recent experience dictates. Twenty-six elements were determined, but owing to space limitations we here report only salient observations and conclusions. A table of abundance and error estimates will be provided on request, and data and interpretations will be presented more fully in a subsequent paper (*2*).

Specimens were split where feasible to provide replicates. Two, 10044 and 10084, were treated more elaborately. The "microgabbro" powder, 10044, was split to yield a representative sample and then sieved with 100-mesh nylon screening. Fine material was split again, and the coarse material was hand-cobbed for separates of plagioclase, pyroxenes, and opaque minerals. The residue from the hand-cobbing, which remains representative of the coarse material to a good approximation, was split to give duplicate samples. The soil specimen, 10084, was sieved with 100-mesh and 325-mesh screening; splits of the origi-

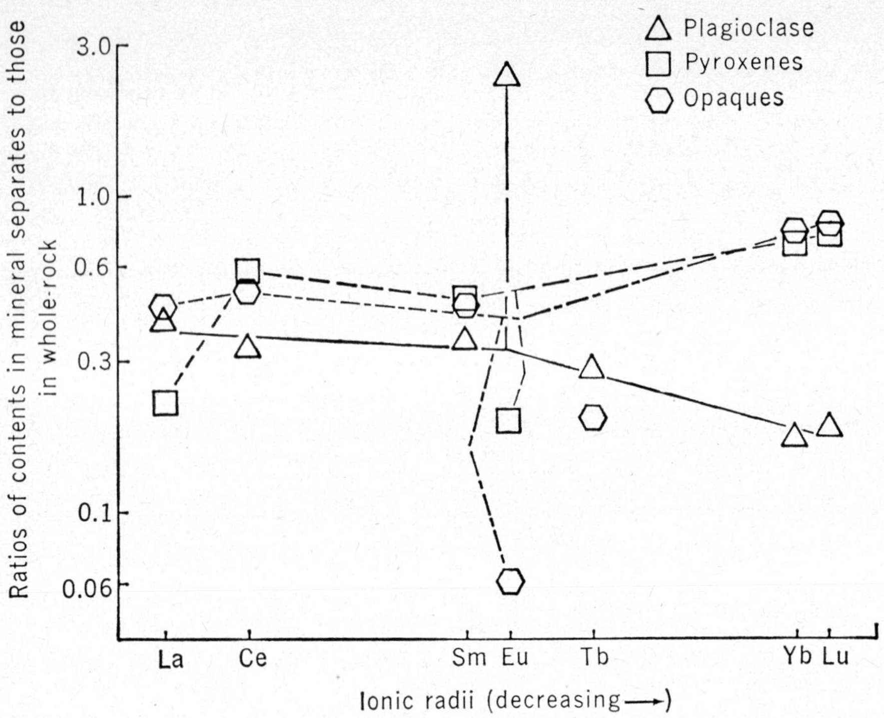

Fig. 1. Rare earth abundances in mineral separates from "microgabbro" 10044. Abundances have been corrected for cross-contamination as outlined in the text and then normalized to whole-rock rare earth contents. Logarithms (base 10) of normalized ratios are plotted versus ionic radii; ordinate values are those of the arguments of the logarithms. Whole-rock abundances would lie along the line at 1.0.

enrichment (Eu/Sm of 1.10 in contrast to 0.16 in the whole rock), to those of terrestrial and achondritic plagioclases (*4*). Both pyroxenes and opaque minerals are depleted in Eu (Eu/Sm of 0.067 and about 0.02). If the major rare-earth-bearing phases among the opaque minerals crystallized before plagioclases did (*5*), europic ion might well have been coherent with other trivalent rare earths in these phases. The Eu-depletion in the opaque minerals thus suggests that that element was present as europous ion at least early in the paragenesis. Depletion of Eu in pyroxenes might reflect crystallization in company with plagioclase, regardless of the oxidation state of Eu in the melt.

Contents of rare earths in bulk specimens are high. La contents range up to 26 ppm and Lu contents to 2.7 ppm. Relative abundances resemble those of Ca-rich achondrites or abyssal basalts (*6*). The contrast with terrestrial continental rocks, where high total rare earth contents are associated with relative enrichment of light rare earths, is striking. All lunar specimens are Eu-depleted by factors of two to three, from comparisons of Eu/Sm ratios with those of chondrites (*7*). This feature may be related to plagioclase fractionation or, more generally, to incoherence of europous ion and trivalent rare earths during magma formation. Lanthanum is also depleted (by about 10% to 30%) in comparison to values predicted from chondritic La/Ce ratios.

nal material and of the intermediate size-fraction were made.

Sodium abundances of bulk specimens are remarkably constant, averaging about 0.33 percent. All of the dispersion about the average value could be accounted for by analytical uncertainties and, in one case, sampling error. It seems unlikely that all specimens were derived from one or a series of closely related magmas, and in any case abundances of other elements do not display comparable constancy. One is led to suspect that some fundamental process, related perhaps to distribution coefficients for Na between solid and fluid phases, has limited the range of Na contents.

Before discussing distribution of elements among minerals separated from 10044, we must estimate contamination of one major "phase" by another (*3*). Sodium is a useful index of plagioclase contamination in the pyroxenes and opaque minerals separates; Al is even more useful in setting a limit on plagioclase in the opaque minerals. Plagioclase constitutes less than 6% of the pyroxenes, less than 2% of the opaque minerals. Chromium contents indicate that there is less than 1% of pyroxenes in the plagioclase, and Si contents limit pyroxenes to less than 15% in the opaque minerals separate. Less than 1% opaque minerals are present in the plagioclase (Cr), less than 6% in the pryoxenes (Ti). The water-white plagioclase grains are the easiest to cobb cleanly, but it is not much more difficult to select pyroxene grains of high purity. It is difficult to detect silicates adhering to opaque grains under the microscope. Consequently, we shall assume that the plagioclase separate is contaminated with 0.5% of each of the other "phases", pyroxenes with 1% of each of the others, and opaque minerals with about 2% plagioclase and 5% pyroxenes.

Under these assumptions, Na contents of plagioclase, pyroxenes and opaque minerals are about 1.09%, 0.05%, and 0.016%, respectively. The latter two estimates are of course strongly dependent on the assumptions stated. Barium is not enriched in the plagioclase separate, within the precision of our data. Rare earth contents of the mineral separates are summarized, normalized to whole-rock abundances, in Fig. 1. Fractional crystallization of a combination of these major phases might leave residual magmas with relative rare earth patterns bowed upward in the middle of the sequence. We observe such patterns in our bulk specimens. Note the similarity of the plagioclase pattern, with a systematic decrease toward heavy rare earths and strong Eu

Rare earth abundances in sieved fractions of 10044 suggest that plagioclase is over-represented in the fine-grained fraction and that there are small grains in this rock in which rare earths are markedly concentrated. A crude estimate indicates that there is about 50% more plagioclase in the fine than in the coarse fractions. Independently, we note from Fig. 1 that as much as two-thirds of the total rare earths other than Eu may be present in minor (presumably late) phases or mineraloids, rather than in the major phases analyzed. Hafnium is distributed in a similar way: it is enriched in fine fractions of 10044, and a mass balance cannot be made using major phases alone. Note also that Hf is not concentrated in pyroxenes (unlike Sc), contrary to expectation.

The "microgabbro" 10045 is impoverished in rare earths and Hf relative to other igneous rocks, and even to breccias and the "soil". Rock 10045 resembles 10044 in these respects and the abundance anomalies may be related to texture. The "soil", which presumably integrates across a wider range of rock types than represented in the Apollo 11 collection, seems to be weighted more toward rocks like 10044 and 10045 than toward 10022, 10071 and 10017. Breccias are not compositionally identical to the "soil" specimens analyzed, but are intermediate toward 10022, etc. Uranium contents range from 0.2 to 0.8 in whole rocks and in sieved fractions and mineral separates and tend to follow rare earths. These relationships should be viewed in the light of accurate geochronological data.

Ratios by mass of Zr/Hf range from about 25 to less than 13 and are low in comparison to terrestrial (and many meteoritic) values (*8*). We can at present offer no explanation for these remarkably low ratios. However, evidence for the dry and chemically reduced character of lunar magmas (*5*) suggests that terrestrial distribution coefficients and coherence relationships may not apply here.

Tantalum contents range from 1 to 2.3 ppm and are comparable to those in many continental crustal rocks (*9*). Tantalum is enriched in opaque minerals from 10044 relative to the whole-rock abundance.

Transition metal contents suggest that there are systematic differences between the igneous rocks, and the breccias and "soil". These trends perhaps display a negative correlation with those for the lithophilic elements noted above. From our Co abundances, one may show that no more than about 3% of the present "soil" can consist of chondritic fragments having Co contents like those of CL chondrites (*10*).

Both from data reported by the Preliminary Examination Team and from our results, it is clear that these rocks have been subjected to complex and unusual differentiation processes. Until we understand better the nature of these processes, it borders on metaphysical argument to compare the lunar rocks with, e.g., chondrites. Consequently, with the exception of a minor portion of our discussion of rare earths, we have eschewed such comparisons.

GORDON G. GOLEŠ, MASUMI OSAWA
KEITH RANDLE, ROBERT L. BEYER
DOMINIQUE Y. JÉROME
DAVID J. LINDSTROM
MARILYN R. MARTIN, SHEILA M. MCKAY
TERRY L. STEINBORN
Center for Volcanology and
Departments of Chemistry and *Geology*,
University of Oregon, Eugene 97403

References and Notes

1. G. E. Gordon, K. Randle, G. G. Goles, J. B. Corliss, M. H. Beeson, S. S. Oxley, *Geochim. Cosmochim. Acta* **32**, 369 (1968).
2. *Geochim. Cosmochim. Acta,* in press.
3. Plagioclase, pyroxene, and opaque minerals in this rock will almost certainly display a range of compositions and may well each comprise several phases *sensu stricto*. Lacking information on the mineralogy of 10044, we here merely try to set reasonable limits on what portions of our "plagioclase" separate are actually pyroxenes and opaque minerals, etc.
4. D. G. Towell, J. W. Winchester, R. V. Spirn, *J. Geophys. Res.* **70**, 3485 (1965); C. C. Schnetzler and J. A. Philpotts in *Origin and Distribution of the Elements* (Pergamon, New York, 1968), pp. 929-938; J. A. Philpotts and C. C. Schnetzler, *Chem. Geol.* **3**, 5 (1968).
5. Lunar Sample Preliminary Examination Team, *Science* **165**, 1211 (1969); D. F. Weill, personal communication.
6. L. A. Haskin, F. A. Frey, R. A. Schmitt, R. H. Smith, in *Physics and Chemistry of the Earth* (Pergamon, New York, 1966), vol. 7, pp. 167-321; J. A. Philpotts, C. C. Schnetzler, H. H. Thomas, *Earth Planet Sci. Lett.* **2**, 19 (1967); F. A. Frey, M. A. Haskin, J. A. Poetz, L. A. Haskin, *J. Geophys. Res* **73**, 6085 (1968); J. A. Philpotts and C. C. Schnetzler, in *Origin and Distribution of the Elements* (Pergamon, New York, 1968), pp. 939-947.
7. L. A. Haskin, M. A. Haskin, F. A. Frey, T. R. Wildeman, in *Origin and Distribution of the Elements* (Pergamon, New York, 1968), pp. 889-912.
8. M. Fleischer, *U.S. Geol. Survey Bull. 1021-A* (1955); J. R. Butler and A. J. Thompson, *Geochim. Cosmochim. Acta* **29**, 167 (1965); E. Mertz and E. Schrage, *ibid.* **28**, 1873 (1964);
9. R. Taylor, in *Physics and Chemistry of the Earth* (Pergamon, New York, 1966), vol. 7, pp. 133-213.
10. R. A. Schmitt, G. G. Goles, R. H. Smith, in preparation.
11. Supported in part by NASA contract NAS 9-7961.

4 January 1970

Isotopic Abundances of Actinide Elements in Lunar Material

Abstract. *The abundances of uranium and thorium were measured mass spectrometrically as 0.59 ± 0.02 and 2.24 ± 0.06 parts per million, respectively, in the fines of Apollo 11 bulk sample. The ratio* $^{235}U/^{238}U$ *was 0.007258 ± 0.000016, in agreement with terrestrial uranium. The following upper limits were set:* $^{236}U/^{238}U \leqslant 3 \times 10^{-9}$*; parts per million of sample,* $^{239}Pu \leqslant 1 \times 10^{-9}$*;* $^{244}Pu \leqslant 9 \times 10^{-11}$*, and* $^{247}Cm \leqslant 1.25 \times 10^{-10}$*. More alpha activity from* ^{227}Th *and its decay products appears to be present than would be in equilibrium with the* ^{235}U *in the sample. The search for other actinide and transactinide nuclides is continuing.*

The purpose of our investigation was to measure the abundance and isotopic composition of thorium, uranium, and any transuranium elements that might also be present in lunar material, using mass spectrometric and radiometric techniques. Among the transuranium elements we made a specific search for ^{244}Pu and ^{247}Cm, two isotopes which had been produced synthetically and whose half-lives had been measured as $(8.18 \pm 0.26) \times 10^7$ (*1*) and $(1.64 \pm 0.24) \times 10^7$ years (*2*), respectively. The abundances of uranium and thorium not only contribute to the total distribution of the elements in lunar material, but are also an important ingredient in calculating the radioactive heat production on the moon. The possible presence of ^{244}Pu or ^{247}Cm, two extinct or nearly extinct nuclides, was interesting because they would have indicated an infusion from a supernova that occurred quite close to the solar system within the last billion years. Another object of our study was to examine the separated actinide and transactinide element fractions by alpha particle analysis and spontaneous fission counting to detect the presence of short-lived nuclides.

Earlier work on terrestrial samples (*1*) showed that it was necessary to work in a laboratory that had never been used for heavy element chemistry; consequently, a new laboratory was equipped for the lunar analysis. All the glassware was carefully leached, and most of the reagents were distilled before use. In some cases special high purity reagents were used without further treatment, but they were tested for Th and U content before being used in the chemical separation procedure.

An alpha count of an 8-mg sample of the fines gave 0.036 ± 0.011 count/min. The single spontaneous fission event observed in 13 days of counting could have been due to counter background.

Two 1-gram samples of fines from the bulk sample (sample No. 10084,75) were dissolved, and known quantities of ^{230}Th (5×10^{-8} g), ^{233}U (7×10^{-8} g), ^{236}Pu (5×10^{-14} g), and ^{242}Cm (2×10^{-14} g) tracers were added. Before use, each of the tracers was carefully purified

Table 1. Micrograms (metal weight) of actinide element per gram of lunar fines.

Element	Sample 1	Sample 2	Sample 3
U	0.600 ± 0.018	0.582 ± 0.018	
Th	2.27 ± 0.06	2.21 ± 0.06	
^{244}Pu	$< 3 \times 10^{-10}$	$< 1 \times 10^{-10}$	$< 9 \times 10^{-11}$*
^{239}Pu	$< 1 \times 10^{-9}$		$< 1 \times 10^{-9}$ *
^{240}Pu	$< 3 \times 10^{-10}$		
^{247}Cm	$< 1.25 \times 10^{-10}$		

* Preliminary results.

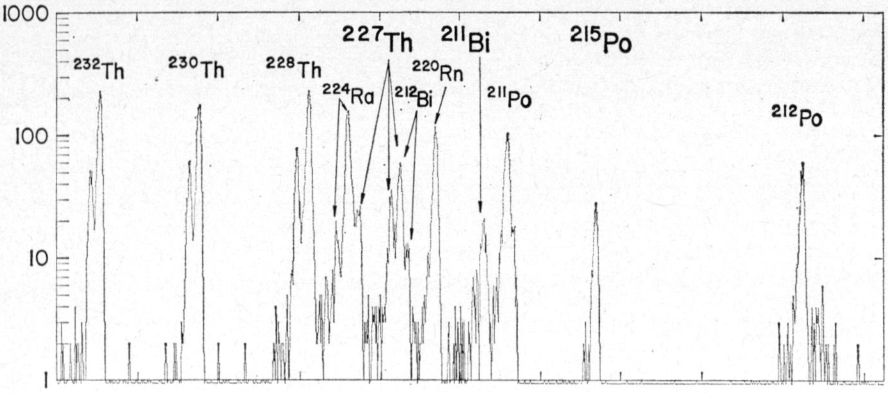

Fig. 1. Alpha pulse analysis of the thorium fraction from lunar fines (not corrected for contribution from reagents.)

and then analyzed by mass spectrometry to determine the isotopic distribution of each of the elements. It was thought that a 1-g sample would be adequate for the determination of the uranium and thorium abundances, and we wanted to conserve our sample for further analysis in the light of the preliminary results obtained from the heavier transuranium element fractions. The samples were chemically separated into thorium, uranium, neptunium, plutonium, curium, and heavier actinide fractions by solvent extraction and ion exchange. Each of the separated actinide fractions was subjected to alpha pulse analysis to check the purity of the fraction. Reagent blanks containing the same tracers were run concurrently with the samples through the same chemical procedure.

The isotopic composition of the separated uranium and thorium fractions was measured in a 30-cm mass spectrometer which offered maximum precision for microgram-sized samples whose isotopic components do not differ by more than 10^4. The plutonium and curium fractions were analyzed with a 250-cm mass spectrometer designed to give maximum sensitivity for detection of actinide elements. The results, corrected for the background introduced by the reagents and by the tracers, are shown in Table 1, columns 2 and 3. The corrections were less than 1 percent for uranium and thorium but accounted for all of the ^{239}Pu observed. The reported errors, based on standard deviations, came from estimates of uncertainties in sampling, counter geometry, alpha counting, and mass spectrometry. No allowance was made for the uncertainties in the half-lives of the ^{233}U (1.62×10^5 yr) and ^{230}Th (8.0×10^4 yr) tracers.

It was almost impossible to observe the alpha particles of the other isotopes of the heavy elements in the various fractions because they were obscured by the larger alpha-counting rates of the ^{230}Th, ^{233}U, ^{236}Pu, and ^{242}Cm tracers that had been added. Consequently, a 10-g sample (No. 3) of the fines was dissolved and only ^{236}Pu tracer added. The plutonium tracer was used to increase the sensitivity of the ^{244}Pu and ^{239}Pu determinations on this larger sample. Sample 3 was also chemically separated into the same element fractions as described earlier, and, in addition to examining the uranium and plutonium mass spectrometrically, all fractions except plutonium are being carefully pulse-analyzed for their alpha particle energies with a 300-mm² solid state detector (Au-Si) and a 1024-channel pulse analyzer set to detect all known natural alpha activities.

Extensive measurements in the 30-cm mass spectrometer showed the ^{235}U/^{238}U ratio to be 0.007258 ± 0.000016, in good agreement with 0.007257 ± 0.0000073 (*3*) found in terrestrial uranium. Another portion of the uranium from sample 3 was analyzed in a special tandem mass spectrometer that can determine neighboring isotopes of widely disparate abundances. This measurement set an upper limit for the ^{236}U/^{238}U ratio of 3×10^{-9}. A small ^{239}Pu peak was perceived, but a background for this sample has not yet been measured. A portion of the plutonium fraction was also examined for ^{244}Pu, but no peak was observed. This phase of the analysis is continuing. Preliminary upper limits are reported for ^{239}Pu and ^{244}Pu in Table 1.

Pulse analysis of the uranium fraction showed alpha particle energies corresponding to ^{238}U, ^{235}U, ^{234}U, and ^{232}U. The ^{235}U/^{238}U ratio was consistent with the ratio observed by the mass measurements. The ^{234}U is a radioactive daughter of ^{238}U and was present in a radioactive equilibrium amount. The ^{232}U is an alpha decay product of the ^{236}Pu tracer that had been added and was also consistent with the amount of tracer added, indicating a good chemical recovery of the uranium fraction.

The thorium fraction was pulse analyzed for 4 days (Fig. 1). The midpoint of the counting occurred about 4 days after the thorium had been purified by ion exchange from uranium, actinium, radium, lead, and virtually all other likely contaminants. The ^{227}Th had been first separated from any ^{227}Ac about 15 days before the pulse analysis. Examination of Fig. 1 shows that (i) the ^{230}Th/^{232}Th ratio is consistent with the ^{238}U/^{232}Th abundances found by mass spectrometer measurements. Thus it can be concluded that there has been no separation of uranium from thorium in the last 2×10^5 years. (ii) The ^{227}Th and its radioactive daughters are about six times more abundant than the equilibrium concentration expected from the known amount of ^{235}U in the sample; however, a reagent blank accounts for approximately one-half of the apparent excess. Actually all the radioactive daughters of ^{227}Th are not present in equilibrium amounts because of the earlier chemical separations. The presence of the easily identifiable ^{215}Po and ^{211}Bi reinforces the assignment of the alpha energy peaks to ^{227}Th and some of its decay products.

In conclusion, the abundances of uranium and thorium are in reasonable agreement with the gamma counting results of O'Kelley and co-workers (*4*). The ^{235}U/^{238}U ratio of the lunar sample is in agreement with meteorites and terrestrial values, indicating that the moon has undergone the same history of nucleogenesis and irradiation as the rest of the solar system. The limit set for ^{236}U implies that the product of cross section and neutron flux was such that each uranium atom in the lunar fines was exposed to a maximum of 4×10^{-22} particle per second if the irradiation was steady and continuous. The absence of ^{244}Pu and ^{247}Cm in lunar material precludes nearby supernovae explosions within the last eon or two. The precise limits depend strongly on the details of the model used in the calculations.

We prefer not to draw any conclusions at this time from the apparent excess of ^{227}Th until further experiments can confirm the initial results.

PAUL R. FIELDS, HERBERT DIAMOND
DONALD N. METTA
CHARLES M. STEVENS
DONALD J. ROKOP, PARKER E. MORELAND
Argonne National Laboratory, Chemistry Division, Argonne, Illinois

References and Notes

1. P. R. Fields, *et al.*, *Nature* **212**, 131 (1966).
2. P. R. Fields, A. M. Friedman, J. Lerner, D. Meta, R. Sjoblom, *Phys. Rev.* **131**, 1249 (1963).
3. Reviewed by G. T. Seaborg, *Encyclopedia of Chemical Elements*, C. A. Hampel, Ed. (Reinhold, New York, 1968), p. 776.
4. Lunar Sample Preliminary Examination Team, *Science* **165**, 1211 (1969).
5. Supported by the AEC and by NASA contract T76536. E. P. Horwitz contributed to the planning and testing of the chemical procedures and C. H. Youngquist provided some of the chemical apparatus.

4 January 1970

Trace Elements and Accessory Minerals in Lunar Samples

Abstract. *Halogen contents similar to those in meteorites are reported in Apollo 11 samples. Concentrations of 0.6 to 13 parts mercury, 0.2 to 0.8 part uranium, 1 to 17 parts lithium, and 1 to 800 parts osmium per million and 0.5 to 1 percent chromium oxide by weight have been measured. In contrast to meteorites and terrestrial rocks a large mercury release below 110°C is observed. Some implications of these results are presented. Fluorapatite, quartz, tridymite, fayalite, and iron-rich, manganese-poor pyroxmangite, previously unreported, have been identified. The small amount of fluorapatite found does not account for the fluorine contents.*

Preliminary estimates of the amounts of F, Cl, Br, I, Hg, Os, U, and Li in several types of lunar material are presented. All elements were measured by neutron activation except F, which was measured separately by photon activation (*1*). In contrast to our previous F measurements, the latter irradiations were monitored by polyvinylfluoride foil sandwiches surrounding samples sealed in fused silica vials and rotated in the photon beam (*2*).

Chlorine, Br, I, U, Te, and Li were measured in one 100- to 200-mg portion of a sample; Hg, Os, Ru, Cr, Sc, and F were measured in another. However, most F measurements reported were obtained on still other portions. Uranium and Te were by-products of the I measurements. The production of ^{18}F from neutron irradiation of Li in silicates (*3*) accounts for our Li results. Mercury was volatilized from samples in a closed system. Complete release of Hg is assumed because all the samples were found to be fused after heating to 1200°C.

The samples measured were representative. They consisted of soil, LRL crushed and homogenized medium-grained rocks, vesicular rock, vuggy rock with ilmenite platelets, and microbreccia. The last three were rock chips from which we removed and crushed small fragments. All sample handling was confined to a glove box with a nitrogen atmosphere.

The data from the halogen measurements are summarized in Table 1. In some experiments the samples were leached about 10 minutes with boiling water, and the separate results are indicated. Up to 80 percent of the Cl and Br may be in water-soluble phases. Duplicate measurements were made on the microbreccia and the soil samples. Chlorine, Br, and Li in the former agree within the present estimates of error; Cl, Br, and Li in the two soil aliquots disagree by an almost constant factor. Sample inhomogeneity would seem to be a reasonable explanation.

The average F/Cl ratio is about 22. The vesicular and vuggy rocks have Cl/Br ratios 3.5 times greater than the ratio of about 51 for the other samples.

Measurement of I was compromised by the high-fission I activities derived from the U in the samples. A refinement of the data will be attempted. The parallel trends appearing in most of the F, Cl, and Br results do not hold for I.

The 140- to 340-ppm F concentrations permit estimates of up to 1 percent of apatite if this mineral is the major site of F as it is in terrestrial rocks. The few tenths percent P_2O_5 required has been observed (*4*), but fluorapatite has been observed rarely and only in one of the rock types. Its presence is discussed below.

Mercury data are summarized in Table 2 and Fig. 1. The Hg volatilized from the samples was significantly different from releases observed in meteorites and terrestrial rocks (*5*): a large release at low temperature ($\leqslant$ 110°C) is observed, and the maximum in the release at 150°C is shifted to 250°C. We were assured by LRL that no Hg pumps or the like were used in their systems. During isothermal heating at 250°C an initial burst occurs, containing 50 to 80 percent of the Hg eventually evolved. In this respect the release pattern is similar to that noted previously for meteorites and terrestrial samples. The large 110°C fraction might be explained as the release of Hg from low-temperature ($\leqslant$ 150°C) traps during thermal cycling on the lunar surface followed by its adsorption on surfaces. The only deviation from the trend noted is in the soil sample 10084-009, which also has the highest relative amount of Hg released above 450°C. Otherwise, the release trends are not inconsistent with the following interpretation. The most degraded material, the soil 10084-019, has the largest low-temperature fraction; it is followed by the microbreccia. The fine-grained rock 10072 has a greater fraction of its Hg in high-temperature traps. The medium-grained igneous rock occupies an intermediate position, possibly because of slower cooling to lower temperatures. The large release ratio for the vesicular rock 10017 can be rationalized as resulting from Hg associated with the vapor phase that formed the vesicules

Table 1. Halogen, lithium, and uranium concentrations in Apollo 11 samples. Except for F, all data on a line are for a single aliquot; subscripts *l* and *s* refer to water extractable and nonextractable portions, respectively.

Sample	F (ppm)	Cl (ppm)	Br (ppm)	I (ppb)	Li (ppm)	U (ppm)
Soil 10084,2	144	2.9	0.051		0.19	
10084,2		$9.9_s + 1.8_l$	$.15_s + 0.08_l$	$625_s + 15_l$	7.2*	0.19
Microbreccia						
10061,22	342	9.6	.15		4.0	
10061,22		$6.8_s + 0.7_l$	$.25_s + 0.01_l$			
10019,29		$9.4_s + 6.4_l$	$.24_s + 0.14_l$	$\leqslant 71_s + 3_l$	$13_s + 0.14_l$	
Medium-grained rock 10044,48	202	$2.6_s + 12_l$	$.076_s + 0.14_l$	$1440_s + 12_l$	0.95*	.17
Medium-grained rock 10047,48		$9.6_s + 4.8_l$	$.11_s + 0.22_l$	$\leqslant 4.9_s + 12_l$	$16_s + 12_l$	
Vuggy rock 10072,24	271	14	.079	>140		.31
Vesicular rock 10017,22	251	15	.080	5.9		.76

*The leach solutions in these experiments were not measured, but little of the Li appears to be water-soluble.

Table 2. Mercury and osmium concentrations (ppb) in Apollo 11 samples.

Sample	Hg		Os
	Total	> 450°C	
Soil 10084,2	5.3	0.060	283
10084,2	0.95	0.22	1.4
Microbreccia 10061,22	1.2	0.042	
Medium-grained rock 10044,48	0.60	0.028	
Vuggy rock 10072,24	5.5	1.1	17
Vesicular rock 10017,22	13	0.09	755

and trapped in the glasslike lining of the voids.

The Hg concentrations range from 0.6 to 13 ppb, the largest being associated with the vesicular rock. These concentrations are significantly lower in general than those in basaltic achondrites (18 to 9000 ppb) (*6*) or in chondrites and approach more nearly the values observed in some terrestrial rocks (*5*).

The Os concentrations (Table 2), as in chondritic and iron meteorites, are in the ppb–ppm range. They are higher than those in a Hawaiian basalt or W-1 and also tend to be higher than the ~ 7.9 ppb reported for the achondrite Nuevo Laredo (*7*).

The Os results could be pertinent to attempts to monitor infall of meteoritic material on the moon by way of the platinum metals (*8*). If these preliminary data are verified, the high value in the igneous rock relative to the soil may imply that the Os is of lunar origin or that the lava flow that produced the rock was associated with the impact and that meteoritic material was incorporated at that time.

Uranium concentrations (Table 1) are in the range indicated by whole-rock counting at LRL (*9*). Two determinations of Cr were made: sample 10084-009 contains 0.54 percent Cr_2O_3 and sample 10017 contains 0.95 percent. These are close to the values reported by LSPET. The Li contents range from 2 to 17 ppm. A value of 13.1 ppm in W-1, not listed in Table 1, was obtained in this work and is in good agreement with literature values (*3*). The Li contents of the samples appear to be similar to or lower than those in achondrites (*3*) and gabbroic rocks. LRL reports similar Li contents (*9*).

Table 3. Electron microprobe analysis of pyroxmangite.

Compound	Average of six grains	Range
SiO_2	46.7	46.3 – 47.1
FeO	46.5	45.9 – 47.7
MgO	0.9	0.3 – 1.2
CaO	5.2	4.7 – 5.6
MnO	0.8	0.6 – 0.9
TiO_2	0.5	0.4 – 0.7
Total	100.6	

Powders of samples 10017, 10072, and 10044 were examined for common rock-forming minerals that are known to contain halogens. In addition, several nonopaque accessory minerals were identified, some of which are reported here. Preliminary identifications based upon optical properties were confirmed by x-ray powder patterns. Chemical analyses were obtained on grain mounts with an electron microprobe.

Sample 10044 contains fluorapatite as euhedral crystals up to 25 μm long and as irregularly shaped areas up to 100 by 25 μm; the mean refractive index is 1.644. Both habits are present as inclusions in a yellow mineral which is an Fe-rich, Mn-poor variety of pyroxmangite (Fe,Mn)SiO_3. The average composition of three selected apatite inclusions is, in percent by weight, as follows: P_2O_5, 39.0; CaO, 52.5; FeO, 1.5; SiO_2, 2.2; F, 3.3; Cl, 0.03. The sum is 98.5; less 1.4 oxygen as F, the sum is 97.1. It is difficult to explain the low summation, for no other elements above atomic number 8 were present in amounts greater than 0.05 percent. Furthermore, it is improbable that much OH was present, because the 3.3 percent F is very close to the maximum of 3.8 percent F for pure fluorapatite. The 2.2 percent SiO_2 in the analysis probably indicates substitution of SiO_4 groups for PO_4 groups and is not due to inclusions of silica minerals which are present within the irregularly shaped areas. The reverse of this type of substitution has been recognized in terrestrial silicates. Although fluorapatite is a common phosphate mineral in terrestrial igneous rocks, it is not known in meteorites (*10*). The bulk of the fluorine in the lunar samples is not located in fluorapatite. This mineral was not found in any of the other rocks examined, even though some of them have higher fluorine contents than the apatite-containing sample 10044. Apatite constitutes approximately 5 percent of the volume of the pyroxmangite grains; thus only 15 of the 202 ppm F in rock 10044 can be accounted for if pyroxmangite constitutes 1 percent of the rock.

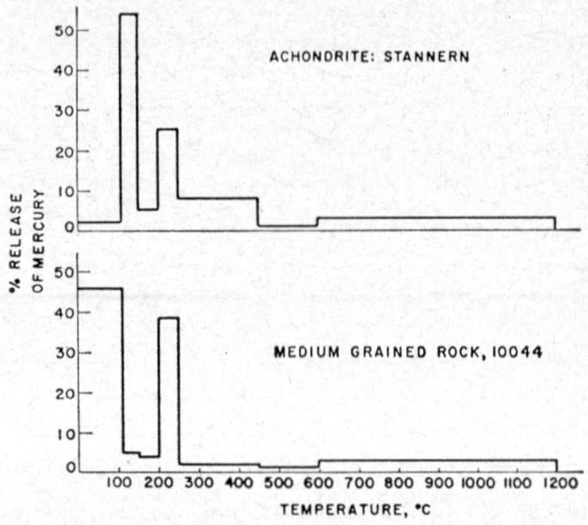

Fig. 1. Mercury release in step-by-step heating of meteoritic and lunar rocks.

The yellow, pyroxmangite-like mineral contains more FeO and less MnO than any known terrestrial mineral. For this reason the lunar occurrence may eventually be designated as a new mineral. The microprobe compositions, corrected for dead time, background, absorption, fluorescence, and atomic number (*11*), are given in Table 3. The formula, based on 18 oxygens, is $(Fe_{4.98}Mg_{0.17}Ca_{0.72}Mn_{0.09}Ti_{0.03})_{5.99}(Si_{5.98}Ti_{0.02})_{6.00}O_{18}$. The optical properties for sodium light are: biaxial (+), $2V = 32 \pm 2°$, $X = 1.745$, $Y = 1.750$, $Z = 1.768$, all values $\pm$ 0.003. Weissenberg photographs (Cu radiation) taken on a single cleavage fragment rotated on each of the three crystallographic axes gave these triclinic cell constants: $a = 6.63 \pm 0.01$ Å, $b = 7.58 \pm 0.01$ Å, $c = 17.43 \pm 0.05$ Å, $\alpha = 114.5 \pm 0.2°$, $\beta = 82.6 \pm 0.1°$, $\gamma = 94.4 \pm 0.1°$. A least-squares analysis of 20 indexed reflcetions from a powder photograph (Cr radiation) of several grains gave: $a = 6.62 \pm 0.01$ Å, $b = 7.54 \pm 0.01$ Å, $c = 17.34 \pm 0.02$ Å, $\alpha = 114.26 \pm 0.04°$, $\beta = 82.81 \pm 0.06°$, $\gamma = 94.36 \pm 0.05°$. X-ray density calculated for a unit cell containing 14 molecules of the basic formula $MSiO_3$ is 3.83. The density determined by the volume displacement of 100 μg of grains in a capillary is 3.68 $\pm$ 0.05.

Sample 10044 also contains trace amounts of fayalite (Fe_2SiO_4) which form narrow bands between pyroxmangite and ilmenite. The formula derived from the uncorrected probe composition is $(Fe_{1.92}Mg_{0.04}Ca_{0.02}Mn_{0.02}Ti_{0.01})_{2.01}Si_{1.02}O_{4.06}$. Olivine of this composition, Fa_{96}, suggests higher oxidizing conditions of formation than those indicated for the finer-grained rocks 10017 and 10072 in which olivine compositions in the range Fa_{30-35} were observed.

Three polymorphs of silica were identified. Tridymite occurs with cristobalite in 10044 and quartz with cristobalite in 10072, both as fine-grained intergrowths. Such associations are characteristic of terrestrial lavas where rapid cooling has delayed the sluggish inversions to account for the coexistence of high- and low-temperature forms of silica. Tridymite and cristobalite are seldom, if ever, found in deep-seated rocks. Nonequilibrium conditions are indicated for 10072, in which silica occurs with olivine of composition Fa_{35}, but silica is compatible with fayalite in 10044. Aside from these equilibrium considerations, the presence of tridymite indicates that not all of the Apollo 11 rocks were subjected to shock pressures in excess of 3 kb.

GEORGE W. REED, JR.
STANKA JOVANOVIC
LOUIS H. FUCHS

Argonne National Laboratory, Argonne, Illinois 60439

References and Notes

1. G. W. Reed, Jr., and S. Jovanovic, *Earth Planet. Sci. Lett.* **6**, 316 (1969).
2. P. Wilkniss and V. J. Linnenboom, *Proceedings of the 2nd Conference on Practical Aspects of Activation Analysis with Charged Particles*, H. G. Ebert, Ed. (EURATOM, 1968), p. 147.
3. M. Quijano-Rico and H. Wanke, *Meteorite Research*, P. M. Millman, Ed. (Reidel, Dordrecht, 1969), p. 132.
4. E. Jarosewich, private communication (1969).
5. S. Jovanovic and G. W. Reed, Jr., *Geochim. Cosmochim. Acta* **32**, 341 (1968).
6. G. W. Reed, Jr., and S. Jovanovic, unpublished work.
7. G. L. Bate and J. R. Huizenga, *Geochim. Cosmochim. Acta* **27**, 345 (1963).
8. J. L. Barker and E. Anders, *ibid.* **32**, 627 (1968).
9. Lunar Sample Preliminary Examination Team, *Science* **165**, 1211 (1969).
10. L. Fuchs, *Meteorite Research*, P. M. Millman, Ed. (Reidel, Dordrecht, 1969), p. 688.
11. J. V. Smith, *J. Geol.* **73**, 830 (1965).
12. We thank the Chemistry Department, Brookhaven National Laboratory, for making its facilities available and especially J. Hudis and Mrs. E. Rowland for their assistance. Carrier and monitor analyses performed by K. Jensen, R. Bane, and I. Fox, Argonne National Laboratory, are gratefully acknowledged. We thank B. Tani and S. Siegel, Argonne National Laboratory, for the single-crystal x-ray studies and J. Sanecki for assistance in the microprobe work. T. E. Bunch, Ames Research Center, analyzed the fluorapatite for fluorine. Supported by USAEC and by NASA contract T-76356.

4 January 1970

Gallium, Germanium, Indium, and Iridium in Lunar Samples

Abstract. *Neutron activation analyses of gallium, germanium, indium, and iridium in eight lunar samples and in meteorites and rocks (including four calcium-rich achondrites and five terrestrial basalts) with similar bulk compositions are reported. Lunar gallium concentrations are remarkably constant at about 5 parts per million, three times higher and four times lower than those in eucritic (calcium-rich) achondrites and terrestrial basalts, respectively. Lunar germanium concentrations range from ≤0.04 to 0.4 part per million. Basaltic achondrites have similar germanium concentrations, whereas those in terrestrial basalts are uniformly higher at about 1.4 parts per million. Concentrations of indium in lunar samples range from 3 to 60 parts per billion, whereas those in terrestrial basalts are 80 to 100 parts per billion, and those in basaltic achondrites 0.4 to 3 parts per billion. Lunar iridium concentrations appear to be in the 0.1 to 10 parts per billion range. The Tranquillitatis samples are distinctly different from calcium-rich achondrites and terrestrial basalts.*

In meteorites containing reduced Ni-Fe, Ge and Ir are heavily concentrated in such phases, Ga is slightly concentrated in the metal relative to the silicate fraction, and In is enriched in the sulfide and silicate phases (*1*). The samples from Mare Tranquillitatis are of basaltic composition, with negligible amounts of reduced Ni-Fe (*2*). Relative to Si, Ga and In abundances in basaltic terrestrial rocks are similar to those in type 1 carbonaceous chondrites. Germanium and Ir are severely depleted in basalts relative to the chondrites. Reasonably good information is available regarding the geochemical behavior of Ga in basalts; Prinz (*3*) indicates that Ga concentrates in the plagioclase and that Ga shows no correlation with the crystallization index, which is an indication of the degree of magmatic fractionation. Very little is known about the geochemical behavior of Ge, In, and Ir in basaltic rocks and magmas, mainly because the concentrations of these elements are at or below the detection limits of emission spectrographic analysis. Thus, to provide a meaningful basis for the interpretation of our analyses of the lunar materials, we initiated studies of the basaltic geochemistry of the elements included in our investigations. We report here concentrations of Ga, Ge, and In in eight Apollo 11 samples and initial investigations of these three elements in terrestrial and meteoritic rocks of basaltic composition. Our Ir data need further refinement and are not tabulated.

The elements Ga, Ge, In, and Ir were determined by a neutron activation scheme involving their sequential radiochemical separation. Gold was also determined in the sequence, but will be discussed elsewhere. The techniques are those outlined previously, with minor changes. After the samples were subjected to radiochemical separation, γ spectra were recorded with NaI scintillation detectors [or occasionally a Ge (Li) ionization detector] in conjunction with a multichannel analyzer.

The precision limits of the method at 68 percent confidence when applied to the analysis of L-group chondrites has been given (*4*) as ± 3.5, 30, 18, and 8 percent for the elements Ga, Ge, In, and Ir. We have not adequately retested the precision during these analyses, but estimate that our Ga (all values), In (values

Table 1. The Ga, Ge, and In concentrations in lunar surface materials.

Sample	Petrolog. class	Ga (ppm)		Ge (ppm)		In (ppb)	
		Replicates	Mean	Replicates	Mean	Replicates	Mean
10084	Fines	4.4, 4.9	4.6	0.34, 0.35	0.34	681, 677	679
10021	Brec	5.7, 5.3	5.5	0.46, 0.31	0.38	14, 30	22 ± 8
10046	Brec	5.6, 5.2	5.4	0.38, 0.34	0.36	25, 6	16 ± 9
10048	Brec	5.1, 5.4	5.2	0.35, 0.29	0.32	56, 65	60
10056	Brec	5.0, 5.1	5.0	0.02, ≤0.04	≤0.04	2.4, 3.5	3.0
10057	CrysA	5.0, 4.9	4.9	0.22, ≤0.08	≤0.2	3.0, 2.9	3.0
10060*	Brec	4.9, 5.5	5.2	0.20, 0.24	0.22	1150, 911	1030 ± 150
10072	CrysA	4.8, 5.1	4.9	0.026, 0.085	≤0.08	49, 54	52

*Ground and split at Manned Spacecraft Center.

above 1 ppb), and Ir (values greater than 2 ppb) limits are about the same as those previously reported. Our Ge (values above 1 ppm) errors appear to be somewhat lower, amounting to limits of about ± 8 percent at 68 percent confidence. Concentrations less than those listed are less precise, and some of our lowest results are listed only as upper limits.

The accuracy of our method can best be judged by comparison of our analyses of USGS standard rocks with those of other research groups. Our earlier Ga data (*4*) on W-1 and a number of L-group chondrites agreed quite well with those of Fouché and Smales (*5*). Our current data on G-2 and BCR-1 average about 15 percent lower than those of Brunfelt *et al.* (*6*), and our PCC-1 value is a factor of two lower than their result. There are no reliable results for Ge on the new USGS rock standards, but our earlier Ge data (*4*) were systematically about 10 percent higher than those of Fouché and Smales (*5*) and Greenland and Lovering (*7*). As discussed by Tandon and Wasson (*4*), our In data are in good agreement with previous results obtained for L-group chondrites and standard rocks, and our current values on the new USGS standards agree well with the concentrations found by Johansen and Steinnes (*8*). We conclude that our systematic errors are, at worst, no greater in magnitude than the random errors discussed above.

In Table 1 are listed Ga, Ge, and In data on eight lunar samples. Also listed is the rock type (*9*). Gallium is remarkably uniform in the lunar samples. In fact, the relative standard deviation of an individual result about the grand mean of 5.1 ppm is only 7 percent, and an appreciable fraction of the scatter is contributed by experimental error. Germanium is much more variable. In the fines and four breccia samples the Ge concentration ranges from 0.2 to 0.4 ppm. In one breccia and in the crystalline rocks the range is from ≤ 0.04 ppm to ≤ 0.2 ppm. The concentration of In is highly variable. In the fines and in sample 10060 the In concentration was very high, on the order of 1 ppm, but these were quite possibly contaminated by the In seal of the rock box and during the homogenization and splitting process at MSC, respectively. Indium contents of the other lunar rocks ranged from 3 to 60 ppm, with no particular correlation with the type of sample. The agreement of replicates was very poor for two of the breccias. This may reflect sampling inhomogeneities, but may also be contamination effects, and there is the possibility that all samples were contaminated, perhaps by the In seal in the lunar rock box. Our data on Ir are currently not publishable, primarily because of incorrect monitor preparation. The values for the lunar samples range from about 0.1 to about 10 ppb.

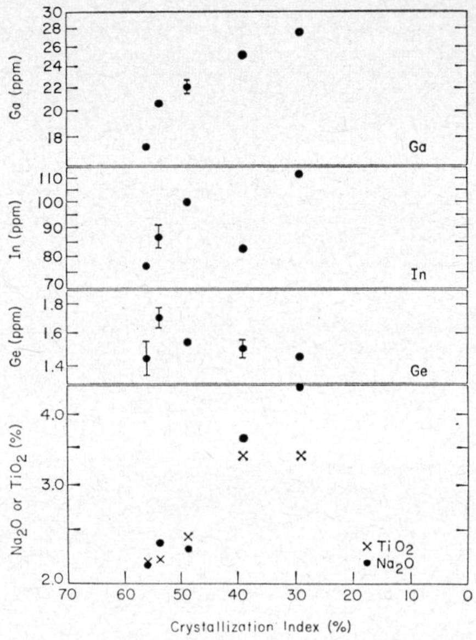

Fig. 1. Chemical data on Hawaiian basalts plotted versus the crystallization index. The Ga and Ge are enriched in the later differentiates, whereas Ge is unaffected by the magmatic fractionation process.

In Table 2 are listed Ga, Ge, and In concentrations for four basaltic achondrites and nine terrestrial rocks, including five basalts and four USGS standard rocks. The basaltic achondrites consisted of three eucrites and one howardite. The eucrites showed very uniform Ga concentrations of about 1.7 ppm, and the howardite a lower value of about 0.9 ppm. These are well below the Ga concentration (5.1 ppm) of the Tranquillitatis samples. Germanium concentrations were low and near our detection limit. All achondrite values were less than or equal to 0.15 ppm, about the same as the lowest values observed in the lunar crystalline rocks. Indium concentrations ranged from 0.4 to 3 ppb, and in three of the four meteorites were lower than the lowest lunar results.

In terrestrial mafic and felsic rocks Ga concentrations were about 20 ppm (four times higher than in the mafic lunar rocks), but much lower values were found in terrestrial ultramafic rocks. Germanium was remarkably uniform in all terrestrial rocks that we investigated and averaged about 1.4 ppm. This is more than an order of magnitude higher than the values in basaltic achondrites and in three of the lunar rocks, and five times higher than the highest lunar Ge concentrations. Terrestrial In concentrations range from 3 ppb in peridotite to about 100 ppb in mafic rocks, a range which is comparable to the range ob-

Table 2. Ga, Ge, and In concentrations in Ca-rich achondrites, terrestrial basalts and other terrestrial rocks. The sources are abbreviated as follows: GAM, G. A. MacDonald, University of Hawaii; MHNP, Muséum d'Histoire Naturelle, Paris; UCLA, University of California, Los Angeles; USGS, United States Geological Survey; and UTü, Mineralogisches Institut, Universität Tübingen.

Sample	Type	Source	No.	Ga (ppm)		Ge (ppm)		In (ppb)	
				Replicates	Mean	Replicates	Mean	Replicates	Mean
Juvinas	Eucrite	UTü		1.8, 1.7, 1.9	1.7	0.06, ≤0.09	≤0.09	1.4, 1.4	1.4
Pasamonte	Eucrite	UCLA	308	1.7, 1.7	1.7	≤0.04, ≤0.01	≤0.04	3.4, 3.0	3.2
Stannern	Eucrite	UTü		1.6, 2.0, 1.6	1.7	0.15, ≤0.02	≤0.15	0.27, 0.45	0.36
Le Teilleul	Howardite	MHNP	711	0.85, 0.93	0.89	0.05, 0.09	0.07	0.88, 0.72	0.80
Waianae, Haw.	Thol. basalt	GAM	C-6	17.2, 17.2	17.2	1.4, 1.4	1.4	85, 69	77
Waianae, Haw.	Thol. basalt	GAM	C-11	20.0, 21.1	20.6	1.9, 1.5	1.7	95, 78	87
Waianae, Haw.	Thol. basalt	GAM	C-20	23.1, 19.8	22.0	1.2, 1.5	1.5	117, 81	99
Waianae, Haw.	Alk. basalt	GAM	C-31	24.1, 26.2	25.1	1.6, 1.4	1.5	87, 78	82
Waianae, Haw.	Hawaiite	GAM	C-178	2.8, 27.2	27.6	1.4	1.4	108, 115	112
Bridal Veil, Ore.	Basalt	USGS	BCR-1	19.8, 19.9	19.8	1.4, 1.5	1.5	98, 94	96
Westerly, R.I.	Granite	USGS	G-2	22.0, 22.0	22.0	1.3, 1.1	1.2	30, 35	32
Cazadero, Cal.	Peridotite	USGS	PCC-1	0.65, 0.60	0.63	1.1, 1.0	1.0	3.4	3.4
Twin Sisters, Wash.	Dunite	USGS	DTS-1	0.46, 4.8		1.1, 1.0	1.0	24	24

served in lunar samples. The terrestrial basalts, however, are distinctly higher in In content than any lunar rocks. The highest Ir values in the lunar samples appear to be about an order of magnitude higher than those observed in basaltic rocks and achondrites.

On the basis of these limited data one can note that the samples collected in Mare Tranquillitatis are distinctly different from the logical comparison materials of Ca-rich (basaltic) achondrites and typical terrestrial rocks of basaltic composition. Our limited conclusions are: (i) Typical basaltic achondrites did not originate in the Mare Tranquillitatis area of the moon; and (ii) the lunar materials of basaltic composition differ from those found on the earth either as a result of differing fractionation processes or differing primitive starting materials, or both. These conclusions have been drawn previously (*2*).

The five Hawaiian rock samples are from the Waianae Volcano on the island of Oahu. Samples 6, 11, and 20 are from the lower member of this volcano [as described by MacDonald (*10*)] and were probably produced during a single stage of magmatic activity. Samples 31 and 178 are from the upper member of this volcano, and may not be as closely related to the former, more basic rocks. Nonetheless, it appears of some interest to consider all these rocks to be related and to attempt to define the behavior of Ga, Ge, and In during the course of magmatic activity. The five samples were chosen to form a suite of related rocks, extending from the most mafic sample, C-6, in order of increasing sample number to the most felsic sample, C-178. To observe possible trends through this suite, we plotted our data against the crystallization index defined by Poldervaart and Parker (*11*) (Fig. 1).

Gallium shows a pronounced enrichment during the course of magmatic fractionation [in disagreement with the data summarized by Prinz (*3*)]. Indium also appears to be enriched, although the C-31 point falls off a smooth curve fitted through the other points. The concentration of Ge does not appear to change appreciably during the fractionation process. Also shown are the concentrations of Na_2O and TiO_2 in the same rocks as reported by MacDonald and Katsura (*10*). These are also enriched in the later differentiates, TiO_2 to about the same degree as Ga and In, and Na_2O to a still greater extent.

One of the intriguing problems regarding the Mare Tranquillitatis samples is their high TiO_2 content. The most plausible explanations of this observation are that the enrichment is the result of a high-temperature volatilization or partial condensation process or the result of magmatic fractionation, or some combination of these possibilities. If volatility is the major cause of the enrichment, one would expect a negative correlation between the contents of TiO_2 and of In, which forms relatively volatile compounds. If magmatic fractionation has established the trends in the lunar samples, our preliminary data indicate that a positive correlation between In and TiO_2 would result. We will search for evidence regarding these possibilities in the data to be published from these proceedings.

PHILIP A. BAEDECKER
JOHN T. WASSON
Department of Chemistry and Institute of Geophysics and Planetary Physics, University of California, Los Angeles

References and Notes

1. P. A. Baedecker and J. T. Wasson, "Gallium (31)," in *Elemental Abundances in Meteoritic Matter*, B. Mason, Ed., in press; "Germanium (32)," *ibid.*; P. A. Baedecker, "Indium (49)," *ibid.*; "Iridium (77)," *ibid.*
2. Lunar Sample Preliminary Examination Team, *Science* **165**, 1211 (1969).
3. M. Prinz, in *Basalts*, H. H. Hess and A. Poldervaart, Eds. (Interscience, New York, 1967), pp. 271–323.
4. S. N. Tandon and J. T. Wasson, *Geochim. Cosmochim. Acta* **32**, 1087 (1968).
5. K. F. Fouché and A. A. Smales, *Chem. Geol.* **2**, 5 (1967).
6. A. O. Brunfelt, O. Johansen, E. Steinnes, *Anal. Chim. Acta* **37**, 172 (1967).
7. L. Greenland and J. F. Lovering, *Geochim. Cosmochim. Acta* **29**, 821 (1965).
8. O. Johansen and E. Steinnes, *Talanta* **13**, 177 (1966).
9. J. Warner, private communication.
10. G. A. MacDonald and T. Katsura, *J. Petrol.* **5**, 82 (1964); G. A. MacDonald, *Geol. Soc. Amer. Mem.* **116**, 477 (1968).
11. A. Poldervaart and A. B. Parker, *Amer. J. Sci.* **262**, 281 (1964).
12. We thank J. Kimberlin, S. Moissidis, S. Dorrance, D. Gondar, W. Simpson, C. Pawlak, and J. Kaufman for assistance. Neutron irradiations were handled at the UCLA and Ames Laboratory reactors by J. Hornbuckle, A. F. Voigt, and their associates. Supported in part by NASA contract NAS 9-8096 and NSF grant Ga-1347.

4 January 1970

Multielement Analysis of Lunar Soil and Rocks

Abstract. *Results for multielement analysis of lunar soil and of seven rocks returned by Apollo 11 are presented. Sixty-six elements were determined with spark source mass spectrography and neutron activation. U.S. Geological Survey standard W-1 was used as a comparative standard. Results indicate an apparent uniformity of composition among the samples. Comparison with solar, meteoritic, and terrestrial abundances reveals depletion of volatile elements and enrichment of the rare earths titanium, zirconium, yttrium, and hafnium. Although there is an overall similarity of the lunar material to basaltic achondrites and basalts, the differences suggest detailed geochemical processes special to the history of this material.*

Chemical analyses of lunar soil and seven rocks from the Apollo 11 mission were performed with spark source mass spectrography (MS) and neutron activation analysis (NAA); this resulted in the determination of approximately 66 elements per sample. Both methods as used for terrestrial samples have previously been described (*1*). The U.S. Geological Survey standard diabase W-1 was again used as a multielement standard, and the same literature values were employed.

For the lunar analyses, a few changes have been made in these methods. One gram of each sample was crushed; then 200 mg was removed for NAA, and 500 mg was removed for MS. For the shorter irradiations in NAA, flux and irradiation times were changed to compensate for the smaller sample size. For the longer irradiations, the samples were inserted in the central thimble of the TRIGA reactor with a resultant flux of 3.5×10^{12} neutron/cm^2-sec. In the case of MS, the 500-mg samples were mixed with equal amounts of graphite (National Carbon Co. spectroscopic powder). Because of small sample size, accurate doping of an internal standard for MS was not possible. Zirconium present in the samples represented a suitable internal standard, and the NAA determination for this element was used in the MS calculations.

Results of the analysis of the eight lunar samples received at our laboratory are given in Table 1, along with the method employed. Results from the method having the better accuracy (as determined by past analyses of terrestrial and meteoritic materials) and the better analytical conditions (higher sensitivity, lower background, and so forth) were used. When these criteria were judged equivalent, an arithmetic average of the results of the two methods was reported. Since these multielement methods possess variations in precision for each element, only average values are given. The relative standard deviation of the determinations by MS is 5 to 25 percent about the mean, whereas that of NAA is 10 percent or less. Accuracy, evaluated by analysis of terrestrial materials, is comparable to or better than the precision. To complete the values for the major elements, Si was determined by atomic absorption spectrophotometry.

The results of both MS and NAA determinations depend on the accuracy of published isotope ratios as applied to the lunar samples. In the determination of lead by MS, the isotope ratios did not correspond to published terrestrial values. Of the four stable lead isotopes, ^{204}Pb was obscured. A summation of

Table 1. Elemental abundances in the lunar samples studied. The results are grouped according to NASA sample type (*2*): A, fine-grained vesicular; B, coarse-grained crystalline; C, breccia; and D, soil. The sample of type D (84-55) is the fine fraction (<1 mm) of soil sample 10002. NAA, neutron activation analysis; MS, mass spectrography; AA, atomic absorption. Numbers used to designate samples are NASA laboratory numbers.

Element	Method	Type A rocks			Type B rocks		Type C rocks		Soil
		20-26	57-32	72-24	58-26	46-23	56-20	60-13	84-55
					Major elements (weight percent)				
Si	AA	19.4	21.5	19.2	18.4	20.6	19.8	18.7	20.2
Al	NAA	5.9	4.0	4.0	5.4	6.2	5.7	6.0	7.3
Ti	NAA	6.1	6.5	6.7	5.8	5.0	5.6	5.1	4.1
Fe	NAA	14.7	15.7	15.4	15.2	13.2	14.8	13.8	12.5
Mg	NAA	5.0	3.7	4.3	3.4	5.5	2.8	4.6	4.6
Ca	NAA	10.3	10.1	10.3	11.0	9.8	11.0	10.4	9.6
Na	NAA	0.28	0.35	0.32	0.32	0.35	0.32	0.34	0.33
K	NAA	0.052	0.22	0.29	0.093	0.17	0.094	0.16	0.11
Mn	NAA	0.20	0.17	0.17	0.19	0.16	0.20	0.16	0.16
Cr	NAA	0.22	0.21	0.24	0.15	0.21	0.14	0.22	0.20
Zr	NAA	0.036	0.056	0.072	0.038	0.062	0.041	0.058	0.039
Ni*	NAA	0.006	0.004	0.003	0.008	0.007	0.005	0.007	0.017
V	NAA-MS	0.0059	0.0040	0.0062	0.0041	0.0068	0.0056	0.0062	0.0078
P	MS	0.070	0.041	0.070	0.024	0.10	0.033	0.060	0.14
					Trace elements (parts per million)				
Li	MS	5	8	14	6	16	16	7	6
Be	MS	2	2.5	4	1.5	6	3	3	4
B	MS	1	4	4	2	9	2	3	2
N	MS	10	20	30	10	70	20	6	30
F	MS	85	70	100	50	220	30	80	66
Cl	MS	150	50	60	50	520	16	150	350
Sc	NAA	85	84	86	87	64	97	70	60
Co	NAA-MS	20	24	28	14	42	15	32	40
Cu	NAA	3.7	5.5	18	7.1	9.7	3.8	11	9.9
Zn	NAA	2.1	2.9	7.0	9.3	30	2.7	25	22
Ga	NAA	3.5	4.7	4.3	4.3	4.9	4.3	5.1	4.6
Ge	MS		1.3	1.1	1.2		1.2	1.4	0.7
As (~)	NAA	0.03	0.04	0.05	0.07	0.05	0.03	0.09	0.07
Se	MS	0.4				0.4		0.9	0.2
Br (~)	NAA	0.1	0.1	0.2	0.3	0.2	0.06	0.3	0.2
Rb	NAA-MS	2.8	4.8	5.7	1.2	3.6	2.0	4.0	4.4
Sr	NAA-MS	170	130	140	180	170	160	180	200
Y	MS	130	210	250	150	190	180	210	150
Nb	MS	36	42	45	47	38	37	45	33
Mo	NAA-MS	0.4	0.4	0.4	0.4	0.7	0.4	0.7	0.7
Pd	MS	0.09	0.1	0.1	0.2	0.1	0.1		0.04
Ag	MS	0.1	0.04	0.6	0.07	0.02	0.2	0.01	0.1
Cd	MS		0.9	1	0.7	0.8	0.9	0.3	0.3
In	MS		0.07	2	0.6	0.08	0.06		0.5
Sn	MS		0.6	0.4	1.2		0.3		0.7
Sb	NAA	0.01	0.005	0.01	0.01	0.005	0.005	0.005	0.005
Cs	NAA-MS	0.2	0.2	0.3	0.3	0.2	0.06	0.2	0.2
Ba	NAA-MS	96	280	300	140	280	100	250	220
La	NAA-MS	11	31	35	16	23	13	24	22
Ce	NAA-MS	34	83	96	45	67	42	62	50
Pr	MS	8.7	22	20	13	20	12	13	9
Nd	NAA-MS	43	84	120	72	70	82	96	46
Sm	NAA-MS	14	24	28	22	20	23	24	18
Eu	NAA-MS	1.6	2.1	2.2	3.0	2.0	2.5	2.0	1.9
Gd	NAA-MS	17	26	31	22	20	24	28	20
Tb	NAA-MS	3.5	5.6	6.8	5.4	4.5	5.4	5.0	3.8
Dy	MS	30	42	45	39	30	40	41	25
Ho	NAA-MS	7	8	10	9	9	9	10	6
Er	MS	19	32	35	36	23	27	30	15
Tm	NAA-MS	1.2	2.3	2.8	2.0	1.6	2.1	1.8	1.2
Yb	NAA-MS	15	26	28	22	20	20	22	12
Lu	NAA-MS	1.5	2.2	2.6	2.3	1.8	1.8	2.0	1.4
Hf	NAA-MS	11	15	18	13	11	11	13	9
Ta	NAA	1.3	1.2	1.8	1.0	1.7	2.2	1.7	1.3
W	NAA	0.13	0.42	0.42	0.36	0.35	0.15	0.35	0.25
Tl	MS								0.2
Pb	MS		3	3	3	2	1.2	3	6
Th	NAA-MS	1.5	4.5	4.8	1.1	2.8	1.4	3.0	2.3
U	NAA	0.14	0.56	0.50	0.20	0.58	0.21	0.60	0.48

* Third decimal place approximate.

the intensities of the remaining three isotopes was used to compute the elemental concentration. In terrestrial basaltic samples, this would lead to a value 2 percent lower than expected, which is well within experimental error.

For the elements Ru, Rh, Te, and I, mass spectrographic determinations were limited by the lack of sensitivity factors. A comparison of the intensities shows that Ru and Te are approximately of the same concentrations as in terrestrial basaltic materials as exemplified by the USGS standard BCR-1; Rh is enhanced by a factor of 2 to 3; and I is depleted by a factor of 2. The elements Re, Os, and Ir were not observed; their detection limits are approximately 0.1 ppm for MS in these samples. Similarly, Hg was not detected by NAA at 0.02 ppm.

The accuracy of elemental determinations is dependent on the amount of possible contaminants. The values for N, F, Cl, Ag, and In may have been influenced by contamination before they were received. The samples were received under a dry nitrogen blanket and prepared for analysis in dry air; this resulted in the possibility of surface N_2 adsorption. On the lunar surface, the samples were sealed in Teflon bags, and the sample boxes were sealed with In-Al alloy O-ring. In addition, a microscopic examination of the samples as received from NASA revealed foreign material in the form of translucent fibers, possibly from the plastic packaging materials.

At the 2σ level of significance, there is neither general sample-to-sample variance, nor is there a sample type-to-type variation of elemental abundances (Table 1). When variations at the 1σ level are examined, division of the samples into two groups, based on the behavior of 13 elements, is possible. For example, if the samples are arranged in order of increasing potassium content, rocks 20, 58, and 56 fall into one group and rocks 60, 46, 57, and 72 fall into another group, with the lunar fine material 84 having intermediate values for these elements but approaching the second group. These groups do not follow the textural types suggested by the Preliminary Examination Team (*2*). However, it should be emphasized that such a compositional classification is possible only if the 1σ variation in values is regarded as significant.

Of the elements studied, there is greatest variation among the eight samples for P, K, Cl, F, In, Zn, and Cu. This variation cannot be attributed to analytical imprecision and must be regarded as reflecting the variation of the samples received by this laboratory. In the case of Zr and Yb, our values are significantly different from the data reported for the same samples by the Preliminary Examination Team. At present, we are unable to assess the source of this discrepancy, apart from commenting on the possible sampling difficulties in the distribution of the samples by NASA.

Comparison of the average lunar elemental abundance data with those of solar (recalculated on a H, He-free basis) meteoritic and terrestrial crust values results in the following observations: In the lunar material, volatile elements such as Sb, As, Br, Cu, Zn and Hg appear depleted. Sodium and potassium are depleted when compared to terrestrial crust and basalts, and they are enriched when compared to solar abundances. The rare earth and refractory elements such as Ti, Zr, Y, and Hf appear enriched. The lunar rare earth abundances, when normalized against La, correspond approximately with solar abundances, are slightly higher than in achondrites, and are considerably higher than those of the terrestrial crust.

With regard to the major elements, the lunar samples more closely resemble basaltic achondrites and, to a lesser extent, terrestrial basalts, with the notable exceptions of Ti and Zr, which are substantially enriched in the lunar material. The trace element distribution in the lunar materials is also comparable to that in achondrites and basalts. Comparison with basalts shows that the trace elements, though dispersed, are evenly distributed about a 1:1 trend line. Comparison with achondrites shows that most trace elements are uniformly enriched in the lunar materials.

It is apparent that the lunar samples are compositionally different from all materials with which they have been compared. Although there is closer similarity to basaltic achondrites and basalts, the lunar material does not match them in all respects. This uniqueness implies detailed geochemical processes special to the history of this material.

GEORGE H. MORRISON
JESSE T. GERARD
A. THOMAS KASHUBA
ESWARA V. GANGADHARAM
ANN M. ROTHENBERG
NOEL M. POTTER
GARY B. MILLER

Department of Chemistry,
Cornell University,
Ithaca, New York 14850

References and Notes

1. G. H. Morrison, J. T. Gerard, A. Travesi, R. L. Currie, S. F. Peterson, N. M. Potter, *Anal. Chem.* **41**, 1633 (1969); G. H. Morrison and A. T. Kashuba, *ibid.*, p. 1842.
2. The Lunar Sample Preliminary Examination Team, *Science* **165**, 1211 (1969).
3. Supported by NASA grant NAS 9-9986.
4 January 1970

Neutron Activation Analysis of Milligram Quantities of Lunar Rocks and Soils

Abstract. *A neutron activation scheme for determining 25 elements in lunar samples weighing 20 milligrams is described and applied to a suite of Apollo 11 lunar materials. Concentrations of titanium, chromium, scandium, tantalum, hafnium, and rare earths are higher than in average basalt, whereas cobalt, nickel, and copper are lower. Chemical variations show groupings of elements possibly associated with the major phases, pyroxene, plagioclase, and ilmenite. The high concentration of "refractory oxides" and low volatile content implies that the raw material for the Apollo 11 samples was condensed from the primitive solar nebula at high temperatures.*

The aim of our program of investigation of lunar material was to develop and employ a neutron activation analysis procedure for the analysis of 20 mg or less of rock or mineral for a large part of the periodic chart. The method we developed is of paramount importance when only small quantities of material are available for analysis. This situation may arise either from a paucity of sample or, more likely, in the analysis of mineral separate fractions from lunar rock and soil. The procedure was adapted from the one described by Perlman and Asaro (*1*).

The general analytical scheme starts with a short neutron irradiation immediately followed by counting with a Ge(Li) diode detector, then proceeding to a longer irradiation and counting with the same system after longer waiting periods. Finally chemical separations of

Table 1. Chemical composition of Apollo 11 lunar rocks and soil. Values in parts per million except Fe and Ti in percentage. A blank indicates that no determination was made. Explanations for the notes in column 2 are as follows: (1) Irradiation B in highly thermalized neutron flux of 4.5×10^{12} neutrons cm^{-2} sec^{-1} for 30 minutes. Counting began 3 hours after removal from pile. Irradiation in type 2 planchet, counting in type 1 planchet. (2) Irradiation B, uranium determined by spark c ounting of induced fission tracks in Lexan clamped on type 2 planchet. (3) Irradiation B, counting began 18 hours after removal from pile. Irradiation in type 2 planchet, counting in type 1 planchet. (4) Irradiation A in core of reactor at thermal neutron flux of 3×10^{13} neutrons cm^{-2} sec^{-1} and large fast neutron component for 3 days. Counting began 8 days after removal from pile. Irradiation and counting in type 1 planchet. (5) Irradiation A, counting began 30 days after removal from pile. Normalization to europium values from irradiation B. (6) Chemical separation and counting as described in text. (7) Probably contaminated.

Element	Note	Vesicular basalt 10020	Gabbro 10044	Breccia 10046	Gabbro 10056	Gabbro 10062	Soil 10084	Breccia 10021	Vesicular basalt 10049	Vesicular basalt 10057
Na	3	2680*	3630*	3720*		2990*	3000*	3470	3800	4050
K	3	1260*	1090*	1200*		230*	1000*	1720	2350	3610
Mn	1	2130*	2290*	1760*	2140*	1510*	1250*	1580	1790	1930
Cu	3	35*	47*	47*		30*	38*	46*	48	60
Dy	1	18*	36*	20*	34*	22*	17*	23	31	37
Eu	3	1.5*	3.0*	1.9*	2.9	1.8*	2.0*	1.8	2.1	2.5
Sm	3	7.6*	18*	13*	17	10*	8.7*	12	18	17
La	4	7.7	14	18	12	12	20			
Yb	4	6.6	9.5	8.7	8.3	6.3	9.5			
Lu	4	0.81	1.3	1.2	1.2	0.87	1.3			
Ce	6	47	86	78	75	48	79			
Fe (%)	5	14	13	12	13	12	14			
Cr	5	2400	1600	1970	1410	1310	2220			
Co	5	21	13	30	14	13	33			
Ni	6	160*		35*	37*		320*			
Sc	5	98	100	71	109	76	80			
Ti (%)	4	5.1	3.7	3.9	3.6	3.9	4.4			
Hf	5	6.6	13	11	11	10	11			
Zr	6	260*	660*	370*			380*			
Ta	5	2.2	2.7	1.8	2.6	1.8	1.8			
U	2	0.25*	0.47*	0.72*	0.45*	0.32*	0.33*	0.39*	0.74	0.48
Ag	6, 7	1.3	0.41*	1.3	0.16*	0.17	0.25			

*Single determination. All others are the average of duplicate determinations.

specific fractions are performed on the sample, and counting is done by either Ge(Li) detector or NaI(Tl) detector, depending on efficiency and purity requirements.

Approximately 250 mg of a lunar rock or soil sample is crushed and ground in an Alundum mortar. Accurately weighed 20-mg aliquots are mixed in a small agate mortar with 80 mg of chromatographic grade cellulose and pelletized with a standard KBr pelletizer under 280 bars pressure. The diameter of the pellet is 1.3 cm, and its thickness is less than 1 mm.

Initially our procedure involved the transfer of the pellet to an accurately weighed high purity (99.9999 percent) aluminum foil planchet just large enough to receive the pellet. The planchet was capped and crimped to form a thin disk. This planchet we shall designate as the type 1 planchet. A standard silicate-matrix material (the "standard pot" made and analyzed by Perlman and Asaro and analyzed by us independently) was treated in identical fashion. In our initial irradiation sequence (called "irradiation A") the samples and standard were placed radially on edge around a cylinder and irradiated. During irradiation the cylinder is rotated around the axis to guarantee a homogeneous flux for all samples and standards.

In irradiation A using type 1 planchets the samples and the aluminum planchets were counted as a whole, and subtraction for the contaminants in the aluminum was made using a blank. The high amounts of ^{24}Na produced in the aluminum, the presence of a large amount of copper and a variable gold contamination in the aluminum foil, as well as the desirability of determining uranium by fission track counting (*2*), led us to develop a new removable container for the pellets. The new planchet is constructed so that an aluminum ring clamps a sheet of Lexan plastic (to detect induced uranium fission tracks) over the pellet seated in a machined slot. This planchet, hereafter designated the type 2 planchet, was used in a second irradiation sequence (irradiation B).

Throughout all the irradiations the cellulose binder, although blackened, maintains its cohesive role, and the pellet can be transferred without loss.

The detailed program of irradiations and countings is given below. All irradiations were performed at the Union Carbide "swimming pool" reactor at Sterling Forest, New York, a 2-hour drive from New Haven, Connecticut, and the counting was done at Yale.

1) The samples are given a 30-minute irradiation in type 2 planchets in a highly thermalized portion of the reactor at a flux of about 4.5×10^{12} neutrons cm^{-2} sec^{-1}. Three hours later, after removal from the pile and after separation of the Lexan, which is etched and spark-counted for induced fission tracks from uranium (*2*), the pellet is transferred to a type 1 planchet and counted for Mn and Dy. Because of excess interference of Ga with the Mn peak in the standard pot, the USGS standard diabase W-1 was used as the standard. Decay for 18 hours is then followed by counting for Na, K, Cu, Sm, and Eu. The Ge(Li) diode detector is used for this part.

2) A long irradiation of the samples in type 1 planchets is next done for 3 days in the core of the reactor with a thermal neutron flux of 3×10^{13} neutrons cm^{-2} sec^{-1} and a large fast neutron component. After 8 days cooling following removal from the pile, the total planchet is counted for short-lived (2- to 4-day half-lives) nuclides. Thirty days after removal from the pile the samples in the planchets are counted again to obtain results for the long-lived nuclides. The Ge(Li) diode detector is used for this part.

3) After the Ge(Li) detector counting sequence is over, the samples are transferred from the type 1 planchet to a nickel crucible, mixed with 500 mg of a solid carrier mixture charge containing the elements sought as oxides, and fused with 10 g NaOH and 10 g Na_2O_2 for half an hour. The fused mass is disag-

gregated in boiling water and centrifuged several times to separate the water-soluble components from the insoluble residue. The residue is dissolved in hydrochloric acid and passed through a column (15 cm long by 3 cm in diameter) containing Dowex-1 resin (50–100 mesh) in chloride form. Two column volumes each of the following acid solutions were passed and collected sequentially: 12*N* HCl carries the rare earths, Ag; 8*N* HCl carries Ti, Zr, Hf; 4*N* HCl carries Co; 0.1*N* HNO_3 carries Hg. Other elements, of course, are also removed from the column with the sought elements. From the appropriate fraction the following elements are separated, purified if necessary, and counted: Silver is separated and counted with the NaI detector, chemical yield is determined and compared with a silver standard; the remaining solution containing Sc and the rare earths is evaporated to dryness on a planchet and counted on the Ge(Li) detector. The Ce/Sc count ratio combined with the previously determined Sc concentration gives the Ce concentration, as compared with the standard pot, without the necessity of yield determination or further purification. The cobalt fraction containing ^{60}Co and ^{58}Co from n,p on ^{58}Ni is purified and counted on the NaI detector, the count ratio $^{58}Co/^{60}Co$ is compared with the standard pot, and the previously determined cobalt concentration gives the results for nickel. The Zr- and Hf-bearing fraction is evaporated to dryness and counted on the Ge(Li) detector, and the Zr/Hf count ratio combined with the previously determined Hf concentration gives the Zr concentration as compared with the standard pot without yield determination or further purification.

Selenium is separated from the water-soluble fraction by passing SO_2 through the acidified solution. Counting is done on the Ge(Li) detector.

The results of our analyses to date are listed in Table 1. Because of probable weighing problems in the irradiation A suite between samples and standards (suspected first by the low Ti results), all the numbers obtained from this irradiation sequence were corrected by the use of data from irradiation B. All values from irradiation A are normalized to the europium determined carefully together with the other short-lived nuclides in irradiation B.

Duplicate results indicate an error of about 5 to 10 percent (coefficient of variations determined from the ranges of a pooled sample). We think that most of this error is due to heterogeneities or sample handling since the counting errors in most cases are lower.

Many of the distinctive chemical properties of the Apollo 11 materials were outlined in the published preliminary results (*3*). We confirm the high rare earth, Ti, Zr, and Cr contents and the low Mn, Co, and Ni contents. In addition, the results given in Table 1 show a deficiency of europium relative to other rare earths as compared with chondrites. Tantalum and hafnium are considerably higher in Apollo 11 site materials than in terrestrial basalts.

The elemental abundances vary among the several Apollo 11 site materials we have examined. Roughly, it appears that three coherent groups exist. One group varying together is Mn, Sc, Ta, Hf, and the rare earths; a second group appears to be Fe, Ti, Co, and Cr; and a third group is Na and U. If we conceive of Apollo 11 material as being composites of different proportions of three dominant phases, pyroxene, ilmenite, and plagioclase, the trace element groupings may be crudely explained as strong associations with these phases. Separated portions must be analyzed before this point can be made explicit.

Although it is premature to present a quantitative model for the chemical makeup of Apollo 11 site materials, several constraints are imposed by the data. The low cobalt and nickel concentrations relative to chromium and scandium, and the low abundance of metallic iron, implies that these components were either not accumulated or were separated out by a secondary process. The high abundances of Ti, Cr, Sc, Ta, Hf, and rare earths imply that the raw material for Apollo 11 material could be a condensate from the solar nebula at temperatures between the higher temperature for the condensation of metallic iron and nickel and the lower temperature for the condensation of sulfides and proton-rich materials, a general model suggested by several people (*4*). The phases and their variations now observed may be due to the redistribution of the elements at the time of impact or as the result of some other magma-producing process.

The analytical procedure we have described is being modified to include an initial x-ray fluorescence scan of the pellet in the type 2 planchet before irradiation. Hence, on the same 20-mg sample we should obtain useful data for Ca, Fe, Ti, Sr, Y, and Zr on the specific sample to be analyzed by neutron activation. Our method is most suitable for small samples and, in particular, may be valuable for the analysis of separated mineral phases and glass beads. Results on a sample as small as 5 mg may be adequate for many problems.

KARL K. TUREKIAN, D. P. KHARKAR
Department of Geology and Geophysics, Yale University, New Haven, Connecticut 06520

References and Notes

1. I. Perlman and F. Asaro, *Archaeometry* **11**, 21 (1969).
2. K. K. Bertine, L. Chan, K. K. Turekian, *Geochim. Cosmochim. Acta*, in press.
3. Lunar Sample Preliminary Examination Team, *Science* **165**, 1211 (1969).
4. J. A. Wood, *Nature* **194**, 127 (1962); J. W. Larimer, *Geochim. Cosmochim. Acta* **31**, 1215 (1967); E. Anders, *Accounts Chem. Res.* **1**, 289 (1968); K. K. Turekian and S. P. Clark, *Earth Planet. Sci. Lett.* **6**, 346 (1969).
5. Research performed under NASA grant 9-8032.

4 January 1970

Elemental Composition of Lunar Surface Material

Abstract. *Elemental abundances, so far obtained, derived from the analysis of Apollo 11 lunar material are reported. Similarities and differences exist between lunar material, the eucritic achondrites, and the augite achondrite Angra dos Reis, the analysis of which is also reported.*

We present here the results of our initial analytical work on (i) lunar fines (sample 10084), (ii) lunar rock (sample 10060,26), (iii) glassy spheres (sample 10085/30/5) received, respectively, on 19 September, 23 October, and 19 November 1969. We also give some data on the elemental abundances in the augite achondrite Angra dos Reis, of which we were able to get 240 mg (BM 63233). In order to give approximate results in an initial survey of as many elements as possible, we used the following methods: spark source mass spectrography (SSM), emission spectrography (E), and x-ray fluorescence spectrometry (XRF). To give more accurate results for a smaller number of elements, we used activation analysis (AA; instrumental, neutron, activation analysis, INAA; radiochemical, neutron activation analysis, RNAA; radiochemical, high-energy gamma, activation analysis, RGAA) and mass spectrometric isotope dilution analysis (MSID). Not all methods were used on the glassy spheres or the meteorite because of the limited amount of sample available.

In the SSM method we used an A.E.I. MS702 instrument. Powdered samples were mixed with graphite previously doped with rhenium as an internal standard and were pressed into electrodes. The electrode surfaces were cleaned by presparking (~ 100 nc), and the analysis was carried out over a wide exposure range (300 to 0.003 nc) at intervals of × 1.5 to × 3. The photographic plates were examined by microdensitometry for the rare earths and by inspection for other elements, results being obtained by the "just detectable line" method. Results recorded for "standard" rocks W-1 and BCR-1 were used to provide empirical corrections for ionization efficiency factors in a number of cases. The accuracy of the method is believed to be within a factor of 3 of the stated values. The cleaved glassy spherule was examined directly by mounting it in an indium support and sparking against a carbon counter electrode. Plates were analyzed visually, and the relative ion exposures were verified by densitometry for several multiisotopic elements. Values for Ti, Mn, and Dy from separate neutron activation analysis were used as internal standards.

In the ES method we used the technique of Webb and Wordingham (*1*); synthetic standards and "standard" rocks (G-1, W-1, AGV-1, BCR-1, DTS-1, T-1, and syenite rock-1) were used for calibration, the spectra being assessed visually and by microphotometry. Coefficient of variation for a single determination was between 5 and 10 percent at concentrations between 20 and 100 parts per million. Accuracy is estimated as ± 20 percent.

In the XRF method we used for major elements a fusion mix consisting of lithium tetraborate, lithium carbonate, and lanthanum oxide (Spectroflux 105) to which had been added 0.8 percent lithium nitrate, the general method being that of Norrish and Hutton (*2*). We measured the minor and trace elements using lithium tetraborate only. Synthetic standards were used in both cases. Comparison of results by these methods was made with published figures on "standard" rocks AGV-1, BCR-1, G-1, and DTS-1, and from this the estimated accuracy is, for major elements (>1 percent) ± 3 percent, minor elements (0.1 to 1 percent) ± 5 percent, trace elements (<0.1 percent) ± 15 percent.

For activation analysis both neutrons (from DIDO, flux 1.3 × 10^{13} neutron cm^{-2} sec^{-1}) and high-energy gamma photons (bremsstrahlung produced from the Harwell electron linear accelerator) were used. For INAA, irradiation and decay periods were 12 seconds and 7 minutes, followed by 5 minutes and 14 hours, respectively, for one set of samples; we used irradiation and decay periods of 2¼ days and 3½ days, respectively, for a different set of samples. Counting was done with two Ge(Li) detectors, a small planar, high-resolution detector (1 kev at 120 kev) for identification of photopeaks, and a larger 30-cm^3 coaxial detector (resolution, 5 kev at 1.32 Mev) for general counting. The associated PDP-8 gamma spectrometer has been described by Pierce *et al.* (*3*); we processed data from this instrument on the main Harwell computer (IBM 360/65), using a specially developed least-squares fitting program which applies corrections for both decay and dead-time. Estimated accuracy is ± 10 percent except where stated. For RNAA we used a flux of 5 × 10^{12} neutron cm^{-2} sec^{-1} generally, but for short-lived radionuclides the higher flux (1.3 × 10^{13}) was used as a pneumatic "rabbit" tube into that flux position. Radiochemical methods used have been described elsewhere (*4*). Estimated accuracy is ± 5 percent. For the fluorine determination by RGAA, samples and standards were irradiated in Al-foil containers for 30 minutes in the bremsstrahlung produced in a tungsten converter ⅛ inch thick from a 10-μa beam of 24- to 28-Mev electrons. Radiochemical separation was achieved by distillation of H_2SiF_6; counting of the 0.51-Mev annihilation radiation (after gamma spectrometry to establish absence of other gamma emitters) was then used for measurement of ^{18}F. Decay of this nuclide was followed for four half-lives as further confirmation of radiochemical purity. "Standard" rocks G-1 and W-1 were also analyzed, and the estimated accuracy is ± 10 percent.

In the MSID methods we used enriched isotopes of the elements concerned, Rb, Cs, Ba, Sr, Ce, Nd, and Sm being determined in the same sample, K being determined in a separate sample. Cation exchange was used to separate the alkali, alkaline earths, and rare earths as three groups, for subsequent mass analysis on a laboratory-constructed instrument with a radius

Table 1. Elements determined by activation analysis* or mass spectrometric isotope dilution, or both.

Atomic number	Element	Concentration of elements: Lunar fines	Lunar rock	Lunar glassy spheres	Meteorite Angra dos Reis
		Percentages			
11	Na	0.32	0.38	0.14	0.03
12	Mg		6	4	
13	Al	7.0	5.2	7.0	5.7
19	K	0.12‖	0.17‖		
20	Ca	~8†	~ 11†	~ 10†	~ 17†
22	Ti	4.7	5.2	4.9	1.7
26	Fe	12.5	14.2	12.3	6.5
		Parts per million			
9	F	74§			
21	Sc	66	70	57	57
23	V		~ 90†	~ 120†	
24	Cr	2300	2800	2300	1700
25	Mn	1630	1620	1610	700
27	Co	~25†	~ 30†	~ 30†	~ 10†
29	Cu		9.1‡		
31	Ga	3.8‡	4.7‡		
32	Ge	0.5‡	0.4‡		
33	As	0.03‡	0.01‡		
37	Rb	3.0‖	4.2‖		
38	Sr	177‖	180‖		
46	Pd		0.006‡		
49	In	2.0¶,0.5¶	0.004‡		
55	Cs	0.12‖	0.19‖		
56	Ba	174‖	212‖		
57	La	19	25	27	10
58	Ce	46‖	56‖		
60	Nd	37‖	45‖		
62	Sm	13‖,14	16‖,16	13	8
63	Eu	1.8‖,2.0	1.9‖,2.6	2.1	2.0
65	Tb	4	6	5	2
66	Dy	23	24	21	12
67	Ho	5	7	5	3.5
70	Yb	12	14	12	6
71	Lu	1.5	1.8	1.4	0.7
72	Hf	11	12	11	3

*Accuracy by INAA, ± 5 percent for all results, except where noted. †Accuracy by INAA, ± 10 percent. ‡Accuracy by RNAA ± 5 percent. §Accuracy by RGAA, ± 10 percent. ‖Accuracy by MSID, ± 5 percent. ¶Contamination of samples suspected.

Table 2. Major elements in lunar fines, expressed as oxides.

Oxide	By XRF (%)	By INAA (%)	By MISD (%)
Na_2O		0.43	
MgO	7.6		
Al_2O_3	13.0	13.2	
SiO_2	43.0		
K_2O	0.15		0.14
CaO	12.2		
TiO_2	7.7	7.8	
Cr_2O_3	0.3	0.34	
MnO	0.2	0.21	
FeO	15.6	16.1	

of 6 inches and a sector of 60° for the two former groups and on an A.E.I. MS5 instrument for the latter group. Details are given elsewhere (*5*). Estimated accuracy is ± 5 percent.

Results obtained by AA and MSID are reported in Table 1; results obtained for the major elements by XRF, MSID, and AA are given in Table 2, and data from some supplementary analyses by SSM, E, and XRF are presented in Table 3. Table 4 lists conservative upper limits for several other elements derived from SSM measurements; some of these, for example, P, S, and Pb, were detected but not measured.

By comparison with the work of the Preliminary Examination Team (PET) (*6*), who used mainly emission spectrography for elemental analysis, and where results for 23 elements and upper limits (for that method) for 30 more were given, we list results so far for 35 elements in Tables 1 and 2, less firm results for 11 more in Table 3, and upper limits (generally at lower levels) for 25 elements in Table 4. Of those we undertook to determine in our original proposal, we have not yet any result for seven elements; on the other hand, we quote results for 18 elements which we did not include in our original proposal. We have not as yet done all the cross-checking of results by our several methods that we would wish to do. (This may explain the conservative accuracy estimates quoted.)

Our results broadly confirm those of the PET, the only significant difference being that for ytterbium (PET reports 1.3 to 7 parts per million; our value is 12 to 14 ppm). We do not know at present whether our samples are taken from those analyzed by the PET or whether they are different. If the samples are different, then our ytterbium figures may simply indicate a larger range for that element in the lunar material rather than an analytical disagreement. The PET comment that "samples are apparently free from inorganic contamination . . . from . . . indium, which forms the seal of the rock box . . ." may no longer hold on the basis of methods of higher sensitivity. Our evidence is not strong but we believe that the disagreement between duplicate indium determinations for lunar fines, coupled with the comparatively high values, indicates contamination. The value for the lunar rock of 0.004 ppm was obtained by irradiating a piece of rock, then leaching it after irradiation with boiling HNO_3-HF, and thereafter proceeding with radiochemical separation on the leach liquor and on the remaining rock as separate samples. The leach liquor contained the equivalent of an additional 0.004 ppm of indium; this result possibly implies that even here some readily removable indium (that is, contamination) was present.

Table 3. Supplementary results by spark source mass spectography, emission spectrography, and x-ray fluorescence spectrometry.

Atomic number	Element	Lunar fines (ppm) SSM	Lunar fines (ppm) E	Lunar fines (ppm) XRF	Lunar rock (ppm) SSM	Lunar rock (ppm) E	Lunar spheres SSM (ppm)	Meteorite Angra dos Reis (ppm) SSM	Meteorite Angra dos Reis (ppm) E
19	K							50	
23	V	50	50			45	40	160	150
28	Ni	170	180		100	150	80	30	50
29	Cu	10			10		10	10	
30	Zn	50			40		20	10	
31	Ga							6	
37	Rb	3					3	1	
38	Sr	200	170	170	200	170	200	100	150
39	Y	100	120	100	100	120	60	40	30
40	Zr		330	320	300	340	200	100	100
41	Nb	10			10		5	5	
56	Ba	200		180	200		150	30	30
58	Ce						40	20	
59	Pr	20			20		20	10	
60	Nd						40	20	
64	Gd	15			20		20	10	
68	Er	10			10		10	5	
69	Tm	2			2		2	1	
90	Th	1			1			1	
92	U	0.4			0.4			0.4	

Table 4. Upper limits by our spark source mass spectrographic method for elements not quantified in lunar fines and rocks.

Atomic number	Element	Upper limit (ppm)
5	B	2
15	P	1500
16	S	500
17	Cl	70
34	Se	20
35	Br	3
42	Mo	10
44	Ru	1
45	Rh	1
47	Ag	1
48	Cd	1
50	Sn	10
51	Sb	1
52	Te	1
53	I	1
73	Ta	3
74	W	1
76	Os	1
77	Ir	1
78	Pt	1
79	Au	1
80	Hg	1
81	Tl	1
82	Pb	10
83	Bi	1

The results for rare earths and barium on the Angra dos Reis meteorite compare quite reasonably with those of Schnetzler and Philpotts (*7*), and our approximate values for potassium, thorium, and uranium are not dissimilar to published values (*8*). Our results for the major elements are in broad agreement with those of Ludwig and Tschermak (*9*), but not unexpectedly our data for sodium and potassium differ markedly from theirs [as also do the results of Muller and Zahringer (*8*) for potassium].

We had some 10 mg of the glassy spheres; one-half sphere we analyzed by SSM; the other whole sphere, after grinding, was analyzed by AA. On grinding the color changed from black to yellow-brown. The analysis shows the spheres to be quite similar to the lunar fines and rocks, except for the sodium content which is distinctly lower in the spheres. The result is very similar to that of Fredriksson and Kraut (*10*) for the analysis of Cachari eucrite glass. The analogy with the arguments about tektites and their probably terrestrial parent material is also obvious.

Several attempts have been made to compare the chemical composition of Apollo 11 lunar meterial with that of known terrestrial and meteoritic silicates. Only the suggestion of terrestrial anorthositic gabbro (*11*) seems to meet the requirements for the major elements, but it fails on the alkali metals and carbon. Several authors have suggested similarities with the basaltic achondrites, especially the eucrites. Turkevich *et al.* (*12*) say, however, that in the lunar material "the high value for titanium and the relatively low value for iron are at variance with the content of these elements in eucrites and other calcium-rich achondrites." The statement about titanium is true, but that about iron is incorrect. Moore County, a eucrite containing 12.4 percent of FeO, comes well within the accuracy range of Turkevich's value of 12.1 percent. Our values for FeO (15.6 and 16.1 percent) on Apollo lunar fines are higher than the latter value and fit well into the eucrite range (*13*).

As a result of these considerations we felt that it would be worthwhile to examine Angra dos Reis, a unique calcium-rich achondrite, which is reported to have the highest titanium content of any meteorite so far analyzed. We analyzed a small sample by SSM and by AA and found that there are some quite interesting similarities to Apollo 11 lunar material. Concentrations of the rare earths, especially, are within a factor of 2 of those in lunar samples, closer than values for any other meteorite. The concentration of europium, which seems very low in lunar material by comparison with the other rare earths, is in this meteorite almost exactly the same as that in the lunar material. (Incidentally, the "normalized to ytterbium" figure for europium in lunar material is lower than for any meteoritic material.) The uranium and thorium abundances in the two materials are also quite similar.

Dissimilarities exist in that for Angra dos Reis, the titanium content is lower [Turkevich *et al.* comment that Sinus Medii material may have lower titanium content than Mare Tranquillitatis material (*12*)], the sodium and potassium content is much lower, and the calcium content is higher than in lunar material. Also the points on the plot of CaO content as a function of Fe/(Fe + Mg) (*14*) are in quite different positions (that for lunar material fitting well with eucrites).

We agree with Turkevich *et al.* (*12*) that "there appears to be no common material (available) on earth that matches in all respects the chemical composition of lunar surface material at Mare Tranquillitatis."

A. A. SMALES, D. MAPPER
M. S. W. WEBB, R. K. WEBSTER
J. D. WILSON
Analytical Sciences Division, Atomic Energy Research Establishment, Harwell, England

References and Notes

1. M. S. W. Webb and M. L. Wordingham, *Harwell Rep. AERE R-5799* (June 1968).
2. K. Norrish and J. T. Hutton, *Geochim. Cosmochim. Acta.* **33**, 431 (1969).
3. T. B. Pierce, R. K. Webster, R. Hallett, D. Mapper, *Modern Trends in Activation Analysis*, J. R. DeVoe, Ed., *U.S. Nat. Bur. Stand. Spec. Publ. 312* (1969), vol. 2, p. 1116.
4. A. A. Smales, T C. Hughes, D. Mapper, C. A. McInnes, R. K. Webster, *Geochim. Cosmochim. Acta* **28**, 209 (1964); A. A. Smales, D. Mapper, K. F. Fouché, *ibid.* **31**, 673 (1967); *Harwell Rep. AERE R-5254* (1967); K. F. Fouché and A. A. Smales, *Chem. Geol.* **2**, 5 (1967); *ibid.*, p. 105.
5. P. Murugaiyan, A. A. Verbeek, T. C. Hughes, R. K. Webster, *Talanta* **15**, 1119 (1968).
6. Lunar Sample Preliminary Examination Team, *Science* **165**, 1211 (1969).
7. C. S. Schnetzler and J. A. Philpotts, *Meteorite Research*, P. M. Millman, Ed. (Reidel, Dordrecht, Holland, 1969), p. 206.
8. H. W. Muller and J. Zahringer, *ibid.*, p. 845; J. W. Morgan and J. F. Lovering, *J. Geophys. Res.* **69**, 1989 (1964).
9. E. Ludwig and G. Tschermak, *Mineral. Petrogr. Mitt.* **8**, 341 (1887).
10. K. Fredriksson and F. Kraut, *Geochim. Cosmochim. Acta* **31**, 1701 (1967).
11. E. Olsen, *Science* **166**, 401 (1969); R. A. Phinney, J. A. O'Keefe, J. B. Adams, D. E. Gault, G. P. Kuiper, H. Masursky, R. J. Collins, E. M. Shoemaker, *J. Geophys. Res.* **74**, 6053 (1969).
12. A. L. Turkevich, E. J. Franzgrote, J. H. Patterson, *Science* **165**, 277 (1969).
13. W. Nichiporuk, A. Chodos, E. Helin, H. Brown, *Geochim. Cosmochim. Acta* **31**, 1911 (1967).
14. B. Mason, *Meteorites* (Wiley, New York, 1962), p. 107.
15. We thank Dr. Moss of the British Museum (Natural History) for the Angra dos Reis sample; Dr. S. O. Agrell of Cambridge University for the lunar glassy spheres; Dr. J. S. Hislop for the fluorine determination by RGAA; and R. Hallett, T. C. Hughes, R. P. Kay, C. A. J. McInnes, C. R. Matthews, A. G. Morton, P. F. Ralph for experimental assistance.

4 January 1970

Abundances of 30 Elements in Lunar Rocks, Soil, and Core Samples

Abstract. *Abundances of 30 elements in seven lunar rocks and soil were determined by instrumental and radiochemical activation analysis. Seven major and minor elements in chips from 27 rocks were determined by instrumental activation analysis. Abundances of ten bulk elements overlap for the breccia rocks and soil samples. All lunar rare earth elements distribution patterns resemble those found in terrestrial abyssal subalkaline basalt, but with Eu depleted by about 60 percent in all lunar samples compared to the adjacent rare earth elements. Precipitation of plagioclase and hypersthene achondritic-like minerals from a melt could account for Eu depletion and the observed distribution of rare earth elements. Abundances of Ti, Al, Ca, Na, and Mn determined by instrumental activation analysis in five core-tube soil samples indicate uniformity for Al and Mn and apparent differences (10 to 20 percent) for Ti, Ca, and Na at 7.8 and 10.5 centimeters as compared to 0 to 5.2 centimeter depths.*

Elemental abundances of ten bulk elements were determined by instrumental activation analysis (IAA) in specimens (about 0.3 to 1 g) of chips, powders, or fine material from five type-B rocks (10017, 44, 45, 47, and 50), two type-C rocks (10019 and 59), and lunar soil (10084), returned from Tranquillity Base (*1*). In collaboration with other investigators (*2*), abundances of seven major, minor, and trace elements were obtained by IAA in chips (about 0.5 to 2 g) from 27 rocks—that is, in two type-A, seven type-B, and 18 type-C rocks. Experiments are in progress for determination of 15 additional minor and trace elements by IAA in these 27 specimens (*2*). The analytical method has been described elsewhere (*3*). Gamma-ray spectroscopy measurements were taken with Ge(Li) and NaI(Tl) detectors.

After IAA, chips (about 0.5 to 1 g) of two type-B rocks (10017 and 50), one type-C rock (10019), and fine soil material (10084) were further subjected to RNAA (radiochemical neutron activation analysis) for determination (*4*) of the 14 rare earth elements (REE), plus Y, Cd, In, K, Rb, and Cs. Abundances of Rb and Cs will be reported elsewhere (*5*). Core-tube specimens of soil were also subjected to IAA for abundance measurements of Ti, Al, Ca, Na, and Mn. Five samples (73 to 146 mg each) were taken at 2.6-cm depth intervals. Further IAA of these core samples are currently in progress for the determination of Cr, Fe, Sc, Co, Ba, La, Sm, Eu, Yb, Lu, Zr, Hf, and Th (*5*).

Chips from rocks 10017, 45, 50, 19, and 59 were split into two smaller chips. Powders from rocks 10044 and 47 were sieved through 100 nylon mesh into a coarse fraction (> 100 mesh) and a fine fraction (< 100 mesh). For 10044, the coarse and fine fraction weights were 0.85 and 1.03 g and for 10047, 0.35 and 0.56 g, respectively. From a total of 5.3 g of lunar soil, three aliquants were split into two unsieved 1-g samples and the third sample of 3.1 g was sieved through 100 and 200 nylon mesh. Weights of the sieved samples were: > 100 mesh, 0.73 g; 100 to 200 mesh, 0.47 g; and <200 mesh, 1.90 g. To check the overall IAA technique, USGS standard rocks W-1 and BCR-1 were simultaneously analyzed with lunar samples.

In general, abundances of Ti, Al, Fe, Mg, and Ca (Tables 1 and 2) are in agreement with those reported (*1*). Values of Na, Cr, and Mn derived by IAA and those of K, by RNAA, are considered superior to those obtained by optical spectrographic measurements (*1*). If a Si value of 20 percent is adopted for lunar soil, then the summation of total elemental concentrations approaches 100 percent, within 2 σ of the individual measurements.

From a comparison of ratios of major and minor elements in the three rock types to soil (Tables 1 and 2), it appears that breccia rocks are most closely related to soil. Within error limits, the chemical compositions of types A and B rocks are identical and distinctly different from type C rocks (Table 2). These data support the conclusion that the closest meteoritic or terrestrial analogs to the composition of matter from Tranquillity Base are the eucrite achondrites and terrestrial anorthositic gabbro (*6*). From rare-gas and shock measurements (*1*), it has been suggested that breccia rocks may be compacted from lunar soils. Because the abundances of 17 lithophilic trace elements, the 14 REE included, are 5 to 20 percent higher in at least two breccia rocks, it is unlikely that breccia rocks were compacted from soil whose composition is comparable to that collected at Tranquillity Base. However, a 10 to 20 percent variation in average abundances of 17 lithophilic trace elements in two 1-g chips from two breccia rocks suggests that a similar variation may exist in Sea of Tranquillity soil; therefore, it is possible that breccia rocks may be derived from soil similar to that at Tranquillity Base.

The elements Na, Al, and Eu are enriched and Ti, Fe, and Mg depleted in the fine fractions relative to the coarse fractions of powders 10044 and 47; Cr and Mn are significantly depleted in the fines relative to the coarse fractions. These measurements are consistent with visual inspection of both fractions; that is, the fine fractions were enriched in light-colored minerals—an indication of an enrichment of plagioclase and a relative depletion of pyroxene and ilmenite minerals.

Total abundances of all 14 REE and Y are 300 ppm in lunar soil, about 307 ppm in chips from the breccia rock 10019, and 279 and 514 ppm in chips from the type-B rocks 10050 and 17, respectively. These total REE plus Y abundances are near to the upper limit of total REE plus Y contents observed in terrestrial continental basalts and gabbros, which average approximately 190 ppm REE plus Y content, with a range from 20 to 530 ppm (*7, 8*). The total REE plus Y in soil and breccia rocks are about 60 and 5 times the total REE plus Y in chondrites and eucrite achondrites, respectively (*9*). However, relative to the REE distribution in chondrites or eucrite achondrites, continental basalts are significantly enriched in the light REE (La to Sm). The REE pattern in soils and breccia rocks, relative to the average pattern in 29 chondritic meteorites (*9, 10*) assumed as the distribution in the primordial solar nebula, resembles that in abyssal subalkaline basalts (*8*). However, the total REE plus Y content in lunar breccia rocks and soil is about three times higher compared to the total REE plus Y content in 12 abyssal basalts, which have an average REE plus Y content of 88 ppm and vary from 40 to 126 ppm (Fig. 1).

Europium is severely depleted in lunar rocks and soil by about 60 percent relative to the adjacent elements Sm and Gd. It seems plausible that plagioclase, which readily incorporates divalent Eu and also accepts to a lesser degree the trivalent

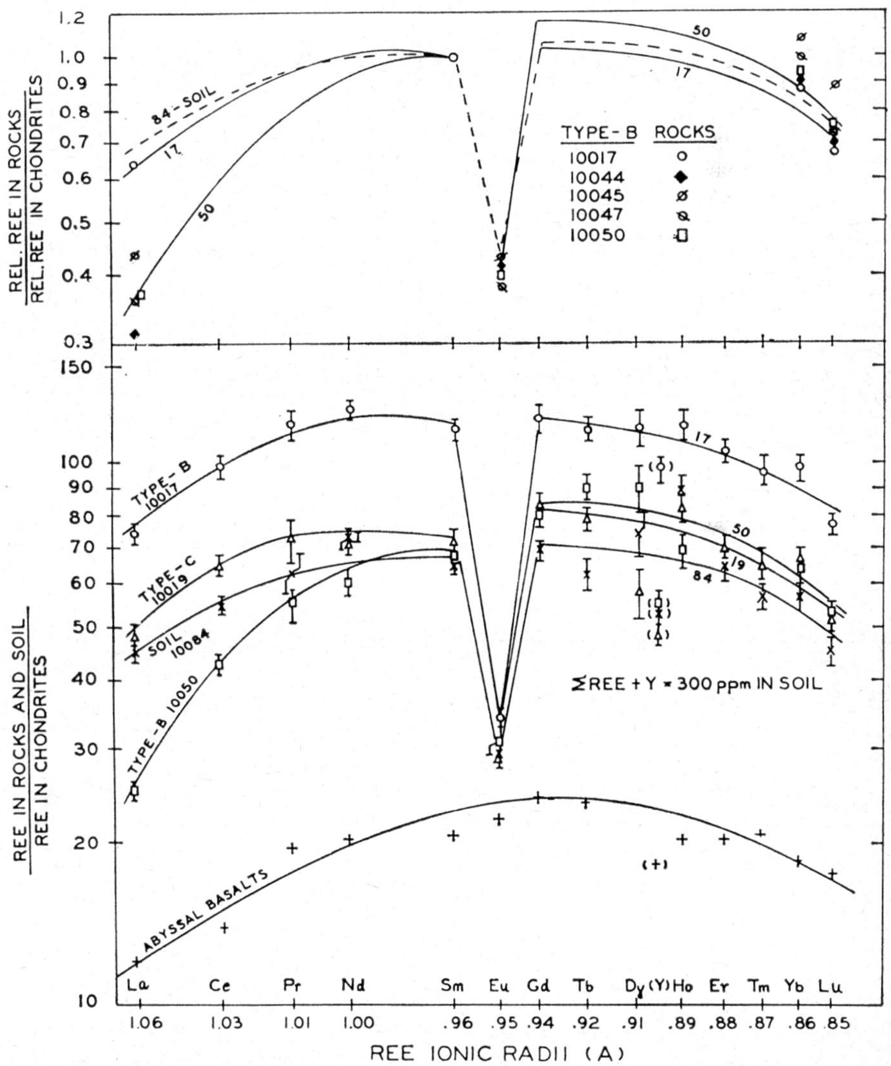

Fig. 1. (Top) Ratios of relative REE (La, Sm, Eu, Yb, and Lu) abundances (Sm ≡ 1.00) in type-B rocks to the same relative REE (Sm ≡ 1.00) abundances in 29 chondrites (*9, 10*). (Bottom) Ratios of absolute REE and Y abundances in two type-B rocks 10017 and 50, one type-C rock 10019, lunar soil 10084, and in 12 terrestrial subalkaline abyssal basalts (*8*) to absolute average REE and Y abundances in 29 chondrites (ppm) (La 0.32, Ce .86, Pr .11, Nd .59, Sm .196, Eu .071, Gd .25, Tb .048, Dy .31, Ho .071, Er .20, Tm .031, Yb .19, Lu .033, and Y 1.9; total REE + Y = 5.2 ppm).

Table 1. Average abundances of major and minor elements in Apollo 11 lunar rocks, soil, and core samples.

Element	Type B: Five rocks (this study)	Type B: Four rocks (others)*	Type C: Two rocks (this study)	Type C: Two rocks (others)*	Type D (soil): This study	Type D (soil): Others*	Ratio (type C / type D) (this study)	Core-tube soil five samples at 2.6-cm depth intervals (this study): Range	Core-tube soil: Average
Si (%)	20.7 ± 0.7	19.0 ± 0.9	18.7 ± 1.1	19.4 ± 0.9	17.5 ± 1.0	20	0.98 ± 0.04†		
O (%)		39.4 ± 1.0‡		41.1 ± 1.0‡		40.8 ± 1.2‡	1.01 ± 0.04†		
Ti (%)	5.9 ± 0.8	5.6 ± 0.8	4.3 ± 0.3	5.3 ± 0.1	3.9 ± 0.3	4.2	1.10 ± 0.12	4.1 to 3.4	3.8 ± 0.3
Al (%)	5.1 ± 0.6	6.2 ± 0.8	6.9 ± 0.1	5.8 ± 0.1	7.1 ± 0.1	6.9	0.97 ± 0.03	7.6 to 7.3	7.4 ± 0.1
Fe (%)	16.2 ± 1.0	14.3 ± 1.1	14.0 ± 0.6	13.6 ± 1.7	12.0 ± 0.1	12.4	1.17 ± 0.06		
Mg (%)	4.3 ± 0.8	4.8 ± 0.9	4.6 ± 0.3	5.0 ± 0.5	4.8 ± 0.1	4.8	0.96 ± 0.06		
Ca (%)	7.7 ± 0.4	7.2 ± 0.2	7.8 ± 0.6	7.9 ± 0.1	7.4 ± 0.2	8.6	1.05 ± 0.07	8.3 to 6.6	7.7 ± 0.7
Na (ppm)	3150 ± 390	4100 ± 500	3440 ± 70	2600 ± 1100	3070 ± 100	4000	1.12 ± 0.05	3800 to 3100	3400 ± 330
K (ppm)	1430(1) ± 70§	1000 ± 600	890(1) ± 40§	1350 ± 220	910(1) ± 50§	1000	0.98 ± 0.07		
Cr (ppm)	2150 ± 680	4200 ± 600	2180 ± 170	2750 ± 350	1780 ± 100	2500	1.22 ± 0.12		
Mn (ppm)	2160 ± 220	3300 ± 900	1750 ± 130	2100 ± 400	1600 ± 10	1750	1.09 ± 0.08	1650 to 1580	1620 ± 30
Total	100.2‖	97.8‖	98.3‖	99.0‖	95.1‖	98.6			

*See (*1*). One σ of this study and others represents one deviation from the mean abundance. †The Si abundances (*14*) of 19.7 ± 0.7 and 20.2 ± 0.2 percent in type C rocks and soil, respectively, and O abundances of 41.1 and 40.8 percent in type C rocks and soil, respectively, were considered more accurate and were used in ratios for these two elements. ‡See (*14*) for O content in 7 type B rocks, 18 type C rocks, and in soil. §Number of rocks or soil samples analyzed for K by RNAA. ‖We assumed O contents of 39.4 percent in type B, 41.1 percent in type C, and 40.8 percent in type D.

REE in a monotonic selection of increasing abundance with increasing ionic radius (*11*), may have precipitated out of a lunar magma and markedly depleted Eu and partially depleted the light REE. Yttrium is similarly depleted on a relative basis in all rocks and soil samples (Fig. 1.) Scandium, with trivalent properties that approximate those of the REE, is enriched by factors of about 7 and about 2 relative to Sc in chrondrites and eucrite achondrites, respectively.

Abundances (ppm) of five representative rare earth elements, La, Sm, Eu, Yb and Lu, and Sc are as follows. For soil, the values are 14.2, 12.6, 2.0, 10.8, 1.5, and 58; for breccia 10019, the values are 14.7, 14.1, 2.1, 12.3, 1.6, and 64; for breccia 10059, the values are 17.8, 16.2, 2.0, 13.6, 1.9, and 67; for type-B rock 10017, the values are 26, 25, 2.7, 21, 2.8, and 90; for type-B rock 10044, the values are 9.3, 18.7, 2.8, 16, 2.1, and 93; for type-B rock 10045, the values are 7.0, 9.8, 1.45, 10.2, 1.45, and 90; for type-B rock 10047, the values are 10.8, 18.7, 2.6, 18, 2.3, and 101; and for type-B rock 10050, the values are 8.4, 14.8, 2.2, 12.9, 1.8, and 93. For type-B rocks we conclude that (i) total REE abundances range within a factor of four; (ii) La is fractionated variably

Table 2. Average elemental abundances in Apollo 11 lunar rocks and soil. The number of rocks analyzed are in parentheses.

Element	Type A rocks: This study (2)*	Type A rocks: Others (4)†	Type B rocks: This study (7)*	Type B rocks: Others (4)†	Type C rocks: This study (18)*	Type C rocks: Others (2)†	Type D soil: This study‡	Type D soil: Others†	Ratio (type A / type B) (this study)	Ratio (type A / type D) (this study)	Ratio (type C / type D) (this study)
Si (%)	18.9 ± 0.8	18.9 ± 1.9	18.7 ± 0.8	19.0 ± 1.0	19.7 ± 0.7	19.4 ± 0.9	20.2 ± 0.2	20	1.01 ± 0.06	0.94 ± 0.04	0.98 ± 0.04
O (%)	38.5 ± 1.2		39.4 ± 1.0		41.1 ± 1.0		40.8 ± 1.2		0.98 ± 0.04	0.94 ± 0.04	1.01 ± 0.04
Ti (%)	6.2 ± 0.5	6.8 ± 0.7	5.9 ± 0.8	5.6 ± 0.8	4.7 ± 0.5	5.3 ± 0.1	3.9 ± 0.3	4.2	1.05 ± 0.17	1.59 ± 0.17	1.21 ± 0.16
Al (%)	4.3 ± 0.6	5.1 ± 0.8	5.5 ± 0.8	6.2 ± 0.8	7.4 ± 0.9	5.8 ± 0.1	7.1 ± 0.1	6.9	0.78 ± 0.16	0.61 ± 0.09	1.04 ± 0.12
Mg (%)	3.9 ± 0.3	4.8 ± 0.7	4.6 ± 1.1	4.8 ± 0.9	4.6 ± 0.6	5.0 ± 0.5	4.8 ± 0.1	4.8	0.85 ± 0.22	0.81 ± 0.07	0.96 ± 0.13
Ca (%)	6.8 ± 0.1	6.9 ± 0.3	7.4 ± 0.9	7.2 ± 0.2	8.1 ± 0.6	7.9 ± 0.1	7.4 ± 0.2	8.6	0.92 ± 0.11	0.92 ± 0.04	1.09 ± 0.09
Na (ppm)	3600 ± 70	4000 ± 700	3180 ± 440	4100 ± 500	3520 ± 160	2600 ± 1100	3070 ± 100	4000	1.13 ± 0.16	1.17 ± 0.04	1.15 ± 0.07
Mn (ppm)	1680 ± 120	2800 ± 800	1830 ± 180	3300 ± 900	1580 ± 130	2100 ± 400	1600 ± 10	1750	0.92 ± 0.12	1.05 ± 0.08	0.99 ± 0.08
Cu (ppm)	12(1) ± 2§	5	11(1) ± 2§	6	13(6) ± 3§	8					

* Chips (1 to 2 g) from rocks of type A were 10022 and 69; type B rocks, 10003, 24, 47, 50, 58, 62, and 71; type C rocks, 10018, 19, 21, 48, 56, 59, 61, 63, 64, 65, 66, 67, 68, 70, 73, 74, and 75. This suite of rock chips was analyzed in collaborative IAA effort by Goles *et al.* (*2*); Si and O were analyzed by Ehmann and Morgan (*14*) and Ti, Al, Mg, Ca, Na, Mn and Cu by our group (R.A.S. and H.W.) Errors are one σ from mean. †See (*1*). ‡The Si and O from (*14*) and remainder of elements of this work. §Number of rocks analyzed for Cu are in parentheses.

with respect to the other light REE—that is, the Sm/La ratio varies from 1.0 to 2.0; (iii) Sm/Eu ratios range from 6.7 to 7.2 for four rocks and 9.3 for one rock, compared to 2.8 in chondritic meteorites (*9, 10*) and to a range of 2.1 to 3.5 in 12 abyssal basalts (*8*); and (iv) Ce abundances are normal, thereby indicating a trivalent oxidation state and suggesting a reducing environment. In brief, from differences of REE patterns in type-B rocks, we conclude that these rocks either were formed from different lava flows or at different depths in a single magma.

If a chondritic-like composition is assumed for the original moon and the REE distribution pattern observed in Apollo-11 rocks is the result of melting and fractional recrystallization of the moon with almost complete concentration (> 90 percent) of the REE in the lunar surface, then a 9-km thick crust may be calculated. Since it is probable that the Ca-rich rocks and soil of the Sea of Tranquillity are abnormally enriched in REE and other lithophilic trace elements, the thickness of the crustal layer will increase inversely with the average enrichment factor. Since complete melting of the moon does not seem possible (*12*), either overall surface heating (for example, a high temperature sun) or localized surface melting resulting from meteoritic impact [Urey (*12*)] may have fractionated only the outer surface of the moon into two or more layers. In any case, for every volume of lunar surface rock that has a composition similar to the Apollo-11 rocks about 60 volumes of the subsurface layer probably are severely depleted in REE and other heavy lithophilic trace elements. The average REE content in this subsurface layer may approximate that found in Ca-poor hypersthene achondrites, such as Johnstown and Shalka, in which the total REE contents (*9*) are considerably depleted relative to chondrites—that is, about 5 to 40 percent of the REE content in chondritic meteorites. For this type of achondrite, the light REE (La to Gd) are normal while the heavy REE (Tb to Lu) are progressively enriched from Tb to Lu, all relative to the chondritic pattern. Material balance arguments suggest that separation of plagioclase and hypersthene achondritic-like minerals (specifically like Shalka) from an original melt, depleted in high temperature minerals, would yield the REE distribution pattern observed in Apollo-11 rocks and soil.

Average Co abundances in lunar soil and two breccia rocks are 31 ± 3 and 38 ± 4 ppm, respectively; in five type-B rocks, the abundances are 23 ± 8 ppm, with a range of 15 to 36 ppm. These Co values, considered superior in analysis, are about two times higher than those previously reported (*1*). No correlation exists between Co and the 14 REE in type-B rocks. If we assume that the average Co content is 700 ppm in ordinary chondritic meteorites, then < 5 percent of chondritic material has been mixed with the soil and rocks in their present state. If the primitive moon had a chondritic composition (about 700 ppm Co content) and was subjected to melting in a reducing environment, it is plausible that Fe metal, Co, and other siderophilic elements would have precipitated out of the magma [Urey (*12*)].

Barium abundances in lunar soil and two breccia rocks are 200 ± 60 and 210 ± 90 ppm, respectively; in five type-B rocks, the Ba abundances are 270 ± 120 ppm, with a range from 70 to 380 ppm. These superior Ba abundances are about three times higher than previous values (*1*). Relative to chrondritic meteorites and subalkaline abyssal basalts, Ba is enriched in lunar rocks and soil by factors of about 50 and 14, respectively (*8, 12*). These enrichment factors correspond to about 60 and 3 for the REE. Since other heavy trace elements (*5*) have been enriched by about the same factor in lunar rocks and soil relative to chondritic meteorites, the hypothesis of an initial chondritic-like lunar surface, followed by melting and fractional recrystallization, becomes more meaningful.

Cadmium abundances of 0.056 and 0.044 ppm in soil and one type-B rock (10017) and an In abundance of 0.005 and 0.001 ppm in one breccia (10019) and one type-B rock (10017) correspond to the general Cd and In amounts in ordinary chondrites and eucrite achondrites (*13*). An In abundance of 0.58 ppm in soil, about ten times that found in carbonaceous chondrites, has been attributed to contamination. These data will be discussed elsewhere (*5*).

Average abundances of three major (Ti, Al, and Ca) and two minor elements (Na and Mn) in five core samples agree with the corresponding elemental abundances in the soil samples that were scooped randomly. For samples from depth intervals of 2.6 cm, no differences were observed for Al and Mn; however, slight variations may exist for Ti (in percent) 4.1 ± 0.1, 4.0 ± 0.1, 4.0 ± 0.1, 3.5 ± 0.1, and 3.4 ± 0.1; for Ca, (in percent) 8.3 ± 0.3, 8.2 ± 0.3, 7.6± 0.3, 7.6 ± 0.3, and 6.6 ± 0.3; and for Na (ppm) 3250 ± 70, 3800 ± 70, 3700 ± 70, 3100 ± 70, and 3140 ± 70 for depths of 0, 2.6, 5.2, 7.8, and 10.5 cm, respectively. Abundance data from 14 other elements in these core specimens will clarify the apparent differences in composition (*5*).

Roman A. Schmitt
Hiroshi Wakita, Plinio Rey
Department of Chemistry and Radiation Center, Oregon State University, Corvallis 97331

References and Notes

1. Lunar Sample Preliminary Examination Team, *Science* **165**, 1211 (1969).
2. G. G. Goles, R. A. Schmitt, W. D. Ehmann, H. Wakita, K. Randle, M. Osawa, J. W. Morgan, "Abundances of major, minor, and trace elements in 27 Apollo-11 lunar rocks," in preparation.
3. R. A. Schmitt, T. A. Linn Jr., H. Wakita, "The determination of 14 common elements in rocks via sequential instrumental activation analysis," *Radiochim. Acta*, in press.
4. A. W. Mosen, R. A. Schmitt, J. Vasilevskis, *Anal. Chim. Acta* **25**, 10 (1961); P. Rey, H. Wakita, R. A. Schmitt, "Radiochemical activation analysis of In, Cd, the 14 rare earth elements and Y in Rocks," in preparation.
5. H. Wakita, R. A. Schmitt, P. Rey, "Abundances of 37 elements in Apollo-11 lunar rocks, soil, and core samples," in preparation. Abundances of V, Zr, Hf, and Th will also be reported.
6. A. Turkevich, E. Franzgrote, J. H. Patterson, *Science* **165**, 277 (1969); E. Olsen, *ibid.* **166**, 401 (1969).
7. L. A. Haskin, F. A. Frey, R. A. Schmitt, R. H. Smith in *Physics and Chemistry of the Earth*, L. H. Ahrens, F. Press, S. K. Runcorn, H. C. Urey, Eds. (Pergamon, New York, 1966), vol. 7, pp. 167–321.
8. F. A. Frey, M. A. Haskin, J. A. Poetz, L. A. Haskin, *J. Geophys. Res.* **73**, 6085 (1968); see A. E. J. Engel, C. G. Engel, R. G. Havens, [*Bull. Geol. Soc. Amer.* **76**, 719 (1965)] for a discussion of the chemical composition of abyssal basalts.
9. R. A. Schmitt, R. H. Smith, J. E. Lasch, A. W. Mosen, D. A. Olehy, J. Vasilevskis, *Geochim. Cosmochim. Acta* **27**, 577 (1963); R. A. Schmitt, R. H. Smith, D. A. Olehy, *ibid.* **28**, 67 (1964).
10. L. A. Haskin, M. A. Haskin, F. A. Frey, T. R. Wildeman, in *Proceedings of the Symposium on Origins and Distribution of the Elements*, L. H. Ahrens, Ed. (Pergamon, New York, 1968), p. 889.
11. J. A. Philpotts and C. C. Schnetzler, *Chem. Geol.* **3**, 5 (1968); J. A. Philpotts, C. C. Schnetzler, H. H. Thomas, *Earth Planet. Sci. Lett.* **2**, 19 (1967); D. G. Towell, J. W. Winchester, R. V. Spirn, *J. Geophys. Res.* **70**, 3485 (1965).
12. H. C. Urey, *Science* **162**, 1408 (1968); H. C. Urey and G. J. F. MacDonald, *Sci. J.* **5**, 60 (1969). See H. C. Urey [*Rev. Geophys.* **2**, 1 (1964)] for a review of the chemical composition of chondritic meteorites.
13. R. A. Schmitt, R. H. Smith, D. A. Olehy, *Geochim. Cosmochim. Acta* **27**, 1077 (1963); R. A. Schmitt and R. H. Smith, see (*10*), p. 283.
14. W. D. Ehmann and J. W. Morgan, *Science*, this issue.
15. Supported by NASA grant NGR 38-002-020 and by NASA contract 9-8097. We thank the Oregon State University TRIGA nuclear reactor group for neutron activations and D. G. Coles, T. Cooper, D. B. Curtis, T. W. Osborn, and R. G. Warren for help in analyses. Dr. V. P. Guinn and J. MacKenzie of Gulf General Atomic provided assistance in Mg analyses.

4 January 1970

Alpha-Particle Activity of Apollo 11 Samples

Abstract. *Nine polished thin sections have been exposed to nuclear track plates, three have been counted by alpha-particle spectrometry, and one has been examined by electron microprobe. Interpretation of the results is in a preliminary stage. Alpha track distribution in the autoradiograph of a breccia forms a network that appears related to the rims of accretionary lapilli composing the breccia. Thorium in a coarse-grained crystalline rock is concentrated in micron-sized, zirconium-rich crystals. Alpha count rates agree with what would be predicted from previously reported thorium and uranium contents of the same rocks, suggesting secular equilibrium for the thorium and uranium decay series.*

Polished, uncovered thin sections of Apollo 11 samples were studied by α-particle autoradiography, α-particle spectrometry, and electron microprobe in an effort to determine the distribution of the α-emitting nuclides, that is, thorium and uranium and their radioactive daughters within the lunar samples. This study was restricted to nondestructive analysis. So far, nine samples have been exposed to nuclear track plates for periods of 2 to 5 weeks. Of these, one track plate has been examined in detail and the others have been scanned in reconnaissance with the petrographic microscope. Alpha-particle spectrometric measurements have been made on three samples. One sample has been analyzed with the electron microprobe.

The autoradiograph of breccia sample number 19,14 has been studied in detail. Figure 1A is a photomicrograph of the sample, and Fig. 1B is a map of the autoradiograph of the sample, showing the distribution of α tracks. Figure 1B shows track density ranging from approximately 300 to 500 tracks per square centimeter for 1 by 1 cm areas. Considering smaller areas, 0.25 by 0.25 cm, the track density varies from approximately 175 to 700 tracks per square centimeter over the area of the sample. This autoradiograph represents a 2-week exposure of the nuclear track plate to the sample.

Groupings of α tracks seen in Fig. 1B indicate areas of thorium or uranium enrichment that are larger than individual rock or crystal fragments, and there appears to be a reticulate pattern to the track distribution. Petrographic examination of the textures of sample 19,14 and other breccias shows that they are composed of ovoid, accretionary lapilli, which have a concentric internal structure. The lapilli generally have a crystal fragment core, surrounded by glass and fine crystal shards and mantled by glass or by glass rimmed by coarse crystal shards. The more prominent lapilli can be seen in Fig. 1A outlined by arcuate chains of crystal fragments. Lapilli range in size from several millimeters to less than 50 μ. Larger lapilli often are composed of aggregates of smaller individually distinguishable lapilli. The matrix between lapilli is the same material which forms the lapilli.

Approximately one third of the breccia is void space, of which we estimate one third is noninterconnected and two thirds are connected. An independent calculation of total void space can be made from the grain density of the fines, 3.1 g/cm^3 (*1*), and the measured bulk densities of 2.0 $\pm$ 0.2 g/cm^3 and 2.02 $\pm$ 0.15 for breccias 18 and 19, respectively (*2*). This calculation indicates that the breccias have 35 percent void space. The bulk densities of the two breccia samples are little more than that of mechanically compacted lunar soil, 1.80 g/cm^3 (*1*), indicating that a minimum of flowage and fracturing is associated with lithification.

Although the breccias contained shocked materials and shock-produced glass, they do not show the textural features characteristic of shock compacted rocks (*3*). The fracture sets, which cut through glass, rock fragments, and matrix without deflection, appear to have formed after induration and probably during excavation of the rock by impact from the original site of deposition. The breccias are bonded together by lightly sintered glass at the

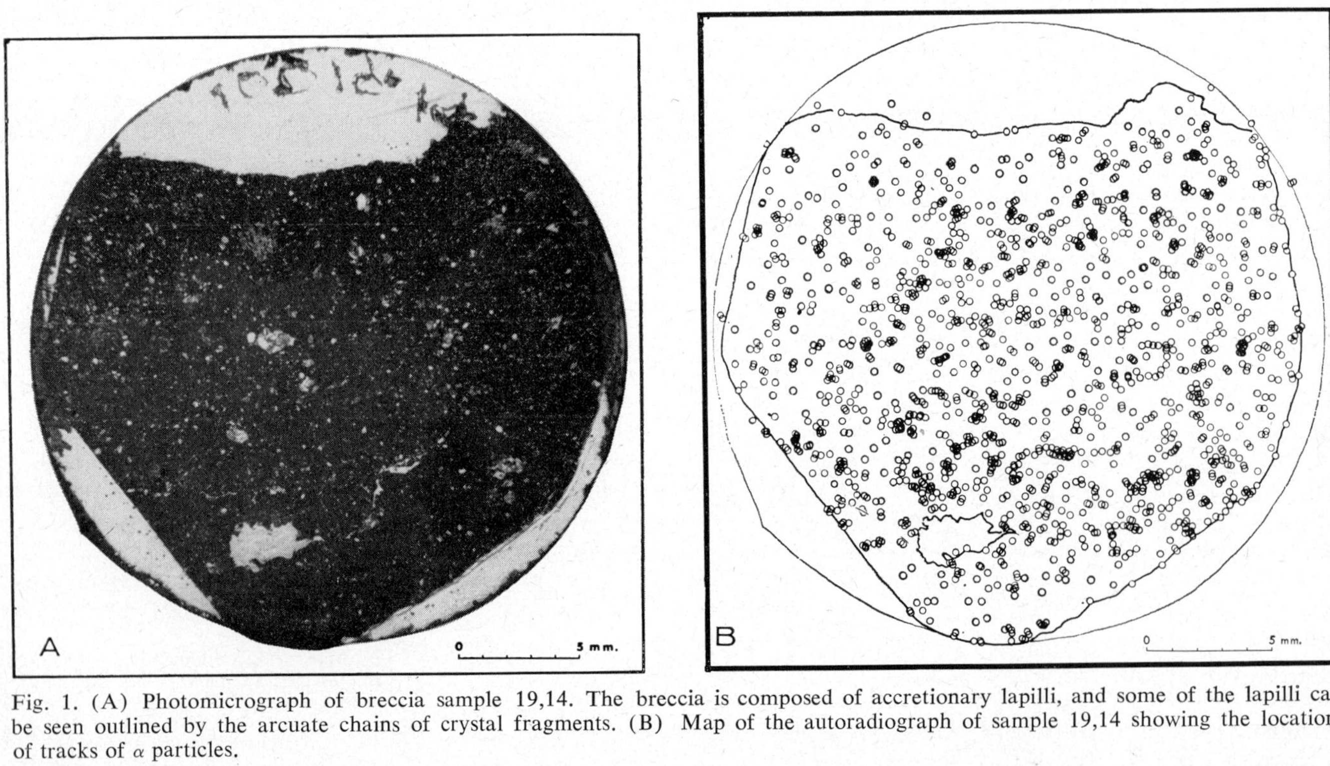

Fig. 1. (A) Photomicrograph of breccia sample 19,14. The breccia is composed of accretionary lapilli, and some of the lapilli can be seen outlined by the arcuate chains of crystal fragments. (B) Map of the autoradiograph of sample 19,14 showing the locations of tracks of α particles.

Table 1. Integral count rates for three lunar samples.

Sample No.	Alpha-particle activity: Measured (count hour^{-1}cm^{-2})*	Alpha-particle activity: Calculated† (disintegrations hour^{-1}cm^{-2})
17,15	0.50 ± 0.06	4.54
19,14	0.14 ± 0.02	1.94
47,42	0.11 ± 0.06	1.33‡

*Counting efficiency is estimated to be 12 percent. † Based on data from (*1*) and (*2*), assuming secular equilibrium obtains in the thorium and uranium decay series. ‡Based on values for type B rock No. 10003.

points of contact with crystals and other glass. This sintering, most apparent in reflected light photomicrographs, suggests that the temperature of the unit was at or above the softening point for some of the glass fragments. The unit was not hot enough (or thick enough), however, to produce densely welded glass or to cause the glass spherules to flow. Cutler and Hendrichsen (*4*) show that highly angular crushed glass sinters much more rapidly than glass spheres of the same equivalent volume. They show that the rate of sintering is inversely proportional to the radius of glass particles at the point of contact and may be five times faster in angular particles than in spheres. It appears, therefore, that the fine-grained angular lunar glass particles could undergo rapid sintering while the larger or more spherical glass particles and crystalline fragments were relatively unaffected. Rock and crystal fragments would be joined by spot welds produced by sintered, micron-sized glass fragments and sintered angular contacts of larger fragments.

Parts of the network of α tracks shown in Fig. 1A appear to coincide with the outlines of lapilli, which suggests that there is some concentration of radionuclides near the periphery of some of the lapilli.

Scanning of the track plates shows no dense clusters of α tracks. However, numerous, micron-sized, zirconium-rich crystals have been seen in sample 58,46, a course-grained crystalline rock. Figure 2 is a backscattered electron picture of the largest zirconium-rich crystals located in that sample. Microprobe analysis indicates that thorium may be concentrated in this crystal.

We calculate, however, that crystals of the size shown in Fig. 2 would not contain sufficient quantities of thorium or uranium to produce concentrations of α tracks on the nuclear track plates that have been exposed.

Alpha-particle spectrometric measurements have been made on polished sections of samples 17,15 (type A), 19,14 (type C), and 47,42 (type B). The samples, ranging from 220 to 352 mm^2 in area and 30 μ thick were placed 3 mm distant from a 300 mm^2 gold-silicon surface barrier detector. Counts were recorded in a 256-channel spectrum. In this configuration, the total counting efficiency for a similar thick source, 260 mm^2 thorium and uranium-bearing synthetic silicate glass, is 12 percent. The integral count rates for the three lunar samples are given in Table 1, with the calculated number of α-particle decays in the sample. The agreement of the measured α activities with those calculated from reported thorium and uranium values indicates that the thorium and uranium decay series are in equilibrium. The thick-source α-particle spectra accumulated from these samples do not contain sufficient counts to permit spectral analysis.

Our conclusions are as follows.

1) Distribution of α tracks from a breccia section indicates a distribution of the α emitters in the rock related to the texture of the breccia, with the radionuclides concentrated at the periphery of some concretionary lapilli.

2) Electron probe examination of a coarse crystalline rock indicates the presence of numerous micron-sized ziconium-rich crystals which may contain concentrations of thorium.

3) Alpha activity of samples of breccia and crystalline rocks is in agreement with reported thorium and uranium values and indicates secular equilibrium obtains in the thorium and uranium decay series.

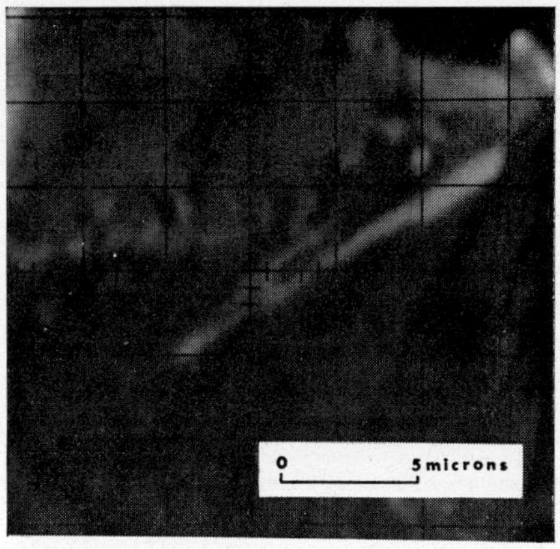

Fig. 2. Backscattered electron picture of zirconium-rich crystal in sample 58,46.

K. A. Richardson, D. S. McKay
W. R. Greenwood, T. H. Foss
Geology Branch, National Aeronautics and Space Administration, Manned Spacecraft Center, Houston, Texas 77058

References

1. Lunar Sample Preliminary Examination Team, *Science* **165**, 1211 (1969).
2. G. D. O'Kelley, J. S. Eldridge, E. Schonfeld, P. R. Bell, personal communication (1969).
3. N. Short, *Science* **154**, 382 (1966).
4. I. B. Cutler and R. E. Henrichsen, *J. Amer. Ceram. Soc.* **51**, 604 (1968).

4 January 1970

Electron-Microprobe Analyses of Phases in Lunar Samples

Abstract. *In fine (type A) and coarse-grained (type B) Apollo 11 lunar volcanic rocks clinopyroxenes are extremely inhomogeneous. Ferrosilite-rich areas in type B rocks have decomposed to submicron vermicular intergrowths of clinopyroxene-fayalite-cristobalite(?). Plagioclase has normal zoning with K_2O up to 0.5 percent in rims. Ilmenites are relatively homogeneous with low MgO (0.1 to 2 percent) and high ZrO_2 (up to 0.26 percent). Metal phase in troilite has <0.02 percent nickel. The breccias (type C) and fines (type D) contain similar phases and aggregates of types A and B together with glassy spherules containing 0.09 to 10.52 percent TiO_2. Rare metal fragments with meteorite-like compositions occur in breccias and fines. Gross similarities between eucrites and Apollo 11 volcanic rocks indicate similar evolutionary environments, but detailed mineralogical differences suggest either separate origins or, if eucrites are lunar, chemical inhomogeneities on the lunar surface.*

In the type A fine-grained volcanic rock studied (10057/67), random microprobe analyses (*1*) of the tiny clinopyroxene grains indicate widely variable Fe (expressed as FeO), Mg, and Ca compositions (Fig. 1). Where larger grains are analyzed the cores are often pigeonitic, whereas the rims are either augite or ferroaugite. Part of the high dispersion apparent in Fig. 1 is due to this zoning of individual grains. One grain analyzed was highly enriched in CaO, with a composition of $Fs_{9.9}$ $En_{1.5}$ $Wo_{88.6}$. The average composition of the clinopyroxene (Table 1) is notable for the appreciable content of Cr_2O_3 and

Table 1. Microprobe analyses of phases in type A lunar volcanic rock 10057/67. Not determined, n.d.

Oxide	Clinopyroxenes		Plagioclase		Ilmenite	
	Range (%)	Average (%)	Range (%)	Average (%)	Range (%)	Average (%)
SiO_2	n.d.	50.91	n.d.	50.04	n.d.	0.09
TiO_2	0.62– 3.35	1.89	n.d.	0.19	51.01–56.02	53.75
Al_2O_3	n.d.	2.07	n.d.	30.49	n.d.	< 0.03
FeO(3)	5.36–32.32	15.42	n.d.	0.82	42.53–46.34	44.69
MnO	n.d.	0.30	n.d.	< 0.04	n.d.	0.15
MgO	0.67–23.83	14.65	n.d.	0.38	0.20– 2.12	1.20
CaO	3.87–37.28	13.88	14.89–16.33	15.89	n.d.	0.21
Na_2O	n.d.	0.09	1.80– 2.35	1.98	n.d.	< 0.04
K_2O	n.d.	0.02	0.19– 0.28	0.24	n.d.	< 0.02
Cr_2O_3	n.d.	0.35	n.d.	< 0.03	n.d.	0.53
ZrO_2	n.d.	< 0.03	n.d.	n.d.	n.d.	0.24
Total		99.58		100.03		100.86

TiO_2, a characteristic property of all rocks collected at the Apollo 11 site (*2*). Random analyses of plagioclase grains show compositions varying from An_{76} to An_{82} with 0.19 to 0.28 percent K_2O and small (but significant) MgO, FeO, and TiO_2 contents (Table 1).

Ilmenite grains are relatively homogeneous, with MgO contents varying from 0.20 to 2.12 percent and significant Cr_2O_3 and CaO contents. Small, but apparently real, Ti excess in structural formulas suggests some Ti^{3+}. The observed MgO contents are consistent with ilmenites' crystallizing in rocks with MgO/FeO ratios around 0.3 (*3*). The high ZrO_2 contents are particularly important and explain the high ZrO_2 contents of these rocks observed in the preliminary investigation (*2*), although rare zircon grains are also present in the lunar volcanic rocks. Walker (*4*) has reported a similar ZrO_2 enrichment in the ilmenite-rich separates from rocks of the Palisades Sill. Other phases present are cristobalite, troilite, and metal, but these are generally too small for analytical resolution.

In the type B coarse-grained igneous rock (10047/11) the most abundant phase present is clinopyroxene (~45 percent), and each grain shows an extreme FeO-MgO-CaO compositional zoning (Fig. 2) with iron enrichments more extreme than are normally seen in terrestrial fractionated basic igneous rocks. In certain areas (~3 percent) of the polished thin section where clinopyroxene grains show extreme iron enrichment, the normally clear clinopyroxene grains suddenly become virtually opaque as the Fs content reaches about 81 molecular percent (see analysis of grain 24 in Fig. 2). Scanning-electron-microscope studies show that these dark areas [probably the unidentified yellow mineral referred to in the preliminary investigation (*2*)] are composed of submicron vermicular intergrowths of three phases which, although they cannot be completely resolved by the microprobe, are best interpreted from the limited data as fayalite, relatively iron-poor pyroxene, and a SiO_2-rich phase (probably cristobalite). The identification of fayalite as the major phase is strengthened by the occurrence on tiny (but resolvable with the microprobe) grains of fayalite (Table 2) along the margins of one such intergrowth area. Scanning-electron-microscope studies indicate that the fayalite grains are continuous with one of the phases in the fine intergrowth. These intergrowths would seem to be best explained as a subsolidus decomposition of extremely ferrosilite-rich pyroxenes after their original metastable (?) crystallization from a rapidly cooling melt, but this can be regarded only as a tentative explanation at present.

The average clinopyroxene analysis calculated in Table 2 is relatively enriched in TiO_2 and Cr_2O_3. Plagioclase (~30 percent) is rather less homogeneous than in the type A rocks and normal zoned grains with An_{90} cores and rims ranging to An_{75} are common (Table 2). Many rims show K_2O enrichment (up to 0.5 percent K_2O) at their outer margins. Ilmenite (~17 percent) is relatively homogeneous (Table 2) with MgO contents ranging from 0.08 to 1.04 percent. Relatively high ZrO_2 contents also occur in the ilmenites from the type B rocks. Cristobalite (~3 percent) is characteristically impure and contains significant amounts of Ti^{4+}, Al^{3+}, Fe^{2+}, Ca^{2+}, and Na^+ ions "stuffed" into its relatively open structure.

The sulfide phase (~1 percent) is troilite (average Fe/S mole ratio, 1.0087) of high purity (Mg, Al, Si, Cr, Zr <0.03 percent; Ti <0.02 percent; Mn <0.04 percent). Tiny blebs of metallic iron (0.60 percent Co) occur in the troilite. This metal is Ni-poor (<0.04 percent) and is also improverished in other important trace elements (Mg, Cr, Zr <0.03 percent; Si <0.02 percent; Ti < 0.01 percent; Mn <0.04 percent). Tiny areas of "acid granophyre" glass, with very variable compositions (Table 2), occur in troilite grains, in one of the

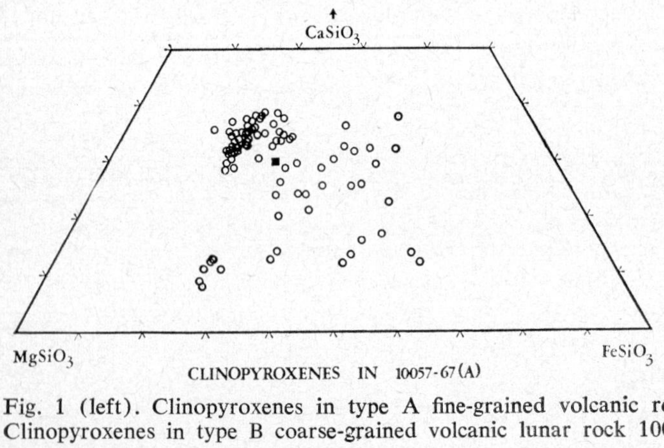

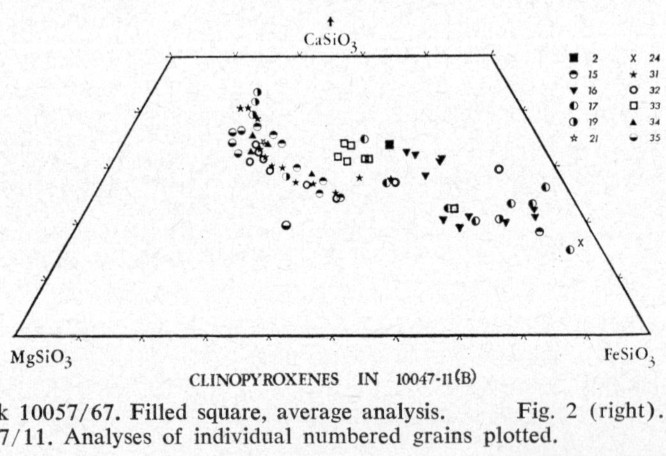

Fig. 1 (left). Clinopyroxenes in type A fine-grained volcanic rock 10057/67. Filled square, average analysis. Fig. 2 (right). Clinopyroxenes in type B coarse-grained volcanic lunar rock 10047/11. Analyses of individual numbered grains plotted.

Table 2. Microprobe analyses of phases in type B lunar volcanic rock 10047/11. Not determined, n.d.

Oxide	Clinopyroxene Range (%)	Clinopyroxene Average (%)	Plagioclase Range (%)	Plagioclase Average (%)	Ilmenite average (%)	Cristobalite average (%)	Fayalite average (%)	Glass range (%)
SiO_2	41.92–51.68	49.08	46.86–50.97	49.26	<0.06	98.32	29.74	62.75–98.43
TiO_2	0.55–2.20	1.19	0.08–0.12	0.09	52.58	0.32	0.09	0.08–0.48
Al_2O_3	0.85–5.53	2.10	30.02–32.98	31.19	0.17	0.95	<0.03	0.86–22.71
FeO(3)	9.49–41.65	20.19	0.45–1.06	0.65	45.75	0.10	68.21	<0.05–1.41
MnO	0.11–0.89	0.39	<0.04	<0.04	0.24	<0.03	0.88	n.d.
MgO	1.07–17.32	12.08	<0.04–0.21	0.14	0.61	<0.02	0.26	<0.04–1.32
CaO	6.07–19.52	14.59	15.32–18.71	17.11	0.08	0.17	0.59	0.05–3.17
Na_2O	0.03–0.14	0.06	0.93–2.23	1.61	<0.04	0.21	n.d.	0.24–0.77
K_2O	n.d.		0.05–0.13	0.09	<0.01	0.01	n.d.	0.21–10.21
Cr_2O_3	0.05–0.58	0.22	<0.03	<0.03	0.29	<0.02	<0.02	<0.03
ZrO_2	n.d.		n.d.		0.12	<0.03	<0.02	n.d.
NiO	n.d.		n.d.		<0.06	n.d.	<0.03	n.d.
Total		99.91		100.18	99.84	100.08	99.77	

fayalite crystals analyzed, and also as separate areas. This glass represents the most highly fractionated residuum left after the crystallization of the major phases.

The type C brecciated rock studied in greatest detail (10046/5) is composed of both A and B rock fragments, fragments of minerals contained in A and B rocks, and spherules and irregular fragments of glass set in a very fine grained matrix part of which appears to be devitrified glass. The clinopyroxene and plagioclase grains have compositions within the ranges found in types A and B. Ilmenite grains occurring as individual grains and as components of types A and B rock fragments have been analyzed. These normally have MgO contents in the range 0.87 to 2.54 percent and in keeping with crystallization in rocks with MgO/FeO ratios about 0.3 (*3*), which is typical for types A and B rocks (*2*). One ilmenite grain crystallizing from an almost opaque glassy (?) fragment had 9.5 percent MgO, which would indicate an MgO/FeO ratio in the glass of about 6 (*3*). The glassy spherules and fragments range from almost colorless, through yellow and brown, to almost opaque (probably as a result of devitrification). Many glasses are inhomogeneous, but what are probably the most abundant glass spherules have a characteristic red-brown color and have a remarkably constant composition both within and between fragments (Table 3), which suggest a common derivation from a single impact-melting event on the lunar surface. The source basic rock resembles the average lunar volcanic rock from the Apollo 11 site with high TiO_2 and Cr_2O_3. The colorless glasses have compositions close to An-rich plagioclase and may represent shock-melted plagioclase grains. The other breccia sample studied (10065/21) was very similar to 10046/5, except that an irregularly shaped metal grain (maximum dimension ~130 μm) containing 93.67 percent Fe, 6.22 percent Ni, and 0.29 percent Co was observed. This composition is very different from the metal phase in the lunar volcanic rocks and most closely resembles a fragment of an iron meteorite.

The sample of fines (10084/53), representing material <1 mm, was split into the following size ranges: >350 μm (15 percent by weight); 250 to 350 μm (1 percent); 150 to 250 μm (11 percent); 100 to 150 μm (10 percent); 70 to 100 μm (10 percent); 80 to 90 μm (15 percent); <80 μm (38 percent). Each of these fractions was then divided into magnetic and nonmagnetic fractions with a permanent magnet, and polished section mounts of all samples obtained in this way were prepared. In the magnetic separation of the >300 μm fraction a rounded metal fragment about 460 μm was separated and found to contain 94.51 percent Fe, 5.27 percent Ni, and 0.48 percent Co. This composition is quite characteristic of an iron meteorite, and this grain would seem to be further evidence of some meteoritic contamination of the lunar surface material. Small grains of anorthite and unusually MgO-rich (up to 8.63 percent MgO) ilmenite are embedded in the margin of the grain, together with another fragment which seems to be a basic glass with 1.47 percent TiO_2. Glassy spherules are common in all size fractions and resemble closely the spherules in the breccias in both color and composition ranges. Many magnetic fractions contain very highly vesicular glassy fragments often containing submicron metal blebs. In the >300-μm fraction plagioclase feldspar grains range in composition from An_{84} to An_{97} and clinopyroxenes show extremely

Table 3. Microprobe analyses of glasses in breccia (10046/5) and fines (10084/53).

Oxide	Glass: Breccia — Red-brown, average (%)	Glass: Fines — Spherules (90–100 μm), range (%)	Glass: Fines — Vesicular and spherules (>300 μm), range (%)
SiO_2	38.92	33.94–51.32	39.08–46.04
TiO_2	10.17	0.12–10.79	0.27–12.05
Al_2O_3	5.31	5.02–30.54	8.25–33.01
FeO (2)	23.33	2.02–23.75	2.43–19.07
MnO	0.33	<0.04–0.35	<0.03–0.28
MgO	13.63	4.56–16.99	3.96–9.48
CaO	7.49	7.21–15.80	10.22–14.80
Na_2O	0.28	<0.04–1.33	0.05–0.63
K_2O	0.06	<0.02–0.68	<0.02–0.69
Cr_2O_3	0.52	<0.03–0.56	0.04–0.36
ZrO_2	<0.04	<0.04–0.07	<0.04–0.07
Total	100.04		

variable FeO-MgO-CaO compositions. Olivines range from Fo_{72} to Fo_{90} and just overlap with olivine compositions reported (*2*) from types A, C, and D materials.

The present work confirms the conclusion of the Preliminary Analysis Team (*2*) that the lunar volcanic rocks crystallized under extremely low partial pressures of oxygen and water. However, the lack of Si in the metal phase or of Ti, Mn, and Mg in the sulfide phase, together with the high FeO contents in the silicate phases, would indicate less severely reducing conditions than prevailed during the evolution of the enstatite chondrites and achondrites. However the gross mineralogies observed immediately invite comparison with the eucritic meteorites. Recent microprobe studies of clinopyroxenes in eucrites have found extreme FeO-MgO-CaO zoning (*5*) similar to that which occurs in the lunar volcanic rocks. Eucrites have crystallized in extremely low partial pressures of O_2 and H_2O and are also characterized by very low alkali abundances just like the lunar volcanic rocks (*2*). But there are also serious differences—for example, the relatively high TiO_2, ZrO_2, and Cr_2O_3 contents of the lunar rocks. These differences are again emphasized by the high TiO_2 content of the clinopyroxenes and ZrO_2, Cr_2O_3 content of the ilmenites in the lunar rocks. Then the lack of Ni in the metal phase in the lunar volcanic rocks is very different from the relatively high Ni content reported in metal from eucrites and howardites (*6*).

If the eucrites are indeed of lunar origin (*7*), then the lunar surface must be chemically inhomogeneous, at least to a degree sufficient to explain the observed differences between the lunar rocks and eucrites. If the K-Ar age of $3.0 \pm 0.7 \times 10^9$ years for the Apollo 11 rocks (*2*) is substantiated then the significantly older Rb-Sr ages ($4.39 \pm 0.26 \times 10^9$ years) for the eucrites (*8*) would also imply different times of formation of various parts of the lunar surface. If, on the other hand, eucrites are not of lunar origin, then those significant similarities between eucrites and moon rocks would at least imply that both evolved in similar, but separate, environments.

N. G. WARE

Department of Geophysics and Geochemistry, Australia National University, Canberra

J. F. LOVERING

School of Geology, University of Melbourne, Victoria 3052, Australia

References and Notes

1. Analyses, using an A.R.L.-EMX microanalyzer, are based on a chemically well-known collection of natural and synthetic minerals. Raw comparative data have then been corrected for absorption [J. Philibert, *X-ray Optics and X-ray Microanalysis* (Academic Press, New York, 1963), p. 379; P. Duncumb and P. K. Shields, *The Electron Microprobe* (Wiley, New York, 1966), p. 284], fluorescence [S. J. B. Reed, *Brit. J. Appl. Phys.* **16**, 913 (1965)], and atomic number [P. Duncumb and S. J. B. Reed, *Natl. Bur. Std. U.S. Spec. Publ. 298*, 133 (1968)]. Experience indicates that the analyses are probably accurate to about 1 percent of the amount present for most major elements.
2. Lunar Sample Preliminary Examination Team, *Science* **165**, 1211 (1969).
3. J. F. Lovering and J. R. Widdowson, *Earth Planet. Sci. Lett.* **4**, 310 (1968).
4. K. R. Walker, *Geol. Soc. Am. Spec. Paper 111*, (1969).
5. A. M. Reid and J. F. Lovering, in preparation.
6. J. F. Lovering, *Nature* **203**, 70 (1964).
7. M. B. Duke and L. T. Silver, *Geochim. Cosmochim. Acta* **31**, 1637 (1967).
8. D. A. Papanastassiou and G. J. Wasserburg, *Earth Planet. Sci. Lett.* **5**, 361 (1969).
9. Polished mounts of material from the fine material (10084/53) were prepared by E. H. Pederson. K. R. Walker, R. B. Hawkins, and R. N. England carried out scanning-electron-microscope studies of submicron phases. Supported in part by A.R.G.C. and Department of Education and Science grants.

4 January 1970

Semimicro Chemical and X-ray Fluorescence Analysis of Lunar Samples

Abstract. *Major and selected minor elements were determined in seven whole rock fragments, five portions of pulverized lunar rock, and the lunar soil. Three different rock types were represented: vesicular, fine-grained basaltic rocks; medium- to coarse-grained, vuggy gabbroic rocks; and breccia. The ranges (in percent) for the major constituents of the lunar samples are: SiO_2, 38 to 42; Al_2O_3, 8 to 14; total iron as FeO, 15 to 20; MgO, 6 to 8; CaO, 10 to 12; Na_2O, 0.5 to 1; K_2O, 0.05 to 0.4; TiO_2, 8 to 13; MnO, 0.2 to 0.3; and Cr_2O_3, 0.2 to 0.4. The high reducing capacity of the samples strongly suggests the presence of Ti(III).*

Thirteen samples weighing a total of 17 g were received for analysis. The samples consisted of lunar fines, seven whole rock fragments, and five portions of rock previously pulverized at the Lunar Receiving Laboratory. Samples were prepared for x-ray spectroscopy (*1*) by fusing 60 mg of the sample with

Table 1. Chemical composition of lunar igneous rocks (in percent by weight).

Constituent	10003	10022		10024	10047
SiO_2	37.8	40.1		39.0	41.3
Al_2O_3	11.0	8.60		9.50	9.80
Fe_2O_3	0.00	0.00		0.00	0.00
FeO	19.8	18.9		18.5	19.0
MgO	7.20	7.74		8.11	6.10
CaO	11.0	10.7		10.0	12.2
Na_2O	0.85	0.91		0.80	0.65
K_2O	0.05	0.30		0.28	0.11
H_2O^-	0.00	0.00		0.00	0.00
Total Ti as TiO_2	12.0	12.2		13.2	10.2
P_2O_5	< 0.2	< 0.2		< 0.2	< 0.2
MnO	0.29	0.25		0.24	0.29
Cr_2O_3	0.26	0.37		0.40	0.22
ZrO_2	< 0.03	0.03		< 0.03	< 0.03
NiO	<0.001	< 0.001		< 0.001	< 0.001
Total	100.25	100.10		100.03	99.87
Total Fe as Fe_2O_3	22.0	21.0		20.5	21.1
R_2O_3	45.8	42.6		44.1	41.7
Constituent	**10049**	**10050**	**10058***	**10058†**	**10062**
SiO_2	41.0	40.9	41.4	41.7	38.8
Al_2O_3	9.50	8.90	10.7	11.8	12.1
Fe_2O_3	0.00	0.00	0.00	0.00	0.00
FeO	18.7	17.3	17.3	18.2	18.3
MgO	7.03	8.03	6.25	6.30	7.21
CaO	11.0	11.3	12.1	11.0	12.0
Na_2O	0.71	0.66	0.79	0.68	0.69
K_2O	0.36	0.05	0.07	0.09	0.07
H_2O^-	0.00	0.00	0.00	0.00	0.00
Total Ti as TiO_2	11.3	12.6	11.1	9.55	10.3
P_2O_5	< 0.2	< 0.2	< 0.2	< 0.2	< 0.2
MnO	0.25	0.27	0.27	0.27	0.27
Cr_2O_3	0.32	0.35	0.21	0.24	0.25
ZrO_2	0.03	< 0.03	< .03	< 0.03	< 0.03
NiO	< .001	< .001	< .001	< .001	< .001
Total	100.20	100.36	100.19	99.83	99.99
Total Fe as Fe_2O_3	20.8	19.2	19.2	20.2	20.3
R_2O_3	42.0	41.4	41.4	43.1	43.6

* Whole rock. † Pulverized.

Table 2. Chemical composition of lunar breccia and soil fines (in percent by weight).

Constituent	10019	10048	10060	311079
SiO_2	41.1	42.2	41.5	42.2
Al_2O_3	13.7	12.9	11.8	14.1
Fe_2O_3	0.00	0.00	0.00	0.00
FeO	15.7	15.7	17.0	15.3
MgO	7.86	7.54	7.52	7.94
CaO	11.9	11.4	11.6	12.1
Na_2O	0.93	0.52	0.78	0.54
K_2O	0.14	0.17	0.18	0.14
H_2O^-	0.00	0.00	0.00	0.00
Total Ti as TiO_2	8.25	8.95	9.15	7.60
P_2O_5	< 0.2	< 0.2	< 0.2	< 0.2
MnO	0.22	0.22	0.23	0.21
Cr_2O_3	0.32	0.31	0.33	0.31
ZrO_2	0.03	< 0.03	0.04	0.03
NiO	0.02	0.02	0.02	0.02
Total	100.15	99.91	100.13	100.47
Total Fe as Fe_2O_3	17.4	17.4	18.9	17.0
R_2O_3	40.3	40.0	40.5	39.3

940 mg of $Li_2B_4O_7$. No heavy absorber was added to the mix. After cooling, the fused glass was brought to 1200 mg with powdered cellulose and then was ground. Three hundred milligrams of the fused mix was pressed into a ½-inch-diameter disc and Na, Mg, Al, and Si were determined with a soft x-ray spectrometer; P, K, Ca, Ti, Mn, Fe, Cr, and Ni analyses were made with an x-ray milliprobe. The rest of the fused mix was pressed into a 1-inch-diameter disc for determination of Si, Al, Fe, Ca, Ti, Mn, Zr, Cr, and Ni.

Chemical determinations were made on all samples for H_2O^-, SiO_2, Na_2O, MgO, R_2O_3, and total reducing capacity. FeO was determined, after catalytic oxidation, on the lunar fines (sample 311079).

The compositional data for the allocated samples are grouped according to rock type and numerical sequence in Tables 1 and 2. The samples listed in Table 1 are type B (gabbro) except for sample 10022, a basalt. Samples 10003, 10022, 10024, 10050, and 10062 were whole rock fragments. Samples 10047 and 10049 were pulverized at the Lunar Receiving Laboratory. Sample 10058 was received as a whole rock fragment and as a pulverized sample.

Chemical compositions of three samples of breccia and one of the lunar fines are given in Table 2. Of the breccias, 10019 was a whole rock fragment, and 10048 and 10060 were received as pulverized samples. These new data support the ranges of ferrous iron, titania, magnesia, and chrome previously reported (*2*). The overall similarity in major element composition for all the samples is apparent. Differences are noted in the minor element contents; for example, the fines and the breccia are characterized by a relatively high NiO (200 ppm) when compared to the igneous rocks (< 10 ppm). In contrast to the previously reported high values for ZrO_2 (*1*, *2*), the new data show that ZrO_2 abundances do not exceed 400 ppm in any of the samples but are more abundant in the breccias and lunar soil fines. As a minor constituent, chromium is relatively abundant (2100 to 4000 ppm Cr_2O_3).

Summations (Tables 1 and 2) for the samples tend to be high and, at first glance, appear to be attributable to metallic iron because it is analytically incorporated in the cited FeO values. Determinations of the total reducing capacity (*3*) of the samples (Table 3) were found to be as much as 4.6 percent higher (equivalent to 1.8 percent additional metallic iron) than can be accounted for by the total iron (as FeO) content obtained by x-ray fluorescence analysis. Optical examination, coupled with calculations based on the assumption that the NiO content is totally associated with meteoritic iron, indicate that the contribution from metallic iron and troilite is no greater than 0.5 percent. The excessive reducing capacity of the returned lunar materials suggests the presence of an element occurring as a major constituent and capable of existing in a lower valence state such as Ti(III). To eliminate the possibility of Ti(III) contributing to the total reducing capacity, a catalytic oxidation was made before a volumetric determination of total iron as FeO. Duplicate volumetric analyses on sample 311079 gave identical results for FeO as those obtained by x-ray spectroscopy.

Table 3. Comparison of total Fe (as FeO) determined by x-ray spectroscopy with total reducing capacity (as FeO) of the lunar materials (in percent by weight).

Sample	Rock type	Reducing capacity	FeO value	Difference (+) as FeO
10003	Gabbro	22.4	19.8	2.6
10019	Breccia	19.4	15.7	3.7
10022	Basalt	21.0	18.9	2.1
10024	Gabbro	22.3	18.5	3.8
10047	Gabbro	19.8	19.0	0.8
10048	Breccia	20.3	15.7	4.6
10049	Gabbro	21.8	18.7	3.1
10050	Gabbro	19.2	17.3	1.9
10058†	Gabbro	19.3	17.3	2.0
10058*	Gabbro	20.3	18.2	2.1
10060	Breccia	20.8	17.0	3.8
10062	Gabbro	20.5	18.3	2.2
311079	Fines	19.4	15.3‡	4.1

* Whole rock. † Pulverized. ‡ The total iron content (as FeO) of sample 311079 was also determined chemically and found to be 15.2_5 and 15.3_6 percent.

The excessive reducing capacity and high summations strongly suggest the presence of Ti(III) in the lunar samples. Investigations continue in this direction.

HARRY J. ROSE, JR., FRANK CUTTITTA
EDWARD J. DWORNIK, M. K. CARRON
R. P. CHRISTIAN, J. R. LINDSAY
D. T. LIGON, R. R. LARSON
U.S. Geological Survey, Washington, D.C. 20242

References and Notes

1. H. J. Rose, Jr., I. Adler, F. J. Flanagan, *Appl. Spectroscopy* **17**, 81 (1963).
2. Lunar Sample Preliminary Examination Team, *Science* **165**, 1211 (1969).
3. The total reducing capacity was determined by dissolution of the samples with 5:3:3 $HF:H_2SO_4:H_3PO_4$ in excess $(NH_4)_3VO_4$ (oxidant) and subsequent back titration of remaining vanadium (V). Two separate determinations of each sample gave identical results.
4. Publication authorized by the director, U.S. Geological Survey.

4 January 1970, revised 6 January 1970

Emission Spectrographic Determination of Trace Elements in Lunar Samples

Abstract. *Eighteen minor or trace elements were detected and determined by emission spectroscopy. Direct d-c arc excitation of powdered samples was used with three variations in the procedure. Thirteen lunar samples consisting of four fine-grained igneous rocks, one medium-grained igneous rock, seven breccias, and one sample of fines were analyzed. The zinc and nickel concentrations in the breccias were approximately one order of magnitude greater than the concentrations of these elements in igneous rocks.*

Thirteen Apollo 11 lunar samples were processed and analyzed spectrographically for 44 elements. The samples we analyzed can be classified into four groups: (i) four samples of type A, fine-grained, vesicular igneous rocks; (ii) one sample of type B, medium-grained vuggy igneous rock; (iii) seven samples of breccias; and (iv) one sample of fines less than 1 mm in grain size from the bulk sample. These sample groups have been described (*1*).

The 12 rock samples consisted of one or more chips and some accompanying

Table 1. Elements detected by emission spectrography. Elemental abundances, in parts per million, are arranged according to rock type and listed in approximate order of volatility in the d-c arc.

Sample*	Zn	Cu	Ga	Rb	Li	Mn	Cr	Co	Ni	Ba‡	Sr‡	V	Be	Nb	Sc	La	Y	Zr
								Rock type A										
57	†	5.7	5.0	4.7	17	2150	2790	30	6.1	440	140	65	3.3	29	99	26	165	635
69	†	8.7	4.9	5.5	18	2390	2760	30	6.7	420	130	72	3.3	20	94	27	164	566
71	†	14	4.8	5.2	17	2230	3060	33	7.0	470	140	78	3.0	24	97	27	162	644
72	†	6.7	4.5	5.0	16	2230	2860	30	6.6	430	130	76	2.7	23	96	25	155	530
								Rock type B										
03	†	6.7	4.7	1.0	9	2580	1860	15	2.7	160	150	82	1.5	21	94	15	113	380
								Rock type C										
18	23	12	4.4	3.6	12	1660	2340	32	197	220	110	60	1.8	25	66	15	97	429
21	24	12	4.6	4.0	13	1770	2480	33	184	270	130	60	2.0	28	72	22	113	424
59	29	21	4.6	3.0	12	1990	2380	33	222	240	120	57	1.7	18	66	19	102	369
61	27	16	5.2	3.4	11	1820	2730	35	241	270	130	60	2.4	21	67	18	103	393
65	23	14	5.0	2.8	12	1925	2390	30	169	260	140	57	2.2	25	69	16	103	390
68	22	12	4.7	3.3	14	2020	2600	33	205	250	130	58	1.9	31	71	21	108	482
73	23	19	3.7	2.1	11	1880	2330	29	199	240	160	66	2.1	14	64	21	89	322
								Rock type D										
84	19	10	3.8	2.7	11	1960	1740	24	185	210	130	50	1.6	18	56	16	81	273

* Last two digits of NASA laboratory number. † Not detected at limit of 4 ppm. ‡ Line width, <5 percent transmission from oscillogram.

fines. Each sample was crushed and pulverized prior to testing. Six of the samples were crushed with a steel diamond mortar prior to grinding to an impalpable powder with an agate mortar and pestle. The remainder of the samples were both crushed and ground in the agate mortar. The sample of fines was quartered twice, and one quarter of the original sample was ground in the agate mortar. A 200-mg portion of each finely ground sample was mixed with 50 mg of graphite powder (*2*); the mixture was ground in an agate mortar and stored in polyethylene capsules.

Three modifications of d-c arc emission spectrography were used: (i) a sample mixed with graphite was excited by an arc at 15 amperes d-c in air with a 3.4-m Ebert spectrograph, and 38 elements in the spectral region between 2300 and 4800 Å were determined (*3*); (ii) a sample mixed with Na_2CO_3 buffer was excited in an argon atmosphere with a 3-m Eagle spectrograph, and nine volatile elements, including Zn, Pb, and Ag, were determined in the second-order spectrum from 2400 to 3650 Å (*4*); and (iii) a sample mixed with a K_2CO_3 buffer was excited in an arc in air with a 3-m Eagle spectrograph, and Cs, Li, and Rb were determined in the spectral region from 6500 to 9000 Å (*5*). The coefficient of variation of the determinations in case (i) is ± 15 percent of the amount present; in cases (ii) and (iii) the coefficient of variation is ± 10 percent of the amount that is present.

Table 2. Elements not detected by any of the three d-c arc techniques used.

Element	Detection limits (ppm)
Ag	0.2
As	4
Au	0.2
B	10
Bi	1
Cd	8
Ce	100
Cs	1
Ge	1
Hf	20
Hg	8
In	1
Mo	2
Nd	100
P	2000
Pb	1
Pt	3
Re	30
Sb	100
Sn	10
Ta	100
Te	300
Th	100
Tl	1
U	500
W	200

Table 3. Average elemental concentrations in type A and type C lunar rocks compared to concentrations in terrestrial rocks (concentrations in parts per million).

Element	Type A	Type C	Basalts (*6*)	Ultramafic rocks (*6*)
Zn	<4	24	105	50
Cu	9	15	87	10
Ga	5	5	17	1.5
Rb	5	3	30	0.2
Li	17	12	17	.X
Mn	2250	1870	1500	1620
Cr	2870	2460	170	1600
Co	31	32	48	150
Ni	7	202	130	2000
Ba	440	250	330	0.4
Sr	135	131	465	1
V	73	60	250	40
Be	3	2	1	.X
Nb	24	23	19	16
Sc	97	68	30	15
La	26	19	15	.X
Y	162	102	21	.X
Zr	594	401	140	45

The results of the analyses are listed in Table 1. The most obvious demarcation in the trace element concentrations occurs between the igneous rocks of types A and B, as compared to the breccias of types C and D. The concentration of nickel is 30 times higher in breccias than in igneous rocks. The concentration of zinc is between five and seven times higher in breccias than in igneous rocks; the magnitude of this difference depends on how much below the limit of detection (4 parts per million) is the concentration in the igneous rocks. In addition, there are several elements for which there are real but smaller differences in concentrations between the igneous and brecciated rocks: (i) Cu is approximately twice as concentrated in the breccias as in the igneous rocks; (ii) Rb and Li are 1½ times more concentrated in type A igneous rocks than in the breccias; (iii) a trend similar to that exhibited by Rb and Li is shown by Sc, La, and Y; (iv) Ba tends to be concentrated in type A igneous rocks, whereas Sr maintains a fairly even concentration in all samples. Several other examples can be noted.

Table 2 lists the elements not detected; also given is the lowest detection limit for the most applicable method of determination for each element.

Comparison (Table 3) of our data with values for terrestrial rocks (*6*) shows that these lunar samples do not closely resemble average terrestrial basalts or ultramafic rocks.

CHARLES ANNELL
ARMIN HELZ

*U.S. Geological Survey,
Washington, D.C. 20242*

References and Notes

1. Lunar Sample Preliminary Examination Team, *Science* **165**, 1211 (1969).
2. Spectroscopic grade graphite powder (United Carbon grade USP 2, mesh size 200) was used as a conducting agent.
3. H. Bastron, P. R. Barnett. K. J. Murata, "Method for the Quantitative Analysis of Rocks, Minerals, Ores and Other Materials by a Powder D-C Arc Technique," *U.S. Geol. Surv. Bull. 1084G* (1960).
4. C. Annell, *U.S. Geol. Surv. Prof. Pap. 575-C* (1967), p. C132.
5. ———, *U.S. Geol. Surv. Prof. Pap. 501-B* (1964), p. B148.
6. K. K. Turekian and K. H. Wedepohl, *Bull. Geol. Soc. Amer.* **72**, 175 (1961).

4 January 1970

Major and Trace Elements and Cosmic-Ray Produced Radioisotopes in Lunar Samples

Abstract. *Analytical data for 42 major and trace elements were mostly obtained by a combination of instrumental and radiochemical neutron activation analyses using both thermal and 14-million-electronvolt neutrons. Excesses of nitrogen and chlorine in the fines, compared with the rocks, are attributed to the solar wind. A striking similarity for contents of seven elements in lunar metal and metal from the calcium-rich achondrite Juvinas was noted. Fractional dissolution was used to separate five radionuclides produced by cosmic-ray bombardment in the fines and rock 57-40. Results for argon-39 from the reaction potassium-39 (n,p), and for argon-37, from the reaction calcium-40 (n,α), seem to require a neutron spectrum containing more neutrons below 2 million electronvolts than the evaporation spectrum or that given by Arnold, Honda, and Lal* (1) *or a strong time dependence for the neutron flux.*

The lunar samples were treated following our standard procedure for the analysis of major and trace elements in meteorites. Pulverized samples (grain sizes ≤ 60 μm) of 250 to 800 mg were first irradiated by 14 MeV neutrons produced by the (d, T) reaction for the determination of O, Si, Mg, and Al. The same samples were then irradiated in the TRIGA research reactor (of the University of Mainz) for 6 hours at a flux of 7×10^{11}n/cm^2 sec, and the γ-ray spectra were measured after various cooling periods by means of a high-resolution Ge(Li) detector. In this way concentrations of the elements Na, K, Ca, Sc, Ti, Cr, Mn, Fe, Co, La, Ce, Sm, Tb, Yb, Lu, Hf, and Ta were determined without dissolving the samples. The reproducibility of these determinations was usually better than 5 percent except for Ca, Ce, and Ta, where the errors are 10 to 20 percent.

Three weeks after their first reactor irradiation, the samples were irradiated for an additional 6 hours at the same flux and radiochemically processed as follows. After addition of carriers, the samples were decomposed under vacuum by Na_2O_2-fusion. The rare gases were purified and xenon, containing the isotope Xe^{135} from the fission of uranium, was isolated for β and β-γ coincidence counting. The melt was then further radiochemically processed for the isolation of individual elements and groups of elements, mainly by solvent extraction techniques. The procedure was similar to one described previously (*2*). Concentrations were determined by γ-ray counting with NaI and Ge(Li) detectors. Chemical yields were determined by reirradiation of samples. The elements determined by this procedure were Li, Cl, Ni, Cu, Ga, Ge, Rb, In, Cs, Pr, Nd, Gd, Dy, Ho, Er, W, Au, Th, and U, with standard deviations less than 10 percent. Portions (about 50 mg) of the samples were set aside before irradiation for the determination of boron by a specially developed fluorescence method (*3*).

Table 1 shows the results of all chemical analyses. Some remarkable features are the high concentrations of Ti (*4*), Zr, and Hf. All rare-earth elements are present in much higher concentrations than in any known type of meteorites. Compared with calcium-rich achondrites, the enrichment amounts to about a factor of 10, except for Eu which seems to be somewhat less enriched. On the other hand, sample 44-32, which shows the lowest concentration of La, has the highest Eu content of all lunar samples analyzed in this laboratory. The concentrations of U, Th, and, to a smaller extent, K are again higher than in any known meteorite. The weight ratio K/U of about 2600 is constant for all our samples and comes very close to that found in achondrites (3000 to 4000). There is a high excess of Cl and especially In in the fine material as compared to lunar rock. Although a possible contamination with In before the distribution of the samples cannot be excluded, the value of 0.75 ppm is only five times higher than the amount of In in the dark phase of the gas-rich meteorite Leighton. A leaching experiment showed that about 80 percent of both Cl and In can be leached out from lunar fines with dilute sulfuric acid (10 percent).

In a similar experiment (*5*), N and Ar were leached from fines in amounts exceeding those from the lunar rock 57-40 by 100 ppm and 1 ppm, respectively. As contamination by atmospheric nitrogen can have been only minor (*5*), we consider this similarity to suggest that not only the rare gases but at least part of the excess N and Cl must be attributed to the solar winds as well.

Some additional interesting information has been obtained from a few milligrams of metal particles, which had been magnetically separated from our sample of lunar fines 84-18. An instrumental neutron activation analysis yielded the following elemental concentrations (values for metal from the calcium-rich achondrite Juvinas are given in parentheses): Fe, 94.6 (96.5) percent; Ni, 4.7 (2.9) percent; Co, 0.52 (0.62) percent; Cu, 340 (150) ppm; Au, 0.83 (0.76) ppm; Ir, 2.6 (0.9) ppm; and W, 24 (21) ppm. Note the surprisingly high concentration of W. We were unable to find any similarly high concentration of tungsten in the metal of several other types of meteorites (calcium-poor achondrite, H and L chondrites). This emphasizes the striking parallelism of the content of trace elements in the lunar metal and that in Juvinas.

From the amount of H_2 evolved during the treatment of a sample of lunar fines with diluted sulfuric acid under vacuum, the amount of metal present in sample 84-18 was estimated to be 0.6 percent by weight. This means that the total amount of Ni (280 ppm) found in this sample (lunar fines) is present in the form of metallic Ni-Fe.

For the examination of cosmic ray isotopes, a technique for fractional dissolution was applied to a sample of the lunar fines (84-18) and the rock (57-40). The fines were processed without crushing; the rock was pulverized to pass a 60-μm sieve. Preliminary experiments on small samples showed hot HNO_3 to be a suitable solvent for the desired fractionation of the target elements of interest. It preferentially dissolves the feldspar, resulting in a fraction enriched in Ca and Al. In addition, in the case of the lunar fines, a considerable fraction of all mineral components was readily dissolved in cold HNO_3. As this effect was not observed for the rock sample, a possible explanation is radiation damage of the surface layers induced by the solar wind or the dissolution of the glass more abundantly present in the fines, or both. The HNO_3 - insoluble residue was decomposed with NH_4HF_2 and H_2SO_4, all steps being performed under vacuum.

Table 1. Concentrations of elements in lunar samples.

Element	Type A (57-40)	Type B (17-33)	Type B (44-32)	Type C (18-22)	Type C (60-16)	Type D (84-18)
			Concentration (%)			
O	40.4	40.7	41.5	40.7	41.4	41.5
Mg	4.2	4.8	3.9	5.0	4.6	4.8
Al	4.0	4.4	6.3	6.1	6.2	6.9
Si	18.9	19.6	20.1	19.6	19.8	19.7
Ca	8.4	8.2	5.1			8.1
Ti	6.5	7.0	6.3	5.5	4.6	4.3
Fe	14.0	14.6	13.3	11.1	12.1	12.0
			Concentration (ppm)			
B	0.8	0.7	1.2			1.03
Li	14.0	23		13.3	10.4	10.4
Na	3000	3470	3580	3920	3600	3150
Cl	12.0	12.2		16.5	15.5	27.1
K	2010	2060	860	1320	1330	1090
Sc	87	86	92	69	70	61
Cr	2160	2310	1300	1900	1820	1830
Mn	1800	1480	2000	1050	1340	1560
Co	25.4	24.5	11.0	24.0	30.0	27.2
Ni	<10					280
Cu	4.3	7.7	4.2			8.2
Ga	5.2	4.2	5.1			4.9
Ge	<1	<1	<1			1.4
Rb	5.2	4.2	5.1			3.0
In	0.0027	0.138				0.75
Cs	0.20	0.12				0.12
La	25	21	12	18	18	15
Ce	107	76	80			63
Pr	9.6	7.8				5.3
Nd	60	58	50	60		47
Sm	12.6	11.9	11.3	8.5	8.7	7.6
Eu	1.80	1.89	2.69	1.68	1.61	1.67
Gd	30	19				18
Tb	5.0	4.6	4.5			2.8
Dy	24	19				17
Ho	5.5	3.8				4.6
Er	12	13				9.5
Yb	16.8	15.6	15.0	11.1	10.9	8.3
Lu	2.15	2.12	1.96	1.56	1.57	1.30
Hf	25.7	23.2	18.3			15.6
Ta	2.0	2.2	2.0	2.1		1.3
W	0.43	0.40	0.24			0.22
Au	0.0016	0.0087	0.0019			0.0021
Th	3.05	3.05	0.76			1.25
U	0.80	0.69	0.28			0.35

After radiochemical purification, ^{22}Na was counted by γ-γ coincidence (NaI), ^{26}Al by high-resolution γ-ray spectroscopy (GeLi), ^{10}Be and ^{36}Cl by β^- counting of BeO and NaCl, and the Ar isotopes by internal gas counting. Table 2 gives the concentrations of the relevant target elements dissolved in the different steps as well as the normalized decay rates. The ^{22}Na and ^{26}Al activities in the total samples were determined before the dissolution as well. For ^{22}Na in the fines this total activity of 61 ± 15 dpm/kg is exactly the same as that obtained by summing the activities in the three fractions; for ^{26}Al, however, the agreement is not as good, the two values being 86 ± 12 dpm/kg and 121 ± 25 dpm/kg. With the exception of ^{39}Ar, the decay rates of all radioisotopes are higher in the fines than in the rock, although to a variable degree. The main reason is the different chemical composition, as discussed below.

In addition to the isotopes listed in Table 2, the ^{14}C content of the fines was measured as 39 ± 5 dpm/kg; a preliminary value for the ^{10}Be decay rate of 27 dpm/kg has been obtained from the cold HNO_3 fraction of the fines. Using the data of Table 2, we calculated the production rates of the radioisotopes on various target elements (Table 3). The high value for the production of ^{39}Ar on K is, as in the case of meteorites (*6*), clearly due to the reaction $^{39}K(n,p)$. The contribution from Ca can be neglected because for the reaction $^{40}Ca(n,2p)^{39}Ar$, the Q value is -8.1 Mev, and the probability that only two

Table 2. Concentrations of relevant elements in lunar fines and rock dissolved in HNO_3, and normalized decay rates.

Element	Lunar fines 84-18: HNO_3 (25°C)	Lunar fines 84-18: HNO_3 (70°C)	Lunar fines 84-18: Residue	Lunar fines 84-18: Total	Lunar rock 57-40: HNO_3 (90°C)	Lunar rock 57-40: Residue	Lunar rock 57-40: Total
			Chemical analyses (percentage by weight)				
O	10.6	3.0	27.1	41.5	4.8	35.6	40.4
Si	3.9	0.8	15.0	19.7	3.0	15.9	18.9
Al	3.92	1.66	1.12	6.9	3.4	1.2	4.6
Mg	1.89	0.14	2.77	4.8	0.4	3.8	4.2
Ca	3.63	1.58	3.61	8.82	2.27	5.60	7.87
Ti	0.49		3.81	4.30	0.33	5.35	5.68
Fe	4.22	0.66	7.12	12.0	1.27	13.03	14.3
K	0.045	0.014	0.050	0.109	0.057	0.144	0.201
			Activity (dpm/kg)				
^{22}Na	32 ± 5	10 ± 2	19 ± 3	61 ± 6	9 ± 2		24 ± 6
^{26}Al	57 ± 13	37 ± 15	27 ± 14	121 ± 25	25 ± 10		70 ± 10
^{36}Cl	6.5 ± 0.28	2.0 ± 0.7	6.13 ± 0.29	14.6 ± 0.8	3.30 ± 0.31	8.85 ± 0.38	12.15 ± 0.5
^{37}Ar	14.4 ± 0.9	4.7 ± 4.3	14.1 ± 0.9	33.2 ± 4.5	—		
^{39}Ar	2.43 ± 0.14	1.18 ± 0.43	3.86 ± 0.20	7.5 ± 0.5	1.99 ± 0.17	7.09 ± 0.23	9.08 ± 0.29

Table 3. Production rates of various radioisotopes in disintegrations per minute per kilogram of target material.

Isotope	Target element					
	Mg	Al	Si	K	Ca	Fe
^{22}Na	100 ± 50	600 ± 200	100 ± 50			
^{26}Al		1000 ± 400	100 ± 50			
^{36}Cl fine					175 ± 8	
^{36}Cl rock					152 ± 9	
^{37}Ar fine					394 ± 18	
^{37}Ar rock						
^{39}Ar fine				4400 ± 500		7.5 ± 1.5*
^{39}Ar rock				3100 ± 500		7.5 ± 2.2*

*The production rate of ^{39}Ar on Ti is assumed to be four times that on Fe, corresponding to a hardness exponent of n equal to 2.5 in the spallation formula.

protons are emitted is extremely small; proton-induced reactions are not possible at all.

Argon-37, on the other hand, is essentially produced from Ca by the neutron-induced reaction $^{40}Ca(n, \alpha)\ ^{37}Ar$. Within the accuracy of the present analysis, ^{36}Cl must be ascribed entirely to Ca as well, although the production rate on Ti and Fe will certainly be comparable to that of ^{39}Ar on these elements.

The ratio of 10 for the production rate of ^{39}Ar from K to that of ^{37}Ar from Ca is about five times higher than the ratio of the effective cross sections for any reasonable choice for the neutron energy spectrum (*1, 6, 7*). Hence, one must either assume that there is a peculiar neutron energy spectrum with an extremely large fraction of the neutrons having energies below 2 Mev, or that the neutron intensity is strongly time-dependent.

Sodium-22 and aluminum-26 cannot be assigned to the different target elements with the same accuracy as the previous isotopes. Nevertheless, it is obvious that for both isotopes Al is the most important target element, which is clearly one reason for the higher decay rates of these isotopes in the fines as compared to the rock. The production of ^{26}Al from ^{27}Al does not imply that the reaction $^{27}Al(n,2n)$ is solely or even predominantly responsible, as the reaction $^{27}Al(p,pn)$ is possible as well.

H. WÄNKE, F. BEGEMANN
E. VILCSEK, R. RIEDER
F. TESCHKE, W. BORN
M. QUIJANO-RICO, H. VOSHAGE
F. WLOTZKA

Max-Planck-Institut für Chemie (Otto-Hahn-Institut), Mainz, Germany

References and Notes

1. J. R. Arnold, M. Honda, D. Lal, *J. Geophys. Res.* **66**, 3519 (1967).
2. R. Rieder and H. Wänke, in *Meteorite Research*, P. Millman, Ed. (Reidel, Dordrecht-Holland, 1969), pp. 76–85.
3. M. Quijano-Rico, in *Dosage des éléments en traces dans les roches et dans d'autres matières naturelles* (Editions du CNRS, Paris, in press).
4. Lunar Sample Preliminary Examination Team, *Science* **165**, 1211 (1969).
5. H. Hintenberger, H. W. Weber, H. Voshage, H. Wänke, F. Begemann, E. Vilcsek, F. Wlotzka, *Science*, this issue.
6. F. Begemann, E. Vilcsek, H. Wänke, *Earth Planetary Sci. Lett.* **3**, 207 (1967).
7. F. Begemann, R. Rieder, E. Vilcsek, H. Wänke, in *Meteorite Research*, P. Millman, Ed. (Reidel, Dordrecht-Holland, 1969), pp. 267–274.
8. Samples were furnished by NASA. Samples were irradiated in the TRIGA research reactor of the Institut für Anorganische Chemie und Kernchemie der Universität Mainz. H. Baddenhausen, A. Balacescu, G. Dreibus, E. Jagoutz, H. Prager, B. Spettel and U. Weidlich provided technical assistance. Supported by the Bundesministerium für Wissenschaftliche Forschung.

4 January 1970

Abundance of Major Elements

Lunar Rock Compositions and Some Interpretations

Abstract. *Samples of igneous "gabbro," "basalt," and lunar regolith have compositions fundamentally different from all meteorites and terrestrial basalts. The lunar rocks are anhydrous and without ferric iron. Amounts of titanium as high as 7 weight percent suggest either extreme fractionation of lunar rocks or an unexpected solar abundance of titanium. The differences in compositions of the known, more "primitive" rocks in the planetary system indicate the complexities inherent in defining the solar abundances of elements and the initial compositions of the earth and moon.*

This report discusses some of the results, limitations, and interpretations of studies of the textures, mineralogy, and chemical composition of lunar samples "gabbro" 10044, gabbroic "basalt" 10057, and the dust (regolith) 10084,28 (*1*). The chemical data (Table 1) were obtained by gravimetric, photometric, and flame photometric techniques. Amounts of alumina were adjusted on the basis of x-ray fluorescent analyses for Cr and Mn in the R_2O_3 group (Table 1).

Analyses of Apollo 11 samples pose problems as difficult as interpretations of the data. The difficulties in our analytical studies resulted from limitations in size of the samples (4 g), time, and the large amounts of titanium.

Interpretations must be made in ignorance of the interrelations of the lunar rock complexes from which the samples came. Our interpretations are based largely upon, and biased by, our understanding of properties and origins of the more "primitive" terrestrial basalts and gabbros, as well as meteorites, especially the "basaltic" achondritic meteorites (*2*).

The three lunar samples consist largely of the minerals pyroxene, ilmenite, and plagioclase. The pyroxene is a pale pink to beige titaniferous augite, and the plagioclase is bytownite. These same minerals with somewhat analogous diabasic textures are common in terrestrial mafic igneous rocks. One major difference is in the proportions of the minerals. The pyroxene is most abundant. It surrounds and interlocks with the plagioclase in diabasic (ophitic) texture. In terrestrial basalts and gabbros, plagioclase commonly equals or exceeds pyroxene in abundance and forms a crystal mesh that envelops the pyroxene. The second striking difference is the abundance of ilmenite in the lunar samples, a reflection of the large concentrations of titanium and iron. Actually TiO_2 is three to five times more abundant in the lunar samples than in the most titaniferous achondritic meteorite, Angra dos Reis (TiO_2, 2.39).

The lunar rocks also are anhydrous and without Fe^{+3} (Table 1). The vesicular (scoriaceous) texture of lunar "basalt" 10057 indicates that this rock originally contained volatiles, but if water or hydroxyl were components of the volatiles, they were expelled. This vesiculation and degassing of the rock would be possible only if the basaltic melt cooled at or very near a dry lunar surface, presumably in a near vacuum.

All terrestrial basalts—and many meteorites—contain some ferric iron and combined water and have formed under higher fluid pressures. In terrestrial basalts both ferric iron and combined water commonly approach or exceed 1 weight percent (Table 2). The most titaniferous terrestrial basalts common to oceanic and many continental volcanoes are also most enriched in K, Na, ferric iron, and combined water (Table 2 and Fig. 1). These are the alkali-olivine basalts that many petrologists

Table 1. Chemical compositions (weight percent) of Apollo 11 regolith and two rocks. Analyst, C. G. Engel. Al_2O_3 values corrected after x-ray fluorescence analyses for Mn and Cr by J. S. Wahlberg (U.S. Geological Survey, Denver, Colorado).

	10044 Gabbro	10057 Vesicular diabase (basalt)	10084,28 Regolith (dust)
SiO_2	42.01	39.79	41.50
TiO_2	8.81	11.44	7.50
Al_2O_3	11.67	10.84	14.31
Fe_2O_3	.00*	.00 ?	.06 ?
FeO	17.98	19.35	15.62
MnO	.24	.20	.22
MgO	6.25	7.65	7.95
CaO	12.18	10.08	11.84
Na_2O	.48	.54	.48
K_2O	.11	.32	.16
H_2O^+	.00	.00	.00
H_2O^-	.00	.01	.01
P_2O_5	.08	.17	.10
Total	99.81	100.39	99.75

* Actual value, —0.08.

Table 2. Compositions (weight percent) of lunar gabbro, average basaltic achondrite, and common terrestrial basalts. The H_2O^+ content in terrestrial oceanic tholeiitic basalt and alkali-olivine basalt due in part to secondary alteration.

	Apollo 11 gabbro 10044	Basaltic achondrite	Oceanic tholeiitic basalt	Alkali-olivine basalt
SiO_2	42.01	48.51	50.01	48.01
TiO_2	8.81	.48	1.37	2.92
Al_2O_3	11.67	13.04	16.18	15.97
Fe_2O_3	.00	1.11 ?	2.32	3.87
FeO	17.98	15.90	7.07	7.56
MgO	6.25	7.87	7.71	5.26
CaO	12.18	11.00	11.33	9.04
Na_2O	.48	.50 ?	2.79	3.73
K_2O	.11	.08 ?	.22	1.89
H_2O^+	.00	.07	.87	1.33
P_2O_5	.08	.19	.13	.42

infer are derived from depths of 50 to 500 km in the earth's mantle. A minority contain from 4 to 7 percent TiO_2, but invariably other major elements such as Si, Al, and P also exceed lunar concentrations (Table 1). Other compositional differences are even more striking. Many alkali-olivine basalts have appreciable ferric iron, combined water or hydroxyl, high Ba, U, Th, Pb, Zr, Th/U, low K/Rb, and, relative to meteorites, more fractionated abundance distribution patterns of rare earths (*2*).

The far-more-abundant tholeiitic basalts of the ocean crust, ridges, and rises actually appear to be the most "primitive" terrestrial basalts. They show many interesting similarities in composition with basaltic achondrite meteorites and have abundance distribution patterns of rare earths like those of many meteorites (*2, 3*). Compared to the lunar samples (Table 2) the oceanic tholeiitic basalts contain approximately the same amounts of Mg, Ca, K, P, U, Th, Y, and Yb, but far less Fe, Ti, Ba, and Zr, and higher concentrations of Si, Al, and Na. Curiously, the one major element ratio common to the most "primitive" terrestrial basalts and meteorites is Al/Ca (0.9–1). But in the lunar "basalt" and "gabbro" this ratio is 0.8 and 0.7, respectively. In many ultramafic lavas extruded from the earth's mantle Al/Ca is as low as 0.6.

Variations in ratios such as Al/Ca, Si/Mg, and Na/K and in amounts of Ti and P seem to emphasize the probability that a variety of igneous processes have fractionated all known planetary rocks. The three major igneous processes are (i) partial melting, (ii) fractional crystallization, and (iii) distillation and volatile transfer (*2*). The effects of widely varying pressures, temperatures, and gravity also are obvious.

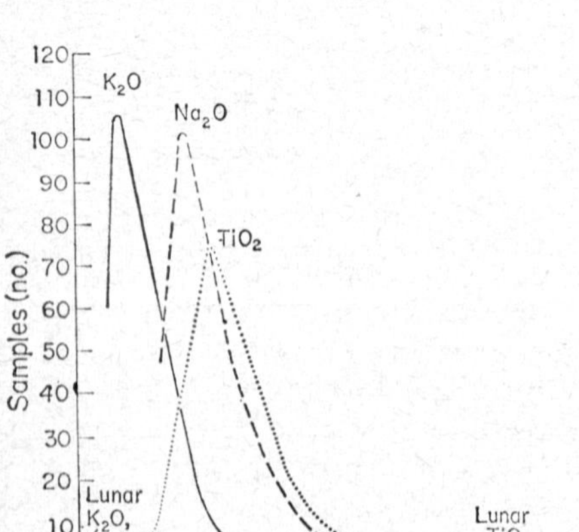

Fig. 1. Frequency distribution of the oxides K_2O, Na_2O, and TiO_2 in alkali-olivine basalts of the oceans. Values for the lunar "gabbro" are plotted as closed triangles on the horizontal coordinate.

Hence, if the earth and moon (and meteorites?) have a common, closely related source and origin, several generalizations follow: (i) available earth rocks are too highly differentiated (fractionated) from their initial source rocks to offer major insights into initial earth, or earth-moon, compositions and processes; (ii) the moon is a rock complex with quite divergent rock types; (iii) lunar and meteoritic matter are products of intermediate and divergent environments and processes in planetary evolution; and (iv) solar and cosmic abundances of elements are more uncertain than ever. Perhaps the most provoking questions are whether current studies of the moon will provide important insights into early earth history, the origin of meteorites, and tektites.

A. E. J. ENGEL
Scripps Institution of Oceanography, University of California, San Diego,
CELESTE G. ENGEL
U.S. Geological Survey, Scripps Institution of Oceanography

References and Notes

1. Lunar Sample Preliminary Examination Team, *Science* **165**, 1211 (1969).
2. A. E. J. Engel, C. G. Engel, R. G. Havens, *Bull. Geol. Soc. Amer.* **76**, 719 (1965); P. W. Gast, *Basalts: A Treatise on Rocks of Basaltic Composition* (Interscience, New York, 1967), vol. 1, p. 325; A. E. J. Engel and C. G. Engel, *The Sea* (Interscience, New York, in press), vol. 4.
3. L. A. Haskin, F. A. Frey, R. A. Schmitt, R. H. Smith, *Phys. Chem. Earth* **7**, 167 (1966).
4. Supported by NASA grant NAS 9-7894.

4 January 1970

Oxygen, Silicon, and Aluminum in Lunar Samples by 14 MeV Neutron Activation

Abstract. *Abundances of oxygen, silicon, and aluminum in 27 lunar rocks and* c *aliquants of lunar soil have been determined by 14 MeV neutron activation. Mean abundances and standard deviations of individual abundances (in weight percent) within each type are: type A (2 rocks), 38.5 ± 1.2 oxygen, 18.9 ± 0.8 silicon, and 4.0 ± 0.4 aluminum; type B (7 rocks), 39.4 ± 1.0 oxygen, 18.7 ± 0.8 silicon, and 5.0 ± 0.6 aluminum; type C (18 rocks), 41.1 ± 1.0 oxygen, 19.7 ± 0.7 silicon, and 6.6 ± 0.5 aluminum; soil (3 aliquants), 40.8 ± 1.2 oxygen, 20.2 ± 0.2 silicon, and 7.2 ± 0.1 aluminum. Oxygen abundances are lower than those in most common terrestrial rocks and are comparable to those found in certain types of stony meteorites. From these results the lunar soil is most similar to the type C lunar rocks.*

The elements oxygen, silicon, and aluminum make up over 60 percent by weight of common terrestrial igneous rocks. The abundances of these major elements may be determined precisely and accurately by means of 14 MeV neutron-activation analysis. In addition, the technique is rapid and essentially nondestructive.

All lunar samples and comparator standards were packaged in heat-sealed polyethylene vials under a dry nitrogen atmosphere (*1*). Samples and standards were irradiated at fluxes of 10^8 to 10^9 neutrons cm^{-2} sec^{-1} provided by a Kaman Nuclear A-1250 14 MeV neutron generator. A single sample transfer system employing nitrogen for the propellent gas was used (*2*). Samples and comparator standards were irradiated separately, and radioactivities were counted sequentially. Corrections for variations in the neutron flux were made by use of a BF_3 neutron monitor. Comparator standards for oxygen and silicon were powders and chunks derived from a piece of pure optical lens quartz. Independent standards of primary standard grade $K_2Cr_2O_7$ were also used for oxygen. Standards for aluminum were NBS 91 standard opal glass and NBS 70a standard feldspar. Samples of USGS standard rock BCR-1 and several basalts furnished by NASA were analyzed simultaneously with the lunar samples (*3*).

The nuclear reactions involved are: $^{28}Si(n,p)^{28}Al$, $^{16}O(n,p)^{16}N$, and $^{27}Al(n,p)^{27}Mg$. Corrections for primary interference reactions and gamma-ray spectral interferences have been considered and applied where required. The most important corrections were applied in the determination of aluminum where the $^{30}Si(n,\alpha)^{27}Mg$ reaction is a primary interference reaction and the $^{56}Fe(n,p)^{56}Mn$ reaction provides a spectral interference. No corrections were applied for the primary interference reactions $^{31}P(n,\alpha)^{28}Al$ in the determination of silicon and $^{19}F(n,\alpha)^{16}N$ in the determination of oxygen. The target nuclides are likely to have relatively low abundances in lunar rocks based on preliminary analyses, but data presented here may be adjusted later, if required, by noting that 0.1 percent fluorine is equivalent to 0.0415 percent oxygen and 1 percent P_2O_5 is equivalent to 0.21 percent silicon. The techniques used, which include a unique multiscaler data-acquisition procedure for oxygen, have been reported (*4*). Application of these new techniques to analysis of the Pueblito de Allende chondrite (*5*) has shown that the standard deviations for six replicate analyses on a given aliquant ranged from ± 0.5 to ± 1.2 percent oxygen and from ± 0.11 to ± 0.24 percent silicon.

All calculations assume terrestrial isotopic abundances for these elements. The same 27 lunar rocks have been analyzed for numerous additional elements by others (*6*).

The most striking feature of these data (Table 1) is the depletion of oxygen in the lunar materials with respect to common terrestrial igneous rocks. Eugster (*7*) tabulated oxygen abundances in common terrestrial igneous rocks and noted that most analyses fell within the relatively narrow range of 40 to 50 weight percent oxygen. Only ultramafic igneous rocks approach the mean oxygen abundances reported here. Eugster has found a good correlation for average compositions of common terrestrial igneous rocks as represented by the equation: %O = 0.194 × %SiO_2 + 35.0 (equation 1). A least squares regression fitted to the pooled lunar rock data of this paper yields the equation: %O = 0.538 × % SiO_2 + 18.2 (equation 2). A null hypothesis test of the significance of the slope yielded t = 5.0 (Student's t test), which is very significant at the 95 percent confidence level. The correlation coefficient for elemental silicon and oxygen is 0.71 (Fig. 1). The silicon and oxygen data, plotted in the manner of Eugster, are shown in Fig. 1. Clearly, any line drawn through points representing the abundances in the three types of lunar rocks is distinct from the line based on terrestrial igneous rocks. A line based on equation 2 does pass close to the points defining the mean abundances in the CH- and CL-group ordinary chondrites.

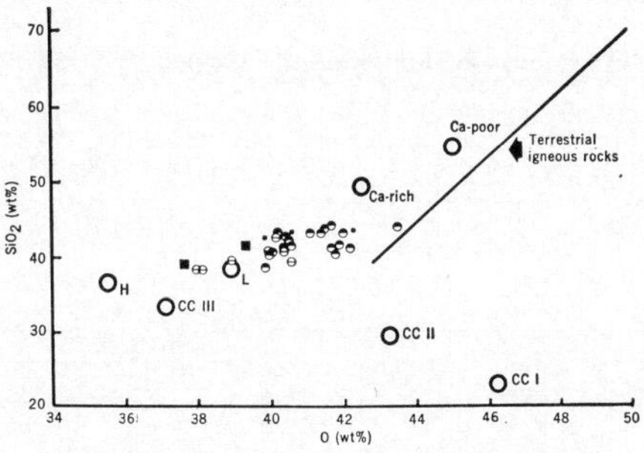

Fig. 1. Relation of silicon as SiO_2 to oxygen in lunar rocks and soil. The line for terrestrial igneous rocks is taken from the tabulation of Eugster (*7*). Large circles represent mean abundances in various types of stony meteorites (*11*). Meteorite identification: CC I, II, and III are carbonaceous chondrites types I, II, and III, respectively; H represents high-iron ordinary chondrites and L represents low-iron ordinary chondrites. Ca-rich and Ca-poor refer to achondrites that are rich and poor in calcum, respectively. (■) type A rock; (○), type B rock; (○), type C rock; (■), soil; and the large circles represent meteorites.

The relative abundance of oxygen atoms to that of silicon atoms (atoms O/atoms Si) for the lunar rocks analyzed is 3.65 and for the lunar soil is 3.55. These values are similar to those found in the ordinary chondrites (CH = 3.9 and CL = 3.7) and to the value of 3.65 estimated for the earth as a whole (*8*). Gault *et al.* (*9*) had estimated this abundance ratio to be 3.1, based on remote analyses of the lunar surface. The corresponding value for the solar atmosphere is 29 (*10*).

In contrast to the situation for silicon and oxygen, both the absolute and atomic abundances of aluminum exhibit appreciable variation among rock types. The relative abundances of aluminum atoms to those of silicon atoms (atoms Al/atoms Si) are 0.22, 0.28, and 0.35 for types A, B, and C, respectively. This ratio for lunar soil is 0.36. Aluminum in lunar rocks is greatly enriched compared to the ordinary chondrites (atoms Al/atoms Si = 0.06), or any other chondrite group. On the basis of these data, the Apollo 11 lunar soil is most similar to type C lunar rocks. Rocks of types A and B would not appear to be the sole contributors to the formation of this lunar soil.

The precise silicon abundance data presented here should be of value in the computation of atomic abundances according to the accepted practice of reporting these relative to Si = 10^6 atoms. Abundances of iron may also be derived from our experimental data (*3*).

WILLIAM D. EHMANN
JOHN W. MORGAN
University of Kentucky, Lexington

Table 1. Oxygen, silicon, and aluminum abundances in lunar rocks by 14 Mev neutron activation. Error limits for individual samples are standard deviations of the means calculated generally on the basis of six replicate analyses for Si and O and two replicate analyses for Al. Error limits expressed for the type means are standard deviations based on values for the individual samples within the set.

Sample	Oxygen (%)	Silicon (%)	Aluminum (%)
Type A			
10022,32	39.3 ± 0.4	19.4 ± 0.1	4.2 ± 0.1
10069,23	37.6 ± 0.1	18.3 ± 0.1	3.7 ± 0.2
Mean	38.5 ± 1.2	18.9 ± 0.8	4.0 ± 0.4
Type B			
10003,36	38.1 ± 0.4	17.9 ± 0.1	5.1 ± 0.1
10024,20	38.9 ± 0.1	18.5 ± 0.1	4.1 ± 0.2
10047,14	40.1 ± 0.1	20.1 ± 0.1	5.5 ± 0.3
10050,29	40.5 ± 0.5	18.4 ± 0.1	5.3 ± 0.2
10058,21	39.9 ± 0.3	19.0 ± 0.2	5.5 ± 0.2
10062,28	38.0 ± 0.2	18.0 ± 0.2	5.3 ± 0.1
10071,21	40.3 ± 0.2	19.1 ± 0.1	4.2 ± 0.1
Mean	39.4 ± 1.0	18.7 ± 0.8	5.0 ± 0.6
Type C			
10018,21	40.1 ± 0.4	20.1 ± 0.1	6.6 ± 0.1
10019,11	39.9 ± 0.2	18.9 ± 0.1	6.5 ± 0.2
10021,22	41.8 ± 0.6	19.6 ± 0.1	6.7 ± 0.1
10048,32	39.8 ± 0.2	18.1 ± 0.1	6.1 ± 0.1
10056,23	41.3 ± 0.3	20.2 ± 0.1	5.7 ± 0.1
10059,30	40.0 ± 0.1	19.1 ± 0.1	6.4 ± 0.1
10060,17	40.3 ± 0.2	20.0 ± 0.1	6.2 ± 0.1
10061,32	41.7 ± 0.2	18.9 ± 0.1	6.7 ± 0.1
10063,05	41.9 ± 0.3	20.3 ± 0.2	6.6 ± 0.2
10064,10	40.5 ± 0.4	19.4 ± 0.1	5.9 ± 0.1
10065,14	41.6 ± 0.7	19.3 ± 0.2	6.6 ± 0.1
10066,05	41.0 ± 0.2	20.2 ± 0.1	6.9 ± 0.1
10067,05	41.6 ± 0.1	20.6 ± 0.1	7.0 ± 0.1
10068,20	40.3 ± 0.3	19.3 ± 0.1	6.3 ± 0.2
10070,05	43.4 ± 0.3	20.6 ± 0.2	7.1 ± 0.1
10074,05	42.1 ± 0.5	19.3 ± 0.1	7.0
10073,21	41.4 ± 0.3	20.5 ± 0.1	7.3 ± 0.1
10075,06	40.4 ± 0.2	19.8 ± 0.3	7.4 ± 0.1
Mean	41.1 ± 1.0	19.7 ± 0.7	6.6 ± 0.5
Soil			
10084,50	42.2 ± 0.5	20.4 ± 0.1	7.2 ± 0.2
10084,50	40.5 ± 0.9	20.3 ± 0.1	7.3 ± 0.1
10084,50	39.8 ± 0.5	20.0 ± 0.2	7.1 ± 0.1
Mean	40.8 ± 1.2	20.2 ± 0.2	7.2 ± 0.1

References and Notes

1. W. D. Ehmann and D. M. McKown, *Anal. Chem.* **40**, 1758 (1968).
2. J. R. Vogt, W. D. Ehmann, M. T. McEllistrem, *Int. J. Appl. Radiat. Isotopes* **16**, 573 (1965).
3. W. D. Ehmann and J. W. Morgan, in preparation.
4. J. W. Morgan and W. D. Ehmann, *Anal. Chim. Acta*, in press.
5. J. W. Morgan, T. V. Rebagay, D. L. Showalter, R. A. Nadkarni, D. E. Gillum, D. M. McKown, W. D. Ehmann, *Nature* **224**, 789 (1969).
6. G. G. Goles and R. A. Schmitt, in preparation.
7. H. P. Eugster, *Handbook of Geochemistry*, K. H. Wedepohl, Ed. (Springer-Verlag, Berlin, 1969), pp. 8C2 and 8E1.
8. B. Mason, *Principles of Geochemistry* (Wiley, New York, ed. 3, 1966).
9. D. E. Gault, J. B. Adams, R. J. Collins, J. Green, G. P. Kuiper, H. Mazursky, J. A. O'Keefe, R. A. Phinney, E. M. Shoemaker, *Science* **158**, 641 (1967).
10. L. H. Aller, *The Abundance of the Elements* (Interscience, New York, 1961).
11. J. R. Vogt and W. D. Ehmann, *Geochim. Cosmochim. Acta* **29**, 373 (1965); *Radiochim. Acta* **4**, 24 (1965); W. D. Ehmann and D. R. Durbin, *Geochim. Cosmochim. Acta* **32**, 461 (1968).

12. Supported by NASA contract NAS 9-8017 and the University of Kentucky Research Foundation.

4 January 1970

Chemical Composition of Lunar Material

Abstract. *Chemical and emission spectrographic analyses of three Apollo 11 samples, 10017-29, 10020-30, and 10084-132, are given. Major and minor constituents were determined both by conventional rock analysis methods and by a new composite scheme utilizing a lithium fluoborate method for dissolution of the samples and atomic absorption spectroscopy and colorimetry. Trace constituents were determined by optical emission spectroscopy involving a d-c arc, air-jet controlled.*

Samples 10017-29 and 10020-30, weighing 16.2 g and 9.8 g, respectively, consisted of rock chips. Sample 10084-132 was an aliquot of the <1-mm fraction of the type D fines, weighing 7.3 g. All samples were crushed to pass a 100-mesh nylon screen. Some metallic fragments were observed in the fines, but an attempt to separate a metal phase from this sample was not successful.

The conventional rock analysis methods used are those described by Maxwell (*1*), except for the determination of total carbon, CO_2, fluorine, chlorine, and total sulfur. A nondistillation method (*2, 3*) with a spectrophotometric finish was used for the simultaneous determination of chlorine and fluorine. An induction furnace was used to liberate the total carbon and sulfur (*4*) as CO_2 and SO_2, respectively, which were then determined titrimetrically, the carbon by nonaqueous titration. The CO_2 released by acid treatment of the sample was similarly determined. The USGS basalt, **BCR-1**, was run simultaneously with the lunar samples to provide a means of evaluating the accuracy of the analytical data. The results are given in Table 1.

The amount of the metallic phase present in 10084-132 was too small for reliable determination. Metallic iron occurring as inclusions in other minerals would not be detected by the methods used to determine metallic iron. It is also possible that some of the titanium is present as Ti(III). Measurement of these possible constituents, and of others contributing to the total reducing capacity of the sample, was attempted by carrying out the decomposition of the sample for the determination of Fe(II) by the modified Pratt method (*1*) in the presence of an added, known amount of Fe(III). Determination of the total iron content of the same solution and subtraction of the added Fe should give an equivalent Fe(II) equal to that obtained by the initial Fe(II) determination, or reveal the presence of some other reducing agent in the solution. The assumption is made that no Fe(III) is present in the sample. The results are given in Table 2; the difference found is expressed as both equivalent Ti_2O_3 and metallic Fe, although it is not possible at this time to say which, if either, is the unknown reducing agent.

Two sample portions were used for analysis by atomic absorption spectroscopy and colorimetry. A 200-mg sample was fused with a fivefold excess of lithium metaborate, the fusion dissolved in dilute hydrofluoric acid, and sufficient boric acid added to provide a solution stable toward glass for at least 2 hours (*5*). Two dilutions were prepared from this, one containing 50 mg of sample and 300 mg of Sr in 100 ml and the other 10 mg of sample and 150 mg of Sr in 100 ml. The more concentrated

Table 1. Chemical analyses (percent by weight) of lunar material [analysts, J. A. Maxwell (conventional) and S. Abbey (atomic absorption spectroscopy, AAS)].

Constituent	BCR-1 (USGS)		10017-29		10020-30		10084-132	
	Conventional	AAS	Conventional	AAS	Conventional	AAS	Conventional	AAS
SiO_2	54.48		40.78	40.14	39.92	41.00	42.28	42.44
Al_2O_3	13.62		8.12	8.17	10.04	9.83	13.76	13.66
TiO_2	2.26		11.71	11.16*	10.72	10.28*	7.35	7.19*
Fe_2O_3	3.77		0.00		0.00		0.00	
FeO	8.65		19.82	19.38†	19.35	19.03†	16.02	15.67†
MgO	3.46		7.65	7.72	7.81	7.77	7.93	7.87
CaO	6.91		10.55	10.99	11.24	11.96	12.00	12.26
Na_2O	(3.36)‡	3.36	(0.51)‡	0.51	(0.37)‡	0.37	(0.42)‡	0.42
K_2O	(1.76)‡	1.76	(0.30)‡	0.30	(0.05)‡	0.05	(0.13)‡	0.13
H_2O^+	0.66		0.00		0.00		0.00	
H_2O^-	0.86		0.01		0.01		0.01	
P_2O_5	0.347	0.35*	0.13	0.15*	0.08	0.07*	0.11	0.10*
MnO	0.165		0.22	0.25	0.24	0.27	0.20	0.21
CO_2	0.055		0.00		0.00		0.03	
F	0.039		0.00		0.00		0.00	
Cl	0.002		0.00		0.00		0.00	
S (total)	0.033		0.22		0.15		0.13	
C	0.00		0.01		0.01		0.00	
Cr_2O_3			(0.36)‡	0.36	(0.40)‡	0.40	(0.33)‡	0.33
Subtotal	100.43		100.39	99.13	100.39	101.03	100.70	100.28
Less O≡S	0.02		0.11		0.08		0.07	
Total	100.41		100.28	99.13	100.31	101.03	100.63	100.28

* Colorimetric, on solutions prepared for atomic absorption spectroscopy (AAS). † Total Fe, expressed as FeO. ‡ AAS values.

Table 2. Reducing substances (percent by weight) in lunar material. N.S., not significant.

Constituent	10017-29	10020-30	10084-132
(1) FeO	19.82	19.43	17.05
(2) Fe_2O_3 equivalent	22.03	21.59	18.95
(3) Total Fe, expressed as Fe_2O_3*	22.01	21.42	17.80
(4) FeO equivalent	19.81	19.28	16.02
(5) Difference (1–4)	N.S.	0.15	1.03
(6) Ti_2O_3 equivalent of (5)		0.15	1.03
(7) Metallic Fe equivalent of (5)		0.04	0.27

* After subtracting Fe(III) added.

Table 3. Minor and trace element content (parts per million) of lunar material (analyst, W. H. Champ).

Constituent	BCR-1 (USGS)	10017-29	10020-30	10084-132
Cr	<20	2200	3000	2300
Zr	<200	410	210	260
Ba	700	230	67	150
Sr	220	100	130	140
V	420	49	120	67
Ni	<20	<20	<20	190
Co	<20	<20	<20	<20
Cu	10	<5	<5	<5
Y	34	160	120	120
Yb	<10	<10	<10	<10
Sc	23	52	78	51
Li*	15	23	12	15
Zn*	120	49	26	47

* By atomic absorption spectroscopy (S. Abbey, analyst). All other determinations by optical emission spectroscopy.

solution was used to determine Si and Al by atomic absorption spectroscopy, with the nitrous oxide–acetylene flame, and K, Mn, and Cr with the air-acetylene flame. All measurements were made with a modified Techtron model AA-3 instrument. The more dilute solution was used to determine total Fe, Mg, Ca, and Na, all with the air-acetylene flame. Details of the method will be given elsewhere. Phosphorus and titanium were also determined photometrically on the main sample solution (*6*, *7*). The results are given in Table 1.

All determinations were based on a comparison of sample solutions with solutions similarly prepared from international reference rock samples. Values assigned to the compositions of the reference samples were based on data prepared by Abbey (*8*) and Govindaraju (*9*). For some elements, it was necessary to add standard solutions of the element to the solutions of the reference samples in order to bring the elemental concentration into the same range as was present in the samples.

An additional 500 mg were decomposed by acid attack and the resulting solution, containing free sulfuric acid and some added hydrochloric acid, was analyzed for lithium and zinc (Table 3) by atomic absorption spectroscopy (*10*). A "standard-addition" technique was used to overcome matrix effects.

The optical emission spectrographic procedure employed a d-c arc, air-jet controlled. Sample portions were mixed with a buffer consisting of alkali salts and graphite powder, at a dilution factor of 7.5, and arced to completion. The spectra were recorded in a 3.4-m Jarrell-Ash grating spectrograph. Because of the general complexity of these spectra, internal standards were not used. Instead, intensity ratios of trace element lines were related to selected iron lines in an "external standard" exposure on the same plate (consisting of Fe_2O_3 treated in the same manner as the samples). Concentrations (Table 3) were obtained from working curves, with corrections as required to correlate with standards of similar composition which were run along with the samples.

JOHN A. MAXWELL
SYDNEY ABBEY
W. H. CHAMP

Geological Survey of Canada, Ottawa 4, Ontario

References and Notes

1. J. A. Maxwell, *Rock and Mineral Analysis* (Interscience, New York, 1968), p. 323.
2. W. H. Huang and W. D. Johns, *Anal. Chim. Acta* **37**, 508 (1967).
3. J. G. Sen Gupta, *ibid.* **42**, 119 (1968).
4. ———, *ibid.*, in press.
5. B. Bernas, *Anal. Chem.* **40**, 1682 (1968).
6. J. A. Maxwell, *Rock and Mineral Analysis* (Interscience, New York, 1968), p. 394.
7. ———, *ibid.*, p. 382.
8. S. Abbey, *Can. Spectrosc.*, in press.
9. K. Govindaraju, personal communication (1969).
10. S. Abbey, *Geol. Surv. Can.*, Paper 67–37 (1967).
11. We thank J. L. Bouvier for the determinations of phosphorus and titanium in the fluoborate solutions, J. G. Sen Gupta for the total sulfur and carbon determinations, and K. A. Church, D. A. Brown, Joanne Crook, and G. A. Bender for their help with the emission spectrographic work.

4 January 1970

Chemical Analyses of Lunar Samples 10017, 10072, and 10084

Abstract. *A crystalline rock, a vesicular rock, and a dust sample returned by Apollo 11 were chemically analyzed by several methods. The compositions of these samples are unlike that of any known rock or meteorite.*

The compositions of lunar samples 10017,20 (a crystalline rock), 10072,20 (a vesicular rock), and of 10084,102 (dust) are given in Tables 1–3. Three 1-g samples were used for each analysis. Wet chemical methods were used in determining the major elements. In order to obtain correct values for some major elements, especially aluminum, it was necessary to determine some trace elements as well. In determining these trace elements, emission spectroscopy, atomic absorption spectroscopy, and neutron activation were used (*1*). The experimentally obtained compositions differ markedly from the composition of all hitherto analyzed rocks and meteorites.

Table 1. Chemical composition of three Apollo 11 moon samples, expressed as oxides. Sample 10017 is crystalline rock; 10072 is vesicular rock; and 10084 is dust. G, gravimetric; C, colorimetric; AA, atomic absorption; T, titration; NA, neutron activity; Q, qualitative; and F, flame.

Oxide	Sample 10017 (%)	Sample 10072 (%)	Sample 10084 (%)	Methods used
SiO_2	40.77	40.20	42.25	G; C
TiO_2	11.82	12.28	7.24	C
Al_2O_3	7.92	7.78	13.83	G; AA
Fe_2O_3	0.0	0.0	0.0	T
FeO	19.79	19.77	15.80	T
Cr_2O_3	0.33	0.36	0.275	AA
MnO	0.22	0.22	0.20	C; NA
MgO	7.74	8.06	7.97	G
CaO	10.58	10.27	11.96	G
Na_2O	0.51	0.52	0.43	F; NA
K_2O	0.29	0.29	0.13	F; NA
P_2O_5	0.18	0.18	0.14	C
H_2O	0.00	0.00	0.00	G
CO_2	0.00	0.00	0.00	Q
Fe met.	0.0	0.0	0.0	Q
Total	100.15	99.93	100.53	

Table 2. Elemental composition of three Apollo 11 moon samples. D, by difference.

Element	Sample 10017 (%)	Sample 10072 (%)	Sample 10084 (%)	Method
Si	19.05	18.78	19.74	G; C
Ti	7.09	7.36	4.52	C
Al	4.19	4.12	7.32	G; AA
Fe	15.38	15.37	12.28	T
Cr	0.226	0.246	0.188	AA
Mn	0.17	0.17	0.155	C; NA
Mg	4.67	4.86	4.81	G
Ca	7.56	7.34	8.55	G
Na	0.38	0.385	0.32	F; NA
K	0.24	0.24	0.108	F; NA
P	0.08	0.08	0.06	C
H	0.00	0.00	0.00	G
O	(40.87)	(40.95)	(41.94)	D

Table 3. Trace elements in three Apollo 11 moon samples. ES, emission spectra.

Element	Sample 10017 (ppm)	Sample 10072 (ppm)	Sample 10084 (ppm)	Methods used
Sc	77	77	59	ES
V	70	82	71	ES
Co	30	30	32	ES; NA
Ni	<5	<5	200	ES; NA; AA
Cu	20	21	13	ES
Zn	47	24	36	AA
Y	160	175	120	ES
Zr	430	460	380	ES
Sm	25	21	18	NA
Eu	2.1	2.0	1.8	NA
Tb	4	3	3	NA
Lu	5	5	3	NA
Yb	19	19	12	NA
Hf	12	12	8	NA
Ta	5	5	3	NA
Th	4.6	4	3	NA

As is apparent from the tables, there is hardly any resemblance with any known rock from the earth or with any meteorite. The almost complete lack of nickel in the rocks is especially peculiar. With 15 percent iron and also the other transition elements of the fourth period well represented, there should be some trace of nickel if the lunar material and the material from which the earth and the meteorites were formed had a common origin. How has the differentiation worked, if the lunar rocks are products of a differentiation, which of course is not proven at all? In any case, the rocks at the Apollo 11 site have neither a basaltic nor a eucritic composition.

H. B. WIIK

Finnish Research Council for Sciences, Helsinki

PENTTI OJANPERA

Geological Survey of Finland, Otaniemi

Note

1. The authors themselves performed the gravimetric, colorimetric, titrimetric, flame photometric, and atomic absorption determinations. We thank the following specialists who contributed to the work: A. Löfgren and A. Savola, who did the emission spectroscopy; P. Kauranen, P. Puumalainen, E. Häsänen, and R. Rosenberg, who did the neutron activation.

4 January 1970

Quantitative Chemical Analysis of Lunar Samples

Abstract. *In three lunar samples the high titanium contents previously reported were confirmed. Water and ferric oxide were absent.*

Three moon samples were analyzed by classical methods (Table 1) (*1*). Determinations for sulfur and chromium were made because of their possible petrologic significance (*2*).

We did not determine metallic iron because we doubt the reliability of published methods and therefore held this determination in abeyance pending discussion with other investigators. Material might be better used in the application of other techniques (for example Mössbauer) to the separated magnetics.

Silica. The procedure was modified to eliminate complications caused by the high concentration of titanium; too much titanium remained with the silica. The procedure (*1*, p. 62) was followed through step 13, the filtrate set aside, 3 g of sodium carbonate added to the silica, and the procedure repeated from step 3 through step 13. The combined filtrates were used in determination of the ammonium hydroxide group which follows.

Table 1. Analyses of moon samples. Values are percentages.

Compound	Olivine basalt 10020-23	Vesicular basalt 10072-32	Lunar fines 10084-131
SiO_2	39.95	40.53	42.20
Al_2O_3	10.19	8.52	14.07
Fe_2O_3	.03	.00*	.00†
FeO	19.14	19.76	15.81
MgO	7.87	7.68	7.93
CaO	11.31	10.42	12.01
Na_2O	.39	.54	.46
K_2O	.05	.27	.12
H_2O^+	.00	.00	.05
H_2O^-	.00	.00	.00
TiO_2	10.52	11.74	7.32
P_2O_5	.07	.14	.08
MnO	.27	.24	.21
S	.18	.24	.14
Cr_2O_3	.38	.35	.28
Subtotal	100.35	100.43	100.68
Less O	.09	.12	.07
Total	100.26	100.31	100.61

*Actual value obtained minus 0.14. †Actual value obtained minus 0.12.

Ferrous oxide. The presence of troilite (FeS) and metallic iron makes results for ferrous oxide uncertain (*1*, pp. 39–42). Hydrogen sulfide should not interfere greatly, because no reducible element is present and most of the hydrogen sulfide is expelled during solution of the sample. Metallic iron is counted as ferrous oxide in the analysis.

Ferric oxide. Ferric oxide is determined by difference after separate determinations of total iron oxide and ferrous oxide. The cumulative errors in both of these determinations are therefore reflected in the value for ferric oxide. The small negative and positive amounts found are probably due to such errors. Within accuracy limits of the method, the samples do not contain ferric oxide.

Combined water. The H_2O^+ reported in sample 10084-131 is probably water adsorbed after the sample reached earth; strongly adsorbed water is not expelled when a silicate is heated at 105°C.

Titanium oxide. Titanium oxide was determined colorimetrically (*1*, p. 70). The precision of the method is probably no better than ±0.1 percent absolute.

Chromium oxide. It is likely that most of the chromium oxide is present in mafic minerals. Sample 10084-131 contained a little black material that might be chromite, but there was no evidence of its presence in the other two samples.

The summation on sample 10084-31 is unacceptably high. The presence of metallic iron, in amounts of about 3 percent, could account for the high total. (Metallic iron is reported as ferrous oxide.) Data from others, however, indicate that 1 percent or less of iron is present. Other factors may contribute to the high total. For example, powder segregation is troublesome in samples containing particles of high density.

LEE C. PECK

VERTIE C. SMITH

U.S. Geological Survey, Denver, Colorado 80225

References and Notes

1. Except where otherwise noted, procedures described in *U.S. Geol. Surv. Bull. 1170* were used.
2. Methods used are described in I. M. Kolthoff and E. B. Sandell, *Textbook of Quantitative Inorganic Analysis* (Macmillan, New York, ed. 3, 1952), p. 724.

4 January 1970

Stable Isotopes, Rare Gases, Solar Wind, and Spallation Products

$^{18}O/^{16}O$, $^{30}Si/^{28}Si$, D/H, and $^{13}C/^{12}C$ Studies of Lunar Rocks and Minerals

Abstract. *The δ ^{18}O of minerals from lunar gabbros and basalts are: plagioclases +6.06 to +6.33), pyroxenes (+5.70 to +5.95), and ilmenites (+3.85 to +4.12). The uniformity of these results indicates isotopic equilibrium in the mineral assemblages. Estimated plagioclase-ilmenite temperatures range from 1150°C to 1340°C. The bulk $^{18}/^{16}O$ and $^{30}Si/^{28}Si$ ratios of these lunar rocks are identical with ratios of terrestrial basalts, but the lunar glass, breccia, and dust are slightly enriched in the heavier isotopes. The lunar hydrogen (formed from solar wind) has a δ D/H of less than —873 per mil and the value may be even lower, as it is probably contaminated with terrestrial hydrogen. The δ ^{13}C of lunar dust and breccia is unusually high relative to reduced carbon in meteorites or on earth.*

The δ ^{18}O values of several Apollo 11 lunar samples are given in Fig. 1. Oxygen isotope analyses were obtained by reacting the samples with fluorine gas at 600°C, following procedures outlined by Taylor and Epstein (*1*). The analyses presented in Fig. 1 can be supplemented by material-balance calculations of whole-rock δ ^{18}O values. The total range of $^{18}O/^{16}O$ in the calculated and measured whole-rock samples from the moon is very small, from δ = +5.7 to +6.3.

Essentially pure mineral separates of plagioclase and pyroxene were analyzed in each case, but the cristobalite was only 80 percent pure, and the ilmenite concentrates typically varied in purity from 70 to 90 percent. The δ ^{18}O values of these impure ilmenites were very accurately corrected for the presence of the impurity (primarily pyroxene), because the volume of SiF_4 gas obtained from each sample was measured. Typically, the corrections for the ilmenite analyses were 0.2 to 0.6 per mil, but the ilmenite concentrate from 10057 was very impure and required a correction of 1.25 per mil. The utility of the SiF_4 correction procedure is illustrated by the fact that, even though the corrections are occasionally large, the final δ ^{18}O values of the ilmenites range only from +3.85 to +4.12 for five different specimens.

All the minerals show a striking uniformity in δ ^{18}O. The clinopyroxenes (δ = +5.70 to +5.95) and plagioclases (δ = +6.06 to +6.33) are practically identical, within experimental error of ± 0.1 per mil. This uniformity and the constancy of the oxygen isotopic fractionations among minerals (given as 1000 ln α, where α is the ratio of $^{18}O/^{16}O$ in mineral A to that in mineral B) suggest that oxygen isotope equilibrium was essentially attained during the crystallization of these mineral assemblages. This is noteworthy in the light of the pronounced chemical zoning in the minerals, particularly the pyroxenes.

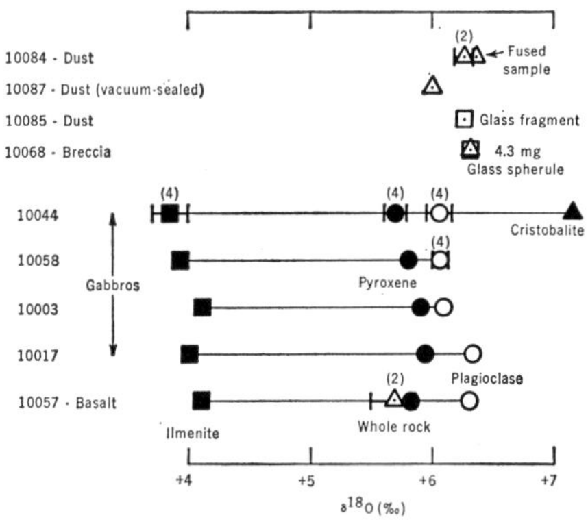

Fig. 1. δ ^{18}O measurements (relative to standard mean ocean water) of lunar rocks and minerals from the Sea of Tranquillity. The numbers in parentheses indicate number of separate analyses, and the error bars represent the average deviation. All other analyses are single determinations.

In a given rock, the order of enrichment of minerals in ^{18}O (ilmenite, pyroxene, plagioclase, free silica) is the same as that observed in many terrestrial igneous rocks and meteorites. However, the ^{18}O fractionations between plagioclase and ilmenite in the lunar gabbros are smaller than in terrestrial gabbros (2.0 to 2.3 versus 3.3 to 4.0). This is in part accounted for by the higher Ca content of the lunar plagioclase, but it also indicates that the $^{18}O/^{16}O$ ratios of coexisting minerals in the lunar gabbros were "frozen in" at a considerably higher temperature than in terrestrial gabbros (*2, 3*).

Although an isotopic temperature calibration curve involving ilmenite has not yet been determined, an estimate of the quartz-ilmenite and plagioclase-ilmenite geothermometer curves can be made on the basis of (i) the plagioclase-magnetite curve used by Taylor (*3*) in estimating "isotopic temperatures" of various igneous rocks; (ii) the plagioclase-H_2O calibration curves of O'Neil and Taylor (*4*); (iii) the quartz-H_2O curve of Clayton *et al.* (*5*); and (iv) the observation that in natural mineral assemblages, 1000 ln α for quartz-ilmenite is equal to 0.92 × 1000 ln α for quartz-magnetite (*6*). The constructed curves are:

$$1000 \ln \alpha_{\text{Q-I}} = 1.35 + 4.56\ (10^6 T^{-2})$$

$$1000 \ln \alpha_{\text{P-I}} = 0.72 + 3.22\ (10^6 T^{-2})$$

where Q-I is quartz-ilmenite and P-I is plagioclase (An_{90})-ilmenite.

The plagioclase-ilmenite temperatures calculated for lunar samples are as follows: 10058, 1243°C; 10044, 1202°C; 10017, 1154°C; 10003, 1338°C; and 10057, 1212°C. When the quartz-ilmenite curve is used, 10044 gives a cristobalite-ilmenite temperature of 1260°C, almost concordant with the 1202°C plagioclase-ilmenite temperature of this rock. The above temperatures can be compared with temperatures of 920°C for a Skaergaard gabbro and 820°C for a Kiglapait gabbro, both of which are compatible with estimates made on these rocks based on the independent Fe-Ti oxide geothermometer of Buddington and Lindsley (*7*). The inference can be drawn that the lunar gabbros crystallized and cooled very rapidly under relatively "drier" conditions than many terrestrial gabbros; if P_{H_2O} had been high, we should expect to observe more evidence of post-crystallization oxygen isotopic exchange.

The bulk $^{18}O/^{16}O$ ratios of the lunar rocks and minerals are indistinguishable from those of terrestrial basalts, the best average δ value for pristine unaltered basalt being +5.9 per mil (*3*). Pyroxenes from most stony meteorites (ordinary chondrites, enstatite chondrites, and achondrites) also have δ ^{18}O values identical with the lunar pyroxenes. However, notable exceptions are the basaltic achondrites, hypersthene achondrites, and mesosiderites, which have δ ^{18}O of +3.7 to +4.4. If the $^{18}O/^{16}O$ ratios of the Sea of Tranquillity samples are characteristic of the rest of the moon, the moon would appear to be more closely related to the chondrites and the earth than to the basaltic achondrites.

The SiF_4 released during reaction of samples with fluorine was collected and analyzed for $^{30}Si/^{28}Si$ in a specially designed, double-collecting mass specrometer for 9-inch radius, having an all-metal sample introduction system (except for the magnetic switching valves). The data are plotted in Fig. 2, along with data for a few meteorites, tektites, and terrestrial igneous rocks. To a first approximation, the lunar gabbros are very similar to terrestrial samples in δ ^{30}Si. Plagioclase is invariably enriched in ^{30}Si relative to coexisting pyroxene, but within experimental error of ± 0.08 per mil the plagioclases from all the gabbros are essentially identical, as are the pyroxenes. Data on terrestrial rocks and meteorites are not sufficient to support any firm conclusions about the δ ^{30}Si variations, but the hypersthene and enstatite achondrite pyroxenes appear to be distinctly lower in ^{30}Si than the lunar pyroxenes, and they are clearly more depleted than the lunar dust, glass, and breccia.

The glass-rich dust and breccia are also clearly richer in both ^{18}O and ^{30}Si than the whole-rock gabbros and basalts. These discrepancies in δ ^{30}Si and δ ^{18}O must be caused by either (i) an ^{18}O- and ^{30}Si-enriched constituent in the dust and breccia that is not derived locally from the nearby gabbros and basalts, or (ii) vapor fractionation of the glass during its formation (presumably mainly by impact melting), with resulting enrichment in both ^{18}O and ^{30}Si. It has been shown experimentally by Walter and Clayton (*8*) that fractional vaporization of SiO_2-rich glass indeed produces

Table 1. Isotopic and concentration data for hydrogen, carbon, and rare gases extracted from samples 10084, 10087, and 10061 from the Sea of Tranquillity.*

Sample fractions	Hydrogen gas			Carbon		Water			Rare gases	
	Conc. (μmole/g)	Deuterium conc. δ D/H	Deuterium conc. ppm	Conc. (ppm)	δ $^{13}C/^{12}C$	Conc. (μmole/g)	Deuterium conc. δ D/H	Deuterium conc. ppm	Conc. (cm^3/g)	H_2 gas/ rare gases (vol. ratio)
				Sample 10084, No. 2 (2.40 g)						
A	None			Trace		6.05	−150	133	None	
B	8.56	−888	18	13.4 (B, C)	+17.8	4.09	−365	100	0.100	4.6
C	12.93	−856	23			0.93 (C, D)	−483	81	0.275	2.5
D	1.45	−615	60	141.7	+16.1				0.140	0.56
Total	22.94	−853	23	155.1	+16.3	11.07			0.515	2.4
				Sample 10084, No. 1 (1.85 g)						
A	None			85 (A, B, C)	+18.6	11.58 (A–D)	−275	114		
B	10.30	−849	24							
C	11.34	−828	27							
D	0.72	−609	61	58						
Total	22.36	−830	27	143		11.58	−275	114		
				Sample 10061 (0.55 g), vacuum-sealed breccia						
A	None			74 (A, B, C)	+10.1	8.9 (A–D)	−412	92		
B	17.86	−882	18						0.044	5.0
C	32.06	−868	21						0.190	2.1
D	Trace			188	+10.9				Trace	
Total	49.92	−873	20	262	+10.8	8.9	−412	92	0.234	2.6
				Sample 10084† ‡, No. 3 (1.71 g)						
A				None		7.83 (A, B)	−329	105		
B	9.40	−850	24	Trace						
				Sample 10087† (~ 2 g)						
B	4.47	−850	24							
C§	20.87	−843	25							
Total	25.34	−844	25							

*Note: δ values for hydrogen relative to standard mean ocean water (SMOW) in per mil. Deuterium concentration of SMOW assumed to be 157 ppm (*10*). δ values for carbon are relative to PDB standard in per mil. Fraction A gases evolved below 500°C, almost wholly water. Fraction B gases evolved between 500° and 550°C, a mixture of CO_2, CO, H_2, H_2O, and rare gases. Fraction C gases evolved between 550° and 650°C, a mixture of CO_2, CO, H_2, H_2O, and rare gases. Fraction D gases evolved between 650° and 1500°C, dominantly CO. †Hydrogen purified through palladium thimble at ~650°C. ‡Lunar dust sample heated in quartz tube to release hydrogen (to avoid contact of released H_2 and H_2O with hot platinum). §Lunar dust exposed to water of δ ≈ −980 per mil prior to extraction of hydrogen fraction C.

an enrichment of ^{18}O in the remaining sample. Whether an analogous process also exists for $^{30}Si/^{28}Si$ cannot be stated until some experiments are done, but it is reasonable to suppose that ^{28}Si would be preferentially boiled off during a high-temperature impact melting.

The isotopic composition and concentration of hydrogen and carbon in the lunar samples are given in Table 1. The results are for different fractions A, B, C, and D extracted from each sample at designated temperature ranges. Note that the first fraction A contains mostly H_2O and the last fraction is primarily CO, resulting probably from reaction of carbon with ilmenite. All samples except lunar dust sample 3 were loaded in an atmosphere dried with P_2O_5.

Two procedures were used to quantitatively separate the hydrogen from the rare gases and other noncondensable gases: (i) the H_2 was converted to H_2O by reaction with hot CuO, the H_2O being isolated and converted to H_2 by reaction with hot U metal, and (ii) the H_2 gas was allowed to diffuse through a palladium thimble at 650°C. The two methods gave similar results. During the combustion of the noncondensable gases, CO_2 was formed and separated from the H_2O. The rare gases remained after all other gases were converted into condensable compounds.

Before gases were extracted from lunar material, about 1 cm^3 of deuterium-depleted H_2 ($\delta = -980$) was run through the extraction procedures. This flushing reduced the errors due to possible contamination with isotopically normal water that may have been present in the extraction line.

The concentration of the H_2 is 22.7 μmole/g for dust and 50 μmole/g for the breccia. The δ D varies from -873 per mil (20 ppm D) to -830 per mil (27 ppm D). It is almost certain that the δ values give only the maximum deuterium concentrations for lunar hydrogen, because the waters extracted from the lunar samples are depleted in deuterium relative to normal terrestrial fresh waters. The data in Table 1 indicate that during the extraction, isotopic exchange occurred between the very light lunar H_2 and the adsorbed H_2O.

We have calculated some material balances: Assuming that the lunar H_2 gas, before contamination, originally had zero deuterium content, the adsorbed water in lunar dust samples (10084) 2 and 1, breccia (10061), and lunar dust (10087) must have had δ values of $+45$, $+57$, $+300$, and -148, respectively. If it is assumed that the contaminating water has a δ value of $-$ 150 per mil, the original lunar hydrogen would have had δ values of -904, -891, -917, and -996, respectively. If the contaminating H_2O has a δ value of -75 per mil (a value reasonable for Houston water vapor), the lunar hydrogen would have δ values of -943, -930, -936, and -1057, respectively. The calculations for sample 10087 are least reliable, because the data in Table 1 represent an incompletely extracted H_2O fraction. If the first condition prevailed (no D in lunar H_2), the contaminating water is unusually concentrated in deuterium. This is possible if the lunar samples were contaminated with evaporated water from, for example, the astronauts' space suits. The presence of some deuterium (about 10 ppm) in the lunar H_2 gives a more reasonable δ value for the contaminating water. Hence, although some deuterium is possibly present in the lunar H_2, we conclude that most of the H_2 is of solar wind origin and is very low in deuterium. Clearly the best procedure to determine the actual δ values for lunar hydrogen is to avoid H_2O contamination, or, if that is impossible, to use H_2O that is highly depleted in deuterium in operations where H_2O might come into contact with the lunar samples. The presence of significant deuterium in solar wind H_2 is very unlikely, considering the mean reaction time of 4 seconds for the reaction: $D + H \rightarrow {}^3He$ (*9*).

Another possibility is that there is present in the lunar rocks some primary water of relatively normal deuterium content in addition to the solar wind hydrogen. This possibility has not been eliminated by our experiments. Further evidence that the lunar H_2 is of solar wind origin is that it is more loosely bound than are the rare gases themselves. The ratio of H_2 to rare gases decreases with progressive heating of the lunar samples (Table 1).

The carbon in the lunar samples appears to be heterogeneously distributed (~145 ppm in the dust and ~260 ppm in breccia), and it has a unique high δ ^{13}C of about + 19 per mil for the lunar dust and about +11 per mil for the breccia. These values are unlike any reduced natural carbon obtained on earth or in meteorites. There are several possibilities that can account for this carbon:

1) Solar-wind carbon is a possible source, inasmuch as there is a correlation between the amounts of carbon and hydrogen in the lunar samples (Table 1). However, if the carbon extracted from the samples is a mixture of lunar carbon and solar-wind carbon, they are both present in essentially similar chemical form. Otherwise, carbon extracted at different temperatures would be expected to have different δ ^{13}C values.

2) If some process involved in the formation of lunar dust and breccia causes fractional loss of carbon, isotopic fractionations can occur; in such a case, we can say little about the ultimate origin of the carbon.

3) Since the possible terrestrial sources of carbon contaminants such as grease and rocket fuel have low δ values, it is unlikely that the δ ^{13}C values of lunar dust are due to terrestrial contamination.

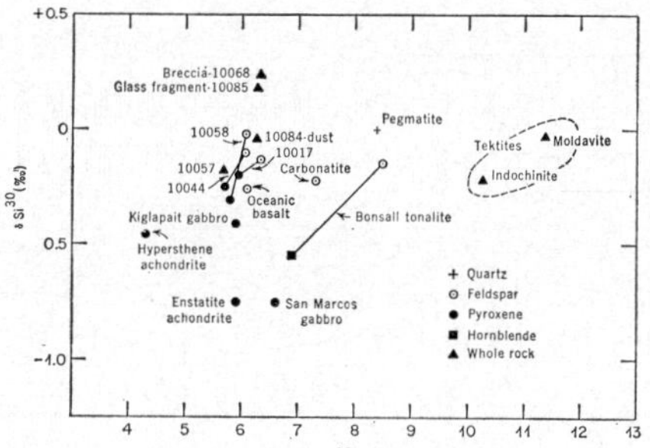

Fig. 2. Graph of δ ^{30}Si (relative to a quartz standard from a pegmatite, Pala, California) versus δ ^{18}O for lunar rocks and minerals. For comparison, some analyses of meteorites, tektites, and terrestrial igneous rocks are also plotted.

SAMUEL EPSTEIN
HUGH P. TAYLOR, JR.

Division of Geological Sciences, California Institute of Technology, Pasadena, California 91109

References and Notes

1. H. P. Taylor, Jr., and S. Epstein, *Bull. Geol. Soc. Amer.* **73**, 461 (1962).
2. ———, *J. Petrol.* **4**, 51 (1963).
3. H. P. Taylor, Jr., *Contrib. Mineral. Petrol.* **19**, 1 (1968).
4. J. R. O'Neil and H. P. Taylor, Jr., *Amer. Mineralogist* **52**, 1414 (1967).
5. R. N. Clayton, J. R. O'Neil, T. Mayeda, unpublished data.
6. G. D. Garlick and S. Epstein, *Geochim. Cosmochim. Acta* **31**, 181 (1967).
7. A. F. Buddington and D. L. Lindsley, *J. Petrol.* **5**, 310 (1964).
8. L. S. Walter and R. N. Clayton, *Science* **156**, 1357 (1967).
9. E. Segrè, *Nuclei and Particles* (Benjamin, New York, 1964), p. 564.
10. H. Craig, *Science* **133**, 1833 (1961).

11. This research was supported by NASA. Some of the laboratory work was done by P. Yanagasawa. Publications of the Division of Geological Sciences, California Institute of Technology, Contribution No. 1706.

4 January 1970

Oxygen Isotope Fractionation between Minerals and an Estimate of the Temperature of Formation

Abstract. *Oxygen isotopic compositions of separated minerals from three type A and four type B rocks are very uniform. The $\delta^{18}O$ values are: plagioclase, 6.20; clinopyroxene, 5.75; ilmenite, 4.45 (parts per thousand relative to Standard Mean Ocean Water). The isotopic distribution corresponds to equilibrium at 1120°C. The isotopic composition of lunar pyroxenes falls within the range for pyroxenes of terrestrial mafic and ultramafic rocks, ordinary chondrites, enstatite chondrites, and enstatite achondrites, but above the range for basaltic achondrites, hypersthene achondrites, and mesosiderites. Glass isolated from the lunar soil has a $\delta^{18}O$ value of 6.2, significantly richer in ^{18}O than the crystalline rock fragments in the soil.*

Oxygen isotope analyses have been done on the separated major mineral phases of seven lunar rocks of types A and B, on whole rock samples of types C and D, and on glass spherules and rock fragments separated from type D.

One-gram samples of the crystalline rocks were crushed in a steel mortar, ground in an agate mortar under acetone, and sieved through stainless steel sieves. For the coarser grained type B rocks, the size fraction between 325- and 500-mesh (25 to 43 μm) was used for mineral separation. For the finer grained type A rocks the fraction passing through the 500-mesh sieve (<25 μm) was used. Mineral separation was accomplished in a few cases by handpicking grains from the coarser fractions, but usually by centrifugation in heavy liquids (bromoform, methylene iodide, and Clerici solutions). Densities of the liquids used are given for each sample in Table 1. Mineral purity was evaluated by x-ray diffraction and optical microscopy, and was >95 percent in all samples, and usually >98 percent.

Oxygen was extracted from 10-mg portions of the separated minerals by the bromine pentafluoride procedure (*1*). Manometric measurement of the oxygen yield gave the quantitative determinations of oxygen content (Tables 1 and 2). Experimental error is estimated as ±1 percent of the amount present. Oxygen was converted to carbon dioxide and the oxygen isotope ratio was measured with a 60°, 15-cm, double-collecting mass spectrometer. Oxygen extraction was carried out in duplicate for all samples, and each gas sample was further analyzed in duplicate. Results of isotope analyses are given in Tables 1 and 2. Standard error of the mean of the duplicate measurements is estimated to be ± .07 per mil.

Oxygen contents of all phases analyzed are in agreement with values calculated from the estimated major-element compositions. No evidence of nonstoichiometry with respect to oxygen has been seen.

The uniformity of isotopic composition from rock to rock, for a given mineral, is very striking. Of the type A and type B rocks, only sample 58(35) shows any systematic departure from the mean, each of its minerals having an ^{18}O content 0.2 per mil lower than the means of the other rocks.

The soil and microbreccia have ^{18}O contents similar to one another, and about 0.5 per mil greater than that of the crystalline rocks. From the approximate modal compositions of the crystalline rocks given in the preliminary reports, their whole-rock $\delta^{18}O$ is estimated to 5.7 ± 0.1 per mil. Fragments of type A and type B rocks were handpicked from the 50- to 100-mesh (150 to 300 μm) fraction of the lunar soil and were found to have isotopic compositions in this range (Table 2). Irregular brown glass fragments (refractive index 1.68) were found to have almost the same composition, and may have been derived by fusion of crystalline rocks.

Glassy spherules, mean diameter 300 μm, were also separated from the soil, and have an ^{18}O content significantly greater than that of the rock fragments, implying addition of an isotopically heavier component or an isotopic fractionation in the process of formation of the spherules. An addition to the soil of 2 percent carbonaceous chondrite-like

Table 1. Oxygen contents and isotopic compositions of minerals from type A and type B rocks. Cr, cristobalite; Pc, plagioclase; Cpx, clinopyroxene; Ol, olivine; Il, ilmenite.

Mineral	Type B rocks			Type A rocks			
	20 (24)	22 (36)	71 (22)	17 (21)	44 (22)	47 (15)	58 (35)
				*Specific gravity**			
Cr	5.74						>2.38
Pc	5.14	2.71–2.81	2.71–2.82	2.70–2.77	2.69–2.77	2.70–2.75	2.71–2.82
Cpx	4.41	3.36–3.44	3.36–3.47	3.33–3.56	hand-pick	hand-pick	3.36–3.52
Ol							
Il	6.19	>4.2	>4.2	>4.3	>4.2	>4.3	>4.2
				Oxygen content (wt. %)			
Cr							53.4
Pc	45.8	46.3	47.2	47.3	46.4	46.5	46.7
Cpx	42.0	42.6	42.1	42.9	42.5	42.0	42.6
Ol	40.5						
Il	33.7	33.9	33.1	33.3	32.0	31.9	32.4
				$\delta^{18}O$†			
Cr	2.70–2.82						7.09
Pc	3.36–3.46	6.31	6.17	6.16	6.18	6.20	6.00
Cpx	3.54–3.65	5.71	5.72	5.81	5.77	5.76	5.53
Ol	>4.2						
Il		4.55	4.47	4.51	4.35	4.29	4.24
				$\Delta^{18}O$‡			
Pc-Cpx	0.45	0.60	0.45	0.35	0.41	0.44	0.47
Cpx-Il	1.32	1.15	1.24	1.29	1.41	1.46	1.28
Pc-Il	1.77	1.75	1.69	1.64	1.82	1.90	1.75

* Values quoted are specific gravities of liquids that bracket the value for the separated mineral. † ^{18}O content in per mil terminology: deviation of $^{18}O/^{16}O$ ratio in parts per thousand from ratio in standard mean ocean water (SMOW). ‡ Oxygen isotope fractionism, $\Delta^{18}O$, defined as 1000 in α, where α is a fractionation factor for two phases, for example, $^{18}O/^{16}O$ in plagioclase divided by $^{18}O/^{16}O$ in ilmenite.

meteorites, as suggested (*2*), would cause an ^{18}O enrichment of about 0.2 per mil. Addition of other known types of meteoritic material would have a negligible effect because of the similarity of their isotopic composition to that of the soil (*3*). Enrichment of ^{18}O in glass resulting from a loss of volatiles in very high-temperature fusion has been shown experimentally (*4*), and may play some part on the moon.

The mean value of the $^{18}O/^{16}O$ ratio in objects of the solar system may be determined in part by the region in the early solar nebula within which the solid particles accreted. There may have been gradients in isotopic composition which are reflected today in differences among the planets, their satellites, and the meteorites and asteroids. Such differences have been observed for various classes of meteorites (*3*). Lacking data for the average oxygen-isotope composition of the moon, perhaps the best we can do at this stage is to compare the isotopic composition of a particular major mineral with that of similar minerals in terrestrial rocks and meteorites. The ^{18}O content of the Apollo 11 pyroxenes is identical with that of pyroxenes from some mantle-derived rocks: garnet peridotites and pyroxenites from kimberlite pipes (*5*). The lunar pyroxenes lie at the high end of the distribution for oceanic basalts and andesites (*5, 6*).

The lunar pyroxenes fall into group II (*3*), which contains the ordinary chondrites, enstatite chondrites, and enstatite achondrites. They are distinctly richer in ^{18}O than the pyroxenes of group I meteorites: basaltic achondrites, hypersthene achondrites, and mesosiderites.

If we speculate that oxygen isotope gradients in the early solar system were in the direction of decreasing ^{18}O outward, as suggested by the data on meteorites, then the oxygen isotope results on Apollo 11 rocks indicate an initial condensation either at the same position as terrestrial material, or somewhat nearer the sun.

Isotopic fractionations between pairs of minerals can be used to determine the temperature of last equilibration of the minerals. In favorable cases, this may simply be the temperature of their initial crystallization. Isotopic fractionations between pairs of major phases are shown in Table 1. For a given mineral pair the observed fractionations are identical, within experimental error, for all seven rocks analyzed: plagioclase-clinopyroxene, 0.45 ± 0.04; clinopyroxene-ilmenite, 1.31 ± 0.08; plagioclase-ilmenite, 1.76 ± 0.06.

Table 2. Oxygen contents and isotopic compositions of various fractions of type C and type D rocks. The $\delta^{18}O$ values are relative to standard mean ocean water.

Sample	Description	O content (wt. %)	$\delta^{18}O$
	Type C		
60 (12)	Whole rock	41.3	6.03
	Type D		
84 (46)	Whole rock	42.6	6.18
84 (46)	Glass spherules	42.2	6.22
84 (46)	Glass fragments	42.3	5.88
84 (46)	Type A fragments	41.7	5.74
84 (46)	Type B fragments	40.6	5.64
84 (46)	"Anorthosite" fragments	44.8	5.89

In principle, any two of these fractionations provide two independent measures of "isotopic temperature." A minimum test of the existence of equilibrium is that these two temperatures agree. Unfortunately, laboratory calibration of isotopic thermometers involving pyroxene and ilmenite has not been done, so that this test cannot yet be carried out rigorously. It is, however, possible to compare the relative magnitudes of two isotopic fractionations with those found within the same mineral assemblage in various terrestrial rocks. This is done in Fig. 1 for the assemblage plagioclase-clinopyroxene-magnetite with analytical data for terrestrial mafic igneous rocks, both intrusive and extrusive (*5, 6*). Because of the dependence of fractionations involving plagioclase on the chemical composition of the plagioclase (*7*), all data have been normalized to a feldspar composition of An_{60}. This amounts to an effective increase in the δ value of lunar plagioclase by 0.16 per mil. Furthermore, the isotopic fractionation between ilmenite and magnetite is very small, but not negligible, at igneous temperatures, so that the ilmenite δ value has been reduced by 0.20 per mil to determine a hypothetical magnetite δ value. The resulting point, representing the average fractionation for the seven Apollo 11 samples analyzed, has been plotted on the plagioclase-pyroxene-magnetite diagram, and falls on the previously determined line through the data points for terrestrial rocks. It is virtually indistinguishable from the results for recent Hawaiian basalts. Hence the evidence for isotopic equilibrium among the major phases is fairly strong.

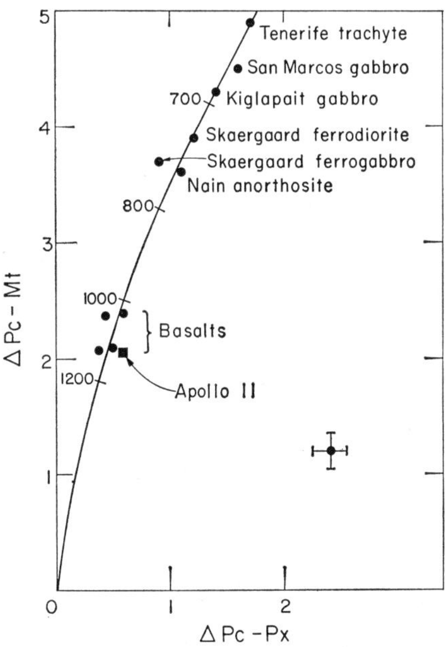

Fig. 1. Plagioclase-clinopyroxene-magnetite "concordancy diagram." Data on terrestrial rocks are from reference (*6*). Temperatures indicated along concordancy line are determined from the plagioclase-magnetite isotopic thermometer (*7, 8*). Pc, plagioclase; Px, pyroxene; Mt, magnetite. Analytical error is indicated by bars on point in lower right-hand corner.

A similar "concordancy" test for the system plagioclase-clinopyroxene-olivine indicates that the olivine of sample 20 (24) was in isotopic equilibrium with the other minerals. Data for terrestrial samples are not available to make this test for cristobalite in sample 58(35), but the magnitude of the cristobalite-plagioclase fractionation of 1.08 is within 0.1 of the value estimated by extrapolation of laboratory calibrations of the quartz-feldspar system to the isotopic temperature inferred for these rocks.

The best estimate of an isotopic temperature can be made from the fractionation between plagioclase and ilmenite. This is possible through the experimental calibration of the plagioclase-magnetite pair (*7, 8*), along with the small ilmenite-magnetite fractionation, estimated on the basis of measurements on terrestrial basalts to be about 0.20 at the temperatures involved. The resulting isotopic temperature is 1120°C, with an analytical uncertainty of ± 30°C. (Somewhat larger systematic errors may exist in the calibration of this isotopic thermometer, particularly since this temperature lies outside the range of the laboratory calibration.)

The isotopic temperature is very similar to those of terrestrial mafic extrusive rocks, which, in turn, are in good agreement with other estimates of extrusion temperatures. Hence it appears that the isotopic temperatures are very close to the initial temperatures of crystallization from a silicate melt. On earth, large differences in isotopic temperatures are found between fine-grained, rapidly quenched volcanic rocks and their coarse-grained plutonic equivalents (*5*, *6*). This has been interpreted as resulting from post-crystallization retrograde isotopic exchange, on a scale comparable to the grain size, during the slow cooling of the plutonic rocks. No terrestrial rock as coarse-grained as the Apollo 11 type B samples has been found with a high isotopic temperature corresponding to its initial crystallization. Apparently the cooling conditions of the lunar samples were such as to prevent significant subsolidus exchange. The absence of water may have been an important factor.

The indistinguishability between types A and B rocks in terms of isotopic compositions or isotopic fractionations suggests that they were derived from magmas of the same isotopic composition, and that in both types the initial isotopic distribution at the time of crystallization has been quenched in, and has subsequently remained unchanged.

NAOKI ONUMA
ROBERT N. CLAYTON
TOSHIKO K. MAYEDA
Enrico Fermi Institute, University of Chicago, Chicago, Illinois 60637

References and Notes

1. R. N. Clayton and T. K. Mayeda, *Geochim. Cosmochim. Acta* **27**, 43 (1963).
2. R. R. Keays, R. Ganapathy, J. C. Laul, E. Anders, G. Herzog, P. M. Jeffery, *Science*, this issue.
3. H. P. Taylor, Jr., M. B. Duke, L. T. Silver, S. Epstein, *Geochim. Cosmochim. Acta* **29**, 489 (1965).
4. L. S. Walter and R. N. Clayton, *Science* **156**, 1357 (1967).
5. G. D. Garlick, *Earth Planet. Sci. Lett.* **1**, 361 (1966); A. T. Anderson, R. N. Clayton, T. K. Mayeda, unpublished data.
6. H. P. Taylor, Jr., *Contrib. Mineral. Petrol.* **19**, 1 (1968).
7. J. R. O'Neil and H. P. Taylor, Jr., *Amer. Mineral.* **52**, 1414 (1967).
8. J. R. O'Neil and R. N. Clayton, in *Cosmic and Isotopic Chemistry*, H. Craig, S. L. Miller, G. J. Wasserburg, Eds. (North-Holland, Amsterdam, 1964), p. 157.
9. Supported by NASA grant NAS-9-7888.

4 January 1970

Water, Hydrogen, Deuterium, Carbon, Carbon-13, and Oxygen-18 Content of Selected Lunar Material

Abstract. *The water content of the breccia is 150 to 455 ppm, with a* δD *from* -580 *to* -870 *per mil. Hydrogen gas content is 40 to 53 ppm with a* δD *of* -830 *to* -970 *per mil. The* CO_2 *is 290 to 418 ppm with* $\delta\,^{13}C = +2.3$ *to* $+5.1$ *per mil and* $\delta^{18}O = 14.2$ *to 19.1 per mil. Non-*CO_2 *carbon is 22 to 100 ppm,* $\delta^{13}C = -6.4$ *to* -23.2 *per mil. Lunar dust is 810 ppm* H_2O *(D = 80 ppm) and 188 ppm total carbon (*$\delta^{13}C = -17.6$ *per mil). The* ^{18}O *analyses of whole rocks range from 5.8 to 6.2 per mil. The temperature of crystallization of type B rocks is 1100° to 1300°C, based on the oxygen isotope fractionation between coexisting plagioclase and ilmenite.*

During the past 20 years a large amount of information on the relative abundance of the stable isotopes of hydrogen, carbon, oxygen, and sulfur in terrestrial and meteoritic materials has been accumulated. Consequently, it was of interest to investigate the relative abundance of these isotopes in lunar materials. In addition, determination of the water and hydrogen contents in lunar samples is important because a knowledge of these contents leads to a better understanding of the development and history of the earth's hydrosphere.

For the deuterium, the water, and some of the carbon analyses, the weighed samples were placed in a platinum crucible and the crucible was then inserted into a vacuum system. The crucible was heated, step by step, to 1350°C and H_2, He, H_2O, CO_2, and SO_2 were collected, separated, and measured, and the isotopic abundances were determined on isotope-ratio mass spectrometers. After melting, the sample was then mixed with approximately half its weight of previously combusted CuO, and the resulting mixture was combusted at 950°C in oxygen at 1 atm pressure. The collected H_2O was analyzed for D/H and the CO_2 for $^{13}C/^{12}C$ and $^{18}O/^{16}O$.

The results of the deuterium analyses are expressed as atomic ratios of deuterium to hydrogen, D/H $\times 10^6$, and also as δD per mil SMOW. The $^{13}C/^{12}C$ analyses are given as $\delta^{13}C$ per mil PDB, whereas the $^{18}O/^{16}O$ ratios are expressed as $\delta^{18}O$ per mil SMOW.

A 9.3-g sample of lunar dust (10084) and 2.7-g piece of breccia (10060-11), which had been sealed in an aluminum vacuum container in Houston until opened for processing in our laboratory, were combusted directly without being melted in vacuum. Another 3.7-g piece of the vacuum-sealed 10060-11 was processed by heating in a vacuum, as described above, to compare results with those for breccia samples 10046-21 and 10046-22 that had been exposed to laboratory atmosphere for several months.

Geoatmospheric contamination was removed by heating the samples to 50°C in vacuum, after which the temperature was raised and the volatiles were collected. No volatiles were emitted from the samples until a temperature of approximately 300°C was reached. Samples were heated at 550°C for several hours and the volatiles processed. Most of the hydrogen, some helium, more than half the H_2O, and a little CO_2 were released at this stage. The temperature was then raised to 950°C and maintained there for several hours, during which time practically all the remaining H_2O, CO_2, and SO_2 and some He were evolved. Finally the crucible was heated to 1350° to 1400°C for 15 minutes to release the remainder of the SO_2, CO_2, and He. The final combustion of the melted sample in oxygen yielded CO_2 and a negligible amount of H_2O. Results of the analysis are given in Table 1.

The hydrogen gas emitted by the samples may have been present as molecular hydrogen trapped in the solid material. Alternately, it may have been generated during heating by the reactions of water with carbon, metallic iron, or ferrous iron present in the silicates. The fact that all the hydrogen was liberated below 550°C, whereas appreciable amounts of water continued to be released above that temperature, would tend to rule out reactions with the reduced forms of iron or with carbon. The possibility still exists that the hydrogen equilibrated isotopically with the water vapor. At 550°C the equilibrium constant of the reaction $H_2O + HD \rightleftarrows HDO + H_2$ is 1.3 (*1*). The hydrogen evolved at 550°C is depleted in deuterium by a factor of 2 to 4 compared with the water evolved at the same temperature—far in excess of the amount to be expected under equilibrium conditions at the temperature of our experiments.

The isotopic compositions of the hydrogen and water are those to be expected for isotopic equilibrium at temperatures from 0° to 200°C, which is well within the lunar temperature range. It is possible that the hydrogen orig-

Table 1. Results of heating and combustion of selected Apollo 11 material. Concentrations are ±20 percent of amount present; D/H × 10^6 = ±20 percent; $\delta^{13}C$ = ±0.2 percent; $\delta^{18}O$ = ±0.2 percent; and δD = ±10 percent. The $\delta^{13}C$ values are per mil deviations from PDB; $\delta^{18}O$ values are per mil deviations from SMOW; δD values are per mil deviations from SMOW. The D/H × 10^6 was calculated from the δD SMOW values, using the relation D/H × 10^6 = .158 (1000 + δD per mil).

			Vacuum heating											Oxygen combustion					Total				
			He		H_2			H_2O			CO_2			H_2O			CO_2		H_2, H_2 + H_2O			Carbon	
Sample No.	Sample type	Weight (g)	cm³/g	$^4He/^3He$	ppm	D/H × 10^6	δD (‰)	ppm	D/H × 10^6	δD (‰)	ppm C	$\delta^{13}C$ (‰)	$\delta^{18}O$ (‰)	ppm	D/H × 10^6	δD (‰)	ppm C	$\delta^{13}C$ (‰)	ppm	D/H × 10^6	δD (‰)	ppm	$\delta^{13}C$ (‰)
10046-21	Breccia	1.704	0.15	4000 ± 300	50	27	−830	455	38	−760	79	+2.3	+19.1	5			102	−23.2	100	33	−790	181	−12.0
10046-22	Breccia	3.574	0.13	4000 ± 200	60	~5	−970	372	66	−580	140	+5.1	+18.8	37			78	−22.7	105	~28	~−820	218	−4.8
10060-11	Breccia	3.712	0.8	2000 ± 200	66	~22	−860	152	21	−870	79*	+3.0	+14.2	10			22	−25.7	83	~46	~−710	>101	−3.2
10060-11	Breccia	2.766	←										→	720	51	−680	113	−10.6	80	51	~−680	113	−10.6
10017-16	Crystal-line	2.028			~1			~25						27			69	−20.0	>6			>69	−20.0
10084-43†	Dust	9.311	←										→	810	80	−490	188	−17.6				188	−17.6

* A small amount of CO_2 released between 550° and 950° C was lost from this sample. † Mineral separation on the lunar dust sample (10084) using heavy liquids and a magnetic separator was unsuccessfully attempted. The separated fractions were then recombined, washed in pure acetone, dried, and combusted as described.

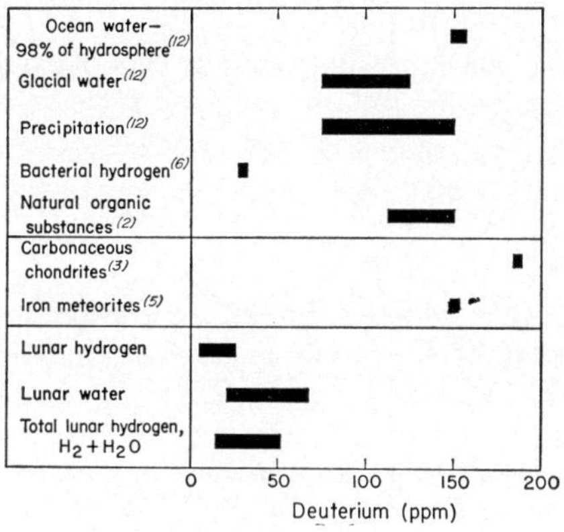

Fig. 1. The deuterium content of hydrogen containing materials from terrestrial, meteoritic, and lunar sources (*13*); D/H was calculated using a value of 158 × 10^{-6} for SMOW.

inates as solar wind protons and then slowly reaches isotopic equilibrium with the water already present in the rocks. The isotopic composition of the hydrogen therefore may not be that of the solar wind. The water may be present as liquid or gas inclusions in the solid phases or as water present in the glass structures.

Although all the samples that we analyzed have had some exposure to terrestrial water vapor, we do not believe that terrestrial water is a serious contaminant in these samples. This opinion is based on several lines of evidence.

1) The water is not evolved below about 300°C. This would tend to rule out adsorbed water.

2) All the breccia samples, both those kept open to the terrestrial environment and those sealed in vacuum, reacted similarly when heated and gave about the same amounts of hydrogen and water with about the same deuterium content.

3) The deuterium abundance of the water is far lower than that to be expected from water picked up from the terrestrial atmosphere by the glass phase present in the rocks. As determined from data on volcanic glasses containing secondary water that is derived from meteoric or atmospheric sources (*2*), the secondary water should range from about 145 ppm D (hydrated in Houston) to about 125 ppm D (hydrated in Denver).

We do not think that the lower water content of the vacuum-sealed sample indicates large-scale contamination by terrestrial water in the other samples because there appears to be great variability in water content between two pieces of the same breccia that had been exposed to the terrestrial atmosphere for several months.

Contamination with unexpended rocket fuel dumped on the lunar surface can be ruled out by the results of Flory *et al.*, which were published in the report of the Lunar Sample Preliminary Examination Team (*3*). Additions of rocket exhaust gases cannot be as easily determined as can those of organic fuel, and these gases still remain a possibility as a contaminant, particularly in the very porous, high-surface-area breccias. The CO_2 and SO_2 results discussed later, however, tend to rule out rocket exhaust contamination.

The piece of vacuum-sealed 10060-11 that was crushed in the laboratory prior to its combustion in oxygen and exposure to the laboratory air for approximately 5 minutes has approximately 70 ppm more water than the total water collected by heating the other vacuum-sealed piece in vacuum. In addition, the water from the combusted sample is slightly enriched in deuterium as compared with the water from the other sample. These differences may be due to the uptake of "heavy" (~ 150 to 200 ppm D) water from the atmosphere during sample handling, or they may represent inhomogeneity in the breccia. From our experience with the several pieces of breccia 10046, we favor the latter explanation—that of sample inhomogeneity.

Figure 1 is a plot of the deuterium abundances of terrestrial, meteoritic, and lunar material. If the analysis given here is of true lunar water, deuterium in the earth is enriched by a factor of about 3 to 5 over that in the moon. If the moon is a sample of the primitive earth, the earth will have lost at least two-thirds of its original hydrogen by escape from the exosphere to outer space.

The total carbon content of the lunar samples is reasonably constant at 150

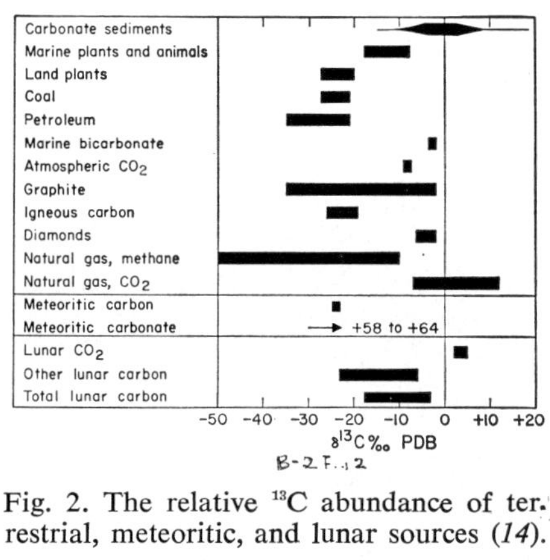

Fig. 2. The relative ^{13}C abundance of terrestrial, meteoritic, and lunar sources (*14*).

Table 2. Oxygen isotope compositions (δ ^{18}O); δ values are per mil deviations from the SMOW standard. Numbers in parentheses refer to number of analyses made.

Sample	Plagioclase	Pyroxene	Ilmenite	Other	Whole rock
Fine grained					
10057-20					5.8 ± 0.1 (2)
Crystalline					
10017-16	6.3 ± 0.1 (2)	5.7 ± 0.1 (2)	3.8 ± 0.1 (2)		5.9 ± 0.1 (2)
10050-34	6.2 ± ? (4)	5.6 ± 0.1 (2)	4.0 ± 0.2 (2)		
Breccia					
10046-21					5.7 ± 0.2 (2)
10060-11					6.2 ± 0.2 (2)
Dust					
10084-43					
Whole rock					6.2 ± 0.1 (3)
Dark glass				6.7 (1)	
Plagioclase (>90%)	6.1 (1)				
Magnetic separate				5.6 ± 0.1 (2)	
Ilmenite (~50%)			5.4 (1)		

± 50 ppm, with a $\delta^{13}C$ of −3 to −20. This is within the range to be expected for average earth carbon (see Fig. 2). Contamination by exhaust gases is again a possibility, but the tenacity with which the CO_2 is held in the samples would tend to rule out this possibility. The CO_2 could have been formed in our laboratory by reduction of the silicates by elemental carbon, but this reaction usually does not begin at temperatures under about 1600°C. The water-gas reaction ($C + H_2O \rightarrow CO + H_2$) is also a high-temperature reaction (~ 900°C). In addition, the evolution of SO_2 is difficult to explain as an artifact of our experimental procedure or as rocket-exhaust contamination. We believe that the CO_2 is truly lunar and may be present as liquid or gaseous inclusions in the solid phases. The carbon extracted by combustion after sample heating and removal of CO_2 is probably present as either carbides or elemental carbon.

The $^{18}O/^{16}O$ ratio of the CO_2 varies from +14.2 to +19.1. These values are much heavier than the $\delta^{18}O$ in the silicates and oxides analyzed below, and this fact is another indication that the CO_2 did not form from the reduction of silicates by carbon. The heavy oxygen may be due to equilibration between the CO_2 and H_2O and the silicates and oxides. It is known that the oxygen of CO_2 in fluid inclusions exchanges with the oxygen present in the host mineral (*4*).

Boato (*5*) has analyzed carbon in both carbonaceous chondrites (−3.7 to −18.8) and ordinary chondrites (−24.5 ± 0.2). These values are in the same range as that determined by our analysis of lunar materials. Clayton (*6*) has found very heavy carbon (+58.6 to +64.4 per mil) in carbonate minerals from Orgueil, a carbonaceous chondrite. This heavy carbon exceeds the range of the carbon in the lunar samples that we have examined.

Oxygen was liberated from the samples by reaction with BrF_5 at 550°C (*7*). The oxygen was converted to CO_2 and analyzed on an isotope-ratio mass spectrometer. The $\delta^{18}O$ values are given relative to the SMOW standard in Table 2.

The lunar whole-rock analyses lie between 5.7 and 6.2 and fall in the 5.5 to 7.0 range found for most terrestrial volcanic rocks (*8*). The lunar pyroxene value of 5.7 is in the 5.3 to 6.3 range found by Taylor *et al.* (*9*) to be characteristic of chondrites and enstatite achondrites. Meteorites which contain anorthitic plagioclase and which have chemical characteristics similar to those of lunar materials (for example, basaltic achondrites and mesosiderites) have δ values for pyroxene which are 1 to 2 per mil lower and thus do not appear to be genetically related to lunar material.

Because mineral separation on the lunar fines was difficult, the analyses of these separates are not particularly meaningful. Since the whole rock is 6.2 and some separates yield lower values, there must be an isotopically heavy component, which may be cristobalite.

From the oxygen isotope fractionations between plagioclase and ilmenite and from the laboratory calibrations of the systems plagioclase-water (*10*) and magnetite-water (*11*), an estimate can be made of the temperature of crystallization of type B rocks. Oxygen yields of 106 to 114 percent (over stoichiometric $FeTiO_3$) were obtained for the ilmenite samples, and the measured δ values of 4.6 (10017) and 4.3 (10050) were corrected to the values shown in the table on the basis that the excess yields were due to pyroxene impurities. The values in the table will have to be raised slightly when the magnesium contents of the ilmenites become known. This uncertainty and the fact that only two crystalline rocks were analyzed in this study places limits on the precision of the isotopic temperature. From our data we estimate the plagioclase-ilmenite fractionation to be between 1.8 and 2.2 per mil, implying a crystallization temperature of 1100° to 1300°C for type B rocks.

IRVING FRIEDMAN
U.S. Geological Survey, Denver, Colorado 80225
JAMES R. O'NEIL
LANFORD H. ADAMI
U.S. Geological Survey, Menlo Park, California 94025
JIM D. GLEASON
KENNETH HARDCASTLE
U.S. Geological Survey, Denver

References and Notes

1. H. E. Suess, *Z. Naturforschung* **5**, 328 (1949).
2. I. Friedman and R. L. Smith, *Geochim. Cosmochim. Acta* **15**, 218 (1958).
3. Lunar Sample Preliminary Examination Team, *Science* **165**, 1211 (1969).
4. R. O. Rye and J. R. O'Neil, *Econ. Geol.* **63**, 232 (1968).
5. G. Boato, *Geochim. Cosmochim. Acta* **6**, 209 (1954).
6. R. N. Clayton, *Science* **140**, 192 (1963).
7. ——— and T. K. Mayeda, *Geochim. Cosmochim. Acta* **27**, 43 (1963).
8. H. P. Taylor, Jr., *Contrib. Mineral. Petrol.* **19**, 1 (1968).
9. ———, M. B. Duke, L. T. Silver, S. Epstein, *Geochim. Cosmochim. Acta* **29**, 489 (1965).
10. J. R. O'Neil and H. P. Taylor, *Amer. Mineral.* **52**, 1414 (1967).
11. J. R. O'Neil and R. N. Clayton, in *Isotopic and Cosmic Chemistry*, H. Craig, S. L. Miller, G. J. Wasserburg, Eds. (North-Holland, New York, 1964), p. 157.
12. I. Friedman, A. C. Redfield, B. Schoen, J. Harris, *Rev. Geophys.* **2**, 177 (1964).
13. C. Bokhoven and H. H. J. Theeuwen, *Koninkl. Ned. Akad. Wetenschap. Proc. Ser. B* **59**, 78 (1956); G. Edwards, *Nature* **176**, 109 (1955); M. I. Krichevsky, F. D. Sisler, I. Friedman, M. Newell, *J. Biol. Chem.* **236**, 2520 (1961).
14. R. E. Zartman, G. J. Wasserburg, J. H. Reynolds, *J. Geophys. Res.* **66**, 277 (1961); S. R. Silverman and S. Epstein, *Bull. Amer. Ass. Petrol. Geol.* **42**, 998 (1958); H. Craig, *Geochim. Cosmochim. Acta* **3**, 53 (1953); R. M. Lloyd, *J. Geol.* **72**, 84 (1964).
15. Publication authorized by the director, U.S. Geological Survey. We appreciate the help of Miss Terry Keith and Dr. Frank Dodge, both of the U.S. Geological Survey, in making the pure mineral separates for ^{18}O analysis.

4 January 1970

Concentration and Isotopic Composition of Carbon and Sulfur in Apollo 11 Lunar Samples

Abstract. *The concentration of carbon and sulfur in six samples ranged between 20 to 200 and 650 to 2300 parts per million, respectively. Carbon was present in gaseous, volatilizable, and nonvolatile forms, and terrestrial contaminants were recognized. Sulfur appeared to exist only as acid-volatile sulfide. The bulk fines contain a high concentration of carbon and a low concentration of sulfur. They are always enriched in the heavier isotope carbon-13 or sulfur-34. The fine-grained basaltic rocks show the reverse relation; lowest carbon, highest sulfide concentrations, and no apparent enrichment in heavy isotopes. The breccias are of intermediate composition.*

The purpose of this study was to differentiate between the various forms of carbon and sulfur present in the lunar material and to measure the $\delta^{13}C$ and $\delta^{34}S$ ratios of these components. The results presented here relate to six samples, five that were submitted directly to UCLA (10084; 1002,54; 10060,22; 10049 and 10057,40) and one bulk fines submitted through Ames Research Center (10086).

Carbon was analyzed in several ways. Total carbon was measured by combustion of 1-g samples in an atmosphere of oxygen at a pressure of about 700 mm-Hg at 1050°C. The gaseous products were passed over copper oxide catalyst, silver wire, and lead oxide (to remove halogens and oxides of nitrogen and sulfur, respectively). The resulting CO_2 was purified from water by distillation through a trap of acetone and dry ice, and its volume was measured on a manometer. It was recognized after preliminary experiments in which smaller samples were combusted that contamination from handling was serious and that the gas volumes obtained were so small, as to make them impractical for purposes of mass spectroscopy.

Table 1. Concentration and isotopic composition of carbon and sulfur.

Sample No.	Carbon Conc. (ppm)	Carbon $\delta^{13}C$* (per mil)	Sulfur Conc. (ppm)	Sulfur $\delta^{34}S$ † (per mil)
	Bulk fines			
10086	143	+ 20.2	680	+ 5.4
10086	170	+ 17.2	640‡	+ 8.2‡
10084	147	+ 18.8	770	+ 4.7
10084	116	+ 19.5		+ 4.4
	Breccia			
10002,54	198	+ 8.4	1070	+ 3.5
10002,54	181	+ 9.2		+ 3.4
10060,22	137	+ 1.6	1120	+ 3.6
10060,22	132	+ 2.7		+ 3.3
	Fine-grained rock			
10049	63	− 18.8	2200	+ 1.2
10049	77	− 21.4		+ 1.3
10057,40	21	− 25.6	2280	+ 1.2
10057,40	11	− 29.8		+ 1.2

* Relative to PDB standard. † Relative to Canyon Diablo standard. ‡ Total sulfur by aqua regia oxidation.

In Fig. 1 the contamination problem in the bulk fines (10084 and 10086) is clearly identified. Samples containing more than 200 ppm of carbon always have $\delta^{13}C$ of less than +15 per mil. These values probably represent mixtures of lunar carbon (+18 to +20 per mil) and common types of terrestrial carbon (−20 to −30 per mil). The data seem to fall into three groups. Those with $\delta^{13}C$ values greater than +17 per mil are considered to be largely or entirely of lunar origin. Those with $\delta^{13}C$ values between +12.5 and +8 have probably been lightly contaminated either at the Lunar Receiving Laboratories or by the lunar module exhaust. The isotopically lightest samples with $\delta^{13}C$ values smaller than −3.5 per mil are considered to have suffered additional contamination in our laboratory. Samples received from the Lunar Research Laboratory, and stored in the original containers (aluminum or steel) in air, yielded values of the intermediate type.

After degassing the sample under simulated lunar–day conditions by heating to 150°C at 10^{-3} torr for 48 hours, a loss in volatile carbon was detected. The total carbon content decreased and isotopic values increased after this treatment. The values in Tables 1 and 2 were obtained on samples treated in this manner.

In order to determine the nature of the carbon, samples were treated chemically or pyrolyzed. The products were captured, measured quantitatively and their $^{13}C/^{12}C$ ratios determined. Some of the results are summarized in Table 2 and will be explained below.

A previously degassed sample (10084) was pyrolyzed under vacuum over three temperature ranges (150 to 250°C, 250 to 500°C, and 500 to 750°C) and the volatile material was frozen out in a liquid nitrogen trap. The products of each pyrolysis step were then volatilized in an atmosphere of oxygen, burnt at 1050°C, passed through the combustion train described earlier, and captured as CO_2 gas. The residue from the pyrolysis experiment was combusted finally, and the released CO_2 also was captured. The total quantity of carbon recovered from the pyrolysis experiment corresponds closely with the total carbon content of the original fines (Table 2); 86 ppm of this carbon was volatilized in this experiment. In a duplicate experiment on the degassed sample of bulk fines 10086, a positive identification of CO_2 in amounts of 48 ppm (minimum), was made by Dr. Sherwood Chang, with a Poropak Q chromatographic column. Thus at least 55 percent of the released volatiles is CO_2, and the remainder is possibly a mixture of carbon monoxide and hydrocarbon decomposition products (neither positively established). It is apparent from the isotope data that material evolved at the lower temperatures may largely consist of such carbon-hydrogen fragments or compounds (as possible contaminants), whereas the gas evolved at the higher temperature is representative of CO_2. The isotopic distribution is in accord with that found for reduced and oxidized carbon compounds (*1*).

When the lunar fines (10086) were treated with hydrochloric and hydrofluoric acids, the residues from duplicate extractions contained 30 and 42 ppm, respectively, relative to the original sample. The $\delta^{13}C$ value and car-

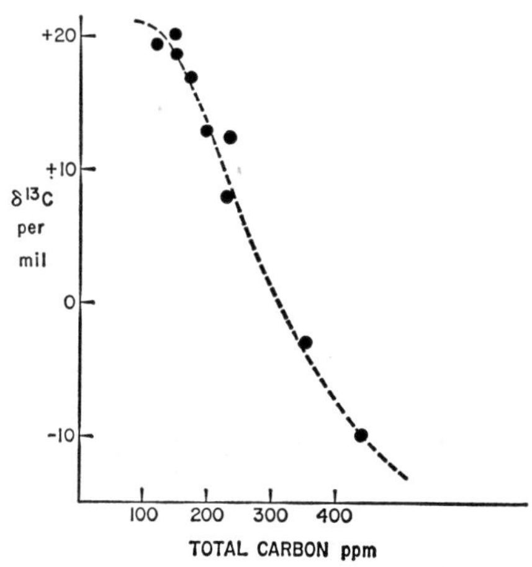

Fig. 1. Relation between the total carbon content of the bulk fines (10084 and 10086) and the carbon isotope abundance.

Table 2. $\delta^{13}C$ of carbon fractions resulting from pyrolysis and hydrolysis.

Experiment	Sample	Carbon content (ppm)	$\delta^{13}C$ (per mil)
Combustion of intact sample	10086	157*	+ 18.7*
	10084	132*	+ 19.1*
Combustion of residue from HCl–HF treatment	10086	30	− 7.9
	10086	42	− 4.3
Pyrolysis† at 150° to 250°C	10084	27	− 25.1
Pyrolysis at 250° to 500°C	10084	19	− 12.0
Pyrolysis at 500° to 750°C	10084	40	+ 23.4
Combustion of residue	10084	59	+ 9.4

* Average of values, in Table 1. † Refers to pyrolysis at the temperature given followed by combustion of volatile products.

bon concentration of this residue are considerably lower than their corresponding values in the residue resulting from pyrolysis. This, therefore, suggests that the nonvolatile carbon from the pyrolysis experiment is composed of two carbon phases, one being soluble in acid. This could be either a carbonate, or carbon dioxide captured within minerals that melt at high temperatures, or a carbide. There was little, if any, chemical evidence for carbonate resulting from acid treatment of the untreated bulk fines, but acid hydrolysis (6*N* HCl) at 105°C for 24 hours yielded measurable quantities of low-molecular-weight hydrocarbons from C_1 to C_4.

Experiments in collaboration with the Ames Research Center on cohenite (Fe_3C from an iron meteorite) indicate that some 20 ppm of carbide carbon would be required to produce the quantities of hydrocarbon gases measured. This carbide would need to have an isotopic composition of $\delta^{13}C$ greater than +20 per mil in order to be balanced by the residual carbon with a $\delta^{13}C$ of −4 to −8 per mil. Although the residue remaining after treatment with hydrochloric and hydrofluoric acids has not been identified, it seems logical to assume it is some form of elemental carbon.

The results of our investigation on carbon compounds can be summarized as follows. The highest concentration of carbon is in the bulk fines and in the breccias. There seems to be a somewhat heterogeneous distribution in the samples. Some of the scatter in the data on total carbon may be due to terrestrial contamination. The indigenous carbon is in three or four possible forms; elemental carbon, metal carbide, carbon dioxide (and possibly carbon monoxide and carbon-hydrogen compounds). The samples with the lowest carbon content are the fine–grained basaltic rocks, which also contained the isotopically lightest carbon. This may indicate that either heavier carbon (as CO_2) escaped during crushing and sample preparation or else that these samples contain only traces of indigenous carbon and are influenced most by terrestrial contamination.

Sulfur was analyzed by reacting samples (approximately 3 g) under vacuum with 85 percent phosphoric acid (10 ml). The reaction mixture was heated at intervals of a few hours to boiling and left standing overnight at 60°C. Evolved hydrogen sulfide was captured in side arms containing a 5 percent solution of silver nitrate and precipitated as silver sulfide. The Ag_2S was separated and then combusted with CuO to evolve SO_2 gas, which was purified by distillation from dry ice in preparation for mass spectroscopy. After the reaction had been completed, the residue was washed to remove excess H_3PO_4 and oxidized with aqua regia. No sulfate could be detected on addition of barium chloride to these solutions. In a separate experiment conducted at the Ames Research Center (*2*), 55 g of lunar sample was extracted with a mixture of methanol and benzene; elemental sulfur could not be detected in this extract. Likewise, no soluble sulfur was found either in the aqueous or acid-hydrolyzed products from this experiment. It is therefore concluded that all the sulfur is in the form of acid-volatile metal sulfides. The concentration of sulfide was inversely related to that of the carbon, highest in the fine-grained basalt and lowest in the bulk fines, the maximum being 2280 ppm as S (or 0.62 percent FeS).

The most surprising result is the enrichment of $\delta^{34}S$ (Table 1). The samples containing the highest sulfur approximate average meteoritic sulfur (*3*), whereas the bulk fines are enriched by 5 per mil or more.

There appears some inconsistency in the results, which have not yet been checked owing to lack of sample. A larger sample (55 g of 10086) on treatment with boiling hydrochloric acid (*2*) yielded H_2S which gave an average measured value for $\delta^{34}S$ of +8 per mil, oxidation of the same fines sample with aqua regia gave a similar $\delta^{34}S$ value of +8.2. However, treatment with phosphoric acid as described above yielded hydrogen sulfide with a $\delta^{34}S$ of only +5.4 per mil. In all cases the total sulfur content varied by only a few ppm. Comparative analyses of HCl and H_3PO_4 treatment of meteorities show $\delta^{34}S$ values for released sulfides that fall within ±1 per mil.

The results of analyses on a large number of meteorites show the isotopic spread rarely to exceed ±3 per mil for a variety of sulfur compounds. Troilite sulfur in iron meteorites, for instance, did not vary beyond the range of ±1 per mil relative to Canyon Diablo troilite.

The isotopic values for $\delta^{13}C$ in the breccias and fines are also anomalously heavy compared to total meteoritic carbon range −4 to −25 per mil (*4*) and to terrestrial carbon with a normal range of +2 to −30 per mil. They are, however, not as enriched in $\delta^{13}C$ as the carbonate phase of meteorites, which appears to lie in the range of +40 to +70 per mil (*5*).

Therefore, it appears that the lunar fines are anomalously enriched in heavy isotopes, relative to the compacted fine-grained rocks, and to terrestrial volcanic rocks and meteorites.

What process is responsible? It must be either one preferentially enriching the heavy isotopes or one preferentially removing the light isotopes. Equilibration fractionation can probably be ruled out in the case of sulfur, as probably only one major phase exists—troilite. In any case, equilibrium effects in a closed system could not account for an enrichment of the total system.

This would therefore suggest addition of heavy isotopes to the fine material and brecciated rocks or removal of light isotopes. In the absence of any evidence for the former (*6*), we prefer to speculate that isotopically light carbon has been preferentially lost from the unconsolidated or weakly consolidated rocks. One possible mechanism for this behavior could be the stripping of carbon and sulfur by protons from the solar wind. Kinetically, this may favor the conversion of ^{12}C to methane and ^{32}S to hydrogen sulfide, which would be lost from the moon's surface.

I. R. KAPLAN
J. W. SMITH*

Institute of Geophysics and Planetary Physics, University of California, Los Angeles

References and Notes

1. Y. Bottinga, *Geochim. Cosmochim. Acta* **33**, 49 (1969).
2. C. Ponnamperuma, E. Barghoorn, I. Breger, S. Chang, A. Duffield, C. Gehrke, B. Halpern, G. Hodgson, H. Holland, R. Johnson, I. Kaplan, K. Keil, K. Krauskopf, K. Kvenvolden, D. Philpott, G. Pollock, W. Schopf, J. Smith, *Science*, this issue.
3. I. R. Kaplan and J. R. Hulston, *Geochim. Cosmochim. Acta* **30**, 479 (1966).
4. G. Boato, *ibid.* **6**, 209 (1954); T. Belsky and I. R. Kaplan *ibid.*, in press.
5. R. N. Clayton, *Science* **140**, 192 (1963); J. W. Smith, I. R. Kaplan, E. Ruth, unpublished data.
6. Lunar module exhaust from a simulated preflight experiment yielded a value for $\delta^{13}C$ of —34.7 per mil.

7. Supported by NASA contract NAS 9-8843. We thank E. Ruth for valuable assistance, and Drs. C. Ponnamperuma, S. Chang, J. Lawless and K. Kvenvolden for collaboration.

* Permanent address: Division of Mineral Chemistry, C.S.I.R.O., P.O. Box 175, Chatswood, New South Wales, Australia.

4 January 1970

Rare Gases, Hydrogen, and Nitrogen: Concentrations and Isotopic Composition in Lunar Material

Abstract. *The concentrations and isotopic abundances of the rare gases have been investigated in fines and three types of rocks. The results obtained from different grain-size fractions and from samples etched to different degrees with nitric and hydrofluoric acids demonstrate the strong concentrations of the solar-wind component in the surface layers of the grains. Exposure ages as well as gas retention ages have been determined in different types of Apollo 11 material. Hydrogen, nitrogen, and other gases have been analyzed by a high-resolution mass spectrometer. As compared with that in terrestrial water, deuterium is depleted by at least a factor of 3 in the investigated type C rocks.*

We have investigated the concentrations and isotopic abundances of the rare gases in fines and in three types of rocks. The results obtained are listed in Table 1.

A systematic decrease of the concentrations of all rare gases has been found with increasing grain size and with increasing loss of material in the etching experiments (Table 1; Figs. 1, 2). The rare gases in the fines are obviously concentrated predominantly in the outer grain layers. Among the light rare gases, Ar shows the greatest and He the smallest decrease of concentration with increasing weight of removed material (Fig. 2), which may be due either to a different distribution or different losses of these nuclides by diffusion or to different penetration depths for He, Ne, and Ar solar wind particles. On the basis of our experience with gas-rich stone meteorites, where the metal fraction invariably shows the least (or no) loss by diffusion, and of a comparison of the $^{4}He/^{20}Ne$ or $^{4}He/^{36}Ar$ ratios in these etched samples with ratios found in the selected iron grains, we believe that preferential loss of the He and Ne by diffusion is responsible. The Kr and Xe concentrations depend similarly on the grain size and on the amount of material removed by etching.

The abundance ratios with one nuclide abundantly formed by spallation and the other abundantly occurring in the solar wind (for instance, $^{3}He/^{4}He$, $^{21}Ne/^{20}Ne$, and $^{38}Ar/^{36}Ar$) show a strong increase with increasing grain size and with increasing degree of etching (Fig. 3).

In the 1 percent residuum of the strongly etched sample the ^{21}Ne is almost purely spallogenic. The amount, together with a production rate of $P_{21} = 0.2 \times 10^{-8}$ cm^3/g (at standard temperature and pressure) times 10^6 years yields an exposure age of $T_{21} = 215 \times 10^6$ years. Furthermore, as the ^4He in this fraction is almost pure solar wind helium, the ^3He can be corrected to obtain the spallogenic component, which is 2.7×10^{-6} cm^3/g at standard temperature and pressure. Hence, using a production rate of $P_3 = 1.3 \times 10^{-8}$ cm^3/g (at standard temperature and pressure) times 10^6 years, we obtain the ^3He exposure age of $T_3 = 207 \times 10^6$ years, in good agreement with the ^{21}Ne exposure age.

The rare gas contents of type A, B, and C rocks are compiled in Table 2. For type C rock 10021.20, similar values have been found as for fines. In type A and B rocks all three components (solar wind, radiogenic, and spallogenic) have been detected. The results for radiogenic ^4H and ^{40}Ar, as well as for spallogenic ^3H and ^{21}Ne, are shown in the table. Exposure ages have been calculated, with production rates $P_3 = 1 \times 10^{-8}$ cm^3/g times 10^6 years and $P_{21} = 0.12 \times 10^{-8}$ cm^3/g

Table 1. Concentrations of rare gas in cubic centimeters per gram at standard temperature and pressure in the bulk material in various grain-size fractions and in acid-etched fractions of fines of sample 10084.18. The concentrations in acid-treated fines are given after 40 percent of the weight has been removed by HNO_3, 93 percent has been removed by HNO_3 and HF, and 99 percent has been removed by acid treatment. The concentrations in a few large grains (>250 μm) selected from the residuum of an acid-etched sample and those in a few milligrams of iron grains selected with a hand magnet are also listed.

Nuclide and ratios	Original fines	Grain-size fractions (μm)			Acid-etched fractions (removed weight in %)			Large grains	Iron grains
		30	30–100	100–250	40	93	99		
^3He	7.88×10^{-5}	1.06×10^{-4}	3.98×10^{-5}	1.80×10^{-5}	3.13×10^{-5}	5.83×10^{-6}	3.07×10^{-6}	2.75×10^{-6}	3.87×10^{-5}
^4He	2.01×10^{-1}	2.78×10^{-1}	1.02×10^{-1}	4.70×10^{-2}	8.72×10^{-2}	1.02×10^{-2}	9.51×10^{-4}	6.84×10^{-3}	9.80×10^{-2}
^{20}Ne	2.21×10^{-3}	3.34×10^{-3}	1.18×10^{-3}	5.85×10^{-4}	5.28×10^{-4}	4.04×10^{-5}	3.14×10^{-6}	3.35×10^{-5}	3.34×10^{-4}
^{21}Ne	5.65×10^{-6}	8.78×10^{-6}	3.42×10^{-6}	1.93×10^{-6}	1.76×10^{-6}	3.59×10^{-7}	4.37×10^{-7}	2.04×10^{-7}	9.44×10^{-7}
^{22}Ne	1.74×10^{-4}	2.69×10^{-4}	2.44×10^{-4}	4.77×10^{-5}	4.31×10^{-5}	3.57×10^{-6}	7.29×10^{-7}	2.82×10^{-6}	2.73×10^{-5}
^{36}Ar	3.71×10^{-4}	6.22×10^{-4}	2.36×10^{-4}	1.63×10^{-4}	3.05×10^{-5}	1.73×10^{-6}	9.46×10^{-7}	1.71×10^{-6}	5.11×10^{-5}
^{38}Ar	7.21×10^{-5}	1.20×10^{-4}	4.72×10^{-5}	3.22×10^{-5}	6.44×10^{-6}	6.96×10^{-7}	5.69×10^{-7}	4.46×10^{-7}	1.03×10^{-5}
^{40}Ar	4.09×10^{-4}	6.48×10^{-4}	3.04×10^{-4}	2.36×10^{-4}	6.60×10^{-5}	4.33×10^{-5}	2.76×10^{-5}	9.09×10^{-6}	7.69×10^{-5}
^{84}Kr	2.11×10^{-7}	3.34×10^{-7}	1.26×10^{-7}	8.87×10^{-8}	2.53×10^{-9}	2.3×10^{-9}	4.96×10^{-9}	1.62×10^{-9}	3.07×10^{-8}
^{132}Xe	2.87×10^{-8}	3.97×10^{-7}	1.82×10^{-7}	1.59×10^{-8}	6.09×10^{-9}	5.7×10^{-10}	1.13×10^{-9}	7.0×10^{-10}	5.5×10^{-9}
$^4He/^3He$	2550	2620	2560	2480	2790	1760	310	2450	2530
$^{20}Ne/^{22}Ne$	12.7	12.4	12.5	12.3	12.2	11.3	4.31	11.9	12.4
$^{22}Ne/^{21}Ne$	30.0	30.6	27.6	24.7	24.5	10.0	1.67	13.8	28.9
$^{36}Ar/^{38}Ar$	5.15	5.18	5.00	5.06	4.74	2.43	1.66	3.83	4.96
$^{40}Ar/^{36}Ar$	1.10	1.04	1.29	1.45	2.16	25	29.2	5.32	1.50

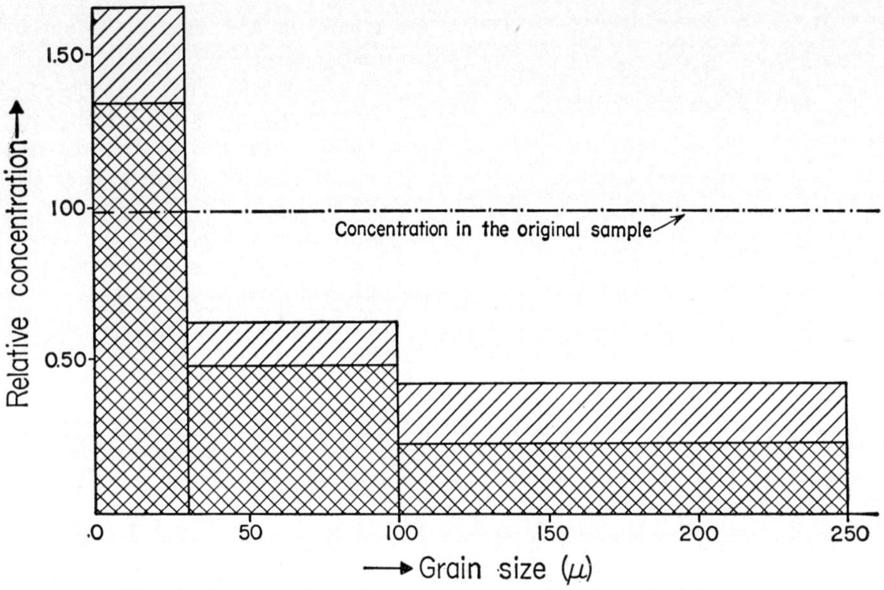

Fig. 1. Concentration of rare gases as a function of grain size.

Table 2. Rare gas concentrations in cubic centimeters per gram in rocks of type A, B, and C.

Nuclide ratios and ages	Type C	Type B			Type A	
	10021.20	10003.11	10017.33	10044.32	10049.20	10057.40
^{3}He	3.27×10^{-5}	1.13×10^{-6}	2.92×10^{-6}	6.37×10^{-7}	2.84×10^{-7}	4.39×10^{-7}
^{4}He	9.05×10^{-2}	2.24×10^{-4}	4.91×10^{-4}	1.49×10^{-4}	8.89×10^{-4}	6.76×10^{-4}
^{20}Ne	1.25×10^{-3}	1.05×10^{-6}	7.75×10^{-7}	1.18×10^{-7}	1.64×10^{-6}	5.49×10^{-7}
^{21}Ne	3.50×10^{-6}	1.29×10^{-7}	4.66×10^{-7}	6.71×10^{-8}	3.15×10^{-8}	5.83×10^{-8}
^{22}Ne	1.00×10^{-4}	2.13×10^{-7}	5.59×10^{-7}	8.38×10^{-8}	1.61×10^{-7}	7.95×10^{-8}
^{36}Ar	1.85×10^{-4}	2.86×10^{-7}	4.65×10^{-7}	9.54×10^{-8}	3.03×10^{-7}	1.16×10^{-7}
^{38}Ar	3.64×10^{-5}	2.20×10^{-7}	6.46×10^{-7}	1.24×10^{-7}	1.02×10^{-7}	8.15×10^{-8}
^{40}Ar	5.63×10^{-4}	2.63×10^{-5}	4.95×10^{-5}	4.47×10^{-5}	6.47×10^{-5}	4.19×10^{-5}
^{84}Kr	8.5×10^{-8}	2.8×10^{-10}	7.2×10^{-10}	2.0×10^{-10}	5.4×10^{-10}	4.2×10^{-10}
^{132}Xe	2.2×10^{-8}	9.8×10^{-11}	3.5×10^{-10}	1.1×10^{-10}	3.1×10^{-10}	1.5×10^{-10}
$^{4}He/^{3}He$	2770	197	168	234	3130	1540
$^{20}Ne/^{22}Ne$	12.6	4.93	1.39	1.41	10.2	6.91
$^{22}Ne/^{21}Ne$	28.4	1.65	1.20	1.25	5.11	1.36
$^{36}Ar/^{38}Ar$	5.08	1.20	0.72	0.77	2.97	1.42
$^{40}Ar/^{36}Ar$	3.04	92	106	496	214	341
^{3}He(spal)		1.1×10^{-6}	2.91×10^{-6}	6.37×10^{-7}	2.30×10^{-7}	4.22×10^{-7}
^{21}Ne(spal)		1.29×10^{-7}	4.64×10^{-7}	6.70×10^{-8}	2.7×10^{-8}	5.69×10^{-8}
^{4}He(rad)	4.5×10^{-3}	1.41×10^{-4}	4.63×10^{-4}	1.44×10^{-4}	7.45×10^{-4}	6.32×10^{-4}
^{40}Ar(rad)	3.7×10^{-4}	2.60×10^{-5}	4.95×10^{-5}	4.46×10^{-5}	6.47×10^{-5}	4.19×10^{-5}
K (ppm)			2060	860	2280	2010
U (ppm)			0.69	0.33	0.81	0.80
Th (ppm)			3.05	1.0		3.05
T_3 (10^6 yr)		110	291	64	23	42
T_{21} (10^6 yr)		108	387	56	22.5	47
T_4 (10^6 yr)			2550	1750	2950	2500
T_{40} (10^6 yr)			2600	3900	3200	2850

Table 3. Exposure ages and gas retention ages of Apollo 11 fines and rocks in million years.

Elements	Fines	Rocks type B			Rocks type A	
	10084	10003.11	10017.33	10044.32	10049.20	10057.40
			Exposure ages			
^{3}He	207	110	291	64	23	42
^{21}Ne	215	108	387	56	22,5	47
			Gas retention ages			
U/Th			2550	1750	2950	2500
K/Ar			2600	3900	3200	2850

times 10^6 years. The ^{3}He and ^{21}Ne exposure ages for individual samples agree within the experimental limits of error, but for different samples they differ by more than an order of magnitude, ranging from 20 to 400 × 10^6 years.

The Th/U ratio has been determined for three of the rock samples, and the average value of Th/U equal to 3.8 has been used in the calculation of the ages of the other samples as well. With the exception of rock type B 10044, where the U-Th/He age is more than a factor of 2 lower than the K/Ar age, both gas retention ages are concordant (Table 2). (The relative abundance of ^{40}K has been determined to agree within 1 percent with the value for a sample of terrestrial reagent potassium.) The Kr and Xe found in type A and B rocks possess a strong spallogenic component.

In addition to the rare gases, the concentrations and isotopic composition of H_2 and N_2 have been measured in a few selected rock specimens. Pulverized samples of 250 mg of the type C rock 10021.20 were heated in a Pt-crucible suspended within a quartz tube. The gases released were pumped continuously by means of a mercury diffusion pump into a Toepler pump to determine the total amount. Typically, 2.6 to 3.0 cm^3/g (at standard temperature and pressure) were given off by the samples upon heating to 1000°C, while additional amounts of less than 0.1 cm^3/g were released when the samples were subsequently heated to the melting point.

A partial analysis by mass spectrometry of the gases from one of the 250-mg samples gave the following results at standard temperature and pressure:

H_2 34.4%; 1.03 cm^3/g
He 17.4%; 0.52 cm^3/g; H/He≈4
CO 13.1%; 0.39 cm^3/g
CO_2 5.0%; 0.15 cm^3/g
N_2 3.5%; 0.106 cm^3/g; H/N≈10
Ar 0.092%; 0.0027 cm^3/g

The residual pressure was due to H_2O, NO, Ne, and minor components. Some isotopic ratios were $^{20}Ne/^{22}Ne$ = 13.2 ± 0.3, $^{36}Ar/^{38}Ar$ = 5.6 ± 0.2, $^{40}Ar/^{36}Ar$ = 2.65 ± 0.10, and $^{13}C/^{12}C$ ≈ 0.012.

To determine the isotopic composition of the hydrogen released from the samples a mass spectrometer (Varian Mat CH5) with a molecular flow inlet system and secondary electron multiplier was used. It permitted complete resolution of the triplet ^{3}He-HD-H_3. Hence, no treatment or operation of the

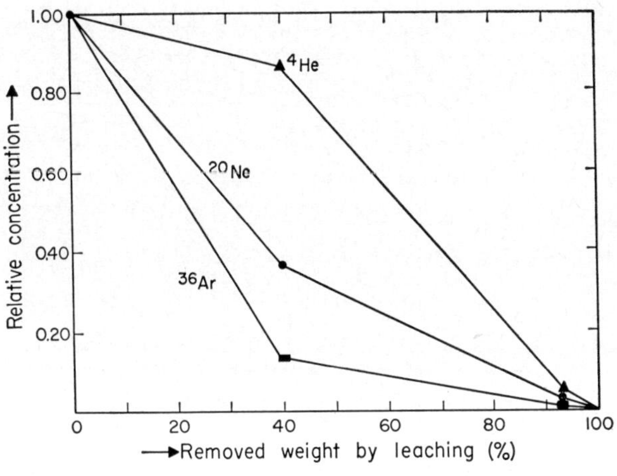

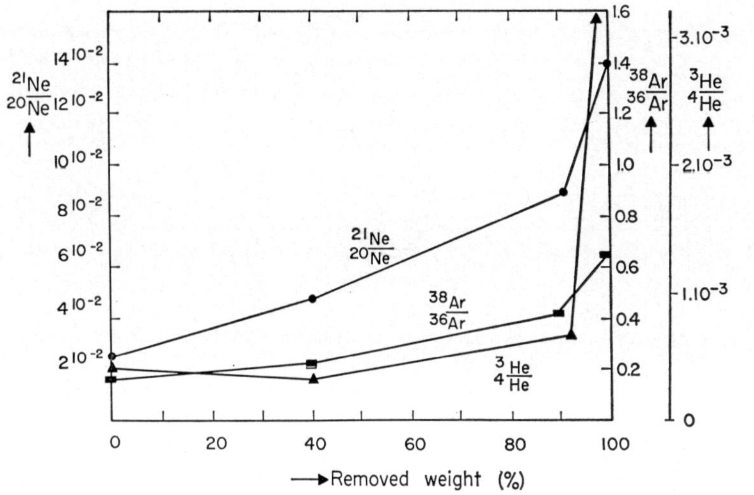

Fig. 2 (left). Relative concentration of rare gases as a function of material removed by leaching. Fig. 3 (right). Abundance ratios of rare gases as a function of material removed by leaching.

gases was required, but the accuracy of the isotopic ratio was still sufficient to measure the rather large deviations from terrestrial standard hydrogen.

Standard hydrogen samples were prepared from distilled tap water, from standard mean ocean water (SMOW) and standard normal ocean water (SNOW) (*2*) by reducing it over uranium at 700°C, and from cylinder hydrogen as well. Their D/H ratios were determined with a mass spectrometer (Varian Mat GD 150) with viscous gas inlet and double collector system. The absolute D/H ratios ranged from 41×10^{-6} for cylinder hydrogen to 156×10^{-6} for SMOW hydrogen (*1*). The standards and the gas samples from two different extractions of 250 mg of rock 10021.20 were alternately run with the CH5 spectrometer. Its resolution was 2500 (10 percent valley definition) at mass number 3. Corrections were made to allow for background and molecular flow. The results obtained from the two sets of measurements were:

$$(D/H)_1 = (1.08 \pm 0.30) \times (D/H)_{H\ cylinder} = (44 \pm 12) \times 10^{-6}$$

$$(D/H)_2 = 1/(3.5 \pm 0.4) \times (D/H)_{SMOW} = (45 \pm 5) \times 10^{-6}$$

The procedure was tested by also determining the $^4He/^3He$ ratio in the Apollo 11 gas as well as in an artificial H_2-^3He-^4He-mixture. The result thus obtained for He from rock 10021.20 is $^4H/^3He = 2660 \pm 200$, in reasonable agreement with independent determinations.

For several reasons the D/H ratio of 45×10^{-6} cannot be expected to reflect the exact solar ratio. It nevertheless shows quite clearly that on the solar surface the deuterium is depleted by at least a factor of three compared with SMOW in accordance with astronomical observations (*2*) and current theories (*3*). For example, the elemental ratios H/He = 4 and H/N = 10 will be even less representative for the solar surface or the solar wind. The abundance ratios of the rare gases have been already altered significantly by diffusion losses alone, and chemical properties will cause additional perturbations. An indication for this was found in a leaching experiment on lunar fines. Upon treatment of a sample with diluted sulfuric acid, less than 10 percent of the total nitrogen (about 100 ppm) evolves as gaseous N_2, while more than 65 percent is leached out in the form of NH_4^+, possibly indicating the existence of nitrides in the lunar material. An additional 15 percent was present as nitrate.

To get an idea about larger anomalies in the $^{15}N/^{14}N$ ratio another gas sample extracted from 250 mg of 10021.20 was run through a gas chromatograph; the isotopic ratio was determined in the nitrogen fraction by the CH5 mass spectrometer. Within the limits of error of this preliminary investigation the $^{15}N/^{14}N$ ratio agreed with that of a terrestrial N_2 sample.

H. HINTENBERGER
H. W. WEBER, H. VOSHAGE
H. WÄNKE, F. BEGEMANN
E. VILSCEK, F. WLOTZKA
Max Planck Institut für Chemie (Otto Hahn Institute), Mainz, Germany

References and Notes

1. The absolute D/H values are based on data for SMOW and SNOW by E. Roth, *International Conference on Mass Spectroscopy*, Kyoto, Japan, 8-12 September 1969.
2. T. D. Kinman, *Mon. Not. Roy. Astron. Soc.* **116**, 77 (1956).
3. W. A. Fowler, J. L Greenstein, H. Hoyle, *Geophys. J.* **6**, 148 (1962); R. Bernas, E. Gradsztajn, H. Reeves, E. Schatzman, *Ann. Phys.* **44**, 426 (1967).
4. We thank C. Müller, G. König, K. F. Roth, and S. Specht for cooperation. The gas chromatographic separation of a nitrogen fraction was performed by E. Jagoutz and K. Schneider. We thank NASA and the Bundesministerium für Bildung und Wissenschaft for making available lunar material for this investigation.

4 January 1970

Isotopic Analysis of Rare Gases from Stepwise Heating of Lunar Fines and Rocks

Abstract. *Highlights of a first effort in sorting out rare gases in lunar material are solar wind rare gases in abundance; variable $^{20}Ne/^{22}Ne$ but constant $^{21}Ne/^{22}Ne$ ratios in fractions of the trapped neon; cosmogenic rare gases similar to those found in meteorites, except for copious ^{131}Xe in one rock but not in another; at Tranquillity Base a rock 4.1×10^9 years old which reached the surface 35 to 65 million years ago, amid soil whose particles have typically been within a meter of the surface for 10^9 years or more.*

This paper describes the results of stepwise heating of three lunar samples. At each of a series of successively higher temperatures, all the rare gases evolved were examined in a glass mass spectrometer. The programmed heating was continued beyond the melting points of the samples until all gas had

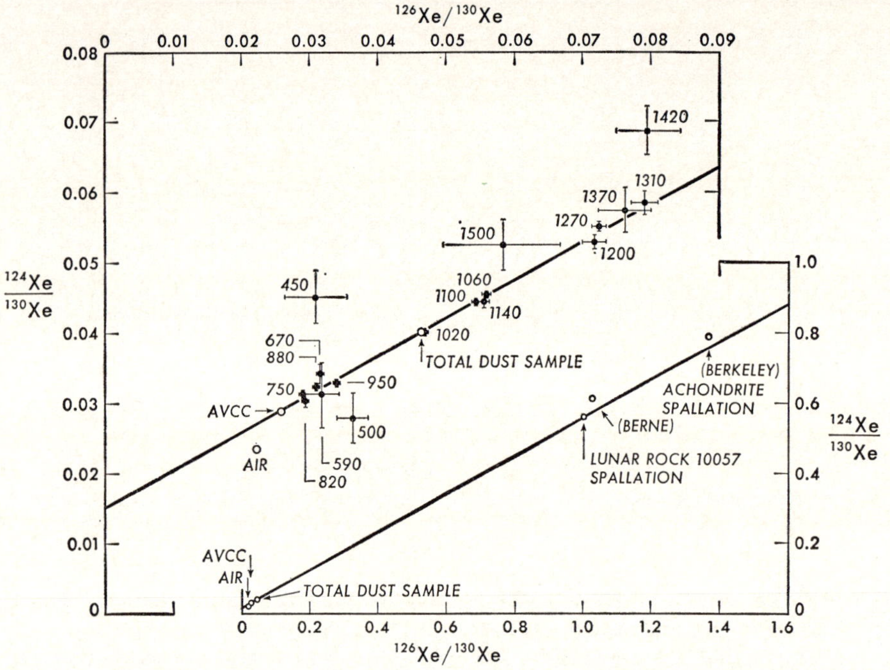

Fig. 1. Correlation of ratios $^{124}Xe/^{130}Xe$ and $^{126}Xe/^{130}Xe$ in lunar fines. The numbers are release temperatures in degrees Celsius for half-hour heatings. These isotopes belong to a two-component system. One component appears to have the AVCC ratios. The other is lunar spallation xenon as deduced from rock 10057. Note the similarity to spallation xenon in achondritic meteorites.

been extracted. Half-gram chips of rocks 57 (fine-grained crystalline) and 44 (coarser-grained crystalline) were studied in one system. A 0.09-g sample of lunar fines (sample 84) was studied in a second system. The lunar material was received at Berkeley in sealed, nitrogen-filled containers. All samples were handled in nitrogen; they were loaded into the vacuum systems without exposure to the atmosphere.

Attention is focused here on resolving the rare gas mass spectra into constituent components. As in meteorite studies, our basic technique has been to plot correlation diagrams for appropriate pairs of isotope ratios in the various temperature fractions from a sample. Two-component mixtures are characterized by straight lines on such a diagram, spanning end points corresponding to the two components. If the end points can somehow be inferred, relative concentrations of the two components can be deduced for each temperature fraction. Departures from a straight line indicate additional components. In particular, concentrations of additional monoisotopic components are easily deduced. The method of analysis should become clear upon discussion of examples.

Xenon in the lunar fines was detected in large concentrations and with an isotopic composition close to that predicted for solar-wind xenon by Marti (*1*) from his studies of xenon in the gas-rich achondrite Pesyanoe. The solar xenon is depleted in ^{134}Xe and ^{136}Xe relative to xenon seen in the carbonaceous chondrites. Our stepwise heating of the fines revealed an additional cosmogenic (cosmic-ray produced) xenon component, as shown in Fig. 1, which is a correlation plot of $^{124}Xe/^{130}Xe$ against $^{126}Xe/^{130}Xe$. The points define a line which spans the compositions of average-carbonaceous-chondrite (AVCC) xenon and spallation-type xenon as deduced from the 1200°C fraction from rock 57. The proximity of the most "southwesterly" points on this plot to the AVCC composition strongly suggests that the solar xenon has the AVCC ratios for the light xenon isotopes, although we have used the composition of the 820°C fraction in the fines for solar xenon in calculations in this paper. All numerical results in this paper are presented in Tables 1 and 2.

Neon both in the fines and in rock 57 is dominated by an abundant trapped component. Figure 2 shows the isotopic composition in the various temperature fractions. Although there is cosmogenic neon present, the spectra are dominated by neon for which the $^{21}Ne/^{22}Ne$ is relatively constant (at a value only slightly greater than the terrestrial value) but the $^{20}Ne/^{22}Ne$ is highly variable. Noncosmogenic neon in any fraction will have a composition somewhere "northwest" of the observed composition on the line joining that point with point *C*. On this basis the $^{20}Ne/^{22}Ne$ ratio for noncosmogenic neon in rock 57 varies from $\leqq 8.3$ to $\geqq 13.9$; similarly, the ratio in lunar fines

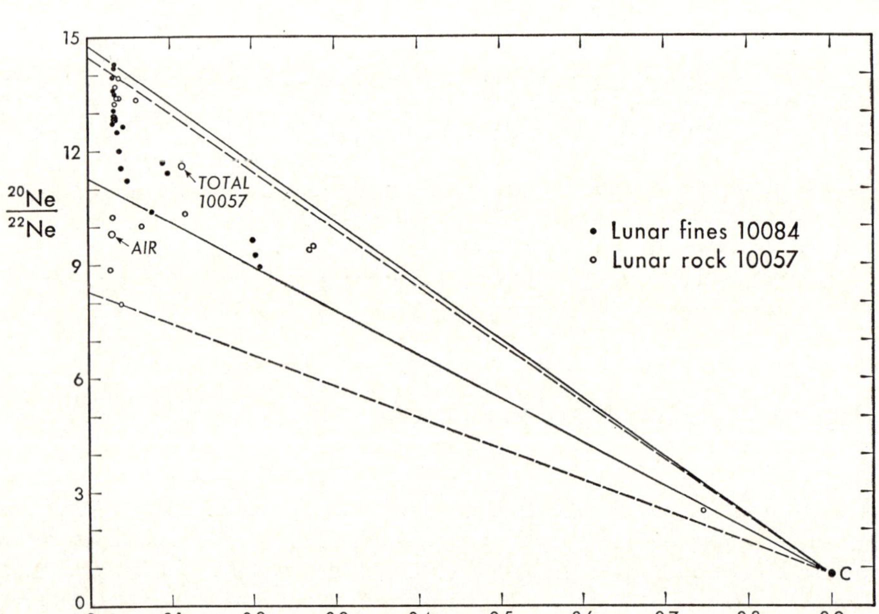

Fig. 2. Isotopic composition of neon fractions released in stepwise heating of lunar fines and rock 10057. The point *C* locates the composition of cosmogenic neon seen in a typical stone meteorite. The noncosmogenic neon composition in any fraction must lie toward the left on the line joining the observed composition with point *C*. Such lines have been drawn for the fractions of extreme composition. The trapped compositions are characterized by large variations in ratio $^{20}Ne/^{22}Ne$ but a nearly constant ratio for $^{21}Ne/^{22}Ne$.

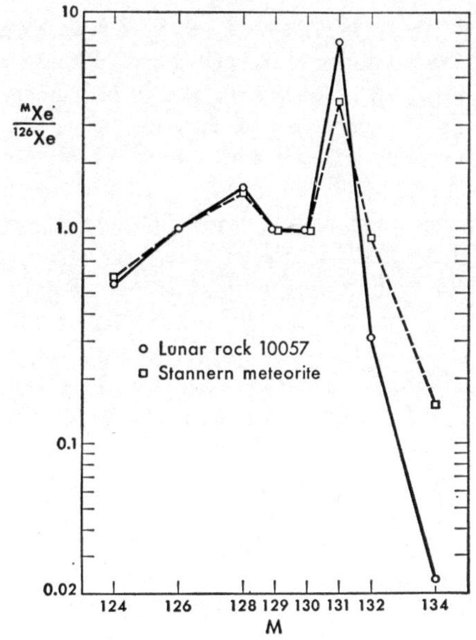

Fig. 3. Comparison of the mass spectrum of cosmogenic xenon in lunar rock 10057 with spallation xenon from the Stannern achondritic meteorite. A large and significant difference is seen for mass 131.

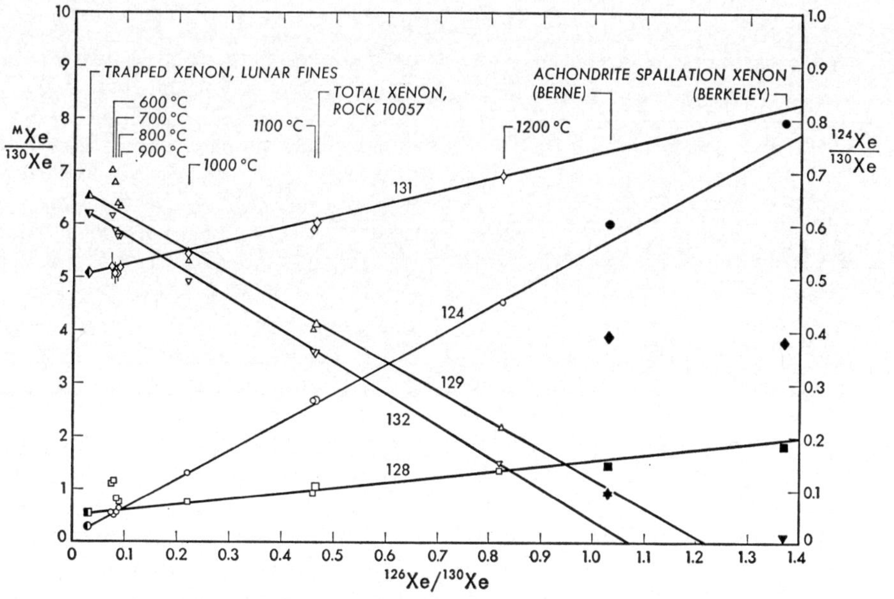

Fig. 4. Graphical analysis of the xenon released in stepwise heating of lunar rock 10057. Small corrections for fission have been made at masses 131 and 132. Isotopes 124, 126, 130, and 132 appear to belong to a simple two-component system (trapped and spallation xenon). Isotope 131 fits a two-component model but with a highly anomalous composition for the cosmogenic component. Isotopes 128 and 129 display excess amounts from 600° to 900°C inclusive.

ranges from ≦11.3 to ≧14.3. Effects of this type are known in stone meteorites (*2*), but have never been adequately explained. Their occurrence in lunar material is one of the most striking results of our study and places constraints upon future theories for their explanation.

Cosmogenic xenon in rock 57 presents a generally consistent picture. The 1200°C fraction is most strongly cosmogenic and has been used to deduce the spectrum. It was assumed that this fraction is a mixture of cosmogenic xenon and xenon with the composition seen at 820°C in the lunar fines. It was further assumed that there is no ^{136}Xe in the cosmogenic fraction. The cosmogenic spectrum is then easily obtained by subtraction. It fits (Fig. 3) the spallation xenon spectrum from the Stannern calcium-rich achondrite, as deduced by Marti *et al.* (*3*), more closely than it fits other published spallation spectra. The most striking difference between the spectra is at mass 131. Rock 57 contains an extraordinary excess of this isotope both in terms of absolute concentration and ratios! The differences at masses 132 and 134, which are small

Table 1. Rare gases from three lunar samples. The gas fractions for the stepwise heatings have been summed. All gas concentrations (boldface) are in units of 10^{-8} cm^3/g (STP). Errors in isotope ratios for total gas are about 1 percent. Errors in concentrations for total gas are about 10 percent.

Isotope	Lunar fines, sample 10084,59			Fine-grained crystalline rock, sample 10057,20					Medium-grained crystalline rock, sample 10044,20
	Total	Solar	Cosmic	Total	Trapped	Cosmic	Fission	Excess	Total
^{3}He/^{4}He	0.00047*			0.000545*†					
[^{4}He]	**2.9×10^7**			**83000†**					
^{20}Ne/^{22}Ne	12.85	12.90		11.58					3.72
^{21}Ne/^{22}Ne	0.0332	0.032		0.1152					0.710
[^{22}Ne]	**41000**	**41000**	>360	**81**	**73**	**8.1**			**10.3**
^{38}Ar/^{36}Ar	0.1925	0.183		0.237					
^{40}Ar/^{36}Ar	1.126	<0.82		39.1					
[^{36}Ar]	**50600**	**50200**	**400**	**388**	**373**	**15**			
^{78}Kr/^{82}Kr	0.0348	0.0315		0.10		0.23			0.164
^{80}Kr/^{82}Kr	0.2064	0.201		0.45		0.68	**0.0043‡**		0.486
^{81}Kr/^{82}Kr						0.0068			
^{83}Kr/^{82}Kr	1.003	0.997		1.06		1.35			1.10
^{84}Kr/^{82}Kr	4.91	4.91		3.57		0.47			2.91
^{86}Kr/^{82}Kr	1.503	1.496		1.02		0			0.786
[^{82}Kr]	**7.6**	**7.5**	**0.10**	**0.034**	**0.023**	**≡0.011**			**0.027**
^{124}Xe/^{130}Xe	0.0400	0.0302		0.265		0.57			0.59
^{126}Xe/^{130}Xe	0.0464	0.0293		0.465		1.03			0.94
^{128}Xe/^{130}Xe	0.530	0.515		1.048		1.61		**0.00031‡**	1.47
^{129}Xe/^{130}Xe	6.28	6.38		4.11		1.03		**0.00027‡**	2.85
^{131}Xe/^{130}Xe	4.98	4.98		6.05		7.50			4.55
^{132}Xe/^{130}Xe	5.97	6.06		3.58		0.32	**0.00008‡**		1.96
^{134}Xe/^{130}Xe	2.20	2.22		1.283		0.020	**0.00012‡**		
^{136}Xe/^{130}Xe	1.78	1.79		1.040		≡0	**0.00013‡**		0.439
[^{130}Xe]	**0.77**	**0.76**	**0.010**	**0.0052**	**0.0029**	**0.0023**			**0.0015**

*No discrimination correction was available for ^{3}He/^{4}He ratio. † Includes estimate for one gas fraction which was lost. ‡ Value is a concentration (not a ratio) for the isotope in the numerator for the row.

Table 2. Exposure ages of two lunar samples. Sample 10087,59 is lunar fines; sample 10057,20 is fine-grained crystalline rock.

Exposure age type	10084,59 (10^6 years)	10057,20 (10^6 years)
^{38}AR	3020	125
^{82}Kr	1600	64
^{130}Xe	720	63

in terms of absolute gas concentration, can perhaps be attributed to fission effects in the Stannern meteorite.

A comprehensive display of the xenon isotope variations in rock 57 is given in Fig. 4, where ratios ^{M}Xe/^{130}Xe are plotted against ^{126}Xe/^{130}Xe. For M equal to 124, 131, and 132, the points fall reasonably well on lines joining the 1200°C composition and the composition seen at 820°C in the lunar fines. Small corrections have been made for a fission contribution at masses 131 and 132 (see the following paragraph). Otherwise, these isotopes, with ^{126}Xe and ^{130}Xe, appear to belong to a two-component system of trapped and cosmogenic xenon. For M equal to 128 and 129, the values observed between 600° and 900°C fall above the lines containing the other points, as would occur if there were an excess of ^{128}Xe and ^{129}Xe being released at these temperatures. The correlation lines in Fig. 4 pass close to the compositions for cosmogenic xenon in calcium-rich achondrites, except at mass 131, where the correlation line observed has the opposite slope from that observed in meteorites.

Fission xenon in rock 57 was indicated, after subtraction of cosmogenic xenon, by higher values of ^{136}Xe/^{132}Xe and ^{134}Xe/^{132}Xe than in the lunar fines. The amount of fission xenon observed is approximately what should accumulate in 4.1 × 10^9 years (see below) of uranium decay in rock 57 (780 ppb uranium).

Argon from the samples is a mixture of radiogenic ^{40}Ar from ^{40}K decay, cosmogenic argon, and trapped argon, which can doubtless be ascribed to the solar wind. Values of the ratio ^{40}Ar/^{36}Ar as low as 0.82 were observed in argon fractions from the lunar fines, and provide an upper limit for this ratio in the solar wind. The much higher values of ^{40}Ar/^{36}Ar in gas fractions from rock 57 have to be attributed to ^{40}Ar from potassium decay. On that basis we compute a K-Ar age of 4.1 ± 0.3 × 10^9 years for rock 57, using the preliminary value (*5*) for the potassium content of the rock.

Krypton exhibits a cosmogenic component in all samples, and an excess of ^{80}Kr in rock 57. Radioactive ^{81}Kr was observed in the highly cosmogenic 1200°C fraction of rock 57, permitting calculation of a ^{81}Kr/^{80,82}Kr age (*6*) of 37 million years for exposure of the sample to galactic cosmic rays. Crude exposure ages (Table 2) were calculated for rock 57 and for the fines from the cosmogenic contents of ^{38}Ar, ^{82}Kr, and ^{130}Xe. We used the production rates from Munk's (*7*) study of the calcium-rich achondrite Nuevo Laredo for this purpose, assuming targets of calcium, strontium, and barium, respectively. These ages for rock 57 exceed the ^{81}Kr/^{80,82}Kr age as they should; the results would agree only if that rock was abruptly shifted to its final location from a completely shielded location. The ^{38}Ar exposure ages are older than the others. For the soil, the ages represent an average for the fine-grained material, but much of the present surface soil material at Tranquillity Base has been within a meter or so of the surface for 10^9 years or so, according to our data.

Although we have not completely analyzed our results from rock 44, one important fact is immediately evident: in rock 44 the large excess of ^{131}Xe in the cosmogenic component is absent.

J. H. Reynolds
C. M. Hohenberg
R. S. Lewis, P. K. Davis
W. A. Kaiser

Department of Physics,
University of California, Berkeley

References and Notes

1. K. Marti, *Science* **166**, 1263 (1969).
2. Reviewed by R. O. Pepin, *Earth Planet. Sci. Lett.* **2**, 13 (1967).
3. K. Marti, P. Eberhardt, J. Geiss, *Z. Naturforsch.* **21a**, 398 (1966).
4. G. W. Wetherill, *Phys. Rev.* **92**, 907 (1953).
5. Lunar Sample Preliminary Examination Team, *Science* **165**, 1211 (1969).
6. K. Marti, *Phys. Rev. Lett.* **18**, 264 (1967); O. Eugster, P. Eberhardt, J. Geiss, *Earth Planet. Sci. Lett.* **2**, 77 (1967).
7. M. N. Munk, *Earth Planet. Sci. Lett.* **3**, 457 (1967).
8. We thank G. McCrory for assistance in the experimental work. Partial support from NASA and from the AEC; AEC code number UCB-34P32-71.

6 January 1970

Solar Wind Gases, Cosmic Ray Spallation Products, and the Irradiation History

Abstract. *The isotopic abundances of the rare gases in the fines are found to be similar to those previously reported for gas-rich meteorites. Relative to the heavy gases, neon and helium are depleted by factors of 2.5 and 10 respectively. Accurate krypton-krypton ages for rocks 10017, 10047, 10057, and 10071 were obtained covering a range of from 47 to 509 × 10^6 years. Varying relative production rates of the krypton and xenon isotopes in these rocks suggest different irradiation conditions and a complex history for at least some of the rocks.*

All 23 stable isotopes of the noble gases He to Xe and Kr^{81} with a half-life of 2.1 × 10^5 years have been determined in the soil (fines ≤ 1 mm) and in four rocks (10017, 10047, 10057, 10071) from the Apollo 11 mission. The results show that we are dealing with gases of different origin. The soil contains large amounts of trapped solar wind gases which almost completely mask other components. In the crystalline rocks, on the other hand, cosmic ray induced spallation and radiogenic gases are predominant. Small amounts of solar wind gases are found too in chips close to the rock surface. No trapped primordial or "planetary" gases were found in the rocks, which suggests that they were completely outgassed at the time of their formation. The relative abundances of the solar wind gases in the soil are given in Table 1. The heavier rare gases Ar to Xe show abundance ratios similar to those found in the Pesyanoe meteorite (*1*), and agree within factors of 2 to 3 with estimates from cosmic abundance calculations. Relative to the heavy gases, Ne and He in the fines are depleted by factors of about 2.5 and 10 respectively as compared with Pesyanoe, a typical gas-rich meteorite. The concentrations of the heavy gases in Pesyanoe are about 300 times lower and there is no evidence of saturation effects. Therefore the agreement between the abundance pattern of Ar, Kr, and Xe in lunar soil and in Pesyanoe is an important fact in the discussion of the origin of these gases.

Table 1. Relative abundances of the rare gases.

	He^4/Ne^{20}	Ne^{20}/Ar^{36}	Ar^{36}/Kr^{84}	Kr^{84}/Xe^{132}
Lunar soil 10084,29	104	6.63	2040	7.8
Pesyanoe (*1*)	388	16.4	2270	7.5
Atmospheric	0.318	0.522	48.5	27.8
Av. chondritic			~200	1.3
Cosmic				
Aller (*7*)	355	72	6760	30
Suess-Urey (*8*)	400	61	4300	27
Cameron (*9*)	1000	13	18,000	13

Table 2. Krypton-krypton ages.

Sample	Kr^{83}/Kr^{81}	Kr_{spall}/Kr^{81}	P_{81}/P_{83}	Kr^{81}-Kr^{83} age (10^6 years)	Kr^{81} ($\times 10^{-12}$ cm³ STP per gram)
		Rocks			
10017,55	2840 ± 160	2840	0.591	509	0.347
10017,56	2597 ± 125	2520	0.588	449	0.367
10047,40	495 ± 22	470	0.604	86.0	0.343
10057,58	424 ± 18	266	0.580	46.8	0.372
10071,32	1739 ± 175	1700	0.606	312	0.420
		Fines			
10084,29	62000 ± 9300	1800	0.60	330	0.40

If we accept trapping and gas retention efficiencies close to unity and a He^4 solar wind flux of 6.3×10^6 atoms per cm² per sec (*2*), we calculate an average irradiation time of about 3×10^5 years for a dust layer of 1 g/cm². The isotopic compositions of the solar wind gases are well within the range of values previously reported for some gas-rich meteorites, except for Ar^{40}. All gases He to Xe contain a small but clear-cut cosmic ray spallation component and therefore do not represent the pure solar wind composition. The magnitude of the spallation components in the soil can be estimated from results obtained from stepwise heating experiments or from the Kr-Kr age of 330×10^6 years (Table 2). The measured Ne^{20}/Ne^{22} ratio of 12.80 is little affected by the spallation component, but in the bulk sample the measured ratios of $He^4/He^3 = 2540$, $Ne^{22}/Ne^{21} = 28.3$, and $Ar^{36}/Ar^{38} = 5.25$ are all slightly too small. The concentration of $Ar^{40} = 3.6 \times 10^{-4}$ cm³ (STP) per gram in the dust is unusually high; the measured ratio Ar^{40}/Ar^{36} is 1.09. From nuclear abundance systematics one would expect a much lower relative abundance of the Ar^{40} isotope in the sun. The low Ar^{40}/Ar^{36} ratio of 0.35 in the meteorite Novo Urei (*3*) also indicates a much smaller value for primordial argon. Less than 30 percent of the Ar^{40} can be accounted for on the basis of the measured K and a K-Ar age of 4.5×10^9 years. A stepwise heating experiment using temperature increments of about 200°C has revealed a variable release of the Ar isotopes. The Ar^{38}/Ar^{36} ratio, however, only increased at higher temperatures when a spallation component became apparent. The lowest ratio $Ar^{40}/Ar^{36} = 0.78$ was found in the 800°C fraction, but gives only an upper limit for the occurrence of Ar^{40} in the solar wind. The true value is probably much lower. The excess Ar^{40} over this upper limit combined with the measured potassium yields a K-Ar age of 4.65×10^9 years. These results suggest the presence of a radiogenic Ar component which is not related to the K in the dust. This is supported by the fact that the release of excess Ar^{40} shows two maxima at 600° and 1200°C.

Inspection of the xenon isotopic composition in the soil reveals large excesses of the light isotopes up to Xe^{131} which can be attributed to spallation, but the neutron-rich isotopes Xe^{134} and Xe^{136} are deficient as compared to both trapped chondritic and atmospheric xenon. The relative abundances of these isotopes are lower than those reported for bulk analyses of the Pesyanoe meteorite, but are rather similar to those reported for the 1000°C fraction of Pesyanoe. From the stepwise heating experiment (Table 3) more information on the composition of solar wind Xe is obtained. The relative excesses on the light isotopes gradually increase with increasing temperature, clearly indicating a spallation component with a large relative yield of Xe^{131}. The relative contents of the heavy isotopes are similar in all temperature fractions, although slightly higher values are observed in the 1000° and 1200°C fractions. This is not expected, because a larger relative spallation contribution to Xe^{132} should tend to decrease the relative abundances of Xe^{134} and Xe^{136}. Fission Xe from spontaneous fission of U^{238} can account for less than 10 percent of the effect. There is no evidence for a large-scale neutron irradiation of solar Xe and Kr such as proposed by Cameron (*4*). As is mentioned above, the gases in the rocks are predominantly spallogenic and radiogenic. We have determined cosmic ray exposure ages for four rocks (Table 2), using the Kr-Kr dating method, which is based on the ratio

Table 3. Concentrations and isotopic composition of xenon in the lunar soil (10084,29), normalized to $Xe^{132} = 100$.

Sample fraction	Xe^{132} ($\times 10^{-8}$ cm³ STP per gram)	124	126	128	129	130	131	134	136
				Bulk samples					
Average	2.05 ± .25	0.632 ± .012	0.725 ± .018	8.75 ± .10	104.5 ± .5	16.66 ± .08	83.5 ± .4	36.93 ± .22	29.90 ± .20
				Temperature fractions					
600°C	0.038	0.484 ± .018	0.474 ± .015	8.29 ± .10	104.9 ± .6	16.33 ± .12	82.4 ± .4	36.90 ± .30	29.95 ± .30
800°C	0.454	0.495 ± .015	0.476 ± .012	8.47 ± .08	105.4 ± .5	16.53 ± .10	82.6 ± .3	36.81 ± .20	29.85 ± .20
1000°C	0.647	0.545 ± .012	0.595 ± .012	8.54 ± .08	104.4 ± .4	16.55 ± .10	82.7 ± .3	37.02 ± .22	30.19 ± .20
1200°C	1.168	0.736 ± .012	0.924 ± .013	9.00 ± .09	104.5 ± .4	16.74 ± .09	84.0 ± .3	37.11 ± .20	30.22 ± .20
1400°C	0.017	0.910 ± .018	1.245 ± .018	9.64 ± .11	107.5 ± .6	17.27 ± .13	88.6 ± .7	36.44 ± .30	29.39 ± .30
Total	2.32 ± .30	0.634 ± .015	0.741 ± .014	8.76 ± .09	104.7 ± .6	16.64 ± .10	83.4 ± .4	37.02 ± .25	30.12 ± .25
Pesyanoe (*1*)	0.0092	0.488	0.504	8.28	103.3	16.40	82.0	37.3	30.7
Atmosphere (*10*)		0.357	0.335	7.14	98.3	15.17	78.8	38.8	33.0

of a stable Kr isotope and the radioisotope Kr^{81}. This dating method is found to be both accurate and best suited for lunar material for the following reasons: (i) the age determination is based on the isotopic composition of one element only and therefore eliminates errors in concentration measurements; (ii) Sr, Y, and Zr, the main target elements for the production of Kr, are strongly enriched in lunar material, making it possible to measure Kr^{81} in samples as small as 0.1 g; (iii) because of both unusual chemical composition and unusual irradiation conditions the production rates of those stable isotopes generally used for age determinations are not well known.

Kr-Kr ages are obtained from the equation (*5*)

$$T_r = \frac{Kr^{83}{}_{spall}}{Kr^{81}} \frac{P_{81}}{P_{83}} \frac{1}{\lambda}$$

where $\lambda = 3.3 \times 10^{-6}$ per year, and $P_{81} = (0.95 \pm 0.05)\ (P_{80} + P_{82})/2$.

The relative production rates of the Kr isotopes are similar for all the investigated rocks, which is indicated by rather constant P_{81}/P_{83} ratios (Table 2). The largest variation in the Kr^{78}/Kr^{83} ratio is 15 percent (between rocks 10047 and 10057). Spallation Kr^{83} in rocks containing solar wind Kr were corrected for solar wind gases by using for the spallation component the lowest Kr^{86}/Kr^{83} ratio of rock-specimen 10017, 55, an interior sample, taking into account a contribution to Kr^{86} from spontaneous fission of U. The $Kr^{83}{}_{spall}/Kr^{81}$ ratio of an outside chip of rock 10017 was found to be 10 percent smaller than that of an inside sample, which may be attributed to a slightly higher Kr^{81} content. However, the relative errors in the Kr contents are about 6 percent.

Do the obtained Kr-Kr ages in fact give the actual time interval of exposure to cosmic rays? As has already been mentioned, there are obvious differences in the relative production rates of certain isotopes which cannot be explained by a varying chemical composition but only by differences in the composition and energy distribution of the interacting cosmic rays. Variations in the relative production of He^3 and Ne^{21} may in part be explained by differences in chemical composition or by possible diffusion losses. One is tempted to ascribe the observed variations in the relative production of the Kr isotopes to differences in the relative abundances of the target elements Sr, Y, and Zr. Large-scale variations in the relative production of the Xe isotopes, however, cannot be explained simply on the basis of chemical differences. Furthermore, the fact that specific relative production rates correlate with each other, such as the Kr^{78}/Kr^{83} and Xe^{126}/Xe^{131} ratios, leads us to conclude that the irradiation conditions of the four investigated rocks did differ from each other. The measured Xe^{131}/Xe^{126} ratios vary by almost a factor of 2 and are on the average also higher than the corresponding ratios found in meteorites. However, Pepin (*6*) found a Xe^{131}/Xe^{126} ratio of ≥ 5.3 in the Estherville meteorite, a value which is within the range covered by our four rocks and the soil sample. The evidence of such large relative Xe^{131} yields, which require a heavily moderated cosmic ray irradiation, suggests that special reactions may be partly responsible for the excess Xe^{131} production. Because of the high capture cross section, Gd is a sensitive indicator for thermal neutrons. We have measured the isotopic compositions of Gd in rock 10017 and in our soil sample. The depletion of Gd^{157} is between 0.1 and 0.4 percent for rock 10017, and for the soil sample we can give an upper limit of 0.2 percent. Because of background interferences in the Gd mass spectrum we can give only limits. This result shows that thermal neutron capture contributes less than 2 percent to the total Xe^{131} production. No neutron resonances are known for Ba^{130} and it seems unlikely that the neutron resonance integral will contribute much to the total cross section.

If the irradiation history of a rock is complex, if the rock has been partially and variably shielded from cosmic rays, then the reported radiation age represents only a lower limit for the actual time of irradiation. Our results indicate such complex histories at least for rocks 10017 and 10057 and for the soil sample.

K. MARTI
G. W. LUGMAIR
H. C. UREY

Department of Chemistry, University of California, San Diego, La Jolla

References and Notes

1. K. Marti, *Science* **166**, 1263 (1969).
2. F. Bühler, P. Eberhardt, J. Geiss, J. Meister, P. Signer, *ibid.* **166**, 1502 (1969).
3. K. Marti, *Earth Planet. Sci. Lett.* **2**, 193 (1967).
4. A. G. W. Cameron, *Icarus* **1**, 13 (1962).
5. K. Marti, *Phys. Rev. Lett.* **18**, 264 (1967).
6. R. O. Pepin, *J. Geophys. Res.* **71**, 2815 (1966).
7. L. H. Aller, *The Abundance of the Elements* (Interscience, New York, 1961).
8. H. E. Suess and H. C. Urey, *Rev. Mod. Phys.* **28**, 53 (1956).
9. A. G. W. Cameron, in *Handbook of Geophysics and Space Environments* (U.S. Air Force Cambridge Research Laboratories, Washington, D.C., 1965).
10. A. O. Nier, *Phys. Rev.* **79**, 450 (1950).
11. We thank K. R. Goldman and M. Meyers for valuable assistance. Supported by NASA contract NAS 9-8107.

5 January 1970

Isotopic Composition of Rare Gases in Lunar Samples

Abstract. *Data from total melt and step-by-step heating experiments on the Apollo 11 lunar samples suggest a close affinity between lunar and meteoritic rare gases. Trapped neon-20/neon-22 ratios range from 11.5 to approximately 15, resembling those for the gas-rich meteorites. Trapped krypton and xenon in the lunar fines and in the carbonaceous chondrites are similar except for an interesting underabundance of the heavy isotopes in both lunar gases which suggests that the fission component found in carbonaceous chondrites is depleted in lunar material. Spallation gases are in most cases quite close to meteoritic spallation gases in isotopic composition.*

Mass spectrometric analyses were carried out for all rare gas isotopes in six lunar crystalline rocks, a breccia, coarse fines sieved from the lunar soil, and the soil.

Gases in the crystalline rocks are in general dominated by radiogenic and spallogenic isotopes. The ^3He radiation ages, on the assumption of a ^3He production rate of 1×10^{-8} cm^3(STP) g^{-1} 10^6 $year^{-1}$, range from $\sim 30 \times 10^6$ (sample 10069,21) to 210×10^6 years (10071,20). U, Th-He, and K-Ar gas-retention ages, which are consistent with the values $\gtrsim 3.5 \times 10^9$ years reported in the preliminary analysis (*1*), as well as Ne, Ar, Kr, and Xe radiation ages, will be discussed in a later publication.

Neon isotopic compositions in the rocks and coarse fines are shown in Fig. 1. The pattern indicates a highly variable mixture of a single spallation composition, perhaps represented in virtually pure form in 10020,47[2], and a trapped neon component with $^{20}Ne/^{22}Ne = 12.7 \pm 0.4$. Neon in 10020,47[2] is essentially identical to that in the C3 and LL3 meteorites Felix and Chainpur, which terminate similar correlation lines for the carbonaceous chondrites (*2*). The striking variability in the proportion of trapped to spallation neon in repeat runs on certain samples may be a consequence of minute contamination by lunar fine material, despite care in handling and cleaning the rock chips. The large relative concentration of trapped Ne in the coarse fines (10085,5), which consisted of 1- to 2-mm fragments of fine- and coarse-grained crystalline rocks, probably derives from direct solar-wind irradiation of the fragment surfaces.

Light isotope compositions of Kr and Xe in four rocks and the coarse fines are shown in Figs. 2 and 3. The Kr points are colinear within the 1σ errors, and the correlation line passes through the composition of atmospheric Kr. The pattern indicates that these samples contain mixtures of an isotopically uniform spallation component and trapped Kr resembling terrestrial Kr. Scatter in the Xe points is somewhat larger; to a first approximation the trend is linear and the same conclusion holds. The trapped Xe component could be either atmospheric or carbonaceous chondrite Xe.

The spallation Kr spectrum, averaged over the four rocks shown in Fig. 2, is $^{78}Kr : ^{80}Kr : ^{82}Kr : ^{83}Kr : ^{84}Kr = 0.26 : 0.67 : \equiv 1.00 : 1.36 : 0.59$, and the spallation Xe spectrum, averaged over the three rocks 10020, 10045, and 10058, is $^{124}Xe : ^{126}Xe : ^{128}Xe : ^{129}Xe : ^{130}Xe : ^{131}Xe : ^{132}Xe : ^{134}Xe = 0.61 : \equiv 1.00 : 1.50 : 1.78 : 0.84 : 3.98 : 0.8 : <0.1$. The spectrums are derived assuming spallogenic ^{86}Kr and $^{136}Xe \equiv 0$. Both are similar to cosmic-ray spallation yields determined for two basaltic achondrites (*3, 4*). Xenon in the fourth rock (10072,21) differs from the achondritic spectrum in showing approximately a factor-2-larger relative production of ^{131}Xe and, to a lesser extent, ^{130}Xe. In this respect the spallation Xe in 10072,21 resembles that found in the silicate fraction of the Estherville mesosiderite, where $^{131}Xe : ^{126}Xe \simeq 7.7$ (*5*). These yields suggest a relatively soft irradiation spectrum.

Four bulk analyses—single-step melt-

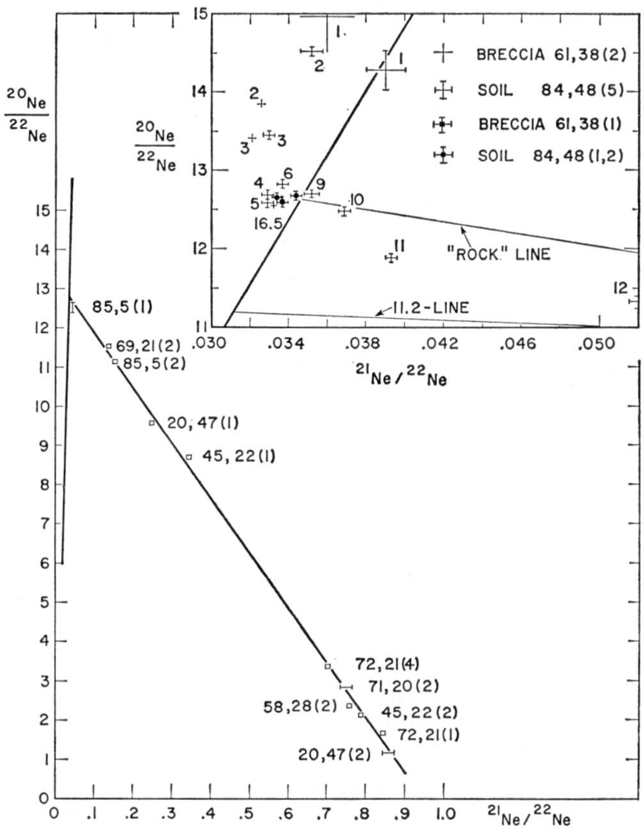

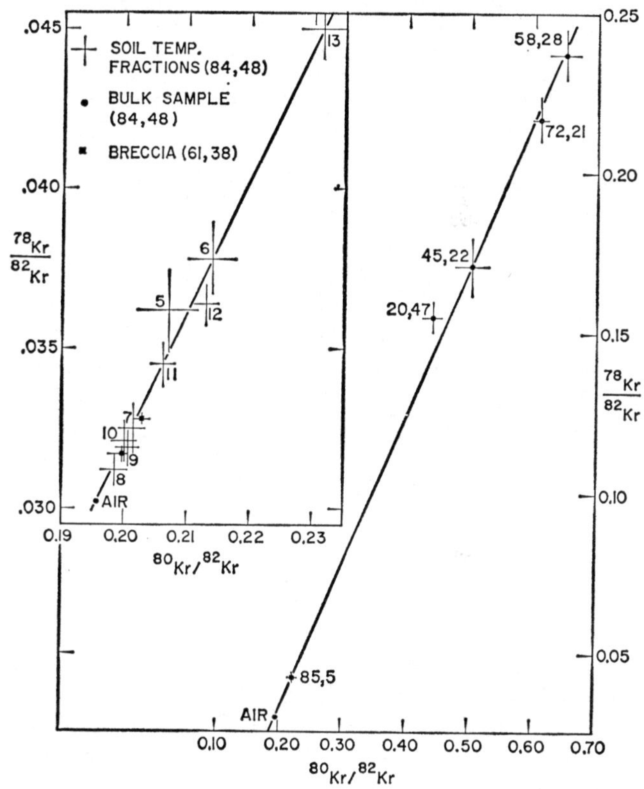

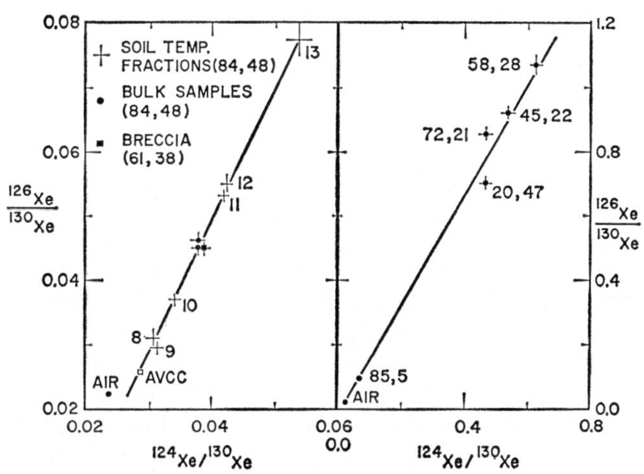

Fig. 1 (top left). Three-isotope correlations in Ne from lunar samples. Crystalline rocks are shown in the main diagram, with bulk and step-by-step heating data on fines and breccia in the insert. Errors are smaller than the plotted symbols except where indicated. Numbers in the insert are out-gassing temperatures in 100°C units; the 1300°C point for the soil lies well to the right of the diagram, slightly above the 11.2 line. Almost all Ne is released in the steps clustering around $^{20}Ne : ^{22}Ne = 12.6$. The heavy diagonal and near-vertical lines are the meteoritic "trend line" (*6*). Fig. 2 (top right). Correlation diagram for the light Kr isotopes. Sample 85,5 in the main diagram is the coarse fines; others are crystalline rocks. Soil and breccia data are given in the insert. Fig. 3 (right). Correlation diagram for the light Xe isotopes, with rock and coarse fine data on the right and soil and breccia data on the left. On the right-hand scales the air and average-carbonaceous-chondrite-composition points are nearly indistinguishable, but on the left plot they are resolved and the correlation line clearly passes through the average carbonaceous chondrite composition.

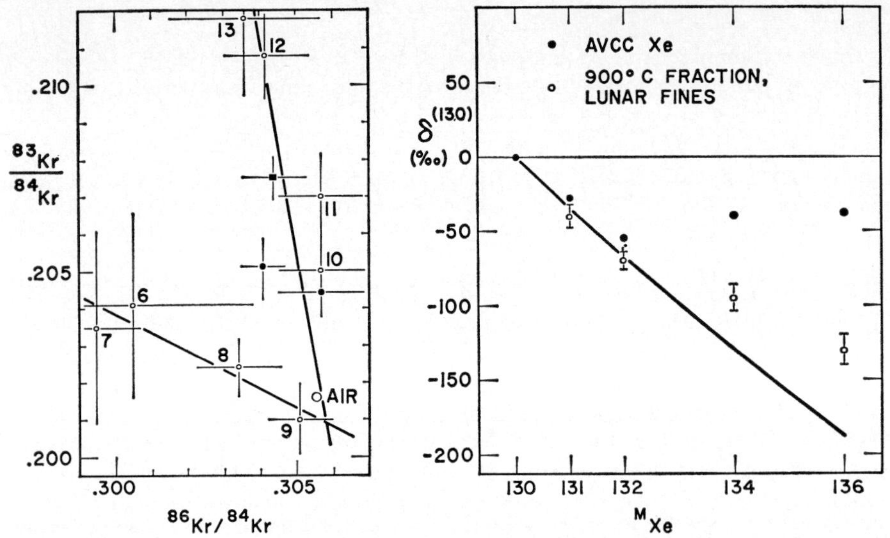

Fig. 4. Heavy Kr and Xe isotopes in the lunar fines and breccia. Filled points on the Kr correlation plot are bulk analyses of soil and breccia, open circles with numbers (100°C units) are step-by-step heating fractions. The lines appear to converge to the air point. All temperature fractions and bulk analyses of the fines and breccia show the general Xe depletion pattern displayed on the right-hand diagram by the 900°C fraction, but here the case is particularly clear because of minimum spallation interference at $^{130-134}$Xe.

ing extractions—were made on rare gases in the lunar fines and the breccia. In addition, step-by-step heating extractions and analysis of all rare gas isotopes in 13 100°C temperature steps were carried out on a sample of the fines.

Values of ^{4}He : ^{3}He = 2400, ^{4}He : ^{20}Ne≃100, ^{20}Ne : ^{22}Ne = 12.66, ^{36}Ar : ^{38}Ar = 5.38, and ^{40}Ar : ^{36}Ar = 1.07 are characteristic of bulk analyses of the soil, but He, Ne, and Ar isotopic ratios show considerable variation during step-by-step heating. The ^{4}He : ^{3}He ratio increases from ~1700 at 100°C to 3050 at 1000°C, then decreases to ~1100 at 1200°C. The ^{36}Ar : ^{38}Ar ratio falls monotonically from 5.77 at 200°C to 4.69 at 1300°C. In both cases the decrease at high temperature ($T \geq$ 1100°C) is due to a relatively large release of firmly sited spallation gases. The ^{40}Ar : ^{36}Ar ratio is of particular interest, for the lunar fines contain about six times more ^{40}Ar than could have derived from *in situ* decay of ^{40}K over 4.5 × 10^{9} years, a difference which suggests that ^{40}Ar must be abundant in the trapped gases. A component which is probably radiogenic ^{40}Ar produced *in situ* outgases preferentially at low temperatures in the step-by-step heating: ^{40}Ar : ^{36}Ar = 14.4 in the 200°C fraction and falls to a minimum of 0.76 at 800°C, the point of maximum ^{36}Ar release and therefore the point at which the trapped gas composition is best represented.

Variations in Ne, Kr, and Xe isotopic compositions in the soil as functions of outgassing temperature are shown in Figs. 1–4. Neon, with very high ^{20}Ne : ^{22}Ne in the low-temperature fractions, is strikingly similar to Ne from step-by-step outgassing of the gas-rich achondrite Kapoeta, except that ^{21}Ne : ^{22}Ne ratios here are generally lower; points tend to lie to the left of the diagonal "trend-line" in the insert in Fig. 1 instead of moving along it as they typically do in meteorites *(6)*. It is clear that the principal Ne component in the lunar fines and breccias is one with ^{20}Ne : ^{22}Ne = 12.6 to 12.7, within error of the composition of neon-B (^{20}Ne : ^{22}Ne = 12.5±0.2), which dominates in the gas-rich meteorites. At high temperatures the neon composition trends toward spallation neon, with the trapped ^{20}Ne : ^{22}Ne ratio approaching 11.2, again a pattern characteristic of gas-rich meteorites (*6*). Except for the lower ^{21}Ne : ^{22}Ne, neon isotopic patterns in the lunar fines and the breccia and in the gas-rich meteorites are virtually identical.

The light-isotope variations in Kr and Xe in the fines, shown in Figs. 2 and 3, form linear arrays on the correlation diagrams. The patterns arise from variable mixing of isotopically uniform spallation and trapped gas components in the various temperature fractions. There is a suggestion in our results that the correlation lines for Kr and Xe in the soil and for Kr and Xe in the crystalline rocks (Figs. 2 and 3) may have slightly different slopes, and therefore that the isotopic compositions of the spallation components may be somewhat different in the two types of material. The apparent shift lies on the margin of experimental error. At the moment, to a first approximation, the soil and rock spallation components are similar.

The step-by-step heating correlation lines in Figs. 2 and 3 pass directly through the isotopic compositions of atmospheric Kr and average carbonaceous chondrite Xe, respectively. In all probability these are the compositions of trapped Kr and Xe in the lunar fines. It is clear that the trapped Xe is not terrestrial, since the air point lies well off the correlation in Fig. 3. If the composition of trapped Xe is assumed to be the average carbonaceous chondrite composition, a simple calculation shows that ~50 percent of the ^{126}Xe in the fines is spallogenic. On the basis of Ba and rare earth abundances in the fines (*7*) and Xe production rate equations developed by Marti *et al.* (*3*), the Xe radiation age of the lunar fines is ≃ 800 × 10^{6} years.

The possibility of atmospheric-type Kr as the trapped component in the fines is strengthened by the heavy isotope correlations shown in Fig. 4. There is evidence here for two spallation components, one appearing in the low-temperature fractions (600° to 900°C) and the other above 900°C. Both correlation lines pass near the composition of terrestrial Kr. The fractionated Kr composition reported to be present in carbonaceous chondrites (*8, 9*) lies well off this plot to the right and does not appear to be a constituent of lunar Kr.

The heavy Xe isotopes in the 900°C fraction of the fines are plotted in the right half of Fig. 4. Less than 0.5 percent of the ^{130}Xe in this fraction is spallogenic. The ordinate of the graph represents per mil deviations from atmospheric composition, normalized to ^{130}Xe. Current studies in this laboratory indicate that ~5 percent of atmospheric ^{136}Xe is fissiogenic. This component has been subtracted from the reference atmospheric Xe in Fig. 4. The heavy isotopes ^{134}Xe and ^{136}Xe are clearly depleted relative to the average carbonaceous chondrite Xe, a finding that suggests a depletion in the fission Xe component known to be present in the carbonaceous chondrites. However, some fission Xe is still present in lunar trapped Xe if our estimate of the composition of fission-free trapped Xe, shown by the heavy curve, is correct. This depletion of fission Xe, coupled with the absence of excess ^{86}Kr in lunar Kr and its definite presence in average carbonaceous chondrite Kr, suggests that excess ^{86}Kr and fission Xe concentrations may correlate, both resulting from the addition of a fissiogenic Kr-Xe component.

R. O. PEPIN
L. E. NYQUIST
DOUGLAS PHINNEY, DAVID C. BLACK
School of Physics and Astronomy, University of Minnesota, Minneapolis 55455

References and Notes

1. Lunar Sample Preliminary Examination Team, *Science* **165**, 1211 (1969).
2. R. O. Pepin, *Earth Planet. Sci. Letters* **2**, 13 (1967).
3. K. Marti, P. Eberhardt, J. Geiss, *Z. Naturforsch.* **21a**, 398 (1966).
4. C. M. Hohenberg, M. N. Munk, J. H. Reynolds, *J. Geophys. Res.* **72**, 3139 (1967).
5. R. O. Pepin, *ibid.* **71**, 2815 (1966).
6. D. C. Black and R. O. Pepin, *Earth Planet. Sci. Letters* **6**, 395 (1969).
7. V. R. Murthy, personal communication.
8. K. Marti, *Earth Planet. Sci. Letters* **3**, 243 (1967).
9. O. Eugster, P. Eberhrardt, J. Geiss, *ibid.* **3**, 249 (1967).
10. Supported by NASA contract NAS 9-8093, ONR contract Nonr-710(58), and NSF grant GA-905. A. O. C. Nier designed the 6-inch (15-cm) double-focusing mass spectrometer used for most of these analyses. We dedicate this paper to the late Rudolph B. Thorness for his 40 years of unparalleled contributions to mass spectrometry.

4 January 1970

Cosmic Ray Production of Rare Gas Radioactivities and Tritium in Lunar Material

Abstract. *The argon radioactivities ^{37}Ar and ^{39}Ar were obtained by vacuum melting from interior and exterior portions of rock 10057 and from a portion of the fines from the bulk sample container. The release of argon and tritium as a function of the temperature was followed for the fine material. A comparison is made of the activities observed in the lunar samples with those expected from the spallation of iron, titanium, and calcium. From these data and the ^{38}Ar content, the cosmic ray exposure age of rock 10057 is deduced as 110 $\times$ 10^6 years.*

A measurement of the stable and radioactive argon isotopes in lunar material can give valuable information on cosmic ray interactions on the lunar surface and can be used to deduce cosmic ray exposure ages of fragmented material resting on the lunar surface. Of particular interest are the stable isotopes ^{36}Ar and ^{38}Ar, and the radioactive isotopes ^{37}Ar (half-life, 35 days) and ^{39}Ar (half-life, 269 years). These isotopes are produced primarily by high-energy bombardment of the relatively abundant elements iron, titanium, and calcium. The primary goal of the investigation discussed here was to measure the ^{37}Ar and ^{39}Ar in surface rocks and in the fine material. The argon is evolved from the solid by vacuum melting and is counted in a small low-level gas proportional counter. By using pulse height analysis the Auger electron spectrum from the electron capture of ^{37}Ar can be easily distinguished from the beta spectrum of ^{39}Ar, and the two isotopes can be measured simultaneously. Since ^{37}Ar has a half-life of only 35 days, the measurements were made at the Lunar Receiving Laboratory during the biological quarantine.

The vacuum melting technique used to evolve and count the argon radioactivities also allows a search to be made for radioactive isotopes of the higher rare gases—namely ^{81}Kr, ^{85}Kr, ^{127}Xe, ^{133}Xe, and ^{222}Rn. The isotope ^{81}Kr (half-life, 2.1 $\times$ 10^5 years) has been observed in meteorites (*1*) and has been attributed to high-energy spallation of Sr, Y, and Zr. Radon-222 arises from the decay of ^{238}U, and a measurement of the ^{222}Rn content can be used as a sensitive measurement of the uranium concentration.

Of special interest is the tritium radioactivity in lunar materials that is also released by vacuum melting. This isotope is produced by high-energy cosmic ray bombardment. The rate of tritium production is an essential quantity needed to determine the tritium-^3He cosmic ray exposure age. Of far greater interest is the possibility that some of the solar flare particles are tritons. Fireman and his associates (*2*) observed tritium in the Discoverer 17 satellite and in a sounding rocket that was exposed to the solar flare of 12 November 1960. An analysis of their measurements indicates that the tritium-to-hydrogen ratio in this event was about 10^{-3} (*2*, *3*). A search for tritium in material exposed on the surface of the moon may afford a unique opportunity to remeasure the tritium content of solar flare particles.

Three separate samples of lunar material were studied; an internal (weight, 7.65 g) and an external (weight, 10.6 g) fragment of rock 10057, and a 10.6-g portion of the fine material (10002,6) from the bulk sample. Among the lunar rocks, the one designated 10057 has been classified type A; this includes the crystalline, fine-grained, vesicular rocks. The external rock fragment and the fine material were processed and measured at the Lunar Receiving Laboratory during the quarantine period, and therefore had to be heat-sterilized. The internal rock chip was measured after the quarantine period had ended, and was not sterilized. The sterilization was conducted in a stainless steel container at a temperature of 120°C for a period of 24 hours. A search was made for ^{37}Ar and ^{39}Ar released from the fine material during the heating and less than 0.6 disintegration min^{-1} kg^{-1} (dpm/kg) was found. The tritium released during the sterilization of the fine material was not determined because it was considered very likely that, if hydrogen were released, it would be absorbed into the stainless steel of the container.

Samples of lunar material in an alumina-lined molybdenum crucible were heated by a radio-frequency induction heater. The gases evolved were collected in a Toepler pump and placed in a tube containing hot vanadium metal. The vanadium at 850°C serves as a getter to remove chemically active constituents. The vanadium was then cooled to room temperature to absorb hydrogen as

Table 1. Volumes of rare gases and hydrogen from lunar material.

Sample	Extraction temperature (°C) and heating period	Rare gases (cm^3/g, STP)	Hydrogen (cm^3/g, STP)
Fines	Sterilization, 120, 24 hr	Not meas.	Not meas.
1000.2,6	120 to 600, 64 min	0.145	0.67
1000.2,6	600 to 900, 73 min	0.068	0.12
1000.2,6	900 to 1200, 92 min	0.003	0.04
1000.2,6	1200 to 1600, 68 min	<0.0005	<0.007
Total	120 to 1600	0.217	0.84
10057,3 (exterior)	120 to 1600	Not meas.	~0.1
10057,3 (interior)	120 to 1600	Not meas.	~0.16

Table 2. Radioactivities of rare gas and tritium.

Sample	Radioactivity (dpm/kg)			Ratio $^{37}Ar/^{39}Ar$
	Tritium	^{37}Ar*	^{39}Ar	
	Fines, sample 10002.6			
Sterilization heating	Not meas.	⩽ 0.6	–	–
Heat I: 120°–600°C	199 ± 12	⩽ 1	–	–
Heat II: 600°–900°C	70 ± 3	6.7 ± 0.7	3.6 ± 0.2	1.84 ± 0.20
Heat III: 900°–1200°C	41 ± 3	11.3 ± 0.8	5.3 ± 0.3	2.14 ± 0.18
Heat IV: 1200°–1600°C	3 ± 2	⩽ 2	–	–
Combined argon heats I and IV	–	0.7 ± 0.2	0.2 ± 0.2	–
Totals	314 ± 13	18.7 ± 1.2	9.2 ± 0.4	2.03 ± 0.16
	Rocks			
10057,3 (exterior)†	224 ± 15	18.5 ± 0.9	11.2 ± 0.3	1.65 ± 0.09
10057,27 (interior)†	214 ± 10	21.6 ± 4.1	13.5 ± 0.4	1.60 ± 0.30

* Values have been corrected for decay to 0000 hours, C.D.T., 21 July 1969. If reentry is taken as the true EOB time, the ^{37}Ar results should be decreased by a factor of 1.076. † Krypton radioactivities were measured for rock 10057: exterior sample, −0.5 ± 0.3; interior sample, 0.0 ± 0.3 dpm/kg.

vanadium hydride. The rare gases were removed from the vanadium tube, their volume was measured, and they were placed in a sample transfer bulb. The vanadium tube was then reheated to 850°C to dissociate the vanadium hydride. The hydrogen was collected with a Toepler pump, measured, and sealed in a second sample transfer bulb. The apparatus for vacuum fusion and extraction of the rare gases and hydrogen from the samples was located in the gas analysis laboratory behind the secondary biological barrier. The sealed glass ampules containing the rare gases and hydrogen were then sterilized and taken through the biological barrier to a purification and counter filling system outside the barrier.

The isolated rare gases were then separated by gas phase chromatography on a charcoal column with helium as the carrier gas. The separated argon, krypton, and xenon were placed in small gas proportional counters for the radioactivity measurements. The counters had a 0.0125-mm tungsten center wire in a cylindrical cathode of zone-refined iron enclosed in an envelope of silica glass. The counter had a thin window at one end to permit energy calibration of the multichannel analyzer with an external ^{55}Fe x-ray source. The active volume of these counters ranged from 0.3 to 0.6 cm^3. The counter was operated inside a ring of anticoincidence gas proportional counters inside a lead shield 10 cm thick. The shield was located 15 meters underground in the Radiation Counting Laboratory. Eight counters could be operated simultaneously, recording the spectra with 256 channel analyzers. The counter voltage was adjusted so that the 2.8 kev Auger electron from the decay of ^{37}Ar would center at channel 104. The counters had a 30 percent resolution (full width at half maximum) for these events. The amplifier was designed so that all events with energy greater than 6.8 kev were stored in channels 250 to 254 (hereafter referred to as "pileup channels"). This arrangement made it possible to distinguish clearly the events arising from the electron capture decay of ^{37}Ar, and the continuous beta spectrum (maximum energy, 570 kev) arising from ^{39}Ar decay. Individual counter efficiencies for ^{37}Ar and ^{39}Ar were measured by filling the counters with argon containing known amounts of these radioactivities. The exact efficiency of the individual counters was measured over the pulse height spectrum and the pileup channels. Their individual distributions depended upon their specific dimensions and gas pressure. In general they exhibited total efficiencies in the range of 50 to 70 percent for ^{37}Ar and ^{39}Ar. To measure krypton, xenon, and radon radioactivities, the counter was adjusted in the same way, so that these decays would be recorded primarily in the pileup channels.

Hydrogen was purified by passing the gas through a palladium metal thimble at 800°C, and the gas was placed in a gas proportional counter; the gas filling was a mixture of 90 percent argon and 10 percent methane. The counter operated at 1 atmosphere and contained less than 0.2 atmosphere of hydrogen gas. Counting was performed in the same system used in the rare gas counting, with the voltage on the counter adjusted so that all events from an externally placed ^{60}Co source occurred in the pileup channels. The cathode was made of copper, and the outer envelope of silica glass. It had an internal diameter of 1.0 cm and was 12.5 cm long.

Preliminary studies (*4*) showed that the fine material contained large quantities of rare gas that probably can be attributed to solar-wind bombardment. We therefore measured the amounts of rare gases and hydrogen evolved from this finely divided material with increasing temperature. As is mentioned above, this sample had been heated previously to 120°C for 24 hours; the quantity of gas released during this period was not measured. The sample was then heated in stages, reaching the maximum temperatures 600°, 900°, 1200°, and 1600°C; the volume of rare gas and hydrogen released at these temperatures, and the periods of time that the fines were maintained at the maximum temperature, are given in Table 1. Detailed studies of the rare gas composition of the fine material show that the rare gas is more than 98 percent helium (*4*, *5*). This gas evolution experiment shows that more than 67 percent of the helium and 80 percent of the hydrogen is removed from this fine material at 600°C. A temperature of 900°C releases the remaining quantities of these gases. We compared the H/He atom ratio of 7.8 measured from the fines to the H/He ratio in the sun. The recently accepted solar value (*6*) is 17, higher by a factor of 2 than the value obtained by heating the fine material. This may be explained by the preferential loss of hydrogen with respect to helium in these finely divided silicates. The volume of hydrogen observed in these measurements would correspond to a hydrogen abundance of 75 parts per million. The two fragments of rock 10057 were heated to 1600°C; they yielded 0.1 and 0.16 cm^3 of hydrogen (Table 1). The larger volume of hydrogen obtained from the interior sample can probably be accounted for

Table 3. Production cross sections for the argon isotopes in the lunar material.

Element	Percentage in sample		Thin target cross section for 600 Mev protons (millibarns)*			
	Fines	Rock 10057	^{37}Ar	^{38}Ar	^{39}Ar	^{42}Ar
Fe	12.4	15.5	5.6	12†	6.3	0.075
Ti	4.2	7.5	11.6	30‡	15.7	0.34
Ca	8.6	7.1	46.7	32‡	2.0	0.039
K	0.10	0.15	47.4	50§	6.3	0
Mn + Cr	0.5	1.3	5§	12§	5§	0.1§

* Absolute errors are in the range of 5 to 10 percent. † Value of Goebel *et al.* (*8*) normalized to the ^{37}Ar and ^{39}Ar values given above for Fe. ‡ Derived from the $^{38}Ar/^{39}Ar$ ratio measured with 380 Mev protons on Ti by Stoenner *et al.* (*9*), and with 3 Gev protons on Ca by A. A. Schaeffer (*10*). § These cross sections are rough estimates from nuclear systematics.

by the fact that this sample had been stored in air for several weeks prior to measurement. We conclude that the hydrogen abundance of the rock was less than or equal to 10 ppm.

The tritium activity released from the fine material in the various temperature intervals is given in Table 2. Sixty-three percent of the tritium was released at 600°C, closely paralleling the evolution of hydrogen (80 percent). The total tritium activity observed in the fine material was approximately the same as that observed in the rock. All of these samples have approximately the same amount of tritium actvity as that usually observed in stone meteorites (*7*)—namely 200 to 700 dpm/kg. It is clear that the high tritium content that one might expect from solar-flare tritons imbedded in an exposed surface is not observed. Perhaps the samples measured were not truly surface samples. The fines were well mixed, and, if there were a high tritium content on the surface, it could be highly diluted with deeper-lying material. The exterior rock sample did have micrometeorite pits on its surface, but it is not at all certain that this surface had been exposed during the last 20 years or so. The interesting observation of Fireman and his associates (*2*) of a triton component in solar-flare particles should be tested with lunar material known to have been exposed in recent years on the surface.

The release of ^{37}Ar and ^{39}Ar radioactivities from the fine material during the stepwise heating shows that the argon evolved almost completely in the temperature interval 600° to 1200°C. Measurements on the combined sample evolved below 600°C and above 1200°C (see Table 2) showed that less than 4 percent of both of these activities was released outside the 600° to 1200°C interval. This result indicates that the argon produced throughout this silicate material with an average particle size of 100 microns (*4*) is strongly retained. The total ^{37}Ar and ^{39}Ar activity measured in this material is approximately the same as that measured in the rock sample. Since the fine material contains large quantities of solar-wind argon, it is impossible to determine the ^{38}Ar produced by cosmic rays, and therefore an effective exposure age cannot be deduced.

The ^{37}Ar and ^{39}Ar activities produced in the fines and the rock were compared with those expected from cosmic ray bombardment. The magnitude of the activities almost exactly equals those observed in recently fallen meteorites (*7*). The meteorite samples studied are regarded as interior pieces of objects, weighing 10 to 100 kg, that receive an isotropic cosmic ray exposure. The surface lunar material might be expected to contain higher activities produced by low-energy solar and galactic cosmic rays, but on the other hand the lunar surface receives only about half the isotropic flux. Apparently the geometric and shielding effects nearly cancel. The ratio of the ^{37}Ar and ^{39}Ar activities can be compared to the ratio expected from cosmic ray bombardment. The cross section for producing these isotopes in thin targets (1 to 5 g cm^{-2}) of iron, titanium, calcium, and potassium at 600 Mev is given in Table 3, along with the chemical analysis of rock 10057 (*4*). On the basis of these data the calculated $^{37}Ar/^{39}Ar$ atom ratio, constant cosmic ray bombardment being assumed, is 3.2 for the fine material and 2.2 for rock 10057. The observed $^{37}Ar/^{39}Ar$ ratio for both of these samples is about 30 percent lower than the estimated cross-section ratios. To obtain an accurate model for the production ratio for these isotopes, a series of bombardments of thick targets of lunar-like material at several energies is required. Experiments along these lines are in progress, but for the present we must use the available information at 600 Mev. To deduce a cosmic ray exposure age based upon the ^{38}Ar content of rock 10057, one must evaluate the relative production rates of ^{38}Ar and ^{39}Ar. Using the cross sections listed in Table 3, we calculate an $^{38}Ar/^{39}Ar$ cross-section ratio of 2.8. The ^{38}Ar content of rock 10057 was measured, by Funkhouser and his associates (*5*), to be 7.1×10^{-8} cm^3 (STP)/g. Using the measured ^{38}Ar and ^{39}Ar production rates and the calculated $^{38}Ar/^{39}Ar$ cross-section ratio, we calculate that rock 10057 was exposed for 110 million years on the lunar surface.

Since the lunar samples are high in Sr, Y, and Zr, it is possible that measurable amounts of ^{81}Kr and ^{85}Kr may be present. A search was made for krypton activities in rock 10057, but only an upper limit of less than 0.5 dpm/kg could be set. From the composition of the rock and an estimate of the production cross sections, we conclude that the level of both ^{81}Kr and ^{85}Kr activities is about 0.2 dpm/kg.

R. W. STOENNER, W. J. LYMAN
R. DAVIS, JR.

Chemistry Department,
Brookhaven National Laboratory,
Upton, New York 11973

References and Notes

1. K. Marti, *Phys. Rev. Lett.* **18**, 264 (1967; O. Eugster, P. Eberhardt, J. Geiss, *Earth Planet. Sci. Lett.* **2**, 77 (1967).
2. E. L. Fireman and J. DeFelice, *Phys. Rev.* **123**, 1935 (1961); E. L. Fireman, in *Proceedings, International Conference on Cosmic Rays, Jaipur* (1963), vol. 1, p. 99; D. Tilles, J. DeFelice, E. L. Fireman, *Icarus* **2**, 258 (1963).
3. S. Biswas and C. E. Fichtel, *Space Sci. Rev.* **4**, 709 (1965).
4. Lunar Sample Preliminary Examination Team, *Science* **165**, 1211 (1969).
5. J. Funkhouser, O. A. Schaeffer, D. D. Bogard, J. Zähringer, *ibid.*, this issue.
6. D. L. Lambert, *Nature* **215**, 43 (1967); until recently a value of 11 for the H/He atom ratio was accepted; see J. R. Taylor, *Quart. J. Roy. Astron. Soc.* **8**, 313 (1967), and I. Iben, *Ann. Phys.* **54**, 164 (1969).
7. R. Davis, Jr., R. W. Stoenner, O. A. Schaeffer, in *Radioactive Dating* (International Atomic Energy Agency, Vienna, 1963), p. 355; E. L. Fireman, J. DeFelice, D. Tilles, in *ibid.*, p. 323; T. Charalambus, K. Goebel, W. Stötzel-Riezler, *Z. Naturforsch.* **24a**, 234 (1969).
8. K. Goebel, H. Schultes, J. Zähringer, *CERN Report 64–12, Geneva.*
9. R. W. Stoenner, O. A. Schaeffer, S. Katcoff, *Science* **148**, 1325 (1965).
10. O. A. Schaeffer, private communication.
11. We acknowledge the assistance of personnel of the Radiation Counting Laboratory and Gas Analysis Laboratory of LRL, where this work was performed. We also are grateful for the use of the cyclotron at the Space Radiation Effects Laboratory, Newport News, Va. Supported by NASA and AEC.

4 January 1970

Inert Gases in Lunar Samples

Abstract. *Sample 10084,40 (fines, less than 1 millimeter) contains substantial amounts of the inert gases. Their concentrations are inversely proportional to particle size; hence the gases appear to be surface-correlated in the soil fragments. The most likely origin of the gas is solar wind or solar cosmic rays. Glass and feldspar are generally poorer in gas than lithic fragments. Ratios of elements in the sample differ significantly from solar values. Ratios of isotopes in the sample are similar to those in meteorites. Argon-40 appears to consist of a radiogenic and a surface-correlated component. An apparent potassium-argon age of* $4.42^{+0.24}_{-0.28}$ *billion years is calculated.*

We have determined (mass-spectrometrically) the concentrations of the inert gases, and in all cases the isotopic compositions of He, Ne, and Ar in the following "splits" of sample 10084,40 ("fines," < 1 mm): 1 and 2, "bulk" of 10084,40 (weights 3.507 and 10.827 mg); 3, lithic fragment, ~800 μm

Table 1. Inert gas contents. Numbers refer to sample description in the text. Units: 3 = 10^{-7}; 4 = 10^{-3}; 20 = 10^{-5}; 21 = 10^{-8}; 22 = 10^{-6}; 36 = 10^{-6}; 38 = 10^{-7}; 40 = 10^{-5}; 84 = 10^{-8}; 132 = 10^{-9}, all cm^3/g at standard temperature and pressure. Numbers in parentheses are cumulative percent errors of the values listed. Only errors which could be experimentally determined are included: weighing error, standard deviation of calibrations, standard deviation of dilutions, standard deviation of peak heights, and errors in blank corrections. We believe that the accuracy of the data is within twice the listed errors. Errors of element ratios such as $^4He/^{20}Ne$ can be computed from this table, but errors in isotope ratios such as $^{20}Ne/^{22}Ne$ cannot.

Sample	Helium		Neon			Argon			Krypton	Xenon
	3	4	20	21	22	36	38	40	84	132
1	923 (1.9)	246 (2.1)	240 (2.1)	640 (3.1)	180 (1.9)	403 (1.8)	754 (1.7)	49.4 (2.0)	18.2 (4.7)	40.3 (9.0)
2	818 (2.4)	225 (2.8)	236 (2.4)	633 (2.7)	180 (2.5)	390 (2.0)	726 (2.0)	48.6 (2.0)	16.4 (3.1)	34.3 (2.6)
3	101 (1.9)	11.7 (1.9)	10.1 (1.7)	150 (1.9)	8.55 (1.7)	17.2 (1.7)	41.2 (1.7)	14.3 (12)	2.3 (10)	10.7 (10)
4	74.6 (1.7)	21.2 (1.9)	28.0 (1.7)	93.5 (3.4)	21.8 (1.7)	86.5 (1.7)	164 (1.7)	15.9 (2.5)	4.66 (8.4)	16.0 (17)
5	80.8 (1.7)	19.1 (1.9)	18.6 (1.7)	93.4 (4.3)	14.6 (1.7)	26.9 (1.7)	56.2 (1.7)	8.72 (5.1)	1.50 (20)	6.3 (20)
6	119 (2.6)	26.1 (1.9)	26.4 (1.7)	143 (3.3)	21.0 (1.7)	50.0 (1.7)	101 (1.7)	13.6 (1.9)	2.71 (15)	7.0 (20)
7	297 (1.8)	75.0 (1.9)	79.3 (1.7)	264 (2.6)	61.1 (1.7)	145 (1.7)	275 (1.7)	24.8 (1.9)	7.4 (10)	19 (20)
8	693 (1.9)	182 (1.9)	202 (1.7)	564 (2.5)	151 (1.7)	333 (1.7)	623 (1.7)	41.2 (1.9)	13.7 (7)	34 (15)
9	368 (1.9)	97.8 (1.9)	129 (1.7)	380 (1.9)	98.8 (1.7)	277 (1.9)	518 (1.7)	34.9 (2.1)	11.9 (10)	34 (15)
10	107 (2.0)	24.2 (1.9)	29.2 (1.7)	107 (6.2)	22.3 (1.9)	36.5 (2.0)	70.7 (1.7)	7.74 (2.6)	2.1 (20)	8.6 (50)
11	7.8 (16)	0.9 (50)	3.34 (5.3)	34 (50)	2.85 (8.6)	5.87 (5.5)	12.0 (6.8)	19.9 (13)		
12	46.2 (4.4)	11.5 (2.2)	26.0 (2.0)	140 (5.3)	20.8 (2.7)	76.9 (1.9)	149 (2.1)	36.4 (2.9)	4.8 (20)	7.6 (50)
13	15.9 (5.6)	3.59 (4.4)	13.2 (2.2)	66 (11)	10.4 (2.8)	14.4 (2.3)	28.7 (2.8)	37.6 (2.8)		
14	11.4 (11)	1.95 (3.6)	1.76 (3.7)	13.7 (15)	1.33 (4.5)	3.31 (3.4)	9.9 (3.7)	2.44 (14)		
15	9.1 (2.6)	2.24 (1.9)	3.02 (1.9)	7.6 (12)	2.24 (2.6)	6.96 (1.9)	13.1 (1.9)	4.19 (5.3)		
16	8.0 (5.2)	2.03 (1.9)	2.16 (1.9)	33.9 (6.2)	1.92 (4.3)	4.45 (4.4)	12.6 (2.7)	7.18 (8.1)		
17	23.5 (2.7)	2.85 (2.6)	3.33 (1.9)	46 (10)	3.02 (4.8)	2.02 (2.8)	3.8 (2.8)	3.15 (6.2)		
18	8.8 (5.3)	2.44 (2.6)	7.59 (1.7)	68.9 (5.2)	6.46 (2.6)	9.19 (1.7)	19.0 (2.6)	6.05 (3.4)		

(0.879); 4, lithic fragment with black fusion crust, ~800 μm (0.704); 5, eight lithic fragments, ~500 μm (0.644); 6, twenty-nine lithic fragments, ~250 μm (0.637); 7, sieve fraction 70 to 105 μm (0.574); 8, sieve fraction < 70 μm (0.409); 9, strongly magnetic portion 70 to 105 μm (0.499); 10, weakly magnetic portion 70 to 105 μm (0.552); 11, brown sphere and spindle, ~250 μm (0.051); 12, three black spheres, ~400 μm (0.175); 13, dark brown glass, ~400 μm (0.171); 14, white "sugary" fragment (feldspar ?), ~550 μm (0.243); 15, lithic fragment with glass and ilmenite, ~800 μm (1.050); 16, four transparent colorless fragments, ~350 μm (0.519); 17, yellow glass, ~600 μm (0.420); 18, green glass, ~500 μm (0.467). The samples were melted in a vacuum furnace, using induction heating. The gases were then cleaned on hot Ti-Zr getters. Absolute amounts were obtained by peak comparison, with carefully metered standards to calibrate the sensitivity of the spectrometer. Preliminary results are listed in Table 1. These results confirm one finding of the preliminary examination team (*1*), namely that the Apollo 11 "fines" contain vast quantities of the inert gases. However, glassy fragments (samples 11–18) are nearly always poorer in gas than lithic fragments of the same size. The same difference is shown by the strongly and weakly magnetic portions of the 70 to 105-μm fraction (samples 9 and 10): the latter contains a substantially greater proportion of clear glassy fragments. The difference could in part be explained by diffusion loss of gas, because the lowest $^4He/^{20}Ne$ ratios occur among samples 11–18. Closer inspection of the data reveals that this cannot be the sole cause—some of the highest $^{20}Ne/^{36}Ar$ ratios are found among the same samples. Perhaps the glassy fragments were formed more recently than the lithic fragments; hence they were exposed to solar wind and solar cosmic rays for a shorter time. Alternatively, the diffusivity of the inert gases in the glasses is so high that gas implanted by solar wind was nearly wholly lost from them, leaving the more firmly implanted solar cosmic-ray ions behind, which might explain the increased $^{20}Ne/^{36}Ar$ ratios. Even so, the ^4He diffusivity in the glasses must be exceptionally high to account for the low $^4He/^{20}Ne$ ratios.

For samples 3–10 (we restrict this discussion to lithic fragments of similar composition and mineralogy), the amount of ^{36}Ar per gram increases in the smaller fragments (Fig. 1), which suggests that the gas is surface-correlated. There is, of course, a possibility that part of the gas is really contained in the very fine, micrometer sized (or less) dust which adheres loosely to surfaces of larger fragments, thus simulating the correlation of Fig. 1. However, an acetone-cleaned sample of fragments in the 105 to 250 μm range agrees closely with the correlation (Fig. 1, point 19). This shows that the trend is real.

Thus, we conclude that most of the inert gases were implanted in the fragments from the solar wind, and from solar cosmic rays of greater energy. We estimate that the average "hemispherical" exposure of a 250-μm fragment to solar wind of present-day intensity was $\sim 10^3$ years (adopted $H^+ = 10^8/cm^2$ sec; $H/^{36}Ar = 10^5$). However, a calculation based on ^{84}Kr gives $\sim 10^4$ years (adopted $H/^{84}Kr = 4 \times 10^9$). These times become much longer if solar cosmic rays were the source of the gas. In addition to direct implantation of solar wind and solar cosmic rays, gas could have been acquired from gas-rich cosmic dust, firmly bonded to the surface of the soil fragments.

Among the principal elemental ratios, $^4He/^{20}Ne$ ranges from 26 to 116, with 13 of 18 ratios >75. These values are substantially below those normally found in gas-rich chondrites or carbonaceous chondrites (*2*), but are similar to the ratios in the howardites (*3*).

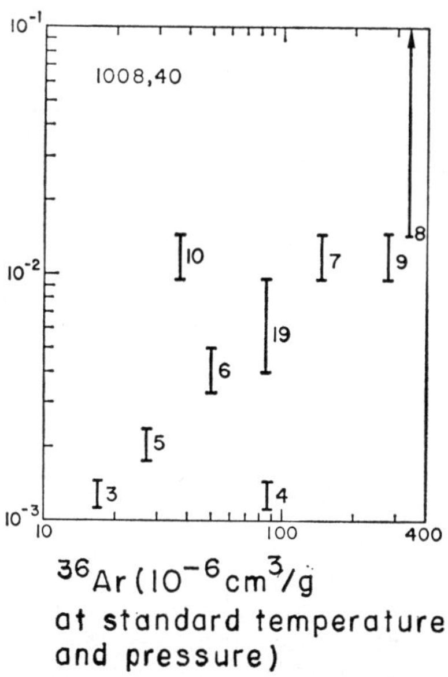

Fig. 1. The ^{36}Ar plotted against inverse of particle size (1/d) for lithic fragments. Glass fragments and feldspar, like the weakly magnetic portion 70 to 105 μm (sample 10) nearly always contain less ^{36}Ar than the lithic fragments.

The $^{20}Ne/^{36}Ar$ ranges from $\sim 3 \simeq 17$, lower than in gas-rich chondrites, but greater than in most carbonaceous chondrites (*2*). The $^{36}Ar/^{84}Kr$ ranges from $\sim 700 \simeq 2400$, similar to the value in the dark portion of the Fayetteville chondrite (*4*), but considerably above the values in carbonaceous chondrites (*2*). Finally, $^{132}Xe/^{84}Kr = 2 \simeq 6$, somewhat above chondritic values (*2*). Certainly, all these ratios are strikingly less than any "solar" or "cosmic" ratios that have been reported. The reasons for this apparent fractionation are not wholly clear; there may be several: preferential diffusion loss of the lighter gases, admixture of meteoritic (chondritic?) material which is known to contain a fractionated component, perhaps a "saturation" effect in the surface for the more abundant, lighter elements, and preferential removal of the lighter gases by sputtering.

A cosmic-ray produced component is present in a number of the samples. We have estimated cosmogenic ^3He_c, $^{21}Ne_c$, and $^{38}Ar_c$ with the following assumptions: "solar" $(^3He/^4He)_s = 3.73 \times 10^{-4}$; $(^{20}Ne/^{21}Ne)_s = 370$; $(^{36}Ar/^{38}Ar)_s = 5.36$; and $(^{36}Ar/^{38}Ar)_c = 0.6$. The amounts of ^3He_c range from ~ 100 to 600×10^{-8} cm^3/g at standard temperature and pressure; $^{21}Ne_c \sim 15 \simeq 120 \times 10^{-8}$ cm^3/g at standard temperature and pressure; and $^{38}Ar \sim 10 \simeq 100 \times 10^{-8}$ cm^3/g at standard temperature and pressure. With one-half of the production rates in Ca-rich achondrites (*5*), the corresponding "exposure ages" are estimated as ranging from tens to hundreds of millions of years. However, the lunar samples may have been exposed to solar cosmic rays as well as to galactic cosmic rays (the effects of solar cosmic rays cannot normally be detected in meteorites because of ablation) such that these "exposure ages" must be considered maximum values. More information on the radioactive species is needed to settle this question more firmly.

The $^3He/^4He$ ranges from $\sim 3.5 \simeq 9 \times 10^{-8}$; the increase is due to ^3He_c as can be seen from concomitant changes in the Ne and Ar isotopic ratios. The $^{20}Ne/^{22}Ne$ ranges from 13.5 to 11.0; $^{21}Ne/^{22}Ne$ from 0.034 to 0.176. On a three-isotope plot (*6*) the points lie near a straight line: $(^{20}Ne/^{22}Ne) = 13.65$ to 14.47 $(^{21}Ne/^{22}Ne)$ (*7*). The isotopic composition of Xe measured in sample 2 is similar, but not identical to that in carbonaceous chondrites. On the basis of $^{132}Xe = 100$, the composition is: 124 (0.48), 126 (0.54), 128 (7.93), 129 (103), 130 (15.7), 131 (81.78), 134 (37.36), 136 (30.86).

In Fig. 2 we have plotted ^{36}Ar against ^{40}Ar in 9 samples (*8*) (most of the glasses and feldspar and samples 3 and 10 do not fall on the correlation, but most lithic fragments do). The points lie close to $^{40}Ar = 1.056\ ^{36}Ar + (7.120 \pm 1.086) \times 10^{-5}$ cm^3/g at standard temperature and pressure. We interpret the correlation as follows: all the samples contain approximately the same amount of K, have about the same apparent K-Ar age, and contain $(7.120 \pm 1.086) \times 10^{-5}$ cm^3/g at standard temperature and pressure radiogenic ^{40}Ar. With K = 0.1 percent by weight (*1*) and terrestrial abundance for ^{40}K, the corresponding K-Ar age is $4.42^{+0.24}_{-0.28}$ billion years. In addition all the samples contain surface-correlated ^{40}Ar with $^{40}Ar/^{36}Ar = 1.056$. This explanation would alleviate the difficulty of embarrassingly high (>7 billion years) K-Ar ages for samples 1 and 2 but it introduces a new puzzle—where does this ^{40}Ar come from? A solar wind origin as postulated for ^{36}Ar seems very unlikely (*9*). The moon itself provides an environment where ^{40}Ar could have been produced by K-decay. Escape of an ^{40}Ar-rich gas from the interior of the moon would make ^{40}Ar available

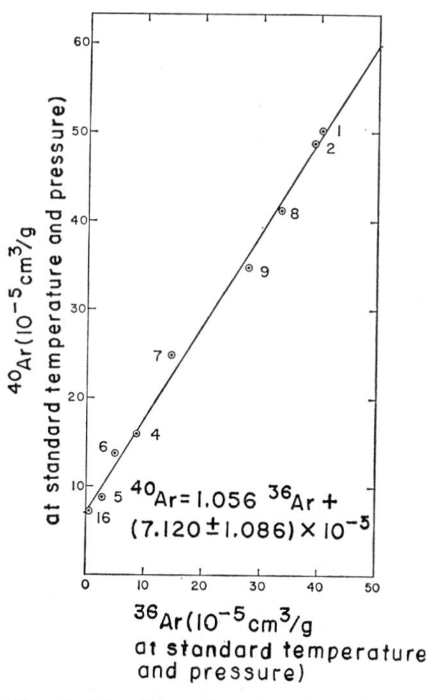

Fig. 2. The ^{36}Ar plotted against ^{40}Ar in nine samples. We conclude that all the samples contain $(7.120 \pm 1.086) \times 10^{-5}$ cm^3/g at standard temperature and pressure of radiogenic ^{40}Ar. With K = 0.1 percent the K-Ar age is $4.42^{+0.24}_{-0.28}$ billion years. The samples also contain surface correlated ^{40}Ar with $^{40}Ar/^{36}Ar = 1.056$.

at the surface (the fraction of ^{40}Ar liberated in the interior need not be large). At the moon's surface this gas must become entrapped in the surfaces of particles. Several mechanisms are possible: shock implantation (*10*), simple dissolution, or "pumping" by elastic collisions with solar wind or solar cosmic-ray ions. We prefer the third possibility because it does not seem to require substantial gas pressures in the moon's atmosphere at any time in the past. An alternative explanation is that ^{40}Ar was acquired from impacting cosmic dust. This requires that the dust is considerably older than the moon, perhaps some 7 to 8 billion years old (*11*).

D. HEYMANN, A. YANIV*
J. A. S. ADAMS, G. E. FRYER
Departments of Geology and Space Science, Rice University, Houston, Texas

References and Notes

1. Lunar Sample Preliminary Examination Team, *Science* **165**, 1211 (1969).
2. E. Mazor, D. Heymann, E. Anders, *Geochim. Cosmochim. Acta*, in press.
3. E. Mazor, E. Anders, *ibid.* **31**, 1441 (1967).
4. O. Müller, J. Zähringer, *Earth Planet. Sci. Lett.* **1**, 25 (1966).
5. D. Heymann, E. Mazor, E. Anders, *Geochim. Cosmochim. Acta* **32**, 1241 (1968).
6. R. O. Pepin, *Earth Planet. Sci. Lett.* **2**, 13 (1967).
7. This equation assumes a two-component mixture. However, Ne in lunar samples may well consist of more than just two components: solar wind, solar cosmic rays (both trapped), and spallation components due to galactic cosmic rays and solar cosmic rays.
8. The point for the acetone-washed sample 19 (105 to 250 μm) which is not shown almost coincides with that of sample 4.
9. D. D. Clayton and W. D. Arnett, private communications.
10. K. Fredriksson and P. DeCarli, *J. Geophys. Res.* **69**, 1403 (1964); R. O. Pepin, J. H. Reynolds, G. Turner, *ibid.*, p. 1406.
11. We thank Drs. E. Anders, D. D. Bogard, R. O. Pepin, and O. A. Schaeffer for helpful discussions and information. Supported by NASA contract NAS-9-7899.
* On leave of absence from the University of Tel Aviv, Israel.

4 January 1970

Trapped Solar Wind Noble Gases, Kr^{81}/Kr Exposure Ages and K/Ar Ages in Apollo 11 Lunar Material

Abstract. *Grain size and etching experiments show that the fine lunar material contains large amounts of trapped solar wind particles. Elemental and isotopic compositions of the noble gases in solar material and in the terrestrial atmosphere are significantly different, except for the* Ar^{36}/Ar^{38} *and the Kr isotope ratios. Exposure ages of two rocks and of the fine material are between 380 and 510* $\times$ 10^6 *years. Feldspar concentrates give K/Ar ages of 3220 and 3300* $\times$ 10^6 *years, significantly higher than the unseparated rock.*

The concentrations and isotopic composition of the noble gases He, Ne, Ar, Kr, and Xe were measured in (i) seven grain-size fractions of fine material (sample No. 10084) (grain sizes between 1 and 150 μm); (ii) four grain-size fractions of ilmenite separated from the fine lunar material (grain sizes between 20 and 110 μm); (iii) four etched samples from an ilmenite grain-size fraction (42 μm); (iv) a number of individual crystals, spherules, and other structural elements separated from fine lunar material; (v) three rock samples from rocks No. 10017 and 10071; and (vi) two feldspar concentrates from rock No. 10071. In addition K and Ba were determined in several of these samples.

Table 1. Comparison of trapped solar gas in lunar ilmenite and terrestrial atmosphere.

Ratio	Trapped solar gas	Terrestrial atmosphere
He^4/Ar^{36}	5300 ± 800;	
Ne^{20}/Ar^{36}	24 ± 3	0.5
Kr^{86}/Ar^{36}	(2 ± 0.3) × 10^{-4}	63 × 10^{-4}
Xe^{132}/Ar^{36}	(1.1 ± 0.2) × 10^{-4}	7 × 10^{-4}
He^4/He^3	2700 ± 100	
Ne^{20}/Ne^{22}	12.90 ± 0.07	9.8
Ne^{20}/Ne^{21}	410 ± 12	338
Ar^{36}/Ar^{38}	5.30 ± 0.05	5.35
Ar^{40}/Ar^{36}	≤0.7.	
Kr isotopes, atmospheric, see text		
Xe isotopes, see text and Fig. 4		

From these results the elemental and isotopic abundances of the trapped noble gases were calculated, and limits on the elemental and isotopic abundances in the solar wind can be inferred. Exposure ages were obtained from Kr^{81}/Kr measurements, and K/Ar ages were derived for the rock samples.

The preliminary examination of the fine lunar material (*1*) already revealed the presence of large amounts of noble gases, presumably of solar wind origin. Other possible sources of these trapped gases (*2*) are energetic solar particles, an ambient atmosphere early in the history of lunar material, or gases released during the impact of comet nuclei. To establish the origin of the trapped gases we measured the noble gas concentration in different grain-size fractions of the bulk material (*3*). The He^4 and Xe^{132} results are given in Fig. 1. A distinct anticorrelation between grain size and trapped noble gas content is evident. Trapped Ne, Ar, and Kr show a similar grain-size dependence.

The experiment was repeated on ilmenite grain-size fractions separated from the fine material. An even stronger anticorrelation between grain size and trapped gas content was observed (Fig. 1 for He^4 and Xe^{132}; Ne, Ar, and Kr show similar dependence). Such a grain size dependence results if the trapped gases are located at the surface of the individual grains and if their concentration per unit of surface area is roughly independent of the grain size.

In a further study of the location of the trapped gases we etched aliquots of the 42 μm ilmenite grain-size fraction with dilute HF to remove a thin layer from the surface of the individual grains. The mean thickness of the etched layer was determined from the amount of Fe and Ti dissolved. The noble gases remaining in the etched grain-size fractions were then measured. While the concentration of the noble gas spallation fraction was essentially unchanged by the HF treatment, the concentration of all trapped gases decreased strongly after removing a surface layer with an approximate average thickness of 2000 Å (Fig. 2). We conclude that the trapped noble gases are located within a depth of 2000 Å below the surface of the individual grains. This is in agreement with the observed grain size dependence of the trapped gas content.

This experimental evidence strongly supports a solar-wind origin of the trapped gases. The range of ions with

Table 2. The K/Ar ages and Kr^{81}/Kr exposure ages of Apollo 11 lunar rocks 10017 and 10071.

Rock No. and fraction	K/Ar ages (× 10^6 years)	Kr^{81}/Kr exposure age (× 10^6 years)
10017 Unseparated	2200 *	510 ± 50
10071 Unseparated	2820 ± 80	380 ± 25
10071 Feldspar CA4	3300 ± 80	
10071 Feldspar CA6	3220 ± 80	

* Ar^{40} concentration without isotopic dilution.

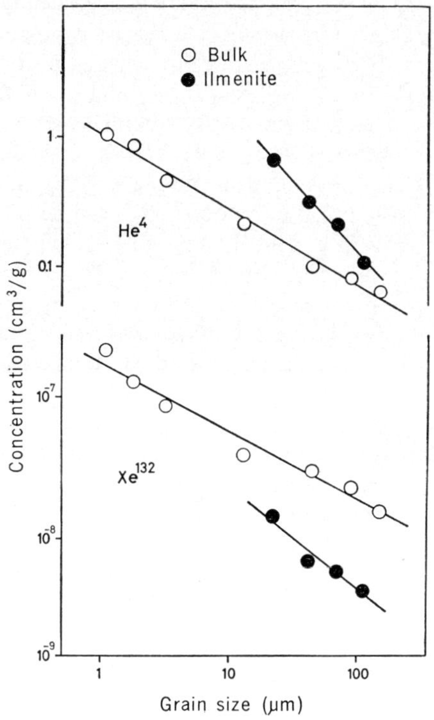

Fig. 1 (left). Grain-size dependency of trapped He and Xe in unseparated Apollo 11 fine material and in ilmenite ($FeTiO_3$) separated from the fine material. Trapped Ne, Ar, and Kr show similar dependencies.

Fig. 2 (right). Trapped He and Xe remaining in the 42-μm ilmenite grain-size fraction after a thin surface layer has been removed from the individual grains by etching with HF. The horizontal axis is equivalent to the thickness of the layer removed from the surface of the individual grains. The two curves for He^4 represent two independent sets of etching experiments. Trapped Ne, Ar, and Kr show similar dependencies.

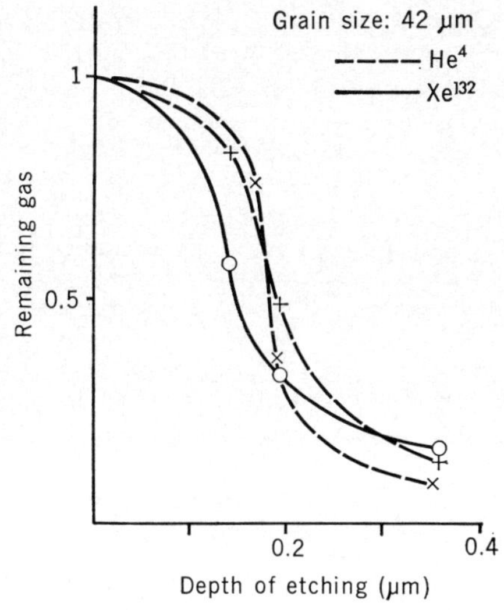

an energy of 1 kev per nucleon, as observed in the present-day solar wind, is of the order of a few hundred to thousand angstroms. Independent experiments have already established that solar-wind noble gas ions indeed reach the lunar surface (*4*).

For ilmenite, the He^4 concentration is inversely proportional to grain size within 10 percent (Fig. 1), that is, the He^4 concentration per square centimeter of surface area is independent of grain size. This requires that either the grains were exposed to the solar wind for the same time interval, independent of grain size, or the exposure was so long that saturation occurred. The observed surface concentrations of 10^{-3} cm^3 He^4/cm^2 correspond to an integrated solar wind flux of 7×10^{17} atom/cm^2. Such an irradiation dose would require an exposure time of approximately 300 years at the present level of the solar wind flux. At these levels of irradiation severe saturation effects have to be expected (*5*), which will most likely influence the relative abundance of the trapped solar wind ions. If we assume that the effective trapping efficiency is reduced to about 10 percent because of saturation, a few hundred million years are needed to obtain the observed helium loading of ilmenite surfaces down to a depth of about 60 cm. This figure compares well with a cosmic ray exposure age of the Apollo 11 sample of fine material of 500×10^6 years, as estimated from $Xe^{126}_{spallation}$ and measured Ba concentrations.

Figure 3 shows the observed range of abundances of trapped gas relative to trapped Ar^{36}. The overall abundance pattern is quite similar to the predicted cosmic abundances (*6*) and to the abundances of trapped gas observed in some gas-rich aubrites (*7–9*). The Ar-Kr-Xe abundances in the bulk fine material and the separated ilmenite are identical. However, the He^4/Ar^{36} and Ne^{20}/Ar^{36} ratios are considerably higher in the ilmenite than in the bulk material. This is best explained by diffusion loss of He and Ne in the bulk material leading to a relative enrichment of trapped Ar, Kr, and Xe. The abundances and isotopic compositions of the trapped gases in the Apollo 11 fine material are not necessarily identical with those of the solar wind. They may have been altered by diffusion losses, systematic differences in trapping probabilities and saturation effects. Diffusion loss and the mass dependency of trapping probabilities should always lead to a depletion of the light gases and isotopes. The highest observed He^4/Ar^{36} and Ne^{20}/Ar^{36} ratios should thus be most representative of the true solar-wind abundances. Since no experimental evidence on the influence of saturation effects on the elemental and isotopic composition is presently available we have to disregard this effect.

The four ilmenite fractions show the least evidence for diffusion loss, and the abundance ratios for the trapped component determined from these mineral concentrates are given in Table 1. The errors correspond to the range of values observed in the four different ilmenite grain-size fractions (except Ar isotope ratios). The necessary corrections for spallation-produced He^3, Ne^{21}, and Ar^{38} were obtained from correlation diagrams and the HF-etched ilmenite fractions. The Ar^{40}/Ar^{36} ratio given is the lowest we observed. A contribution of radiogenic argon from the decay of K^{40} cannot be excluded, and the ratio Ar^{40}/Ar^{36} of 0.7 must be considered as upper limit for the Ar^{40}/Ar^{36} ratio of the trapped gas. All grain-size fractions of the bulk material give Ar^{40}/Ar^{36} ratios higher than 0.7. In the finer grain-size fractions the measured K concentration is not high enough to account for this Ar^{40} excess in terms of *in situ* decay.

All ilmenite fractions have a sizable spallation Kr content, and the very precise evaluation of the trapped Kr isotopic composition which is required to distinguish between atmospheric Kr and AVCC Kr (*10*) is difficult. The bulk grain-size fractions contain relatively little spallation Kr, and correlation diagrams indicate that the trapped Kr in these fractions is more similar to atmospheric Kr than to AVCC Kr.

All bulk and ilmenite grain-size fractions show spallation Xe. From suitable correlation diagrams we can show that the trapped Xe cannot have atmospheric composition. The correlation lines for the light Xe isotopes Xe$^{124,\,126,\,128,\,130}$ virtually pass through the points corresponding to AVCC Xe, and we conclude that the relative abundances of these isotopes in the trapped gas can be compatible with AVCC Xe within 2 percent. We shall use the AVCC isotopic composition for these isotopes for the further discussion. The abundance of the heavy isotopes Xe$^{131,\,132,\,134,\,136}$ relative to Xe^{130} is significantly lower than in atmospheric or AVCC Xe (spallation correc-

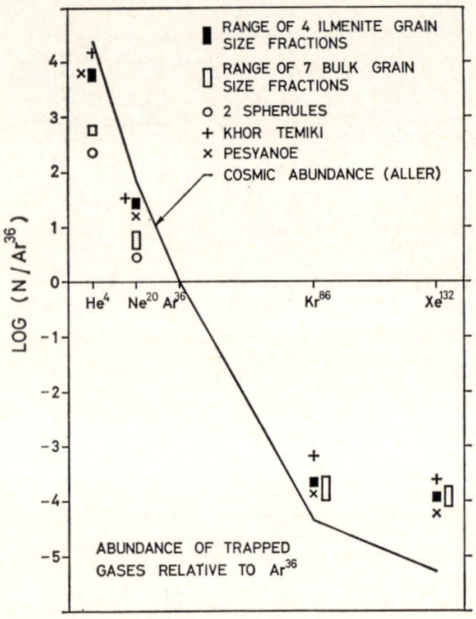

Fig. 3. Trapped gas abundances relative to Ar^{36} in bulk grain size (1 to 150 μm) and ilmenite grain size (20 to 110 μm) fractions separated from Apollo 11 fine lunar material. The two spherules separated from the fine lunar material had diameters of 0.2 mm. Other data from (*6–9*).

tion applied). From six fractions with the smallest spallation we obtain the δ values given in Fig. 4.

The mass dependence of the absolute abundances of noble gases in the trapped gases is less steep than the predicted cosmic abundances (*6*) (Fig. 3). We cannot yet decide whether this is due to saturation effects in the trapping of the solar wind or to mass discrimination in the solar wind or whether this reflects the true chemical composition of the solar wind reservoir.

For the discussion of the solar-wind noble gas isotopic compositions we will assume that these ratios represent to a

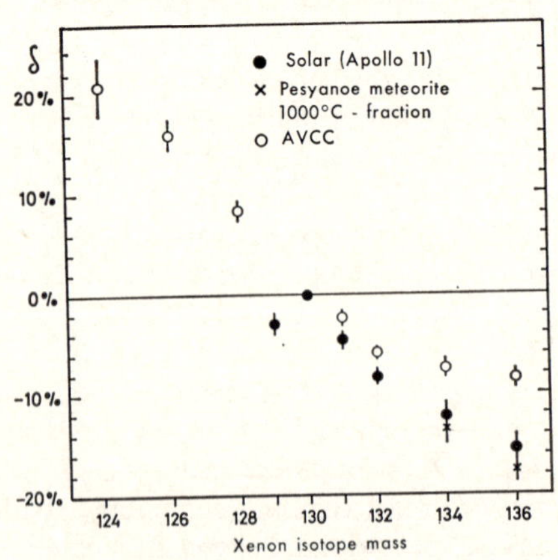

Fig. 4. Isotopic composition of trapped solar wind Xe in Apollo 11 fine material, of AVCC Xe (*10*) and of trapped Xe in the Pesyanoe meteorite (*9*).

$\delta = [(Xe^M/Xe^{130})/(Xe^M/Xe^{130})_{atm} - 1]$ is the relative deviation from the atmospheric composition.

good approximation the true isotopic compositions prevailing at the time of formation of the solar system. The lower terrestrial Ne^{20}/Ne^{22} and Ne^{20}/Ne^{21} ratios are in excellent agreement with a mass dependent fractionation process (for example gravitational escape), terrestrial Ne representing the remaining few per mil of the original gas. If we calculate the overall fractionation factor from the Ne^{20}/Ne^{22} ratio, then the terrestrial Ne^{20}/Ne^{21} ratio predicted from the solar wind Ne^{20}/Ne^{21} ratio is rather high compared with the observed terrestrial value. This may indicate that the terrestrial atmosphere contains some Ne^{21} from spallation reactions or (α,n) reactions on O^{18} (*11*). Terrestrial and solar Ar^{36}/Ar^{38} ratios agree within the present experimental uncertainty.

The difference between the abundance of the heavy isotopes in AVCC Xe and solar Xe could be explained by a fission component in AVCC Xe. The required fission spectrum would be $Xe^{131} : Xe^{132} : Xe^{134} : Xe^{136} = 0.74 : 1.03 : 0.81 : 1.0$ with $Xe^{136}_{fission}/Xe^{136} = 0.08$ in carbonaceous chondrites. A possible long-lived isotope from the "island of stability" near $N=184$ might be the progenitor (*12*). However, if fission Kr is indeed present in AVCC Kr (*10*), we would have $(Xe^{136}/Kr^{86}{}_{fission}) \sim 10$, a ratio much too low for the fission of super heavy nuclides. The large differences observed between solar wind and atmospheric Xe are difficult to explain as the mass fractionation factors proposed are very hard to realize, especially because virtually no fractionation is observed in Kr and Ar. The assumption of a completely different origin for terrestrial Xe might be required to explain the Xe isotope anomalies, with AVCC Xe and terrestrial Xe originating from different nucleosyntheses.

The noble gases extracted from lunar rocks 10017 and 10071 are essentially mixtures of pure spallation noble gases and radiogenic He^4 and Ar^{40}. Compared to the fine material the trapped gas content is very low. The abundance patterns of the spallation noble gases are similar to the well known spallation yield spectra observed in meteorites. The K/Ar ages and Kr^{81}/Kr exposure ages (*13*) are given in Table 2.

The K/Ar ages of the two feldspar fractions are distinctly higher than the age of the whole rock, reflecting the limited significance of total rock K/Ar ages. The feldspar fractions show He^3 diffusion loss and we cannot yet decide on the basis of this data alone, whether their K/Ar ages are significantly affected by prolonged diffusion losses or shock, or whether they represent the time of crystallization and rapid cooling.

The exposure ages are an order of magnitude lower than the K/Ar ages. Hence crystallization and ejection of the rocks must have occurred at different times. It cannot be decided yet whether the difference in exposure age of the two rocks reflects variable shielding conditions during exposure, or whether the two rocks have been ejected at different times.

From rock 10071 we calculate the following spallation production rates per million years in units of $10^{-12}cm^3$ STP/g: $He^3 = 7200$; $Ne^{21} = 1000$; $Ar^{38} = 1300$; $Kr^{83} = 2.7$; and $Xe^{126} = 0.44$.

P. Eberhardt
J. Geiss, H. Graf
N. Grögler
U. Krähenbühl
H. Schwaller
J. Schwarzmüller
A. Stettler

Physikalisches Institut, University of Berne, 3000 Berne, Switzerland

References and Notes

1. Lunar Sample Preliminary Examination Team, *Science* **165**, 1211 (1969).
2. As in meteorite research we use the term "trapped gases" for the gases which are not the result of *in situ* nuclear processes (radioactive decay, spallation reactions, neutron capture, and so forth) (*8*).
3. The grain size and etching experiments are similar to those made with the meteorite Khor Temiki (*7*).
4. F. Bühler, P. Eberhardt, J. Geiss, J. Meister, P. Signer, *Science* **166**, 1502 (1969).
5. H. C. Lord, *J. Geophys. Res.* **73**, 5271 (1968).
6. L. H. Aller, *The Abundance of the Elements* (Interscience, New York, 1961).
7. P. Eberhardt, J. Geiss, N. Grögler, *Tschermaks Mineral. Petrog. Mitt.* **10**, 535 (1965); *J. Geophys. Res.* **70**, 4375 (1965).
8. O. Eugster, P. Eberhardt, J. Geiss, *J. Geophys. Res.* **74**, 3874 (1969).
9. K. Marti, *Science* **166**, 1263 (1969).
10. The abbreviation AVCC is used to denote the average Kr and Xe isotopic composition observed in carbonaceous chondrites. The AVCC Kr shows a 0.3 percent per mass unit systematic depletion of the light isotopes relative to atmospheric Kr and perhaps a small fission Kr component. The AVCC Xe shows isotopic anomalies up to 20 percent [D. Krummenacher, C. M. Merrihue, R. O. Pepin, J. H. Reynolds, *Geochim. Cosmochim. Acta* **26**, 231 (1962); O. Eugster, P. Eberhardt, J. Geiss, *Earth Planet. Sci. Lett.* **3**, 249 (1967)].
11. G. W. Wetherill, *Phys. Rev.* **96**, 679 (1954).
12. E. Anders and D. Heymann, *Science* **164**, 823 (1969); M. Dakowski, *Earth Planet. Sci. Lett.* **6**, 152 (1969).
13. K. Marti, *Phys. Rev. Lett.* **18**, 264 (1967); O. Eugster, P. Eberhardt, J. Geiss, *Earth Planet. Sci. Lett.* **2**, 77 (1967).
14. We thank M. Bill, R. Küpfer, P. Bochsler, H. Debrunner, R. Giovanoli, H. Hofstetter, V. Horvath, E. Lenggenhager, A. Schaller, U. Schwab, and H. Wyniger for their help. Supported by the Swiss National Science Foundation (grants NF 2. 73. 68 and NF 5079).

4 January 1970

Gas Analysis of the Lunar Surface

Abstract. *The rare gas analysis of the lunar surface has lead to important conclusions concerning the moon. The large amounts of rare gases found in the lunar soil and breccia indicate that the solar atmosphere is trapped in the lunar soil as no other source of such large amounts of gas is known. The cosmogenic products indicate that the exposure ages of the 17 lunar rocks measured vary from 20 to 400 million years with some grouping of the ages. The most striking feature is the old potassium-argon age which for the 14 rocks analyzed varies from 2.5 to 3.8 billion years. It is concluded that Mare Tranquillitatis crystallized about 4 billion years ago from a molten state produced by a large meteorite impact or volcanic flow.*

The rare gases offer important clues to the history of the lunar surface layers. Variations in isotopic abundances and absolute amounts of the noble gases indicate that a separation from the primordial gas component took place early in the moon's evolution such that subsequent additions of rare gases are rather easily traced. We now have evidence that the processes that add rare gases to the moon include radioactive decay and solar wind, solar flare and cosmic ray implantation, and nuclear reactions induced by the higher energy particles. Interpretation of rare gas measurements leads to surface exposure ages, the time of crystallization of rocks, and insight into the composition of the solar surface.

Most of the gas analyses were made with a 6-inch, 60° magnetic deflection mass spectrometer. A number of the "gassier" samples were analyzed on a less sensitive 5-inch, 180° instrument.

Standard techniques of gas extraction, chemical purification with hot titanium, and separation were employed (*1*). All samples were initially degassed at 125° to 150°C for 5 to 24 hours prior to introduction into the vacuum system. Subsequent adsorption of atmospheric rare gases was considered negligible in most samples after we observed the small amount of gases released at 400° to 450°C in several of the samples.

The 6-inch spectrometer was operated at a sensitivity of 2×10^{-10} cm^3/mv for He, Ne, Ar and 3×10^{-13} cm^3/mv for Kr and Xe. The 1650°C-furnace blank comprised most of the background and it normally amounted to $< 5 \times 10^{-9}$ cm^3 ^{3}He, 1×10^{-7} cm^3 ^{4}He, 9×10^{-9} cm^3 ^{20}Ne, 7×10^{-8} cm^3 ^{40}Ar, 6×10^{-12} cm^3 ^{84}Kr, and 2×10^{-12} cm^3 ^{132}Xe. Long term reproducibility determined from daily standards, was within $\pm$ 5 percent for all five noble gases. Isotopic ratios were corrected for mass spectrometer discrimination and are accurate to $\pm$ 2 percent for the gases He, Ne, and Ar and less than 1 percent for the gases Kr and Xe. The accuracy in the absolute amount of each gas is estimated to be within $\pm$ 20 percent.

The high gas content of the fines and breccia allowed the use of milligram size samples. Only interior portions of crystalline rocks were analyzed and these were mechanically cleaned of contaminating fines; nevertheless, some degree of solar contamination was unavoidable. Most samples of the crystalline rocks measured ranged from 10 to 100 mg.

Because of the large amounts of solar gases in the fines and breccia (Table 1), surface exposure and K-Ar ages could only be calculated for the crystalline rocks (Table 2) plus two lithic fragments handpicked from two different breccias. Both the radiation and K-Ar ages are depicted (Fig. 1). Corrections for determining spallation gases were applied in the normal manner using solar ratios measured in the fines and breccia. No atmospheric ^{40}Ar was assumed present in the calculation of radiogenic Ar, and the solar correction was negligible in most all cases.

The breccia and fines contain extraordinarily large amounts of noble gases which are derived predominantly from the solar wind. Compared to relative cosmic abundances, depletion of the major elemental isotope decreases with increasing atomic weight, as would be expected, invoking loss by diffusion and sputtering-saturation mechanisms. The average relative abundances of ^{4}He, ^{20}Ne, ^{36}Ar, ^{84}Kr, ^{132}Xe in both the fines and breccia samples, normalized to ^{36}Ar (100), are 42,000; 660; 100; 0.054; 0.020, respectively. Both the absolute and relative amount of gas varies by a factor of three, even within a single sample, yet the isotopic abundances vary only slightly. The ratio ^{4}He/^{3}He is about 2800, in agreement with estimates of solar abundances (*2*). The average value for ^{20}Ne/^{22}Ne is 12.4 and for ^{22}Ne/^{21}Ne, about 30.7 (corrected for spallation ^{21}Ne). Both these ratios are similar to those in gas rich meteorites. The high, yet variable, ^{20}Ne/^{22}Ne ratio, relative to the terrestrial atmosphere, might be attributed to ^{23}Na (p,α) ^{20}Ne reactions generated by solar flares in the surface layers of the lunar fines. Shock or some similar process apparently tends to homogenize the noble gases throughout the fines and breccia on a small scale as indicated by temperature release patterns. Yet millimeter-size lithic fragments removed from the breccia contain only small amounts of cosmic noble gases.

The breccias returned by Apollo 11 generally contain more gas than the fines. The isotopic ratios are similar with one notable exception: ^{40}Ar/^{36}Ar is about 1.1 in the fines, but averages 2.2 in the breccia samples. Theoretical

Table 1. Rare gas content of fines and breccia (concentration in units of 10^{-8} cm^3/g).

Sample	Helium		Neon			Argon			^{84}Kr	^{132}Xe
	^{4}He	4/3	^{20}Ne	20/22	22/21	^{36}Ar	40/36	36/38		
				Fines						
10010	11,000,000	2430	200,000	12.4	31.6	35,000	1.1	5.20	21	10
10010	19,000,000	2540	313,000	12.8	29.4	34,000	1.2	5.17	20	4.1
				Breccia						
10018	24,000,000	2650	480,000	13.4	27.8	72,000	2.1	5.16	35	12
10021	37,000,000	3300	750,000	12.4	30.6	100,000	2.1	5.13	28	10
10023	25,000,000	2570	350,000	12.7	29.3	52,000	2.4	5.15	24	11
10027	15,000,000	2930	320,000	12.2	29.9	67,000	2.2	5.08	41	12
10048	21,000,000	2600	350,000	11.1	28.6	55,000	2.1	5.22	31	12
10061	47,000,000	3400	710,000	12.1	30.0	89,000	1.8	5.14	48	11
10068	25,000,000	2780	330,000	12.2	29.0	42,000	2.7	5.22	20	7.5

calculations for elemental synthesis predict an $^{40}Ar/^{36}Ar$ ratio of less than 0.01 percent (*2*), thus the large amounts of ^{40}Ar observed must be of radiogenic origin, not produced by *in situ* decay of ^{40}K but adsorbed and/or injected into the fines and breccia in characteristically different amounts. Perhaps the breccia were not derived locally, but from fines with a higher $^{40}Ar/^{36}Ar$ ratio in another site. The $^{36}Ar/^{38}Ar$ ratio is a remarkably constant 5.28 ± .06.

The major Kr and Xe component in the fines and breccia is ascribed to a solar wind origin. The isotopic composition among these samples varied only slightly and showed an average composition for xenon 124/126/128/129/130/131/132/134/136 of 0.57/0.65/8.5/104/16.5/82.5/≡100/37.1/30.4, and an average composition for krypton 78/80/82/83/84/86 of 0.72/4.10/20.4/20.4/≡100/30.8. The lighter isotopes of Kr and Xe show a variable excess in the fines and breccia over the Kr and Xe trapped in carbonaceous chondrites. Xenon-134 and -136 appear depleted by several percent in the solar xenon component from that observed trapped in carbonaceous chondrites. This particular xenon isotopic pattern in lunar fines and breccia is consistent with the trapped plus fission two-component xenon found in temperature release studies of carbonaceous chondrites but shows values for the 134/132 and 136/132 ratios lower than any temperature release performed to date on these chondrites (*3*). The $^{129}Xe/^{132}Xe$ ratio in the solar component is the same as that found in typical carbonaceous chondrites.

The breccias have many of the characteristics of gas rich meteorites. It is not unreasonable, then, to suggest that gas rich meteorites were formed in a manner similar to the lunar breccia; that is, on a surface of a large body irradiated by solar and cosmic particles, with a certain amount of surface mixing, and then lithified by shock generated by meteoritic impact.

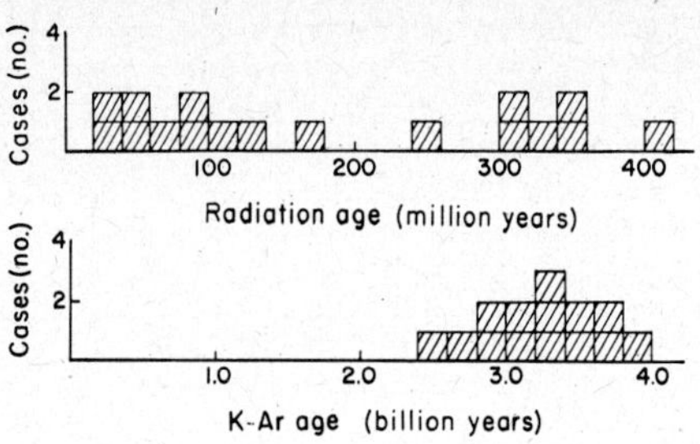

Fig. 1. Histograms of radiation ages (upper) and K-Ar ages (lower) of crystalline rocks.

All the crystalline rocks measured contain noble gases produced by spallation reactions. The amount of the spallogenic gases depends on the time the sample was exposed to cosmic rays, whereas the relative isotopic abundance reflects both the chemistry and the degree of radiation shielding of the sample. Elemental isotopic abundances of 15 crystalline rocks show an average value of 9 for $^{3}He/^{21}Ne$ and 0.9 for $^{21}Ne/^{38}Ar$.

Assuming 2π-geometry for the lunar surface rocks and, therefore, a production rate of 1.0×10^{-8} cm^3 per million years for ^{3}He, then absolute radiation ages can be assigned to the different samples. Production rates, relative to ^{3}He, have been derived for ^{21}Ne and ^{38}Ar from excitation functions applied to the average chemical composition of the crystalline rocks. The variation in the measured $^{3}He/^{21}Ne$ ratio appears to be dependent upon the effective energy spectrum rather than chemical composition. The ^{38}Ar is generated primarily from Ca, and the production rate is also a strong function of radiation hardness. The calculated $^{21}Ne/^{38}Ar$ ratio is rather uncertain because of limited cross-section data for Ca. Nevertheless, exposure ages determined from the calculated production rates of ^{21}Ne and ^{38}Ar agree within 20 percent of the ^{3}He based radiation age in almost all of the 17 samples measured.

Production rates of ^{21}Ne based on the measured ^{22}Na content of three rocks (*1*) agree well with calculated values as does a ^{3}He-T age on a crystalline rock that was analyzed for tritium (*4*). This fact, the general concordancy of the ^{3}He, ^{21}Ne, ^{38}Ar ages, and the concordant K-Ar and U-He ages (discussed later) imply that no widespread diffusive gas loss has occurred for the spallogenic gases in the crystalline rocks.

The radiation ages (Fig. 1) range from 20 to 400 million years and fall into two rather amorphous groups at about 100 and 350 million years. Spallation gases in lithic fragments from two breccia samples indicate radiation ages of approximately 350 million years. Thus, the breccia and the rocks were scattered on the surface of Mare Tranquillitatis by few, but distinct, events, most probably meteoritic impact. The estimated thickness of the regolith in the vicinity of the Apollo 11 landing site (*1*) and an assumed age of the basement rock of about 4 billion years leads to an average rate of "soil" formation of 1 to 2 mm per million years. In addition, the fact that even the rocks with the youngest surface ages exhibited rounding and that no rocks were found with radiation ages greater than about 400 million years again implies a combined erosion and surface coverage rate of 1 to 2 mm per million years Some degree of turnover in the top meter of the regolith must be assumed in order to provide lunar rocks older than a few tens of million years that still remain uncovered.

Table 2. Rare gas content of crystalline rocks (concentrations in units of 10^{-8} cm^3/g).

Sample	^{3}He	^{4}He	^{20}Ne	^{21}Ne	^{22}Ne	^{36}Ar	^{38}Ar	^{40}Ar	^{84}Kr	^{132}Xe
10017	310	58,000	110	47	56	37	47	4100	0.080	0.040
10018 *	420	260,000	5,100	76	460	690	230	6100	0.51	0.17
10020	110	20,000	72	14	20	20	19	1800	0.046	0.031
10022	350	63,000	210	52	67	59	50	5700	0.13	0.063
10024	260	50,000	220	32	76	75	68	4100	0.46	0.39
10032	150	15,000	750	16	73	98	32	7400	0.071	0.041
10044	82	34,000	250	8.8	27	36	15	2500	0.094	0.046
10047	96	25,000	210	15	30	43	19	1500	0.090	0.036
10049	25	77,000	73	2.6	8.7	11	5.0	5600	0.059	0.021
10050	440	80,000	900	51	110	150	73	2500	0.35	0.21
10057	44	64,000	110	5.7	14	19	9.7	4800	0.020	0.013
10058	58	21,000	64	7.1	12	15	12	3700	0.015	0.007
10061 *	350	95,000	13,000	74	1000	2300	620	8200	1.58	0.46
10069	28	69,000	38	4.0	7.0	6.0	5.4	6500	0.007	0.005
10070	110	98,000	790	11	68	110	32	2400	0.071	
10071	320	73,000	177	37	52	36	34	6100	0.050	0.029
10072	190	75,000	235	26	44	41	36	7600	0.033	0.012

* Lithic fragments handpicked from breccia.

The lunar crystalline rocks show a wide isotopic variation in krypton and xenon, all characterized by excesses of xenon 124–132 and krypton 78–84 relative to ^{136}Xe and ^{86}Kr, respectively. These excesses are attributed to cosmic-ray spallation reactions in the rocks, mainly on Ba, Sr, Y, and Zr. The spallation spectrum for the various rocks are all similar and are (normalized to $^{126}Xe \equiv 1$ and $^{83}Kr \equiv 1$): 124/126/128/129/130/131/132/134 ≡ 0.55 ≡ 1/1.5/1.7/1.07/6.3/0.9/0.07, and 78/80/82/83/84 ≡ 0.2/0.48/0.76 ≡ 1/0.38. These spectra are quite similar to the spallation spectra found in achondrites except for a much higher ^{131}Xe and a somewhat lower ^{84}Kr. The higher 131 yield is probably an effect of low energy protons as barium is the major target for spallation Xe production, and 730-mev proton spallation of barium does not show this high ^{131}Xe yield (*5*). The amounts of ^{126}Xe and ^{83}Kr in the rocks vary by almost an order of magnitude among samples because of the differing exposure histories, yet these amounts correlate reasonably well with the spallation ^{38}Ar. The slight isotopic excesses of the lighter isotopes of Kr and Xe in the lunar fines and breccia over that in carbonaceous chondrites is also probably due to spallation reactions. Not only is the amount of excess ^{126}Xe in different breccia variable, but in amount it is only slightly greater than the spallation ^{126}Xe found in those crystalline rocks with the longest exposure history.

Radiogenic Ar and He are apparent in all crystalline rocks examined. K-Ar ages were obtained for 14 rocks and fragments for which K was known (*1*). Potassium was determined by emission spectroscopy but not from aliquots used for noble gas analysis. The K-Ar ages, including two from lithic fragments from two breccia samples, range from 2.5 to 3.8 billion years (Fig. 1). The U-He ages are subject to large errors because of the solar wind correction to the measured ^{4}He and the uncertainty in the U-Th content derived from the K values (*1*); nevertheless, most of the ages are concordant within this error with the K-Ar data. Thus, the spread in the K-Ar ages is probably real to some extent, and not just caused by variable leakage of radiogenic Ar.

Mare Tranquillitatis, then, probably first crystallized about 4 billion years ago from a large molten body produced either by meteoritic impact or volcanism. Subsequent large impacts or volcanic events would further mold the mare to its present form and produce younger rocks. Despite the time spread of approximately 1 billion years in volcanic events, it appears that the features of the moon at Mare Tranquillitatis were well delineated at least 2 billion years ago.

J. G. FUNKHOUSER
O. A. SCHAEFFER
State University of New York, Stony Brook
D. D. BOGARD
NASA Manned Spacecraft Center, Houston, Texas
J. ZÄHRINGER
Max-Planck-Institut für Kernphysik Heidelberg, Germany

References and Notes

1. Lunar Sample Preliminary Examination Team, *Science* **165**, 1211 (1969).
2. D. D. Clayton, *Principles of Stellar Evolution and Nucleosynthesis* (McGraw-Hill, New York, 1968).
3. J. H. Reynolds and G. Turner, *J. Geophys. Res.* **69**, 3263 (1964); H. Funk, F. Podosek, M. W. Rowe, *Geochim. Cosmochim. Acta* **31**, 1721 (1967); M. W. Rowe, *ibid.* **32**, 1317 (1968).
4. R. Davis, private communication.
5. H. Funk, *Earth Planet. Sci. Lett.* **2**, 215 (1967).
6. We wish to acknowledge helpful discussions with D. Clayton (Rice University) regarding nuclear reactions, especially proton reactions on sodium to produce neon. We thank W. Hart, W. Hirsch, D. Moore, C. Polo, L. Simms, R. Wilkin of Brown & Root-Northrop (MSC) and G. Barber, R. Warasila of SUNY, Stony Brook, for their assistance in the analyses. Supported by NASA contract NAS 9-8820.

5 January 1970

Solid State Studies of the Radiation History of Lunar Samples

Abstract. *Particle track densities up to* $> 3 \times 10^9$ *per square centimeter have been measured in different samples. Rocks 17, 47, 57, and 58 have VH* ($Z > 22$) *galactic cosmic ray ages of 11, 14, 28, and 13* $\times 10^6$ *years, respectively. Rock 57 has a calculated erosion rate of* $\lesssim 10^{-7}$ *centimeter per year. Near-surface track versus depth data in rock 17 can be fit with solar flare particles that have a differential energy spectrum* αE^{-3}; *lunar samples can be used to study the history of solar activity. The uranium in the crystalline rocks occurs principally in small regions* <10 *to* ~ 100 *micrometers in size. The (low) thermoluminescence of the fines increases with depth in core 10004. With one possible exception, x-ray studies have not shown pronounced radiation damage effects. The total energy release upon heating is small up to 900°C and occurs in three broad regions.*

We have measured particle tracks, thermoluminescence, x-ray distortions, U distributions, and energy release patterns in lunar samples. Table 1 gives a summary of track data. The lower track densities were measured with an optical microscope and the higher densities with a Cambridge Mark IV stereo-scan electron microscope. A series of cross calibrations (Fig. 1) have convinced us that the geometrically shaped, deep surface holes normally counted in the stereo-scan are essentially equivalent to the long tracks seen in the optical microscope after etching.

Track densities versus depth were measured in rocks 17 and 57. For the following reasons we believe that the tracks found at $\gtrsim 5$ mm from an external surface are due primarily to VH nuclei ($Z > 22$) from galactic cosmic rays: (i) the uranium concentrations are too low to give an appreciable spontaneous fission track density; (ii) the tracks are frequently anisotropically distributed and have a length distribution characteristic of VH tracks studied in meteorites; (iii) long tracks attributable to VVH nuclei ($z \gtrsim 31$) are observed with an abundance of $\sim 10^{-3}$ of the total density; and (iv) deeper than 5 mm, there is an accord between theory and experiment for the depth profile of galactic tracks.

Using feldspar data from interior portions we calculate the following VH exposure ages: (i) 10017, 11 $\pm$ 1 $\times 10^6$ years, (ii) 10047, 14 $\pm$ 4 $\times 10^6$ years, (iii) 10057, 28 $\pm$ 5 $\times 10^6$ years, (iv) 10058, 13 $\pm$ 2 $\times 10^6$ years. The quoted errors are statistical. Greater uncertainties arise from the lack of precise sample position in the rocks. Our calculations are based on previous theoretical treatments (*1*) using a modified low energy spectrum (*2*). Because all the surfaces, including "tops" and "bottoms," have high densities compared with densities of interior samples, we use a "rolling stone model" in which the irradiation is uniform on all sides.

The VH exposure ages are all approximately equal and are in striking contrast to the proton spallation ages that can be inferred from the rare gas data reported in the preliminary examination. From both ^{3}He and ^{21}Ne data, rock 17 has the oldest spallation age, probably $\gtrsim 200 \times 10^6$ years. Rock 57, in contrast, has the youngest spallation age, probably $\gtrsim 40 \times 10^6$ years.

The general similarity of the VH ages could be the result of an equilibrium process such as erosion or thermal annealing of tracks. However, the considerable difference in the ages of rocks 17 and 57 argues against this view.

The simplest interpretation of the low ratio of VH to spallation ages in rocks 17, 47, and 58 is to assume that these rocks were buried at least 15 cm (17) and 11 cm (47, 58) below the surface during most of their spallation age. If this explanation is correct, we would expect more isotope production by secondary nuclear cascade products in these rocks than in rock 57.

With the method outlined by Price *et al.* (*3*), we found that the accord between the VH and spallation ages in rock 57 can be used to set an upper limit of $\lesssim 0.1$ cm/10^6 years for the surface erosion rate of this rock. This value is in accord with the surface track data discussed next.

Starting at about 5-mm depth, the track density rises abruptly as the surface is approached, reaching values $>3 \times 10^9$ per square centimeter. The VVH/VH ratio is $\sim 10^{-3}$. The densities are highly variable from one surface grain to the next and frequently show strong gradients within a single grain. The profiles of density versus depth in the first 30 μm of certain grains are quantitatively similar to the profiles reported by Pellas *et al.* and by Lal and Rajan (*4*) for grains in the meteorite Kapoeta. The extrapolated galactic spectrum can account for <5 percent of these surface tracks, and we attribute them to either solar flares (most likely) or to a demodulated galactic spectrum (less likely) very different from that currently observed in periods of low solar activity.

We have fitted out 0- to 5-mm depth data using the following differential energy spectra (*5*): $S_1 = 8 \times 10^{12} E^{-3}$ p/cm^2 10^6 years sr/Mev/amu, $S_2 = 1.6 \times 10^{11} E^{-2}$, and $S_3 = 1.9 \times 10^{14} E^{-4}$. These spectra are normalized to give an integral VH flux of 10^5 p/cm^2 year for an energy ≥ 20 Mev/amu, averaged over a solar cycle. Within broad errors due to sampling difficulties, the results in rock 17 can be fit with the spectrum S_1 between 1 and 5 mm. Below 1 mm this spectrum predicts higher densities than are observed and does not agree with the gradients measured in surface grains. However, agreement can be restored by taking an erosion rate of slightly less than 0.1 cm/10^6 years and by assuming that the rock was lightly covered with a thin layer of dust. The spectrum S_3 definitely does not fit the data.

Most crystals removed from the <1-mm fraction of the fines and from the breccias show high track densities similar to those observed in surface crystals from rocks. Some also show edge effects and the tracks are probably produced primarily by solar flares. If rock erosion proceeds gently by the flaking off of individual surface grains, which have already received a solar flare irradiation, then we would conclude from the track densities observed in the fines that crystals are removed (and subsequently covered over) at the rate of $\sim 2 \times 10^{-7}$ cm/year or one layer of 100 μm crystals every 5×10^4 years. The core results suggest a possible depth dependence from which average turnover rates could be calculated; however, more data are needed. The large amount of rare gas in the fines cannot be accounted for by stopping flare particles and must be due to the solar wind.

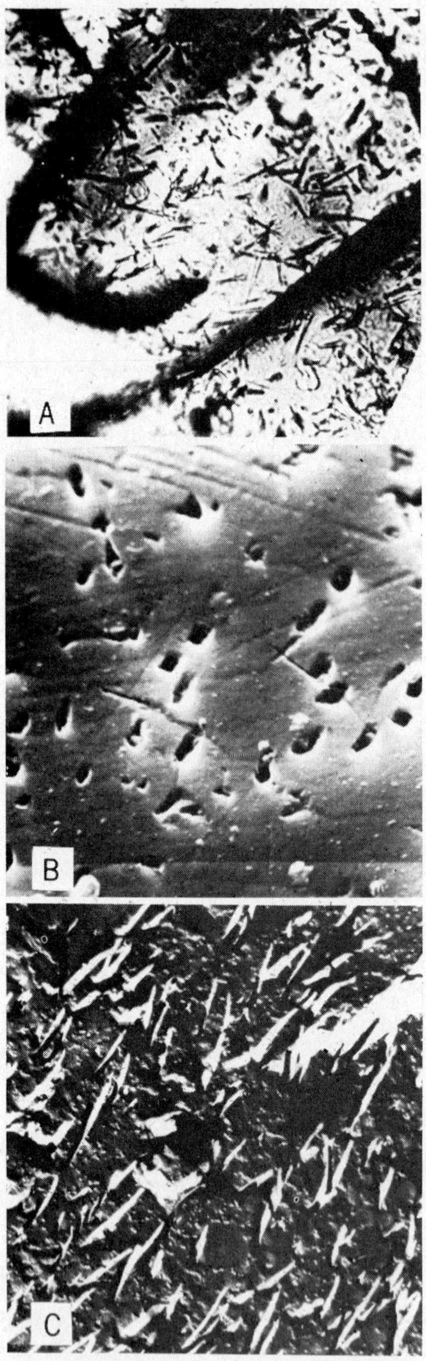

Fig. 1. Etched fossil tracks in feldspar as seen with different types of instruments: (A) in the optical microscope, tracks appear as long dark lines; (B) in the stereo-scan electron microscope, the tracks show as deep, geometrically shaped holes; (C) plastic replicas viewed with a transmission electron microscope prove that the holes seen in the stereo-scan at high track densities are not superficial surface pits. Agreement is found between these three methods when measurements are made on the same crystal (not shown here).

Track densities measured in the internal portions of glass spherules are highly variable, with ~ 80 percent of the spherules having densities too low to measure. Tracks are difficult to observe on the external surfaces, which are frequently covered with holes of a quite different origin as well as secondary glass splatters, impact pits, and a variety of other interesting features. The differences in track densities from one spherule to the next are probably due both to different radiation histories and to different track retentivities. Track annealing temperatures (1 hour) in spherules of different composition range from ~ 100°C to >400°C, suggesting the use of spherules as temperature monitors. Tracks that contain spherules frequently have an edge effect similar in profile to that observed in surface grains. The edge effect is generally uniform, which indicates an isotropic irradiation. In some spherules the track pits fall into two size classes, with the larger resembling fission tracks. If this were proved to be the case, some spherules would have ages of $\sim 10^9$ years. Dating of the formation time of some spherules is likely to prove possible, particularly in the case of nonhomogenized spherules which contain small inclusions with high uranium concentrations.

Uranium maps of individual grains and rock fragments were made by registering thermal neutron-induced fission tracks either in external mica detectors or in the annealed grains themselves (*6*). In all the materials, with the exception of some glass spherules that have uniform distributions of U in the range 150 to 500 ppb, the U occurs in small regions ranging in size from <10 to $\sim$ 100 μm and with concentrations from 1 to $\sim$ 100 ppm. About one-

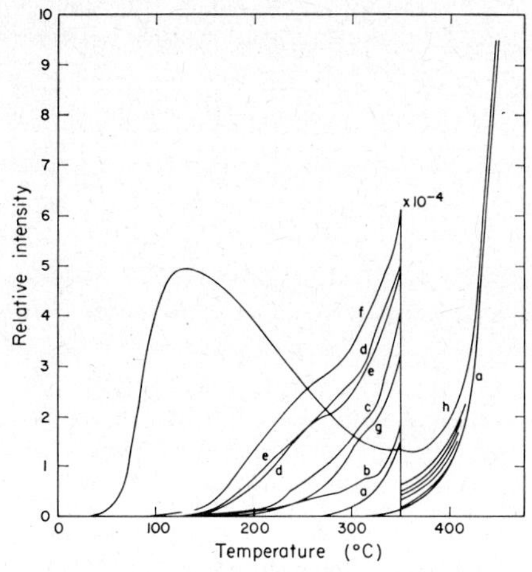

Fig. 2. Thermoluminescence (TL) glow curves in unseparated lunar fines from core 10004. Light output is measured as a function of temperature in samples heated rapidly in argon gas. The sample is then cooled, irradiated with ^{90}Sr β rays, and reheated to calibrate its response. The TL output increases with increasing depth. The high temperature β-ray response is the same for all these samples. (a) background; (b) core tube, top; (c) core tube, 3 cm from top; (d) core tube, 6 cm from top; (e) core tube, 9 cm from top; (f) core tube, 12 cm from top; (g) lunar fines; (h) core tube, top, 1 hour.

third of these regions are associated with discrete objects that range in color from clear white to black. Apart from inclusions, the average U concentrations in feldspars, pyroxenes, and ilmenites were measured as <50 ppb by one method and, in fewer samples, as <5 ppb by a second, more sensitive method. Powder samples of the bulk fines and of the breccias 46 and 23 had uranium concentrations of 550, 610, and 620 ppb, respectively. We conclude that the uranium resides principally in several different small grain phases. These phases, which are likely linked to trace elements such as Zr and P, would have uranium concentrations high enough to date by the fission track method.

The x-ray measurements were made with Debye-Scherrer, Laue, and precession camera techniques to identify minerals and to look for crystal distortions characteristic of radiation damage. In addition to the major minerals, we found a total of six red grains ranging in size from 20 to 200 μm in samples from core 10004. Two small grains were found in the surface sample, three found at 6 cm deep, and another at 9 cm. From their space group and lattice spacings, they were identified as alpha aluminum oxide. Most of the grains from the fines and breccias showed no obvious crystalline distortion. In some crystals the Laue spots formed continuous streaks similar to those we have previously observed in samples of feldspar and pyroxene irradiated with heavy ions (7). However, contrary to the minerals irradiated with heavy ions, these distortions did not anneal at temperatures of 900°C. Lattice parameter measurements were made on the fine material before and after annealing at 900°C. These measurements showed less than a 10^{-3} fractional change in spacings for both pyroxenes and feldspars above 200-mesh in size. This is consistent with what we would predict from our track studies on lunar materials. Pyroxenes below 400-mesh in size, where the effects of solar wind and solar flare particles should be most easily detected, show a change in spacing of up to 0.5 percent upon annealing. This effect may be due to radiation, although other explanations are possible.

Thermoluminescence (TL) was measured in an argon atmosphere to 500°C with an EMI9635QB photomultiplier. The natural TL is low and is not saturated in any sample in spite of the large radiation doses on the moon.

The TL output of lunar fines increases with depth in core 10004 (Fig. 2). From these data we infer that the average depth of bulk lunar fines was ~2 cm. The natural and β-ray sensitivities of the oriented breccia 46 are similar to the fines, but no clear-cut depth effect is seen.

Both fines and breccias show a large TL peak at ~ 120°C when irradiated with ^{90}Sr β rays. Partial annealing of β-irradiated samples gives curves similar to the natural TL curves. Heating of a sample to >800°C for 8 hours fails to modify the subsequent β response. A low temperature peak is also present in a surface sample of rock 57, but it is missing in an interior sample. It is also absent in both external and internal samples of rock 47. The ratio of natural TL to β-ray TL was ~5 times greater at the surface than in the interior of rock 57 but not in rock 47.

Handpicked grains from rock 47 show that its TL output is due primarily to feldspar. The results indicate that separation of feldspars should greatly enhance the TL signal in the fines.

Since the ionization does not increase with depth, it is tempting to associate the core results with the temperature gradient in the lunar surface. Detailed kinetic studies, as well as low-temperature irradiations and dose rate studies, are in progress to test this hypothesis.

Measurements of total energy release were made with a Dupont model 900 differential analyzer. When heated in air to 900°C, the fines and breccias give

Table 1. Summary of fossil track results. These results do not include the surface crystals which have densities that range above 3×10^9 tracks per square centimeter and which depend critically on the position with respect to the surface.

Sample	Density range (tracks/cm²)	Average density (tracks/cm²)
	Crystalline rocks	
10047 int. ~5 mm from surface	5×10^6 to 1.7×10^7 (12 feldspars)	$(1.2 \pm 0.4) \times 10^7$
	4×10^6 to 8.4×10^6 (13 pyroxenes)	$(5.6 \pm 0.4) \times 10^6$
10058 int. ~12mm from surface	7 to 7.7×10^6 (2 feldspars)	$(7.4 \pm 0.6) \times 10^6$
	4.4 to 6.4×10^6 (8 pyroxenes)	$(5.6 \pm 0.3) \times 10^6$
10057	12 positions in vertical section give monotonic variation from 1.5×10^7 at center to $> 10^9$ at surfaces	
10017	12 positions in vertical section give monotonic variation from 3.0×10^6 at center to $> 10^9$ at surfaces	
	Breccia	
10046 ~15mm from surface	$\sim 10^9$ (2 feldspars); many others with reaction layer; out of 5 dark spherules, one with $\sim 2 \times 10^7$ (small pits), another with $\sim 5 \times 10^5$ (large pits)	
	Fines	
Olivines (5)	1.6×10^7 to 2.3×10^7	$(1.8 \pm 0.2 \times 10^7)$
Feldspars (30)	1.5×10^7 to $> 10^9$	
Pyroxenes (30)	6×10^6 to $> 5 \times 10^8$	
Spherules (36)	Six dark spherules with densities that ranged from 2×10^7 to 1.2×10^8 (small pits); three of them with $\sim 10^5$ (large pits); other spherules contain only shallow pits.	
	Coarse fraction of fines	
Pyroxenes from 2-mm fragment (2)		$\sim 5 \times 10^6$
	Core tube (4 crystals for each location)	
Surface	10^9 to 2×10^9	
6 cm below surface	6×10^8 to 10^9	
12 cm below surface	2.6×10^7 to 3.5×10^8	

three weak exotherms: (A) 250° to 450°C, < 4 cal/g; (B) 450° to 710°C, ≲ 20 cal/g; and (C) 750° to 850°C, 13.5 cal/g. Peaks A and B are suppressed when heated in N_2. Peak A varied from <0.5 cal/g to ~5.7 cal/g in different core samples. Peak C is relatively sharp and has two subpeaks. Rock 57 shows no DTA peaks. The absence of peak C in rock 57 suggests that peak C may be associated with the glass in the fines. This would be consistent with two other observations: (i) at ~800°C, several euhedrally shaped glassy grains were observed by us to burst explosively; and (ii) x-rays showed reversion of a glassy spherule to a feldspathic crystalline material after heating at 900°C. From the DTA data we conclude that there is no large stored energy release due to the recombination of lattice defects produced by the lunar radiation environment.

Perhaps our most interesting result is the observation of solar flare effects. From studies of tracks in internal grains from breccias and in the surfaces of crystalline rocks that have been buried at different times in the past, lunar samples can be used to study the history of solar activity.

G. CROZAZ
U. HAACK
M. HAIR, H. HOYT, J. KARDOS
M. MAURETTE, M. MIYAJIMA
M. SEITZ, S. SUN, R. WALKER
M. WITTELS, D. WOOLUM

Laboratory for Space Physics, Washington University, St. Louis, Missouri

References and Notes

1. R. L. Fleischer, M. Maurette, P. B. Price, R. M. Walker, *J. Geophys. Res.* **72**, 331 (1967); M. Maurette, P. Thro, R. Walker, R. Webbink, *Meteorite Research: Proceedings*, P. M. Millman, Ed. (Springer-Verlag, Berlin, 1969), p. 41.
2. G. M. Comstock, C. Y. Fan, J. A. Simpson, *Astrophys. J.* **146**, 51 (1966).
3. P. B. Price, R. S. Rajan, A. S. Tamhane, *J. Geophys. Res.* **72**, 1377 (1967).
4. P. Pellas, G. Poupeau, J. C. Lorin, H. Reeves, J. Audouze, *Nature* **223**, 272 (1969); D. Lal and R. S. Rajan, *ibid.*, p. 269 (1969).
5. B. McKibben and E. C. Stone, personal communication.
6. K. C. Condie, C. S. Kuo, R. M. Walker, V. Rama Murthy, *Science* **165**, 57 (1969).
7. M. Seitz, M. C. Wittels, M. Maurette, R. M. Walker, H. Heckman, *J. Radiat. Res.*, in press.
8. This work would have been impossible without the assistance of P. Swan. We also wish to thank M. Marberry for her invaluable help. Much of the laboratory work was done by A. Jones. This work was supported in part by contracts from NASA and the McDonnell Douglas Corporation. Dr. Crozaz is Chargé de Recherches du Fonds National Belge de la Recherche Scientifique, on leave; Dr. Haack is on leave from University of Göttigen, Germany; Dr. Maurette is on leave from Institut de Physique Nucléaire, Laboratoire de Spectrométrie de Masse, Orsay, France; Dr. Miyajima is on leave from Waseda University, Japan; and Dr. Wittels is on leave from the U.S. Atomic Energy Commission.

4 January 1970

Tritium and Argon Radioactivities in Lunar Material

Abstract. *Tritium and argon radioactivities, attributable to galactic and solar cosmic-ray interactions, were measured in lunar soil and in three lunar rocks. The tritium in the soil, 325 ± 17 disintegrations per minute per kilogram, is slightly higher than that in the rocks, 212 to 250 dpm/kg. For two rocks, the tritium was combined with the helium-3 in order to calculate exposure ages of 375 ± 40 and 205 ± 25 million years. The argon-37 radioactivities, 21.0 to 27.2 dpm/kg, and the argon-39 radioactivities, 12.1 to 16.4 dpm/kg, are slightly higher than those in stony meteorites. Higher exposure ages were obtained from the argon isotopes than from tritium and helium-3. On the basis of the known galactic cosmic-ray flux and the known cross section, at least half of the observed radioactivities are produced by solar cosmic rays.*

The amounts of tritium and argon radioactivities in lunar material are of interest because they give information about galactic and solar-source mechanisms and because they help determine exposure ages. These ages are related to the erosion history of the lunar surface.

The apparatus used for the argon measurements in lunar materials is similar to that for argon measurements in meteorites except for the improved proportional counters, whose characteristics are given in a recent article (*1*). For tritium, counter and other improvements made determinations possible in samples as small as 1 g.

Lunar rocks 72, 17, and 61, of petrological types A, B, and C, respectively, and a sample of soil fines, which have been described by the Lunar Sample Examination Team (*2*), were analyzed. The samples were melted under vacuum in the presence of argon carrier at 1600°C for 2 hours. The evolved gases were reacted over vanadium foil at approximately 800°C. The noble gases were removed from the vanadium foil at room temperature, and their volumes were measured. The helium and neon were separated by freezing the argon and heavier noble gases on charcoal at liquid-nitrogen temperature. Carrier krypton was added to the charcoal. The argon was separated from the krypton at Dry-Ice temperature and placed in low-level proportional counters with volumes between 0.54 and 0.75 cm^3. The argon pressure in the counters was standardized at 1.4 atm. The argon yields were between 95 and 99 percent. Methane (10 percent) was added to the argon. The counters were removed from the extraction system and counted in a low-level counting system where the backgrounds were 1 count/day or less in 2.8 ± 0.6 kev channels of a 100-channel analyzer. The 2.2 to 3.4 kev energies cover the ^{37}Ar peak to 1/10 maximum. The backgrounds were less than 4 count/day above the 4.0-kev channel where the ^{39}Ar was counted. The counting efficiency for ^{37}Ar in different counters varied from 41 to 45 percent; and for ^{39}Ar, from 14 to 20 percent. A thin quartz window at one end of the counter permits the passage of the 5.9 kev x-ray from an ^{55}Fe source used to adjust the energy scale. Calibration curves with known amounts of ^{37}Ar and ^{39}Ar for a typical counter are given in another publication (*1*). Figures 1 and 2 show the argon counting data, together with backgrounds, from the 9.9-g lunar soil sample for 9061 minutes between 16 and 23 September 1969, and from the 10.0-g sample of lunar rock 17 for 12,936 minutes between 3 and 13 October 1969. The argon samples were counted for approximately 2 months, and the ^{37}Ar activity decayed with a 35-day half-life.

The krypton was removed from the charcoal at 0°C and put into an argon-type counter to which methane had been

Table 1. Radioactivities and gas contents from lunar samples. The only krypton data obtained were from soil 84-24 (dpm/kg): ^{81}Kr, 0.14 ± 0.10; ^{85}Kr, < 1.

Sample	Weight (g)	Tritium (dpm/kg)	Hydrogen (cm^3/g)	Helium (cm^3/g)	^{39}Ar (dpm/kg)	^{37}Ar* (dpm/kg)
			Apollo 11			
Soil 84-24	(9.9)	325 ± 17	1.2	0.26	12.1 ± 0.7	27.2 ± 2.2
Rock 17-14	(10.0)	219 ± 7	0.47	<5 × 10^{-3}	16.4 ± 0.9	21.0 + 2.0
Rock 72-11	(9.4)	234 ± 10	0.76	<5 × 10^{-3}	15.8 ± 1.0	25.7 ± 2.0
Rock 61 surface	(1.49)	235 ± 15	2.5	~ 0.3		
Rock 61 interior	(0.96)	231 ± 10	1.4	~ 0.4		
			Meteorite(1)			
Sprucefield L4 (nonmagnetics)	(21)	290 ± 30		<5 × 10^{-3}	7.7 ± 0.4	12.6 ± 1.4

*Extrapolated to 21 July 1969.

added. The electronics were adjusted so that the 5.9 kev x-ray from the ^{55}Fe source was peaked in channel 30 of the 100-channel analyzer. The 11.9 kev peak from ^{81}Kr should then appear in channel 61. The counter efficiency for ^{81}Kr was assumed to be the same as for ^{37}Ar, and for ^{85}Kr, to be 30 percent above 4.0 kev.

The hydrogen was removed from the vanadium foil while heated, and the volume of the released hydrogen was measured. The hydrogen counter consisted of a copper cathode and a 0.002-inch Therlo center wire in a quartz envelope that had a thin window at one end; its total volume was 42 cm^3. The electronics were adjusted so that the 100-channel analyzer scanned energies up to 19.3 kev. With 400-torr pressure of P-10 gas (commercially available counting gas containing 90 percent argon and 10 percent methane) and 200-torr pressure of hydrogen, the counting efficiency for tritium between 1 and 19.3 kev was 60 percent. The resolution for the 5.9 kev source was 20 percent. The background was 0.14 count/min. The tritium spectrum does not consist of a sharp peak; however, it has a different shape from the background spectrum.

Table 1 summarizes the results. For comparison, the radioactivities in the nonmagnetics of the recently fallen Sprucefield meteorite, an L4 chondrite, were measured with the same equipment. The radioactivities in the lunar material are within a factor of 2 of those in the meteorite. There are, however, interesting differences between the lunar samples and also between lunar samples and meteorites.

The tritium in the soil was 325 ± 17 dpm/kg; in rocks 17 and 72, it was 219 ± 7 and 234 ± 10 dpm/kg. A chip from the surface of rock 61, which on the basis of pit count is thought to be the top side, had 235 ± 15 dpm/kg; an interior sample of the same rock had 231 ± 10 dpm/kg. The higher content in the soil cannot be caused by the small differences in chemical composition between the soil and rocks. The helium content of the soil was 0.26 cm^3/g, in accord with the mass spectrometer results (*2*); this content was attributed to the solar wind. The hydrogen in the soil was 1.2 cm^3/g, which may in part be solar-wind hydrogen. However, the difference between the tritium in the soil and that in the rocks should not be attributed to solar-wind tritium until other possibilities are eliminated. On the basis of the small difference in the tritium contents of the surface and interior sample of rock 61, one could set an upper limit of 0.5 triton/cm^2 sec for tritium implanted by solar wind, if the surface sample was the top side. However, tritium from solar flares, which would not be implanted at the surface, might be the source of the difference. There is a possible difference in the average depths of the soil and rock material that could account for the tritium difference. More detailed studies of tritium versus depth are necessary before the excess tritium in the soil can be interpreted.

The tritium contents of rocks 17 and 72 can be combined with their spallation ^3He contents of $(3.2 \pm 0.2) \times 10^{-6}$ and $(1.9 \pm 0.1) \times 10^{-6}$ cm^3/g measured at the Lunar Receiving Laboratory (*3*) to obtain cosmic-ray exposure ages. The exposure age of rock 17 is 375 ± 40 million years, and that of rock 72 is 205 ± 25 million years, with the ^3He production rate assumed to be twice the present tritium decay rate. These ages are simply interpreted as the time since the rocks have been thrown out from large impact craters. There are two possible complications to this simple interpretation: some ^3He may be lost by diffusion, and the production of ^3He may not have been constant because of changes in the amount of surrounding material with time. The rocks have been eroded by small-particle bombardment,

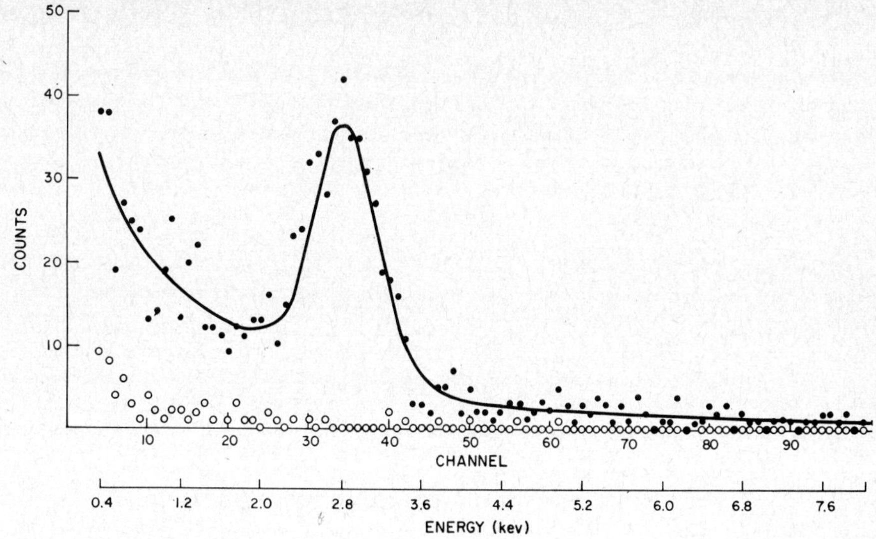

Fig. 1. Argon radioactivity for 9.9 g of lunar soil. Solid line, curve from calibrations with ^{39}Ar = 12.1 dpm/kg and $(^{37}Ar)_0$ = 27.2 dpm/kg. Solid circles, counts per channel for 9061 minutes 57 to 64 days after 21 July 1969 (there were 123 counts beyond channel 99). Open circles, background counts per channel for 11,426 minutes (there were 11 counts beyond channel 99).

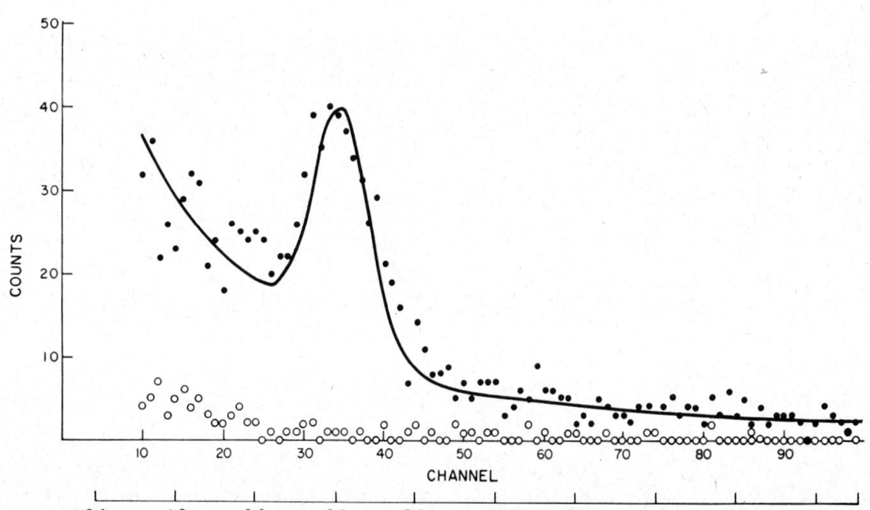

Fig. 2. Argon radioactivity for 10.0 g of lunar rock 17. Solid line, curve from calibrations with ^{39}Ar = 16.4 dpm/kg and $(^{37}Ar)_0$ = 21.0 dpm/kg. Solid circles, counts per channel for 12,936 minutes 74 to 84 days after 21 July 1969 (there were 178 counts beyond channel 99). Open circles, background counts per channel for 13,078 minutes (there were 22 counts beyond channel 99).

as indicated by minute pits on the rounded surfaces (*2*). The soil may have been in a state of flux because of this bombardment and because of internal stresses. Consequently, the rocks could have been buried in the soil at various times in the past. It is therefore of interest to obtain exposure ages from the argon isotopes.

The ^{39}Ar activity in the soil was 12.1 ± 0.7 dpm/kg, and those in rocks 17 and 72 were 16.4 ± 0.9 and 15.8 ± 1.0 dpm/kg. The ^{37}Ar activity in the soil was 27.2 ± 2.2 dpm/kg, and those in rocks 17 and 72 were 21.0 ± 2.1 and 25.7 ± 2.0 dpm/kg. An exposure age could be obtained from combination of spallation ^{38}Ar and the argon radioactivities if the relative production rates were known. To a first approximation, the ^{38}Ar production rate equals that of ^{37}Ar plus ^{39}Ar. The ^{38}Ar produced by cosmic rays in rocks 17 and 72, as measured at the Lunar Receiving Laboratory (*3*) is $(50 \pm 10) \times 10^{-8}$ and $(33 \pm 5) \times 10^{-8}$ cm^3/g, respectively. The ^{38}Ar to the ^{37}Ar plus ^{39}Ar exposure age of rock 17 is 640 ± 160 million years, and that of rock 72 is 410 ± 80 million years. These ages are 30 percent or more higher than the ^3He to ^3H ages and indicate either that some ^3He has been lost or that the amounts of material surrounding the rocks have changed during the past 400 million years.

The tritium and argon radioactivities in meteorites are attributed to the interactions of galactic and solar cosmic rays with the material. It is reasonable to attribute these radioactivities in lunar material to the same mechanism. Since the flux of galactic cosmic rays at 1 A.U. is known (*4*) and the depth of the lunar samples is fairly small (less than 25 g/cm^2), one can estimate what fraction of the radioactivities are attributable to galactic cosmic rays. The flux of nucleons with energy greater than 1000 Mev is 0.8 ± 0.1 nucleon/cm^2 sec and that with energies between 400 and 1000 Mev is 0.35 ± 0.1 nucleon/cm^2 sec averaged over the solar cycle. The total tritium-production cross section in oxygen is 38 millibarns, independent of the bombardment energy above 400 Mev (*5*). For heavier elements, the tritium cross section increases with atomic weight A as approximately $A^{2/3}$ for particles above 1000 Mev and is roughly independent of atomic weight for particles between 400 and 1000 Mev (*6*). With these fluxes and cross sections, we obtain a dpm/kg, which is less than half the tritium production rate of 85 ± 15 amount observed. Similar considerations for the ^{39}Ar production from iron and titanium lead to the conclusion that only one-quarter or less of the ^{39}Ar observed in the lunar rocks is produced by galactic cosmic rays. It therefore appears that at least half the observed radioactivities are produced by solar cosmic rays.

EDWARD L. FIREMAN
JAMES C. D'AMICO, JAMES C. DEFELICE
Smithsonian Astrophysical Observatory, Cambridge, Massachusetts 02138

References

1. E. L. Fireman and R. Goebel, *J. Geophys. Res.*, in press.
2. The Lunar Sample Preliminary Examination Team, *Science* **165**, 1211 (1969).
3. D. D. Bogard, private communication.
4. W. R. Webber, *Handbuch Phys.* **46**(2), 181 (1967); P. Meyer, *Ann. Rev. Astron. Astrophys.* **7**, 1 (1969).
5. E. L. Fireman and F. S. Rowland, *Phys. Rev.* **97**, 780 (1955); L. A. Currie, W. F. Libby, R. L. Wolfgang, *ibid.* **101**, 1557 (1956).
6. T. A. Kirsten and O. A. Schaeffer, *Interactions of Elementary Particle Research in Science and Technology*, C. L. Yuan, Ed. (Academic Press, New York, in press).

5 January 1970

Particle Track, X-ray, Thermal, and Mass Spectrometric Studies of Lunar Material

Abstract. *Particle tracks in Apollo 11 samples are dominantly of cosmic ray and solar origin: primary galactic and solar flare particles, likely spallation recoil tracks, and possible solar-wind heavy particles. The energy spectrum of iron-group nuclei is inferred from track density gradients in surface layers, and a limit of* $<< 10^{-7}$ *centimeter per year is deduced for the surface erosion rate. From cosmic ray tracks in rock and core samples it is clear that the lunar soil is stirred often during each few million years. X-rays reveal augite, anorthite, olivine, ilmenite, troilite, nonmeteoritic iron, and assorted glasses, but no major structural damage. Hydrogen, helium, and other gases in the fines are compatible with expected solar wind ratios.*

In addition to being of direct crystallographic and thermodynamic interest, phase identification is vital to proceeding with particle track studies. X-ray, electron microprobe, and optical, chemical etching, and standard mineral separation techniques were used. Debye-Scherrer patterns revealed that the dominant crystalline phases in the lunar fines (sample 2,84) are pyroxene and feldspar. Selected (diameter, >50 μm) crystals from the coarse fines (sample 2,85) and from the interior of rock 44 (sample 44,47) were examined by Laue transmission and x-ray precession camera techniques. They revealed anorthite, pyroxene, olivine of 30 (±5) percent fayalite, and ilmenite. The crystals were generally of good quality without major structural defects. Twinning occurs in some larger crystals, an example of deformation twinning in ilmenite being visible in Fig. 1. Microprobe studies of this sample (rock 2,85,4) revealed the same phases plus troilite and free iron (nonmeteoritic, since nickel was not detectable). The pyroxene was determined to be augite. Specific crystals identified by the microprobe are indicated in Fig. 1. Comparison of this section with a thin section in both transmitted and reflected light provided us with a means of rapid visual identification of most of the crystalline phases of interest for track etching (*1*). Glass was readily identifiable by its response to etching of particle tracks in hydrofluoric acid (*2*).

Our objectives are, first, to identify the variety of particle tracks present in lunar materials; second, to use these tracks to learn about the history of the moon; and finally, to learn about the solar and galactic cosmic rays that impinge upon the moon. Possible track origins in extraterrestrial materials have been described (*3, 4*), and examples have been given of the identification of several varieties in meteorites (*5–7*). In this work we report the observation of tracks thought to be solar flare particles, spallation recoil tracks in glass, and possible solar-wind heavy nuclei in a crystalline phase.

The most abundant tracks found in this study are from heavy iron-group nuclei (Fig. 2a). They have been identified by criteria we have noted elsewhere (*3, 5*): track densities decreasing toward the interior (rocks 3,44, and 17); preferred orientation of tracks (rocks 3,44, and 5-2.6,1,3); approximately equal track densities in adjacent crystals of different mineral types (rocks 2-85,4; 3; 17; and 44); and length distribution peaked at ~ 10 μm, but including shorter and longer tracks (rocks 3,44, and 5,2.6,1,3) and present in crystals (anorthite and augite) having too little uranium ($\leq 10^{-10}$ weight fraction) to have supplied a significant contribution.

A second, abundant type of track is found in glass (Fig. 2b); it consists of abundant, extremely short tracks ranging in length from 2 μm down to unresolvable dots that can be removed by heating, and is similar in appearance to tracks produced in terrestrial basaltic glass by high-energy proton bombardment. We identify these as tracks caused by spallation recoil nuclei.

A third variety of track (Fig. 2c) appeared in a single unidentified crystal that was removed from the upper few thousand angstroms of the top of rock 17, etched, and viewed in the electron microscope. The tracks appear as a high density ($\sim$4 to 5 $\times$ 10^{10}/cm^2) of shallow tracks that are $<$ 2000 Å in depth and are extremely low-energy particles. One likely possibility is that these are heavy (mass $\sim$ 200) solar-wind particles which Walker (*4*) has noted would produce particle tracks much like those of energetic α-recoil nuclei (*8*). Since we estimate a 2-million-year cosmic ray exposure age for the surface of rock 17, and since the flux of solar-wind proton is $\sim 10^8$ cm^{-2} sec^{-1}, the abundance of heavy solar-wind particles relative to protons would be $\sim 2 \times 10^{-11}$, provided our identification is correct. For comparison, the lead-to-hydrogen ratio in the solar photosphere is also $\sim 2 \times 10^{-11}$ (*9*).

The only fission tracks that we have positively identified are seen radiating from localized regions in some augites, it being likely, but unproved, that there is a distinct, separate phase where the fissionable nuclide is segregated. We have not yet attempted to date such a sample, but we have an upper-limit age of 35 million years for one glass fragment, computed on the assumption that all tracks longer than 2 μm are from spontaneous fission of uranium-238.

We examined 15 individual particles of augite, anorthite, and glass from the top and from depths of 2.6, 5.2, 7.8, and 10.5 cm in sample 5, one of two cores taken on the Apollo 11 mission. For soil that had been undisturbed over its whole life as a surface layer, there should be a monotonic decrease, by a factor of at least 20, in going from depths of a few millimeters to 10.5 cm (*3*) in material of the observed (*10*) density. Between depths of zero and a few millimeters, a further decrease would have occurred, of an amount which we determine here to be a factor of $\sim$100. Instead of showing a regular decrease, the track densities are chaotic. In both augite and anorthite, variations from less than 10^6/cm^2 to nearly 10^9/cm^2 occur at the top, bottom, and middle regions of the core, a clear indication that the material has been thoroughly stirred. One possible explanation of this behavior would be malfunction of the coring device, which allowed the sample to be shaken and mixed during transport. Because of the strong, apparently electrostatic cohesiveness of the soil, however, it would be extremely difficult to produce such complete mixing. We therefore conclude that this mixing was caused by natural forces on the moon. Since 5 of the 15 particles examined had track densities $\gtrsim 10^8$/cm^2, almost surely at

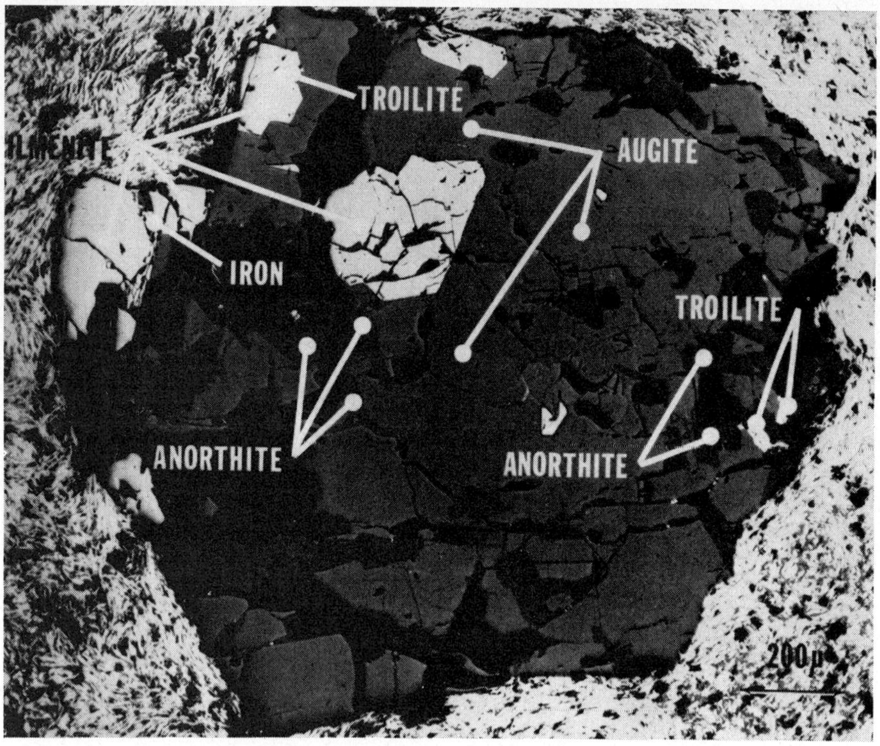

Fig. 1. Section of rock 2,85,4 with positively identified phases indicated. The mineral composition of this sample is typical of the "crystalline" (*10*) rocks used in this work. This sample spent at least 2 million years close to the lunar surface.

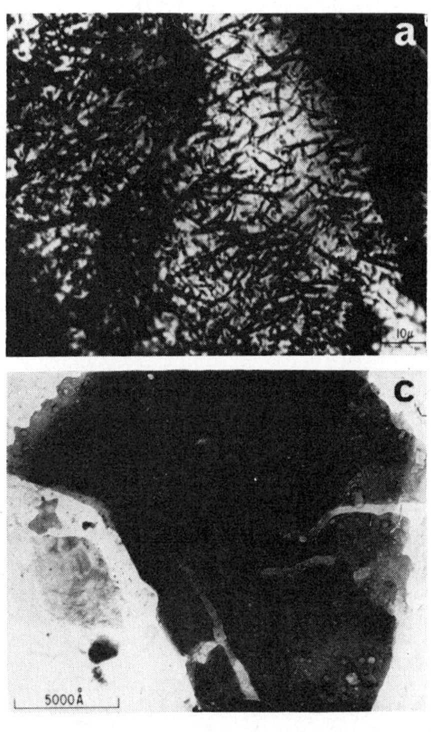

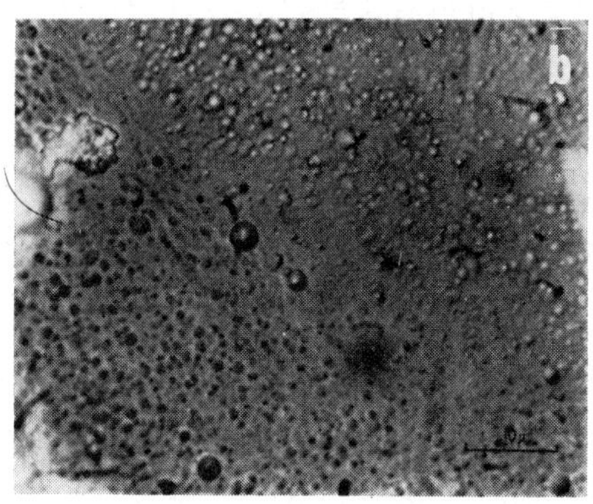

Fig. 2. Examples of cosmic-ray-related tracks. (a) Tracks of low-energy iron-group primary cosmic ray nuclei of energy 0.1 to 15 Mev per nucleon in an augite crystal (5-2.6,1,3). (b) Spallation recoil nuclei in a small glass spherule (5-7.8,1,2). (c) Possible heavy solar-wind particles in an unidentified surface crystal [rock 17, SD(R)-4].

some time in the past they were in the top millimeter of the soil, and this indicates that stirring has been frequent indeed. The distribution of the track densities observed indicates that the residence time of the soil in the top 10.5 cm of the surface is approximately 15 million years; thus the soil, as fine particles, is at least that old, and, in the probable event that the depth of thoroughly stirred soil exceeds 10.5 cm, is proportionately older. From a statistical model we estimate that the top layer has been remixed every 10,000 to 100,000 years.

Each of these samples shows evidence of having been turned over during its life at the surface of the moon. Rock 17 has high cosmic ray track densities at both the top and bottom surfaces, with lesser densities in the interior. The gradients, plus the density value of $\sim 1.6 \times 10^6/cm^2$ at a position 3 cm from the rock surface, indicate (*3*) that each side of this sample spent ~ 2 million years directly exposed to space at the lunar surface—a total of 4 million years. From radioactivity measurements (*10*) it was inferred that this rock spent >100 million years near the surface of the moon. Hence we can conclude that, during most of this time, it was covered over with soil to estimated depths of >10 cm.

Sample 44 contains tracks of anisotropic orientation a few millimeters below the surface and isotropic tracks at its center. Although one can conceive of a single position during cosmic ray bombardment which would allow the isotropy at the center, the more likely origin is multiple, random reorientation of the sample. The estimated total cosmic ray exposure time at the lunar surface is 2.5 million years.

Rock 46 is the only one for which there is clear photographic evidence as to which side was up on the moon. We find a glass-lined pit of probable impact origin at the bottom, indicating that at one time the position was inverted.

The variation, with depth, of cosmic ray track densities in lunar pyroxenes and feldspars has, folded within it, the cosmic ray energy spectrum for iron-group nuclei down to very low energies. Lunar surface materials, in contrast to the outer layers of meteorites, can be found with zero ablation; thus tracks are present from particles of energies extending down to those corresponding to the shortest resolvable lengths, $\sim 10^{-5}$ to 10^{-6} cm.

Our data were obtained on anorthite and augite from rock 17 and from a core sample (5-2.6,1,3) of augite that at one time was a surface crystal, as evidenced by the steep track gradient across it, viewed by means of both light and electron microscopy.

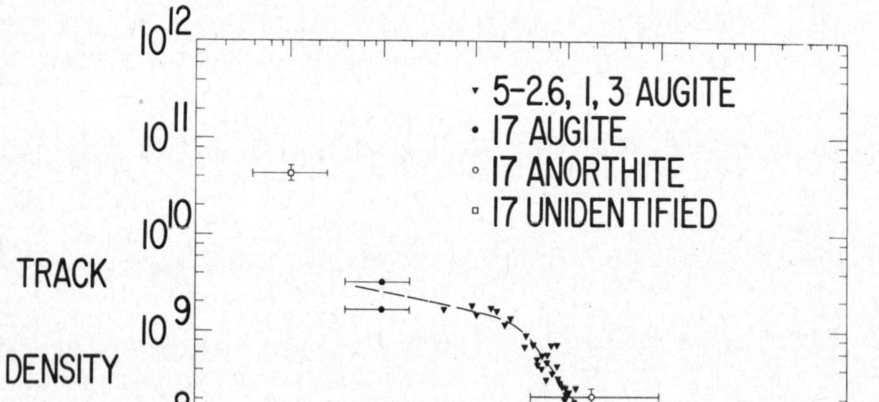

Fig. 3. Track density plotted against depth for samples 17 and 5-2.6,1,3. The data for sample 5-2.6,1,3 are renormalized to those for sample 17 at a depth of 100 μm. Except for the datum at 1000 Å, the electron micrograph results are reduced by a factor of 3, a reduction that corresponds to the observed increase in track count due to the resolution of shorter (that is, non-iron) tracks.

Surface samples plucked from rock 17 on a plastic replica, etched, and viewed in the electron microscope showed tracks of very low-energy particles; interior samples showed tracks of higher-energy cosmic rays. A composite curve for track density plotted against depth is given in Fig. 3. Except for the point at ~ 1000 Å, which, as noted above, may well be from solar-wind particles of mass ~ 200, the tracks are dominantly those of iron-group nuclei.

For cosmic ray tracks of range 10 to 120 μm, the track densities are closely similar to those seen in some of the crystals of gas-rich meteorites (*7*) that are believed to have been irradiated with iron-group nuclei at the beginning of the solar system. Our results support that interpretation and give, for more recent time, the ratios between the low and intermediate energy fluxes. Approximate differential flux ratios φ/φ (200 to 300 Mev per nucleon) are 76, 32, 17, 10, 1, and 1 for the energy intervals 0–0.5, 0.5–5, 5–10, 10–15, 50–100, and 100 200 Mev per nucleon, so the differential flux is constant from 300 to 50 Mev per nucleon and then rises steeply at lower energies.

The observed track density gradient in the rocks is satisfactorily accounted for by adding to the known galactic spectrum a solar flare spectrum of $\exp(-R/R_0)$, with reference rigidity $R_0 \approx 100$ Mv, it being assumed that present-day fluxes from both sources have been maintained over the last several million years. On the basis of this assumption, the rate of erosion of the rock surfaces would have been much less than 10^{-7} cm/yr.

Preliminary experiments on lunar fines by thermal-analysis, micrographic, and mass-spectrometric techniques have been carried out, to study the physical behavior and stability of this material at high temperatures and the resultant evolution of gases.

The following general observations and preliminary conclusions can be drawn. Hydrogen, helium, neon, argon, nitrogen, and trace levels of krypton and xenon evolve on heating in a pumped vacuum. Also observed are CO_2, CH_4, H_2S, HCl, C_6H_6, and numerous other breakdown products, pointing to significant contamination.

The preliminary indicated level of hydrogen is higher, by a little more than 1 order of magnitude than that of helium, but higher by approximately 3 orders of magnitude than the levels of neon and argon. Nearly all the hydrogen evolved is driven off by the time a temperature of 800°C is reached, whereas heavier gases continue to evolve in significant amounts to higher temperatures, owing to their slower diffusion from the bulk of the solid.

Results of thermogravimetric analysis show that, unless oxygen-bearing impurities are held to very low levels during heating in the range 600° to 1200°C, net weight gains of the sample can actually occur while the hydrogen, nitrogen, inert gases, and other contaminants are still evolving. This indicates that the lunar dust is received in a not fully oxidized state. This result is compatible with the observation of free, nonmeteoritic iron discussed above, and with some of the other phases observed.

The heating of the fines in high-purity helium reproducibly produced melting, with some gas evolution, at a temperature of approximately 1165°C. Upon cooling this produced the characteristic porous, dark glassy phase observed abundantly in some of the received samples. Thus it is clear that melting of the powder, probably by meteoritic impact, is the cause of this structural feature, but that temperatures well over 1100°C are required.

R. L. Fleischer, E. L. Haines*
R. E. Hanneman, H. R. Hart, Jr.
J. S. Kasper, E. Lifshin
R. T. Woods

General Electric Research and Development Center, Schenectady, New York 12301

P. B. Price

University of California, Berkeley

References and Notes

1. R. L. Fleischer, P. B. Price, R. M. Walker, *Annu. Rev. Nucl. Sci.* **15**, 1 (1965); *Science* **165**, 383 (1965).
2. R. L. Fleischer and P. B. Price, *J. Geophys. Res.* **69**, 331 (1964); *Geochim. Cosmochim. Acta* **28**, 755 (1964).
3. R. L. Fleischer, P. B. Price, R. M. Walker, M. Maurette, *J. Geophys. Res.* **72**, 331 (1967).
4. R. M. Walker, *Proceedings Clermont-Ferrand Track Conference* (1969), vol. 2 p. VII-2.
5. R. L. Fleischer, P. B. Price, R. M. Walker, M. Maurette, G. Morgan, *J. Geophys. Res.* **72**, 355 (1967).
6. R. L. Fleischer, P. B. Price, R. M. Walker, *ibid.* **70**, 2703 (1965); *Geochim. Cosmochim. Acta* **32**, 21 (1968).
7. D. Lal and R. S. Rajan, *Nature* **223**, 269 (1969); P. Pellas, G. Poupeau, J. C. Lorin, H. Reeves, J. Audouze, *ibid.*, p. 272.
8. W. H. Huang and R. M. Walker, *Science* **155**, 1103 (1967).
9. L. H. Aller, *Stars and Stellar Systems*: vol 6, *Stellar Atmospheres*, G. P. Kuiper, Ed. (Univ. of Chicago Press, Chicago, 1960), p. 156.
10. Lunar Sample Preliminary Examination Team, *Science* **165**, 1211 (1969).
11. We acknowledge the able experimental assistance of R. B. Bolon, H. C. Couch, W. R. Giard, A. S. Holik, E. F. Koch, L. Osika, G. P. Schacher, and C. Van Buren. Samples of rock 17 were taken with the cooperation and help of J. R. Arnold, D. Lal, M. Maurette, and R. M. Walker. Proton irradiations at the Princeton-Pennsylvania accelerator were carried out with the aid and cooperation of H. Allen and F. Homan. This work was supported in part by NASA under contract NAS 9-7898.
* Permanent address: Jet Propulsion Laboratory, Pasadena, California.

4 January 1970

Rare Gases in Lunar Samples: Study of Distribution and Variations by a Microprobe Technique

Abstract. *The rare gas distribution in lunar soil, breccias, and rocks was studied with a micro-helium-probe. Gases are concentrated in grain surfaces and originate from solar wind. Helium-4 concentrations of different mineral components vary by more than a factor of 10 apart from individual fluctuations for each type. Also grains with no detectable helium-4 exist. Titanium-rich components have the highest, calcium-rich minerals the lowest concentrations. The solar wind was redistributed by diffusion. Mean gas layer thicknesses are 10, 6, and 5 μm for helium, neon, and argon respectively. Lithic fragments in breccias contain no solar gases. Glass pitted surfaces of crystalline rocks contain about 10^{-2} cubic centimeter of helium-4 per square centimeter. Etched dust grains clearly show spallogenic and radiogenic components. The apparent mean exposure age of dust is approximately 500 × 10^6 years, its potassium-argon age is approximately 3.5 × 10^9 years. Cavities of crystalline rocks contain helium-4, radiogenic argon, H_2, and N_2.*

One of the first important observations of the Preliminary Examination Team (*1*) was the occurrence of large amounts of He, Ne, and Ar in lunar fines and breccias. Their relative abundances and the isotopic composition resemble the solar abundances as well as the abundances of light primordial rare gases. The gases could be due to trapped solar wind, to trapped primordial gases, or to stopped solar flare particles. The distribution of these gases on a microscale provides the clue to their origin. Enrichment at the surface of grains points to solar wind, whereas a homogeneous distribution is expected for solar flare particles or for primordial gases. Diffusive fractionation of light and heavy rare gases yields information about the thermal history of lunar samples and about turnover rates. A split of the components could yield cosmic ray exposure ages and radiogenic ages. Also, the original orientation of rock surfaces can be inferred from gas concentration profiles.

The combination of a microprobe with a He mass spectrometer has previously been applied to gas-rich meteorites (*2*). Samples are bombarded by a 35 kev electron beam of 1 μa with a beam width of ~ 10 μm. Helium-4 in situ liberated by the electron beam was registered by a 60° mass spectrometer simultaneously with elements like Ti, Fe, Ca, Cr, and Mn. Helium profiles

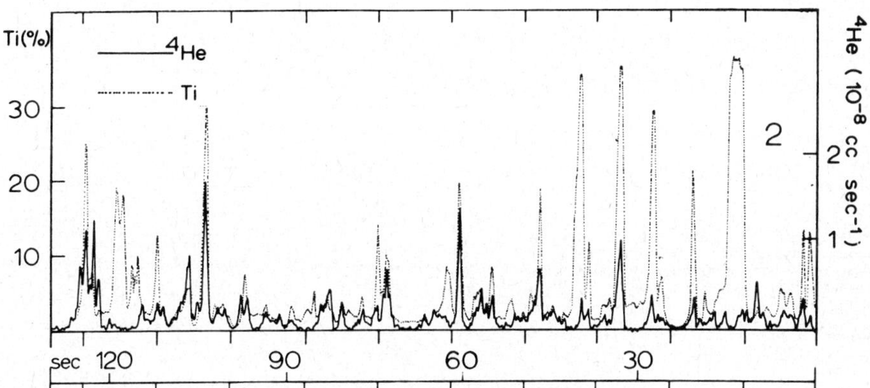

Fig. 1. Correlation of ^{4}He and Ti in lunar breccia 10019.13.

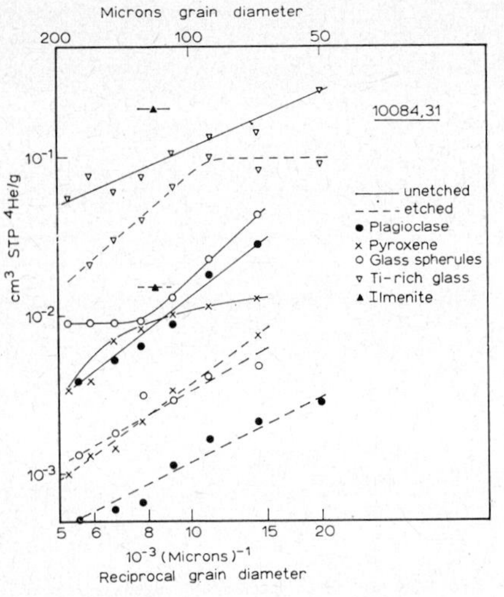

Fig. 2. Average ^{4}He concentrations plotted against reciprocal diameter of 800 handpicked dust particles before and after etching.

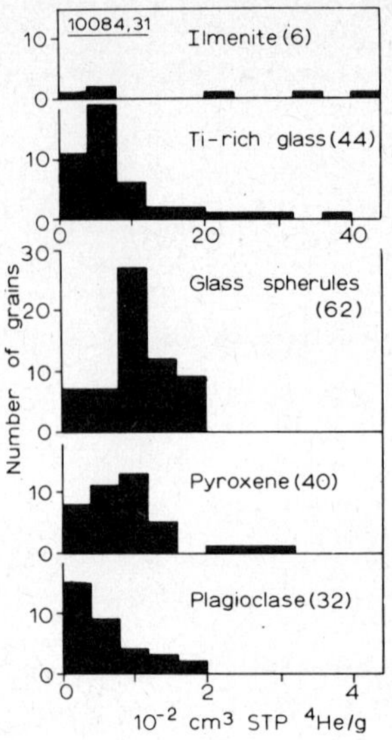

Fig. 3 (left). Fluctuation of ^{4}He concentration in individual grains of separated dust components 120 ± 20 μm in diameter.

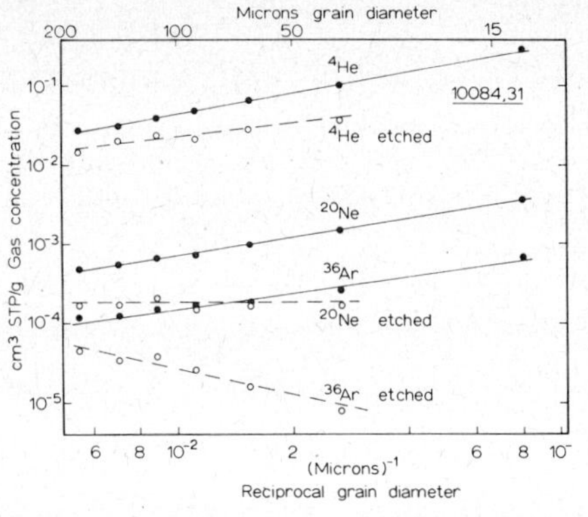

Fig. 4 (above). Concentrations of ^{4}He, ^{20}Ne, and ^{36}Ar plotted against reciprocal grain size diameter of unseparated lunar dust before and after etching.

related to the variation of these elements were obtained. The sample holder was provided with a hat to yield efficient gas collection, with a resulting sensitivity of ~ 10^{-8} cm^3 of ^{4}He per millivolt per second. Calibration was made by means of a standard leak and by comparison with statically measured test samples. The technique has been applied in a scanning mode (16 μm/sec) to polished breccias and rocks as well as to surfaces of unpolished rocks. A special mechanism kept the beam in focus even for irregularly shaped samples. The high gas retentivity of the minerals required a larger beam, which in turn caused a larger beam diameter. Therefore the local resolution was sometimes less than the average grain diameter. Hence, to supplement the helium profiles with data about well defined minerals, 800 handpicked grains of five major constituents of lunar dust and of different grain sizes were singly analyzed. Half the 800 grains were etched with a HF/H_2SO_4 mixture to remove the outer 2 μm of each grain. Etching velocities were determined under the microscope. Etching times varied between 8 seconds (glassy fragments) and 2 minutes (ilmenite). All grains were singly mounted and bombarded with the electron beam, and ^{4}He was analyzed together with Ti or Fe. Each grain was degassed by a single shot. During bombardment, the irregularly shaped crystals were melted to spherules. From their diameters, precise volumes were determined under the microscope. To supplement the results and to check the calibration of the dynamic measurements, seven grain size fractions of unseparated lunar dust were analyzed for He, Ne, and Ar in a static mass spectrometer. Also, an etched portion (adjusted to etch 2 μm of the pyroxene) was measured.

To analyze gases included in cavities of crystalline rocks, chips were broken in the high vacuum system without heating. Non-noble gases were rapidly scanned with a mass filter, rare gases were measured mass spectrometrically.

A ^{4}He-Ti scan typical for breccias is shown in Fig. 1. It reveals a very inhomogeneous He distribution. Helium is concentrated in spots of some 10 μm in diameter, mostly in very fine grained areas; larger grains and especially small crystalline inclusions are poor in ^{4}He. A strong correlation between Ti content and He is found. The content of Fe, Cr, and Mn is also correlated with He, but to a lesser extent and mostly involving Fe. On the other hand, Ca and He are inversely related. These findings can be generalized from hundreds of scans across breccias 10019.13, 10046.-19 and 10061.26. The mean peak height for He-rich spots corresponds to gas concentrations of the order of 3 cm^3 of ^{4}He per gram.

The averaged ^{4}He concentrations for each of seven grain size fractions and five mineral components of handpicked dust grains are shown in Fig. 2. Clearly, the He content decreases by a factor of about 10 from ilmenite to Ti-rich glass fragments, to pyroxene, and to plagioclase. It is also seen that He concentrations increase toward smaller grain sizes nearly linearly with the reciprocal diameter. Figure 2 also shows the He reduction due to etching of the outer 2 μm for the same five components and seven grain sizes. Amounts of He are generally lowered by factors of 2 to 10; nevertheless, even for etched minerals the relation between He and reciprocal diameter essentially remains.

The statistical distribution of He content in single grains of the same size is shown for each component in Fig. 3. The histogram applies to the 120 ± 20 μm fraction; similar histograms were obtained for all other sieve fractions. Variations are large for each type but largest for the Ti-rich components.

Figure 4 shows the results of the measurements on sieve fractions of unseparated dust. The linearity between gas concentration and reciprocal diameter is valid for He as well as for Ne and Ar. Going from 90 μm to 13 μm radius, the ^{4}He/^{20}Ne ratio slightly increases, from 58 to 77. The same is true for the ratios ^{20}Ne/^{36}Ar (3.9 to 5.8) and ^{4}He/^{36}Ar (230 to 440). After etching these ratios increase, pronouncedly for smaller grain sizes. The isotopic ratio ^{4}He/^{3}He (2800), ^{22}Ne/^{21}Ne (32 to 23), ^{20}Ne/^{22}Ne (13.1 to 12.5), ^{36}Ar/^{38}Ar (5.4 to 5.2), and ^{36}Ar/^{40}Ar (1.0 to 0.6) decrease only very little with increasing grain size. Etching reduces ^{4}He/^{3}He by about 8 percent, ^{20}Ne/^{22}Ne by about 15 percent, and ^{22}Ne/^{21}Ne by a factor of about 2 for all grain sizes. The ^{38}Ar/^{36}Ar ratio in etched samples decreases with grain size down to 3.7 for the 25 to 50 μm fraction. The ^{40}Ar/^{36}Ar ratio increases after etching, especially for small grain sizes, where it reaches a maximum value of 6.6.

Scans across polished sections of the crystalline rock 10057.22 liberated less He than was detectable by our method, even with the beam intensity increased to 25 μa. Two other specimens (10057.80 and 10085.41) had pitted surfaces and were most likely exposed to the solar wind also. Many scans going from a freshly broken surface to the pitted surface clearly showed the rare gas implantation on the pitted surface (Fig. 5.) The quantity of He per square centimeter is $\sim 10^{-2}$cm^3. Repeated scans perpendicular to the pitted surface revealed an average thickness of $\sim$ 10 μm for the He loaded layer (Fig. 5). Again, if ilmenite grains form the surface, the He concentrations are highest. For penetration depths of about 10 μm, an average ^{4}He concentration of 3 cm^3 per gram results.

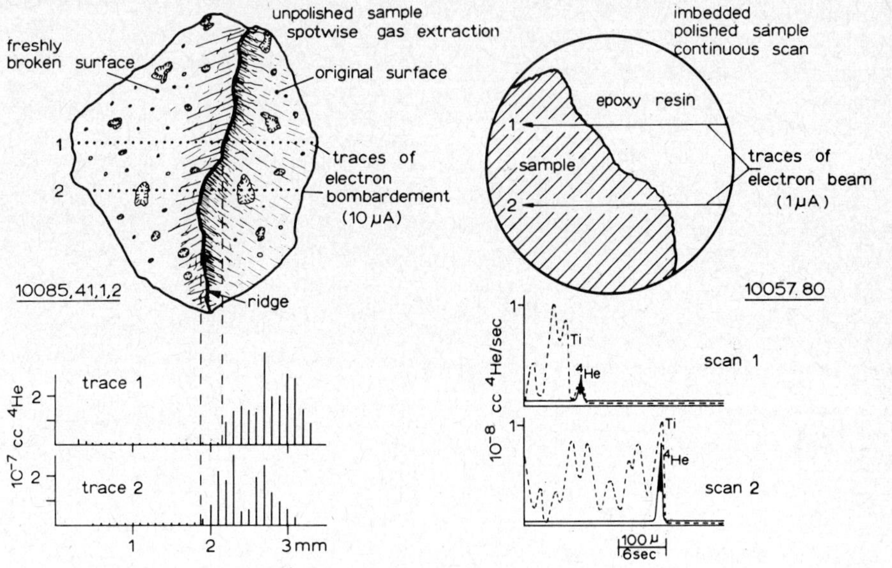

^{4}He distribution in surface rocks

Fig. 5. Distribution of ^{4}He at the surface of crystalline rocks 10057.80 and 10085.41.

The analysis of gases included in cavities of the crystalline rock 10057.22 gave the following results (in 10^{-8}cm^3): In a 56 mg piece, N_2, 700; H_2, 5; ^{4}He, 3; ^{40}Ar, 1.4; ^{40}Ar/^{36}Ar = 700. In a 50 mg piece: ^{4}He, 0.7; ^{40}Ar, 0.63; H_2 and N_2 not detected. The Ar is clearly radiogenic, probably because of degassing after rock crystallization.

The foregoing results show that the rare gases consist of three components. The main contribution seems to come from trapped solar wind particles. There are also small radiogenic and spallogenic components.

A solar wind origin is suggested by the proportionality of the gas concentrations with the reciprocal diameter of the grains, both in mineral components and in unseparated sieve fractions. Since the concentrations increase with the surface to volume ratio, the gases must be concentrated in the outermost surface layers. This idea is supported by the composition of the gases, which agrees with the so-called solar abundances. All gas bearing grains must have been at the uppermost surface at one time or another. In a similar way, the surfaces of glass pitted crystalline rocks were loaded with gases while they were exposed to the sun, but the interior was always shielded. Since breccias are composed of dust and lithic fragments, it is natural that the matrix contains solar gases while they are undetectable in lithic fragments. Mineral grains without detectable gas content most likely were never exposed to the surface or were degassed at a later stage. The correlation of He with Ti and to some extent with Fe, Cr, and Mn and the anticorrelation between He and Ca may be explained by different diffusion constants or gas adsorption properties of different minerals. The fact that grains of the same mineral differ substantially in their gas concentration indicates different exposure times or partial outgassing.

The mean thickness b at which the gas of the surface is reduced by a factor of e can be calculated from a simple mathematical treatment. Assuming an exponential decrease of gas concentration toward the interior of the grains, it is possible to infer b from the measured degree of gas reduction due to etching; b is 10 μm for He, 6μm for Ne, and 5 μm for Ar. These different b values indicate that diffusion is responsible for the fact that b_{He} is larger than the mean penetration depths of solar wind He (0.05 μm). The mean layer thickness b_{He} is in good agreement with the value directly measured at the surface of crystalline rocks (Fig. 5). Applying b = 10 μm to the He measurements, one ends up with the same surface concentration of 10^{-2} cm^3 of ^{4}He per square centimeter as found before. This concentration is much smaller than the product of solar wind flux and exposure age, because of shielding and possibly surface saturation. For the smallest sieve fraction the apparent b_{Ar} value is smaller than for larger grain sizes. This is still unexplained; it may be that some particular mineral is enriched in the smallest fraction.

Stepwise heating experiments (*3*) yield an average diffusion constant for ^{4}He at 97°C in lunar dust of $D/a^2 \sim 3 \times 10^{-2}$ sec^{-1}. A rough estimate for the time required to fill the interior of a 100 μm grain yields $\sim 10^{3 \pm 1}$ years. Assuming a continuous turnover, the average exposure age of lunar surface material yields a rate of $\sim$ 1 m per 100 $\times$ 10^6 years. This would correspond to a solar wind penetration time of $\sim$ 10^4 years per 100 μm. More detailed diffusion experiments should be undertaken.

The location of the solar wind particles in the surface makes it easy to remove them together with the surface layer so that spallogenic and radiogenic components become relatively enriched in the remaining sample. The spallogenic component can clearly be distinguished from the ^{22}Ne/^{21}Ne and ^{36}Ar/^{38}Ar ratios in the etched dust fractions. Assuming a production rate of 0.15 $\times$ 10^{-8} cm^3 of ^{21}Ne per gram per 10^6 years gives an apparent exposure age of 500 $\times$ 10^6 years. An ^{38}Ar exposure age could be estimated only for the smallest sieve fraction. That age is also about 500 $\times$ 10^6 years. The apparent ^{3}He exposure age (200 $\times$ 10^6 years) is uncertain because of high ^{4}He content and possible diffusion losses. The ^{21}Ne and ^{38}Ar exposure ages are considered as upper limits, since spallogenic gases could be partially inherited.

The etched small grain size fractions show a pronounced enrichment of ^{40}Ar. Assuming an ^{40}Ar/^{36}Ar ratio of 1 in the solar wind, the radiogenic ^{40}Ar content amounts to ~ 4.2×10^{-5} cm^3 per gram of dust. If we assume an average K concentration of 1000 parts per million, a K-Ar age of 3.5×10^9 years results.

T. KIRSTEN
F. STEINBRUNN, J. ZÄHRINGER
Max-Planck-Institut für Kernphysik, Heidelberg, Germany

References and Notes

1. Lunar Sample Preliminary Examination Team, *Science* **165**, 1211 (1969).
2. J. Zähringer, *Earth Planetary Sci. Lett.* 1, 20 (1966).
3. O. A. Schaeffer *et al.*, private communication and *Science*, this issue.
4. We thank W. Gentner, A. Haidmann, P. Horn, E. Jessberger, H. W. Müller, H. Richter, D. Storzer, and H. Weber for their assistance in this work.

4 January 1970

Pattern of Bombardment-Produced Radionuclides in Rock 10017 and in Lunar Soil

Abstract. *A large number of radionuclides have been measured as a function of depth in lunar rock 10017 and in bulk fines. Data are reported on ^{10}Be, ^{22}Na, ^{26}Al, ^{36}Cl, ^{49}V, ^{53}Mn, ^{54}Mn, ^{55}Fe, ^{56}Co, ^{57}Co, and ^{59}Ni and on upper limits for ^{46}Sc, ^{48}V, ^{51}Cr, and ^{60}Co. The results for several nuclides show striking evidence of excess surface production attributable to solar flare particles. Data for short-lived species, ^{56}Co, ^{57}Co, ^{54}Mn, ^{55}Fe, and ^{22}Na, appear consistent with fluxes from known recent events. Long-lived species demonstrate the existence of solar flare protons and alphas at least for the last 10^5 to 10^6 years, at fluxes comparable to those now observed.*

Measurements of the concentration of short- and long-lived radionuclides in meteorites have been useful in unraveling the history of the high-energy radiation which produced them and of the target objects (*1*, *2*). Similar studies on lunar surface materials have the same potential usefulness, with some special advantages. A lunar sample has suffered no ablation, and in principle its recent orientation and depth are known. The planet which they sample is a single, intensely studied one at a known orbital position.

We attempted to study processes of gardening by impact or other processes on the lunar surface, through times of exposure and depth variations, and to decipher the fossil record of solar and galactic high-energy particles in the earth-moon region. Since the longest half-lives available are less than 3×10^6 years, and since measured bombardment ages are generally much longer than this (*3*, *4*), it appears that the samples so far received are, for our present purposes, relatively fixed targets. Thus we have generally concentrated on the objective of using the rock and soil samples to study the high-energy particle emission of the sun on a time scale up to millions of years. Nevertheless, a combination of radionuclide, rare gas, and track data can still give much information on turnover rates.

We received two lunar samples. First, 100 g of bulk fines (10084,16), <1 mm size, was received on 12 September (52 days after lunar landing). The second sample, two chunks of rock 10017 (a coarse-grained rock of mass 980 g) was received on 3 October. Figure 1a shows a drawing of this rock, with an outline of the approximate position of the specimens. The rock had two plausible surface orientations; most photographs made in the preliminary examination were in what we now consider a position inverted relative to its orientation on the moon (*5*).

Samples were subjected to grinding (rock), magnetic separation (rock and dust), and leaching (dust). They were dissolved in $HF\text{-}HNO_3$, with carrier addition where appropriate, in a Teflon still under flowing N_2. Chlorine and Si were recovered in the distillate. The residue from the distillation was heated with H_2SO_4 in a Pt dish to remove HF; HCl was added, and Fe was extracted. The "main chemistry" was completed by anion- and cation-exchange column steps. This main chemistry was carried out, with variations, for two 50-g portions of soil and for six subdivisions of the rock (Fig. 1b). The resulting fractions, containing usually one element of interest, sometimes two or three, were purified chemically and radiochemically.

Chemical analyses for calculation of chemical yields and for following the progress of separations and recoveries were carried out by atomic absorption spectrometry. Agreement was good between these analyses and those available to us from others, with a few exceptions. In some cases the analytical data of others were used for final yield calculations. Overall yields were usually in

Table 1. Summary of counting results (dpm/kg). Limits (2σ) in bulk fines sample: ^{7}Be, 250; ^{46}Sc, 9; ^{48}V, 38; ^{51}Cr, 45; and ^{60}Co, 5.

Nuclide	Half-life	Rock 10017					Bulk fines	
		T4DF (0 to 4 mm)	T4U (4 to 12 mm)	T4I (12 to 30 mm)	T3SI (60 mm)	Houston RCL*	This work	Houston RCL
^{10}Be	2.5×10^6y	16 ± 2		16 ± 2			15.7 ± 1.6	
^{22}Na	2.6 y	76 ± 10	43 ± 6	37 ± 6			55 ± 7	51 ± 9
^{26}Al	7.4×10^5y	133 ± 15	70 ± 10	57 ± 9	31 ± 5	39 ± 7	108 ± 17	120(± 19)
^{36}Cl	3×10^5y	12 ± 5	16 ± 2	16 ± 2	63 ± 12	73(± 13)	17.0 ± 1.6	
^{49}V	330 d						7.4 ± 2.0	
^{53}Mn	~ 2×10^6y	107 ± 21	45 ± 13	52 ± 10				
^{54}Mn	303 d	33 ± 17	34 ± 22	10 ± 11		34(± 13)	18 ± 10	28 ± 9
^{53}Mn and ^{54}Mn			98 ± 15	88 ± 10			82 ± 11	
^{55}Fe	2.6 y	420 ± 44	94 ± 27	93 ± 30	97 ± 24		195 ± 22	
^{56}Co	77 d	125 ± 20	< 16	< 8		26(± 6)	44 ± 6	40 ± 10
^{57}Co	270 d	5.6 ± 2.6	< 1.7	< 2.4			< 1.8	
^{59}Ni	8×10^4y	< 4					2.6 ± 0.8	

* Houston 10017 data were measured using the whole rock (*14*).

the range of 80 percent, varying with the element, though in some cases of low specific activity (^{26}Al and ^{55}Fe) only a fraction of the available material was purified and counted. In some cases small amounts of radioactive impurities were seen in the counting, and where necessary these samples were recycled.

Several detector systems were used, depending on the decay scheme of each species and the demands of schedule. The most important were: (i) the Evans counter (E), a β-γ system developed from a design by N. Bhandari, consisting of a small Geiger counter (external sample) in the well of a 7.5 by 7.5 cm NaI(Tl) crystal; (ii) the Loosli counter (H), also modified from a design by Bhandari (*6*), a small internal pressurized proportional counter in a 10 by 10-cm NaI(Tl) well crystal, for x-ray, x-γ, and β-γ measurements; (iii) the Tanaka counter (T) (*7*), a Geiger-plastic scintillator package for β-counting and spectrometry; (iv) the Lal counter (L), from a design by Rajagopalan (*8*), a pair of Geiger counters with fixed absorber between, also for β-counting; (v) the Delany counters (D), internal x-ray counters of large area, 60 and 160 cm^2; and (vi) a 40-cm^3 solid-state Ge(Li) crystal (SS) in an annulus of NaI(Tl) for gamma counting.

Samples were counted on Cu or plastic holders, with the use of a Millipore deposition technique or electroplating, both of which gave extremely smooth and uniform deposits. Counting standards were prepared from standardized solutions. Where possible, the form and thickness of samples and standards were approximately matched. Beyond this, self-absorption corrections were calculated by standard methods. Backgrounds were taken where possible with the same or similar chemical species, carried through the same chemistry as the samples. In several cases, samples were counted on more than one system, generally with concordant results. All these procedures were carried out more or less completely several times on terrestrial materials before arrival of the sample.

The counting results to date are summarized in Table 1. The errors quoted include all known sources: counting statistics, chemical yield, calibration, and others. Ratios, especially of the same nuclide in different samples, may be more accurate. Table 1 also shows the results obtained at Houston for comparable materials, where available. In general the agreement is good.

Table 2. Comparison of observed and calculated activities in Bruderheim and Harleton meteorites and in rock 10017 (units dpm/kg).

Nuclide	Bruderheim			Harleton		Rock 10017 (12 to 30 mm)	
	Obs.	Calc. $\alpha = 400$	Calc. S(10)	Obs.	Calc. $\alpha = 600$	Obs.	Calc. $\alpha = 600$
^{55}Fe	340	140	280	$\leqslant$ 180	92	93	37
^{54}Mn	100	56	92	38	41	10	16
^{53}Mn	85	56	68	45	42	52	17
^{49}V	34	25	25	20	22	7*	9
^{36}Cl	7.5	6.5	7.2	7.0	6	16	9
^{26}Al	60	54	65	45	41	57	33
^{22}Na	90	88	94	64	67	37	29
^{10}Be	19	24	24	21	23	16	15

* Estimated from bulk fines.

The gentle acid leaching of the dust sample, before the main chemistry, dissolved a large fraction of the Ni and Co in the sample, with a little Fe and other elements. A small amount of Cl was isolated and showed no ^{36}Cl activity. The tabulated activities of ^{56}Co and ^{59}Ni are based on chemical yield for the undissolved portion; we assume that Fe was the only significant target.

The ratios of activities in surface samples to those in deep samples (Table 1) are of special interest. For all accurately measured nuclides the ratio of sample T4DF to T4I is substantially greater than 1. It is particularly large for ^{55}Fe, and in the extreme case of ^{56}Co no activity is observed below the top sample. A comparison of fines and deep samples shows a similar but smaller effect. We infer that the mean depth of the dust below the undisturbed surface is of the order of 2 cm. All these data seem to show the importance of solar proton irradiation near the surface.

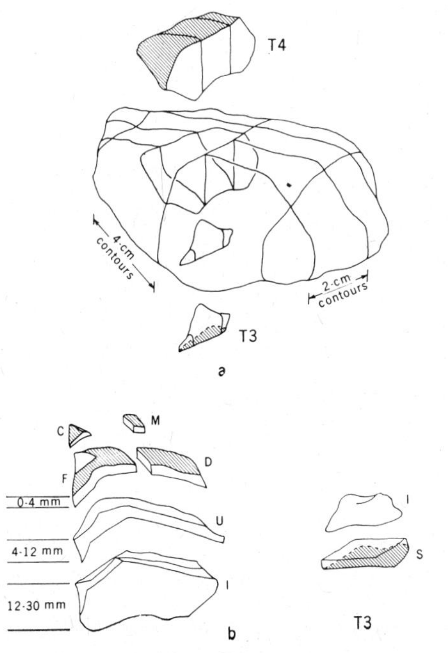

Fig. 1. (a) Rock 10017, as oriented by counting data, showing the location of our two samples. (b) Exploded drawing of samples T4 and T3, showing location and orientation of subsamples divided by us.

The ^{59}Ni is significant as showing the presence of solar α-particles. It is produced uniquely by ^{56}Fe (a,n) ^{59}Ni.

For the production of radionuclides in these materials, then, we must consider two sources: the galactic cosmic radiation and solar particles. The galactic production can be calculated on several models (*1*, *2*). It is reasonable to assume for this component the same constancy with time observed in meteorites (*1*).

We have used a model developed from that of Arnold *et al.* (*9*). Primary and secondary spectral shapes were chosen, usually of the form $\kappa/(\alpha + E)^{2.5}$. Smaller values of α correspond to a softer spectrum (more secondaries, greater depth). The absolute intensity is related to that at the surface by assuming that particles >1 Bev move linearly with a mean attenuation length of 155 g/cm^2, and normalizing the spectra above that point. The value of α must decrease smoothly with depth. Excitation functions are taken from experiment, analogy, or reference (*9*, *10*) for each product nuclide and important target nuclides. Integration gives the production in disintegrations per minute per kilogram of lunar material.

The model was tested with data for the Bruderheim and Harleton chondrites (*1*). A value of α=400 Mev was found to give the best fit for Bruderheim

Table 3. Short-lived nuclides in 0 to 4 mm layer from solar proton data: Experiment versus calculation.

Nuclide	Half-life	Expt. (dpm/cm²)	Calculated (dpm/cm²)
^{56}Co	77 d	0.18	0.13*, 0.36†
^{57}Co	270 d	0.01	0.004*, 0.01†
^{54}Mn	303 d	~0.02	0.01*, 0.02†
^{55}Fe	2.6 y	0.5	0.5
^{22}Na	2.6 y	0.06	0.12

* See (*15*). † See (*16*).

(Table 2); for Harleton the best relative and absolute fit was for $\alpha = 600$. A low energy tail, of the AHL S(10) (*9*) form, gave improved agreement for ^{55}Fe, but made little change otherwise. Since for primaries $\alpha = 1$ Bev, and since even at the top surface secondaries must be important, we chose a value of $\alpha = 700$ at zero depth, falling to $\alpha = 500$ for the deepest samples of rock 17.

With this spectrum normalized for 2π bombardment as above, we then computed the expected galactic production for the deep sample T4I (Table 2). It must be emphasized that this procedure, while consistent with available physical information, is in effect a method of interpolation, in which the soundness of the model is determined empirically in meteorites, and the results are translated to the different composition and geometry of the lunar samples. We believe the results are not more than 30 percent in error on the average.

It can be seen that some species are well accounted for by galactic production. Others show a marked excess.

Solar particles have lower energies than galactic cosmic rays. They occur during flares in short bursts, with spectral shapes generally well represented for our purposes in exponential rigidity form (*10*). Since secondary particles are not important at these energies, we adopt a different model for production calculations. Particles are assumed to lose energy almost entirely by ionization, and production is integrated along the path, giving results in dpm/cm² for a given total flux above 10 Mev and mean rigidity R_0. For ^{56}Co, a product of the low energy ^{56}Fe(p,n) ^{56}Co reaction, and the observed low mean rigidity (~50 MV), all the production occurs in the outer 4 mm. Thus we need not calculate angular dependence or interior spectral shape. For higher energy products and higher R_0 a more elaborate program is used to calculate dpm/cm² for given depth ranges.

For short-lived products we have compared our data with predictions from known flares (Table 3). It can be seen in particular that the results for ^{56}Co are well accounted for by flares—mostly by the event of 12 April 1969. The other short-lived species are also present in roughly the calculated amounts; somewhat less for ^{22}Na.

From the activities of the long-lived species ^{36}Cl, ^{26}Al, ^{53}Mn, and ^{10}Be, we can see that solar protons have been present for the past million years in amounts comparable to those observed recently. Only the high-energy product ^{10}Be fails to show a large excess over galactic production (Table 2). But before we can make definite conclusions about the flux and rigidity of past solar events we must examine the excess production observed in the deep samples. At present, without detailed rare gas and nuclear track information and precise ^{53}Mn measurements by activation analysis, we can explain our data in two ways. Model (A) assumes that there has been no gardening or erosion for rock 10017 over 10^7 years or so. This would lead us to believe in a mean flux of solar protons similar to the presently observed one (of order 30/cm²sec), but higher mean energy or rigidity, $R_0 \geqslant 150$ MV, over millions of years, and perhaps in different fluxes in different periods. Model (B) assumes that the higher interior values are those characteristic of production at deeper levels in the moon and that the exposure of the present surface is relatively recent ($\leqslant 5 \times 10^5$ years age). This leads to a higher flux and lower rigidity in the period of exposure to produce the observed surface excess. Thus in either case the mean solar proton flux appears to have been as high as (or even higher than) that observed at present.

Several other topics remain to be discussed. The comparison with rare gas data shows a complex radiation history, discussed by Marti *et al.* (*4*). The Gd isotope data (*4, 11*) will give an independent integral of the neutron flux. Nuclear track data (*12, 13*) confirm the location of the upper rock surface. The lunar sample results have implications for meteorite studies, and for cosmic dust influx rates, both as deduced from limits on erosion rates, and for comparison with ^{26}Al data in sediments.

We conclude that: (i) Our data on short-lived nuclides in the lunar surface can be explained by known bombarding fluxes: the cosmic radiation and solar flare particles. (ii) The long-lived nuclides show a similar pattern. Hence we conclude that solar flares, and presumably sunspots, have been occurring for at least the last 10^5 to 10^6 years. (iii) While details of flux and mean rigidity depend on data not yet available, the mean solar particle flux appears to have been comparable to or perhaps greater than that observed at present.

JULIAN P. SHEDLOVSKY

National Center for Atmospheric Research, Boulder, Colorado 80301

MASATAKE HONDA

University of Tokyo, Tokyo, Japan

ROBERT C. REEDY
JOHN C. EVANS, JR.
DEVENDRA LAL
RICHARD M. LINDSTROM
ANTHONY C. DELANY
JAMES R. ARNOLD
HEINZ-HUGO LOOSLI
JONATHAN S. FRUCHTER
ROBERT C. FINKEL

Department of Chemistry, University of California at San Diego, La Jolla

References and Notes

1. M. Honda and J. R. Arnold, *Handbuch der Physik* 46/2 (Springer-Verlag, Berlin-Heidelberg, 1967), pp. 613–632.
2. T. A. Kirsten and O. A. Schaeffer, in *Interactions of Elementary Particle Research in Science and Technology*, L. C. L. Yuan, Ed. (Academic Press, New York, 1969).
3. Lunar Sample Preliminary Examination Team, *Science* **165**, 1211 (1969).
4. K. Marti, G. W. Lugmair, H. C. Urey, *ibid.*, this issue.
5. We thank E. Schonfeld (private communication) for information which suggested the correct orientation and hence led to much improved sampling of rock 10017. Our experimental program was however designed to verify this conclusion. We also received help from R. M. Walker, R. L. Fleischer, P. B. Price, and M. Maurette in this matter.
6. N. Bhandari, *Nucl. Inst. Meth.* **67**, 251 (1969).
7. E. Tanaka, S. Itoh, T. Maruyama, S. Kawamura, T. Hiramoto, *Int. J. Appl. Radiation Isotopes* **18**, 161 (1967).
8. D. Lal, G. Rajogopalan, Rama, in *Radioactive Dating and Methods of Low-Level Counting* (Int. Atomic Energy Agency, Vienna, 1967).
9. J. R. Arnold, M. Honda, D. Lal, *J. Geophys. Res.* **66**, 3519 (1961).
10. D. Lal and V. S. Venkatavaradan, *Earth Planet. Sci. Lett.* **3**, 299 (1967).
11. A. L. Albee, D. S. Burnett, A. A. Chodos, O. J. Eugster, J. C. Huneke, D. A. Papanastassiou, F. A. Podosek, P. G. Russ II, H. G. Sanz, F. Tera, G. J. Wasserburg, *Science*, this issue.
12. G. Crozaz, M. Hair, P. Hoyt, J. Kardos, M. Miyajima, M. Maurette, M. Seitz, R. Walker, M. Wittels, D. Wollum, *ibid.*, this issue.
13. R. I. Fleischer, E. L. Haines, R. E. Hanneman, H. R. Hart, Jr., J. S. Kasper, E. Lifshin, R. T. Woods, P. B. Price, *ibid.*, this issue.
14. G. D. O'Kelley, J. S. Eldridge, E. Schonfeld, P. R. Bell, *ibid.*, this issue.
15. J. N. Kinsey and F. B. McDonald, preliminary data.
16. J. Hsieh and J. Simpson, preliminary data.

17. Supported by NASA contract NAS 9-7891.

4 January 1970

Cosmogenic and Primordial Radionuclides in Lunar Samples by Nondestructive Gamma-Ray Spectrometry

Abstract. *The ^{7}Be, ^{22}Na, ^{26}Al, ^{44}Ti, ^{46}Sc, ^{48}V, ^{51}Cr, ^{54}Mn, ^{56}Co, ^{57}Co, ^{60}Co, ^{40}K, ^{238}U, and ^{232}Th were measured in lunar fines and portions of three rocks. Major production of cosmogenic radionuclides is due to solar protons, thus their concentrations are far different than those in meteorites. Surface exposures of the rocks and fines are long compared with the 0.74 million year half-life of ^{26}Al. Lunar fines show substantially higher concentrations of low energy reaction products. The ratios of thorium to uranium are extremely constant at 3.8, which indicates very little geochemical differentiation and are in good agreement with a common nucleosynthesis for lunar and earth materials.*

The relative and absolute radionuclide concentrations of lunar samples are far different from those observed in meteorites, which reflects substantial differences in the chemical composition of the lunar surface and in the energy spectrum of incident cosmic rays. Whereas meteorites are exposed to a somewhat similar but less intense cosmic ray spectrum, the isotopes produced by solar cosmic rays are largely lost by ablation during entry of the meteorite through the earth's atmosphere. Thus, while the cosmogenic isotopes observed in meteorites result mainly from the high energy galactic component, those on the lunar surface are produced largely from the solar component of the cosmic ray spectrum.

Preliminary gamma ray spectrometric measurements made during sample quarantine at the Lunar Receiving Laboratory provided approximate radionuclide concentrations of lunar samples (*1*). The objectives of the present work were to perform very precise measurements of the radionuclide content of lunar surface rocks and fines and interpret these observations in terms of cosmic ray exposure and the history of the specimens.

All of the radionuclide measurements were made by nondestructive gamma ray spectrometric techniques. The lunar samples were maintained in a dry nitrogen atmosphere in thin containers of stainless steel, or aluminum-lined lucite for the counting measurements. The counting was performed on anticoincidence shielded multidimensional gamma ray spectrometers (MDGRS) (*2*), employing large NaI(Tl) detectors and on a Ge(Li) spectrometer (*3*). The MDGRS which employed either two 28 cm diameter by 15 cm thick NaI (Tl) crystals or two 23 cm by 20 cm thick NaI(Tl) crystals offer a very high sensitivity for the measurement of those radionuclides which decay by emission of several gamma rays or by emission of a positron and a gamma ray [^{22}Na, ^{26}Al, ^{44}Ti, ^{46}Sc, ^{48}V, ^{56}Co, ^{232}Th (^{208}Tl) and ^{238}U(^{214}Bi)]. The anticoincidence shielded Ge(Li) spectrometer which viewed the sample with two 70 cm^3 detectors provided an improved sensitivity for the measurement of radionuclides which emit only one photon per disintegration (^{7}Be, ^{51}Cr, ^{54}Mn and ^{57}Co) and served to verify the measurements of the more abundant cascade photon emitters (^{22}Na and ^{26}Al). The concentrations of the radionuclides in the lunar samples were based on 10,000 to 20,000 minute count intervals on both types of analyzers. Standard mockups with the same size, shape, physical density, and electron density as each lunar rock and containing known amounts of each radionuclide of interest were prepared from a mixture of plaster of paris, iron powder, and water (*4*). Mockups of the lunar fines were prepared in mixtures of 5 percent Al_2O_3 and 95 percent dunite to provide the same density as the fines. The standard radioisotopes were first pipetted into the Al_2O_3, dried, and after mixing, the labeled Al_2O_3 was homogeneously mixed with the dunite to form the mockup. Other types of mockups prepared in powdered basalt, and dunite with 25 percent Fe were shown to have essentially identical counting efficiencies to those in the dunite (*5*), thus substantiating the adequacy of the mockups.

Inhomogeneous mockups where the activity was all located in the top 25 percent produced only a 5-percent effect on the total counting efficiency but could easily be determined by differences in the ratios of the coincidence photons from ^{22}Na and ^{26}Al which were deposited in the two NaI(Tl) crystals.

The standard radionuclide solutions used in preparing mockups were all calibrated by counting them in a precisely controlled geometry on a Ge(Li) diode for which counting efficiency versus photon energy had been very accurately established (*4*). The radio-

Table 1. Cosmogenic and primordial radionuclides in lunar materials. The $\pm$ values are the standard deviations associated with all counting statistics. Rock A(57), sample 10057,30; rock B(17), sample 10017,37; rock B(3), sample 10003,25; fines 1 (84), sample 10084,41; fines 2 (84), sample 10084,113-2; fines 3 (84), sample 10084,113-1. Values for rows 1 (^{7}Be) through 11 (^{60}Co) are expressed as disintegrations per minute per kilogram; row 12 (K) is weight percent; rows 13 (Th) and 14 (U) are ppm; weights of sample and of rock are expressed in grams. Values prefaced with $<$ represent 2σ of the gross counting rate in the photopeak area.

Radionuclides	Rock A(57)	Rock B(17)	Rock B(3)	Fines 1 (84)	Fines 2 (84)	Fines 3 (84)
^{7}Be				$<$ 112		
^{22}Na	43 $\pm$ 2	45 $\pm$ 2	49 $\pm$ 2	64 $\pm$ 3	63 $\pm$ 2	
^{26}Al	84 $\pm$ 2	80 $\pm$ 2	75 $\pm$ 2	131 $\pm$ 4	137 $\pm$ 4	
^{44}Ti	$<$ 2					$<$ 1.5
^{46}Sc	11 $\pm$ 2	13 $\pm$ 2	8 $\pm$ 2	11 $\pm$ 1		
^{48}V				$<$ 10		
^{51}Cr	$<$ 62			$<$ 63		
^{54}Mn	41 $\pm$ 8	46 $\pm$ 17	60 $\pm$ 7	24 $\pm$ 3		
^{56}Co	19 $\pm$ 9					53 $\pm$ 10
^{57}Co	$<$ 3.9	$<$ 5.0	$<$ 4.3	1.3 $\pm$ 1.6		
^{60}Co	$<$ 1	0.8 $\pm$ 0.6	$<$ 1	0.3 $\pm$ 0.8		
K	0.23 $\pm$ 0.02	0.26 $\pm$ 0.01	0.046 $\pm$ 0.006	0.11 $\pm$ 0.01	0.11 $\pm$ 0.01	
Th	3.6 $\pm$ 0.1	3.2 $\pm$ 0.1	1.08 $\pm$ 0.05	2.2 $\pm$ 0.1	2.3 $\pm$ 0.1	
U	0.95 $\pm$ 0.05	0.86 $\pm$ 0.05	0.29 $\pm$ 0.02	0.54 $\pm$ 0.03	0.56 $\pm$ 0.04	
Sample weight	230	115.5	117.2	999	327.0	270.8
Rock weight	897	971	213*			

* Weight uncertain.

Table 2. Calculated production rates of ^{26}Al from galactic cosmic rays in lunar material. Rates are expressed in disintegrations per minute per kilogram, based on work of Fuse and Anders (*6*). Samples are identified fully in Table 1.

Material	Al	Si	Fe	Ca	Ti	Total 4π	Total 2π	Observed	Calculated percent of observed
Fines (84)	32.9	62.0	0.3	0.6	0.1	95.9	48.0	134	36
Rock A (57)	27.6	52.1	0.3	0.5	0.2	80.7	40.4	84	48
Rock B (17)	25.2	58.0	0.3	0.5	0.2	84.2	42.1	80	53
Rock B (3)*							42.1	75	56

* Chemical composition not available but assumed to be the same as for rock 17.

nuclide content of the standard mockups were thus known within ±2 to 3 percent. The instruments used for the lunar sample measurements were calibrated by counting duplicates or triplicates of the mockups to a precision of ± 1 to 2 percent. The radionuclide concentrations in the lunar samples were calculated by direct comparison of their counting rates with those of mockup standards after application of appropriate background and Compton corrections.

A serious problem in the measurements of cosmogenic radionuclides that is not encountered in meteorite measurements resulted from the high concentrations of Th and U in the lunar material. The U and Th which were 10- to 50-fold higher in the lunar material than in meteorites produced a high Compton scatter "background" over the entire coincidence spectrum.

The radionuclide concentrations observed in samples of three lunar rocks and in lunar fines are summarized in Table 1, with standard deviations as low as ± 3 percent for ^{22}Na and ^{26}Al. One of the more striking observations of the measurements is the large differences between ^{22}Na and ^{26}Al concentrations of the rocks and fines (Table 1). This large range was not evident in the LRL preliminary measurements (*1*), perhaps because of the large statistical uncertainties in their measurements. In addition, concentrations of these radionuclides are much higher than one would have predicted from a 2π isotropically incident galactic cosmic ray flux. The expected galactic production of ^{26}Al, after buildup of secondaries, was determined for the chemical compositions (*1*) of the lunar samples (Table 2), with the method of Fuse and Anders (*6*) for stone meteorites. Since the calculated concentrations listed for the 2π isotropic flux include the contribution from secondary buildup, these values should be considered as upper limits, as should the accompanying values for the percent galactic contribution. Extensive measurements in near surface samples from the meteorite Saint Séverin (*7*) provided the first direct proof that ^{22}Na and ^{26}Al in near surface samples are almost twofold lower than in deeper regions of a large meteorite. Thus, the galactic contributions may be as little as one-half that listed in Table 2. The low ^{22}Na and ^{26}Al concentrations in the Denver meteorite (*8*) are apparently due to a lack of secondary flux buildup because of its small size (Table 3). The range in the concentrations of ^{22}Na and ^{26}Al in lunar samples shown in Table 1 is not accounted for by differences in chemical composition and is apparently due to differences in shielding. Unfortunately, documentation of the orientation of the rock on the lunar surface, the relationship of our specimens to the original rocks, and of the sampling depth of the lunar fines is not presently available to help explain the observed concentration differences. The ^{26}Al concentrations in the three rock samples range from 56 to 63 percent of that in the lunar fines while the ^{22}Na concentrations in the three rocks are also low, ranging from 67 to 77 percent of that in the fines. Since both of these radionuclides are produced by low energy reactions, a closer average distance to the surface which the fines apparently occupied could explain their higher concentrations.

The ratios of ^{26}Al to ^{22}Na are two- to threefold higher than those normally present in meteorites. This high ratio is apparently due to a low ^{22}Na production rate. The threefold higher aluminum content of the lunar samples (*1*) than of meteorites (*9*) would favor ^{26}Al production relative to ^{22}Na from the low energy solar protons. Also, Mg which is a major target element for ^{22}Na production is about threefold lower in the lunar samples and would further limit ^{22}Na production. The ^{26}Al to ^{22}Na ratios in lunar samples bear a direct relationship to exposure age and could be employed for estimating the surface exposure age in a manner similar to that used in the age dating of meteorites. However, since the relative production rates of ^{26}Al and ^{22}Na for the low energy solar component is not well known, caution should be exercised in applying this dating technique either with this isotope pair or for other cosmogenic pairs. Regardless

Table 3. Comparison of cosmogenic radionuclides in lunar sample with those in meteorites. Meteorite A_1 is an 11-cm deep sample from the Allende meteorite (*10*); meteorite A_2 is a sample from > 25 cm deep in the Allende meteorite (*10*); meteorite B is the Denver meteorite (*8*) and meteorite C is a near-surface sample from the Saint Séverin meteorite (*7*).

Isotope	Activity (dpm/kg)								Ratio of isotope to ^{22}Na							
	Lunar sample				Meteorite				Lunar sample				Meteorite			
	Rock			Fines					Rock			Fines				
	(57)	(17)	(3)	(84)	A_1	A_2	B	C	(57)	(17)	(3)	(84)	A_1	A_2	B	C
^{22}Na	43	45	49	66	69	81	61	56								
^{26}Al	84	80	75	134	57	61	43	36	1.95	1.78	1.53	2.03	0.83	0.75	0.70	0.64
^{44}Ti	<2			<1.5					<0.05			<0.02				
^{46}Sc	11	13	8	11	6	9			0.26	0.29	0.16	0.17	0.09	0.11		
^{48}V				<10		21						<0.12		0.26		
^{51}Cr	<62			<63		54			<1.9			<0.95		0.67		
^{54}Mn	41	46	60	24	99	129		48	1.0	1.0	1.2	0.36	1.4	1.6		0.86
^{56}Co	19			53				4	0.59			0.80				0.07
^{57}Co	<3.9	<5.0	<4.3	1.3		32		28	<0.12	<0.11	<0.09	0.02		0.40		0.50
^{60}Co	<1	0.8	<1	0.3	41	128	<4	11	<0.03	0.018	<0.02	0.04	0.59	1.6	<0.07	0.20

of any variations in the ^{26}Al and ^{22}Na production rates with energy, depth, solar cycle, and so forth, it is obvious that the lunar rocks and fines have surface exposure ages which are comparable to or much longer than the half-life of ^{26}Al. If one assumed that the production ratios of $^{26}Al : ^{22}Na$ were the same in the lunar samples, and that the $^{26}Al : ^{22}Na$ ratio of about two which was observed in both the lunar fines and in rock 57 (Table 3) represents a saturation value, then rocks 17 and 3 would indicate lunar surface exposure ages of 2.2 and 1.5 million years, respectively. However, in view of known differences in $^{26}Al : ^{22}Na$ production rates with energy it appears more likely that all of the samples are at saturation and that the ratio differences are due to shielding.

The ^{54}Mn concentrations in the lunar samples show an even wider range than ^{22}Na and ^{26}Al and do not appear to show any correlation with them (Tables 1 and 3). The main target element for ^{54}Mn production is iron and the much lower concentrations of iron in the lunar materials than in chondrites helps explain the generally lower ^{54}Mn concentrations. The efficient production of ^{54}Mn in lunar materials requires more energetic protons than the production of either ^{22}Na or ^{26}Al. The lunar fines show a much lower ^{54}Mn concentration than the rocks which is not explained by their somewhat lower iron content. It appears that while the solar protons greatly enhance the production of ^{22}Na and ^{26}Al, they contribute relatively little to ^{54}Mn production; thus one observes the low ^{54}Mn in lunar fines and the higher concentrations in rocks where significant secondary flux buildup may occur.

The titanium content of the lunar samples is about 50 times that in meteorites and one would therefore expect cosmogenic isotopes from this target element to be more prevalent. The ^{46}Sc which is produced from titanium with a cross section of 80 mb at 45 Mev (*5*) was present at concentrations of 8 to 13 disintegrations per minute per kilogram in the lunar rocks and fines. Considering that only a 2π isotropic flux is incident on the lunar surface, these values are high compared with 6 to 9 disintegrations per minute per kilogram in interior samples of the Allende meteorite (*1*) (Table 3) where iron is the principal target element. From ^{46}Sc concentrations in meteorites (*10, 11*), the 2π lunar irradiation effect, and limited secondary flux buildup in the lunar samples, a lunar surface ^{46}Sc concentration of 2 to 4 disintegrations per minute per kilogram would have been expected. The threefold higher ^{46}Sc concentration is attributed mainly to the high titanium content of the samples plus the additional solar protons in a sufficiently high energy range for its effective production. The ^{44}Ti concentrations in rock 57 and in the lunar fines sample were <2 and <1.5 disintegrations per minute per kilogram. The cross section for production of ^{44}Ti from Ti is low (*5*), comparable to that from iron, whereas the threshold is rather high. Therefore, ^{44}Ti concentrations comparable to, or lower than the observed 1.4 and 2.0 disintegrations per minute per kilogram in the Harleton and Bruderheim meteorites (*12*) would be expected.

The isotopes ^{56}Co and ^{48}V are both produced by low energy (p,n) reactions on Fe and Ti, respectively. They have almost identical excitation functions (*5*) and are thus excellent indicators of the solar particle flux intensity. Cobalt-56 concentrations of 53 and 19 disintegrations per minute per kilogram in the lunar fines and rock 57, respectively, were observed. These concentrations are more than ten times what one would predict from the reported 4 disintegrations per minute per kilogram in meteorites (*7*), indicating that >90 percent of the ^{56}Co originates from solar proton reactions. Because of the short half-life of ^{48}V (16.1 days) and the 2- to 3-month delay in sample receipt only a less than value, <10 disintegrations per minute per kilogram, in lunar fines could be measured. Based on the nearly identical nature of the excitation functions for ^{56}Co and ^{48}V formation (*5*), we would predict a ^{48}V concentration at the time of sampling of <8 disintegrations per minute per kilogram.

The ^{57}Co concentrations in the lunar samples are about tenfold lower than in meteorites. The principal target elements Fe and Ni are about two- and tenfold lower, respectively, in lunar material and this is evidently responsible for the very low ^{57}Co values.

The very low ^{60}Co concentrations (Table 1) in the lunar materials are to be expected from the low content of stable cobalt of the lunar surface which is about thirtyfold lower than that in meteorites. Only less than values were obtained for ^{7}Be (53 days) and ^{51}Cr (28 days). Beryllium-7 is a high energy reaction product and has not yet been observed in meteorites. Chromium-51 had decayed by about fivefold at the time of sample receipt and was thus well below its expected and measurable concentration.

As indicated by the very high concentration of low energy products such as ^{56}Co and ^{26}Al, the solar proton component contributes heavily to their production, and a depth profile measurement of such radionuclides would provide a basis for determining surface exposure age, accretion rate, and long and short term variations in cosmic ray flux.

Interest in the K, U, and Th content of the moon lies in its radiogenic heating together with its history in relationship to that of the earth, meteorites, tektites, and other solar system bodies. Our measurements of the natural radioactivity due to K, U, and Th are in general agreement with preliminary measurements (*1*), but they provided smaller error limits and showed a much narrower range in Th : U ratios. Our measurements extend the K : Th weight ratio to cover a range of 1600 to 3000 and reduce the Th : U weight ratio range to less than 10 percent. The lunar sample Th : U ratios of 3.8 ± 0.2 are in good agreement with the calculated $^{232}Th : ^{238}U$ value of 3.8 to 0.3 (*13*) for the present-day solar system and with the ratios in chondrites (*9*), and terrestrial crustal rocks (*14*). These constant Th : U ratios, which would be expected from their very similar ionic radii and identical outer electron configurations, add strong support for cessation of nucleosynthesis of lunar matter at the same time as that of the earth's and meteorite's.

The K, U, and Th concentrations of the lunar materials are considerably different than those observed in other extraterrestrial and major terrestrial materials. Compared with stony meteorites (*9*), the U and Th in lunar samples are some tenfold higher while the K ranges from comparable to fivefold higher concentrations. The lunar material has U and Th concentrations that are similar to some tektites (*15*), but its K content is about tenfold lower. Compared with the average crustal composition of the earth, the lunar materials are about two- to fourfold lower in U and Th and tenfold lower in K. Of the major terrestrial materials the lunar samples most closely resemble basalt (*11*) with similar U and Th concentrations but about a fourfold lower K content. The lunar fines represent somewhat of an integrated sample, that reflects major lunar surface processes, and since their primordial radionuclide concentrations are similar to those of the lunar rocks, the total meteorite contribution to the lunar fines must not have been large enough to significantly alter its composition. The lunar materials thus appear rather unique in their primordial radionuclide composition, which indicates a substantially different geochemical history than that of known earth and other extraterrestrial materials.

RICHARD W. PERKINS
LOUIS A. RANCITELLI
JOHN A. COOPER
JAMES H. KAYE
NED A. WOGMAN
Battelle Memorial Institute, Pacific Northwest Laboratories, Richland, Washington 99352

References and Notes

1. Lunar Sample Preliminary Examination Team, *Science* **165**, 1211 (1969).
2. N. A. Wogman, D. E. Robertson, R. W. Perkins, *Nucl. Instr. Meth.* **50**, 1 (1967); N. A. Wogman, *et al., ibid.* **74**, 197 (1969).
3. J. A. Cooper, in preparation.
4. R. W. Perkins, L. A. Rancitelli, N. A. Wogman, J. A. Cooper, *Semiannual Technical Progress Report to NASA Manned Spacecraft Center,* 1 December 1968 through 1 June 1969.
5. R. W. Perkins, L. A. Rancitelli, J. A. Cooper, J. H. Kaye, N. A. Wogman, in preparation.
6. K. Fuse and E. Anders, *Geochim. Cosmochim. Acta* **33**, 653 (1969).
7. K. Marti, J. P. Shedlovsky, R. M. Lindstrom, J. R. Arnold, N. G. Bhandari, *Meteorite Research* (Springer-Verlag, New York, 1969), p. 22.
8. L. A. Rancitelli, in preparation.
9. H. C. Urey, *Rev. Geophys.* **2**, 1 (1964); A. E. Ringwood, *ibid.* **4**, 113 (1966).
10. L. A. Rancitelli, R. W. Perkins, J. A. Cooper, J. H. Kaye, N. A. Wogman, *Science* **166**, 1269 (1969).
11. M. Honda and J. R. Arnold, *ibid.* **143**, 203 (1964).
12. J. P. Shedlovsky, P. J. Cressy, Jr., T. P. Kohman, *J. Geophys. Res.* **72**, 5051 (1967).
13. W. A. Fowler and F. Hoyle, *Ann. Phys.* **10**, 280 (1960).
14. B. Mason, *Principles of Geochemistry,* (Wiley, New York, ed. 3, 1966), p. 45.
15. G. R. Tilton, *Geochim. Cosmochim. Acta* **14**, 323 (1958).
16. We thank D. R. Edwards, J. G. Pratt, and J. H. Reeves of this laboratory for their aid in the preparation of the standards and the acquisition of data. The very-high-sensitivity instrumentation that made this work possible was developed in programs sponsored by the U.S. Atomic Energy Commission.

4 January 1970

Elemental Compositions and Ages of Lunar Samples by Nondestructive Gamma-Ray Spectrometry

Abstract. *A gamma-ray spectrometry system with low background was used to determine the radioactivity of crystalline rocks, breccias, and fine material. Nuclides identified were* ^{40}*K,* ^{232}*Th,* ^{238}*U,* ^{7}*Be,* ^{22}*Na,* ^{26}*Al,* ^{44}*Ti,* ^{46}*Sc,* ^{48}*V,* ^{52}*Mn,* ^{54}*Mn, and* ^{56}*Co. Concentrations of K, Th, and U ranged between 480 and 2550, 1.01 and 3.30, and 0.26 and 0.83 parts per million, respectively. Concentrations of thorium and uranium were those of terrestrial basalts, while the potassium concentrations were near values for chondrites. Products of low-energy nuclear reactions showed pronounced concentration gradients at rock surfaces. Concentrations of K and of* ^{22}*Na determined here were combined with concentrations of rare gases to estimate gas-retention ages and cosmic-ray exposure ages with ranges of 2200 to 3200 and 34 to 340 million years, respectively, for three rocks.*

Lunar samples from Apollo 11 were analyzed by nondestructive, gamma-ray spectrometry at the Lunar Receiving Laboratory (LRL), Houston, Texas (*1*). The preliminary data have now been refined and additional samples also have been analyzed.

The low-background, gamma-ray spectrometer used in these studies has been described (*1*). Because the samples were viewed by two detectors, gamma-ray spectrums could be recorded either as noncoincident, or singles, data if an event occurred in one detector only, or as a gamma-gamma coincidence event if a signal was produced in both detectors simultaneously. The data acquisition system included a coincidence-anticoincidence logic circuit and dual 12-bit analog-to-digital converters interfaced to an on-line, digital computer. The computer was programmed to record singles data from each detector at a resolution of 256 channels, gamma-gamma coincidence data in a folded matrix with a resolution of 127 by 127 channels but requiring only 8192 storage locations, and a sum-coincidence spectrum of 256 channels. All of these spectrums were used in the detailed analysis of each sample.

It was necessary to use a variety of sample containers (three types of stainless steel and two types of polypropylene) with different attenuation and scattering properties for gamma rays. In addition, variations in radioactive contamination caused uncertainties in the background corrections. The preliminary examination was made especially difficult because it was not known in advance of the mission that some of these containers were to become necessary. Spectrum libraries for calibration in some cases had to be acquired during the active mission period, which reduced the time available for measuring lunar samples. After the sample quarantine was lifted, a number of small samples were measured which had been exposed to atmosphere, and so did not require special containers.

Before receipt of lunar samples, response functions of 19 expected nuclides were collected. Extended sources of electrolytically reduced iron powder containing 10^4 to 10^5 disintegrations per minute of each nuclide were prepared, measured, and compiled into a library of spectrums for the preliminary examination.

Potassium, thorium, and uranium standards were calibrated by weighing. Potassium standards were prepared from reagent-grade potassium chloride, and for some sources from potassium enriched in ^{40}K. Certified standards of thorium and uranium were obtained from the U.S. Atomic Energy Commission, New Brunswick Laboratory. All other radionuclides were standardized by absolute gamma-ray spectrometry to an overall uncertainty of less than 5 percent.

Final calibrations were performed through use of replicas filled with radionuclide standards which reproduced closely the bulk and electronic densities of the lunar samples. The replicas consisted of 0.08-mm aluminum-foil shells hand-formed around a plaster cast of each lunar sample, reinforced with epoxy cement and filled with iron powder. Electronic densities were calculated from the gross chemical composition (*1*) and matched in the replicated standards by use of a mass of iron powder about 5 percent in excess of the lunar sample mass. The bulk density was reproduced by the addition of 3-mm polyurethane spheres (0.016 g/cm^3). Such exact replicas were prepared for the four crystalline rocks measured. Lunar fines were simulated with mixtures of powdered iron and dunite. Nine replicas of each lunar sample were filled with a known portion of each radionuclide found in the preliminary examination. A spectrum library was collected for each set of replicas and used for final analysis of the five samples. The breccia samples were analyzed with cyclindrical standards used in the preliminary examination.

The concentration of each nuclide

Table 1. Gamma-ray analyses of whole rocks and fines from Apollo 11. Values for short-lived nuclides have been corrected for decay to 0000 hours, C.D.T., 21 July 1969. TI, tentatively identified.

Measurement	Sample number (type)*							
	10057 (A)	10072 (A)	10003 (B)	10017 (B)	10018 (C)	10019 (C)	10021 (C)	10002 (D)
Weight (g)	897	399	213	971	211.5	234	157	301.5
Bulk density (g/cm³)	2.73 ± 0.14	2.37 ± 0.24	2.88 ± 0.35	3.00 ± 0.15	2.0 ± 0.2	2.02 ± 0.15		1.55 ± 0.05
K (ppm)†	2550 ± 130	2300 ± 120	480 ± 25	2430 ± 120	1420 ± 70	1200 ± 60	1600 ± 80	1100 ± 60
Th (ppm)†	3.30 ± 0.20	2.80 ± 0.17	1.01 ± 0.06	3.25 ± 0.18	2.30 ± 0.20	1.90 ± 0.19	2.50 ± 0.25	1.92 ± 0.10
U (ppm)†	0.79 ± 0.06	0.76 ± 0.06	0.26 ± 0.03	0.83 ± 0.07	0.60 ± 0.09	0.43 ± 0.06	0.54 ± 0.08	0.49 ± 0.04
^{26}Al (dpm/kg)	75 ± 8	73 ± 8	74 ± 8	73 ± 8	108 ± 16	101 ± 15	110 ± 15	120 ± 12
^{22}Na (dpm/kg)	41 ± 4	46 ± 5	41 ± 4	39 ± 4	55 ± 8	47 ± 7	55 ± 8	51 ± 5
^{44}Ti (dpm/kg)	TI	TI						TI
^{46}Sc (dpm/kg)	10 ± 2	8 ± 2	13 ± 3	13 ± 3	13 ± 4	10 ± 3	13 ± 4	8 ± 2
^{48}V (dpm/kg)			12 ± 9	11 ± 7				
^{52}Mn (dpm/kg)			35 ± 20				33 ± 21	
^{54}Mn (dpm/kg)	32 ± 6	20 ± 4	35 ± 7	33 ± 7	38 ± 10	28 ± 9	21 ± 6	28 ± 7
^{56}Co (dpm/kg)	31 ± 8	40 ± 10	43 ± 10	26 ± 7	33 ± 10	35 ± 10	50 ± 15	40 ± 7
^{7}Be (dpm/kg)	TI		TI	TI				TI

*Classification according to (1). † Standardization for assay of K, Th, and U with reference to terrestrial isotopic abundances. Equilibrium of Th and U decay series also assumed.

was determined from a resolution of the complex gamma-ray spectrum by the method of least squares (2) with the Univac 1108 computer complex at the Manned Spacecraft Center. Each of the four types of spectrums [singles from both NaI(Tl) detectors, sum coincidence, and two-parameter] were analyzed separately. Data from the two-parameter array were selected by adding adjacent planes in four regions of interest; the energy interval of each resulting spectrum was restricted to give the optimum discrimination against interferences.

In the preliminary analyses the extended sources described above were used to accumulate the spectrum libraries. When the counting sample did not match the particular reference library, an interpolation was required. The agreement between these preliminary results and the final results obtained with replicas was generally within 10 percent.

To check the experimental method, a test sample was prepared in a replica of sample 10017 containing K, Th, U, and ^{26}Al at the same concentrations found in the preliminary examination. The test sample was measured for a time similar to that employed during the acquisition of lunar sample data (1428 minutes). Analysis of the test sample on the basis of the exact library yielded observed concentrations that differed from the added concentrations by less than the statistical errors of the determination (0.9 percent for K, to 7 percent for ^{26}Al).

Because of the complex operations involved in handling lunar material, preparation of samples for gamma-ray spectrometry was slow during the quarantine period, and analysis of the first sample could not begin until 4 days after samples arrived at the LRL. Thus, radioactive species with half-lives of less than a few days were undetectable. In addition, the intense interferences from the gamma rays of the Th and U decay series in the samples made detection of weak gamma-ray components difficult. The precision of the measurements on these samples could have been improved by longer counting times; however, since these analyses were part of the preliminary examination at the LRL, time was limited. All samples were measured in at least two separate experiments, and total counting times were from 1000 to 3000 minutes per sample.

The results are summarized in Table 1. Twelve radioactive species were identified, some tentatively. The nuclides of shortest half-life identified were ^{52}Mn (5.7 days) and ^{48}V (16.1 days). The concentrations listed in Table 1 represent averages for the whole sample (usually a rock). Thus, if the sample is inhomogeneous, data obtained on a small portion may not be directly comparable to the concentrations of Table 1. The errors assigned in Table 1 include, in addition to the statistical errors of counting, estimates of possible systematic errors due to uncertainties in the detector efficiency calibrations.

Bulk densities for seven samples were obtained from the sample weights and the volumes of the replica shells fabricated as described above and are presented in Table 1. The crystalline rocks, breccias, and fines show a distinct correlation between bulk density and sample type.

The breccias and fines are very similar in their chemical compositions, as evidenced by their similar concentrations of K, Th, U, and cosmogenic radionuclides. This substantiates the hypothesis that the breccias are compacted mostly from the fine material. With the exception of sample 10003, the crystalline rocks appear to form

Table 2. Radionuclide concentrations in lunar samples as a function of depth. Concentrations for "Top" and "Interior" for 10017 were determined by an instrumental method and are only approximate. See text. Values for whole rock are from Table 1.

Nuclide	Sample 10017				Sample 10057	
	Top	Interior	Bottom	Whole rock	Top	Whole rock
^{26}Al (dpm/kg)	95 ± 15	65 ± 10	50 ± 7	73 ± 8	115 ± 9	75 ± 8
^{22}Na (dpm/kg)	43 ± 6	33 ± 6	30 ± 5	38 ± 4	63 ± 7	41 ± 4
^{56}Co (dpm/kg)	44 ± 18		10 ± 11	26 ± 7	90 ± 25	31 ± 8
K (% by weight)	0.235 ± 0.005	0.240 ± 0.005	0.240 ± 0.005	0.243 ± 0.012	0.258 ± 0.010	0.255 ± 0.013

Table 3. Estimates of gas-retention ages and cosmic-ray exposure ages for crystalline rocks. Gas contents from (*9*). Errors quoted for ages are combined analytical errors of gas contents and radioactivity determinations.

Sample	Gas content (10^{-8} cm³/g)		Age (10^6 year)	
	^{40}Ar	^{21}Ne	K-Ar	Exposure
10017	4080 ± 200	47 ± 3	2200 ± 150	340 ± 37
10057	5200 ± 260	5.0 ± 0.3	2500 ± 180	34 ± 4
10072	7600 ± 380	25 ± 1	3200 ± 230	150 ± 17

a group with general chemical similarities. However, sample 10003, a coarsely crystalline rock resembling the terrestrial gabbro, differs significantly from the others in its concentrations of K, Th, and U. This suggests either that this rock came from another region of the moon or that it is the product of an igneous process different from that which formed the other crystalline rocks examined.

Potassium concentrations range from just below to three times the average potassium concentration of chondrites (0.085 percent). Such low K abundances are also characteristic of some oceanic gabbros and basalts (*3*). The Th and U contents are near those of terrestrial basalts, and the ratio of Th to U is constant at the terrestrial value of about 4. The abundance of radioactive elements (K, Th, and U) in the surface materials is much greater than that inferred for the mean content of radioactive elements in the moon (*4*), which suggests that the surface materials have been enriched in radioactive elements with respect to the whole planet. The ratio of K to U for the samples investigated here varies from about 1800 to 3200. This range of values is unusually low, both by comparison with chondrites (45,000) and with terrestrial rocks (10,000). Since this ratio is not readily changed by terrestrial igneous processes (*5*), it may be inferred that the average ratio (~2700) observed here does not vary greatly over the whole moon. This suggests that the moon, like the earth, is depleted in potassium relative to chondrites (*6*).

In addition to analyses of whole rocks, attempts were made to study radionuclide concentration gradients, since it was expected that solar flare protons could produce high concentrations of some nuclides at the surfaces of lunar materials (*7*). If nuclides decay by emission of coincident radiations, the presence of a concentration gradient on the surface of such an extended sample can be detected by variations in the intensities of the symmetrical energy zones of the two-parameter matrix. An instrumental method for deriving surface concentrations at upper and lower surfaces of a sample was developed, and a replica containing a steep surface gradient of ^{26}Al was used to derive a calibration factor. An exterior, 234-g portion of 10017 was counted in a special experiment to determine whether the surface had been exposed to solar proton bombardment. The area of the 1.46-Mev peak of ^{40}K was used as a guide for centering the sample between the detectors. These instrumental results, combined with conventional analyses of a fragment from the bottom of 10017 and a 7.6-g, 4-mm-thick specimen from the top of 10057, are shown in Table 2.

Although errors were large, concentration gradients were observed for ^{56}Co, ^{26}Al, and ^{22}Na. Because of the low threshold energy for the production of ^{56}Co, the concentration gradients for this nuclide are particularly steep. This suggests that solar proton activation is the principal source of the ^{56}Co, especially the flare (*8*) of 12 April 1969. The data of Table 2 suggest that ^{26}Al and ^{22}Na are produced both by solar and by cosmic-ray protons. Concentrations determined at the tops of the rocks agree approximately with values obtained for fine material. Since the fines were collected close to the surface, this agreement suggests that the instrumental method for distinguishing between the tops and bottoms of rocks is valid. The presence of such concentration gradients makes difficult a precise comparison between some data in Table 1 and data from fragments of the same rocks.

Information obtained by gamma-ray spectrometry may be combined with gas analysis data to provide estimates of crystallization and cosmic-ray exposure ages. In Table 3 are shown gas retention ages derived from the potassium analyses on three crystalline rocks (Table 1) combined with radiogenic ^{40}Ar concentrations obtained by Schaeffer *et al.* (*9*). Estimates of cosmic-ray exposure ages were made by the ^{22}Na-^{22}Ne method. From similar work on meteorites (*10*), it was assumed that the cross sections for the direct production of ^{22}Na and ^{22}Ne were equal. Specific activities of ^{22}Na from the present work were used in combination with Ne concentrations obtained by Schaeffer *et al.* (*9*). The ^{22}Ne of spallation origin was estimated (*9*) to be 10 percent higher than the concentration of spallogenic ^{21}Ne. The ages obtained are in good agreement with ^{3}He exposure ages determined by Schaeffer *et al.* (*9*) but are subject to considerable uncertainty in the production rates chosen for ^{22}Na and ^{22}Ne. Variations in cosmic-ray exposure ages of a factor of 10 are consistent with the concept of a continual series of disturbances of the lunar surface due to impact.

G. Davis O'Kelley
James S. Eldridge
Oak Ridge National Laboratory, Oak Ridge, Tennessee 37830
E. Schonfeld
P. R. Bell
NASA Manned Spacecraft Center, Houston, Texas 77058

References and Notes

1. Lunar Sample Preliminary Examination Team, *Science* **165**, 1211 (1969).
2. E. Schonfeld, *Nucl. Instr. Meth.* **52**, 177 (1967).
3. For example, C. G. Engel and R. L. Fisher, *Science* **166**, 1136 (1969).
4. D. L. Anderson and R. A. Phinney, *Mantles of the Earth and Terrestrial Planets*, S. K. Runcorn, Ed. (Interscience, London, 1967), p. 113.
5. F. Birch, *Bull. Geol. Soc. Amer.* **68**, 483 (1958); G. J. Wasserburg, G. J. F. MacDonald, F. Hoyle, W. A. Fowler, *Science* **143**, 465 (1964).
6. P. W. Gast, *History of the Earth's Crust*, R. A. Phinney, Ed. (Princeton Univ. Press, Princeton, N.J., 1968), pp. 15–27.
7. D. B. Ebeoglu and K. M. Wainio, *J. Geophys. Res.* **71**, 5863 (1966); D. Lal, R. S. Rajan, V. S. Venkatavaradan, *Geochim. Cosmochim. Acta* **31**, 1859 (1967); G. D. O'Kelley, S. Kupferman, K. A. Richarson, unpublished work.
8. D. K. Bailey, personal communication to J. A. Arnold.
9. O. A. Schaeffer, J. Funkhouser, D. D. Bogard, J. Zähringer, personal communication and this issue.
10. G. Spannagel and C. Sonntag, in *Radioactive Dating and Methods of Low-Level Counting, Proceedings of a Symposium, Monaco, 2–10 March 1967* (International Atomic Energy Agency, Vienna, 1967), p. 231.
11. The authors gratefully acknowledge contributions to the work reported here by V. A. McKay, R. T. Roseberry, T. F. Sliski, R. E. Wintenberg, K. J. Northcutt, W. Portenier, and M. K. Robbins. We thank J. R. Arnold, P. W. Gast, and O. A. Schaeffer for helpful discussions. This work was carried out under Union Carbide's contract with the U.S. Atomic Energy Commission through interagency agreements with the National Aeronautics and Space Administration.

4 January 1970

General Mineralogy

Mineralogy and Petrology of Some Lunar Samples

Abstract. *Chemical analyses and norms of four samples are presented which confirm original estimates of low silica, unusual abundance of titania, and low oxidation state of the rocks. Accounts are given of mineralogy and petrology of fine- and coarse-grained igneous rocks and microbreccias with emphasis on chemical composition of individual minerals and glasses. The glasses are either spheres that scatter widely around the composition of lunar basalts or coating glasses that approximate basalts and microbreccias in composition.*

Rocks returned by the Apollo 11 mission are all part of the lunar regolith; it may be inferred that some are volcanic rocks originating at no great depth, while others, the microbreccias, form part of a surface layer possibly differing from the dusty maria surface only in degree of compaction. Some components of the regolith show evidence of shock metamorphism and melting; many may have originated at a considerable distance from the point at which they were collected.

The compositions of individual phases within this material have been determined with a Cambridge "Geoscan" electron probe operating at an accelerating potential of 20 kv. The experimental procedure and treatment of data have been described *(1)* and all analyses have been referred, either directly or indirectly, to the primary standards listed by these authors. Bulk rock compositions have been determined with classical wet chemical techniques.

Chemical analyses *(2)* reveal low silica values, comparatively low alumina values, and unusual abundance of titania (Table 1); iron exists in a very low state of oxidation. This may be explained either by the presence of some of the titanium in the titanous state or by interference in the ferrous iron titration, caused by the presence of sulfides. The metallic iron values accepted are based on values obtained by Bancroft from Mössbauer spectra on samples 1, 2, and 3; it is assumed at this stage that sample 4 has the same metallic iron content as analysis 1.

An attempt was made to determine the metallic iron content of these samples by subjecting them to attack by a stream of dry chlorine gas (free of HCl) in a tube furnace heated to 350°C, following the method of Moss and Hey *(3)* for the treatment of meteorites. The volatilized $FeCl_3$ was then determined by means of a silver reductor and standard ceric sulfate solution. A correction was made for the presence of FeS which is also attacked under these conditions (Table 1). These figures appear to be distinctly too high, especially so in analysis 2. The high results are thought to be due to attack of some other mineral phase present, probably an iron-titanium oxide. Tests were made on a Norwegian ilmenite which proved to be completely decomposed by the chlorine treatment. It is unlikely that the ilmenite present in the lunar samples could have been so completely broken down, for much higher results for Fe would have been obtained. It is more likely that a spinel is involved. Investigation of this matter is proceeding.

The SiO_2 was determined by dehydration with HCl and volatilization with HF. Any SiO_2 in solution after this stage was recovered from the pyrosulfate fusion of the ammonia precipitate. The Al_2O_3 was obtained by difference from the ammonia precipitate. Total iron was determined by titration with ceric sulfate solution after treatment in a silver reductor. The FeO was determined by the modified Pratt method. The MnO was obtained colorimetrically after oxidation with KIO_4. The MgO was precipitated as magnesium ammonium phosphate and ignited to $Mg_2P_2O_7$. The CaO was precipitated as oxalate and ignited to oxide. The Na_2O and K_2O were determined with an EEL flame photometer. Total H_2O was determined by the Penfield method with $PbCrO_4$ as a flux. The almost constant value of H_2O+ is significant; it must be held in the glass and would appear to be original water, since the samples have never been subjected to a damp atmosphere for any significant length of time. The H_2O- was obtained from the loss in weight at 105°C. The TiO_2 was estimated colorimetrically with H_2O_2. The P_2O_5 was determined colorimetrically by the vanadomolybdate method. The Cr_2O_3 was determined by the diphenyl carbazide method *(4)*. Sulfur was precipitated as $BaSO_4$ after Na_2CO_3 fusion and extraction of the cake with water, allowance being made for reagent blank. Fluorine was determined by the Sen Gupta modification of the method of Huang and Johns *(5)*. Trace elements on these rocks are reported by Brown *(6)*.

Detailed examinations were made of thin sections from four igneous rocks: 03/44 (type B); 17/50 (type B, but transitional to type A); 45/34 (type A); and 58/23 (type A), probably the coarsest grained rock of the entire Apollo 11 suite. In addition, grain mounts of the ferromagnesian mineral fractions from 44/43 were examined, and small xenoliths of type A rocks, some hardly altered, others thoroughly recrystallized, were studied in the breccia 65/15.

The grain size of the igneous types ranges from that of a fairly coarse-grained basalt to that of the coarsest dolerite with plagioclase crystals up to 3 mm in length. Type B rocks are holocrystalline; there is a little acid glass in the residuum of 17/50 (Table 2, analysis 13), and rather more in 45/34. Three minerals, clinopyroxene, bytownite, and ilmenite (in decreasing order of abundance) are prominent phases in all the rocks; of these, clinopyroxene predominates absolutely. In 45/34 there are up to 5 percent phenocrysts and microphenocrysts of olivine (Fa_{29}), which are enclosed by clinopyroxene and sometimes sharply jacketed by it, suggesting a reaction relation similar to that seen occasionally in terrestrial basanitoids. In

03/44 and 58/23, coarse patches of cristobalite are found in the residuum, sometimes intergrown with a more sodic plagioclase or with little patches of an alkali feldspar. In 03/44, 45/34, and 58/23 the pyroxene is intergrown with or mantled by an iron-rich pyroxenoid. This is very unevenly distributed and has continued to crystallize until a very late stage, for it is sometimes intergrown with the cristobalite. This mineral is also described *(7)*. Accessories are a brown spinel (picotite) that occurs as tiny octahedra enclosed in the olivine as in alkali basalts, an opaque chrome titanospinel (Table 2, analysis 7) associated with the ilmenite (Table 2, analysis 8), a few needles of rutile in the residuum, and slender prisms of apatite. Other accessories are troilite, chalcopyrite, and metallic iron, and a late spinel, probably ulvöspinel, though the composition of this phase has not been confirmed. In the heavy residue of 44/43, one grain of an alloy of composition roughly $Cu_{70}Zn_{30}$ (brass) was found. Contamination was suspected but an additional grain of apparently the same material which has not yet been analyzed had troilite and feldspar firmly attached to it.

Textures range from subpoikilitic in 03/44 and 58/23 to intergranular in 17/50 and 45/34 where lath-shaped plagioclases separate areas composed mainly of overlapping equant pyroxene grains, pyroxenoid, and ilmenite. The texture of 17/50 recalls that of the perpendicular feldspar rock of the Skaergaard intrusion. In 17/50 and 58/23 Universal stage studies suggest some degree of preferred orientation in the plagioclase.

In general a single clinopyroxene phase is present, which in the coarse dolerites display remarkably little variation in $2V$ (46° to 42°), but probe data indicate zoning from about $Ca_{35}Mg_{45}Fe_{20}$ to a subcalcic ferroaugite near $Ca_{28}Mg_{13}Fe_{60}$. A remarkable feature in these coarse-grained rocks is the prominence of subcalcic augite. In the basalt 45/34 and transitional rock 17/23 a different trend emerges characterized by a large variation in $2V$; the early pyroxene, $2V = 48°$, $Ca_{44}Mg_{41}Fe_{15}$ is more lime-rich and is zoned continuously from $2V = 42°$ to 20°, OAP / / (010). From probe data the final pyroxene of 45/34 has a composition $Ca_{15}Mg_{54}Fe_{31}$ which is remarkably magnesian. In this rock the pyroxene fractionation appears to be following the tholeiitic quench trend *(8)*. Analyses of pyroxenes (Table 2, analyses 1 and 2) indicate a striking reduction in the alumina and titania contents and a slight increase in manganese with fractionation. At a moderately late stage pyroxene ceased to crystallize, hence iron-rich pigeonite and ferroaugite are absent.

Exsolution lamellae were not observed optically but x-ray studies *(9)* indicate that they are present on a submicroscopic scale. A possible explanation of the abundance of subcalcic augite and of the absence of visible exsolution lamellae in such coarse-grained rocks may lie in the titania content of the pyroxenes depressing the solvus. Orthopyroxene is absent except possibly in a very fine-grained granulitic xenolith in one of the breccias where the original pyroxene has unmixed into two phases, one of which has low birefringence.

The pyroxenoid has the fractured appearance in thin section typical of minerals of the wollastonite group; it has also their fibrous character and a marked tendency to grow as curved crystals. Mosaic textures of interlocking grains and undulose extinction, even where the crystals are not bent, may indicate a monoclinic parent. A deeper yellow color peripherally and an associated variation in birefringence are evidence of zoning. The mineral is triclinic optically, has perfect (110) and ($1\bar{1}0$) cleavages and a poorer one on (100). Twins on (100) are as common as in clinopyroxenes; in such cases the optic axial planes of the two subindividuals intersect at an angle of 5°. The angle $2V_z$ ranges from 42° to 47°, and $Z \wedge [001] = 43°$ to 45°. The optic axial plane passes within 5° of [001]. An apparently single crystal from 58/23 was found to be zoned continuously from $Wo_{29}En_{42}Fs_{29}$ to $Wo_{14}En_5Fs_{81}$ (Fig. 1A). The

Table 1. Analyses of lunar rocks and C.I.P.W. norms. 1. Apollo 11, 84/14 fines; Fe 1.22 by Cl method. 2. Apollo 11, 45/24 type A; Fe 1.78 by Cl method. 3. Apollo 11, 44/39 type B; Fe 1.23 by Cl method, revised estimate of metallic Fe 0.20 percent. 4. Apollo 11, 60/25 type C; Fe 1.18 by Cl method.

	1	2	3	4
		Analyses of lunar rocks		
SiO_2	42.16	39.60	42.46	41.96
Al_2O_3	13.60	9.51	10.21	11.85
Fe_2O_3	nil	0.06	nil	nil
FeO	15.34	19.12	17.60	16.51
MnO	0.20	0.28	0.28	0.23
MgO	7.76	8.10	5.96	7.63
CaO	11.94	11.07	12.25	11.38
Na_2O	0.47	0.36	0.48	0.49
K_2O	0.16	0.05	0.11	0.20
H_2O+	0.05	0.15	0.10	0.11
H_2O-	nil	nil	0.01	0.04
TiO_2	7.75	11.10	9.18	9.02
P_2O_5	0.05	0.04	0.04	0.07
Cr_2O_3	0.30	0.47	0.21	0.31
S	0.12	0.14	0.18	0.15
F	nil	nil	nil	nil
Fe	0.60	0.20	0.60	0.60
Total	100.50	100.25	99.67	100.55
Less S = O	0.03	0.04	0.05	0.04
Total	100.47	100.21	99.62	100.51
Total Fe as Fe_2O_3	17.56	21.59	20.85	18.90
		C.I.P.W. norms		
Q	0.46	0.78	4.34	1.87
Or	0.95	0.30	0.65	1.18
Ab	3.98 } 39.46	3.05 } 27.54	4.06 } 30.09	4.15 } 34.88
An	34.53	24.19	25.38	29.55
Wo	10.18	12.72	14.67	11.05
En	5.60 } 19.98	6.81 } 25.04	6.89 } 29.17	6.03 } 21.70
Fs	4.21	5.51	7.61	4.62
Hy (En	13.73 } 24.04	13.36 } 24.18	7.96 } 16.76	12.97 } 22.91
Hy (Fs	10.31	10.81	8.80	9.94
Cr	0.44	0.69	0.31	0.46
Il	14.72	21.08	17.44	17.13
Mt	nil	0.09	nil	nil
Ap	0.12	0.09	0.09	0.17
Py	0.22	0.26	0.34	0.28
Fe	0.60	0.20	0.60	0.60
Total	100.04	99.94	99.13	99.98

explanation of the appearance of the pyroxenoid is not obvious, but extremely low pO_2 might suppress the formation of fayalitic olivine. The appearance of the pyroxenoid in place of iron-rich pyroxenes must imply that here we have in some way passed beyond the stability field of pyroxenes, possibly by way of a monoclinic-triclinic inversion associated with increasing iron and decreasing calcium and magnesium contents. Titania and manganese contents are significant and must have a bearing on the genesis of the mineral which seems to be related to ferrosilite III. Crystals from 44/43 examined by electron diffraction prove to be unmixed into approximately equal amounts of two phases with similar structures. These may represent an iron wollastonite and ferrosilite III.

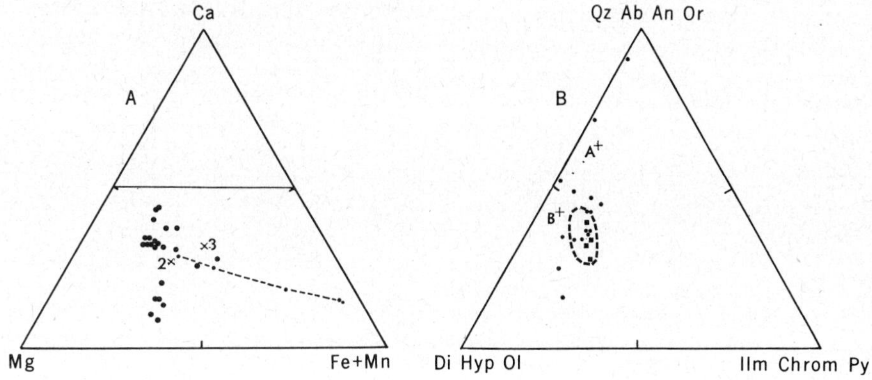

Fig. 1. (A) Atomic percentages of Ca, Mg, and FeMn in pyroxene and pyroxenoid. Zoning of pyroxene-pyroxenoid grain indicated by dashed line. × 2, × 3, Normative pyroxenes from analyses 2 and 3 of Table 1. (B) Plot of normative constituents of glasses. Squares represent lunar rocks 84/14, 60/25, 44/39, 45/24; dotted lines represent area enclosed by welding or coating glasses; circles represent individual glass spheres; +A is olivine tholeiite, Mid-Atlantic Ridge; and + B is olivine plus clinopyroxene-enriched dolerite.

Plagioclase crystals are flattened on (010); twinning is mostly on the albite, carlsbad, and albite-carlsbad laws; some pericline twinning was observed and an acline A twin determined in 17/50. In 58/23 a banate twin was observed enclosed in the pyroxenoid. Euler angles determined on 17/50 demonstrate it to be virtually unzoned plagioclase An_{78}, in 58/23 the composition is An_{88}, and in 45/34 where the feldspar is zoned compositions ranging from An_{70} to An_{59} were encountered. In 58/23 the late plagioclase intergrown with cristobalite has a composition of An_{76}; it is associated with little rectangular patches of an optically negative alkali feldspar.

In the coarser rocks two iron-titanium oxide phases are present, and these are always closely associated; in 45/34 ilmenite only was observed. Analyses of two ilmenites and a spinel from 85/48 are presented in Table 2 (analyses 8, 9, and 7). These ilmenites carry distinctly more magnesium than do those from 45/35 but are otherwise similar, and all are notable for titanium being consistently in excess of its stoichiometric proportion; it is likely that some is in the titanous state. The spinel is rich in chromium and titanium and carries considerable magnesium and aluminum; again some titanium may be trivalent. Exsolution of a chromium-rich phase is present in some of the coarser ilmenites.

As the norms of the analyzed samples must correspond closely with the modes it is remarkable that all show the presence of normative quartz as the silica values are so very low. Other features are the exceptionally low soda, and very high chromium in relation to the high Fe/Mg ratios. The intimate association of metal and sulfide phases indicates liquid immiscibility in a magma which must already have undergone considerable differentiation. The remarkable chemical features are clearly portrayed in Fig. 1B where comparisons are made with terrestrial basaltic types. In these surface or near-surface lunar rocks it is likely that soda, alumina, and silica may have escaped with the volatile phase; the reduced effect of gravity may account for the extreme fractionation *in situ* observed in the ferromagnesian silicates of the coarser rocks. The interstitial glass (Table 2, analysis 13) indicates the possibility of a high silica, high potassium fractionation trend.

Microbreccias 46/48, 48/42, 60/28, 65/15, and parts of 85/4 and 85/5 contain four principal components—angular fragments of basaltic igneous rocks or their individual minerals, glassy spheres, meteoritic material, and a cementing glass.

The igneous fragments show a wide variety of quench textures; individual minerals may be normal or may display shock effects ranging from the granulitization of ferromagnesian minerals and twinning of ilmenite to maskelynitization and ultimately fusion of plagioclase. In some cases the latter has recrystallized anew to feathery, twisted aggregates, in others to panidiomorphic crystals set in glass.

Some small fragments occur with

Table 2. Analyses of lunar samples. 1, Augite, 45/35/1; 2, pigeonite, 45/35/1; 3, pyroxenoid from microbreccia, 65/15.1; 4 and 5, plagioclase, 60/28; 6, maskelynite, 85/4/10.2; 7, spinel, 85/4/14.2; 8, ilmenite, 17/12; 9, geikielo-ilmenite, 85/4/10.3; 10, tachylitic glass, 85/4/5; 11, glass sphere, 85/30/4; 12, coating glass, 85/4/17a; 13, interstitial glass, 17/12 (high total due to loss of alkalis during measurement of silica).

	1	2	3	4	5	6	7	8	9	10	11	12	13
SiO_2	48.00	52.06	46.09	50.06	43.59	44.90	.40	.05	.05	40.81	41.61	41.08	78.92
TiO_2	5.91	1.35	1.22	—	—	.02	25.12	54.20	56.30	10.73	7.21	8.34	—
Al_2O_3	4.93	.85	.75	30.60	35.38	34.78	6.23	.05	1.64	10.20	8.95	13.07	11.88
Cr_2O_3	.70	.25	.13	—	—	—	21.40	.60	.34	.32	.48	.30	—
FeO	8.49	19.15	37.53	.61	.32	.21	40.71	43.94	32.39	19.51	21.74	17.65	2.88
MnO	.09	.35	.61	—	—	—	.34	.43	.34	.23	.24	.23	—
MgO	13.27	18.61	1.66	.26	.07	.05	5.72	.71	9.63	6.11	8.50	8.11	.08
CaO	19.88	6.89	10.40	15.05	19.44	18.92	.01	.10	.44	11.92	9.97	9.93	2.10
Na_2O	—	—	.08	2.14	.38	.60	—	—	—	.34	.24	.40	.56
K_2O	—	—	—	.20	.06	.08	—	—	—	.17	.11	.16	6.86
Total	101.27	99.51	98.47	98.92	99.24	99.56	99.93	99.98	101.08	100.34	99.05	99.27	103.28

microporphyritic plagioclase (An_{95}) in a granulitic mosaic of plagioclase, olivine with or without clinopyroxene, and geikielo-ilmenite (Table 2, analysis 9) plus rutile, suggesting remelted rocks but not necessarily resulting from shock.

The glassy spheres range from 2 mm to ~ 1 μm in diameter, the larger ones being found in 85/4 and 85/5. They are predominantly spherical but ellipsoidal and dumbbell-shaped forms also occur, suggesting oscillation of liquid drops in flight. Some are colorless, others pale green, yellow, or brown. The smaller ones are nonvesicular but, with increase in size, vesicles become more common; in one example a centrally disposed vesicle was seen to occupy some 25 percent of the total volume. Some spheres are wholly glassy apart from very occasional droplets of metallic iron which sometimes project from, or are concentrated at, the surface and may possibly have been incorporated during flight. A number contain partially resorbed relicts of any of the minerals of the igneous rocks. Others have devitrified or partially crystallized and occasionally resemble chondrules in texture but differ in their mineralogy (ilmenite with or without rutile, olivine, plagioclase).

In some specimens metallic iron droplets (with 1 to 3 percent nickel) with or without troilite are more common and occur in a cellular or contorted flow pattern similar to that found in the glass which welds together some fragments of microbreccia in the lunar fines. The presence of minute droplets at the surface of one sphere was also revealed by the use of the scanning electron microscope which showed many particles 0.1 to 0.5 μm in diameter with an atomic number considerably greater than that of the matrix. Further examination in the electron microscope-microanalyzer *(10)* indicated that the Ni, S, and Ti contents of these droplets were low and that the iron content was at least 70 percent. Two small spheres ~ 60 μm in diameter contained within the microbreccia (60/28) show a "plum-pudding" structure of uniformly sized and distributed droplets and crystallites, possibly of ilmenite, which suggests a cooling rate intermediate between that which produces the very fine droplets and that which gives the chondritic texture.

The composition of some glassy spheres is plotted in Fig. 1B; measurements at different points on the same sphere vary only slightly and though the

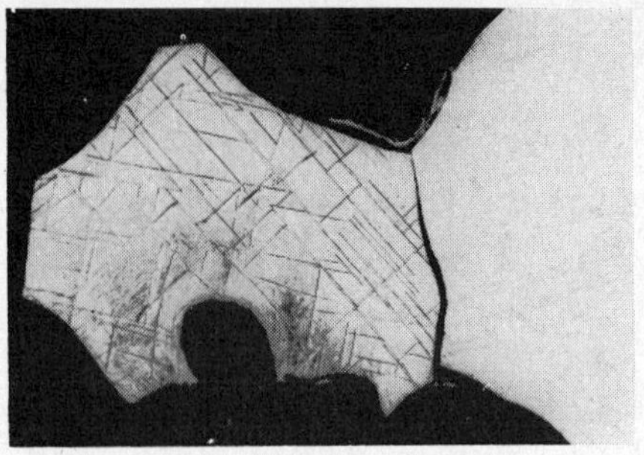

Fig. 2. Fragment of iron meteorite from 85/4 showing kamacite with Neumann bands and α-ε iron transformation structure (× 110).

analyses concentrate around the composition of lunar basalt and microbreccia it is much more variable than that of the coating glasses described below. This variation is probably best explained as the result of impact fusion of basaltic rocks. Alternatively, those close to lunar basalts in composition may have arisen by lava fountaining in the same way as the spheres found associated with Peles hair, or any forced discharge of hot, dust-laden, incandescent gas such as occurs in the furnaces of coal-burning power stations.

A second type of glass in the microbreccias is found molded around preexisting rock, mineral, or glassy fragments as irregular seams from 3 to 200 μm in thickness. This glass is probably of the same origin as that which coats and welds together 1- to 5-mm fragments of microbreccia occurring in the lunar fines (85/4). It is vesicular and with a complex flow pattern marked by slight differences in color and refractive index and by trails of immiscible droplets of metallic iron (Ni < 3 percent) and subordinate troilite. The maximum size of these droplets is of the order of 10 μm, but, in the transmission electron microscope, particles down to 20 Å diameter are visible. In total, the droplets probably amount to less than 0.2 percent of the glass and the ratio of iron to troilite is higher than in the igneous rocks, suggesting that sulfur vapor from the dissociation of troilite is responsible in part for the vesicular nature of this glass. The finely divided metal probably accounts for the dark body color of the glass which may be olive, pale brown, or pale yellow in thin section.

The composition of a typical glass is given (Table 2, analysis 12), and in Fig. 1B it is seen that these coating or welding glasses are close in composition to the microbreccias and basalts. They may be therefore considered as their direct fusion products or the introduced quench products of a basaltic magma.

In 85/4, 85/30 some glass coatings are up to 2 mm thick differing from those above only in the absence of a complex flow pattern. Their contact with basaltic rocks simulates trachylitic chilled edges but the discordant nature of the glass-ilmenite phenocryst contacts precludes such an origin. Their composition is close to that of basalt and again they must be regarded as direct fusion or introduced quench products of a basaltic magma. One fragment of which the outer surface was preserved showed several microimpact craters: these were studied with the scanning electron microscope and, in general, consisted of a central smooth crater surrounded by features controlled by steep radial and shallow concentric fractures.

Finally, occasional angular or wire-like fragments of meteoritic iron are observed in both microbreccias and fines. These are dominantly kamacite (about 7 percent Ni : 0.5 percent Co) although one fragment (Ni 7 percent) with well-developed Neumann bands and weakly developed α-ε transformation structure, includes small drops of zoned taenite (core Ni 9 percent, rim Ni 14 percent) (Fig 2).

S. O. Agrell, J. H. Scoon
I. D. Muir, J. V. P. Long
J. D. C. McConnell, A. Peckett
Department of Mineralogy and Petrology, University of Cambridge, England

References and Notes

1. T. R. Sweatman, J. V. P. Long, *J. Petrol.* **10**, 332 (1969).
2. J. H. Scoon (Cambridge Univ.) performed the chemical analyses of the rocks.
3. A. A. Moss, M. H. Hey, D. I. Bothwell. *Mineral Mag.* **32**, 802 (1961); A. A. Moss, M. H. Hey, C. J. Elliot, A. J. Easton, *ibid.* **36**, 101 (1967).
4. E. B. Sandell, *Colorimetric Determination of Traces of Metals* (Interscience, New York, 1944), p. 196.
5. J. G. Sen Gupta, *Anal. Chim. Acta* **42**, 119 (1968).
6. G. M. Brown, C. H. Emeleus, J. G. Holland, R. Phillips, *Science*, this issue.
7. A. T. Anderson, Jr., A. V. Crewe, J. R. Goldsmith, P. B. Moore, R. C. Newton, J. E. Olsen, J. V. Smith, P. J. Wyllie, *ibid.*, this issue.
8. I. D. Muir and C. E. Tilley, *Geol. J.* **4**, 143, (1964).
9. P. Gay, G. M. Bancroft, M. G. Bown, *Science*, this issue.
10. P. Duncumb, *The Electron Microprobe* (Wiley, New York, 1966), pp. 490–499.
11. I. D. Muir is responsible for the petrography of the igneous rocks.
12. We thank the Cambridge Scientific Instrument Company for use of the scanning electron microscope, P. Duncumb (Tube Investments Researeh Laboratory) for use of the electron microprobe microanalyzer, and our colleagues in this department for assistance.

4 January 1970

Petrologic History of Moon Suggested by Petrography, Mineralogy, and Crystallography

Abstract. *Opaque mineral compositions indicate that the fugacity of oxygen is approximately 10^{-13} (earth basalts, 10^{-10}). Experiments under reducing conditions suggest that the crystallization range is approximately 1140° to 1070°C. Iron-rich pyroxmangite, fayalite, and hedenbergite occur in microgabbro. Ferropseudobrookite rimmed by ilmenite containing rutile and Cr-spinel lamellae occurs in ferrobasalt. Plagioclase vitrophyres in breccia can explain highland Surveyor VII analysis. We suggest crystal-liquid differentiation of out-gassed convecting moon with growing Fe-rich core, olivine-pyroxene mantle, plagioclase-rich dynamic crust underlain by nonspherical, inversely stratified ferrobasalt. Impact-breaking or convection-thrusting of crust releases fraction rich in Fe and Ti. Scanning electron microscopy of glass balls reveals minute depressions consistent with micrometeorite impact.*

Type A ferrobasalts 22, 57, and 72, type B microgabbros 44 and 47, type C breccia 61, and type D 84 and 85 soil (*1*) yielded optical and electron microscope data, x-ray diffraction patterns, and electron microprobe analyses. We concentrated on evidence of crystal-liquid differentiation, particularly in the breccia which contains plagioclase- and olivine-vitrophyres as well as fragments of ferrobasalts.

Petrography and mineralogy of basalts and microgabbros. Ferrobasalts are vesicular containing sporadic olivine and ferropseudobrookite phenocrysts within fine-grained augite with intergrown pigeonite, ilmenite, plagioclase, and devitrified, microvesicular residual glass. Microgabbros lack phenocrysts and contain coarse-grained augite (strongly zoned), Fe-rich pyroxmangite, ilmenite, plagioclase, low-cristobalite, and a symplectitic residuum of fayalite and hedenbergite composition. Ferrobasalt fragments in breccia contain more residual glass than individual ferrobasalts, while some are entirely devitrified glass without microphenocrysts,

Ferrobasalts and microgabbros petrographically and chemically resemble coarse segregation veins or schlieren in terrestrial basalts formed from Fe-Ti-rich residual liquid flowing into cracks and crystallizing (*2*, *3*). Figure 1 shows compositions of selected minerals and glasses.

Ilmenite in ferrobasalt occurs as rims to ferropseudobrookite phenocrysts (Gi 13), early crystals lining vesicles (Gi 11), later tabular crystals (Gi 6) in areas rich in ilmenite and pyroxene, and finally as plates (Gi 3 to 0.3) restricted to plagioclase-glass rich regions [Gi, mole percent $MgTiO_3$]. The very low Fe_2O_3 content increases slightly with crystallization. Ilmenite in microgabbro ranges from center of large crystal (Gi 3; Cr_2O_3, 0.7; MnO, 0.3 weight percent) to tiny crystal associated with residuum fayalite (Gi 0.4; Cr_2O_3, 0.05; MnO, 0.6 weight percent).

Ferropseudobrookite ($FeTi_2O_5$) megacrysts in ferrobasalts 22 and 72 and ferrobasalt fragments in breccia 61 are rimmed invariably by ilmenite containing lamellae of rutile and Cr-rich spinel.

Ulvospinel occurs in the residuum of microgabbro 44.

Troilite ($Fe_{1.02}Cu_{0.00}Ni_{0.00}S$) occurs as subspherical interstitial grains in basalt and microgabbro enclosing tiny metallic iron particles ($Fe_{0.97}Ni_{0.003}$) yielding bulk composition of approximately $Fe_{1.1}S$. The complex texture of some iron suggests unmixed oxide, carbide, or phosphide, or all three. The invariable troilite-iron assemblage suggests an Fe-S liquid immiscible with the silicate.

Subcalcic augite is predominant in both ferrobasalt and microgabbro. Except for lower Ca, the zoning (Fig. 1A) parallels that of terrestrial basaltic pyroxenes (*4*) and ranges from hourglass to irrational planar sector zoning. The subcalcic augite of ferrobasalt has a smaller composition range than for the microgabbro, and is further distinguished by containing small patches of pigeonite. The microgabbro pyroxene (Fig. 1A) zones discontinuously into almost Mg-free metasilicate, which may be iron-rich pyroxmangite.

Subcalcic augites in ferrobasalt contain (weight percent) 6 to 1.5 Al_2O_3, 4.1 to 1.1 TiO_2, 0.15 to 0.26 MnO (no data for Fe-rich specimens), and 0.6 to 0.1 Cr_2O_3 (no data for Fe-rich specimens) as compared with 2 to 0.6 Al_2O_3, 1.7 to 0.9 TiO_2, 0.3 to 0.7 MnO, and 0.3 to 0.1 Cr_2O_3 in microgabbro.

Pyroxmangite is the unidentified yellow mineral of the microgabbro (*1*). Single crystals from sample 44 yielded: triclinic *a* 6.61, *b* 7.54, *c* 17.3Å; α 114.3°, β 82.7°, γ 94.6° (Table 1, analysis A). With stereomicroscopy, yellow pyroxmangite clearly abuts cinnamon pyroxene; in thin section, distinction is difficult. On earth, pyroxmangite has Fe/(Fe + Mn) ratio below 0.6; however terrestrial Fe-rich pyroxmangite might have been confused with pyroxene.

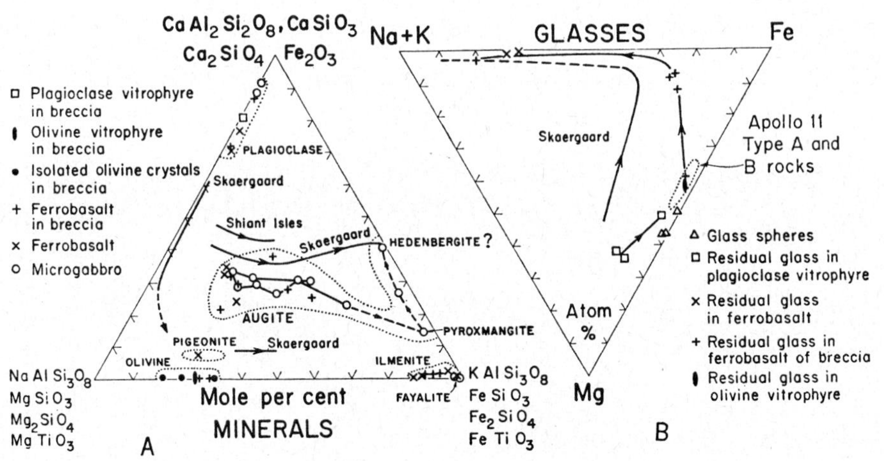

Fig. 1. (A) Compositions of minerals from Apollo 11 rocks. End members and minor element data which apply to different minerals are: plagioclase [$CaAl_2Si_2O_8$, $NaAlSi_3O_8$, and $KAlSi_3O_8$ (mole percent), and 0.5 to 1.3 FeO, 0.2 MgO, and 0.1 TiO_2 (weight percent)], pyroxenes and pyroxmangite ($CaSiO_3$, $MgSiO_3$, $FeSiO_3$, 0.15 to 0.7 MnO, 0.3 to 0.1 Cr_2O_3, 8 to 1 Al_2O_3, 4 to 1 TiO_2, data for augites only; Al and Ti are highest in augites from ferrobasalts), olivine (Ca_2SiO_4, Mg_2SiO_4, Fe_2SiO_4, 0.00 to 0.03 Ni, 0.14 to 0.29 Mn, 0.8 Mn in fayalite), ilmenite (Fe_2O_3, $MgTiO_3$, $FeTiO_3$, 0.5 to 0.0 Al_2O_3, 0.8 to 0.3 Cr_2O_3, 0.4 to 0.9 MnO). Comparison curves for Shiant Isles and Skaergaard pyroxenes and plagioclases from (*4*, *8*). (B) Iron, magnesium, and alkali diagram of glasses from Apollo 11 rocks. Arrows on lines of liquid variation point toward increasing solidification. Comparison curve for estimated Skaergaard liquids from (*8*). For relations of plagioclase vitrophyre and ferrobasalt see petrogenesis section. Composition of glass beads indicates volatile loss of Na and K.

Table 1. Microprobe analyses mentioned in text (weight percent). The analyses are: A, pyroxmangite, rock 10044; B and D, devitrified glasses in fragments of ferrobasalt in breccia 10061; C, devitrified glass in ferrobasalt 10022; E, devitrified glass (90 percent) in plagioclase vitrophyre in breccia 10061; F, glass (70 percent) in olivine vitrophyre in breccia 10061; and G, synthetic glass used for melting experiments.

Oxide	Analysis						
	A	B	C	D	E	F	G*
SiO_2	45.6	56	62	73	46	54	43
TiO_2	0.3	2	0.35	1.2	0.6	1.0	10
Al_2O_3	1.1	19	23	13	24	17	10
FeO	45.9	14	3	2.4	5	10	19
MnO	0.6	0.3		0.2			0.4
MgO	0.6	1	0.1	0.2	0.7	4	6.5
CaO	5.9	8	9	2.1	15	14	9.7
Na_2O	0	1.4	1.7	0.5	1	0.1	0.5
K_2O	0	0.9	2.3	5.5	0.06	0.02	0.16
P_2O_5	0	0.3	0.09	0.3	0.02	0.02	

*Also Cr_2O_3 0.47 and ZrO_2 0.16.

Olivines (69 to 72 mole percent Mg_2SiO_4) from two ferrobasalt fragments in breccia 61 have (Fo_{69-72}), Mn 0.25, Ca 0.19, and Ni 0.00 to 0.01 (weight percent). The Mn and Ca contents fall in the range of terrestrial volcanic olivines, but the Ni content is less (*5*).

Plagioclase in ferrobasalt grows into hollow, square tubes (extreme hopper structure) which enclose devitrified liquid and other phases. Bytownite up to $An_{87}Or_{1.5}$ is typical, but adjacent to rhyolitic residual glass $An_{71}Or_2$ occurs.

Devitrified glass in ferrobasalts varies in quantity and composition from that of ferrobasalt itself to "andesite", to "anorthosite", and to "rhyolite" (Table 1, analyses B, C, and D).

Symplectic residuum in microgabbros consists of fayalite and metasilicate of composition near hedenbergite (Fig. 1A).

Low-cristobalite which is spatially associated with the residuum of the microgabbro and the rhyolitic composition of residual glass in ferrobasalt emphasize the silica saturation of Apollo 11 rocks evident in their norms ($\sim$10 percent SiO_2).

Although the vesicularity of ferrobasalts indicates past presence of gases, none of the rocks shows obvious evidence of alteration.

Breccia contains ferrobasalt and minerals at various states of shock including melting (i) glass beads—clear to red brown; (weight percent) SiO_2 35 to 42, TiO_2 0.6 to 10, Na_2O <0.1, and K_2O <0.02; glassy to spherulitic; some bubbles; rare crystal ghosts and small blebs of Fe metal; (ii) irregular fragments of vesicular glass—clear to yellow, brown, or green; glassy to devitrified; many vesicles; anhedral crystal relics; blebs of Fe (<4 percent Ni) surrounded by halos of Fe-poor glass; rare microlites; (iii) glass rims on rock fragments—clear to yellow-brown; similar to (ii) but fewer Fe blebs and more devitrified: origin; melting by external heating possibly involving hot, impact-generated gas; (iv) plagioclase vitrophyres—clear, almost nonvesicular, euhedral or dendritic plagioclase microphenocrysts; metal absent; composition (Table 1, analysis E); (v) olivine vitrophyres—pale-brown, nonvesicular, skeletal olivine; no metal; composition of 70 percent residual glass (Table 1, analysis F).

Soil has similar constituents to breccia, showing similar x-ray and microprobe data for minerals. Fragments of meteoritic iron are rare.

Glass balls show complex surface features under hard-vacuum, scanning electron microscopy—for example (i) minute, rubble-filled, circular, lipped depressions consistent with micrometeorite impact (Fig. 2), (ii) dimpled (golf-ball) surface, and (iii) deformed surface with small angular fragments sticking to surface. The phenomena are consistent with spheres of varying rigidity and stickiness falling onto soil. Surface tension shaping of fiery rain and collision deformation and breakage ex-

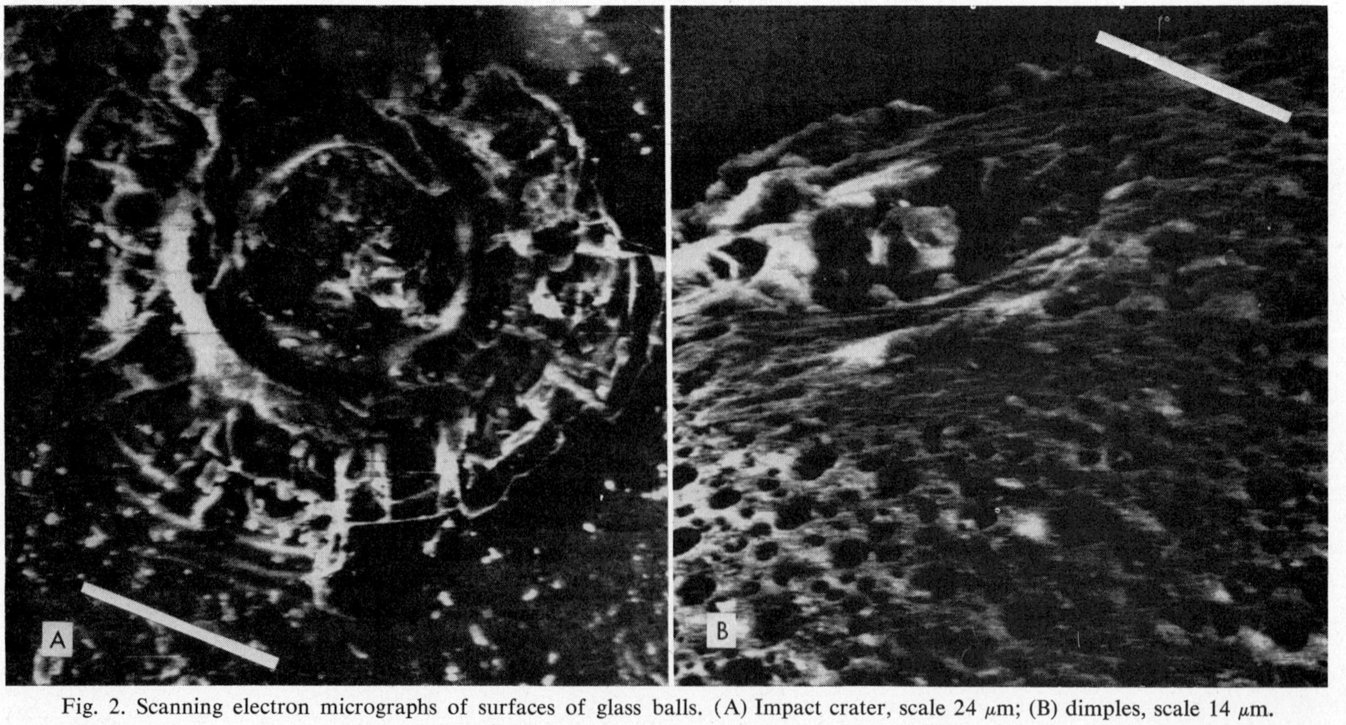

Fig. 2. Scanning electron micrographs of surfaces of glass balls. (A) Impact crater, scale 24 μm; (B) dimples, scale 14 μm.

plain many features. Some balls contain cavities.

Melting experiments. Synthetic ferrobasalt glass (Table 1, composition G) was heated in stoppered iron capsules in evacuated silica tubes from 1½ to 12½ hours and then water-quenched. The results showed that at 1235°, 1185°, and 1155°C, it was all glass; at 1125°C, 70 percent glass, the rest being ilmenite and pyroxene; at 1100°C, 30 percent glass, the rest being ilmenite, pyroxene, plagioclase; and at 1075°C, <10 percent glass. Small Fe metal beads pervade the glass. Ilmenite has <1 percent Fe_2O_3. The low liquidus temperature ($\sim$1140°C) and narrow melting range are similar to those in basaltic segregation veins (*3*) and are suggestive of a differentiation origin of ferrobasalt. The mineral sequence agrees with textural interpretation of Apollo 11 ferrobasalts.

Petrogenesis. Outgassing of the moon probably was more extensive than that of the earth because of the lack of visible alteration and low Na_2O content of Apollo 11 ferrobasalt and microgabbro. Although the more calcic nature of Apollo 11 than terrestrial basaltic plagioclase might result from incorporation of Al in high-pressure phases of the earth's mantle not found on the moon, the sodic nature of chondritic plagioclase suggests volatile loss of Na from the moon as evidenced by the extremely low Na_2O and K_2O contents of the glass beads. The extreme zoning of pyroxenes may result from a paucity of catalyzing volatiles.

Reduction of the moon was greater than that of the earth but similar to ordinary chondrites (*6*) since the oxygen fugacity of the ferrobasalts and microgabbros, as estimated from compositions of iron silicate, oxide, sulfide, and metal phases with the use of standard thermochemical data, is about 10^{-13} at 1100°C compared to 10^{-10} for earth basalts (*7*). The following data support a greater reduction of the moon: Low Ni in olivine (greater reduction to metal) and low P_2O_5/K_2O ratio (greater solution of P in metal).

Differentiation of the moon occurred of a type similar to that of the earth except for differences due to the moon's extreme reduction and volatile outgassing as well as smaller mass. Variation in liquid composition with crystallization of ferrobasalt is similar to that of the Skaergaard intrusion (*8*) except for lower alkalis (Fig. 1B). In ferrobasalt, a sulfide melt, ilmenite and augite separate early from silicate melt and are joined by plagioclase only after semisolidification as evidenced by fragments in breccia of glassy ferrobasalt containing only euhedral ilmenite and pyroxene, blebs of sulfide and devitrified glass.

Plagioclase vitrophyres (in breccia) contain 70 percent normative plagioclase and probably are fragments of highland material because their composition is close to that of Surveyor VII (*9*). Because they have much more soda than glass beads (which are almost certainly impact fused ferrobasalts, microgabbros and their minerals) and lack blebs of Fe metal, the plagioclase vitrophyre fragments probably were not superheated on a small (grain size) scale. Crystallization of plagioclase vitrophyres produces glasses which trend toward ferrobasalt composition on an alkali, iron, magnesium (AFM) diagram (Fig. 1B) but which have less TiO_2 and more Na_2O.

The ferrobasalts and plagioclase vitrophyres are complementary: the delayed appearance of plagioclase in ferrobasalts (density, 3.54 g/cm^3) corresponds to the super-abundance of plagioclase in the plagioclase vitrophyres (density, 3.0 g/cm^3). The TiO_2/K_2O ratio ranges from 6.7 to 10 in three plagioclase vitrophyres but is 28 in ferrobasalt suggesting either addition of Ti to or loss of K from ferrobasalt. Tentatively we explain these facts by gravitative separation of plagioclase (2.8 g/cm^3), augite (3.4 g/cm^3), ilmenite (4.7 g/cm^3), and liquid (3.1 g/cm^3) followed by melting of liquid plus plagioclase mush and liquid plus augite plus ilmenite mush yielding plagioclase vitrophyre and ferrobasalt liquids respectively.

Breccia cohesiveness (*1*) may result from thermal sintering from hot gas and ash clouds. Impact rather than volcanic origin of the cloud is indicated since vesicular, glassy rims require high superheating. Thermal sintering is favored over impact compaction because of preservation of thin-walled vacuoles in some spherules.

Proposed petrologic history of the moon. Our data and petrogenetic interpretations suggest, in the light of previous studies (*10*), the following model for the evolution of the moon. (a) Initially the moon was melted sufficiently for Mg-rich olivine and pyroxene to settle towards the center, perhaps around an iron-rich core. Major chemical fractionation took place by crystallization within the interior and by loss of volatiles from the surface. (b) Iron content and density of the convecting liquid mantle of the moon increased as successive shells were formed. Convection-thrusting and impact-breaking continually disrupted a thin primitive crust rich in olivine and pyroxene. (c) After about 80 percent of the initial mass had crystallized (to within about 130 km of the surface), plagioclase commenced to form and the iron and titanium content and density of the remaining liquid rapidly increased. Plagioclase flotation created the first substantial crust—now the lunar highlands. (d) Gravitational pull of some body in the solar system caused concentration of the dense liquid on the present nearside of the moon and antipodal bulge of light crust on the farside. (e) Ilmenite crystallized and sank, forming TiO_2-rich cumulates at the base of the liquid layer, which now amounted to <5 percent of the lunar mass. (f) Impacting bodies (*11*) continued to destroy the thin nearside crust exposing large areas (seas) of ferrobasalt liquid and creating additional liquid by impact melting of cumulates. Impacting bodies of the same energy would form only large, deep craters on the thick farside crust. (g) After total solidification, local remelting possibly caused explosive volatilization and ash flow emplacement of breccia deposits.

Several tests of this model are possible. (i) The proposed density stratification, especially the inverse stratification, could be tested for by a large seismic array. (ii) A search could be made for cumulate rocks and primitive olivine-pyroxene crust. (iii) A higher heat flow on the nearside would follow from concentration of radioactive elements in the residual liquid.

A. T. ANDERSON, JR.
Department of the Geophysical Sciences, University of Chicago, Chicago, Illinois 60637

A. V. CREWE
Department of Physics, Enrico Fermi Institute, Chicago

J. R. GOLDSMITH
P. B. MOORE, J. C. NEWTON
Department of the Geophysical Sciences, University of Chicago

E. J. OLSEN
Department of the Geophysical Sciences and *Field Museum of Natural History, University of Chicago*

J. V. SMITH
P. J. WYLLIE
Department of the Geophysical Sciences, University of Chicago

References and Notes

1. Lunar Sample Preliminary Examination Team, *Science* **165**, 1211 (1969).
2. H. Kuno, *J. Petrol.* **6**, 302 (1965).
3. C. E. Tilley, H. S. Yoder, Jr., J. F. Schairer, *Carnegie Inst. Wash. Yearbook* **62**, 77 (1963).
4. G. M. Brown, *Mineral. Mag.* **31**, 511 (1957); R. J. Murray, *Geol. Mag.* **91**, 17 (1954).
5. T. Simkin and J. V. Smith, *J. Geol.*, in press.
6. R. F. Mueller, *Geochim. Cosmochim. Acta* **28**, 189 (1964).
7. R. F. Fudali, *ibid.* **29**, 1063 (1965); I. S. E. Carmichael and J. Nicholls, *J. Geophys. Res.* **72**, 4665 (1967); A. T. Anderson, Jr., *Amer. J. Sci.* **266**, 704 (1968).
8. L. R. Wager and W. A. Deer, *Medd. Grønland* **106**, 1 (1939); L. R. Wager, *J. Petrol.* **1**, 364 (1960).
9. A. L. Turkevich, E. J. Franzgrote, J. H. Patterson, *Science* **162**, 117 (1968).
10. P. D. Lowman, Jr., *J. Geophys. Res.* **74**, 495 (1969); R. T. Reynolds and A. L. Summers, *ibid.*, p. 2495; Y. Nakamura and G. V. Latham, *ibid.*, p. 3771.
11. G. P. Kuiper, *Proc. Nat. Acad. Sci. U. S.* **40**, 1096 (1954).

12. We thank M. Isaacson and D. Johnson for electron microscopy, R. Zechman and O. Draughn for microprobe operation, G. A. Desborough for sulfide standards, the Clayton and Hafner groups for sample cooperation, and R. Banovich, A. Devitt, and I. Baltuska for technical help. We thank J. Warner of NASA for excellent cooperation. Supported by NASA grant NAS 9-8086. We thank NSF (grants G-1658, GA-4420, and GA-1656), ARPA, Union Carbide Corporation, Herz Foundation, and University of Chicago for basic and specific support.

4 January 1970

Electron Microprobe Analysis of Lunar Samples

Abstract. *Plagioclase feldspar, clinopyroxene, and ilmenite in a polished thin section of a type A crystalline rock were analyzed. The clinopyroxene grains are compositionally variable, and both high Ca and low Ca phases are present. The plagioclase is compositionally homogeneous. The ilmenite is chemically homogeneous except for occasional, small areas of high local chromium concentration. Accessory minerals are: apatite (containing Cl, F, Y, and Ce), troilite, and metallic iron. Glassy spherules from the lunar soil are for the most part similar in composition to the crystalline rocks; however, some appear to have been monomineralic. The crystalline rock has apparently formed by relatively rapid cooling of a silicate melt under conditions of low oxygen partial pressure. Many components of the soil appear to have formed by meteoritic impact.*

We present here electron microprobe and petrographic analysis of Apollo 11 samples of a crystalline rock and separates from a soil sample. A polished thin section of sample 10017-16 was examined microscopically. The sample is an equigranular, crystalline rock in which most crystals are about 0.5 to 1.0 mm in largest dimension. In transmitted light, four major phases can be recognized: (i) clinopyroxene, which occurs as short subhedral crystals; (ii) plagioclase, occurring as anhedral lath-shaped crystals; (iii) opaque minerals, dominantly euhedral ilmenite; and (iv) a partly crystalline, very fine-grained intergranular mesostasis containing several phases. A few vesicles, generally circular and about 1 mm in diameter, constitute about 5 percent by volume of the sample.

The modal mineral composition of the sample, determined by point counting (816 points, 200 mm^2 area), was (in percent by volume): clinopyroxene, 49.7; plagioclase, 18.0; opaque minerals, 23.9; and mesostasis, 8.3. The subophitic texture of the sample and the modal mineral composition are typical of the type A crystalline rocks described by the Preliminary Examination Team (PET) (*1*).

The clinopyroxene is colorless to very pale reddish-brown. The crystals do not exhibit definite pleochroism or color-zoning, and no reaction rims were observed at the contacts with other minerals. There is an irregular cleavage in the clinopyroxene grains. The extinction angle $Z \wedge c$ is about 45°, thus indicating an augite. The plagioclase is apparently unzoned. The pyroxene grains are slightly fractured, but neither pyroxene nor plagioclase shows any evidence of shock metamorphism. Distinctive shock features such as intense fracturing, "planar features," or isotropization of plagioclase were not observed. In view of this lack of evidence for shock metamorphism of the major minerals, it seems clear that the mesostasis cannot be shock-melted glass and that it probably represents residual crystallization of the melt from which the sample formed.

Although sample 10017-16 does not contain olivine, it has a mineral composition consistent with the chemistry of other samples (*1*) and is representative of the material at the Apollo 11 landing site.

The plagioclase phase of sample 10017-16 appears generally homogeneous both between grains and within individual grains. The average composition of the plagioclase is SiO_2, 49 percent; CaO, 16 percent Al_2O_3, 33 percent; total iron as FeO, 1 percent; TiO_2, 0.2 percent; K_2O, 0.2 percent; Na_2O, 2 percent; and anorthite, 76 percent.

In contrast, the clinopyroxenes in the sample appear heterogeneous both between grains and within grains. Most of the measured compositions center around 30 atom percent $FeSiO_3$, 30 atom percent $CaSiO_3$, and 40 atom percent $MgSiO_3$. Some pyroxene crystals show bands and regions of higher atomic number due to higher iron concentrations (see Fig. 1). These areas appear to be a second pyroxene phase with low calcium concentration (pigeonite?) at the boundaries of the prevalent pyroxene phase (Table 1).

Table 1. Representative composition of various phases in crystalline rock 10017 (type A).

Sample	Weight percentages as oxide									Sum
	Na	Mg	K	Al	Si	Ca	Ti	Mn	Fe as FeO	
Pyroxene		14.6		3.3	49.6	15.8	2.3		15.1	100.7
Pyroxene		15.9		3.6	49.6	17.0	2.7		9.9	98.7
Pyroxene		12.4		2.2	49.4	14.0	1.7		19.5	99.4
Pyroxene		20.8		0.9	52.0	4.1	0.8		21.5	100.2
Ilmenite		0.91		0.63	0.11		52.8	0.34	46.1	100.9
Ilmenite		0.88		0.71	0.12		51.4	0.33	46.5	100.0
Spherule (130-3)	0.2	5.0	0.1	12.5	42	14	7.4	0.18	16.5	98.4
Spherule (L-8)	0.3	11	0.1	13	46	12	2.5	0.14	15	100
Spherule (122)	0.5	13	0.16	4.5	37	8	10	0.24	24.5	97.9
Spherule (128)	3.0	13.5	0.06	7.4	43	10	0.5	0.19	21.2	96.2
Spherule (129)	0.7	5.4	0.1	24.5	47	16	0.5	0.07	5.7	100

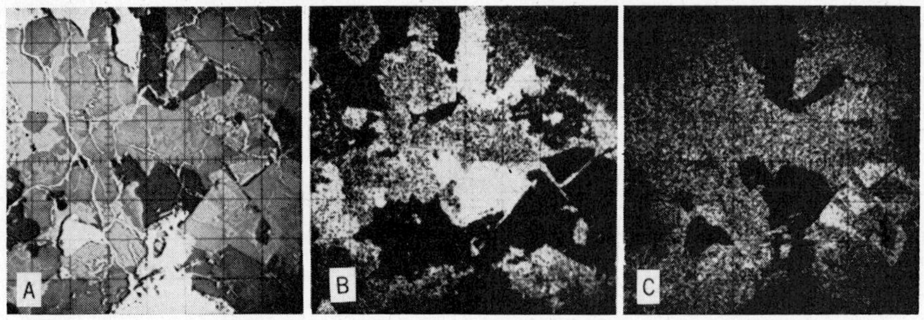

Fig. 2. Electron microprobe scan pictures of sample 10017, type A, showing silica-rich areas mixed with the plagioclase. (A) Target current scan: The darkest areas are ilmenite, the middle-gray areas are pyroxene, and the light-gray areas are plagioclase. (B) Fe K x-ray scan: The brightest areas are the ilmenite, and the next brightest areas are the pyroxene; (C) Si K x-ray scan: The bright areas in the lower right are due to SiO_2.

Several apatite grains were identified. The mineral was found to contain small concentrations of Cl, F, Y, and Ce.

One part of the section contained a number of very silica-rich areas mixed with the plagioclase. The distribution of these areas is shown in the electron microprobe scanning pictures (Fig. 2).

Ilmenite is the dominant opaque phase. The grains, except for small areas of increased Cr content, are homogeneous within grains and between grains. Representative compositions of the ilmenite are shown in Table 1. The ilmenite is virtually stoichiometric in Fe and Ti, thus indicating little or no solid solution of Fe_2O_3 or Ti_2O_3. Chromium is present but has not yet been determined quantitatively. Very small areas of high local Cr concentration occur within the ilmenite grains and near their edges. As the Cr content increases, the Ti concentration decreases markedly while the Fe level is relatively unchanged.

Small grains of troilite were occasionally observed. The troilite contained no detectable impurities, and the composition is apparently stoichiometric FeS. Metallic Fe particles 5 mm or smaller in diameter are either associated with or are included in the troilite grains (Fig. 3). The two particles shown, an iron grain and a troilite grain, occur in the area of interstitial mesostasis. Around these grains K, Al, and Si show high concentration, but Na and Ca are much lower than in the plagioclase in this rock section.

The lunar soil contains numerous distinctive glass particles, generally spherules believed to represent droplets of shock-melted material (*1*, p. 1219). These particles were concentrated from 4 g of lunar soil by mineral separation with heavy liquid techniques, a procedure which allowed almost complete recovery of glassy spherules greater than 150 μm in diameter. A total of 107 spherules were recovered. These ranged in size from 35 μm in diameter up to an oval-shaped body 1280 μm long by 1041 μm wide. Approximately 50 percent of the spherules are 125 to 500 μm in diameter.

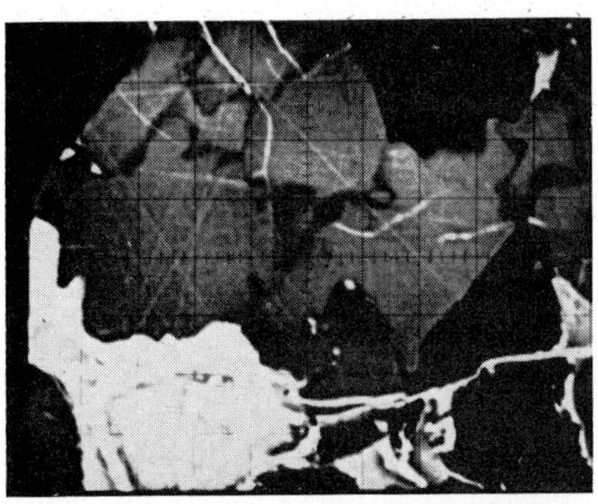

Fig. 1. Electron microprobe scan pictures of sample 10017 showing areas of higher iron concentrations in the pyroxenes. The very bright area is plagioclase, and the larger light-gray area consists of pyroxene. The darker gray bands and regions show substantial increases in iron concentration.

Nearly 95 percent of the glassy particles are spherical or oval in shape. The remainder are dumbbell-, teardrop-, or rod-shaped. Most (~80 percent) of the spherules are opaque black or translucent wine-red, and most of the remainder are brown or green. The surfaces of the spherules range from smooth to rough and pitted.

Many spherules have what appears to be a metallic coating or beads on their surface, or both. Scanning electron microscope photographs of one such spherule (320 μm in diameter) show that portions of the surface are covered with small flattened beads up to approximately 25 μm in width. Electron probe microanalysis shows that the beads are predominantly iron (95 $\pm$ 2 percent), with 3.5 to 4 percent approximately 0.3 percent Co and 0.15 percent P homogeneously distributed.

Ten spherules were analyzed by electron probe microanalysis. The compositions of some of the spherules included in Table 1 are similar to those reported for lunar rocks and soil. These spherules were mainly deep red or brown in color. One of these spherules contained over 15 metallic inclusions less than 10 μm in diameter consisting of iron containing 1 to 2 percent Ni. The amount of TiO_2 in the host glass was anomalously low (2.5 percent).

The remaining three spherules had different composition from that of the majority (Table 1). One spherule, No. 122, contained anomalously low concentrations of Al_2O_3 and high concentrations of FeO (24.5 percent) and TiO_2 (10 percent). Another, No. 129, has a composition close to that of plagioclase. The third, No. 128, is characterized by low amounts of Al_2O_3 and high concentrations of FeO and MgO.

Four angular glass particles were analyzed. These were for the most part isotropic, ranged from translucent to opaque, and contained weakly, birefringent lath-like inclusions. All four angular fragments were heterogeneous, with compositions similar to that of the lunar rocks and soil. The presence of Ni-Fe spherules in the glassy spherules is strong evidence for impact origin.

Two types of metals were investigated: (i) metallic areas associated with particles separated magnetically from soil sample 10084 and (ii) a metal fragment extracted from a breccia rock 10046-18a.

Several particles were extracted from

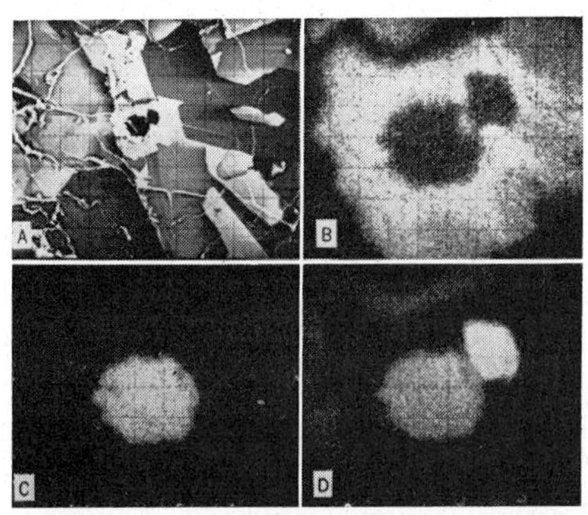

Fig. 3. Electron microprobe scan pictures of sample 10017 showing a troilite and an iron bleb and the distribution of K around these blebs. (A) Sample current scan; (B) K x-ray scan; (C) S x-ray scan; and (D) Fe x-ray scan.

the soil with a hand magnet. Some are melted, spherical in shape, and composed mainly of glass. Others are vesicular and complex in shape and contain large cavities and fragments of crystalline rock.

Scanning electron microscopic examination showed that metallic areas occur both within and on the surface of the soil particles. The metal content varied greatly from one particle to another. The metal inclusions in the soil particle are small ($< 30\ \mu m$) and often rounded. They are swathed in troilite, and their microstructure indicates shock-hardening.

Two metal areas, less than 10 μm in the largest dimension, were analyzed. One contained 13 percent Ni, 85 percent Fe, approximately 0.9 percent Co, and 0.4 percent P. The other contained 10 percent Ni, 85 ± 5 percent Fe, approximately 0.6 percent Co, and 0.35 percent P.

Another rounded metallic phase in a frothy complex soil particle is composed of essentially Fe and 0.4 percent Co. Nickel and P were not detected.

The metallic sample from the breccia is irregularly shaped and about 0.7 mm in its longest dimension. In a polished section, it exhibits a eutectic-like structure consisting of a very fine dispersion of phosphide and metal with lesser amounts of sulfide and carbide.

Extensive electron microprobe analysis of this particle shows the following. The metallic area surrounding the eutectic structure consists mainly of Fe with about 2 percent Ni, 1 ± 0.2 percent P, and no Co. The eutectic-like area contains on the average 6 ± 1 percent Ni and 1 to 1.5 percent P with small amounts of S and C. Electron microprobe scans show a strong correlation between Ni and P in the eutectic with the highest Ni region adjacent to the eutectic area. The carbides contain about 6.5 percent Ni and up to 8 percent P. The sulfides also contain Ni in solution. The amount of Ni in the phosphides [$(Fe\text{-}Ni)_3P$] was 10 percent as contrasted to that of the adjacent metal with 6 ± 1 percent Ni and less than 1 percent P.

Our findings bear out the suggestion made by PET (*1*) that the terrain from which these samples were collected is internally generated igneous rock overlain by an impact-formed breccia with some meteoritic material.

ISIDORE ADLER
LOUIS S. WALTER
PAUL D. LOWMAN, BILLY P. GLASS
BEVAN M. FRENCH, JOHN A. PHILPOTTS
NASA Goddard Space Flight Center, Laboratory for Space Physics, Greenbelt, Maryland 20771
K. J. F. HEINRICH
National Bureau of Standards, Gaithersburg, Maryland
J. I. GOLDSTEIN
Metallurgy Department, Lehigh University, Bethlehem, Pennsylvania

References and Notes

1. Lunar Sample Preliminary Examination Team, *Science* **165**, 1211 (1969).
2. We thank F. Wood, Jr., W. Kouns, A. Doan, and C. Inman for their assistance.

4 January 1970

Mineralogical and Petrological Investigations of Lunar Samples

Abstract. *Fragments of igneous rocks and breccias, and one coarse-grained rock with thin sections, have been studied. Minerals found include pyroxene, plagioclase, olivine, ilmenite, troilite, ulvöspinel, native iron, cristobalite, tridymite, alkali feldspar, apatite, and quartz. Textures are described and interpreted. Among features revealed by optical, microprobe, x-ray diffraction, and electron microscope methods are extreme zoning and unmixing in pyroxene grains, compositional variations in ilmenites, and effects of shock metamorphism. Some trace elements were determined by x-ray fluorescence analysis.*

Work we have carried out so far has been primarily on particles and dust (10085,2) from the 1- to 10-mm fraction of the bulk sample, and on rock 10044 (rock specimen 10044/43, covered section 10044/42, polished section 10044/41). In this report specimens from 10085,2 are given the prefix B followed by our own laboratory serial number.

In the igneous rocks grain size has been found to vary from one rock fragment to another, and associated with varying grain size there tend to be characteristic textures. No phenocrysts have been observed in any rocks. The coarsest rock examined (10044/43) has equant pyroxenes up to 2 mm across, skeletal-to-acicular ilmenites up to 2 mm, and plagioclase laths up to 1.5 mm in length. With decreasing grain size, the ilmenites change from skeletal (up to 0.5 mm across) through acicular grains and a dendritic variety to equant ilmenites (~ 0.1 mm) in the finest rock. These changes are paralleled by the plagioclases. The pyroxenes, on the other hand, are anhedral and subophitic. Within the range outlined, six distinct textural types have been seen: (i) medium-grained equant (up to 2 mm): 10044/41, B4; (ii) acicular plagioclase and ilmenite with anhedral pyroxene of medium to fine grain size (up to 0.5 mm): B3, B8, B9; (iii) fine-grained rock (up to 0.15 mm) with characteristic sets of thin subparallel ilmenite lamellae: B1; (iv) granular fine-grained rock (up to 0.1 mm): B5; (v) rock characterized by plumose intergrowths of plagioclase and pyroxene (0.25 mm) resembling chondritic texture: B18a; (vi) vesicular glass with rounded ilmenites: B18b. The principal minerals in types i to v are the same, although the proportions vary (see Table 1).

The modal analyses of the different igneous rock types (Table 1) show a remarkable consistency in pyroxene content, an inverse relation between plagioclase and ilmenite, and an increase in ilmenite content with decreasing grain size. A very high content of ilmenite is notable in all of the rocks.

Among breccias two distinct and several intermediate types can be distinguished. The first consists of fragments of igneous rocks, single mineral grains, and a matrix of fine-grained dust. Ilmenites in these breccias frequently show evidence of shock. The second type of breccia is predominantly glassy, containing angular fragments of low-reflectivity glass in a glassy matrix of slightly higher reflectivity. Specks of nickel-iron up to 40 μm in diameter have been seen. This breccia may be the product of more severe shock metamorphism. A spherule from within it has the composition $Fe_{94.0}\ Ni_{4.8}\ Co_{0.9}$, which is similar to that of a spherule from impactite glass from Lake Bosumtwi Crater, Ghana (*1*). No oxides or sulfides are present in the breccia.

Intermediate between these two types are breccias consisting of varying proportions of glass as well as mineral and rock fragments. Vesicles have been observed in many of these breccias. Flow textures occur in some of the glassy portions. Native iron occurs both as minute spherules and as larger (up to 50 μm) fragments with serrated rims.

Ilmenites in various rock types were surveyed. They display minor variations in TiO_2:FeO ratios within single rock types ranging from 60:40 to 50:50 (molecular proportions) and are free of Fe_2O_3. Single grains, however, are homogeneous, with one exception, where part of a grain has distinctly lower reflectivity.

The ilmenites investigated are free of any large-scale exsolution or replacement phenomena. However, minute lamellae (up to 10 μm) of a gray mineral occur in some ilmenites parallel to (0001).

The effects of shock metamorphism on ilmenite are represented by sets of twinning lamellae (sometimes bent and folded) following mostly one, but occasionally two, directions. Lamellar ilmenites are widespread in the breccias, but they also occur in some igneous rocks.

Ulvöspinel usually occurs intergrown with ilmenite in the medium-grained igneous rocks. Quantitatively, it is insignificant (less than 0.01 percent of the rock). Its presence does, however, support the assumption of an oxygen-deficient environment of formation.

The particle size of the dust-like fraction from the bulk sample was suitable for transmission electron microscopy. About 50 percent of the particles (both spherical and angular) gave only diffuse electron scattering, a result that indicates a glassy nature. Glassy spherules have diameters down to about 1000 Å, and many of them have adhering crystalline particles. Ilmenite grains were recognizable as thin (0001) plates, often with hexagonal outline and characteristic electron diffraction pattern. These showed an extremely high density of dislocations as compared with ilmenites from within rock 10044/43, and a higher degree of stacking disorder could also be inferred from the diffraction pattern. The increased dislocation density and stacking disorder are probably associated with a higher degree of shock metamorphism.

Several black, glassy-looking spheres with diameter approximately 0.1 mm were found among the fines. X-ray diffraction analysis of these indicated the presence of glass and some α iron, either as cores or adhering to the surface. They are probably similar to the spherules found within the glassy breccia. X-ray fluorescence analysis of 0.5 g of the dust gave the following concentrations: Nb, 16 ppm; Zr, 265 ppm; Y, 140 ppm; and Sr, 215 ppm. One colorless crystal from the "fines" was identified as α quartz.

In rock 10044/43 the principal phases which can be seen in transmitted light are clinopyroxene, opaque minerals, plagioclase, cristobalite, and tridymite (mode in Table 1); apatite and K-feldspar are also present. In reflected light some areas exhibit a fine-grained, simplectite-like texture. The dominant opaque phase is ilmenite, but there are also troilite, native iron, and ulvöspinel.

Table 1. Modal analyses by volume. The mode of thin section 10044/42 was determined by the conventional method. Other modes were determined by superimposing a transparent grid on reflected-light photographs.

Sample	B4	B8	B3	B9	B1	B5	10044/41	10044/42
Plagioclase	24.2	34.8	33.9	36.2	26.8	24.4	33.1	34.1
Pyroxene	64.2	50.5	48.8	48.5	48.5	47.0	47.3	46.4
Ilmenite *	9.8	14.7	18.3	15.3	21.7	25.4	14.4	12.3
Cristobalite and tridymite							5.2	6.3

* Ilmenite here includes pyrrhotite, native iron, and ulvöspinel.

The pyroxenes as compared with those in most terrestrial rocks are remarkably free of inclusions of oxides and sulfides. Many grains show shadowy extinction and regions with different birefringence but similar refractive index, probably indicating misorientation at grain boundaries; similar subgrain boundaries have been observed by electron microscopy (Fig. 1). Other grains are zoned both in color and birefringence, and electron-probe analyses show these to have Mg-rich cores and Fe-rich margins ranging from augite to ferroaugite. The $2V_\gamma$ for different grains ranges between 37 and 48½ ° and some show simple (100) twinning. On the margins of some pyroxene grains crystals occur devoid of Mg and with unusually low Ca content (Fig. 2).

X-ray diffraction studies of some pyroxene single crystals showed them to be unmixed into monoclinic Ca-rich and Ca-poor components (*2*), the corresponding diffraction spots from each component being joined by streaks indicating either strain or a continuous range of composition. For each component $a \sin \beta$ and b were determined in each of three grains, and their compositions (Fig. 2) were derived on the basis of the data of Brown (*3*) and Viswanathan (*4*). The unmixing was confirmed by electron microscopy which showed (001) lamellae approximately 500 Å apart. Electron diffraction showed the existence of stacking faults as well as other disorder. A lamellar structure is seen in many grains in immersion oil under high power.

A separated pyroxene fraction (density 2.85 to 3.5 g/cm³) gave an x-ray powder pattern with moderately broad peaks yielding cell parameters $a \sin \beta = 9.33_5$, $b = 8.94_5$ Å, and suggesting an average composition of Ca:Mg:Fe = 31:42:27 (*3*).

The plagioclase is twinned on albite, pericline, and Baveno laws; it is slightly zoned. The $2V_\alpha$ ranges from 80 to 84°. A few crystals of a plagioclase separate with density lower than 2.85 g/cm³ were melted; the refractive index of the glass was 1.564 (±0.001), corresponding to An_{88}. Plagioclases from a light fraction of the above separa-

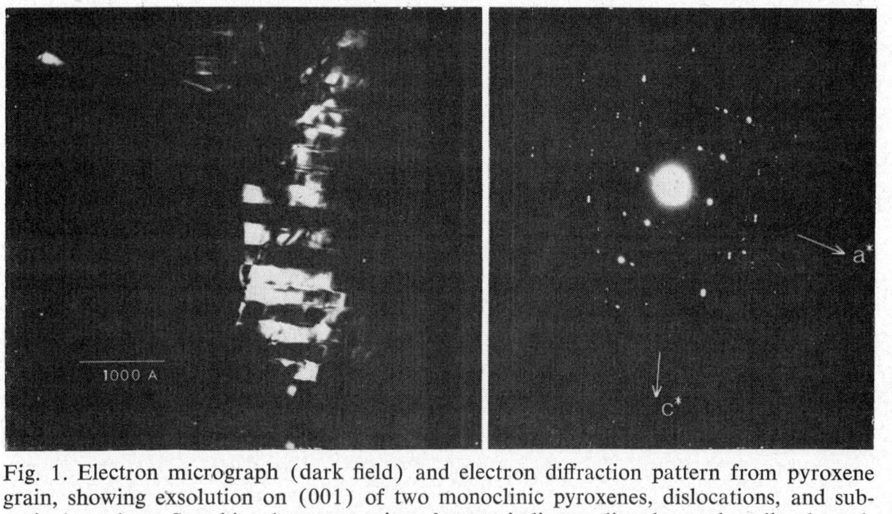

Fig. 1. Electron micrograph (dark field) and electron diffraction pattern from pyroxene grain, showing exsolution on (001) of two monoclinic pyroxenes, dislocations, and subgrain boundary. Streaking between pairs of spots indicates disorder at lamellar boundaries, and streaking parallel to a^* indicates stacking disorder on (100).

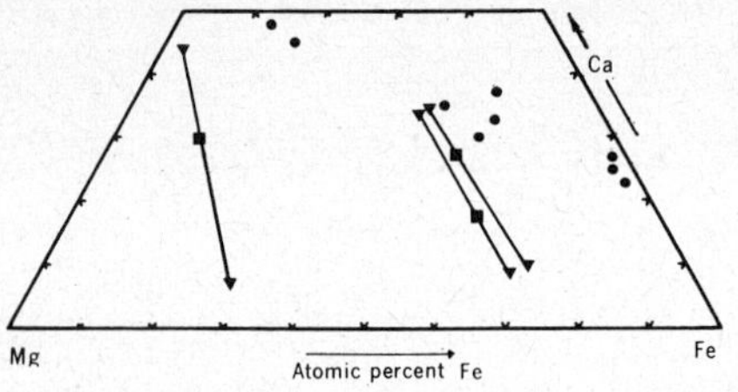

Fig. 2. The pyroxene quadrilateral. Circles, electron-probe analysis from a single grain; triangles, unmixed pairs of pyroxenes determined by single-crystal x-ray diffraction; squares, estimated bulk composition of pairs.

tion gave a glass with a refractive index 1.550 (±0.001) corresponding to An_{74} (*5*). Electron-probe analyses suggested a range from An_{72} to An_{89} within single grains.

The ilmenite in this rock is relatively free of shock lamellae and has a low density of dislocations and stacking faults. Twinning lamellae are rare, and there is no sign of compositional zoning or inhomogeneities within grains. The TiO_2:FeO ratio is 60:40, but small variations from grain to grain are observed. X-ray powder diffraction from an ilmenite separate gave cell parameters $a_{hex} = 5.08$, $c_{hex} = 14.06 \pm 0.01$ Å.

Troilite is the second most abundant opaque phase (~0.1 percent). It usually contains minute blebs of native iron. Electron-probe analyses show the blebs to be virtually free of Ni and Co, in contrast to the iron spherules in the glassy breccia and to meteoritic iron.

Minor amounts of ulvöspinel (<0.01 percent) occur in association with ilmenite and native iron. Textural evidence suggests that spinel may have formed after ilmenite (a postcrystallization "thermal event"). This possibility is further supported by the occasional occurrence of troilite on minute cracks and fissures.

A small amount of apatite in hexagonal prisms has been seen optically, and its presence has been confirmed by microprobe. Cristobalite (mean r. i. 1.487) with characteristic twinning and tile structure occurs in euhedral crystals and filling interstices, and the presence of acicular tridymite (mean r. i. 1.472) in parallel growth with pyroxene was confirmed by x-ray single-crystal diffraction. A few crystals of K-feldspar were identified optically and by microprobe.

The yellow mineral referred to in the Preliminary Examination Team report (*6*) has been observed, and its cell parameters are being determined. A small amount of purplish mineral with hexagonal habit and a dark brown mineral have also been seen, but the minerals have not been identified.

X-ray fluorescence analysis of 0.5 g of rock 10044 gave the following concentrations: Nb, 41 ppm; Zr, 465 ppm; Y, 215 ppm; Sr, 250 ppm; and Cr, 1180 ppm. Semiquantitative analyses showed that the chromium is present in approximately equal amounts in the light pyroxene fraction and the ore fraction, with lesser amounts in the dense pyroxene fraction.

We also made trace element determinations on 0.5-g samples of rocks 10045 and 10060, which gave Nb, 28 and 38 ppm; Zr, 300 and 565 ppm; Y, 190 and 170 ppm; Sr, 185 and 225 ppm; and Cr 3010 and 2165 ppm.

The lobate embayed textures of the ilmenites and the arrangement of the plagioclase and pyroxene within the lobes suggest that ilmenite was the first phase to begin to crystallize. It was probably soon followed by pyroxene, which shows clear centers to the crystals but plagioclase enclosed in the outer parts. Tridymite is involved in intergrowths with pyroxene and plagioclase and in at least one case with K-feldspar, so it probably crystallized fairly late.

The compositions and mutual relation of the opaque minerals can be used to put limits on fS_2 and fO_2.

For the general evolution of moon rocks we would conjecture that, since lunar gravity and thermal gradient are both low, there has been no crust-mantle recycling. Low-melting silicates may have formed an original crust later to be partially covered by the basic rocks now reported. These may have originated in a deep magma chamber floored by phases in the system Ni-Fe-S.

J. C. Bailey, P. E. Champness
A. C. Dunham, J. Esson
W. S. Fyfe, W. S. MacKenzie
E. F. Stumpfl, J. Zussman
*Department of Geology,
The University,
Manchester M13 9 PL, England*

References and Notes

1. A. El Goresy, in *Shock Metamorphism of Natural Materials*, B. M. French and N. M. Short, Eds. (Mono, Baltimore, 1968), pp. 531-53.
2. M. G. Brown and P. Gay, *Am. Mineral.* **44**, 592–602 (1959).
3. G. M. Brown, *ibid.*, **45**, 15-38 (1960).
4. K. Viswanathan, *ibid.* **51**, 429–42 (1966).
5. J. R. Smith, *Carnegie Inst. Wash. Yearbook* No. 56 (1957), p. 216.
6. Lunar Sample Preliminary Examination Team, *Science* **165**, 1211 (1969).
7. We thank the Department of Mineralogy and Petrology, University of Cambridge, for mineral separates and loan of a spindle stage, D. L. Hamilton for melting the plagioclases, and J. E. Chisholm for help with x-ray work.

4 January 1970

Mineralogy and Deformation in Some Lunar Samples

Abstract. *Observations on the mineralogy and deformation in samples of crystalline rocks, breccias, and fines from Tranquillity Base provide evidence for magmatic and impact processes. Overall homogeneity, igneous textures, and absence of xenoliths in the crystalline rocks indicate derivation from a common titanium-rich magma by internal, anorogenic volcanism rather than by impact. Crystallization conditions allowed strong compositional variation in pyroxenes, olivine, and plagioclase and the growth of a new mineral, the iron analog of pyroxmangite. Subsequently, impact produced breccias containing shock-deformed crystals and glasses of varying compositions.*

The preliminary examination (*1*) has shown that the Tranquillity Base rocks have undergone igneous and deformational processes in many ways comparable with terrestrial and meteoritic phenomena. The major minerals can be assigned to known rock-forming mineral groups, but their compositional variation, association, and textures demonstrate relationships which complement and extend petrological experience gained from investigations of terrestrial basic rocks and meteorites and from synthetic studies. To this end we have examined four crystalline rocks, seven breccias, and a sample of fines (*2*) by optical, x-ray, and electron microprobe methods.

The major minerals of the crystalline rocks and lithic fragments are, in common order of abundance, clinopyroxene, plagioclase, and ilmenite. Slight differences in mineral proportions between most rocks indicate a general crystal fractionation process, which in a few cases has led to essentially bimineralic plagioclase-ilmenite, clinopyroxene-ilmenite, and plagioclase-pyroxene rocks. In general, strong compositional zoning and grain-to-grain compositional variation reflect incomplete equilibration between minerals and liquid during crystallization.

Clinopyroxene shows extreme variation in composition (Fig. 1A), with single grains having domains of varying Ca/Mg and Mg/Fe ratios. The majority of the compositions plot in a band across the pyroxene quadrilateral from approximately $Ca_{40}Mg_{43}Fe_{17}$ to $Ca_{25}Mg_{5}Fe_{70}$. Large parts of this trend are commonly present within systematically zoned grains, and Fig. 1B shows the variation along a traverse of 0.8 mm.

The unidentified yellow mineral (*1*) has been studied in detail in 10047. It occurs as discrete grains associated with cristobalite and as marginal growths on clinopyroxene. Unlike the clinopyroxene, it is homogeneous; the composition, averaged from six grains, is $(Fe_{0.85}Ca_{0.13}Mg_{0.02})SiO_3$. Optical parameters are $2V_z = 32°$ to $38°$; $n_X = 1.750$, $n_Y = 1.752$, and $n_Z = 1.768$. The x-ray powder pattern has 64 measured lines, with the seven strongest at (spacing, relative intensity) 3.10 (6), 2.93 (10), 2.68 (9), 2.62 (6), 2.58 (6), 2.19 (5), and 2.16 Å (6). Unit cell dimensions, based on precession and Weissenberg data, and a least squares refinement of powder data are $a = 7.55$, $b = 17.38$, $c = 6.62$, $\alpha = 82.69$, $\beta = 94.53$, $\gamma = 114.34$. These data show that it is a new mineral (phase A), a triclinic pyroxenoid-type polymorph with *Siebenerketten* analogous to pyroxmangite and synthetic "iron-rhodonite" (*3*).

The continuation of the clinopyroxene trend beyond the calcium minima of normal terrestrial differentiated basaltic magmas toward $Ca_{25}Mg_{5}Fe_{70}$ and the crystallization of phase A are indicative of unusually high silica activity (*4*).

Plagioclase, like clinopyroxene, shows large compositional variations over submillimeter distances. The overall range of composition is An 100 to An 60. The range detected in a fine-grained rock, 10020, is An 95 to An 67, with variation mainly from grain to grain, more calcic compositions being associated with olivine and blocky ilmenite. In a coarser rock, 10047, a similar range occurs as normal zoning in single grains. Comparison of compositions and $2V$ determinations shows correlation with high structural state curves (*5*).

Olivine occurs in some rocks as mainly large, early grains, commonly enveloped by clinopyroxene. In 10020,24 core compositions are consistently about Fo 65 and rims are zoned to about Fo 55. Comparison of these core compositions with the Fe/(Fe + Mg) ratio of the rock indicates a supercooling effect before olivine nucleated. Single crystal fragments in the breccias extend the compositional range to Fo 35.

An iron-titanium-chromium mineral (phase B) of the spinel group occurs within or in close proximity to olivine. It is slightly birefringent, has a cell edge of 8.415 ± 0.005 Å, and a compositional formula of $Fe_{1.0}(Cr_{0.60}Ti_{0.56}Fe_{0.27}Al_{0.28}Mg_{0.19})O_4$. Most grains show marginal enrichment of iron and titanium at the expense of chromium; some have rare lamellae of ilmenite and rutile.

Tabular to lamellar ilmenite appears to have a composition close to its

Table 1. Electron microprobe analyses of some glasses from two breccias. Color of glasses in plane-polarized light: 1, yellow-brown; 2, light yellow; 3, light yellow; 4, yellow-brown; 5, colorless; 6, semiopaque.

Oxide	Sample 10065,21			Sample 10018,27		
	1*	2*	3	4	5	6
SiO_2	38.3	42.9	43.9	38.1	40.9	41.4
TiO_2	13.3	5.9	5.3	6.6	1.0	0.3
Al_2O_3	7.0	13.8	15.5	15.0	25.5	18.9
Cr_2O_3	0.4	0.3	0.2	0.2	0.1	0.15
FeO†	20.0	13.5	12.6	15.9	6.8	6.4
MnO	0.25	0.15	0.15	0.2	0.1	0.1
MgO	8.9	9.0	7.8	9.6	8.4	19.4
CaO	10.2	12.3	12.5	13.7	16.1	11.5
Na_2O	0.5	0.8	0.8	0.3	0.5	0.3
K_2O	0.05	0.05	0.1	0.01	0.01	0.01
Total	98.9	98.7	98.85	99.61	99.41	98.46

* Analyses 1 and 2 from different parts of the same glass spherule. † Total iron as FeO.

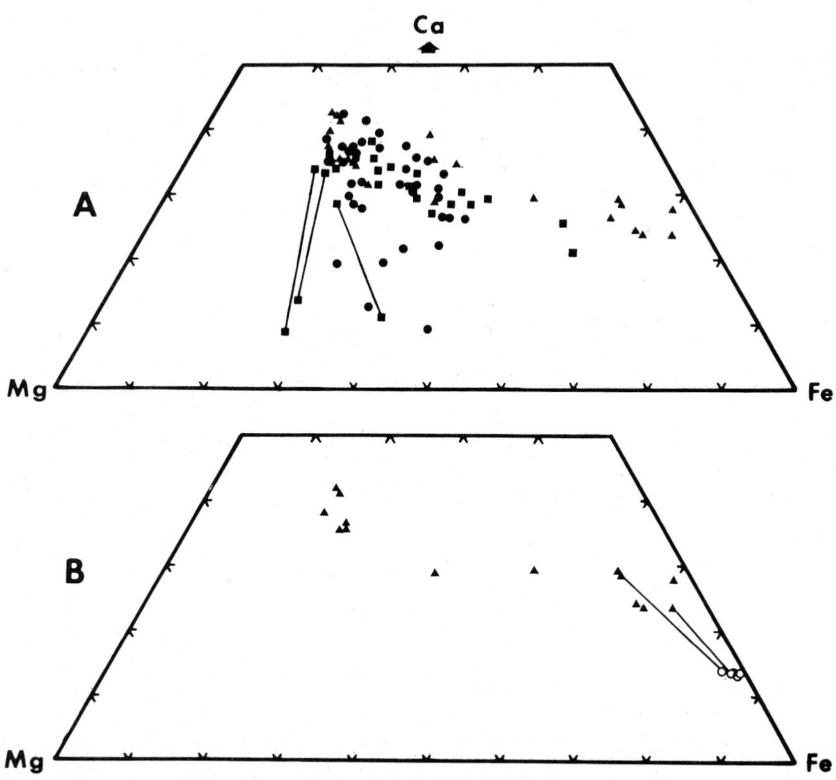

Fig. 1. (A) Plot of pyroxene compositions in terms of Ca, Mg, and Fe atomic percent for three crystalline rocks: 10017 (■), 10047 (▲), and 10069 (●). Tie lines join domains of different compositions coexisting within a single grain. (B) Plot of compositions in terms of Ca, Mg, and Fe atomic percent for one traverse across a single grain in rock 10047, and of the coexisting pyroxenoid, phase A (○). The core has an average composition $Ca_{38}Mg_{43}Fe_{19}$ and the rim $Ca_{25}Mg_{5}Fe_{70}$. Tie lines join phase A adjacent to the clinopyroxene.

ideal formula but slightly enriched in titanium. This suggests that the element is partly in the trivalent state (*6*), which would imply a very low oxygen fugacity in the magma. Some crystals show twinning, and some contain fine lamellae of rutile and a Cr-bearing mineral, possibly phase B. The x-ray powder pattern of the ilmenite is normal with no extra lines. Unit cell parameters are a_0 = 5.088 ± 0.005 Å and c_0 = 14.055 ± 0.005 Å.

Minor opaque phases are troilite containing traces of Ni, Ti, and Cr, and associated blebs of native iron with trace Ni.

Low cristobalite occurs in both fine- and coarse-grained rocks, including those containing olivine. Although optically clear grains contain minor amounts of Ca, Al, Ti, and Fe (K and Mg were not detected), the x-ray powder pattern (a_0 = 4.970 ± 0.005 Å, c_0 = 6.931 ± 0.005 Å) is in close agreement with that of the NBS standard (*7*). Crystallizing with the cristobalite are potassium-barium feldspars; apatite; at least two zirconium-bearing phases, one of which is an iron-titanium-zirconium silicate (phase C) occurring as minute red-brown flakes; and in some cases a largely isotropic mesostasis rich in potash and silica. Observed in the breccias are orthopyroxene close to En 70; single grains of pigeonite; and a few meteorite fragments of nickel-iron containing up to 17 percent Ni.

The breccias and fines are in addition the source of strongly deformed crystals, glasses, and recrystallized fragments which we interpret as the products of shock metamorphism. Tensional fractures, a common feature of the larger fragments in the breccias and, to a lesser extent, of the crystalline rocks, are taken to be the result of weaker shock stresses.

However, other styles of deformation occur which we do not consider to be shock induced; in some cases stresses accompanied crystal emplacement and growth, whereas in a few fragments postcrystallization strain may have been imposed at tectonic rates. The deformation attributed to crystallization stress is most clearly seen in pyroxenes in some coarser rocks. It is commonly developed as an undulous extinction ranging over as much as 30°. In many grains the boundaries between areas of differing extinction position are sharp and coincide with color and compositional domain boundaries. In others variation in extinction position, color,

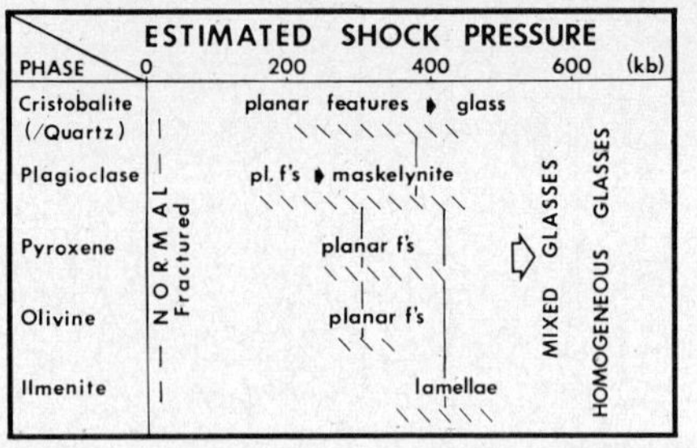

Fig. 2. A generalized scheme of shock metamorphism for the main minerals in breccia samples with estimated shock pressures. Observations have been made on normal and fractured phases, on mixed and homogeneous glasses, and in the regions indicated by diagonal patterns. Vertical lines link phases in polycrystalline fragments.

and composition takes place over a zone 10 to 25 μm wide. This zone may also be the site of narrow (1 to 3 μm), curved or wavy, closely spaced features, possibly fine kinks, which differ slightly in refractive index from the host crystal. The close correlation of these features with compositional variation in the pyroxenes suggests that they result from strains induced during crystallization. This conclusion gains support from the absence of comparable deformation in coexisting plagioclase and cristobalite which are more sensitive to postcrystallization stresses.

The distinct shock deformation features seen in single and multiple mineral fragments in the breccias and fines have been assigned to a scheme of progressive shock metamorphism (Fig. 2) on the basis of the degree of development of planar features and changes in refractive index and birefringence.

Most common among the shocked minerals are clinopyroxene fragments averaging 50 μm across with nearly normal optics but containing one or two sets of planar features. The latter are up to 2 μm wide and 10 μm apart, have distinctly lower refractive indices than their host, and lie close to the zone $[\bar{h}11]$ (h = 1, 2, 3).

Plagioclase exhibits a comparable development of one to four sets of planar features. In most cases this development is accompanied by variable extinction and decrease in refractive index and birefringence toward the isotropic state, maskelynite. Silica phases, originally either cristobalite or possibly quartz, also contain multiple sets of planar features; large single grains with as many as eight sets of planar features have been measured. Such grains show variations in mean refractive index and birefringence, some domains being isotropic, and others have an apparent $2V_z$ about 35°. These grains coexist with completely isotropic maskelynite.

Also coexisting with maskelynite are more strongly deformed clinopyroxenes with as many as four planar sets and a fine, mosaic undulous extinction. Similar deformation states are seen in olivine, orthopyroxene, and pigeonite. Ilmenite at this stage is not noticeably deformed, though commonly fractured and twinned. A few grains enclosed in glass are strongly disrupted and show fine striations between twin lamellae.

The glasses are considered the products of intense shock metamorphism (Fig. 2). They commonly contain schlieren of variable refractive index and show great variation in shape, composition (Table 1), vesicularity, degree of recrystallization, and character of inclusions. Some compositions (analyses 1 to 4) are comparable with the reported analyses (*1*); others (analyses 5 and 6) are significantly different and may be derived from bimineralic assemblages.

We have not recognized tectosilicates with well-developed weak shock features as are common in large (>1 km) impact craters (*8*); the shocked phases observed are probably the products of the small craters in the immediate vicinity of Tranquillity Base (*9*).

On the other hand, fragments of strongly crushed plagioclase-orthopyroxene rocks and orthopyroxene with broad kink bands resemble tectonites rather than shocked rocks, suggesting that either tectonic or crater-related faulting has occurred.

The presence of possible tectonites and the indications from the bimineralic rocks and the more homogeneous mineral fragments of crystal fractionation and slow cooling at depth, point to the operation of internal lunar processes. However, the clearest evidence for an active moon comes from the fine-grained crystalline rocks. Their overall textural homogeneity and freedom from xenoliths is in strong contrast to impactites of similar grain size (*10*). We therefore favor generation of magma by anorogenic volcanism from a common titanium-rich source rather than by impact melting.

The crystallization and the absence of distinct phenocrysts indicate a simple magmatic history. Cooling conditions were generally rapid and characterized by supercooling, high silica activity and low fO_2, fS_2, and fH_2O. The unusual mineral assemblages and metastable phases so produced owe their preservation to the apparently anhydrous environment.

J. A. V. DOUGLAS
Geological Survey of Canada, Ottawa, Ontario
M. R. DENCE
Observatories Branch, Department of Energy, Mines and Resources, Ottawa, Ontario
A. G. PLANT, R. J. TRAILL
Geological Survey of Canada, Ottawa

References and Notes

1. Lunar Sample Preliminary Examination Team, *Science* **165**, 1211 (1969).
2. Crystalline rocks: 10017,23; 10020,24; 10047,25; 10069,29. Breccias: 10018,27; 10023,9; 10046,53; 10060,20; 10061,39; 10065,21; 10073,25. Fines: 10085,14.
3. J. H. Whiteley and A. F. Hallimond, *J. Iron Steel Inst.* **99**, 199 (1919).
4. D. H. Lindsley and J. L. Munoz, *Amer. J. Sci.* **267A**, 295 (1969).
5. J. R. Smith, *Amer. Mineral.* **43**, 1179 (1958).
6. J. B. MacChesney and A. Muan, *ibid.* **46**, 572 (1961).
7. H. E. Swanson, M. I. Cook, E. H. Evans, J. H. deGroot, *Nat. Bur. Stds. Circ. 539* **10**, 48 (1960)
8. P. B. Robertson, M. R. Dence, M. A. Vos, *Shock Metamorphism of Natural Materials* (Mono, Baltimore, 1968), p. 433.
9. Apollo 11 Preliminary Science Report, *NASA SP-214* (1969).
10. M. R. Dence, *Ann. N.Y. Acad. Sci.* **123**, 941 (1965); F. C. Taylor and M. R. Dence, *Canadian J. Earth Sci.* **6**, 39 (1969).
11. E. C. T. Chao, *Shock Metamorphism of Natural Materials* (Mono, Baltimore, 1968), p. 135; W. v. Engelhardt and D. Stöffler, *ibid.*, p. 159.
12. Technical assistance of colleagues at the Geological Survey of Canada is gratefully acknowledged.

4 January 1970

Mineral Chemistry of Lunar Samples

Abstract. *Glass spherules, glass fragments, augite, ferroaugite, titanaugite, pyroxmangite, pigeonite, hypersthene, plagioclase, potassium feldspar, maskelynite, olivine, silica, ilmenite, TiO_2, "ferropseudobrookite," spinel, ulvöspinel, native iron, nickel-iron, troilite, and chlorapatite were analyzed with the electron microprobe. There are no indications of large-scale chemical differentiation, chemical weathering, or hydrous minerals. Contributions of meteoritic material to lunar surface rocks are small. Rocks with igneous textures originated from a melt that crystallized at or near the surface, and oxygen fugacities have been low. Shock features indicate that at least some surface material is impact-produced.*

Polished thin sections of loose surface fines < 1 mm in diameter (10084-97), breccias (10019-22, 10059-27, 10067-8, 10068-34), and igneous rock (10045-29) were studied by optical and electron microprobe techniques. Some of us are coinvestigators on other projects and studied breccia 10061-37 (T.E.B.) and loose surface material 10086-3 (K.K.).

Sample 10084-97 consists largely of rock fragments, feldspar, pyroxene, ilmenite, and glass spherules and glass fragments, with lesser amounts of pyroxmangite, olivine, ulvöspinel, and metallic nickel-iron. Breccias have essentially the same major phases and, in addition, accessory phases (pyroxmangite, pigeonite, hypersthene, maskelynite, olivine, silica, TiO_2, "ferropseudobrookite," spinel, native iron, metallic nickel-iron, troilite, chlorapatite, and silica-rich glass). Loose surface material and breccias contain a variety of rock fragments, many of which resemble in their mineralogy and chemistry the igneous rocks. However, there are also a variety of lithic fragments that are apparently not represented in the hand-specimen-sized samples. Fragments in sample 10059-27 consist largely of plagioclase [An_{95} (95 mole percent anorthite)] and minor amounts of olivine [Fo_{66} (66 mole percent forsterite)], augite [$Fs_{15.3}En_{45.6}Wo_{39.1}$ (15.3 mole percent ferrosilite, 45.6 mole percent enstatite, 39.1 mole percent wollastonite)], pigeonite ($Fs_{30.3}En_{61.4}Wo_{8.3}$), and ilmenite (6 percent MgO by weight) resembling terrestrial anorthositic gabbros. This occurrence, in association with Ti-rich mafic rocks, is noteworthy because terrestrial anorthositic rocks also tend to be associated with Ti-rich provinces. Fragments in sample 10019-22 consist of plagioclase (An_{95}), olivine (Fo_{81}), plagioclase glass, fine-grained lath-like feldspar and pyroxene embedded in the glass, and spinel A (Table 1). One inclusion in sample 10061-37 is largely maskelynite (An_{91}), clinopyroxene ($Fs_{16.3}En_{41.4}Wo_{42.3}$), and orthopyroxene ($Fs_{37.6}En_{58.5}Wo_{3.9}$).

Sample 10045-29 is a medium-grained ophitic igneous rock containing augite, plagioclase, and ilmenite; ferroaugite, titanaugite, pigeonite, potassium feldspar, olivine, silica, native iron, troilite, and chlorapatite are present in accessory amounts.

The following is a summary of the observed phases and their compositions in the samples studied:

Calcium-rich pyroxenes contain TiO_2 (0.67 to 9.3 percent by weight) and Al_2O_3 (0.58 to 9.7 percent). All pyroxenes with more than 2 percent TiO_2 are referred to as titanaugite (*1*). Most grains contain Cr_2O_3 (0.1 to 0.8 percent), and the amount of Cr_2O_3 increases with increasing MgO content. Manganese monoxide ranges from 0.06 to 0.54 percent; K_2O and Na_2O rarely exceed 0.10 percent each.

Pyroxenes exhibit minor to large within-grain and grain-to-grain compositional variabilities. Major elements in larger grains vary widely across grains and even broad areas (10 to 15 μm or more) do not generally show repetitions of patterns consistent with fine-scale exsolution, although well-defined exsolution of orthopyroxene or pigeonite from augite, and augite from hypersthene have been observed in a few grains. Typical patterns across large pyroxene grains show that FeO varies

Table 1. Electron microprobe analysis of less common minerals. Compositions are given in percentages by weight.

Mineral	SiO_2	Al_2O_3	FeO	CaO	Na_2O	K_2O	Cr_2O_3	TiO_2	MnO	MgO	ZrO_2	Total
					Silicates							
Potassium feldspar	64.6	19.9	0.60	0.72	0.33	13.1						99.25
Silica	96.9	1.03	0.35	0.52	0.05	0.25						99.10
					Oxides							
Ulvöspinel		1.88	62.2	0.03			3.5	32.7	0.22	0.04		100.57
Spinel A		68.0	3.33				2.16			25.9		99.39
Spinel B		6.0	46.3				19.3	23.0		4.3		98.9
"Ferropseudobrookite"		1.48	14.7	0.32			1.94	71.4	0.07	8.7	0.05	98.66

Table 2. Electron microprobe analyses of glass spherules in percentage by weight. Analyses of angular glasses are similar in composition and are not given. Parenthetical numbers indicate the number of spherules analyzed.

Oxide	Clear (15)	Green (10)	Orange to brown (20)	Red to black (22)
SiO_2	45.5	44.2	40.1	37.2
Al_2O_3	27.0	21.0	14.8	5.2
Cr_2O_3	0.12	0.17	0.32	0.61
TiO_2	0.45	1.81	7.6	10.7
FeO	3.8	7.9	15.3	24.7
MgO	6.9	8.6	8.4	13.1
MnO	0.8	0.15	0.22	0.30
CaO	15.0	14.7	12.9	7.7
Na_2O	0.24	0.41	0.14	0.27
K_2O	0.02	0.39	0.05	0.04
ZrO_2	0.02	0.04	0.07	0.06
P_2O_5	0.04	0.14	0.04	0.05
Total	99.17	99.51	99.94	99.93

sympathetically with MgO and that FeO and MgO vary antipathetically with CaO; TiO_2 varies sympathetically with Al_2O_3, and TiO_2 and Al_2O_3 vary antipathetically with SiO_2. However, patterns have been observed where (i) CaO varies sympathetically with FeO and MgO; (ii) CaO varies antipathetically with FeO, but MgO shows no relation to either; (iii) FeO varies sympathetically with MgO, but CaO shows no relation to either; and (iv) CaO, FeO, and MgO show variations unrelated to one another, and TiO_2 always varies sympathetically with Al_2O_3.

Analyzed pyroxenes range from augite to ferroaugite to ferrohedenbergite (or pyroxmangite); a few grains conform to subcalcic augite and ferroaugite compositions. Figure 1 appears to define a compositional trend similar to that of the diopside-ferropigeonite series (*2*). However, these pyroxenes may have originated in nonrelated parent rocks and the compositional trend may be fortuitous.

Calcium-poor pyroxenes are pigeonite that is always in association with augite or titanaugite in lithic fragments, and hypersthene that is always in association with exsolved augite. Both are low in Al_2O_3, Cr_2O_3, and TiO_2 by comparison with calcium-rich pyroxenes, but they contain higher amounts of MnO.

Plagioclase (anorthite or bytownite, sometimes labradorite) is usually somewhat zoned, and within one rock or fragment type it may range in composition by about ± 20 percent An. In five lithic fragments of sample 10059-27, it ranges from An_{64} to An_{90}. The most common plagioclase in the surface fines and breccias is An_{93}. Potassium feldspar was found in small patches in sample 10045-29 (Table 1). Maskelynite of An_{91} occurs in sample 10061-27. Other maskelynite grains are also similar to unshocked plagioclase in composition, and no particular depletion of alkalis was noted.

Olivine is sometimes zoned and ranges in composition from Fo_{48} to Fo_{80}. It is remarkable because of its consistently high CaO content (~ 0.5 percent) and minor amounts of TiO_2 and Cr_2O_3.

Silica occurs in small (<10 μm) grains not suitable for precise quantitative microprobe analysis. However, one grain about 25 μm in diameter was found in sample 10045-29 and analyzed (Table 1); the analysis indicated large amounts of elements other than Si and O. Identification of the phase by optical or x-ray diffraction methods was not possible; the composition suggests that it is not quartz, but probably tridymite.

Ilmenite analyses show the presence of Al_2O_3 (0.1 to 0.3 percent), MnO (0.5 to 0.6 percent), MgO (0.8 to 6.2 percent), CaO (<0.02 to 0.3 percent), Cr_2O_3 (0.5 to 1.3 percent), and V_2O_3 (up to 0.2 percent) besides the major oxides FeO and TiO_2.

Titanium dioxide occurs as small grains (<20 μm) in ilmenite and appears to be pure TiO_2; accurate quantitative analysis was prevented because of the small grain size.

"*Ferropseudobrookite*," $(Mg,Fe)Ti_2O_5$, is a new mineral which was discovered in samples 10059-27 and 10067-8. It occurs in small grains (<20 μm) in contact with and enclosed by ilmenite. The chemical composition (Table 1) is intermediate between that of the theoretical compounds $FeTi_2O_5$ and $MgTi_2O_5$. Insufficient amounts of the phase in the sections at hand have thus far prevented x-ray diffraction studies. A survey of other sections is presently being made in an attempt to locate a grain large enough for x-ray work.

Spinel A is nearly pure $MgAl_2O_4$ and occurs in a lithic fragment in sample 10019-22 that consists largely of plagioclase, plagioclase glass, olivine, and pyroxene (Table 1). An unusual spinel B containing large amounts of Cr_2O_3 and TiO_2 was found in sample 10059-28 (Table 1). It resembles a spinel from the meteorite Cachari which was probably produced by shock melting of ilmenite and chromite, followed by rapid cooling (*3*).

Ulvöspinel was found in two 30-μm grains in sample 10084-97 (Table 1).

Native iron is usually intergrown as blebs or droplets in troilite and is essentially pure Fe. One grain in sample 10045-29, however, has about 0.1 percent Ni.

Metallic nickel-iron commonly occurs

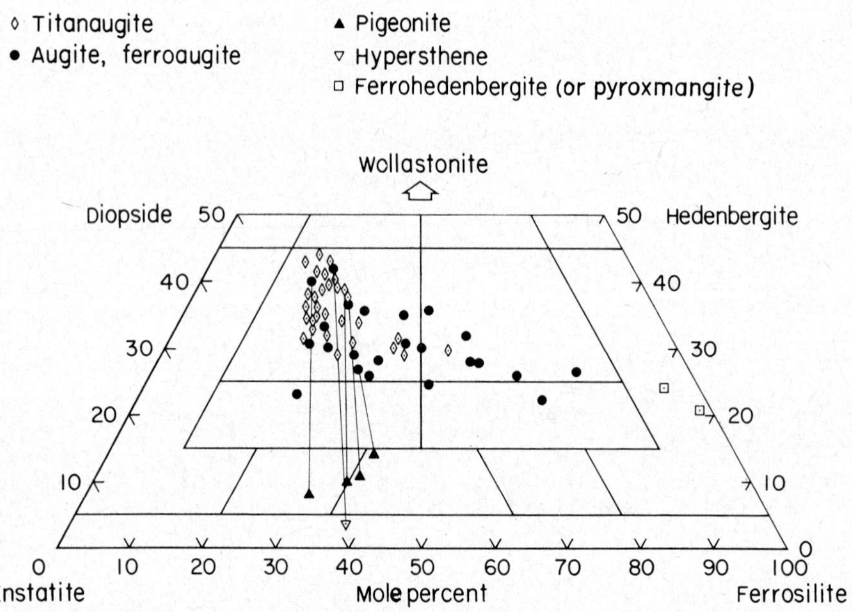

Fig. 1. Composition of 54 fairly homogeneous pyroxenes from breccias plotted in the ferrosilite-enstatite-wollastonite diagram (in mole percent).

as small (<20 μm) droplets in glass spherules and glass fragments or as small grains in the fine-grained breccia matrix. Nickel contents range from 0.5 to 19.8 percent; Co ranges from 0.35 to 0.94 percent.

Troilite occurs as an intergrowth with native iron or as a separate phase. Minor amounts of Mn (0.4 percent) and Cr (0.03 percent) are common.

Chlorapatite was identified with the electron microprobe; however, the grain size is too small (<10 μm) to allow accurate quantitative microprobe analysis.

Glass spherules and fragments are similar in composition for any given color (Table 2). The amounts of SiO_2, Al_2O_3, and CaO decrease as the color of the spherules darkens, whereas the amounts of Cr_2O_3, TiO_2, FeO, and MgO increase; other oxides show no definite trends.

Igneous glass occurs in minute patches <30 μm in diameter of silica-rich glass which range in SiO_2 from about 72 to 84 percent. A typical analysis is: SiO_2, 78.1 percent; TiO_2, 1.11 percent; Al_2O_3, 10.6 percent; FeO, 0.96 percent; MnO, 0.04 percent; MgO, 0.07 percent; CaO, 5.1 percent; Na_2O, 1.24 percent; K_2O, 3.4 percent; P_2O_5, 0.19 percent; Cr_2O_3, 0.05 percent; ZrO_2, 0.04 percent; total, 100.90 percent.

The following tentative conclusions may be drawn from the data presented herein: (i) There are no indications of chemical weathering or the presence of hydrous minerals. (ii) The presence of troilite, native iron, and the less oxidized end members ulvöspinel (Fe_2TiO_4), rather than magnetite (Fe_3O_4), in the magnetite-ulvöspinel series; ilmenite ($FeTiO_3$), rather than hematite (Fe_2O_3), in the hematite-ilmenite series; "ferropseudobrookite" ($FeTi_2O_5$), rather than pseudobrookite (Fe_2TiO_5), in the "ferropseudobrookite"-pseudobrookite series indicates that the oxygen fugacities of the lunar rocks are low relative to those of terrestrial rocks. (iii) Contributions (by weight) of meteoritic material to lunar surface rocks appear to be small. (iv) The texture of sample 10045-29 and many lithic fragments is typical of rapidly cooled rocks and is analogous to terrestrial volcanic or hypabyssal rocks. (v) The ilmenite content of all samples is higher ($\sim$ 15 to 20 percent by weight) than that of most terrestrial rocks and is similar in bulk composition to that of meteoritic ilmenite. (vi) Shock features indicate that at least some of the loose surface material and breccias are impact-produced. (vii) Metallic nickel-iron present in glass spherules and glass fragments is interpreted as being of meteoritic origin.

KLAUS KEIL, MARTIN PRINZ
Department of Geology and Institute of Meteoritics, University of New Mexico, Albuquerque 87106

T. E. BUNCH
Space Sciences Division, NASA Ames Research Center, Moffett Field, California 94035

References and Notes

1. K. Yagi and K. Onuma, *Hokkaido Univ. J. Fac. Sci.* 14, 463 (1967).
2. H. Kuno, *Amer. Mineral.* 40, 70 (1955).
3. T. E. Bunch and K. Keil, unpublished data.
4. We made electron microprobe analyses at the University of New Mexico and at the NASA Ames Research Center, using techniques described elsewhere [K. Keil, *Fortschr. Mineral.* 44, 4 (1967)]. We thank G. Conrad, J. Erlichman, R. Geitgey, F. Busche, and J. Rasho for assistance in the electron microprobe work, and R. Taylor and Mrs. R. Taylor, Mrs. J. Etheridge, and Miss M. Busch for assistance in the data reduction. Supported in part by NASA contract NAS9-9365.

4 January 1970; revised 6 January 1970

Petrographic, Mineralogic, and X-Ray Fluorescence Analysis of Lunar Igneous-Type Rocks and Spherules

Abstract. *Three lunar rocks show almost identical mineralogy but grain sizes that vary from basaltic to gabbroic. Clinopyroxene is zoned from augite to subcalcic ferroaugite compositions and is accompanied by decrease in Cr, Al, and Ti. Plagioclase is zoned from 93 to 78 percent anorthite. Olivine (68 percent forsterite) is present in one rock and apatite is rare. Cristobalite, ilmenite with Ti-rich lamellae, ulvöspinel (often Cr-rich), troilite, and kamacite are low in trace elements. Glassy spherules are of basaltic or feldspathic (92 percent anorthite) composition but contain abundant iron spheres of taenite composition (13 percent Ni). Four rock analyses by x-ray fluorescence show affinity with terrestrial basalts but with anomalous amounts of Ti, Na, Cr, Zr, Y, Rb, Nb, Ni, Cu, and Zn.*

Three polished thin-rock sections (17/50, 45/35, 58/23) were provided for nondestructive optical and electron microprobe analysis, and three 2-g samples of rock powders (44/39, 45/24, 60/25) for nondestructive x-ray fluorescence analysis of minor elements. A 5-g sample of 1-mm to 1-cm heterogeneous fragments (85/3) yielded different rock types, which were prepared for microprobe analysis, and 2 g of 1- to 3-mm size was powdered and analyzed for major and minor elements by x-ray fluorescence.

The three main rock types are very similar in mineralogy and have no terrestrial or meteoritic analogs. They are closest to terrestrial tholeiite basalt and meteoritic eucrite, but the high titanium and reduced state of each lunar rock make such comparisons superficial. The textures and mineral assemblage

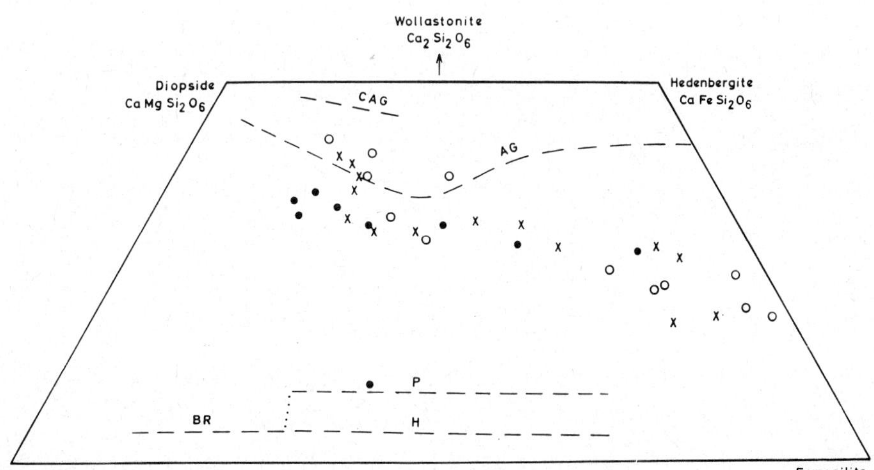

Fig. 1. Pyroxene compositions in relation to the trends (*2*) for terrestrial igneous pyroxenes (broken lines: CAG, calcic augite; AG, augite; BR, bronzite; H, hypersthene; P, pigeonite series). Symbols refer to the three igneous-type rocks (Table 1): closed circles, A; crosses, B; open circles, C. The pale yellow phase (crystallographically a pyroxenoid according to S. O. Agrell *et al.*, this issue) is present as the most iron-rich compositions in B and C. Only one pigeonite composition was obtained, but more measurements are needed.

favor crystallization from a silicate-rich melt, without crystal sorting, and from a lava mass near the surface. The grain size varies from basaltic, with pyroxene in excess of plagioclase nuclei (17/50), to diabasic (45/35) and gabbroic (58/23), with ophitic texture in which nuclei proportions are reversed. The coarser rocks could be blocks derived either from shallow subsurface plutons or from the centers of thick lava masses. Preferred flow or sedimentation orientation of grains was not observed. Range values are given because grain sizes and proportions vary too much for a dependable statistical analysis on a limited volume.

Clinopyroxene, plagioclase feldspar, ilmenite, cristobalite, troilite, ulvöspinel, and kamacite, in decreasing order of abundance, are ubiquitous. Olivine and chrome spinel are only in section 45/35, and apatite in section 58/23. For the three rocks we find clinopyroxene (47 to 59.4 percent, 0.1 to 2 mm); plagioclase (25.1 to 37.2 percent, 0.04 to 1 mm); ilmenite (10.5 to 14.5 percent, 0.1 to 3 mm); olivine (0 to 4.6 percent, 0.6 mm); cristobalite (0.2 to 5.1 percent, 0.1 to 1 mm); troilite (0.3 to 1.9 percent, maximum of 130 μm); kamacite (0.05 to 0.4 percent, maximum of 50 μm); trace of apatite (150 by 10 μm needles). Ulvöspinel (50 to 200 μm) and exsolution lamellae in ilmenite (maximum width 5 μm) were not counted before coating.

A Cambridge Instruments Company Geoscan Mk2 electron microprobe was used for all mineral composition analyses (Table 1). Standards of natural and synthetic minerals were used, and correction procedures followed Boyd (*1*). At a beam size of 2 μm (15 kv), grains below 5 μm gave poor resolution.

Pale brown clinopyroxene is strongly zoned from augite to subcalcic ferroaugite. Pale yellow rims and interstitial patches are chemically, but not necessarily crystallographically, subcalcic ferroaugite. The 35 plotted points (Fig. 1) show a distinctive band trend, and the full range is spanned by some crystals. The one-clinopyroxene trend may be indicative of much higher crystallization pressures than the cristobalite stability would suggest, although pyroxene metastability should be recognizable from x-ray diffraction, cryptoperthitic intergrowth patterns. Minor elements are fractionated, as between terrestrial Ca-rich and Ca-poor phases (*2, 3*). The amount of Ti is fairly low, and Al is comparable with the amount in low-pressure tholeiitic pyroxenes. Broad (?strain) lamellae, short 5-μm-thick lamellae of another translucent phase (too oblique for analysis), and hourglass zoning require further analysis.

Plagioclase is unusually calcic (93 percent anorthite) for a water-poor basalt (*2*) but the more sodic rims (to 78 percent anorthite) complement the pyroxene rims as indicators of crystallization from a cooling liquid. Olivine is very low in nickel. Although magnesian (65 to 70 percent forsterite), it is apparently compatible with cristobalite since only augite forms rare, probable reaction rims. The mosaic of fractures in cristobalite is attributed to inversion to the twinned low form already recorded (*4*). A similar shock mosaic in quartz (Lac Couture, Quebec) is discounted here in view of the sparseness of shock lamellae in the plagioclase. Al is the only important trace element, and an upper pressure limit of 5 kb is likely. Areas in the gabbroic rock suggest alteration associated with the cristobalite and low-temperature plagioclase and pyroxenic rims.

Ilmenite is far more abundant than in basaltic or achondritic analogs. Melanocratic Skaergaard ferrodiorites (*5*) are more ilmenitic but are distinctly cumulative and are rich in alkalis. The grains vary much in Mg, and exsolution lamellae are enriched in $FeTi_2O_5$ (7 percent) relative to the host. Ulvöspinel as discrete, often vermicular and cruciform crystallites is not intimately associated with ilmenite. Cr is clearly concentrated in the spinel phase, and one crystal contains 22 percent Cr_2O_3 (Cr : Ti = 1 : 1).

Troilite generally seems to be a late phase and appears either as irregular patches or spheres. Minor elements found in meteoritic troilites (*6*) are traces here. Kamacite (0.5 percent Ni) invariably occurs as spheres within the troilite. The low nickel is reflected in the rock analyses and contrasts with the dust and microbreccia nickel (Table 2). The iron : troilite ratio (approximately 1 : 6) is slightly higher than for the Fe-FeS eutectic at low pressure (*7*) but either this proportion is raised by Ni (*8*) or O (*9*), or there is 2-liquid equilibrium.

The tiny apatite needles were only confirmed qualitatively for P.

A pure plagioclase rock fragment (3 mm^3) contains about 3000 crystals and could not therefore be from a gabbroic specimen. It is strong evidence

Table 1. Mineral compositions (percentages by weight) by electron microprobe analysis. Minor phase data are given in text. Samples are (A) basalt type 17/12, (B) diabase type 45/35, (C) gabbro type 58/23, and (D) large-spherule rims from fines 85/3. The number of preferable quality analyses is given in parentheses in the sample column. Compositions were averaged over points and crystals. Three dots in a column indicate that an element was sought but, if present, was below 0.05 percent.

Sample	Si	Al	Fe	Mg	Ca	Na	Mn	Ti	S	Ni	Co	Cr
					Clinopyroxene cores							
A(5)	23.5	1.4	9.5	9.5	11.1	. . .	0.2	1.4				0.4
B(5)	22.6	2.4	9.6	8.6	12.2	. . .	0.2	1.9				0.5
C(6)	22.0	1.0	11.6	7.6	12.0	. . .	0.3	1.1				0.2
					Clinopyroxene rims							
A(3)	22.8	0.8	22.5	5.3	9.8	. . .	0.3	0.7				0.2
B(5)	20.7	0.9	26.9	2.5	6.8	. . .	0.3	0.7				0.1
C(5)	21.1	0.4	28.0	2.0	6.9	. . .	0.4	0.6				0.1
					Plagioclase cores							
A(4)	22.9	17.0	0.4		11.5	1.5		0.1				
B(5)	21.4	17.9	0.3		12.9	0.6		0.1				
C(3)	21.7	18.8	0.3		12.7	1.0		0.1				
D(5)	21.3	18.4			13.2	0.3		0.1				
					Olivine							
B(13)	18.2	. . .	21.0	19.8	0.2		0.3	0.1		. . .		0.1
					Cristobalite							
A(1)	45.8	0.8	0.1		0.2	0.1		0.3				
C(2)	45.6	0.4	0.1		0.1	0.1		0.2				
					Ilmenite							
A(2)	. . .	. . .	35.5	0.7	. . .		0.4	32.7				0.5
B(1)	0.1		35.3	0.7	0.1		0.4	33.4	. . .	. . .	. . .	0.4
					Chrome spinel							
B(1)	0.2		37.4	2.9	0.1		0.4	13.3	. . .	. . .	. . .	15.0
					Ulvöspinel							
C(1)	0.1	0.8	49.3	. . .	0.1		0.2	22.9	. . .	. . .	. . .	0.1
					Troilite							
A(1)	0.1		60.2	. . .	. . .		0.1	0.1	36.6	. . .		0.1
C(1)	. . .	0.1	59.2	. . .	. . .		0.1	. . .	39.2	. . .	. . .	. . .
					Iron							
A(1)	0.6	. . .	96.4	. . .	0.2		0.1	0.1	1.0	0.5	0.5	0.3
C(1)	0.1	. . .	98.5	. . .	. . .		. . .	. . .	. . .	. . .	0.8	0.1
D(1)	0.2		84.1	. . .	. . .		0.1	0.1	0.1	13.4	0.5	. . .

Table 2. Minor elements (in ppm) by x-ray fluorescence analysis of rock and fines samples. Samples: (A) dust-size fines, 85/3; (B) gabbro-type, 44/39; (C) diabase-type, 45/24; (D) microbreccia type, 60/25. Mineralogy and petrology of 44/39 by J. Zussman and of 60/25 by S. O. Agrell, and major element chemistry (including Cr) of 44/39, 45/24, and 60/25 by J. Scoon, this issue. The following elements were also determined but were close to or below the detection limit (5 ppm): U, Th, Bi, Pb, Sb, Sn, Cd, Ag, and Mo. Only sample 45/24 contained detectable Sn (15 ppm). Standard deviations for sample A are given in parentheses in column 2.

Element	Samples			
	A*	B†	C†	D†
Ni	117 (3.1)	4	4	162
Cu	16 (2.5)	4	6	6
Zn	19 (0.9)	3	3	30
Rb	3 (1.0)	1	1	5
Sr	159 (1.8)	206	145	164
Y	124 (1.6)	147	85	127
Zr	351 (3.1)	366	219	393
Nb	15 (1.1)	21	12	16
Ba	257 (15.7)	82	44	162

* Thirty-two determinations. † Two determinations.

for crystal fractionation near the lunar surface, with magnetic sinking or floating of 92 percent anorthite.

A polished slice (with a diameter of 2 mm) from across the largest spherule in the fines sample shows almost complete quenching to plagioclase laths (92 percent anorthite), the glass trace being of the same composition. Minor taenite (Table 1) and troilite rims are the only other phases; thus the dry liquidus temperature would have been about 1500°C. Metallic fragments (1 to 5 μm) concentrated in the hollow spherical core during sawing are rich in Cu, Ni, and Co, but their original location could not be determined. Seven 1-mm spherules are of vesicular glass, which varies from calc-feldspathic to Ti-rich basaltic and includes subangular ilmenite, and abundant small spheres (1 to 13 μm) of taenite (13 percent Ni) sometimes rimmed by a thin sheath of troilite. Absence of detectable exsolution suggests taenite freezing above 650°C. The high Ni content of the spherules' iron phase and the high iron: troilite ratio are in contrast to the phase-assemblage of the volcanic rocks and indicate meteoritic origins for part of each spherule. The calcic normative feldspar and high Ti in the rest of each spherule indicates melting of the lunar surface volcanic debris.

Table 3. Major oxides and norm of dust-size fines sample analyzed by x-ray fluorescence. Total iron was calculated as FeO except for the trace of Fe_2O_3 estimated from wet chemical data on comparable rocks by Scoon (this issue). The metallic Fe value is not known.

Oxide	* Mean (%)	S.D.	Mean norm	
			CIPW	%
SiO_2	42.13	0.123	Qz	1.7
Al_2O_3	13.64	0.071	Or	1.0
Fe_2O_3	0.06		Ab	4.6
FeO	15.29	0.063	An	34.7
MnO	0.21	0.002	Wo	8.9
MgO	7.38	0.076	En	18.6
CaO	11.32	0.041	Fs	14.1
Na_2O	0.54	0.065	Mt	0.1
K_2O	0.16	0.003	Ilm	14.7
TiO_2	7.69	0.038	Ap	0.3
P_2O_5	0.10	0.005	Tr	0.9
Cr_2O_3	0.33	0.001	Chr	0.5
S	0.31	0.004	FR†	0.67
Total	99.16		AR‡	0.86

* Thirty-two determinations. † FR, iron ratio (FeO/MgO + FeO). ‡ AR, anorthite ratio by weight (An/An + Ab + Or).

A Philips-1212 automatic x-ray fluorescence spectrometer was used for four rock analyses. Major and trace elements were obtained from a prepared briquette (*10*) for the fines sample (Tables 2 and 3) but the other three rocks were loaned to us by another investigator (Table 2) before his analysis for major elements; it was therefore necessary to present them to the x-ray beam as powders mounted on 6-μm Mylar film in sealed chambers. The trace element data (Table 2) could not, therefore, be based on so many determinations and could not be subjected to statistical analysis. For trace elements, calibration curves were drawn from special glasses (*14*) synthesized relative to preliminary data (*4*) for matrix and spiked accurately with the trace elements being sought. Interelement interference precluded analysis for Sc, V, and Co. For major elements, reliance was placed on a glass standard (*14*) with high Ti, in conjunction with international rock standards (*10*).

The fines analysis could represent a mean either of fragments of larger Apollo 11 rocks or of the regolith at Tranquillity Base. The regolith mean is favored because of the marked chemical correlation (Table 2) between fines and microbreccia and because of chemical distinction from the two volcanic rocks (see Ni, Zn, Rb, Ba). Nickel, in particular, illustrates the contribution of metallic meteoritic material to the fines (about 0.1 percent) and suggests that the microbreccia is of indurated regolith fragments. The volcanic rocks show low Ni, Zn, Cu, Rb, and high Y, Nb, Zr compared with most terrestrial basalts, but Zr is lower and Nb higher than reported earlier (*4*). Major and other minor elements (Table 3) of the heterogeneous sample illustrate features common to other Apollo 11 rocks in comparing with terrestrial basalts except for low SiO_2 and Na_2O and extremely high TiO_2, fairly high Cr_2O_3, and a combination of high AR and FR ratios (see Table 3, footnote).

The most remarkable aspect of the lunar rock chemistry is the high titanium content. Eucritic meteorites (*11*), like terrestrial basalts (*12*), do not have this feature. Most Apollo 11 rocks examined are believed to be volcanic, with textures and mineral zoning indicative of crystallization from a slowly cooled liquid. Troilite and kamacite crystallized at low oxygen fugacities which, assuming a quartz-fayalite-iron type buffer (*13*), would be about 10^{-13} bar at 1200°C and low total pressures. Low Ni (even by basaltic standards) and the pyroxene assemblage have no meteoritic analogs. The igneous rocks could be fractionated basalt equivalents if alkalis had been lost during the extensive vesiculation and if Ti had increased through crystal fractionation, but the pyroxene evidence does not support Mg-Fe fractionation between separated crystals and liquid. The contribution from meteoritic material is confined to the spherules, and then only to the nickeliferous iron parts.

G. MALCOLM BROWN
C. HENRY EMELEUS
J. GRENVILLE HOLLAND
ROY PHILLIPS

*Department of Geology,
University of Durham, England*

References and Notes

1. F. R. Boyd, *Carnegie Inst. Wash. Yr. Book* **67**, 327 (1968).
2. G. M. Brown, in *Basalts*, H. H. Hess and A. Poldervaart, Eds. (Interscience, New York, 1967), vol. 1, p. 103.
3. F. R. Boyd and G. M. Brown, *Carnegie Inst. Wash. Yr. Book* **67**, 83 (1969).
4. Lunar Sample Preliminary Examination Team, *Science* **165**, 1211 (1969).
5. L. R. Wager and G. M. Brown, *Layered Igneous Rocks* (Oliver and Boyd, Edinburgh, 1968), table 5.
6. B. Mason, *Geochim. Cosmochim. Acta* **30**, 365 (1966).
7. P. R. Brett and P. M. Bell, *Carnegie Inst. Wash. Yr. Book* **67**, 198 (1969).
8. G. Kullerud, *ibid.* **62**, 175 (1963).
9. A. J. Naldrett, *ibid.* **66**, 419 (1968).
10. J. G. Holland and D. W. Brindle, *Spectrochim. Acta* **22**, 2083 (1966).
11. J. A. Wood, in *The Solar System*, B. M. Middlehurst and G. P. Kuiper, Eds. (Univ. of Chicago Press, Chicago, 1963), vol. 4, p. 337.
12. V. Manson, in *Basalts*, H. H. Hess and A. Poldervaart, Eds. (Interscience, New York, 1967), vol. 1, p. 215.
13. H. P. Eugster and D. R. Wones, *J. Petrol.* **3**, 82 (1962).
14. We thank the Natural Environment Research Council (U.K.) for financial aid and the whole of the Geology Department, University of Durham, for support in many ways. Special technical assistance was provided by G. Randall (polished sections), G. Dresser (photographs), A. Carr (electronics), R. Lambert (chemistry), F. Frost and G. H. Gale (computer programs), and the Department of Mineralogy and Petrology, Cambridge (sectioning and distribution of rock sections). High-quality synthetic glasses were made as x-ray fluorescence standards, based on the preliminary Apollo 11 data, by the Pilkington Research Laboratory (Lathom, England).

4 January 1970

Lunar Anorthosites

Abstract. *Sixty-one of 1676 lunar rock fragments examined were found to be anorthosites, markedly different in composition, color, and specific gravity from mare basalts and soil breccias. Compositional similiarity to Tycho ejecta analyzed by Surveyor 7 suggests that the anorthosites are samples of highlands material, thrown to Tranquillity Base by cratering events. A lunar structural model is proposed in which a 25-kilometer anorthosite crust, produced by magmatic fractionation, floats on denser gabbro. Where early major impacts punched through the crust, basaltic lava welled up to equilibrium surface levels and solidified (maria). Mascons are discussed in this context.*

We prepared and studied thin sections of 1676 rock fragments (diameter range, 1 to 5 mm) from the Apollo 11 bulk sample (*1*). In almost all cases the rocks are fine-grained enough for fragments of this size to constitute representative samples. We found the following proportions of rock types, evidently a mixture from many sources: soil breccias, 52.4 percent; basalts, 37.4 percent; glasses, 5.1 percent; anorthositic rocks, 3.6 percent; others (including less than 0.1 percent of recognizable meteoritic debris), 1.5 percent.

Of these types, only the anorthosites were totally unexpected. These differ from the dark, titanium-rich basalts in appearance, composition, and density. They are light gray or white rocks with compact granular-to-loose ophitic textures composed of equant, xenomorphic-to-subidiomorphic crystals. The dominant mineral is calcic plagioclase. There are true anorthosites (plagioclase, > 90 percent), gabbroic anorthosites (plagioclase, > 77 percent, < 90 percent), and anorthositic gabbros (plagioclase, < 77 percent). Of these the gabbroic anorthosites are the most common. Chemically they are all distinguished by low Ti and high Ca and Al contents (Table 1). Their densities are between 2.8 and 2.9 g/cm^3 (*2*). Most of these particles have suffered shock metamorphism. The following descriptions are based on relatively unaltered specimens.

True anorthosites are scarce in the sample. They consist of tightly packed anorthite polyhedra with traces of olivine and opaque minerals. The anorthite crystals (An_{96-98}) are typically untwinned and homogeneous. These anorthosites occur as clots within a variety of gabbroic anorthosite (Fig. 1) which contains small olivine crystals (Fa_{49}) lying along the contacts between anorthite polyhedra.

Fig. 1. Gabbroic anorthosite (37-1) with clots of true anorthosite. Olivine crystals (high relief) outline anorthite polyhedra in the gabbroic areas. The rock is believed to be an igneous cumulate, but the fine-grained polygonal texture may have formed by recrystallization. Oblique nicols. Width of field, 750 μm.

The true anorthosites are probably adcumulates, a variety of igneous cumulate (*3*) from which the intercumulus liquid has been expelled by continued growth of the primary precipitate crystals. The associated gabbroic anorthosites are probably mesocumulates, cumulates in which some of the mafic intercumulus liquid crystallized as olivine between the growing anorthite crystals. This intercumulus olivine is distinctly richer in Fe than cumulus olivine, described below.

There are more complex examples of anorthosites and gabbroic anorthosites, but all reflect cumulative origins. Some contain plagioclase crystals which are both complexly twinned and zoned. Oscillatory zoning has been observed. Olivine and, rarely, clinopyroxene (pigeonite) join plagioclase as primary precipitates in some of the gabbroic anorthosites. Troilite, titaniferous chromite, ilmenite, and kamacite occur as accessories.

The anorthositic gabbros contain subidiomorphic plagioclase, olivine, and, rarely, clinopyroxene crystals in a mafic groundmass. They probably represent (or approach) orthocumulates, cumulates (*3*) in which the intercumulus liquid has crystallized between loosely packed, primary precipitate crystals. Thus their bulk compositions approach that of the contemporaneous magma. The intercumulus liquid in some anorthositic gabbros crystallized as large (1 mm) pigeonite crystals enveloping the primary olivine (Fa_{27-30}) and anorthite crystals. In terms of most elements, such gabbros trend toward the bulk compositions of lunar basalts, but because the ophitic pigeonite contains little Ti, a major discrepancy remains.

The anorthosites and anorthositic gabbros are products of fractional crystallization under plutonic, hypabyssal, or stagnant volcanic conditions. They are more calcic than corresponding terrestrial rocks—another manifestation of the low alkali content of the Apollo 11 sample. Of the two types of terrestrial anorthosites (*4*), the Adirondack type and the stratiform type, the latter, which occur in layered intrusions, are the more calcic and display textures of a more certainly cumulative origin. On compositional as well as textural grounds, therefore, the lunar anorthosites and anorthositic gabbros more closely resemble the stratiform-type terrestrial anorthosites. By terrestrial standards

Table 1. Bulk compositions of anorthositic particles, determined by averaging the results of seven to ten randomly placed, defocused-beam, electron microprobe analyses. a, Anorthositic gabbro (9–8); b, anorthosite-gabbroic anorthosite (37–1).

Oxide*	Percentage, by weight	
	a	b
SiO_2	46.0	45.4
TiO_2	0.3	tr
Al_2O_3	27.3	33.8
Cr_2O_3	0.2	tr
FeO	6.2	2.8
MnO	0.1	0.1
MgO	7.9	1.7
CaO	14.1	17.5
Na_2O	0.3	0.4
K_2O	tr	tr
NiO	tr	0.0
SO_3	0.1	0.0
Total	102.5	101.7

* Oxygen is calculated stoichiometrically, on the assumption that all Fe is divalent.

they are fine-grained rocks, but this fine grainedness may have been caused more by the absence of volatiles than by rapid cooling. There is also the likelihood that many of the fine-grained granular rocks have been recrystallized.

We doubt that there is a close genetic relationship between these rocks and the more common titanium-rich basalts. If they are consanguineous and related by local fractional crystallization, we would expect to see some continuity between compositions of the intercumulus material in the anorthositic rocks and the titanium-rich basalts. We do not find continuity; these two materials are distinctly different in Ti content. If the anorthosites and basalts are related by fractional crystallization, the crystallization was on a scale large enough to bury and hide all but these two of its products.

Cratering activity is constantly exchanging and mixing material on the lunar surface, so it is likely that not all of our rock fragments derive from the Sea of Tranquillity. Some component of the sample, however small, must have come from the lunar highlands, which begin only 42 km south of Tranquillity Base. Indeed, light-colored rays extending northward from the highlands can be seen on the mare surface near Tranquillity Base (*5*). The albedo and elevation of the highlands lead us to expect that highlands fragments, if present in our sample, would be lighter in color and lower in density (*6*) than mare basalts. Of the rock types we observed, anorthosites are the most obvious candidates. In Table 2 the bulk analysis of the anorthositic gabbro particle of Table 1 is recast in atomic percentages and compared (*7*) with the Surveyor 7 α-scattering analysis of ejecta from the highlands crater Tycho (*8*). The agreement is good enough to make a highlands origin for our anorthosites seem very likely.

O'Keefe (*6*) has argued that the lunar topography is approximately isostatically adjusted. Highlands terrain stands ~ 3 km higher, on average, than the maria surfaces (*9*). If materials of density ~ 2.9 g/cm^3 (anorthosite) and ~ 3.3 g/cm^3 (basalt) comprise the highlands and maria, respectively, these density differences must persist to a depth of ~ 25 km to compensate for the mass of highlands material that stands above the maria surfaces. Material beneath the 25-km level cannot be much denser than ~ 3.3 g/cm^3, in view of the overall density of the moon, and is unlikely to be less dense, because this would constitute a density reversal beneath the maria. The simplest structural model consistent with these requirements would be an anorthositic crust 25 km thick "floating" on gabbro of density 3.3 g/cm^3 (Fig. 2). Where major planetesimal impacts punched through this crust in early times, basaltic material (then still liquid, or possibly melted by the impacts) welled up to positions of hydrostatic equilibrium and solidified, forming the maria.

It is possible that lunar mascons (*10*), which constitute a minor but interesting deviation from lunar isostasy (*6*), can be understood as an artifact of this lava-filling process. Basic lavas contract on crystallization and increase in density by about 10 percent (*11*), so the equilibrium surface of a lava lake penetrating the lunar crust would stand about 2 km lower after crystallization than before (Fig. 2). The important point to note is that, after a lava-filled mare solidified and dropped to level II (a state of hydrostatic equilibrium), any additional lava that could still find its way to the surface would be capable of refilling the mare basin to level I. Level I would continue to be the free liquid surface in hydrostatic equilibrium with subcrustal magma, in spite of the (solid) equilibrium nature of level II.

Table 2. Comparison of the chemical composition of an Apollo 11 anorthositic gabbro fragment with the composition of Tycho ejecta (see *8*).

Element	Atomic percentage	
	Anorthosite	Tycho
C		<2
O	60.9	58 ± 5
Na	0.2	<3
Mg	4.1	4 ± 3
Al	11.3	9 ± 3
Si	16.2	18 ± 4
P, S, K	5.4	6 ± 2
Ca, Ti, Cr, Mn, Fe, Ni	2.0	2 ± 1

If a liquid zone underlay the solidified mare, addition of such an overburden would probably drive the whole system down to a new position of overall equilibrium. If the mare was supported beneath, however, extrusion of lavas from surrounding subcrustal zones onto the mare surface would constitute a genuine addition of mass to that area of the moon and would give rise to a positive gravity anomaly. Complete filling of the zone between level I and level II with lava of density 3.3 g/cm^3 (solidified), which is hydrostatically possible, would give rise to a gravity anomaly of 276 mgal (*12*). Since the observed gravity anomalies are little more than half this great (*10*), it

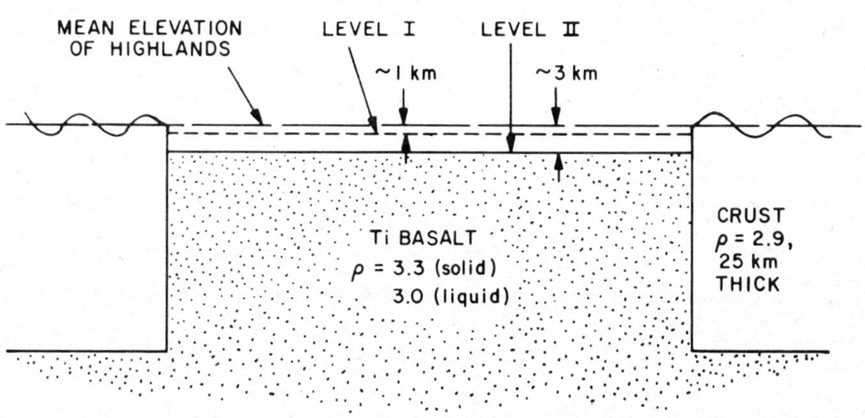

Fig. 2. Schematic representation of a lunar crustal section, showing equilibrium surface levels for mare basalt when molten (level I) and when solid (level II).

is not necessary to postulate 100 percent efficiency in lava overfilling or in the support of overfilled maria by the material underlying them.

Formation of an anorthositic layer on the moon seems to require magmatic fractionation. Anorthite crystals (ρ = 2.76 g/cm^3) will float in dense mafic magmas, similar to the mare basalts; in principle, a cumulate crust could have formed by flotation in the postulated lunar system. At best such a simple model can form only the framework of the moon's obviously complicated geologic history, however; observational details not explained by it include the small grain size of the anorthositic rocks, the low Ti content of even the most mafic anorthositic gabbros, and the higher Fe content of olivines in some of the anorthositic rocks (Fa_{49}) than of olivines in the basalts ($Fa_{\sim 30}$). Formation of a 25-km anorthositic layer would require igneous activity on a grand scale, possibly the early melting of a substantial fraction of the moon. The concept of a hot moon has been out of favor in recent years (*13*), but evidence discrediting it cannot be considered conclusive at this time.

JOHN A. WOOD
JOHN S. DICKEY, JR.
URSULA B. MARVIN
BENJAMIN N. POWELL
Smithsonian Astrophysical Observatory, Cambridge, Massachusetts 02138

References and Notes

1. Optical, x-ray diffraction, and microprobe analysis techniques were used. Detailed results are reported in Smithsonian Astrophys. Obs. Spec. Rep. 307.
2. Densities of three particles, determined by suspension in Clerici solutions of known density, were 2.88, 2.88, and 2.84 g/cm^3.
3. L. R. Wager, G. M. Brown, W. J. Wadsworth, *J. Petrol.* **1**, 73 (1960).
4. A. F. Buddington, *Geol. Soc. Amer. Mem. 7* (1939).
5. *Lunar Sample Information Catalogue—Apollo 11* (NASA Manned Space Center, Houston, 1969), p. 20.
6. J. A. O'Keefe, *Science* **162**, 1405 (1968).
7. Table 1 contains results of all the bulk analyses of anorthositic material we performed. Of fragments a and b of Table 1, the anorthositic gabbro fragment (a) is more nearly representative of the 61 anorthositic fragments studied.
8. R. A. Phinney, J. A. O'Keefe, J. B. Adams, D. E. Gault, G. P. Kuiper, H. Masursky, R. J. Collins, E. M. Shoemaker, *J. Geophys. Res.* **74**, 6053 (1969).
9. D. L. Lamar and J. McGann, *Icarus* **5**, 10 (1966).
10. P. M. Muller and W. L. Sjogren, *Science* **161**, 680 (1968); ———, unpublished work.
11. S. P. Clark, *Handbook of Physical Constants* (Geological Society of America, New York, 1966), p. 93.
12. The infinite slab approximation was used for this calculation: g_z (mgal) = 4.19 ρ (g/cm^3) L, where L is slab thickness, in kilometers.
13. H. C. Urey, *Science* **162**, 1408 (1968); also earlier papers dealing with the moon.
14. This work is supported by NASA grant NAS 9-8106. We are grateful to J. F. Bower and W. A. Frank for their assistance.

4 January 1970

Special Mineral Studies

Experimental Petrology of Lunar Material: the Nature of Mascons, Seas, and the Lunar Interior

Abstract. *One-atmosphere melting data show that Apollo 11 samples are near cotectic. Melting relations at pressures up to 35 kilobars show that clinopyroxenite or amphibole peridotite are possible lunar interiors. Mascons cannot be eclogite; they may be ilmenite accumulate. Hot lunar surface material will boil off alkalis.*

Sample 10017.51 was studied in open iron capsules (1 to 40 hour runs) at elevated pressure; 10084.16 in sealed $Ag_{70}Pd_{30}$ below 1200°C, sealed Pt (10 minute runs) above. At atmospheric pressure $Ag_{70}Pd_{30}$ was used at fugacity $fO_2 \sim 10^{-8}$; molybdenum at $fO_2 \sim 10^{-12.5}$; runs were 4 to 8 hours.

Figure 1 summarizes results obtained during the study of two specimens from Tranquillity Base.

1) Clinopyroxene is the liquidus phase over a wide range of pressures and is succeeded by garnet at high pressures in the dust. If these materials were to be regarded as dry primary magmas, clinopyroxenite is the most likely source rock and residuum at depths less than 700 km in the moon. Gabbro is an alternative source for the dust, but not the rock sample in the range 0 to 100 km. Eclogite is an alternative source at depths greater than 350 km.

2) Clinopyroxenite is acceptable as a model of the lunar interior in terms of density and absence of change to a denser phase at less than 50 kb. An eclogite source is not acceptable on account of its high density. See paragraph 3.

3) At temperatures > 1000°C metamorphism to rutile-ilmenite-quartz-eclogite occurs at ~ 300 km. This eclogite contains iron-rich garnet, ilmenite, and rutile and probably has a density greater than 3.7 g cm^{-3}. If the moon is hot, or ever has been, its uniform density of 3.34 g cm^{-3} excludes the Tranquillity Base materials as mean lunar composition.

4) Mascons are not due to eclogite formation in mare material on or near thermal gradients required to explain the formation of lavas. See paragraph 8.

5) The dust sample is rich in Al_2O_3, indicated by the high-temperature anorthite liquidus at low pressure and early garnet entry at high pressure compared with the rock sample. The TiO_2 content, 6.25 percent, is lower than in all rock specimens reported previously (*1*).

6) Preliminary quenching data on synthetic compositions (Tables 1 and 2) are summarized in Figs. 2 and 3. Liquidus temperatures are comparable to those of similarly iron-enriched earth basalts (*2*). The solidus/liquidus temperature interval is narrow (60° to 90°C). Olivine, two pyroxenes, anorthite, and two titaniferous oxides appear at the liquidus in the range covered by the 13 compositions.

7) These features characterize compositions lying close to and surrounding a cotectic liquid composition. Liquid of this composition will be generated by advanced crystal fractionation at low pressure or by initial fusion of suitable ilmenite gabbro.

8) When the quenching data are considered together with the topographic appearance of the lunar seas, they point strongly to the conclusion that the rocks collected by Apollo 11 came from the upper surface of a huge lava lake and that they probably represent liquid extensively modified by near-surface fractionation.

9) Accumulation of a titaniferous oxide at the base of lava lakes of this size could account for mascons. Figure 1 shows that, even in compositions with titaniferous oxide as liquidus phase at low pressure, fractionation of the dry liquids at high pressure is accomplished by the separation of phases poorer in TiO_2 than the liquid, thus enriching the liquid in TiO_2. High-pressure TiO_2 enrichment beyond the limit at which titaniferous oxide will become the liquidus phase at low pressure is therefore clearly possible. Those lava lakes with sufficient previous TiO_2 enrichment will possess high-density basal concentrations of ilmenite, ulvöspinel, or pseudobrookite, 300 m of which on a mare-wide scale may be sufficient to account for the mascons. A 10-km-deep lava lake with 10 percent TiO_2 in the initial liquid could provide such a basal mass concentration.

10) Rhythmic layering in a well-fractionated lava lake could produce a

Table 1. Compositions of the nine synthetic mixes made up with 1 mole anorthite, 1 mole augite, 2.5 moles hypersthene, ½, 1 or 1½ moles ilmenite, saturated or with 1 mole quartz or olivine. Interspersed are some analyses of lunar rocks (*1*).

Component	Mix 1	Mix 2	Sample 10050 B	Mix 3	Mix 4	Sample 10002 D	Mix 5	Sample 10017 B	Sample 10020 A	Mix 6	Mix 7	Mix 8	Sample 10021 + 10061 C	Sample 10022 A	Mix 9
SiO_2	43.5	40.5	38	37.9	44.9	43	41.3	40	38	38.3	48.5	44.8	41.5	43	41.7
TiO_2	3.8	7.2	9	10.1	4.6	5.6	8.5	10	12	11.7	4.3	7.9	9.3	11	11.1
Al_2O_3	9.8	9.2	11	8.5	11.7	13.5	10.8	11	11	10.0	11.0	10.2	11.5	7.7	9.4
FeO	19.8	21.7	20	23.4	16.6	13	19.0	19	18	21.1	15.5	17.9	17.5	21	20.0
MgO	12.2	11.4	10	10.6	9.3	10	8.5	8.5	8	7.9	8.7	8.0	8.2	6.5	7.4
CaO	10.8	10.1	10	9.4	12.9	11	11.9	10	10	11.0	12.1	11.2	11	9	10.4

structure having a multiplicity of density variations, with the possibility of unusual seismic transmission properties.

11) What source rocks could yield a primary magma whose low-pressure fractionation leads to the cotectic situation observed? Volatile-poor materials which simultaneously meet this requirement and the density requirements are restricted to two types: spinel- or garnet-poor magnesian peridotites and clinopyroxenites. Peridotite is probably unacceptable on the grounds of the minor element geochemistry, which is anomalous by comparison with terrestrial rocks thought to be derived from dry peridotites. The required magnesian peridotite is, in the absence of a core, irreconcilable with a chondrite model of mean lunar composition. A clinopyroxenite moon would go far toward explaining features of the minor element geochemistry.

12) Alternatively, if the moon contains combined water, substantial volumes of iron-rich amphiboles, perhaps with olivine and pyroxene, may be present. This material could accommodate a carbonaceous chondrite composition within the density requirements. Plausible amphibole compositions may perhaps be found among the lherzite layers of the Pyrenean peridotites (*3*). If it can be shown that such amphibole-bearing ultramafic rocks yield partial melts close to the amphibole component in composition, many problems of major and minor element composition in the Tranquillity Base specimens would be answered.

13) Experiments were carried out at 2 kb, 1050°C, with water pressure equal to total pressure, at the quartz-fayalite-magnetite buffer (*4*) on nine synthetic compositions (Table 1) and the dust sample. The primary phase volume of olivine, which barely intersects the three siliceous compositions at atmospheric pressure, expands with increasing water vapor pressure so that olivine is encountered during the crystallization of all samples.

14) The assemblage olivine-clinopyroxene-liquid (no orthopyroxene)

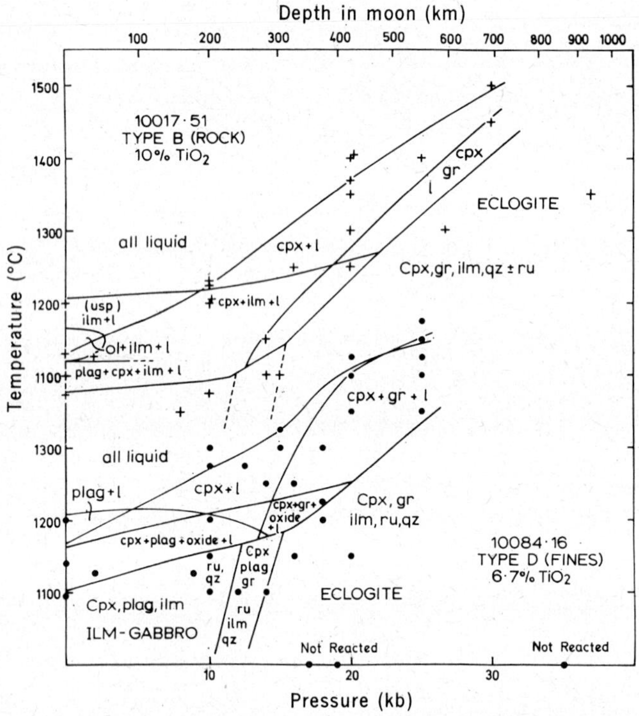

Fig. 1 (left). Interpretation of experiments on two lunar samples. Note displaced temperature scale. Abbreviations used in the figures: an, anorthite; cpx, clinopyroxene; suffix c, C lattice (augitic), suffix P, primitive lattice (pigeonitic); gr, pyralspite garnet; ilm, ilmenite; l, liquid; mgt, magnetite; ol, olivine; plag, calcic plagioclase; qz, quartz; ru, rutile; usp, ulvöspinel or other nonmagnetic opaque cubic spinel; psbr, pseudobrookite.

Fig. 2 (below left). Variation of liquidus phase with TiO_2 and SiO_2 at two oxygen activities in the nine compositions in Table 1, shown by heavy boundaries and large lettering. The second phase to crystallize is that found in the adjacent liquidus field except where labeled otherwise in smaller lettering. Broken lines delineate the fields of different second crystalline phase. The shaded boundary marks the boundary of saturation with a titaniferous oxide at the liquidus.

Fig. 3 (below right). Variation of liquidus phase with the compositions of Table 2 plus 2, 5 from Table 1, plotted as SiO_2 against Al_2O_3 (left hand figures) and SiO_2 against FeO + MgO (right hand figures), the upper pair being for the more oxidizing conditions.

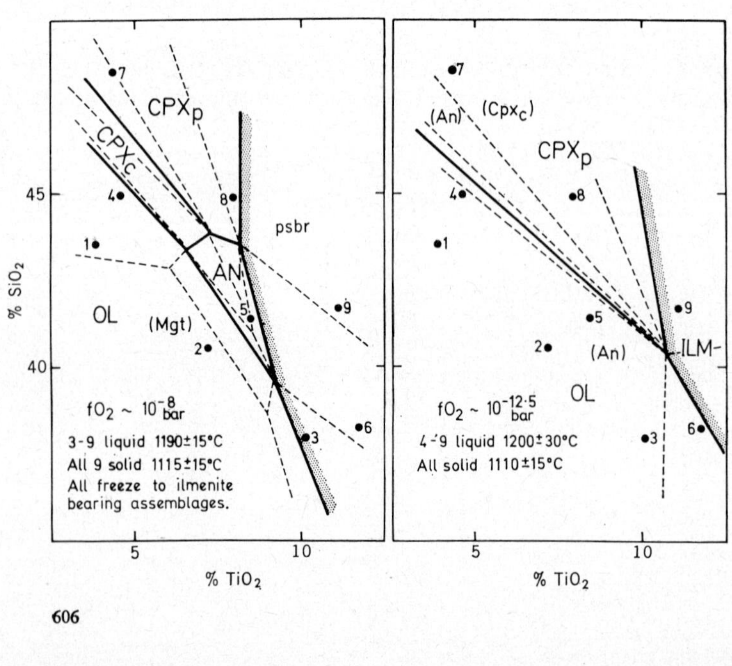

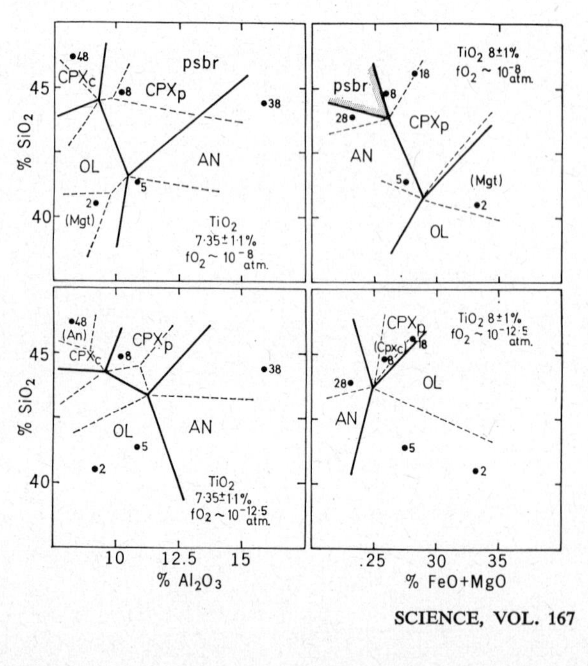

Table 2. Compositions of four further mixes made up by modifying mix 8 as indicated, 1 mole shift in each case.

Component	Mix 48	Mix 8	Sample 10037 D	Mix 38	Mix 18	Mix 28
SiO_2	46.2	44.8	43	44.5	45.6	43.8
TiO_2	6.5	7.9	7	6.2	7.1	9.0
Al_2O_3	8.3	10.2	13	15.9	9.1	11.4
FeO	17.1	17.9	16	14.0	18.8	16.7
MgO	8.4	8.0	8	6.3	9.2	6.5
CaO	13.6	11.2	12	13.1	10.0	12.6
	8 + Diopside			8 + Anorthite	8 + Hypersthene	8 − Hypersthene

was observed in the wet melting experiments, accompanied by a deep brown amphibole in some charges, including the dust. The appearance of amphibole supports the hypothesis set out in paragraph 12. Assuming that amphibole is not the liquidus phase, fractionation must lead to enrichment of TiO_2 in residual liquids. Such liquids will, of course, be saturated with a titaniferous oxide phase on loss of volatiles and reduction in pressure. See paragraph 9.

15) An alkali-enriched synthetic composition (*5*) held near its liquidus in a vacuum furnace at 1200°C, 10^{-5} torr, for 14 days lost 3.4 percent in weight. Analysis after the experiment (*5*) showed a reduction of total alkalis from 3.6 to 2.2 percent.

Volcanic rocks on the lunar surface may have cooled for some days at temperatures above the solidus under vacuum conditions, with loss of the alkalis and other volatile oxides. Therefore the material collected by Apollo 11 may be the residue of fractional sublimation superimposed upon the effects of fractional crystallization. All geochemical conclusions based upon the present composition must take into account the much evolved character of the lunar surface material.

MICHAEL J. O'HARA
GORDON M. BIGGAR
STEPHEN W. RICHARDSON
Grant Institute of Geology,
West Mains Road, Edinburgh
EH9 3JW, Scotland

References and Notes

1. Lunar Sample Preliminary Examination Team, *Science* **165**, 1211 (1969).
2. C. E. Tilley, H. S. Yoder, Jr., J. F. Schairer, *Carnegie Inst. Wash. Year B.* **65**, 260 (1966).
3. A. Lacroix, *Compt. Rend.* **165**, 381 (1917).
4. D. R. Wones and M. C. Gilbert, *Amer. J. Sci.* **267-A**, 480 (1969).
5. Initial composition as percent by weight of sample, with result after 14 days at 1200°C, 10^{-8} atm, in parentheses. SiO_2 50.13 (51.53), Al_2O_3 13.99 (14.30), TiO_2 5.48 (5.36), FeO 12.34 (n.d.), MgO 5.52 (5.74), CaO 8.97 (9.03), Na_2O 1.42 (0.47), K_2O 2.15 (1.70). We are indebted to M. J. Saunders for these analyses.
6. C. E. Ford and B. G. Jamieson carried out the experiments at 2 kb, and the latter was responsible for many calculations in connection with the synthetic compositions. We are indebted to Nuclear Enterprises Ltd., who provided a new x-ray fluorescence spectrometer for analysis of iron loss and molybdenum gain on 10-mg experimental samples.

4 January 1970

Petrogenesis of Lunar Basalts and the Internal Constitution and Origin of the Moon

Abstract. *Petrographic and electron-microprobe studies combined with high pressure-temperature investigations of phase relationships in average Apollo 11 basalt and possible source material show that the lower parts of maria may be composed of eclogite (density 3.74 grams per cubic centimeter), thus explaining the existence of mascons. The Apollo 11 basalt was probably formed at depths of 200 to 400 kilometers by a small degree of partial melting from pyroxenitic source material [FeO/(FeO + MgO) = 0.25, Al_2O_3 ~ 4 percent, CaO ~ 3 percent]. This composition may be representative of the lunar interior and yields the observed mean lunar density and moment of inertia. Present data are in conflict with fission, binary planet, and capture hypotheses of lunar origin but are consistent with Ringwood's (1966) precipitation hypothesis.*

Over 400 complete electron-probe analyses and another 400 partial analyses of minerals and glasses from Apollo 11 rocks and dust have been made. The data show exceptionally strong crystal-liquid fractionation within small volumes of single rocks, but little fractionation between different rocks, suggesting a rather direct derivation of lunar basalts from their source regions without extensive fractionation en route. Apollo 11 basalts are enriched in the "incompatible" trace elements U, Th, Ba, rare earths, Y, Zr, and Ti by factors of about 50 over their chondritic abundances (*1*). This suggests that the basalts may have been derived by a small degree (~ 3 percent) of partial melting of source material within the lunar interior (*2*). We have explored this hypothesis further by an experimental investigation of phase equilibria in average Apollo 11 basalt over the ranges 0 to 30 kb and 1000° to 1600°C. Corresponding experiments have also been carried out upon a possible source material of the Apollo 11 basalt.

The starting material was a synthetic glass possessing the average composition of Apollo 11 basalts (Fig. 1). The experimental methods used are as described elsewhere (*3*) except that charges were contained in iron capsules. After the charges were held long enough to reach equilibrium at the desired pressure and temperature, they were quenched and examined by optical, x-ray, and electron-probe methods. Results are shown in Figs. 1–3. Subsolidus results are discussed first.

The pressure required to transform the low pressure "gabbroic" assemblage (clinopyroxene + feldspar + ilmenite) to an eclogitic assemblage (titaniferous garnet + clinopyroxene + rutile) is suprisingly small—12 kb at 1100°C. This is due to the low sodium and silicon and high iron content of Apollo 11 basalt (*3*). The transition interval between gabbro and eclogite fields is also narrow. Our experiments, combined with other extensive results (*3*), indicate a gradient of 20 bars/°C for the transformation. This gradient implies that at zero pressure Apollo 11 basalt would pass into the eclogite stability field at 500°C, a temperature at which silicate reaction rates are still substantial. This has a bearing on the origin of lunar mascons. It is possible that the deeper regions of maria cooled sufficiently slowly to transform to eclogite. However, the near-surface flows cooled rapidly (as is shown by lack of fractionation) and were metastably preserved with basaltic mineral assemblages. If the deeper regions of some maria are composed of eclogite, with a calculated density of 3.74 g/cm³, a simple explanation of the origin of lunar mascons is provided.

The probable range of internal lunar temperature distributions given by

Table 1. Comparison of source regions of terrestrial (oceanic) basalts and Apollo 11 basalt (based on this paper and on *1, 2, 3, 5, 7, 10,* and *12*).

Parameter	Earth	Moon
Chemical composition	Dominantly Mg-Fe silicates Less than 5 percent each Al_2O_3 and CaO	
Mineralogy	Dominantly olivine	Dominantly pyroxenes (Mg-rich)
FeO/(FeO+MgO) (mole %)	12	25
Oxygen fungacity at 1200°C	10^{-8} to 10^{-9}atm	$10^{-13.5}$atm
Siderophile elements (Ni, Cu, Ga)	Apparently depleted in moon as compared to earth	
Fractionation of volatile elements, Rb, K, Na, Zn, Pb*		
Rb/Ba	0.10	0.03
K/Ba	93	11
K/U	10,000	3,000
Na/Al	0.22	0.06
Zn/Mn	.07	.01
Pb/Ba	.06	.02

* To offset the effects of crystal-liquid fractionation upon absolute abundances, comparison is made on the basis of abundance ratios in which the abundance of each volatile element is compared to the abundance of a relatively nonvolatile element which behaves similarly during crystal-liquid fractionation processes in relevant silicate systems (*10*).

several authors (*4*) is shown in Fig. 2. Note that they lie in the eclogite field throughout. The Apollo 11 basalt therefore cannot represent the mean lunar composition, for it is far too dense. It follows that Apollo 11 basalt is a differentiate from less dense, more primitive material. This conclusion is supported by an interpretation of trace-element abundances (*1, 2*).

Crystal-liquid equilibria (Fig. 1) show three distinct pressure regimes. In the low pressure regime (0 to 3kb, 0 to 60 km depth), liquidus phases are olivine (Fo_{74}) and karooite [$(Fe_{0.5}Mg_{0.5})Ti_2O_5$]. Apollo 11 basalt cannot have formed by partial melting from a mineral assemblage consisting dominantly of these silica-deficient minerals, for they are too dense ($\rho \geqslant 3.5$ g/cm³) to be representative of the lunar interior ($\rho = 3.34$ g/cm³). There is also other evidence related to the strength of the lunar crust and the depths, mass distributions, and thermal regimes of the maria which shows decisively that Apollo 11 lavas must have been derived from much deeper within the moon (*5*).

In the high pressure regime ($\geqslant$25 kb, $\geqslant$550 km) near-liquidus phases are garnet and clinopyroxene. In principle, therefore, Apollo 11 magma could have formed by a few percent of partial melting of a compositionally similar garnet-clinopyroxene assemblage below 550 km. However, on the basis of the phase boundaries shown in Fig. 2, this assemblage is too dense (>3.5 g/cm³) to be representative of the lunar interior. If the experimental data on the gradient of the gabbro-eclogite transition are ignored, a lunar model consisting of a 350-km shell of gabbro [Fe/(Fe+Mg) = 0.25, ρ = 3.0 g/cm³] passing downward into eclogite (ρ = 3.55) might be considered. This model could provide a correct mean lunar density; however, it would be in sharp conflict with the moon's moment of inertia, yielding a value of 0.385 for I/MR^2 compared with the observed value of 0.402 ± .002 (*6*). Detailed analysis along these lines demonstrates conclusively that the moon's mean chemical composition cannot remotely resemble that of basalt (*5*).

Consider now the formation of magmas in the intermediate pressure regime (10 to 20 kb, 200 to 400 km; Fig. 1). Here the liquidus phase is typically a subcalcic clinopyroxene (average composition Fs, 21 percent; En, 64 percent; Wo, 15 percent) containing ~ 4 percent Al_2O_3 and ~ 7 percent CaO. Occasionally, orthopyroxene is found (Fs, 20 percent; En, 74 percent; Wo, 6 percent). Further experiments, in which 10 percent $MgSiO_3$ was added to synthetic lunar basalt, followed by refusion, showed that orthopyroxene became the liquidus phase in this pressure interval. Clearly, the Apollo 11 basalt is saturated or nearly saturated with orthopyroxene between 10 and 20 kb. Accordingly, this basalt could have formed by a small degree of partial melting of an assemblage consisting of orthopyroxene + subcalcic clinopyroxene at depths of 200 to 400 km in the lunar interior. The experiments do not establish the relative proportions of orthopyroxene and subcalcic clinopyroxene in the source region. An interpretation of Ca/Al partitions (*5*) together with evidence of a relatively unfractionated rare earth abundance pattern (*1*) strongly suggests that orthopyroxene predominates over clinopyroxene. Additional evidence (*5*) indicates that some olivine could be

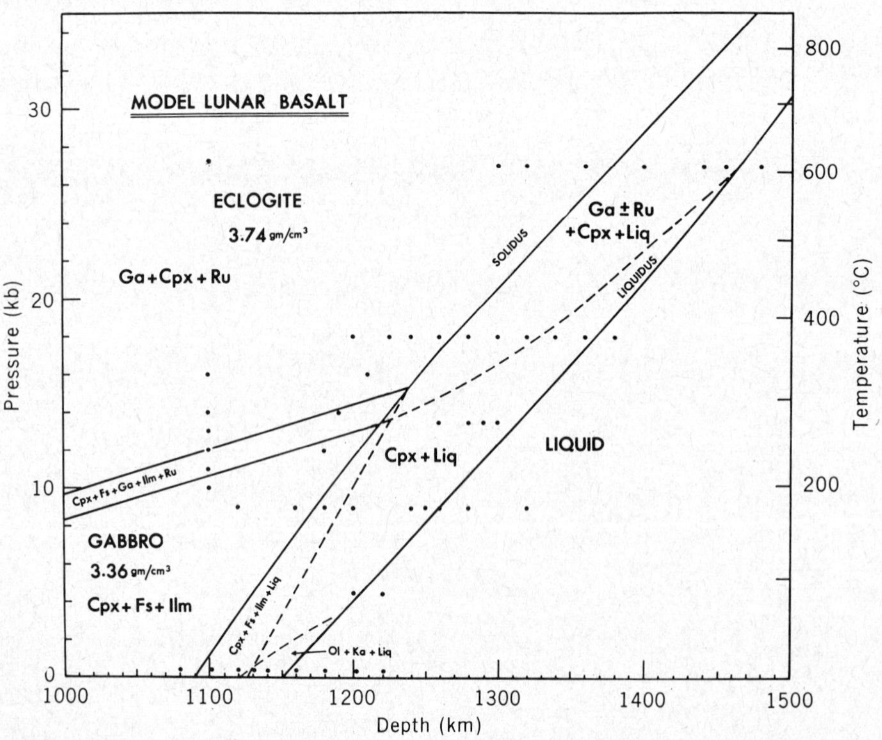

Fig. 1. Stability fields of mineral assemblages and melting equilibria in average Apollo 11 basalt at high pressures and temperatures. Each dot represents a separate experiment. Composition of basalt was as follows (percent by weight): SiO_2, 41.2; TiO_2, 10.2; Al_2O_3, 10.5; Cr_2O_3, 0.56; FeO, 18.0; MgO, 7.9; MnO, 0.34; CaO, 10.5; Na_2O, 0.51; K_2O, 0.15; ZrO_2, 0.14. Key: Cpx, clinopyroxene; Fs, feldspar; Ga, garnet; Ilm, ilmenite; Ru, rutile; Ka, Karooite; Ol, Olivine; Liq, liquid.

present in the source regions, although this is not required. Quartz can be excluded as a component of the source region (*5*).

Next we ask whether a dominantly pyroxenite source region composed of orthopyroxenes and subcalcic clinopyroxenes as found on or near the liquidus of the Apollo 11 basalt between 10 and 20 kb can explain the moon's mean density and moment of inertia. Mineral stability fields and densities, based on the results of 42 high pressure and temperature experiments for such a model lunar pyroxenite are shown in Fig. 3, together with the probable maximum range of lunar internal temperature distributions. Calculations show that for reasonable temperature distributions this model provides a mean density and value of I/MR^2 very similar to those of the moon (*5*). However, agreement is not perfect, and small corrections to the model composition are needed. To match the density, FeO must be reduced by about 1 percent and Al_2O_3 by about 1 or 2 percent. The Al_2O_3 reduction has the effect of increasing the pressures required for the stability of the garnet pyroxenite assemblage, and is also necessary to achieve agreement between calculated and observed moments of inertia.

Our petrologic investigation (*7*) has revealed many close similarities in mineralogy, petrochemistry, and oxidation states between Apollo 11 basalt and basaltic achondrites, previously suggested to be of lunar origin (*8*). The present experimental investigation has shown that Apollo 11 basalt could have formed by a small degree (~ 3 percent) of partial melting of a pyroxene-rich assemblage at depths of 200 to 400 km.

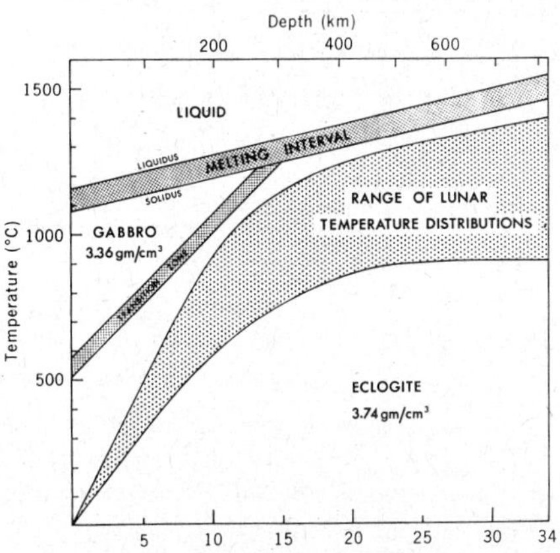

Fig. 2. Gabbro, eclogite, and liquid fields for average Apollo 11 basalt (composition given in Fig. 1) in relation to probable range of internal temperature distributions in the moon.

It is possible that basaltic achondrites and howardites have formed by a larger degree (10 to 30 percent) of partial melting from similar source material. A small degree of partial melting would cause large enrichments of incompatible elements (U, Th, Ba, Zr, Y, rare earths) in the magma, whereas with a large degree of melting (achondrites) the concentration of incompatible elements would be much smaller.

Comparisons between source regions of Apollo 11 basalt and terrestrial basalt are summarized in Table 1. Principal similarities are that both source regions are composed dominantly of Mg-Fe silicates with lesser Ca and Al and that both are characterized by strong depletions of relatively volatile metals, among them Na, K, Rb, Pb, Zn, when compared to the primordial abundances. Superimposed on these overall similarities are some major differences in mineralogy, chemical composition, the FeO/(FeO+MgO) ratio, and oxidation state. The Apollo 11 basalt source region has been depleted in Na, K, Rb, Pb, and Zn compared to the terrestrial basalt source region by factors of 3 to 10. Apollo 11 basalts on the average also display marked depletion of the siderophile elements Ni, Cu, and Ga compared to terrestrial basalt (*1*).

These differences between lunar and terrestrial basalts and between their source regions preclude the moon's having once been fissioned directly from the earth's mantle, for fission requires that lunar basalts be essentially similar to terrestrial ones. Most versions of the binary planet hypothesis require that the silicates in the earth's mantle and moon are similar and invoke physical fractionation processes during primary accretion to account for the excess iron now present in the earth's core. This hypothesis implies that the basalts formed by partial melting from the identical silicates in the moon and in the earth's mantle would display similar composition ranges: the compositional differences between lunar and terrestrial basalts are inconsistent with this hypothesis. The capture hypotheses are also unsatisfactory. They do not explain the low density (*9, 10*) of the moon or the differences in oxidation states and compositions of lunar and terrestrial basalts and their source regions.

The traditional hypotheses of lunar origin are clearly untenable, and alternative hypotheses must be seriously examined. We call attention to the hypothesis of Öpik (*11*), based on study of lunar craters and tidal evolution, that the moon was formed by the coagulation of a "sediment-ring" of planetesimals, somewhat analogous to the rings of Saturn only much more massive. A further hypothesis developed by Ringwood (*10*) offers an explanation of the origin and composition of the sediment ring. This maintains that in the later stages of accretion of the earth a massive primitive atmosphere was developed and that the temperature was sufficient to evaporate selectively a substantial proportion of the silicates falling upon the earth. The primitive atmosphere was subsequently dissipated as the sun passed through a T-Tauri phase, and, on cooling, the silicates were precipitated to form an assemblage of planetesimals orbiting the earth. This hypothesis provides a framework for interpreting and explaining the low density of the moon and the other inferred compositional differences between earth and moon. Some of these—fractionation of volatile elements, difference in redox state, the pyroxene-rich nature of lunar interior—

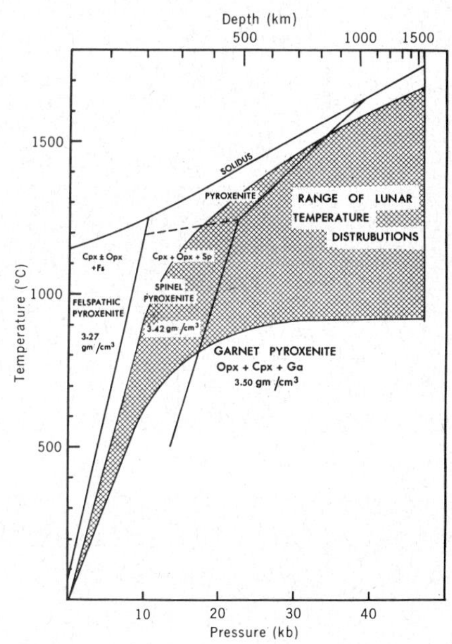

Fig. 3. Diagram showing stability fields of mineral assemblages displayed by a pyroxenite in relation to the probable range of lunar internal temperature distributions. The pyroxenite represents a possible source material for Apollo 11 basalt. The increase in density with increasing pressure across the boundary of pyroxenite (± spinel) and garnet pyroxenite fields is gradual, and the maximum theoretical density of 3.50 g/cm³ for the garnet pyroxenite requires pressures exceeding 50 kb. This is a preliminary diagram based upon the results of 42 experimental runs upon the model pyroxenite combined with data from a closely related system (*13*). Composition of pyroxenite was as follows (percent by weight): SiO_2, 53.1; TiO_2, 1.0; Al_2O_3, 5.0; Cr_2O_3, 0.4; FeO, 13.5; MgO, 22.5; MnO, 0.4; CaO, 4.0; Na_2O, 0.1.

were indeed specifically predicted by Ringwood in 1966 (*10*). The "precipitation" hypothesis is a close relative of the fission hypothesis, since it maintains that the material now in the moon was ultimately derived from the earth, not from the mantle but from the earth's massive primitive atmosphere.

A. E. RINGWOOD, E. ESSENE

Department of Geophysics and Geochemistry, Australian National University, Canberra

References and Notes

1. Lunar Sample Preliminary Examination Team, *Science* **165**, 1211 (1969); B. Chappell, personal communication.
2. A. E. Ringwood, in preparation.
3. D. H. Green and A. E. Ringwood, *Geochim. Cosmochim. Acta* **31**, 767 (1967); *Contrib. Mineral. Petrol.* **15**, 103 (1967).
4. G. J. F. MacDonald, *J. Geophys. Res.* **64**, 1967 (1959); B. J. Levin, *The Moon* (Academic Press, New York, 1960), p. 157; P. E. Fricker, R. T. Reynolds, A. L. Summers, *J. Geophys. Res.* **72**, 2649 (1967).
5. A. E. Ringwood and E. Essene, in preparation.
6. H. C. Urey and G. J. F. MacDonald, in *Physics and Astronomy of the Moon*, Z. Kopal, Ed., ed. 2, in press.
7. E. Essene and A. E. Ringwood, in preparation.
8. M. Duke and L. Silver, *Geochim. Cosmochim. Acta* **31**, 1637 (1967).
9. H. C. Urey [*Physics and Astronomy of the Moon*, Z. Kopal, Ed. (Academic Press, New York, ed. 1, 1962), p. 481] has attempted to reconcile the low density of the moon with a capture origin. However, his theory is based upon the proposition that the moon is a "primitive object" with the same iron/silicon ratio as the solar photosphere. Recent redeterminations of *f* values for iron [T. Garz, M. Kock, J. Richter, B. Baschek, H. Holweger, A. Unsöld, *Nature* **223**, 1254 (1969)] have shown that previous estimates of the iron abundance in the photosphere were much too low, and the iron/silicon ratio is now believed to be similar to that occurring in chrondritic meteorites, in the earth, and, by inference (*10*), in Mars and Venus. Urey's hypothesis is therefore no longer tenable.
10. A. E. Ringwood, *Geochim. Cosmochim. Acta* **30**, 31 (1966).
11. E. J. Öpik, *Astron. J.* **66**, 60 (1961).
12. A. E. Engel, C. G. Engel, R. G. Havens, *Bull. Geol. Soc. Am.* **76**, 719 (1965).
13. D. H. Green and A. E. Ringwood, *Earth Planet. Sci. Lett.* **3**, 151 (1967).

14. We thank A. Major, W. Hibberson, and N. Ware for assistance in the experimental work.

4 January 1970

Crystallization of Some Lunar Mafic Magmas and Generation of Rhyolitic Liquid

Abstract. *During crystallization of lunar crystalline rocks 10022 and 10024, augite changes in composition almost continuously from titanaugite (~$Ca_{36}Mg_{47}Fe_{17}$ with TiO_2 ~3 percent) to a very iron-rich variety ($Ca_9Mg_5Fe_{86}$), pigeonite changes from $Ca_9Mg_{66}Fe_{25}$ to $Ca_{10}Mg_{51}Fe_{39}$, and olivine changes (in 10022) from $Mg_{71}Fe_{29}$ to $Mg_{41}Fe_{59}$, whereas plagioclase stays as bytownite. These compositional variations, as well as the textural relations, may be explained by rapid crystallization of undercooled magmas. The residual liquids found as mesostasis are rhyolitic, which suggests that fractional crystallization of some lunar mafic magmas can generate rhyolitic magmas. Melting experiments were made on crystalline rocks to determine liquidus temperatures and crystallizing phases.*

Crystalline rocks 10017,20 and 10024, 23 [type B (*1*)] and 10022,41 and 10072,33 (type A) were examined with microscope and microprobe analyzer. This paper reports mainly the analytical results for 10022 and 10024. Major minerals of 10022 are titanaugite-subcalcic ferroaugite, pigeonite, ilmenite, plagioclase, olivine, troilite, native iron, and apatite. Those of 10024 are the same as those of 10022 except that olivine is absent and there are relatively large cristobalite crystals. All the rocks have many interstices filled with mesostasis, that is, fine-grained aggregates of potassium-rich phase and plagioclase with or without ilmenite, iron-rich clinopyroxene, nearly pure SiO_2 phase, troilite, native iron, apatite, and a Zr- and Ti-rich mineral (possibly $ZrTiO_4$). The minerals of 10017 and 10072 are similar to those of 10024 and 10022, respectively. The terms dolerite and microgabbro (or dolerite pegmatite) would be appropriate for 10022 and 10024, respectively. However, peculiar textures such as the radial and fine intergrowth of augite and plagioclase and of augite and pigeonite are unlike those in the terrestrial dolerites and microgabbros.

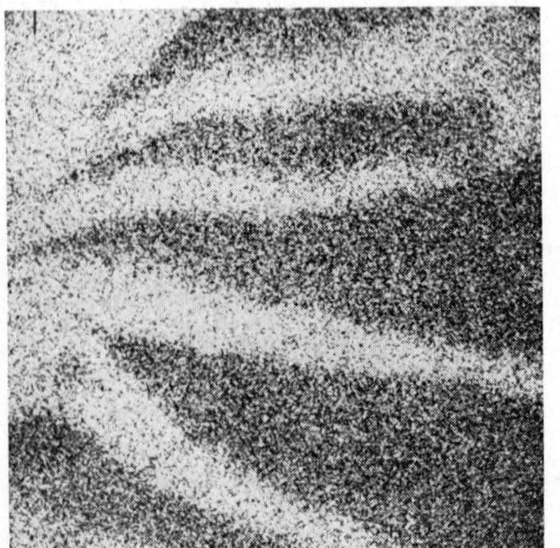

Fig. 1. Photomicrograph of the Ca Kα emission of one type of intergrowth of augite (light areas) and pigeonite (dark areas) in 10024. One edge of the photograph corresponds to 80 μm.

In the pyroxenes augite-subcalcic ferroaugite predominates over pigeonite in amount. No orthopyroxene was found. Pigeonite is closely associated with augite and often occurs as rounded or irregularly elongated inclusions in augite. One type of the intergrowth of both the clinopyroxenes is shown in Fig. 1. Similar intergrowth of needle-like augite and plagioclase is often observed. These intergrowths are indicative of rapid growth of clinopyroxenes and plagioclase. Some of the very fine intergrowth of augite and pigeonite may be a result of the rapid exsolution of metastable subcalcic augite.

More than 50 complete analyses of clinopyroxenes have been made with the electron-probe microanalyzer (*2*). The greater parts of the augite-subcalcic ferroaugite crystals are titaniferous augite with TiO_2 $2 \sim 3$ percent by weight; however, marginal parts of many crystals abutting on mesostasis are very rich in Fe and poor in Ca, Mg, and Ti. The plot of clinopyroxene analyses in the Ca-Mg-Fe diagram (Fig. 2) shows that the compositional variations are very large, with a trend toward ferrosilite. The most iron-rich one, which is yellowish in thin section, is close to ferrosilite, its composition being outside the compositional areas of subcalcic ferroaugite and ferropigeonite so far given (*3*). The large part of the compositional variation of clinopyroxene is observable in a single crystal, as shown in Fig. 2. Continuous scanning from the center to the margin of strongly zoned crystals with the Kα emission of Ca, Mg, Fe, Ti, and Al has shown that the zoning is essentially continuous, although steep changes in Ca and Fe are observed within the distance of 10 μm in the marginal zones of some crystals. The compositional range of pigeonite is not large compared with those of augite-subcalcic ferroaugite. Some of the tie lines between coexisting augite and pigeonite are drawn in Fig. 2. The partition coefficient K_D of Mg-Fe (*4*) between the augite and pigeonite is $0.90 \sim 1.01$, with an average of 0.96, indicating the relatively high temperature of equilibration. The compositional

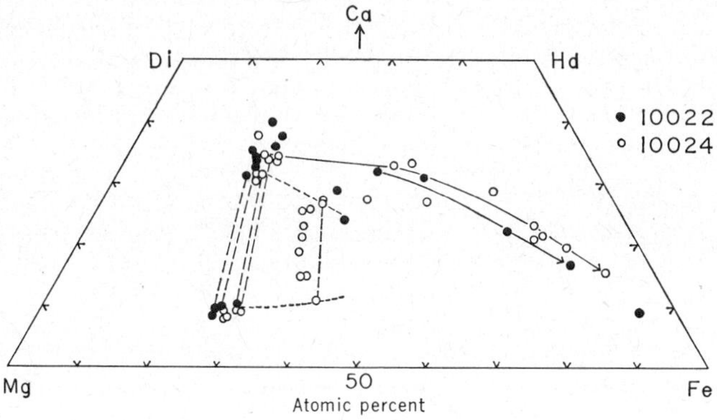

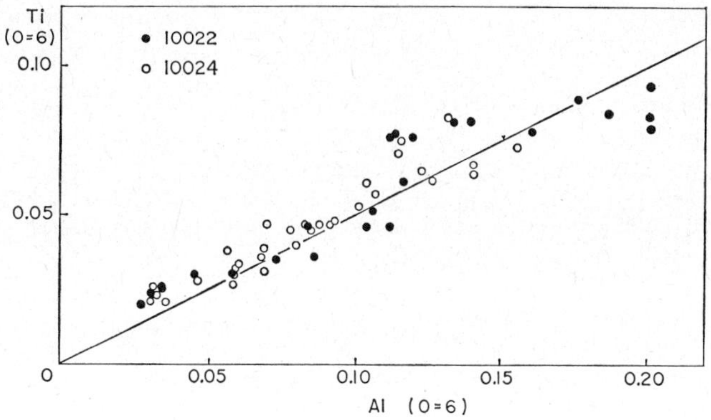

Fig. 2 (left). Ca-Mg-Fe plot of the analyzed clinopyroxenes in 10022 in 10024. Solid lines indicate the ranges of compositional zoning within single crystals, dashed lines are tie lines between coexisting augite and pigeonite, and dotted lines indicate a possible miscibility gap between augite and pigeonite in 10024. Fig. 3 (right). Al-Ti relations of the analyzed clinopyroxenes in 10022 and 10024. Solid line indicates the ratio Ti/Al = ½.

gap between augite and pigeonite is considerably smaller than those found in the terrestrial plutonic rocks. The clinopyroxenes which occur as needle-like crystals intergrown with plagioclase are poor in Ca and fall in the area intermediate between augite and pigeonite (Fig. 2). These subcalcic augites may have formed by rapid crystallization. They are most probably metastable because their compositions fall within the miscibility gap. The relatively iron-poor titanaugites in 10022 have higher Al and Ti and lower Si than those in 10024. The Al-Ti relations of the analyzed clinopyroxenes are shown in Fig. 3. Augites crystallized from the more silica-undersaturated magmas have higher Al and Ti and lower Si content (5), and the relative concentration of silica in the original magma of 10022 may therefore have been lower than that of 10024 when augite crystallized.

Olivine is found in 10022 and 10072. The compositional range of olivine in 10022, on the basis of the 11 microprobe analyses, is $Fo_{71}Fa_{29} \sim Fo_{41}Fa_{59}$. This compositional zoning is observable in a single crystal. Olivine has slightly higher Mg/Fe ratios than the contacting pigeonite and augite.

Four ilmenite crystals have been analyzed. They are close to stoichiometric $FeTiO_3$ but contain small amounts of Mg, Cr, Mn, and Zr and show little zoning.

Nine plagioclase crystals have been analyzed. Their compositions are between $An_{73}Ab_{27}$ and $An_{81}Ab_{19}$ in 10022 and $An_{78}Ab_{22}$ and $An_{80}Ab_{20}$ in 10024; that is, they are all bytownite. Many other plagioclase crystals were examined by scanning and found to be essentially homogeneous. In some crystals, however, weak zoning to sodic bytownite was observed at the margins. Those which occur in the mesostasis are also sodic bytownite.

Semiquantitative analysis shows that a relatively large native iron crystal in a troilite of 10024 contains 0.6 percent Ni by weight.

Mesostasis is dark in transmitted light. The grain size of the phases in the mesostasis varies from 1 to 20 μm. These phases, identified by the microprobe, are described above. The potassium-rich phase, which is most probably glass, has the following composition (percent by weight): SiO_2 75 ~80; Al_2O_3, 10; CaO, 1; Na_2O, < 0.5; K_2O, 9. From the textural relations the mesostasis is considered to have crystallized in the latest stage of crystallization of the rocks; that is, it represents the residual liquid. The microprobe analysis with a large beam size (20 μm) was made to obtain the bulk composition of the mesostasis. Because of the pres-

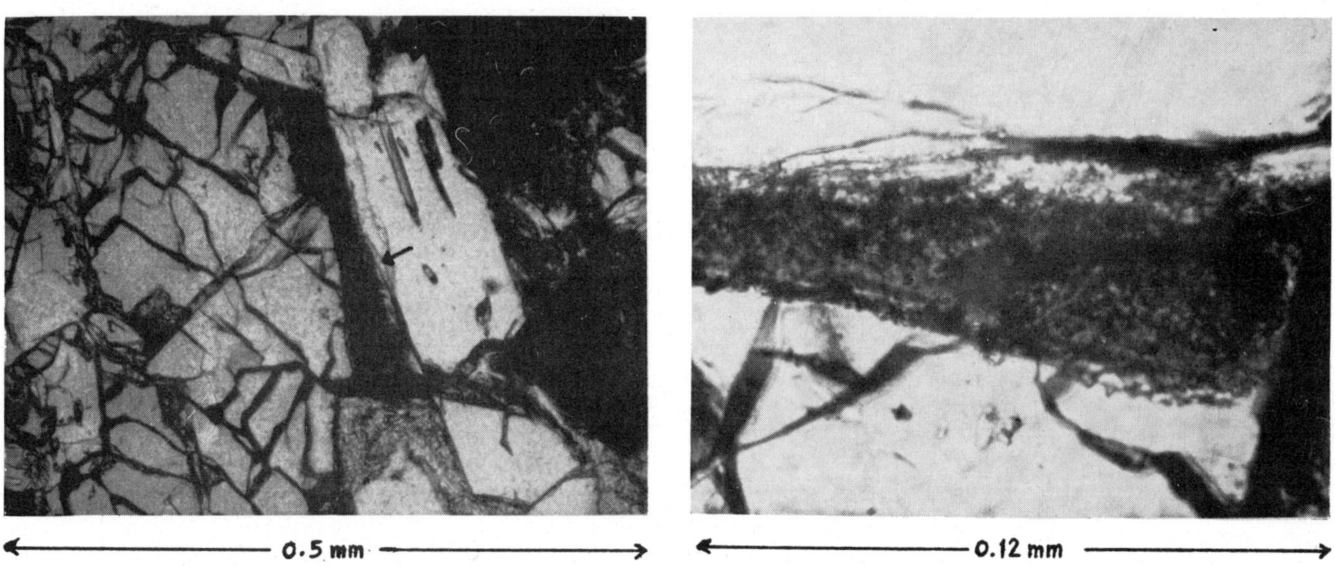

Fig. 4. Melting relations of a lunar crystalline rock (type A); (left) 0.5 mm; (right), 0.12 mm (horizontal dimensions).

ence of the potassium-rich glass, most of the mesostasis have high potassium contents, although the relative amount of the glass is different in different mesostasis. One of the 10 analyses is as follows (percent by weight): SiO_2, 71; TiO_2, 1.1; Al_2O_3, 11.5; FeO, 6.1; MnO, 0.1; MgO, 0.2; CaO, 2.4; Na_2O, 0.7; K_2O, 6.6; total, 99.7. It is of interest that the mesostasis is granitic or rhyolitic in composition, though the sodium content is considerably lower and the iron content is higher than in most terrestrial granites. Absence of K-feldspar in the mesostasis indicates that the temperature of the solidification of the mesostasis was at least above the plagioclase-K-feldspar-silica quaternary minimum.

The rapid-growth texture and the presence of metastable clinopyroxenes suggest that most of the rock crystallized rapidly. During the crystallization of the magmas, clinopyroxenes and olivine change considerably in composition, whereas plagioclase changes very little and stays as bytownite. These features are similar to those observed in the chilled zones of the prehistoric Makaopuhi lava lake studied by Evans and Moore (*6*), although the compositional variation of augite is narrower and that of plagioclase is wider in Makaopuhi than in 10022 and 10024. Evans and Moore suggest that the narrow compositional range of plagioclase in the chilled zones as compared with those in the slowly cooled zones is due to the relatively rapid crystallization of the undercooled magma. In view of the cooling history in the Makaopuhi lava lake, it is suggested that the lunar rocks 10022 and 10024 were also formed by crystallization of undercooled magmas. In this case crystallization would have been rapid. During the crystallization, fractional crystallization of ferromagnesian minerals took place, and the residual liquids were enriched in silica and potassium and other minor elements such as phosphorus, sulfur, and zirconium. Thus, rhyolitic liquids were formed in the last stage of the crystallization. The presence of native iron indicates that the oxygen pressure was low during the crystallization.

The melting experiments were made at 1 atm and at high pressures. A fine-grained crystalline rock of type A which was separated from the fines 10085,13 and numbered 10085,13-1 and a type B rock, 10017,66, were used. The bulk chemical compositions, determined by wet chemical method, are as follows, in percent by weight for 10017, 66 and 10085,13-1, respectively: SiO_2, 40.03 and 40.16; TiO_2, 11.61 and 11.93; Al_2O_3, 8.04 and 7.50; Fe_2O_3, 0.0 and 0.0; FeO, 20.46 and 20.37; MnO, 0.24 and 0.26; MgO, 8.17 and 8.09; CaO, 10.26 and 10.17; Na_2O, 0.49 and 0.53; K_2O, 0.30 and 0.29; H_2O, 0.0 and not detected; P_2O_5, 0.17 and 0.15; Cr_2O_3, 0.34 and 0.42; NiO, 0.00 and not detected; totals, 100.11 and 99.87 (*7*).

The melting experiments at atmospheric pressure were made by the quenching method under controlled oxygen partial pressure of 10^{-8} to 10^{-13} atm. Pt-40-percent-Rh and Ag-40-percent-Pd crucibles were used as sample containers. In some of the runs Fe-saturated Pt-40Rh crucibles were used to prevent iron loss from the sample. The high-pressure experiments were made with the tetrahedral-anvil type apparatus described by Akimoto *et al.* (*8*). The lunar sample, which was ground finely in an agate mortar, was put in molybdenum capsule. Temperature was measured with Pt/Pt-13-percent-Rh thermocouple. The phases after the runs were identified with both the microscope and electron-probe microanalyzer.

Melting relations of rock 10085,13-1 are given in Fig. 4. At 1 atm chromite and chromian $(Mg,Fe)Ti_2O_5$ are on the liquidus in the runs made with the Pt-40Rh crucible; however, only chromite is on the liquidus in the run made with the Fe-saturated Pt-40Rh crucible. In the former runs, the crucible was found to dissolve considerable amounts of iron from the samples. This may explain the crystallization of chromian $(Mg,Fe)Ti_2O_5$ near the liquidus temperatures. The runs made with the Fe-saturated Pt-40Rh crucible show that the liquidus temperature of the present lunar rock is slightly above 1200°C. In the run made with the Ag-40Pd crucible at 1160°C, a small amount of chromian $(Mg,Fe)Ti_2O_5$ was observed with clinopyroxene, as in the runs made with the Pt-40Rh crucible. Since the Ag-40Pd crucible is believed to dissolve much less iron than the Pt-40Rh crucible, the chromian $(Mg,Fe)Ti_2O_5$ might be a phase crystallizing from the liquid of lunar rock composition. On the basis of the runs made with the untreated Pt-40Rh crucible, the sequence of crystallizing phases is chromian $(Mg,Fe)Ti_2O_5$ and chromite, Ca-rich clinopyroxene, olivine, ilmenite, plagioclase with lowering temperature. At 1125° and 1130°C both chromian $(Mg,Fe)Ti_2O_5$ and ilmenite crystallized, whereas at 1100°C ilmenite but no chromian $(Mg,Fe)Ti_2O_5$ crystallized. Since chromite and chromian (Mg,Fe) Ti_2O_5 were not found in rock 10085, 13-1, the rock would have crystallized at temperatures below 1125°C if it crystallized near the surface of the moon.

At high pressures the liquidus phases for the 10085,13-1 composition are different from those at 1 atm: At 20 and 30 kb, Ca-rich clinopyroxene and rutile crystallize slightly below the liquidus, and at 40 kb Ca-rich clinopyroxene is on the liquidus. It is noted that no chromite and chromian $(Mg,Fe)Ti_2O_5$ crystallize at high pressures. On the basis of the high-pressure runs it is suggested that the original magma of the rock 10085,13-1 could be formed by partial melting of materials containing clinopyroxene and rutile in the depths of the moon. Of course other hypotheses involving formation of the rock by the fractional crystallization of magmas with lower Fe/Mg ratio are also possible.

IKUO KUSHIRO
YASUO NAKAMURA
HIROSHI HARAMURA
Geological Institute, University of Tokyo, Tokyo, Japan
SYUN-ICHI AKIMOTO
Institute for Solid State Physics, University of Tokyo

References and Notes

1. Lunar Sample Preliminary Examination Team, *Science* **165**, 1211 (1969).
2. Japan Electron Optics Laboratory Co., model JXA-5 with 40° take-off angle. Standards: natural albite (Na), adularia (K), and hematite (Fe); synthetic wollastonite (Ca); synthetic pure oxides (Mg, Al, Si, Ti, Cr, Mn, and Ni). Accelerating voltage, 15 kv; electron beam diameter, 1 ~ 2 μm unless otherwise noted; five to ten spots for each analysis. The relative intensities were collectively corrected by the method of A. E. Bence and A. L. Albee, *J. Geol.* **76**, 382 (1968).
3. H. H. Hess, *Am. Mineral.* **26**, 515 (1941).
4. R. Kretz, *J. Geol.* **69**, 361 (1961).
5. I. Kushiro, *Am. J. Sci.* **258**, 548 (1960); M. J. LeBas, *ibid.* **260**, 267 (1962).
6. B. W. Evans and J. G. Moore, *Contrib. Mineral. Petr.* **17**, 85 (1968).
7. Analyst, H. Haramura.
8. S. Akimoto, H. Fujisawa, T. Katsura, *J. Geophys. Res.* **70**, 1969 (1965).
9. This study was carried out as a memorial to the late Hisashi Kuno, who had planned the study and had been appointed a Principal Investigator. We are indebted to T. Katsura and M. Nishikawa for their help in the melting experiments, to Mrs. Kuno, who kindly offered a financial support and to R. Kubo, T. Tatsumi, M. Ozima, and other members of the University of Tokyo for their suggestion and help. A grant from the Ministry of Education is also appreciated.

4 January 1970

Iron-Titanium Oxides and Olivine from 10020 and 10071

Abstract. *A new mineral (approximately $Fe_{0.5}Mg_{0.5}Ti_2O_5$) related to the pseudobrookite series has been discovered in section 10071,28. Electron-probe analyses for this mineral, a coexisting ilmenite, and a chromian ulvöspinel-ilmenite assemblage in section 10020,40 indicate crystallization under highly reducing conditions. Analytical and optical absorption studies of the olivine in 10020 show it to contain unusually high Cr (1400 parts per million) probably as Cr^{2+}.*

Petrographic and electron-probe studies carried out primarily on rocks 10020 and 10071 have revealed a variety of opaque minerals whose compositions differ in considerable detail from terrestrial counterparts. Reflection microscopy indicates that ilmenite whose composition closely approaches $FeTiO_3$ is by far the most abundant primary opaque phase, forming 10 to 20 percent by volume of the crystalline rocks (Table 1). Magnesian ferropseudobrookite, chrome spinel, and chromian ulvöspinel have been found in accessory amounts. The magnesian ferropseudobrookite is sufficiently distinctive in chemical composition to merit consideration as a new mineral. Troilite, containing blebs and subhedral crystals of metallic iron, has been observed in all samples as a late-stage precipitate.

Three grains of magnesian ferropseudobrookite, all mantled by ilmenite, were discovered in section 10071,28. One of these grains (Fig. 1) was analyzed for 12 elements with the electron probe (Table 2) and found to be essentially $Fe_{0.5}Mg_{0.5}Ti_2O_5$ with minor amounts of Cr, Al, Mn, and Ca. The mineral is pale gray and shows medium to strong reflection anisotropy in oil. Its white-light reflectivity in air is in the range 15 to 18 percent, and it is softer than ilmenite. The three grains are rectangular in outline with longest dimensions from 100 to 300 μm, whereas their ilmenite overgrowths are 100 to 200 μm thick. These grains and their overgrowths appear to be primary.

The electron-probe analysis (Table 2) shows that the magnesian ferropseudobrookite is intermediate in composition between $Fe^{2+}Ti_2O_5$ (ferropseudobrookite) and $MgTi_2O_5$ (karrooite). Both $Fe^{2+}Ti_2O_5$ and $MgTi_2O_5$ have been synthesized, but a solid solution series between them has not been experimentally demonstrated.

The formula of pure pseudobrookite is $Fe^{3+}{}_2TiO_5$, and solid solution of $MgTi_2O_5$ or $Fe^{2+}Ti_2O_5$ with this phase is accomplished by the coupled substitution (Mg, Fe^{2+})-Ti for Fe^{3+}-Fe^{3+}. Both $Fe^{2+}Ti_2O_5$ and $MgTi_2O_5$ have been found as subordinate components in various pseudobrookite-like minerals in terrestrial occurrences (*1*). An example of such a mineral is kennedyite ($Fe^{3+}{}_2MgTi_3O_{10}$), a phase found in the Karroo basalts (*2*). Kennedyite is isostructural with pseudobrookite and contains 28.77 percent Fe_2O_3 and 60.33 percent TiO_2. The MgO content of kennedyite (6.45 percent) is similar to that of the lunar phase (7.70 percent), but the apparent absence of Fe_2O_3 and higher TiO_2 content in the lunar phase makes its occurrence distinctive.

A low content of Fe_2O_3 in the lunar magnesian ferropseudobrookite is obviously essential for its recognition as a new mineral. A technique for the analysis of Fe^{3+} with the electron probe is being developed (*3*), but its application to this group of minerals will require further work. There are several reasons, however, that lead to the strong presumption that the concentration of Fe^{3+} is very small in this lunar phase.

If the electron-probe analysis of the magnesian ferropseudobrookite is recalculated to an ionic formula where $O = 5$ (Table 2), Ti is found to equal 1.97, in comparison with the maximum possible value of 2.00. Inasmuch as half of any Fe^{3+} present must replace Ti, there is obviously not much possibility of significant Fe^{3+}. Moreover native Fe was observed as a phase in specimen 10071. As discussed below, the presence of native Fe and the compositions of other solid solutions indicate that these rocks crystallized under highly reducing conditions.

A spinel has been discovered in section 10020,40, whose composition is as unique as that of the magnesian ferropseudobrookite described above. Unlike the magnesian ferropseudobrookite, however, the spinel is a solid solution between two well-established minerals, ulvöspinel (Fe_2TiO_4) and chromite ($FeCr_2O_4$).

Grains of spinel in section 10020,40 are generally euhedral with dimensions from 100 to 200μm, and they form about 10 percent by volume of the Fe-Ti oxides. These spinels are opaque, opti-

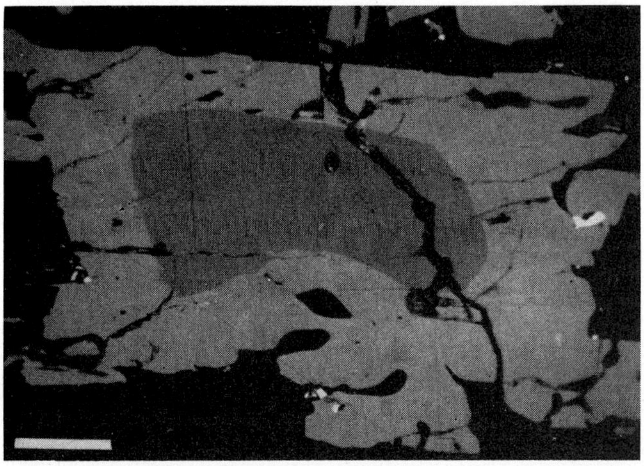

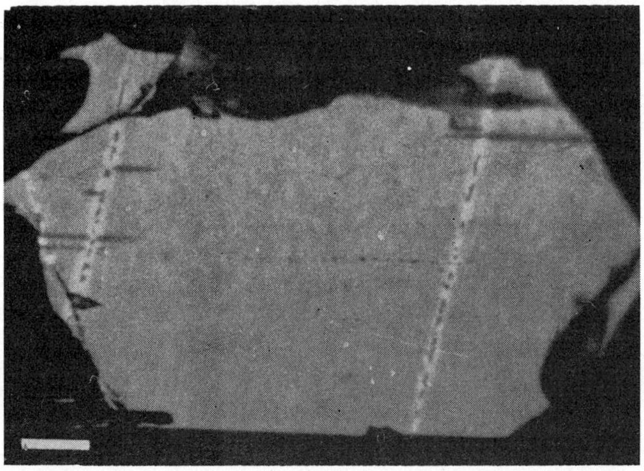

Fig. 1 (left). Magnesian ferropseudobrookite (dark gray) mantled by an overgrowth of ilmenite (pale gray). Note the fine lamellae of rutile (pale white) in the ilmenite and also the associated troilite (bright white) along the grain boundaries (scale 50 μm). Fig. 2 (right). An ilmenite crystal (medium gray) with two thick lamellae of exsolved rutile (pale white). Lensoidal rods of spinel (dark gray) occur in the ilmenite, whereas irregular blebs occur within the rutile. The crack is infilled with troilite (bright white) (scale 10 μm).

Table 1. Modal analyses measurements were made on areas of thin sections selected for minimum density of voids. The average area measured was 32 mm^2.

Section	Pyroxene	Opaque	Plagioclase	Olivine	Cristobalite	Other	Hole or vesicle
10003,37	51.7	18.2	29.0		0.3 (?)	0.2	0.5
10020,31	54.8	17.1	21.4	4.8	0.9 (?)		0.9
10020,40	52.9	13.5	28.5	2.9	1.7 (?)	<0.1	0.6
10057,35	50.9	15.7	19.2		<0.1	3.3	10.8
10071,28	58.3	14.7	21.7		0.8	2.7	2.0
10072,36	59.4	14.8	20.4		0.2 (?)	3.7	1.5

cally homogeneous, tan in color, and weakly anistropic with white-light reflectivity values from 18 to 20 percent. Many of them are mantled by ilmenite, in contrast to occurrences in terrestrial lavas, where all similar overgrowths are observed to be titanomagnetite. Spinel inclusions in olivine and pyroxene in 10020,40 are not mantled. In terrestrial rocks Cr-spinel is one of the earliest minerals to crystallize, whereas in meteorites it is always later than olivine and usually later than pyroxene (*4*). These spinels (Table 1) have compositions unlike any previously reported, and for purposes of comparison may be regarded either as Cr-rich titanomagnetites (23.5 percent Cr_2O_3) or Ti-rich chromites (20.9 percent TiO_2). Terrestrial titanomagnetites contain between 0.01 percent and 0.07 percent Cr_2O_3 (*5*). Chromites generally have rather small concentrations of TiO_2, although both rutile (*4*, *6*) and ilmenite (*4*, *7*) have been observed as exsolution intergrowths. Among terrestrial chromites those richest in TiO_2 are usually found in basaltic lavas and range between 1.3 and 2.5 percent TiO_2 (*8*). An unusual occurrence in the Bushveld, however, has a reported TiO_2 content of 12.8 percent (*9*). A meteoritic chromite analyzed by Buseck and Keil (*6*) contains 2.9 percent TiO_2. A complete solid solution exists between magnetite (Fe_3O_4) and chromite ($FeCr_2O_4$) and also between magnetite (Fe_3O_4) and ulvöspinel (Fe_2TiO_4). Although a solid solution series has not been experimentally demonstrated between Fe_2TiO_4 and $FeCr_2O_4$, our analysis of the chromiferous ulvöspinel in rock 10020 strongly suggests that such a series exists, confirming the suggestion of Legg (*7*). The absence of alteration or exsolution products in the spinel is indicative of relatively rapid cooling and low oxygen fugacity.

Table 2. Electron-probe analyses of iron-titanium oxides and olivine.

Substance	Section 10020,40: Olivine	Section 10020,40: Chromian ulvöspinel	Section 10020,40: Ilmenite	Section 10071,28: Ilmenite	Section 10071,28: Magnesian ferropseudobrookite
Percent by weight					
SiO_2	37.5				
TiO_2	0.09	20.9	53.9	52.8	73.4
Al_2O_3	0.05	8.61	0.11	0.15	1.62
Cr_2O_3	0.21	23.5	1.30	0.59	2.15
FeO*	25.5	42.1	41.9	44.2	15.3
MnO	0.30	0.25	0.47	0.32	0.08
MgO	36.5	4.23	2.59	2.06	7.70
CaO	0.33	0.03	0.07	0.01	0.01
V_2O_3	<0.01	0.4	<0.3	<0.3	<0.5
CoO	<0.03	<0.03	<0.03	<0.03	<0.02
NiO	<0.01	<0.01	<0.01	<0.01	<0.01
CuO	<0.01	<0.01	<0.01	<0.01	<0.01
ZnO	<0.01	<0.01	<0.01	<0.01	<0.01
ZrO_2	<0.10	<0.10	<0.10		
Total	100.5	100.0	100.3	100.1	100.3
Atomic proportions					
	O = 4	O = 4	O = 3	O = 3	O = 5
Si	0.989				
Ti	0.002	0.541	0.995	0.987	1.969
Al	0.001	0.349	0.003	0.004	0.068
Cr	0.004	0.638	0.025	0.012	0.061
Fe	0.561	1.213	0.860	0.919	0.456
Mn	0.007	0.007	0.010	0.007	0.007
Mg	1.433	0.217	0.095	0.076	0.410
Ca	0.009	0.001	0.002		
V		0.011			
Totals	3.006	2.977	1.990	2.005	2.966

*Total Fe as FeO.

Ilmenite is generally euhedral, although coarse skeletal crystals with trapped pyroxene, metallic iron, and troilite are present in specimen 10047. Fine skeletal ilmenite in large concentration is also found in partly devitrified glass spheres and fragments in the breccias. Ilmenite is generally homogeneous, but some grains contain lamellae of rutile <10 μm thick oriented along $\{01\bar{1}2\}$ and lensoidal rods of spinel along {0001}. The secondary spinel and rutile invariably coexist within ilmenite, and they are often intimately associated (Fig. 2). Both spinel and a second generation of ilmenite have been observed within the rutile, suggesting that the rutile was iron-enriched when it crystallized. The textures of these intergrowths suggest that the phases are interrelated by a complex rather than a simple exsolution process with possibly one phase triggering nucleation and growth of the other. Glass or silicate inclusions and contacts with metallic iron or troilite appear to have promoted exsolution. Such intergrowths in ilmenite have not been reported in terrestrial rocks but have been observed in meteorites (*6*). Growth of rutile in ilmenite is generally the result of oxidation, but Ramdohr (*10*) has suggested that reduction could also cause the generation of rutile by the breakdown of ilmenite in environments of low oxygen fugacity. However, the reaction

$$2FeTiO_3 \rightarrow 2TiO_2 + 2Fe + O_2$$

liberates free iron, and this was not observed in the lunar ilmenite. The rutile intergrowths have a texture suggesting exsolution rather than alteration, inasmuch as they are oriented parallel to $\{01\bar{1}2\}$ and have well-defined contacts that pinch at the intersection of two or more lamellae.

Electron-probe microanalyses of two optically homogeneous ilmenite grains (Table 2) in rocks 10020,40 and 10071,28 show the presence of considerable MgO and Cr_2O_3. Ilmenite from the Farmington chrondrite (*6*) in which rutile and chromite were observed as exsolution products has a similar composition but contains significantly lower Cr and higher Mg.

Sparse crystals of olivine in rock 10020 range up to 1.4 mm in their largest dimension and have the appearance of phenocrysts. Nevertheless, most olivine crystals in this rock grade down in grain size to 0.036 mm, which is equivalent to the coexisting pyroxene, feldspar, and opaque minerals. The olivine forms 3 to 5 percent by volume of the rock (Table 1). It has proved ideal for optical absorption and x-ray studies and for electron-probe analysis because of its coarse grain size and relative homogeneity and because lattice distortions due to shock are absent.

The 10020 olivine has been analyzed for 14 elements with the electron probe, and of these Mg, Fe, Si, Ca, Mn, Cr, Al, and Ti were found in significant concentration (Table 2). A dozen or more grains of the olivine in section 10020,40 were included in the analysis, and the average composition is $Fo_{72}Fa_{28}$ (mole percent) with a range in composition of approximately ±3 mole percent Fo. The Cr_2O_3 content of the olivine (0.21 percent by weight) is higher by a factor of 2 than that of any known terrestrial olivines, and the NiO content (<0.01 percent by weight) is considerably lower than that of most terrestrial olivines of igneous origin. The latter usually contain 0.01 percent by weight or less Cr_2O_3, with the exception of olivine crystals that form inclusions in natural diamond; these olivine inclusions have been found to have Cr_2O_3 contents ranging up to 0.1 percent by weight (*11*). A common factor in the crystallization of olivine in a lunar lava and in a terrestrial diamond is a reduced oxygen fugacity, suggesting that the chromium may be present as Cr^{2+}. More direct evidence in support of this suggestion is cited below.

Preliminary microscopic measurements of the crystal field absorption spectra were made on a single crystal picked from a chip of sample 10020 and on olivines in section 10020,40. The spectra exhibit broad bands in the near infrared, similar to those observed for terrestrial olivines close to this composition (*12*). Specifically, absorption bands were observed at approximately 9000 and 10,600 Å, corresponding to Fe^{2+} in octahedral coordination. When the crystal was rotated about the *a* axis, pleochroism was observed when the *b*-*c* plane was made parallel with the plane of polarization of the substage nicol. Minimum absorption with maximum resolution occurred approximately parallel to *b*. These phenomena are well known in olivines (*13*), but differences were observed in that the weak band at 6300 Å usually attributed to Fe^{3+} in octahedral coordination is absent and a weak, poorly resolved band at about 10,400 Å is present. We believe this band is caused by a trace of Cr^{2+}. No near-infrared bands for Cr^{3+} were observed. Weak absorption was measured at both ends of the visible region. Maximum transmission occurred between 4000 and 5000 Å, characteristic of Cr^{2+} (*14*). The absence of Cr^{3+} was confirmed by the fact that no band occurred in that region. Divalent chromium $[d^4(t_{2g})^3(e_g)^1]$ is peculiar in that it causes severe Jahn-Teller distortion of an octahedral site (*15*) and thereby is destabilized relative to Cr^{3+}. If chromium were incorporated as Cr^{2+} during crystallization, it would imply an extremely reducing environment.

The x-ray diffraction of an olivine crystal picked from 10020 is consistent with the accepted space group of *Pbnm* for terrestrial olivines. The unit cell parameters are $a = 4.76 \pm 0.01$, $b = 10.284 \pm 0.004$, $c = 6.018 \pm 0.003$ Å, consistent with the composition determined by electron probe.

A number of features observed in these samples that distinguish them from terrestrial rocks appear to result from highly reducing conditions during primary crystallization. Outstanding features are (i) the presence of metallic iron, (ii) unusual oxide phases rich in ferrous iron, including magnesian ferropseudobrookite and chromian ulvöspinel, and (iii) exceptionally high Cr in the olivine, believed to be present as Cr^{2+}.

S. E. HAGGERTY, F. R. BOYD
P. M. BELL
L. W. FINGER, W. B. BRYAN
Geophysical Laboratory, Carnegie Institution of Washington, Washington, D.C. 20008

References and Notes

1. D. G. W. Smith, *Amer. Mineral.* **50**, 1982 (1965).
2. O. von Knorring and K. G. Cox, *Mineral. Mag.* **32**, 676 (1961).
3. A. L. Albee and A. A. Chodos, Abstract of paper presented at Geological Society of America Meeting, Atlantic City, New Jersey, part 7, 1 (1969).
4. P. Ramdohr, *J. Geophys. Res.* **68**, 2011 (1963).
5. W. A. Deer, R. A. Howie, J. Zussman, *Rock-Forming Minerals* (Wiley, New York, 1962), vol. 5; I. S. E. Carmichael, *Contrib. Mineral. Petrol.* **14**, 36 (1967).
6. P. R. Buseck and K. Keil, *Amer. Mineral.* **51**, 1506 (1966).
7. C. Legg, *ibid.* **54**, 1347 (1969).
8. J. Babkine, F. Conquéré, J. Vilminot, K. D. Phan, *Bull. Soc. Franc. Mineral. Crist.* **88**, 447 (1965); S. E. Haggerty, thesis, University of London (1968).
9. J. J. Frankel, *South African J. Sci.* **38**, 152 (1942).
10. P. Ramdohr, *Carnegie Inst. Wash. Year Book* **63**, 217 (1964).
11. H. O. A. Meyer and F. R. Boyd, *ibid.* **68**, in press.
12. R. G. Burns, thesis, University of California, Berkeley (1965); W. B. White and K. L. Keester, *Amer. Mineral.* **51**, 774 (1966).
13. R. G. Burns, in preprint of *The Application of Modern Physics to the Earth and Planetary Interiors,* S. K. Runcorn, Ed. (Wiley, New York, 1969).
14. The visible and near-infrared spectra of a pure chromous olivine (Cr_2SiO_4) provided by G. Healy were measured as a comparison for us by W. White.
15. L. E. Orgel, *An Introduction to Transition-Metal Chemistry: Ligand Field Theory* (Wiley, New York, 1960).
16. We thank NASA for financial support with the contract NAS 9-9988.

4 January 1970

Opaque Minerals of the Lunar Rocks and Dust from Mare Tranquillitatis

Abstract. *The opaque minerals in the lunar rocks 10047, 10050, 10057, 10059, 10068 and in the dust were studied. Rock 10047 contains ilmenite as a main component besides Al-bearing chromian ulvöspinel, a new Ti, Fe, Zr, Y, Ca silicate, troilite, cobaltian* α *Fe, minor hafnian baddeleyite and dysanalyte. Dysanalyte contains Ca, Ti, and Fe as major elements and shows concentrations (rare earth elements together* ≈ *10 percent) of La, Ce, Pr, Nd, Sm, Eu, Gd, Tb, Dy, Ho, Er, Tm, Hf, Y, and Zr plus traces of Nb, Ba, and Na. The chemistry of ilmenite in the igneous rocks shows no appreciable variations. A new mineral of the solid solution series* $MgTi_2O_5$*-*$FeTi_2O_5$ *has also been encountered in the polymict breccias. The microbreccias and the fines contain meteorite debris indicating an impact origin.*

Three igneous rocks (10047, 10050, 10057), two polymict microbreccias (10059, 10068) and 5 grams of lunar dust were studied in reflected light and with the electron microprobe. The dust was separated under acetone to six fractions. Of those fractions only some dozens of grains > 200 μm especially the magnetic ones and those denser than 4.03 were thoroughly studied. Hence we

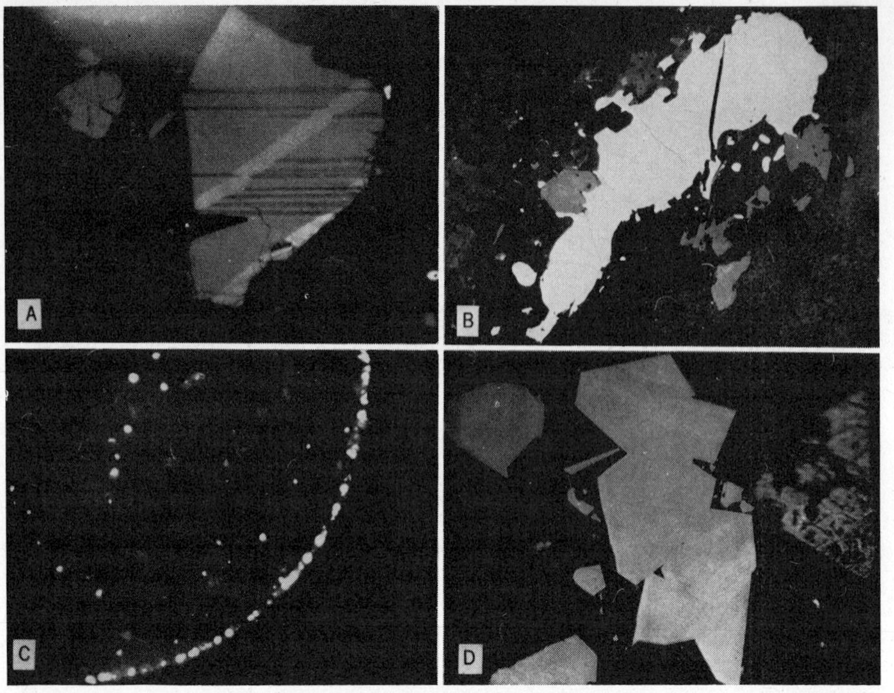

Fig. 1. (A) Ilmenite (in the breccia) containing lamellae of "anosovite" parallel to (0001) and two broad lamellae of rutile (with internal reflections) (oil immersion, × 300). (B) Troilite overgrown by a baddeleyite crystal. The groundmass consists of pyroxene, fayalite, and K-rich glass (oil immersion, × 150). (C) A globule of homogenized glass (some spherules of iron) with a rim of iron spherules migrated to the surface (oil immersion, × 125). (D) A new spinel mineral similar to magnetite, idiomorphic and zoned (oil immersion, × 150).

will be presenting data on breccias and dust which appeared to be important with respect to their mineralogy and observations which seemed to be peculiar. Electron microprobe analyses were carried out on many coexisting minerals with emphasis on the opaques and semiopaques. Observations on transparent components will also be discussed. Geologically, it is evident that the crystalline igneous rocks here represent the primary rock components of the original lunar surface. These rocks are fresh, and show no alterations except on their outermost surface due to numerous impacts of micrometeorites and to the effect of the solar wind. In contrast, the breccias and dust are of polygenic origin and contain products of many geological events, mainly meteorite impacts. All stages of impact metamorphism starting from stress twinning up to indications of complete melting of ilmenite, rutile, and the silicates were encountered in the breccias and the dust. It seems that the sum of influences of small meteorites is minute compared to those caused by bigger ones. Very large impacts create peak pressures and temperatures causing partial to complete melting and evaporation of huge amounts of the target rocks and tremendous shock effects in the surrounding rocks. The meteorite itself, depending on its impacting mechanics, may be partially or completely evaporated.

The impacting bodies are mostly either stony or iron meteorites. Stony meteorites, being brittle and more fusible than iron meteorites hardly survive such impacting. Only once did we observe a globule comparable with a real chondrule. In the breccias some chromite grains—probably relicts of chondrites—were also encountered. On the other hand we observed relicts of iron meteorites in the form of molten or condensation globules and splinters of nickel-iron in the breccias and dust. Considering the very low content and the small quantity of iron in the igneous rocks, this iron cannot be mistaken for meteoritic iron. Only in the case of very small grains which could not be analyzed with the probe is the recognition between autochthonic and meteoritic iron impossible. In the case of troilite the situation is more complex.

The crystalline igneous rocks and rock splinters in the breccias have been briefly described (*1*). With some restrictions regarding the high TiO_2 content, those rocks are comparable to doleritic feldspar basalt, high in SiO_2 and thus poor in olivine. In general, the high TiO_2 content is not exceptional for such rocks; some basalts with > 4 percent TiO_2 are known. The mineral assemblage is the same, consisting of clinopyroxene, calcic feldspar (An_{75}–An_{98}), ilmenite, cristobalite, tridymite, and small amounts of troilite and cobaltian iron. Troilite and iron form eutectic intergrowths. Small amounts of potassium-rich glass (in part, two immiscible phases) are common. Besides the main components we observed the following accessories which were not observed before: ulvöspinel, a new Fe, Ti, Zr, Y, Ca silicate, hafnian baddeleyite, dysanalyte with high concentrations of rare earths, a member of the $MgTi_2O_5$-$FeTi_2O_5$ solid solution series and an unknown fibrous mineral. Fayalite has been observed in myrmekitic intergrowths with clinopyroxene and K-rich glass. Hematite occurs as fine lamellae within a few ilmenite grains in the shocked breccias. The optical identification of hematite was confirmed by microprobe analyses. Rutile exsolution lamellae are quite rare. Only one small grain of native Cu was observed.

Ilmenite is always an early mineral like plagioclase. The crystals are identical to those in many basalts, gabbros and norites; the form being somewhat skeletal, due to preferential growth of the (0001) face. No exsolutions could be observed with highest magnification, as in most terrestrial basalts. In the crystalline rocks twin lamellae are present but quite rare. In similar rock splinters, in the breccias and dust, twinning of ilmenite is abundant but without magnetite at twin crossings. The composition does not vary considerably in the different rocks. Ilmenite in rock 10057 has the average composition (%): TiO_2, 61.7; FeO, 36.6; MgO, 0.62; MnO, 0.52; Al_2O_3, 0.26; Cr_2O_3, 0.26. Coarser grains are slightly enriched in Cr_2O_3. The formation of colorless rutile in broad to thin sharp lamellae parallel to $(22\bar{4}3)$ of ilmenite is mostly restricted to the splinters in the breccias and dust and is probably caused by shock.

Ulvöspinel was observed in every rock specimen examined and in the rock splinters of the breccias as well. It occurs mostly as tiny grains with or without ilmenite as inclusions in the pyroxene. The optical properties are typical but its recognition is rather difficult due to optical similarties to ilmenite. In part it replaces ilmenite. Loose skeletons appear to be young and may belong to the latest crystallization in the interstices. Quantitative electron microprobe analysis of numerous grains revealed ulvöspinel to be rich in Al, Cr, and Mn (57.8 percent FeO, 38.2 percent TiO_2, 3.29

percent Al_2O_3, 0.89 percent Cr_2O_3, 0.67 percent MnO). Ulvöspinel in the crystalline rocks has the formula:

$$(Fe^{+2}_{14.07}Mn^{+2}_{0.17}Ti^{+3}_{1.70})_{15.94}(Ti^{+4}_{6.66}Al^{+3}_{1.13}Cr^{+3}_{0.21})_{8.00}O_{32}.$$

A dark blue grayish mineral with low reflectivity, very probably a new member of the $MgTi_2O_5$-$FeTi_2O_5$ solid solution series was observed as thin sharp lamellae parallel to (0001) of ilmenite (Fig. 1A). This gray mineral is quite similar to some lamellae observed (*2*) in the mesosiderite Vaca Muerta. The same mineral occurs in the rock splinters in the breccias, mostly as coarse inclusions in ilmenite, and is partly replaced by the latter. Quantitative electron microprobe analysis revealed: 75.6 percent TiO_2, 11.9 percent FeO, 8.12 percent MgO, 1.87 percent Al_2O_3, and 1.81 percent Cr_2O_3. Accordingly the formula is:

$$(Ti^{+4}_{14.99}Al^{+3}_{0.61}Cr^{+3}_{0.40})_{16.00}(Mg^{+2}_{3.36}Fe^{+2}_{2.76}Ti^{+4}_{0.80})_{6.92}O_{40}.$$

Baddeleyite, ZrO_2: Regarding the exceptionally high ZrO_2 content of the rocks we looked for a Zr-bearing mineral (in vain) in rock 10057. In 10047,13 we found baddeleyite partly in crystalline grains (Fig. 1B). Zircon, with similar properties, was not observed. In the breccias and in a heavy concentrate from the dust, large clusters of baddeleyite were encountered. The U and Th contents of baddeleyite are quite low, which explains the high transparency and the lack of metamictization. Baddeleyite is quite rich in Hf (≈ 2 percent) and contains other minor elements, for example, rare earths. The mineral forms thick, tabular, untwinned grains, which is quite different from terrestrial occurrences.

A new mineral A was first mistaken for kennedyite (a mineral not rare in acid basalts) due to optical similarities. Electron microprobe analysis indicated, however, that the mineral is a Ti, Fe, Zr silicate with lower concentrations of Ca and Y and minor amounts of Mn, Gd, Nd, Dy, Hf, Ce, Al, and Na. The new mineral, very probably a basic silicate, occurs in clusters of presumably orthorhombic crystals. Its color is gray to dark gray without noticeable anisotropism. The mineral is opaque with a reflectivity similar to Ti_3O_5 or ludwigite.

Dysanalyte: This mineral, or at least a member of the perovskite group, was discovered in the coarse crystalline rock 10047,13 during the probe analysis of baddeleyite with which it is intimately intergrown. It forms round, in part perfectly crystallized, grains with distinct purple internal reflections. Electron microprobe analysis revealed the major elements Ca, Ti, and Fe with high concentrations of La, Ce, Pr, Nd, Sm, Eu, Gd, Tb, Dy, Ho, Er, Tm, Hf, Y, and Zr besides traces of Nb, Ba, and Na. The mineral differs from members of the pyrochlore family due to the low Nb content and its high reflectivity. Another species, mineral B, rich in Ce and Ca (probably a member of the apatite group) was also observed. A mineral C, with a form similar to titanomagnetite but much lower in reflectivity was observed in the breccias. It is a two-phase unmixing product containing hexagonal, spindle-shaped lamellae in an isometric matrix.

In the dust we observed a mineral D occuring in idiomorphic grains intergrown with ilmenite, rutile, and baddeleyite. Mineral D is enormously hard (probably >9) and shows some internal reflections.

Native copper was observed only once in the coarse crystalline rock 10047,13. This is not unusual if we remember that native copper occurs in many basalts.

Troilite and iron are minor components in the igneous rocks and form mainly eutectic intergrowths. The troilite is stoichiometric in composition with traces of Ti (0.07 to 0.3 percent). The iron is nearly Ni free but contains up to 0.7 percent Co.

The breccias (10059,25 and 10068, 25) consist essentially of five groups of components: (a) fragments of the above described rock; (b) foamy, pumice-like fragments sometimes more abundant than the former; (c) glassy globules of two types; c_1 not homogenized, full of schlieren; c_2 completely homogenized, but partly devitrified; (d) globules of metal or troilite or both; and (e) splinters of meteorites, mostly nickle-iron.

Group (a) differs from the igneous rocks described above only through indications of shock metamorphism (for example, formation of rutile and hematite and stress twinning in ilmenite). Group (b) consists of material of different origin, partly fused to glass but not evaporated. It represents prevalently lunar material as observed from the nondigested remnants—mainly ilmenite and anorthite. During fusion plentiful gas formed many gas bubbles, yielding pumice-like structures. Minute iron globules are quite abundant, more abundant than in the igneous rocks, and very probably originated from evaporated nickel-iron of an impacting meteorite. Globules of c_1 have variable refractive indices, since they are full of schlieren. Globules of c_2 type may have variable refractive indices between 1.45 and 1.90. As mentioned above, c_2 globules are clear, without inclusions or bubbles. Such homogeneous glasses could only be formed in short times by condensation from vapor. Strong variation in reflectivity indicates remarkable variation in chemistry. The presence of evenly distributed fine iron dust supports the idea that globules are condensed from a silicate cloud evaporated following a meteorite impact.

The rose or blue colored pearls or globules (*1*) are composed of acidic glass with the lowest refractive index. Those colors are caused by colloidoptic effects (like colors of rose quartz or blue quartz) of finest devitrification products. Practically opaque basic glasses with high *n* show much stronger devitrification into different components below the resolution power of the microscope.

A peculiar single observation may be added here. A large type c_2 globule contains, besides a few tiny iron spherules in its center, thousands of iron spherules at the outermost layer of the glass globule (Fig. 1C). Dr. R. Bloch (Bersheba) (*3*) gave a possible explanation. The fact that iron spherules form a complete layer without touching each other may indicate that this phenomenon is probably a frozen flotation process. The lyophobe, equally loaded spherules of iron swim up quickly in the lyophile silicate melt. The equal loading of the iron spherules prevents their coagulation.

Group (d) globules of iron or troilite or both are partly identical to the condensation spherules of Meteor Crater in Arizona (*4*). Often they contain (as in Meteor Crater) thin films of schreibersite. In any case they are products of large iron meteorite impacts. Obviously the relative quantities Fe : FeS : Fe_3P will vary. Sometimes Fe : FeS is nearly 1 : 1 or 1 : >1, a ratio very rare in iron meteorites (except Soroti), but in stony meteorites such a ratio is common. Nearly always Ni : Fe is high, so that a genetic relationship to lunar iron is precluded.

Group (e), splinters of meteorite iron, disrupted by the impact explosion at any stage, are rare. We observed such splinters only three times in the breccias and often in a concentrate of heavy minerals from the dust. Sometimes they reach a size of 1 mm or more. After these explanations, the dust merits only a short discussion. Only the relative abundance of the groups (a) to (e) vary.

The magnetic and heavy fractions,

however, contain some constituents observed only in minute quantities in the breccias. First, the above-mentioned splinters of iron and globules of iron and troilite, and so on are different. Baddeleyite and dysanalyte are abundant in this fraction and may be studied in detail. Last we discovered well-developed octahedral crystals, distinctly zoned, in all properties (polishing, reflectivity, color, up to details of the natural faces) very similar to magnetite; very probably a new spinel (Fig. 1D).

The iron spherules in the dust also have their pecularities. Some spherules contain a radiating structure of troilite and silicate inclusions (in the form quite similar to enstatite chondrules). In others the ratio Fe : FeS and their intergrowth varies strongly. One splinter of schreibersite was observed with idiomorphic kamacite crystals.

PAUL RAMDOHR, AHMED EL GORESEY
Max Planck Institut Für Kernphysik, Heidelberg, Germany

References

1. Lunar Sample Preliminary Examination Team, *Science* 165, 1211 (1969).
2. P. Ramdohr, *Monatsber. Deut. Akad. Wiss. Berlin* 7, 12 (1965).
3. R. Bloch, personal communication.
4. P. Ramdohr, *Chem. Erde* 25, 3, 259 (1966).

4 January 1970

Search for Magnetite in Lunar Rocks and Fines

Abstract. *Magnetite crystals larger than 2 micrometers are absent from rocks and fines. Smaller opaque spheres in the fines can tentatively be identified as magnetite. Their concentration is not higher than 1 × 10^{-6} particle per particle smaller than 1 millimeter. In the fines from the sampling site, the contribution of material similar to type 1 carbonaceous meteorites is insignificant, either because it never existed, or because it was evaporated or comminuted by impact or was diluted by indigenous material. Other magnetite habits typical of carbonaceous meteorites or possibly of cosmic dust or comets were also sought without success—such as rods, platelets, framboids, spherulites, and idiomorphic crystals.*

The magnetite in Wiik's type 1 carbonaceous meteorites shows well defined crystal habits, some of which are exclusively found in these meteorites (*1*). These habits are easily recognizable under the reflecting microscope; their properties emerge sufficiently and thus the specific particles are quickly discernible, even when in small proportions in mixtures.

This mineral may thus be used as a path-finder for Orgueil-like materials, either indigenous or foreign. The discovery on the moon of materials belonging to this type is very important for the theories of the moon, planets, meteorites, comets, and cosmic dust (*2*). We thus undertook a search for the typical magnetite crystals in lunar materials, with the fines representing a mixture of indigenous and foreign materials and the hard rocks providing a good basis of comparison with materials of lunar origin.

Our results do not prove any theory and must be considered as part of a series dealing with sampling sites with various histories.

We studied rocks 10060,40 and 10072,31 and fines 10084,96. The rocks were analyzed by photonic microscopy (transmitted and reflected light), electron microprobe, x-ray diffraction, and scanning electron microscopy. An original method of mounting and polishing small quantities of micron-sized particles was devised (*3*). Control of contamination was attained by mounting the fines in the glove box, filtering all fluids on Millipore, and running blanks concurrently with all the experiments.

The only oxide in sample 10072,31 that could superficially be confused with magnetite was ilmenite. It occurs as idiomorphic laths (sections of thin platelets) and as larger poikilitic monocrystals (0.1 to 1 mm) with an idiomorphic tendency. Optical properties are those which may be observed and measured on pure ferriferous members in terrestrial rocks. A decisive property for differentiating ilmenite from magnetite is the pleochroism in air and oil (*4–6*). Table 1 summarizes the data obtained for lunar and terrestrial ilmenites, as well as terrestrial magnetite. The ilmenite in this hard rock is never twinned. Two other oxides were found: rutile as fine exsolutions and rare chromite.

Rock 10060,40 is very friable, made up of a loose conglomerate of all the rock and mineral fragments found in the fines.

From fines sample 10084,96, transparent polished sections were prepared as follows: (i) The powder was sieved at less than 1 mm; (ii) the size fractions obtained by sieving under a CCl_4-jet were 1 to 0.250 mm, 0.250 to 0.100 mm, 0.100 to 0.044 mm, and less than 0.044 mm; (iii) the heavy liquid centrifugates obtained from each size fraction had densities (g/cm^3) of $d<2.8$, $2.8<d<3.3$, and $d>3.3$. Magnetic separations were tried without noticeable success. The calculated concentration factor for the heaviest fraction of grain smaller than 0.044 mm is about 30.

Several free opaque or half-opaque minerals were found: ilmenite, sulfides, metallic iron, hematite, and goethite (?). Sulfides and iron are often altered into transparent, red, or brown low-reflecting oxides, which seem to be at least partially due to artifacts.

Ilmenite occurs in the fines as irregular fragments with conchoidal fractures. It is transparent under intense illumination (different from magnetite) and often twinned. Spheres of ilmenite have never been found, but opaque spheres of very low reflectivities are frequent (chromite and glass of various composition?). No one opaque mineral whose optical and morphological properties could fit in with those of magnetite has been found. Because the number of scanned heavy particles comes to at least 1×10^6, the concentration of typical magnetite crystals does not exceed 1 in 3×10^7 particles smaller than 1 mm.

Table 1. Reflectivity of magnetite and ilmenite. Not determined, N.D.

Sample	Reflectivity at 546 nm (%) (with maximum absolute error)	
	In air	In oil (n_{546} = 1.517)
SiC standard	20.61 ± 0.01	7.51 ± 0.01
Magnetite from Bisperg, Sweden	19.2 ± 0.5	6.9 ± 0.4
Ilmenite from Ilmen Mts., U.S.S.R. (R_0)*	20.1 ± 0.7	7.5 ± 0.7
Three Mg-free ilmenites (R_0)		
Miask, U.S.S.R.†	19.9	N.D.
Ilmen Mts, U.S.S.R.†	20.1	N.D.
Bodenmais, Germany†	19.5	N.D.
Ilmenite from lunar sample 10072,31 (R_0)	19.3 ± 0.5	6.9 ± 0.2

* See (*5*). † See (*6*).

The only objects that could possibly correspond to magnetite are small opaque spheres of steel-gray shine and with reticulated surface. Diameters do not exceed 1 or 2 μm. Their concentration, measured in the total powder and in the concentrates, is less than 1×10^{-6}.

Ilmenite has been found everywhere with a series of habits and dimensions in broken pieces of crystalline rocks. Some fragments have their ilmenites strongly cataclazed and twinned, whereas others show no shock effects at all. The only other oxides found are rutile, chromite, and pseudobrookite (?), all associated with ilmenite.

One of the most frequent varieties of glass is that with spherical inclusions of iron-nickel, twinned ilmenite, bubbles, and flow structures. The ilmenite crystals have their usual fracture form, but with a little rounding of the sharp edges. They have therefore not been completely melted, as opposed to iron and glass *(7)*.

Minerals from conglomeratic fragments are made up of all the minerals, rock fragments, and spheres in the fines. The ilmenites are often twinned.

About 50 oxide grains from rock 10072,31 and about 100 from the fines 10084,96 were analyzed by electron microprobe for Fe, Ti, Mg, Cr, and Mn. We did not find any mineral that had Fe as a major element, and that had less Ti than ilmenite. Two high Ti minerals have been found associated with ilmenite: rutile and a gray mineral similar to pseudobrookite *(7)*. A few percent of Mg and traces of Mn are generally present in the ilmenites. These elements do not vary very much in relation to habits.

Under the scanning electron microscope the vuggy rock 10072,31 shows black opaque minerals of two basic forms: hexagonal platelets and octahedra. As both are undoubtedly ilmenite, the last pseudo-form can immediately be interpreted as a combination of a trigonal rhombohedron $(10\bar{1}1)$ (?) with the basis (0001) *(8)*. Etch pits are absent from both habits, but pseudo-octahedra show a strong tendency to grow as skeletons.

A powder diagram of a pseudo-octahedra group gave all the ilmenite lines and no others *(9)*, which confirms the homogeneity observed under the photonic microscope. X-ray diffraction patterns obtained from the heavy fraction showed ilmenite to be the sole constituent.

Concerning the absence of magnetite in crystalline rocks, it has been shown that high concentrations of Ti and low oxygen pressures prevent the separation of a magnetite phase *(10)*.

It is difficult at this time to interpret the negative results in relation to the specific problems of carbonaceous meteorites. If the absence of magnetite is confirmed in other indigenous materials, the interpretation of later magnetite finds in the lunar fines from other sites will be made easier.

The total absence in the fines of magnetite crystals typical of carbonaceous meteorites confirms the uniqueness of these materials and suggests that one or more very selective phenomena are active. Several hypotheses may be considered. (i) Indigenous lunar rocks of composition similar to type 1 carbonaceous meteorites are absent from the sampling site. The preservation of ilmenite included in rocks, glass, and conglomerates shows that if such magnetite-bearing rocks had ever existed, this mineral could have withstood impacts at both low and high velocities. (ii) Meteorites of type 1 impinging on the moon with cosmic velocities are completely volatilized. But one may point out that the occurrence of just slightly rounded ilmenite in vesicular glass (see above) and of almost undisturbed magnetites in basalts which have undergone nuclear shocks *(11)* testifies to a good preservation of meteoritic magnetites. (iii) Carbonaceous meteorites of type 1 are rare objects in the solar system. Mixed with very high quantities of indigenous material, the typical crystals could not be found in the fines, even after being concentrated 30 times.

The absence of magnetites of typical habits, and of magnetite in general, shows that Orgueil-like materials are probably not common rocks on the moon's surface. This result gives us a comparison term for the study of other sampling sites, where the specific phenomena responsible for the appearance of Orgueil-like materials could have occurred, such as falls of comets or cosmic dust-balls.

JACQUES JEDWAB
ALAIN HERBOSCH
Laboratoire de Géochimie, Université Libre de Bruxelles, 50, Avenue Roosevelt, 1050 Brussels, Belgium
ROLAND WOLLAST
GEORGES NAESSENS
NICOLE VAN GEEN-PEERS
Laboratoire de Chimie des Solides, Université Libre de Bruxelles

References and Notes

1. J. Jedwab, *Earth Planet. Sci. Lett.* **2**, 440 (1967); L. H. Ahrens, *Origin and Distribution of the Elements* (Pergamon, Oxford, 1968), p. 467.
2. B. Nagy, *Geol. Foren. Stockholm Forsch.* **88**, 235 (1966); J. W. Larimer, *Geochim. Cosmochim. Acta* **32**, 1187 (1968); E. J. Öpik, *Irish Astron. J.* **8**, 185 (1968).
3. J. Jedwab, M. Defleur-Schenus, A. Herbosch, in preparation.
4. P. Ramdohr, *Die Erzmineralien und ihre Verwachsungen* (Akademie-Verlag, Berlin, 1950) p. 658.
5. K. von Gehlen and H. Piller, *N. Jahrb. Mineral. Monatsh.* **1965**, 97 (1965).
6. B. Cervelle, *Bull. Bur. Rech. Géol. Minières. Ser. II* **6**, 3 (1967).
7. A. El Goresey, *Geochim. Cosmochim. Acta* **28**, 1881 (1964).
8. C. Palache, H. Berman, C. Frondel, *Dana's System of Mineralogy* (Wiley, New York, 1944), vol. 1, p. 528.
9. L. G. Berry and R. M. Thompson, *X-Ray Powder Data for Ore Minerals* (Geological Society of America, New York, 1962), p. 175.
10. A. F. Buddington and D. H. Lindsley, *J. Petrol.* **5**, 310 (1964).
11. N. M. Short, *Goddard Space Flight Center Report X-664-69-117* (1969).
12. We thank W. De Keyser, who made the microprobe accessible for this work, and Mrs. Blankoff-Scarr, who helped in the translation.

4 January 1970

Quantitative Optical and Electron-Probe Studies of the Opaque Phases

Abstract. *Ilmenite, chrome-titanium spinel, ulvöspinel, troilite, native iron, iron-nickel alloy, and native copper are present. In addition mackinawite is provisionally identified. Three additional opaque phases are present but not identified. Modal analysis indicates that the breccia is enriched in iron relative to ilmenite and troilite, and the high nickel content of this iron suggests that it is largely of meteoritic origin. The bulk composition of liquids corresponding to iron/troilite droplets in the rocks indicates an oxygen fugacity no greater than $10^{-15.5}$ and a sulfur fugacity of 10^{-6} bar. Complete melting of rocks produced a glass containing complex iron/troilite globules and skeletal ilmenites.*

Optical and electron-probe studies of opaque phases in samples 45.35.5, 48.42, 58.23, 72.46, 84.5, and 85.4 indicate the presence of ilmenite, chrome-titanium spinel, ulvospinel, troilite, native iron, iron-nickel alloy, and native copper. In addition mackinawite is tentatively identified. Two phases, one with very low reflectance, the other with higher reflectance, form fine lamellas in

Table 1. Modal analyses.

Rock type and number	Ilmenite	Troilite	Iron
A (45.35.5)	9.66	0.132	0.028
A (72.46)	12.8	0.210	0.013
B (58.23)	14.2	0.066	0.027
C (48.42)	3.03	0.026	0.022*

* This includes iron-nickel alloy.

ilmenite. These have not yet been identified. A fine flamelike exsolution in troilite is not yet identified.

Ilmenite in rocks of type A (basaltic, specimens 45.35.5 and 72.46) exhibits two modes. The first mode comprises randomly oriented needles up to 1 mm long. They are enclosed within pyroxene crystals but occasionally show slightly indented margins against randomly oriented feldspar laths. The second mode comprises subangular grains molded on feldspar laths but intergrown with pyroxene and represents the major amount of ilmenite in the rock. Small amounts of chrome-titanium spinel are associated with this ilmenite mode. Ilmenite in sample 45.35.5 carries a fine-grained lamellar exsolution parallel to $[10\bar{1}1]$ of a brighter, whiter anisotropic phase. This exsolution is sometimes accompanied by a second phase comprising a very dark, finer lamellar phase parallel to [0001]. The brighter phase may occur alone in ilmenite, but the darker exsolution is always associated with the brighter phase and never occurs alone. Textural relationships of these exsolution phases suggest that they developed simultaneously. Well-defined occasional cracks are present in the ilmenites and in the rock as a whole and slightly offset the brighter exsolution lamellas. A hairline crack in ilmenite carries a bright phase which is considered to be troilite. This development is also noted in sample 72.46, where the crosscutting bright phase runs out along the crack a small way into the surrounding silicates.

Ilmenite in rocks of type B (doleritic, specimen 58.23) consists mainly of coarsely crystalline, near-rectangular crystals up to 1 mm long partially molded on feldspar and intergrown with pyroxene. Ilmenite also occurs rarely intergrown with ulvöspinel in fine-grained, late-stage euhedral crystals embedded in pyroxene. These crystals have textural features characteristic of subsolidus unmixing and have a core of what in polished section appears to be a graphic intergrowth of feldspar and pyroxene. Although these textures appear to be of a subsolidus nature, substantial oxidation would be required for their formation. It is likely therefore that the simultaneous crystallization of the cubic and rhombohedral phases has taken place. These complex crystals are associated with troilite and are surrounded by a well-defined reaction rim in the enclosing pyroxene which appears generally lighter in polished section than the main mass of pyroxene. The ilmenites in the above mentioned rock types are generally untwinned although occasional needlelike twins are observed.

In breccia specimen 48.42 ilmenite occurs as single relict grains or enclosed within rock fragments. The coarse fines, sample 85.4.16, contain chips of rocks of the two primary igneous varieties having ilmenite textures similar to those described above. One fragment of type A rock lacks the second mode of subangular crystals described above; in this case the first mode of acicular crystals is very well developed. Another fragment of type A rocks shows a fine-grained equigranular texture with subangular ilmenites.

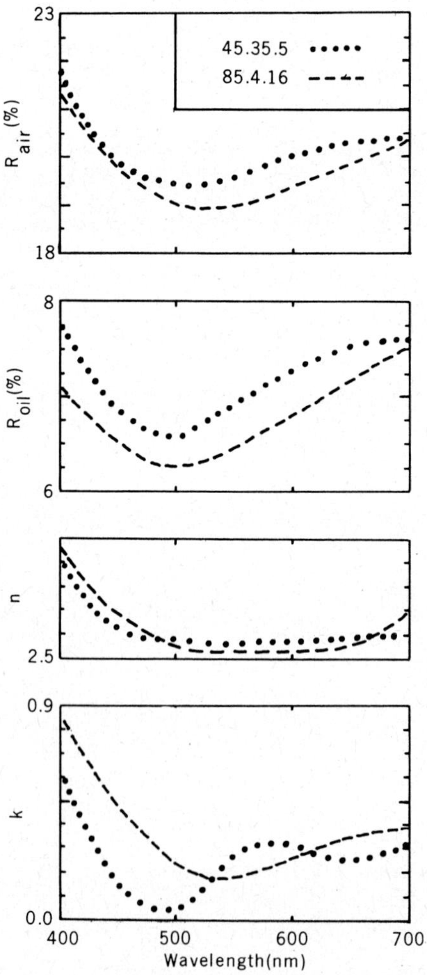

Fig. 1. The reflectance, R, in air and oil, refractive index, n, and absorption coefficient, k, in the visible spectrum for the ordinary vibration direction of ilmenite from samples 45.35.5 and 85.4.16. The immersion oil used is Cargille type "A" at 22.5°C.

Table 2. Analyses of metals and alloys. Sample 10045·35 is a type A rock in which iron occurs intergrown with troilite. Sample 10048·42 is a breccia; in this rock iron with a high nickel content occurs alone and may have a meteoritic origin, whereas the iron with high cobalt content is associated with troilite and is more likely to be derived from the igneous rock types.

Metal	Sample			
	10045·35	10045·35	10048·42	10048·42
Fe	99.19	99.50	89.10	98.12
Co	0.65	0.50	0.58	1.23
Ni	0.00	0.00	10.32	0.64
Total	99.84	100.00	100.00	99.99

In a glassy fragment from coarse fines, sample 85.4.14, ilmenite occurs as skeletal crystals; troilite and native iron are the only other crystalline phases in this glass. One rock fragment from coarse fines, sample 85.4.16, carries an ilmenite grain with multiple curved twin lamellas which coincide with actual displacements in the grain. This ilmenite crystal is on the margin of the rock fragment. Other ilmenites within the fragment appear normal, but optical studies (Fig. 1) of one of these crystals indicate that they too may be affected by shock metamorphism. Oxide distribution in the fine fines, sample 84.5, is very similar to that in the breccia described above.

Troilite is present as an accessory phase in all the different types of material. It is usually observed intergrown with native iron. In the case of one relatively large droplet in sample 72.46, the iron has migrated to form a continuous rim on one side of the sulfide droplet. An unidentified flamelike exsolution of possible pentlandite is observed within troilite at the contact with iron. In the same rock another relatively large troilite droplet devoid of iron has a rim of spinel where the troilite is molded on an early ilmenite lath. In general, in this rock the troilite occurs intergrown with hexagonal grains of iron which together form ragged patches enclosed within silicates or alternatively have smooth contacts molded on ilmenite. Composite iron/troilite ovoid droplets occur enclosed within ilmenite. Where this occurs calculations indicate that the composition of the liquid from which the droplet crystallized contained 68 percent Fe, which corresponds closely to the composition of the eutectic at 988°C in the system Fe-S at 1 bar (*1*). The oxygen fugacity for a liquid of this composition would be less than $10^{-15.5}$ and the sulfur fugacity would be 10^{-6} at 1050°C (*2*). However, the absence of iron and the presence of spinel in association with troilite in the large droplet described above indicate local variation in the oxygen and sulfur fugacities of the sulfide liquid. The oxygen and sulfur fugacities at 1050°C for magnetite, pyrrhotite and liquid are $10^{-10.04}$ and 1 bar respectively (*2*). The smaller of the sulfide liquid droplets are more likely to be buffered by the host silicate magma. The fine-grained iron/troilite droplets therefore probably indicate the very low fugacities of oxygen and sulfur for the silicate magma.

In sample 45.35.5 native copper occurs in association with troilite and iron as a small segregation in troilite at a troilite/ilmenite contact. This is further evidence of the low oxygen and sulfur fugacities since the native copper appears to have crystallized as a primary phase.

In sample 72.46 the ragged boundaries of the troilite/iron composite droplets enclosed within silicates contrast strongly with the smooth contacts against ilmenite. A fine-grained disseminated bright phase in silicates bordering troilite globules may represent an iron residuum from the breakdown of troilite. This sulfur loss may be related to the process responsible for natural cracks in the rock. In one case iron has migrated along a crack and cut a composite iron/troilite droplet enclosed within ilmenite. In another case possible troilite is present in a crack crosscutting ilmenite. These features may best be explained by a metamorphic event which resulted in loss of sulphur after the rock had solidified and which occurred while it was hot.

In the glasses of sample 85.4.14, grain 12, troilite and iron occur intergrown in spherical globules (*3*). The variety of textures observed may be accounted for by the phase relations in system Fe-S below 1600°C (*1*). Liquids range in composition from pure iron to troilite. This implies (*1*) temperatures over 1535°C at one atmosphere for the iron sulfide liquids in this sample. In general the smaller the globule the greater the probability that it consists entirely of iron. A diffusion of very fine globules showing a flow pattern occurs in glass in sizes down to the resolution limit of the microscope. Each iron/sulfide globule tends to act as a complete system not in equilibrium with adjoining globules from which it may differ radically in bulk composition. Iron globules give evidence of shearing in glass where iron forms infilling in a zone of dilation, thereby indicating that some glass suffered mechanical deformation shortly after solidifying. A strongly anisotropic phase is noted forming fine spots and bands within troilite in a troilite/iron globule in a chondrule-like body from sample 85.4.14, grain 7. This phase is tentatively identified as mackinawite.

Modal analyses for the principal opaque phases in rocks are given in Table 1. The primary igneous rocks (types A and B) have comparable amounts of ilmenite, troilite, and iron, whereas the breccia (type C) exhibits depletion of ilmenite and troilite relative to the primary igneous rocks. Iron modes for the breccia are comparable with those of primary igneous rocks. However, the presence of nickel in amounts ranging from 0.6 precent up to 10.3 percent in the iron in the breccia suggests that this iron has a diverse origin.

The low modal amounts of ilmenite and troilite in breccia relative to primary rock types suggest that ilmenite and troilite of the primary igneous rocks are destroyed in the glass-forming process. The presence of skeletal ilmenite crystals and composite iron/troilite globules in glass from coarse fines, sample 85.4, supports this conclusion. The glass-forming process may also involve some loss of sulfur. Ilmenite, troilite, and iron are unlikely to be preferentially preserved or destroyed in this process. However, material of iron or iron-nickel composition is enriched relative to ilmenite and troilite in the breccia by a factor of 4 as compared with the abundance of iron in the primary igneous rocks. It would appear, therefore, that approximately 75 percent of the iron or iron-nickel alloy in the breccia could have a meteoritic origin.

Reflectance measurements (*4*) in air and oil have been made on selected ilmenites to determine the reflectance, R, absorption coefficient, k, and refractive index, n, in the wavelength range 400 to 700 nm (Fig. 1). The shocked ilmenite described above in sample 85.4.16 exhibits simple absorption of the ordinary ray, whereas the unshocked ilmenite from sample 45.35,5 displays a curve with a maximum at about 590 nm, which may be indicative of more complex bonding.

The following phases were analyzed by electron probe: ilmenite, chrome-titanium spinel, ulvöspinel, native iron, native copper, iron-nickel alloys and troilite. All data are corrected by computer programs written by Aucott (*5*). The nickel content of iron in breccia varies between 0.6 and 10.3 percent, whereas nickel is low, or lacking, in iron in the igneous rocks. Cobalt is often present up to 0.65 percent in these cases. Table 2 presents analyses for some phases. Ilmenites in all rock types are unzoned and carry appreciable but variable amounts of Mg, Al, Si, Ca, Cr, and Mn.

Primary igneous rocks studied have a simple mineralogy as far as opaque minerals are concerned. The opaque mineral assemblages have crystallized under conditions of very low oxygen and sulfur fugacities. Similar opaque mineral assemblages are present in glasses, but in this case equilibrium has not been established. Optical data indicate that ilmenites in the fines are affected by shock metamorphism. The fines contain some metal of meteoritic origin.

P. R. SIMPSON
S. H. U. BOWIE
Institute of Geological Sciences, 64-78 Gray's Inn Road London, W.C.1. England

References and Notes

1. G. Kullerud and H. S. Yoder, *Econ. Geol.* **54**, 4 (1969).
2. A. J. Naldrett, *J. Petrology* **10**, 2 (1969).
3. R. Brett, *Amer. Mineral.* **52**, 721 (1967); A. El Goresey, *Shock Metamorphism of Natural Materials* (Mono, Baltimore, 1968), p. 531.
4. P. R. Simpson, in preparation; D. R. Curry, Analytical & Ceramics Unit Report No. 37, Institute of Geological Sciences, London (1969).
5. J. W. Aucott, unpublished electron-probe correction programs, Institute of Geological Sciences, London.
6. We thank M. P. Jones (Imperial College of Science and Technology) for the use of a Quantimet Image Analysing Computer for the modal analysis.

4 January 1970; revised 7 January 1970

Lunar Troilite: Crystallography

Abstract. *Fine, euhedral crystals of troilite from lunar sample 10050 show a hexagonal habit consistent with the high-temperature NiAs-type structure. Complete three-dimensional counter intensity data have been measured and used to confirm and refine Bertaut's proposed low-temperature crystal structure.*

The rare mineral troilite is formed under strongly reducing conditions and has previously been found in nature only in serpentinized rocks and meteorites. Troilite was early recognized as a very minor but generally occurring constituent of the material returned by Apollo 11 from the Sea of Tranquillity.

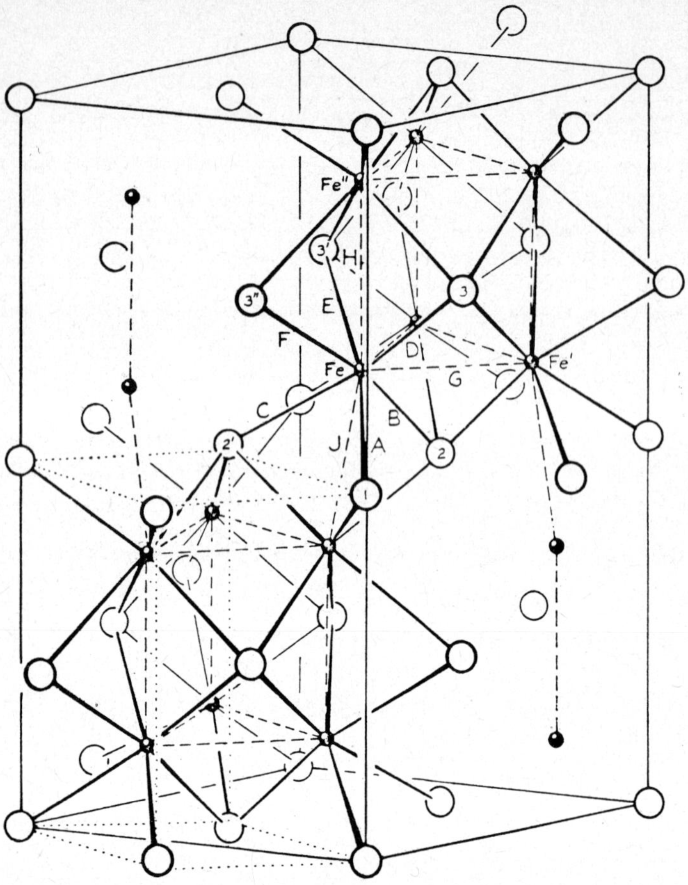

Fig. 1. Crystal structure of troilite. Dashed lines, shortest Fe-Fe distances; dotted lines at lower left, the NiAs-type subcell. Interatomic distance vectors are listed in Table 2.

Table 1. Structure and thermal parameters for lunar troilite. Standard errors are shown in parentheses in terms of the last significant digits.

Atom	Site	Structure parameters *x*	*y*	*z*
Fe	12(*i*)	0.3797(3)	0.0551(3)	0.1230(1)
S_1	2(*a*)	0	0	0
S_2	4(*f*)	⅓	⅔	0.0199(4)
S_3	6(*h*)	0.6652(9)	−0.0031(8)	¼

Atom	Thermal parameters β_{11}	β_{22}	β_{33}	β_{12}	β_{23}	β_{31}
Fe	0.0100(4)	0.0087(3)	0.00167(4)	0.00049(3)	−0.0001(2)	0.0000(2)
S_1	0.0051(11)	$= \beta_{11}$	0.0025(3)	$= \beta_{11}/2$	0	0
S_2	0.0064(6)	$= \beta_{11}$	0.0018(2)	$= \beta_{11}/2$	0	0
S_3	0.0060(8)	0.0051(8)	0.0019(1)	0.0027(7)	0	0

Table 2. Interatomic distances and thermal ellipsoids in lunar troilite. Standard errors are shown in parentheses in terms of the last significant digits.

To atom	Interatomic distances for Fe (vectors labeled in Fig. 1) Vector	Distance (Å)
S_1	A	2.565(2)
S_2	B	2.379(3)
S_2'	C	2.504(3)
S_3	D	2.359(4)
S_3'	E	2.416(4)
S_3''	F	2.721(4)
Fe′(2)	G	2.919(3)
Fe″	H	2.947(3)
Fe‴	J	2.984(3)
Fe		3.669(3)
Fe		3.796(3)

Atom	Thermal ellipsoids, rms axes (Å) *u*	*v*	*w*
Fe	0.104(3)	0.109(2)	0.116(2)
S_1	0.083(14)	$= u$	0.131(8) ($\parallel c$)
S_2	0.092(8)	$= u$	0.114(6) ($\parallel c$)
S_3	0.083(7)	0.091(7)	0.116(4) ($\parallel c$)

So far, troilite is the only sulfide positively identified in these lunar samples.

Troilite is found in the massive lunar material as thin stringers in ilmenite, or small blebs up to 0.3 mm in size, which are commonly finely polycrystalline. The masses of troilite often contain small particles of metallic iron. The mineral also sometimes appears on the walls of cavities, where it forms bright metallic yellow patches, rounded as though solidified from a molten droplet. In lunar sample 10050, two fine, euhedral single crystals were found implanted on pyroxene crystals that lined a vug. One of these crystals was removed for single crystal study, the results of which are reported below.

The crystals are well developed, with bright faces that give good optical signals. The crystal studied was the upper part of a somewhat barrel-shaped habit. The forms *m*, *p*, and *s* are those that would be expected for the simple NiAs-type crystal unit cell that prevails above 140°C. (Miller indices are $\{10\bar{1}0\}$, $\{10\bar{1}1\}$, and $\{10\bar{1}2\}$, respectively, for the high-temperature cell and $\{11\bar{2}0\}$, $\{11\bar{2}2\}$, and $\{11\bar{2}4\}$ for the low-temperature cell.) The crystal gave sharp, undistorted Buerger precession diffraction patterns and was very well suited to crystal structure study. Another euhedral crystal taken from lunar sample 10047 proved to be an aggregate of subparallel crystals ranging in orientation over 5°.

X-ray powder diffraction patterns (Debye-Sherrer) and single-crystal, x-ray goniostat measurement give *d*-spacing and unit cell data that are in close agreement with those of Haraldsen (*1*). The unit cell is hexagonal, space group $P\bar{6}2c$, with $a = 5.962 \pm 0.002$ Å, $c = 11.750 \pm 0.003$ Å, and a cell content of 12 FeS. This cell represents a superstructure of the simple NiAs-type structure, as originally found by Hägg and Sucksdorff (*2*), based on small displacements of the Fe and S atoms from their ideal positions. The chemistry is assumed to be very close to stoichiometric. Experimental work (*3*) has shown that, up to very high temperatures, for the phase $Fe_{1-x}S$, x does not take any measurable negative value. Other investigators have found that lunar troilite may contain up to a total of 1 percent Co, Ni, and Ti in solid solution, but the crystal studied here has not been analyzed. The density calculated for pure FeS is 4.841 g/cm³.

With a Picker automatic diffractometer, diffraction intensities were measured with MoKα radiation by 2θ scan to a limit of $(\sin\theta)/\lambda = 1.25$, encompassing 1135 independent reflections. Of these, 600 gave values more than twice the standard error based on counting statistics, and these were used for the structure analysis. The low proportion of observed reflections is characteristic of the systematic intensity weighting of a superstructure lattice of this type. The data were processed in the usual way for Lorentz and polarization corrections, but no account was taken of absorption or extinction. Although absorption is moderate (linear coefficient = 135.2 cm^{-1}), it is ignored in this study because the crystal is small (~0.15 mm). The structure parameter overdetermination is quite large (600 intensities/6 parameters), and the final standard error for these parameters is satisfactorily low.

The refinement was carried out by least squares analysis (*4*). The mode is anisotropic, and the program automatically controls the parameter dependencies imposed by the symmetry. A scale factor, 6 structure parameters, and 14 thermal parameters were allowed to vary. The data were treated as F values with unit weights. The starting parameters were those of Bertaut (*5*), which gave $R = 0.29$; convergence was complete after four cycles. Anomalous dispersion effects were included, but, without absorption corrections, they did not significantly affect the final result. The final unweighted conventional reliability factor was $R = 0.084$. The structure factors are listed in a table not included here (*6*).

Bertaut (*5*) previously solved the problem of the nature of the distortion of the ideal NiAs structure inherent in troilite. Using visually estimated Weissenberg data obtained with CoKα radiation from a synthetic crystal, he determined the space group and a structure consistent with the intensities of the superstructure reflections. Without considering the thermal motions, he derived his parameters by least squares analysis of 31 of these data, obtaining $R = 0.195$. The present analysis of lunar troilite wholly confirms Bertaut's structure proposal and now measures the bond lengths with a standard error of the order of 0.004 Å. The final parameters are listed in Table 1.

The structure is characterized, as Bertaut (*5*) found, by a moderate drawing together of the Fe atoms into triangular groups normal to the c axis (Fig. 1). The hexagonal close-packed sulfur framework remains essentially undistorted, except for a small displacement of S_2 (0.14 Å) parallel to the c axis away from the center of the Fe triangles. The Fe-Fe distances are apparently longer than they are in pyrrhotite, where they may range down to 2.84 Å. In fact, atomic displacements may be present in pyrrhotite also (as well as ordered vacancies), so that this estimate of Fe-Fe distance may be too low, and the distances may actually be comparable to those in troilite. The shortest Fe-Fe distance in troilite is in the horizontal triangular groups (2.920 Å), where the octahedra are edge-shared (3.43 Å in pyrrhotite), rather than in the c direction as might be expected, through the shared octahedral face (2.947 to 2.985 Å). The FeS_6 octahedron is rather severely distorted, with six different Fe-S bond lengths ranging from 2.359 to 2.721 Å. The dimensions of the thermal ellipsoids of the atoms are given in Table 2. The ellipsoid for Fe is nearly isotropic and for the S atoms is slightly elongated along the c direction. Absorption corrections may be needed to improve the reliability of the latter result.

Finally, the neutron powder diffraction study of troilite by Andresen (*7*) should be mentioned. In order to account for his neutron diffraction intensities, he found it necessary to adjust Bertaut's structure parameters for Fe to $x = 0.383 \pm 0.005$ and $y = 0.050 \pm 0.005$. These values are within one standard deviation unit of the ones in this study. Andresen further found that the magnetic moments of the Fe atoms are oriented parallel to the c axis in an antiparallel arrangement, with all moments in one horizontal layer pointed in one direction, opposite to those in the next layer. At the transition temperature of 140°C, where the superstructure disappears, the magnetic structure also changes, with the moments now directed normal to the c axis. The magnetic ordering then decreases with temperature increase to the Néel temperature of 327°C.

HOWARD T. EVANS, JR.
U.S. Geological Survey,
Washington, D.C. 20242

References and Notes

1. H. Haraldsen, *Z. Anorg. Allgem. Chem.* **246**, 169 (1941).
2. G. Hägg and I. Sucksdorff, *Z. Physik Chem.* **B22**, 444 (1933).
3. G. Kullerud, in *Researches in Geochemistry*, P. Abelson, Ed. (Wiley, New York, 1967), vol. 2, p. 286.
4. Least squares analysis, bond lengths, and errors were calculated with a computer program written by Dr. L. Finger, Geophysical Laboratory, Washington, D.C.
5. E.-F. Bertaut, *Bull. Soc. Franc. Mineral. Crist.* **79**, 276 (1956).
6. This table may be obtained upon request from the author (H.T.E.).
7. A. F. Andresen, *Acta Chem. Scand.* **14**, 919 (1960).
8. Supported by NASA contract T-75406.

6 January 1970

Opaque Minerals in Lunar Samples

Abstract. *Microscope study and electron microprobe analysis of lunar rocks and soil show that ilmenite, troilite, and native iron are accompanied by trace amounts of ulvöspinel, titanochromite (new mineral name), an unidentified Ti-Fe oxide, and a complex Zr-Y silicate. The assemblage requires a strongly reducing environment. Textures and modal proportions show that the rocks present are not a differentiation series. The restricted nature of the opaque mineral assemblage suggests a narrow range of composition for the materials from which the parent liquids of the rocks were generated. Textural variety must reflect differences in cooling rates, probably related to depths of formation.*

Opaque minerals in four samples of lunar igneous rocks (10022, 10044, 10049, and 10058), in one fragmental rock (10060), and in rock fragments separated from a sample of the less than 1 mm fraction of the lunar soil (10084) were studied by microscope and electron microprobe. Principal attention was given to the crystalline materials, but opaque minerals in glass fragments and spheres were examined briefly.

In all four igneous rocks and in all holocrystalline rock fragments in the other samples, the dominant opaque mineral is ilmenite, 10 to 17 percent by volume in various rocks. Troilite containing globules of iron is the next most abundant mineral, ranging from 0.3 to 1.3 percent by volume. The troilite-iron ratio in 10044, as indicated by results of image analysis, is roughly 35 to 1. Though minor in amount, troilite and native iron are present in all holocrystalline rocks and rock fragments and are almost uniformly disseminated.

Other minerals thus far found, all in trace amounts, are the following: (i) An aluminous ulvöspinel, disseminated in 10044 and 10058 and found

Table 1. Chemical compositions of minerals in lunar samples. The samples are: 1. 10022, ilmenite. 2. 10049, ilmenite. 3. 10044, ilmenite. 4. 10058, coarse ilmenite. 5. 10058, ilmenite intergrown with ulvöspinel. 6 to 8, Ilmenites in various rock fragments from lunar soil. 9 to 11, From single rock fragment in lunar soil—homogenous ilmenite (9); titanium-iron oxide (11), rimmed by ilmenite (10). 12. Ulvöspinel. 13. Titanochromite 2. 14. Titanochromite 1.

Sample	Oxides							
	TiO_2	Al_2O_3	Cr_2O_3	FeO*	MnO	MgO	CaO	Total
1	53.6	0.11	0.63	45.0	0.34	2.0	0.08	101.8
2	53.0	0.32	0.45	43.8	0.34	0.82	n.d.	98.7
3	52.9	0.14	0.27	45.8	0.37	0.93	n.d.	100.4
4	52.9	0.06	0.33	47.7	0.40	0.22	0.13	101.7
5	52.1	0.17	0.04	47.2	0.33	0.15	0.0	100.0
6	51.1	1.7	2.0	44.5	0.03	0.88	0.23	100.4
7	51.8	0.14	0.7	44.0	0.32	1.4	0.48	98.8
8	52.9	0.2	1.0	42.4	0.31	2.5	0.14	99.5
9	53.5	1.6	1.2	41.1	0.33	3.5	0.11	101.3
10	55.6	0.07	0.87	38.9	0.33	4.1	0.18	100.5
11	72.3	2.0	1.4	17.0	0.02	6.0	0.13	98.8
12	33.2	2.21	0.47	64.4	0.30	0.16	0.02	100.8
13	20.6	5.2	20.9	49.8	0.36	1.1	n.d.	98.0
14	17.4	10.1	26.6	39.5	0.30	5.0	n.d.	98.9

* Total iron as FeO.

also in a single rock fragment from lunar soil; (ii) titanochromite 1, a Ti-rich chromite found in one rock fragment from lunar soil; (iii) titanochromite 2, a second Ti-rich chromite found in a single rock fragment from the soil; (iv) an unidentified titanium-iron oxide, found in 8 fragments from lunar soil; and (v) a Fe-Ti-Zr-Y silicate disseminated in two lunar rocks (10044, 10058).

Compositions of oxide minerals were determined by microprobe analysis (Table 1). The analytical procedure of Bence and Albee (*1*) was used. Reflectivities of oxide minerals are given in Table 2. Accuracy of the procedure for major elements is considered to be 1 to 5 percent of the amounts present (*2*), but summations and calculations of atomic proportions for minerals reported here suggest an accuracy closer to the lower figure.

Ilmenite exhibits a variety of crystal habits, and much additional analytical work is needed to determine its range in composition. Data thus far obtained suggest that ilmenite in each of the four igneous rocks (samples 1 to 5, Table 1) is uniform in composition and that individual crystals are homogeneous. Compositions, except that of 10022, are those of common ilmenites; that is, elements other than Ti and Fe are present only in trace amounts. Some ilmenites from rock fragments in lunar soil, however, show Al_2O_3, Cr_2O_3, or MgO in excess of 1 percent (samples 6 to 10, Table 1). Analyses 9 and 10 are for ilmenite in a rock fragment in which the mineral occurs both as homogeneous crystals (sample 9, Table 1) and as rims on the unidentified Ti-Fe oxide (sample 11, Table 1). Both differ from normal ilmenite in MgO or Cr_2O_3, or both.

Ulvöspinel is known in earth rocks only from fine-scale intergrowths with titanomagnetite; hence its reflectivity (Table 2) has not previously been reported. In 10044 and 10058, it is always intergrown with ilmenite, but homogeneous areas as large as 120 by 50 μm appear in the sections. Its genetic relations to ilmenite are unclear. Textures can be interpreted to indicate either simultaneous crystallization with ilmenite or replacement of ulvöspinel by ilmenite. The indicated formula for ulvöspinel in 10058 (sample 12, Table 1) is $(Fe_{1.98}Mn_{.01}Mg_{.01})(Ti_{.90}Al_{.09}Cr_{.01})O_4$, remarkably close to pure Fe_2TiO_4.

The two other spinels reported above, like ulvöspinel, occur in relatively coarse-grained rocks, in crystals 220 μm or less in diameter. They differ in composition but are both essentially Ti-rich chromites. A preliminary search of the literature indicates that no comparable terrestrial spinel is known. The name titanochromite is accordingly proposed for members of the spinel group, such as those described above, that are rich in Ti, Fe, and Cr. Reflectivities, crystal habits, and hardness are within the range characteristic of members of the spinel group. Both titanochromites show incipient alteration to ilmenite along octahedral planes or grain margins. The indicated formulas are: Titanochromite 1 $(Fe_{1.15}Mn_{.01}Mg_{.26}Ti_{.45}Al_{.41}Cr_{.72})O_4$; titanochromite 2 $(Fe_{1.53}Mn_{.01}Mg_{.06}Ti_{.57}Al_{.22}Cr_{.61})O_4$.

Accessory mineral (iv), essentially a titanium-iron oxide, has been found only in rocks that appear to contain two generations of ilmenite, an earlier generation composed of relatively large, well-formed crystals, and a later one composed of thinly platy to skeletal crystals. In each of eight rock fragments, one or more grains (120 μm or less in diameter) consist of crystals of the oxide marginally replaced and therefore rimmed by ilmenite. The unknown thus appears to be an early-formed phase that reacted with the parent liquid to form ilmenite. Both the ilmenite of the rims (sample 9, Table 1) and that of the homogeneous crystals (sample 10, Table 1) are of abnormal composition. The unknown mineral is bluish gray and distinctly anisotropic (Table 2), with A_r equal to 1.0° in air and 1.9° in oil (n = 1.515). The reflectivity is in the range exhibited by terrestrial spinels, not that of the TiO_2 group of minerals, but the anisotropism suggests nonisometric symmetry, and calculation to a spinel formula shows a 20 percent cation deficiency. Structural

Table 2. Reflectivities (in percent; accuracy ± 0.2 percent) of ilmenite and spinels in lunar samples.

Wave-length (nm)	Ilmenite (10049)		Ulvö-spinel	Titano-chromite 1	Titano-chromite 2	Ti-Fe oxide	
	$R\omega$	$R\varepsilon$				R_1	R_2
450	18.9	15.8	15.1	14.0	14.8	14.1	15.2
470	18.6	15.6	14.7	13.8	14.8	14.0	15.0
500	18.5	15.4	14.6	13.6	14.8	13.8	14.7
520	18.5	15.4	14.8	13.4	14.8	13.7	14.5
546	18.7	15.7	15.1	13.6	15.0	13.4	14.4
589	18.9	16.1	15.6	13.7	15.1	13.3	14.3
620	19.2	16.6	16.1	13.9	15.2	13.2	14.2
640	19.5	17.0	16.4	14.0	n.d.	13.0	14.1

work is clearly needed if larger crystals can be found.

Accessory mineral (v) is found in blades and clusters of blades sparsely disseminated in 10044 and 10058. The largest crystal is 120 μm in length. The mineral is gray in reflected light, isotropic, and semiopaque. It has a reflectivity of 12.9 ± 0.2 percent at 546 nm. Dispersion of the reflectivity is negligible. The approximate composition indicated by microprobe analysis is (in percent) SiO_2 14, TiO_2 20, ZrO_2 16, Y_2O_3 5, Al_2O_3 1, FeO 43, and CaO 1.

Troilite has not been analyzed. In 10044, ten metallic globules in troilite gave 0.6 to 0.9 percent Co, averaging 0.7 percent. In 10058, ten globules gave 0.7 to 1.3 percent O, averaging 0.8 percent. No Ni was detected.

The total opaque mineral assemblage in the lunar samples examined has features in common with both terrestrial rocks and meteorites. It differs, however, from assemblages in terrestrial rocks in the association of ilmenite with troilite and native iron and in the presence of the other Ti-rich oxides here described. It differs from assemblages in meteorites in the high content of ilmenite, in the apparent absence of Ni in iron, and the presence of Ti oxides other than ilmenite.

The various holocrystalline rocks and rock fragments show a wide range of average grain size and a variety of textural fabrics. Average grain size of ilmenite ranges from 5 to 300 μm in various rocks. Fabrics range from subophitic to microgranular to combinations of tabular and skeletal ilmenite. Four features of this display appear significant. (i) Ilmenite and troilite plus iron are constantly associated in all holocrystalline rocks, regardless of grain size or fabric. (ii) No correlation between modal proportions of these minerals and grain size or fabric is apparent. (iii) Textural features that are characteristic of cumulates, such as igneous lamination, are absent. (iv) Textural relations of ilmenite suggest that it has formed in the early to middle stages of crystallization of the host rocks. Taken together, these features set narrow limits on the interpretation of the high Ti content of the samples. Specifically, there is no evidence of a differentiation series leading to concentration of ilmenite in special rock types. The appearance of the unknown Ti-Fe oxide as an early and transitory phase is too minor a feature to be of much significance, although it indicates local fractional crystallization involving the Ti-Fe content of the parent liquid. It is thus necessary to conclude that the crystalline rocks closely reflect the bulk compositions of the parent liquids and that these liquids had a limited compositional range. The striking variety of textural fabrics must be assigned to differences in rates of cooling probably related to depths of formation.

In terrestrial igneous rocks, comparable or higher concentrations of titanium minerals are found in both intrusive and extrusive rocks. Intrusive rocks are those of the Precambrian gabbroanorthosite complexes and those of the layered or zoned mafic intrusives. Concentrations in these rocks are variously ascribed to separation of immiscible liquids, to fractional crystallization and gravitative settling of early-formed Ti-Fe minerals, or to concentration of Ti and Fe in late residual liquids. The textural features and modal proportions of the lunar rocks are contrary to the operation of any of these mechanisms. The high concentration of titanium in the lunar rocks must therefore go back to the stage at which the parent liquids were generated, by melting of preexisting materials. Either these materials were already rich in Ti, or the process of melting was selective, and the resulting liquid or liquids became enriched in the process.

Lavas relatively rich in TiO_2 have been reported from certain basalts from Iceland (*3*), Hawaii (*4*), and north-central Oregon (*5*); all basalts reported to be of oceanic origin. There is a genetic resemblance between these and lunar rocks, in that high Ti content appears to reflect bulk compositions of parent liquids. The Ti-rich rocks of Hawaii are regarded as special differentiates of tholeiitic magmas, whereas there is no suggestion of such parental magmas in Apollo 11 rocks. The alkalic basalts of central Oregon are reported (*5*) to contain up to 11.5 percent opaques, mostly ilmenite, and Robinson regards these as derived from primary alkali basalt magma. Both the Hawaiian and Oregon basalts are alkali-rich, whereas Apollo 11 rocks are reported (see *6*) to be notably deficient in alkalis.

An appraisal of the chemical system in which the Apollo 11 rocks crystallized must rest on consideration of both opaque and transparent minerals, but some conclusions can be drawn from the opaque mineral assemblage alone. The constant association of ilmenite, troilite, and native iron suggests that these minerals are an equilibrium assemblage, one that must have formed in a strongly reducing environment with low oxygen fugacity. The absence of magnetite is consistent with this conclusion.

The reported absence (see *6*) of hydroxyl-bearing minerals and the presence of cristobalite instead of quartz suggests an anhydrous system crystallizing at high temperatures. Taken in sum, the chemical parameters seem significantly different from those that characterize igneous systems on earth.

Rock fragments from the lunar soil include numerous glass spheres and fragments, glass-welded fragmental material, and holocrystalline rock fragments with fusion crusts. Metallic globules are present in many glass fragments and spheres. The total mineralogy of these is very complex and has been given only preliminary investigation. The globules, however, are of special interest because, unlike the globules in the holocrystalline rocks, some of these contain, in addition to iron or iron plus troilite, phosphorus and nickel in significant amounts. The possible derivation of some of these fragments from meteoritic materials similar to those found on earth obviously requires investigation (*7*).

EUGENE N. CAMERON
Department of Geology and Geophysics, University of Wisconsin, Madison 53706

References and Notes

1. A. E. Bence and A. L. Albee, *J. Geol.* **76**, 382 (1968).
2. C. R. Knowles, J. V. Smith, A. E. Bence, A. L. Albee, *ibid.* **77**, 439 (1969).
3. A. Johannsen, *Petrography* (Univ. of Chicago Press, Chicago, 1938), vol. 3, p. 261.
4. G. A. Macdonald, *Mem. Geol. Soc. Amer.* **116**, 477 (1968).
5. P. T. Robinson, *Contrib. Mineral. Petrol.* **22**, 349 (1969).
6. Preliminary Science Report, Apollo 11, *NASA SP 214*, 123 (1969).
7. My attention has been called to a report by J. J. Frankel [*S. African J. Sci.* **38**, 152 (1942)] of a chromite containing about 12 percent TiO_2.
8. I wish gratefully to acknowledge the invaluable counsel of E. D. Glover in microprobe analysis and the skill and care of D. M. Fadness in preparation of sections of lunar samples. Supported by NASA grant NAS 9-9987.

4 January 1970

Diffraction and Mössbauer Studies of Minerals from Lunar Soils and Rocks

Abstract. *Constituents of lunar soils and rocks were studied by powder and single-crystal x-ray diffraction. In addition to identification of minerals, including rare amphibole, mica, and aragonite, a detailed study of the important rock-forming minerals of the plagioclase, pyroxene, and olivine groups has begun. Mössbauer spectra were recorded from lunar soils, ground rock samples, and separates of iron-bearing minerals. The proportions of iron-bearing minerals were estimated from computer-fitted areas for the bulk samples. The* Fe^{2+} *in the lower-density fraction of pyroxene was ordered, whereas that of the higher-density fraction was disordered.*

Powder and single-crystal x-ray diffraction procedures permit not only the identification of the components of lunar soils and rocks but also detailed study of the structural state, exsolution, alteration, and so forth, of individual minerals to provide data on the conditions of formation. Some qualitative observations are reported on plagioclases, pyroxenes, and olivines; the main work is still in progress.

Plagioclase, clinopyroxene, olivine, and ilmenite were identified as the predominant minerals in fines 84-11 and 85-30; various forms of glassy material, including spherules, are also present. A diffraction pattern from one of these spheres showed very weak powder rings superposed upon scattered isolated maxima. Devitrification appears to have produced at least two crystalline components of different grain size; there are similarities between the weak powder pattern and that of a sphene-like structure, though the identification is unproven. Small amounts of native iron were also present in these fines, very occasionally with apparently crystalline faces; the only specimen of this kind examined proved to be a fine aggregate with a poor degree of preferred orientation. During the preliminary optical examination, two small crystals of micaceous appearance were selected for further examination; initial single-crystal study of one of these is consistent with a mica, but the polymorphic type has not yet been established.

Single crystals of the plagioclases, pyroxenes, and olivines were examined in detail. Two plagioclase crystals were examined to ascertain the structural type and, if possible, the structural state. Both showed transitional anorthite structures (*1*) with slight differences in the intensities and diffuseness of type (*c*) reflections; without detailed analytical data on the crystals, it is impossible to assess the structural state for either. Three pyroxene crystals were studied by the methods of Bown and Gay (*2*) for submicroscopic exsolution; all three were unmixed with a host augite structure, and variable proportions of an exsolved pigeonite structure sharing the (001) plane. There is some variation in the angular separation of the augite and pigeonite components, and in all three patterns weak streaks join augite and pigeonite spots; these features suggest that the exsolution process may be incomplete. In one crystal there is a second orientation of exsolved pigeonite with the *c* axis of the pigeonite almost parallel to that of the augite, similar to that described for the pyroxene from the gabbro-picrite 4526 from the Skaergaard intrusion (*3*); for all three specimens, class (*b*) reflections with $h + k$ odd for the pigeonite structures are slightly diffuse (*4*). Single-crystal patterns from three olivine crystals are typical of forsteritic specimens with no evidence of oxidation, alteration, or other such effects.

Material from four rocks described as type A (*5*) is under investigation; plagioclase, pyroxene, ilmenite, and olivine have been identified together with small amounts of native iron and a fragment (from 17-50) of what appears to be a metallic alloy (cubic, *F*-lattice, $a \sim 3.68$ Å).

Detailed studies of three plagioclase crystals from two of these rocks reveal interesting variations in the subsidiary reflections. Very weak and diffuse type (*c*) reflections were detected for only one crystal (from 45-24), and this reflection class was apparently absent from both the other crystals (from 45-35). For 45-24, type (*b*) reflections are weak and diffuse and apparently split; the separation of these type (*e*) reflections for this unusual intermediate plagioclase structure was measured to give δ_c as approximately 166°, corresponding to a composition of about $An_{72\text{-}73}$ (*6*). For one of the crystals from 45-35, very weak diffuse single type (*b*) reflections were just visible, but they could not be observed on normal exposures for the other crystal. Type (*f*) reflections were not present on the patterns from any of the crystals. These plagioclase crystals must be in an intermediate-high structural state.

All four pyroxene crystals so far examined from this rock type have

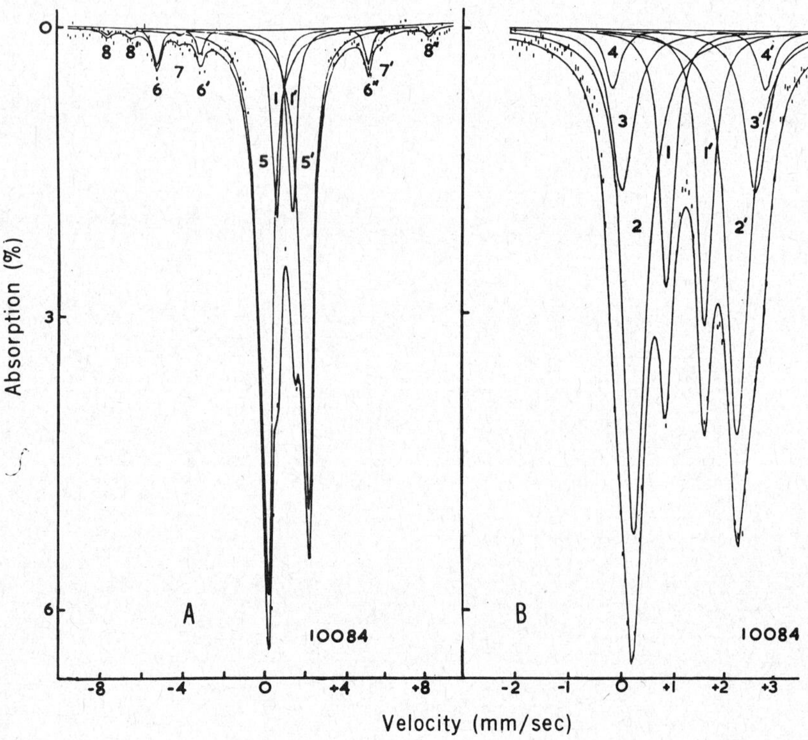

Fig. 1. Mössbauer spectra of 84-14 at room temperature. (A) High-velocity scan; (B) low-velocity scan. The numbering of peaks 1, 2, and so forth, is explained in the text.

shown predominantly pigeonite structures [twinned on (100) for one crystal] with class (*b*) reflections slightly diffuse. Three of them have indications of incipient augite exsolution in weak streaks associated with pigeonite spots, whereas the fourth also has weak maxima showing augite sharing (001) planes and possibly a second orientation almost parellel to the *c* axis of the host pigeonite; again the angular separation of the *z* axes for the pigeonite host and the (001) augite lamellae is less than that usually observed for intergrowths of this kind. Three olivine crystals (from 45-24, 45-35, and 72-47) give simple forsteritic patterns.

Crystals from two rocks described as type B (*5*) are under examination. Troilite and crystals of a yellow mineral thought to be a pyroxenoid comparable to pyroxmangite have been identified in 44-43; investigations of this material are incomplete, but some crystals show effects that may be associated with exsolution phenomena. In 58-23, a rare crystal of an amphibole from a vug (approximately a = 9.85 Å, b = 18.20 Å, c = 5.30 Å, and β = 104½°) has been studied, and a pink lath-like fragment showed a random aggregate of aragonite crystallites.

Three plagioclase crystals from 58-23 show transitional anorthite structures with some variation in the diffuseness and intensity of type (*c*) reflections; there may be some slight differences in the single type (*b*) maximums, but it is not possible to assess the structural states of these feldspars. Two pyroxene crystals from the same rock show pigeonite exsolving on (001) planes from an augite host structure; the amount of pigeonite is different in the two crystals, but both have a slight diffuseness of the class (*b*) reflections. The angular separation of augite and pigeonite is that normally observed in such intergrowths, but there are still weak streaks joining augite and pigeonite maximums.

Iron-bearing minerals have been found to give characteristic Mössbauer spectra (*7*) which make it possible to identify such minerals in a bulk rock sample; semiquantitative estimates of the relative proportions of these minerals can be made from the areas of peaks (*8*). Furthermore, since the parameters are sensitive to small differences in cation site symmetry, the Mössbauer effect has been widely used to measure accurately intracrystalline site populations in amphiboles and pyroxenes (*9*). Studies of this kind have been carried out on the fines 84-14 and 85-4, ground rock from 44-39 and 45-24, and mineral separates (olivine, ilmenite, and pyroxene) from 44-43.

Table 1. Percentages of iron-bearing minerals. These preliminary percentages are calculated on the assumption that olivine contains 20.0 percent Fe^{2+} by weight, that pyroxene contains 10 percent Fe^{2+}, and that the other minerals correspond to their ideal chemical formulas. The percentage of non-iron-bearing components is taken from the C.I.P.W. norm derived from chemical analysis by J. H. Scoon. These final values are expected to have semiquantitative significance, despite these and other assumptions.

Mineral	Percent of total Mössbauer area		Weight percent of total sample	
	84–14	45–24	84–14	45–24
Ilmenite	19.7	26.9	4.2	7.4
Pyroxenes	67.6	60.8	53.1	60.6
Olivine	4.4	6.1	1.8	3.0
Iron	5.8	2.1	0.6	0.2
Troilite	⩽1.1	⩽2.0	⩽0.1	⩽0.2
Magnetite	1.4	2.1	0.2	0.2
Noniron			40.0	28.4

Typical room temperature spectra of 84-14 at high and low velocity scans are shown in Fig. 1; qualitatively similar spectra were obtained for the other three bulk samples. Computed peaks have been assigned as follows (*10*): 1 and 1′ as ilmenite; 2, 2′, 3, and 3′ as pyroxene (mainly augite and subcalcic augite); 4 and 4′ as olivine; 5 and 5′ as an overlap of pyroxene and olivine peaks; 6, 6′, and 6″ as metallic iron; 7 and 7′ as troilite; 8, 8′, and 8″ as magnetite (with some Ti). Center shifts (C.S.), quadrupole splittings (Q.S.), and the magnetic peak positions are generally in good agreement with those of corresponding terrestrial minerals, and these parameters are consistent for all samples. Ilmenite peaks (Q.S. = 0.70 mm/sec and C.S. = 1.17 mm/sec) are in excellent agreement with those reported previously (*11*), whereas the Q.S. value (2.86 mm/sec) for olivine is slightly lower than that obtained in earlier work (*7*) for similar compositions. The Q.S. parameters for augite at 295°K are 2.00 mm/sec (inner lines) and 2.50 mm/sec (outer lines); these are similar to values for synthetic and natural terrestrial augites (*12*). Resolution of the pyroxene peaks is markedly improved at 80°K; parameters for the disordered pyroxene (Fig. 2B; see final paragraph) are very similar to those for the augite. The iron peak positions are in excellent agreement with the well-known peak positions for pure iron and indicate that there is little, if any, replacement of iron by nickel (*13*).

Ilmenite peaks (1 and 1′) are distinctly asymmetric, suggesting that there are appreciable quantities of another iron-bearing mineral present, one peak of which strongly overlaps 1′. The line widths of the pyroxene peaks for 85-4 and 84-14 are much greater than those from the rock samples; this broadening probably reflects a wider variation in chemical content for the soil pyroxene.

Table 1 lists the calculated percent-

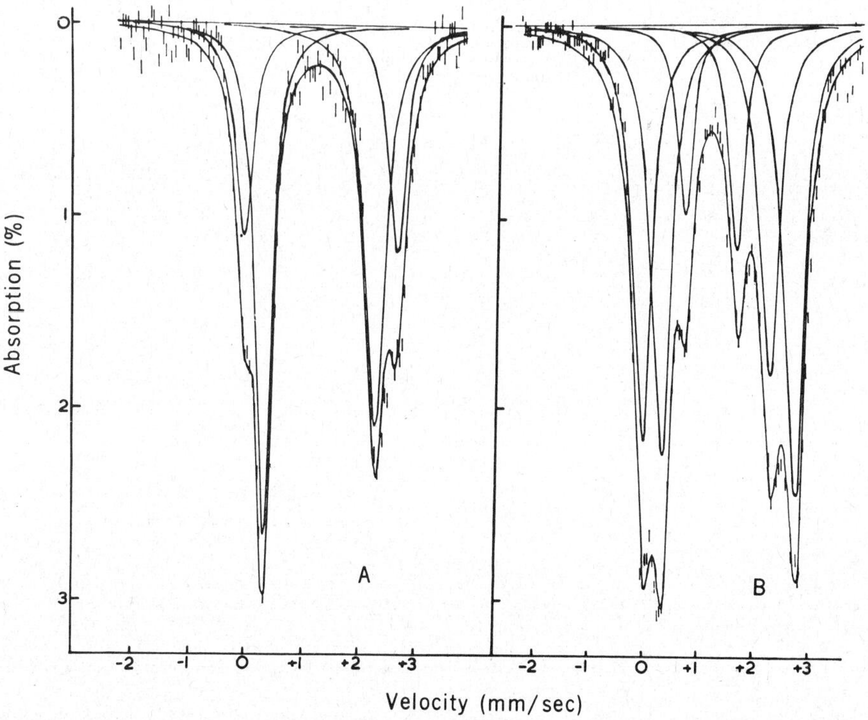

Fig. 2. Mössbauer spectra of two pyroxene fractions from 44-43 at 80°K. (A) Lower-density fraction; (B) higher-density fraction.

ages of iron-bearing minerals derived from the Mössbauer areas for 84-14 and 45-24; the assumptions and possible errors in these values will be discussed elsewhere. However, it is apparent that the iron content of 84-14 is much greater than that for 45-24, whereas 45-24 has appreciably more ilmenite, pyroxene, and olivine than 84-14.

Mössbauer spectra at 80°K of two pyroxene separates from 44-43 are shown in Fig. 2; the inner peaks in B are due to ilmenite. Fraction A (density 2.85 to 3.50) is ordered ($area_{inner}/area_{total} = 0.67$), whereas fraction B (density 3.70 to 3.90) is essentially disordered ($area_{inner}/area_{total} = 0.52$). This surprising result agrees with the optical, x-ray, and microprobe work reported here and elsewhere that this rock contains both a clinopyroxene (A) and a pyroxenoid (B) of the pyroxmangite type. For B, the line widths of the inner peaks are greater than those of the outer peaks (0.48 and 0.40 mm/sec, respectively), and the room-temperature areas are not consistent with those obtained at 80°K. Although this evidence could be interpreted as implying there are more than two cation sites with appreciably different Q.S. values, it could also be due to a lack of homogeneity caused by exsolution.

P. GAY
Department of Mineralogy and Petrology, University of Cambridge, Cambridge, England

G. M. BANCROFT
Department of Chemistry, University of Cambridge

M. G. BOWN
Department of Mineralogy and Petrology, University of Cambridge

References and Notes

1. M. G. Bown and P. Gay, *Z. Krist.* **111**, 1 (1958).
2. ———, *Amer. Mineral.* **44**, 592 (1959).
3. ———, *Mineral. Mag.* **32**, 379 (1960).
4. ———, *Acta Cryst.* **10**, 440 (1957).
5. Lunar Sample Preliminary Examination Team, *Science* **165**, 1211 (1969).
6. P. Gay, *Mineral. Mag.* **31**, 21 (1956).
7. G. M. Bancroft, A. G. Maddock, R. G. Burns, *Geochim. Cosmochim. Acta* **31**, 2219 (1967).
8. E. L. Sprenkel-Segel and G. J. Perlow, *Icarus* **8**, 66 (1968).
9. D. Virgo and S. S. Hafner, *Mineral. Soc. Amer. Spec. Paper 2*, 66 (1969); G. M. Bancroft and R. G. Burns, *ibid.*, p. 137.
10. The assumptions and constraints used to compute these spectra will be discussed elsewhere.
11. T. C. Gibb and N. N. Greenwood, *J. Inorg. Nucl. Chem.* **39**, 947 (1969); the C.S. values quoted here are relative to stainless steel. To convert to sodium nitroprusside 0.16 mm/sec must be added.
12. G. M. Bancroft, A. Turnock, P. G. L. Williams, in preparation.
13. C. E. Johnson, M. S. Ridaut, T. E. Cranshaw, P. E. Madsen, *Phys. Rev. Lett.* **6**, 450 (1961).
14. We thank K. O. Rickson for assistance in taking the diffraction photographs.

4 January 1970

Lunar Clinopyroxenes: Chemical Composition, Structural State, and Texture

Abstract. *Single-crystal x-ray diffraction, microprobe, optical and electron optical examinations of clinopyroxenes from Apollo 11 lunar samples 10003, 10047, 10050, and 10084 show that generally the crystals are composed of (001) augite-pigeonite intergrowths in varying ratios. Transmission electron micrographs reveal abundant exsolution lamellae, many only 60 Å thick. In addition to the phase inhomogeneities, primary chemical inhomogeneities are clearly demonstrated. There are reciprocal relationships between calcium and iron and between* Ti^{4+} + *2Al and* R^{2+} + *2Si. Our evidence suggests that a chemically inhomogeneous subcalcic* C2/c *augite was the only primary pyroxene from which pigeonite later exsolved.*

The lunar clinopyroxenes, augite and pigeonite, from rocks 10003,38 10047,27, 10050,32, and soil 10084,100 collected by the Apollo 11 astronauts were examined by standard petrographic microscopy, by the Buerger single-crystal x-ray precession technique, by transmission and scanning electron microscopy, and by electron-microprobe analyses of single crystals mounted in epoxy cement and polished to expose a plane surface.

Examination of the clinopyroxenes with the petrographic microscope shows intercumulus textures and, within the grains, complex patterns of color varying from very light pink to fairly dark reddish brown. The characteristic properties of the lunar clinopyroxenes are the "sectorial" extinction and the apparent warping or bending of the (110) cleavage planes. These effects are caused by lattice rotation of two or more blocky, sometimes pie-shaped, "domains" contained within a crystal. Precession photographs of crystals containing such domains reveal that they are offset from one another by rotation about a common *b* axis (*1*). The rotation is generally 1° to 5°, and often a one-to-one correlation can be made between the number of domains seen optically in a crystal and the number observed in its *h0l* diffraction pattern. Color variations occur across the domain boundaries. The "sectorial" extinction so commonly displayed by the pyroxenes when examined in plane polarized light is caused by this offset domain structure.

Single-crystal x-ray diffraction patterns were made of 25 crystals or crystal chips. Almost all the crystals consisted of augite plus pigeonite oriented on (001). The approximate amount of pigeonite contained in the host augite was determined by comparing the relative intensities of the *h0l* reflections (*2*). The ratios of augite to pigeonite in some of these single crystals are as follows: rock 10003, 4 : 1, 7 : 3, 1 : 1, 1 : 1, 2 : 3, 1 : 4; rock 10047, 17 : 3, 13 : 7, 1 : 0, 7 : 3; rock 10050, 19 : 1, 19 : 1, 3 : 7, 17 : 3, 1 : 0, 17 : 3, 3 : 2, 50 : 1; rock 10072, 17 : 3; soil 10084, 13 : 5, 17 : 3, 17 : 3.

Cell parameters obtained from measurements of precession patterns can be summarized as follows: augites, *a* 9.68 to 9.76 Å, *b* 8.90 to 8.95 Å, *c* 5.25 to 5.28 Å, *β* 105.8° to 106.8°; pigeonites, *a* 9.68 to 9.76 Å, *b* 8.90 to 8.94 Å, *c* 5.21 to 5.24 Å, *β* 108.8° to 109.1°. The augites have space group *C2/c*, and the pigeonites, $P2_1/c$. The reflections violating *C2/c* symmetry in a crystal

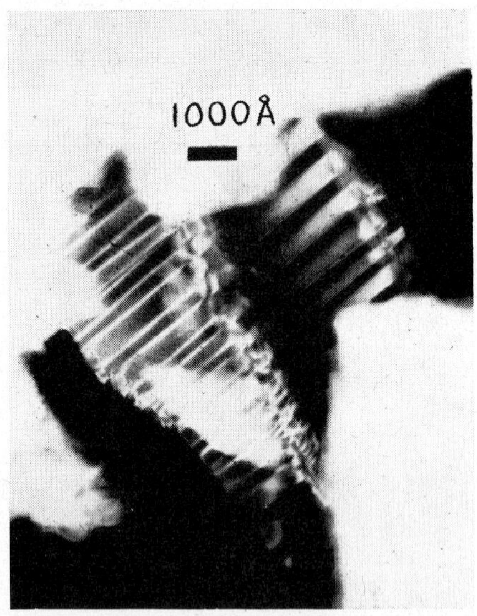

Fig. 1. Transmission electron photomicrograph of a small crystal fragment of clinopyroxene from rock 10047. The narrow lamellae to the left of the photograph (thin white bands) vary in thickness from 60 to 100 Å and are interpreted as being pigeonite unmixed on (001) from augite host (wide, dark bands 200 to 500 Å thick). Coalescence of the thin pigeonite lamellae into much thicker lamellae may be occurring to the right of the photograph (light bands approximately 400 Å thick). The discontinuity in the center of the grain trending 11 to 5 o'clock may be a domain boundary with the left domain becoming more augitic, the right, more pigeonitic.

from rock 10003 with augite-pigeonite ratio 1 : 4 are found to be diffuse, similar to those observed by Morimoto and Tokonami (*3*) for pigeonite from andesite, Isle of Mull, Scotland. With one exception, the pigeonites included in the augite hosts are oriented with the (001) planes and *a* and *b* axes in common with those of the host. Pigeonite oriented on (100) of augite was observed within a prismatic crystal growing into a vug of rock 10072,33.

Only occasionally are pigeonite exsolution lamellae observed with the petrographic microscope. Transmission electron microscopy, however, reveals a complex array of (001) pigeonite lamallae, some only 60 Å thick (Fig. 1). Generally the lamellae are thinner than 1000 Å, so that identification by observations made with the petrographic microscope is difficult. The pigeonite contains more iron than the augite host and is thus denser and more opaque to electrons. We therefore assume that the less transparent pigeonite lamellae appear in the electron photomicrographs as gray to white zones and the more transparent augite areas appear as dark gray to black zones. Usually the small crystal fragments show abundant exsolved pigeonite at the crystal margins, and on some margins there appears to be more pigeonite than augite host. We have found no direct evidence for primary pigeonite in the rock samples studied, with the possible exception of crystal 10072-1 where a (100) overgrowth of primary pigeonite on augite may have occurred.

We conclude that the crystal fragments containing more pigeonite than augite originated from a much larger crystal which was initially a subcalcic augite. This primary augite unmixed a large amount of pigeonite and then underwent partial *intra*granular recrystallization such that parts of the grain became pigeonitic and others, augitic. A chip from such a grain, containing mostly pigeonite, could be misinterpreted as being a primary host phase. Such *intra*granular recrystallization textures are known in terrestrial rocks, for example: the inversion of pigeonite to orthopyroxene, and the unmixing and segregation of cummingtonite and actinolite to opposite ends of a grain that was originally a very subcalcic actinolite (*2*).

Additional evidence for unmixing of about 50 weight percent pigeonite from primary subcalcic augite appears in the *h0l* diffraction pattern of crystal 10003-8. This diffraction pattern shows the crystal to be composed of a single domain and indicates a present 1 : 1 ratio of augite to pigeonite. The x-ray reflections are generally sharp, but a diffuse streaking parallel to the *c** direction appears along a line connecting the equivalent *h0l* reflections of the two clinopyroxenes. For example, there is an intense streak between the 10,0,0 reflections of pigeonite and augite. This diffuse streaking is thought to be due to the diffraction effects produced by "islands" or "zones," some perhaps only a few unit cells in volume, of clinopyroxenes possessing a variety of compositions and possibly structural states. Such streaking would not be expected from primary intergrowth of augite and pigeonite but would be the expected result of incomplete exsolution and structural transformation over a range of temperature. Another interesting feature of the pigeonite unmixing is that those crystals which display multiple domains do not necessarily have the same amount of pigeonite in each domain. For example, crystal 10050-8 contains three domains, divided into augite-pigeonite ratios as follows: 3 : 7, 7 : 3, and 1 : 1.

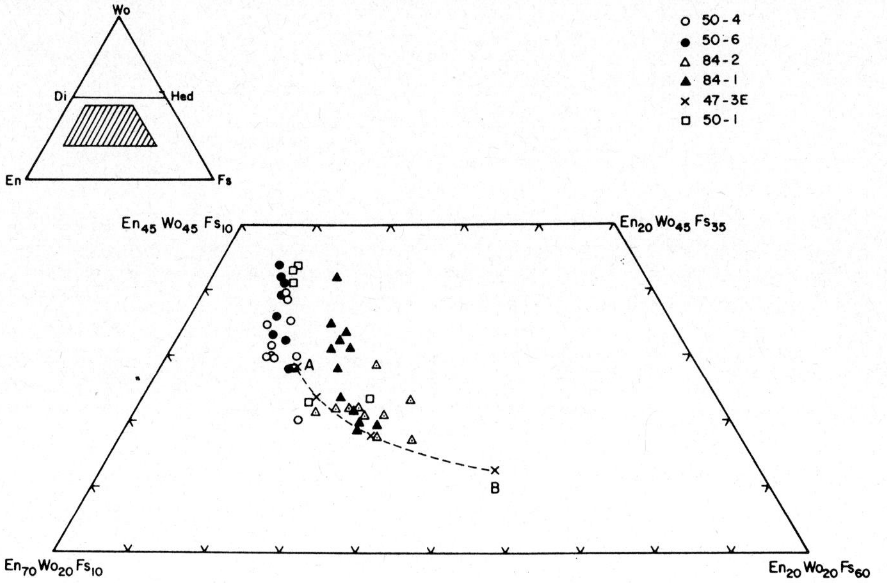

Fig. 2. Plot of electron-microprobe chemical analyses of six lunar clinopyroxenes on a part of the pyroxene quadrilateral. Separate analyses were made at several points on each crystal.

Fifty electron-microprobe chemical analyses were made at various points on six single crystals previously examined by optical and x-ray methods (*4*). The results of these analyses are plotted in the pyroxene quadrilateral (Fig. 2). This plot ignores Al, Ti, Na, Cr, and Mn, which together correspond to about 3 mole percent of the clinopyroxene components, but this should not appreciably affect the usefulness of the diagram. The augite-pigeonite ratios for the crystals analyzed are: 10050-4, 19 : 1; 10050-6, 1 : 0; 10084-2, 13 : 5; 10084-1, 17 : 3; 10047-3E, 13 : 7; 10050-1, 19 : 1. Knowledge of the amount of unmixed pigeonite in the grains enables us to separate the primary chemical inhomogeneities appearing in the single crystal from secondary chemical inhomogeneities due to pigeonite unmixing from the primary clinopyroxene.

Four of the crystals analyzed show a striking distribution of compositions which trend nearly normal to the En-Fs join when plotted on the quadrilateral. The variation of composition for these crystals is as follows: 10050-6, $En_{45}Wo_{42}Fs_{14}$ to $En_{47}Wo_{34}Fs_{19}$; 10050-4, $En_{45}Wo_{40}Fs_{16}$ to $En_{48}Wo_{30}Fs_{21}$; 10084-1, $En_{41}Wo_{41}Fs_{18}$ to $En_{45}Wo_{29}Fs_{26}$; 10050-1, $En_{43}Wo_{42}Fs_{15}$ to $En_{43}Wo_{32}Fs_{25}$. Three of these crystals (10050-1, -4, -6) contain very little pigeonite, so that the analytical results reflect primary chemical inhomogeneities. Crystal 10050-6 is composed of six domains of unequal size rotated from one another by a few degrees. The chemical composition of this crystal varies from domain to domain.

The vertical trend demonstrated by these nearly pigeonite-free crystals is thought to be due to a concentration of iron into the *M*2 site at the expense of calcium, with only small change in total Mg content. This enrichment of Fe over Ca in *M*2 is believed to be related to the persistent crystallization of An-rich plagioclase. Ordinarily, in layered intrusive igneous rocks, we would expect plagioclase to trend from An_{80-70} at the early stages of crystallization to An_{40-30} at the later stages. The plagioclases in these rocks, however, are persistently calcic, ranging from approximately An_{85} to An_{73} (*5*). It appears that the primary *C*2/*c* clinopyroxenes crystallizing in these rocks became more and more depleted in Ca as they grew because of depletion of Ca in the melt by the continued precipitation of An-rich plagioclase. Magnesium tends to remain in the *M*1 site, and enrichment of Fe in the ordinarily Ca-rich *M*2 site must therefore occur. Once the concentration of Fe into *M*2 becomes significantly high, the crystallization trend can turn toward the ferrosilite corner of the pyroxene quadrilateral by Fe replacing Mg in *M*1.

Crystals 10084-2, 10047-3E appear to be lowest in Ca (Fig. 2). The scatter for crystal 10084-2 seems somewhat random. Crystal 10047-3E shows a continuous variation of Si, Fe, Ca, Mg, Ti, Al, Cr, and Mn from one end of the grain to the other (Fig. 3). Four complete analyses were made along the path of the traverse (dashed line, Fig. 2). The composition varies continuously from $En_{47}Wo_{34}Fs_{19}$ at the calcic end to $En_{37}Wo_{26}Fs_{36}$ at the other end. Unmixed pigeonite is undoubtedly concentrated into the more subcalcic portions of the crystal. Chemical formulas for the four analyzed positions (marked by *X*, Fig. 2) are, in order of decreasing Ca content, as follows: $Ca_{0.671}Na_{0.005}Mn_{0.010}Fe_{0.374}Mg_{0.921}Ti_{0.053}Cr_{0.011}Al_{0.084}Si_{1.871}O_{6.000}$, $Ca_{0.628}Na_{0.006}Mn_{0.011}Fe_{0.427}Mg_{0.0923}Ti_{0.047}Cr_{0.009}Al_{0.076}Si_{1.873}O_{6.000}$, $Ca_{0.569}Na_{0.006}Mn_{0.013}Fe_{0.525}Mg_{0.879}Ti_{0.039}Cr_{0.007}Al_{0.066}Si_{1.896}O_{6.000}$, $Ca_{0.518}Na_{0.006}Mn_{0.015}Fe_{0.709}Mg_{0.735}Ti_{0.032}Cr_{0.006}Al_{0.060}Si_{1.918}O_{6.000}$.

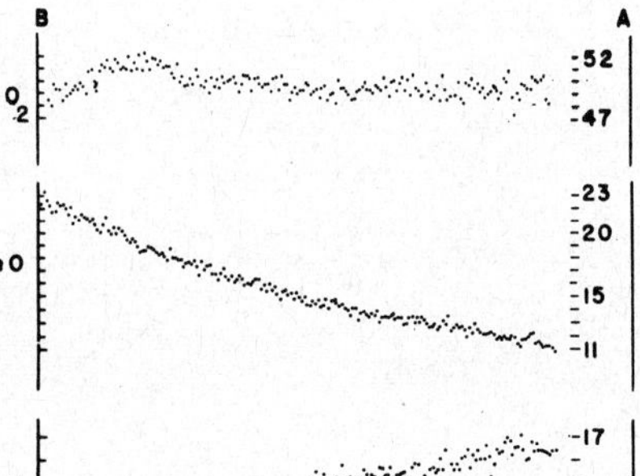

Fig. 3. Chemical trends in weight percent oxides for SiO_2, FeO, CaO, MgO, TiO_2, Al_2O_3, MnO, Cr_2O_3, and Na_2O obtained from an electron-probe traverse at 1 micron intervals from one end of crystal 10047-3E to the other end approximately parallel to the *c* axis.

The six crystals are generally lower in Ca and richer in Fe at the margins; the Ca and Fe vary nearly inversely in probe scans across the grain. The six crystals also show a nearly exact coupling of Ti and Al with Si and R^{2+} which can be interpreted as replacement of $R^{2+} + 2Si^{4+}$ by $Ti^{4+} + 2Al^{3+}$. Thus one Ti^{4+} would replace Mg, Mn, or Fe^{2+} in the *M*1 site, and charge balance would be maintained by simultaneous substitution of two Al^{3+} for two Si cations in the tetrahedral sites. This nearly exact coupling with Ti^{4+} suggests that little or no Ti^{3+} is present, for otherwise we should observe an octahedral Ti^{3+}–tetrahedral Al couple.

The change from low to high Ti+Al is observed from a probe scan that crosses a color boundary, the high Ti and Al being associated with an intense reddish brown color. In turn, these sharp color boundaries often appear to be associated with the boundaries of the adjacent domains. The TiO_2 and Al_2O_3 contents each vary from about 1.5 weight percent in the nearly colorless areas to about 3 weight percent in the reddish brown areas. This distribution of Ti and Al found in the lunar pyroxenes has been found in certain terrestrial pyroxenes (*6*). The coupled relationship between Ca and Fe appears to be largely independent of the Ti and Al content, so that color is not necessarily related to pigeonite content.

The chaotic phase and chemical inhomogeneities displayed by these lunar clinopyroxenes suggest rapid crystallization from a melt of changing composition. We believe that a single *C*2/*c* clinopyroxene began to crystallize on the liquidus and above the augite-pigeonite solvus. We know from the studies of Smyth (*7*) and Prewitt, Papike, and Ross (*8*) that at high temperatures the low-temperature $P2_1/c$ pigeonite changes to the *C*2/*c* structure type, a fact which suggests a potential closing of the augite-pigeonite miscibility gap. Whether we call the primary *C*2/*c* lunar pyroxene subcalcic augite or calcic pigeonite may then be merely academic. The extreme chemical inhomogeneities found in the lunar clinopyroxenes may be due partly to the presence of a very flat solidus, such that small changes in melt composition would have an appreciable effect on the composition of the solid. The primary chemical inhomogeneities are further complicated by the subsolidus unmixing of two clinopyroxenes. The multiple domain structure is believed to be related to complex conditions of high-temperature crystal growth rather than to any later shock effects.

MALCOLM ROSS
U.S. Geological Survey, Washington, D.C. 20242

A. E. BENCE
Department of Earth and Space Sciences, State University of New York, Stony Brook 11790

E. J. DWORNIK, JOAN R. CLARK
U.S. Geological Survey, Washington, D.C. 20242

J. J. PAPIKE
Department of Earth and Space Sciences, State University of New York, Stony Brook 11790

References and Notes

1. For the purposes of this paper a "domain" will mean a portion of a pyroxene crystal that does not possess rotational offset.
2. M. Ross, J. J. Papike, K. W. Shaw, *Mineral. Soc. Amer. Spec. Pap. No. 2* (1969), p. 275.
3. N. Morimoto and M. Tokonami, *Amer. Mineral.* **54**, 725 (1969).
4. Analyses were reduced by the Bence-Albee method [A. E. Bence and A. L. Albee, *J. Geol.* **76**, 382 (1968)] with analyzed pyroxene monitors interspersed throughout the analysis of the unknowns. Analyses 1–30 were made for Na, Mg, Al, Si, Ca, Ti, Cr, Mn, and Fe, and sum to 99±1 weight percent. Analyses 31–50 were made for Mg, Al, Si, Ca, Ti, and Fe. The sums on these analyses run about 0.5 weight percent lower due to neglecting Na, Mn, and Cr contents.
5. D. B. Stewart, personal communication.
6. A. E. Bence, unpublished data.
7. J. R. Smyth, *Earth Planet. Sci. Lett.* **6**, 406 (1969).
8. C. T. Prewitt, J. J. Papike, M. Ross, *Amer. Mineral.*, in press.
9. Research supported by NASA. We acknowledge with gratitude the initiation of this study by our co-investigator, Dr. Jeffrey Warner, MSC, NASA, and the assistance of Mrs. Judith A. Konnert in some of the x-ray studies.

4 January 1970

Compositional Zoning and Its Significance in Pyroxenes from Three Coarse-Grained Lunar Samples

Abstract. *The calcium-rich pyroxenes in lunar samples 10047, 10058, and 10062 show pronounced sectoral and radial compositional variations which correlate with sharp to gradual variations in color and optical properties. The pyroxenes apparently grew as nearly euhedral crystals from melts of approximately the same composition as that of the samples. The coupled substitutions determined across sector boundaries suggest that Al is predominantly in the tetrahedral site and that Ti is predominantly quadrivalent. The pyroxene differentiation trend (unknown in terrestrial pyroxenes) is toward extreme enrichment in the ferrosilite molecule. The iron-enriched portions of the pyroxene grains may have grown with a triclinic pyroxenoid structure.*

The three type B lunar samples studied (10047, 10058, and 10062) are all ilmenite-rich gabbros with conspicuous reddish brown pyroxene. Sample 47 [54 percent, by volume, pyroxene (Px), 0.7 mm median grain diameter; 27 percent plagioclase (Pc), 0.2 mm; 15 percent ilmenite (Ilm), 0.3 mm; and 4 percent cristobalite (Cr), 0.4 mm] is very similar modally and in grain size to sample 58 (55 percent Px, 0.4 mm; 24 percent Pc, 0.2 mm; 15 percent Ilm, 0.3 mm; 6 percent Cr, 0.3 mm). Sample 62 is much finer grained and contains olivine (Ol) instead of cristobalite (50 percent Px, 0.15 mm; 26 percent Pc, 0.1 mm; 20 percent Ilm, 0.1 mm; 4 percent Ol, 0.25 mm). Sulfide, containing rounded blebs of native iron, is a minor accessory in all three samples; grains of ulvospinel were identified in sample 58, and chromite (in the olivine and ilmenite) and rutile (in the ilmenite) in sample 62. An unidentified Zr mineral, with essential Fe, Ti, Si, and Ca, and significant Y, was also found in sample 58.

The pyroxenes are typically pink, subhedral, and in ophitic or subophitic intergrowth with plagioclase. Prismatic cleavage is rare; the irregular, roughly parallel fractures present represent, in some grains at least, the (001) parting. Zoning is conspicuous in all pyroxenes, manifest in plane polarized light by a sharp to gradual change in color from pink to near colorless. Discontinuous peripheral zones on some grains in sample 47 are distinctly honey-yellow; similar material may also be present, in lesser amount, in sample 58 (Fig. 1, zone *T*). The zoning is emphasized under crossed polarizers, with both sharp and gradual extinction changes coinciding with the plane-light color boundaries. The zone distribution approximates an hourglass pattern in some grains, but in others it appears exceedingly complex, with irregular variations and repetitions. The possibility of occurrence of strain shadows cannot entirely be excluded. The pyroxenes in sample 62 are particularly fractured, with highly irregular extinction patterns; single pyroxene "areas" appear to be composed of a number of separate zoned individuals. Compositional irregularities in probe traverses of these pyroxenes suggest the presence of submicroscopic Ca-poor pyroxene exsolution lamellae.

Optical properties, understandably, show considerable variation. Although those of sample 58 have been studied most intensively, there is considerable overlap between the samples. $2V_z$ measurements on sample 58 vary from 40.5° to 47.5°; $Z \wedge c$ could not be determined precisely but apparently varies more continuously than does 2V. Refractive indices on 13 grains range from Ny = 1.702 to Ny = 1.742, birefringence 0.23 to 0.33. Yellow border phase of sample 47 has $2V_z$ = 35°, Ny = 1.750 to 1.755. Efforts to obtain "average" unit cell parameters for the pyroxene in sample 58 resulted in unrealistic values with prohibitively large errors after least-squares refinement. Because of the prevalent zoning, similar attempts on the pyroxene in other samples were abandoned. X-ray powder photographs of the yellow "pyroxene" border phase (handpicked) in sample 47 show it to be significantly different from the pink pyroxene core, suggesting a different (and possibly lower) symmetry for this yellow phase. Qualitatively, the yellow border phase of sample 47 appears equivalent in composition to the outermost zone in sample 58 (Fig. 1, zone *T*).

The compositional variations along microprobe traverses *a* and *b* of one sample 58 pyroxene (Fig. 1) are qualitatively illustrated in Fig. 2. Discontinuities in composition occur at points *B*, *C*, and *D* (Fig. 2a), and these discontinuities correlate exactly with the extinction discontinuities at *B*, *C*, and *D* of Fig. 1. Points *A* and *E* of Fig. 2 mark the position of the two outermost extinction discontinuities in Fig. 1, but do not coincide with sharp chemical changes. The composition, best represented by the Fe, Mg, and Ca profiles, varies continuously from points *B* and *D* toward the edges of the crystal, and, with the exception of the slight discontinuity at *C*, is nearly constant between points *B* and *D*. The area of continuous change of composition correlates with a continuous change of extinction angle; conversely, along the line between *B* and *D* no change of extinction angle was observed. Figure 2b shows a second profile, nearly perpendicular to the first

Table 1. Results of eight analyses of pyroxene.

Substance	1	2	3	4	5	6	7	8
				Oxides				
MgO	2.29	16.71	15.13	15.28	16.57	15.39	1.85	14.60
Al_2O_3	0.86	1.58	3.06	2.84	1.66	1.22	0.78	2.78
SiO_2	47.09	51.56	49.40	49.85	51.36	50.78	46.53	49.76
CaO	8.04	15.81	18.66	17.89	16.08	14.53	7.16	20.78
TiO_2	0.72	1.54	2.61	2.54	1.74	1.22	0.78	2.92
Cr_2O_3	.06	0.18	0.28	0.20	0.15	0.10	.04	0.28
MnO	.67	.24	.19	.22	.26	.32	.61	.21
FeO	41.07	12.76	10.45	10.88	13.27	16.56	43.42	9.86
Total	100.8	100.4	99.8	99.7	101.1	100.1	101.2	101.2
				Cation proportions				
Si	1.96	1.92	1.85	1.87	1.91	1.92	1.95	1.85
Al	0.04	0.07	0.14	0.13	0.07	0.06	0.04	0.12
Ti	.02	.04	.07	.07	.05	.04	.03	.08
Cr	.002	.005	.008	.006	.004	.003	.001	.008
Mn	.02	.01	.01	.01	.01	.01	.02	.01
Mg	.14	.93	.85	.85	.92	.87	.12	.81
Fe	1.43	.40	.33	.34	.41	.53	1.52	.31
Ca	0.36	.63	.75	.72	.64	.59	0.32	.83

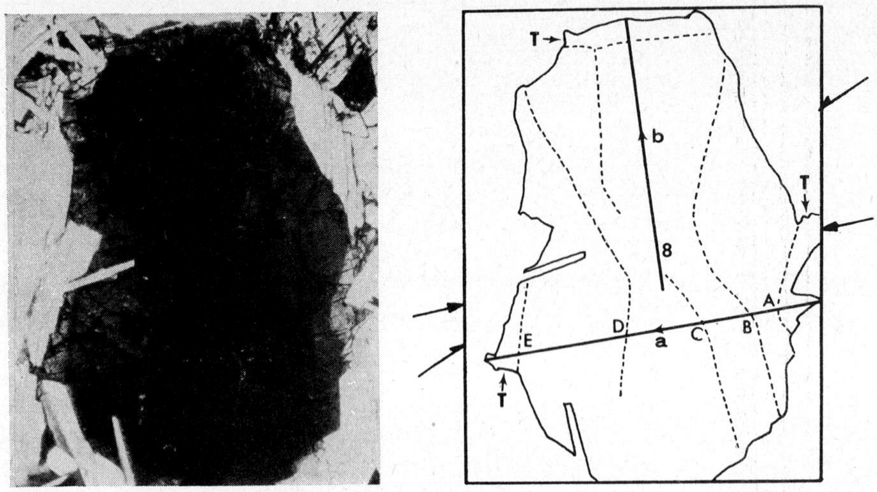

Fig. 1. (Left) Pyroxene crystal from sample 10058; crossed polarizers. (Right) Outline of the crystal at left, showing zone boundaries (broken lines). Heavy lines locate microprobe traverses *a* and *b* (traverse *a* is approximately 1 mm long). Letters *A* through *E* mark positions of discontinuous change of extinction angle. Double-headed arrows parallel (001) parting; single-headed arrows parallel trace of optic axial plane. Zones designated *T* are referred to in text.

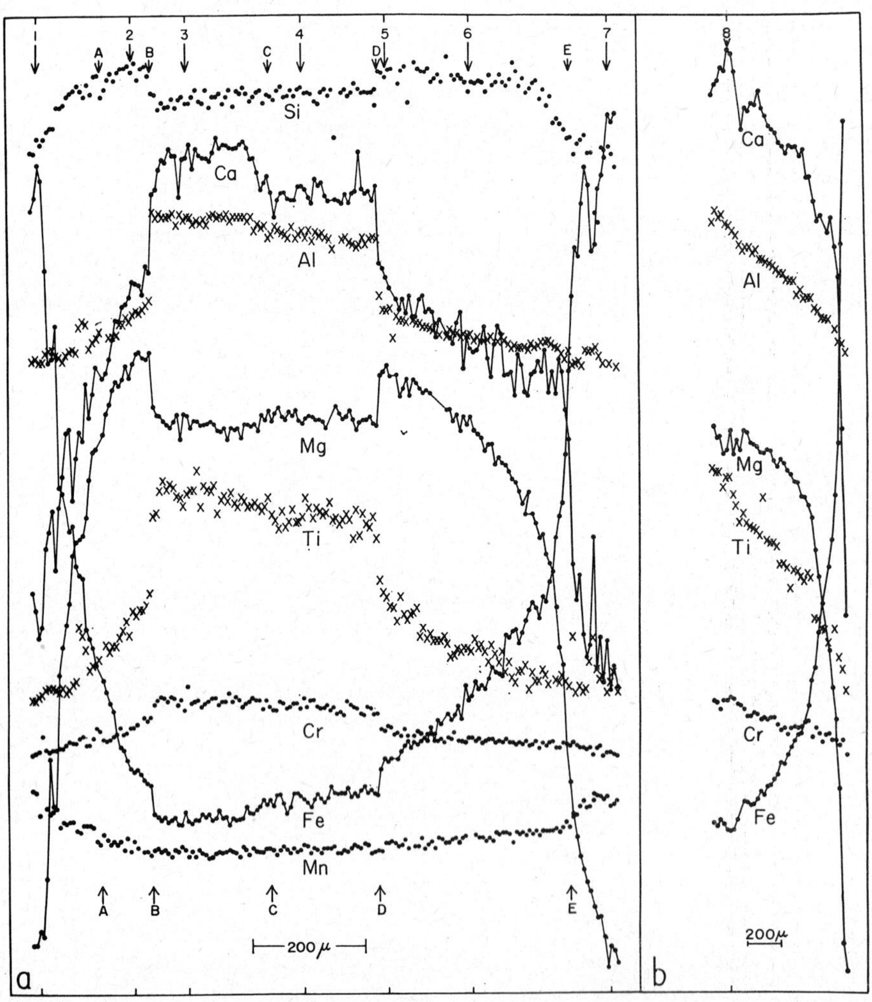

Fig. 2. (a) Variations of the elements Ca, Mg, Fe, Mn, Ti, Cr, Al, and Si along traverse *a* of Fig. 1. (b) Variations of the elements Ca, Mg, Fe, Ti, Cr, and Al along traverse *b* of Fig. 1; point 8 is at the center of the crystal. The letters *A* through *E* mark positions of discontinuous changes of extinction angle. The numbers 1 through 8 mark the horizons where the analyses in Table 1 were determined. Electron beam diameter is less than 1 μ; analysis interval (a) is 6 μ; (b) is 20 μ. Vertical scale (not shown) is counting rate.

traverse (Fig. 1, right) and extending from the center of the crystal to the edge. Again, pronounced continuous compositional zoning correlates with a continuous change of the extinction angle.

Table 1 shows total analyses for the points indicated in Fig. 2; the analyses are based on the correction factors of Bence and Albee (*1*), relative to almandine garnet (Fe), synthetic diopside (Ca, Mg, Si), jadeite (Al), ilmenite (Ti), chrome metal (Cr), and spessartine garnet (Mn). Wavelength profiles did not indicate the presence of more than 0.03 percent by weight of any other element. The analyses total close to 100 percent; but the estimated errors are about 2 percent of the amount of the major elements present and about 5 percent for the minor elements, with the exception of Al which may be low by 10 to 20 percent. The formulas shown in Table 1, calculated on the basis of six oxygen atoms to the unit cell, clearly indicate tetrahedral substitution of Al for Si. The substitution of Ti in the M sites, in an amount which is approximately half of the Al substitution of Si, indicates the coupled substitution (Ti + 2 Al) for (2 Si + 1 divalent cation). This coupled substitution implies that the titanium is predominantly quadrivalent. The large possible error in the Al determination arises from the fact that the Al standard (jadeite) has Al in sixfold rather than fourfold coordination. Because the spectrometer was peaked on the jadeite, the correction would increase the atomic proportion of Al to a number closer to twice the titanium content.

The hourglass pattern of areas with coherent optical extinction (Fig. 1) and the discontinuous changes of composition between such areas suggest that the pyroxenes of lunar sample 10058 are sector-zoned. The sector-zoning phenomenon has been described elsewhere in detail for staurolite (*2*) and has been observed (*3*) in titaniferous augite. The profiles of Fig. 2 are remarkably similar in geometric form to those reported for sector-zoned staurolite (*2*). The analyses between the edges and points *B* and *D* (Fig. 2) are within the (110) sectors, and those between *B* and *D* may be within the (001) sector. The small discontinuity at *C* implies another sector, besides the (001) sector, between points *B* and *D*. A number of different sectors are possible in the lunar pyroxenes because pyroxene typically grows with more than two forms.

The nearly constant composition between *B* and *D* and the continuous change of composition to the edges from *B* and *D* and within the (001) sector (Fig. 2b) are consistent with the sector-zoning interpretation. The constant composition section of the profile is nearly parallel to the (001) crystal face (Fig. 1), which would imply that that part of the profile was parallel to a growth surface which could be expected to have a constant composition; on the other hand, the other profiles are oblique to the growth surfaces and therefore would show any continuous changes of composition with growth.

The major chemical difference between sectors, higher Ti and Al and lower Si in the (001) sector relative to the (110) sectors, is also consistent with the sector-zoning interpretation: On a growing (001) face the M and T sites are exposed simultaneously, whereas on a growing (110) face either an M site or a T site is exposed. The simultaneous exposure of the two sites on (001) facilitates a coupled substitution because a local charge balance can be achieved, whereas a coupled substitution on the (110) face requires a local charge imbalance during the time between additions of growth layers. This type of mechanism is discussed elsewhere for staurolite (*4*). A mechanism to explain the discontinuity of Ca between sectors is not immediately obvious. The sector-zoning model implies, of course, that the lunar pyroxenes were nearly euhedral during most of their growth.

Seven other profiles exhibiting chemical features similar to those of Fig. 2 were made on the pyroxene grain of Fig. 1 and other pyroxene grains in sample 10058. However, they did not all exhibit sharp discontinuities in all elements. Several showed sharp discontinuities of the multivalent cations Ti, Si, and Al, but were diffuse in Fe, Mg, and Ca. It is here suggested that there were originally sharp discontinuities in all the elements but that during the time since the pyroxenes crystallized localized diffusion led to local obliteration of the sharp discontinuities. At some places, the diffusion was only by the divalent cations, but at others all the cations may have been involved. A localized diffusion pattern could lead to a very irregular optical appearance of the pyroxenes.

The very pronounced continuous zoning of Fe, Mg, and Ca (Fig. 2) was noted in pyroxene grains in all three samples. This type of zoning pattern has been reported for garnet (*5*) and suggests that the pyroxenes, as they grew, depleted the matrix (melt) in Mg relative to Fe. The pronounced zoning suggests that the composition of the reservoir from which the pyroxenes grew was the same as that of the rock in which they now occur. This interpretation precludes the possibility of differentiation by physical separation of pyroxene from a large magma chamber. The edge content of Mg in the pyroxene of sample 58 was calculated on the basis of the Rayleigh fractionation model (*5*), and compares favorably to the measured edge content.

The essential characteristic of the compositional variation in these lunar pyroxenes is one of extreme enrichment in the ferrosilite molecule. Analyses 1 and 7 (Table 1), representing the outermost, iron-enriched zones, plot in an area of the pyroxene quadrilateral in which there are no terrestrial equivalents (*6*) and where the stable existence of a single pyroxene phase at low pressures is "forbidden," according to the recent experimental work of Lindsley and Munoz (*7*). These analyses are, nevertheless, almost identical with the most ferrosilite-rich metasilicate phases obtained by Bowen *et al.* (*8*) in the system $CaO–FeO–SiO_2$: 80 $FeSiO_3$, 20 $CaSiO_3$ in monoclinic hedenbergite solid solutions stable below 980°C; and 76 $FeSiO_3$, 24 $CaSiO_3$ in triclinic β-wollastonite solid solutions stable above that temperature. Lindsley and Munoz (*7*), however, found that the maximum amount of ferrosilite that could occur in stable solid solution with either hedenbergite or wollastonite is closer to 60 percent. They attribute the discrepant experimental results of Bowen *et al.* (*8*) to either incomplete reaction or metastable equilibrium. Optically, the yellow border phase of sample 47 pyroxene appears homogeneous; reconnaissance probe study shows that it is similar to the border phase in sample 58, represented by analyses 1 and 7 (Table 1). The x-ray powder pattern of the border phase, however, is distinct from that of the core pyroxene, and, although not yet positively determined, it may represent a triclinic pyroxenoid modification. The closest natural occurrence of this phase is the Skaergaard ferrohedenbergite, deemed to have inverted from the ferriferous β-wollastonite form (*9*). This inversion rendered original single crystals into a mosaic of smaller clinopyroxene grains with slightly different orientation. A similar process may be responsible for the unusual complex patchwork texture of the pyroxenes in sample 62, although the possibility of shock deformation cannot be excluded.

If a ferriferous calcium metasilicate with pyroxenoid structure is indeed present, whether or not it is a metastable phase (*7*), then probably the solidus for these rocks was above 980°C (*8*, p. 213). The 1933 prediction of Bowen *et al.* (*8*, p. 270) "that somewhere wollastonites with relatively high amounts of $FeSiO_3$ should be found" will have been surprisingly fulfilled.

R. B. HARGRAVES
L. S. HOLLISTER
G. OTALORA

Department of Geological and Geophysical Sciences, Princeton University, Princeton, New Jersey

References and Notes

1. A. Bence and A. Albee, *J. Geol.* **76**, 382 (1968).
2. L. Hollister and A. Bence, *Science* **158**, 1053 (1967).
3. J. Preston, *Amer. Mineralogist* **51**, 1227 (1966); A. L. Smith and I. S. E. Carmichael, *ibid.* **54**, 909 (1969).
4. L. Hollister, *ibid.*, in press.
5. ———, *Science* **154**, 1647 (1966).
6. W. A. Deer, R. A. Howie, J. Zussman, *Rock Forming Minerals* (Wiley, New York, 1963), p. 5.
7. D. H. Lindsley and J. L. Munoz, *Amer. J. Sci.* **267-A**, 295 (1969).
8. N. L. Bowen, J. F. Schairer, E. Posnjak, *ibid.* **26**, 193 (1933).
9. G. M. Brown and E. A. Vincent, *J. Petrol.* **4**, 175 (1963).
10. We thank N. Dorety, T. Forseman, C. Kulick, D. Roy, R. Warren, and J. Wolinski (Princeton University) for assistance in preparing the samples for analysis and for assistance in the reduction of the data.

4 January 1970

Crystallography of Some Lunar Plagioclases

Abstract. *Crystals of calcic bytownite from type B rocks have space group* $I\bar{1}$ *with* $c \approx 14$ *angstroms. Bytownite crystals from type A rocks are more sodic and have space group* $C\bar{1}$, $c \approx 7$ *angstroms. Cell parameters of eight bulk feldspar separates from crystalline rocks indicate that the range of angle gamma is about 23 times the standard error of measurement, and its value might be useful for estimation of composition. Cell parameters of seven ilmenites are close to those of pure* $FeTiO_3$.

Bulk bytownite separates were obtained from eight crystalline rocks by flotation of pulverized rock in bromoform (specific gravity 2.87). Suitable single crystals were removed from the chips and the separates before grinding, and single-crystal x-ray diffraction patterns were obtained by the Buerger precession method with MoKα radiation (50 kv, 20 ma). The $0kl$, $h0l$, and $h1l$ nets of crystals from rocks 10003, 10047, 10050, 10071, and 10072 were recorded with exposures of about 160 hours on Ilford G film. Crystals from type B rocks (10003, 10047, 10050) give strong "a" reflections ($h+k$ even, l even) and weak, but fairly sharp, "b" reflections ($h+k$ odd, l odd), an indication that these feldspars have $c \approx 14$ Å and space group $I\bar{1}$. Crystals from type A rocks (10071, 10072) show only "a" reflections, indicating $c \approx 7$ Å and space group $C\bar{1}$. All of the plagioclase crystals examined were twinned on the albite law. In every case the cell parameters determined from single-crystal measurements were indistinguishable from those given by the powder-diffraction refinements of bulk plagioclase samples from the corresponding rocks (Table 1). The crystals examined can therefore be considered representative of the plagioclases in these rocks.

Powdered bytownite was combined with an internal standard (either CaF_2, $a = 5.4620$ Å, or BaF_2, $a = 6.1971$ Å), and each diffraction pattern was recorded once with CuKα radiation on a diffractometer run continuously from 60° to 13° 2θ. The diffractometer output was collected for successive intervals of 0.005° 2θ and recorded on paper tape as counts per 0.005° 2θ. The position of each diffracted line was reduced from the raw data by a computer program (*1*). The cell parameters given in Table 1 were then obtained by least-squares methods (*2*). Lines affected by ilmenite or clinopyroxene interference were deleted.

For seven samples the cell parameters of ilmenite were refined (Table 2) and were found to be close to those of pure $FeTiO_3$ (*3*), which suggests that these rocks crystallized under reducing conditions (*4*). Good values for the cell parameters of the clinopyroxenes could not be obtained because the diffraction lines were broad. Some zoned bytownite crystals were observed optically in thin sections, but the diffraction lines for each sample were sharp, showing that the proportion of each sample that differs from the bulk is very small. The standard errors listed in Tables 1 and 2 reflect the internal consistency of line positions from one particular pattern of each sample; experience (*5*) indicates that the precision of such cell-parameter determinations is at least twice the stated standard errors. Plagioclase patterns from type C breccias (10021, 10048, 10061, 10073) and type D dust (10084) show peaks that are too broad and weak for refinement.

Corresponding cell parameters for the eight crystalline samples range only 4 to 11 times the associated standard errors, except for the values of the angle γ, which have a range about 23 times the associated standard errors. The separa-

Table 1. Crystallographic data and estimated compositions for eight Apollo 11 plagioclases.

Rock and chip No.	03, 38	50, 32	47, 27	20, 41	57, 36	72, 33	71, 29	49, 22
Rock type	B	B	B	A	A	A	A	A
Space group	$I\bar{1}$	$I\bar{1}$	$I\bar{1}$			$C\bar{1}$	$C\bar{1}$	
Cell parameters								
a (Å)	8.192	8.178	8.187	8.167	8.174	8.178	8.175	8.176
Standard error	0.003	0.002	0.003	0.003	0.003	0.002	0.003	0.002
b (Å)	12.892	12.872	12.882	12.861	12.876	12.871	12.860	12.864
Standard error	0.003	0.002	0.003	0.003	0.003	0.003	0.004	0.003
c (Å)	7.093*	7.092*	7.098*	7.093	7.096	7.097	7.095	7.096
Standard error	0.002*	0.001*	0.001*	0.001	0.002	0.002	0.002	0.001
α (°)	93.24	93.28	93.37	93.46	93.34	93.39	93.25	93.40
Standard error	0.03	0.02	0.03	0.02	0.03	0.03	0.04	0.02
β (°)	116.03	115.95	116.05	115.92	116.06	116.09	116.04	116.07
Standard error	0.02	0.02	0.02	0.02	0.02	0.02	0.03	0.02
γ (°)	91.02	90.95	90.81	90.54	90.46	90.45	90.41	90.33
Standard error	0.02	0.03	0.03	0.03	0.03	0.03	0.04	0.03
Cell volume (Å^3)	671.3*	669.5*	670.7*	668.3	669.3	669.3	668.6	668.8
Standard error	0.2	0.2	0.2	0.3	0.2	0.2	0.3	0.2
Number of lines refined	24	31	22	22	30	31	24	25
Standard error, °2θ, CuKα_1	0.019	0.018	0.016	0.017	0.023	0.022	0.023	0.016
Estimated chemical composition (weight percent An)	85	83	81	78	75	75	74	73
Calculated $\Delta 2\theta$, (131)–($1\bar{3}1$), °2θ, CuKα_1	2.22	2.20	2.18	2.09	2.00	2.01	1.92	1.97
Γ, °2θ, CuKα_1 (*6*)	1.37	1.35	1.28	1.11	0.94	0.95	0.79	0.85
B, °2θ, CuKα_1 (*6*)	0.77	0.74	0.78	0.74	0.81	0.82	0.82	0.82

* True value is twice the tabular value for space group $I\bar{1}$.

Table 2. Crystallographic data for seven Apollo 11 ilmenites.

Rock and chip No.	03, 38	50, 32	47, 27	20, 41	72, 33	71, 29	49, 22
Cell parameters							
a (Å)	5.088	5.087	5.088	5.086	5.086	5.087	5.088
Standard error	0.001	0.001	0.001	0.001	0.001	0.001	0.001
c (Å)	14.085	14.081	14.085	14.070	14.068	14.068	14.079
Standard error	0.007	0.002	0.002	0.004	0.005	0.008	0.005
Cell volume (Å^3)	315.8	315.6	315.8	315.2	315.2	315.2	315.6
Standard error	0.1	0.1	0.1	0.1	0.1	0.2	0.1
Number of lines refined	8	9	10	9	7	7	8
Standard error, °2θ, CuKα_1	0.019	0.013	0.009	0.018	0.012	0.021	0.012

tion in degrees 2θ of the 131 and $1\bar{3}1$ lines depends strongly on the value of γ, and varies by 0.3° 2θ for CuKα radiation in the samples studied. When the compositions of the feldspars in Table 1 are known independently, the 131-$1\bar{3}1$ separations given in Table 1 can be calibrated and may then prove useful in estimating feldspar compositions in other similar samples.

The cell parameters of calcic plagioclases vary slightly with change of composition and with Al/Si order. From investigations utilizing methods based on the indices of refraction (*6*) and on cell parameters and line positions (*7*) of terrestrial plagioclases, it appears that the lunar plagioclases in type B rocks are from 5 to 10 percent by weight more calcic than those in type A rocks, but that within each rock type the plagioclase composition varies relatively little. The lunar plagioclases do not differ in any obvious ways from terrestrial plagioclases that are chemically equivalent. If the compositional estimates in Table 1 are accurate, values of the parameters Γ and B (*7*) suggest that the plagioclases in type B rocks probably have relatively high Al/Si disorder compared to those in terrestrial intrusive igneous rocks.

DAVID B. STEWART
DANIEL E. APPLEMAN
J. STEPHEN HUEBNER
JOAN R. CLARK

U.S. Geological Survey, Washington, D.C. 20242

References and Notes

1. U.S. Geological Survey programs 9208 CURVE and 9208 PEAKS written for this research by M. S. Hellman.
2. H. T. Evans, Jr., D. E. Appleman, D. S. Handwerker, Program, Am. Cryst. Assoc. meeting, Cambridge, Mass. (1963), p. 42.
3. G. Shirane, S. J. Pickart, R. Nathans, Y. Ishikawa, *J. Phys. Chem Solids* **10**, 35 (1959).
4. See R. W. Taylor, *Am. Mineral.* **49**, 1016 (1964); D. H. Lindsley, *Ann. Rept. Director Geophys. Lab.* **62**, 60 (1963); A. H. Webster and N. F. H. Bright, *J. Am. Ceram. Soc.* **44**, 110 (1961).
5. T. L. Wright and D. B. Stewart, *Am. Mineral.* **53**, 38 (1968).
6. F. Chayes, *Am. J. Science*, Bowen volume, 85 (1952).
7. J. V. Smith and P. Gay, *Mineral. Mag.* **31**, 744 (1958); H. -U. Nissen, in press.

8. We thank Mary E. Woodruff, U.S. Geological Survey, for assistance in mineral separation and data processing. Supported by NASA. Publication authorized by director, U.S. Geological Survey.

4 January 1970

Petrology of a Fine-Grained Igneous Rock from the Sea of Tranquillity

Abstract. *All phases in a thin section of sample 10022 have been analyzed by electron microprobe. Augite grains show strong iron enrichment in the outer 15 to 20 microns. Pigeonite cores occur within augite grains. The plagioclase has an anorthite content of between 73 and 81 mole percent and is high in Si and low in Al compared to stoichiometric feldspar. Residual phases include microcrystalline Fe-rich "pyroxene," plagioclase, K-rich alkali feldspar, silica, and rare areas rich in P and Zr with concentrations of Ba, Y, and rare earth elements. The density, viscosity, and crystallization history of the lava of sample 10022 are discussed.*

Results reported here are based primarily upon electron probe microanalyses (*1*) of a fine-grained vesicular igneous rock, NASA sample 10022,28. Phase designations are based principally on chemical composition and do not imply direct knowledge of structure. The rock has an igneous texture, and our interpretations assume that it crystallized from a melt. Average compositions of analyzed phases are presented in Tables 1 and 2. Volume percentages of the major phases have also been determined on the basis of 2000 random point identifications in transmitted and reflected light microscopy.

Skeletal ilmenite crystals up to several hundred microns long are distributed evenly throughout the section, some having delicate branches less than 10 μm wide. Euhedral ilmenite grains are enclosed in some augites; this suggests that ilmenite crystallized first. In section some ilmenite grains have rounded cavities filled with fine crystals of the other phases. The ilmenite exhibits no zoning or exsolution that can be resolved with the probe. Plagioclase crystals are often hollow and appear tubular in thin section, a habit characteristic of relatively rapid growth. The "tubes" are filled with an assortment of pyroxene, ilmenite, additional plagioclase, and residual phases. Plagioclase crystals occasionally transect the outer rims of augite. Many plagioclases grade into microcrystalline or glassy material of quartzofeldspathic composition. Figure 1a (detail) shows the range of composition in 19 plagioclase grains. There is not much systematic zoning of individual grains. In the average plagioclase of Table 2 Si − (Na + K) = 2.10 and Al − (Ca + Fe + Mg) = 0.82 (based on eight oxygens), whereas these parameters are 2.00 and 1.00, respectively, in stoichiometric feldspars. Excess Si and a deficiency of Al have also been detected in terrestrial feldspars (*2*). The Ti and Fe (assumed to be Fe^{2+}) content is higher in the average lunar plagioclase of rock 10022 than in most terrestrial plagioclases.

Clinopyroxene is the most abundant phase and occurs in approximately equant grains up to several hundred microns in diameter. Most grains are composite, containing cores of pigeonite (10 to 100 μm) surrounded by augite. Pigeonite compositions (Fig. 1b) cluster close to the area where terrestrial pyroxene trends show inversion from orthopyroxene to pigeonite. There is a sharp composition break between the pigeonite and augite (Fig. 1, b and c). In contrast to the lack of zoning in the pigeonite, there is a continuous increase of Ca relative to Mg in the main body of the augite, followed by a sharp increase of Fe and a decrease of Mg and Ca in the outer 20 μm. The minor elements Ti, Al, and Cr are enriched in augite relative to pigeonite whereas Mn shows a reversed distribution. Titanium, Al, and Cr decrease in the augite rims whereas Mn increases. The smooth composition trends in the main body of the augites are similar to trends observed in plutonic terrestrial pyroxenes and may reflect growth from an ample liquid reservoir under near equilibrium conditions, whereas the abrupt zoning of the rims may be due to rapid changes of local residual liquid composition. Some augite rims appear to grade into material of Fe-rich pyroxene composition associated with other residual phases. The profile labeled *ar-rp* (Fig. 1b) extends from an augite rim into such a residual phase area. X-ray area scans across pigeonite-augite boundaries often reveal irregular borders suggestive of resorption of pigeonite and compatible with a peritectic reaction between pigeonite, augite, and liquid. Several grains of olivine (10 to 20 μm) are included in augite crystals. It seems likely that minor olivine crystallized before augite. Troilite is present as small interstitial blebs 5 to 30 μm in diameter. Its composition is close to that of stoichiometric FeS, and it has a lower content of Ti than some chondritic troilites have, in spite of the high Ti content of the rock. This may be related to conditions that contribute to a greater amount of oxidation or to the late crystallization of troilite after most of the Ti had been incorporated into ilmenite. Subspherical grains of iron ($\leqslant 5$ μm) are enclosed in some troilites with a texture suggestive of immiscible liquids. The Ni content of the iron averages 0.9 percent by weight with a maximum content of 1.4 percent observed.

As magma crystallization proceeded to the final stages, it was probable that certain constituents of the melt would solidify in phases other than the major phases described above. In terrestrial lavas it is usual to find residual constituents in the form of glass plus fine-grained matrix minerals. We have not identified glass, but a systematic search for residual phases has revealed patches of interstitial material which we interpret to be the crystallization product of a highly differentiated last-stage liquid. The overall average composition of this interstitial material is given in Tables 1 and 2; the analysis was also calculated in terms of normative components, and the results show the composition to be quite siliceous. The residual material is further resolved by probe analysis into the following fractions: (i) Areas of Fe-rich pyroxene composition, often adjacent to augites. Analyses of such areas are plotted as triangular symbols in Fig. 1b, and the average is given in Tables 1 and 2. Some of these "pyroxene" compositions are thought to be unstable in the pyroxene structure at low pressures (*3*). Within the resolution of the probe, these areas represent a homogeneous phase, but they may not have true pyroxene structure. (ii) Nearly pure silica, believed to be a primary igneous phase crystallized from a silica-saturated residual melt strongly differentiated from the initial ultrabasic

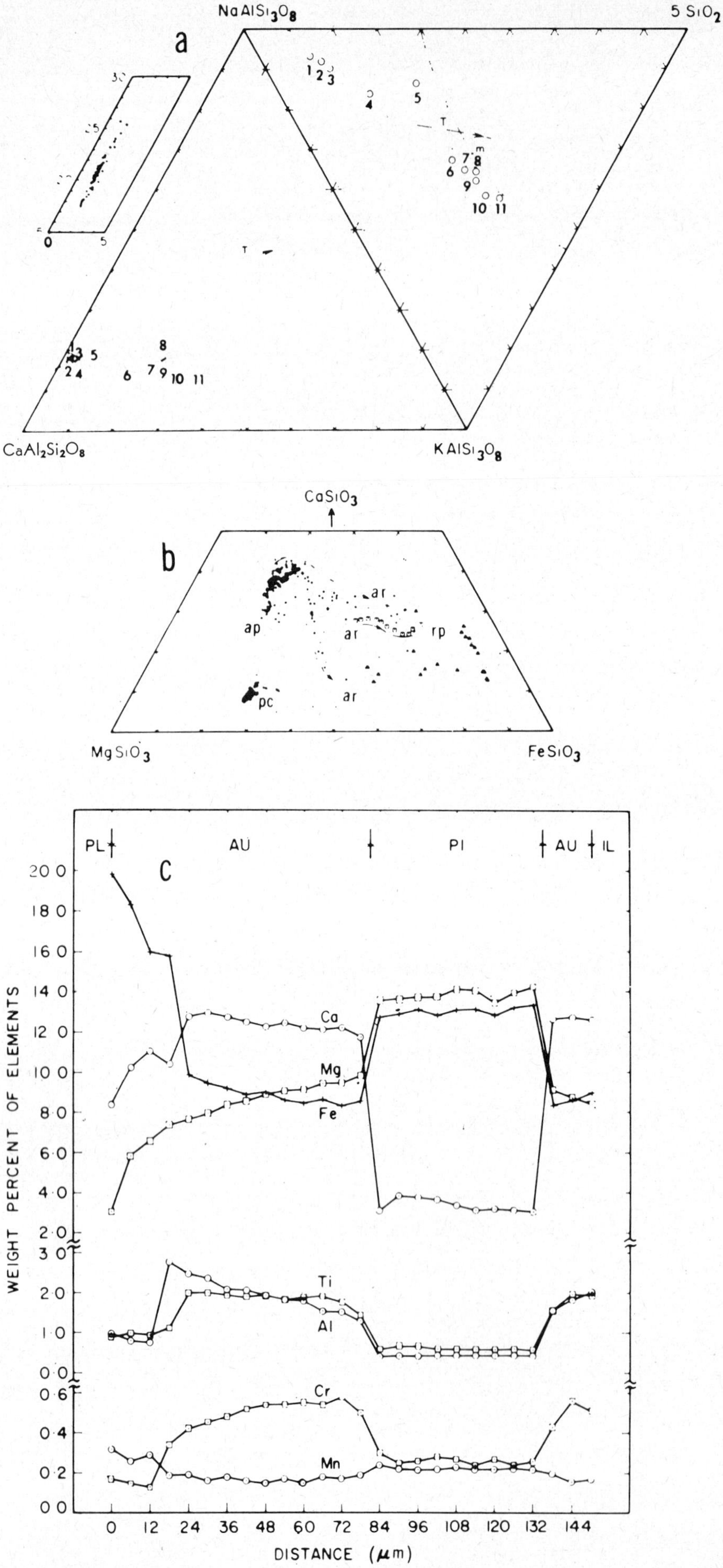

Fig. 1. Plagioclase, pyroxene, and residual phase compositions. (a) Small filled circles (detail) represent the results of 48 plagioclase analyses; Nos. 1 through 11 represent a profile from a plagioclase grain through a residual material of quartzo-feldspathic composition (anorthite + albite + orthoclase + quartz $\geqslant$ 88 percent by weight of total); T = Tasmanian dolerite salic trend (*4*); m = albite-orthoclase-quartz minimum at a water pressure of 500 kg/cm^2. (b) Pigeonite cores, *pc;* profiles from augite-pigeonite boundary to augite rim, *ap-ar;* profile from augite rim through residual "pyroxene," *ar-rp;* residual "pyroxene" analyses, triangles. (c) Profile across composite pyroxene; *PL,* plagioclase; *AU,* augite; *PI,* pigeonite; *IL,* ilmenite.

Table 1. Average compositions of phases in sample 10022 by electron probe microanalysis. Totals are given as oxides except for troilite and iron. Assumed oxidation states of elements are shown.

Phase	No. of spots-No. of grains	Elements (percentage by weight)												Totals
		Si^{4+}	Ti^{4+}	Al^{3+}	Cr^{3+}	Fe^{2+}	Mg^{2+}	Ca^{2+}	Mn^{2+}	Ni^{2+}	Na^{+}	K^{+}	S^{2-}	
Ilmenite	30-9	0.16	31.9		0.37	35.6	0.75	0.21	0.28					101.78
Olivine	7-2	17.2	0.19	0.10	0.15	22.1	20.3	0.35	0.23					100.47
Pigeonite	54-7	24.7	0.61	0.50	0.28	13.2	13.8	3.33	0.23		0.017			100.05
Augite	149-12	22.7	1.68	1.83	0.43	11.2	8.22	11.6	0.20		0.07			100.10
Plagioclase	30-19	23.8	0.10	16.0		0.66	0.20	10.9			1.75	0.21		100.39
Troilite	14-11	0.20	0.33		0.03	63.2		0.22		0.02			36.1	100.53
Iron	5-5	0.21	0.32		0.02	97.1			0.02	0.89			0.17	98.73
Residual														
Average	114-5	27.5	0.44	9.04	0.03	6.81	0.80	6.92	0.08		0.86	1.62		99.94*
"Pyroxene"	20-4	21.3	0.85	0.86	0.07	29.8	2.54	5.88	0.41		0.01			100.10
Silica	12-4	46.7	0.19	0.35	0.01	0.43	0.08	0.08			0.04			101.69
P-rich area	1-1	11.5	0.23	1.7		6.1	0.24	22.1			0.13	0.64		99.27†

*Includes 0.25 percent by weight BaO. †See text for additional elements which are included in total.

liquid. (iii) Rare areas rich in P and Zr which are associated with notable enrichment in exotic elements. In Table 1 we list the results of an analysis of an area rich in P which contains in addition to 9.9 percent by weight P (in percentage by weight): Y = 1.5, Ba = 0.22, La = 0.54, Ce = 1.4, Pr = 0.37, Nd = 1.5, Sm = 0.42, Gd = 0.57, Dy = 0.48, plus detectable amounts of Cl. Separate Zr concentrations (up to 18 percent by weight) with detectable amounts of Y, Ba, La, and Ce (but no P) have also been found in the residual material. We interpret these occurrences to be due to microcrystalline apatite and possibly zircon only partially resolved with the probe. (iv) Patches resembling eutectic microstructures often gradational with plagioclase grains. The average composition of these areas is represented by the salic portion of the formula for the average residual material in Table 2. A profile (4-μm steps) starting at the edge of a plagioclase grain and extending into this salic residue is illustrated in Fig. 1a (Nos. 1 through 11). The trend shows strong K and Si enrichment outward from the plagioclase, thus reflecting the progressive differentiation of the residual liquid. In Fig. 1a these residual liquid trends may be compared to those observed in the salic constituents of a Tasmanian dolerite (*4*). In both cases the trend, projected from the anorthite apex, approaches the ternary minimum in the albite-orthoclase-quartz system. While the terrestrial trend is defined by the salic constituents of a large differentiated complex, we are here dealing with several microns in one thin section. Although the microstructure is too fine to be completely resolved with the probe ($\leqslant$2 to 3 μm), it is possible with a small probe to partially resolve it into Ca-, K-, and Si-rich areas, which we interpret to be microcrystalline intergrowths of plagioclase, alkali feldspar, and silica, that is, microgranodiorite. Whether comparable igneous differentiation occurred on a larger scale on the moon remains to be seen.

Experimental work on a synthetic equivalent [11 major elements as given in (*5*)] of sample 10022 has been carried out in an Ar-atmosphere furnace with iron filings in the hot zone as an oxygen getter. Results given here are for runs held for 10 hours at the desired temperature, followed by quenching to 0°C in less than 2 seconds. Analyses of quench products are made by microprobe, x-ray diffraction, and microscopy. Ilmenite crystallizes at a liquidus temperature of 1300° ± 25°C, and there is an interval of approximately 100°C before pyroxene begins to crystallize. Plagioclase crystallizes at a somewhat lower temperature than pyroxene but is not equilibrated with the liquid in 10 hours, as indicated by a wide range of compositions. The equilibrium partition of Si, Al, Na, and Ca between plagioclase and liquid has been developed into a geothermometer (*6*). We have considered the temperatures at which the average plagioclase [An_{79} (79 mole percent anorthite)] would equilibrate with liquids from sample 10022 remaining after all of the ilmenite (21 percent by weight) and pigeonite (5 percent by weight), and varying amounts of augite, have been removed. Accordingly, the beginning of plagioclase crystallization is calculated to have occurred at between 1075° and 1175°C, corresponding to prior crystallization of 0 to 40 percent by weight augite. Experimental and theoretical arguments thus lead to a general view of the crystallization of sample 10022

Table 2. Average chemical formulas of phases in sample 10022.

Phase	Formula	Percentage by volume	Density (g/cm3)	Percentage by weight
Ilmenite	$(Fe_{7.54}Mg_{.37}Ca_{.06}Mn_{.06})_{8.03}(Ti_{7.86}Cr_{.08}Si_{.07})_{8.01}O_{24}$	15	4.8	21
Olivine	$(Mg_{1.34}Fe_{.64}Ca_{.014}Mn_{.0066}Cr_{.0047})_{2.01}(Si_{.986}Al_{.006}Ti_{.006})_{.998}O_4$			
Pigeonite	$(Fe_{2.09}Mg_{5.02}Ca_{.74}Mn_{.037}Na_{.0065}Cr_{.048}Al_{.060})_{8.00}(Si_{7.78}Al_{.10}Ti_{.11})_{7.99}O_{24}$	5	3.4	5
Augite	$(Fe_{1.82}Mg_{3.08}Ca_{2.64}Mn_{.033}Na_{.028}Cr_{.075}Al_{.29})_{7.97}(Si_{7.36}Al_{.33}Ti_{.32})_{8.01}O_{24}$	50	3.4	50
Plagioclase	$(Ca_{.744}Na_{.208}K_{.014}Fe_{.032}Mg_{.022})_{1.020}(Si_{2.32}Al_{1.62}Ti_{.0059})_{3.95}O_8$	29	2.7	23
Troilite	$(Fe_{1.00}Ti_{.007}Si_{.007}Ca_{.005})_{1.02}S$	1*	4.8	1
Iron	$Fe_{.997}Ni_{.0091}S_{.0025}Ti_{.0059}Cr_{.0003}Mn_{.0001}Si_{.0051}$			
Residual				
Average †	$23.5(Wo_{23}Fs_{60}En_{16}Rh_1) + 59.7(Or_{20}Ab_{18}An_{61}Cn_1) + 16.3\ SiO_2 + 0.5\ TiO_2$			
"Pyroxene"	$(Fe_{5.36}Mg_{1.04}Ca_{1.47}Mn_{.075}Na_{.0044}Cr_{.01}Al_{.16})_{8.12}(Si_{7.62}Al_{.16}Ti_{.18})_{7.96}O_{24}$			
Silica	$(Si_{11.9}Al_{.091}Cr_{.002}Fe_{.054}Ca_{.013}Mg_{.023}Ti_{.028}Na_{.01})_{12.1}O_{24}$			

*Includes iron and all others. † Formula based on one cation per pyroxene and feldspar molecule.

which confirms the textural relations.

The density and logarithm of viscosity of silicate melts are approximately linear functions of composition, and published density and viscosity data for simple silicate systems allow one to calculate these properties for natural lavas with accuracies comparable to those of current measurements (*7*). We calculate the density of sample 10022 liquid to be 2.93 to 2.96 g/cm^3 between 1400° and 1250°C. The viscosity varies from 6 to 27 poises over the same temperature range, which is comparable to the viscosity of glycerin at room temperature. The low viscosity [the viscosities of olivine basalts (*8*) range at least an order of magnitude higher] is related to low concentrations of Si and Al and makes it likely that, for comparable thermal histories, lunar lavas, in spite of lower gravity, are capable of faster and more extensive flow than their terrestrial counterparts. In view of the large areal extent of plateau-type lavas on earth, it is not difficult to imagine lunar lavas flowing, topography permitting, over areas comparable to mare basins. The pronounced fluidity might also result in the formation of long lava tubes which could form sinuous rills when the upper crusts are breached. The large density contrast between ilmenite and the liquid, the long ilmenite crystallization interval, and the low viscosity of the liquid would favor ilmenite-liquid separation wherever lavas were ponded to any depth. The igneous rocks of the Sea of Tranquillity may owe their high ilmenite content to such a mechanism, and mascons may be related to concentrations of dense, early-crystallizing solids in the central portions of the basins of the deeper circular maria. In any case, such reasoning leads one to expect that the ilmenite (and therefore Fe and Ti) content may be quite variable over the lunar surface.

DANIEL F. WEILL
I. STEWART MCCALLUM, YAN BOTTINGA
MICHAEL J. DRAKE, GORDON A. MCKAY
Center for Volcanology, University of Oregon, Eugene 97403

References and Notes

1. Standard analytical conditions: Applied Research Labs electron microprobe x-ray analyzer; 15 kv; sample current, 30 to 50 × 10^{-9} amperes; 1- to 2-μm beam; 10-second counting interval. When convenient, Ti through Ni were analyzed at 20 kv. The counting interval was increased for minor elements, and a larger beam was used in alkali-rich areas. Drift, dead time, background, absorption, generation, and fluorescence corrections were applied to all analyses. Details of all analyses are available on request.
2. K. Perry, Jr., *Lithos* **1**, 201 (1968).
3. D. H. Lindsley and J. L. Munoz, *Amer. J. Sci.* **267-A**, 295 (1969).
4. I. McDougall, *Bull. Geol. Soc. Amer.* **73**, 279 (1962). Salic trends are taken from the interpretation by I. S. E. Carmichael, *Quart. J. Geol. Soc. London* **119**, 95 (1963).
5. Lunar Sample Preliminary Examination Team, *Science* **165**, 1211 (1969).
6. A. M. Kudo and D. F. Weill, *Contrib. Mineral. Petrol.*, in press.
7. Y. Bottinga and D. F. Weill, abstract, *Trans. Amer. Geophys. Union* **50**, 676 (1969); in preparation.
8. Kani, *Proc. Imp. Acad. Tokyo* **10**, 29, 79 (1934).
9. We thank R. Freeman, H. Howard, and E. Oas for technical assistance. Supported by NASA contract NAS 9-8071 and NASA grant NGR 38-003-009.

4 January 1970

High-Voltage Transmission Electron Microscopy Study of Lunar Surface Material

Abstract. *The internal substructures of a type B sample have been examined at high magnification and compared with terrestrial rocks. Selected ultrathin sections were prepared from these multiphase materials by an ion-thinning technique and examined in a 1-Mev electron microscope, with complementary optical analyses. The structures in the ilmenite and plagioclase indicate that the lunar material has undergone plastic deformation by dislocation movement and possibly microtwinning, with subsequent recovery. The pyroxene exhibits complex lamellar structures of submicron spacing. These various observations are consistent with the optical microscopy evidence for distortion and recovery and identify the processes involved.*

The use of transmission electron microscopy to study the fine-scale substructure in thin foils of metals and some ceramic materials during the past decade has been of considerable value in elucidating various structure-history and structure-property relationships (*1*). However, with few exceptions, this technique has not been applied to geological materials because of the difficulty of preparing suitable electron-transparent foils from multiphase nonmetallic solids. Previous electron microscopy studies of geological importance appear to have been restricted to a few single-phase rock-forming minerals (*2, 3*), with thin foils usually prepared by mechanical cleavage. The latter method has the major disadvantages of uncertainty as to the location of the fragment in the original specimen and the possibility of introducing artifacts.

Although the potential of ion-bombardment thinning to prepare foils for electron microscopy was recognized some 20 years ago (*4*), only recently has apparatus capable of producing glow-discharge plasma beams that are stable over long periods of time been perfected (*5*). Such apparatus can successfully thin metals and ceramics and has now been shown to be suitable for a number of geologically important materials (*6*). However, the foil thickness of a few thousand angstroms required for conventional (100-kv) transmission microscopy results in a delicate and friable specimen which is easily broken and destroyed in handling. Consequently, the recent advent of high-voltage microscopy offers a major advantage for the study of geological materials because the specimen thickness can be increased to the order of 1 μm, which facilitates both preparation and handling. Furthermore, the higher voltage permits increased resolution by reduction of chromatic aberration, increased precision of selected-area electron diffraction, and reduction of the possibility of specimen damage in the microscope from ionization and beam heating effects.

The combination of ion thinning and high-voltage microscopy has particular value for the study of the substructure of lunar material, in that wastage and loss during specimen preparation of the limited amount of available material is minimized. The present paper describes the application of these new techniques to the study of the substructure of a sample of type B lunar rock (Preliminary Examination Team Report sample 10029-1) with the objectives of comparing the principal characteristics with those of analogous terrestrial material to assist in elucidating the mechanical and thermal history of the lunar material. This study is the first in a program of electron microscopy observations of the internal structure of lunar surface material brought back from the Apollo 11 and 12 missions.

The specific techniques used for preparing and examining the 10029-1 sample were developed on specimens of terrestrial rocks chosen to simulate the lunar material as closely as possible. The initial series of simulation specimens included several Hawaiian basalts, together with pyroxenite and anorthosite. After the release of chemical analyses and

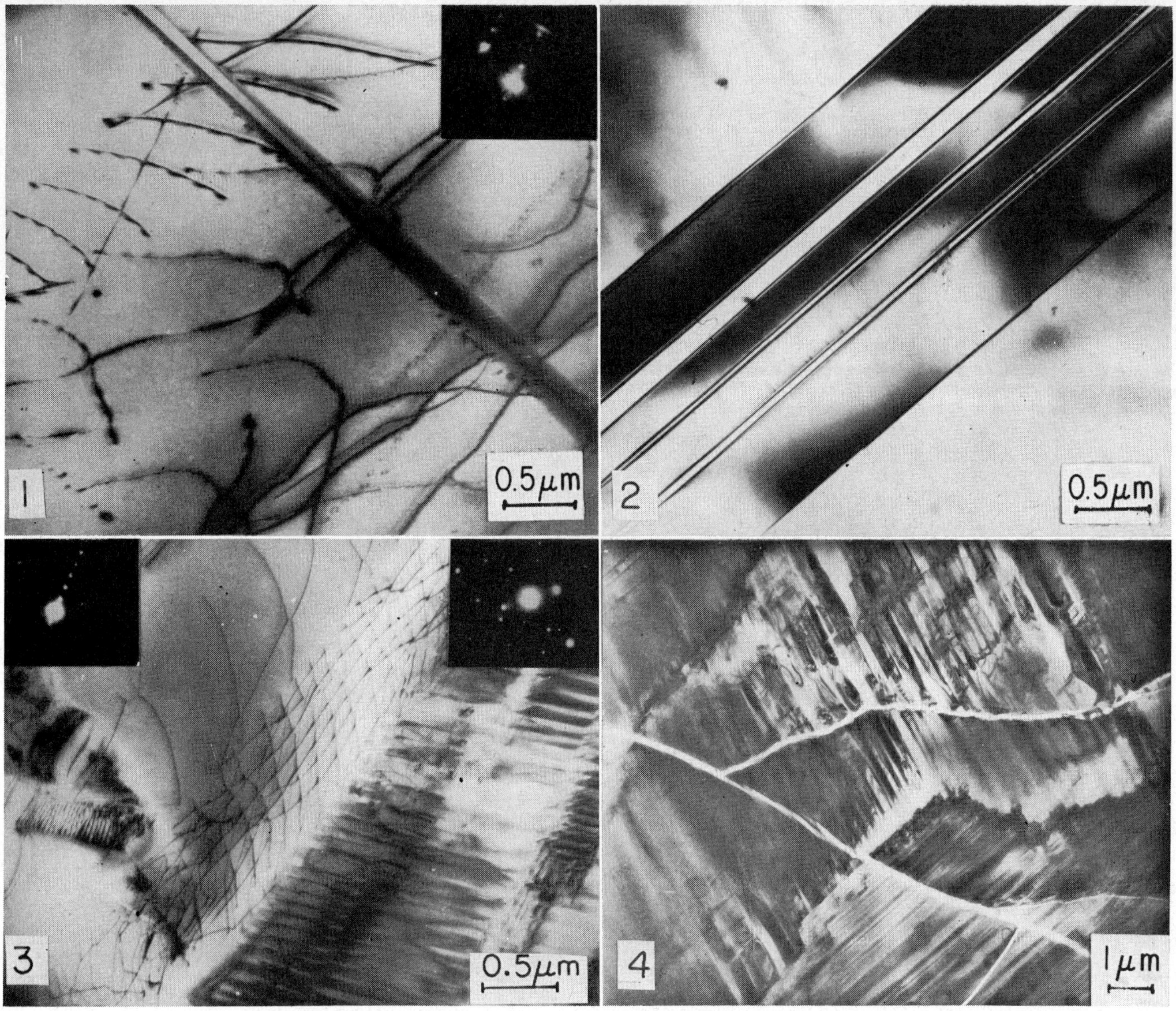

Fig. 1. Transmission electron micrograph of ilmenite, showing deformation twin and slip dislocations. Apollo 11 sample 10029-1; 800 kv; × 19,600. Fig. 2. Transmission electron micrograph, showing deformation twins in plagioclase. Apollo 11 sample 10029-1; 800 kv; × 19,600. Fig. 3. Transmission electron micrograph, showing interface between pyroxene and plagioclase phases. Apollo 11 sample 10029-1; 800 kv; × 19,600. Fig. 4. Transmission electron micrograph of pyroxene, showing characteristic substructure and cracking. Exsolution lamellae occur along several intersecting crystallographic orientations. Apollo 11, sample 10029-1; 800 kv; × 5880.

petrographic information in the Apollo 11 Preliminary Examination Team report, specimens of diabase (Ironwood, Michigan, and Beaver Bay, Minnesota) were examined as more closely resembling the lunar material. The examination sequence finally adopted to correlate large-area studies at low magnification by optical petrography with high magnification electron petrography is the following. Standard thin sections of a given specimen are examined in the optical microscope to identify the phases present, their textural relationships, and evidence for deformation or chemical zoning. Areas of interest for more detailed observation are then selected and 3-mm diameter discs cored out (with the features of most interest in their central region) ultrasonically. Each disc is then mounted on an alumina ring (approximately 0.25 mm thick) to provide mechanical support for direct insertion in the electron microscope and other handling. The disc is next reduced in its central region to a thickness of about 1 μm by bombardment with argon-ion plasma beams from both sides. In this procedure, the specimen is positioned in a high-vacuum chamber between two plasma guns and rotated about an axis at an angle of 75° to the plasma beams. With the guns operated at 6 kv and a beam current of 50 μA, a suitably thin section (indicated by the development of perforations) is prepared in 15 to 20 hours. Although the resulting specimens are too thin to show interference colors between crossed polarizers, a careful photographic comparison is made before and after thinning (with a large-scale montage or map of the complete specimen made at the latter stage) to permit exact identification in the electron microscope of specific features examined optically. The specimen is then transferred to a specially designed holder and examined by transmission electron microscopy. The high-voltage machine used is that at the 1-Mev facility completed recently at the U.S. Steel Research Center (*7*). Experiments with the various geological materials mentioned earlier have shown that no radiation damage under the high-voltage electron beam is apparent in any of the minerals observed. The only mineral in which such damage has been found to occur to date is quartz (*3*). On the basis of these experiments, an optimum operating voltage of 800 kv was selected for the examination of the lunar material, with a specimen beam current of 1 μA. Apertures to permit the selection of areas down to 0.1 μm in diameter at the specimen for electron diffraction were used in conjunction with a two-circle goniometer stage fitted with an anticontamination device to permit prolonged observation for diffraction-contrast experiments.

The three major minerals in the medium-grain-sized 10029-1 rock sample are clinopyroxene, calcic-plagioclase, and ilmenite, in order of the volume fraction present and in agreement with the preliminary analysis for type B rocks given in the Preliminary Examination Team Report. Optically, the major features of the lunar samples are (i) a more general cracking (irregular and curved cracks and cleavage) of all minerals than is normally found in terrestrial rocks (this is consistent with mild shock damage from meteorite impact), and (ii) a wide variety of features in the pyroxenes, including growth twins with [001] axis and planar composition surfaces parallel to (100) planes (common in terrestrial rocks), marked misorientations within many of the single grains (subgrains and kinking, with kink boundaries nearly perpendicular to [001]), and lamellar structures or striations defined by zones of different interference colors. The latter features are of variable orientation but appear close to the (001) plane and parallel to the *a*-axis—they are not fully resolvable optically but could be interpreted as deformation bands, twins, or arrays of slip dislocations from pencil glide. Despite the optical evidence of deformation, the general absence of visible "damage" in the subgrains suggests that some recovery occurred subsequent to or coincidental with the deformation.

The rates of ion thinning of these major minerals were found to differ sufficiently (decreasing in the order plagioclase, pyroxene, ilmenite) that the specimens were best examined at successive stages of thinning for large-area observation of desired phases in the electron microscope. These differences exit in both lunar and terrestrial materials, but the controllability of the thinning process appears to be easier for the lunar specimens (this may be due to the absence of hydrated phases—a particularly prominent structural difference between the lunar rock and the diabase). The optically opaque ilmenite was observed in electron transmission to contain substantial arrays of slip dislocations and some microtwinning (Fig. 1), together with an occasional subgrain. The latter probably originated in the original crystal formation, since the dislocation arrays show little evidence of rearrangement other than the existence of small loops (Fig. 1). In contrast to the ilmenite, the principal feature of the plagioclase is extensive microtwinning (Fig. 2). The twins, at least some of which are of the deformation type, appear in groups or bundles and may correspond to the "lamellar twinning" parallel to the long dimension of the lath and plate-shaped crystals observed optically. The less extensive arrays of dislocations in the plagioclase exhibit rearrangements, including the development of networks corresponding to low-angle boundaries (Fig. 3), indicative of recovery. None of the occasional compositional zoning observed optically was found, nor any fine-scale exsolution structures, such as have been noted in recent electron microscopy studies of K-rich feldspars (*8*).

The clinopyroxene differed from both the other minerals in that no evidence of slip dislocations was found in electron transmission. The principal structural feature observed was an extensive and complex pattern of parallel striations on a submicron scale (Fig. 3). A more illustrative example of this structure in a single pyroxene grain is shown in a low-magnification composite electron micrograph (Fig. 4). Selected-area electron diffraction measurements failed to detect any misorientation between individual striations or across the "domain" boundaries shown in the micrograph (*9*). The appearance of the striations changes markedly as the diffraction contrast conditions are altered (compare Figs. 3 and 4), and any dislocations present may be obscured by this structure. The domains are on a much smaller scale than the subgrains observed optically, and none of the latter have yet been observed in the areas examined. The terrestrial pyroxene in the diabase contained more complex and irregular striated structures. The optical measurements made on the lunar pyroxene are consistent with the composition of pigeonite or its extension into the augite field (subcalcic augite). In terrestrial volcanic rocks, pigeonite is commonly zoned between Ca-rich and Ca-poor regions; some evidence for such zoning was observed in the lunar sample. Electron-microprobe evidence for a fine-scale Ca-zoning in bronzite was recently reported (*10*). In view of these indications, it appears likely that the striations in the lunar rock arise from a submicron scale exsolution and that these fine striations are responsible for the "lamellar structures" observed optically. Additional features observed were precipitate particles in both the plagioclase and pyroxene (of the order of a micron in size and as yet unidentified), and some evidence of highly localized "black-spot" radiation damage in the plagioclase.

It is concluded from the above results that the lunar sample 10029-1 has been lightly deformed and simultaneously or subsequently heated sufficiently to permit recovery. This electron microscopy evidence is consistent with the optical observations and provides more specific knowledge of the structural features. This preliminary study also shows that high-voltage microscopy provides a significant addition to the field of petrography.

S. V. Radcliffe
A. H. Heuer
Division of Metallurgy and Materials Science, Case Western Reserve University, Cleveland, Ohio
R. M. Fisher
Bain Fundamental Laboratory, U.S. Steel Research Center, Monroeville, Pennsylvania
J. M. Christie
Department of Geology, University of California, Los Angeles
D. T. Griggs
Institute of Geophysics and Planetary Physics, University of California, Los Angeles

References and Notes

1. P. B. Hirsch, A. Howie, R. B. Nicholson, D. W. Pashley, M. G. Whelan, *Electron Microscopy of Thin Crystals* (Butterworths, London, 1965).
2. S. Amelinckx, *The Direct Observation of Dislocations* (Academic Press, New York, 1964), p. 487; H. U. Nissen and W. Bollmann, *Proc. Intern. Congr. Electron Microscopy, 6th, Kyoto, 1966*, p. 591.
3. A. C. McClaren, J. A. Retchford, D. T. Griggs, J. M. Christie, *Phys. Stat. Solidi* **19**, 631 (1967).
4. G. Induni, *Helv. Phys. Acta* **20**, 463 (1947); R. Castaing, *Rev. Met.* **52**, 669 (1955).
5. M. Paulus and F. Reverchon, *J. Phys. Radium* **22**, 103A (1961).
6. N. J. Tighe, *Ultrafine Grain Ceramics* (Syracuse Univ. Press, New York, in press); H. W. Green and S. V. Radcliffe, in preparation; H. W. Green, A. H. Heuer, S. V. Radcliffe, in preparation.
7. R. M. Fisher and J. S. Lally, *Proc. European Conf. Electron Microscopy, 4th, Rome, 1968*.
8. W. Bollmann and H. U. Nissen, *Acta Cryst.* **A24**, 546 (1968); J. D. C. McConnell, *Phil. Mag.* **19**, 221 (1968).
9. Evidence for the expected misorientation associated with the striations is being sought through Kikuchi-line diffraction analysis.
10. R. W. Kistler, J. D. Obradovich, E. D. Jackson, *J. Geophys. Res.* **74**, 325 (1969).
11. Supported through the Manned Spacecraft Center, NASA, Houston. We are appreciative of the cooperation of E. C. T. Chao, U.S. Geological Survey, in facilitating the selection of an appropriate lunar specimen. We thank particularly the following colleagues for their advice and assistance: H. W. Green and J. Norton (Case Western Reserve University); J. S. Lally, A. Szirmae, G. P. Wray, and W. R. Duff (U.S. Steel Research Center) and J. de Grossé, University of California, Los Angeles.

8 January 1970

Fine Particles, Glasses, and Shock Effects

Silicate Liquid Immiscibility in Lunar Magmas, Evidenced by Melt Inclusions in Lunar Rocks

Abstract. *Examination of multiphase melt inclusions in 91 sections of 26 lunar rocks revealed abundant evidence of late-stage immiscibility in all crystalline rock sections and in soil fragments and most breccias. The two individual immiscible silicate melts (now glasses) vary in composition, but are essentially potassic granite and pyroxenite. This immiscibility may be important in the formation of the lunar highlands and tektites. Other inclusions yield the following temperatures at which the several minerals first appear on cooling the original magma: ilmenite (?) liquidus, 1210°C; pyroxene, 1140°C; plagioclase, 1105°C; solidus, ~1075°C. The glasses also place some limitations on maximum and minimum cooling rates.*

Ever since certain silicate systems were found to exhibit liquid immiscibility, there have been numerous attempts to apply this process to explain certain rocks. In 1927 Greig (*1*) showed that the unusual compositions and unreasonably high temperatures made immiscibility unlikely as a process in the origin of igneous rocks. Roedder (*2*) then described silicate immiscibility in the system leucite-fayalite-silica, and in the parent quaternary system K_2O-FeO-Al_2O_3-SiO_2, at more geologically reasonable temperatures and compositions. Recent work has also resurrected this hypothesis for certain syenite-gabbro associations (*3*).

Although normal, primary silicate melt inclusions are not rare in lunar rocks; other more unusual primary inclusions provide abundant evidence of immiscibility. Such evidence can be found readily in loose grain mounts, but is most obvious in thin sections. All 58 thin sections (or probe mounts) of 15 crystalline rocks (numbers 03, 17, 20, 22, 24, 29, 44, 45, 47, 49, 50, 57, 69, 71, and 72) showed such evidence under brief examination, although it was relatively sparse in sections from rocks 20 and 71. The fragments of crystalline rocks in many of the sections of breccias examined (33 slides from 11 rocks) showed similar evidence, as did most sections of 29 fragments from the lunar soil (sample 85) (*4*).

Normal silicate glass inclusions, representing the trapping of a *homogeneous* melt, were found in plagioclase, ilmenite, pyroxene, and olivine. Those in olivine were by far the largest and most abundant per unit of mineral volume. As in similar melt inclusions in earth rocks, failure to nucleate stable phases frequently prevented the attainment of equilibrium on cooling.

Most plagioclase crystals are remarkably free of inclusions, but a very few, such as 20-29 (*5*), contain in their cores one large inclusion or groups of amoeboid inclusions of dark brown glass of high index of refraction. In addition to the glass, there is usually a bubble (from shrinkage on cooling) and one or more opaque masses, presumably ilmenite. These inclusions were probably trapped by the nucleation and rapid growth of a skeletal plagioclase crystal in a portion of melt otherwise free of plagioclase nuclei at the time. The proportion of ilmenite appears to be approximately equal to that in the rock.

Many ilmenite crystals are skeletal, so most of the apparently unconnected blebs of silicate seen in them are not actually isolated. Some blebs of silicate melt were isolated by the growth of ilmenite, however, yielding large inclusions. On cooling, these generally have formed brown glasses, either isotropic or containing a mass of minute, feathery birefringent crystals, presumably pyroxene. A few also contain large crystals of tridymite.

Some growth phenomenon has resulted in the trapping of melt inclusions in many, if not most, of the larger olivine crystals. These inclusions vary from a small fraction of a micrometer (in secondary, healed fractures) (*6*) to 655 μm, and the present phase assemblage is, to a degree, a function of this size. The smallest are all glass, or have nucleated only a shrinkage bubble or one or two opaque phases (Fig. 1a). The larger ones are generally more crystalline, and contain up to six phases (Figs. 1b and 2); in effect, these present us with a sample of the magma and a documented model of its future evolution, starting at the time of trapping.

A series of heating experiments were performed on the inclusions in olivine. Individual tiny grains of olivine containing suitable multiphase inclusions were held in iron foil envelopes *in vacuo* for hours or days at various temperatures and then quenched and reexamined (Fig. 2). These experiments, to be reported in full elsewhere, provided the following data on the homogenization temperatures: ilmenite (*7*) liquidus, 1210°C; joined by pyroxene, 1140°C; joined by plagioclase, 1105°C; solidus, ~1075°C. As these runs were all made in "containers" of olivine, they represent liquids saturated with olivine. There are small amounts of volatile constituents in these inclusions, as more pyroxene crystals are present in inclusions open to the furnace vacuum than in adjacent, completely sealed inclusions in the same grain, at near-liquidus temperatures.

Heating experiments that result in the formation or melting of daughter minerals in such inclusions in olivine permit us to place some limits on both the maximum and the minimum cooling rates (*8*). This is complicated by what appear to be large differences in the ease of nucleation of the various phases, and the present data are partly contradictory, but the epitaxial nature of the daughter minerals in the multiphase inclusions (Figs. 1 and 2) suggests that the original cooling from the temperature of trapping to somewhat below 1100°C must have been very slow. The frequent lack of nucleation of pyroxene in the glass remaining after crystallization of ilmenite and plagioclase and the presence of dendritic crystals in those that did nucleate suggest that further cooling of the magma containing these phenocrysts was rapid. It may be possible to select laboratory cooling rates that will duplicate these textures in previously homogenized inclusions, but this has not yet been achieved.

When the rocks were approximately 90 to 98 percent crystallized, as a mush of ilmenite, olivine, plagioclase, and pyroxene, the residual silicate melt between the crystals, still very mafic in composition, started to form minute immiscible globules of a second, high-silica, high-potassium melt. The main evidence for this *heterogeneous* mixture of two silicate melts lies in the following facts:

1) Residual masses of high index glass that is brownish, clear, and isotropic or in part devitrified occur in the angular interstices between late plagioclase crystals and contain various sizes of almost spherical blebs of colorless, isotropic, low index glass and occasionally a vapor bubble. The larger of such low index blebs in turn have minute spherical masses of high index glass toward their centers. Similar examples of the reverse situation, the angular interstices between plagioclase crystals being filled with low index glass containing spherical masses of high index glass, were found in several rocks, particularly 03-37; 50-31; and 72-36 (Fig. 7b). Similar phenomena are frequently seen effectively in two dimensions in the form of exceedingly thin platelike inclusions between subparallel plagioclase crystals. In some of these the high index glass has crystallized to a small amount of an opaque phase (ilmenite?), and a much larger amount of a fine-grained or single-crystal, birefringent, high-index phase, presumably pyroxene, that contains a row of circular masses of colorless, low index glass (Fig. 5a).

2) The compositions of the two liquids, as determined by electron microprobe (to be detailed elsewhere), do not differ greatly from certain immiscible liquid pairs found in the synthetic system leucite-fayalite-silica (*3*). The nature of the lunar samples made accurate analysis difficult, and the compositions vary significantly from rock to rock, but most cluster closely about the following weighted averages of 37 complete microprobe analyses (high- and low-silica glasses, respectively): SiO_2 76.1, 47.8; Al_2O_3 11.7, 3.2; FeO 2.5, 31.4; MgO 0.3, 2.3; CaO 1.9, 11.2; Na_2O 0.4, 0.1; K_2O 6.6, 0.3; TiO_2 0.5, 3.7. In composition these glasses are essentially potassic granite and pyroxenite.

3) A synthetic charge, closely approximating a mixture of equal portions of the two glasses, fused in a pure iron container *in vacuo* at 1450°C, formed a homogeneous melt (n of glass 1.569 ± 0.005). Portions of this glass held in pure iron *in vacuo* in the range 1135° to 1075°C formed two immiscible liquids, one yellow brown, with high n, the other colorless, with low n, with or without crystals of silica (cristobalite?) and pyroxene (Fig. 8).

4) The most common, and obvious, evidence for immiscibility is found in the low-index glass inclusions in the outer margins of many late plagioclase crystals. This feature is present in all the rocks, including the fragments in the breccias, but is relatively rare in rocks 44, 50, and 71. In section these inclusions appear as rows of closely spaced, rounded or elongate blebs, with an almost regular spacing of 5 to 15 μm between centers, and almost always near a contact with pyroxene (Figs. 5b and 6). A few are connected with an external residual film of glass lining a vug, but many are found to be completely within a single plagioclase crystal (Fig. 6, a and b). They may make a zone about 20 μm wide that starts abruptly (presumably from the onset of immiscibility) at a growth zone near the edge of the crystal and extends to just a few micrometers from the contact with pyroxene. Only rarely are these blebs as large as 15 μm in diameter. The larger ones tend to bifurcate, and almost always contain 1- to 2-μm specks with a high index of refraction (pyroxene or high index glass?) on their surfaces and centers (Fig. 6, a and b). The occasional chance section

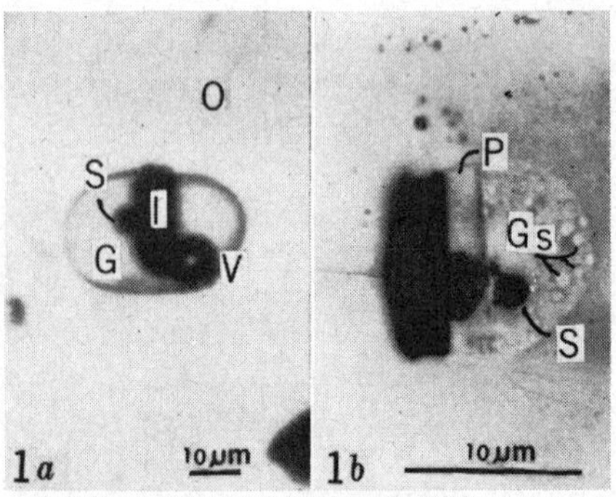

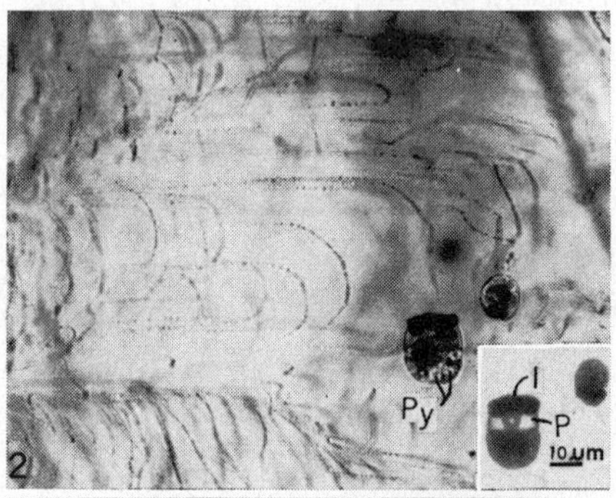

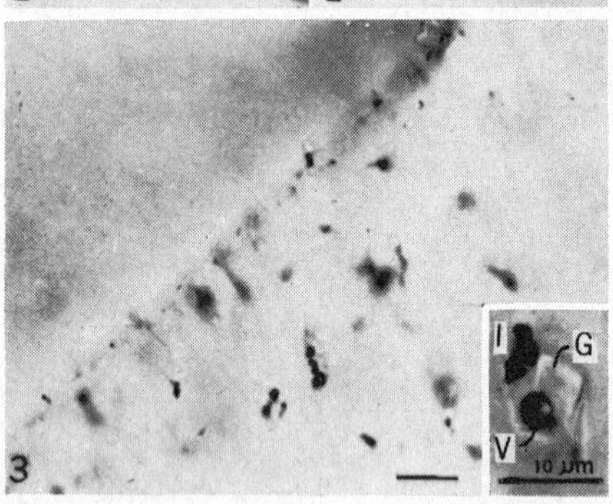

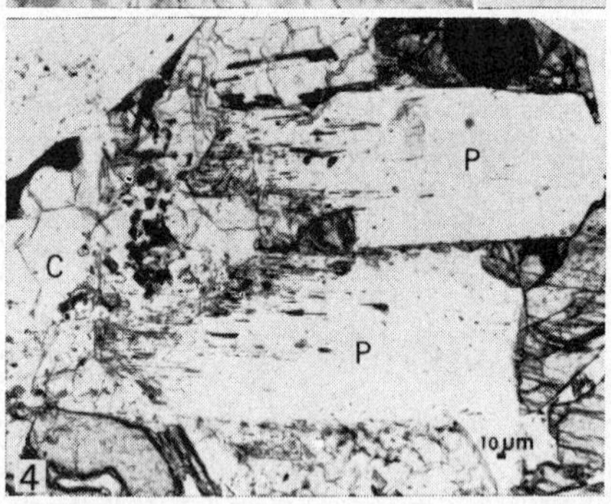

Fig. 1. (a) Unusually large inclusion of glass (G) in olivine (O) with daughter minerals ilmenite (I) and sulfide (?) (S), and shrinkage bubble (V). Slide 71-28. (b) Similar but also contains daughter crystal of plagioclase (P), and minute globules of much lower-index, high-silica glass (Gs). With rare exceptions, the plate-like ilmenite and plagioclase daughter crystals are always arrayed parallel to (100) of the enclosing olivine—a case of "internal epitaxy." The sulfide (?) grain (S) is spherical. Slide 45-32. Fig. 2. Pair of inclusions in olivine, similar to Fig. 1, before (inset) and after 2 days at 1110°C. The plagioclase has melted, and the almost opaque devitrified glass has recrystallized to a few coarser pyroxene crystals (Py) in clear glass. A series of dislocation loops (?), almost invisible in the original grain, have become decorated with minute, unidentified grains. Loose grain A-8 from olivine phenocryst, rock 20–41. Fig. 3. Pyroxene single crystal, in part inclusion-free (upper left), with a sharply delineated zone rich in glass inclusions (lower right). The upper left shows uniform extinction, but that of the lower right is blocky and irregular. Individual inclusions in such zones (inset) may have glass (G), opaque daughter crystal (I, ilmenite?), and shrinkage bubble (V). Slide 72-36. Fig. 4. Plagioclase crystals (P), rich in inclusions, near a mass of late, interstitial cristobalite (C), ilmenite, apatite, and high-silica glass (at left). Slide 47-7576.

parallel to the pyroxene-plagioclase contact shows them to be circular in cross section. Gelatin-slide mounts of crushed fragments show indices of refraction in the range 1.486 to 1.515 for these glasses. Many of the microprobe analyses were of such blebs in plagioclase.

5) A sharply delineated zone consisting of large numbers of tiny colorless silicate glass inclusions of very low index of refraction occurs near the margins of some pyroxene crystals (Fig. 3). In those rocks having strongly zoned pyroxenes, these inclusions start at the zone of most rapid increase in birefringence. Many of these inclusions contain an opaque daughter crystal, but most are so small that they have not even nucleated a bubble on cooling. Unlike the inclusions of mafic melt in olivine, not even the larger ones show any evidence of crystallization or devitrification to form silicates from this glass.

6) Large numbers of inclusions occur in both pyroxene and plagioclase within about 100 μm of most vugs and miarolytic cavities, and adjacent to or at the contact with *late* interstitial crystals of tridymite, cristobalite, and apatite (*9*). In pyroxene these inclusions are round or vermicular; in plagioclase they are generally thin tubes parallel to the length of the laths, and contain birefringent, high-index crystals (presumably pyroxene) as well as blebs of low-index glass (well displayed in rock 47; Fig. 4). In pyroxene these may be seen even though no other evidence of zoning is visible. Early crystals of tridymite and cristobalite, which are sharply euhedral against single-crystal pyroxene, plagioclase, and ilmenite (particularly in rocks 44 and 47), have no such inclusion zones in the minerals along their borders.

7) Inclusions of dark greenish or yellowish brown, high-index glass were found in late cristobalite. Some of these occur as globules in low-index glass (Fig. 7b). Others are large, spherical to elongate, curving blebs with indices of refraction much higher than the low-index glasses described above, and most of the larger ones contain birefringent, higher-index crystals, presumably pyroxene, that are either equant, solid, single crystals or long feathery dendrites. A few of the blebs contain minute blebs of colorless, very low-index glass within the high-index glass (*10*).

8) Almost perfectly circular masses of cristobalite in late, inclusion-rich plagioclase were found in or near some vugs (47-16). These presumably represent the crystallization, essentially in place, of round blebs of high-silica glass.

9) Three inclusions were found in ilmenite, from three different rocks, each consisting of dark brown, high-index glass and colorless, low-index glass.

10) Tiny blebs of very low-index glass (maximum 0.3 percent by volume) were found on the walls of three small undevitrified glass inclusions in olivine from rocks 20 and 45 (Fig. 1b). These are believed to represent immiscibility in metastable glass, in that pyroxene should have grown in these inclusions.

The immiscibility seems to occur too late in the crystallization to be helpful in explaining the odd bulk composition of these rocks, but, if the samples from the Sea of Tranquillity are at all representative of the maria, it is possible that this same splitting has occurred on a large scale in the early history of the moon, and that, as a result of the rising of such low-density liquid globules, the lunar highlands may be, in effect, partly "granitic" in composition. If so, the 30- to 100-fold increase in potassium should result in gross differences in heat flow and surface radioactivity. Any such surface mass would also greatly simplify the problem of a possible lunar origin for tektites, as the compositions of the siliceous glasses are obviously very much closer to tektites than are the compositions of the bulk rocks returned by Apollo 11.

EDWIN ROEDDER

U.S. Geological Survey, Washington, D.C. 20242

PAUL W. WEIBLEN

Department of Geology and Geophysics, University of Minnesota, Minneapolis 55455

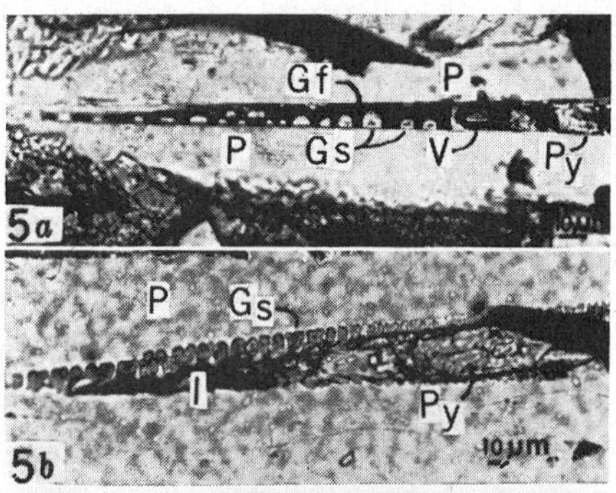

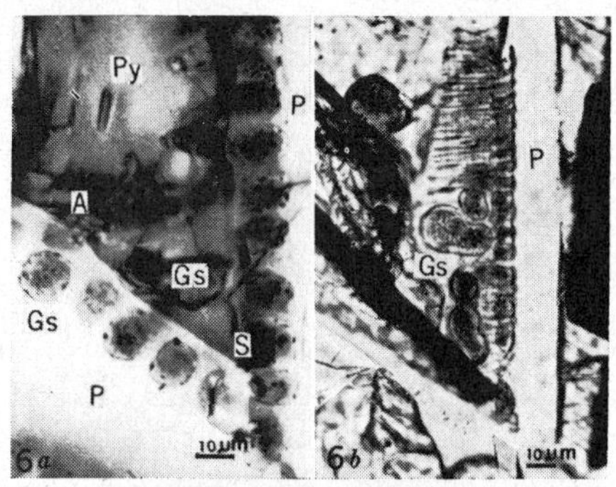

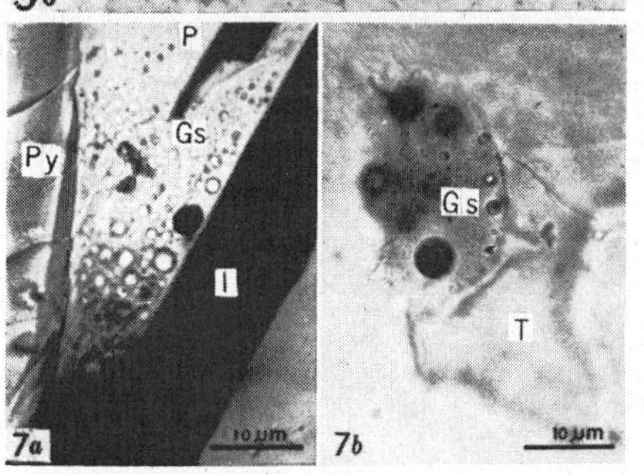

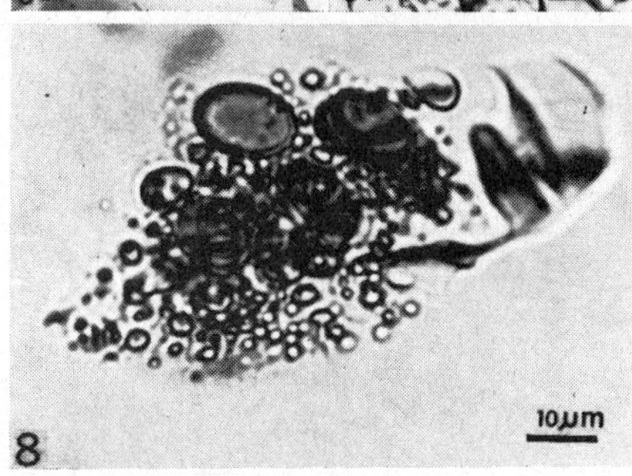

Fig. 5. (a) Wedge-like inclusion between two parts of single plagioclase crystal (P), now consisting of coarse pyroxene (Py), shrinkage bubble (V), isotropic (but devitrified?) low-silica, high-iron glass (Gf), and blebs of high-silica glass (Gs). Slide 71-32. (b) Similar wedge-like inclusion in plagioclase (P), but consisting of coarse pyroxene (Py), opaque mineral (I), and teeth-like row of high-silica glass blebs (Gs). Slide 72-46. Fig. 6. (a) Completely isolated blebs of high-silica glass (Gs) embedded in plagioclase single crystals (P). Each bleb has numerous much smaller blebs of high index (high-iron glass?), mainly on the interface with plagioclase. The pyroxene (Py) is strongly zoned, compositionally, and contains many blebs of high-silica glass (Gs), apatite (A), and sulfide (S) near its contact with plagioclase. Slide 72-36. (b) Somewhat similar, with mass of blebs of high-silica glass (Gs) all embedded in a single crystal of plagioclase, crystallographically continuous with the inclusion-free part (P). The center bleb is one of the largest such inclusions found. All show many minute, high-index blebs. Slide 45-32. Fig. 7. (a) Spheres of dark, high-index, high-iron glass in colorless, low index, high-silica glass (Gs). The glasses interfinger with plagioclase (P), between ilmenite (I) and pyroxene (Py); the largest sphere (dark) has devitrified, but the others are isotropic. Slide 72-36. (b) Similar, but with spheres of high-index glass (now devitrified and dark except for one small, bright, isotropic one), in a rounded mass of high-silica glass (Gs) in tridymite (T). Gelatin-slide mount of grain from rock 50-32, in 1.479 oil. Fig. 8. Synthetic melt, previously homogeneous, after 30 minutes at 1135°C. This grain of colorless glass, $n = 1.513$, is mounted in a matching oil, and contains many blebs of brownish, high-index glass plus a few crystals of silica (cristobalite?). The bar in the lower right of each photograph is 10 μm long.

References and Notes

1. J. Greig, *Amer. J. Sci.* (ser. 5) **13**, 1 (1927); *ibid.* **15**, 375 (1928). The term is used here to mean immiscibility between two silicate melts. There is ample evidence of immiscibility between silicate melts and: liquid iron; liquid sulfides; dense, supercritical CO_2 [E. Roedder, *Amer. Mineral.* **50**, 1746 (1965)]; and dense NaCl-H_2O fluids [E. Roedder and D. S. Coombs, *J. Petrol.* **8**, 417 (1967)].
2. E. Roedder, *Amer. Mineral.* **36**, 282 (1951); *Geol. Soc. Amer. Bull.* **64**, 1466, 1554 (abstracts) (1953).
3. A. R. Philpotts and J. A. Philpotts, *Geol. Soc. Amer. Abstracts 1969* (part 7), 176 (abstract) (1969).
4. Some of the inclusion types found raised still unanswered questions, owing to the small inclusion size, inclusion rarity, and hence lack of optimum orientation for observation, the necessary prohibition of destructive testing with these sections, and the obvious lack of time. As a result, many of the findings are too tentative to present here, but we believe the evidence for silicate immiscibility is conclusive.
5. During the examination of many of these rocks at the Lunar Receiving Laboratory, the individual thin sections or probe mounts of each rock were identified by the appropriate five-digit rock number followed by a dash and a six-digit section container number (for example, 10072-64-7668). When these sections were released to the principal investigators, NASA replaced these container numbers with new, two-digit section numbers such as 10072, 46). To aid other investigators in checking the specific sections examined, we give the container numbers where section numbers are not currently available, or both when there is ambiguity. We have omitted the "100" of the sample number and the "64" of the container number, and changed the comma to a hyphen to avoid ambiguity in serial listing. For some of these sections, we understand that a complete tabulation of container numbers and their equivalent section numbers cannot be made until the sections are returned to NASA.
6. A few very small secondary inclusions in olivine in samples 03-37 and 20-29 contained only gas, and this gas may be CO_2. Only destructive tests or the discovery of larger inclusions will permit more positive identification. Small amounts of noncondensable gases were found in some of the bubbles in glass fragments from samples 84 and 85, but most bubbles held only a vacuum.
7. All that is known about the identity of the phase at the liquidus is that it is the last portion to dissolve of an opaque phase remaining in the position of the original daughter ilmenite crystal. It could also be the ferropseudobrookite found at the liquidus in one synthetic lunar magma (R. Tuthill and M. Sato, personal communication).
8. Similar but very rapid vacuum heating experiments on minute fragments and spheres of glass from the lunar soil (sample 84) show that the dark glasses, with indices above 1.70, devitrify throughout within a few seconds in the range 1000 to 1100°C, but under these same conditions colorless glasses of index <1.60 may develop only a thin surface film of crystals. Such devitrification experiments may permit placing semiquantitative limits on the cooling history of some lunar materials, as similar devitrification textures are seen in some of the fragments of the breccias and soil, including well-developed chondrule-like features in some former glass spheres.
9. This interstitial occurrence of high-potassium glass and apatite may explain some of the biological growth effects reported for lunar soil. The silica crystals occur as close to olivine as a few hundred micrometers in several rocks, indicating nonequilibrium.
10. As a result of the need for multiple impregnations with epoxy (diluted with toluene) and the use of several epoxy resins with different indices of refraction in the preparation of these sections, there are many examples of apparent immiscibility (for example, filling of inclusion shrinkage bubbles with epoxy where they are cut by the section surface, or filling of the bubbles left in vugs by shrinkage of the previous impregnation), and of real immiscibility (residual globules of one epoxy component in another). These artifacts (and several other types) occur in cracks within grains, between grains, and, most confusingly, in the microvugs where evidence of actual magma immiscibility is most apt to be found.

11. Much of the preliminary work was done at the Manned Spacecraft Center, Houston. We are indebted to the personnel of the Lunar Receiving Laboratory for courtesies extended during this hectic period.

4 January 1970

Petrology of Unshocked Crystalline Rocks and Shock Effects in Lunar Rocks and Minerals

Abstract. *On the basis of rock modes, textures, and mineralogy, unshocked crystalline rocks are classified into a dominant ilmenite-rich suite (subdivided into intersertal, ophitic, and hornfels types) and a subordinate feldspar-rich suite (subdivided into poikilitic and granular types). Weakly to moderately shocked rocks show high strain-rate deformation and solid-state transformation of minerals to glasses; intensely shocked rocks are converted to rock glasses. Data on an unknown calcium-bearing iron metasilicate are presented.*

Evidence of impact metamorphism is widespread in the lunar sample. The fine material and breccias contain abundant small fragments of shocked crystalline rocks and mineral grains and particles of shock-melted rocks; only in the large rock fragments are the shock effects negligible. Moreover, some fragments have a history of multiple impact (for example, fragments of older breccia in younger breccia, and spherules of shock glass coated by different shock glass).

In order to study the effects of shock in the returned lunar samples, it is necessary to know in detail the rock textures and mineralogic variations characteristic of the unshocked crystalline rocks. On the basis of rock modes, mineralogy, and textures, we have classified these rocks into two suites: a dominant ilmenite-rich suite and a subordinate feldspar-rich suite. All the large samples and the bulk of the smaller rock fragments and mineral grains in the fines and breccias are from the ilmenite-rich suite; rocks of the feldspar-rich suite are found only as small fragments in the fines and breccias.

In the ilmenite-rich rocks, the major mineral constituents are clinopyroxene, calcic plagioclase, and ilmenite. Olivine is a common minor constituent, and cristobalite, troilite, and native iron are accessory and trace constituents of nearly all rocks. Interstitial glass is present in many rocks. The rocks are dominantly volcanic; both the presence of vesicles and vugs in all large samples and the rock textures suggest that they crystallized from magmas at or near the lunar surface (*1*).

Most of the ilmenite-rich rocks have one of three types of texture: (i) intersertal, (ii) ophitic-subophitic, and (iii) hornfels (*2*).

Intersertal rocks (Fig. 1A) have abundant interstitial glass that contains skeletal plagioclase grains, fine myrmekitic intergrowth of plagioclase and glass, and minute grains of pyroxene. Olivine forms sparse phenocrysts marginally reacted to clinopyroxene or small rounded cores in pyroxene grains. Ilmenite grains are subhedral. Clinopyroxene grains tend to be subhedral, but locally they form subophitic patches partly enclosing ilmenite. Patches of micrographic intergrowth of pyroxene and plagioclase fill some interstices. Grain size is typically fine and ranges from about 0.05 to 0.2 mm.

Ophitic rocks (Fig. 1B) are characterized by ophitic to subophitic intergrowth of clinopyroxene and plagioclase; plagioclase forms subhedral laths, and pyroxene is anhedral and partly encloses plagioclase. Olivine occurs as phenocrysts or as partly resorbed cores in pyroxene grains, and the pyroxene grains may be pigeonitic or contain cores of pigeonite. Interstitial myrmekitic intergrowth of plagioclase with glass is common. Grain size is highly variable and ranges from about 0.05 to 1 mm.

Hornfels (Fig. 1C) are characterized by reduced size of pyroxene and ilmenite grains and growth of coarse poikiloblastic grains of plagioclase. The ilmenite grains are equant rather than bladed; the pyroxene grains are equant and subhedral, or rounded, or form a mosaic of minute grains. The rocks may be highly inequigranular; ilmenite and pyroxene range from about 0.02 to 0.15 mm, and poikiloblastic plagioclase may be up to 3 mm across. Locally, primary rock texture is preserved in relict patches.

In nearly all the ilmenite-rich rocks, troilite (containing minute droplets of native iron) forms irregular patches or globules between grains. Minor cristobalite (locally intergrown with tridymite) fills interstices. In the unmetamorphosed rocks, pyroxene shows feathery, patchy, banded, hourglass, or polygonal extinction and color variation, and these features tend to disappear with increasing degree of thermal metamorphism. Apatite occurs as euhedral inclusions in, or myrmekitically intergrown with, pyroxene. Rutile and chrome-titanium spinel are trace and accessory constituents, occurring both as exsolution lamellae in ilmenite and as single grains in the rocks.

Tables 1 and 2 present mineral proportions and compositions of the major silicate minerals from typical ilmenite-rich rocks. These minerals are relatively homogeneous in the intersertal and fine-grained ophitic rocks and relatively heterogeneous in the hornfels and the coarse-grained ophitic rock. Compositions of interstitial material have also been investigated. Interstitial glass is potassium-rich: semiquantitative microprobe analyses of three spots in glass in the intersertal rock indicate minimum K_2O contents of 6.0, 6.6, and 5.4 percent. Plagioclase that forms skeletal grains in interstitial glass or is myrmekitically intergrown with glass is relatively enriched in sodium; semiquantitative microprobe analyses suggest compositions of about 70 to 73 percent anorthite.

The feldspar-rich rocks have characteristic modal compositions, mineralogy, grain sizes, and textures much different from those of the ilmenite-rich rocks. The essential minerals are calcic plagioclase, clinopyroxene, and olivine. Modes of two specimens are presented in Table 1; content of plagioclase is much higher and content of opaque and oxide minerals is much lower than in ilmenite-rich rocks. Grain size is typically very fine; the maximum we have observed in our samples is 0.15 mm.

Most of the undeformed feldspar-rich rocks are of two textural types: (i) poikilitic and (ii) granular. In poikilitic rocks (Fig. 1D), pyroxene is poikilitic and interstitial, plagioclase grains have euhedral shapes bordering pyroxene, and olivine forms either equant rounded grains enclosed by pyroxene or poikilitic interstitial grains between and enclosing plagioclase grains. Textures are similar to those of terrestrial cumulates (*3*), and at least one of us (E.D.J.) considers some of these rocks to be the product of crystal accumulation. In granular rocks (Fig. 1E), plagioclase forms a mosaic of lath-shaped grains that commonly define a rough foliation, and olivine and pyroxene occur as small equant rounded grains along boundaries between the plagioclase laths.

Table 1. Rock modes.

Minerals	Ilmenite-rich rocks: Intersertal	Fine-grained ophitic	Coarse-grained ophitic	Hornfels	Feldspar-rich rocks: Poikilitic	Granular
Olivine		7	0.5		6	19 (olivine and clinopyroxene)
Clinopyroxene	53	39	48.5	48	7	
Plagioclase	23.5	32	36	27	86.5	80.5
Cristobalite	2	4	0.5	1.5		
Myrmekite	3	6		2		
Glass	4.5			6.5		
Ilmenite and spinel	13	11	14	14	0.5 (ilmenite, spinel, sulfides and native iron)	0.5 (ilmenite, spinel, sulfides and native iron)
Sulfides and native iron	1	1	0.5	1		
Similar to:	10072	10020	10047	10017		

Table 2 presents compositions of the major silicate minerals of the poikilitic rock shown in Fig. 1D. In this specimen and in the three other feldspar-rich rocks that were analyzed, olivine and pyroxene are richer in magnesium and plagioclase is richer in calcium than in any of the ilmenite-rich rocks studied. The oxide minerals we have identified with the microprobe (*4*) are ilmenite, rutile, chrome spinel, and magnesian ferropseudobrookite, and the proportion of ilmenite relative to other oxides is much lower than in the ilmenite-rich rocks.

Two types of shocked material are common in the breccias and fine material of the returned lunar sample and may be classified (*5*) as follows: (i) Weakly to moderately shocked. Minerals show high strain-rate de-

Table 2. Electron microprobe analyses of silicate minerals (percent by weight).

Oxides	Ilmenite-rich: Intersertal (10084, 12-12)	Fine-grained ophitic (10084, 12-2)	Coarse-grained ophitic (10029)*		Hornfels (10084,12-27)*		Feldspar-rich poikilitic (10084, 12-61)
			Olivine				
SiO_2		35.6	31.6	29.8			37.4
FeO	None	29.4	44.4	55.4	None		25.2
MgO		33.9	19.6	11.1			37.2
CaO		0.33	0.34	0.44			0.06
MnO		0.28	0.54	0.53			0.18
Total		99.51	96.48	97.27			100.04
Fo		66.7	44.0	26.3			72.5
			Clinopyroxene				
SiO_2	45.2	45.1	49.1	46.2	47.5	49.9	51.4
TiO_2	4.96	4.43	1.55	1.66	1.53	2.22	1.58
Al_2O_3	5.24	3.81	1.49	0.83	1.57	2.34	1.64
FeO	10.24	15.2	15.3	32.3	19.6	11.71	8.40
MgO	12.93	10.53	14.6	6.51	10.54	16.4	17.8
CaO	20.5	19.0	16.3	11.38	17.5	17.6	19.0
MnO	0.17	0.30	0.30	0.59	0.28	0.18	0.09
Total	99.24	98.37	98.64	99.47	98.52	100.35	99.91
Ca	44.1	41.7	33.6	24.9	36.9	35.5	37.7
Mg	38.7	32.2	41.8	19.9	30.9	46.0	49.2
Fe	17.2	26.1	24.6	55.2	32.2	18.5	13.1
			Plagioclase				
SiO_2	50.9	44.4	44.2	47.7	47.9	51.3	43.8
Al_2O_3	28.5	34.1	34.7	31.6	31.8	29.5	34.8
Fe_2O_3	1.14	N.D.	N.D.	N.D.	0.59	1.37	0.17
CaO	14.4	18.2	18.6	15.6	16.4	14.2	18.8
Na_2O	2.51	0.89	0.81	2.62	1.79	2.61	0.71
K_2O	0.24	0.05	0.05	0.18	0.19	0.60	0.07
Total	97.69	97.64	98.36	97.70	98.67	99.58	98.35
An	74.9	91.6	92.4	75.9	83.5	75.0	93.6

* Different grains from the same rock. N.D. not determined.

Table 3. Selected electron microprobe analyses of mineral and rock glasses. Indices of refraction were determined by interference microscopy.

Component	Anorthite A-1	Feldspathic A-6	? A-10	Olivine-rich B-6	Titaniferous B-14	Titaniferous C-1	Titaniferous C-9
SiO_2	45.4	46.0	51.0	43.5	41.2	40.1	41.1
TiO_2	0.1	0.4	2.0	0.4	5.0	6.5	9.2
Al_2O_3	34.5	24.6	12.1	6.9	16.9	13.6	10.3
Total Fe as FeO	0.3	5.9	14.4	22.1	12.1	15.0	18.9
MgO	0.2	8.0	8.5	17.4	8.8	7.6	6.8
CaO	19.0	15.1	9.9	8.4	13.3	11.9	10.5
Na_2O	0.8	0.2	0.2	0.2	0.1	0.1	0.5
K_2O	0.1	0.1	0.3	0.1	0.1	0.1	0.2
MnO		0.1	0.2	0.3	0.2	0.2	0.3
Total	100.4	100.4	98.6	99.3	97.7	95.1	97.8
Index of refraction							
	1.5701	1.5942	1.6164	1.6534	1.6500	1.6705	1.7001

formation features and solid-state transformation to glasses [thetomorphic glasses (*5*)]. (ii) Intensely shocked. Rocks are melted by shock and form glasses that contain inclusions of shocked minerals and minute spherules of meteoritic nickel-iron. Rocks that show intermediate degrees of shock metamorphism are rare.

Most weakly shocked feldspar-rich rocks are intensely granulated and brecciated. The samples consist of angular fragments of feldspar-rich rock, plagioclase, olivine, and pyroxene in a finely granulated matrix of the same minerals; many of the larger fragments are strongly sheared and deformed. In most of the specimens there is little conclusive evidence that the brecciation is shock induced, but a few show features diagnostic of shock [planar features (*5*) in plagioclase, fine lamellar twinning parallel to (001) in clinopyroxene (*6*)]. Moderately and strongly shocked feldspar-rich rocks are rare.

Ilmenite-rich rocks show somewhat different shock effects. In weakly shocked rocks, texture is generally preserved and intense shear and brecciation are uncommon. Grains of all minerals are strongly fractured and cleaved (Fig. 1F). Pyroxene shows fine lamellar mechanical twinning (*7*), and ilmenite shows fine lamellae that may be mechanical twins. Olivine shows weak extinction variation and faint possible deformation lamellae. In some samples, pyroxene has poorly defined kink bands, and some plagioclase shows lamellar mechanical twinning. In slightly more shocked rocks, in plagioclase, planar features appear and become abundant, birefringence and indices of refraction decrease, and the plagioclase is converted in patches to thetomorphic glass (Fig. 1G). Rock texture, however, is perfectly preserved. In moderately shocked rocks (Fig. 1H), plagioclase is wholly converted to glass. The grains retain their gross lath shapes, but in some rocks they are slightly bent and show internal streaks of varying indices of refraction that suggest small-scale flow. Pyroxene is finely granulated, and the frequency of twinning appears less than in weakly shocked rocks. Ilmenite shows more abundant and more closely spaced fractures than in weakly shocked rocks.

The particles of glass found in the fines and breccias can be classified into mineral glasses or rock glasses depending on their chemical compositions (Table 3). The mineral glasses are clearly formed by shock, as thermal melting is fractional and should not produce glass of the composition of single minerals. The rock glasses are also shock-melted, for they contain nickel-iron spherules, shocked mineral fragments, and flow bands, and they are highly heterogeneous.

At present, the only mineral glass we have identified is of plagioclase composition. It occurs mostly in colorless angular chips or cleavage fragments and only rarely in spherules. Where recrystallized, the fragments are white in color, transparent to opaque.

Fragments of a single clear glass chip of thetomorphic plagioclase (92 percent anorthite) were annealed for 2 to 24 hours in air or in evacuated capsules at 850°, 875°, 900°, and 950°C and were air quenched to study variation of its optical and crystalline properties. Annealing at 850° for 2 hours reduced the index of refraction from 1.5688 to 1.5674 ± .0002, an indication that this anorthite glass is a dense glass with a high pressure history (*8*). The same glass recrystallized into a spherulitic to mosaic mat of anorthite crystals after heating at 900°C for 2 hours; at 875°C, recrystallization did not take place after 8 hours but occurred after 24 hours. Heating in air and in evacuated capsules produced no distinguishable differences. These crude experiments produced recrystallized plagioclase glass with textures similar to those in many fragments in the fines and breccias.

Rock glasses occur as angular fragments, spherules, oblate spheroids, dumbbells, and teardrops and also as spatter on surfaces of fragments and vesicular rinds wrapped around small fragments. Their colors are pale green, yellow, orange, brown, garnet red, and black.

The indices of refraction of the plagioclase mineral glass and rock glasses correlate well with their chemical compositions (Table 3). Indices vary directly with iron and titanium content (reflecting the amount of normative ilmenite

Fig. 1. Textures of unshocked and shocked crystalline rocks. Scale in each is 0.3 mm. (A) Photomicrograph of intersertal ilmenite-rich rock in plane light. Interstices between pyroxene and ilmenite grains are filled by glass (gray, granular appearing) that contains skeletal plagioclase grains (left center, top left). (B) Photomicrograph of fine-grained ophitic ilmenite-rich rock in plane light. Olivine (right center) forms phenocrysts, and subhedral plagioclase (clear) is enclosed by ophitic clinopyroxene. (C) Photomicrograph of hornfels ilmenite-rich rock in light passed through crossed Nicol prisms. Fine pyroxene and ilmenite grains are enclosed by coarse poikiloblastic plagioclase. (D) Photomicrograph of poikilitic feldspar-rich rock in plane light. Pyroxene (left, top center) and olivine (bottom right) are interstitial and poikilitic and enclose subhedral plagioclase grains (clear). (E) Photomicrograph of granular feldspar-rich rock taken in light passed through crossed Nicol prisms. Pyroxene and olivine form small grains along boundaries between plagioclase laths. (F) Photomicrograph of weakly shocked ilmenite-rich rock taken in light passed through crossed Nicol prisms. Plagioclase (right center) and pyroxene (left) are strongly fractured, and pyroxene (center) shows fine lamellar twinning parallel to (001) (bottom center) and (100) (top center). (G) Photomicrograph of plagioclase grain (center) partly converted to thetomorphic glass, taken in light passed through crossed Nicol prisms (rotated slightly from perpendicular). The birefringent parts of the grain show planar features (left center). (H) Photomicrograph of moderately shocked ilmenite-rich rock in plane light. Ilmenite (center) and pyroxene (bottom right) are intensely fractured, and laths of plagioclase glass (right center, left) are slightly bent.

present) and inversely with aluminum content (reflecting the amount of plagioclase present). Among the samples studied, there is a tendency of indices of rock glasses to cluster around $N = 1.59$ (rich in plagioclase) and $N = 1.66$ (rich in normative ilmenite). This grouping may indicate that these glasses correspond to the two suites of rocks described earlier.

There is also a general correlation between color and the index of refraction or composition of the glasses. Feldspar-rich glasses range from colorless to pale green to green. Glasses of intermediate composition are gray-green or yellow, and glasses high in iron and titanium are brown to garnet red.

Many rock glasses show recrystallization or devitrification textures. They range from textures typical of annealing to textures typical of igneous crystallization in a melt. Features indicating that these glasses were originally of impact origin are inclusions of shocked or recrystallized shocked minerals and meteoritic nickel-iron spherules.

An unknown transparent yellow mineral from sample 10047 has been identified. It is a calcium-bearing iron metasilicate isostructural with pyroxmangite (*9*). Electron microprobe analysis: 46.8 percent SiO_2, 0.5 percent TiO_2, 0.3 percent Al_2O_3, 44.6 percent FeO, 0.8 percent MgO, 6.0 percent CaO, trace Na_2O, and 0.8 percent MnO. Chemical formula: $(Fe_{.81}Ca_{.14}Mg_{.03}Mn_{.02})(Si_{.98}Ti_{.01}Al_{.01})O_3$. This mineral is biaxial positive with optical constants $\alpha = 1.753–1.756$, $\beta = 1.755–1.758$, $\gamma = 1.766–1.767$, birefringence 0.013–0.011, $2V = 35–40°$. It is triclinic, space group $P\ \bar{1}$, with $a = 6.623 \pm .001$ Å, $b = 7.543 \pm .002$ Å, $c = 17.354 \pm .006$ Å, $\alpha = 114.34 \pm .02°$, $\beta = 82.72 \pm .02°$, $\gamma = 94.51 \pm .02°$, $Z = 14$. Strong lines of yellow mineral are: *d* 2.934, *hkl* (014), intensity 100; 3.09 (021) 35; 4.68 (110) 35; 6.56 (100) 25; and 2.672 ($0\bar{2}6$) 25. Calculated specific gravity is 3.82 (x-ray), 3.80 (Gladstone-Dale). The measured specific gravity (flotation technique) is 3.76.

E. C. T. CHAO
ODETTE B. JAMES
JEAN A. MINKIN
JUDITH A. BOREMAN

U.S. Geological Survey, Washington, D.C. 20242

E. D. JACKSON
C. B. RALEIGH

U.S. Geological Survey, Menlo Park, California 94025

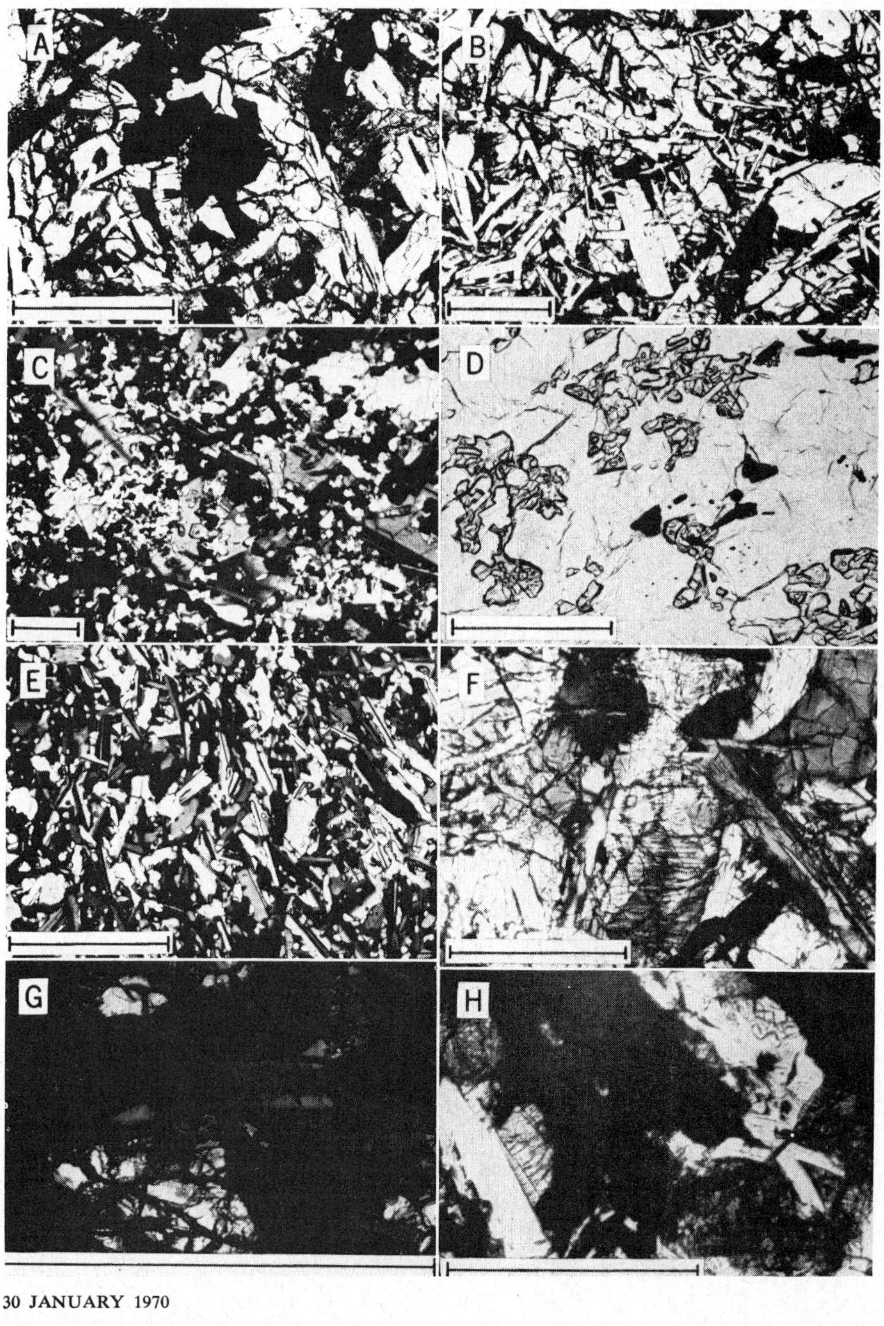

References and Notes

1. Lunar Sample Preliminary Examination Team, *Science* **165**, 1211 (1969).
2. Similar to textures of reheated basalts in Hawaiian lava lakes (R. Fiske and T. Wright, unpublished data).
3. E. D. Jackson, in *Ultramafic and Related Rocks*, P. J. Wyllie, Ed. (Wiley, New York, 1967), p. 20.
4. Electron microprobe analyses of grains of these oxides in the sample shown in Fig. 1D follow:

	Ilmenite	Rutile	Chrome spinel	Magnesian ferro-pseudo brookite
TiO_2	55.8	98.7	3.89	72.0
Al_2O_3	0.11	0.26	14.1	0.97
Cr_2O_3	0.33	0.27	51.5	1.26
FeO	31.4	0.00	20.8	11.3
MgO	10.6	0.10	9.09	11.1
MnO	0.24	0.00	0.20	0.01
Total	98.48	99.33	99.58	96.64

5. This qualitative classification of degree of shock is based on diagnostic criteria of shock or impact metamorphism established from studies of terrestrial shocked materials [E. C. T. Chao, in *Researches in Geochemistry*, P. H. Abelson, Ed. (Wiley, New York, 1967), vol. 2, p. 204; E. C. T. Chao, *Science* **156**, 192 (1967); E. C. T. Chao, in *Shock Metamorphism of Natural Materials*, B. M. French and N. M. Short, Eds. (Mono Book, Baltimore, 1968), p. 135.
6. O. B. James, *Science* **166**, 1615 (1969).
7. In pyroxene, mechanical twinning is dominantly parallel to (001) and subordinately parallel to (100) (determined by Universal stage and x-ray measurements). X-ray studies also indicate shock produces asterism and fading of higher-order reflections in pyroxene.
8. E. C. T. Chao and P. M. Bell, *Carnegie Inst. Wash. Yr. Book* **67**, 126 (1969); P. M. Bell and E. C. T. Chao, *ibid.*, in press.
9. F. Libau. *Acta Cryst.* **12**, 177 (1959).
10. The petrographic and electron microprobe data were obtained primarily from polished thin sections of materials from the allocated 1-mm to 1-cm chips (10084,12). We thank N. G. Benjamin for preparing these polished thin sections. We also thank George Desborough for electron microprobe analyses of the glasses. Work was supported by NASA contract T-75412.

4 January 1970

Lunar Soil: Size Distribution and Mineralogical Constituents

Abstract. *The lunar soil collected by Apollo 11 consists primarily of submillimeter material and is finer in grain size than soil previously recorded photographically by Surveyor experiments. The main constituents are fine-grained to glassy rocks of basaltic affinity and coherent breccia of undetermined origin. Dark glass, containing abundant nickel-iron spheres, coats many rocks, mineral, and breccia fragments. Several types of homogeneous glass occur as fragments and spheres. Colorless spheres, probably an exotic component, are abundant in the fraction finer than 20 microns.*

Material finer than 1 mm constitutes approximately 90 percent of the two core tubes returned by Apollo 11 and a similar proportion of that of the bulk sample box. Our sample consisted of 15 g of submillimeter material from the bulk sample box, necessarily studied without orientational context, on which we have made size analyses and morphological and mineralogical descriptions which bear on the petrology of the source rocks and the mode of development of the soil. Our work was accomplished on a split of half the original sample, and various analyses were made on smaller splits, ranging from 300 mg to about 2 g.

The size distribution was determined by wet and dry sieving (wet sieving below 67 μm) and by Coulter electronic counter for the fraction finer than 125 μm. In the sieve determinations, all fractions were weighed before and after sieving and the data corrected for small losses of material. The Coulter counter has the advantage of operating on a total sample. Particles suspended in electrolyte are drawn through an aperture and the resistivity change, as electrolyte is displaced in the aperture, is amplified and counted for individual particles. The sample was run using a 200 μm aperture, was allowed to settle to remove the coarser constituents, and was then run using a 50 μm aperture to obtain good resolution of the distribution down to 1 μm.

Distributions obtained by the two methods are consistent and give a median grain size of 62 μm and a modal grain size of 20 μm for the total sample. To obtain the distribution for the whole sample, we assumed that our sample had the same proportion of submillimeter material as the core tubes studied in the preliminary examination (*1*). The distribution is compared in Fig. 1 with those of a residual desert soil and a glacial till, in which mechanical weathering has been the dominant comminution process. The lunar sample is not as poorly sorted as the residual soil but resembles the glacial till in size distribution. However, the lunar soil is asymmetrically skewed to finer grain size, whereas the glacial till is skewed toward coarser grain size. The lunar soil is notably deficient in material finer than 15 μm, containing less than either of the terrestrial examples.

Figure 2 compares the size distribution for our sample with photographic determinations made for Apollo 11 in the preliminary examination (*1*) and from Surveyor photography of other uplands and maria sites. The distribution is consistent with the interpretation of Apollo 11 photographs and is similar in slope to other maria sites studied by Surveyor. The data consistently indicate that the Apollo 11 site has a greater abundance of fine material than other sites that have been photographed.

The mean density of the submillimeter fraction is estimated to be 3.01 g/cm^3 on the basis of an examination of its suspension in a density gradient (methylene iodide–dimethyl formamide) from 2.0 to 3.32 g/cm^3. The range of grain densities is from less than 2.2 to greater than 3.32 g/cm^3. The mean density is less than the density of the larger rock fragments, which range from 3.2 to 3.4 g/cm^3 (*1*), and is less than the mineral density of the core tubes, 3.1 g/cm^3 (*1*). The lower density is due to increased amounts of glass in the submillimeter fraction. A large amount of the breccia and glass fraction can be removed from the sample by a hand magnet. The nonmagnetic fraction is much denser than the magnetic fraction, about 70 percent of this material being denser than 3.32 g/cm^3.

There are four principal constituents of the soil:

1) Rock fragments, which can be placed in three groups on the basis of grain size: equigranular, 0.1 to 0.5 mm; equigranular, 0.05 to 0.1 mm; wholly crystalline or partly glassy, with grain size generally less than 0.05 mm. The coarser rock fragments consist primarily of clinopyroxene and feldspar, blocky ilmenite (rarely with rutile lamellas) and cristobalite, with minor troilite, which contains blebs of metallic iron with about 0.5 percent nickel. The very fine-grained to glassy material varies in texture. Fragments in polished section

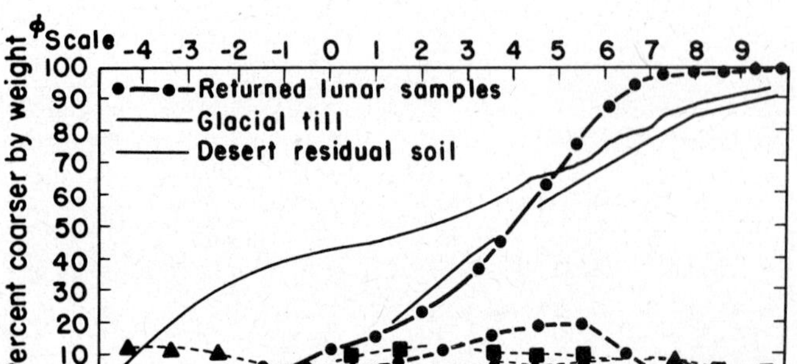

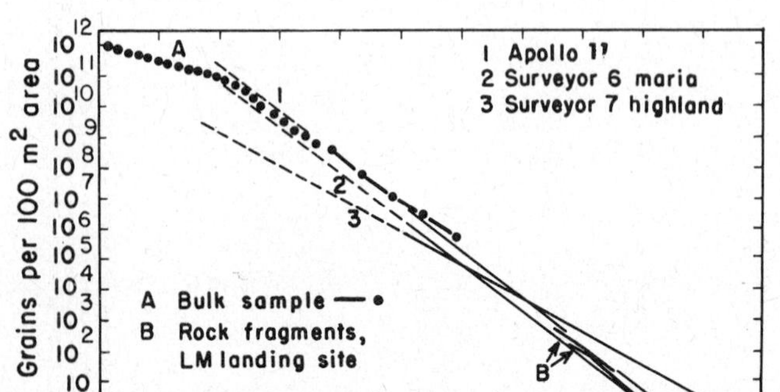

Fig. 1 (left). Cumulative size distributions and histograms for lunar soil, residual desert soil, and glacial till. Fig. 2 (right). Comparison of size distribution for lunar fines and distributions determined photographically for Apollo 11 and by two Surveyor experiments. Our data have been converted to area by assuming particles to be spherical and of constant density.

can be characterized by the form of the ilmenite, which occurs as blebs or as skeletal crystals of different growth stages. The silicates vary from wholly crystalline to glassy, commonly with pyroxene skeletal crystals interspersed with lamellas of glass rich in silicon, calcium, and aluminum, and poor in iron and titanium. Sulfides are not abundant, nor is metal, and these fragments are not separated into the magnetic fractions.

Broken crystal fragments make up approximately 10 to 15 percent of the submillimeter fraction. Crystal fragments as large as 1 mm are not common, suggesting that medium- and coarse-grained rocks are not generally abundant. Olivine is present in the crystal fragments but was not observed in the rock fragments. Many of the crystal fragments are conspicuously deformed; some of them are translucent and polycrystalline, giving stretched spot patterns and line patterns in x-ray diffraction studies of individual grains.

From the observed abundance of definable rocks and broken crystal fragments, the crystalline components of the lunar soil represent near-surface igneous rocks of basaltic affinity in the proportions: very fine to glassy varieties, 50 percent; fine-grained equant varieties, 30 percent; fine- to medium-grained varieties, 20 percent.

2) Breccia fragments are a principal constituent of the soil, constituting approximately 25 percent of the 500 to 1000 μm fraction and a larger proportion of finer size fractions. These fragments are typically gray, with irregular shapes and surface textures. Many of them are quite porous with many reentrant surfaces; others are compact, hard aggregates including crystal, rock, and irregular glass fragments in a matrix of glasses of variable composition similar to that shown in analyses 1 and 2, Table 1. The rock fragments, vesicular glass fragments, and glass spherules observed in the breccia indicate that it originated as a surficial debris layer. It does not contain as much highly magnetic glass as the total soil sample and apparently represents a rock type of complex history, rather than simply fused soil.

3) Dark, magnetic glass, varying in color from dark olive brown to black in large grains but generally colorless to brown in very thin flakes, is a major constituent of the soil (20 to 25 percent). It occurs in a variety of shapes, including irregular masses, commonly coated with fine particles, surficial coatings on breccia and other rock fragments, and as incipient and well-formed spherules. The glass is inhomogeneous, containing partially resorbed crystals of feldspar and ilmenite. It is generally similar in composition to the bulk soil sample (see analyses 3 and 4, Table 1). The glass contains abundant metal spheres, with nickel contents ranging from about 4 to 15 percent, which are submicroscopic to a few micrometers in diameter. The intensity of color appears to be related to the abundance of tiny inclusions. Flow structures are present, and in some instances strings of metal spheres are concentrated along flow bands. The glass is typically vesicular, but the vesicles are not always related to the surface. Where it covers breccia or crystalline material, cavities tend to be concentrated at the contact of glass and rock, suggesting that cavities are due to surface tension effects as liquid flowed imperfectly over the rock. Surfaces may be covered by cavities or by nodes, which give the surface a lumpy appearance. The nodular surface texture of the melted material is possibly due to preferential fusion of fine-grained particles, which stick and fuse to the glass surface.

Table 1. Analyses of glasses from the submillimeter fraction. Analyses performed by electron microprobe; data are uncorrected for absorbtion or fluorescence: 1, irregular glass common in the fraction finer than 10 μm; 2, amber glass sphere; 3, olive-brown glass sphere; 4, black opaque glass sphere; 5 and 6, dark amber glass spheres; 7, deep red-brown glass; 8, pale green glass sphere; 9, colorless glass sphere, 8 μm diameter; 10, pale green sphere; 11 and 12, pale green spheres.

Oxide	1	2	3	4	5	6	7	8	9	10	11	12
SiO_2	45.2	45.5	43.0	38.1	34.9	37.7	41.9	44.5	49.5	55.4	47.0	41.7
TiO_2	2.0	3.1	6.5	10.4	9.8	10.4	9.3	0.1	0.5	0.9	0.2	0.4
Al_2O_3	29.4	14.0	11.7	4.5	13.5	5.0	8.5	14.6	28.0	1.4	24.9	6.5
Cr_2O_3	<0.1	0.1	0.2	0.4	0.2	0.5	0.3	0	0	0	<0.1	0.4
FeO	6.4	12.0	14.0	23.8	15.8	23.6	17.4	15.0	2.5	10.4	4.0	14.7
MgO	3.1	8.0	7.2	11.9	8.3	12.3	6.9	9.6	2.0	16.7	7.5	23.6
CaO	13.0	12.0	10.6	7.8	14.8	8.2	11.0	11.8	14.0	20.1	13.7	6.8
Na_2O	<0.1	<0.1	0.4	0.9	0	0.1	0.4	0.5	<0.1	0	0	0
Total	99.1	94.6	93.7	97.6	97.1	97.9	97.1	96.1	96.5	105.2	97.3	94.1

The morphology and composition of the glass suggest that it represents very briefly melted material. The presence of nickel-rich metal spherules indicates that temperatures were locally and briefly above the melting temperature of the rock and suggests that the glass originated as the melted ejecta from relatively nearby meteorite impacts, predominantly on soil.

4) Transparent glass of a variety of colors, including amber, deep red brown, pale green, yellow and colorless, makes up a conspicuous but relatively small proportion of the soil. These glasses are homogeneous, in contrast to the brown and black spheres associated with the glass described in 3). They rarely contain incompletely melted ilmenite grains. Most occur as irregular fragments and as spheres, rods, and dumbbells. The proportion of spheres and related forms increases with decreasing grain size, reaching approximately 4 percent in the fraction finer than 10 μm. The proportion of spheres by color varies also with grain size: amber spheres are dominant in most size ranges but are most prevalent in the fractions between 37 and 250 μm; yellow and green spheres are generally about 10 percent of the sphere population at all grain sizes; and the colorless, or very pale spheres, are highly concentrated in the fraction finer than 20 μm.

Electron microprobe analyses of glasses are given in Table 1. Most of the glass spheres are not simply melted mineral grains but some type of composite. Many amber spherules are similar in composition to the bulk rock (*1*). The lighter-colored spheres tend to have less iron and titanium, greater magnesium/iron ratios, and higher calcium and aluminum contents. The amber color appears to be related to high titanium content (compare analyses 7 and 8). Because the compositions and size distribution of colorless spherules are distinctive, we consider these to be exotic constituents of the sample, and it is tempting to compare them with the sample analyzed at Tycho by Surveyor 7 (*2*), which has similar characteristics.

There are three obvious differences in the soil as a function of grain size: (i) The proportion of magnetic material, consisting largely of glass, increases markedly with decreasing grain size, from 50 percent at 500 μm to 75 percent at 37 μm. (ii) The proportion of identifiable crystalline rock fragments decreases and the proportion of crystal fragments, including intergrowths of minerals, increases with decreasing grain size; however, the finest-grained to glassy rocks are identifiable in the range below 37 μm. (iii) The composition of the soil is more homogeneous at finer grain size, containing more glass and glass of a more limited compositional range. Most of the glass in the fraction finer than 10 μm is of a low-titanium, low-iron, high-aluminum type similar to that shown in analysis 1, Table 1 (about 60 percent), with 20 percent of glass similar to the amber glass (analysis 2). Feldspar, pyroxene, and ilmenite in approximately equal proportions make up the remainder. In coarser size fractions, low-titanium glass is less abundant and more variable in composition, and the variety of glasses and crystals is greater (*3*). The low-titanium glass has some compositional affinities with a low-melting fraction of the bulk rock. The relatively high abundance of ilmenite in the fines probably is due to its resistance to melting during formation of impact glasses.

The following summarizes our interpretation of the lunar soil: (i) It has been formed primarily by comminution of fine-grained, near-surface basaltic rocks and breccia of uncertain origin. (ii) Melting is a conspicuous feature tending to cause consolidation of the soil into clumps, and it may be a partial explanation of the deficiency of the sample in very fine-grained material. (iii) Metallic spherules in the glass, colorless and pale spheres, and possibly some other minor constituents are exotic material, but their abundance is certainly less than 0.1 percent of the total sample.

MICHAEL B. DUKE, CHING CHANG WOO
MELVIN L. BIRD, GEORGE A. SELLERS
ROBERT B. FINKELMAN
U.S. Geological Survey,
Washington, D.C. 20242

References and Notes

1. *Apollo 11, Preliminary Science Report, NASA SP-214* (National Aeronautics and Space Administration, Washington, D.C., 1969), size frequency of core tubes, p. 128; photographic determinations, p. 47; density of rocks, p. 124-126; density of soil, p. 115; chemical composition, p. 132.
2. A. L. Turkevich, E. J. Franzgrote, J. H. Patterson, *Science* **162**, 117 (1968).
3. The pyroxene assemblage in the soil is variable, containing predominantly subcalcic augite and augite, with minor pigeonite. A few grains low in calcium are probably orthopyroxene (with higher Mg/Fe than most of the augite). Some of the yellow grains left undescribed by the preliminary examination team are hedenbergitic pyroxene and fayalitic olivine. One grain of an opaque mineral, occurring as an inclusion in olivine, is a chrome-titanium spinel, with composition FeO, 47.6; MgO, 4.6; CaO, 0.1; Al_2O_3, 5.3; TiO_2, 24.5; and Cr_2O_3, 18.7 (all percent by weight).
4. The contributions of the following people were essential: R. Fuhrman, C. Merrill, P. Rosenfield, and R. Commeau. Publication authorized by the Director, U.S. Geological Survey.

4 January 1970

Mineralogy and Petrology of Coarse Particulate Material from Lunar Surface at Tranquillity Base

Abstract. *Five grams of coarse fines (10085,11) contains 1227 grains, mostly mafic holocrystalline rock fragments, microbreccia, and glass spatter and agglomerates with less abundant anorthosite fragments and regularly shaped glass. The crystalline lithic fragments in the coarse fines and microbreccias represent a closely related suite of gabbroid igneous rocks that have a wider range of modal analyses and textures than seen in the larger crystalline rock samples returned by Apollo 11. Petrographic evidence of shock metamorphism is common, and the abundant glass is almost all shock-produced. None of the glass observed is similar to tektite glass.*

This work is based chiefly on a nondestructive study of sample 10085,11, a 4.92-g sample of coarse fines (1 cm to 1 mm), and work with two microbreccia petrographic thin sections 10046,65 and 10068,36. In addition, some of the coarser grains from a 15.28-g sample of fines (less than 1 mm) 10084,80 have been examined, and preliminary studies have been made of thin sections of the following samples: 10020, 10045, 10048, 10049, 10062, 10065, 10072.

Sample 10085,11 contained 1227 grains as follows: mafic holocrystalline rock fragments, 585; microbreccia, 395; glass spatter and agglomerates, 204; regularly shaped glass (spherules, dumbbells, and so forth), 6; anorthosite and other light-colored rock fragments, 37. Of the holocrystalline rock fragments 428 were unshocked or only mildly shocked and 157 grains were moderately to strongly shocked, as judged by microfracturing and vitrification of feldspars. Of the microbreccia fragments 135 have portions of their surface covered with glass.

Mafic holocrystalline rock fragments are dominantly gabbroid with the basic mineralogy of clinopyroxenes, ilmenite, calcic plagioclase, and accessory olivine. Relative proportions of these minerals and textures show considerable variation, and there is a wide range of shock effects. Glass is very rare, except for probable impact metamorphism products. Preliminary measurements on optical properties show a relatively wide range for the clinopyroxenes which is consistent with titaniferous augite, normal augite, ferroaugite, subcalcic augite, and pigeonite: for example, $2V_z$ of 22 grains (U-stage and spindled) ranges from 21° to 48°, with dispersion of $2V_z$ $r>v$ dominant with grains with axial angles greater than 40° and $r<v$ dominant in grains with axial angle between 28° and 38°. Values for $2V_z$ of 34° and 46° with opposite dispersion of $2V$ occur in a single rock fragment. Badly fractured grains with undulatory extinction illustrate apparent shock damage. Plagioclase optics of a few grains fit low-temperature plagioclase, An_{85-90} ($n_\beta = 1.578 \pm 0.001$, $2V_x = 78 \pm 2$), but zoning has not been checked in detail and a wider range is indicated. Olivine optics place it in the range Fo_{75-80} ($n_\beta = 1.696$ to 1.713, each $\pm\ 0.003$; $2V_x = 87° \pm 3°$).

A considerable range of crystallization textures is displayed in the various crystalline rock types in the breccias, but most are familiar basaltic and gabbroid varieties. Grain sizes range from 0.01 to 1 mm. The more common relations are ophitic and subophitic pyroxene molded onto euhedral to subhedral laths of plagioclase, both intergrown with platy to irregular ilmenite. Pyroxene is also commonly intergranular euhedral or anhedral. In a few rocks subhedral plagioclase mantles rod-shaped cores of pink pyroxene (Fig. 1, a and b). This texture with similar mineralogy is known in terrestrial rocks such as the Patuki Volcanics (Permian) from South Island, New Zealand (Fig. 1c) (*1*). Except for two fragments, no porphyritic textures were seen. One exception is a microporphyritic felsite with distinctly fluidal (trachytic) texture consisting of subaligned seriate phenocrystic laths in a similarly aligned matrix of microlites and possibly minor glass (Fig. 2). A second porphyritic relation occurs in an ilmenite-rich basaltoid rock in which a 0.1-mm olivine grain is set in an intergranular matrix of 0.01-mm grains of pyroxene, ilmenite, and sparse plagioclase. Rock sample 10045 shows similar olivine phenocrysts in a diabasic matrix. The most striking exceptions to common terrestrial textures are: (i) an even-grained

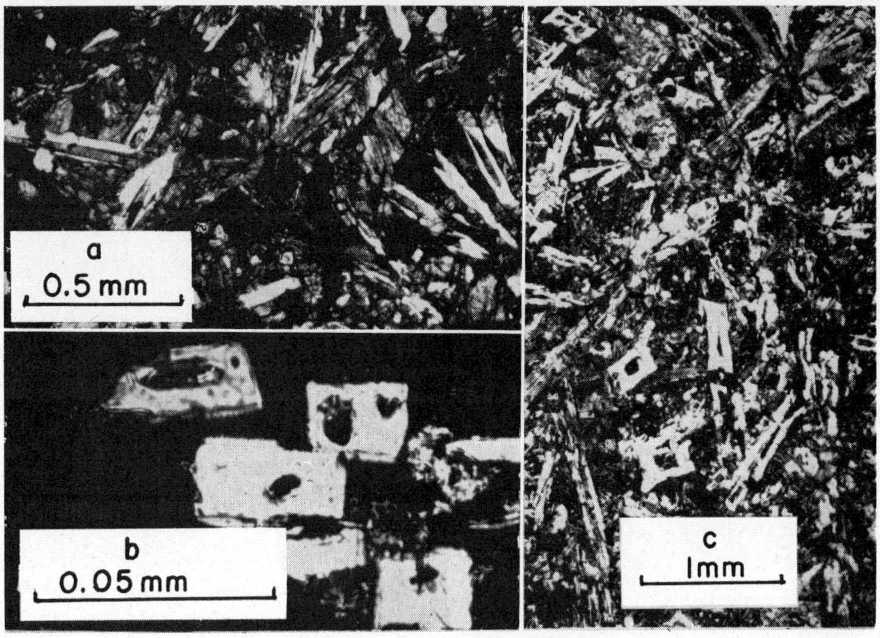

Fig. 1. Calcic plagioclase mantling pink clinopyroxene in mafic holocrystalline lithic fragments in Tranquillity Base sample 10046. Plane polarized (a) and crossed polarizers (b). (c) Identical texture in terrestrial rock with similar major mineralogy from the Patuki Volcanics (Permian), South Island, New Zealand (plane polarized light).

(0.05 to 0.2 mm) idiomorphic mixture of stubby pyroxene (colorless cores with narrow pink margins) and platy ilmenite set in a very fine matrix of fibrously intergrown pyroxene and plagioclase(?) dusted with opaque granules, and (ii) "ophitic" plagioclase enclosing subequant grains of pyroxene and ilmenite.

Monomineralic polycrystalline plagioclase rock fragments occur in sample 10085,11 and also in several thin sections of microbreccias. Other fragments contain sparse olivine or pyroxene. Most appear to be, or have been, relatively coarse-grained and to be derived from an anorthositic assemblage. Many, if not most, show severe granulation of plagioclase with relics of coarser grains grading into an extremely fine matrix. Plagioclase colors of the larger grains range from medium to light gray, gray-green, and white. This rock type is not represented in the larger specimens returned by Apollo 11. The microporphyritic fragment mentioned earlier is too fine-grained for certain determination of the feldspar composition by optical methods, but a few of the phenocrysts appear to be optically positive, suggesting calcic plagioclase and a rock name of "andesite" or "volcanic anorthosite." This rock may be from surface volcanic flows in the Sea of Tranquillity; however, the possibility that the anorthosite originated from the lunar highlands cannot be overlooked. The other light-colored rock fragments appear to be very fine-grained and shocked particles similar in mineralogy to the mafic holocrystalline rock fragments, but with very low ilmenite contents.

The microbreccia fragments have a wide range of induration from easily friable to very hard. These are mostly typical microbreccia fragments with a fine-grained groundmass composed mostly of thetomorphic and fused mineral and rock glass. Many fragments contain regularly shaped glass particles and generally light-colored lithic fragments. Thin sections of most microbreccias, such as 10068,36 contain fragments of preexisting microbreccias. Three of the microbreccia fragments in 10085,11 contain reddish-brown glass that appears to have been produced *in situ*. Although positive petrographic evidence has not been found, we believe that the microbreccias are shock-welded pieces of the lunar regolith. One large fragment of mafic holocrystalline rock has glass that is apparently derived from it. This glass is reddish-brown to brown and contains numerous microlites, and has a mean *n* of 1.552, ranging from 1.550 to 1.554. The microbreccias contain abundant evidence of shock metamorphism and multiple shock.

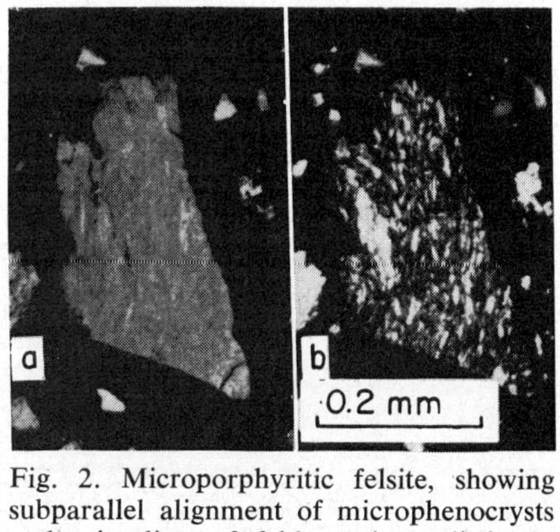

Fig. 2. Microporphyritic felsite, showing subparallel alignment of microphenocrysts and microlites of feldspar in a distinctly trachytic texture. (a) Plane polarized light; (b) under crossed polarizers.

Agglomerates and glass spatter consist chiefly of vesicular, irregularly shaped brown to gray and black glass masses commonly including and cementing together numerous mineral grains and small polymineralic rock fragments from the regolith. The refractive indices of these glasses have a narrow range compared with the glass spherules, and they are mostly translucent, containing abundant, dust-size mineral grains and microlites. These glasses are generally very similar in their optical properties, color, and morphology and may have originated from the same nearby impact event, or from similar events in similar parent materials.

Regularly shaped glass in the form of spherules, dumbbells, teardrops, and modifications of these forms are relatively rare in the coarse fines, but are abundant in the finer sizes and breccias (Fig. 3, a–d). These range in color from almost colorless to gray-green, green, yellow, orange-brown, to deep wine red. The refractive indices generally correlate well with the increasing depth of color if there are not abundant microlites or evidence of devitrification. Refractive indices are: light-green homogeneous glass spherule, $n = 1.580 \pm 0.002$; grass-green grain with surface flow grooves, $n = 1.601 \pm 0.002$; brown heterogeneous glass spherule, $n = 1.608$ to 1.620 ± 0.002; dark red to red-brown glass fragment, average $n = 1.716 \pm 0.002$. Surfaces of spherules range from highly reflective and regularly spherical to completely dust-covered and irregular. The dust-covered surfaces of glass spherules have mineral and glass particles up to 10 μm, but mostly less than 2 μm, stuck in their surfaces. Some flattened spherules have dust adhering only to one of the flattened sides as if the spherule had struck the regolith before it was completely solid, and was able to stick dust to the side that impacted the surface, the impact slightly flattening the spherule. The surfaces of some spherules, as well as of some mineral grains, have excellently preserved hypervelocity im-

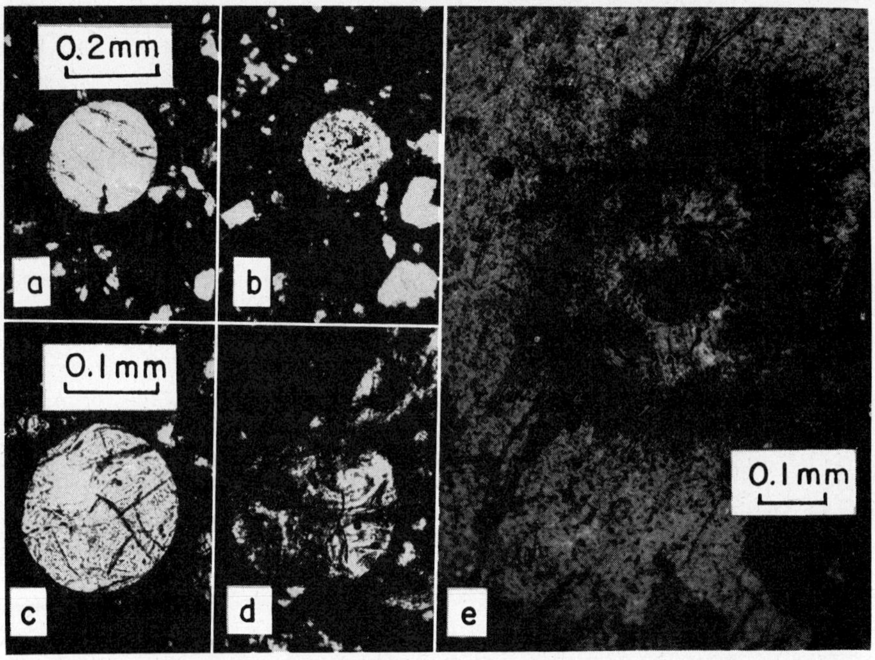

Fig. 3. (a) Homogeneous glass spherule, yellow to slightly greenish-yellow, fractured, holohyaline with no mineral inclusions or devitrification (10065). (b) Slightly irregular orange glass spherule with abundant mineral inclusions (10065), scale same as in (a). (c) Yellowish-brown glass spherule with flow structure, fractured with incipient devitrification and few small mineral inclusions (10048). (d) Mixed glass spherule of yellowish-green and orangy-brown glass with contorted flow structure and moderate devitrification (10065), scale as in (c). (e) Hypervelocity impact crater on surface of dark-red glass fragment (10085,11), incident light. (a–d) Plane polarized light.

pact craters (Fig. 3e). These range in size from a few tenths of a millimeter down to the limit of resolution of the binocular microscope and probably represent the impacts of very small meteoroids and high-velocity secondary particles. Some glass fragments are parts of spherules that appear to have been split apart by one of these impacts. The morphology of these small impact craters is remarkably uniform with an approximately hemispherical central crater lined with glass, a slightly raised lip, grading outward into a highly fractured and crushed ring syncline surrounded by a wide zone of thin, platelike spalls and an outer zone of redeposited ejecta commonly having a ray-like structure. One glass spherule has been observed that has a light, silvery-colored metallic coating on a portion of the surface, possibly nickel-iron. None of the glasses yet examined are similar to tektites in refractive index or general petrography. Almost all of the glass spherules contain spherical vesicles, and these have a wide range in size and abundance. In general, the heterogeneous and mixed glass spherules with obvious flow structure have more vesicles than the homogeneous spherules with little or no flow structure.

In the initial processing of sample 10085,11, the material was washed with acetone to remove much fine dust adhering to the larger grains. The fine material has a mean grain size of less than 5 μm, but a diffractogram demonstrated that clinopyroxene, calcic plagioclase, and ilmenite are the major constituents in this size range, similar to the larger particles. No fragments of meteoritic material, stony or iron, have been recognized in the coarse fines from sample 10085,11.

Coarse fines from Tranquillity Base contain fragments of a closely related suite of igneous rocks that have a wider range of modal analyses and textures than that seen in Apollo 11 samples larger than 1 cm (*2*). The textures of many of the mafic holocrystalline rocks are similar to those of terrestrial volcanic and shallow intrusive rocks. Fragments of anorthosite occur in the coarse fines and microbreccias, but this rock type is not represented in the larger rocks returned by Apollo 11. Anorthosite may originate from pods or crystal cumulates in the surface material at or near Tranquillity Base, but may have come from the lunar highlands. Petrographic evidence of shock metamorphism and meteoritic impact is much more common in the holocrystalline rock fragments in the coarse fines than in the larger holocrystalline rock samples. Almost all of the glass in the samples examined has apparently been produced by the impact of meteoroids or high-velocity secondary particles. None of the glass yet examined is similar to tektite glass. This observation tends to reinforce strongly the previous conclusions of many workers that tektites do not originate from the moon (*3*).

ELBERT A. KING, JR.
MAX F. CARMAN, JOHN C. BUTLER
Department of Geology,
University of Houston, Houston, Texas

References and Notes

1. M. F. Carman, *Trans. Roy. Soc. N.Z.* 1968, 6, 8, 91 (1968).
2. Lunar Sample Preliminary Examination Team, *Science* 165, 1211 (1969).
3. E. A. King, Jr., *Guidebook, Meteoritical Society, 32nd Annual Meeting* (1969).
4. We thank the staff of the Lunar Receiving Laboratory and the Lunar Sample Preliminary Examination Team for documenting the relationship of our sample to the total sample returned by Apollo 11 and for providing background information. Sponsored by the National Aeronautics and Space Administration under contract NAS 9-10373.

4 January 1970

High Crystallization Temperatures Indicated for Igneous Rocks from Tranquillity Base

Abstract. *Complex intergrowths of troilite (FeS) and iron in the igneous rocks from Tranquillity Base contain 8.4 percent native iron by volume. The intergrowths were derived from an initially homogeneous sulfide liquid that separated immiscibly from the magma at 1140°C or above. Textures show that the sulfide liquid formed late in the crystallization and cooling history of the igneous rocks and after the major ilmenite and pyroxene had formed.*

Two textural varieties of igneous rocks were among the rocks returned from Tranquillity Base. They have been arbitrarily called type A, "a fine-grained vesicular" rock, and type B, "a medium-grained vuggy" rock (*1*). Two samples of type A rocks (10062 and 10072) and one of type B (10050) have been examined in the present study. The mineralogical compositions are similar in all three; 10050 and 10072 contain predominantly pyroxene, calcic plagioclase, and ilmenite with minor amounts of cristobalite; 10062 has the same major minerals, but instead of cristobalite contains a minor amount of olivine. Each of the rocks contains in addition small amounts of both pyrrhotite (variety troilite, FeS) and native iron. It is with the petrological significance of these last two phases that this paper is concerned.

X-ray powder diffraction of a fragment of pyrrhotite from rock 10050 yielded a pattern identical with that of troilite from Del Norte County, California (*2*). Electron-microprobe analyses of pyrrhotites in both 10050 and 10072, using carefully prepared synthetic FeS as a standard, gave an identical composition of FeS. The pyrrhotite has the ideally stoichiometric composition FeS and is therefore the variety troilite. A search for other minor elements such as Cu, Ni, Co, and Mn did not reveal any above the detection limits of approximately 0.1 percent.

Metallic iron was identified as a phase by optical examination on polished surfaces, by x-ray powder diffraction, and by microprobe analysis. The iron in rocks 10050 and 10072 contains detectable amounts of both Co and Ni in solid solution, with Co more abundant than Ni. Lacking reliable analysis standards, it is possible only to specify that the Co content is approximately 1 percent.

The polished surfaces available for textural examination differed considerably. A 1-in.^{2} (6.45-cm^2) polished section of 10072 was available, whereas for 10062 and 10050 only 0.25- and 0.15-in.^{2} surfaces, respectively, were examined.

Troilite is present in the vesicles of rock 10072, where it occurs as rare euhedral to subhedral crystals. The vesicular troilites overgrow minerals of the rock matrix forming the vesicle walls and clearly formed later than the body of the rock. They presumably involve vapor-phase movement into the vesicles.

Within the matrix of the rocks, the troilite is never euhedral and does not occur as individual grains but is invariably intergrown with iron. Where apparently separate grains of either iron or troilite were observed, further grinding and polishing showed them to be the partly exposed projections of larger two-phase intergrowths. Iron is always the minor phase of the two, and is present in rounded to subrounded grains scattered through the troilite. Some grains have 4- or 6-sided shapes suggestive of sections through single crystals (see Fig. 1A). Although it is not clear whether or not all the iron grains in a given intergrowth have a single orientation, it was observed that the troilites were monocrystalline aggregates that extinguished as a single unit under crossed nicols.

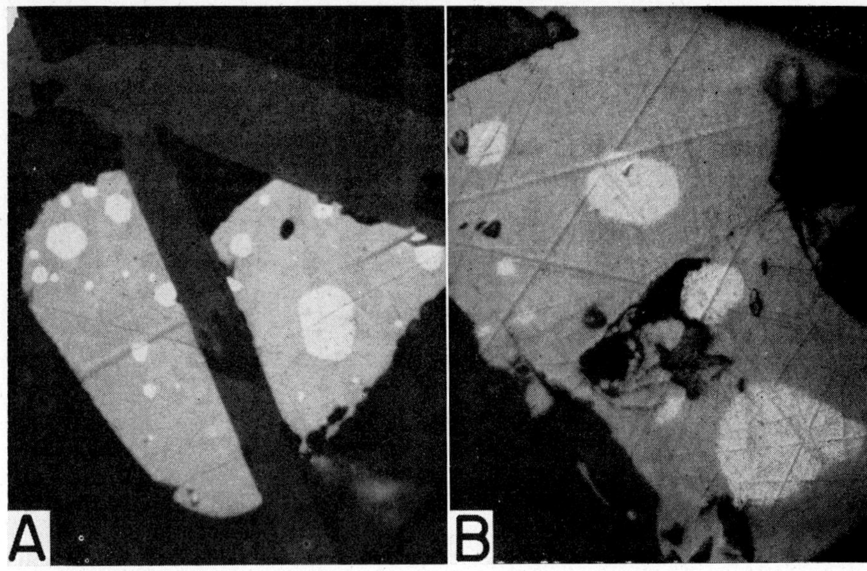

Fig. 1. (A) Rock 10072. Intergrowth of troilite (light gray) and iron (white) enclosing a branching arm of a skeletal ilmenite crystal (dark gray). Notice the rough, six-sided shapes of some iron grains. The troilite is a mono crystal. Width of field of view, 0.01 mm. (B) Rock 10050. Intergrowth of troilite and iron. Lines are scratches remaining from polishing. Width of field of view, 0.02 mm.

The troilite-and-iron intergrowths were apparently one of the last features to form in the rocks. They often enclose the branching limbs of the skeletal crystals of ilmenite (Fig. 1A) or are intergrown with bladed pyroxene crystals (in 10072) in such a way as to suggest formation after the pyroxene had grown. Though such textural criteria are rarely unambiguous, the troilite-and-iron intergrowths give every indication of having crystallized late in the cooling and crystallization of the silicate magma.

Visual examination of the polished surfaces suggests a constant composition for the intergrowths. Because the intergrowths rarely exceed 200 μm in diameter, it was not possible to analyze the volumetric content of iron directly. The polished surfaces of individual grains were therefore photographed and the exposed areas of iron and troilite were measured planimetrically. In the absence of specific evidence of the third dimension, it was assumed that the surface measurements yielded a reasonable estimate of the volume percentages. For rock 10072, 24 intergrowths were measured, and for 10062, 11. The observed range in 10072 is 3.3 to 14.6 percent iron by volume; in 10062 from 4.4 to 13.3 percent iron by volume. The average of 24 measurements in 10072 is 8.4 percent iron by volume with a standard deviation of 2.3 percent, and the average of 11 measurements in 10062 is 8.3 percent with a standard deviation of 3.3 percent. If iron contents are grouped in 2 percent brackets and their frequency is plotted (Fig. 2), a narrow distribution pattern emerges for 10072. The distribution of values from 10062 does not show a clear pattern, but one might hardly be expected on the basis of 11 measurements.

Too few intergrowths were measured in rock 10050 to yield either a meaningful average composition or a meaningful distribution. It is clear from a visual examination, however, that the iron content differs little from those in 10062 and 10072 (see Fig. 1B).

The ubiquitous intergrowth of troilite and iron in a constant ratio precludes the possibility of their having co-crystallized from the magma, or of the iron's having been formed by post-magmatic desulfuration of pyrrhotite grains. Loss of sulfur would lead to a concentration of iron grains along the rims of the troilite grains and would probably give inhomogeneous intergrowth compositions. Had the troilite and iron grown

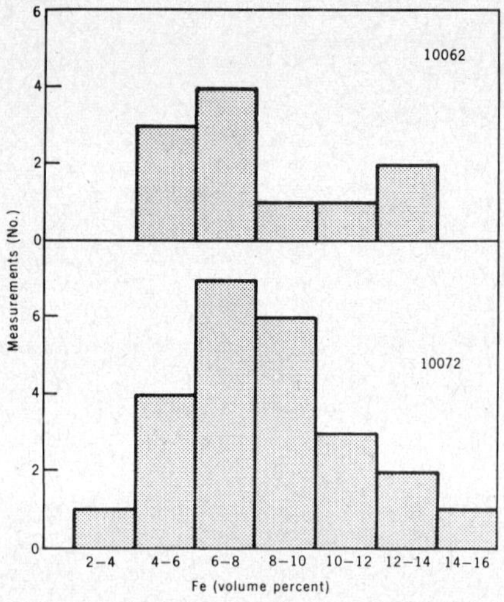

Fig. 2. Frequency distribution of measured iron contents for troilite-and-iron intergrowths in rocks 10062 (top) and 10072 (bottom).

as separate crystalline phases from the magma, we should expect to sometimes find them as independent phases—or if as intergrowths because of unknown nucleation effects, to observe a wider spread in intergrowth compositions.

The constant intergrowth composition clearly suggests the breakdown of an initial homogeneous phase. Despite extensive work in the system Fe-S (*3*), no solid phases have ever been found between FeS and Fe, nor does troilite or iron reach the intergrowth composition by solid solution. The homogeneous parent phase must therefore have been an iron sulfide liquid which separated immiscibly from the silicate magma.

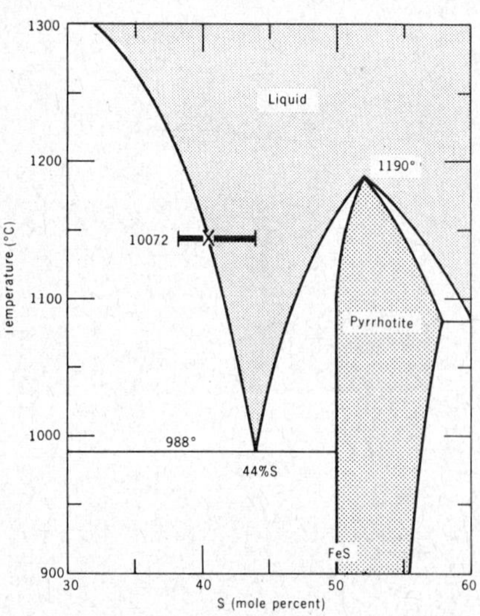

Fig. 3. Portion of the phase diagram for Fe-S after Hansen and Anderko (*3*). The composition of the troilite-and-iron intergrowth in 10072 (X) falls at 1140°C on the liquidus. The bar graph shows the standard deviation of measurement of the intergrowth.

If volumes are converted to weights on the basis of a density of 7.87 g/cm^3 for iron and 4.79 g/cm^3 for troilite, the minimum temperature at which the intergrowth composition could be a homogeneous liquid is 1140°C (Fig. 3). The first solid to separate below 1140°C would be iron, in agreement with the observation that some of the iron grains have crystal outlines.

The density of a sulfide liquid with the composition of the intergrowth would be approximately 5 g/cm^3. The density of a silicate magma with the bulk composition of rocks 10062 or 10072 would not have exceeded 3.5 g/cm^3. The immiscible sulfide liquid should therefore have separated gravitatively. That it did not do so, but instead remained evenly dispersed through the rock, means either that the magma was so rapidly quenched there was insufficient time for settling or that the sulfide liquid did not separate until the magma was already sufficiently crystallized to prevent settling. Rapid quenching is probable for the fine-grained vesicular rocks, 10062 and 10072, but not for the coarser-grained type B rock, 10050. The second circumstance must therefore be the applicable one. This is in accord with the previously mentioned textural evidence, that the troilite-and-iron intergrowths formed late in the crystallization history, after ilmenite and pyroxene had started crystallizing. Most of the crystallization relations of magmas with the composition of Tranquillity Base rocks must therefore be above 1140°C.

BRIAN J. SKINNER

Department of Geology and Geophysics, Yale University, New Haven, Connecticut 06520

References and Notes

1. Lunar Sample Preliminary Examination Team, *Science* 165, 1211 (1969).
2. L. G. Berry, and R. M. Thompson, "X-Ray Powder Data for Ore Minerals: The Peacock Atlas," *Geol. Soc. Am. Mem. 85* (1962).
3. M. Hansen, and K. Anderko, *Constitution of Binary Alloys* (McGraw-Hill, New York, 1958).
4. Supported by NASA contract NAS 9-8075.

4 January 1970

Morphology and Related Chemistry of Small Lunar Particles from Tranquillity Base

Abstract. *Glass spherules show multiple high-velocity impact craters and are coated with small particles including glass, plagioclase, clinopyroxene, ilmenite, olivine, chromite, rock fragments, and frozen droplets of iron, nickel-iron, and troilite. These spherules passed through an impact cloud of hot fragmental material, condensing iron-rich vapor and high-velocity projectiles. Breccia contains concentric, accretionary lapilli units and appears to be a sintered deposit from a hot lunar base surge generated by impact.*

Lunar fines and breccias were examined by optical microscope, scanning electron microscope with nondispersive x-ray detector, and electron microprobe. Prior to this detailed examination all chip-sized (4 to 10 mm) material from the bulk fines was examined in order to determine rock types and their relative proportions prior to distribution to the principal investigators. Of 122.3 g of chips, 35.2 percent are breccia, 36.4 percent are fine-grained crystalline rock, 15.4 percent are medium coarse-grained crystalline rock, 0.8 percent are anorthosite, and 12.2 percent are glass and glass-coated rock. The amount of actual glass is estimated to be about 6 percent. Crystalline rock fragments therefore constitute about 60 percent of the nonglass fragments, and breccia constitutes the remaining 40 percent. Because of the large number of chips and the lack of apparent sampling bias, we consider this to be a relatively unbiased estimate of the overall proportion of these two rock types at the Tranquillity Base site.

Our examination of the fines and breccia confirmed the general description provided by the Preliminary Examination Team (*1*). The majority of the fines result from mechanical disaggregation and from the production of glass by meteorite impact. The composition of the fines is generally similar to that of the breccia and will be discussed in detail in a later paper.

Small glass spherules, which occur in the fines and breccia, provide a record of some of the processes associated with lunar surface impact. Many of the spherules are dark, nearly opaque glass and are covered with smaller particles. A typical spherule of this type is shown in Fig. 1A (insert). This spherule is a titanium- and iron-rich glass and is covered with several different types of material. The microstratigraphy of the spherule indicates a history of melting, capture, and accumulation. The innermost material on the glassy skin of the spherule is a partial coating of micrometer- and submicrometer-sized frozen droplets of iron, nickel-iron, and troilite. The nickel-iron droplets (Fig. 1, A and C) contain 5 to 6 percent nickel. Most droplets contain detectable amounts of nickel but a few contain less than the limit of detection for our nondispersive x-ray system, which is about 2 percent for nickel in iron. The droplets range in size from about 0.05 to about 30 μm. The average size is about 1 μm. In one region on the spherule the droplets are so numerous that they form a coating over the glass skin. Smooth indentations made in the glass by the droplets indicate that the glass was still molten or soft during the capture of the droplets. The glass was also vesiculating during this period, and many vesicle openings ranging from 0.1 to 10 μm in diameter were preserved as the glass cooled. Troilite droplets are relatively rare and are mixed with nickel-iron droplets. Figure 1B is a mixture in which the troilite forms a partial coating over and a matrix between a complex of nickel-iron droplets. There is apparently no mixing between the two phases.

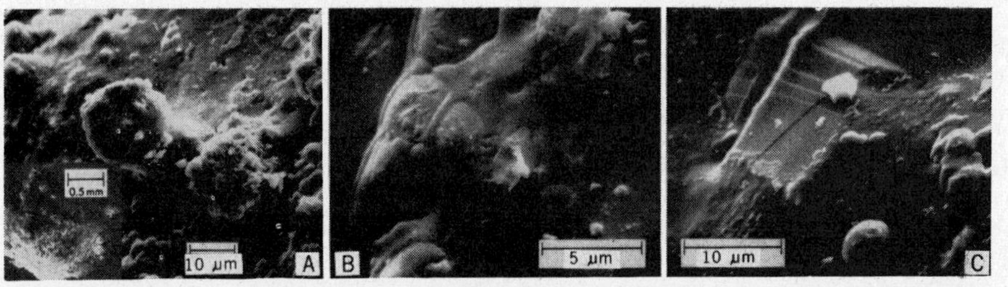

Fig. 1. Scanning electron microscope photographs of coated glass spherule. (A) (insert) Overall view; (A) nickel-iron droplets (right) and large glass splash droplet (center); (B) troilite coating and matrix surrounding nickel-iron droplets; (C) partly buried ilmenite grain, tiny glass splash droplets (on ilmenite), and larger nickel-iron droplets (bottom).

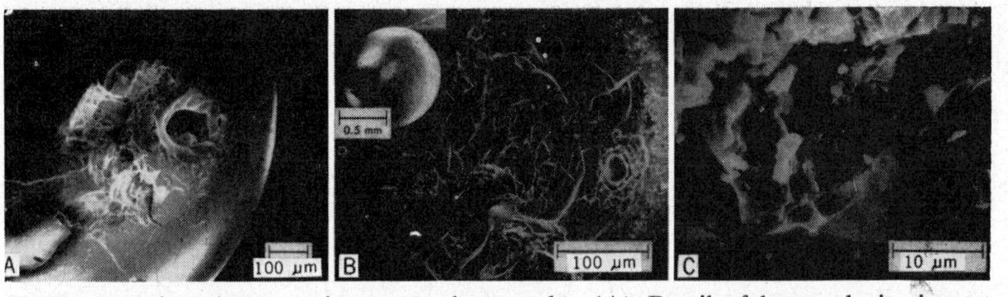

Fig. 2. Scanning electron microscope photographs. (A) Detail of hypervelocity impact crater; (B) (insert) overall view of pitted spherule; (B) detail of spalled zone showing cone of percussion (lower center) and subsequent small hypervelocity impact (right); (C) broken vesicle surface in glass fragment coated with potassium-rich material, possibly vapor condensate.

After the deposition of the droplets had occurred on the large spherule, many small chips of glass and mineral fragments were also captured. Some of this captured material was partially imbedded in the molten glass as shown by the ilmenite grain (Fig.1C). The lower left portion of the grain is completely covered with glass. The contact between the glass and ilmenite clearly shows that the impact of the ilmenite caused some of the molten silicate to splash out. Hundreds of tiny (0.1 μm) glass spherules sprayed out by the impact can be seen on the ilmenite.

Figure 1A shows a glass spherule about 20 μm in diameter which is necking out from the larger spherule as a rebound from an impact of a particle into this molten spherule. Also visible in this picture are numerous partly coalesced nickel-iron droplets.

As this large glass spherule cooled, it continued to capture a variety of material which became partly imbedded in the surface. These particles were welded to the glass spherule and to each other and could not be removed by ultrasonic cleaning. Glass fragments predominate, but grains of plagioclase, clinopyroxene, olivine, and chromite were also identified. Many captured glass spherules show a capture and accumulation sequence on their own surfaces. Some of the large fragments on the surface of the spherule consist of microbreccia and contain a wide variety of particles which were accumulated and partly welded together before the fragment was captured by the large spherule. The material forming these microbreccia fragments has a softened or sintered appearance.

The microstratigraphy, as preserved on this and many similar spherules, indicates that they passed in a molten state through an impact plume or cloud and encountered and captured a wide assortment of ejecta material during that passage. The part of the cloud encountered first contained molten nickel-iron, iron, and troilite droplets having a mean diameter on the order of 1 μm. Some of the iron droplets may represent a condensation aerosol from an iron-rich vapor generated by the impact of an iron meteorite. Similar droplets are found in terrestrial impact-produced glass (*2*).

Another type of glass spherule is characterized by a surface relatively free from captured material. This surface typically shows small impact craters which reflect a wide range of impact velocities up to and including hypervelocity. Figure 2, A and B, shows a variety of features associated with this type of glass spherule. Figure 2A shows a hypervelocity impact which has melted a central region and radially and concentrically fractured a larger region. The impact vented into a vesicle beneath the surface and displaced a large segment of glass outward. Circular structures in the fracture zone are vesicles exposed by the spallation of the spherule skin. Figure 2B is an overall view (insert) and a closeup of another glass spherule showing a large circular spalled area resulting from impact. A small cone of percussion is visible in the lower part of the picture. At some later time a hypervelocity impact occurred and melted a small crater near the edge of the larger crater. Other craters ranging from those caused by multiple hypervelocity impact to those caused by simple lower velocity impact are present on this spherule.

These and similar spherules indicate an environment of numerous high-velocity particles. It is possible, although it does not seem likely, that these spherules remained exposed directly at the lunar surface long enough to allow all of the impact craters to be formed by a random rain of micrometeoroids. Alternatively, it might be more likely that all of the impact pits on any one glass spherule were closely associated in time and were in fact part of the same major event. Gault *et al.* (*3*) have shown that the velocity of secondary material may exceed that of the primary impacting object. It is possible that the impact pits on a glass spherule are all caused by the secondary (or tertiary, or so on) projectiles from a single impact of a large meteorite on the lunar surface. These projectiles could be expected to exhibit a wide range of velocities including hypervelocities, and this is reflected in the variation in the types of pits. By inference some or all of the pits on lunar rocks also may result from impacts in one or more impact plumes, and therefore the relative densities of pits on rocks may not be a satisfactory indicator of length of exposure on the lunar surface.

Gradations between the completely covered and smooth spherules were also found. In several cases the two types were welded together, and both may have been formed in the same event. Nickel-iron droplets were found on both types of spherules.

A considerable variation existed in the composition of the glass particles in the fines and breccia. Some single glass spherules were homogeneous; others contained zones or flow bands of different compositions. This is not surprising in view of the number of different ways in which glass spherules were formed and the possible variations in temperatures and mixing. At least six mechanisms produce lunar spherules. These include (i) expansion and tearing apart of large masses of molten glass formed toward the center of major impacts, (ii) the breakup of impact-produced liquid jets into droplet trains, (iii) splash and rebound from objects hitting molten glass, (iv) drag of splattered glass over hard nonwetted surfaces, (v) condensation from a vapor, and (vi) vesiculation of impact-produced or volcanic magma. The production of glass spherules by vesiculation may be a major lunar mechanism and occurs when segments of broken bubble walls are pulled together by surface tension (*4*).

A typical homogeneous spherule in breccia 10068,32 contains 41.8 percent SiO_2, 14.7 percent Al_2O_3, 7.0 percent TiO_2, 12.2 percent CaO, 8.5 percent MgO, 14.0 percent FeO, 0.07 percent K_2O, and 0.17 percent Na_2O. This breccia contains two unusual glass fragments which are similar in texture and contain about 30 percent euhedral to subhedral mostly equant olivine crystals averaging 20 to 50 μm in diameter. The glass matrix contains 45.2 percent SiO_2, 15.5 percent Al_2O_3, 0.3 percent TiO_2, 9.9 percent CaO, 13.5 percent MgO, 13.6 percent FeO, 0.2 percent K_2O, and 1.4 percent Na_2O. The normally zoned olivine ranges from 85 to 92 (mean, 88) mole percent forsterite. The fragment may be a partly quenched glass resulting from the impact melting of pyroxene and plagioclase, or it may contain chondritic material.

A chrome titanium spinel of unusual composition was analyzed in a small fragment of crystalline rock in breccia 10068,32. This opaque phase contains 41.5 percent FeO (total iron), 24.5 percent TiO_2, 21.4 percent Cr_2O_3, 4.6 percent MgO, 6.5 percent Al_2O_3, and 0.3 percent SiO_2. Several small (15 to 30 μm) grains of this phase are present and are homogeneous at the scale of the electron beam (1 μm).

A slightly botryoidal coating (Fig. 2C) on the wall of a broken vesicle in a breccia glass fragment has a very high concentration of silica and contains about 5 percent K_2O. This coating may constitute a condensate from a silicate vapor formed when the glass was heated by impact. Condensed silicate vapor has been observed during volcanic eruptions (*5*), has been produced experimentally (*4, 5*), and has been produced in nuclear explosions (*6*). Material formed in this way may have complex and variable composition and morphology, and our search for such material in the lunar fines is continuing.

Fisher and Waters (*7*) have discussed the possible characteristics of lunar base surges. The lunar breccias show many features which could be explained if they were the deposit from a hot base surge produced from a large impact event. Turbulence in the base surge could build the observed accretionary rims (*8*) on nuclei of glass spheres and crystal fragments by rolling and tumbling action. Coarse layering in some of the breccias (*1*) corresponds to the bedding which is often characteristic of base surge deposits. After the flow had stopped, sintering of the finest glass fractions in the still hot deposit lithified the matrix which bonded the accretionary lapilli to each other. Partial devitrification of some glass fragments took place during cooling. Additional heat or thicker deposits could produce a more densely welded rock with a dense crystal-bearing glass as the final product in a sequence analogous to the formation of terrestrial welded ash flow tuffs. Sintering of lunar base surge deposits may proceed at temperatures which produce no sintering in the earth's atmosphere. Air interacts with gold particles (*9*) and MgO particles (*10*) to inhibit sintering by comparison with that which takes place in a vacuum. A similar vacuum enhancement of lunar glass sintering might be expected.

DAVID S. MCKAY
WILLIAM R. GREENWOOD
DONALD A. MORRISON
NASA Manned Spacecraft Center, Houston, Texas 77058

References and Notes

1. Lunar Sample Preliminary Examination Team, *Science* **165**, 1211 (1969).
2. L. J. Spencer, *Mineral. Mag.* **23**, 387 (1933).
3. D. E. Gault, W. L. Quaide, V. R. Oberbeck, *Shock Metamorphism of Natural Material*, B. French and N. Short, Eds. (Mono, Baltimore, 1968), pp. 87–99.
4. K. A. Richardson and D. S. McKay, in preparation.
5. B. Vonnegut, R. K. McConnell, Jr., R. V. Allen, *Nature* **209**, 445 (1966).
6. D. E. Rawson, *Shock Metamorphism of Natural Material*, B. French and N. Short, Eds. (Mono, Baltimore, 1968), pp. 626–627.
7. R. V. Fisher and A. C. Waters, *Science* **165**, 1349 (1969).
8. See K. A. Richardson, D. S. McKay, W. R. Greenwood, T. H. Foss, *Science*, this issue, for details of breccia structures and formation.
9. K. H. Olsen and G. C. Nicholson, *J. Amer. Ceram. Soc.* **51**, 669 (1968).
10. P. E. D. Morgan and E. Scala, *Bull. Amer. Ceram. Soc.* **44**, 301 (1965).
11. We are grateful to G. Ladle for his assistance in maintaining and operating the instruments used in this study.

4 January 1970

Mineralogy and Petrography of Lunar Samples

Abstract. *The lunar samples consist largely of augite, calcic plagioclase, and ilmenite. Olivine is a minor constituent of some rocks, as is cristobalite. Other minerals present in small amounts include tridymite, chromite, kamacite, taenite, and troilite. The principal rock types can be broadly grouped into ilmenite basalts and breccias. Except for their high ilmenite content, the lunar rocks resemble the calcium-rich achondritic meteorites (eucrites and howardites) in composition and structure. Evidence of a meteoritic increment in the lunar soil is provided by the presence of nickel-iron particles in glass and breccia, and the occurrence of metal-troilite spheroids; the breccias contain occasional silicate aggregates that resemble meteoritic chondrules. The lunar fines contain 325 parts of water-soluble calcium per million.*

We have confirmed the presence of plagioclase, pyroxene, ilmenite, olivine, cristobalite, tridymite, chromite, kamacite, taenite, and troilite in the lunar samples. Microprobe analyses show a range of plagioclase compositions from An_{70} to An_{100}, individual grains being zoned with increasing An content to-

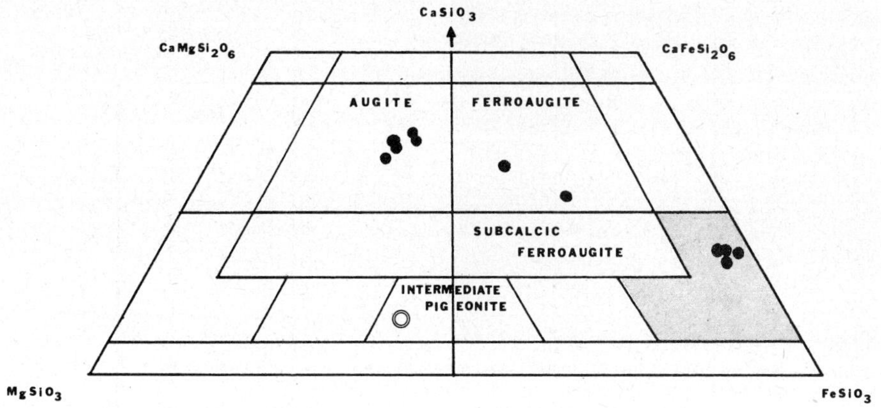

Fig. 1. Augite and ferroaugite compositions in lunar crystalline rock 10047, and the composition of the yellow pyroxenoid mineral in the shaded field, a compositional region for which natural pyroxenes were previously unknown. Also shown (open circle) is an intermediate pigeonite grain, found in crystalline rock 10003.

wards the center. Pyroxene is also zoned and shows a range of composition, falling in the fields of augite, ferroaugite, and pigeonite; some grains show exsolution lamellae of hypersthene or clinohypersthene. Microprobe analyses of pyroxenes in one rock (10047), recalculated in terms of the $CaSiO_3$ (Wo), $MgSiO_3$ (En), and $FeSiO_3$ (Fs) components, range from $Wo_{40}En_{36}Fs_{24}$ to $Wo_{28}En_{21}Fs_{51}$ (Fig. 1). Associated with the pyroxene in 10047 is a yellow mineral (*1*); microprobe analyses of this mineral show that its mean composition is $Wo_{18}En_4Fs_{78}$, and J. V. Smith (*2*) has identified it as a pyroxenoid mineral with the pyroxmangite structure. Olivine occurs in small amounts in some of the lunar rocks and has a composition near Fa_{30}. Ilmenite is universally present, usually as idiomorphic platy crystals. Cristobalite is interstitial to the other minerals and is evidently a late crystallization. Rare grains of tridymite were found in a low-density fraction of the lunar soil. Chromite is present as an accessory mineral. Trace amounts of kamacite and troilite are present in the crystalline rocks, frequently as globules in the mesostasis. Taenite, probably of meteoritic origin, occurs as metal particles in the breccias and lunar soils; these metal particles contain up to 30 percent nickel.

The lunar crystalline rocks (types A and B) (*1*) form a mineralogically uniform group. Although they range widely in grain size, they consist of the three principal minerals augite, plagioclase, and ilmenite in order of decreasing abundance. Modal analyses of five samples (10003, 10020, 10047, 10049, 10050) gave very small variations. Augite ranges from 47.7 to 53.4 volume percent and averages around 50 percent. The range in plagioclase and ilmenite is larger: 24.4 to 36.9, and 9.8 to 18.5 percent, respectively. The average grain size ranges from about a few hundredths of a millimeter up to nearly 2 mm. By terrestrial standards, this includes rocks ranging from basalts (less than 0.5 mm) to dolerites (less than 2 mm) and approaches the grain size of gabbros (greater than 2 mm). This grain-size comparison suggests that the finest grained lunar rocks correspond to quenched surfaces of flows, or ejected cinders or bombs, and the coarser grained samples correspond to flow interiors or perhaps to the interior of thin dikes or sills. Perhaps the simplest hypothesis is that all the crystalline rocks are fragments of a single flow blown to their present sites by meteorite impact.

The coarser grained rocks (type B) show a clear-cut crystallization sequence. Ilmenite is one of the first minerals to crystallize, and appears to be followed shortly by augite. When present, olivine crystallized largely before augite, and is commonly mantled by it. The crystallization of ilmenite and augite precedes and overlaps that of plagioclase. In general, the sequence follows the well-known Bowen reaction series. Progressive crystallization eventually produced residual liquids enriched in FeO and SiO_2 and depleted in CaO, MgO, and Al_2O_3 compared to the parent liquids. From these residual liquids, cristobalite and the peculiar Fe-Ca pyroxenoid crystallized, with possibly apatite and other phases.

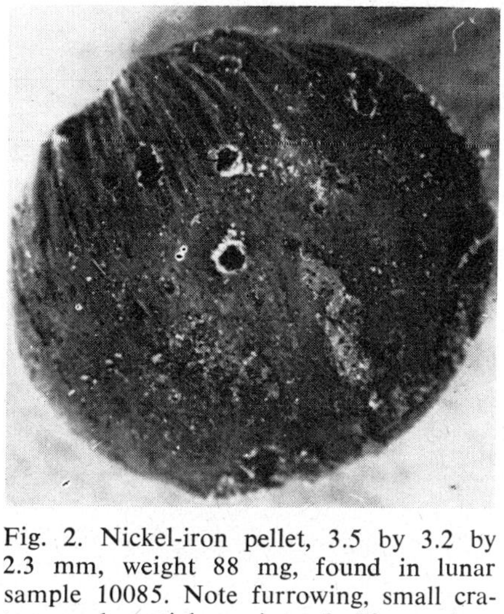

Fig. 2. Nickel-iron pellet, 3.5 by 3.2 by 2.3 mm, weight 88 mg, found in lunar sample 10085. Note furrowing, small craters, and partial coating of silicate glass.

The lunar rocks are comparable in some regards to terrestrial basalts. As in the lunar samples, plagioclase and augite are the principal phases in terrestrial basalts. However, ilmenite is rarely abundant, and normally there is more plagioclase than augite. The high content of pyroxene plus ilmenite in the lunar samples justifies calling them ilmenite melabasalts or ilmenite meladolerites, following the nomenclature of Johannsen (*3*). Although the lunar igneous rocks can be classified by a terrestrial scheme, there remain fundamental chemical and mineralogical differences between the lunar rocks and common terrestrial basalts (*1*). Compared to tholeiitic basalts of comparable grain sizes, the lunar rocks contain very little mesostasis or glass. They appear to have come from melts that are very difficult to quench, and thus perhaps of quite low viscosity. Among terrestrial basalts, low-silica magmas, particularly those of the alkaline basalt family, are similarly difficult to quench even when rapidly cooled.

One of the most striking features of the coarser grained lunar rocks is a well-developed ophitic texture, that is, a texture composed of large pyroxene crystals enclosing a number of plagioclase crystals. This is the commonest texture of terrestrial intrusive basaltic rocks. Although the texture appears to indicate early crystallization of plagioclase followed by augite, it has more often been taken to indicate simultaneous crystallization of the two phases (*4*).

There are certain analogies between the iron-bearing basalts of Disko Island, Greenland, and the lunar basalts. The main similarity is the occurrence of kamacite in both. As in the lunar samples, the kamacite in the Disko basalts commonly occurs in interstitial areas and is associated with sulfides. Some of the Disko basalts have high FeO contents and crystallized under low oxygen fugacities, as did the lunar basalts. Eventually one may find the Fe-Ca pyroxenoid in the Disko basalts. We have examined one of these basalts but find the most iron-rich pyroxenes to be ferriferous pigeonite.

Among meteorites, there are close analogies in composition and structure between the lunar basalts and the Moore County eucrite, a calcium-rich achondrite consisting of calcic plagioclase, pyroxene, and tridymite, with accessory chromite and ilmenite; however, this meteorite contains only a little titanium.

The lunar breccias are composed of rock, mineral, and glass fragments in a comminuted matrix of the same materials; irregular fragments of nickel-iron and of silicate aggregates resembling chondrules are present in small amount. They range from friable to hard and completely indurated. Effects of shock metamorphism are pronounced. For example, plagioclase grains are typically chalky white in hand specimen, and in thin section are seen to be partly or wholly converted to maskelynite. The breccias appear to be shock-lithified lunar soil and contain much glass which is highly variable in color, refractive index, and composition. Some of this glass is present as transparent orange-brown spheres that are superficially similar to chondrules in meteorites. Urey (*5*) has proposed that chondrules (and hence chondritic meteorites) originate on the moon, impacts on the lunar soil having produced considerable quantities of molten droplets. However, the Apollo 11 site cannot be the source of presently known chondritic meteorites, because the chemical and mineralogical compositions of the lunar samples are quite different from those of these meteorites. However, the structure of the breccias is very similar to that of the howardite meteorites (calcium-rich achondrites) such as Kapoeta (*6*), and they have comparable compositions, except for the high titanium content. Duke and Silver (*7*) have suggested a lunar origin for the calcium-rich achondrites.

The lunar soil (material <1 mm diameter) comprises fragments of basalts and breccias, single mineral grains, and much glassy material. Some fragments of a type not seen in larger pieces were noted, in particular a fine-grained rock consisting of subequal amounts of anorthite (An_{95}) and colorless glass, this glass having the approximate composition (weight percent); SiO_2, 43; Al_2O_3, 13; TiO_2, 1.3; FeO, 13; MgO, 17; CaO, 8; Na_2O, 0.6. The x-ray powder photographs of different size fractions from

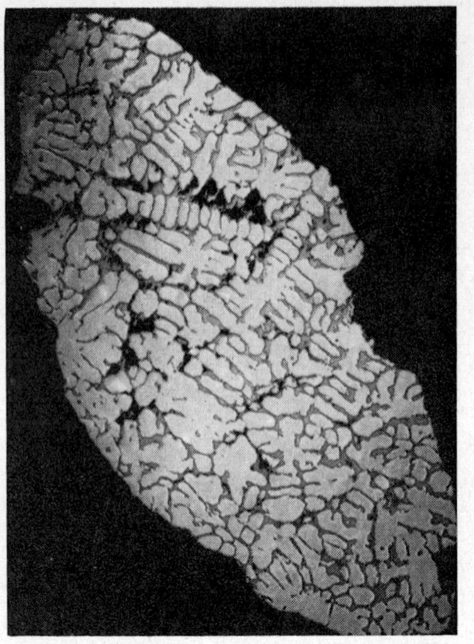

Fig. 3. Polished section of nickel-iron pellet, showing dendritic aggregate of taenite (white) and troilite (gray); length of section is 2.5 mm.

the lunar soil show uniform proportions of augite, plagioclase, and ilmenite throughout. The finest component (<10 μm) consists largely of glass shards, and electron micrographs show that these are vesicular on a submicron scale, individual vesicles being as small as 100 to 200 Å in diameter.

We sieved our specimen of lunar fines (10084) and obtained more than 2 g of −325-mesh material. Since this material was destined for eventual mineral separation, we decided to extract it first with triply distilled water and to determine the water-soluble cations. A control experiment without sample yielded less than 1 ppm of the cations determined. A sample of the Allende meteorite was extracted for comparison. The values obtained [lunar sample, Allende sample (in ppm)] were : K, <5, <5; Na, 21, 288; Mg, 16, 77; Fe, <5, <5; and Ca, 325, 95. The high amount of extractable calcium is noteworthy. We do not know what anions balance this calcium, but we do not believe that they are halogens. Sulfate or carbonate are distinct possibilities.

Evidence for a meteoritic increment on the lunar surface is provided by the nickel-iron fragments in the breccias and the occurrence of nickel-iron pellets in the soil, such as that illustrated in Figs. 2 and 3. This pellet has a lensoid form with one surface (here referred to as the upper surface) with a greater radius of curvature than the other. This upper surface has a series of furrows, evidently produced by shrinkage during solidification; since these furrows are confined to the upper surface, this indicates that the upper half cooled more slowly than the lower half, presumably because the latter was in direct contact with the lunar surface and lost heat by conduction rather than by radiation. The upper surface shows a partial polish, probably the result of dust abrasion, whereas the lower surface has a rough matte-like texture.

The surface of the pellet is cratered, more abundantly on the upper than on the lower half. These craters were evidently formed by the impact of small silicate particles, the remains of which form a glassy coating on their interior. There are at least three types of craters, each with raised or overturned rims; with increasing ratios of depth to diameter they may be described as (i) saucer craters; (ii) bowl craters; (iii) flask craters (in which a tubular neck leads into a spherical cavity).

A density determination of this pellet gave a value of 7.36; this would indicate a bulk composition of about 85 weight percent nickel-iron and 15 percent troilite, but these figures are approximate because of the presence of some silicates on the surface. The internal structure is a dendritic intergrowth of taenite and troilite. Microprobe analyses show a variation from 14 to 16 percent nickel in the taenite, with the higher nickel values at the margins of the dendrites. The troilite contains a small and variable amount of nickel, usually 0.1 to 0.5 percent, sometimes as high as 0.9 percent. Inclusions of chromite are present in the troilite, and a small amount of silicate. At the interface of the taenite and the troilite there is a narrow rim of a distinct and separate nickel-iron phase, visible only after etching. This phase has sharply defined edges, and has a variable nickel content, usually 32 to 36 percent but occasionally as high as 45 percent.

The taenite in this pellet has a pseudo-martensitic structure, which may be interpreted either as shock-induced or as the result of rapid cooling. Buchwald (*8*) has produced similar structures by quenching a synthetic nickel-iron alloy with 15 percent Ni and 0.2 percent P from 1100°C. The structure of the taenite-troilite intergrowth is remarkably similar to that described by Begemann and Wlotzka (*9*) from the Ramsdorf meteorite, a highly shocked chondrite; they also produced similar structures by melting pieces of the Pantar chondrite and cooling in a vacuum.

From these data we infer the following history for this pellet. Its material was originally part of a meteorite that impacted the moon. The impact produced molten nickel-iron droplets which fell into the lunar soil. Solidification took place quickly (within minutes), but cooling was more rapid on the lower than on the upper surface. After the pellet solidified it was hit by high-velocity particles, thereby producing craters. Some abrasion and polishing by drifting dust has also occurred.

B. MASON, K. FREDRIKSSON
E. P. HENDERSON, E. JAROSEWICH
W. G. MELSON, K. M. TOWE
J. S. WHITE, JR.
Smithsonian Institution,
Washington, D.C.

References and Notes

1. Lunar Sample Preliminary Examination Team, *Science* **165**, 1211 (1969).
2. J. V. Smith, personal communication.
3. A. Johannsen, *A Descriptive Petrography of the Igneous Rocks* (University of Chicago Press, Chicago, 1937), vol. 3.
4. F. Walker, *J. Geol.* **65**, 1 (1957).
5. H. C. Urey, *Icarus* **7**, 350 (1967).
6. B. Mason and H. B. Wiik, *Amer. Museum Novitates* **2273**, 10 (1966).
7. M. B. Duke and L. T. Silver, *Geochim. Cosmochim. Acta* **31**, 1637 (1967).
8. V. F. Buchwald, *Acta Polytechnica Scand.* **51**, 7 (1966).
9. F. Begemann and F. Wlotzka, *Geochim. Cosmochim. Acta* **33**, 1351 (1969).
10. We are grateful to D. B. Ballard and H. Yakowitz of the U.S. Bureau of Standards and D. Massey of the U.S. Geological Survey for assistance and advice during this investigation. Supported by NASA contract 9-8016.

4 January 1970; revised 7 January 1970

Phase Chemistry, Structure, and Radiation Effects in Lunar Samples

Abstract. *Phase chemistry, structure, and radiation effects were studied in rock, breccia, and soil samples. The regolith apparently developed in the final stages of accretion and was modified by later impact processes and radiation weathering. Exposure ages indicate transfer of buried igneous rock fragments to the near surface late in lunar history. With a few exceptions igneous rock fragments, soil, and breccia share the same distinctive chemistry, probably acquired before accretion of the moon. The igneous rocks texturally resemble basaltic achondrites, and the soil and breccias contain glassy spheres analogous to chondrules.*

Polished thin sections from three igneous rocks (10024, 10017, and 10057) and one breccia (10018), plus rock fragments, glass, and mineral grains from the soil (*1*) were examined optically and analyzed by electron microprobe. Samples of soil and of the surface and interior of 10017 were examined by x-ray diffraction, electron diffraction, and transmission and scanning electron microscopy.

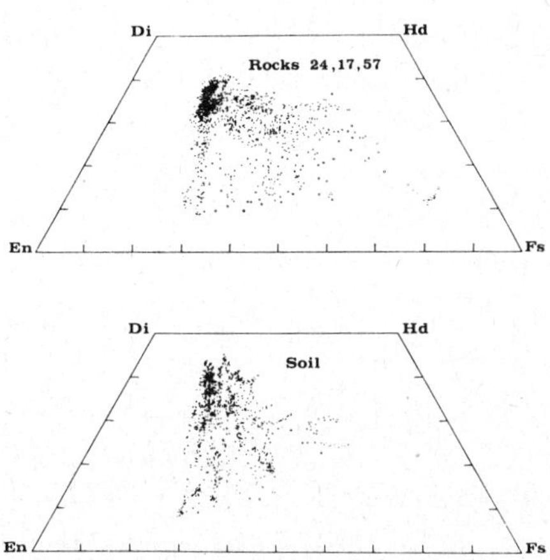

Fig. 1. Ranges of compositions measured in pyroxene grains from igneous rocks and soil. The trend toward high iron values reflects zoning, and the spread in calcium values is largely due to exsolution.

Analytical work was directed toward establishing the major element composition of pyroxene, feldspar, and ilmenite. Other phases analyzed are discussed briefly, but no effort was made to identify all phases present in each assemblage.

Ilmenites in the rocks and soil have essentially the same composition ($Mg_{.04}Fe_{.93}Ti_{.99}Cr_{.02}Al_{.01}O_3$) with no zoning or exsolution. Plagioclase composition ($Or_1Ab_{20}An_{78}$) is also remarkably similar in the various igneous rocks. Plagioclase grains in the soil have a wider range of compositions (An_{51} to An_{96}) with an average composition ($Or_1Ab_{13}An_{86}$) more calcic than that of the igneous rocks. Potassic feldspar is present in minor amounts in rocks and soil.

In both rocks and soil the dominant pyroxene is an augite with minor amounts of Al_2O_3 and TiO_2. A combination of zoning and exsolution, with lamellae too narrow to be resolved by microprobe techniques, results in extreme compositional heterogeneity within individual pyroxene grains (Fig. 1). The range of pyroxene compositions, the average composition ($Wo_{31}En_{44}Fs_{25}$), and the clustering of analyses around $Wo_{36}En_{46}Fs_{18}$ are essentially the same for the igneous rocks and the soil. Both rocks and soil contain troilite, commonly as small rounded interstitial grains. Metallic iron with 0.4 to 1.1 percent nickel by weight occurs as tiny blebs within or adjacent to sulfide.

The fine-grained interstitial material, which apparently was the last liquid fraction to crystallize, contains a variety of phases, including silica, potassic feldspar, calcium phosphate, zirconium silicate, and zirconium oxide. Partial analyses were made of some of these to determine the location of minor elements present in anomalous proportions (*2*). About two-thirds of the Zr occurs as silicate and oxide. The remainder is largely in ilmenite, ranging from 300 to 3000 ppm with large variations from grain to grain and within single grains. The extensive and variable Zr substitution in ilmenite is consistent with high temperatures of formation and rapid cooling. Zr/Hf atomic ratios (150–250) in the high Zr phases fall in the upper range of terrestrial values. A major portion of the yttrium is in the phosphate. Y/Yb atomic ratios (150–250) in the high centrates from rock 10017 and the soil are comparable to meteorite and terrestrial ratios and considerably below initial estimates (*2*). Barium is concentrated in the potassic feldspar.

Rock fragments from the soil and breccias have the same distinctive bulk chemistry as the igneous rocks, while displaying a range of crystallite sizes. Some chemically dissimilar rocks sampled in the coarser soil fraction (1 mm to 1 cm) include fragments that are almost wholly feldspar (anorthosite) and one with a high proportion of calcic plagioclase (An_{95}) coexisting with olivine (Fo_{80}) and orthopyroxene

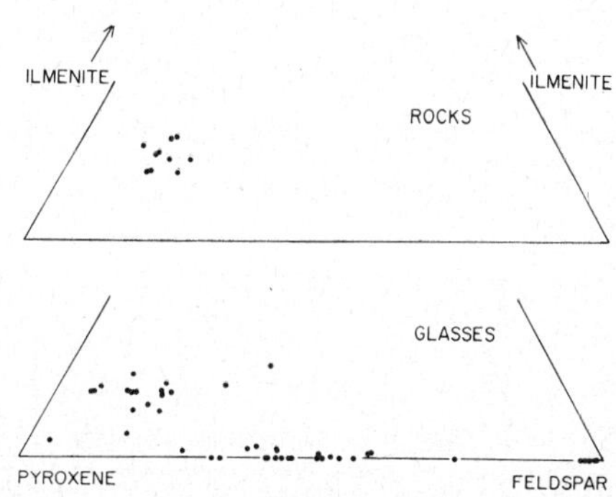

Fig. 2. The composition of Apollo 11 rocks (*2*) and glasses from the soil, expressed in terms of normative pyroxene, feldspar, and ilmenite. Over the entire range of glass types these three normative minerals can account for virtually the total composition (average 95 percent).

($Wo_{4.5}En_{71}Fs_{24.5}$) containing exsolved blebs of clinopyroxene ($Wo_{36.5}En_{47}Fs_{16.5}$). Olivine ($Fo_{67}$), orthopyroxene ($Wo_3En_{80}Fs_{16}$), maskelynite, graphite (one grain, about 1 mm), and two phyllosilicates (discussed below) were also found in the soil.

A major component of the soil is glass. Flow structure and bubbles are common in the darker glasses, which are richer in Fe, Mg, Ti, and Cr. The lightly colored glasses are generally devoid of flow structure and bubbles, and approach feldspar in composition. All the glass types occur both as spheres and as angular fragments, and are more abundant in the fine fraction of the soil. Many of the spheres are coated by largely crystalline excrescences.

The glass compositions could be produced by melting together various proportions of the minerals in the igneous rocks (Fig. 2). There are apparently no glasses with high normative ilmenite, and crystalline ilmenite occurs within glass grains. The trend toward high normative feldspar values (Fig. 2) suggests that these compositions are favorable for formation and preservation of glass. The normative feldspar calculated from the glass analyses is, for many samples, more calcic than the igneous rock feldspar, suggesting loss of alkalis.

The effects of exposure on the lunar surface were studied on grains from the soil and on a fragment of rock 10017 that had approximately 1 cm^2 of the original upper surface. The surface was readily distinguished from the interior by its dull appearance and by cosmic ray particle track densities ($2–7 \times 10^7$ cm^{-2} at the surface compared with $3.5–7 \times 10^6$ in the interior). The surface has an abundance of minute cracks, and near-surface grain boundaries are filled with loose material.

On the surface of rock 10017 we found hexagonal platelets about 0.5 to 1 μm wide showing a hexagonal diffraction pattern (a_0 = 5.3 Å, b_0 = 9.2–9.3 Å on orthogonal axes). Similar material and also hexagonal platelets with different morphology and unit cell size (a_0 = 5.1–5.2 Å, b_0 = 8.89–9.00 Å) were observed in the soil. Both phases show only two-dimensional diffraction on tilting and become amorphous on prolonged exposure to the electron beam. These materials are almost certainly phyllosilicates. Controls indicate that they are not talc from the gloves used at NASA–MSC or contaminants from our laboratory. The possibility that these

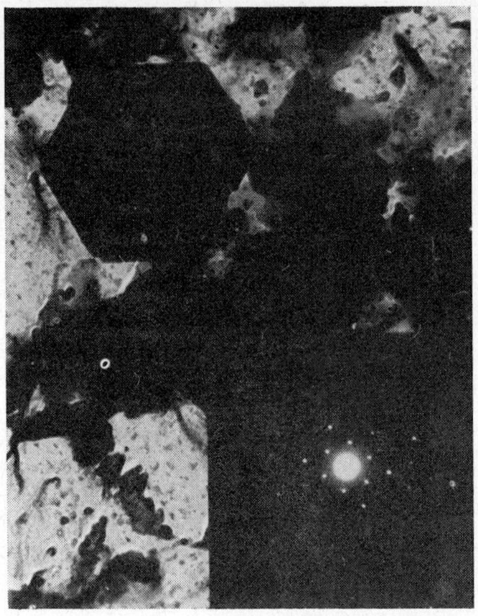

Fig. 3. Electron transmission image and electron diffraction pattern of a hexagonal flake (a_0 = 5.15 Å; b_0 = 8.90 Å) from the lunar soil ($\times$ 10,000).

phases result from contamination from other sources or from reaction of radiation-damaged material with the terrestrial atmosphere cannot yet be eliminated. A most interesting possibility is that the phyllosilicates did form on the moon. One possible process is hydroxylation of silicates by proton bombardment (*3*) followed by crystallization; another is silicate-water interaction on the moon.

The Apollo 11 samples all share the same distinctive chemistry. As impact processes have played a role in the formation of the soil, it probably contains material from a much wider area than the astronauts could sample directly. The unusual composition consequently is not a local phenomenon but is typical of a large area.

The igneous rocks appear to be a single generation of material crystallized from a wholly liquid melt. The homogeneous melt cooled rapidly at first with little or no removal of early formed phases, and the cooling rate diminished with time allowing development of zoning and then exsolution in the pyroxene. The rocks formed at high temperatures (probably 1200° to 1500°C), low pressures (near surface), and low oxygen fugacities. No rock fragments have been found with compositions comparable to the residual liquid fraction (that is, enriched in alkalis, Ba, Y, rare earths, Zr, Si, and P). Although a wide range of grain sizes was observed, neither very rapidly cooled glassy rocks nor truly plutonic rocks were found.

If the melt from which the Apollo 11 igneous rocks crystallized was derived from material akin to the common meteorites or terrestrial rocks, then the differentiation process must have been unlike terrestrial crustal processes. Reduction and removal of a metallic fraction might explain the near absence of nickel and the noble metals, and loss of volatiles at very high temperatures may explain the low alkali content and concentration of the more refractory elements. Serious problems remain, however, in trying to explain the concentrations and proportions of the group 3b and 4b elements. The unique nature of the differentiation process, and the absence of rocks with related compositions, that is, members of a differentiation sequence, coupled with physical evidence such as the high density of the lunar surface, suggest that the Sea of Tranquillity rocks attained their distinctive chemical composition prior to lunar accretion, and that this composition must differ from that of the lunar interior.

The soil is a loose aggregate of rock fragments similar to the major igneous rocks, mineral fragments from such rocks, and a variety of glasses. As several glass fragments contain spherules of nickel-iron (some with associated phosphide and sulfide) analogous to the spherules found in terrestrial impactites, it is apparent that the soil formed in part by melting and comminution of the igneous rocks by meteorite impact. The paucity of meteorite material in the soil suggests that most of the impacting material was not meteoritic. If much of the soil was formed in the terminal stages of accretion, then the impacting material may have had a composition like that of bulk Apollo 11 material. The original regolith appears to have developed early in lunar history and has been modified by later impacts and only to a small extent by radiation weathering.

Basement rocks such as 10017 have been transported, presumably by impact, to the near surface late in lunar history. Rock 10017 has a much lower exposure age ($10 \pm 5 \times 10^6$ years), as determined from counts of particle tracks, than the soil from the same area ($\geq 10^9$ years). Preliminary studies of the distribution of grains with high particle track densities indicate that there is some vertical mixing in the soil, but the process is extremely slow.

The glass spheres resemble meteoritic chondrules, particularly in the breccias, where some of the spheres have crystallized. Compositionally these spheres are quite unlike chondrules, but they do illustrate a mechanism whereby chondrule-like bodies can develop. The glass beads probably formed by melting different proportions of minerals from a preexisting rock. Thus the bulk composition may vary considerably, but certain element ratios (reflecting phase compositions in the parent rock) will be preserved. This suggests a possible explanation of one of the puzzling characteristics of ordinary chondrites, where chondrules vary in bulk composition but contain olivine and pyroxene of fixed compositions.

There are many similarities between the Apollo 11 samples and basaltic achondrites, despite the obvious differences in bulk composition and the nature of the pyroxenes. Textures are remarkably similar, as are the development of zoning and degree of exsolution, apparent cooling rates, depth of burial and oxygen fugacities, the range of textures for similar bulk compositions, and the common occurrence of monomict breccias. The lunar surface may not have the appropriate composition to generate basaltic achondrites, but the necessary environments and processes probably did exist.

In the terminal stages of lunar accretion the material being added had a very distinctive chemical composition. Local melting formed a shallow liquid silicate pool that cooled rapidly. Impacts destroyed the uppermost surface and formed a soil layer over the igneous rocks very early in lunar history. There has been no major exchange of material between the surface and the deep interior, and the moon retains its accretionary stratigraphy. Later impacts and radiation effects have modified but not destroyed the imprint of these early processes.

GUSTAF ARRHENIUS, SAARA ASUNMAA
JAMES I. DREVER, JOEL EVERSON
RAY W. FITZGERALD, JANE Z. FRAZER
HITOSHI FUJITA, JEFFREY S. HANOR
DEVENDRA LAL, SAMUEL S. LIANG
DOUGLAS MACDOUGALL, ARCH M. REID
JOHN SINKANKAS, LAUREL WILKENING
University of California, San Diego,
La Jolla 92037

References and Notes

1. The term soil is used here for unconsolidated material of less than 1 cm diameter.
2. Lunar Sample Preliminary Examination Team, *Science* **165**, 1211 (1969).
3. E. J. Zeller and L. B. Ronca, *Icarus* **7**, 372 (1967).
4. Support from NASA grant NAS 9-7892.

4 January 1970

Mineralogy, Petrology, and Surface Features of Lunar Samples 10062,35, 10067,9, 10069,30, and 10085,16

Abstract. *The primary rocks are a sequence of titanium-rich basic volcanics, composed of clinopyroxene, plagioclase, and ilmenite with minor olivine, troilite, and native iron. The soil and microbreccias are respectively loose and compacted mixtures of fragments and aggregates of similar rocks, minerals, and glassy fragments and spheres. Impact events are reflected by the presence of shock metamorphosed rock fragments, breccias, and glasses and their resulting compaction to form complex breccias, glass-spattered surfaces, and numerous glass-lined craters. Chemistry of the glasses formed by the impact events is highly variable, and the high iron and nickel content of a few moundlike features suggests that at least some of the projectiles are iron and nickel-rich meteorites.*

Samples 10062,35; 10067,9; 10069,30; and 10085,16 have been analyzed. Samples 10062,35 and 10069,30 are thin sections of rocks; sample 10067,9 is a thin section of a microbreccia, and sample 10085,16 is a 10-g portion of lunar soil.

Sample 10062,35 is composed of radiating lathlike and acicular calcic-plagioclase crystals that enclose irregular grains of clinopyroxene and olivine. Skeletal crystals up to 1.5 mm of ilmenite occur throughout the section, and irregular patches of troilite with rounded exsolution blebs of iron are common. The plagioclase and clinopyroxene have a patchy extinction. The ilmenite has two types of exsolution features; the first occurs as bluish-gray elongated blebs, probably a spinel, and the second as fine whitish-gray lamellas of rutile. The overall texture of the section is very similar to a rapidly quenched silicate liquid of low viscosity. The mineralogy indicates that the liquid had the composition of a titanium-rich olivine basalt.

Sample 10069,30 is composed of euhedral to subhedral clinopyroxene and stubby subhedral to lathlike ilmenite in a matrix of poikilitic calcic-plagioclase. Small amounts of troilite with small exsolved blebs of native iron occur in irregular and rounded patches. The clinopyroxene crystals are either zoned or show a patchy extinction, suggesting compositional variation. In many crystals the cores contain numerous inclusions that are less than 1 μm in diameter. The plagioclase crystals

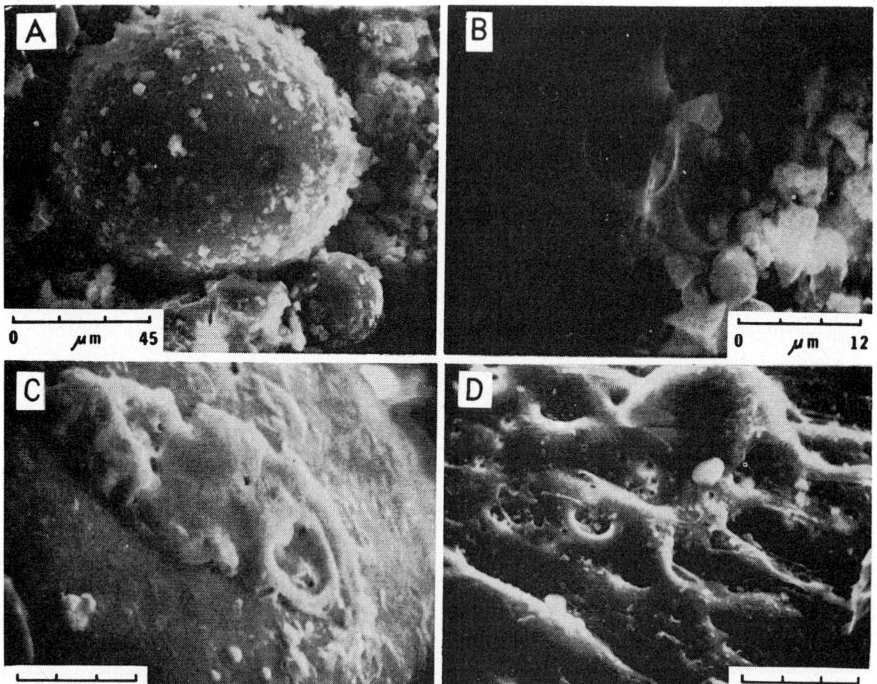

Fig. 1. Scanning electron micrographs (*2*). (A) Sphere with adhering dust fragments in a typical angular microbreccia matrix; (B) vesicular glassy fragment in microbreccia; (C) irregular glass bleb on glass sphere; and (D) ropy glass spatter on glass.

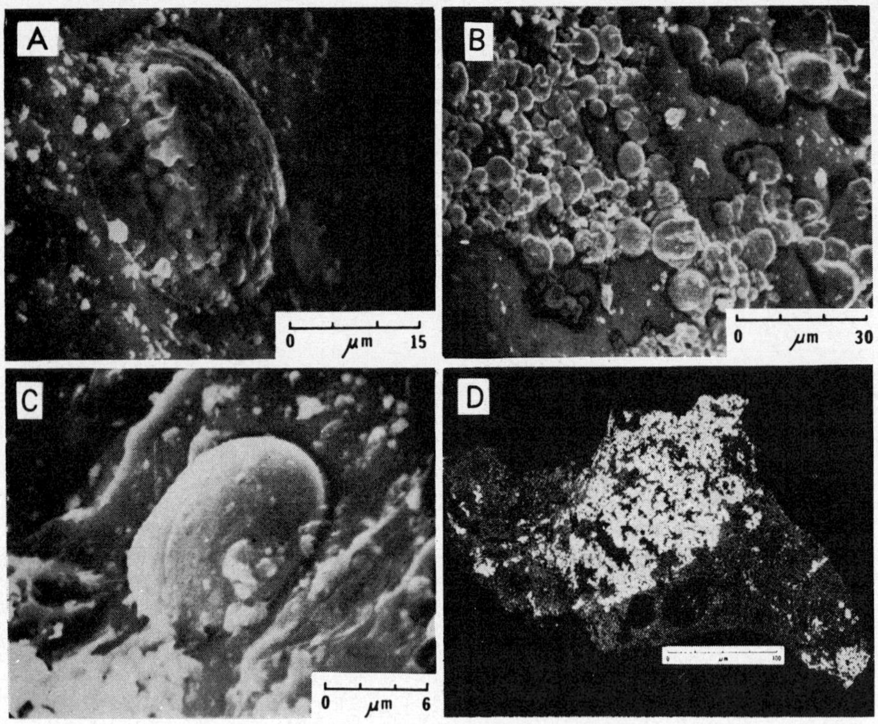

Fig. 2. Scanning electron micrographs (*2*). (A) Textured mound on glass sphere. (B) Magnetic iron-nickel mounds on glass fragment. (C) Magnetic iron-nickel mound surrounded by depressed rim on glass sphere. Note distortion of scanning picture by magnetic mound. (D) Electron microprobe iron scan of glass fragment shown in (B).

occur both as poikilitic crystals that enclose clinopyroxene and ilmenite grains or as radiating laths; their birefringence is very low, and the undulatory extinction is suggestive of mild shock metamorphism. The mineralogy, texture, and a few large vesicles indicate that this rock is a fine-grained vesicular basalt. The textures indicate that both clinopyroxene and ilmenite occur as primary minerals in a plagioclase-rich liquid. Patches of radiating "quench" plagioclase suggest rapid chilling, and minor shock effects indicate a later bombardment by meteoritic material.

Sample 10067,9 is a microbreccia. It is a compacted composite of rock, mineral, and glass fragments and fine lunar dust. The rock fragments may be divided into three types. Most common are fragments of basaltic rocks of varying grain size made up of mixtures of clinopyroxene, calcic-plagioclase, and ilmenite. Also abundant are aggregates of calcic-plagioclase, clinopyroxene, and ilmenite with characteristic feathery or herringbone textures reminiscent of quench products. In addition fragments of breccia occur. Mineral fragments of calcic-plagioclase, clinopyroxene, and ilmenite are common, while smaller amounts of troilite and native iron are present. Small glass spheres and fragments of glass varying in color from colorless to brownish orange are scattered throughout the slide. In many cases small fragments of rocks and minerals are imbedded in the spheres and glasses. The very fine-grained matrix was not identifiable in thin section, but x-ray diffraction data confirmed the presence of plagioclase, ilmenite, clinopyroxene, and abundant glass. A number of rocks and mineral fragments indicate varying degrees of shock from shattered fragments to irregular glassy patches. Scanning microscopy shows local welding of fragments by silicate glasses, which result either from shock metamorphism or from admixed droplets in the breccia.

Sample 10085,16 is a portion of the lunar soil. The soil is composed of a mixture of rock, mineral, and glass fragments and glass spheres (see Fig. 1, A and B) much like the microbreccia (sample 10067,9). Aggregates of rocks, minerals, and glasses are common. The rock fragments are microbreccias, vesicular basalts, gabbros, and microgabbros. The lithology, texture, and mineralogy of the microbreccias are similar to those of sample 10067,9. The igneous rocks are composed of varying proportions of calcic-plagioclase, clinopyroxene, and ilmenite. The coarser-grained rocks vary from anorthosites to gabbros, while the finer-grained rocks are basaltic in their mineralogy. Mineral fragments are abundant and are composed of the three primary minerals calcic-plagioclase, clinopyroxene, and ilmenite. Minor amounts of olivine, troilite, and native iron are present. The glass spheres and fragments occur in many forms, and range from hollow to vesicular to solid. The glass spheres varied in color from colorless through green, brown, and wine-red to opaque which suggests variation in chemical composition. Semiquantitative electron-microprobe analysis of glass spheres showed that they are chemically heterogeneous; approximate weight percentage ranges are: SiO_2, 40 to 50; Al_2O_3, 2 to 18; TiO_2, 3 to 15; Cr_2O_3, 0.00 to 0.08; total iron as FeO, 3 to 18; MgO, 2 to 9; CaO, 1 to 12; Na_2O, 0.1 to 0.7; MnO, 0.03 to 0.1; and NiO, 0.00 to 0.3.

An important aspect of the interpretation of the petrogenesis of the Apollo 11 material has been the examination of surface textures of the rocks and glasses with a scanning electron microscope. The fine-grained igneous rocks, breccias, and mineral fragments often are covered with thin films of glass, or may have a fine frosted surface (Fig. 1D). The frosted surface was shown to be either the result of intense fracturing or a glass coating. In addition, small craters from 20 to 2000 μm, often glass-lined, occur on the surface of many of the tiny rock fragments. The craters are often surrounded or underlain by a whitish margin which is composed of fractured particles of the host. In contrast, the vesicules of the rocks contain highly reflective euhedral mineral faces.

The glass-covered surfaces and spheres have three major types of features: vesicules (Figs. 1B, 1D, and 2D), craters (Fig. 3), and mounds (Figs. 1C and 2). The vesicules are usually spherical or subspherical in shape and often coalesce into irregular networks (Fig. 1D). Occasionally perfectly hemispherical vesicules are seen. The vesicules probably formed by the expansion of gas bubbles from the silicate glass, and thin fragile shells of the original bubbles may be seen in a number of cases.

There appear to be three types of craters: a hole surrounded by a flared lip (Fig. 3A); a raised "crown" of glass surrounded by a circular ring with radiating sets of conchoidal fractures (Fig. 3B); and a deep glass-lined hole with a "splashed" rim surrounded by radiating sets of fractures (Fig. 3, C and D), and, at greater distances, very fine radiating rays of splashed glass. It is inferred that the different crater features result from variations in the plasticity of the target with the target becoming increasingly brittle in the sequence described. It is probable that some of the impact phenomena occur in the particulate cloud that results from the impact event.

The glass surfaces and spheres are often covered with small moundlike features which sometimes are small droplets of glass and sometimes highly magnetic features (Figs. 1C and 2, A–C). They may have smooth or rough surfaces and in some cases are surrounded by a narrow depression or moat (Fig. 2, A and C).

These small moundlike features coat many surfaces, giving them a finely dusted appearance. In addition, some surfaces are splashed with irregular blebs or stringers of ropy glass (Fig. 1D). These and the moundlike glass features are often of a different color than the substrate.

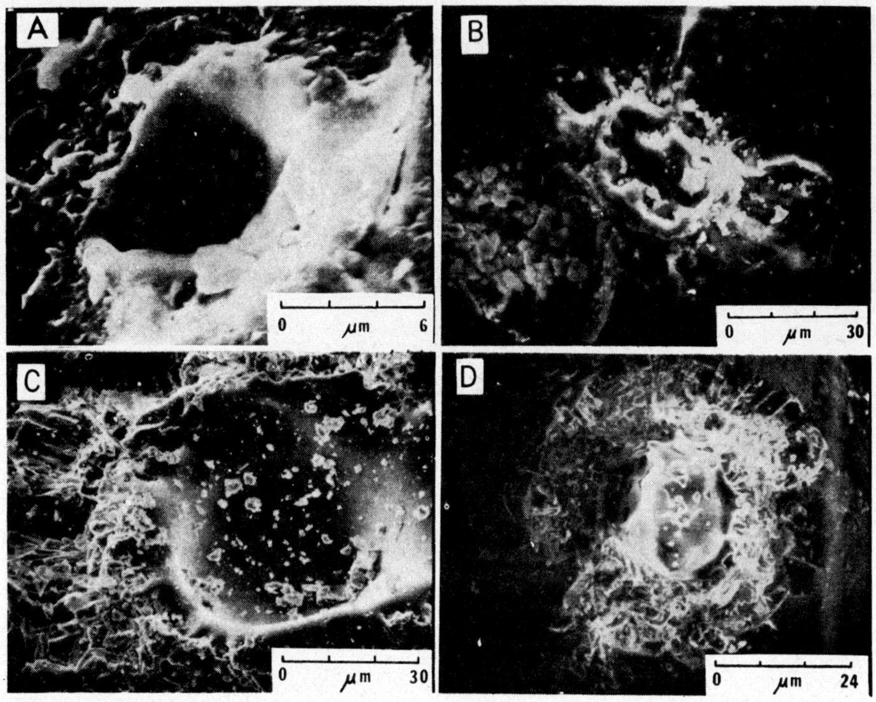

Fig. 3. Scanning electron micrographs of impact structures (*2*). (A) Glass-lined crater with flared lip on glass sphere. (B) Glass-lined crater with raised "crown" of glass surrounded by a circular ring with radiating sets of conchoidal fractures on glass sphere. Note hemispherical pit lower left. Small particles welded to surface of pit. (C) Glass-lined hole with "splashed" rim surrounded by radiating sets of fractures on glass fragment. (D) Glass-lined hole with "splashed" rim surrounded by radiating sets of fractures on glass-spattered surface.

Semiquantitative electron-microprobe analyses of the cratered and moundlike features indicate that the glass within and surrounding the cratered features is essentially isochemical, which suggests that the impacting particle has been expelled and that the raised lips and crowns are essentially a remelted glass of the area surrounding the impact. However, the moundlike features and ropy glasses have a distinctly different chemistry from the underlying substrate. For example, Fig. 2D shows the distribution of iron on a glass fragment spattered with the small moundlike magnetic features shown in Fig. 2B. In these moundlike features, iron is the major element, nickel is minor with cobalt, magnesium, aluminum, phosphorus, and silicon in lesser amounts.

Our data support the conclusions of the Preliminary Examination Team (*1*) that the primary rock types at the Apollo 11 landing site are basic volcanic rocks. The textures indicate crystallization in a low pressure environment where crystallization has been rapid, and in those cases where quench textures are observed, almost instantaneous. The main phase assemblage is fairly simple, being composed primarily of clinopyroxene, calcic-plagioclase, and ilmenite with minor amounts of olivine, troilite, and native iron in a few rocks. No secondary or hydrous minerals were noted. Superimposed on their primary history, later events are reflected by the presence of shock-metamorphed minerals; fragmentation of the rocks and the resulting compaction to form breccia; the presence of glass-spattered surfaces and the presence of numerous craters, often glass-lined, on their surface. The breccias and soil contain similar rocks and minerals degraded by the secondary processes following their formation.

These data suggest that the main process responsible for the degradation of the igneous rocks is the rain of fragments, both from primary and secondary impact events. All of the surface features observed for the Apollo 11 samples are compatible with phenomena associated with meteoritic events. The numerous small glassy spheres, shells, and the glass mounds are probably relicts of the spattered molten beads that are ejected during the impacting event. The chemical data also indicate that in the cratered features the impacting particle was probably completely ejected while the moundlike glassy features and irregular glassy surface blebs are the result of foreign droplets of liquid that have splashed onto the surfaces. The range of cratered features, moatlike surrounds of the mounds, and fragments adhering to glassy surfaces indicate collisions between particles while some were still plastic. The local melting resulting from the impact event is probably responsible for most of the splattered glass. The presence of the moundlike features that are rich in iron and nickel suggests further that iron-nickel meteorites played a role in the formation of the Apollo 11 samples.

JAMES L. CARTER
University of Texas at Dallas (formerly Southwest Center for Advanced Studies), P.O. Box 30365, Dallas 75230

I. D. MACGREGOR
Department of Geology, University of California, Davis 95616

References and Notes

1. Lunar Sample Preliminary Examination Team, *Science* **165**, 1211 (1969).
2. Uncoated surfaces.
3. Supported by NASA contract NAS 9-8766. We thank W. Brown, T. Lyon, D. Martin, J. Pangborn, J. Toney, and R. Wittkopp for technical assistance. Contribution No. 127, geosciences division.

4 January 1970

Lunar Glasses and Micro-Breccias: Properties and Origin

Abstract. *Impact has been an important rock-forming process on the moon. Electron- and ion-probe analyses of major and minor elements show that most glasses and chondrule-like particles formed by shock melting of various proportions of mainly pyroxene, plagioclase, and ilmenite. This is the first direct evidence that chondrule-like molten droplets can form in impact events. Welding and shock lithification resulted in rocks texturally similar to chondrites but composition rules out the moon as source for chondrites. Impact craters on a nickel-iron sample evidence the importance of secondary impacts by accelerated lunar matter.*

Glass fragments and particles occur in a wide variety of shapes and colors in the lunar soil and breccias (*1*). The glasses could be traced back to melting of the main minerals in the Apollo 11 lunar materials. Furthermore, all the crystalline rocks and their minerals could be found in the soil samples as well as in the "micro-breccias" (*1*). We have concentrated our work on rock fragments, and especially glasses, in the soil samples and breccias 21, 61, and 73, although major and trace elements in the various phases of several crystalline rocks have been analyzed.

Major elements were determined in some 60 glasses ranging from spheres, and irregular fragments, to vesicular "crust" adhering to rock fragments and semidevitrified veins in breccias. In general no correlation was found between morphology or size and chemical composition. However, the color in the thin sections changes from colorless to brown to opaque with increasing concentration of Fe and Ti. Silica with a few exceptions ranges from 33 to 48 weight percent; Na_2O from <0.1 to 0.4 percent which indicates some over all loss of these elements. Also, norm calculations for the glasses and the rocks (*1*) indicate more olivine in the former. For Si this is consistent with Walter's (*2*) studies of volatilization of glasses, and although he did not observe a preferential loss of Na the more reduced condition of the lunar rocks may explain this discrepancy.

A strong and consistent anticorrelation of Al_2O_3 with TiO_2 is apparent (Fig 1). On the other hand TiO_2 and FeO show positive correlation. It thus appears clear that most glasses formed by shock melting and quenching of different proportions of ilmenite, pyroxene and plagioclase, the principal minerals in the crystalline rocks. The possibility that some glasses represent material condensed from hot, high density gases cannot be excluded. However, many particles are definitely not such condensates (Fig. 2). One sphere with almost no TiO_2 or Al_2O_3 has pyroxene compositon and may be of meteoritic origin, and two with low TiO_2 and intermediate Al_2O_3 may be mixtures of lunar and meteoritic materials. In Fig. 1 the spheres from a breccia fragment (in 85) appear to form a distinct family. However, only brown, transparent spheres were analyzed in this sample while the spherules in breccias 21 and 61 (Fig. 1) range from colorless through green, yellow and brown to opaque.

Ion-probe analysis (*3*) of trace elements in one brown sphere, a clear individual fragment, two different parts of "crust" on a breccia fragment, and a clear fragment embedded in the same breccia show only small variations and a general similarity with the micro-breccias and soil (*1*). The ranges for some of the elements analyzed were, expressed in atoms per 10^6 Si: Li 107 to 182, B 90 to 230, Na 1.93 to 2.28 $\times 10^4$, Sr 320 to 548, Zr 253 to 357, and Ba 146 to 177 (with exception for the sphere which has high iron and 82 Ba/10^6 Si). Some elements such as V, Cr, and Mn show stronger variation and are positively correlated with iron content.

Figure 2 shows a composite, unusually large (0.4 mm) spherule which graphically illustrates the process of glass formation. Part of the sphere is glass, but some of the original material is still crystalline, severely shocked pyroxene, plagioclase, and ilmenite; the latter mineral is partly dissolved in the glass giving darker colors locally.

The glass spheres may be called chondrules, that is quenched silicate droplets. A few directly resemble glassy chondrules described from the Chainpur chondrite (*4*) in that they have skeletal olivine (Fa_{24-26}) in a glass depleted in Mg and Fe but enriched in Si, Al. The chemical differentiation is weaker, however, than in the Chainpur chondrule, indicating less reducing conditions. Lithification by welding and/or shock has produced rocks texturally resembling types II and III carbonaceous chondrites. The lunar breccias are also, texturally, somewhat similar to terrestrial impactites and some ignimbrites. Still they can be more favorably com-

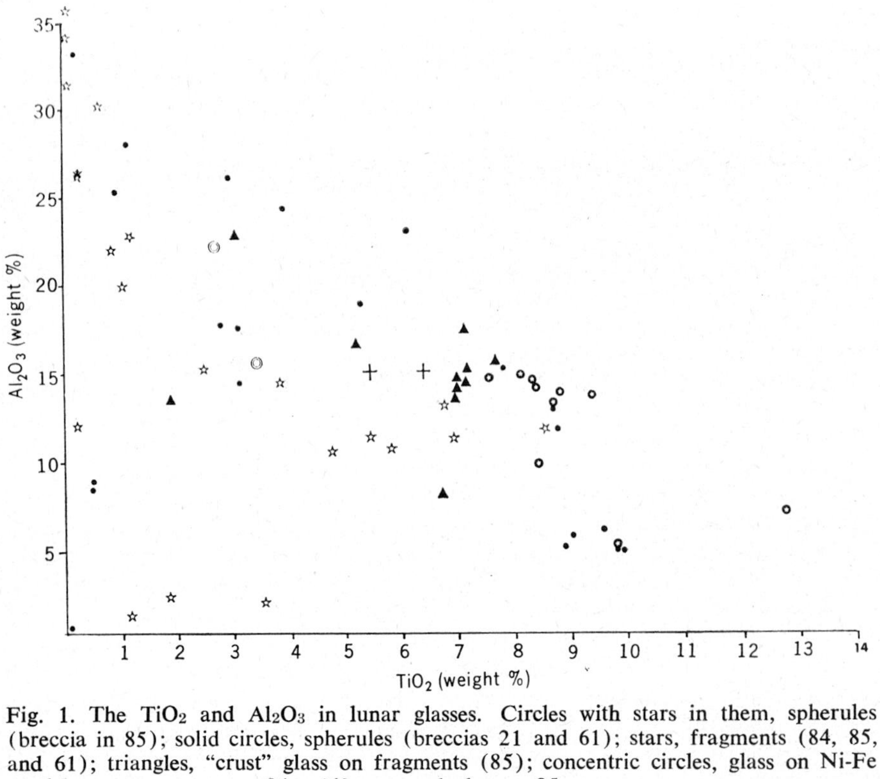

Fig. 1. The TiO_2 and Al_2O_3 in lunar glasses. Circles with stars in them, spherules (breccia in 85); solid circles, spherules (breccias 21 and 61); stars, fragments (84, 85, and 61); triangles, "crust" glass on fragments (85); concentric circles, glass on Ni-Fe particle; crosses, average 84, <43 μm, and glasses, 85.

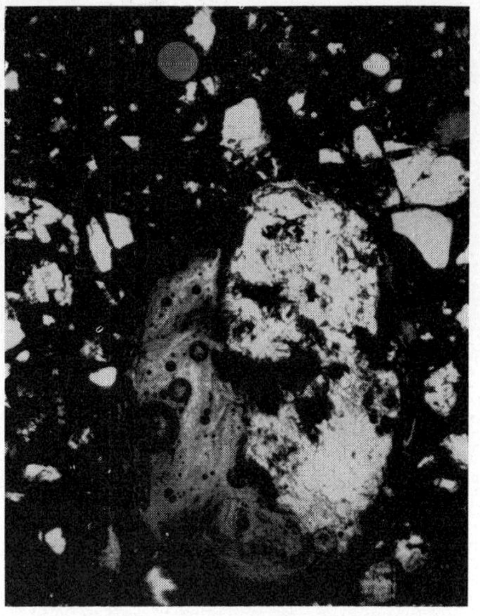

Fig. 2. The left part plus the entire rim of the 0.4-mm long, oval bead is yellowish glass with darker flow lines. The right, lighter part is severely shocked pyroxene and plagioclase. The ilmenite, center, is partly dissolved in the glass causing brown coloring. At top center is a typical, brown glass spherule.

pared to chondrites, although their composition apparently eliminates the moon as a source (*5*) for any type of chondrite.

Our soil sample 84,22 was somewhat finer, that is, 50 percent < 43 μm than reported (*1*). In the size fraction 0.25 to 0.5 mm we found, among 596 grains studied, 37 percent igneous rock fragments similar to types A and B (*1*), 6 percent other igneous rock types, 38 percent breccias, 10 percent "bubbly" glasses with crystallites and/or inclusions, and 9 percent clear glasses including spherules. The fraction 70 to 100 μm is similar and x-ray diffraction analysis (*6*) showed little variation with grain size. Bulk analysis of the fines (<43 μm) in 84 and of randomly selected glasses from 85 (*6*) are almost identical with the soil (*1*). However, preliminary electron-probe analysis of the matrix in breccias 21 and 61 indicates lower Ti/Al, Ti/Si, and Fe/Si ratios than in any bulk samples.

Among the rock fragments are several with low contents of opaque minerals, especially ilmenite. Some are also rather coarse grained and may be similar to rocks apparently present in the Apollo 12 samples. Sample 85,17 (*6*) contained a 4-mm, lens shaped, apparently meteoritic, nickel-iron particle (see cover). Its surface shows craters caused by high velocity impact of high TiO_2 lunar matter which has been transformed to glasses similar to those described above (Fig. 1). Apparently secondary impacts may play an important role in modifying the lunar surface.

Most minerals in the soil are similar to those in the crystalline rocks. However, we have also found unusually pure anorthite (< 0.1 Na_2O), and a pyroxene displaying 10 to 20 μm-wide exsolutions. The host pyroxene is $En_{43}Fs_{15}Wo_{42}$ with 1 to 3 percent TiO_2 and Al_2O_3; the lamellae are $En_{64}Fs_{36}$. The metal and sulfide grains particularly in the glassy fragments have more Ni (up to 12 and 2 percent, respectively) than do the crystalline rocks. Whereas the latter may have been reduced out *in situ*, the metal in the glassy fragments and mostly in the breccias is probably of meteoritic origin.

The crystalline rocks have a narrow range in chemistry (*1*), mineralogy and modal composition (*2*) but vary in grain size and texture. We have analyzed individual minerals in rocks 3, 20, 22, 47, and 50. However, consultation with several colleagues indicates that, with a few exceptions, our data have no exceptional trends or implications. An Fe-rich pyroxene-like mineral, $En_4Fs_{78}Wo_{17}$, was found in 47 (and in the soil), and a trace of pigeonite, $En_{53}Fs_{39}Wo_8$, in rock 3. Olivines in 20 and in a type B fragment were zoned with sharp increase in Fe by about 15 percent in the outer 30 to 40 μm of grains 0.2 to 0.4 mm in size. The centers were rather homogeneous with Fa_{30}. The CaO content is relatively high ~ 0.4 percent and Cr enriched to about 0.15 weight percent in the center parts. Metal in 3 and 47 is low in Ni but has ~ 0.5 percent Co. We learned (*7*) that the metal in 22 had about 1 percent Ni; we found 0.6 and 1.7 percent Ni, and 1.2 and 2.6 percent Co, respectively, in two grains both minute exsolutions in troilite (Ni, Co, Mn, Cr, < 0.05 percent).

Some of our ion-probe trace element analysis in individual phases in one type B rock fragment are summarized (Fig. 3). It must be emphasized, however, that large variations have been observed. For example, 1 ppm (atomic) Ba and 100 ppm K were found in plagioclase from type B while a grain in the soil had ~ 1500 ppm Ba and 500 ppm K. Similarly B and Ba vary by a factor of 40 or more, and Zr by a factor of 20, in

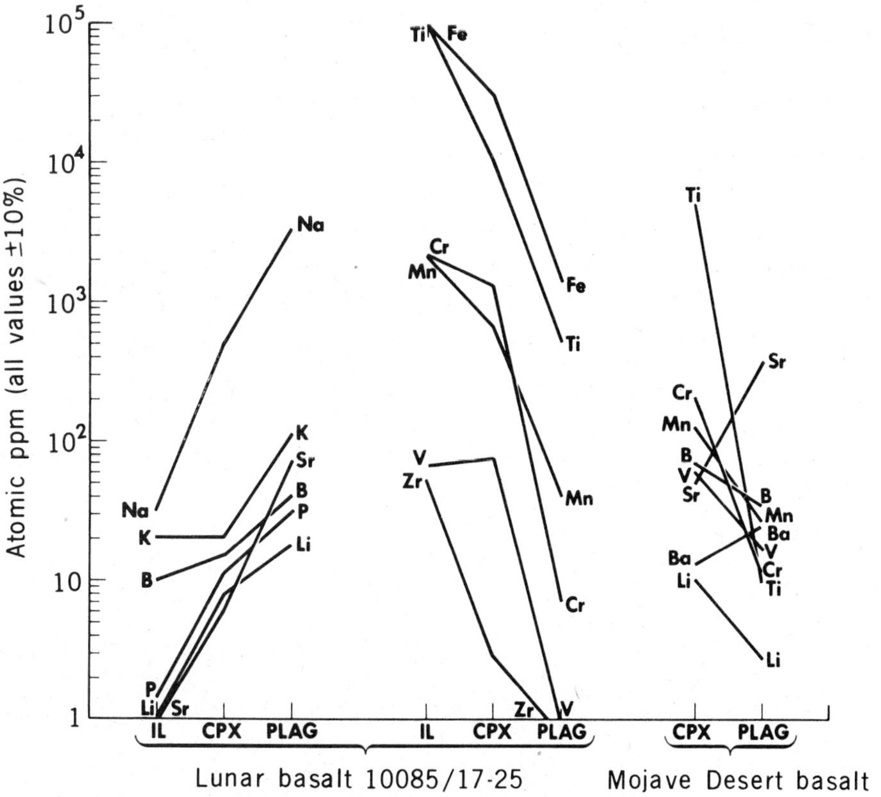

Fig. 3. Trace element distribution in the main minerals of a lunar rock, type B, and a terrestrial basalt. Elements in the lunar sample showing similar trends are grouped together for convenience. IL, ilmenite; CPX, clinopyroxene; PLAG; plagioclase.

otherwise similar pyroxenes from different rocks. This indicates that these minerals have different sources, that is parent magmas.

The origin of the parent magma for the crystalline rocks is uncertain. It may have been derived by partial fusion at depth caused by radiogenic heating or perhaps by melting triggered by impact of very large bodies. In any event, the textures are typical of igneous rocks which cooled mainly under static conditions. It is of course possible that the magma from which they crystallized flowed a long distance before crystallization began, as is the case in many terrestrial basalts. The high TiO_2 contents and other composition characteristics of the lunar crystalline rocks indicate that either the processes which produced them or their parent materials were quite unlike terrestrial processes and parent materials. If the great age (*1*) of these rocks holds, one can either presume that the moon was accreted from already differentiated material, or that early in the moon's (and earth's) history large numbers of impacts were a major geological agent creating superheated melts over considerable areas. Similar origins have been suggested for chondrites, that is, that they were created by impacts on relatively small, low density (compared to the moon) bodies (*8*).

K. FREDRIKSSON, J. NELEN
W. G. MELSON, E. P. HENDERSON
Smithsonian Institution,
Washington, D.C. 20560
C. A. ANDERSEN
Hassler Research Center,
Applied Research Laboratories,
Goleta, California 77058

References and Notes

1. Lunar Sample Preliminary Examination Team, *Science* **165**, 1211 (1969).
2. L. S. Walter, *Geochim. Cosmochim. Acta* **31**, 2043 (1967).
3. C. A. Andersen, *J. Mass Spec. Ion Phys.* **2**, 61 (1969); *ibid.*, in press.
4. K. Fredriksson and A. M. Reid, *Science* **149**, 856 (1965).
5. H. Wänke, *Z. Naturforsch.* **21a**, 93 (1966).
6. B. Mason, K. Fredriksson, E. Jarosewich, W. G. Melson, E. P. Henderson, J. S. White, Jr., *Science*, this issue.
7. O. Weill, personal communication.
8. K. Fredriksson, *Trans. N.Y. Acad. Sci.* **25**, 756 (1963).
9. We thank G. Moreland who prepared a number of outstanding polished thin sections without which much of this work would hardly have been possible. Also we are indebted to a number of principal investigators and colleagues who openly and generously discussed their results. J. Hinthorne assisted with the ion probe analyses and B. J. Fredriksson with electron probe data evaluation. Supported in part by NASA, contract NAS 9-8015.

5 January 1970

Deformation of Silicates from the Sea of Tranquillity

Abstract. *Plagioclase and olivine crystals in the crystalline rocks from the Sea of Tranquillity show little or no evidence of either static or dynamic deformation. The large disorientations in many of the pyroxene crystals are commonly consistent with slip on the system* T = {100}, t = [001], *but these distortions are not due to plastic flow. They are ascribed to rapid growth and quenching phenomena as deduced from studies of chondrules and of quenched natural and experimentally produced melts. Some of the silicates in the breccias and regolith show evidence of shock deformation, from mild to intense, as indicated by pervasive fracturing, shock lamallae, and partial transformation of pyroxene and plagioclase crystals to glass.*

Recent static experiments on silica and silicates have demonstrated that the plastic flow mechanisms vary with temperature, with strain rate, and, for some materials, with pressure. The slip mechanisms observed at the higher temperatures and lower strain rates have been successfully correlated with those operative in natural slow deformations. The primary objective of the present investigation of silicates returned from the Sea of Tranquillity was to determine the presence or absence of such deformation features. Given a representative sample from various provinces of the lunar surface, such studies should discern the importance of tectonic activity during the history of the moon. In addition, we have examined the lunar material for evidence of dynamic deformation due to the impact of foreign bodies. In our studies to date we have used primarily optical microscopy and x-ray Laue and precession methods.

Plagioclase in euhedral and subhedral crystals is a common constituent of both the fine- and coarse-grained igneous rocks but shows little or no evidence either of static or dynamic deformation. Thin plagioclase laths are commonly bent slightly, but such bending is also evident in thin laths from experimentally and naturally quenched terrestrial basalts. Olivine crystals in the fine-grained igneous rocks show no deformation whatever. Most of the pyroxenes (chiefly augite), however, show large distortions within individual grains. The disorientations resemble, in many respects, those produced by both static and dynamic deformation of pyroxenes and will be discussed in more detail below.

Three types of disorientations are observed in the pyroxene grains: (i) uniform sweeping undulatory extinction (Fig. 1a) over angles ranging to 100°; (ii) discontinuous changes in extinction with subboundaries separating zones of uniform extinction—somewhat similar to kink bands or polygonization in crystals; and (iii) more irregular disorientations in which the grains are composed of mosaics of irregularly shaped blocks having different orientations. Universal stage measurements of the variation in crystal orientations in grains showing disorientations of types i and ii generally indicate that an axis near [010] (rotation axis) is common to the various subgrains with [001] nearly normal to zones of undulatory extinction or subgrain boundaries. If the bending is interpreted as being due to plastic flow, the slip system T (slip plane) = {100} and t (slip direction) = [001] is inferred from the measurements. The slip system {100} [001] has been observed in clinopyroxenes deformed experimentally at high temperatures (*1*).

Six small grains extracted from both the crystalline rocks and regolith have been examined by x-ray Laue and precession methods. All of these crystals except one show some disorientation or mosaic structure; the disorientations differ in degree, but the rotational patterns are similar. In each crystal, rotations are greatest about the [010] axis, as is shown by a representative precession photograph in Fig. 2. Most of the Laue and precession photographs also show complexities that are not yet fully understood.

The more irregular disorientations (type iii) do not exhibit any simple crystallographic relationship between subgrains. For the simplest of these, the "hour glass" structure, the various zones tend to share the [010] axis, but the boundaries between zones are not, in general, rational. Optic angle-measurements suggest that the disorientations generally are not related to changes in

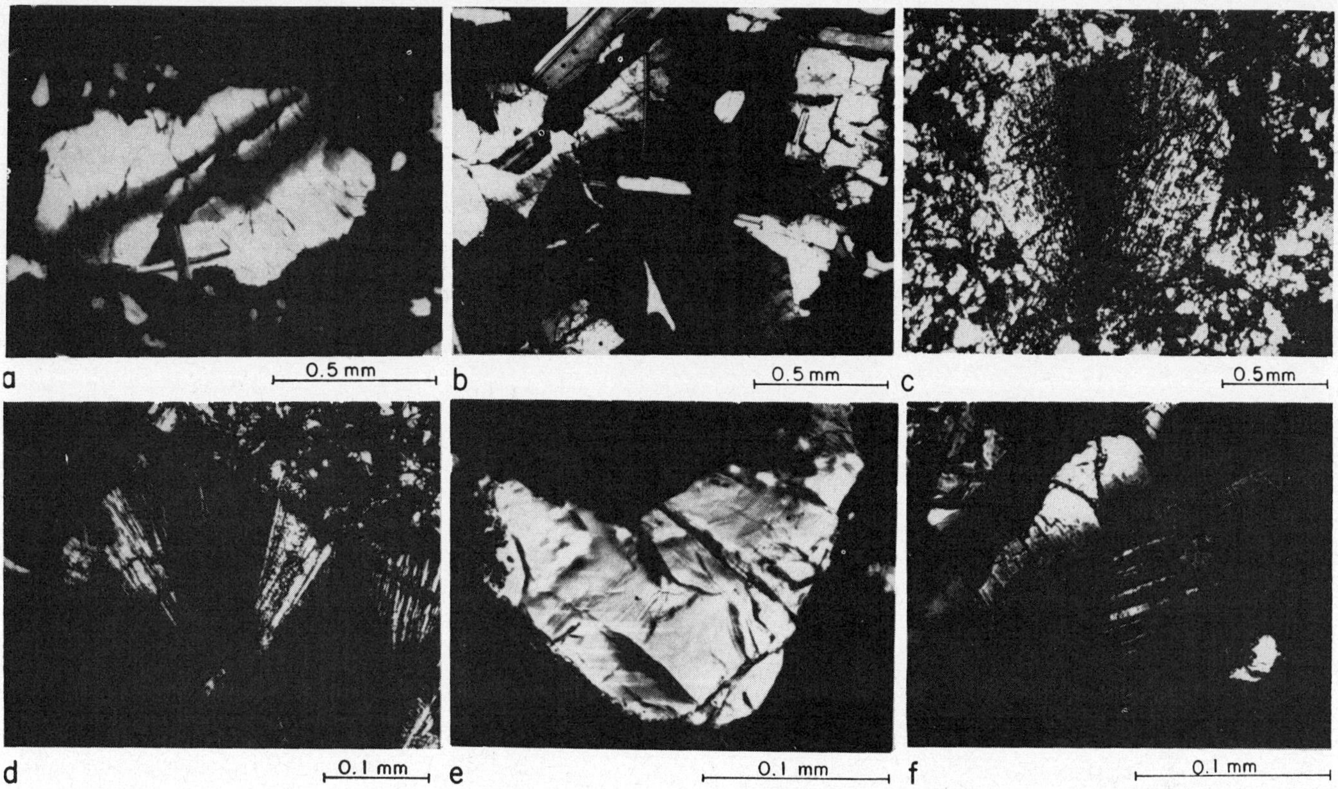

Fig. 1. Photomicrographs taken between crossed polarizers. (a) Pyroxene showing strong undulatory extinction from beccia 10046. (b) Distorted pyroxene grain containing plagioclase laths in rock 10047. Note the thin vertical plagioclase lath in highly distorted area of pyroxene near top of photograph. (c) Pyroxene from Pantar olivine-bronzite chondrite. (d) Radiating fibers of enstatite in rapidly quenched experimentally melted pyroxene rock. (e) Shock lamellae (light, east-west linear features) in pyroxene from lunar regolith sample 10085. (f) Deformation twins (light northeast-trending bands) in pyroxene in fine-grained crystalline rock in lunar regolith sample 10085.

chemical composition. That is, the optic angle varies as much within a single optically homogeneous zone as it does from zone to zone. Confirmation of this conclusion has come from electron-microprobe determinations of Ca, Mg, and Fe distributions in two adjacent highly disoriented zones having the same optical angle (*2*). The variations in these elements are large within each zone but show no significant differences between zones. Such chemical inhomogeneities within small domains may well contribute to the complexities observed in the x-ray photographs. The mosaic structure in this class of pyroxenes (type iii) is distinctly different from the mosaic structure produced in pyroxenes experimentally shocked to high pressures (*3*).

Several lines of evidence indicate that the disorientations in the pyroxenes are not due to plastic deformation. First, there is no evidence of deformation in the other silicates; in particular, olivine, which deforms more readily than pyroxene under both static and dynamic conditions, is totally undeformed. Second, the dominant mode of plastic deformation of monoclinic pyroxenes is twin gliding; only a few broad growth twins were observed in the crystalline rocks. Finally, straight, thin plagioclase laths such as the vertical lath near the top of Fig. 1b are undisturbed in highly bent regions of the pyroxene. Such laths should be uniformly bent or broken if the distortions in the pyroxene are the result of plastic deformation.

Studies of pyroxene chondrules in chondritic meteorites, of similar structures in rapidly quenched experimentally produced melts of pyroxene-bearing rocks, and of rapidly quenched terrestrial basalts suggest that the disorientations in the pyroxenes may be accounted for by rapid growth during quenching and subsequent annealing. Figure 1c shows radiating fibers of pyroxene in a chondrule from the Pantar olivine bronzite chondrite, a structure typical of pyroxene chondrules in many chondrites. A similar radiating structure was observed (Fig. 1d) in a natural eclogite melted experimentally at 1100°C, 15 kb confining pressure at a constant strain rate of 10^{-4}/sec and then quenched rapidly. In that specimen, the enstatite fibers are rotated about [010], as is true also of some of the pyroxene chondrules measured. However, the rotation axes in some of the radiating chondrules are not rational as might be expected from their more complex three-dimensional geometry (*4*). Undulatory extinction very similar to that shown in Fig. 1a has been observed in an augite crystal in molten basalt withdrawn from Kilauea Iki in mullite tubes and quenched rapidly. Again, the axis of rotation is [010], and [001] is normal to zones of undulatory extinction. Thin laths of plagioclase, some of which are bent slightly, as in the lunar samples, were also observed in those specimens.

We envision the growth process giving rise to disorientations in the pyroxenes in the coarse-grained rocks as shown schematically in Fig. 3. The dominantly pyroxene melt interstitial to the plagioclase-ilmenite network is quenched rapidly, the fibers or platelets nucleating on the framework and growing rapidly preferentially nearly parallel to the {001} plane (Fig. 3, left). This process gives rise to the preferred optical and crystallographic relations shown in Fig. 3 (center),

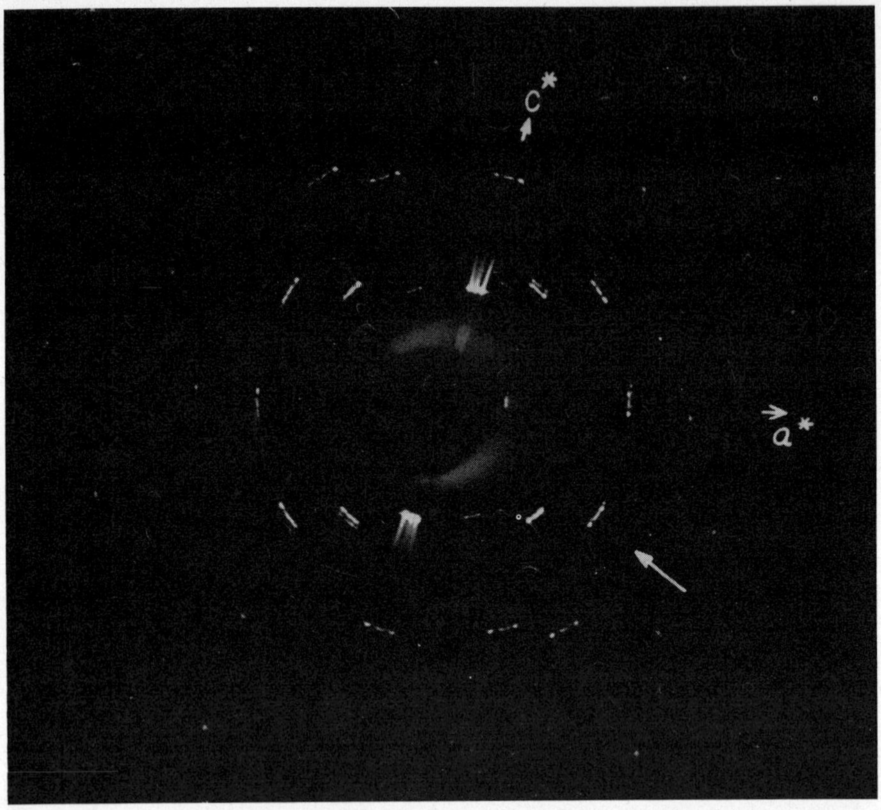

Fig. 2. Precession photograph of *hol* net of a pyroxene grain selected from sample 10084 (fines). Slight displacements of weaker sets of spots in directions nearly normal to the arrow indicate complex structural domains. Zr-filtered Mo radiation (45 kv/20 ma), 66-hour exposure.

although the relations are generally not so simple, as is revealed by the x-ray precession studies. Subsequent annealing permits the fibers or platelets to coalesce into subgrains separated by walls of dislocations as shown in Fig. 3 (right). Study of grains showing the uniform undulatory extinction using phase contrast illumination reveals the presence of many such low-angle subboundaries.

The irregular disorientations in the pyroxenes arise as a natural consequence of the growth scheme proposed. Thus, the "hourglass" structure is explained as having nucleated at the central point of the "hourglass" with radiating growth along oppositely directed conical surfaces. The more irregular mosaics may be the result of sections cut through the pyroxenes at large inclinations to the cone axes. Such irregular disorientations are quite common in thin sections of chondrules and are due chiefly to the orientation of the thin section relative to that of the chondrule (*4*).

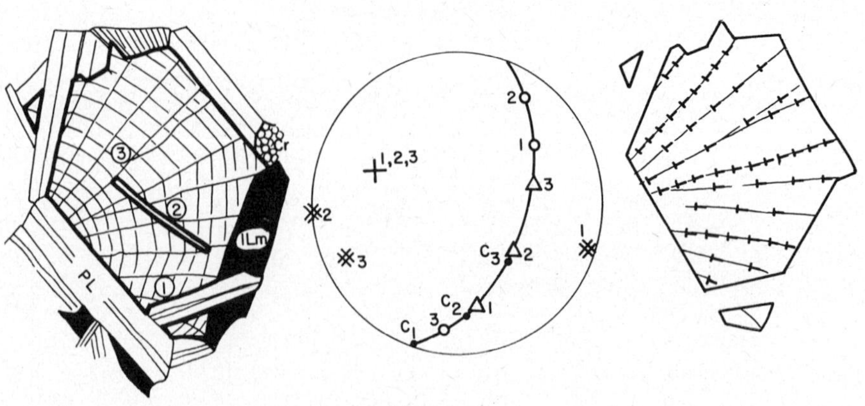

Fig. 3. Schematic diagram showing proposed origin of uniform disorientations in the pyroxene crystals. Pyroxene-rich melt confined in plagioclase-ilmenite framework crystallized rapidly during quenching (left), giving idealized rotations shown in center stereogram. Subsequent annealing causes coalescence of crystallites separated by dislocation walls.

The silicates in the lunar breccias and regolith are derived mainly from the crystalline rocks and hence show all of the structures described above. In addition, some of them show evidence of mild to intense deformation due to shock by impact of foreign bodies. The evidence for such deformation is discussed in detail by others and will be treated only briefly here.

Pervasive fracturing in chips of crystalline rock within the breccias and regolith and angular fragments of silicates gives evidence of mild shock (*3*, *5*). Shock lamellae (glass) parallel to {001} in some pyroxene (Fig. 1*e*) and plagioclase grains indicate moderate shock deformation, as may mechanically induced pericline twins in plagioclase. Deformation twins parallel to {001} were observed in some of the pyroxenes (Fig. 1*f*); twins of this orientation have been produced in static experiments at low temperature or high strain rates or both (*1*) and by shock deformation experiments (*6*). Slight undulatory extinction in two olivine grains produced by slip parallel to [100] and the {001} twins in the pyroxenes may be ascribed either to static or to dynamic plastic deformation. However, the absence of such deformation in the crystalline rocks suggests that these features were produced by shock, although slip parallel to [001] is most common in shocked olivine (*3*). A few of the plagioclase and pyroxene crystals appear to be partially transformed to glass, suggesting intense shock (*5*).

We conclude that all of the features in the silicates in the crystalline rocks studied by us can be ascribed to growth, quenching, and annealing processes; there is no substantive evidence of static plastic deformation. Shock deformation due to impact, varying from mild to intense, has been operative and may well be the primary disintegrating process giving rise to the lunar regolith.

N. L. Carter
I. S. Leung
H. G. Ave'Lallemant
Department of Geology and Geophysics, Yale University, New Haven, Connecticut

References and Notes

1. D. T. Griggs, F. J. Turner, H. C. Heard, *Geol. Soc. Am. Mem.* **79**, 39 (1960); C. B. Raleigh, J. L. Talbot, *Am. J. Sci.* **265**, 151 (1967).
2. L. Fernandez, personal communication.

3. N. L. Carter, C. B. Raleigh, P. S. DeCarli, *J. Geophys. Res.* **73**, 5439 (1968).
4. K. Fredriksson, *Trans. N.Y. Acad. Sci.* **25**, 756 (1963).
5. E. C. T. Chao, *Shock Metamorphism of Natural Materials* (Mono, Baltimore, 1968), p. 135.
6. C. B. Sclar, personal communication.
7. We are indebted to L. Fernandez for determining the distributions of several elements in a disoriented pyroxene, E. Roedder for use of his photograph published in our Fig. 1a and to B. Skinner for the loan of specimens from Kilauea Iki. Supported by NASA contract NAS 9-9937.

4 January 1970

Shock Metamorphism in Lunar Samples

Abstract. *Indications of shock metamorphism produced by pressures up to the megabar region have been observed in the fine material and the breccias, but very rarely in the coarser fragments of crystalline rocks. These indications are deformation structures in plagioclase and pyroxene, diaplectic plagioclase glasses, and glasses formed by shock-induced melting of lunar rocks. Two sources of shock waves have been distinguished: primary impact of meteorites and secondary impact of crater ejecta. There are two major chemical types of shock-induced melts. The differences in chemistry may be related to impact sites in mare and highland areas.*

The following shock-induced transformations produced by solid-state reactions and melting have been observed in breccias (59, 60, 65, and 27) and fine material (85, 84), besides weaker effects of plastic deformation and fracturing probably due to shock: (i) deformation structures in plagioclase and pyroxene; (ii) diaplectic feldspar glasses; and (iii) glasses produced by shock-induced melting.

Plagioclase grains with lamellae of low index of refraction and low or no birefringence, which might be classed with those from terrestrial meteorite craters and shock experiments, are very rare in the samples studied. For the most part they display only one set of lamellae that are isotropic or of low birefringence. Apparently, the conditions (duration of shock waves, composition of plagioclase, and texture of lunar rocks) do not favor the formation of these structures. This is confirmed by the absence of deformation lamellae in the shocked rock fragment described below.

Very few of the pyroxene grains found in the fine material exhibit multiple sets of closely spaced lamellae similar to those that are well known from shocked quartz.

In the fine material and in the breccias, colorless and isotropic grains of plagioclase composition occur which exhibit no indications of fusion, such as vesicles or flow structures. Some of them show straight grain boundaries and cleavage. Microprobe analyses and refractive index measurements were made on 14 grains, isolated from samples 84,106; 85,25; and 85,26 (Table 1). The densities of six grains were also determined. Figure 1 demonstrates that the refractive indices of these glasses are distinctively higher than those of normal glasses produced by fusion. Likewise, the density is higher than that of fused plagioclase. These observations show that the glasses are not fusion products but are diaplectic glasses (*1*) formed in the solid state by shock waves with amplitudes between about 300 kb and about 500 kb. They have been produced by meteoritic impact on plagioclase-bearing crystalline rocks of the lunar surface which were broken during excavation into fragmental mineral grains. Fragments of such rocks with diaplectic plagioclase glass (very seldom with small isotropic alkali feldspar inclusions) in its original paragenesis with pyroxene and ilmenite are rarely found in the fines and breccias. Diaplectic plagioclase glasses are also known from terrestrial impact craters (*2*), from the Shergotty meteorite (*3*), and from shock wave experiments (*3*). To check further on the diaplectic nature of the lunar glasses, annealing experiments were carried out on various fragments of a large grain (Fig. 2). Annealing at 700° and 800°C increased the index of refraction. The beginning of recrystallization was observed at 900°C, and complete recrystallization at 1000°C. The same increase of refraction on annealing was observed on diaplectic plagioclase glass of the Manicouagan crater (*4*), whereas fused plagioclase shows no change on annealing at temperatures below the transformation point.

Table 1. Refractive index n_D, density d, and anorthite (An) content of lunar diaplectic plagioclase glasses.

No.	An (mole percent)	n_D	d
1	74	1.5620	2.650
2	75	1.5651	
3	75	1.5675	2.653
4	83	1.5702	
5	86	1.5709	
6	86	1.5720	
7	86	1.5760	2.673
8	88	1.5747	2.684
9	88	1.5762	
10	89	1.5716	
11	89	1.5772	
12	89	1.5797	2.684
13	91	1.5802	2.684
14	93	1.5807	

Glasses in the fine material and in the breccias occur as (i) bodies of regular spheroidal, ellipsoidal, dumbbell, or teardrop shape, (ii) irregular fragments, and (iii) vesicular coatings on fragments of breccias and crystalline rocks.

1) The spheroidal, ellipsoidal, dumbbell, and teardrop glass bodies form about 10 percent of all glass in the fine material. Their sizes are in the range between 2 mm and 0.3 μm. Red, brownish, and yellow glass bodies are most frequent. About half as many are colorless or greenish. Violet-brownish-colored spherules are rarer. The glasses are generally homogeneous. Some contain vesicles. Schlieren and inclusions of mineral grains (plagioclase, pyroxene), often partially fused, are rarer. Very small spherules of metallic Fe occur in some glasses. Apparently the very regular shape of these bodies and their rotational symmetry resulted from an equilibrium between surface tension and inertial forces acting on liquid drops rotating and falling in the lunar vacuum. Divitrification is rare, but a few devitrified spherules have been found which resemble orthopyroxene chondrules of chondrites. The homogeneity of most of these glasses indicates their initially high temperatures. The rarity of devitrification indicates the predominantly rapid rates of cooling.

2) There are colorless and greenish irregular fragments and about as many red, yellow, and brown pieces. Violet-brownish colors are rarer. Some of these glasses are homogeneous. Glasses containing vesicles, mineral fragments, and flow structures are more frequent than in group 1. Minute Fe spherules are rather frequent in this kind of glass. Devitrified glass fragments with dendritic and skeletal crystals have been frequently found in the breccias.

Microprobe analyses of 13 rounded glass bodies (group 1) and 26 glass fragments (group 2) showed a broad

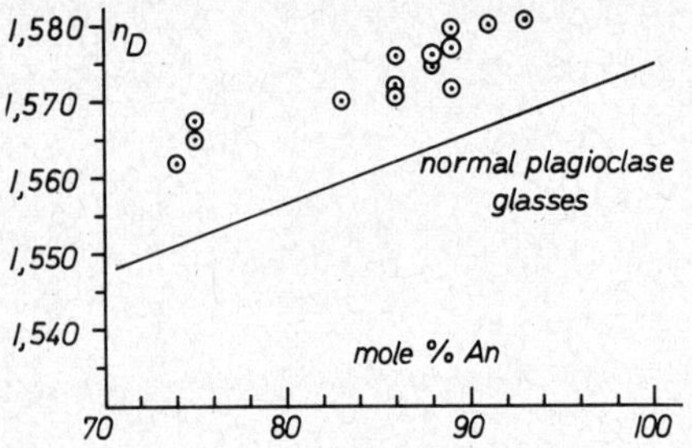

Fig. 1. Refractive indices of lunar diaplectic plagioclase glasses. (Solid line) Refractive indices of fused plagioclase. [After Barth (5)]

variability of chemical composition, with the SiO_2 content ranging from 33 to 50 percent. Each individual piece of glass can be interpreted as a fusion product of a particular mixture of the lunar rock-forming minerals plagioclase, pyroxene, and ilmenite, in some cases with admixtures of free SiO_2, olivine, or even metallic Fe.

The chemical heterogeneity of the lunar glass bodies and fragments distinguishes lunar fine material and breccias from terrestrial pyroclastic rocks. The individual glass pieces constituting the latter are all of essentially the same chemical composition. Because of their heterogeneity, it is concluded that the lunar glass bodies and fragments did not originate in large and homogeneous lava pools. Instead, they must have been formed, individually, by fusions of relatively small rock volumes.

Notwithstanding the chemical differences between the individual glasses, at least two major chemical types can be distinguished.

Type 1: colorless to green glasses. Average composition (weight percent; 13 microprobe analyses): SiO_2, 46; TiO_2, 0.8; Al_2O_3, 24; FeO, 7; MnO, 0.08; MgO, 8; CaO, 13; Na_2O, 0.6; and K_2O, 0.1. Average norm: plagioclase, 67; pyroxene, 21; olivine, 9; and ilmenite, 1.5.

Type 2: yellow, brown, red, and violet glasses. Average composition (13 microprobe analyses): SiO_2, 40; TiO_2, 9; Al_2O_3, 11; FeO, 19; MnO, 0.2; MgO, 10; CaO, 11; Na_2O, 0.5; and K_2O, 0.1. Average norm: plagioclase, 32; pyroxene, 35; olivine, 16; and ilmenite, 17.

Generally, spherules and the other regular bodies of both types contain less SiO_2 than the corresponding fragments, so that calculation of the norm results in relatively high percentages of olivine and, in the case of two green spherules, even of metallic Fe (2 and 4 percent). Metallic Fe occurs in many lunar glasses as minute Fe spherules. There are two possible explanations for their formation: (i) admixture of meteoritic Fe, condensing from the vapor phase, produced by the impacts of Fe-bearing meteorites, or (ii) preferential volatilization of SiO_2 (or SiO) from a rock-melt heated to very high temperatures in the lunar vacuum.

3) Coatings of vesicular glass are frequent on breccia and rock fragments occurring in breccias and fine material. The glasses of this kind so far investigated belong to type 2. Microprobe analyses of some glassy coatings showed that the crust was not formed by fusion of the coated rock and has taken up, at most, minor amounts of it. The glass crusts may be considered in most cases projectiles ejected from meteorite craters as splashes of melt or small rock fragments which have impacted rock fragments on the lunar surface, causing progressive shock metamorphism within them. This could be best demonstrated by a small gabbroidic fragment ($10 \times 6 \times 4$ mm). In the immediate vicinity of the glass crust (up to 0.3 mm thick), within a zone of 2.5 mm maximum thickness, all plagioclase is transformed into isotropic diaplectic glass. A very narrow transition zone (maximum thickness, 0.2 mm) with partially isotropic plagioclase separates the outer zone from the interior of the rock, where the plagioclase shows normal birefringence.

The boundary between diaplectic glass and birefringent plagioclase represents a shock wave amplitude of about 300 kb. The small extension of the transition zone indicates a rapid decay of the shock wave along its path into the rock. Therefore, the thickness of the impacting projectile must have been rather small, probably not more than 2 mm.

We conclude that all types of glasses have been produced by shock-induced fusion of lunar rocks due to meteoritic impacts, presumably at pressures up to the megabar region. Some parts of the melt have been dispersed as fine droplets which are preserved as regularly shaped bodies of rotational symmetry. Larger masses have been broken into fragmental grains, and still other parts of impact melts are found as crusts adhering to the surface of rocks.

The obvious division of the glasses into two major chemical types indicates

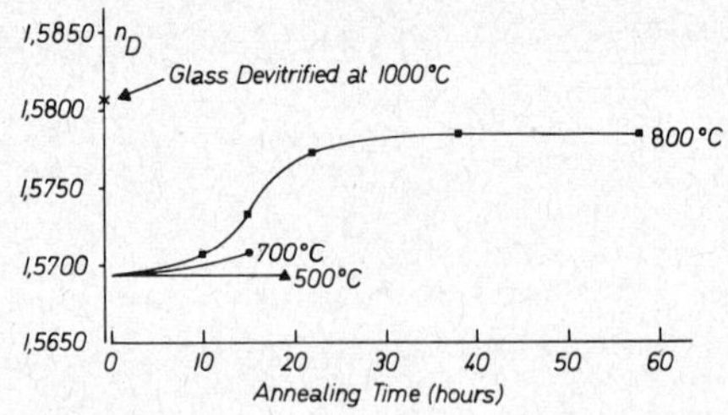

Fig. 2. Results of annealing experiments with lunar diaplectic plagioclase glass (anorthite, 86 mole percent).

that, at the Apollo 11 landing site, material occurs which has been produced by meteoritic impact events in at least two areas of the lunar surface which differ in chemical composition. The yellow, brown, red, and violet glasses of high TiO_2 and FeO content (type 2) have a composition similar to that of mare material represented by the alpha-scattering experiments of Surveyor 5 and 6 and the rocks of Apollo 11. They may be products of impacts within the Sea of Tranquillity area. The colorless and greenish glasses of higher SiO_2 and Al_2O_3 and lower TiO_2 and FeO content (type 1) correspond to the highland material north of Tycho analyzed by Surveyor 7. We suggest that these glasses came from an impact site at the highlands.

The two types of shock-melted material may perhaps be connected with the two ray systems in the vicinity of the Apollo 11 landing site. The north-northeast-trending ray, which is perhaps related to Tycho or another crater on the highlands, may have delivered the colorless and greenish glasses of type 1.

W. von Engelhardt
J. Arndt
W. F. Müller, D. Stöffler
Mineralogisch-Petrographisches Institut, University of Tübingen, Tübingen, Federal Republic of Germany

References and Notes

1. W. von Engelhardt and D. Stöffler, in *Shock Metamorphism of Natural Materials*, B. M. French and N. M. Short, Eds. (Mono, Baltimore, 1968), pp. 159–168.
2. See *Shock Metamorphism of Natural Materials*, B. M. French and N. M. Short, Eds. (Mono, Baltimore, 1968).
3. D. J. Milton and P. S. DeCarli, *Science* **140**, 670 (1963).
4. T. E. Bunch, A. J. Cohen, M. R. Dence, in *Shock Metamorphism of Natural Materials*, B. M. French and N. M. Short, Eds. (Mono, Baltimore, 1968), pp. 509–518.
5. T. F. W. Barth, *Feldspars* (Wiley, New York, 1969), p. 167.
6. We acknowledge financial support from the "Ministerium für Bildung und Wissenschaft" of the Federal Republic of Germany.

4 January 1970

Impact Metamorphism of Lunar Surface Materials

Abstract. *Silicate grains from Tranquillity Base have shock-induced features ranging from internal fragmentation through complete disruption of the lattice to thermal melting. Half the crystalline grains with diameters less than 125 micrometers have features of shock equivalent to those produced in the laboratory at pressures greater than about 40 kilobars. One quarter have features indicative of pressures greater than 90 kilobars. These properties together with great quantities of melt glass and aluminum-26 in the fine-grained material are indicative of repeated shock-loading by meteoritic bombardment over long periods of time.*

Samples of granular material with diameters less than 1 cm (10084,99; 10085,18) and thin sections of breccias (10023, 10060, 10061, 10065, 10073) and igneous rocks (10020, 10047) were studied by means of x-ray, microscopic, and electron microprobe techniques. In addition, fine material (10084,113) and igneous rocks (10003, 10017, 10057) were analyzed by means of γ-ray spectrometric methods.

All samples are part of a continuum of particles and blocks. The cumulative number of particles (N) in the fine material can be plotted as a power function of grain diameter (D), $N = KD^{-\alpha}$, where α equals 3 over the size range of 15 to 250 μm. Constituent particles consist of mineral grains, igneous rock fragments, glass grains, glass-bonded mineral aggregates, and breccias in proportions which are a function of grain size. Glass in discrete grains is the principal component of particles with diameters less than 5 μm, and igneous rock fragments are the main constituent in fractions with diameters greater than 500 μm. Glass-bonded mineral aggregates and mineral and igneous rock fragments in equal numbers are prevalent in intermediate sizes.

Mineral grains are of the same composition as components of the igneous rocks. Electron microprobe analyses of plagioclase range from anorthite $(An)_{63-97}$, but most grains are in the range An_{80-90}. Clinopyroxenes range in composition from augite and titanaugite to ferroaugite. A few grains of pigeonite, minor quantities of olivine and cristobalite, and large amounts of ilmenite are present. The mineral grains are fragments of igneous rocks. Glass-bonded aggregates and breccias, however, are aggregated components of previously comminuted material. Thus, the particulate distribution can be described simply, $N = KD^{-\alpha}$, but the process which produced it is complex, including both comminution and aggregation.

Impact has played a major role in producing and modifying the regolith (*1*). Its importance is particularly evident when considering the extent to which mineral grains have been metamorphosed by shock. Modification of plagioclase has been extensive, ranging from simple bending to complete disruption of the lattice. Undulatory extinction is evident in most grains. A number have bent cleavage planes, faulted twin lamellas, and planar features; a high proportion have been internally fragmented, fragmentation ranging from simple breaking into large mosaic blocks to extensive internal milling. The increase in internal fragmentation is accompanied by a decrease in index of refraction and birefringence.

Many optically isotropic grains are probably plagioclase vitrified by shock, as suggested by their indices of refraction. Two such grains from breccias 10061 and 10065 were analyzed and have the compositions An_{91} and An_{94}, characteristic of the plagioclase of igneous rock samples.

Clinopyroxene grains also show evidence of metamorphism by shock, but to a lesser degree than does the plagioclase. Many have bent and faulted cleavage planes, and a significant proportion have been internally fractured to various degrees. A few have been vitrified by shock.

Internal fragmentation of plagioclase grains is so commonplace that it has been used as a basis for evaluating the extent to which the crystalline material has been metamorphosed by shock. Single crystals of oligoclase were shock-loaded at various pressures, a ballistic range being used to provide standards for correlating the degree of internal fragmentation with pressure. The extent of fragmentation is revealed by the size of mosaic blocks and by the degree of asterism on Debye-Scherrer x-ray pho-

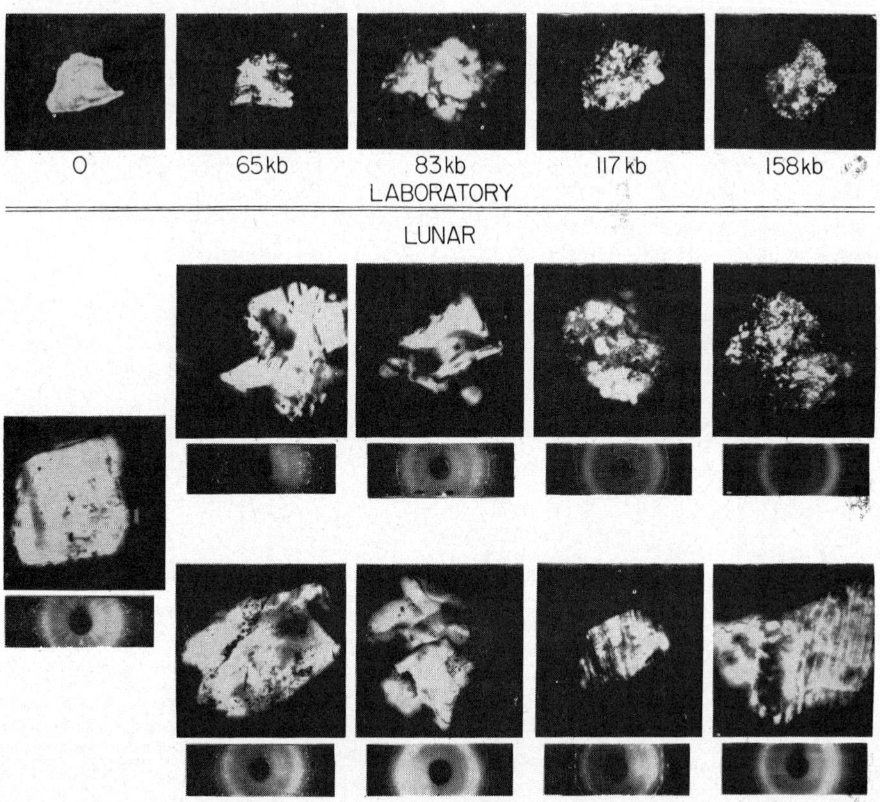

Fig. 1. Progressive shock-induced internal fragmentation of plagioclase as evidenced by reduction in optical domain size and an increase in x-ray asterism. Photomicrographs in the first row are of plagioclase shock-loaded to known pressures. Photomicrographs and front reflection portions of Debye-Scherrer x-ray patterns in the second and third rows are of plagioclase from the regolith. Grains illustrated have diameters between 63 and 125 μm.

tographs (*2*). Single crystals subjected to pressures between 30 to 40 kb are internally fragmented with mosaic blocks having average dimensions of 40 μm. Individual blocks have undulatory extinction and have been slightly rotated from their original lattice orientations, but relict cleavage and twin planes are recognizable. The x-ray diffraction pattern is simply more spotty than that of a single crystal. As pressures increase, domain size decreases (Fig. 1).

Original structural features are lost and blocks become more highly milled. X-ray patterns become more spotty with spots defining diffraction lines. Grains subjected to pressures of about 90 kb have typical powder patterns and mosaic blocks with average dimensions of 5 μm. At 150 kb, the mosaic is evenly granular with micron size domains with sutured boundaries.

Debye-Scherrer patterns of 120 plagioclase grains from the 63- to 125-μm fraction of the lunar fines were analyzed, and dimensions of the mosaic blocks in each grain were determined by means of a petrographic microscope with a micrometer eyepiece. Photomicrographs and front reflection x-ray patterns of selected grains are shown in the second and third rows of Fig. 1. Grains in the second row illustrate progressive internal granulation. Those shown in the third row also show progressive fragmentation, but the structural collapse has been controlled by the original lattice. The domains have a preferred orientation with x-ray patterns ranging from lens-shaped streaks to pinching and swelling rings. Interpretations of the pressure history based on x-ray asterism (*2*) indicate that 51.5 percent of the grains have been shocked to pressures greater than 40 kb, and 21.5 percent were subjected to pressures greater than 90 kb. Dimensions of mosaic blocks indicate similar histories: 49.5 percent were subjected to pressures greater than 40 kb, and 24 percent have witnessed pressures greater than 90 kb.

The breccias contain the same constituents in the same proportions as the granular debris. They have the same distributions of grain size and probably represent compressed clots of the regolith. The percentages of crystalline material subjected to pressures greater than 40 kb in four breccias range from 47.5 to 54.6 according to domain size criteria. The breccias thus appear to have been produced by shock waves of low-pressure amplitude which did not significantly increase the proportion of shock-damaged material.

Coarse grains in the breccias are igneous rock fragments. Many constituent minerals in the fragments have sharp extinction, and most have but slight undulatory extinction. Most of the fragments bear no evidence of shock deformation supporting the hypothesis that most of the breccias were formed by shock waves of low-pressure amplitude. Large igneous rock fragments contain little evidence of shock deformation. The proportion of shock-damaged material in the lunar regolith is an inverse function of grain size. This is compatible with the observation that the more highly shocked components of single impact events are concentrated in the more finely comminuted products and with the fact that multiple impact reduces grain size.

Glass in the fine debris and in the breccias occurs in various forms and colors. Angular grains are most common. Spheroidal grains are present in about the same proportion in each size class (1 to 2 percent). Colors range from colorless through shades of green and brown to red, reflecting differences in chemical composition as indicated by a range in refractive indices from 1.46 to 1.74 and a corresponding range in densities from less than 2.582 to 3.128. Electron microprobe analysis of representative samples confirms the correlation of color and composition. Clear glasses have high percentages of silica and alumina, and low percentages of iron, titania, and magnesia. The reverse is true of red glasses.

Some of the glass is of shock vitrification origin and retains the form of its parent mineral, but most solidified from a melt. Glass from glassy aggregates is often vesicular and sometimes has flow banding. Its average specific refractive energy (*3*) of 0.214 compares closely with the value of 0.219 calculated from the chemical analysis of the lunar fines (*4*). It may have been produced by melting of this widespread material. Percentages of major oxides in the glass scatter widely when plotted as a function of silica percentages, unlike samples from related volcanic rocks. A first generation volcanic origin is improbable. Rather, the compositions of the glasses can best be explained as representing different proportions of shock-melted mineral components of the lunar surface. Melting of surface materials by solar glazing may be theoretically possible (*5*), but evidence of shock pressures in the range below those required to melt silicate rocks, 600 to 1000 kb (*6*), is so extensive that no other process of melting is considered necessary to account for the quantities of glass observed.

Many features have been interpreted here as originating through meteoroid impact. Appropriately, nickel-iron grains with diameters less than 25 μm are present in all breccia and granular samples. One 2-mm fragment of polycrystalline kamacite (5.6 percent Ni) with a few included grains of schreibersite and a single grain of bronzite was found.

Materials of the regolith have witnessed a variety of shock pressures. Many grains bear no evidence of shock deformation, but the proportion that do is very significant when it is considered that debris from a single impact crater which exhibits shock damage is of the order of 1 percent of the total mass ejected (*7*). The regolith at Tranquillity Base, therefore, must have been subjected to extensively repeated shock-loading and reworking by impacting projectiles. The extent of the working of the surface is not surprising, however, in view of the very ancient solidification age of surface rocks (*5*). The very high content of ^{26}Al($T = 7.4 \times 10^5$ years) in the top few centimeters of the fine debris (14 dpm/kg) suggests that these materials have been near the surface for a minimum of several half-lives, and that time required for the surface to have been mixed to a depth of 1 m may have been of the order of tens of millions of years.

WILLIAM QUAIDE
THEODORE BUNCH
ROBERT WRIGLEY

Space Sciences Division, NASA, Ames Research Center, Moffett Field, California 94035

References

1. V. R. Oberbeck and W. L. Quaide, *Icarus* **9**, 446 (1968).
2. F. Horz and W. L. Quaide, *Abst. Meteoritical Soc., 32nd Ann. Meet.* (1969).
3. J. H. Gladstone and T. P. Dale, *Phil. Trans. Roy. Soc. London Ser. A* **153**, 37 (1864).
4. Lunar Sample Preliminary Examination Team, *Science* **165**, 1211 (1969).
5. T. Gold, *ibid.*, p. 1345.
6. D. Stoffler, *Contr. Mineral Petrol.* **16**, 51 (1967); E. C. T. Chao, *in Shock Metamorphism of Natural Materials,* B. M. French and N. M. Short, Eds. (Mono, Baltimore, 1968), p. 135.
7. W. L. Quaide and V. R. Oberbeck, *Earth Sci. Rev.* **5**, 255 (1969).

4 January 1970

Evidence and Implications of Shock Metamorphism in Lunar Samples

Abstract. *Lunar microbreccias and loose regolith materials contain abundant evidence of shock metamorphism related to crater-forming meteorite impacts. Diagnostic shock effects include (i) planar features in a silica phase and feldspars, and lamellae in clinopyroxene, (ii) thetomorphic feldspar glass, (iii) heterogeneous glasses of rock and mineral composition, (iv) distinctive recrystallization textures, and (v) characteristic changes in crystal structure as indicated by x-ray diffraction analysis and measurements of refractive index. The microbreccias are produced from regolith materials (ejected from craters) by shock lithification. Some feldspar-rich fragments may represent ejecta introduced from nonlocal sources, such as the lunar highlands.*

The most conspicuous features on the moon (and Mars) are circular depressions covering much of the exposed surface. Many planetologists regard them as meteorite impact craters, but others argue that they are dominantly volcanic forms. I have previously pointed out (*1*) that the origin of most lunar craters will ultimately be established by evidence from rock materials directly associated with these structures. If significant proportions of such rocks display signs of shock metamorphism, then correlation with impact events can be safely assumed (*2*).

Studies of ejecta from a small nuclear explosion crater in basalt (Danny Boy event) (*3, 4*) indicate the following effects to be specific indicators of shock damage in crystalline rocks of basic composition: (i) planar features in feldspar; (ii) feldspar glass thetomorphs (maskelynite); (iii) dispersion of fragmented (and variably decomposed) mafic minerals within shock-melted feldspar glass, thus leading to glasses of heterogeneous composition; (iv) intense alterations of crystal structure, revealed by measurements of refractive index, x-ray diffraction asterism, and infrared absorption. However, considering the relative resistance of rocks rich in mafic minerals to shock damage at lower (<300 kb) pressures (*5*), I concluded (*3*) that only 1 to 2 percent of a basaltic target rock surrounding a meteorite impact point would experience sufficiently high pressures to produce recognizable shock-metamorphic effects.

These observations apply directly to the lunar samples. I did not identify diagnostic shock effects in four crystalline rocks (types A and B) from the Apollo 11 site. If such samples were ever shocked, they must represent fragments originally located in the low-pressure region at varying distances from each impact point prior to ejection from nearby craters. Abundant shock-metamorphic effects were noted in lunar soil and microbreccias, thus indicating that these materials (i) contain variably shocked ejecta from positions closer to impact points of primary craters formed in the crystalline lavas of Mare Tranquillitatis, and (ii) are probably shocked further by repeated impacts or "gardening" during buildup of the lunar regolith (*6*). These effects, similar to those in shocked Danny Boy basalt and in nonbasaltic rocks from some terrestrial impact structures, are here documented.

Lamellar structures. Possible planar features (shock lamellae) (*7*) were detected in only six individual feldspar fragments during examination of 14 polished sections from 12 microbreccia samples. Planar features in two silica grains (phase unknown) (Fig. 1a) in a section from sample 10060-30 are definitely like those commonly observed in shocked terrestrial quartz. These grains displayed four and five discrete sets of planar features, respectively, when rotated on the universal stage. Numerous close-spaced, crisscrossing lamellar structures in two other fragments (sample 10046-12) (Fig. 1b), identified by microprobe analysis as plagioclase [An_{70-75} (70 to 75 mole percent anorthite)], meet some of the criteria for planar features defined by Carter (*8*). The fragments are untwinned, composite crystals, having a reddish-brown translucency and low birefringence typical of some highly shocked feldspars from terrestrial impact sites. The lamellae, however, also resemble certain quench crystallization structures. Poles to these lamellae, measured by universal stage techniques and plotted on a stereogram, tend to align in a girdle roughly normal to the *c*-axis.

The general scarcity of planar features in lunar feldspars, considering the wide range of shock damage observed in the fragmental samples, is consistent with studies of terrestrial samples (*5*) which indicate that these shock lamellae develop less readily in feldspars from basic igneous rocks than from quartz-bearing silicic rocks.

One clinopyroxene fragment (sample 10065-21) displayed three sets of closely spaced lamellae (deformation twins?) (Fig. 1c) which are oriented subparallel to (001), (100), and ($\bar{2}$21) planes. Under crossed Nicol prisms, this fragment shows a pronounced blotchy, granulose texture, characteristic of thermal recrystallization of heavily shocked mafic minerals in meteorites (*9*). Another clinopyroxene fragment (sample 10060-30) contains well-developed kink bands associated with a type of mosaic structure noted in mafic minerals (*9*).

Isotropization. Glassy phases that retain the initial shapes of shock-isotropized crystals (thetomorphs) were rare in the lunar samples I examined. An excellent example of thetomorphic feldspar glass occurs in a type B crystalline rock fragment (Fig. 1d) in microbreccia in which original feldspar laths are completely isotropic.

In another fragment (sample 10060-39) internal flow lines within feldspar thetomorphs with distorted crystal shapes, a marked lowering of birefringence in pinkish pyroxene, and a "blurring" of pyroxene and ilmenite boundaries into feldspar glass attest to more advanced breakdown of crystal structures accompanying the first stages of fusion. A single fragment from the soil, consisting of several broad isotropic zones within an irregular network of slightly birefringent pigeonite (giving a weak x-ray diffraction pattern), may be representative of thetomorphic pyroxene glass.

Shock-produced glasses. Numerous spheres and irregular fragments of glassy materials occur in the microbreccias and soil. Such glassy bodies resemble shock-melted masses found in breccia deposits at terrestrial impact craters. Many lunar glass fragments contain highly shocked crystalline inclusions. Relict textures within some fragments indicate partial fusion of granular materials or individual crystals, or selective isotropization of specific phases. Many glassy fragments exhibit heterogeneous compositions in the regions scanned by microprobe. Some glasses show Ni-Fe concentrations, thus supporting derivation by shock pressures and temperatures generated during impact with iron meteorites.

Quench crystals. Certain fragments within the microbreccias contain distinctive reticular, radial, or skeletal crystal groups indicating quench-crystal growth during rapid cooling of shock-melted glasses. Some fragments consist entirely of such crystals whereas glassy phases remain in others.

Recrystallization. In the microbreccias and soils, many small crystalline fragments, presumably once thetomorphs or shock-melted glasses, possess microtextures comparable to devitrified or recrystallized phases found in strongly shocked rocks from terrestrial impact craters. Feldspar-rich fragments, particularly, show a variety of characteristic textures (Fig. 1e)—small interlocked laths, new tiny grains with sutured interfaces, sheath-like crystal bundles—similar to those observed in feldspar at Clearwater Lakes, Manicouagan, Rieskessel, and other impact structures (*7*).

Chemical changes. Feldspars in 12 fragments from sample 10046-12 were analyzed by electron microprobe for their Ca, Na, and Si contents. Feldspars in two unshocked type B fragments have both CaO and Na_2O contents equivalent to An_{80-85}, whereas extensively recrystallized feldspars in two shocked type B fragments show apparent anorthite contents (expressed by CaO only) between An_{63} and An_{68}. Three single fragments of recrystallized feldspar glass have An_{73}, An_{80}, and An_{91} contents, respectively. A colorless glass spherule of approximate plagioclase composition analyzes as An_{77}. In all shocked feldspar masses, CaO loss is moderate (1 to 3 percent less than values for unshocked type B feldspar masses) but the Na_2O content drops from between 1.4 and 1.7 percent [Ab_{14-18} (14 to 18 mole percent albite)] to between 0.4 and 0.6 percent (Ab_{4-6}) in three grains with planar features, to values in the clear glass near 0.1 percent (Ab_{1-2}). The silica content in all shocked feldspar masses decreases, except in one recrystallized type B fragment.

X-ray diffraction asterism. Shock damage in lunar materials was also detected by rotation of apparent single-crystal grains in an x-ray beam. This technique produces diffraction patterns on film that range from spots to streaks to arcs (asterism), depending on the degree of polygonization, mosaicism, or other crystal structure disorientations developed by increasing shock pressures (*10*). Diffraction patterns for grains (three each) of plagioclase, clinopyroxene, and ilmenite from an unshocked type B crystalline rock (sample 10047-9) photographed in MoKα radiation consisted of narrow spots.

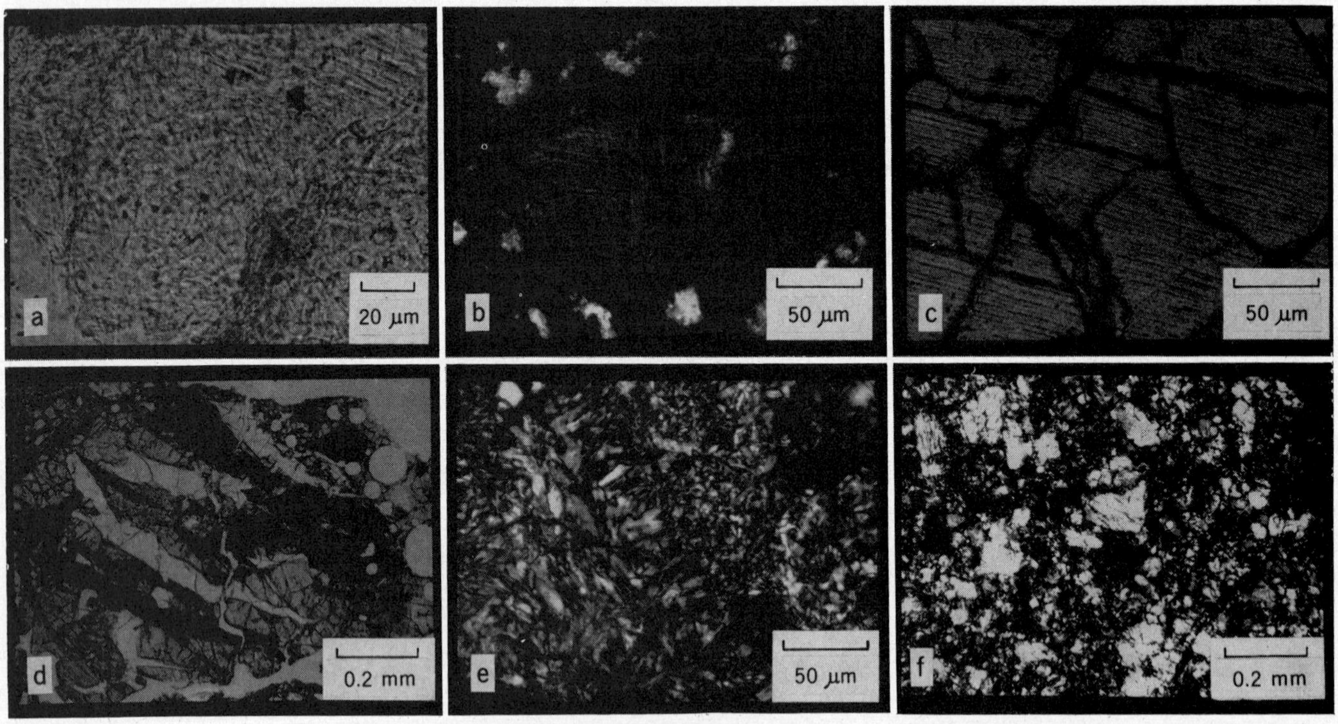

Fig. 1. (a) Three sets of planar features in a silica (quartz?) fragment in microbreccia sample 10060-30. Two other sets become visible when the section is rotated on the universal stage. Plane polarized light. (b) A small fragment of highly shocked plagioclase (An_{75}) containing closely spaced lamellar structures interpreted to be planar features. Most of the grain is a single crystal with five distinct sets of lamellae. A second crystal (upper right) follows a different orientation and has three sets of shock lamellae. Plane polarized light. (c) Detailed view of a single clinopyroxene fragment showing closely spaced lamellae (west-northwest) oriented subparrallel to (100). Two other distinct sets become visible as the section is rotated on the universal stage. Plane polarized light; under crossed Nicol prisms, the grain shows a prominent recrystallization texture. (d) Fragment of a type B crystalline lava rock, now in microbreccia (sample 10065-21). Feldspar laths (light-gray) are completely isotropic (maskelynite) when viewed in cross-polarized light. Pyroxene grains (darker gray) show reduced birefringence and a characteristic shock-induced mosaic extinction pattern. Note glass with vesicles (upper right) encrusting the fragment. Plane polarized light. (e) Portion of a completely recrystallized feldspar (glass?) grain from microbreccia sample 10065-21. This texture is identical to that observed in highly shocked and recrystallized feldspar crystals in rocks from the impact structure at Clearwater Lakes, Canada. Crossed Nicol prisms. (f) Part of a 15-mm wide fragment in microbreccia sample 10046-12. About 85 percent of the shattered individual grains are plagioclase (An_{93}); the remainder are mostly light-green olivine. Individual grains are randomly oriented; this suggests that this fragment is a (possibly shock-lithified) granular aggregate. Crossed Nicol prisms.

Several grains from lunar soil produced similar, sometimes slightly elongated, spot patterns. Most other crystals from the soil exhibited patterns of elongate streaks or continuous arcs. Microscope examination indicates that some grains are crystalline aggregates formed in the parent rocks or by later recrystallization. However, many feldspar, clinopyroxene, and ilmenite grains were originally single crystals until conversion, by shock, to assemblages of displaced mosaic blocks.

Refractive indices. Several strongly shocked feldspar crystals (unrecrystallized), whose asterism patterns consist of coalesced streaks, have N_X values as low as 1.554 to 1.562, equivalent to An_{45-65}. Clear to green-tinted, feldspar-rich glass gave N_D values from 1.56 to 1.59, depending on the amount of Fe-Mg contamination.

Clinopyroxenes that exhibit reduced birefringence and "cloudy" areas can also have anomalously low indices of refraction resulting from shock damage. Several grains picked from the soil met all conditions of a general test for shock metamorphism in lunar materials (*3*, p. 92). These grains show pronounced asterism, and one (sample 10084) contains possible planar features aligned near $(\bar{1}00)$, $(\bar{1}10)$, and $(\bar{2}10)$ planes. The clear crystalline region in that grain has $N_X = 1.618 \pm 0.002$ (compared with 1.70 to 1.72 for unshocked pyroxenes), whereas cloudy areas (decomposition products?) have values between 1.588 and 1.608.

Shock lithification. Comparison of fragment types and degrees of shock damage between lunar soils and microbreccias indicates that the latter are the consolidated (coherent) equivalent of regolith material. The thickness of rubble around the Apollo 11 site appears insufficient to have produced microbreccias by overburden compaction (*11*); the absence of water-bearing phases seemingly rules out fluid-controlled cementation. Although, conceivably, the soil could be derived from the disruption of microbreccia units, a more plausible explanation is that these microbreccias formed by an impact-related compaction process operating on loose, surficial rubble.

Earlier, I predicted the existence of cohesive breccias formed by shock lithification as possible lunar surface rocks (*12*). Similar materials occur as ejecta from terrestrial cratering explosions in loose alluvium. In that shock-produced "instant rock," coherence results from grain-interlocking and bonding by simple compression at lower (100 to 200 kb) pressures and from melting of the clay-silt fractions at higher pressures.

The less cohesive microbreccias may develop by simple compressive bonding during subsequent impacts on the lunar soil. More highly consolidated specimens, however, are held in a glassy matrix, readily visible where polished sections are thin-ground, that forms by compression melting of the "dust" and other fines that predominate in the soil.

A critical minimum thickness of regolith must be exceeded in order that significant amounts of "instant" microbreccia be produced at a given site. Microbreccias should be especially common in the lunar highlands, where ejecta deposits may exceed a thickness of 1 to 2 km (*13*).

Light-colored fragments. Small, whitish, granular masses are infrequent constituents of lunar microbreccias and soils. Some were identified as highly shocked olivine or clinopyroxene crystals. Others are feldspar-rich (70 to 90 percent) aggregates (Fig. 1f) in which plagioclase (An_{90-95}) shows random dispersion of crystal axes indicative of polycrystalline rock rather than shattered single grains. I suggest that those fragments are either fractured anorthosite (resembling the Serra de Mage meteorite) or shock-lithified feldspar rubble, perhaps introduced as ejecta from distant sources, including the lunar highlands (*14*).

In summary, the hypothesis that impact cratering has shaped and modified the moon's surface is supported by the widespread evidence of shock metamorphism in fragmental materials at the Apollo 11 site. The regolith there is largely a local accumulation of ejecta, derived mainly by cratering of underlying crystalline lavas, from which microbreccias were produced by shock lithification during later impacts. This regolith cannot have been a pyroclastic deposit prior to alteration by impact metamorphism, because of the broad range, rather than constant levels, of shock damage in fragments within individual soil and microbreccia samples.

NICHOLAS M. SHORT

Laboratory for Meteorology and Earth Sciences, NASA Goddard Space Flight Center, Greenbelt, Maryland 20771

References and Notes

1. N. M. Short, *Amer. Astronaut. Soc. Sci. Technol. Ser.* **14**, 17 (1967).
2. B. M. French, in *Shock Metamorphism of Natural Materials*, B. M. French and N. M. Short, Eds. (Mono, Baltimore, 1968), p. 1.
3. N. M. Short, *Mod. Geol.* **1**, 81 (1969).
4. O. B. James, *Science* **166**, 1615 (1969).
5. N. M. Short, *Bull. Geol. Soc. Amer.*, in press.
6. D. E. Gault, *Trans. Amer. Geophys. Union* **50**, 219 (1969); A. H. Marcus, *Johns Hopkins Univ. Tech. Rep. 122* (1969).
7. See pertinent papers (pp. 243, 413, 433, 453, 475) in *Shock Metamorphism of Natural Materials*, B. M. French and N. M. Short, Eds. (Mono, Baltimore, 1968).
8. N. L. Carter, *ibid.*, p. 453.
9. Assuming the applicability of criteria developed primarily for olivine, as reported by N. L. Carter, C. B. Raleigh, P. S. DeCarli, *J. Geophys. Res.* **73**, 5439 (1968).
10. F. Dachille, P. Gigl, P. Y. Simons, in *Shock Metamorphism of Natural Materials*, B. M. French and N. M. Short, Eds. (Mono, Baltimore, 1968), p. 555; F. Horz, abstract, program of the 32nd annual meeting of the Meteoritical Society, Houston, Texas, 1969.
11. V. R. Oberbeck and W. L. Quaide, *Icarus* **9**, 446 (1968).
12. N. M. Short, *Science* **154**, 383 (1966); ———, in *Shock Metamorphism of Natural Materials*, B. M. French and N. M. Short, Eds. (Mono, Baltimore, 1968), p. 219.
13. ———, in preparation.
14. E. Olsen, *Science* **166**, 401 (1969); A. L. Turkevich, E. J. Franzgrote, J. H. Patterson, *Science* **162**, 117 (1969).
15. Supported by the Planetology Branch, Laboratory for Space Physics, and by the Laboratory for Meteorology and Earth Sciences, Goddard Space Flight Center, Greenbelt, Maryland. I thank Dr. B. P. Glass, Dr. P. D. Lowman, C. Kouns, and F. Wood for assistance in sample analyses.

5 January 1970

Shock-Wave Damage in Minerals of Lunar Rocks

Abstract. *The mineral fragments that constitute the Apollo 11 microbreccia and fines show a number of shock-induced microstructural effects including multiple twinning on (001) in clinopyroxene and incipient development of lattice-controlled discontinuities in olivine. These structures coupled with the effects of shock-induced melting as manifested by spherules and angular fragments of glass and their partly to completely crystallized equivalents indicate that Apollo 11 materials were subjected to weak to moderate shocks with associated peak pressures of the order of 100 to 200 kilobars.*

The results of preliminary examination of the Apollo 11 lunar rocks and fines (*1*) indicated that many of them had been subjected to one or more shock events presumably produced by meteorites striking the lunar surface. The evidence for this conclusion was based largely on the occurrence of (i)

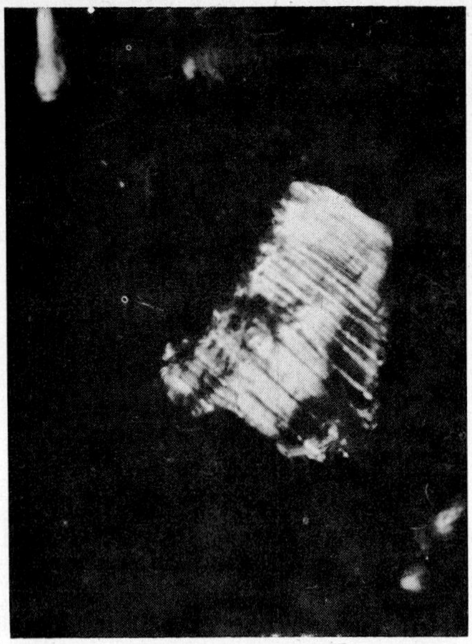

Fig. 1. Photomicrograph of clinopyroxene fragment in microbreccia (10060) showing multiple twinning parallel to (001). × 530. Crossed Nicols.

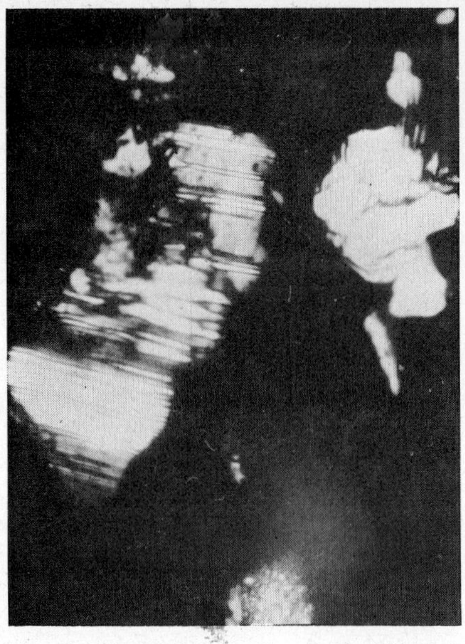

Fig. 2. Photomicrograph of diopside experimentally shock loaded at about 200 kb showing shock-induced multiple twinning parallel to (001). × 530. Crossed Nicols.

microbreccia which appears to be shock-lithified ejecta from impact events, (ii) angular fragments of microbreccia within a matrix of microbreccia, (iii) spherules, broken spherules, and angular fragments of glass of variable composition, (iv) glass-splashed exposed surfaces of microbreccia, (v) nickel-iron spherules, and (vi) undulatory extinction of birefringent minerals in microbreccia and fines. Small-scale shock-induced microstructural damage in minerals of the rocks and fines was studied in detail by light and electron microscopy. Additional evidence supporting the concept of multiple shock events on the lunar surface was found.

In general, the minerals of the fine- and medium-grained crystalline igneous rocks show little or no evidence of shock-wave damage. A striking feature of the minerals that constitute the microbreccia and fines, however, is their extreme heterogeneity in terms of the degree of shock recorded in coexisting fragments. so that in a highly magnified microscopical field of view one may observe fragments that were structurally damaged by high transient pressures, fragments that show high-temperature melting phenomena ascribable to shock processes, and fragments that are apparently unshocked. Such variability is characteristic of materials from terrestrial impact craters (*2*) and surface or near-surface experimental explosion craters (*3*).

Many of the plagioclase fragments in the microbreccia and fines are shattered and now consist of a number of contiguous subunits. Some show microfaulting, as revealed by offsets of twin lamellae, and others are highly strained, as shown petrographically by undulatory extinction effects. Correspondingly, many of the clinopyroxene fragments have been shattered into subunits and appear highly bent around axes roughly normal to the *c* axis as revealed by the traces of the prismatic cleavage planes. Some clinopyroxene grains are microfaulted as revealed by offsets in the prismatic cleavage traces and, rarely, by offsets in what appear to be exsolution lamellae of orthopyroxene(?) parallel to (100). In addition, many of the clinopyroxene fragments are highly strained as shown by undulatory extinction effects. On the other hand, the extent of mechanical damage recorded in the olivines appears, in general, to be much less, although a small percentage of the olivine fragments in the microbreccia and fines show abundant microfractures and undulatory extinction. All these features are indicative of shock but are not conclusive.

Many clinopyroxene fragments in the microbreccia show multiple narrow twin lamellae parallel to (001) (Fig. 1). It has been experimentally demonstrated on this program that multiple lamellar twinning parallel to (001) is developed in diopside (variety alalite, Mussa Alp, Piedmont, Italy) by shock-loading fine particulate material at about 200 kb (Fig. 2). Some of the unshocked diopside particles are twinned parallel to (100), so that the superimposed shock-induced basal-plane twinning produces a distinctive quadrille or microcline-like structure. The minimum peak pressure required to produce this structure, however, is not known, but it has just been reported (*4*) that augite in weakly shocked basalt from a nuclear explosion crater at the Nevada Test Site shows shock-induced multiple twinning on (001). It must be noted that, although rare, twinning of clinopyroxene on (001) can be developed by tectonic processes

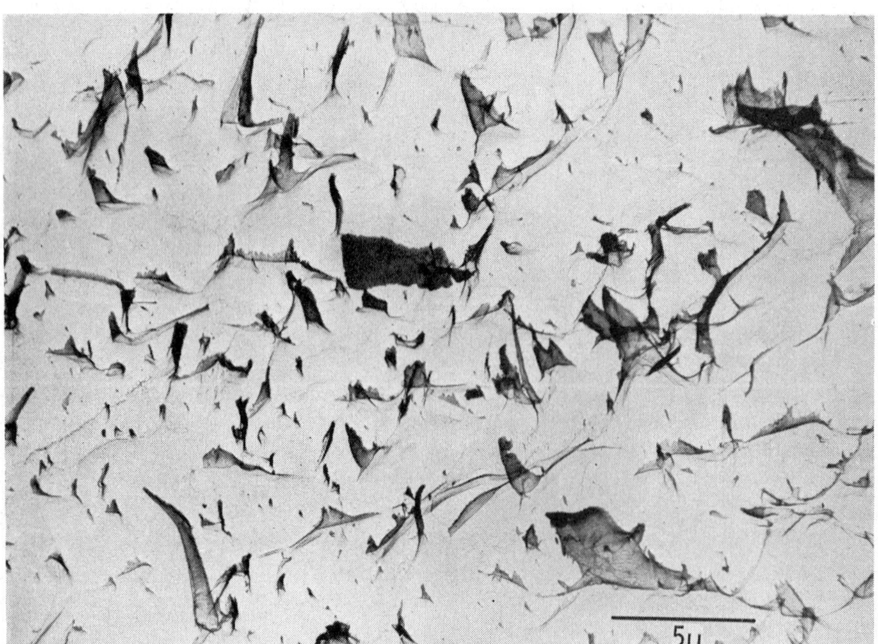

Fig. 3. Electron micrograph of replica of etched polished surface of olivine grain in crystalline fragment from fines showing incipient development of shock-induced planar features interpreted as lattice-controlled fractures.

and has been produced experimentally at low strain rates (*5*). Nevertheless, the evidence strongly suggests that the multiple basal-plane twinning observed in clinopyroxene fragments in Apollo 11 material is the result of shock.

Although some of the olivine fragments in the microbreccia and fines show irregular microfracturing and undulatory extinction, only a few grains were found that showed any development of planar or lamellar features considered to be indicative of shock. One olivine from a crystalline igneous rock fragment in the fines showed in the electron microscope (Fig. 3) the incipient development of lattice-controlled discontinuities of the type that are highly developed in particulate olivine samples experimentally shock loaded at about 200 kb (*6*). Such planar features, interpreted as microfractures, constitute excellent evidence of shock-wave damage. Several olivine grains were observed to contain one set of optically resolvable thin lamellae, which appear to be similar to those described in olivine from shocked basalt [fig. 8D in (*4*)]; another olivine grain in a microbreccia (10061) showed some recrystallization (development of tiny new grains) similar to the well-developed recrystallization effects in dunite experimentally shock loaded to 1 Mb (*7*).

In a thin section of microbreccia (10060,20) are three grains of a colorless untwinned mineral with a low birefringence of section which show several close-spaced sets of planar features that resemble those observed in shocked quartz (*8*) and shocked plagioclase (*9*). These grains show no cleavage and are biaxial positive with $2V$ in the range 25° to 32°. They are either shocked and strained quartz or shocked plagioclase in which the optical indicatrix has been changed markedly, or, possibly, shocked clinopyroxene. Chao (*2*, p. 146) has reported that shocked quartz may be biaxial positive with an optic angle as high as 28°, and Bunch (*10*, p. 425) has found that shocked labradorite may have an optic angle as much as 20 percent smaller than its unshocked compositional equivalent. Evidently these grains should be checked with the electron microprobe, inasmuch as the possible occurrence of quartz could have significant petrological implications.

Each of two magnetic particles from the fines was found to contain an angular grain of metallic iron containing about 5 percent nickel as revealed by the electron microprobe. The grains are probably kamacite fragments from a metallic meteorite. When etched with Nital, one grain developed what appear to be Neumann bands typical of lightly shocked iron meteorites [fig. 3a in (*11*)]; the other shows an exceedingly fine microstructure near the resolution limit of the light microscope which may represent the $\alpha \rightleftarrows \varepsilon$ transformation as recorded in meteorites shocked between 100 and 200 kb [fig. 3b in (*11*)]. Electron micrographs of the latter show a fine granular structure with a tendency toward a "woven cloth" appearance.

Shock-induced thermal effects are manifested in the microbreccia and fines by spherules and angular fragments of glass of variable composition. In addition, many of the spherules and angular fragments show the development of quench crystals, which are typically skeletal or dendritic and are enclosed in a matrix of glass or devitrified glass. Finally, less common are fragments of partly melted pyroxene and plagioclase enclosed in brown and colorless vesiculated glass, respectively.

Although the crystalline igneous rocks show little or no effect of shock, selected mineral fragments in the microbreccia and fines show shock-induced structural damage and shock-induced thermal effects. These phenomena suggest that the Apollo 11 rocks were subjected to relatively weak to moderate shocks with associated peak pressures of the order of 100 to 200 kb.

CHARLES B. SCLAR
Department of Geological Sciences, Lehigh University, Bethlehem, Pennsylvania 18015

References and Notes

1. Lunar Sample Preliminary Examination Team, *Science* **165**, 1211 (1969).
2. E. C. T. Chao, in *Shock Metamorphism of Natural Materials*, B. M. French and N. M. Short, Eds. (Mono, Baltimore, 1968), p. 135.
3. N. M. Short, *ibid.*, p. 185.
4. O. B. James, *Science* **166**, 1615 (1969).
5. C. B. Raleigh and J. L. Talbot, *Amer. J. Sci.* **265**, 151 (1967).
6. C. B. Sclar, final report, NASA contract NSR 36-002-062 (1968); *Trans. Amer. Geophys. Union* **50**, 219 (1969).
7. N. L. Carter, C. B. Raleigh, P. S. DeCarli, *J. Geophys. Res.* **73**, 5439 (1968).
8. W. v. Engelhardt and W. Bertsch, *Contr. Mineral. Petrol.* **20**, 203 (1969).
9. P. B. Robertson, M. R. Dence, M. A. Vos, in *Shock Metamorphism of Natural Materials*, B. M. French and N. M. Short, Eds. (Mono, Baltimore, 1968), p. 433.
10. T. E. Bunch, *ibid.*, p. 413.
11. M. E. Lipschutz, *ibid.*, p. 571.
12. Supported by NASA contract NAS 9-8084.

4 January 1970

Cathodoluminescence Properties of Lunar Rocks

Abstract. *Calcic plagioclase is the dominant luminescent mineral in crystalline rocks and breccias. Minor amounts of cristobalite and tridymite are also luminescent, as are trace grains of potassium feldspar. Two types of intergrowths of potassium feldspar with a silica phase, possibly quartz, were found in the breccias. Luminescence spectra of plagioclase show significant similarities to, and differences from, spectra of terrestrial plagioclase. Shock damage in the breccias is reflected in systematic changes in the plagioclase spectra, thus giving evidence of disordering on the angstrom scale. Associated extinction patterns seen between crossed Nicol prisms give evidence of disordering on the micrometer scale.*

Luminescence petrography, the microscopic study of rock luminescence in thin section, is a significant extension of petrographic technique. Studies of terrestrial rocks have shown that this technique can provide information unobtainable by any other method (*1*). By means of auxiliary equipment one can measure the spectra of luminescing regions as small as 40 μm in diameter. Such measurements are important in the work reported here.

Crystalline rocks. The crystalline igneous rocks studied (samples 10020 and 10045) are intermediate in grain size between rocks of types A and B (*2*). Both specimens are similar and contain cristobalite and olivine. The dominant luminescent mineral is calcic plagioclase which has an abundance of ≈ 25 percent. Its luminescence is blue to yellowish-blue. Cristobalite (≈ 1 percent abundance) emits blue luminescence and, like terrestrial quartz, shows development of red-emitting luminescence centers under electron bombardment. Tridymite ($\ll 1$ percent abundance) is disseminated throughout as pink luminescing crystals 10 to 50 μm in diameter and is often closely associated with plagioclase. Pyroxene (≈ 45 percent abundance), olivine (≈ 5 percent abundance), and several unidentified iron-bearing trace minerals are nonluminescent, as are terrestrial mafic minerals containing more than a few percent iron. Apatite and zircon, although brightly luminescent in terrestrial rocks, were not found.

The plagioclase luminescence shows a range of intensity and color. Central regions of many laths are more yellow and more intensely luminescent than peripheral regions. Some small plagio-

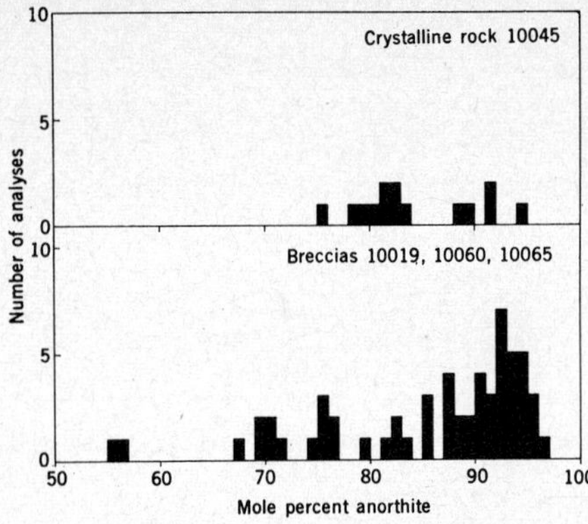

Fig. 1. Summary of data from microprobe analysis of calcic plagioclase. Selection was not strictly random. The crystalline rock specimens range from An_{89} to An_{97} in bright luminescing regions, grading to An_{75} in dull rims or grains. The low values of anorthite (An_{56}) in breccias were found in lithic fragments. The dominant range of compositions is very similar in crystalline rocks and breccias.

clase crystals show dull luminescence only, like the rims of larger laths. Microprobe analysis shows that the zoning of luminescence correlates with normal compositional zoning. The brightest luminescing central regions are calcium-rich [$\approx An_{95}$ (95 mole percent anorthite)], and the amount of Ca diminishes to An_{75} in the dullest rims or grains (Fig. 1). This suggests preferential partitioning of activator ions in the crystalline phase so that they become impoverished in the melt as crystallization proceeds.

Terrestrial feldspar specimens from diverse localities show a wide range of luminescence color, although the spectra of such specimens show a surprising degree of regularity. Virtually all of the varied color effects are the result of the presence or absence of, and the relative intensities of, three broad peaks which are characteristic of the spectra of terrestrial feldspar. The luminescence spectrum of a terrestrial plagioclase An_{85}, where all three peaks are well developed, is shown in Fig. 2A. To facilitate comparisons with lunar specimens, experience with a suite of terrestrial feldspars will be briefly summarized.

Luminescence spectra of 47 diverse terrestrial feldspars were measured. Composition as determined by microprobe analysis shows 20 of them to be alkali feldspar, 20 to be plagioclase, and 6 to be approximately albite; one of them has mixed composition (calcic alkali). The relative intensity of the three peaks varies widely among specimens and suggests that three distinct luminescence centers are involved. The green peak ($\approx$ 5590 Å) does not appear in the spectra of alkali feldspar but is present in the 16 plagioclase spectra where anorthite exceeds 11 mole percent. It probably results from a divalent activator ion proxying for calcium in the plagioclase structure. The position of the green peak is not sensitive to composition within the plagioclase series. The blue peak ($\approx$ 4500 Å) shows some variability in position and shape. Neither feature correlates with composition, and the causes are unknown. In some cases more than one luminescence center may be involved in the blue emission. In any case, a definite blue emission band is found in 45 of the 47 specimens. The two exceptions are both alkali feldspar. The position of the red-infrared peak ($\approx$ 7100 Å) is somewhat sensitive to composition. It shows a range of about 300 Å and correlates roughly with the mole percent albite without regard to whether the feldspar is calcic or potassic. The red-infrared peak of alkali feldspar is usually less intense than that of plagioclase. Potassium-rich specimens usually exhibit a broad blue peak and a small but definitely observable red-infrared peak. In three cases alkali feldspar shows no red-infrared peak at all. In the other 44 specimens the red peak is clearly present.

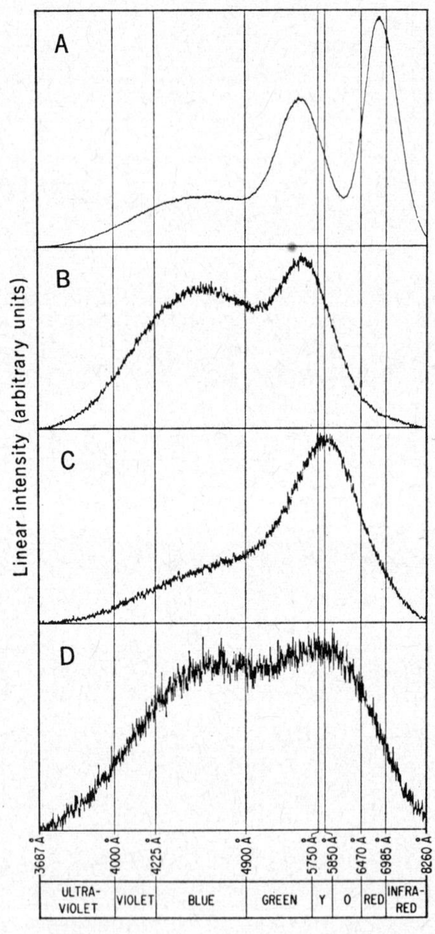

Fig. 2. Raw spectra for comparison. (A) Terrestrial plagioclase An_{85}. (B) Lunar crystalline rock plagioclase. The red-infrared peak is absent. (C) Plagioclase grain from lunar breccia showing intermediate degree of shock damage. The green plagioclase peak is shifted and broadened. The luminescence of this specimen appears orange. (D) Maskelynite grain; note the nonlinear abscissa. Correction of the spectra for phototube spectral response and for monochromator resolution will increase the shifts evident in C and D.

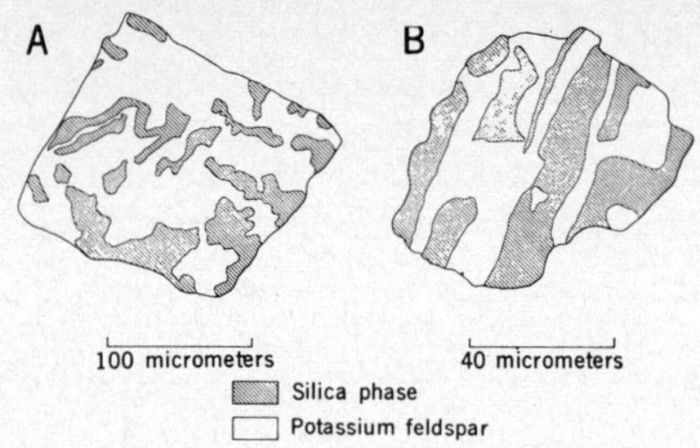

Fig. 3. Two types of intergrowth between potash feldspar and silica phases, possibly quartz. (A) Pink luminescing silica; (B) dully luminescent silica.

A typical spectrum of plagioclase from lunar crystalline rock is shown in Fig. 2B. The blue and green peaks are present, but the red-infrared peak is missing. The spectra of over 100 grains of plagioclase have been measured in lunar crystalline rocks and breccias, and no evidence of a red-infrared peak was found in any of them. The reason is not known.

Luminescence petrography of the crystalline rocks showed no evidence of complex or inverted zoning, nor was there any evidence of secondary hydrothermal or aqueous mineralization.

Breccias. Breccias studied (samples 10019, 10060, 10065) consist of fragments of minerals, rocks, and glass of several types (*2*). Lithic fragments are abundant. In sample 10060 we identified 50 fragments between 0.2 and 2 mm in size. Constituent grain sizes of lithic fragments vary from a few micrometers in aphanitic clasts to 1-mm phenocrysts, but most fall between 0.02 and 0.2 mm. Monomineralic fragments are less than 0.6 mm in size and cover a range of 0.01 to 0.6 mm. The lithic fragments, some of which are well rounded, are generally basaltic. Fragments recognizable as clasts of lunar crystalline igneous rocks are common, but the lithic fragments appear to encompass a somewhat wider range of basaltic rock types than the crystalline specimens available for study. For example, some are porphyritic and others have felty textures. In some the plagioclase is anhedral. The modal composition is also more variable than in the crystalline rocks. Plagioclase is dominant in some fragments, and opaques are rare. Some clasts appear to be breccia fragments but are composed mainly of plagioclase, whereas opaques and mafic minerals constitute only about 10 percent. Some fragments contain plagioclase as sodic as Ab_{44}.

The dominant luminescent phase both in lithic fragments and mineral grains is calcic plagioclase, and most fragments are similar in composition to plagioclase in the crystalline rocks (Fig. 1). Luminescent tridymite and cristobalite are also present, whereas luminescent apatite and zircon are not.

The crystalline igneous rocks and closely associated rock types are probably the source of most of the clastic and glassy fragments found in the breccias. However, the breccias may also contain grains or fragments from a distant source. Considerable effort was expended in a search for alkali feldspar.

Two distinct types of alkali feldspar were found. Both consist of intergrowths on a fine scale between potassium feldspar and a silica phase, possibly quartz (Fig. 3). Careful microprobe analysis indicated that the potassium feldspar composition in each type was stoichiometric. One type (Fig. 3A) has pink luminescing silica similar in color to that of the tridymite of the crystalline rocks. Several grains of this type were found. The second type (Fig. 3B) has a nonluminescent or a dully luminescent silica phase, and only one such grain was found in three thin sections of breccia. Both show spectra typical of potassium feldspar. Because such intergrowths are common in terrestrial acidic rocks, it is tempting to speculate that similar rocks exist on the moon. The argument is equivocal at best, however, because such intergrowths do point to differentiation but say nothing about the scale.

Numerous isotropic or weakly birefringent grains of stoichiometric plagioclase composition were found in the breccia. Most show dull reddish luminescence. We interpret these to be maskelynite, that is, plagioclase disordered by shock waves. Single grains showing both birefringence and normal plagioclase luminescence on one end, and isotropism and dull reddish luminescence on the other, were found. The reddish luminescence results from both shifting of the green peak toward longer wavelengths and broadening of the peak. This results in increased relative red emission (Fig. 2D). Intermediate degrees of disordering are reflected in

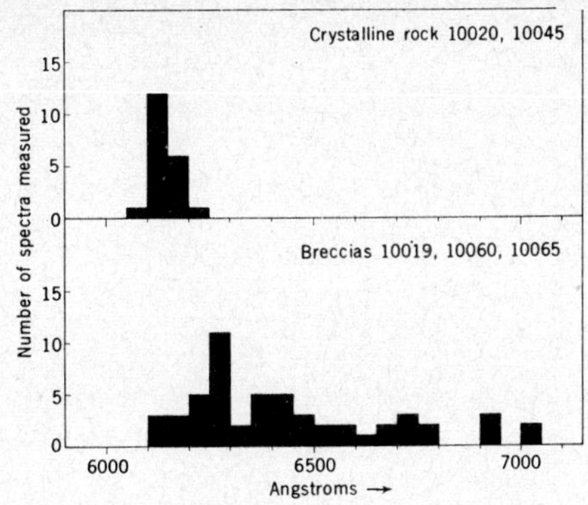

Fig. 4. Distribution of upper half intensity points of plagioclase spectra. Maxima were taken from the green to the yellow peak.

the luminescence spectra. Something extraordinary is affecting the spectra of plagioclase in the breccias, as can be readily seen from the following comparisons. Average positions of the green peaks were calculated for 16 samples of terrestrial plagioclase, 21 plagioclase grains from the lunar crystalline igneous rocks, and 48 plagioclase grains from lunar breccias. The average values and extremes are: 5590^{+120}_{-65} Å, 5590^{+160}_{-65} Å, and 5735^{+495}_{-230} Å. Notice not only the extraordinary average value for the breccias but also the increased breccias which just permits overlap. Both shifts and broadening can be reflected by plotting the upper half intensity position (Fig. 4). This graphically shows that a continuous range of effects is involved in plagioclase from the breccia, a result in stark contrast with the plagioclase of the crystalline rocks. Correction of the raw spectra will enhance the result. Grains that show intermediate degrees of disordering (Fig. 2C) may appear yellow or orange in luminescence and frequently show mosaic extinction patterns when viewed between crossed Nicol prisms. Similar phenomena have been reported in shock-damaged specimens, but such features in and of themselves might not be sufficient evidence to support the interpretation of shock damage (*3*). Strangely, the breccia specimens studied show very little of the fracturing and lamellar and planar features diagnostic of shock damage (*4*). This same observation was made by Short for experimentally shock-lithified sand (*5*). The association of evidence of disordering on the micrometer scale, as shown by the mosaic extinction patterns, and evidence for disordering on the angstrom scale, as shown by luminescence, strengthens our conclusion that these features result from shock damage.

Our interpretation of peak shifts, peak broadening, and decrease of luminescence intensity with lattice disordering is consistent with the general theory of solids. Luminescence spectra depend in part on the nature of the luminescence center but also on the nature of the solid because excited states are displaced in energy by the crystal field of the solid. Disorder in the crystal lattice results in crystal field perturbations. Local variations in the crystal field result in an altered or broadened distribution of excited-state energies. Anisotropic distortions in the crystal field might produce peak shifts, whereas isotropic disordering would certainly lead to broadened emission bands. Nonradiative transitions would also tend to become more probable as order decreases.

In addition to the above process, for which evidence is ubiquitous, some grains suggest the production of luminescence centers by shock damage. These show enhanced luminescence in the blue end, and reduced intensity in the red end, of the spectrum. Enhancement of luminescence in the blue end of the spectrum has also been observed in shocked anorthosite from Manicouagan, Canada (*6*), and has been qualitatively observed in quartz in a graded suite of shock-damaged specimens (*7*).

On the whole, the breccias seem to attest to the actuality of the process postulated by Short: the initial comminution of consolidated rocks by meteorite impacts and subsequent shock lithification of the rubble so produced (*5*). In contrast, the crystalline igneous rocks studied showed no evidence of shock damage.

ROBERT F. SIPPEL
ALEXANDER B. SPENCER
Mobil Research and Development Corporation, Dallas, Texas 75211

References and Notes

1. J. V. P. Lond and S. O. Agrell, *Mineral. Mag.* **34**, 318 (1965); J. V. Smith and R. C. Sterstrom, *J. Geol.* **73**, 627 (1965); R. F. Sippel and E. D. Glover, *Science* **150**, 1283 (1965); R. F. Sippel, *Rev. Sci. Instrum.* **36**, 1556 (1965); *J. Sediment. Petrol.* **38**, 530 (1968).
2. Lunar Sample Preliminary Examination Team, *Science* **165**, 1211 (1969).
3. F. Dachille, J. Fauth, V. Vand, abstract from the program of the annual meeting of the Geological Society of America, 1964; N. M. Short, *J. Geophys. Res.* **71**, 1195 (1966).
4. E. T. C. Chao, in *Researches in Geochemistry* P. H. Abelson, Ed. (Wiley, New York, 1967), vol. 2, p. 204; T. E. Bunch, A. J. Cohen, M. R. Dence, in *Shock Metamorphism in Natural Materials*, B. M. French and N. M. Short, Eds. (Mono, Baltimore, 1968), p. 509.
5. N. M. Short, *Science* **154**, 382 (1966).
6. We thank T. E. Bunch for supplying these specimens.
7. N. M. Short kindly supplied these specimens showing graded shock from the Sedan nuclear explosion of July 1962.
8. We thank C. M. Wall for diligent and careful microprobe work and Dr. W. L. Medlin for helpful consultation.

5 January 1970

Mössbauer Studies

Mineralogy and Composition of Lunar Fines and Selected Rocks

Abstract. *Mineralogical descriptions and both wet chemical analyses and microprobe analyses are given of the glasses and crystalline components of the lunar fines and of the minerals in microgabbros (samples 10050 and 10047). The principal minerals described are various clinopyroxenes, plagioclase, olivine, low cristobalite, low tridymite, ilmenite, iron-nickel, iron, schreibersite, cohenite, troilite, and a new CaFe pyroxenoid. Descriptions are given of small craters produced by hypervelocity particle impact on glass and iron-nickel fragments in the fines. The rounding of grains in the fines and of surface rocks is attributed to mechanical abrasion and not to cratering.*

Approximately half of the lunar fines (<1 mm) consist of particles below 37 μm in size. Wet chemical analyses of the bulk fines, of the <37-μm fraction, of an abundant scoriaceous glass component, and of the microbreccia rock show almost identical compositions (Table 1, analyses 1 to 4). Analyses of the microgabbro and vesicular basalt (Table 1, analyses 5 and 6) show somewhat higher FeO and TiO_2 and less Al_2O_3, with a higher K_2O content in the basalt. These differences probably arise because the two rock analyses are not representative of the rock suite from which the fines, the glass, and the microbreccia were derived.

Glasses in the fines occur as spheroids and other rounded forms and as angular or abraded fragments, which vary from colorless ($d \leq 2.6$ g/cm^3; n, 1.46 to 1.60), pale yellow, yellow-brown, tan, and green shades ($d \sim 2.7$ to 2.8, $n \sim 1.59$ to 1.65), dark brown and reddish brown ($d \sim 2.9$ to 3.1, $n \sim 1.65$ to 1.75) to almost opaque ($d \sim 3.1$ to 3.3, $n \sim 1.75$). A dark brown scoriaceous glass (Table 1, analysis 3) shows color banding and contains minute spherical iron inclusions and unfused inclusions of clinopyroxene, ilmenite, plagioclase, and lithic fragments. Plagioclase glasses (61 to 94 percent anorthite) and compositions between anorthite and the scoriaceous glass are less abundant. Figure 1 shows glass compositions that range from plagioclase to glasses higher in FeO and TiO_2 than the scoriaceous material. They can be formed by varying admixtures of plagioclase, clinopyroxene, and ilmenite.

The crystalline phases in the fines

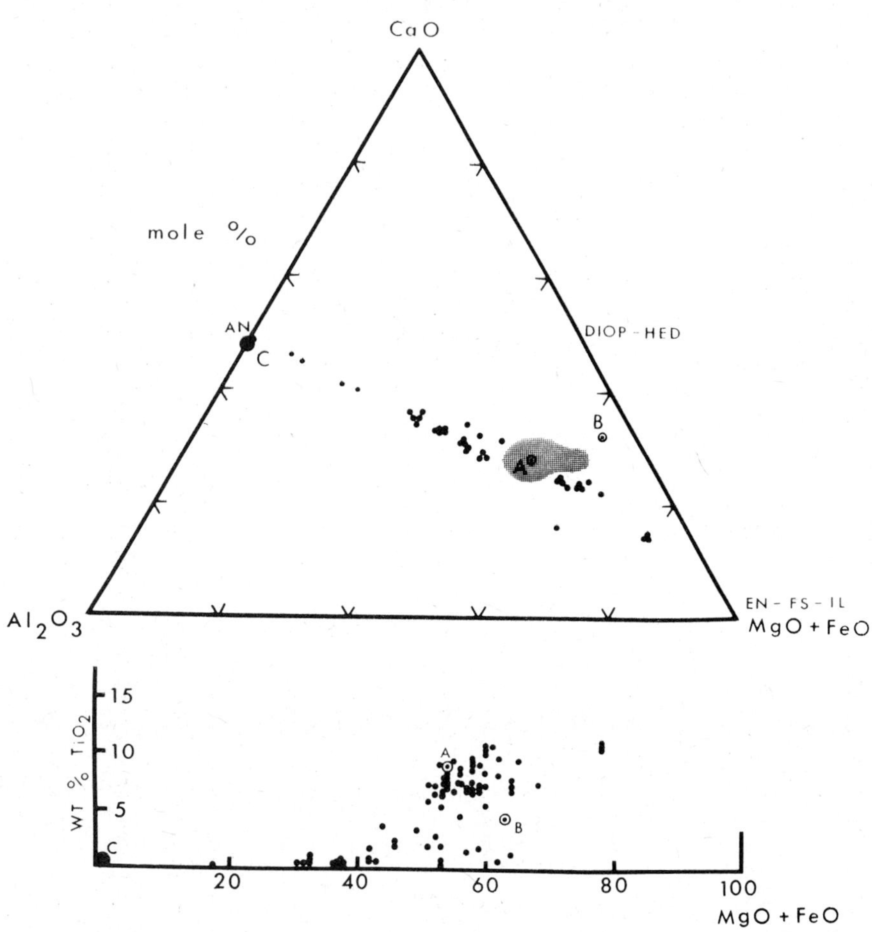

Fig. 1. (Top) Electron probe analyses (J. C. Drake, analyst) of glasses in lunar fines in terms of molecular percentages of CaO, Al_2O_3, and (MgO + FeO). Each dot represents the analysis of a single glass spherule, dumbbell, or angular fragment. The shaded area represents 48 analyses. Point A is a plot of the scoriaceous glass analysis in Table 1 (analysis 3) and point B is a plot of the two clinopyroxene analyses in Table 2 (analyses 1 and 2). Point C is an average of 15 plagioclase glasses (from 61 to 94 percent anorthite). In general, the density of the glasses increases from left to right, from $d \leq 2.6$ to $d \sim 3.25$. Abbreviations are An (anorthite), Diop (diopside), Hed (hedenbergite), En (enstatite), Fs (ferrosilite), Il (ilmenite). (Bottom) Percentage by weight of TiO_2 in the glasses as a function of (MgO + FeO).

Table 1. Wet chemical analyses of bulk fines (< 1 mm), glass in fines, and rock fragments in coarse fines (1 cm to 1 mm).

Component	Average fines* No. 1	Fine fines† No. 2	Scoriaceous glass‡ No. 3	Microbreccia§ No. 4	Fine-grained gabbro‖ No. 5	Vesicular basalt¶ No. 6
SiO_2	41.77	41.81	41.84	41.69	41.65	40.60
TiO_2	7.42	7.20	7.09	7.35	8.79	11.28
Al_2O_3	13.68	14.64	14.00	13.52	11.00	8.05
Cr_2O_3	0.36	0.31	0.30	0.36	0.34	0.33
FeO (total)	15.98	15.21	15.78	16.05	18.20	20.09
MnO	0.23	0.20	0.23	0.23	0.25	0.24
MgO	8.38	7.55	8.18	8.40	7.80	7.90
CaO	11.68	12.59	12.03	11.76	11.22	10.57
Na_2O	0.41	0.43	0.40	0.42	0.41	0.44
K_2O	0.13	0.14	0.14	0.14	0.12	0.24
P_2O_5	0.12	0.1	0.1	0.12	0.12	0.15
NiO	~0.01	~0.01	~0.01	0.01		
ZrO_2	0.05	0.05	0.05	~0.06	~0.1	~0.1
S	0.1	0.1	0.1	~0.1	~0.1	~0.2
− S = O	−0.05	−0.05	−0.05	−0.05	−0.05	−0.1
Total	100.20	100.24	100.15	100.16	100.05	100.09

* Particle size < 1 mm; sample weight, 1.16 g. † Particle size < 37 μm; sample weight. 1.01 g. ‡ Mixture of approximately 60 percent dark brown glass and 40 percent enclosed or adhering clinopyroxene, plagioclase, and ilmenite; sample weight. 0.52 g. § Sample weight, 1.05 g. ‖ Approximately 50 percent tan clinopyroxene, 35 percent plagioclase, 15 percent ilmenite, and rare olivine and cristobalite; trace of troilite with iron blebs; sample weight, 1.07 g. ¶ Approximately 50 percent tan clinopyroxene, 30 percent plagioclase, 20 percent ilmenite, and traces of troilite with iron blebs; sample weight, 1.06 g.

Table 2. Wet chemical analyses of clinopyroxene (Nos. 1 and 2), plagioclase (Nos. 3 and 4), and ilmenite (No. 5) in lunar fines and fine-grained gabbro. Clinopyroxene No. 1 and plagioclase No. 3 from lunar fines; clinopyroxene No. 2, plagioclase No. 4, and ilmenite No. 5 from fine-grained gabbro (10050,37).

Component	Clinopyroxene		Plagioclase		Ilmenite
	No. 1*	No. 2*	No. 3†	No. 4‡	No. 5§
SiO_2	48.2	49.2	45.1	46.3	
TiO_2 (total)	3.38	2.58	0.05	0.05	52.6
Al_2O_3	4.65	3.55	35.1	34.0	
Cr_2O_3 (total)	0.53	0.46	0.01	0.01	0.51
FeO (total)	12.45	12.62	0.42	0.99	45.3
MnO	0.23	0.24	0.03	0.03	0.37
MgO	15.1	16.0	0.3	0.3	0.83
CaO	15.25	15.20	18.0	17.1	
Na_2O	0.12	0.10	0.85	1.05	
K_2O			0.07	0.09	
ZrO_2	~0.02	~0.02			~0.3
Total	99.93	99.97	99.9	99.9	99.9

Number of ions on the basis

	Of 6 oxygens				*Of 8 oxygens*				*Of 3 oxygens*	
Si	1.808	2.00	1.844	2.00	2.082	3.99	2.135	3.98		
Al	0.192		0.156		1.910		1.848			
Al	0.014		0.001							
Ti	0.095		0.073		0.002		0.002		0.994	
Cr	0.016		0.013						0.010	
Fe	0.391		0.395		0.016		0.038		0.952	
Mn	0.007	1.99	0.008	2.00	0.001		0.001		0.008	0.99
Mg	0.844		0.893		0.020	1.01	0.020	1.00	0.031	
Ca	0.613		0.610		0.890		0.845			
Na	0.009		0.007		0.076		0.094			
K					0.004		0.005			
Zr									0.002	
Mg	45.7		47.0		An 91.7		An 89.5			
Fe	21.1		20.8		Ab 7.8		Ab 9.9			
Ca	33.2		32.1		Or 0.4		Or 0.5			

* Both samples contain approximately 0.5 percent (by volume) ilmenite as inclusions. This ilmenite was not deducted from the analysis. Weight: sample 1, 0.36 g; sample 2, 0.45 g. † This sample (weight, 0.27 g) contained 4 percent by weight impurity (0.35 percent ilmenite and 3.65 percent clinopyroxene). These impurities were substracted and the analysis was recalculated to 100 percent. TiO_2 and MgO values were obtained by electron probe. ‡ The sample (weight, 0.36 g) contained 2.3 percent by weight impurity (0.46 percent ilmenite and 1.84 percent clinopyroxene). These impurities were subtracted, and the analysis was recalculated to 100 percent. TiO_2 and MgO values were obtained by electron probe. § The sample (weight, 0.22 g) contained 8.5 percent by weight adhering clinopyroxene and plagioclase. The analysis was recalculated to 100 percent after subtraction of these phases. Electron probe analyses showed absence of Al.

include (in decreasing abundance) clinopyroxenes, plagioclase, ilmenite, olivine, tridymite, cristobalite, iron-nickel, iron, troilite, schreibersite, and cohenite. The pyroxenes are highly inhomogeneous and vary from augite, in part with exsolved pigeonite, to subcalcic augite, pigeonite, ferroaugite, and titanaugite. A wet chemical analysis of a bulk, average sample of the pyroxene is given (Table 2, analysis 1). Figure 2 shows probe analyses of clinopyroxene grains. The titanaugite in Fig. 2 represents a brown, almost opaque type containing ilmenite inclusions. A wet analysis of bulk anorthite from the fines corresponds to 92 percent anorthite (Table 2, analysis 3). The total variation in anorthite content, determined by probe analyses of individual grains, ranges from 90 to 98 percent anorthite. Ilmenite is the main accessory mineral. Olivine is yellowish green and transparent, and varies from approximately 61 to 76 percent forsterite. Low cristobalite occurs rarely as microgranular grains with n 1.486 and birefringence ~0.002. Low tridymite occurs sparingly as grains with irregular extinction and inversion twinning. The indices of refraction (α 1.470, β 1.472, γ 1.476) are relatively low. Isotropic silica glass (n 1.462), probably derived from tridymite, was tentatively identified. Wavy intergrowths of colorless glasses with $n \sim 1.462$ and colored, high index glasses were observed. Iron-nickel (~3 to 14 percent Ni, mostly 6 to 7 percent Ni) occurs rarely as spheroidal and discoidal bodies, as cubo-octahedrons, and as fragments of meteoritic iron. Iron (<0.5 percent Ni) occurs as tiny spheres, cubical crystals, and thin films in glass, and as grains intergrown with plagioclase and pyroxene. The iron-nickel is intergrown with schreibersite, cohenite, and troilite.

Microgabbro 10050,37 was studied in detail, and wet analyses of the clinopyroxene, anorthite, and ilmenite from this sample are given in Table 2 (analyses 2, 4, and 5). Compositionally, the pyroxenes are extremely variable. Figure 2 shows the bulk pyroxene analysis and the compositional variations from augite, with pigeonite, to iron-rich pyroxenes. The outer edges of large grains, and smaller clinopyroxene grains interstitial to plagioclase, have the most Fe-rich compositions. The range in TiO_2 content of the pyroxenes is from 0.6 to 5.1 percent by weight, and Al_2O_3 ranges from 0.5 to

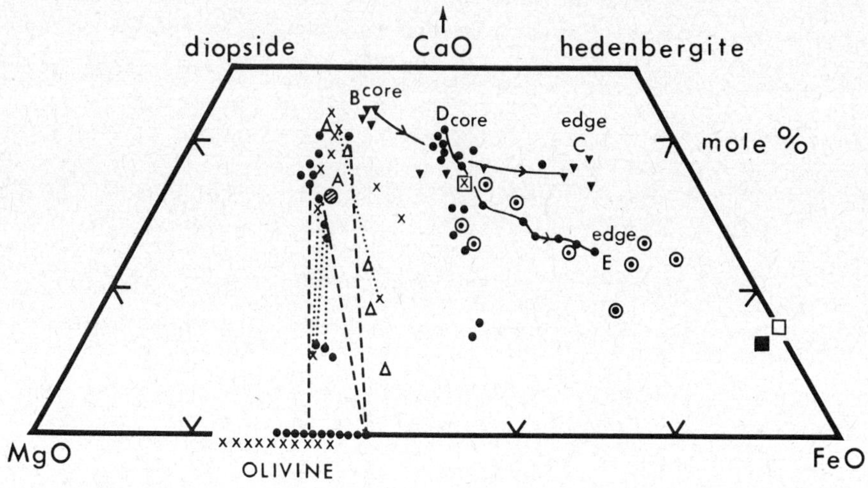

Fig. 2. Pyroxene, olivine, and pyroxenoid compositions in rocks and fines in terms of molecular percentages of CaO, MgO, and FeO. All compositions, except point *A*, are electron microprobe determinations (C. Klein and J. C. Drake, analysts). Point *A* represents the bulk analyses of the two clinopyroxenes in Table 2 (analyses 1 and 2). The symbols represent pyroxenes, olivines, and pyroxenoid in different assemblages: (solid circle) olivine and pyroxene in rock 10050,37 (fine-grained gabbro); (solid circle in ring) fine-grained, interstitial, yellowish tan pyroxene in 10050,37; (inverted solid triangle) pyroxenes in rock 10062,34 (gabbro); (open triangle) pyroxenes in rock 10045,32 (vesicular ilmenite basalt); (cross) pyroxene and olivine in fines; (cross in square) titanaugite in fines (TiO_2, 7.0 percent by weight); (solid square) yellow phase in rock 10047 similar to synthetic Ca-Fe pyroxenoid (C. W. Burnham and D. H. Lindsley, personal communication; their synthetic composition is shown by the open square). Dotted lines connect coexisting augite and pigeonite in fines and rocks. Dashed lines connect coexisting augite and olivine. Lines *B–C* and *D–E* represent probe traverses across single pyroxene grains, from core to edge (traverse length, approximately 250 μm).

6.0 percent by weight. The average plagioclase is 90 percent anorthite; the total range is from 84 to 94 percent anorthite. The anorthite probably contains some Fe in its structure. The accessory ilmenite is the main host mineral for Zr (Table 2, analysis 5). The olivines, from 58 to 70 percent forsterite, occur as rounded or bleb-like grains in clinopyroxene. Low cristobalite occurs sparingly as flattened octahedrons that are oriented microgranular inversion pseudomorphs after high cristobalite. Low tridymite occurs as inversion pseudomorphs after crystals of high tridymite, which are tabular on (0001). Both minerals show inversion twinning. Troilite, an accessory mineral, shows exsolution blebs of iron. An unidentified yellow-brown mineral containing Fe and probably Cl that decomposes under the electron beam was noted.

In microgabbro 10047 a new yellow mineral (with α 1.752, β 1.758, and δ 1.767, faintly pleochroic) consisted of CaO 6.0, FeO 47.2, MgO 0.9, MnO 1.0, Al_2O_3 0.2, TiO_2 0.5, SiO_2 45.0 percent by weight (total 100.8), corresponding to $Ca_{13}Mg_3Fe_{84}$. It has a composition similar to, and an x-ray powder pattern identical with, that of a synthetic pyroxenoid of composition $Ca_{15}Fe_{85}$ (*1*).

Impact craters produced by hypervelocity particles and ranging up to about 1 mm in diameter are common on spherules and fragments of the glass and iron-nickel in the fines. The craters have a complex morphology: in nickel-iron, a surface rim and two inner subcraters of different diameter; in glass, two subcraters, the inner with a rim of glass, a zone of granulated glass, and an outer shallow fracture crater. These impact craters are viewed as a transient event, produced while the material was in free flight in the explosion cloud of a meteorite impact in which the hypervelocity particles were themselves generated.

Well-rounded, abraded grains of glass, of plagioclase, of pyroxene, and of lithic fragments are common; they resemble terrestrial detrital sands. These features, and the rounding of large rock fragments on the lunar surface, are attributed to abrasion attending meteorite impacts. The surface material also was impacted by hypervelocity particles at the same time.

CLIFFORD FRONDEL
CORNELIS KLEIN, JR.
JUN ITO, JOHN C. DRAKE
Department of Geological Sciences, Harvard University, Cambridge, Massachusetts 02138

Reference and Note

1. C. W. Burnham and D. H. Lindsley, personal communication.
2. Supported by NASA contract NAS-9-8100. We thank C. W. Burnham and M. Campot for assistance.

4 January 1970

Mössbauer Spectrometry of Lunar Samples

Abstract. *Nuclear gamma resonance measurements for the nuclide ^{57}Fe in lunar material were made in transmission on lunar fines and in scattering on intact lunar rock chips. No appreciable amount of ferric iron was detected. Resonances were observed for ilmenite in all samples. Strong resonances attributed to ferrous iron in silicates, including pyroxenes and, in some samples, glasses and olivine, were also present. Metallic iron, alloyed with nickel, and troilite were also detected in the lunar fines. Differences in the spectra of various samples of lunar material and their significance are discussed.*

This paper reports on the application of Mössbauer spectrometry (*1-3*) to lunar material. Mössbauer spectrometry is appropriate for the study of solid materials that contain appreciable concentrations of elements that exhibit nuclear gamma resonance. These elements include iron, one of the major elements in the material returned to earth by Apollo 11 (*4*), and which was detected in each remote chemical analysis at Surveyor sites on the moon (*5*).

This technique has proved especially useful for the identification and characterization of iron-bearing minerals and mineraloids, determination of the phase distribution of the iron, and determination of the oxidation state of the iron in the sample. It involves the measurement of nuclear hyperfine spectra of iron atoms in the lunar material by determination of the energy spectrum of the resonance absorption of nuclear gamma radiation transmitted through

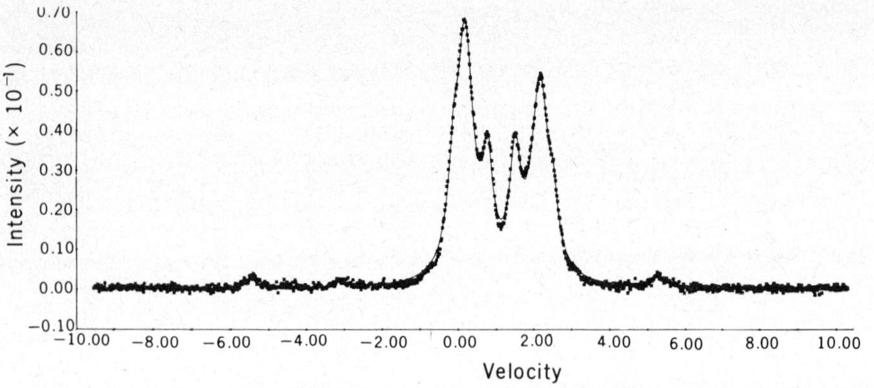

Fig. 1. Mössbauer hyperfine spectrum for unsieved documented sample lunar fines (specimen 10087,4) with least-squares computer fit (measured in transmission).

samples, or alternatively, by detection of the energy spectrum of resonantly absorbed gamma radiation from radiation reemitted from the surface of lunar specimens. These Mössbauer hyperfine spectra contain quantitative information on all major compounds containing significant amounts of iron in the sample. The method can distinguish readily among many important mineral phases, but is relatively insensitive to phases containing only trace amounts of iron.

Two general types of instruments have been used: transmission spectrometers and a scattering spectrometer. In each case, a collimated beam of gamma radiation from a radioactive source of ^{57}Co having a source strength in the 10 to 100 mc range was directed at the lunar sample. Proportional counters of several different designs were used to detect the transmitted or backscattered gamma radiation. In each case, the radioactive source was driven with a controlled periodic motion in order to provide the resonantly emitted gamma radiation with Doppler shifts of appropriate magnitude to cover the energy range of hyperfine splittings for the ^{57}Fe nuclei present in the sample. Multiscaling electronics operated in synchrony with a constant acceleration mode transducer motion provided for the recording of the number of gamma rays transmitted or scattered at each intermediate value of the Doppler shifting velocity. Mössbauer hyperfine spectra measured in this manner have permitted the analyses of terrestrial rocks for iron-bearing minerals, total iron content, and ferric to ferrous ratios. For some terrestrial samples, further useful information concerning grain size, shock exposure, radiation damage, and thermal history has been obtained (*1–3, 6*).

The lunar fines examined included material from the documented sample and a variety of sieve size fractions from the bulk sample. Figure 1 is characteristic of the spectra obtained for lunar fines. It was measured using a specimen of approximately 70 mg of unsieved fines from the documented sample. The lunar dust covered an area of approximately 2 cm^2 corresponding to a sample thickness of about 35 mg/cm^2 between the beryllium windows. Other samples of fines ranging in sample thickness between 31 mg/cm^2 and 73 mg/cm^2 are in use. These also were completely enclosed in sample holders between thin beryllium or lucite windows. In most cases, the samples of fines in the holders have been maintained in a dry nitrogen atmosphere throughout the period of investigation.

The spectrum of Fig. 1 indicates that iron is abundant in the lunar fines and that it is distributed in a number of different phases (*1*). This sample spectrum is clearly different from the spectra of meteorites (*2, 3, 7*) and from the spectra of terrestrial rocks which have been examined using this technique (*1*). The most prominent peaks (near zero velocity and +2.1 mm/sec) largely originate from resonant absorption by high spin ferrous iron in the silicate phases (including iron-bearing glasses and monoclinic pyroxenes) (*1–3, 6, 8*) and by Fe^{2+} in minor opaque phases. Strong peaks are also present from ilmenite (these being located in the vicinity of +0.7 mm/sec and +1.4 mm/sec) (*1, 9*). At higher velocities, three of the six peaks characteristic of metallic iron indicate its presence in the fines (*3, 7*). In some samples of fines, additional low intensity peaks in the high velocity range indicate the presence of troilite (*3, 10*).

Standard techniques of spectral analysis, including least-squares computer fitting of the spectra (*1–3*), have been used for the quantitative interpretation. Some variations both in total iron and in the distribution of iron among mineral phases have been observed between the fines of the documented sample and the fines of the bulk sample, and between different sieve size fractions of the bulk sample; however, all samples of fines which we have examined give grossly similar spectra. No major systematic trends in phase distributions were apparent in the sieve frac-

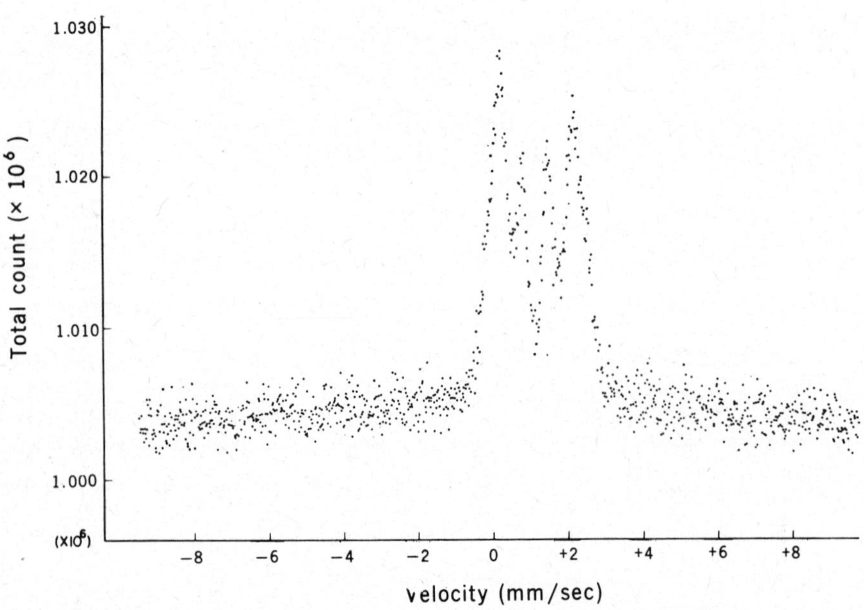

Fig. 2. Mössbauer hyperfine scattering spectrum for an interior surface of lunar rock 10020,20 (a dense fine grained crystalline rock with scattered vesicles).

tions from the Lunar Receiving Laboratory. In a typical sample of fines, about 73 percent of the total iron appears to be in the silicate phases, about 17 percent in ilmenite, and about 5 percent in metallic iron. Two percent or less of the total iron is typically present in troilite. We find no evidence for a significant amount of ferric iron. While our estimates of the total iron content are only approximate, we find an average of 12 percent, in good agreement with the 12.4 percent already reported (*4*) as typical for type D fines. We also find some evidence of a trend toward higher total iron content in the finer sieve fractions of the fines.

The spectral contribution from the metallic iron, which typically represents of the order of 0.5 weight percent of the fines, shows that the iron is alloyed, and the line positions and widths may be interpreted as indicating a variable composition containing some few percent nickel (*3*, *7*). This suggests a probable meteoritic origin for a significant portion of the metallic iron present in the fines. The silicate phases in the fines are rather complex, including both iron-bearing glasses of varying composition and a wide range of pyroxene compositions. These factors contribute to the exceptional width of the strongest manifest lines in the vicinity of zero velocity and +2.1 mm/sec. The presence of strong absorption attributable to iron-containing glasses is compatible with direct observations of the presence of much glassy material having high indices of refraction in the fines (*4*).

Comparison of the spectra of lunar fines exposed to air with those maintained under dry nitrogen, and with fines originally under vacuum, gave no evidence for any major changes in the lunar fines as a result of exposure to the atmosphere over a time scale of weeks, in agreement with previous observations using other techniques (*4*).

Additional measurements made on lunar fines cooled to liquid nitrogen temperature have, in general, verified the mineral phase identifications already made on the basis of measurements performed at room temperature. None of the major iron-bearing mineral phases appears to undergo a magnetic transition between 300°K and 80°K; this is in accord with the properties of the identified phases, and suggests that the contribution from certain other phases in the opaque assemblage, such as ulvöspinel (if stoichiometric), is reasonably small in these samples.

Prominent features of this material can be observed in the rock spectra (Figs. 2 and 3). Spectral differences are considerably more pronounced among the rocks than among the many samples of fines which we have examined.

Figure 2 shows a scattering spectrum measured on an interior surface of a fine grained crystalline rock (specimen 10020,20). About 25 percent of the iron in this rock appears to be in ilmenite, with the remainder mainly in the silicate phases, which include pyroxenes, some olivine and some iron-bearing glass. Our spectra show that less metallic iron is present in the type A and type B lunar rocks (including this rock) than in the fines or the breccia. Application of spectral smoothing techniques reveals evidence for some troilite.

Initial scattering measurements made on the obverse side of this rock chip (believed from circumstantial evidence to be from the bottom outer surface of the rock) show spectra grossly similar to those of the interior, which indicates a general similarity in mineralogical composition between the surface and the interior. More detailed studies of exposed surfaces are continuing, in order to examine the surface layers for radiation effects and secondary mineral formation.

The ilmenite peaks of specimen 10003,22 (Fig. 3) are relatively less intense for this rock than for the finer grained specimens, which supports the early observation (*4*) of a seemingly lower abundance of ilmenite in coarser grained rocks. There is no evidence for olivine in this rock, and iron-bearing glasses appear to be largely absent. The silicate spectrum in this rock is thus dominated by the pyroxene contributions, which exhibit relatively narrow line widths and have outer lines with unusually low relative intensities. This is indicative of appreciable ordering of the ferrous ions in the cation sites of these lunar pyroxenes, which is of special interest, as the cation distributions in pyroxenes reflect temperatures and pressures of mineral formation as well as shock exposure (*6*, *8*). While much more detailed study is required for an analysis of the thermal history of this material, our results may be interpreted as indicating a slow cooling history for this rock.

Spectra measured for the vesicular lunar basalt (rock 10022) show that about one-third of the iron in this rock is in ilmenite, which exceeds the amount in any other sample we have

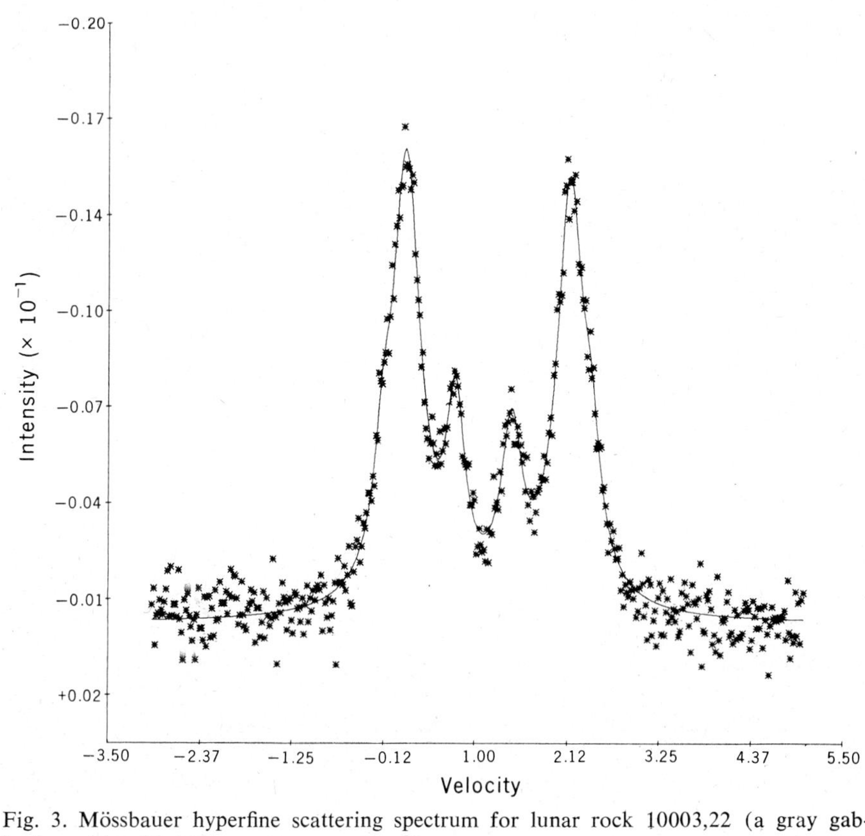

Fig. 3. Mössbauer hyperfine scattering spectrum for lunar rock 10003,22 (a gray gabbroic rock), with least-squares computer fit.

examined. The silicate phases contribute wide lines, indicating that this rock contains iron-bearing glasses in addition to pyroxenes, which is compatible with its volcanic character (*4*).

Initial scattering measurements on breccia 10065 produced spectra somewhat similar to those of the fines, as would be expected in view of the probable origin of lunar breccia as shock lithified fines (*4*); spectral differences include considerably reduced intensity in the ilmenite regions. Magnetically ordered material is present in this breccia and it seems to be primarily metallic iron (*3*, *7*).

Ferric iron is largely absent from the lunar material. At most a few percent of the iron is in the Fe^{3+} state. All of the rocks and fines from Apollo 11 which we have examined appear to have an appreciable iron content and grossly similar chemical and mineralogical compositions with the individual differences as remarked upon, as evidenced by their Mössbauer spectra.

C. L. HERZENBERG
D. L. RILEY
IIT Research Institute, Chicago, Illinois

References and Notes

1. C. L. Herzenberg, in *Mössbauer Effect Methodology*, I. J. Gruverman, Ed. (Plenum Press, New York, in press), vol. 5; ——— and D. L. Riley, in *Developments in Applied Spectroscopy*, E. L. Grove, Ed. (Plenum Press, New York, in press), vol. 8.
2. A H. Muir, Jr., in *Mössbauer Effect Methodology*, I. J. Gruverman, Ed. (Plenum Press, New York, 1968) vol. 4, p. 75.
3. E. L. Sprenkel-Segel and G. J. Perlow, *Icarus* **8**, 66 (1968); E. L. Sprenkel-Segel and S. S. Hanna, *Geochim. Cosmochim. Acta* **28**, 1913 (1964).
4. Lunar Sample Preliminary Examination Team, *Science* **165**, 1211 (1969).
5. A. L. Turkevich, E. J. Franzgrote, J. H. Patterson, *ibid*. p. 277; *ibid*. **160**, 1108 (1968); *ibid*. **162**, 117 (1968); J. H. Patterson, E. J. Franzgrote, A. L. Turkevich, W. A. Anderson, T. E. Economou, H. E. Griffin, S. L. Grotch, K. P. Sowinski, *J. Geophys. Res.* **74**, 6120 (1969).
6. S. S. Hafner and D. Virgo, *Science* **165**, 285 (1969); R. W. Dundon and L. S. Walter, *Earth Planet. Sci. Lett.* **2**, 372 (1967); G. M. Bancroft and R. G. Burns, *ibid* **3**, 125 (1967).
7. L. D. Lafleur, C. D. Goodman, E. A. King, *Science* **162**, 1268 (1968).
8. G. M. Bancroft, A. G. Maddock, R. G. Burns, *Geochim. Cosmochim. Acta* **31**, 2219 (1967).
9. T. C. Gibb, N. N. Greenwood, W. Twist, *J. Inorg. Nucl. Chem.* **31**, 947 (1969); G. Shirane, D. E. Cox, W. J. Takei, S. L. Ruby, *J. Phys. Soc. Japan* **17**, 1598 (1962); Magnitudes of Doppler shift velocities are reported with respect to the center of the iron metal spectrum at room temperature, in accordance with the recommendation of the Mössbauer effect standards committee.
10. S. Hafner and M. Kalvius, *Z. Kristallogr.* **123**, 443 (1966).
11. We thank many colleagues for their contributions to this program, in particular R. B. Moler, S. I. Baker, and P. I. Flinn. We thank S. S. Hafner, A. H. Muir, and N. N. Greenwood for communicating results of their Mössbauer studies on Apollo 11 samples prior to publication. Supported by NASA contract NAS 9-8083.

4 January 1970

Mössbauer Effect and High-Voltage Electron Microscopy of Pyroxenes in Type B Samples

Abstract. *Site occupancy numbers for ferrous iron, magnesium, and calcium at the M1 and M2 sites in lunar clinopyroxenes are estimated from nuclear gamma-ray resonant absorption spectra of* ^{57}Fe. *The cation distribution is ordered; calcium and magnesium prefer M2 and M1, respectively. The distribution corresponds to an equilibrium at a temperature lower than 680°C. Crystals cleaved and sectioned by diamond-knife ultramicrotomy were examined by high-voltage (200 kv) electron microscopy and diffraction. Uniform 300- to 600-Å-wide bands that correspond to single crystal domains were found. Correlation of the bands with magnetic ordering at low temperatures is considered.*

The crystallization and subsequent cooling history of pyroxenes is reflected in their exsolution phenomenon and in the cation distribution over structural sites. Therefore, studies of the domain structure in pyroxene crystals and of the intrinsic site occupancy in the distinct phases are important (*1*).

Clinopyroxenes separated from the type B specimens 10003 and 10044 were crushed to a grain size of approximately 50 μm and further separated into fractions of different Mg,Fe,Ca compositions by means of heavy liquids. Each fraction yields a distinct resonant absorption spectrum (Fig. 1). The nuclear quadrupole doublets due to Fe^{2+} at the M1 and M2 sites are generally well resolved. The spectra of two fractions from 10044 are shown in Fig. 1. Isomer shifts and nuclear quadrupole splittings (peak to peak separation) of the clinopyroxene from rock 10044 and those of a terrestrial augite with similar chemical composition and an iron-rich orthopyroxene are given in Table 1. The widths at half peak height are approximately 0.36 mm/sec.

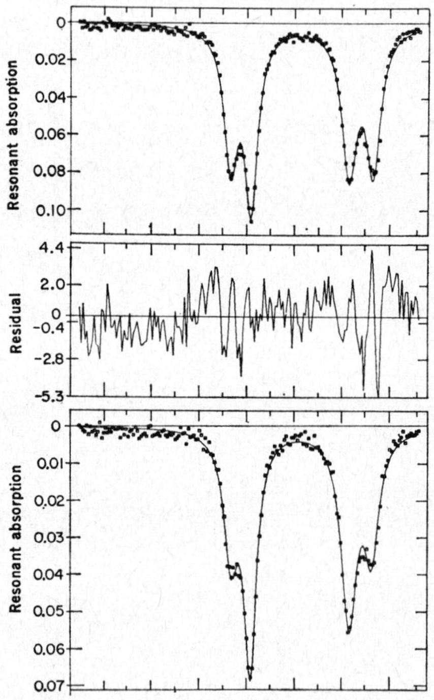

Fig. 1. Resonant absorption spectra of ^{57}Fe in clinopyroxene from specimen 10044. Upper spectrum: iron-rich fraction 10044-P3. Lower spectrum: magnesium-rich fraction 10044-P2. The solid line is a least squares fit (13 variables) to the uncorrected data. Absorbers were held at 77°K. Outer peaks: Fe^{2+} at M1; inner peaks: Fe^{2+} at M2.

The x-ray diffraction powder data indicate that the 10044 pyroxene consists almost exclusively of augite, whereas some fractions from 10003 also include pigeonite. However, measurements of isomer shifts and quadrupole splittings at 77°K in terrestrial pyroxenes with compositions within the enstatite-diopside-hedenbergite-ferrosilite quadrilateral show that the shifts and splittings are only slightly affected by small changes in Mg,Fe,Ca composition or by changes in crystal symmetry, for instance from space group $C2/c$ to $P2_1/c$. The data of Table 1 are consistent with those observations. Therefore, the resolution of the M1 and M2 doublets, for instance in the case of separates from rock 10003, is not surprising, and the hyperfine data of ^{57}Fe can be interpreted in terms of average occupancy numbers.

Average compositions and site occupancy numbers of the two chemically distinct 10044 fractions are shown in Table 2. The Mg,Fe,Ca distribution is principally an ordered one, M2 being almost exclusively occupied by Ca and Fe^{2+}, M1 being occupied by Mg and excess Fe^{2+} (Table 2). No significant Fe^{3+} peaks could be detected.

Estimates of the stability field of the observed cation distribution can be made from heating experiments. Specimens of clinopyroxene 10044-P2 were heated at 675°C for 76 hours and at 1000°C for 17 hours. A significant change in the relative doublet intensities was observed (Table 2). Therefore, the cation distribution in the natural 10044 clinopyroxene corresponds to an equilibrium temperature lower than 675°C.

Studies of the cation distribution in rapidly cooled volcanic orthopyroxenes (*2*) and kinetic studies of the cation exchange process (*3*) suggest that partial ordering of cations in lunar clinopyroxenes may proceed rapidly during initial quenching from a high temperature but complete order is only attained after significantly long geological times at low temperatures. Disordered cation distributions are attained by initial rapid cooling. For comparison some data on the terrestrial Kakanui augite (*4*) are given in Table 2. The site occupancy for this volcanic sample is considerably complicated by the presence of other cations (*5*) but the significantly higher relative proportions of Mg in the M2 sites is consistent with a more rapid cooling history in

Table 1. Nuclear quadrupole splittings and isomer shifts of Fe^{2+} in lunar and terrestrial pyroxenes at 77°K.

Sample	Quadrupole splitting (mm/sec)		Isomer shift referred to metallic iron (mm/sec)	
	M1	M2	M1	M2
10044-P2	2.92	2.06	1.29	1.26
10044-P3	2.96	2.03	1.29	1.26
10044-P2 heated 675°C	2.91	2.04	1.28	1.26
10044-P2 heated 1000°C	2.92	2.05	1.28	1.26
Kakanui-Augite	2.80	2.09	1.26	1.25
Orthopyroxene XYZ	3.10	2.04	1.29	1.26

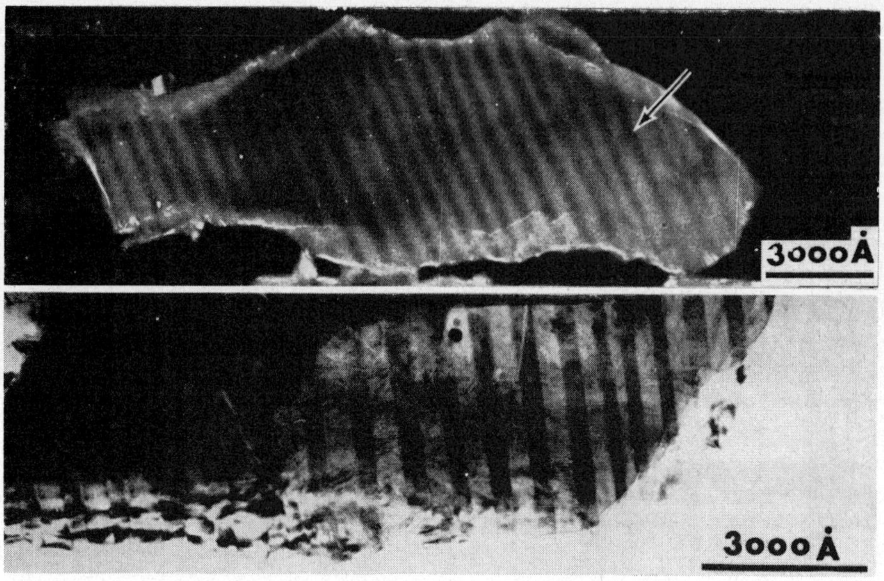

Fig. 2. High-voltage, dark field electron micrograph of cleaved clinopyroxene 10044 (above) and high-voltage electron micrograph of iron-rich clinopyroxene 10044-P3 showing uniform dense bands which are parallel (001) and irregular granules (below).

comparison with the augite from rock 10044. The change of site occupancy in lunar augite is comparable to the change in ordered orthopyroxenes heated at the same temperature. The slightly smaller change in the lunar specimen is indicative of a somewhat larger Gibbs free energy difference for the Mg,Fe^{2+} exchange reaction between M1 and M2 at 1000°C.

Pyroxenes cleaved and sectioned by diamond knife ultramicrotomy (*6*) and mounted directly on thin film specimen grids were examined by high voltage (200 kv) electron microscopy under conditions of increased penetration power and decreased specimen damage (*7*).

In crystals from fraction 10044-P3, regularly spaced, dense bands (Fig. 2, dark areas) with widths of 300 to 600 Å were recorded in approximately 2000 plates. These uniform bands appear to be single crystal domains and are oriented with their long axis in the plane of the crystalline layers, approximately normal to crystallographic *c*. Intrinsic lattice spacings of 2.5 Å (probably d_{002}) can be detected within the bands parallel to their long axis.

The bands resemble electron-optical images of magnetic domain walls as seen in thin layers of magnetic materials (*8*). They are predominantly seen in iron-rich (10044-P3) crystals but are absent in magnesium-rich (10044-P2) crystals and in the terrestrial orthopyroxene XYZ. In this connection resonant absorption spectra of specimen 10044-P3 taken at low temperatures do reveal magnetic ordering, below a Néel point of approximately 20 to 30°K. However, it is interesting to note that iron-bearing chain silicates are generally not mag-

Table 2. The Fe^{2+} distribution data over M1 and M2 sites in lunar and terrestrial pyroxenes. The Fe^{2+} distribution numbers were determined from the ^{57}Fe resonant absorption spectrums. The site occupancy was calculated from the distribution numbers assuming a preference of Ca for M2. Al, Ti, and Mn are ignored; suggested values are Al = 0.03, Ti = 0.03 and Mn = 0.02 (*10*). En, enstatite; Fs, ferrosilite; Wo; wollastonite.

Sample	Molecular composition	Fe^{2+} distribution numbers		Site occupancy M1			Site occupancy M2		
		M1	M2	Mg	Fe	Ca	Mg	Fe	Ca
10044-P2	$En_{0.44}Fs_{0.20}Wo_{0.36}$	0.36	0.64	0.86	0.14	0	0.02	0.26	0.72
10044-P3	$En_{0.36}Fs_{0.34}Wo_{0.30}$	0.46	0.54	0.68	0.32	0	0.03	0.37	0.60
10044-P2 heated to 675°C		0.38	0.62	0.85	0.15	0	0.03	0.25	0.72
10044-P2 heated to 1000°C		0.40	0.60	0.84	0.16	0	0.04	0.24	0.72
Kakanui-augite*		0.34	0.66	0.72	0.06	0	0.19	0.12	0.62
Orthopyroxene XYZ	$En_{0.12}Fs_{0.85}Wo_{0.02}$	0.447	0.534	0.24	0.76	0	0.02	0.94	0.02

*0.90 Mg, 0.21 Fe, 0.61 Ca, 0.33 Al, 0.09 Na, 0.02 Ti per six oxygen atoms (*5*).

netically ordered, even at very low temperatures, especially if the amount of diamagnetic cations (magnesium, calcium) substituting for iron at the octahedrally coordinated positions is larger than 25 percent as in the 10044-P3 augite (*9*). We believe that the observed ordering is due to iron-iron clustering in the single crystal domain bands as revealed by electron microscopy (Fig. 2). In the domains between the bands, clustering of Mg,Ca seems likely. A statistical estimate shows that the total volume of the bands and interband domains within each crystal is about the same. Thus, a speculation on their respective chemical compositions can be made. If one assumes the extreme case that the ferrous ions are entirely located within the bands one is led to the pigeonite composition $En_{0.19}Fs_{0.68}Wo_{0.13}$ for the bands and an almost pure diopside of the composition $En_{0.53}Wo_{0.47}$ for the interband domains (molecular percent). The speculated band composition is consistent with the Fe^{2+} distribution data for 10044-P3 in Table 2. If one assumes complete ordering of Mg and Ca in the bands, the Fe^{2+} site occupancy in M1,M2 is 0.62, 0.73, respectively. This occupancy could certainly produce magnetic ordering at low temperature due to M1-oxygen-M2 superexchange.

The electron microscopic studies suggest an unexpected complexity in the intrinsic atomic organization of the clinopyroxenes from sample 10044. While these may indicate implications on the history of the crystals with respect to temperature and pressure, more detailed investigations will be needed for definite conclusions and their relevance.

H. FERNÁNDEZ-MORÁN
STEFAN S. HAFNER
MITSUO OHTSUKI, DAVID VIRGO
Departments of Geophysical Sciences and Biophysics (Pritzker School of Medicine), University of Chicago, Chicago, Illinois 60637

References and Notes

1. S. S. Hafner and D. Virgo, *Science* **165**, 285 (1969).
2. D. Virgo and S. S. Hafner, *Amer. Mineral.* **55**, (Jan.-Feb. 1970).
3. ———, *Mineral. Soc. Amer. Spec. Pap.* **2**, 67 (1969).
4. B. Mason, *N.Z. J. Geol. Geophys.* **9**, 474 (1966).
5. J. R. Clark, D. E. Appleman, J. J. Papike, *Mineral. Soc. Amer. Spec. Pap.* **2**, 31 (1969).
6. H. Fernández-Morán, *Exp. Cell Res.* **5**, 255 (1953); *Ind. Diam. Rev.* **16**, 128 (1956); *J. Biophys. Biochem. Cytol. Suppl.* **2**, 29 (1956).
7. H. Fernández-Morán, *Proceedings of Sixth International Congress of Electron Microscopy* (Maruzen, Tokyo, 1966), pp. 13, 27, and 147; H. Fernández-Morán, *The Neurosciences*, F. O. Schmitt, Ed. (Rockefeller Univ. Press, New York, 1967), p. 281.
8. J. Silcox, *Phil. Mag.* **8**, 7 (1963).
9. G. K. Shenoy, G. M. Kalvius, S. S. Hafner, *J. Appl. Phys.* **40**, 1314 (1969).
10. A. T. Anderson, personal communication.
11. Mössbauer spectroscopy was aided by Mrs. L. Century, Miss B. Janik, and Mr. H. P. Weber and supported by NASA grant NAS-9-8080. Electron microscopy was aided by Mr. C. L. Hough, Mr. C. Weber, and Mr. G. Bowie for the photographic reproduction; Mrs. V. Iglesias, Misses A. Hibino and M. Hanaoka, Mr. R. Vicario, Mr. H. Krebs, and Mr. G. Arcuri for specimen preparation; Miss S. Rowe for editorial assistance; and Mrs. S. Erikson for administrative assistance. It was supported by the Pritzker Fund, the L. Block Fund, and the Otho Sprague Memorial Fund of the University of Chicago and NASA grant 14-001-012 and NIH grant GM-13243-05.

4 January 1970

Mössbauer Spectroscopy of Moon Samples

Abstract. *Lunar bulk sample 10084,85 (< 1 mm size dust), and samples from rocks 10017,17 (fine grained, vesicular), 10046,17 (breccia), 10057,59 (fine grained, vesicular, top surface), 10057,60 (fine grained, vesicular, interior), and 10058,24 (medium grained, not vesicular) have been investigated by ^{57}Fe Mössbauer spectroscopy. Iron metal and the Fe^{2+} minerals ilmenite, pyroxene, troilite, and iron containing glass have been identified. An iron line of sample 10084,85 (originally sealed in nitrogen) showed no significant intensity change when the sample was exposed to air. The antiferromagnetic transition in several lunar ilmenites at 57° ± 2°K corresponds to stoichiometric $FeTiO_3$. Magnetically separated 10057 showed troilite and some metallic iron.*

Mössbauer spectroscopy has been used to investigate the iron-containing phases in lunar bulk and rock samples. In addition to identifying and analyzing the iron-containing minerals, we looked for reactions with air and made a more detailed study of the magnetically ordered minerals. A search for unusual charge states of Fe which might be expected to be unstable in air was given first priority, followed by a test of the reactivity of the lunar material with air.

The Mössbauer apparatus included two electromechanical systems and a constant velocity mechanical system. The sources were ^{57}Co in Cu. All spectra are shown relative to zero velocity at the center of the metallic iron spectrum. The samples were received (from NASA) sealed in nitrogen in nesting plastic containers.

The initial material, designated 10084,85 was from the < 1 mm size fraction of the bulk sample. Immediately upon receipt, it was placed in a glove box which was flushed with nitrogen 3 times before the sample was opened and sieved. A portion (approximately 40 mg/cm^2) of the fraction < 74 μm was loaded into a vacuum cryostat (Dewar) also in the glove box.

Using an electromechanical system, we obtained spectra with ~ 10^6 count/channel for the velocity ranges ± 11 and ± 3.4 mm/sec at room temperature, 80°K and 5°K. All high-velocity scans showed weak lines of Fe metal. The rest of the absorption could in all cases be accounted for by Fe^{2+} in ilmenite, in pyroxene M1 and M2 sites (*1*), and in glass. A low-velocity scan at room temperature is shown in Fig. 1*a*. There was no indication either of unusual Fe charge states or of Fe^{3+} ions.

The intensity of one of the outer iron lines was then monitored with the constant-velocity spectrometer. After counting for 1 day to establish a reference, we exposed the sample to air. Counting continued for 4 days with no observable change, and we concluded that no significant reaction of the metallic phase with air occurred at room temperature. Subsequently all samples were handled in air. Measurements repeated on this sample after about 2 months exposure to air still indicate no reaction.

Approximately 330 mg of the < 74-μm dust from sample 10084 was separated into three fractions with heavy liquids (specific gravity, 3.3 and 4.25). Under a microscope the light fraction (about 250 mg) appeared to contain most of the glass and less than or about 10 percent pyroxene. The medium fraction (about 65 mg) appeared to be enriched in pyroxene, but ilmenite and some glass were also apparent. The heavy fraction (about 15 mg) was enriched in ilmenite, and considerable pyroxene was apparent.

Examination of the fractions with a magnet showed that most of the glassy material was strongly attracted, while less than half the material in either of the two heavier fractions was attracted. Mössbauer spectra showed that the Fe metal lines were relatively enhanced for the light glassy material compared with those in the spectra of the two heavier fractions. Therefore, we find a strong tendency for the Fe metal in the dust

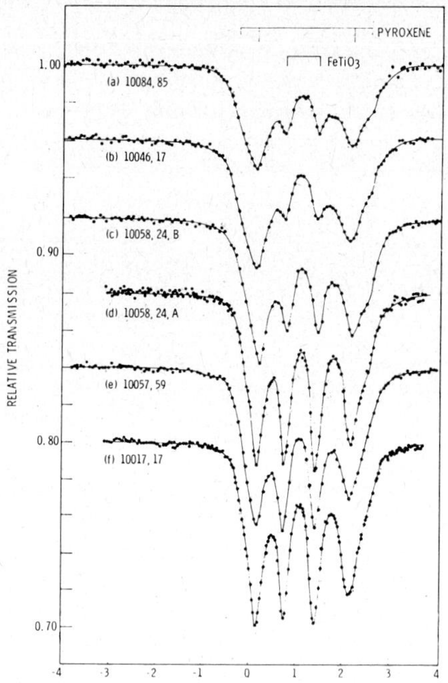

Fig. 1. Room-temperature Mössbauer spectra of lunar dust (*a*) and rock samples (*b–f*). Curves *b–f* have been displaced for clarity.

to be intimately associated with glass.

Approximately 100-mg portions of chips from four rocks, Nos. 10017,17 (fine grained, vesicular), 10058,24 (medium grained, not vesicular), 10057, 59 (fine grained, vesicular), and 10046, 17 (breccia), were crushed to fine powders. Mössbauer spectra were taken at room temperature with both wide and narrow velocity scans on the electromechanical systems. The narrow velocity spectra are shown in Fig. 1, curves *b–f*. Sample 10058 consisted of two fragments, and separate absorbers were made from parts of each. As indicated by Fig. 1, curves *c* and *d*, sampling differences within a given rock (medium grained) can be important. A few grains from each rock also were mounted for microscopic examination.

The Mössbauer spectra of the breccia 10046 were almost identical with those of the bulk material 10084. In the case of rock 10057, we had both a surface piece 10057,59 and an interior piece 10057,60. Although the top of the surface piece had visible glazing of glass, finely fractured "sugary" feldspar, and other physical differences showing surface damage, the absorber prepared by grinding up a fragment about 2 mm thick gave a Mössbauer spectrum indistinguishable from that of the interior piece, thus indicating negligible penetration of surface damage.

The central parts of the spectra of the rocks, excluding the breccia 10046, show mainly ilmenite lines and pyroxene M1 and M2 site lines. However, preliminary spectrum-stripping analysis (*2*) also suggests the presence of small amounts of glass. Again there is no evidence of unusual charge states or of Fe^{3+} ions. Only a very small amount of olivine could be present in any of these rocks.

Although the pyroxene M1 and M2 site lines are not well resolved at room temperature, differences between the various rocks are apparent (especially after spectrum stripping). This may indicate chemical differences between the rocks and also provide information about the thermal or shock histories.

In the wide velocity scans of the rocks, metallic Fe was observed only in the breccia. However, weak absorption at the troilite FeS positions was definitely seen in the rocks 10017 and 10057 and possibly in 10058. The presence of pyrrhotite could not be ruled out on the basis of these data alone.

Relevant information was obtained from microscopic examination of the grain mounts. The breccia 10046 and the dust contained Fe metal (i) enclosed in glass spheres, (ii) associated with troilite blebs, (iii) as occasional irregular subangular grains, and (iv) as very fine (micrometer size) particles throughout the fine breccia matrix. Rocks 10017, 10058, and 10057 all contained troilite blebs often enclosing blebs of metallic iron. Metallic Fe was occasionally seen lining cracks in ilmenite and silicates. In sample 10058 and within rock fragments in the breccia, occasional exsolution lamellae of a phase more reflective and lighter in color were seen in the ilmenite.

On the basis of the above results we decided to investigate the stoichiometry and homogeneity of the ilmenite and to clarify the nature and association of the magnetically ordered phases. The temperature and sharpness of the antiferromagnetic transition in ilmenite of rocks 10017 and 10058 and the heavy fraction of the bulk sample 10084 were measured by the following technique. The constant-velocity spectrometer was set at the position of the higher energy line of the ilmenite quadrupole split doublet, and the counting rate was monitored as a function of temperature from about 20°K to about 80°K. Normalized results are shown in Fig. 2, together with data for a reference sample of terrestrial ilmenite (*3*) from Ashe County, North Carolina.

Approximately 500 mg of the interior portion of rock 10057 was finely powdered and forced to flow in an acetone suspension past a magnet (about 1 kilogauss). A total of 7.4

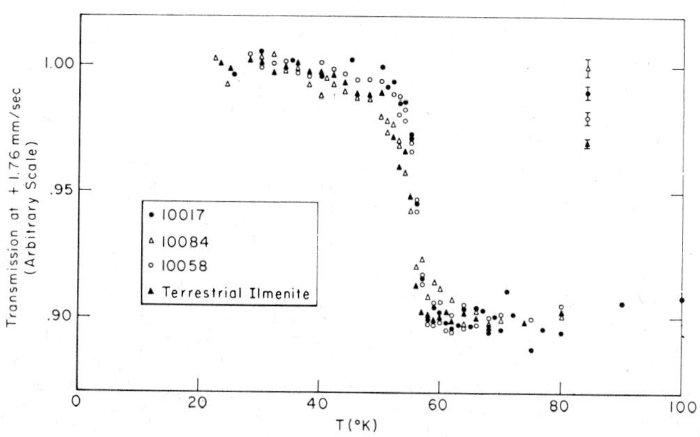

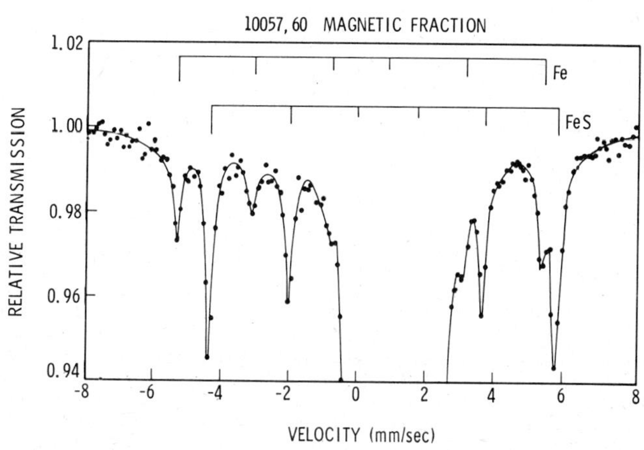

Fig. 2 (left). Normalized curves of counting rate plotted against temperature at + 1.76 mm/sec, from which the antiferromagnetic transition temperature of ilmenite is inferred. Fig. 3 (right). Room-temperature Mössbauer spectrum of magnetically separated material from lunar rock 10057.

mg of magnetic material was recovered. A spectrum of this material at room temperature is shown in Fig. 3. Relatively strong troilite lines and weaker Fe metal lines are easily visible. There is no evidence for pyrrhotite. The complete spectrum (untruncated) indicates an increase in the ratio of ilmenite to pyroxene but is otherwise similar to the spectrum of the unseparated rock.

To see whether the Fe metal plus FeS in the rock could possibly be sufficient to account for the Fe in the dust we obtained ratios of Fe metal and FeS to total Fe by summing the areas in appropriate groups of channels in the spectra. Metallic Fe plus FeS constitute about 11 percent of the total Fe in the dust and only about 5 percent of the total Fe in the rocks. On rock 10057 we found Fe metal and FeS individually to be about 1 and 4 percent of the total Fe. In the dust Fe metal is greater than or about 7 percent of the total Fe.

Assuming our samples are representative, we summarize our present conclusions as follows.

1) The original magma from which the rocks (at least 10057) crystallized was in a net-reduced state so that the assemblage of Fe metal + troilite formed on solidification with no pyrrhotite. The Fe metal lining cracks in ilmenite and silicates can be accounted for by the low temperature of the eutectic in the Fe-FeS system. It can be assumed that the major constituents of the rock were solid and that the ilmenite had already cracked because of contraction stresses before this Fe-FeS solution froze, causing some of the excess Fe to diffuse into the cracks in the ilmenite.

2) There is considerable excess of Fe metal in the dust over that in the rock or over that which could be formed by reducing all the FeS in the rocks. Therefore, a large fraction of the Fe metal in the dust must either result from reduction of Fe in the silicates during formation of the glass or be of meteoritic origin or perhaps both. Reduction of the silicates is possible with the right temperature-time relationship and could be enhanced if the surface being melted has been saturated with solar-wind hydrogen. If the different density fractions of the dust were analyzed for total Ni and the results compared with the Fe metal determined by the Mössbauer measurements, a semiquantitative estimate of the relative importance of the processes could be made.

3) The ilmenite of all samples seems to be homogeneous, well ordered, and very nearly stoichiometric $FeTiO_3$. The magnetic transition of rocks 10017 and 10058 is very sharp, and within the temperature accuracy of $\pm$ 2°K it corresponds to the values reported for synthetic $FeTiO_3$ (*5*). The transition for the dust is slightly broadened on the low-temperature side, perhaps because of disorder induced either by shock or by quenching from a high temperature. On the basis of available data (*5*), hematite (Fe_2O_3) in solution is expected to lower the magnetic transition temperature of ilmenite at a rate of about 2.8°K per mole percent. Therefore, we conclude that the ilmenite in our samples must contain less than 0.5 mole percent Fe_2O_3, and, since considerably more than this is soluble, the observed exsolution lamellae must not be Fe_2O_3. Since the original magma appears to have been highly reduced, we suggest the possibility that the lamellae might even be a phase containing Ti_2O_3.

4) As determined from bulk sample measurements, there appears to be no significant reaction of the metallic iron with laboratory air at room temperature over the time scale of about 2 months. Also, no other time-dependent spectrum changes were observed. There is no evidence for Fe^{3+} or unusual charge states.

A. H. MUIR, JR.
R. M. HOUSLEY, R. W. GRANT
M. ABDEL-GAWAD, M. BLANDER
Science Center, North American Rockwell Corporation, Thousand Oaks, California 91360

References and Notes

1. B. J. Evans, S. Ghose, S. Hafner, *J. Geol.* **75**, 306 (1967); G. M. Bancroft, A. G. Maddock, R. G. Burns, *Geochim. Cosmochim. Acta* **31**, 2219 (1967).
2. A. H. Muir, Jr., in *Mössbauer Effect Methodology*, I. J. Gruverman, Ed. (Plenum, New York, 1968), vol. 4, p. 75.
3. Stoichiometric within about 1 percent by chemical analysis.
4. C. H. Shomate, *J. Amer. Chem. Soc.* **68**, 964 (1946); J. J. Stickler, S. Kern, A. Wold, G. S. Heller, *Phys. Rev.* **164**, 765 (1968).
5. Y. Ishikawa, *J. Phys. Soc. Japan* **17**, 1835 (1962).
6. We thank W. B. Norris, K. G. Rasmussen, M. R. Bloombaum, and D. H. Hengstenberg for technical assistance; and Dr. C. L. Herzenberg and Prof. S. S. Hafner for communicating the results of their Mössbauer studies on Apollo 11 samples before publication. Supported in part by NASA contract NAS 9-10208 with Space Division, North American Rockwell Corporation, Downey, California 90241.

4 January 1970

Magnetic and Electrical Properties

Magnetic Properties of Lunar Samples

Abstract. *A breccia sample (10023) from the moon was found to have a strong and fairly stable remanent magnetization. If this sample was not magnetized by local fields in the spacecraft or in the lunar receiving laboratory, it must have been magnetized on the moon. This could have happened in a variety of ways, such as cooling through the Curie temperature, by continuous thermal cycling, or by impact, but all of these require the presence of a magnetic field. Such a field could have been of internal origin in the moon, or it could have been a residual effect from the earth's magnetic field at a time when the moon and the earth were much closer together. Thermomagnetic studies identify the presence of iron with about 1 percent nickel (igneous), iron with about 5 to 10 percent nickel (meteoritic), iron with about 33 percent or more nickel (meteoritic), and ilmenite.*

We studied the magnetic properties of a solid breccia sample (10023) and a powder sample (10084). Preliminary investigation of this solid sample and others was carried out with a fluxgate magnetometer (*1*). This breccia sample was found to have a magnetic remanence of 7.7 $\times$ 10^{-3} emu/g and an induced magnetization of 7.6 $\times$ 10^{-3} emu/g in a field of 0.92 oersted (*1*).

Magnetic studies of two major areas are reported. The first is a detailed investigation of the natural remanent magnetization (NRM). This property should reveal something about the magnetic history of the moon because any NRM present in the samples would be the result of exposure to a magnetic field. The moon does not at present have a significant field (*2*). In an orbiting vehicle 300 km above the surface it was shown that the magnetic moment of the moon was less than 4 $\times$ 10^{20} cgs (the present value for the earth is approximately 8 $\times$ 10^{25}) and that the regional surface field must be less than 16 γ (1 γ = 10^{-5} oersted). As a result the presence of a significant and stable remanent magnetization would suggest either that the moon once had an intrinsic magnetic field of its own, much larger than the present one, or that it was in the vicinity of another body, such as the earth, which had a significant field.

The second area of study is to consider the application of magnetic techniques to identify the amount and types of various iron-bearing minerals. This approach could positively identify very minute amounts of such minerals as magnetite, iron oxyhydroxides, siderite, pyrite, and many others (*3*). In the particular context of the lunar samples as discussed in the preliminary examination, it is clear that the major role to be played is the identification of iron and iron-nickel minerals. The preliminary report suggests also that small amounts of troilite (FeS) and large amounts of ilmenite are present in the igneous rock.

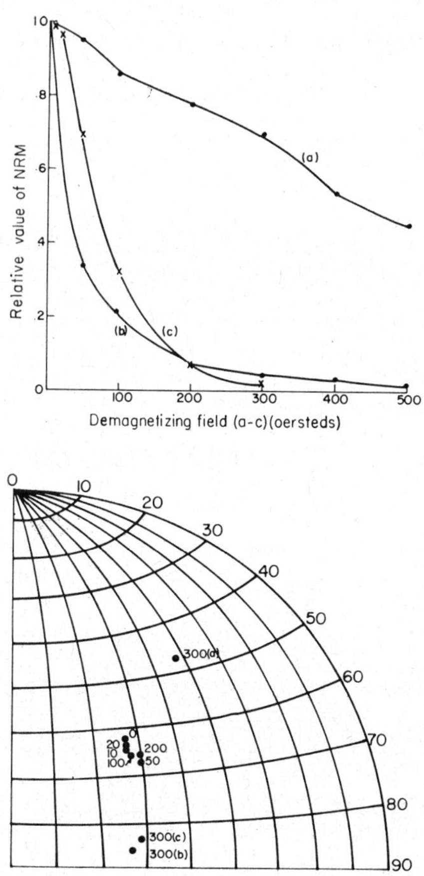

Fig. 1. (Top) Alternating current demagnetization curves of natural remanent magnetization. (a) Oxidized basalt, Flagstaff, Ariz.; (b) unoxidized basalt, Flagstaff, Ariz.; (c) lunar breccia sample (10023). (Bottom) Stereogram of directions of magnetization on demagnetization (a–c).

Measurements of the NRM of the breccia sample show that the sample has a strong magnetization of 2.8 $\times$ 10^{-3} emu/g. The direction of this magnetization is of course not known with respect to north but it is pointing up at about 60° from the horizontal as defined by the upper surfaces. The sample was subjected to alternating field demagnetization in fields up to 300 oersteds. The decay of the magnetization is quite similar to that for thermoremanent magnetization (TRM) in typical basaltic rocks with unoxidized magnetite grains a few microns in size (*4*). The direction of magnetization is essentially unchanged during the demagnetization process although some scatter is present in repeat measurements at 300 oersteds (Fig. 1a). This type of demagnetization curve is not typical of rocks with a highly stable magnetization but it is typical of many rocks generally considered adequate for paleomagnetic studies. There is no doubt that this specimen acquired its remanent magnetization in the presence of a magnetic field. It is possible that this was a field of several hundred oersteds in the spacecraft or the lunar receiving laboratory but this seems unlikely and if similar results are found on many samples this possibility will be even less likely.

The following pertinent minerals were observed in the breccia sample using a reflecting microscope. (i) Iron is present in particles a few microns in size, generally associated with troilite. This iron is probably of igneous origin and has a Curie temperature of 760°C suggesting the presence of a few percent

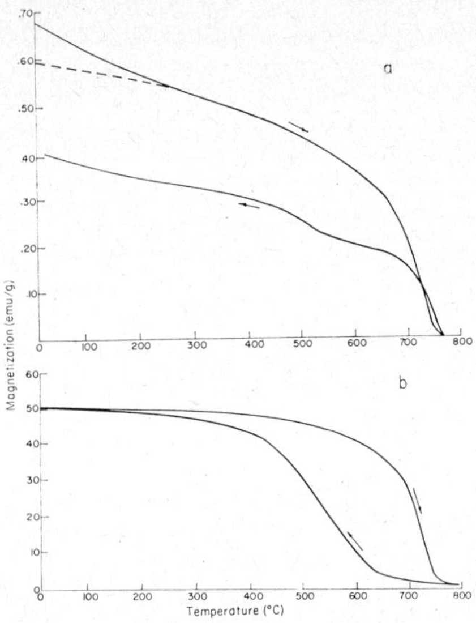

Fig. 2. Saturation magnetization versus temperature curves (at 3000 oersteds). (a) Surface powder showing presence of iron and two iron-nickel phases (10084,90). (b) Particle of iron-nickel separated from breccia sample 10023. Apparent thermal hysteresis is due to phase change in kamacite.

of impurities—perhaps 1 or 2 percent nickel or silicon. (ii) In the breccia, ragged particles of iron-nickel are observed. These have a range in grain size from 1 or 2 μm up to 50 μm or more. The particles are believed to be α-phase (body-centered), and on heating they change to the nonmagnetic γ-phase (face-centered) at 750°C and on cooling revert to the α-phase at 650°C. (iii) Troilite (FeS) is abundant but it cannot be detected magnetically. If it were nonstoichiometric it would give a ferrimagnetic Curie temperature of 320°C. The fact that none is seen magnetically means that it is antiferromagnetic and stable on heating both in air and in vacuum. (iv) Ilmenite ($FeTiO_3$) is present in great abundance as reported (*1*). It shows lamellar twinning parallel to the 0001 plane. As will be seen from the low temperature magnetic data, it is nearly stoichiometric. (v) Very small amounts of a deep gray phase are found, intergrown in the ilmenite grains. The composition of this is not known, although it is possible that it is chromite.

Very few studies of the magnetic properties of meteorites and tektites have been made in the past and the available literature is quite limited. Stacey *et al.* (*5*) examined a series of chondritic meteorites and found that many of them do have a significant remanent magnetization carried by α-phase iron-nickel which has 5 to 6 percent nickel (kamacite). They suggest that this means that there was an ancient field in the parent body of the meteorites. As they point out, this phase loses its magnetization on heating at 750°C where a phase change from the α form to the γ form occurs on heating. On cooling the reverse phase change takes place at a lower temperature giving an apparent thermal hysteresis.

This irreversibility is characteristic of iron-nickel alloys with less than 30 percent nickel. Pure iron undergoes a phase change from the α-phase to the γ-phase at 910°C. Alloying with nickel causes the transition to occur at lower temperatures and it reverts to the α-phase on cooling at still lower temperatures unless cooling is at a very slow rate. Bozorth (*6*) gives a graph of the effective temperature of this change on cooling.

In our experiments on the powdered sample (10084) of surface fines, a similar phenomenon was found. A vibrating sample magnetometer (Princeton Applied Research, FM-1) was used to measure the magnetization while heating the samples (Fig. 2a). The heatings were done in a vacuum of somewhat less than 1 μm because heating in air caused complete oxidation of the ferromagnetic phase. Even at the pressures used, it is seen that some oxidation took place since the measurements are not fully repeatable. This was the result of some oxidation of the fine particles, since subsequent tests done by measuring the magnetization before and after heating to 800°C in a furnace with a pressure of less than 3×10^{-5} mm Hg showed that no such breakdown occurred.

In the heating curve of Fig. 2a, three distinct phases can be identified. The first of these has a Curie temperature of about 760°C and is essentially repeatable. This phase is an iron phase, and on heating up to 800°C it undergoes no phase change. The Curie temperature is however distinctly less than that of pure iron (770°C), and from studies of the dependence of Curie temperature on composition this phase must contain a few percent of an impurity such as 1 or 2 percent of nickel or silicon. This phase is believed to be due to the iron found in the igneous rocks (*1*). A second phase is present in the powders which on cooling appear to have a Curie temperature of about 570°C. This type of curve is better illustrated in Fig. 2b, which is a temperature versus magnetization curve for a single iron-nickel particle taken from the breccia sample. On heating, the sample loses its magnetization at about 750°C and reacquires it at about 600° to 650°C on cooling. This curve is completely repeatable and indicates the presence of an iron-nickel alloy as considered above for meteorites. The graph from Bozorth (*6*) suggests that 5 to 7 percent nickel is present in this phase which is typical of the kamacite of iron-nickel meteorites. This phase is not present in the igneous rocks (*1*) but is present in both the powders and the breccias.

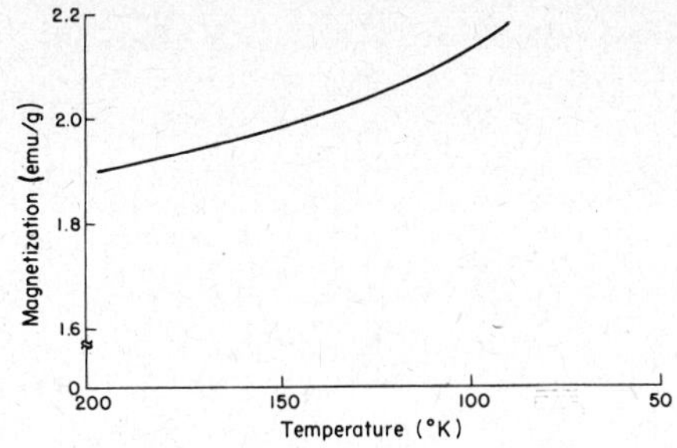

Fig. 3. Saturation magnetization versus temperature curve (at 5800 oersteds) of lunar powder sample (10084,90) at low temperatures.

A third phase is indicated in the curve of Fig. 2a (above the dashed line). On heating and cooling there is a distinct change in the magnetization at about 200°C but the temperature is not well defined. Results similar to this have been discussed by Hoselitz (*7*) who shows that a γ-phase of iron-nickel gives just this type of magnetic response. Since the temperature at the lunar surface probably decreases to about 100°K during the lunar night, this phase must be γ-phase and remain in this state down to this temperature. This requirement can be met by an iron-nickel alloy with a Curie temperature of 200°C and with 33 percent or more nickel. This is probably due to the presence of meteoritic taenite which is common in iron meteorites (*8*). The lack of a distinct Curie temperature may be due to the presence of a range of nickel concentrations.

A similar set of experiments was done at low temperature with the same vibrating magnetometer by making observations down to liquid nitrogen temperature (77°K). On cooling to this low temperature an additional magnetic phase appears (Fig. 3). The shape of the curve is that of a paramagnetic substance; however it will not fit the standard $1/T$ relation. It does however fit a relation of the form $1/(T + \theta)$ where $\theta = 64$°K. This is the relation expected above the Néel temperature for an antiferromagnetic substance such as ilmenite which is known to occur abundantly in these samples. The Néel temperature for ilmenite is reported as 68°K; thus the ilmenite present must be nearly stoichiometric, because solid solution with hematite would tend to make the ilmenite ferrimagnetic and to raise the characteristic Néel temperature.

Hysteresis loops were run on several of the powdered samples (10084) in an attempt to add to our understanding of the magnetic properties. From these observations it is possible to determine the ferromagnetic content assuming that the value of saturation magnetization quoted for iron holds for this material [217.8 emu/g (*6*)]. These results indicate that the powdered material has about 0.3 to 0.5 percent of iron and iron-nickel present in the form of ferromagnetic particles. As pointed out by Senftle *et al.* (*9*), the shape of spherical iron particles in tektites controls the susceptibility up to fields of 6000 oersteds. Our measurements were made in fields of 7500 oersteds. The saturation remanent magnetism is about one-tenth the saturation magnetization, and this suggests that there is a considerable fraction of the iron material present in the form of particles a few microns or so in size. Iron in larger grain sizes does not have such a high remanence. In fact, these data are much like those reported for basalts (*4*) in which particles of magnetite of about this size were found. The bulk coercive force is typically about 100 to 200 oersteds, which again is a value considerably higher than that of larger iron particles and which suggests the presence of some grain fractions with a fairly high coercive force.

The weak field susceptibility was measured on a bridge (Geophysical Specialities MS 3-B) in fields of about 1 oersted. This gave a value of 1.13×10^{-3} emu/g for the powder sample and a value of 1.1×10^{-3} emu/g for the breccia sample.

Several magnetic phases have been identified in this study but it is clear that only the iron or iron-nickel particles, or both, are responsible for carrying the large and fairly stable remanent magnetization found in the sample. These particles are of sufficiently small size as observed both magnetically and optically to be the carriers of the NRM but it is still not known how the magnetization was acquired. If we assume that the magnetization is not caused by exposure to large man-made fields, the following possibilities need to be considered.

1) It is a thermoremanent magnetism (TRM) acquired when the specimen cooled in the presence of a magnetic field. The precise field strength required is not known but it is likely to be within an order of magnitude of the present field of the earth. The novel possibility that it was acquired at the time of the phase change from γ to α needs to be considered as well as the possibility that it is a normal TRM.

2) It is a remanent magnetization caused by the thermal cycling found at the lunar surface in the presence of a nonrandom applied field. This possibility can be further amplified by consideration of the work of other investigators (*10*). If this is a real possibility the effect should attenuate very rapidly with depth beneath the surface.

3) It may be related to the shock process that formed the breccias. Magnetostriction could well be important in this material, and a preferred alignment could be expected if a magnetic field were present at the time of stressing. Unless the shock process itself somehow generates a local magnetic field no remanence would be expected. In general, shock processes reduce the NRM which is already present (*11*).

It is interesting to note that all of these mechanisms require the presence of a magnetic field so that such a remanent magnetism can be formed, and it is tentatively concluded that the moon did have a magnetic field at some time in its history or that these rocks became magnetized when the moon was very much closer to the earth than at present. If the effect found is limited to the breccias, it may be either because no field was present when the igneous rocks were formed or because the shock or the presence of iron-nickel particles increases the magnetic stability of the breccias relative to the igneous rocks.

D. W. STRANGWAY
Department of Physics, University of Toronto, Toronto 181, Canada

E. E. LARSON
Department of Geological Sciences, University of Colorado, Boulder 80302

G. W. PEARCE
Department of Physics, University of Toronto

References and Notes

1. Lunar Sample Preliminary Examination Team, *Science* 165, 1211 (1969).
2. N. F. Ness, K. W. Behannon, C. S. Scearce, S. C. Cantarano, *J. Geophys. Res.* 72, 5769 (1967).
3. D. W. Strangway, B. E. McMahon, J. L. Bischoff, in *Hot Brines and Heavy Metal Deposits*, E T. Degons and D. A. Ross, Eds. (Springer-Verlag, New York, 1969), pp. 460–473.
4. D. W. Strangway, B. E. McMahon, E. E. Larson, *J. Geophys. Res.* 73, 7031 (1968).
5. F. D. Stacey, J. F. Lovering, L. G. Parry, *ibid.* 66, 1523 (1961).
6. R. M. Bozorth, *Ferromagnetism* (Van Nostrand, Princeton, N.J., 1951).
7. K. Hoselitz, *Ferromagnetic Properties of Metals and Alloys* (Clarendon Press, Oxford, 1952).
8. J. A. Wood, *Meteorites and the Origin of Planets* (McGraw-Hill, New York, 1968).
9. F. E. Senftle, A. N. Thorpe, R. R. Lewis, *J. Geophys. Res.* 69, 317 (1964).
10. D. J. Dunlop and G. F. West, *Rev. Geophys.* 7, 709 (1969); Y. Shimizu, *J. Geomag. Geoelec.* 11, 125 (1960); L. Neel, *Advan. Phys.* 4, 191 (1955).
11. R. B. Hargraves and W. E. Perkins, *J. Geophys. Res.* 74, 2576 (1969).
12. Early parts of our studies were supported by NASA grant NGR 22-009-176. For the past year and a half the work has been supported by the National Research Council of Canada. We thank W. Mottley and D. Redman for laboratory assistance.

4 January 1970

Magnetic Properties of Lunar Dust and Rock Samples

Abstract. *Determinations on 20- to 80-milligram portions of the rock samples and the −150 mesh fraction of the lunar dust show pronounced Curie points between 680° and 780°C. Remanent intensities of five rock fragments vary from 8.4×10^{-5} to 0.30×10^{-5} emu/gram. Upon demagnetization, two of the samples had only viscous magnetization and two other samples had stable magnetizations with remanent coercivities in excess of 50 oersteds. Partial thermal demagnetization suggests that these apparently stable moments may have been acquired in a magnetic field in excess of 1500 gammas.*

Current theories regarding the origin of the earth's magnetic field require the presence of a conducting fluid core (*1*, *2*). Consequently, it is of considerable interest to know whether the moon ever had an appreciable magnetic field. If it had one in the past, the implication would be that the interior was at one time a conducting fluid or that the moon was within the earth's magnetic field at the time the rock cooled. Studies of the magnetic properties of five Apollo 11 lunar rock samples, as well as the dust, show that some of the samples have a stable magnetic moment that could have been acquired as a result of cooling in the presence of a magnetic field. The samples studied are (i) a fragment of rock sample 10022, a vesicular basalt with spherical vesicules; (ii) a fragment of rock sample 10069, a fine-grained basalt with irregular vesicules; (iii) a fine-grained gabbro from dust sample 10085,16 fragment 2; (iv) a vesicular basalt fragment from dust sample 10085,16, which is similar to sample 10022 fragment 4; and (v) a fine-grained basalt fragment from dust sample 10085,16, which is similar to sample 10069; and (vi) the −150 mesh fraction of dust sample 10084.

Thermomagnetic analyses were made on the −150 mesh portion of each sample by sealing 10- to 100-mg portions in evacuated quartz glass tubes which were placed in a magnetic force balance in a high gradient field of about 2 kilo-oersteds, and heated at a rate of 1° to 2°C per second to a temperature of up to 800°C. The results of these analyses are shown in Fig. 1. The most prominent Curie point in the dust samples is between 750° and 770°C, suggesting the presence of native iron with very low nickel content. The dust sample also yields a prominent Curie point in the vicinity of 180°C after it has been heated to 800°C. This transition is also present in the initial heating curve but is only evidenced by a very slight change in slope near 200°C. When the sample is cycled below 500°C, the heating and cooling curves are very similar, and the 200°C transition remains very obscure. When cycled to 650°C, this transition becomes more prominent and when cycled to 720°C, the 200°C transition becomes quite pronounced. When the dust is heated beyond 600°C, the heating and cooling curves are quite different, suggesting that a chemical change has occurred within the sample.

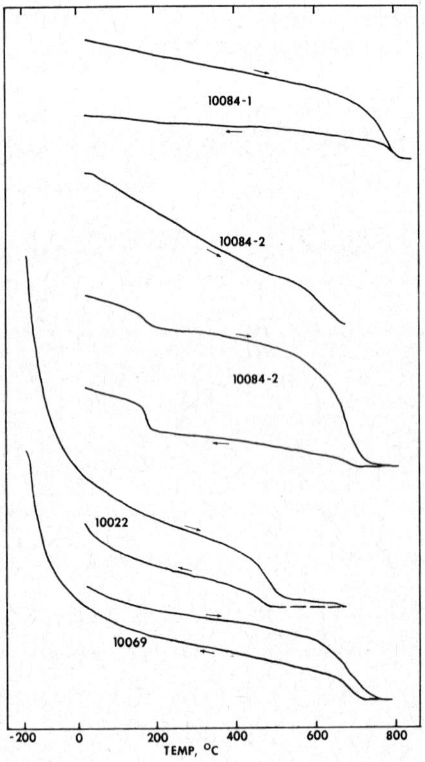

Fig. 1. Selected thermomagnetic analysis curves for lunar dust and rock samples. Vertical scale is arbitrary. Arrows indicate direction of temperature change.

Small portions (approximately 75 mg) of each of the larger rock samples (10022 and 10069) were also subjected to thermomagnetic analysis. The results of these experiments are also shown in Fig. 1 and are in general very similar to those for the dust samples. These samples were also cooled to −196°C in order to check for any ilmenite Curie points. For sample 10069 a Curie point near −170°C seems indicated, but sample 10022 shows no evidence of a Curie point above −196°C. High-temperature Curie points for sample 10069 occur at 520°C and in the range 680° to 752°C. These are not reproducible during cooling; instead they are replaced by one at 690° to 700°C and another at 115°C. In sample 10022, the high-temperature Curie points are at 485°C and at > 680°C and upon cooling at 465°C.

All of the above observations are not characteristic of normal basalts found on the earth. Instead they have many similarities to those reported by Stacey *et al.* (*3*) from chondritic meteorites. This suggests that the major magnetic constituents in these samples are iron with a low nickel content, perhaps a pyrrhotite or troilite(?), and an abundant ilmenite phase with very little, if any, Fe^{3+}.

Initial measurements of the natural remanent magnetization (NRM) of the three rock fragments from the dust sample 10085,16 and the two rocks 10022 and 10069 yielded specific intensities ranging from 8.41 × 10^{-5} emu/g to 0.297 × 10^{-5} emu/g. Two of the small samples had previously been exposed to an external field of about 60 oersteds during petrologic examination in a scanning electron microscope. Consequently, their NRM might be expected to be anomalous. The NRM moments of the three samples that had not been exposed to large external fields are 2.46 × 10^{-5}, 1.92 × 10^{-5} and 0.297 × 10^{-5} emu/g.

All samples were subjected to alternating field demagnetization with peak

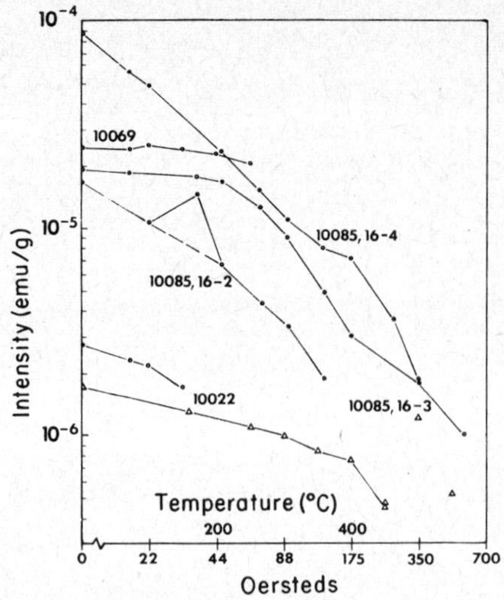

Fig. 2. Progressive alternating field (circles) and thermal demagnetization curves (triangles) for rock samples studied. Note use of log scale.

fields varying from 11 to 570 oersteds, and the results are shown in Fig. 2. Samples 10069 and fragment 3 from dust sample 10085,16 have a stable NRM and have remanent coercivities in excess of 44 oersteds. Both these samples appear to be of the same rock type. Fragments 2 and 4 from the dust sample (the ones that had been in the scanning electron microscope) show only an IRM and no stable NRM. One is a fine-grained gabbro and the other a fragment of vesicular basalt similar to 10022. Upon alternating field demagnetization, they show an almost linear decrease on a log-log scale. Sample 10022 is more stable than fragments 2 and 4, yet it does not show the stability evidenced by fragment 3 and sample 10069.

After being demagnetized in an alternating field of 33 oersteds, sample 10022 was further demagnetized by progressive thermal demagnetization in a vacuum at intervals of 50°C. The results of this demagnetization are shown in a log-linear plot at the bottom of Fig. 2. In a second heating after the 250°C and 350°C demagnetization steps, the sample was allowed to cool in an applied field of 1000 gammas for the first 100°C and then to cool the rest of the way in the absence of any applied field (*4*). The ratio of the intensity lost in the temperature interval to that gained during cooling in the 1000-gamma field is 1.88 and 1.57 for the intervals 150° to 250°C and 250° to 350°C, respectively. Progressive thermal demagnetization showed an increasing rate of decrease in remaining moment up to 450°C, as is characteristic of samples with distributed thermal blocking temperatures. Above 450°C, no reliable demagnetization results could be obtained. The cause of this instability is not known, but it may be the result of exceeding the pronounced Curie point at 485°C (Fig. 1). The moment due to the phase with the higher Curie point may be unstable, or it may be that it is completely overridden by the moment acquired during cooling through the 465°C transition in a small residual field (10 to 20 gammas).

After alternating field demagnetization, fragment 3 of the dust sample was heated in vacuum to 450°C and allowed to cool in a field of 500 gammas. No additional detectable moment was acquired during this process. This sample was then heated in vacuum to 750°C and allowed to cool in an applied field of 1000 gammas. Again no detectable moment was acquired, indicating that it acquired its NRM in a field in excess of 20,000 gammas or that it was acquired by some means other than cooling through the Curie point. Alternatively, an irreversible chemical change may have taken place in the sample so that remanence can no longer be acquired in the same way the initial NRM was acquired.

Thus the evidence from fragment 3 of the dust sample allows no estimate of the possible range of values for the moon's field in the past. The partial thermal demagnetization of sample 10022 suggests that the sample may have acquired its initial NRM in a field of 1500 to 2000 gammas. However, the marked instability at higher temperatures leaves this conclusion somewhat in doubt; it is subject to verification by apparently more stable samples such as sample 10069.

Thermomagnetic analysis of the Apollo 11 dust and rock samples studied indicates that the major magnetic minerals are ilmenite, native iron, and possibly pyrrhotite. Of these, the native iron and the pyrrhotite (troilite?) are the dominant carriers of the natural remanence in these rocks. The natural remanences are low in all these samples in comparison with similar rocks from the earth. The stability of the NRM is low in three of the samples studied but is high in the other two. If the NRM of these samples is interpreted as being a TRM acquired by passage through the Curie point in the presence of a magnetic field, a field in excess of 1500 gammas is suggested.

CHARLES E. HELSLEY
University of Texas at Dallas,
(formerly Southwest Center
for Advanced Studies),
P.O. Box 30365, Dallas 75230

References and Notes

1. W. V. R. Malkus, *Science* **160**, 259 (1968).
2. For a summary of the origin of the earth's field see J. A. Jacobs, *The Earth's Core and Geomagnetism*, (Macmillan, New York, 1963).
3. E. Thellier and O. Thellier, *Ann. Geophys.* **15**, 285 (1959).
4. F. D. Stacey, J. F. Lovering, L. G. Parry, *J. Geophys. Res.* **66**, 1523 (1961).
5. Supported by NASA contract NAS 9-8767. Contribution 126, Geosciences Division.

4 January 1970

Magnetic Studies of Lunar Samples

Abstract. *The remanent magnetism of a lunar type C breccia sample includes a large viscous component with a time constant of several hours, and a high coercivity remanence, possibly acquired by impact processes on the lunar surface. Ilmenite(?) and metallic iron in breccias, and ferrous and metallic iron in glass beads separated from lunar fines (type D) were identified by high-field and low-temperature experiments. The iron appears to occur in a wide range of grain sizes including the single domain and multidomain states.*

The purpose of magnetic studies of Apollo 11 lunar materials is to ascertain whether the materials possess a stable remanent magnetization that was originally acquired on the moon, and to compare the magnetic properties of individual glass beads in the lunar fines with those of tektites. These magnetization studies could indicate past lunar magnetic environments and thus be of value in theories concerning the moon's earlier history. Intrinsic bulk and single-particle magnetic properties are also of interest with respect to the magnetic mineralogy of the lunar material.

Magnetization of type C breccia sample number 10059,24 in high applied magnetic fields was measured with a vacuum magnetic balance of the Faraday type (*1*). Results obtained for magnetization versus applied field to 8400 oe and magnetization (in 2000 oe) versus temperature between −180°C and 830°C are given in Fig. 1A. The magnetization curve rises very sharply at low fields and gradually approaches a saturation value near 2.7 emu/g. The thermal curves show a well defined Curie temperature at 775°C, and a suggestion of a second Curie or Néel point slightly below −180°C. These are interpreted as due to reported metallic iron and ilmenite (*2*), with a Curie point of 770°C and a Néel point of −205°C, respectively (*3*). About one-half the original room-temperature magnetization was lost during the heating experiment, but subsequent prolonged heating at 800°C did not further change the shape of the cooling curve, which indicated that the changes took place during the first heating cycle. There is also a small inflection in the cooling curve near 250°C; no interpretation is given for its origin.

Similar high field experiments were made on individual glass beads (0.1 to 1.5 mg) selected from lunar fines type D sample number 10084,86. A quartz-spring balance with the sample in a dry helium atmosphere (*4*) was used for these studies. The low field (< 3000 oe) magnetization at room temperature varied from 10 to 100 $\times$ 10^{-4} emu/g, and saturation magnetization varied from nil to 1.2 emu/g. Magnetization (in 3400 oe) versus temperature measurements indicated a Curie point of 770°C, but unlike the breccia, identical heating and cooling curves with a shape resembling the cooling curve of Fig. 1A (without the inflection at 250°C) were obtained. The magnetization versus applied field curve of individual glass beads are very different from those of the breccia: some were linear up to about 5000 to 7000 oe and reached saturation values of about 1 emu/g at 9000 oe. Other individual glass beads did not exhibit saturation magnetization in high fields at room temperature, indicating that essentially all the iron is in the paramagnetic state. Low temperature measurements verified this fact and also showed that in those glass beads that had an appreciable saturation magnetization, only a small part of the iron was in the metallic form with the major part in the paramagnetic state. In some respects these magnetic properties are very similar to those of tektites (*5*); moreover, electron microprobe analysis for total iron detected essentially the same concentrations of ferrous and metallic iron that are indicated magnetically; these concentrations are also similar to tektites. These similarities therefore suggest that small iron spherules in the beads are the carriers of the magnetization (5).

The rapid rise and gradual saturation in the plot of breccia magnetization versus applied field suggests, as in the case of the single-bead experiment, the presence of grains (spherules) of varying size, that range from the multidomain through single domain states. About one-half the magnetization of the breccia is destroyed on heating; in contrast, the beads have reproducible heating-cooling curves, thus implying that up to one-half the iron in the breccia may be within glass beads while the remainder is "unprotected," free to react with other substances. The exact nature of this reaction, which takes place in an atmosphere near 10^{-5} torr, is not known. It is clear, however, that the iron is not being oxidized to a cubic ferrite such as magnetite. The saturation magnetization of iron ferrites is generally about half that of metallic iron, so that their effect should be readily evident in the cooling curve if they were the alteration product.

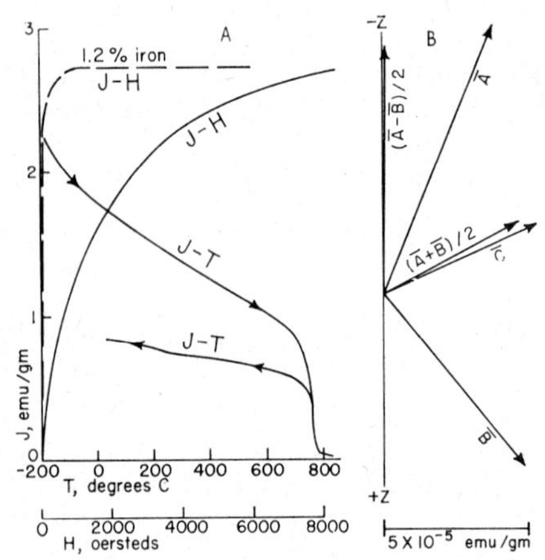

Fig. 1. (A) Magnetization of sample 10059, 24 in high applied fields. The magnetization versus applied field data (J–H) (lower abscissa) were made at 20°C. The magnetization versus temperature (J–T) data were obtained in an applied field of 2000 oe at a pressure of 2 to 3 $\times$ 10^{-5} torr (data below 20°C were obtained on a 103-mg sample in dry nitrogen at atmospheric pressure). Heating and cooling rates were 10°C per minute. For comparison, the curve for 1.2 percent iron in large grain sizes is shown by the dashed curve. (B) Results of Traînage experiments; ($\bar{A}$), magnetization vector after alignment for 8 hours with the earth's magnetic field directed parallel to the −Z sample axis; ($\bar{B}$), similarly for 8-hour alignment of earth's field and +Z axis; ($\bar{C}$), remanence vector after tumbling 8 hours in the earth's field. Within a few degrees, all vectors lie in the same plane.

To summarize these intrinsic magnetic studies, it appears that the magnetic mineralogy of the breccia consists mainly of metallic iron (1.2 percent by weight) in a wide range of grain sizes including single and multidomain states, and ilmenite that is antiferromagnetic at room temperature. Probably about half the iron is in the form of spherules within glass beads while the remainder reacts to form nonmagnetic or very weakly magnetic products upon heating to the Curie temperature of iron.

Remanent magnetization measurements of breccia sample 10059,24 (14.25 g) with spinner magnetometers (6) suggested that the magnetization may have changed during the experiments or that the measurements were adversely affected by anisotropy of susceptibility (7). Additional measurements after the sample rested in the earth's magnetic field for varying lengths of time (Traînage experiments) showed that the remanence did indeed change. Moreover, removal of the earth's field during measurement in the magnetometer improved the internal consistency of the data, which also suggests that the sample is magnetically anisotropic. Thereafter, all measurements were made as quickly as possible in null field without letting the sample come to rest during the measurement sequence.

Traînage experiments showed that viscous remanent magnetization (VRM) was acquired rather quickly in the earth's field but became essentially "saturated" after 4 to 5 hours. Moreover, this VRM could be removed by tumbling the sample in the earth's field for similar lengths of time. These results are summarized in Fig. 1B. The vector $(\bar{A} - \bar{B})/2$ is the VRM acquired in the earth's field and has a value of 8.3×10^{-5} emu/g. The "stable" remanence, $(\bar{A} + \bar{B})/2$, is 5.3×10^{-5} emu/g and is close in direction and intensity to the remanent vector measured directly by removal of VRM by tumbling in the earth's field.

Partial alternating field (AF) demagnetization experiments (Fig. 2) were all repeated at each step to ascertain the amounts of "anomalous" magnetization added by the process (8), and the samples were left tumbling in the earth's field for several hours between AF treatment and measurement of remanence. Progressively larger amounts of anomalous magnetization forced termination of the experiments at a peak AF of 400 oe.

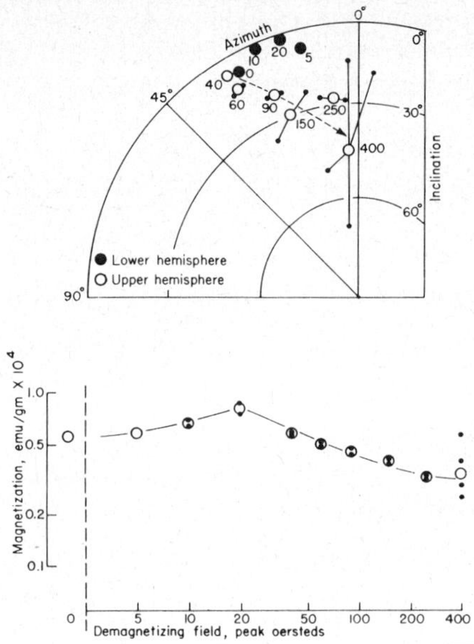

Fig. 2. Partial demagnetization experiments. The upper diagram shows changes in direction of the remanence during demagnetization as points on a Lambert equal-area projection, and the lower diagram shows the corresponding changes in intensity. Large symbols are mean values and lines connect individual values with their mean.

Within the experimental accuracy, these partial AF demagnetization data suggest the presence of three components of remanent magnetization with different coercivities in addition to the large VRM, which was removed by tumbling after each AF treatment. The first component has maximum coercivities of the order of 20 to 40 oe with irregular directions. It is probably an isothermal remanent magnetization (IRM) acquired during sample processing. The second component has high coercivities ranging up to at least 400 oe since the resultant vector was still moving after demagnetization at this peak field strength. The third apparent component—the one the resultant was approaching as the second component was erased—may be interpreted in two ways. It may be a very high coercivity remanence, or it may be an apparent remanent component generated by large susceptibility anisotropy. The measurement data after the 400-oe AF treatments are only consistent within about 10 percent so that the remaining remanence is very likely in part due to anisotropy. Such an argument cannot be invoked for the component with coercivities to 400 oe, however, since it responds linearly to the AF treatment. Its minimum strength—the vector difference of the resultant after 40-oe AF treatment and that after 250 to 400 oe —is about 3.5 × 10^{-5} emu/g.

In summary, the breccia contains a rather complicated remanence. It acquires a VRM of about 8 × 10^{-5} emu/g if left in the earth's field for a few hours. There is a probable IRM component of about 2 × 10^{-5} emu/g intensity with coercivities to 40 oe. Another component, with coercivities ranging over 400 oe, has a minimum intensity of 3.5 × 10^{-5} emu/g, and there also possibly exists a very high coercivity remanence with a maximum intensity of 3 × 10^{-5} emu/g. However, this last component is at least in part only an artifact of susceptibility anisotropy. The wide range of remanence coercivity is consistent with the previous inference of a wide range of iron grain sizes including single and multidomain magnetic states.

The remanent component with coercivities up to at least 400 oe is quite interesting in that no process is known by which the sample could have acquired this remanence after it left the lunar surface. The relatively slow decrease of remanence with increasing AF treatment is not characteristic of an IRM acquired in a large magnetic field. Thus, it is likely that this remanence is of a kind that results from the presence of a magnetic field plus some energy change such as cooling from high temperatures, chemical change, shock, and so on. Which process is most likely depends, of course, upon the mode of genesis of the breccia itself. If the rock acquired its magnetization by cooling from high temperatures (or by chemical action), then a relatively permanent field in the lunar vicinity at some time in the past might be postulated. On the other hand, shock induced magnetization by lunar impact could take place essentially instantaneously, and ionization processes at the time of impact might provide the necessary magnetic field. It would be desirable to study artificial magnetizations of these types in the breccia; unfortunately the large VRM, anisotropic properties, and anomalous magnetization acquired during AF treatment in sample 10059,24 preclude useful studies of this type.

RICHARD R. DOELL
C. SHERMAN GROMMÉ
U.S. Geological Survey, Menlo Park, California 94025
A. N. THORPE
F. E. SENFTLE
U.S. Geological Survey, Washington, D.C. 20242

References and Notes

1. R. R. Doell and A. Cox, in *Methods in Palaeomagnetism*, D. W. Collinson, K. M. Creer, S. K. Runcorn, Eds. (Elsevier, Amsterdam, 1967), p. 159.
2. Lunar Sample Preliminary Examination Team, *Science* **165**, 1211 (1969).
3. S. P. Clark, Ed., *Handbook of Physical Constants, Geol. Soc. Amer. Mem. 97*, (1966).
4. F. E. Senftle, M. D. Lee, A. Monkewicz, J. Mayo, T. Pankey, *Rev. Sci. Instr.* **29**, 429 (1958).
5. F. E. Senftle, A. N. Thorpe, R. R. Lewis, *J. Geophys. Res.* **69**, 317 (1964).
6. R. R. Doell and A. Cox, *U.S. Geol. Surv. Bull. 1203-A* (1965).
7. ———, in *Methods in Palaeomagnetism*, D. W. Collinson, K. M. Creer, S. K. Runcorn, Eds. (Elsevier, Amsterdam, 1967) p. 196.
8. ———, *ibid.*, p. 241.
9. We thank E. Mankinen and S. Sullivan for assistance with the laboratory measurements.

4 January 1970

Magnetic Properties of Lunar Samples

Abstract. *The magnetic properties of samples of rock, fines, and magnetic separate from the fines from Apollo 11 have been measured. Native iron, or possibly nickel-iron, of submicroscopic particle size is the most important constituent, with minor contributions from ilmenite, paramagnetic iron minerals, and other iron-titanium oxides. The remanent magnetization of a sample of the microbreccia rapidly acquires a viscous magnetization and does not appear to have a significant stable remanence. The crystalline sample has a weak natural remanence showing some stability.*

The samples we investigated were fines (10084,13), a magnetic separate from fines (10084,135), a 17-g sample of the microbreccia rock (10046,46), and an 11-g sample of crystalline rock (10017,64).

Washing followed by sorting of the fines under the microscope showed (i) glass, including spheres, dumbbell shapes, and broken fragments, of brown, green, and colorless hue with the rare brown fragments containing opaque cruciform and square-section crystallites 10 μm across; (ii) purplish pyroxene with refractive index β 1.698 and $2V = 44°$, suggesting $Ca_{37}Mg_{38}Fe_{25}$ (atomic percent); and (iii) opaque grains.

The salient feature in the microbreccia is the variability of texture in different parts of the sample. Combined study of thin and polished sections showed: (i) Ilmenite in a wide variety of forms, from crystals 50 by 20 μm through laths and patterned growths to tiny interphase rods. It makes up about 16 percent by volume. One crystal of ilmenite showed a core of lower reflectivity than the bulk of this mineral, and investigation with the electron-probe microanalyzer showed the presence of titanium-rich pseudobrookite. (ii) Two types of sulfide, making up about 2 percent volume. One of the types is probably troilite or pyrrhotite. (iii) Pyroxene, the chief translucent phase. (iv) A few grains of olivine ($2V = 90°$, suggesting $Mg_{86}Fe_{14}$). (v) An intergrowth of unknown nature, making up about 38 percent by volume of the section studied, opaque and yet of low reflectivity. Ilmenite and sulfide occur as tiny interphase rods and granules within the intergrowth.

The identity of the pyroxene has been confirmed by x-ray powder diffraction. No iron particles have been definitely identified microscopically in any of the material.

The ^{57}Fe Mössbauer spectra of samples of the fines were obtained at 290°, 78°, and 4.2°K. The predominant feature of all spectra was a series of overlapping quadrupole-split doublets centered at the Fe^{2+} chemical-isomer-shift position of typical silicate minerals. In addition, about 3 to 5 percent of the total resonance appeared as a magnetically split six-line spectrum with a hyperfine field of 330 ± 5000 oersteds attributable to free iron or a dilute alloy of nickel in iron. It is also probable that ~10 percent of the iron in the sample is present as Fe^{3+}. Typical

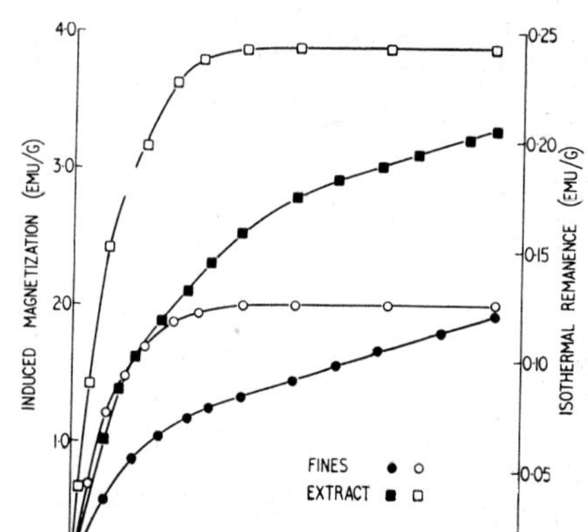

Fig. 1. Curves of J_i–H (left-hand scale, full symbols) and J_r–H (right-hand scale, open symbols) for fines and magnetic separate.

Table 1. Mössbauer data for ^{57}Fe at 78°K. Chemical isomer shift relative to sodium nitroprusside, $Na_2[Fe(CN)_5NO]$; $2H_2O$ taken as zero.

Chemical isomer shift (mm/sec)	Quadrupole split (mm/sec)	Assignment	Resonance (% by area)
0.90	1.0	Fe^{3+}	10
0.26	0	Fe (metal) or Fe/Ni alloy	4
1.45	1.1	Ilmenite	20
1.52	2.0	Tektites, orthopyroxenes, and site 2 of clinopyroxenes	66
1.57	2.9	Olivines and site 1 of clinopyroxenes	

computer-fitted data and assignments are given in Table 1. The presence of ilmenite was confirmed by the disappearance of the paramagnetic quadrupole doublet in the spectrum at 4.2°K (below the Néel temperature).

The Mössbauer spectrum of the magnetic separate was similar to that of the fines except that the percentage of the resonance due to metallic iron increased to ~7. No evidence for iron sulfides or Fe_3O_4 was obtained either from the Mössbauer spectra or from x-ray powder diffraction patterns. This indicates that their concentration, if they were present, was less than 1 percent of the iron in the sample.

Figure 1 shows the dependence of induced (J_i) and isothermal remanent (J_r) magnetization on the applied field, H. The J_i–H curve may be interpreted as a constituent saturating at about 3000 oersteds, together with a nearly paramagnetic constituent apparent above 4000 oersteds, with susceptibility about 0.12×10^{-3} emu g^{-1} $oersted^{-1}$. The susceptibility value is too large to be accounted for by the iron present, even if all the iron is in paramagnetic form; thus superparamagnetic material may be present. However, there is some suggestion of slight curvature in the upper part of the curve, and the whole curve may also be explained by superparamagnetism (*1*). The evidence suggests that native iron is the important magnetic constituent ($J_s \sim 220$ emu g^{-1}) in the fines and separate, and the J_i–H curves indicate the presence of the order of 1 and 2 percent of iron, respectively. Saturation remanence is reached in a field of about 3000 oersteds (0.13 and 0.24 emu g^{-1}, respectively), and the coercivity of remanence in the fines is 400 oersteds.

Figure 2 shows the reversible susceptibility (χ_r) of the fines and the separate in the temperature range −196° to 800°C; the samples were contained in an evacuated quartz tube and χ_r was measured in an a-c field (1500 hz) of 2.5 oersteds r.m.s. The variation of χ_r at low temperature is small and no discontinuities occur, so the existence of large-grained magnetite is ruled out. Above room temperature no discontinuities indicative of titanomagnetites or magnetite are present, and the Curie points of both the fines (778° ± 3°C) and the separate (775° ± 3°C) correspond closely to that of pure iron (782° ± 3°C) measured on the same instrument. The curve for iron falls much more steeply toward the Curie point than do those of the fines and the separate; this steepness is explained by the demagnetizing effects in isolated grains of high intrinsic susceptibility (*1*), and these effects seem to be less important in the lunar material. This is confirmed by the large amount of iron (~8 percent) required to be present on the basis of the observed χ_r of the fines (2.4×10^{-3} emu g^{-1} $oersted^{-1}$). The lower susceptibility after heating may be ascribed either to oxidation of iron particles or to growth of particles (by coalescence) above the critical size for superparamagnetism.

The thermomagnetic curve in a high field (10,000 oersteds) again shows a high Curie point of about 775°C, with a shape more similar to that expected from a paramagnetic or superparamagnetic component—that is, a slow decrease toward the Curie point without marked discontinuities; there is evidence that physical or chemical changes commence at about 500°C. This is confirmed in the course of

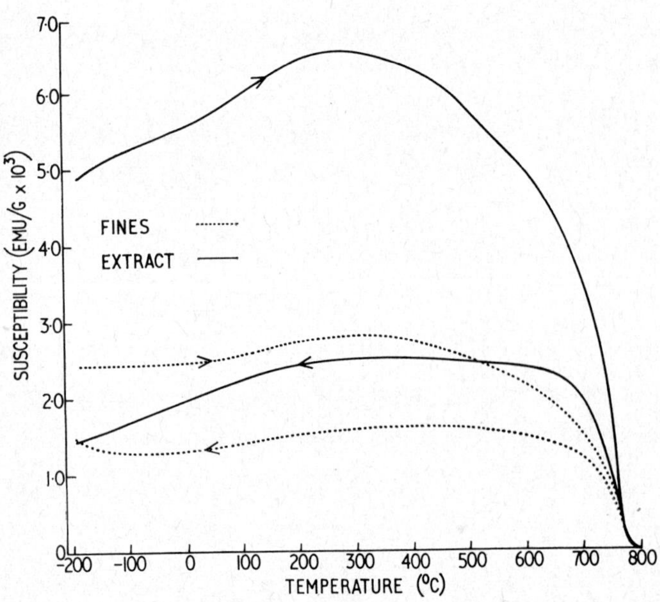

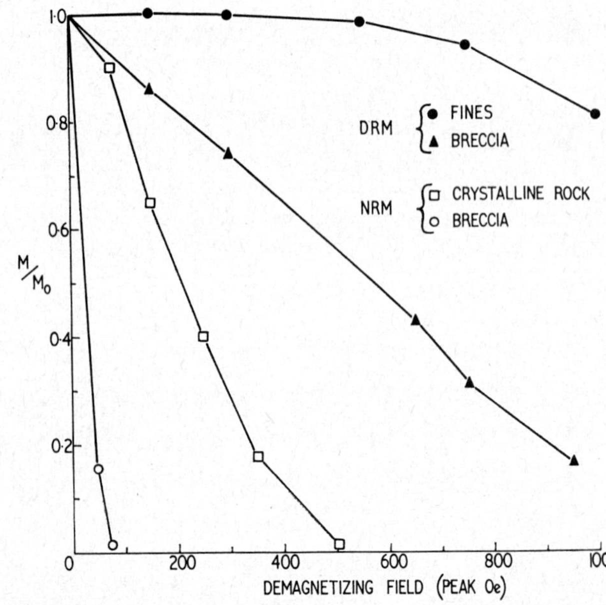

Fig. 2 (left). Reversible susceptibility of fines and magnetic separate plotted against temperature. Fig. 3 (right). Stability of NRM in breccia and crystalline rock, and of artificial DRM in fines and crushed breccia.

partial thermoremanent magnetization (PTRM) experiments (*1*), in which the TRM acquired in 100°C intervals up to 800°C in a field of 0.42 oersted was measured. The importance of iron is again indicated by the fact that the maximum PTRM was acquired between 800°C and 700°C, and measurement of susceptibility after each temperature interval shows that changes occur in the material after 500°C. Smaller PTRM's are acquired in all temperature intervals; these may be due to blocking temperature effects or small amounts of iron-titanium oxides.

Measurement of high field anisotropy (rotational hysteresis) of the fines on the torque magnetometer (*2*) shows a curve of which part is characteristic of iron, together with a peak at about 2500 oersteds, similar to that found in measurements on titanomagnetites in terrestrial rocks; this peak is absent in the separate, but appears after heating in vacuum to 800°C. These results suggest the presence of titanomagnetites with spinel structure in the fines, but the absence of such titanomagnetites in the separate is puzzling; there is as yet no evidence of significant quantities of iron-titanium oxides, other than ilmenite, under the microscope.

When first measured, the breccia sample had a natural remanent magnetization (NRM) of 62×10^{-6} emu g^{-1}, which was reduced to about 1.0×10^{-6} emu g^{-1} after demagnetization in a peak field of 75 oersteds (Fig. 3). The initial NRM was clearly a viscous remanent magnetization (VRM) acquired in the earth's field before the sample was received in Newcastle, and further tests showed the rapid growth of VRM in a weak field (0.16 oersted) to be logarithmic with time, with a viscosity coefficient S of 22×10^{-6} emu g^{-1}. The fines gave a similar result with $S = 27 \times 10^{-6}$ emu g^{-1}. The ease with which the breccia sample acquires stray magnetizations made further demagnetization difficult, and consistent results were not obtained. At present it may be stated that any "hard" remanence has intensity less than 1.0×10^{-6} emu g^{-1}, and probably less than 10^{-7} emu g^{-1}. Measurements on the Princeton Applied Research spinner and also on the astatic magnetometer on an approximately equidimensional chip of the breccia showed its susceptibility anisotropy to be less than 0.5 percent; its bulk susceptibility (2.9×10^{-3} emu g^{-1}) is close to that of the fines.

The presence of any grains possessing a hard magnetization was investigated in the fines and in a sample of crushed breccia by allowing a suspension of each in acetone to settle out in a weak field (~1.0 oersted); after the acetone was evaporated off, the resulting depositional remanent magnetization (DRM) was measured and subjected to alternating field demagnetization. The stability of the DRM is shown in Fig. 3. It is seen that there are particles of extremely high stability in the fines and of somewhat less stability in the breccia.

Preliminary measurements on the crystalline rock sample indicate an NRM of intermediate hardness. An initial NRM of 5.6×10^{-6} emu g^{-1} was reduced to 2.0×10^{-7} emu g^{-1} in a demagnetizing field of 500 oersteds (Fig. 3); this is clearly not a VRM acquired in a few weeks or days. The growth of VRM in the rock in a field of 2.4 oersteds over a period of 9 days suggests that the measured NRM could not have been acquired in any reasonable magnetic field or time on the moon, if the growth is logarithmic. The only other measurement made on this rock was of reversible susceptibility; a value of 7.7×10^{-5} emu g^{-1} $oersted^{-1}$ was obtained.

Although so far no native iron has been definitely identified under the microscope, it appears to dominate the magnetic properties of the fines and breccia. However, a range of submicroscopic particle sizes can account for the results described in this report; iron particles less than about 160 Å in diameter will be superparamagnetic, those between about 160 and 320 Å will be single domains, and larger ones will be multidomain grains. Superparamagnetism can account for the anomalously high initial susceptibility and for the shape of the J_i–H curve, single domains are presumably responsible for the stable DRM observed in the fines and breccia, and these and larger grains carry the isothermal remanence. Viscous remanence may be carried by multidomain grains as well as by those on the margin of superparamagnetism. The magnitude of the observed properties is explained by the presence of between 0.1 and 1.0 percent of native iron; this amount is in agreement with the Mössbauer results (~0.5 percent). It is possible that any nickel present is associated with the iron. If all the nickel present in the fines (250 ppm) is assigned to 0.5 percent of nickel-iron, then the latter will contain about 5 percent nickel, a somewhat lower concentration than that usually found in iron meteorites.

The remaining iron is contained in the abundant ilmenite (~15 percent) and in the paramagnetic pyroxenes and olivines, with possibly a small proportion in a spinel phase. Although the paramagnetic forms of iron will contribute to the magnetic properties in an applied field and the spinel phase will contribute to all properties, the native iron is the dominant magnetic constituent.

It is unlikely that the breccia has any stable remanence, and it appears that the "hard" magnetic particles that it and the fines contain are randomly oriented single domains. The NRM of the crystalline rock requires further investigation; it is not a recently acquired VRM, nor does it appear to be a simple VRM acquired logarithmically in any reasonable lunar field. The magnetometer results from Apollo 12 can be interpreted by supposing that the bedrock in the maria is uniformly magnetized to about 10^{-5} emu g^{-1}. (The monthly thermal cycling in the presence of the component of the solar wind field along the moon's rotation axis requires study.) Failing this, the NRM might be a TRM acquired $3{\text -}5 \times 10^{9}$ years ago; it is possible that the moon has a small iron core in which dynamo action was once more vigorous than it is now.

S. K. RUNCORN
D. W. COLLINSON
W. O'REILLY
A. STEPHENSON
Department of Geophysics and Planetary Physics, School of Physics,
N. N. GREENWOOD
Department of Inorganic Chemistry, School of Chemistry,
M. H. BATTEY
Department of Geology, The University, Newcastle-upon-Tyne, England

References and Notes

1. T. Nagata, *Rock Magnetism* (Maruzen, Tokyo, 1962).
2. D. W. Collinson, K. M. Creer, S. K. Runcorn, Eds., *Methods in Paleomagnetism* (Elsevier, Amsterdam, 1967).
3. We are indebted to the following for valuable discussion and assistance with measurements: K. M. Creer, R. G. J. Strens, D. H. Tarling, A. Manson, and P. Redman, Department of Geophysics and Planetary Physics, School of Physics; J. M. Jones and P. C. Jones, Department of Geology, and A. T. Howe, Department of Inorganic Chemistry, School of Chemistry. Support from the Natural Environment Research Council is acknowledged.

4 January 1970

Magnetic Properties of Lunar Sample 10048-22

Abstract. *The natural remanent magnetization (3.7 × 10^{-4} electromagnetic units per cubic centimeter) and the susceptibility per cubic centimeter (6.3 × 10^{-13}) of an 18.5-gram breccia specimen were determined with instrumentation and techniques currently used in paleomagnetism. The relatively low magnetic stability of the rock in the earth's field and in alternating demagnetizing fields precludes considering it as a reliable carrier of paleomagnetism. A magnetic balance study yields an unusually high Curie temperature (750°C) which is possibly diagnostic of metallic Fe containing less than 5 percent nickel. The estimated relative abundance of the iron in the sample is about 0.5 percent.*

The identification of a stable component of thermoremanent magnetization in a sample of lunar material could have important implications relative to the past history of the moon's interior. On this basis, an analysis was made of the natural remanent magnetization and of the magnetic properties of sample 10048-22. More specifically, the stability of its natural remanent magnetization was studied and compared with that of an anhysteritic component artificially introduced in the sample. Its magnetic susceptibility and its thermomagnetic properties were also determined in an attempt to identify the ferromagnetic constituents of the sample.

The sample studied consists of an 18.5-g, irregularly shaped portion of the breccia 10048. Because of its irregular shape and of the relatively high friability of its surface, it was necessary to mount the sample in a special lucite container that could be accurately positioned and oriented in some of the apparatus. In order to design the lucite container, to determine the volume of the sample, its centroid and its specific weight, replicas of the sample were cast with various materials. As a by-product of this preparation it was found that the bulk density of the sample is 2.4.

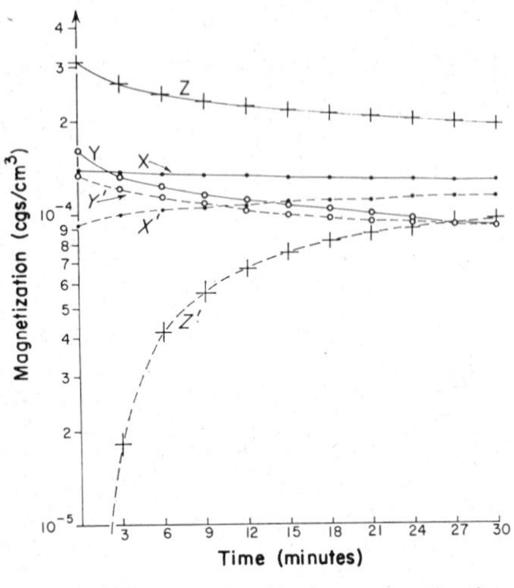

Fig. 1. Decay of viscous magnetization acquired in earth's field in 1 hour with Z axis of sample 10048 pointing upward (solid lines) and downward (dashed line).

The volume susceptibility of the sample was determined with a bridge-type susceptibility meter (*1*) in a field of 0.5 oe and of 1 khz. The value of 6.3 × 10^{-3} cgs units thus obtained would be considered relatively high for most terrestrial rocks and could be diagnostic of an unusual ferromagnetic mineral.

The remanent magnetization of the sample set in a null-field space was determined with a three-magnet astatic magnetometer (*2*). Prior to the original measurements, the sample was allowed to rest in the earth's magnetic field with its Z axis (the axis of the cylindrical lucite container) pointing upward along the vertical. On account of the relatively high magnetic susceptibility of the sample, its centroid was kept a minimum distance of 7.5 cm from the magnet system during the measurements. The magnetization was then determined ten consecutive times at intervals of 3 minutes. It was noted (Fig. 1) that the intensity of its Z component decreased substantially during the 30 minutes required for the series of measurements. The sample was then replaced in the earth's field with its Z axis pointing downward. This resulted in a reversal of the magnetization along the Z axis, but the soft component of magnetization thus introduced in the sample again gradually decreased during 30 minutes (Fig. 1). This experiment points to the presence of a low-coercivity ferromagnetic mineral in the sample.

The sample was then exposed to a series of alternating field cleaning treatments, a standard procedure in most paleomagnetic laboratories. The apparatus used for these tests has been described (*3*).

The peak intensity of the demagnetizing field was gradually raised to 800 oe in steps of 15, 25, or 50 oe. The net result of this treatment was that the original magnetization of the sample (3.7 × 10^{-4} emu/cm³) decreased to 4.5 × 10^{-5} emu/cm³ after the treatments in the higher fields. Of particular interest is the variation in the relative orientation of the magnetization in the sample. The inclination changed systematically after the treatments in demagnetizing fields below 250 but very erratically after the treatments in higher fields (Fig. 2). In comparison with the behavior of a typical stable terrestrial rock to which a similar treatment was applied, it is clear that the magnetization of the lunar sample is relatively unstable, either due to the mechanism of its formation or to the nature of its carrier.

A further set of alternating field demagnetization treatments was carried out after the sample was exposed for 5 minutes in a 0.5-oe constant field on which an 880-oe alternating field (60 hz) was superposed. The 3.5 × 10^{-3} emu/cm³ magnetization then present in the sample was reduced to 2.5 × 10^{-5} emu/cm³ after the cleaning treatment in 700 oe. This experiment suggested

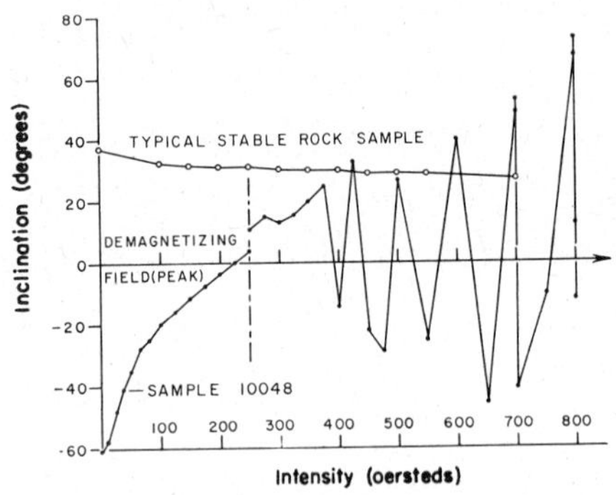

Fig. 2. Variation of magnetic inclination with demagnetization treatment for a typical stable terrestrial rock sample (open dots) and for sample 10048 (filled dots).

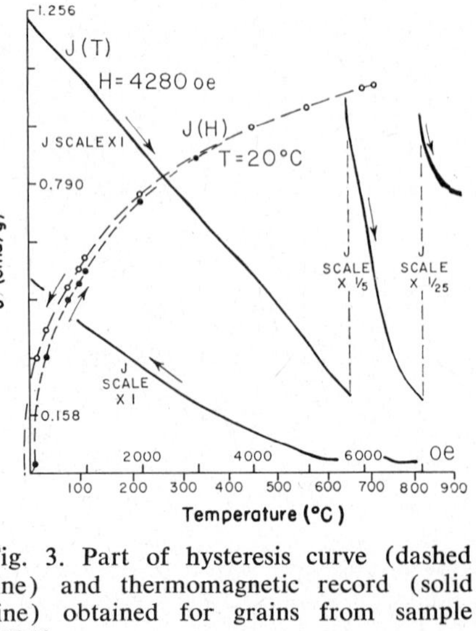

Fig. 3. Part of hysteresis curve (dashed line) and thermomagnetic record (solid line) obtained for grains from sample 10048.

that the instability of the sample magnetization should be attributed to the nature of its carrier rather than to the mechanism of its formation.

To investigate the nature of the carrier(s) of remanent magnetization in the sample, small grains detached from its surface were examined with a recording magnetic balance (*4*). The investigation was carried out with a sample weighing approximately 20 mg and under an air pressure of 3×10^{-3} mm-Hg.

Figure 3 shows the dependence of the magnetization (J) on the applied field (H) at 20°C and the change in J (H = 4280 oe) during heating to 900°C and subsequent cooling. The J(H) curve shows that the original sample approached magnetic saturation at about 4000 oe and that its coercive force was approximately 70 oe. Such a low coercive force is in accord with the previously noted tendency of the sample to acquire a large component of viscous magnetization when exposed for a few minutes in a field as low as the earth's magnetic field (Fig. 1). Moreover, it suggests that there is no major component of high micro-coercivity in the sample.

The J(T) curve shows a monotonic decrease in magnetization from −180°C to the Curie temperature at 750° ± 8°C (only partly shown on Fig. 3). The high Curie temperature is diagnostic of a mineral that is uncommon in terrestrial rocks. The cooling part of the J(T) curve suggests that a substantial transformation of the original mineral took place at high temperature.

The Curie temperature of the original ferromagnetic constituent at 750° ± 8°C and the transformation of this constituent into one with a Curie point of 580°C and with lower saturation magnetization were interpreted as follows: (i) the original ferromagnetic constituent was relatively pure metallic Fe with perhaps as much as 5 percent Ni, this estimate being based on standards given by Kneller (*5*). In this context, it is noteworthy that the Fe in meteorites normally contains more than 5 percent Ni; (ii) the estimated abundance of fine-grained Fe in the sample is about 0.5 percent; (iii) it is possible that virtually all the metallic Fe was oxidized to Fe_3O_4 at higher temperatures. This is in keeping with the Curie temperature of the new mineral and with the halved saturation moment of the sample at 20°C after heating.

In the light of the above magnetic data it is not possible to state conclusively whether a substantial magnetic field witnessed the last cooling of sample 10048 on the lunar surface. It would thus be presumptuous to speculate further on the original configuration of the moon's interior along this avenue on the basis of this sample.

A. LAROCHELLE
E. J. SCHWARZ
Geological Survey of Canada, Ottawa, Canada

References and Notes

1. K. W. Christie and D. T. A. Symons, *Geol. Surv. Can. Paper 69-14* (1969).
2. A. Larochelle and K. W. Christie, *Geol. Surv. Can. Paper 67-28*, (1967).
3. A. Larochelle and R. F. Black, *Can. J. Earth Sci.* 2, 684, (1965).
4. E. J. Schwarz, *Geol. Surv. Can. Paper 68-37* (1968).
5. E. Kneller, *Ferromagnetismus* (Springer-Verlag, Berlin, 1962).
6. We thank K. W. Christie and G. N. Freda for assistance in the preparation of the sample and in making some of the measurements.

5 January 1970

Search for Magnetic Monopoles in the Lunar Sample

Abstract. *An electromagnetic search for magnetic monopoles of the minimum size predicted by Dirac, or of any larger magnitude, has been performed on 8.37 kilograms of lunar surface material. No monopole was found. This experiment sets new limits on the production cross section for monopoles and on their occurrence in cosmic radiation.*

For several years now, the hunt has been on for particles that would interact with the magnetic field just as electric charges interact with the electric field, acting as a source for the field and being accelerated by it. These particles, called monopoles, would be stable. They would have a magnetic charge measured by an integer ν, the Dirac charge 3×10^{-8} emu being used as a unit (*1*). Their existence would give credence to the only known explanation for the extraordinarily accurate phenomenon of charge quantization (*2*). According to a recent theory (*3*), they would be the most fundamental particles, the building blocks of the universe. However, no such particle or combination with a net nonzero magnetic charge has ever been found (*4–6*).

In view of the negative results of these experiments (*4–6*), the lunar surface is considered to be the most likely hiding place for monopoles, whether they belonged to the primary cosmic rays or were produced in the collision of a high-energy cosmic ray particle with a nucleon of the lunar surface. In either case, the lunar material would slow the monopole down and trap it. The reasoning that favors the lunar sample involves its great age, 3 to 4×10^9 years, and the small depth to which the surface has been churned during the long period of time. These two factors give the lunar surface the longest known exposure to cosmic rays. Furthermore, the absence of both an atmosphere and a magnetic field on the moon allows the fate of a monopole after it has been slowed down to be assessed with more certainty than it could be on the earth.

Our detection technique relies on the electromotive force induced in a coil by a moving monopole. As in previous work (*7*), the sample was transported along a continuous path threading the windings of a coil. In this experiment the coil was made of superconducting material and was short-circuited by a superconducting switch. A small current was stored in the superconducting loop before a sample was run. If a sample containing a monopole had been run, the induced electromotive force would have modified this current. After each sample had been circulated 400 times, the superconducting switch was opened and a signal proportional to the current in the loop was transferred electrically out of the cryostat, amplified, and finally recorded on an oscilloscope. A real magnetic charge would have been detected as a difference between the signal obtained when the switch was opened and the one normally observed when the opening of the switch interrupted the "standard current" that had been introduced as an overall check on the apparatus. A zero magnetic charge therefore corresponded to a nonzero standard signal. This technique assured us that the equipment was working at all times.

An overall calibration was obtained from a long solenoid in which a known change of current simulated the "missing term" in Maxwell's equations—the one describing the contribution of a "magnetic current density." A statistical study of our signals shows that the measurement of the magnetic charge was affected by a 1 standard deviation error of about ⅛ of a Dirac unit, when a ride of 400 passes was given the sample. Therefore, the smallest monopole compatible with Dirac theory was expected to produce an 8 standard deviation signal. There are reasons to believe the smallest actual charge would have twice the Dirac value (*8*), and this would correspond to twice as big a signal.

The lunar surface material analyzed in this experiment consisted of 28 individual samples. One sample was composed of three rocks (NASA 10022-1, 10023-1, and 10024-3) weighing all together 213 g. The remaining 27 samples were all fines from the bulk sample (NASA 10002). The individual samples of fines ranged in weight from 261 to 356 g, and weighed all together 8.13 kg.

The measured magnetic charge of each sample was consistent with zero, and statistically incompatible with the hypothesis that the absolute value of the magnetic charge was as large as, or larger than, a single Dirac unit of magnetic charge. We can therefore set upper limits on the number of monopoles present in the primary cosmic rays *and* on the number of monopoles produced by high energy cosmic ray particles interacting with nucleons of the lunar surface material. We quote our results at the 95 percent confidence level, including a correction of 10 percent to the monopole density to take into account the possibility that any individual sample may have contained paired monopoles of opposite charge.

The actual values of both upper limits depend upon unknown properties of the hypothetical magnetic monopole—namely, its charge, its mass, and all the parameters that determine its range inside the lunar material before it comes to rest. Therefore we express our results as a function of n, a parameter which relates the approximate range R, in grams per square centimeter, to the kinetic energy E, in Gev, by $R = 0.1\ E/n^2$. For low velocities, when the monopole loses energy by ionization only, n in this formula is the magnetic charge ν measured in Dirac units. At higher velocities, the effective value of n is expected to increase with E due to bremsstrahlung.

The values of both upper limits depend also upon the assumption we make regarding the depth, D, to which the lunar surface has been churned. We have represented our results in Figs. 1 and 2 for an assumed exposure time of 3×10^9 years and for two mixing depths, (i) 5 cm (solid curve), which represents effectively no mixing depth, and (ii) 100 cm (dashed curve).

In Fig. 1, we plot our upper limit for the flux of monopoles per square centimeter per second, per steradian, in the cosmic rays. The curves are displayed as a function of the kinetic energy of the monopole with n and D as parameters. Curves A and B represent upper limits known from the most extensive previous searches for monopoles in cosmic rays. Curve A results from examination of deep ocean deposits (*5*), and curve B results from an analysis of tracks in obsidian and mica (*6*). The results of the most extensive search carried out in the earth's atmosphere are given by curve C (*9*).

The production of monopoles by proton-nucleon interactions depends upon the monopole pair-production cross section, σ. In Fig. 2 we have plotted the upper limit for σ, as it results from our experiment, as a function of the monopole mass for different values of D and n. The flux of primary cosmic rays above the energy E, in Gev, was assumed to be $1.4 \times E^{-1.67}$ $cm^{-2}\ sec^{-1}\ sr^{-1}$ (*4*).

The incident proton was assumed to lose 40 percent of its energy at each proton-nucleon interaction (*9*). The monopole pair-production cross section was assumed to be constant above the threshold energy for monopole pair production. Curves A and C represent the limits for σ as known from previous work (*5* and *9*, respectively). Curve D comes from a search for monopoles in a meteorite (*10*) as interpreted in reference (*4*). Curve E corresponds to the most extensive accelerator study made to date (*11*).

The search for monopoles in the lunar sample of Apollo 11 resulted in the finding that there was neither an unpaired north or south monopole in any of the 28 samples studied. This result sets upper limits on the presence of monopoles both in the primary cosmic rays and in the proton-nucleon interactions, without any assumption concerning the migration of the monopoles through matter under the influence of the magnetic field. If the lunar mixing depth is less than 10 meters over a

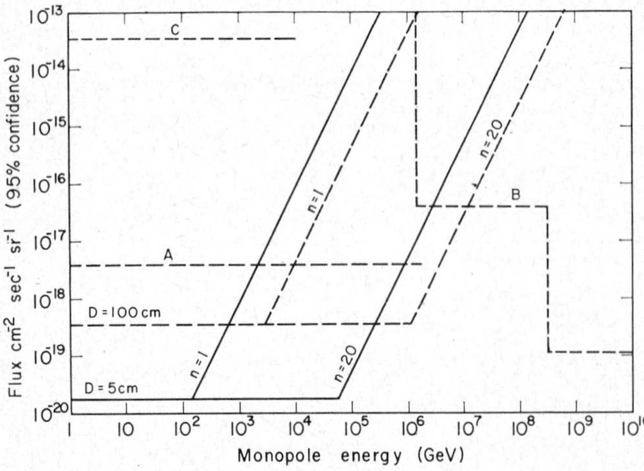

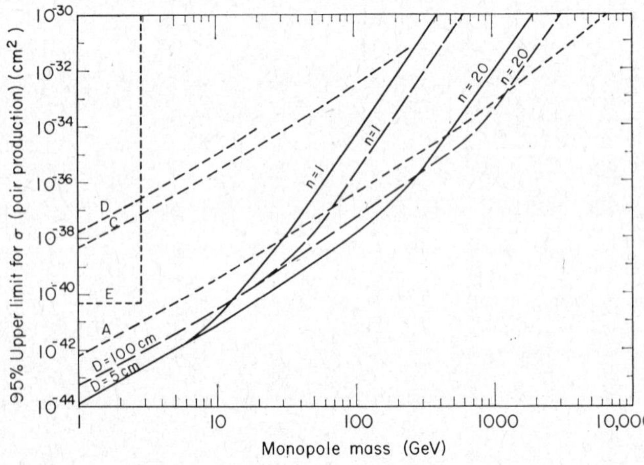

Fig. 1 (left). Ninety-five percent confidence level upper limit on the flux of monopoles as a function of monopole energy. The solid and dashed curves for D=5 and 100 cm are from this work. The parameters n and D are defined in the text. Data for curyes: A, reference (*5*); B, reference (*6*); C, reference (*9*). Fig. 2 (right). Ninety-five percent confidence level upper limit on the monopole pair-production cross section in proton-nucleon collisions. The solid and dashed curves for D=5 and 100 cm are from this work. The parameters n and D are defined in the text. Data for curves: A, reference (*5*); C, reference (*9*), D, reference (*10*); E, reference (*11*).

period of 3 × 10^9 years, our upper limits are lower than any previous values except in high ranges of mass and energy, as shown in Figs. 1 and 2.

LUIS W. ALVAREZ
PHILIPPE H. EBERHARD
RONALD R. ROSS
Lawrence Radiation Laboratory and Space Science Laboratory, University of California, Berkeley 94720
ROBERT D. WATT
Stanford Linear Accelerator Center, Stanford, California

References and Notes

1. P. A. M. Dirac, *Phys. Rev.* **74**, 817 (1948); *Proc. Roy. Soc.* **A133**, 60 (1931).
2. L. J. Fraser, E. R. Carlson, V. W. Hughes, *Bull. Amer. Phys. Soc.* **13**, 636 (1948) (paper E 17); J. G. King, *Phys. Rev. Lett.* **5**, 562 (1960).
3. J. Schwinger, *Phys. Rev.* **173**, 1536 (1968); *Science* **165**, 757 (1969).
4. For a systematic review of experimental searches for Dirac monopoles up to 1968, see E. Amaldi "On the Dirac magnetic poles," in *Old and New Problems in Elementary Particles*, G. Puppi, Ed. (Academic Press, New York, 1968), p. 20. Two more recent searches are reported in references (*5* and *6*).
5. R. L. Fleischer, H. R. Hart, I. S. Jacobs, P. B. Price, W. M. Schwarz, F. Aumento, *Phys. Rev.* **184**, 1393 (1969).
6. R. L. Fleischer, P. B. Price, R. T. Woods, *ibid.* **184**, 1398 (1969).
7. L. W. Alvarez, A. J. Schwemin, R. G. Smits, R. D. Watt, Semi-annual Report, UCRL 11466, October 1964, p. 6 (unpublished).
8. L. W. Alvarez, Physics Notes 479, Lawrence Radiation Laboratory, 1963 (unpublished); J. Schwinger, *Phys. Rev.* **144**, 1087 (1965).
9. W. C. Carithers, R. Stefanski, R. K. Adair, *Phys. Rev.* **149**, 1070 (1966).
10. V. A. Petukhov and M. N. Yakimenko, *Nucl. Phys.* **49**, 87 (1963).
11. E. Amaldi, G. Baroni, A. Manfredini, H. Bradner, L. Hoffman, G. Vanderhaeghe, *Nuovo Cimento* **28**, 773 (1963).
12. We are indebted to Roscoe Byrns for the design of our cryogenic equipment, to John Taylor and Leo Foley for the design and test of many parts of the apparatus, to Maurilio Antuna, Robert Gilmer, and Hans Stellrecht for the construction of the electronic equipment, to Egon Hoyer for the design of the sample containers, and to all the members of the team that worked on this experiment in Berkeley and Houston. We acknowledge the help we have received from the staff of the Lunar Receiving Laboratory and the Lunar Sample Analysis Planning Team. Work done under NASA contract NAS 9-8806.

5 January 1970

Magnetic Properties of the Lunar Crystalline Rock and Fines

Abstract. *Magnetic measurements have shown that nondiamagnetic minerals in a lunar crystalline rock of type B are (free Fe^{2+} in paramagnetic pyroxenes) : (antiferromagnetic $FeSiO_3$) : (antiferromagnetic $FeTiO_3$) : (ferromagnetic iron) = 4.3 : 7 : 20 : 0.08 in weight percentage. The abundance of ferromagnetic Fe in the lunar fines is about 7.5 times its abundance in the crystalline rock. The natural remanent magnetization of the crystalline rock of 7.5 × 10^{-6} emu/g in intensity may not be attributable to its thermoremanent magnetization.*

Mineralogical and chemical compositions of a lunar crystalline rock of type B (10024,22) were examined through a scanning x-ray microprobe to identify magnetic minerals responsible for magnetic properties of the rock sample. Electron microscope scannings and high-power optical microscope surveys were also made to clarify the mutual mineralogical relationships among the magnetic minerals. Main magnetic minerals in the sample have been identified as (i) pyroxenes [(Fe, Mg, Ca)SiO_3], (ii) almost stoichiometric Fe pyroxenes ($FeSiO_3$), (iii) almost stoichiometric ilmenites ($FeTiO_3$), and (iv) almost pure metallic irons. Fine grains of metallic irons are usually included in troilites (FeS).

The magnetization of the rock was measured in magnetic fields from 0 to 30 kilooersteds in a temperature range from 4.2° to 300°K and also in another range from 300° to 1090°K in a highly controlled non-oxygen atmosphere. The results, summarized in Fig. 1, clearly show that the magnetism of the sample consists of four phases: (i) a paramagnetic phase, which may be attributable mostly to free Fe^{2+} in pyroxenes, (ii) an antiferromagnetic phase of almost stoichiometric Fe pyroxene represented by a Néel point at 41°K, (iii) an antiferromagnetic phase of almost stoichiometric ilmenite represented by a Néel point at 57°K, and (iv) almost pure metallic irons represented by a Curie point at about 1040°K. The scanning x-ray microprobe analysis indicated that the metallic irons include such small portions of Ni (<0.2 percent) and Co (<1.2 percent) that the observed Curie point is very close to the established value (1043°K) of pure iron. Troilites surrounding these metallic irons contain 57.5 percent Fe by weight on the average. Since very little Mn^{2+} (~0.2 percent) and almost no Fe^{3+} have been detected in the rock sample, the paramagnetism of the rock can be attributed to Fe^{2+}. Because we know (i) the paramagnetic behavior of free Fe^{2+} in Fe-Mg-Ca oxide minerals such as pyroxene (*1*), (ii) antiferromagnetism of Fe pyroxene (*2*), (iii) antiferromagnetism of ilmenite (*3*), and (iv) ferromagnetism of pure iron (*4*), we attempted a quantitative simulation of the dependence of magnetization of the lunar rock on temperature. The individual referred dependences of phases i to iv on temperature and the simulated magnetization curve are illustrated in Fig. 1, where the assumed weight percentages of the four phases are free Fe^{2+}:$FeSiO_3$:$FeTiO_3$:Fe = 4.3:7:20:0.08. The simulated curve is in good agreement with the observed plots except for temperatures around 50°K. This discrepancy around 50°K may be due to a certain unclarified antiferromagnetism of (Fe, Mg, Ca)SiO_3, which is close to $FeSiO_3$ in chemical composition. The weight percentages of FeO and TiO_2 in the rock are evaluated from the simulation model to be 17.7 percent for FeO and 10.5 percent for TiO_2, which are in approximate agreement with the results of chemical analysis. Further, the simulated curves of magnetization versus the magnetic field for the range of $H = 0 \sim 30$ kilooersteds at various temperatures are in good agreement with observed data. Thus the model of magnetic composition of the lunar rock seems to be in good agreement with all results of its chemical and petrological analyses.

The intensity of natural remanent magnetization of the lunar rock (10024, 22) amounts to 7.50 × 10^{-6} emu/g. By an a-c demagnetization procedure of every 3-oersted step from 0 to 12 oersteds, the intensity of remanent magnetization is reduced to 88 percent of its initial value, and the scattering of its direction measured at each demagnetization step is within 13° and 9°, respectively, with respect to its declination and inclination. The observed stability may suggest that the remanent magnetization is not simply attributable to a contamination caused by the earth's magnetic field after the arrival of the sample at the earth's surface. However, the intensity and direction of the remanent magnetization are widely different in five separated pieces that were cut from the sample; that is, the remanent magnetization is not homogeneous at all. There is almost no doubt that the remanent magnetization is due to ferromagnetism of fine metallic irons. Experimentally, the thermoremanent magnetization of the sample is saturated with respect to temperature at about 800°C. The specific intensity of the saturated thermoremanent magnetization acquired in a unit magnetic field is 2.90 × 10^{-4} emu/g

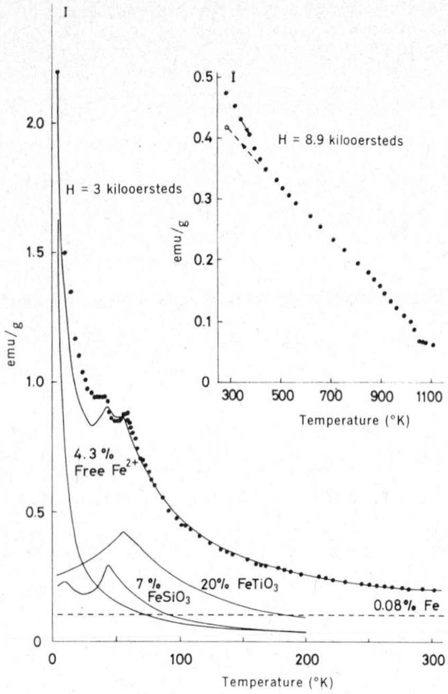

Fig. 1 (left). Magnetization (I) versus temperature (T) for lunar crystalline rock 10024, 22. Solid circles are observed values; unbroken lines are the simulated curve and the basic curves of the dependences of magnetization of Fe^{2+}, $FeTiO_3$ on temperature; the broken line is the basic dependence of the Fe particle on temperature.

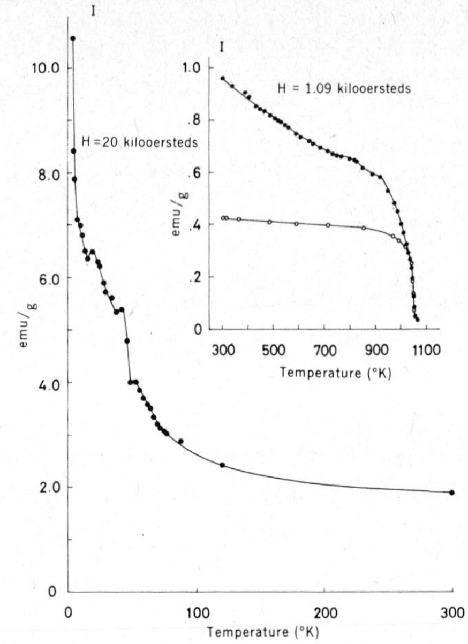

Fig. 2 (right). Magnetization (I) versus temperature (T) for lunar fines 10084,89. Solid circles are observed values before heating to 800°C; open circles are observed values after heating to 800°C.

oersted^{-1}. If the observed natural remanent magnetization is assumed to be the thermoremanent magnetization, the magnetic field during the cooling period must be about 2600 γ. This value seems hardly probable. Moreover, the heterogeneity of the remanent magnetization is incompatible with the homogeneity of the thermoremanent magnetization. The saturated magnetization and the saturated isothermal remanent magnetization of the ferromagnetic component of this rock at room temperature are 1.55×10^{-1} emu/g and 1.52×10^{-3} emu/g, respectively; these values are about 2×10^4 and 2×10^2 times as large, respectively, as the natural remanent magnetization.

Although the cause of the observed natural remanent magnetization has not yet been identified, it seems likely that the remanent magnetization might be caused by (i) an instantaneous local magnetic field generated by some kind of lightning, (ii) the piezomagnetic effect caused by meteorite impacts in the presence of a weak solar wind magnetic field, or (iii) a simple heterogeneous distribution of fine single-domain particles of metallic irons.

The same sequence of analyses was used for lunar fines (10084,89). Both x-ray microprobe and magnetic analyses have shown that magnetic minerals in the lunar fines are basically the same as in the lunar crystalline rock 10024,22; that is, that magnetic minerals in the fines are paramagnetic pyroxenes, antiferromagnetic Fe pyroxenes, antiferromagnetic ilmenites, and metallic irons (see Fig. 2). However, the saturation magnetization of the ferromagnetic component (metallic iron) of the lunar fines at room temperature amounts to 1.17 emu/g, which is about 7.5 times as large as the same magnetic parameter of the crystalline rock. The curve of magnetization versus temperature in Fig. 2 also indicates that the abundance of antiferromagnetic ilmenite is less and the abundance of antiferromagnetic and paramagnetic pyroxenes is more in the lunar fines than in the crystalline rock.

Analyses of the magnetization versus magnetic field curves at various temperatures have shown that about 20 percent of metallic iron in the crystalline rock and about 50 percent of metallic iron in the fines is superparamagnetic and in the form of very fine grains, which are less than 100 Å in mean diameter and which are superparamagnetic at temperatures above room temperature. These fine grains seem to be easily oxidized to become wüstite (FeO) at high temperatures, even in a high vacuum chamber. The remaining metallic irons are considered to be of fairly large size, and their magnetization curve is subjected to the shape demagnetization effect. For instance, five pieces of 92 spherules separated from 150 mg of the lunar fines are iron spherules with diameters that range from 20 to 100 μm and with Fe contents that range from 85 to 90 percent (±5 percent).

T. NAGATA, Y. ISHIKAWA
H. KINOSHITA, M. KONO, Y. SYONO
University of Tokyo, Tokyo, Japan
R. M. FISHER
U.S. Steel Corporation Research Center, New York 10006

References and Notes

1. T. Nagata, *Rock Magnetism* (Maruzen, Tokyo, 1961), pp. 76–78.
2. A. Sawaoka, S. Miyahara, S. Akimoto, *J. Phys. Soc. Japan* **25**, 1253 (1968).
3. Y. Ishikawa and S. Akimoto, *ibid.* **13**, 1298 (1958); J. J. Stickler, S. Kern, A. Wold, G. S. Heller, *Phys. Rev.* **164**, 765 (1967).
4. R. M. Bozorth, *Ferromagnetism* (Van Nostrand, Princeton, N.J., 1951), pp. 53–55 and 160–180.

4 January 1970

Magnetic Resonance Properties of Some Lunar Material

Abstract. *Paramagnetic resonance spectra of Apollo 11 fines and rocks were measured at 9 and 35 gigahertz and at 4°, 80°, and 300°K. At both frequencies the material has an intense absorption at* g = 2, *with a line width of ~950 gauss. Fe ions with strong exchange interactions produce this resonance. A comparison of the resonance absorption due to* Fe^{3+} *showed that the energy of the crystal field interaction was ~0.1 per centimeter.* Mn^{2+} *was identified in several samples, and an absorption at* g = 1.89 *was tentatively attributed to* Ti^{3+}. *The nuclear magnetic resonance spectra of* ^{27}Al *had a distribution of asymmetry parameters* η *ranging from 0.25 to 0.75 and had nuclear quadrupole coupling constants* e^2qQ/h *of approximately 3 megahertz.*

Elements present in lunar material (Apollo 11) with the highest relative abundance, which would be expected to produce electron paramagnetic resonance absorption (EPR), are Ti, Cr, Mn, and Fe (*1*). Ti^{3+}, Mn^{2+}, and Fe^{3+} have the highest probability of being detected (*2*). Exposure of rocks and minerals to energetic photons and particles can introduce paramagnetic defect states, the concentration of which may be a measure of the integrated flux of radiation (*3*). The valence states of ions such as Fe^{3+} and Mn^{2+} may also be altered by radiation (*4*). The characteristics of nuclear magnetic resonance (NMR) spectra are determined by interactions between nuclear magnetic moments and nearby electrons localized on the nucleus, on surrounding atoms, and in a metal by the conduction electrons (*5*). In lunar material (Apollo 11) the nuclei that are present in amounts sufficient to be detected by usual experimental techniques (*5*) are ^{1}H, ^{23}Na, ^{27}Al, and ^{29}Si.

Samples from rocks 10065 and 10062 and from fines were encapsulated in containers (*3*), which were either sealed with a cap or sealed to an ion pump in the vacuum chamber (F201) in the Lunar Receiving Laboratory. EPR and NMR measurements could be made with the material in the container, and samples were therefore relatively uncontaminated by terrestrial atmosphere. The pressure in containers with attached ion pumps has been $\leq 10^{-8}$ mm Hg since their removal from F201. Samples from rocks 10046, 10047, and 10057 encapsulated in air and from 10046 and 10005 encapsulated in N_2 were received. EPR and NMR measurements were repeated on 10062 and 10065 samples after exposure first to N_2 and then to air. Magnetic separation of fines failed, owing to the attraction of most of the particles by the magnet. Subsequently, mineral separations were made in air based on gross visual differences seen through a microscope with × 30 magnification. Part of sample 10047-49 was also separated in this fashion into three components: (i) fragments with an amber color in transmission, (ii) lustrous, opaque, black fragments with conchoidal surfaces, and (iii) colorless, white, or transparent fragments.

EPR measurements were made at 9 and 35 GHz at 4°, 80°, and 300°K. Resonance lines are designated by $g = h\nu/\beta H$, where H is the value of the magnetic field at which dI/dH, the derivative of the absorption, changes sign, ν is the spectrometer frequency, β is the Bohr magneton, and h is Planck's constant. NMR measurements were made over a range of resonant frequencies from 8 to 16 MHz by means of a Varian wide-line NMR spectrometer with a Varian "proton free" probe and a time-averaging computer.

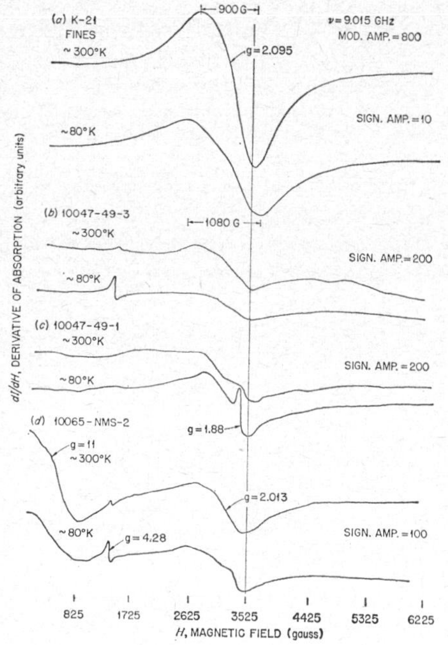

Fig. 1. Effect of temperatures on resonance spectra of several lunar samples.

^{1}H was not detectable in the probe and container for a 90-trace readout from the time-averaging computer for a 500-second trace time.

A spectrum characteristic of as-received fines and of 10065 (encapsulated in F201, exposed to N_2 and then to air) and many other rock fragments is shown in Fig. 1a. For this line $2.01 < g < 2.11$ and the width ΔH of the line measured between the two inflection points of the derivative curve is $900 \leq \Delta H \leq 980$ gauss at both 9 and 35 GHz when measured at room temperature. At 9 GHz, ΔH increased from 900 to 1080 gauss when the temperature was decreased from 300° to 80°K (compare the two curves in Fig. 1a). There was also a twofold decrease in peak-to-peak height. Use of the equation $I \sim \Delta H^2 h_{pp}$, where h_{pp} is peak-to-peak height, gives the intensity of the absorption at 80°K as ~12 percent less than at 300°K. At 4°K it was not possible to make direct comparisons of intensity with the measurements at 80° or 300°K. On the basis of a comparison of spectrometer sensitivities at the two temperatures, the intensity of the detected resonance was at least 100 times less than the resonance detected at 300°K.

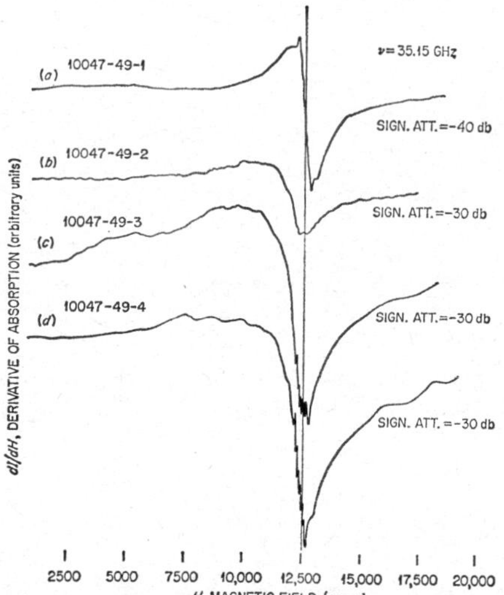

Fig. 2. Electron paramagnetic resonance spectra of mineral separates from rock 10047-49. (a) Amber, transparent, ~3 mm³ (sample 1); (b) lustrous, black, opaque, conchoidal surfaces, ~3 mm³ (sample 2); (c) colorless, white, transparent [~10 percent (A) and (B) material], 3 mm³ (sample 3).

Resonance spectra at 9 GHz of two fractions of 10047-49 are shown in Fig. 1 (c, b). At 300°K the $g = 2.1$ resonance in 10047-49-2 and 10047-49-3 decreased in intensity with decreasing temperature, whereas in 10047-49-1 the absorption in this region of g value increased. These samples differed in other respects. A resonance in 10047-49-3 at $g = 4.28$, quite weak at 300°K, showed a fourfold increase in intensity at 80°K. This resonance was not detected in either 10047-49-1 or 49-2. Resonance absorption at $g = 1.89$ in 10047-49-1, with a temperature dependence similar to the $g = 4.28$ resonance in 10047-49-3, was not detected in either 10047-49-3 or 49-2.

A nonmagnetic piece of 10065 has a spectrum (Fig. 1d) that is quite different, except in the region of $g = 2$, from the spectrum of the magnetic fraction. The resonance at $g = 2$ was similar in width, shape, and temperature dependence to the resonance in Fig. 1a. There was an intense absorption at $g = 11$ and a weak one at $g = 4.28$ at 300°K. At 80°K the resonance at $g = 11$ decreased twofold in intensity and the resonance at $g = 4.28$ increased fourfold. The change in the shape of the absorption at $g = 1.89$ with decrease in temperature, which can be seen by comparing the two curves in Fig. 1d, is probably due to a resonance similar to the one in 10047-49-1.

The spectra of the three fractions of 10047-49 measured at 35 GHz and 300°K are all characterized by relatively intense absorption in the region of $g = 2$. Sample 10047-49-1 has an almost symmetric absorption in this region (Fig. 2a). Although not well resolved in the figure, the central part contains six equally spaced lines with a spacing of 90 gauss and $g = 2.00$. This six-component spectrum has better resolution in the curves for 49-3 and the powder (Fig. 2, c and d). It was not detected in 49-2. The broad asymmetric shape of the $g = 2.1$ absorption in 49-3 is similar to the resonance in amber-colored fragments separated from the fines sample (for example, amber spheres).

All the described resonances had an intensity that was proportionate to incident microwave power and shapes that were independent of power when it was <0.3 watt.

Room temperature NMR spectra of ^{27}Al ($I = 5/2$) from a sample of 10062 and of 10065 were recorded in the dispersion mode. Two of the spectra are shown in Fig. 3 (b and c) with the spectrum of the probe, sample container, and ^{27}Al in $AlCl_3$, which was used as a reference for ν_0, in Fig. 3a. Line shapes were due to second-order quadrupole broadened ($1/2 \leftrightarrow -1/2$) transitions (*5*) with a distribution of nuclear quadrupole coupling constants and of asymmetry parameters. These spectra are characterized by nuclear quadrupole coupling constants of several megahertz. An estimate of the asymmetry parameters η was obtained by comparison with ^{27}Al resonance from powdered samples of microcline ($\eta = 0.21$, $e^2qQ/h = 3.22$ MHz), and albite ($\eta = 0.62$, $e^2qQ/h = 3.29$ MHz) (*6*) and was found to range from 0.25 to 0.75. Spectra of ^1H from 10062 and 10065 and from 5 grams of fines were not obtained in reproducible form for integration times >30 hours. Spectra of ^{23}Na, ^{29}Si, and $^{47,49}Ti$ were not detected for integration times $\sim$10 hours.

The most probable source of the intense resonance that is characteristic of the fines and breccia samples is Fe^{3+}, but other types of resonance from Fe (such as ferromagnetic resonance) may also contribute to the resonance absorption. The frequency-independent g value and the temperature dependence of width and intensity of this characteristic line are evidence that these properties are determined by strong magnetic interactions between the paramagnetic ions. Such interactions could occur in metallic particles of iron whose diameter is smaller than a domain (superparamagnetism) (*7*) or in regions of an oxide of iron where there is a high concentration of paramagnetic Fe^{3+}. No resonance absorption would be expected in pure FeO under the condition of these experiments, but partly reduced Fe_2O_3 has an intense resonance with characteristics similar to the lunar material resonance (Fig. 1a).

Fe^{3+} in crystal fields of symmetry lower than cubic, in which the energy of interaction of the d^5 electrons with the crystal field is of the same order of magnitude as $h\nu$, where ν is the frequency of measurement, may have strong resonance absorptions at $g \sim 10$, $g \sim 6$, $g = 4.28$, and $g \sim 0.8$ (*8*) depending on the symmetry of the crystal field. Absorption at $g \sim 10$, $g = 6$, and $g = 4.28$ has been observed in many compounds (*2*) and in silicate glasses (*9*). On the basis of these data and analyses, the $g = 11$, $g = 4.28$ absorptions shown in Fig. 1 are due to Fe^{3+}. The increase in absorption at high field observed in 10047-49-3 may be due to absorption from Fe^{3+}, which in some crystal fields would have an absorption peak at $g \sim 0.8$. The decrease in intensity of the absorption at $g = 11$ in 10065-NMS-2 and at high fields in 10047-49-3 is consistent with this assignment. Since these absorptions may occur in an excited electronic state (depending on the nature of the crystal field interactions), a decrease in temperature decreases the population of electrons in the states with a resultant decrease in absorption. Measurements at 35 GHz show that most of the resonance absorption is in the neighborhood of $g = 2$ and that none is observed at $g \sim 11$ or $g \sim 4.28$. Hence, we can conclude that the crystal field interactions are much less than $h\nu$, when $\nu = 35$ GHz (1.2 cm^{-1}).

Cr^{3+} and Ti^{3+} in various crystals have resonance absorptions at $g < 2.0$ (*2*). Cr^{3+} is usually observed at 300°K, whereas Ti^{3+} is usually observed only at temperatures ≤ 80°K. On the basis of the relatively high abundance of Ti as compared with Cr in lunar material (Apollo 11) and the temperature dependence of the $g = 1.89$ absorption, we tentatively attribute this absorption to Ti^{3+}. The congruence of spacing and g value of the six-line set in 10047-49-1, 49-3, and powder with the six hyperfine lines of the ($1/2 \leftrightarrow -1/2$) electronic transition of Mn^{2+} in many silicate minerals (*4*) is positive identification of Mn^{2+} in these samples. This assignment is also supported by the relative abundance of Mn ($\sim$0.2 percent). The width and shape of these lines are similar to those of Mn^{2+} in amorphous compounds, such as tektites (*4*).

EPR absorption due to singly charged oxygen vacancies has been found in several silicate minerals after irradiation with energetic protons (*3*). Lunar surface material, if exposed to a sufficiently large flux of solar protons, particularly energetic flare protons, might be expected to exhibit a similar absorption. A flux of $\sim 10^{16}$ cm^{-2} of 40-Mev cyclotron protons at $T \sim 100$°C during irradiation is sufficient for the production of enough centers in $\sim$3 mm^3 of sample of a mineral to be easily detected. In none of the white and colorless transparent fragments separated from our fines sample, or in 10047-49-3 or 10047-49-1, with the possible exception of one piece, was this absorption observed. In view of (i) the small distance of penetration of protons ($\sim$3 mm for 40-Mev protons incident on olivine), (ii) the relatively low flux per year of protons with energy ≥ 10 Mev (of the order of 10^{10} cm^{-2}/year) (*10*), (iii) the relatively short exposure time reported for lunar material (Apollo 11) ($\sim 10^7$ to $\sim 10^8$ years), and (iv) the small probability of obtaining material from within 0.5 cm of the surface due to mixing during collection and return, this negative result is not unexpected.

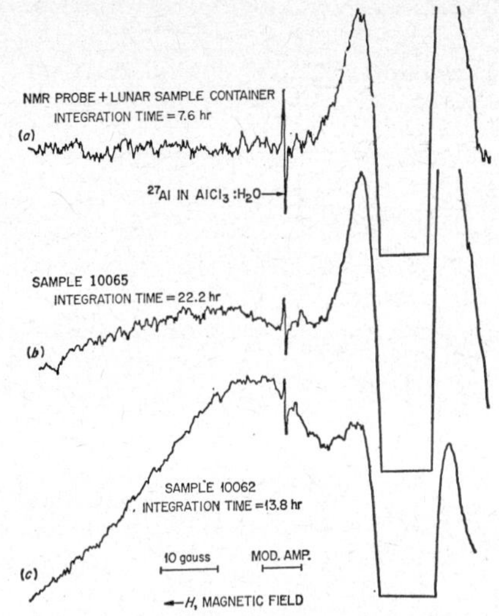

Fig. 3. Nuclear magnetic resonance spectra of ^{27}Al in lunar material (Apollo 11) at $\nu_0 = 16$ MHz recorded in the dispersion mode. The two samples whose measurements are shown in (a) and (b) were of approximately equal volume.

R. A. WEEKS, A. CHATELAIN
J. L. KOLOPUS

*Solid State Division,
Oak Ridge National Laboratory,
Oak Ridge, Tennessee 37830*

D. KLINE

*State University of New York,
Albany 12203*

J. G. CASTLE

*University of Alabama,
Huntsville 35807*

References and Notes

1. Lunar Sample Preliminary Examination Team, *Science* **165**, 1211 (1969).
2. W. Law and E. L. Offenbacker, *Solid State Phys.* **17**, 136 (1965).
3. R. A. Weeks and J. L. Kolopus, Paramagnetic Resonance Spectra of Some Silicate Minerals, "Semi-Annual Progress Report for Period Ending December 31, 1968," (*ORNL-CF-69-3-5*, contract NASA-MSC-76458).
4. R. A. Weeks and J. L. Kolopus, Paramagnetic Resonance Spectra of Some Silicate Minerals, "Semi-Annual Progress Report for Period Ending June 30, 1968," (*ORNL-CF-68-8-38*, contract NASA-MSC 76458).
5. A. Abragam, *The Principles of Nuclear Magnetism* (Oxford Univ. Press, London, 1963); C. P. Slichter, *Principles of Magnetic Resonance* (Harper & Row, New York, 1963).
6. P. Hartmann, thesis, University of Zurich, Switzerland (1963).
7. C. P. Bean and J. D. Livingston, *J. Appl. Phys. Suppl.* **30**, 1205 (1959).
8. H. H. Wickman, thesis, University of California (*UCRL Rep. 11538*).
9. E. Lell, N. J. Kreidl, J. R. Hensler, *Progr. Ceram. Sci.* **4**, 59 (1966).
10. E. J. Zeller, L. B. Ronca, P. W. Levy, *J. Geophys. Res.* **71**, 4855 (1966).

11. Research sponsored by the U. S. Atomic Energy Commission under contract with Union Carbide Corporation and by NASA contract MSC-T-76458. We thank H. S. Story (State University of New York at Albany) for his comments on the NMR results, M. M. Abraham (Oak Ridge National Laboratory) for his comments on the EPR results, S. Hafner (University of Chicago) for his comments on the NMR measurements, and Orson Anderson (Lamont-Doherty Geological Observatory) for the loan of his ~5-gram sample of fines.

4 January 1970

Optical and High-Frequency Electrical Properties of the Lunar Sample

Abstract. *Reflectivity and polarization laws for the powder sample and its spectrum are close to the mean for the lunar maria. Solid samples show a marked absorption feature at 1 micron. The low albedo appears to be due to a surface coating on dust grains rather than to volume absorption. The high-frequency electrical properties resemble those of a fine powder made from typical dense terrestrial rocks and are consistent with previous estimates from ground-based radar observations. The differential mass spectrum is almost constant from 100-micron particles down to 0.1-micron particles; most particles are smaller than 0.3 micron. Their shapes disclose a variety of processes of generation.*

Lunar dust and rock chip samples have been analyzed in the lunar laboratory of the Cornell Center for Radiophysics and Space Research; our concern has been with the optical and electrical properties of the sample and their relation to those known for the lunar surface as a whole, and with the questions surrounding the origin of the lunar dust. Four salient points have emerged.

1) The optical scattering law and polarization properties of a surface of lunar dust generally correspond closely to these properties as observed for the moon as a whole. The rock chip sample shows a strong absorption feature at 1 μm which is not prominent in the lunar scattered light. It is probable, therefore, that most of the lunar surface is covered with a material similar to the powder that was investigated.

2) The dielectric constant is within the range that had been estimated for the moon as a whole by radar methods.

3) The particle size distribution indicates that the differential mass spectrum as a function of radius is almost constant from 100 μm down to 1000 Å. The shapes of the particles indicate a variety of sources. Some have the sharp edges that are characteristic of fracture; others are rounded, indicating processes of melting or condensation. Some cannot readily be attributed to either of these mechanisms.

4) The darkness of the lunar dust is mainly due to dark surface deposits on the grains, probably metallic, rather than to absorptivity of the bulk material.

The optical scattering law as a function of phase angle and the optical polarization law were measured with the same instrument that had been used for measuring many sample powders in the past and in the same manner (*1*). The lunar powder proved to resemble, both in appearance and in the measured optical properties, the lunar maria as observed from the earth and the terrestrial powders previously proposed (*1*) as being most closely representative of the moon. These powders also proved to be similar under optical microscope examination. The particle size was similar, the great majority of the particles being less than 10 μm in diameter. The adhesion of the small particles to each other indeed created the "dendritic growth" appearance under the microscope that has been given the name "fairy castles." It appears that the large part of the pronounced lunar opposition effect—that is, the brightness surge toward zero phase—can be attributed to the shadows cast by this lacy surface structure.

Figures 1 and 2 summarize the optical properties of the Apollo 11 samples. Each data point represents the mean of several observations of different portions of a sample, and the measurements repeated very well. In Fig. 1 the photometric phase function of the lunar dust sample is generally steeper than the mean lunar case (*1*) for phase angles less than 15°, but the difference is very small. The curve for polarization plotted against phase angle (Fig. 1) also demonstrates the similarity of the dust sample to the moon as a whole, but, again, there are minor differences; the crossover from negative to positive polarization occurs at a lower phase angle, and polarization in the positive branch is greater.

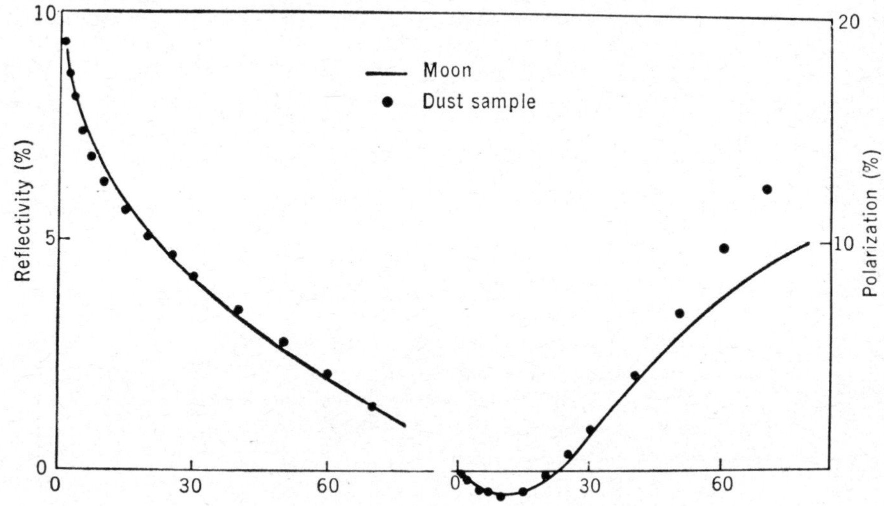

Fig. 1. The dependence of reflectivity and polarization on phase angle at wavelength of 5600 Å and at normal viewing. [Curve for the moon from Wattson and Danielson (*3*)]

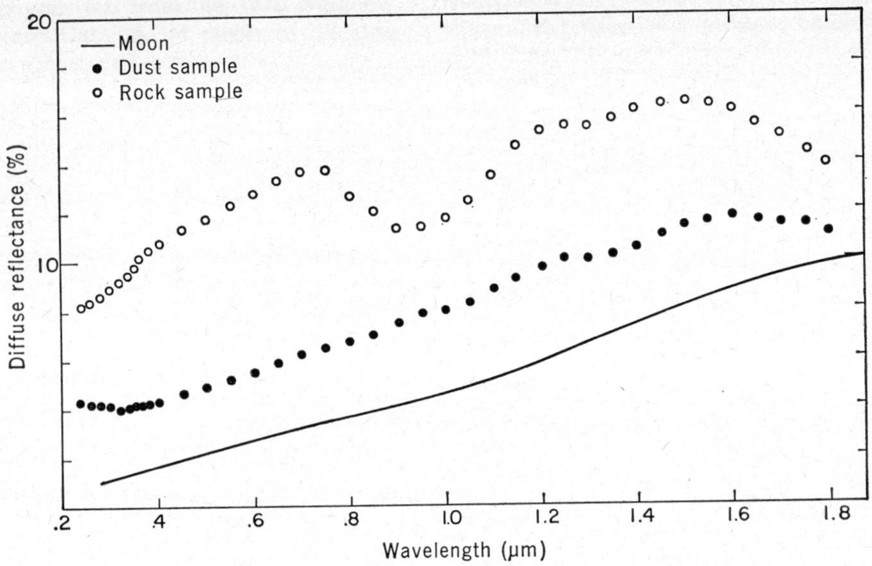

Fig. 2. The spectral reflectance of Apollo 11 lunar samples. [Curve for the moon from Hapke (*1*), with arbitrary normalization of reflectivities]

The normal albedo of the dust sample at 5600 Å was measured as 10.2 ± 0.2 percent. This value is in close accord with the value 9.96 percent for the Apollo 11 site, as derived from Apollo 10 orbital photography (*2*). Moreover, in the hemispherical reflectance measurements performed on a Cary 14 spectrophotometer, the albedo values of the dust sample in the visible and near-infrared were similar to lunar maria values obtained from earth-based observations. Both spectra are featureless, climbing steadily in albedo from ~ 0.3 to 1.5 μm (Fig. 2). Lunar rock chip samples were also measured on the spectrophotometer, and a strong absorption band, not present in the powder sample, appeared near 1 μm. A weak band in this region had previously been suggested, from earth-based observations of the moon (*3*, *4*). Further details of the optical properties of the Apollo 11 samples will be presented in the near future (*5*).

Rock powders in the size range of a few microns tend to be very light in color. Most rocks have too low an opacity to absorb much of a light ray, which is generally scattered out of the surface after having traversed only a few microns of material. Trying to account for the very low albedo of the lunar surface has been a long-standing problem, in view of the indication, from the optical scattering and polarization properties, of a very small particle size, since even the darkest rocks tend to be quite light when powdered. We have previously, in this laboratory, undertaken sputtering experiments with kilovolt protons and with α-particles on powdered rock surfaces, which have indicated darkening. It has been suggested that this darkening was due to the deposition of reduced metals, perhaps chiefly iron, on the surface as a result of the dissociation by the sputtering process, the partial escape of the oxygen, and the slowness of surface recombination limited by diffusion.

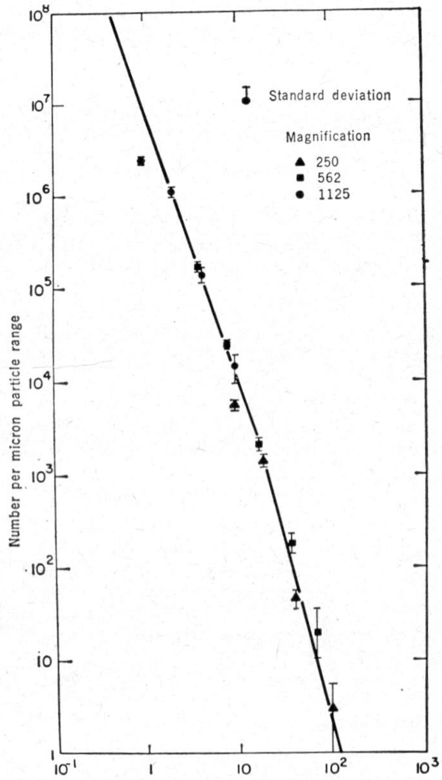

Fig. 3. The differential particle size distribution for the bulk sample.

Metallic surface coatings of as little as 30 Å can provide much opacity but would make only an insignificant contribution to the bulk chemical composition. We have seen strong evidence for such coatings, but we have not yet been able to analyze them adequately. Whether they are indeed the result of sputtering or of other metal evaporation (vacuum plating) processes, or whether perhaps they are just the reduction by the hydrogen of the metallic surface layers produced by solar wind, is not yet clear. However, we have had the following indications of the presence of metallic layers.

We observed under the microscope that some larger particles in the size range 50 to 200 μm that could be found in the lunar soil sample had a metallic appearance, sometimes over only a certain part of their surface. Some particles could be clearly seen as translucent glass in which a well-defined area appeared metallic. One sphere, for example, looked like a honey-colored glass from one side but like a steel ball from the other. When treated with the common acids (hydrochloric and nitric) that attack metals, the metallic appearance was generally reduced but not completely removed. Hydrofluoric acid generally tended to remove the metallic appearance entirely, even before a visible erosion of the particle had taken place.

For most of the material an optical examination is not feasible because the particles are too small. Nevertheless, when the same acids were applied to a microscopic sample of fine powder it quickly turned very much lighter, almost white. It seems likely, therefore, that in the finer material a metallic surface coating is also normally present, and responsible for the low albedo.

The measurement of a particle size distribution for such small grains is not an easy matter. The cohesion of the grains prevents analysis of the smaller sizes by sieving, as was pointed out by the preliminary investigators (*6*). We have used three techniques: (i) making microscope slides of the powder mixed into a transparent varnish and smeared out into a thin layer, which permits us to count the particles with an oil emersion microscope down to sizes of about 2 μm; (ii) constructing a water sedimentation column in which the descent of particles as small as 1 μm can be photographically registered; (iii) determining the size distribution of the smallest particles from the scanning

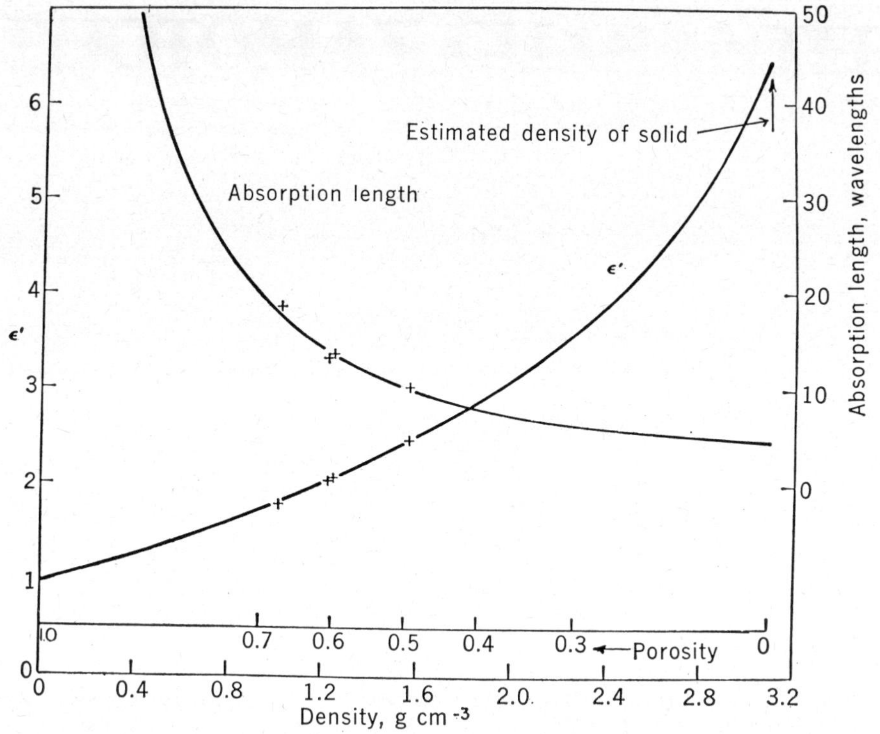

Fig. 4. The dielectric constant and absorption length of the bulk sample at 450 Mhz as a function of the powder density. The solid curves are the Rayleigh formula (7). The estimated density of solid is from Lunar Sample Preliminary Examination Team (6).

electron micrographs referred to below. The three methods give consistent results (Fig. 3).

The detailed shapes of particles can be seen to a resolution of 300 Å in numerous scanning electron microscope pictures that were taken to see whether the origin of the material was revealed by the particle shapes.

Our studies indicate that various different effects have been active in producing the fine material. Some particles are spherical and rounded, suggesting condensation from a vapor or freezing of a liquid in free fall. Others are sharp-edged and angular, undoubtedly the result of fracture. They lack, in general, any obvious indication of a crystalline structure, as neither cleavage planes nor preferred angles are seen. It would appcar that most of the fractured material is amorphous, or, if any of it is crystalline, that the size of the crystals is below the limit of resolution.

The spherical or compact round particles seen are less frequent but may form a continuous sequence from the 100-μm range down to very small sizes. The great majority of particles in the 10- to 1-μm size range have, however, more intricate shapes that are not readily understood. There are many rounded surfaces, and yet the particle as a whole is not compact. Elongated objects with rounded ends, surfaces where the sense of the curvature often changes, rough spots occurring in smooth surfaces, and various other features argue against any single explanation—liquid droplets, condensation, or fracturing. Additional processes such as erosion by sputtering, partial melting, and partial evaporation must be considered, and scanning electron microscope study of these mechanisms is needed before all the responsible processes can be identified.

Measurements were made by means of the technique used for determining the electrical properties of terrestrial rock powders (7). The dielectric constant (ε') and loss tangent of lunar dust at several stages of compaction were measured at 450 Mhz. The measurement in each case included a measurement of the density of the sample, and the porosity was calculated from the quoted specific gravity of the rock of which the powder is composed (6). The dielectric constant and the absorption length (Fig. 4) are consistent with the values deduced from ground-based radar and radiometric observations, respectively. As with terrestrial rock powders, the dielectric constant and loss tangent as a function of porosity follow the Rayleigh mixing formula and, by extrapolation, suggest a permittivity for the solid rock of the same composition as the lunar dust which is near the average for dense terrestrial rocks (about 7). The permittivity is about 3 for the dust at a typical "loose packing" porosity of 0.6. The absorption length at the same porosity, in this sample, is about 10 wavelengths.

T. GOLD
M. J. CAMPBELL
B. T. O'LEARY
Center for Radiophysics and Space Research, Cornell University, Ithaca, New York 14850

References and Notes

1. B. Hapke, *Science* 159, 76 (1968).
2. R. L. Wildey and H. A. Pohn, *Astrophys. J.* 158, L129 (1969).
3. R. B. Wattson and R. E. Danielson, *ibid.* 142, 16 (1965).
4. R. G. Tull, *Icarus* 5, 505 (1966).
5. B. O'Leary and F. Briggs, in preparation.
6. Lunar Sample Preliminary Examination Team, *Science* 165, 1211 (1969).
7. M. J. Campbell and J. Ulrichs, *J. Geophys. Res.* 74, 5867 (1969).
8. This work was done under NASA contract NAS9-8018. We are grateful to Corning Glass Works for the assistance given with the electron micrography. We thank Paul Shapshak, F. Briggs, and J. Winters.

5 January 1970

Magnetic Resonance Studies of Lunar Samples

Abstract. *Electron spin resonance searches at 9.5 gigahertz on several fines samples and portions of several rocks have yielded signals whose lineshapes and temperature dependences show that the samples are principally ferromagnetic in nature. Proton magnetic resonance searches at 60 megahertz of these samples have not revealed any signals ascribable to water or any other types of hydrogen in concentrations greater than 0.0001 percent by weight contained in narrow lines (5 oersteds wide or less) and 0.01 percent by weight in wide lines (as wide as 100 oersteds).*

We are conducting magnetic resonance studies of the lunar samples returned by the Apollo 11 mission in order to (i) determine by nuclear magnetic resonance (NMR) the total proton concentration along with the distribution of any protons detected between water, hydrated minerals, organic

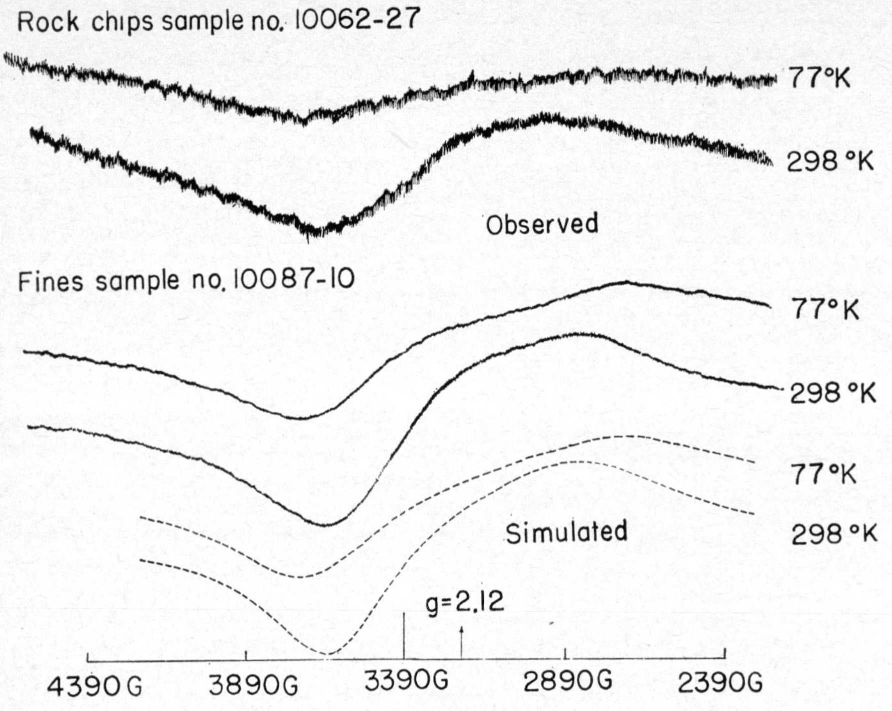

Fig. 1. Observed and computer-simulated ferromagnetic resonance spectra for the lunar samples.

protons, and solar wind protons; (ii) study by NMR any other magnetic nuclei present at significant concentrations to be satisfactorily examined by NMR; and (iii) search for electron spin resonance (ESR) signals which might arise from radiation damage to inorganic or organic components, or both, of lunar rock and soil. We describe here our initial ESR searches for species with unpaired electrons and our NMR studies directed towards the detection of protons in the lunar samples.

The ESR studies were carried out with a conventional 9.5 Ghz Varian V-4500 spectrometer fitted with 100 khz modulation, a rectangular cavity, and variable temperature provisions (down to 4°K). The NMR spectrometer consisted of a Varian 30-cm magnetic system (14 kilooersteds) whose power supply was swept by a digitally generated voltage ramp. The latter was either provided by frequency-to-voltage conversion from a digitally swept HP-5100 Frequency Synthesizer or from a HP-5480 Signal Averager. In all cases the sweep ramp was advanced synchronous with the HP-5480 address advance by an HP-2115 Controller. Repetitive linear sweeps up to 200-oersted wide could be generated. The detection system consisted of a 60 Mhz Varian V-4311 rf unit and cross coil probe [constructed of low proton materials and possessing a 15-mm (inner diameter) low proton-containing insert] whose output was passed to a PAR Model HR-8 Lock-in Amplifier. The modulation frequency was 205 hz; modulation levels between 1 millioersted and 50 oersteds were employed.

Our samples consisted of several portions of fines (sample 10087) and portions of rock 10017 (external and internal sample), rock 10046 (external and internal sample), and rock 10062 (external and internal sample). For ESR, samples in quartz tubes (2 to 3 mm outer diameter) were employed. The NMR samples were studied in quartz tubes with outer diameter of 15 mm. Samples not sealed at the Lunar Receiving Laboratory were handled and stored under an atmosphere of dry nitrogen.

In the ESR experiments, a rather broad signal centered at $g = 2.12 \pm 0.05$ was detected in both the rock chip sample 10062-27 and in the fines sample 10087-10 (Fig. 1). For the latter, the signal intensities were roughly an order of magnitude stronger than those observed for the rock chip sample. In both cases, no temperature dependence of the intensities was noted. The lineshapes of these signals were studied over the range of 77° to 298°K. Computer simulation of the lineshapes (*1*) employing a model consisting of ferromagnetic centers of the order of 1 μm in diameter distributed throughout the samples appears to fit the observed signals. The g value of 2.12, as well as the first order anisotropy constant ($2k_i/Mz$) of 400 oersteds obtained at room temperature and 500 oersteds at 78°K, suggest that these ferromagnetic centers are probably metallic iron.

Our magnetic resonance searches at 60 Mhz on all our lunar samples have thus far failed to detect any signals that we would attribute to protons. The instrumentation and the low proton probe we are employing is capable of detecting and has been calibrated to measure proton contents down to the 0.0001 weight percent (1 ppm) contained in resonance lines 5 oersteds wide and the 0.01 percent (100 ppm) contained in resonance lines as wide as 100 oersteds. We have studied a variety of model samples containing protons. They consisted of H_2 gas at various pressures between 760 mm and 7.6 mm, H_2 gas admixed with paramagnetic NO, H_2 gas in silica (*2*), terrestrial rocks and soils, and samples of various terrestrial materials admixed with known amounts of protons. In every case, proton NMR signals were easily detected.

In the initial rare gas analyses of a number of Apollo 11 samples (*3*), seven samples exhibited ^4He contents in the range of 0.01 to 0.8 scm^3/g (volume of 1 cm^3 of gas at 1 atm and 273°K per gram of sample). For a different fines sample than we have, the preliminary gas analysis (*3*) indicated possible H_2 and ^4He contents of 0.8 and 0.2 scm^3/g, respectively. In view of the ratio of ^1H to ^4He in the solar wind (*4*) and if these resident gas contents arise from the solar wind, it is not unreasonable to expect the occurrence of proton contents of the order of 1 to 3 scm^3/g in some lunar surface rock and soil samples. As discussed above, our instrumentation can readily detect gaseous H_2 equivalent to 0.01 scm^3/g of rock or soil if it gives rise to a resonance line 10 oersteds wide or less. The linewidths of gaseous H_2 samples are 0.6 oersted wide or less. Gaseous H_2 trapped in silica (*2*) exhibits a linewidth of 1.4 oersteds. Based on our experiences with terrestrial rock and soil samples, we expected that any protons in water molecules or on carbon atoms in the lunar samples might exhibit linewidths of 1 to 10 oersteds. We have detected such protons in terrestrial samples down to the 10 ppm level. In any case, we did not expect to detect the residual proton content of sample 10017 because of its low He content (3×10^{-4} to 11×10^{-4} scm^3/g) (*3*).

Because aluminum was found to the extent of 4.1 to 6.9 weight percent in all Apollo 11 samples thus far analyzed (*3*), we searched for ^{27}Al resonances in our samples but our results were ambiguous. All the terrestrial samples containing Al examined exhibit strong ^{27}Al signals.

Our failures to detect proton magnetic resonance signals could indicate that the proton contents of our samples are very low or that the resonances of any resident protons are too broad to be detected under the conditions employed thus far. This latter situation could very well prevail based on the ESR experimental detection of ferromagnetic centers discussed above and the large magnetic susceptibilities of the Apollo 11 samples found in the initial magnetic measurements (*3*).

STANLEY L. MANATT
DANIEL D. ELLEMAN
ROBERT W. VAUGHAN
SUNNEY I. CHAN
FUN-DOW TSAY
WESLEY T. HUNTRESS, JR.

Space Sciences Division, Jet Propulsion Laboratory, and Division of Chemistry and Chemical Engineering, California Institute of Technology, Pasadena 91103

References and Notes

1. C. Kittel, *Phys. Rev.* **73**, 155 (1948).
2. These materials have been described by S. P. Faile, J. J. Schmidt, and D. M. Roy [*Science* **156**, 1593 (1967)]; we wish to thank P. C. Lauterbur for making portions of these materials available to us.
3. Lunar Sample Preliminary Examination Team, *Science* **165**, 1211 (1969).
4. The ratio of ^{1}H to "He" in the solar wind is of the order of 20; for example, see M. Neugebauer and C. W. Snyder, *J. Geophys. Res.* **71**, 4469 (1966).
5. We thank D. B. Nash, M. M. Neugebauer, E. M. Bollin, P. C. Lauterbur for their many helpful discussions. This paper is contribution No. 3998 from the Division of Chemistry and Chemical Engineering, California Institute of Technology. Sponsored under NASA contract No. NAS 7-100.

4 January 1970

Other Physical Properties

Thermoluminescence of Lunar Samples

Abstract. *Appreciable natural thermoluminescence with glow curve peaks at about 350 degrees centigrade for lunar fines and breccias and above 400 degrees centigrade for crystalline rocks has been recognized in lunar samples. Plagioclase has been identified as the principal carrier of thermoluminescence, and the difference in peak temperatures indicates compositional or structural differences between the feldspars of the different rock types. The present thermoluminescence in the lunar samples is probably the result of a dynamic equilibrium between acquisition from radiation and loss in the lunar thermal environment. A progressive change in the glow curves of core samples with depth below the surface suggests the use of thermoluminescence disequilibrium to detect surfaces buried by recent surface activity, and it also indicates that the lunar diurnal temperature variation penetrates to at least 10.5 centimeters.*

Thermoluminescence (TL) studies on samples from two medium-grained crystalline rocks (type B), two breccias (type C), and lunar fines (type D, less than 1 mm) that included material from the bulk sample and from core No. 1 were undertaken to determine properties of lunar material and to investigate possible uses of TL for the study of lunar geology (*1*).

Between 5 and 20 mg of material, in powdered form, was mounted in silicon oil on a small aluminum disk and heated at a rate of 20°C/sec by focused infrared radiation from below. The thermoluminescence was measured by a photomultiplier tube with a Cs_3Sb cathode, and the glow curve and the area under the glow curve were recorded as a function of temperature on an x-y-y′ recorder. The instrument was calibrated with a ^{14}C-activated light source of known luminous flux, which made it possible to read the height of a TL peak in lumens and the amount of TL (the area under the glow curve) in lumen-sec. Samples were irradiated with x-rays of 14-kev effective energy, and the doses were monitored with an integrating dosimeter.

The TL measurements on the fines (sample 10084) from the bulk sample box yielded a glow curve with a single weak peak at about 350°C. In order to identify the TL carrier, a portion of the fines was sieved and separated into fractions with different relative proportions of the major mineral constituents, by means of heavy-liquid and magnetic techniques. The natural TL was measured on portions from nine fractions, including the untreated fines. All of these natural glow curves have a single peak at about 350°C. The same samples were then exposed to 5000 rad of x-radiation, and their "artificial" glow curves were measured 3 minutes after exposure. The artificial glow curves have nearly identical shapes, with prominent peaks at 125° and 185°C and evidence of less prominent peaks at about 260° and 350°C. The ratio N/A was used to compare the relative TL sensitivities of the nine fractions, where N is the area per milligram under the natural glow curve from 300° to 350°C and A is the area per milligram under the artificial glow curve from 85° to 350°C. Although the areas differ by more than an order of magnitude (N from 0.44 to 17.6 nlumen-sec/mg and A from 6.0 to 283 nlumen-sec/mg), N/A is nearly constant for all fractions, and has a value of 0.063 ± 0.006 (S.D.). The similarity of the natural glow curves, the similarity of the artificial glow curves, and the constancy of N/A show that a single mineral is responsible for the TL. Both N and A increase as the amount of feldspar in the nine fractions increases, which indicates that plagioclase is the primary TL carrier.

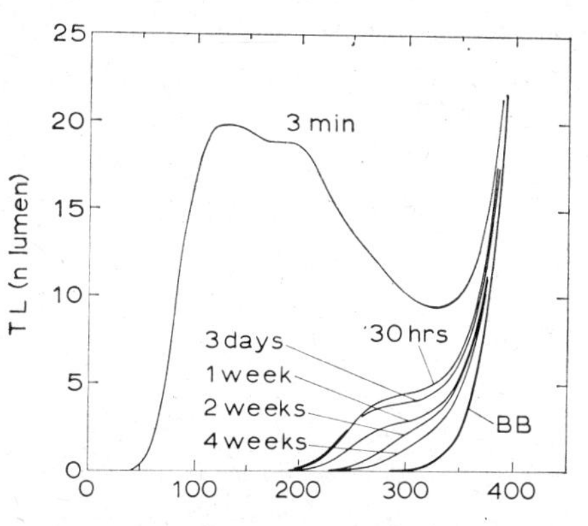

Fig. 1. Glow curves, showing the decay of artificial TL (5000-rad) as a function of time at 100°C for the 45- to 75-μm fraction of the lunar fines. *BB*, Blackbody curve.

In order to observe the change in TL as a function of time and temperature, the following decay experiments were performed on material from the fines: (i) decay of 5000-rad artificial TL in air at room temperature (≈22°C); (ii) decay of 5000-rad artificial TL in vacuum at 100°C; and (iii) decay of natural TL in vacuum at 100°C. At room temperature, the 125°C peak is almost entirely gone within a few days, but measurable TL in the region above 150°C remains even after a period of 4 weeks. At 100°C the artificial TL of the 125°C- and 185°C-peaks is completely gone by 30 hours (Fig. 1). The 260°C peak is prominent after 1 week at 100°C, but after 4 weeks it has greatly decayed so that the shape of the glow curve closely resembles that of the natural glow curve. There is noticeable decay of the natural TL with storage at 100°C; by 4 weeks the area under the natural glow curve between 300° and 350°C is reduced by approximately one-third.

Although difficult to measure accurately because of mutual interference, the half-lives of the various peaks can be estimated from the decrease in area under the glow curves and the change in peak heights with time. The 125°C and 185°C peaks appear to have half-lives of less than 2 weeks at room temperature and will be of little value for lunar geologic studies. At 100°C ambient temperature, the 260°C peak decreases with an apparent half-life of about 2 weeks, and the 350°C peak

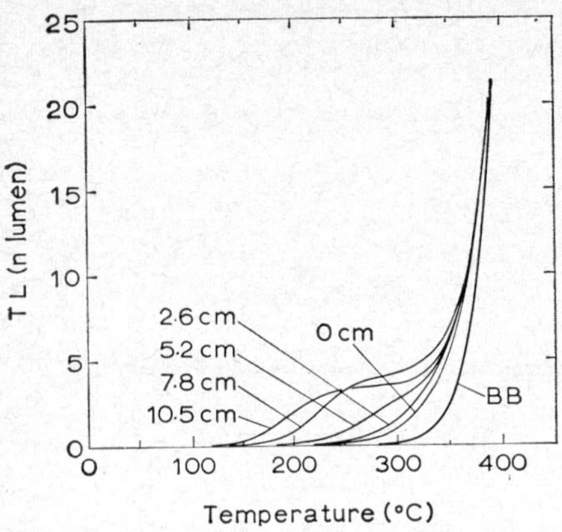

Fig. 2. Glow curves on samples from core No. 1 (10005), showing the change in natural TL with depth below the lunar surface. *BB*, Blackbody curve.

decreases with a half-life of about 1 to 2 months. These half-lives were compared with values calculated from a simple exponential model (*2*) with

$$t_{1/2} = (\ln 2/s) \exp (E/kT)$$

where

$$s = (qE/kT_p^2) \exp (E/kT_p)$$

The activation energy E is approximately 25 kT_p, q is the heating rate, T_p is the temperature of the glow curve peak, k is Boltzmann's constant, and T is the isothermal decay temperature. For isothermal decay at 100°C this approximation yields $t_{1/2}$ values for peaks at 260°C and 350°C of about 1 day and 11 months, respectively; the corresponding values of $t_{1/2}$ for decay at 22°C are 18 years and 60,000 years. The model and the limited experimental data are in reasonably good agreement inasmuch as they predict a rather short lifetime for both peaks at 100°C. In addition, the model predicts that energy storage in the form of TL from ionizing radiation is highly temperature-dependent. A change of only 20°C in the isothermal decay temperature will change the half-life of a peak by more than a factor of 10.

The short lifetime and high degree of temperature-dependence of TL indicate that lunar surface temperatures have had a significant influence on the observed TL. Thus the glow curves from material collected near the surface, as the Apollo 11 samples were, will retain only an imperfect record of the radiation history of the samples, and it seems doubtful that TL studies on such material can result in meaningful radiation exposure ages. Finally, the apparent great age of the Apollo 11 samples (*1*) and the short half-life of TL in the lunar surface environment suggest that the TL observed in a near-surface sample is probably the result of a dynamic equilibrium between the growth of TL due to radiation and its decay in the lunar thermal environment.

The glow curves of untreated material from core No. 1 show a systematic change with depth below the lunar surface (Fig. 2). Comparison of the 5000-rad artificial glow curves further shows that the differences observed in the natural TL are not due to differences in either mineral content, mineral type, or susceptibility to radiation. The TL of the sample from the top of the core is nearly identical to the TL of untreated fines from the bulk box. With depth, there is an increase in the total amount of TL and the glow curves shift progressively to the left (lower temperatures). This systematic change is probably a result of the rapid decrease in mean temperature with depth below the lunar surface. Near the surface, where temperatures may exceed 100°C during the lunar day, only high-temperature TL remains. The daytime temperatures to which the material is subjected decrease rapidly with depth, resulting in slower decay rates and the preservation of more, and lower-temperature, TL. Because the glow curves in Fig. 2 shift continually to lower temperatures with depth, we conclude that the lunar diurnal temperature variations penetrate to a depth of at least 10.5 cm. It is also interesting to note that from 0 to 7.8 cm there is a steady increase in TL in the region of the glow curve above about 250°C, but that at 10.5 cm there appears to be a reversal in this trend. If real, this reversal may be caused by attenuation of solar radiation with depth.

It seems likely that the TL observed in the core samples is a result of an equilibrium between TL acquisition and TL loss, which suggests that TL disequilibrium may be useful in detecting disturbances in the lunar regolith. For example, if a surface was buried by a few centimeters of material such as might be ejected during the formation of a nearby crater, the normal progression of glow curves with depth would be disturbed. This disturbance would remain until equilibrium conditions were reestablished. The minimum length of time that such an anomalous glow curve pattern would remain detectable can be estimated from the time required to produce glow curves similar to those found in the core samples. Preliminary estimates, with decay neglected, suggest that a radiation dose of at least 10^4 rad would be required. The radioactive decay of U, Th, and K within the material (*1*) at present contributes about 0.3 rad/yr. The contribution from solar and cosmic radiation depends on depth and would be about 6 rad/yr at a depth of 10 cm (*3*). This suggests that a detectable TL anomaly at a depth of 10 cm would remain for more than 10^3 years after burial. Thus TL studies on core tube samples from the lunar regolith may be a means of detecting, and in the future perhaps of dating, recent surface activity that might otherwise go unnoticed.

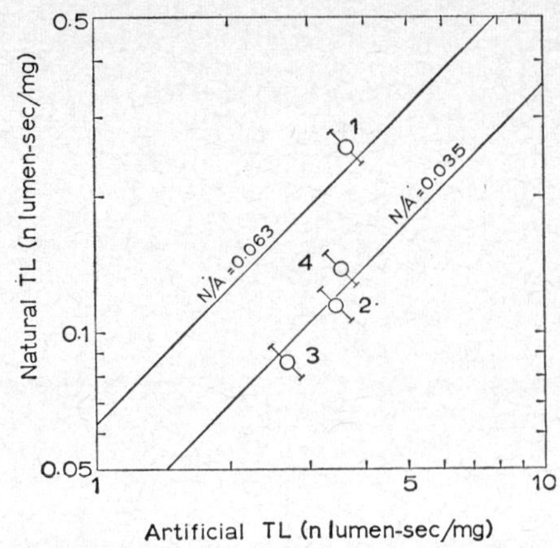

Fig. 3. Natural TL (300° to 350°C) plotted against artificial TL (85° to 350°C) for samples from breccia 10048: 1 = top; 2 = 4 mm below top; 3 = interior; 4 = bottom. The bars indicate estimates of the standard deviation of precision at the two-thirds confidence level.

The natural glow curve and the 5000-rad artificial glow curve were measured on material ground from the surfaces of chips from the top, interior, and bottom of breccia samples 10046 and 10048. The shapes of these curves indicate that the TL carrier in the breccias is identical to the TL carrier in the fines. There was no progressive change in the glow curves with depth, such as was found in the core samples. This is probably because of the lack of significant thermal gradients in the rocks. There is also no significant trend with depth of the ratios N/A except for the sample scraped from the uppermost surface of breccia 10048, which has a natural TL about twice that of the TL from the other three samples, as indicated by the N/A values (Fig. 3). This indicates a higher radiation dose, possibly from low-energy solar radiation, which loses most of its energy within the upper millimeter or less. The TL measurement from the uppermost surface of breccia 10048 was lost because of instrument malfunction, and no significant differences were detected in the interior and bottom samples.

Studies of the crystalline rocks (10003 and 10044) are still in progress, but it is evident that the TL carrier (plagioclase) in these rocks is different from the TL carrier in the breccias and the fines. There is more TL, by several orders of magnitude, than there is in the fines; there is a natural peak above 400°C, with no evidence of a peak at 350°C; and the artificial glow curves are unlike those of the fines. There also appear to be differences between the glow curves of the two crystalline rocks, although these differences are not as pronounced as those between the crystalline rocks and the fines and breccias. These differences in TL characteristics are probably caused by dissimilarities either in composition or in the degree of shock metamorphism.

Finally, it is of interest to consider the amount of energy represented by TL in the lunar material. This has been estimated for the fines and breccias by taking the entire area under the glow curves (in lumen-sec), converting it to units of energy through use of the conversion factor for monochromatic green light at 5550 Å, and assuming that the light is emitted from the sample equally in all directions. The values are 8×10^{-9} cal/g for the fines and 2×10^{-9} cal/g for the breccias. Although only rough estimates, these data indicate that TL energy stored in the Apollo 11 samples is exceedingly small.

G. BRENT DALRYMPLE
RICHARD R. DOELL

U.S. Geological Survey, Menlo Park, California 94025

References and Notes

1. Lunar Sample Preliminary Examination Team, *Science* **165**, 1211 (1969); "Apollo 11 Preliminary Science Report," *NASA Spec. Publ. No. 214* (1969).
2. J. T. Randall and M. H. F. Wilkins, *Proc. Roy. Soc. London Ser. A* **184**, 366 (1945); L. I. Grossweiner, *J. Appl. Phys.* **24**, 1306 (1953).
3. J. A. Ryan, in *The Lunar Surface Layer*, J W. Salisbury and P. E. Glaser, Eds. (Academic Press, New York, 1964), p. 265.
4. We thank B. Myers for assistance with the laboratory measurements. Publication authorized by the director, U.S. Geological Survey.

4 January 1970

Luminescence and Thermoluminescence Induced by Bombardment with Protons of 159 Million Electron Volts

Abstract. *We have observed a red luminescence, qualitatively similar to that of enstatite achondrites, in unsorted fines, their separated mineral phases, and rock chips. The energy efficiency of the plagioclase fraction is approximately 1 percent. At −196°C the effect is enhanced by a factor between 1 and 2. All fractions except ilmenite exhibit blue thermoluminescence with a glow peak near −135°C and an energy efficiency ≈ 4×10^{-6}. Unlike the thermoluminescence of terrestrial and meteoritic material, it is nonrepeatable even after annealing at 200°C. Similar thermoluminescence is seen in rock chips, but in unsorted fines it is masked by the opaque fractions.*

There are two reasons for studying the luminescence of lunar rocks. First, the results may be used, alone or in combination with results from terrestrial and meteoritic rocks, as indicators of specific crystalline defects, hence giving information on the thermal and chemical history of the lunar surface.

Table 1. Luminescent and thermoluminescent parameters of lunar rocks.

Material	Energy efficiency of direct luminescence		Thermoluminescence	
	−196°C	+100°C	Glow peak (°C)	Efficiency
Unsorted fines	1.97×10^{-3}	1.13×10^{-3}		
Plagioclase	1.07×10^{-2}	6.66×10^{-3}	−134	4.32×10^{-6}
Pale pyroxene	7.16×10^{-3}	4.12×10^{-3}	−135	0.72×10^{-6}
Dark pyroxene	3.97×10^{-3}	2.80×10^{-3}	−135	4.09×10^{-6}
Ilmenite	2.54×10^{-3}	1.83×10^{-3}		
Surface chip	Optical depth not known		−132	Optical depth not known
Interior chip			−145	

Second, the existence of a class of transient lunar phenomena has often been ascribed to luminescent effects, and studies of the energy efficiency of such processes, in conditions approximating those found on the moon, are clearly relevant to this interpretation.

Using techniques similar to those we have employed previously (*1*), we have measured the luminescence of samples from Apollo 11 when irradiated in the extracted proton beam of the synchrocyclotron at the Atomic Energy Research Establishment, Harwell. Our beam flux of about 4×10^8 protons cm^{-2} sec^{-1} gave rise to a local energy density of ionization of 2×10^{15} ev g^{-1} sec^{-1}, which is insufficient to cause radiation damage during the measurements. This energy density is comparable with that resulting from the solar wind on the lunar surface and is a factor of 10^7 less than the value obtained with low-energy, electrostatically accelerated protons (*2*). Spectra were observed with a set of seven wide-band interference filters and an EMI 9558 photomultiplier tube. The specimens could be cooled to liquid nitrogen temperature and heated to 200°C. The phototube current and thermocouple voltage were recorded on a two-pen recorder.

The specimens studied were two unsorted aliquots of the < 1-mm fines, sample 84; four separated mineral phases from these fines; and two rock chips, a surface and an interior piece, from sample 17. The mineral separation was performed by washing and sieving, followed by magnetic separation of the fraction in the size range 40 to 150 μm. Modal analysis of the four phases yielded the following compositions: (i) plagioclase 88 percent (with 10 percent colorless glass and 2 percent pyroxene composites); (ii) pale pyroxene 68 percent (with 17 percent plagioclase, 13 percent glass, and 2 percent ilmenite); (iii) dark pyroxene 66 percent (with 26 percent ilmenite composites, 4 percent plagioclase and 4 percent glass); (iv) ilmenite 80 percent (with 16 percent pyroxene composites, 2 percent plagioclase, and 2 percent glass). Fractions B and C each contain about 3 percent of unidentified olivine. Note that this size range contains only about 5 percent of glassy fragments and spherules.

The six powder specimens were mounted on aluminum disks using nonluminescent silicone grease. The amount of plagioclase and pale pyroxene was insufficient to completely cover the

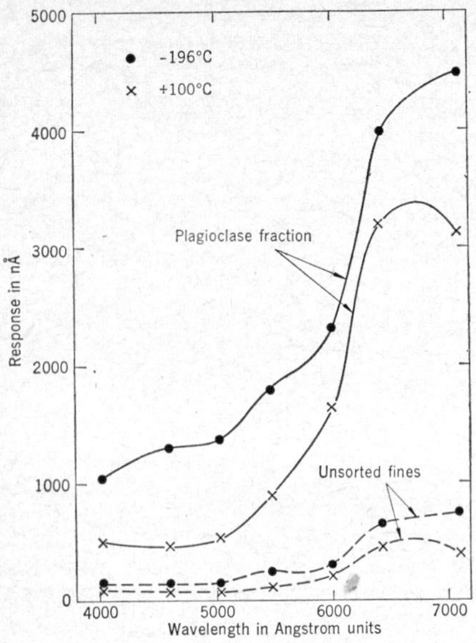

Fig. 1. Luminescent spectra of the unsorted fines and the plagioclase fraction, at −196° and +100°C.

beam target area and allowance was made for this in calculating absolute efficiencies. Spectra for the powder specimens were taken at −196°, 20°, and 100°C, the first and last corresponding roughly to lunar night and day temperatures. Thermoluminescence was looked for after irradiation at −196°C, using a heating rate of approximately 0.5°C sec^{-1}. We also searched for natural thermoluminescence in the unirradiated unsorted fines. No effect was seen on heating from room temperature to 200°C.

The rock chips were mounted similarly. Spectra were taken at 20°C and −196°C; thermoluminescence spectra were also recorded, though the temperature scale of these spectra is unreliable owing to the thickness (≈4 mm) of the chips and the consequent temperature difference between irradiated face and thermocouple junction.

All constant temperature spectra were recorded twice, with an average reproducibility of ±3 percent. Using our knowledge of the gain of the phototube, its spectral response, the transmission of the filters and other optical elements, and the geometry of the system, we were able to deduce from the phototube output current the total number of photons emitted per second per unit wavelength interval, $N(\lambda)$. From this result and the bandwidth of the filters, the total energy emitted as visible light, E, was calculated. We estimate the systematic standard error of $N(\lambda)$ to be ±14 percent and of E to be ±17 percent.

In a subsidiary experiment we measured the optical depth of our powder samples and found it to be 0.0099 ± 0.0003 g cm^{-2} for the unsorted fines and 0.0078 ± 0.0006 g cm^{-2} for the ilmenite. The other specimens were transparent and thinner than their optical depth.

We then calculated E_0, the energy deposited within the optical depth by the proton beam. The energy efficiency ε of the direct luminescence is given by E/E_0 and of the thermoluminescence by

$$\left(\int_0^{t=t_1} E\,dt\right) \Big/ \left(\int_0^{t=t_0} E_0\,dt\right)$$

where t_0, t_1, are the times of irradiation and heating, respectively. Estimating the systematic error of E_0 to be ±12 percent, we find the systematic error in ε to be ±20 percent.

The spectra for direct luminescence exhibit the following features. There is a pronounced red enhancement at all three temperatures, the ratio red/blue being about seven at 20° and 100°C, and varying from 2½ (pale pyroxene) to 4½ (unsorted fines and ilmenite) at −196°C. The spectra of all specimens have the same general shape and vary significantly in their absolute brightness. In Fig. 1 we show the responses of plagioclase (the brightest fraction) and the unsorted fines, at −196° and 100°C. As is generally the case with terrestrial and meteoritic specimens, the responses at 20° and 100°C are very similar. The unsorted fines and the ilmenite fraction have almost identical spectra, presumably because the opaque ilmenite grains mask the brighter response of the transparent minerals. In Table 1 we show the energy efficiency ε summed over the visible spectrum using the expressions above. We give ε for each specimen at 100° and −196°C. The ratio is seen to vary from 1.39 to 1.75.

As a check on our calculations we also measured the luminescence of a granite specimen that had previously been compared with a commercial phosphor of known efficiency. The results were in agreement with the previous measurements and confirm that the efficiencies approach 1 percent for the plagioclase fraction.

All specimens except the ilmenite and the unsorted fines showed pronounced blue thermoluminescence, peaked sharply in the region 4350 to 4850 Å, at temperatures below −120°C. The glow peak for the plagioclase fraction is shown in Fig. 2. There was no significant thermoluminescence in any

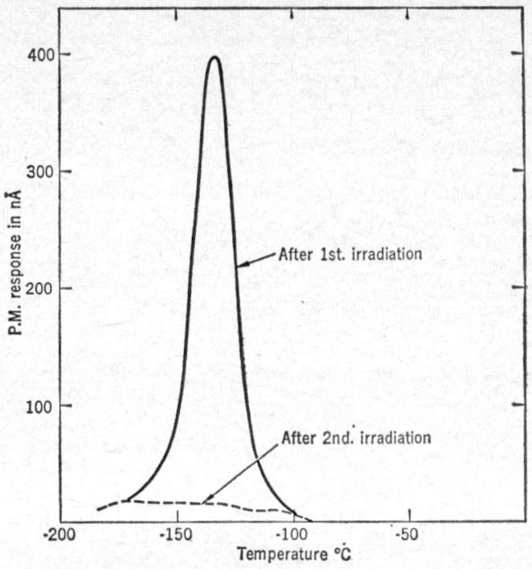

Fig. 2. Thermoluminescence of the plagioclase fraction, measured in the wavelength band 4350 to 4850 Å.

other wavelength band nor at higher temperatures. Specimens were heated to 200°C and held at this temperature for several minutes. A very interesting effect was then noted: it was not possible to reproduce the blue glow peaks after a second irradiation at low temperature. This effect is shown in Fig. 2, where in place of the sharp peak we see only a small effect not very different from the background.

This phenomenon renders precise calculation of the energy efficiency rather difficult, as this quantity will be determined eventually by the number of luminescent centers available rather than by the period of irradiation. The values of ε given in Table 1 are therefore lower limits (since the same glow curve could perhaps have been produced by a shorter irradiation).

The rock chips exhibited similar thermoluminescence, and it is interesting to note that, though our knowledge of heating rates is poor, the glow peaks for the interior and the surface chip have very different efficiencies. We do not yet know whether this phenomenon reflects a real difference in radiation history or is due, for instance, to different minerals' being fortuitously exposed.

As with many other properties of the lunar rocks, these spectra are not precisely matched by any terrestrial counterpart. It is nevertheless interesting to note that the red luminescence seen in enstatite achondrite meteorites [Khor Temiki (*1*), Bustoe (*3*), Bishopville (*3*)] is qualitatively very similar and is roughly of the same magnitude. The similarity in the blue is strikingly improved if one assumes an equal admixture of terrestrial vesicular basalt, spectra of which show, at −196°C, the plateau near 5000 Å characteristic of all our lunar specimens. The enstatite achondrites do not show such bright blue thermoluminescence as do the lunar rocks. They have been studied by ourselves [Khor Temiki (*1*)], and by Sun and Gonzales [Cumberland Falls (*4*)], and we both find a red glow peak near 0°C. Sun and Gonzales saw a blue glow peak near −120°C, but we found this peak to be much less intense than the red one, which does not appear at all in the lunar rocks.

Presumably the red luminescence shown by all our specimens is due either to a mineral phase common to all fractions or to the ubiquitous presence of an activator that is increasingly quenched by the larger concentrations of iron in the more magnetic fractions. In view of the modal analysis above, showing a moderately clean separation, we incline to the second suggestion.

The blue thermoluminescence might be ascribed to the same activator were it not for its complete absence from the ilmenite fraction. As regards its nonrepeatability, such an effect is observed in some crystalline phosphors, particularly alkali halides, in which thermoluminescent recombination occurs at *F* centers that are thermally stable up to the annealing temperature. However, we had not previously seen this effect in any of our terrestrial or meteoritic specimens. The phenomenon is obviously highly relevant to the problem of transient lunar phenomena, many of which have recurred in the same areas of the moon's surface.

Two lines of thought have been suggested. Either the surface is irradiated at night and glows on heating during the day, or subsurface layers at temperatures near − 50°C (*5*) are sporadically exposed and heated and give off as light the energy accumulated during many years of irradiation (by high-energy particles). We can now state, at least for the Sea of Tranquillity, that the energy efficiency of thermoluminescence is too low for the first effect to be seen and that the glow appears at a temperature too low for the second hypothesis to hold. Moreover, repeated thermoluminescence of the same material is unlikely as local temperatures never exceed 200°C.

J. A. EDGINGTON

Queen Mary College, University of London, England

I. M. BLAIR

Atomic Energy Research Establishment, Harwell, England

References and Notes

1. I. M. Blair and J. A. Edgington, *Nature* **217**, 157 (1968).
2. D. B. Nash, *J. Geophys. Res.* **71**, 2517 (1966).
3. C. J. Derham, J. E. Geake, G. Walker, *Nature* **203**, 134 (1964).
4. K. H. Sun and J. L. Gonzales, *ibid.* **212**, 23 (1966).
5. We thank Dr. W. J. French and Dr. A. B. Poole of the Geology Department, Queen Mary College, University of London, for assistance in separating and identifying the various mineral phases. We also thank Dr. J. A. Bastin for discussions particularly on lunar temperatures.

5 January 1970

Luminescence, Electron Paramagnetic Resonance, and Optical Properties of Lunar Material

Abstract. *Dust samples have been found to luminesce weakly under proton excitation, but not under ultraviolet. Damage, recovery, and heating effects have been investigated. Chips of breccia show luminescence, from white inclusions only, under ultraviolet and protons. Some rock chips show general luminescence, mainly from plagioclase. No natural or excited thermoluminescence has been found for dust or chips. The electron paramagnetic resonance spectrum shows the same broad Fe^{3+} dipole resonance for dust and for some chips; other chips show no response. The polarization characteristics of dust are found to be identical to those of the Sea of Tranquillity, independently of proton damage. Chips show characteristics unlike any part of the lunar surface.*

Dust sample 10084.6 showed no visible luminescence under ultraviolet excitation at room temperature, or when cooled in liquid nitrogen; two lamps emitting radiation at mainly 3650 and 2537 Å, respectively, were used. No infrared emission (beyond 7300 Å) was detected by a Ag-AgO-Cs photomultiplier or by a PbS cell when dust at either temperature was irradiated by intense visible light (cutoff at 6000 Å) from a quartz/iodine lamp.

Under excitation by 60-kev protons, weak luminescence with no afterglow was observed; the spectrum of this luminescence was scanned by a grating spectrophotometer with successive trialkali and Ag-AgO-Cs photomultipliers (Fig. 1). The emission is nearly white, with a low green peak at about 5600 Å and a lower blue one at about 4500 Å. The efficiency is very low; for fresh dust it is about an order of magnitude less than for Gnadenfrei meteorite (*1*) or standard basalt BCR-1 (*2*) and is estimated at about 10^{-5}. Basalt gave a peak of emission at about 7300 Å, and this was not found for the lunar dust.

The albedo, with a fresh MgO screen taken as unity, was about 7 percent when a tungsten lamp at 2850°K and a trialkali photomultiplier were used; for basalt it was about 40 percent. Heating the dust to about 760°C in argon (as described later) doubled both the luminescence efficiency under proton excitation and the albedo. The emission spectrum changed (Fig. 1) and so did the diffuse-reflection spectrum (Fig. 2), the dust now appearing brown in color. There is some x-ray evidence that heating produces some structural annealing, and this may be the cause of the increased efficiency. Dust heated to 960°C in argon appeared slightly darker than the 760°C sample, and its luminescence emission spectrum (Fig. 1) was slightly more like that for the fresh dust.

Damage by proton irradiation caused

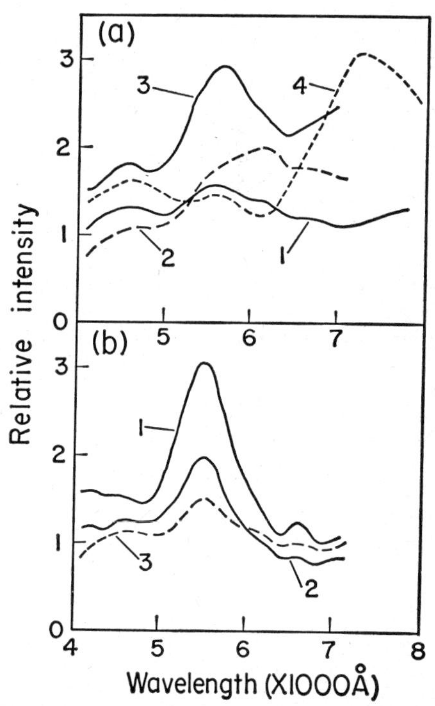

Fig. 1. Luminescence emission spectra with 60-kev proton excitation. (a) 1. Lunar dust 10084.6, fresh; 2. after heating to 760°C in argon; 3. after heating to 960°C in argon; 4. basalt BCR-1. (b) 1. Light fraction from rock 10044.43—mainly plagioclase. 2. Rock chip 10058.37. 3. Rock chip 10057.54. The intensity scales of the separate spectra are unrelated.

a rapid initial falloff in luminescence efficiency (typically, by a factor of 2½ in the first 5 minutes at 1 μa/cm² followed by a decreasing rate of falloff). However, the decrease was less rapid than for basalt, possibly suggesting that the lunar dust is already significantly damaged throughout (as discussed later); furthermore, there was no detectable change in visual appearance or in albedo, even after 8 hours of such damage. Heating the damaged dust to 600°C, for example, caused a recovery to double the "undamaged" efficiency and turned the dust brown. There was some indication of recovery without any color change after heating to only 150°C, but this may have been due to accidental disturbance of the dust surface. The dust which had been turned brown by heating was found to darken slowly with proton damage; perhaps after prolonged damage it would revert to the color of the fresh lunar dust.

All our rock chips were inspected with a microscope, under ultraviolet radiation of mainly 3650 or 2537 Å, separately. Most of the chips are breccias and showed no general luminescence; under 3650 Å excitation, several of them showed some faint blue-white luminescence from white inclusions. Under 2537 Å excitation, the effect was the same, plus some fainter white luminescence from clear inclusions. Only 10058.37,38, a coarse-grained rock, showed general luminescence except from its black ilmenite regions.

Under excitation with 60-kev protons, the effect was similar but more striking; there was negligible overall luminescence from the breccias, but bright blue-white or white luminescence from many white or glassy-looking inclusions, including those observed under ultraviolet. Fragments of some of the white inclusions were removed in an attempt to identify them by inspection with a polarizing microscope and by refractive index measurements (*3*). They were very soft (probably because they were highly fractured) and complex in structure. One large, white (blue-white luminescent) inclusion was found to be (probably) mainly cristobalite; other inclusions were probably complex mixtures of clinopyroxene and plagioclase, sometimes in the form of very small crystals embedded in glass. The coarse-grained rock 10058 luminesced brightly from all except its ilmenite regions, the luminescent parts apparently consisting of plagioclase with some pyroxene; its spectrum was scanned (Fig. 1). No difference was detected between the external chip 10058.37 and the internal one 10058.38. For comparison, a separated light fraction (probably 95 percent plagioclase) from another rock, 10044.43 (*4*), was scanned (Fig. 1). The efficiency was high (about 10^{-3}) and the spectrum very similar to that for 10058.37,38 except for a somewhat higher green peak—probably because in the rock the plagioclase peak is diluted by whiter pyroxene emission. Microprobe analysis of the separated plagioclase showed a high Sr content (800 ppm) and a trace of Yt.

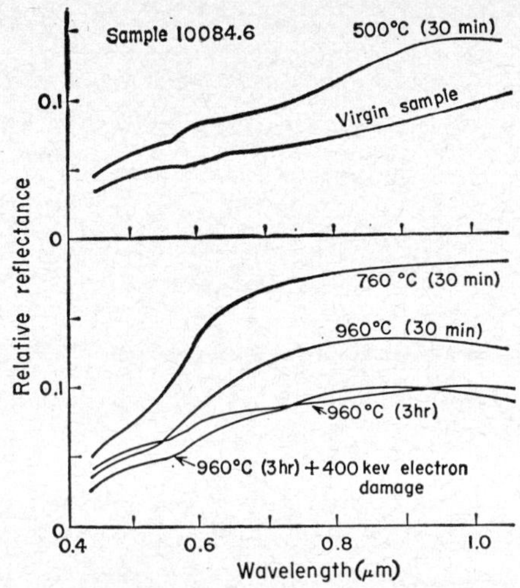

Fig. 2. Diffuse-reflection spectra of lunar dust 10084.6, measured at 18°C before and after various thermal treatments.

To look for thermoluminescence, samples were heated in a vacuum at a uniform rate (1°C per second) from 20°C to above 500°C and viewed through a chopper disk by cooled photomultipliers covering the range 0.35 to 1.1 μm. No detectable emission was obtained from either virgin samples or those subjected previously to ultraviolet irradiation, x-rays, 30-kev ions, or 400-kev electrons. Experiments with excitation by ultraviolet radiation at 77°K likewise gave no detectable thermoluminescence. This correlates with the absence of photoluminescence in most samples. Samples with small luminescent inclusions are still being studied.

Diffuse-reflection spectra were measured at room temperature before and after each thermoluminescence experiment and after excitation by ionizing radiation. Significant changes in body color of all dust samples are observed after they are heated in a vacuum or in an inert atmosphere to temperatures just above 200°C. The changes are even more marked after heating to higher temperatures; at 760°C, the dust assumes a medium brown color that is much nearer that of virgin samples when treatment at 900° to 960°C is carried out. No change in weight was found after such treatments and the microscopic appearance, apart from color, was unchanged. There was a slight tendency to sinter. The overall albedo, increased by heating to 760°C, then decreased with further rise in temperature. The change in body color is associated with a loss of broad absorption bands in the red and near-infrared regions; these absorptions return if treatment is carried out at 900° to 960°C. Electron bombardment of the latter samples does not produce significant changes, even for large doses (Fig. 2). However, proton irradiation does have an effect, as described above. Studies of x-ray diffraction patterns of powders from virgin and treated samples (960°C) indicate structure development after heating, due perhaps to the annealing out of crystalline defects. More investigations of these changes are in progress, including the detailed study of separated components.

The X-band (3 cm) equipment was used for measurement of the electron paramagnetic resonance spectra of samples at room temperature. Samples of dust showed a very broad but strong resonance of the dipole interaction broadened type with maximum at about the free spin value (*g* value slightly greater than that for free spin); this spectrum was relatively unaffected by thermal treatment involved in successive thermoluminescence experiments. However, heating at 900° to 960°C produced a strong, relatively unresolved, spectrum across the 6-kgauss field width of the usual spectrum. The above resonance spectrum for dust samples (10084.6) was also found for all rocks chips (10057.54,55; 10057.53,54,55; 10023.7,8; 10059.35,36; 10020.42; and for core samples 10004.31,32). However, in the case of clear fragments, 10058.37,38, no detectable signal was obtained. Observed signals are most likely due to Fe^{3+} ions in the components. The electron paramagnetic resonance results correlate with data on paramagnetic susceptibility given in the Lunar Receiving Laboratory preliminary report.

Since large changes in albedo and in reflection spectra can be obtained at temperatures not much greater than 200° to 250°C it seems unlikely that the lunar surface, or layers near to the surface, have been at such temperatures since their original formation or deposit. Formation at high temperatures must have been followed by relatively rapid cooling since only minutes at a few

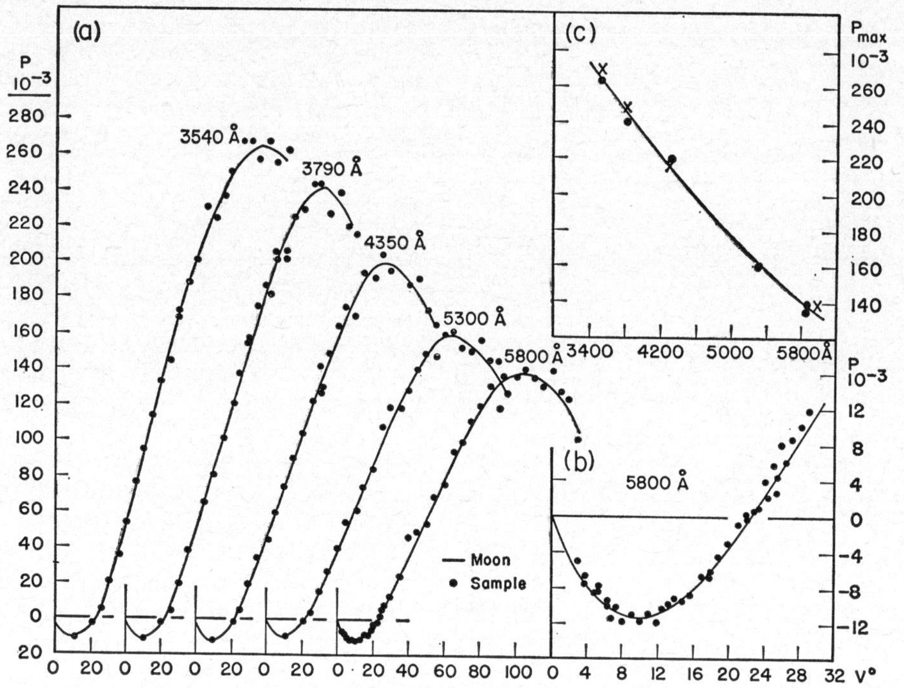

Fig. 3. Polarization curves for lunar dust 10084.6 and for Sea of Tranquillity. (a) Complete curves. (b) The negative branch, on a larger scale. (c) The change of P_{max} with wavelength. Crosses, Sea of Tranquillity; circles, lunar sample.

hundred degrees centigrade are needed to produce changes in the reflection spectra and hence in the body color. In view of the marked effects of heating to 900°C or so on the electron paramagnetic resonance spectrum, which suggests magnetic ordering, it would seem that the known magnetic properties of the lunar samples are also related to a rapid cooling process.

The polarization of light scattered by a surface gives a powerful indication of its texture and optical properties (*5*). We have investigated both dust and rock chips, and these gave different results.

The variation of degree of plane polarization *P* with angle between incident and emergent light *V* has been investigated for part of lunar dust sample 10084.6 and found to be identical with that obtained by telescopic measurements (*6*) of the Sea of Tranquillity. The results are shown in Fig. 3, where polarization is given as negative when in the plane of the incident and emergent rays, and as positive when at right angles to this (the only two cases found). The agreement between the lunar and sample results is within the experimental error of both of them; for example, the shape of the characteristic negative branch plotted in Fig. 3b agrees with the lunar curve to within 10^{-3} in *P*, providing the surface of the sample is roughened. Smooth dust gives a slightly shallower negative branch. The agreement is also good for the maximum of positive polarization, as regards the angle at which it occurs and particularly as regards the way its height, P_{max}, changes with wavelength as shown in Fig. 3c. Five filters were used with peaks from 3520 to 5850 Å.

It had already been shown, by comparison with terrestrial samples (*5*), that the characteristics of the lunar curve, as represented by the shape of the negative branch and by the wavelength variation of P_{max}, could only be produced by a fine, dark, roughened powder. The main significance of the agreement between the telescopic measurements and our lunar sample ones is that it suggests that the optical properties of the exposed surface layer may be taken as also representing the subsurface material which constitutes the bulk sample. This could not previously be confirmed. That the surface layer was not changed by its exposure to solar radiation was confirmed by attempts to damage a lunar dust sample by irradiation; prolonged bombardment with 60 kev protons produced no change in either the visual appearance of the dust, or in its polarization characteristics or albedo (measurements were made after 2 hours at about 1 μa/cm^2; the appearance was still unchanged after 7 hours). This may be because the dust is already damaged to saturation throughout the depth sampled, each individual grain having at some time spent a prolonged period on the exposed surface.

One of the heated dust samples referred to earlier was investigated. After proton bombardment it darkened perceptibly, but no change in its polarization curve was observed.

Surface chips were investigated from two rocks as different in appearance as possible.

1) Chip 10057.54 from the top (and presumably exposed) surface of a very dark grey rock. This showed a negative branch nearly as deep as for the dust sample and extending to $V = 19°$; P_{max} was about 350×10^{-3} and occurred at $V = 130°$. Its height was very sensitive to the position of the sample, and the value of P_{max} could be moved up and down over a range of at least 50×10^{-3}. The wavelength variation of P_{max} also changed with position. Photographs taken with a scanning electron microscope of part of the same region of this chip revealed a complex surface grain structure down to a scale comparable to the wavelength of light; this, combined with the darkness and opacity of the grains, probably accounts for the strong negative branch.

2) Chip 10058.37 from the surface of a rock containing much light-colored material. This showed no negative branch; P_{max} was about the same as for the previous rock. These results are consistent with the much larger amount of light material present.

The characteristics of these two extreme types of rock chip are both sufficiently different from those observed for any part of the lunar surface to suggest that exposed rocks have little influence on the optical properties of the moon, as observed from the earth. However, information about the characteristics of these exposed rocks is important in explaining measurements made for asteroids, which may be composed of the same materials but are too small to have sufficient gravitational attraction to retain a dust layer (*7*).

J. E. GEAKE
University of Manchester Institute of Science and Technology, Manchester, England

A. DOLLFUS
Paris Observatory, France

G. F. J. GARLICK, W. LAMB
University of Hull, England

C. WALKER
University of Manchester Institute of Science and Technology

G. A. STEIGMANN
University of Hull, England

C. TITULAER
Paris Observatory, France

References and Notes

1. J. E. Geake and G. Walker, *Geochim. Cosmochim. Acta* **30**, 929 (1966). Meteorite samples on loan from the British Museum (Natural History), to whom we are grateful.
2. F. J. Flanagan, *ibid.* **31**, 289 (1967).
3. We thank Professor W. S. MacKenzie for help with these identifications.
4. We thank Prof. J. Zussman for the loan of this sample, and for the data.
5. A. Dollfus, *Physics and Astronomy of the Moon*, Z. Kopal, Ed. (Academic Press, New York, 1962), chap. 5.
6. ——— and E. Bowell, final report U.S. Air Force contract 61(052)-914, June 1969.
7. We thank I. Robinson, A. Manwaring, and R. J. Fryer for assistance with the experimental work, and to the Science Research Council for grants (Manchester and Hull). We thank the Textile Technology Department, U.M.I.S.T., for the use of their Stereoscan miscroscope, and J. Sparrow for operating it.

6 January 1970

Luminescence of Apollo 11 Lunar Samples

Abstract. *Luminescence measurements were made of four lunar rocks, two terrestrial rocks (granite and gabbro), and one terrestrial mineral (willemite) by comparing the spectral curves with the curve of a barium sulfate standard. Efficiencies with 3000 angstrom excitation were* $\leqslant 6 \times 10^{-5}$ *for the lunar samples,* $\leqslant 8 \times 10^{-5}$ *for gabbro of very similar composition to the lunar samples, approximately* 10^{-4} *for granite, and approximately* 2×10^{-2} *for willemite. If these are typical values for other ultraviolet excitation wavelengths, the Apollo 11 site appears to contribute little to the observed lunar luminescence.*

On the basis of lunar eclipse observations, Link *(1)* proposed that the light of the moon contained a luminescence component in addition to the reflected solar component. Later observations, especially by means of the Fraunhofer line-depth technique, seemed to confirm this idea *(2–5)*. Our objective is to evaluate these observations and understand the luminescence behavior of lunar lithologic materials by measuring the luminescence spectra and efficiencies of the lunar samples with the chief excitation sources in the space environment and by comparing the results with those of similar measurements of terrestrial rocks and minerals. The excitation sources of interest are ultraviolet, x-rays, protons, and electrons.

We give here some results of our measurements of the Apollo 11 samples with ultraviolet excitation. A Hanovia 2500-watt high pressure xenon-mercury lamp was the source of ultraviolet. The light, before striking the samples, passed through a Gaertner quartz prism monochromator with a Corning 7-54 filter at the exit slit to cut out wavelengths longer than 4000 Å. The light from the sample passed through a Jarrell-Ash 0.25-m grating spectrometer, either with or without a Corning 3-74 filter at the entrance slit to cut out wavelengths shorter than 4000 Å, and finally to an EMI 9558Q photomultiplier tube. The photomultiplier output was recorded by a Moseley 2D-2 X-Y recorder.

Initially, some runs were made on terrestrial comparison samples with a Hunter hydrogen capillary glow discharge lamp as a source. This lamp, which concentrates the glow discharge within a small capillary, is used as our source of hydrogen Lyman-alpha (1216 Å). The light in this case irradiated the sample directly with only a magnesium fluoride window intervening. Preliminary results appeared to show no significant luminescence. The luminescence excited by Lyman-alpha appears to be less than that excited by the middle and near ultraviolet and will require a different experimental set-up to be detected and measured. For this reason, subsequent runs on the lunar and terrestrial samples were made with maximized irradiation intensity with the Hanovia lamp. A wavelength band of about 200 Å width at half maximum, centered at about 3000 Å, was used.

The samples were mounted on a 12-position turntable inside a vacuum chamber. Six samples from four lunar rocks were included: 10044,38 (exterior) and 10044,53 (interior) from a coarse-grained igneous rock, 10048,36 (interior) and 10048,37 (bottom) from a breccia, and 10057,45 (interior) and 10022,55 (surface) from two fine-grained igneous rocks. The exposed surface area ranged from a little less than 1 to a little less than 2 cm^2. Terrestrial comparison samples consisted of granite (El Capitan formation, Mariposa County, California), gabbro (near San Marcos, San Diego County, California), and willemite (Franklin Furnace, New Jersey), the latter a highly luminescent mineral. The remaining three turntable positions were occupied by a barium sulfate surface to serve as a reflectance standard, a thermopile for absolute energy calibration of the excitation beam and the system, and an aluminized mirror for alignment of the system.

The luminescence efficiencies at 3000 Å excitation for all samples except the willemite proved to be too low to give a distinct, unambiguous luminescence spectrum. The curve for willemite (Fig. 1) shows the spectrum obtainable from a material of high efficiency, in this case about 2×10^{-2}. The other eight samples gave curves of the type shown in Fig. 2 recorded for lunar sample 10057,45. In these cases, at the maximized intensity of irradiation and the high sensitivities required to sense any luminescence, the effects of scattered light and of the small, nonzero transmission of the filter made

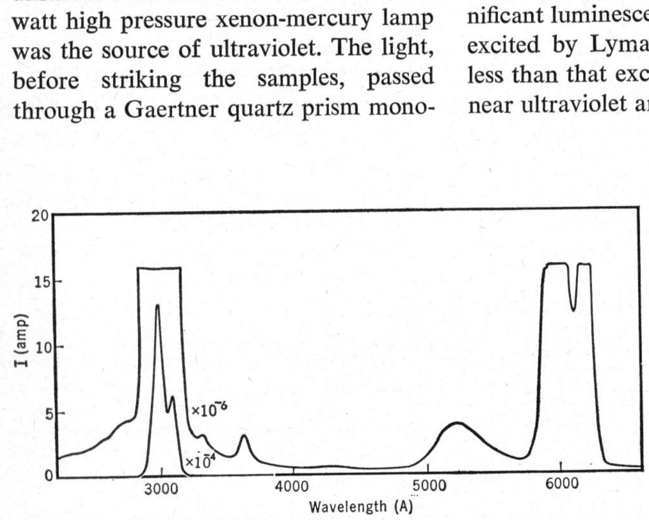

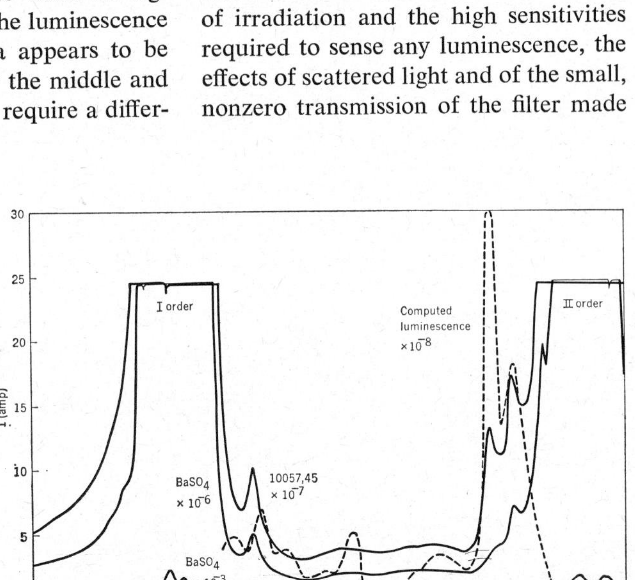

Fig. 1 (above). Luminescence of willemite with 3000-Å excitation. Fig. 2 (right). Luminescence of lunar sample 10057,45 compared with barium sulfate standard (3000-Å excitation).

the signal background limited. To extract the luminescence signal, shown as the computed curve of Fig. 2, therefore, it was necessary to use the technique of comparing the sample curve with that of the barium sulfate reflectance standard.

The barium sulfate has a reflectance of very nearly 100 percent over the entire wavelength band investigated and is not known to have any significant luminescence in this band. Its curve, therefore, represents the spectrum of the light exciting the sample, and the ratio of the sample curve to that of barium sulfate is the sample reflectance. Any increase of apparent reflectance of the sample can be the result of either reflectance changes or luminescence or both. Comparison of such curves from all samples showed increases, and therefore possible luminescence, in the neighborhood of 5300, 5500, 3500 to 4100, 4300 and 4970 Å, in order of decreasing intensity. The 3500- to 4100-Å band, especially, is a region in which we have found granite and gabbro reflectance values to remain constant or to decrease, so that in this region, at least, the case for luminescence is strengthened. However, the effects are small and could arise from other than luminescence. In this event, the calculated efficiences represent an upper limit for the samples for the conditions of this particular set of measurements.

The efficiencies are as follows: all lunar samples, $\leqslant 6 \times 10^{-5}$; granite, $\sim 10^{-4}$; gabbro, $\leqslant 8 \times 10^{-5}$; and willemite, $\sim 2 \times 10^{-2}$. The values represent total efficiencies; that is, the ratio of energy in the luminescent band to energy in the band incident upon the sample. Despite the uncertainty, the values shown are consistent in that the lunar samples, which are very similar to gabbro in composition, are also similar in apparent luminescence efficiency.

The outstanding feature of these results is the low values of the efficiencies for the lunar samples and the two terrestrial rocks. They are comparable to the terrestrial rock efficiencies that we and Nash measured with protons and that we measured with electrons and x-rays; they are also much lower than those we estimated for terrestrial rocks with the 1800- to 2200-Å band by a less refined filter method (*6, 7*). In our earlier study, with a 1 percent efficiency for the 1800 to 2200 Å band, we calculated a level of luminescence, 5 percent of the reflected background in the 2200 to 2500 Å band, that was the average observed around 3950 Å by Grainger (*4*), who made the most accurate line-depth measurements of the moon. The efficiencies we measured in this study of the Apollo 11 lunar samples are more than two orders of magnitude lower; in a similar calculation they would yield a level of luminescence that would be undetectable by the line-depth technique. If our further studies with excitation at other ultraviolet wavelengths and with proton, electron, and x-ray excitation show no greater efficiencies than these, the Apollo 11 landing area can be considered to contribute little or nothing to the luminescence determined by the line-depth technique. Grainger's observations did not include the Apollo 11 landing area; if correct, they imply higher luminescence efficiencies on other areas of the moon. Terrestrial gabbros, which are similar in composition to the Apollo 11 rocks, are found in both this and previous studies to possess low luminescence efficiency as compared to granites. Therefore, other lunar areas, especially in the highlands, could contain rocks of higher efficiency than those from Apollo 11. In this connection, note that the Surveyor 7 alpha-scattering analysis measured less than half as much iron, a luminescence-quenching element, at the Tycho site than was measured at the Surveyor 5 and 6 mare sites (*8*). Finally, it is possible that certain excitation mechanisms that exist in the space environment, such as simultaneous irradiation with ultraviolet and charged particles, may result in higher efficiencies and await laboratory testing. For these reasons, samples from many areas of the moon need to be studied in order to characterize the luminescence of the lunar rocks and to understand its significance.

NORMAN N. GREENMAN
H. GERALD GROSS
Space Sciences Department, McDonnell Douglas Astronautics Company–Western Division, Santa Monica, California 90406

References and Notes

1. F. Link, *Compt. Rend.* **223**, 976 (1946).
2. N. A. Kozyrev, *Izv. Crimean Astrophys. Obs.* **16**, 148 (1956).
3. J. Dubois, *Rozpravy Cesk. Akad. Ved* **69**, 1 (1959).
4. J. F. Grainger, *Astron. Contrib. Univ. Manchester.* **104**, 1 (1963).
5. H. Spinrad, *Icarus* **3**, 500 (1964).
6. N. N. Greenman, V. W. Burkig, H. G. Gross, J. F. Young, Douglas Aircraft Co. Rept. SM-48529 (1965).
7. D. B. Nash, *J. Geophys. Res.* **71**, 2517 (1966).
8. A. L. Turkevich, W. A. Anderson, T. E. Economou, E. J. Franzgrote, H. E. Griffin, S. L. Grotch, J. H. Patterson, K. P. Sowinski, NASA SP-184 (1969).
9. Supported by NASA contract NAS9-7966. We wish to thank M. W. Wegner, W. M. Hansen, and A. D. Pinkul for their assistance in these studies.

4 January 1970

Luminescence and Reflectance of Tranquillity Samples: Effects of Irradiation and Vitrification

Abstract. *Luminescence measurements of Tranquillity samples indicate that energy efficiencies for excitation by protons and ultraviolet are in the range 10^{-6} or below; natural and induced thermoluminescence is even weaker. If these samples are typical, lunar surface luminescence cannot occur at reported levels. Comparison of proton luminescence spectra from the exterior and interior of rocks and fine fragments provides evidence of solar wind impingement on the moon's surface. Spectral reflectance and albedo measurements of fresh rock powders before and after both laboratory proton irradiation and fusion indicate that vitrification may be an important mechanism of lunar darkening.*

Experiments simulating solar wind proton irradiation and solar ultraviolet irradiation on the moon's surface were performed to determine the emission properties and energy efficiencies of luminescence and thermoluminescence as well as the effects of irradiation and vitrification on spectral reflectance (0.23 to 2.5 μm) and albedo. Samples examined (*1*) represent each of the three rock types and fine soil collected (*2*).

Samples were handled and prepared in dry N_2. This was later omitted when experiments showed no detectable effects on any of our measurements from exposure to air. Proton and ultraviolet irradiations were conducted in vacuum at 10^{-6} torr. Thermoluminescence and spectral reflectance measurements were made with instruments flushed with dry N_2.

Irradiations were carried out with 1-, 2-, and 4-kev protons (to simulate solar wind) at fluxes in the range 3 to 10×10^{13} protons cm^{-2} sec^{-1} (~ 5 to 15 μa cm^{-2}) for doses of up to 4×10^{19} protons cm^{-2} (equivalent to 2×10^4 years of solar wind) per sample, and simultaneously and independently with

0.25 to 0.43 μm ultraviolet at solar equivalent intensity (~ 10^5 ergs cm^{-2} sec^{-1}). Luminescence spectra and energy efficiency (*3*) in the range 0.25 to 0.72 μm were measured with a calibrated grating monochromator and a photomultiplier with S 20 response. Rapid spectral scanning capability allowed complete luminescence spectra to be recorded every 10 sec in order to measure the rapid changes in relative spectral intensity exhibited by silicate luminescence during initial proton excitation (*4*).

Proton-excited luminescence was observed in all rock and soil samples. Initial spectra from both fresh crystalline rock material and freshly crushed soil fragments show two emission bands of nearly equal intensity in the blue at 0.40 μm and yellow at 0.55 μm. With continued irradiation the spectra decrease in intensity and change in relative spectral distribution with disappearance of the yellow band and shift in the peak position of the blue band to 0.46 μm. This behavior is similar to that observed for terrestrial and meteoritic silicates (*4*) and for convenience will be referred to as luminescence decay. The resulting spectra will be referred to as decayed spectra.

Energy efficiencies at initial laboratory proton excitation are in the range 1 to 6 × 10^{-6} for bulk samples and 2.4 × 10^{-5} for a plagioclase mineral separate; these values are low in the range of values determined for terrestrial and meteoritic silicates (*5*). Efficiency for a given sample decreases with increasing proton dose, proton energy density, and sample temperature. Most of the luminescence from lunar samples is produced by the plagioclase mineral component. Since all samples examined contain plagioclase in varying amounts, all spectra are similar in shape. Typical spectra at initial excitation by 2.0-kev protons are shown in Fig. 1 along with corresponding values of integral energy efficiencies. From sample to sample, efficiency increases with increasing plagioclase and decreasing Fe and glass concentrations. Crushed artificial glass, made by fusing (*6*) a small portion of sample 10020, yields no detectable proton-excited luminescence.

The luminescence data obtained, if typical for the whole moon, confirm that it is impossible for solar wind or flare proton bombardment to excite luminescence from rocks and soil on the moon's sunlit surface that would be detectable from the earth. Figure 2 illustrates the energy relationship between incident proton intensity, luminescence intensity, and required brightness levels to account for reports of lunar luminescence (*7*) by various observational techniques. Note that even at laboratory proton intensities, which are ~ 10^5 times greater than maximum solar intensity, the required brightness is too low by approximately two orders of magnitude. At lower proton intensity the luminescence energy efficiency increases (Fig. 2), but the luminescence brightness decreases; at equivalent solar wind or solar flare intensity the extrapolated luminescence efficiency of lunar material is in the range 0.1 to 1.0 percent, but the luminescence brightness would be too low for observation by five orders of magnitude. Luminescence efficiency of 100 percent would be insufficient to account for reported observations, as pointed out in previous studies (*8*).

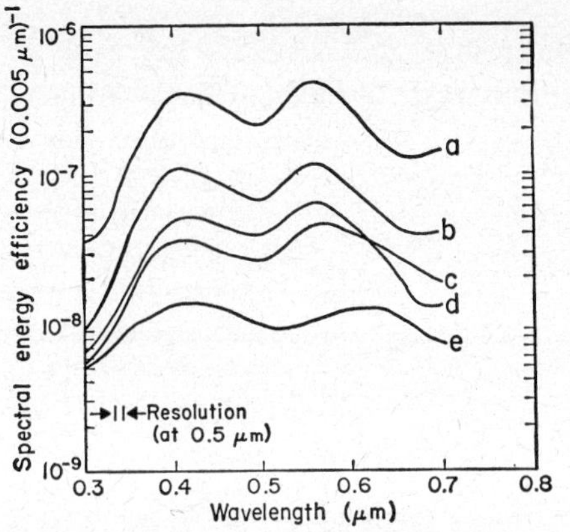

Fig. 1. Proton-excited luminescence spectra for Tranquillity samples at initial exposure to laboratory irradition. 2.0-kev proton excitation, 15μa cm^{-2} (~ 10^{14} protons cm^{-2} sec^{-1}). Samples powdered to < 50 μm. Pressure 2 × 10^{-6} torr. Temperature 50°C. Curve *a*, plagioclase from medium-grained igneous sample 10058, integral energy efficiency 2.4 × 10^{-5}. Curve *b*, medium-grained igneous sample 10058, integral energy efficiency 5.6 × 10^{-6}. Curve *c*, 0.15 to 1.0-mm crushed fines, integral energy efficiency 2.5 × 10^{-6}. Curve *d*, fine-grained igneous sample 10020, integral energy efficiency 3.0 × 10^{-6}. Curve *e*, 1.0-mm uncrushed fines, integral energy efficiency 0.9 × 10^{-6}.

The exterior surfaces of lunar crystalline rocks, the microbreccia, and the uncrushed fines tend to yield decayed luminescence spectra and low efficiencies at initial laboratory proton irradiation relative to the spectra from fresh interiors of crystalline rocks and freshly crushed fines. This is especially evidenced by absence or relative weakness of the yellow (0.55 μm) band in the initial spectra. The top exterior surfaces of 10058, 10020, and 10057 and the bottom of 10058 show decayed spectra, while the bottom of 10020 gives a fresh spectrum.

The decayed proton luminescence spectra suggest that these particular crystalline rock surfaces, breccia materials, and fines have been exposed to low-kev ion irradiation on the moon's surface, presumably from the solar wind. Comparison of natural and laboratory decayed spectra indicates that the incident solar wind doses required to produce the natural decayed spectra are > 10^{16} protons cm^{-2}. The luminescence spectra observed in the laboratory therefore provide direct evidence for impingement of solar wind ions on the moon's surface and may be a qualitative measure of the minimum time of exposure of lunar material surfaces to the solar wind. The quantitative aspects of this concept, however, require further study. The fact that fines uniformly give decayed spectra suggests sufficient tilling of the lunar regolith at least to the depth sampled to permit solar wind exposure of all fine particles, top and bottom, to > 10^{16} ions cm^{-2}; but all rocks (e.g., 10020) were not uniformly exposed in the process.

Ultraviolet-excited luminescence was undectable in all samples examined, putting an upper limit on the energy efficiency at about 10^{-6}. Likewise, simultaneous ultraviolet and proton irradiation was no more effective in exciting luminescence than the proton irradiation alone. If these results are typical for the moon, solar ultraviolet irradiation alone or together with the solar wind cannot yield detectable lunar luminescence in rocks and soil.

Natural thermoluminescence (*9*) below 120°C was absent in all samples examined; some natural thermoluminescence above 120°C was observed in sample 10058. Low-level induced thermoluminescence was achieved by x-ray exposure of crystalline material (10020 and 10058); storage efficiencies (*3*) are in the range 10^{-12} to 10^{-11} for the temperature range 23° to 500°C. No thermoluminescence could be induced with x-rays in soil or breccia. Ultraviolet exposure equivalent to 4 days of solar-ultraviolet exposure induced no detectable thermoluminescence in sample 10058. Proton irradiation (1 × 10^{18} protons cm^{-2} at 2 kev and 5 × 10^{16} protons cm^{-2} at 4 kev) yielded no induced thermoluminescence. The pos-

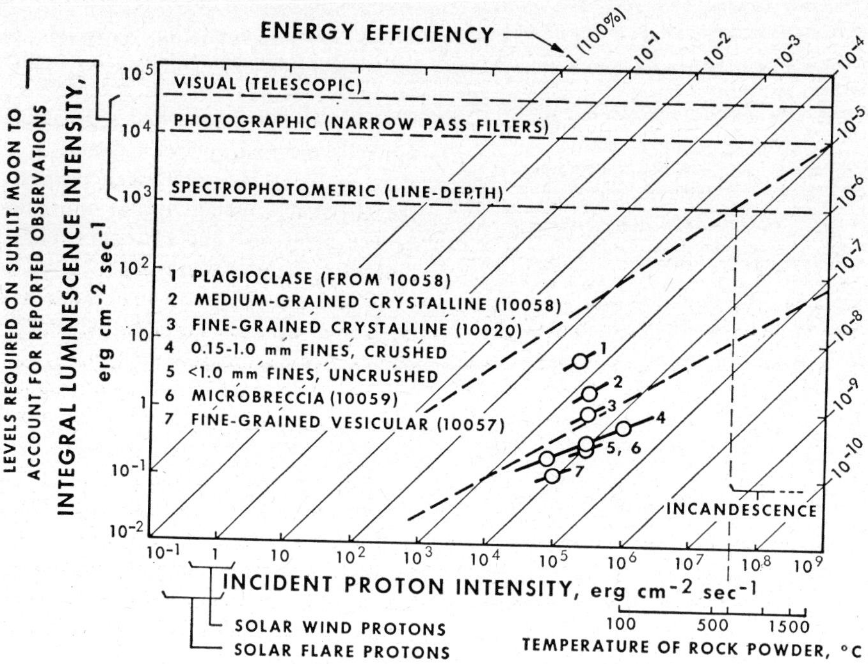

Fig. 2. Summary of the solar-proton excited lunar luminescence detection problem. Light solid diagonal lines indicate energy efficiency. Dashed diagonal lines show upper and lower bounds of luminescence efficiency trends for terrestrial and meteorite silicate rocks under various proton intensities (*5*). Intense proton flux causes rock material to incandesce before luminescing bright enough for detection on the sunlit moon from earth (*5*). Solar flare protons, although higher in energy per ion, amount to lower total energy flux than solar wind protons.

sible occurrence of detectable thermoluminescence on the moon near the lunar dawn terminator has been suggested (*10*). The present data indicate that the storage efficiency of lunar material is many orders of magnitude too low to yield levels of emission that would be detectable on the lunar surface against a background of either direct solar illumination or earthshine.

Spectral reflectance and albedo (*11*) were measured to determine how proton irradiation experiments affect the reflecting properties of lunar material and if such irradiation could alter the reflectance of fresh crushed rocks and glass to that of average lunar soil. Much of the soil was found to consist of glass, which presumably formed in part from crystalline parent material (judging from glass-lined pits and glass spatters on crystalline rocks). To determine the effects of irradiation on the reflectance and albedo of fresh glass, a representative sample of lunar crystalline material (10020) was fused (*6*). Glass and parent material were both ground to < 50 μm and subjected to a 2-kev proton irradiation of 4 × 10^{19} protons cm^{-2} for crystalline material and 2 × 10^{19} protons cm^{-2} for the glass. Reflectance spectra before and after irradiation are shown in Figs. 3 and 4. The reflectance of crystalline material has a prominent minimum near 1.0 μm, other subtle structures near 1.25 and 2.2 μm, and a suggestion of structure around 1.7 μm. In glass there are broad minimums at ~ 1.02 and 1.8 μm. By comparison, fine soil material shows subtle structure near 0.95 μm. In the visible, glass shows a pronounced enhancement in the red compared with unfused material. Elsewhere the spectra are relatively featureless. In all spectral regions the reflectance of glass is less than that of the original crystalline sample.

The experimental proton irradiations have reduced the spectral reflectance of both crystalline and artificial glass materials in the infrared below about 1.5 μm and uniformly in the visible; indicated changes in the ultraviolet reflectance of glass are within experimental uncertainty. The irradiation reduced the albedo of the crystalline material from 20 to 17 percent, and that of the glass from 8 to 5 percent; vitrification alone changes the albedo from 20 to 8 percent for the sample studied.

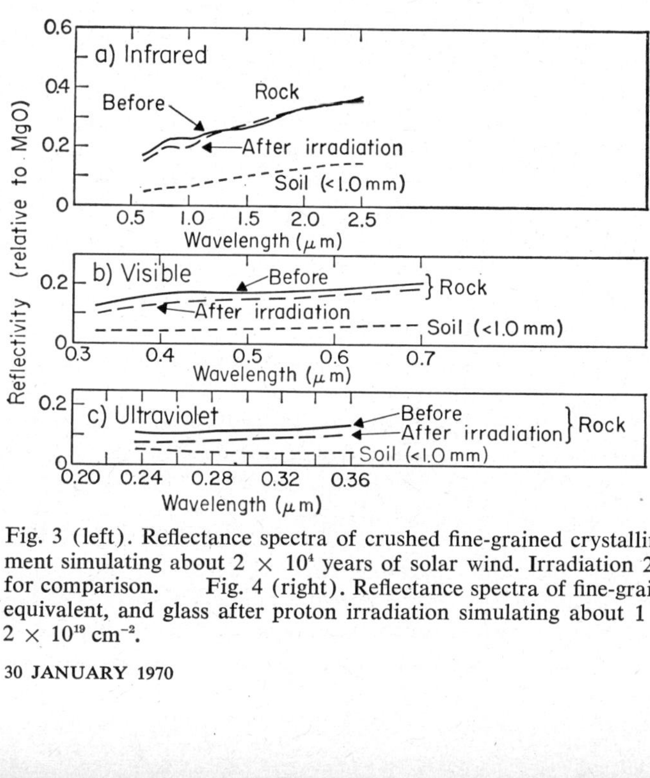

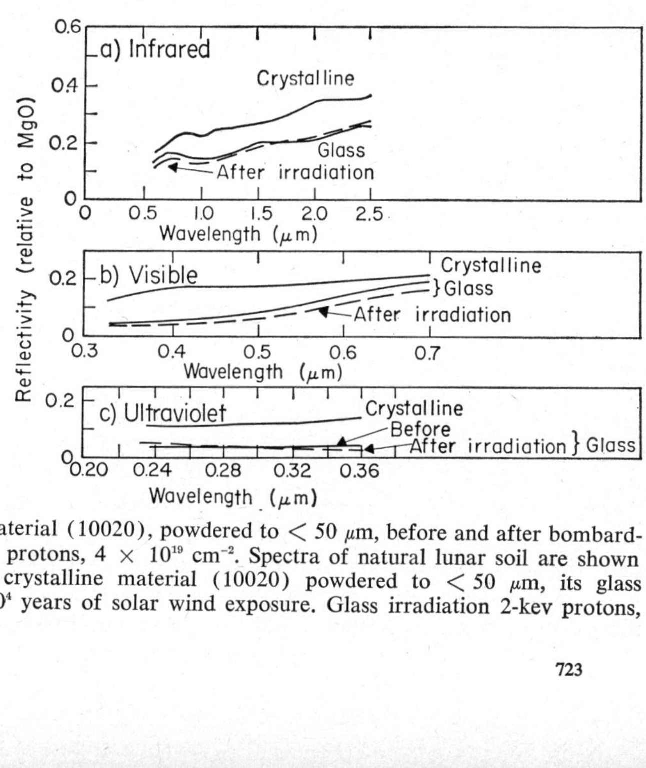

Fig. 3 (left). Reflectance spectra of crushed fine-grained crystalline material (10020), powdered to < 50 μm, before and after bombardment simulating about 2 × 10^4 years of solar wind. Irradiation 2-kev protons, 4 × 10^{19} cm^{-2}. Spectra of natural lunar soil are shown for comparison. Fig. 4 (right). Reflectance spectra of fine-grained crystalline material (10020) powdered to < 50 μm, its glass equivalent, and glass after proton irradiation simulating about 1 × 10^4 years of solar wind exposure. Glass irradiation 2-kev protons, 2 × 10^{19} cm^{-2}.

The cause of experimental proton irradiation darkening is not understood, but contamination is suspected to play a role (*12*). We are thus cautious in asserting that proton irradiation produces significant lunar darkening. Alternatively, vitrification may be an important darkening process. Except in the ultraviolet, neither the experimental irradiation dose nor the artificial vitrification alone produces material with the reflectance properties of lunar soil.

DOUGLAS B. NASH
JAMES E. CONEL
RAYMOND T. GREER*

Jet Propulsion Laboratory, Pasadena, California 91103

References and Notes

1. Lunar Receiving Laboratory rocks numbered 10058 (medium-grained igneous), 10020 (fine-grained igneous), 10057 (fine-grained vesicular), 10059 (microbreccia), and fines < 1.0 mm. All samples were received packaged in N_2, rock samples as 1- to 1.5-g chips from top exterior, bottom exterior, and interior of each rock.
2. Lunar Sample Preliminary Examination Team, *Science* **165**, 1211 (1969).
3. Energy efficiency is defined as the ratio of integral luminescence energy to applied proton energy. Thermoluminescence storage efficiency is defined here as the ratio of total thermoluminescence energy observed to total applied x-ray, ultraviolet, or proton energy prior to trap saturation.
4. D. B. Nash, *J. Geophys. Res.* **71**, 2517 (1966); J. E. Geake and G. Walker, *Geochim. Cosmochim. Acta* **30**, 929 (1966).
5. D. B. Nash, *ibid.*; D. B. Nash, in *Thermoluminescence of Geological Materials*, D. J. McDougall, Ed. (Academic Press, New York, 1968), p. 587.
6. Glass was fused in dry nitrogen using a platinum vessel after slow heating to drive off adsorbed atmospheric constituents.
7. Z. Kopal and T. W. Rackham, *Icarus* **2**, 481 (1963); J. F. Grainger and J. Ring, *Proc. Intern. Space Sci. Symp.*, W. Priester, Ed. (North-Holland, Amstersdam, 1963), p. 989; H. Spinrad, *Icarus* **3**, 500 (1964); T. B. McCord, *J. Geophys. Res.* **72**, 2087 (1967).
8. E. J. Flam and R. E. Lingenfelter, *Nature* **205**, 1301 (1965); D. B. Nash, *ibid.* (1966); E. P. Ney, N. J. Woolf, R. J. Collins, *J. Geophys. Res.* **71**, 1787 (1966).
9. Thermoluminescence was measured using an RCA 6199 photomultiplier and a heating rate of 80°C min^{-1}.
10. K. H. Sun and J. L. Gonzalez, *Nature* **212**, 23 (1966); M. Sidran, *J. Geophys. Res.* **73**, 5195 (1968).
11. Hemispherical spectral reflectance was measured with a Beckman DK-2A spectroreflectometer modified to accept horizontal, uncovered samples. Albedo, defined here as the bidirectional integral reflectance (0.4 to 0.7 μm) measured with vertical illumination and phase angle of 15°, was determined with a goniophotometer. All measurements were made relative to fresh smoked MgO.
12. D. B. Nash, *J. Geophys. Res.* **72**, 3089 (1967).
13. We thank Clay Seaman, Warren Rachwitz, Fraser Fanale, Elsa Abbott, Edgar Bollin, and Charles Fisher for their assistance. This research under NASA contract NAS 7-100.
* National Research Council Resident Research Associate appointed to the Jet Propulsion Laboratory. Present address: Department of Nuclear Engineering, Iowa State University, Ames 50010.

4 January 1970

Thermal Radiation Properties and Thermal Conductivity of Lunar Material

Abstract. *The thermal radiation properties were measured for lunar fines and chips from three different lunar rocks. Measurements for the fines were made at atmospheric pressure and at a pressure of 10^{-5} torr or lower. The directional reflectance was obtained over a wavelength range of 0.5 to 2.0 microns for angles of incidence up to 60 degrees. The bidirectional reflectance—the distribution of reflected light—was measured for white light angles of illumination up to 60 degrees. The thermal conductivity was measured over a temperature range 200 to 400°K under vacuum conditions.*

Directional reflectances of Apollo 11 lunar fines and rock chips are presented for white light and for wavelengths from 0.5 to 2.0 μm. The measurements include the directional reflectance (*1*), $\rho(\psi)$, and the normalized bidirectional reflectance (*2*), $\rho_b(\psi, \theta)$. The directional reflectance subtracted from unity gives the absorptance as a function of angle of illumination; the bidirectional reflectance indicates the distribution of reflected light as a function of angles of illumination and viewing. Both quantities are needed for calculations of thermal energy balance.

A second thermophysical property, the thermal conductivity of the fines, k, was measured under vacuum conditions for temperatures from 205° to 404°K. This property, like the reflectance, is needed for thermal energy balance calculations.

The directional reflectance was obtained with a sample center-mounted in an 8-inch diameter integrating sphere reflectometer. The sphere coating was magnesium oxide. The sphere system is constructed so that the sample can be held in a horizontal position—a necessity for powders—while the sphere and external optics can be rotated so that angles of illumination or viewing up to approximately 75° can be obtained. The system can be evacuated to pressures in the 10^{-7} torr range.

The integrating sphere was operated in the reciprocal mode, that is, the sample was illuminated by diffuse light from the sphere walls. The directional reflectance $\rho(\theta)$ was obtained as the ratio of intensity when the sample was viewed to intensity when the wall was viewed (*1*). This measurement is equivalent (*1*) to illuminating the sample at an angle of incidence ψ equal to the angle of viewing θ. For the total or white light measurements a 1000 watt tungsten-iodine lamp (DXW) with a reflector was used. The detector was a Kipp-Zonen CA-1 thermopile. The spectral results were obtained with a Perkin-Elmer 112U spectrometer and a lead-sulfide cell detector.

Measurements were made on samples 10084,68 (powder), 10047,23 (basalt) (*3*), 10048,21 (breccia), and 10057,29 (crystalline). The spectral results are shown in Fig. 1 only for angles of 10° and 60°, but measurements for angles of viewing of 10°, 20°, 30°, 45°, and 60° were also obtained. All samples received were packaged in dry nitrogen.

The directional reflectance of the lunar powder shows no absorption bands over the spectral range studied. Sample 10084,68,1 was mounted in the integrating sphere and the system evacuated to pressures below 10^{-5} torr. Sample 10084,68,2 was run at atmospheric pressure 2 months after No. 1 sample and No. 2 had been resealed and stored in the original shipping container for these 2 months. The mea-

Table 1. Total directional reflectance of Apollo 11 material. The source was a 1000 watt DXW tungsten-iodine lamp.

Angle of viewing ψ	Sample: No. 10084,68,1 Vacuum run	No. 10084,68,1 Atmospheric run	10057,29,1	10047,68,2	10048,21,1
10°	0.102	0.098	0.137	0.118	0.162
20°	0.101	0.096	0.163	0.126	0.166
30°	0.117	0.103	0.145	0.124	0.168
45°	0.115	0.113	0.135	0.128	0.175
60°	0.148	0.135	0.129	0.129	0.183

surements obtained in a vacuum are higher than those at 1 atm. The difference could possibly be caused by the surface roughness or texture, contamination, or material packing. These measurements point out the need to study a vacuum packaged sample (*4*) so that the clean surface can be investigated before it is exposed to the atmosphere. Under these conditions surface texture would be the same. The reflectance increases with increasing angle of viewing (θ), and this corresponds to what one would expect of a dielectric material.

Total directional reflectances were obtained for the powder under vacuum and dry nitrogen atmospheres. The vacuum results are higher for these measurements than for the atmospheric pressure results as in the case of spectral measurements. The results are given in Table 1. Here we can see that the reflectance increases as the angle of viewing increases, as is common for dielectrics.

Sample 10047,23,2 shows a reflectance spectrum most similar to the lunar powder (Fig. 1, B). There are no major absorption bands in the spectrum but possibly a minor one around 1 μm. This rock chip, as well as the others, was run under atmospheric conditions. Photographs of all rock chips were taken for documentation, and the area used for study on each sample has been recorded.

It should be noted here that the measurement of the reflectance on the rock chips as a function angle is somewhat questionable because of the irregular shape and condition of the surface. However, for those rock chips where it was reasonable to make these measurements, they were obtained.

The remaining rock chip measurements in Fig. 1, C and D show very pronounced absorption bands. Sample 10084,21,1, a breccia type, has a minimum reflectance centered at 1.0 μm and an apparent band centered beyond 2.0 μm. The reflectance spectrum of sample 10057,29,1, a crystalline type of material with a cinder-like appearance, also has a strong band at 1.0 μm but has no other strong bands. However, the data suggest a weak band around 1.7 μm. A second sample, 10057,29,2, was also tested. Its reflectance over the wavelength range was between 10 and 20 percent below that of 10057,29,1. These measurements again show the strong band at 1 μm but a much weaker one, if any, at 1.7 μm. The difference in reflectance is most likely due to the greater density of surface holes and craters of 10057,29,2.

The bidirectional reflectance measurements were made with a goniometric system designed so that the sample could be viewed and illuminated while under vacuum conditions. The complete system including the optics is described elsewhere (*5*). The measurements were obtained for wavelengths of 0.6, 0.8, 1.0, 1.5, and 2 μm and with tungsten-iodine lamps of 650 and 1000 watt outputs. Because of some unexpected problems with equipment no vacuum measurements will be reported here. Bidirectional infrared measurements are being obtained.

Two types of bidirectional measurements were made, one where we held fixed the angle of viewing θ and varied the angle of illumination ψ, and one where we held the angle of illumination fixed and varied the angle of viewing. For the fixed θ mode of operation, the area of illumination varies as $1/\cos \psi$; in the fixed ψ mode, the area of viewing varies as $1/\cos \theta$. For both modes the detector output can be interpreted as an intensity measurement. Because it is very difficult to obtain absolute bidirectional reflectances, the data have been normalized by the specular read-

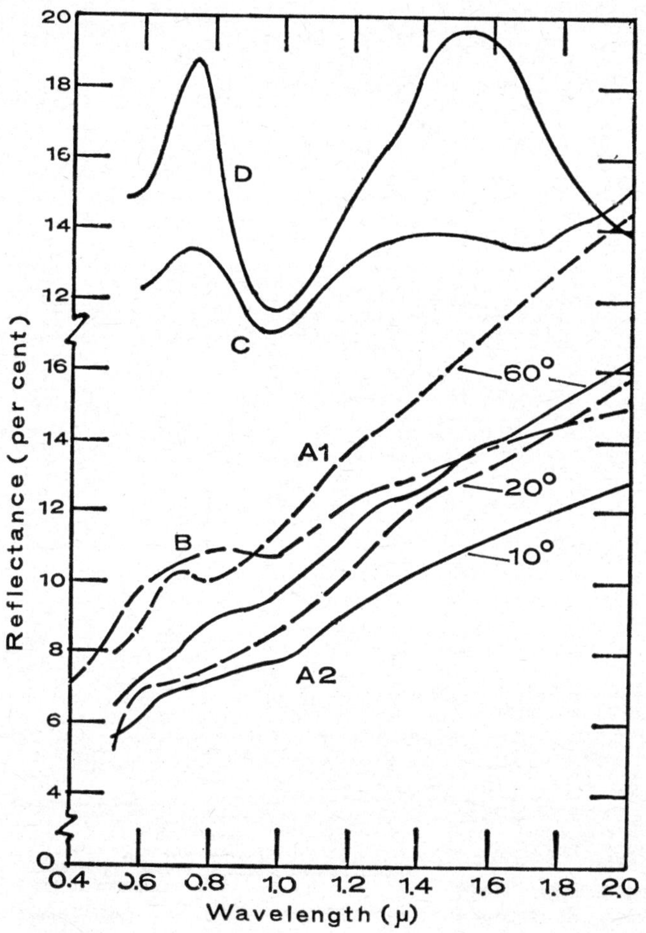

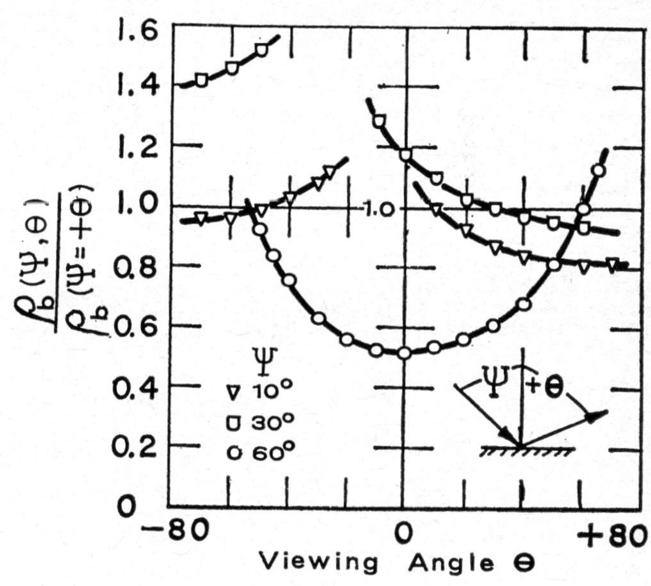

Fig. 1 (left). Reflectance spectra of lunar fines and rock chips. A1 (dashed lines), powder, sample 10084,68,1, vacuum run ψ = 10°, 60°. A2 (solid lines), powder, sample 10084,68,2, atmospheric run ψ = 20°, 60°. B (dash-dot line), basaltic type, sample 10047,23,2, ψ = 10°. C, crystalline type, sample 10057, 29,1, ψ = 10°. D, breccia type, sample 10048,21,1, ψ = 10°. Note ordinate for C and D shifted. Fig. 2 (above). Normalized bidirectional reflectances for lunar fines sample 10084,68,2 as a function of angle of viewing with the angle of illumination fixed at 10°, 30°, or 60°. The light source was a 1000 watt tungsten-iodine lamp.

ing where the angle of illumination ψ equals the angle of viewing θ. The angles were scanned automatically and the output of the detector was recorded on a strip chart recorder.

In the fixed θ mode, the blockage of the area of view of the detector by the primary illuminating mirror was too large to be acceptable, so the measurements were not available for this paper. Figure 2 shows the normalized bidirectional reflectance results for the fixed ψ mode. For such a presentation a Lambertian surface would be a horizontal line. The lunar powder shows strong back-scatter in the direction of illumination and a much weaker forward-scatter (if any) in the specular direction. The spectral measurements obtained show the same general characteristics but do indicate a slight dependence on wavelength.

The thermal conductivity of sample 10084,68,2 was measured with a line heat source technique. A Teflon cell was constructed, the sample volume being approximately 25 by 13 by 13 mm. A Nichrome heating wire 0.203 mm in diameter was used as the line heat source as well as a resistance thermometer to measure the powder temperature at the wire. An iron-constantan 36-gage thermocouple, placed 0.2 mm from the wire, was used to monitor the temperature of the powder as well. The thermal conductivity cell was placed in a vacuum chamber and surrounded by a blackened copper shroud. The shroud could be heated or cooled so that we could control the average powder temperature. Pressures of 10^{-3} torr and lower were generally used when data were obtained. The thermal conductivity of a powder is, in general, independent of pressure for pressures below 10^{-2} torr.

The general scheme for obtaining data was to heat or cool the sample to the desired temperature. When equilibrium was reached, a steady heat flux from the wire was started by electrical heating. A temperature-time recording was then obtained for the wire. From the heat input and the temperature-time characteristics the thermal conductivity of the sample can be calculated (*6*).

Sample 10084,68,2 was loaded and settled by tapping lightly onto the cell. The measured density was 1.265 g/cm³. Only three data points over the temperature range used are presented here. The thermal conductivity in watts per meter per degree Kelvin was found to be 1.71×10^{-3} (205°K); 2.07×10^{-3} (299°K); and 2.42×10^{-3} (404°K). The data would seem to indicate a dependency on temperature, but because of experimental uncertainties in the measurements no comment can be made at this time concerning temperature effects.

RICHARD C. BIRKEBAK
CLIFFORD J. CREMERS

Department of Mechanical Engineering, University of Kentucky, Lexington 40506

JAMES P. DAWSON

Scientific Specialties Corporation, Houston, Texas 77058

References and Notes

1. The directional reflectance: If a surface is illuminated by a beam of light at an angle ψ from the surface normal with energy $de_i(\psi)$ and the reflected light is collected over the hemispherical space $de_r(\psi)$ the directional reflectance is $\rho(\psi) = de_r(\psi)/de_i(\psi)$; or the surface can be illuminated hemispherically with diffuse light $e_{i,h}/\pi$ while the reflected intensity $I_r(\theta)$ is collected in a specific angular direction (θ) in a small solid angle. The directional reflectance is $\rho(\theta) = I_r(\theta)/[e_{i,h}/\pi]$. It can be shown then $\rho(\psi) = \rho(\theta)$.
2. The bidirectional reflectance $\rho_b(\psi, \theta)$ is expressed as the intensity of reflected light in the direction (θ) divided by the incident light per unit time and surface area contained within a solid angle $d\Omega_i$ in the direction ψ, $\rho_b(\psi, \theta) = dI_r(\psi, \theta)/de_i(\psi)$. Intensity is defined as $dI(\theta) = de(\theta)/d\Omega\cos\theta$.
3. This descriptive terminology was taken from *Apollo 11 Preliminary Science Report* (NASA SP-214, 1969).
4. A vacuum handling system was built by NASA but was not ready for the Apollo 11 mission.
5. C. B. Neville, Jr., thesis, University of Kentucky (1970).
6. J. H. Blackwell, *Can. J. Phys.* 34, 412 (1956).
7. We thank C. Neville, C. Allen, E. Yates, E. Hoover, and W. Buchholtz for assistance and support with the design and construction of the equipment, and C. Cremers, M. Birkebak, and T. Birkebak for assistance with data reduction. Supported by NASA grants NAS9-8098 and NGR 18-001-026.

4 January 1970

Elastic Wave Velocities of Lunar Samples at High Pressures and Their Geophysical Implications

Abstract. *Ultrasonic measurement of P and S velocities of Apollo 11 lunar samples 10020, 10057, and 10065 to 5 kilobars pressure at room temperature shows a pronounced increase of velocity (as much as twofold) for the first 2 kilobars. The travel times predicted from the velocity-depth curve of sample 10057 are consistent with the results of the Apollo 12 seismic experiments. At pressures below 200 bars, the samples are highly attenuating; for both P and S waves, the value of Q is about 10.*

Measurement of the velocities of elastic waves in lunar samples has two objectives: to provide the basic data necessary for interpreting lunar seismograms, and to study properties of rocks which were formed in extraterrestrial environments. Seismic data combined with laboratory data provide the most direct clues to the understanding of the lunar interior. In addition, mechanical, thermal, and chemical processes near the lunar surface may be inferred from the properties of the rocks. We present here the P- and S-wave velocities measured up to 5 kb pressure on the Apollo 11 samples. Comparison will be made with the results obtained in the Apollo 12 seismic experiments.

The samples were provided by NASA in the form of a rectangular cylinder [1 by 1 by 2 cm; see Table 1 and (*1*)]. Samples 10020 and 10057 are dense crystalline rocks with intrinsic densities of 3.2 to 3.4 g/cm³ (*2*). Sample 10057 is the largest and the most uniform, though probably weakly to moderately shocked, rock returned on Apollo 11 (*1*); it is similar to basalt. Sample 10020 is a fine-grained igneous rock containing plagioclase, pyroxene, and minor amounts of olivine and cristobalite. Sample 10065 is a fine-grained breccia (terrestrial analog, microbreccia) and has a bulk density of 2.35 g/cm³.

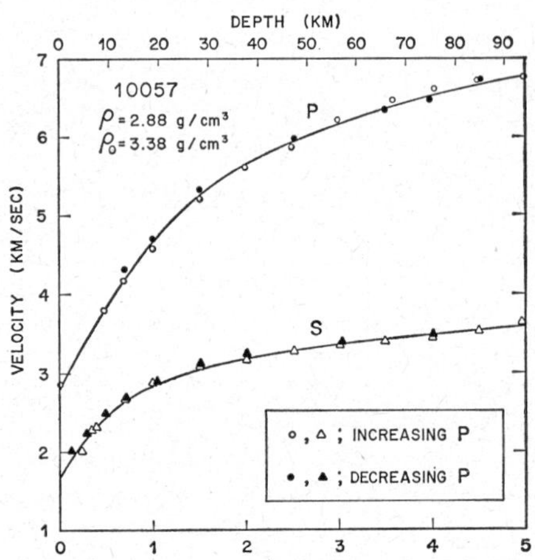

Fig. 1. The P and S velocities of sample 10057 as a function of pressure; ρ is bulk density and ρ_0 is estimated intrinsic density. The upper scale gives the depth in the moon converted from the pressure.

The standard pulse-transmission technique of Birch (3) with 1-Mhz barium titanate transducers for both P and S waves was used. High-pressure measurements were made in a simple piston-cylinder high-pressure cell with petroleum ether as the pressure medium. Samples were jacketed by copper foil or rubber tubing, or encapsulated with Sylgard. The average values of the results obtained by several runs with different sample-transducer assemblies are listed in Table 1. The estimated accuracy is 2 to 3 percent for P waves and 5 percent for S waves. Because of the small size of the samples and the high attenuation, these accuracies are considerably lower than those normally achieved. The example given in Fig. 1 is typical of the behavior of the velocity with pressure of lunar rocks. The velocity increase of about twofold for the first 2 kb pressure increase is much more pronounced than in earth rocks.

Table 1. Bulk density and velocity (in km/sec) of samples.

Sample	Wave	Pressure (kb)						
		0.0	0.5	1.0	2.0	3.0	4.0	5.0
10020	P	3.50	4.80	5.55	6.30	6.80	7.00	7.20
$\rho = 3.18$ g/cm^3	S	2.20	2.88	3.25	3.67	3.87	4.00	4.08
10057*	P	2.82	3.80	4.65	5.62	6.18	6.52	6.78
$\rho = 2.88$ g/cm^3	S	1.70	2.45	2.82	3.20	3.39	3.50	3.62
10065	P	1.50	2.90	3.50	4.05	4.30	4.40	4.50
$\rho = 2.34$ g/cm^3	S	1.05	1.70	2.00	2.28	2.42	2.65	2.78

*This sample is densely pitted. The estimated density is 3.38 g/cm^3 (see 2).

The elastic properties of the Apollo 11 crystalline rocks appear to be representative of the material in the Sea of Tranquillity. A second set of observations may be obtained from the data of Surveyor 5, recorded about 25 km northwest of the Apollo 11 landing site. Simulated Surveyor 5 lunar rocks were synthesized (4, 5) with the chemical composition given by the alpha backscattering experiment. Because the velocities of elastic waves measured on the simulated rocks (ρ = 2.8, V_P = 6.2, and V_S = 3.5) closely approximate the intrinsic properties of the Apollo 11 rocks (extrapolated to zero pressure to eliminate the effects of microcracks), neither having been corrected for porosity, we believe that our results apply to the entire area of the Sea of Tranquillity.

The densities and the P-wave velocities at 10 kb extrapolated from the data of sample 10057 fall on the curve of mean atomic weight of about 23 in Birch's (6) plot of velocity, density, and mean atomic weight. This value of mean atomic weight agrees reasonably well with that calculated from the chemical composition (5) of the Apollo 11 samples, 23.0 to 23.7. This agreement suggests that Birch's diagram can be used for the interpretation of the lunar interior.

The extremely low velocities at normal pressures are probably due to the presence of microcracks between grains; these cracks may have formed either by the large temperature fluctuations of the lunar surface or by shock metamorphism. Microscopic examination of the thin sections revealed cleavage fractures but no through-going fractures.

The lunar samples have a very high attenuation (low Q) at low pressures for elastic waves. Although we could not measure precisely the value of Q, we estimated its order of magnitude by measuring the ratio of the amplitude of the signal through the sample to that through a steel test piece having the same size and shape as the sample. If we assume that the effect of Q dominates over those of geometrical ray spreading and internal reflections and that the steel has a much higher Q than the sample, then the value of Q can be calculated approximately from the amplitude ratio a by $a = \exp[-\pi f t/Q]$ where f is frequency and t is travel time through the sample. At low pressures ($P < 200$ bars), the values of Q are of the order of 10 for both P and S waves for all the samples. At high pressures, all the samples, and in particular 10020, showed an appreciable increase in Q. It was not possible, however, to determine the value of Q because the assumptions made above are not valid for high-Q samples.

The fact that two of the samples (10020, 10057) have densities nearly equal to the bulk density of the moon, 3134 g/cm^3, is noteworthy; rock samples collected, at random, on the earth's surface would represent the density of neither the bulk earth nor the mantle. One implication of this finding is that the vertical differentiation process in the moon may have been much less extensive than that in the earth; the structure of the moon may be relatively homogeneous, without well-defined crustal layers. This idea can be tested by comparing our laboratory data with the results obtained in the Apollo 12 seismic experiment. If the dense samples 10020 and 10057 are representative of the lunar material, we can predict, ignoring the temperature effect, the depth variation of seismic waves in the moon

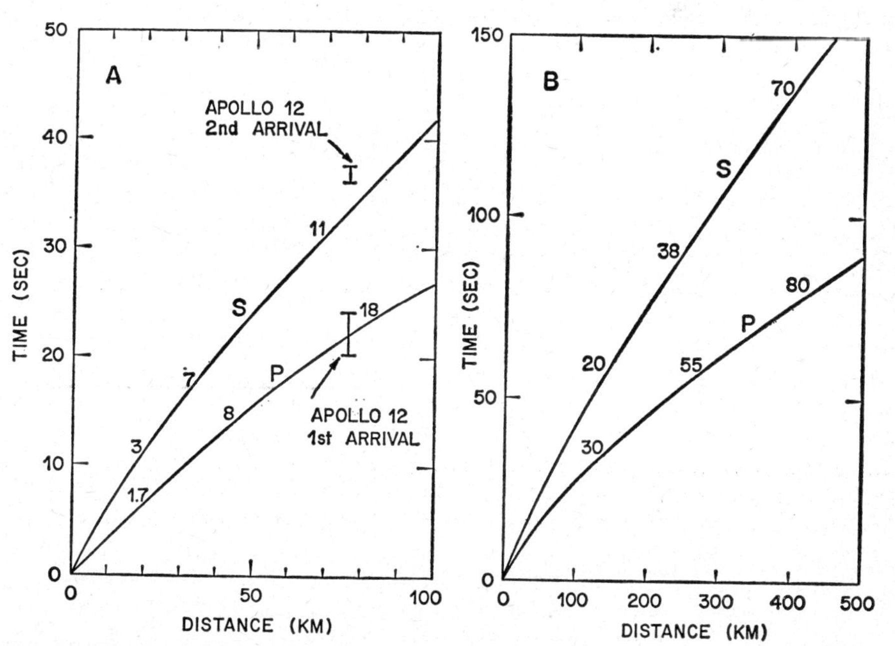

Fig. 2. The travel-time curve for the moon predicted from the velocity-depth curve of 10057 (Fig. 1), the most uniform sample. The times of the first and second arrivals in the Apollo 12 seismogram (7) are shown in A. The numbers on the curve indicate the depth (in km) of the penetration of the ray. B shows the travel-time curve to a larger distance.

(see Fig. 1), from which the travel-time curve of seismic waves can be predicted as shown in Fig. 2. In the Apollo 12 seismogram which recorded the impact of the ascent module, two distinct arrivals were detected (*7*); the first at 20.1 to 24.1 seconds and the second at 37.5 seconds after the origin time. The epicentral distance is 75.9 km. As is shown in Fig. 2, these arrival times agree remarkably well with the predicted P and S arrival times. Although the second arrival has not been confirmed as S, our proposed undifferentiated moon model is at least compatible with the seismic results. This agreement also suggests that the special textures (probably microcracks) of the samples also prevail at depths down to 20 km or so. These microcracks may have important bearings on the physical processes that have taken place on the lunar surface, and therefore deserve more extensive investigation.

If the samples 10020 and 10057 are representative of the lunar material, an efficient wave guide must exist near the lunar surface because of the sharp velocity increase with depth. Calculations of surface-wave dispersion curves made for the structure predicted from the velocity-pressure curve of the sample 10057 show a relatively constant group velocity (velocity, 1.57 km/sec) of Rayleigh waves over a period range 0 to 5 seconds. If future seismic experiments confirm this dispersion character, the evidence for an undifferentiated moon will be strengthened. If, on the other hand, significant deviation from the predicted dispersion curves is found, it will indicate the existence of structural heterogeneity at depths due to temperature, phase change, and compositional change.

HIROO KANAMORI*, AMOS NUR
D. CHUNG, D. WONES, GENE SIMMONS
Department of Earth and Planetary Sciences, Massachusetts Institute of Technology, Cambridge 02139

References and Notes

1. NASA Lunar Sample Information Catalog, Apollo 11, Manned Spacecraft Center, Houston (1969).
2. The bulk density of any of these rocks is the ratio of its mass to bulk volume. We also estimated the intrinsic density of sample 10057, which contains numerous voids. Assuming that the ratio of the total cross-section area of the voids on the surface to the total surface area of the sample is equal to the porosity, we obtained a porosity of 17.4 percent and intrinsic density of 3.38 g/cm^3. Sample 10020 has few visible voids; the intrinsic density probably does not exceed the bulk density by more than 10 percent.
3. F. Birch, *J. Geophys. Res.* **65**, 1083 (1960).
4. Full details on synthesis and other physical properties of the simulated Surveyor 5 rocks are to be reported elsewhere by Chung, Horai, Simmons, and Wones. The mineralogical composition of the simulated rocks closely approximates that of the Apollo 11 rocks. The composition (in mole percent) of the Apollo 11 rocks and the simulated rocks are, respectively, for plagioclase, 25 to 40 and 30 to 44; for pyroxene, 46 to 56 and 45 to 60; for olivine, 0.6 to 3 and none; for ilmenite, 10 to 17 and small; and for free silica, 5 to 11 and 6 to 10. The composition of the simulated rocks was based on the recent analysis of Surveyor 5 [see *Science* **165**, 277 (1969)] as well as analyses given for A-22 and B-50 (*5*).
5. Lunar Sample Preliminary Examination Team, *Science* **165**, 1211 (1969).
6. F. Birch, *J. Geophys. Res.* **66**, 2199 (1961).
7. G. Latham, personal communication, December 1969.
8. We thank Ki-iti Horai for his help, Lotfi Mohsen for the porosity measurements, and William Brace for valuable suggestions. We benefitted from discussions with Frank Press, Nafi Toksoz, and Ralph Wiggins concerning various seismological problems. Gary Latham made available to us some of the results of the Apollo 12 seismic experiments. Supported by NASA contract NAS 9-8102.
*On leave from Earthquake Research Institute, University of Tokyo.
4 January 1970

Infrared and Thermal Properties of Lunar Rock

Abstract. *The infrared absorption properties of lunar rock throughout the range 2 to 2000 micrometers were investigated and, in addition, direct measurements of specific heat and thermal conductivity of rock samples were made. The results suggest that pure radiation is an important, if not dominant, process in heat flow in the lunar surface layer. A new method for determining the mean conductivity of this layer gives somewhat lower values than earlier earth-based measurements. There is also evidence to suggest that, at depths of about 10 meters, the rock is still of a porous and fragmental nature.*

There is a strong connection between the infrared and thermal properties of the lunar surface layer. The infrared emissivity is a controlling factor on the rate at which heat leaves the moon's surface. It is also apparent that, within the regolith, a considerable fraction of heat transfer takes place by radiation rather than by conduction, and measurements of the particle size and infrared absorption in the 2- to 200-μm range now make determination of the relative efficiencies of the two processes possible. Similar measurements at longer wavelengths, together with microwave results from earth-based telescopes, now give the thermal conductivity of the lunar surface regolith. Previously the conductivity could be estimated only from measurements at 8 to 14 μm.

An increase of temperature with depth in the top few meters of the lunar surface is implied by the longer-wavelength ($\gtrsim$ 5 cm) earth-based microwave measurements; now that we have direct measurements of the thermal conductivity of the rock, the infrared and thermal measurements can be used to determine the actual thermal gradient in the layer and to compare this gradient with that predicted from internal radioactive heating.

Thermal sintering as well as mechanical shock may be involved in the formation of the breccia; continuous measurement of the thermal conductivity of a heated sample should be a means of watching the breccia formation process at an enhanced rate.

We describe here infrared and conductivity measurements now in progress. The infrared transmission throughout the range 2 to 2000 μm has been measured for the type D material with (i) double-beam spectrometers with Nernst sources and Golay detectors (range, 2 to 50 μm); (ii) Fourier transform spectrometer with a mercury source and Golay detector (range, 50 to 1000 μm); and (iii) grating spectrometer of the type described by Bloor *et al.* (*1*) with a mercury lamp source and Golay detector (range, 200 to 2000 μm).

In addition, a check at 1300 μm has been made with a 1.5-m Cassegrain telescope with the sun as source and an InSb photoconducting detector. A similar spot measurement at 338 μm is being made with a cyanogen laser with a Golay cell as a detector. The fines were compressed to a density of 1.8 g/cm^3. In each case we measured the attenuation or extinction coefficient μ defined by the equation

$$\frac{I_x}{I_0} = e^{-\mu\rho x} \qquad (1)$$

where I_x is the intensity after passage through a thickness (x) of material, I_0 is the incident intensity, and ρ is the sample density. Below 50 μm, measurements were also made with crushed fines in a transparent substrate, and, in this way, the coefficient of pure absorption K_λ was determined. Because of the tendency for the dust to shake down and form macroscopic cracks, the far infrared measurements were made with the powder spread evenly over a horizontal Mylar (melinex) window. These measurements will be described elsewhere (*2*).

Measurements of the thermal conductivity of types A (fine grained crystalline), C (breccia), and D (fines) materials are being made with a method in which a thermal step function is supplied to the sample by a straight electrically heated wire. In the case of the rock types A and C, comparison is being made with materials of known conductivity. From the diffusion equation for heat flow

$$\frac{\delta T}{\delta t} = \frac{k}{\rho c}\Delta^2 T \qquad (2)$$

which relates the temperature T to the specific heat c, thermal conductivity k, and density ρ, it is seen that, for systems of the same boundary conditions, the temperature at any point will only depend on $kt/\rho c$. These measurements will be described elsewhere (*3*).

We have also made measurements of specific heat and are, at present, measuring the Hall effect and electrical conductivity of samples. Since both the electrical and thermal conductivities are dependent on interparticle contacts, we have made a number of measurements of the shapes and distribution of sizes of particles of the type D material, and have examined its microcrystalline morphology using electron-microscope techniques.

The infrared absorption and scattering coefficients of the type D sample are shown in Fig. 1. In the middle infrared (5 to 50 μm), these measurements show features associated with the Si–O bond. At the longer submillimeter wavelengths the absorption coefficient of the fines varies very much more rapidly with wavelength than would be expected from an extrapolation of the microwave data obtained from earth-based telescopes in the millimeter wavelength region. At the long wavelength limit of our measurements there is an octave of overlap (1000 to 2000 μm) with the microwave measurements. The microwave results essentially give the parameter

$$\delta = \frac{1}{K_\lambda}\left(\frac{\omega c}{2\rho k}\right)^{1/2}$$

where K_λ is the absorption coefficient at wavelength λ, ω the angular frequency corresponding to the lunar diurnal period, c the specific heat, ρ the density of "average" lunar material, and k its thermal conductivity. Our measurement of K_λ of the fines at 1400 μm, together with previously obtained values of the microwave parameter (*4*), give a value of the conductivity $k = 2.5 \times 10^{-6}$ cal cm^{-1} $°K^{-1}$ sec^{-1} for the lunar surface layer.

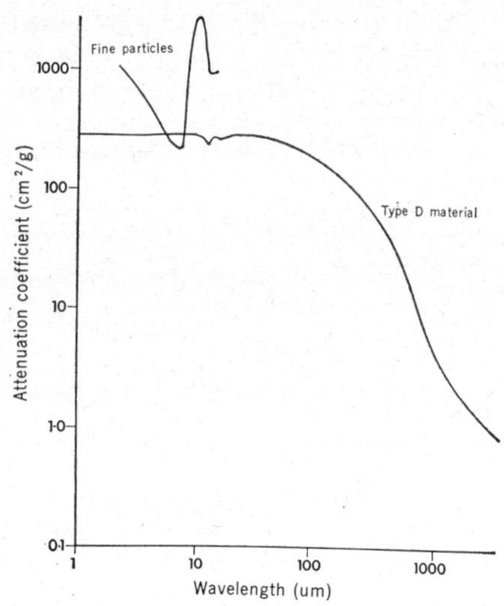

Fig. 1 (above). Absorption properties of type D material in the infrared range.

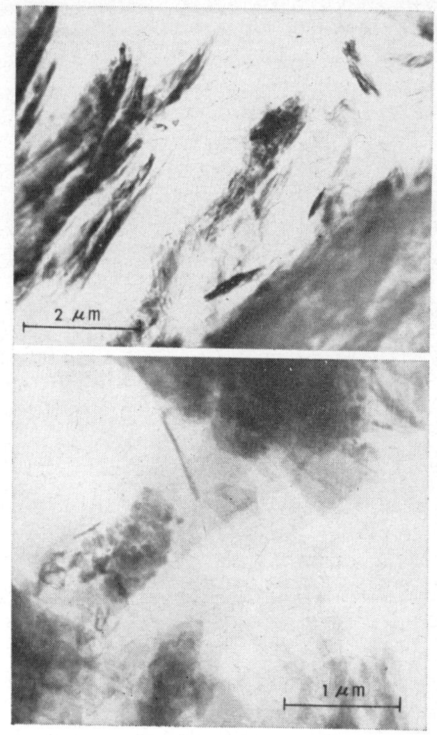

Fig. 2 (right). Electron micrographs of type D material.

Thermal measurements indicate a specific heat of 0.20 ± 0.02 cal g^{-1} $°K^{-1}$ for the breccia (sample 10065). The thermal conductivity of the fines seems to be somewhat higher than that found by the indirect method just described. The conductivity is dependent on the degree of packing, and we are at present studying this effect. For the crystalline (type A) rock and the breccia, the conductivities have provisionally been found to be in the range 0.5 to 2.0 $\times 10^{-3}$ cal cm^{-1} $°K^{-1}$ sec^{-1}, the conductivity of the breccia being lower than that of the type A material.

The electron micrographs, which were made to investigate interparticle contacts, suggest a predominance of lamellar crystals up to 500 Å thick (Fig. 2).

The mean conductivity of the lunar surface layer 2.5×10^{-6} cal cm^{-1} $°K^{-1}$ sec^{-1} determined here is somewhat lower than that deduced from earth-based lunation measurements at 8 to 14 μm. The difference is most likely to be explained by the fact that the mean measured midnight temperatures are weighted strongly in favor of those areas with high thermal conductivity. This explanation is reinforced by the fact that the values of conductivity obtained from eclipse observations at 8 to 14 μm are in good agreement with our determination.

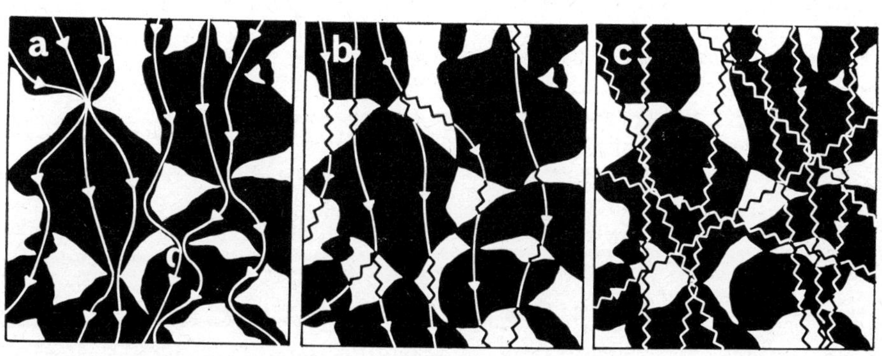

Fig. 3. Possible modes of heat transfer in type D material. (a) Pure conduction, (b) conduction-radiation, and (c) pure radiation. The curved line denotes the photon path. The solid line denotes the line of heat flow (normal to isothermal). ■ Solid material; □ void (vacuum).

In considering the heat flux coming from the lunar interior, our direct measurements of the thermal conductivity of the rock types A and C can be compared with the variation of mean disk temperature with wavelength (*5, 6*). If we assume that the region from 1 to 20 m below the surface is composed of these rocks (A and C), then a heat flux of about 3×10^{-5} cal cm^{-2} sec^{-1} is implied—a result two orders of magnitude greater than that predicted by Jaeger (*7*) and MacDonald (*8*) from radioactive heating of chondritic material. This suggests that, even at depths of about 10 m, the lunar material has a mean thermal conductivity considerably lower than that of the solid samples we have investigated, and this in turn implies that, at these depths, the crust is still likely to be fragmental and porous.

The possible mechanisms for heat flow in the uppermost layer of the lunar surface fines are shown in Fig. 3. An analytic expression for the pure conduction process is possible only if the nature of the particle contacts is known, but expressions for the conduction-radiation (k_{rc}) and pure radiation (k_r) processes have been given by Troitski (*9*) and Clegg *et al.* (*10*)

$$k_{rc} = 4\,\sigma\alpha T^3 l\left(\frac{\varepsilon}{2-\varepsilon}\right) \qquad (4)$$

$$k_r = \frac{16}{3}\,\sigma\,\frac{T^3}{K(\bar{T})} \qquad (5)$$

where l is the mean spacing between particles, α the void ratio, σ Stefan's constant, ε the particle emissivity, and $K(\bar{T})$ the Rosseland mean absorption coefficient. From our measurements of absorption coefficients throughout the infrared wavelength region we can determine this mean absorption coefficient and the calculation shows that the pure radiation process is in general more effective than the conduction-radiation process. The spectroscopic results given in Fig. 1 demonstrate that the effects of scattering and reflection of the infrared radiation must also be considered in a model for heat transfer in the type D material. Since the breccia has a higher conductivity than the fines, it is clear that we cannot ignore the pure conduction process especially where the fines are well packed.

J. A. BASTIN
P. E. CLEGG
G. FIELDER

Physics Department, Queen Mary College, London University, London, England

References and Notes

1. D. Bloor, T. J. Dean, G. O. Jones, D. H. Martin, P. A. Mawer, C. H. Perry, *Proc. Roy. Soc. London Ser. A* **260**, 510 (1961).
2. P. E. Clegg, A. B. Poole, W. J. French, in preparation.
3. E. L. G. Bowell, G. Fielder, D O. Gough, S. J. Pandya, in preparation.
4. J. A. Bastin and A. E. Gear, *Proc. Roy. Soc. London Ser. A* **296**, 348 (1967).
5. J. E. Baldwin, *Monthly Notices Roy. Astron. Soc.* **122**, 513 (1961).
6. V. S. Troitski, *Proc. Roy. Soc. London Ser. A* **296**, 366 (1967).
7. J. C. Jaeger, *Nature* **183**, 1316 (1959).
8. G. J. F. MacDonald, *J. Geophy. Res.* **64**, 1967 (1959).
9. V. S. Troitski, *Nature* **213**, 688 (1967).
10. P. E. Clegg, J. A. Bastin, A. E. Gear, *Monthly Notices Roy. Astron. Soc.* **133**, 63 (1966).
11. We thank E. Bowell of the University of London Observatory for the measurements of thermal conductivity. We also thank W. J. French and A. B. Poole of the Geology Department and E. H. Andrews of the Department of Materials, Queen Mary College, for the near-infrared and electron-microscope measurements.

4 January 1970; revised 8 January 1970

Thermal Diffusivity and Conductivity of Lunar Material

Abstract. *The thermal diffusivity and conductivity of type C lunar samples returned by Apollo 11 are lower and less dependent on temperature than those of type A samples. The thermal properties of both types are lower than the corresponding properties of normal terrestrial rocks.*

The thermal diffusivity of four Apollo 11 lunar specimens, 10020, 10046, 10057, and 10065, with bulk densities 2.99, 2.21, 2.88, and 2.36 g/cm^3 respectively, was measured over the temperature range –130°C to +150°C by the modified Ångstrom method (*1, 2*). All of these samples were rectangular prisms, 1 by 1 by 2 cm. The thermal diffusivity κ may be determined from either the phase lag or the amplitude decay with distance of a periodic temperature wave. Geometry B of Kanamori *et al.* (*2*), in which the temperature wave propagates through the specimen in the direction perpendicular to the long axis of the prism, was used. The d-c component of the temperature wave heated the samples about 100°C above the ambient temperature, which was controlled by either liquid nitrogen or solid carbon dioxide in order to obtain the desired sample temperature.

Results of the investigation are summarized in Fig. 1. The temperature variation of thermal diffusivity is almost identical for samples 10020 and 10057 and for samples 10046 and 10065, but is quite distinctive for each of the two groups. The difference is attributed to the differences in texture and the composition of the specimens. Samples 10020 and 10057 are fine-grained vesicular crystalline igneous rocks, type A according to the classification of the Preliminary Examination Team (*3*). Samples 10046 and 10065 are breccias (type C). The presence of microcracks, larger cracks, and glass in the type C samples is probably the cause of the lower thermal diffusivity and smaller temperature dependency.

In many solids, the reciprocal of thermal diffusivity varies almost linearly with temperature *t*. Coefficients of a linear relationship were determined by least-squares from our data. For type A

$$\kappa^{-1} = (0.314 \pm 0.159) \times 10^2 + (0.378 \pm 0.051)t$$

and for type C

$$\kappa^{-1} = (0.545 \pm 0.207) \times 10^2 + (0.648 \pm 0.068)t$$

where κ is in cm^2/sec and *t* in degrees K. These relationships are illustrated in Fig. 1.

In order to obtain the thermal conductivity, $K = \kappa\rho c_p$, and the thermal inertia $\gamma^{-1} = \sqrt{K\rho c_p}$ from the diffusivity, we calculated the heat capacity at 273°K, c_p, for samples 10020 and 10046 from the mineral composition (*4*) and the thermodynamical data on minerals (*5*). Taking the values of c_p as 0.174 cal/g °C for both samples 10020 and 10046, we find, for type A, $K = 3.87 \times 10^{-3}$ cal/cm sec °C and $\gamma = 22$ cm^2 $sec^{1/2}$ °C/cal and for type C, $K = 1.66 \times 10^{-3}$ cal/cm sec °C and $\gamma = 40$ cm^2 $sec^{1/2}$ °C/cal. These estimates of γ are substantially smaller than those obtained from infrared and passive microwave data on the lunar surface (*6*). It must be mentioned that the thermal diffusivity measurements reported here were made in air at atmospheric pressure. Under lunar surface conditions the materials with con-

Table 1. Attenuation of diurnal temperature variation of the moon with depth.

Amplitude °C	Depth (cm)	
	Type A*	Type C†
150	0	0
10	67	52
1	124	96
0.1	180	140

* $\kappa = 0.75 \times 10^{-3}$ cm^2/sec. † $\kappa = 0.45 \times 10^{-3}$ cm^2/sec.

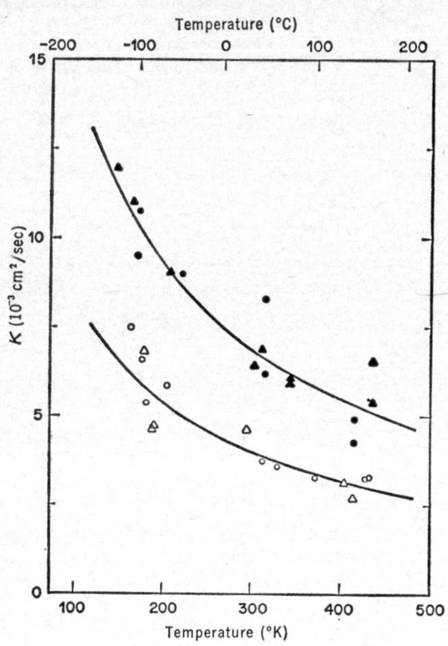

Fig. 1. Thermal diffusivity plotted against temperature for Apollo 11 lunar materials. Closed circles, sample 10020; closed triangles, 10057 (type A). Open circles, 10046; open triangles, 10065 (type C).

siderable porosity (*7*) should exhibit lower thermal diffusivities. An even lower thermal diffusivity is to be expected for the lunar soil, type D material.

The equatorial lunar surface temperature is known to vary about 150°C in amplitude with a period of 2.5×10^6 sec. The measurement of heat flow from the interior of the moon in a shallow borehole, nominally 3 m, expected on Apollo 13 (*8*) depends on a rapid decrease of amplitude of the diurnal wave. For lunar material with thermal diffusivities as high as those found for rocks in air in this investigation, 0.75 and 0.45 $\times$ 10^{-3} cm²/sec, corresponding respectively to the materials of type A and type C, the characteristic depths (*9*) for the diurnal wave are 120 and 155 cm. In Table 1, the depths below the flat lunar surface at which the amplitude of the diurnal variation of temperature, 150°C, becomes 10°C, 1°C, and 0.1°C are shown. Almost surely our values may be taken as upper limits for the diffusivity of lunar surface material, and a reliable thermal gradient for heat-flow determination may be measured in a 3-m hole. However, the presence of boulders scattered through the lunar soil of very low diffusivity may prove to be troublesome.

KI-ITI HORAI, GENE SIMMONS
HIROO KANAMORI, DAVID WONES
Department of Earth and Planetary Sciences, Massachusetts Institute of Technology, Cambridge 02139

References and Notes

1. H. Kanamori, N. Fujii, H. Mizutani, *J. Geophys. Res.* **73**, 595 (1968).
2. H. Kanamori, H. Mizutani, N. Fujii, *J. Phys. Earth* **17**, 43 (1969).
3. Lunar Sample Preliminary Examination Team, *Science* **165**, 1211 (1969).
4. Modal analysis of sample 10020 shows (600 points counted) pyroxene, 43.6 percent; ilmenite, 26.5 percent; plagioclase, 23.6 percent; olivine, 3.7 percent; void, 1.7 percent; and other, 0.7 percent. The sample 10046 shows (1571 points counted) pyroxene, 16.8 percent; opaques, 8.6 percent; plagioclase, 4.7 percent; unidentified, 3.5 percent; glass, 2.9 percent; and matrix (<40 μm), 63.5 percent. In calculating the heat capacity of sample 10046, we approximated the opaques by ilmenite and the matrix by 30 percent glass, 55 percent pyroxene, and 15 percent plagioclase.
5. R. W. Goranson, in "Handbook of Physical Constants," F. Birch, J. F. Schairer, H. C. Spicer, Eds., *Geol. Soc. Am. Spec. Pap. No. 36* (1942); R. A. Robie and D. R. Waldbaum, "Thermodynamic Properties of Minerals and Related Substances at 298.15°K (25°C) and One Atmosphere (1.013 Bars) Pressure and at High Temperatures," *U.S. Geol. Surv. Bull.* 1259 (1968).
6. V. D. Krotikov and V. S. Troitskii, *Sov. Phys. Uspekhi* **6**, 841–71 (1964); V. S. Troitsky, *Proc. Roy. Soc. London Ser. A* **296**, 366–98 (1967).
7. Porosity of sample 10057, obtained from the point count of cavities on the surfaces of the sample, is 0.174. (see H. Kanamori, A Nur, O. Chung, D. Wones, G. Simmons, *Science*, this issue.)
8. M. Langseth, Jr., A. Wechsler, E. Drake, G. Simmons, *Science*, in press.
9. The characteristic depth given by $2\pi/\sqrt{\omega/2\kappa}$ is the distance at which the amplitude of a periodic temperature variation with frequency ω is reduced by a factor of exp $(-2\pi)=0.0019$.
10. Supported by NASA contract NAS9-8102.

4 January 1970

Compressibilities of Lunar Crystalline Rock, Microbreccia, and Fines to 40 Kilobars

Abstract. *The compressibilities of three lunar samples were studied at room temperature from 0 to 40 kilobars. The samples were a fine-grained vesicular crystalline rock (type A), a microbreccia (type C), and fines (type D). All samples were porous. The microbreccia and fines were quite compressible at all pressures; the compressibility of the crystalline rock was somewhat less, being 8.4 megabar*$^{-1}$ *at 1 atmosphere and 1.5 megabar*$^{-1}$ *at 35 kilobars. Some porosity appeared to remain in the samples at all pressures. Thus the pressure-volume data derived from these samples may be representative of porous surface and near-surface material in the vicinity of the Apollo 11 landing site but may not be representative of lunar material at depth.*

The pressure-volume properties of three lunar materials were studied to 40 kb at room temperature. These materials were sample 10017,69, a fine-grained vesicular crystalline rock (type A): sample 10048,52, a microbreccia (type C), and sample 10084,78, fines (type D). The method consisted of measuring bulk volume changes under nearly hydrostatic pressure (*1*). If any of the samples can be considered to be representative of lunar material at depth, the pressure-volume data will be useful in calculating lunar density as a function of radius. The data can also be used to perform numerical calculations of meteoritic impact processes. Finally, they provide basic information on the lunar surface material such as compressibility and porosity, and qualita-

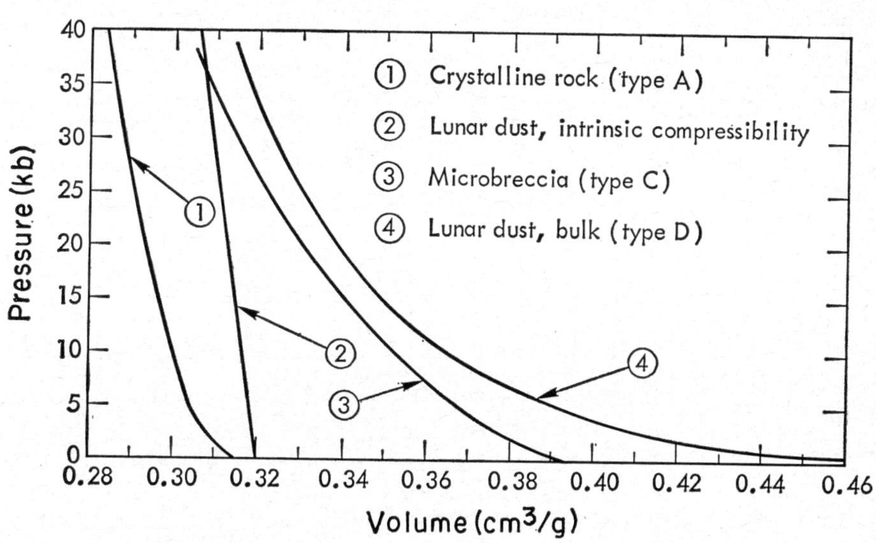

Fig. 1. Pressure versus specific volume for lunar materials.

tive information about the strength of the constitutive minerals can be inferred.

Compressibilities (Fig. 1) of these materials are typical of porous terrestrial rocks (*1*). For example, the change in volume of the crystalline type A rock (Fig. 1, curve 1) is 9.8 percent at 40 kb, which is similar to that of many terrestrial igneous rocks. This sample contained more than 5 percent initial porosity, and some porosity was present at the highest pressures. The compressibility at 35 kb was 1.5 Mb^{-1}, which is much higher than intrinsic compressibilities calculated at this pressure for a plagioclase-clinopyroxene-ilmenite rock of reported composition (*2*). These are estimated to be about 0.9 Mb^{-1} at 35 kb. The higher compressibility of the lunar rock is due to residual porosity.

If near-surface lunar rock is similar to this crystalline rock, one can predict P-wave velocities near the lunar surface. If one assumes a Poisson's ratio of 0.25 and uses the initial compressibility of 8.43 Mb^{-1}, the predicted velocity of P-waves is 2.6 km/sec.

The lunar dust is much more compressible, as would be expected for an unconsolidated powder (Fig. 1, curve 4). The intrinsic compressibility for the fine dust (Fig. 1, curve 3) was determined by mixing equal volumes of dust and tin powder; the mixture was hot-pressed to remove all porosity. After correcting for the effects of tin, we obtained the intrinsic pressure-volume curve of the dust which can be represented by $-\Delta V/V_0 = ap - bp^2$ with $a = 1.28$ Mb^{-1}, $b = 5.5$ Mb^{-2}, and $V_0 = 0.3197$ cm^3/g. Note that some porosity remains in the dust even at 38 kb, as is also the case for quartz-feldspar sands and gravels (*3*). The strength of the individual lunar dust minerals and glasses thus seems to be of the same order of magnitude [40 to 50 kb at a confining pressure of 30 kb (*4*)] as quartz and feldspars.

The microbreccia (type C) rock is also very compressible at all pressures (Fig. 1, curve 3), and it is certainly not a well-compacted rock. Its compressibility is 17 Mb^{-1} at 1 atm and 3.6 Mb^{-1} at 35 kb.

Inelastic effects were noted in all three lunar materials. Some of the porosity was permanently removed upon unloading to 1 atm; these were 4.5 percent for the crystalline rock, 12.8 percent for the microbreccia and 27 percent for the fines. However, at low pressure some porosity was recovered, as is also the case for porous terrestrial igneous rocks (*1, 3*).

In conclusion, all the samples, including the crystalline rock, contained porosity up to pressures of 40 kb. The samples may be representative of surface and near-surface material in the vicinity of Tranquillity Base. If the lunar interior is hot, one would not expect porous rock to be present at high pressures and temperatures, and in that case the lunar crystalline rock, for example, could not be used as a model for interior material. However, if the moon is cold, then a porous rock such as the crystalline rock may be a model for at least part of the lunar interior. Data for lunar rocks from many sites must be obtained to construct a reasonable model of the interior of the moon.

D. R. STEPHENS
E. M. LILLEY

University of California, Lawrence Radiation Laboratory, Livermore 94550

References and Notes

1. D. R. Stephens, E. M. Lilley, H. Louis, *Int. J. Rock Mech. Mineral. Sci.*, in press.
2. Lunar Sample Preliminary Examination Team, *Science* **165**, 1211 (1969).
3. D. R. Stephens and E. M. Lilley, in *Shock Metamorphism of Natural Materials*, B. M. French and N. M. Short, Eds. (Mono, Baltimore, Md., 1968), p. 51.
4. J. M. Christie, H. C. Heard, P. N. LaMori, *Amer. J. Sci.* **262**, 26 (1964).
5. Work performed under the auspices of the U.S. Atomic Energy Commission.

4 January 1970

Sound Velocity and Compressibility for Lunar Rocks 17 and 46 and for Glass Spheres from the Lunar Soil

Abstract. *Four experiments on lunar materials are reported: (i) resonance on glass spheres from the soil; (ii) compressibility of rock 10017; (iii) sound velocities of rocks 10046 and 10017; (iv) sound velocity of the lunar fines. The data overlap and are mutually consistent. The glass beads and rock 10017 have mechanical properties which correspond to terrestrial materials. Results of (iv) are consistent with low seismic travel times in the lunar maria. Results of analysis of the microbreccia (10046) agreed with the soil during the first pressure cycle, but after overpressure the rock changed, and it then resembled rock 10017. Three models of the lunar surface were constructed giving density and velocity profiles.*

Glass Spheres. Sound velocities and Poisson's ratio, σ, of glass spheres from the lunar soil (300 to 600 μm diameter) were measured with the resonance sphere technique (*1*). Densities were measured using heavy liquids and also by measuring volume and mass. The results are shown in Fig. 1, along with data of tektites (*1*), and some artificial lunar glasses (*2*) with composition close to the lunar soil. Glass spheres with low density are colorless or lightly colored and may have low TiO_2 and FeO content; high density glass is dark brown, red or black, and may contain large amounts of TiO_2 and FeO.

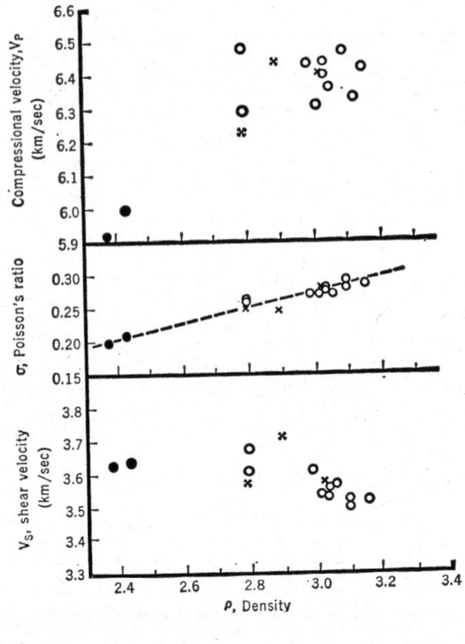

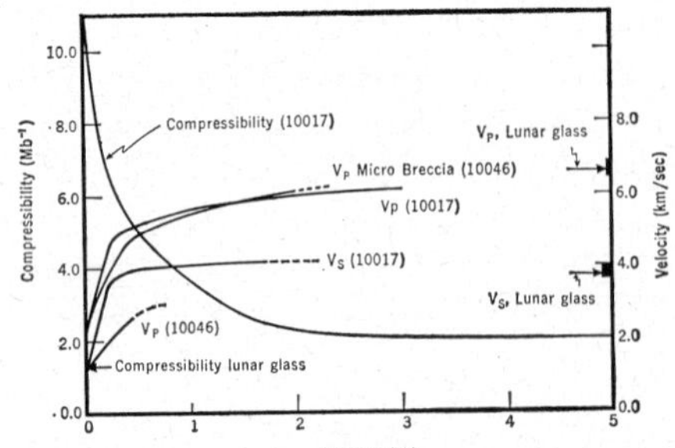

Fig. 1 (left). Sound velocities and Poisson ratio for ten lunar soil glass spheres, two tektites and three artificial glass spheres, whose composition is similar to Apollo 11 lunar soil and rocks. Solid circles, lunar glass; open circles, tektite; crosses, pseudo lunar glass. Fig. 2 (above). A composite of pressure data showing β of rock 10017, V_P and V_S for rock 10017, V_P for the microbreccia 10046 (first run and final run), and V_P, V_S, and β for the glass spheres.

Although the density ranges from 2.79 to 3.15, the compressional velocity, V_P, remains about the same, while the shear velocity changes slightly. This is consistent with the known fact that the density of rocks and minerals, but not the velocity, increases with the addition of Fe and Ti. This is also true of the artificial lunar glass. There seems to be a useful correlation, showing that σ increases with ρ from sample to sample; σ may be an indicator of Ti and Fe content. We anticipate a correlation between σ and the index of refraction. We wondered if values of V_S and V_P for lunar glass could be considered as typical of lunar rocks in an idealized state (no porosity or cracks). We applied the fourth-power relationship between the moduli and the density for minerals and glasses (*3*). The assumption of a smaller volume (95 percent) for the idealized crystal, with no shift in bulk modulus and a slight shift in shear modulus, leads us to conclude that V_P for the idealized Apollo 11 lunar rock is 6.5 to 7.0 km/sec at P=0, with V_S about 4 km/sec.

The Debye temperature of the glass beads is about 530, computed from V_P and V_S. The data from the glass beads indicate they are not like tektites, although it may turn out that they resemble microtektites (*4*).

Compressibility. Static compressibility of the fine grained vesicular rock 10017 was measured using the strain gage method of Brace (*5*). Measurements of linear compressibility were made in three perpendicular directions at pressures up to 2 kbar and in one direction to 6 kbar. The volume compressibility, β, is typical of that observed for many terrestrial rocks; it decreases rapidly with pressure from a very high value within a kilobar (Fig. 2). This effect is due to the closing of cracks (*5, 6*).

Above 0.5 kbar, linear compressibility is nearly isotropic, indicating little preferred orientation of minerals. Below that pressure, however, some anisotropy was found, which suggests a preferred orientation of cracks. Such a preferred orientation could have been produced by a variety of mechanisms, and in particular by impact or thermal shock.

The value of β at high pressure is surprisingly high, closer to acidic than to basic igneous rocks (*6*). At 5 kbar β is 1.98 Mb^{-1} and decreases steadily with P. This high value is probably due to porosity associated with nearly spherical vesicles, which will affect β even at very high pressures [since they are not readily closed (*7*)]. An estimate of the intrinsic

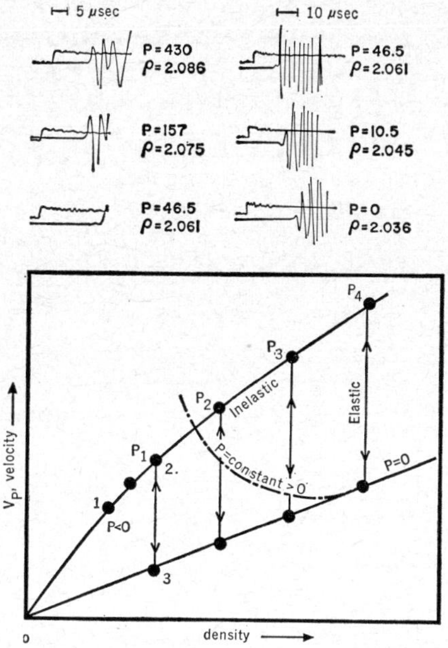

Fig. 3. V_P, compressional velocity of the lunar soil. The top curve is inelastic compression; the bottom curve the extrapolated value of V_P plotted against ρ at P=0; the straight lines, the loci of elastic compressions with P, from different porosity states; the dashed lines, curves of constant P. Typical arrival time patterns are traced above the curve.

β, calculated from the results on the glass beads, is indicated in Fig. 2.

Thus, the presence of open cracks explains the high β and low V_P of 10017 and 10046 near P=0. This argument also may explain the low velocities inferred from the LEM impact (*8*), during the Apollo 12 seismic experiment. Because of the low gravity, open cracks (and their effects) will reach to far greater depths on the moon than on earth. Furthermore, the anomalous low velocities observed at low pressure are not observed on earth, because the cracks are filled with water (*9*). On the moon, however, because water is probably not present, *in situ* velocities in the upper 20 km can be expected to be low compared to the intrinsic values of rocks.

Sound Velocities of Rocks 10017 and 10046. The sound velocity of rocks 10017 and 10046 (a microbreccia) was measured versus pressure with a modified Birch technique (*10, 11*). The data obtained are relative changes in the acoustic delay time. These, together with the sound velocities measured at ambient pressure (V_P=1.84 km/sec for 10017, 1.2 km/sec for 10046) were used to compute the velocity pressure curves shown in Fig. 2. The shear velocity was calculated from the V_P and β data.

For each sample, we found a sharp increase in velocity with applied pressure. The initial run on the microbreccia yielded the smallest initial rate of change of velocity and the final value of V_P at high P was about 3 km/sec. After pressure cycling to 7 kbar, in a subsequent run V_P increased faster with P and attained a value at high P double that in the initial run and became similar to that of rock 10017. The initial slope, however, remains less than for rock 10017. The change is most likely due to alteration of the pore structure by pressure. We feel, therefore, that in the initial run on 10046 the effect of equant pores dominated both the change in velocity and the final velocity at high P. After cycling through 7 kbar, the microstructure was irreversibly altered, and cracks developed at the expense of

Table 1. Three models of the velocity profile of the upper surface of the lunar maria. Compressional velocity, V_P; shear velocity, V_S (km/sec). Density, ρ (g/cm³). Depth below surface, Z (km). Three models corresponding to lunar soil (1), broken rock (2), and solid rock (3).

Property	Pressure (kbar)					
	0	0.05	0.1	0.5	1.0	1.5
			Number 1			
ρ	2.20	2.21	2.22	2.23	2.24	2.25
V_P	1.07	1.20	1.40	2.06	2.65	3.20
V_S	0.67	0.73	0.80	1.08	1.44	1.75
Z	0.0	1.5	2.8	13.7	27.2	39.3
			Number 2			
ρ	3.10	3.10	3.11	3.12	3.15	3.17
V_P	1.81	1.89	1.97	2.46	2.98	3.38
V_S	1.01	1.05	1.10	1.38	1.67	1.89
Z	0.0	1.0	2.0	9.8	19.5	29.0
			Number 3			
ρ*	3.10	3.10	3.10	3.10	3.10	3.10*
V_P	1.80	2.55	3.30	5.20	5.60	5.80
V_S	1.20	1.75	2.30	3.90	4.10	4.12
Z	0.0	1.0	2.0	9.9	19.6	29.0

*Density does not change because only thin cracks are closing.

equant pores. This is evidenced by the facts that: (i) the velocity at P=0 changed from an initial value of 1.2 to 2.2 km/sec, with little change in density; (ii) the final V_P-P curve is similar to that of rock 10017. Because of the presence of many equant pores, it is unlikely that V_P of about 6 km/sec represents the intrinsic values for these rocks. The idealized rock should have a higher value, perhaps closer to that of the glass spheres.

V_P of Lunar Fines. Compressional sound velocities were measured at various pressures and compactness on the soil, by a technique (*12*) that incorporates the pulse transmission method (*11*). Sample length and density were determined throughout the experiment.

Behavior of V_P, as a function of compaction (density), is shown schematically in Fig. 3. The soil compacts inelastically, along the locus P_1-P_4. If at point 2 pressure is released, the density changes very little, but V_P drops elastically (point 2, to point 3 at P=0).

The line marked "P=0" represents the locus of values of V_P (ρ, P=0) generated by a series of elastic and inelastic pressure cycles. Curves of constant pressure (P > 0) have the general shape shown by the dashed line. For the soil along an elastic curve V_P drops from 2.23 (ρ=2.08, P=430 bars) to 0.67 km/sec (ρ=2.03, P=0).

For P=0, V_P increases with ρ along the zero isobar, and importantly, has a value of 1.1 km/sec at ρ=2.2 comparable to the zero pressure values of the microbreccia rock 10046 (V_P=1.2, ρ= 2.2). From the observed $\Delta P/\Delta\rho$ along the elastic curve, the bulk modulus, K, of the lunar soil is found. Significantly at P=0, the value of K for the soil and 10046 are close (12 kbar and 19 kbar, respectively). The soil and microbreccia are mechanically similar.

The Velocity Profile in the Lunar Surface. Three simple models were calculated from our data giving velocity versus pressure at shallow lunar depth (Table 1). Model 1 assumes a soil-microbreccia mixture under self-compaction. Model 2 assumes a breccia of "igneous" rock 10017 under self-compaction. The variation of K with P as measured for the soil and 10046 (first pressure run) is used. However zero pressure values of V_P, ρ, and K are taken from 10017. This "igneous heterogeneous" model corresponds to a regolith of brecciated and broken up rock blocks.

Model 3 is derived from V_P and β of the rock 10017 and represents the case of a homogenous half space, in which cracks simply close under pressure. All three are dry models, that is, there are no pore fluids.

EDWARD SCHREIBER*
ORSON L. ANDERSON
NAOHIRO SOGA†
NICHOLAS WARREN
CHRISTOPHER SCHOLZ

Lamont-Doherty Geological Observatory of Columbia University, Palisades, New York

References and Notes

1. N. Soga and O. L. Anderson, *J. Geophys. Res.* **72**, 1733 (1967).
2. The artificial lunar glasses were prepared by C. Greene (Alfred University) and given to the authors for these measurements.
3. O. L. Anderson and J. E. Nafe, *J. Geophys. Res.* **70**, 3951 (1965); N. Soga and O. L. Anderson, *Proc. VII International Congress on Glass*, article 37 (Gordon and Breach, New York, 1966).
4. J. A. O'Keefe, *J. Geophys. Res.* **74**, 6795 (1968).
5. W. F. Brace, *ibid.* **70**, 391 (1965).
6. F. Birch, *Geol. Soc. Amer. Mem.* **97**, 97 (1966).
7. J. B. Walsh and W. F. Brace, *Felsmechanik und Ingenieurgeologie* (Springer-Verlag, Vienna, 1966), vol. 4.
8. G. Latham, personal communication.
9. G. Simmons and A. Nur, *Science* **162**, 789 (1968).
10. F. Birch, *J. Geophys. Res.* **65**, 1083 (1960).
11. P. Mattaboni and E. Schreiber, *ibid.* **72**, 5160 (1967).
12. N. Warren, *ibid.* **74**, 713 (1969).
13. Supported by contract NAS-98790. The sequence of the authors was chosen by lottery. Lamont-Doherty Geological Observatory contribution No. 1443.
* Also at Queens College, C.U.N.Y.
† Presently at Department of Space Science, Rice University, Houston.

6 January 1970

Apollo 11 Drive-Tube Core Samples: An Initial Physical Analysis of Lunar Surface Sediment

Abstract. *Two drive-tube core samples were obtained at Tranquillity Base. Fines include much glass, are unweathered, medium gray, loose, nonstructured, very weakly coherent, and demonstrate both accumulation and mixing in a waterless vacuum environment. In contrast to chemical weathering characteristic on the earth, lunar alteration processes are primarily mechanical. We infer that environmental processes of the lunar surface may be expressed as follows:* R (regolith) = f(cl, p, r, t, b, a, . . .), *in which climate* (cl) *is constant and the time* (t)-*dependent processes of bombardment* (b) *and accumulation* (a) *assume significance unparalleled on the earth because of their effects on parent material* (p) *and relief* (r).

Two drive-tube core samples were collected successfully late during Apollo 11 exploration on the surface of the moon by Neil A. Armstrong and Edwin E. Aldrin, 20 July 1969. These cores thus provide the first samples of unconsolidated lunar surficial deposits and their subsurface characteristics.

During sampling, both drive tubes penetrated the first few inches easily, but required hammering to penetrate more deeply (Fig. 1) (*1*). The drive-tube samplers consist of an aluminum outer barrel, or tube, equipped with a disposable steel bit with inward taper at one end and a disposable aluminum handle at the other; the bit was replaced after use with a cap to protect and to retain the collected sample. An extractable aluminum split-tube assembly was housed inside the barrel to receive the sample as the tube was driven, and to permit later removal of the sample in undisturbed condition. This assembly consists of an anodized aluminum tube split lengthwise, surrounded by a Teflon sheath previously heat-shrunk into place to bind halves of the split-tube in position. A Teflon follower and steel spring are driven back inside the tube as the sample enters, and remain in place at the top of the sample to support it during trans-

Fig. 1. Collection of Core Tube No. 1, approximately 3.5 m NNW of LEM, showing position relative to solar wind composition experiment in background. Note slant of tube and use of hammer required by Aldrin to drive it. [NASA photo AS11-40-5964 by Neil A. Armstrong]

port. Maximum possible sample size is determined by inner dimensions of the split aluminum tube, 31.75 cm in length and 1.95 cm in diameter. Both drive tubes were opened in the Biological Preparations Laboratory of the Lunar Receiving Laboratory on 28 July, in a sterile dry nitrogen environment. After preliminary sketching, description, and photography of each core, the cores were split lengthwise by removing half of each core by trowelling. The material, consisting of 22.39 g from Core No. 1, and 26.73 g from Core No. 2, comprised the prime biological sample. Particle size distribution of each half-core removed was obtained by sieving; bulk densities were estimated by calculation; and the remaining core halves were stored.

Core No. 1 (sample 05,0) is 10 cm long (Fig. 2); Core No. 2 (sample 04,0) is 13.5 cm long. Both appeared to be essentially undisturbed in regard to their primary characteristics. Core No. 2 is faintly stratified; the other is not. Both cores include very small particles with highly reflective surfaces, and the matrix of both cores is nearly uniform medium gray (N3 to N4) *(2)* with a low-chroma brownish hue (10YR 3/1 to 4/1) that is most evident under tungsten light. Texturally, the cores appeared to be a silty fine sand or loam with admixed angular rock fragments, glass spherules, and aggregates of glass up to 3 or more millimeters in diameter; average grain size is about 0.11 mm. Core No. 2 included a large subrounded glass aggregate of 1.2 cm maximum diameter (Fig. 3). The core material is loose, weakly cohesive, and consists of single grains except for minor aggregates of glass. When exposed in the opened core-tube liner, the material clearly had bearing strength insufficient to support available soil penetrometers.

When probed with a small spatula, the material disintegrated particle by particle or formed extremely fragile units of subangular blocky shapes (in the case of tiny slump blocks 2 to 4 mm across) or crumb-like shapes (when less than 2 mm in diameter). Because similar ephemeral units could be reproduced experimentally, it is not demonstrable whether these shapes (Fig. 2) were original structures comparable to the peds of terrestrial soils. A single band or layer of slightly higher value and chroma about 0.5 cm thick was observed in Core No. 2, 6.2 to 6.5 cm below its original surface. Upper and lower boundaries of this band were sharply defined, but it was not macroscopically different in texture. No systematic distribution of coarse particles (> 1 mm) was observed.

Data from Surveyor experiments, and observations and photographs made by Aldrin and Armstrong, record a surface veneer of material with high albedo, visibly disturbed by the astronauts' footprints. No trace of light-colored material was found at the top of either core sample (note Fig. 2), although scattered small ellipsoidal patches of fine incoherent material of lighter gray color (N7) and dimensions up to slightly more than 1 mm were found in both cores. These light patches may be fragments of a disrupted layer, now mixed with glass spherules and rock fragments in the fine matrix of the cores.

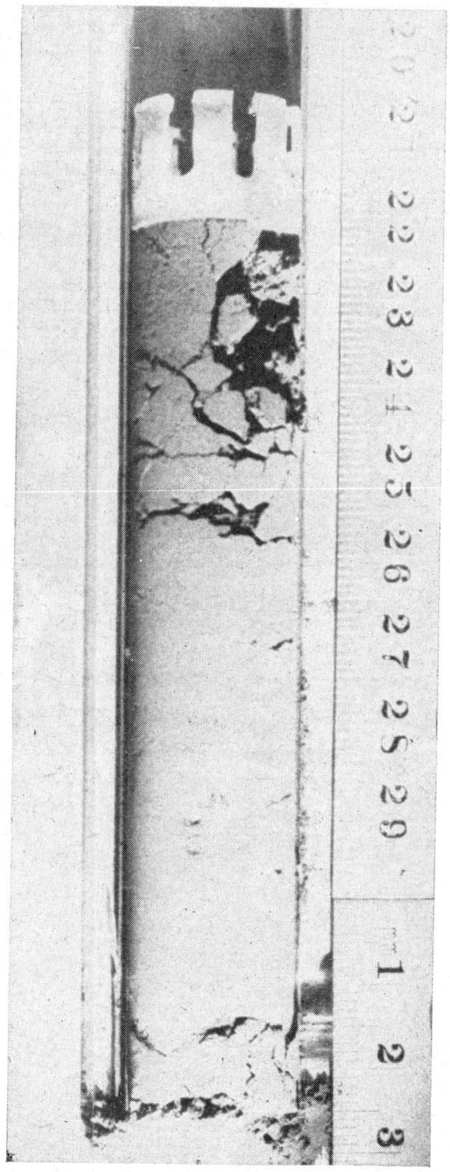

Fig. 2. Contents of Core Tube No. 1 (sample 05,0) before dissection. Note (i) void space near top beneath Teflon follower, possibly resulting from irregular lunar surface, (ii) lack of light-colored surface veneer, and (iii) fracture pattern of core material, showing weak coherence despite otherwise loose consistence and single-grain structure except for weak aggregates of glass. Scale is in centimeters. [NASA photo S-69-45049]

Quantitative measurements of grain size distribution and bulk densities of the core samples were limited in their accuracy by restricted working conditions in the environmental cabinets in which the samples were handled. Bulk densities of 1.66 ± 0.03 g/cm^3 and 1.54 ± 0.03 g/cm^3 (*3*, p. 1219) were calculated by using a nominal diameter of 2.00 cm, whereas blueprint specifications call for an interior diameter of 1.95 cm.

Assuming that the diameter may range from 1.95 to 2.00 cm, the calculated densities would be 1.70 ± 0.04 g/cm^3 and 1.58 ± 0.04 g/cm^3, respectively. More precise calculation is impossible because of undetermined compression by the funnel-shaped bit of the core tubes, and because halves of the split-tube liners of both core tubes had slipped out of register. Whether this occurred during or after sampling is not known, and effective diameter of the core tube during sample collection cannot, therefore, be established.

Similarly, difficulties limit the size range for which sieve analyses made in the nitrogen cabinets (*3*, figure 11, p. 1219) are reliable. The small sample size limits accuracy of any percentage determined for particles coarser than 1 mm, and mechanical difficulties beset sieve measurements in the range below about 65 μ.

Mechanical analyses made outside the LRL cabinets consistently have been different from those made inside them during the quarantine period of Apollo 11. Subsequent analyses show both a larger concentration of fines than was found earlier, and a range of material to diameters far smaller than was first estimated.

Two possibilities for these discrepancies are evident: (i) material sieved inside the cabinets may have failed to pass meshes larger than actual particle diameters because of static electricity; and (ii) weak aggregates of lunar material may continue to break down with continued sieving. The first possibility is probable; the second is supported by much evidence. During microscopic examination, other members of the Lunar Sample Preliminary Examination Team repeatedly observed and measured fine aggregates, especially of glass, including fragments only a few microns or even less than a micron in diameter (*4*). Such fragments not only demonstrate the presence of a very finely comminuted fraction, highly glassy in composition, but also make clear the great difficulty of obtaining reliable sieve analyses on any consistently reproducible basis which does not break down all aggregates. Unfortunately, the actual mechanical properties of lunar surficial deposits are determined by sediment in aggregated rather than disaggregated form.

Microscopically, Apollo-11 fines are about half glass, with plagioclase, clinopyroxene, ilmenite, and olivine. Freshness (except for shock metamorphism), limited devitrification, lack of hydration of even the finest glass fragments, and unoxidized free iron particles and spheres (*3*, p. 1220) indicate a waterless vacuum environment for the samples throughout their history. This lack of chemical weathering is reflected by the almost neutral colors of the cores. The near absence of any effective chemical alteration processes on the lunar surface is further illustrated by the similarity in initial chemical analyses of 23 Apollo-11 samples, including fines (*3*, p. 1221 and table 1, p. 1222), which suggest that the surficial debris are simply fragmented or remelted versions of the three major rock types collected there.

In almost all terrestrial weathering environments, physical and chemical weathering not only proceed concurrently, but generally shift toward dominance of chemical weathering as decrease in particle size results in greater surface-to-volume ratios and thus in greater reactivity per unit volume. Clearly this is of little importance in the case of Apollo-11 fines, which lack clay minerals but instead have anhydrous glass in even the finest size fractions.

The influence of mechanical processes, however, appears to control features of the cores, which demonstrate both accumulation and mixing. Accumulation is recorded by the subtle layering of Core No. 2; mixing, by the variation in composition and morphology of glass, presence of rock fragments of mixed lithology, and the incoherent gray inclusions of both cores. Thus, the lunar regolith at Tranquillity Base is not homogeneous, but varies three-dimensionally as shown by the cores and by photographs of the surface. Obviously, the dominant mechanical process responsible for both mixing and accumulation of regolith debris on the unprotected lunar surface is bombardment. Apollo-11 data unquestionably show evidence of shock metamorphism to be abundant in the fines, and bombardment has resulted in development of craters, minute pitted rock surfaces, shattering and mixing of the regolith, scattering of materials, shock metamorphism, melting, vaporization, condensation, and addition of glass and extralunar material to surficial deposits (*3*, pp. 1226–1227).

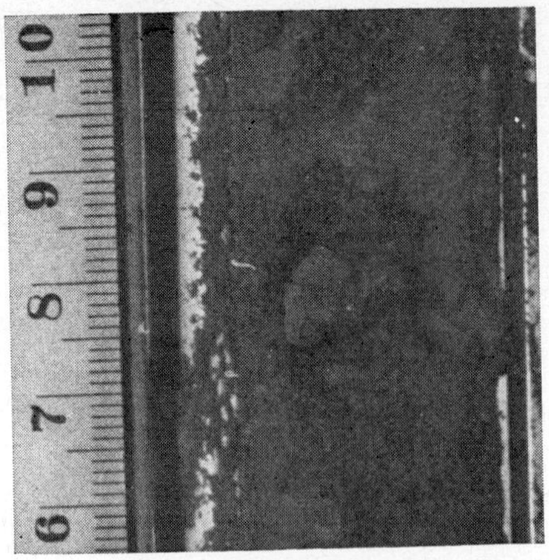

Fig. 3. Large subrounded aggregate of glass, found during dissection of Core Tube No. 2 (sample 04,0). Scale in centimeters indicates approximate depth below lunar surface. Note contrast in size with fine texture of matrix. [NASA photo S-69-45536]

The contrast between lunar and terrestrial surface environments and resulting characteristics of surficial deposits may be expressed through comparison of the Apollo-11 cores with analogous terrestrial materials, especially soil profiles.

Geologically, weathering profiles are a primary criterion for establishing relative chronologies in Quaternary deposits, and—if recognizable—eventually could be of similar use on the moon (*5*). Absence of such profiles is as important as their presence, for macroscopic comparison of morphology of the cores with terrestrial soil profiles permits a qualitative estimate of the relative importance and nature of weathering or alteration processes affecting lunar surficial deposits.

Pedologically, the terrestrial weathering environment has been summarized by Jenny's (*6*) classic formula expressing the relationship between the solum and soil-forming factors acting upon it:

$$S = f\,(cl, o, r, p, t, \ldots)$$

where S = soil and the independent variables recognized are climate (cl), organisms (o), relief (r), parent material (p), and time (t). Absence of any trace of organic life in Apollo-11 samples eliminates organic influence as a factor in the lunar surficial environment. Similarly, the lack of an effective atmosphere to serve as a variable dispersing medium for insolation and moisture reduces climate for any particular spot on the lunar surface to a constant range of diurnal temperature extremes. Thus relief, parent material and time clearly are the factors in the terrestrial equation having effective variability on the lunar surface.

Even lacking the terrestrial influences of climate and organisms, development of alteration profiles or zonation of sediments on the lunar surface is expectable as a result of at least the following agencies: (i) mixing, with other lunar or extralunar materials; (ii) bombardment by radiation and by atomic particles as well as by meteorites; and (iii) gravity movement. Cosmic ray and atomic bombardment may provide alteration profiles not macroscopically evident (none were recognized in these cores), and gravity movement probably is a dependent rather than independent process. But because Apollo-11 cores (and, more recently, those of Apollo 12) (*7*) display layering, mixing, and evidence of bombardment, recognition of these characteristics permits construction of a revised formula to express environmental factors which influence lunar surficial debris:

$$R = f\,(cl, p, r, t, a, b, \ldots)$$

in which R = regolith and the new variables introduced are accumulation (a) and bombardment (b). Because climate (cl) is essentially constant, the time (t)-dependent, mechanical processes of bombardment and accumulation assume significance unparalleled on earth because of their effects on parent material (p) and relief (r). Thus, in contrast to earth, lunar alteration processes are primarily mechanical rather than chemical.

ROALD FRYXELL

Laboratory of Anthropology, Washington State University, Pullman 99163

DANIEL ANDERSON
DAVID CARRIER
WILLIAM GREENWOOD
GRANT HEIKEN

Manned Spacecraft Center, Houston, Texas 77058

References and Notes

1. For original location of the samples at Tranquillity Base, see Fig. 3-16 of E. M. Shoemaker *et al.*, in *Apollo 11 Preliminary Science Report*, NASA, SP-214 (1969), p. 52.
2. Color notations follow the system of *Munsell Soil Color Charts* (Munsell Color Co., Baltimore, 1954).
3. The Lunar Sample Preliminary Examination Team, *Science* **165**, 1211 (1969).
4. E. C. T. Chao, C. Frondel, R. E. Wilcox, personal communication (1969).
5. Geology Group Report, in *NASA 1965 Summer Conference on Lunar Exploration and Science*, NASA SP-88 (1965), pp. 95–98.
6. H. Jenny, *Factors of Soil Formation* (McGraw-Hill, New York, 1941), pp. 1–19, especially p. 16.
7. The Lunar Sample Preliminary Examination Team, *Science*, in press.
8. We thank T. H. Foss, C. Frondel, J. Lindsay, and R. E. Wilcox for helpful discussions or for critical review of the manuscript.

4 January 1970

Spectral Reflectivity of Lunar Samples

Abstract. *Twelve rock chips and two samples of fines all have electronic absorption bands in diffuse reflected light between 0.32 and 2.5 micrometers. Major bands occur between 0.94 and 1.00 micrometer and at 2.0 micrometers, and arise from* Fe^{2+} *in clinopyroxene and to a lesser extent in olivine. A band at 0.95 micrometer and other details of curve slope and shape for the lunar surface fines match McCord's telescopic curve for an 18-kilometer area that includes the Apollo-11 site. Results confirm mineralogical predictions based on telescopic data and support the feasibility of obtaining mineralogical information by remote reflectivity measurements.*

Our investigation had the following objectives: (i) To search for electronic or other absorption bands in the visible and near-infrared portions of the spectrum and to relate any bands to the mineralogy of the samples; (ii) to investigate parameters other than mineralogy that might affect the frequencies or depths of bands; and (iii) to relate laboratory spectra and sample mineralogy to earth-based telescopic spectra of the moon and to evaluate the feasibility of obtaining mineralogical information by remote reflectivity measurements.

The basis for interpretation of absorption bands in silicates between 0.3 and 2.5 μm was developed through the application of crystal field theory to mineralogy (*1*). Transmission spectra of single oriented crystals, with polarized light, have led to further refinements in band assignments (*2*). Electronic bands are produced by transition elements, notably iron, in various valence and coordination states. Band frequencies are sensitive to distortion of *d*-orbital shells of transition metals by neighboring anions. Because metal-oxygen distances differ for most minerals, the absorption band frequencies can be used for identification of minerals that exhibit bands.

Absorption bands also appear in diffuse reflectance spectra of minerals and their powders. The feasibility of making mineral identifications based on reflectivity curves of minerals and rocks has been discussed (*3*). Interpretations of bands in lunar and planetary reflectivity curves have been made on the basis of the above arguments (*3, 4*). The reflectivity measurements of the lunar samples provide the first opportunity to test the validity of mineralogical interpretations that were based on reported absorption bands in the telescopic curves. Major uncertainties have been whether the lunar surface has been altered in an unknown way and whether the lunar reflectivity curves can be correctly interpreted in terms of terrestrial mineral and rock curves.

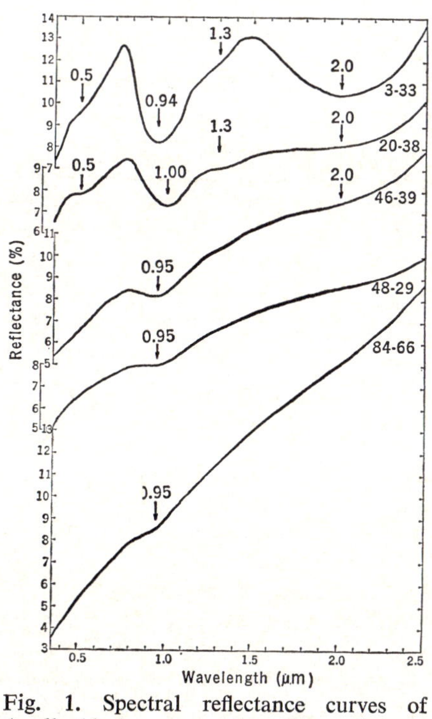

Fig. 1. Spectral reflectance curves of Apollo-11 lunar samples. The reference is MgO. Samples 3-33 and 20-38 are crystalline rocks; 46-39 and 48-29 are breccias; 84-66 is surface fines.

We made reflectivity measurements of type A, crystalline rock containing olivine; type B, crystalline rock without olivine; type C, breccia (two samples); and type D, surface fines (two samples) (*5*). Each sample of rock and of breccia consisted of three chips taken from the top, interior, and bottom parts. One sample of fines was packaged in a vacuum container. All other samples were kept in an N_2 atmosphere except for a small fraction of fines that was deliberately exposed to air.

Two spectroreflectometers were used, a Cary-14 RI and a Beckman DK-2A. The instruments and the measurement procedures will be discussed elsewhere (*6*). Both instruments employ MgO-lined integrating spheres. Freshly smoked MgO and sandblasted gold were used as reflectivity standards.

Broad absorption bands occur in the diffuse reflectance spectra of all of the lunar samples investigated. There are significant differences in band frequencies and band depths among the samples. All observed bands are attributed to electronic transitions in iron and titanium. No vibrational bands were observed. Conspicuously absent are the OH^- or H_2O bands at 1.4 and 1.9 μm that occur in many terrestrial minerals and rocks.

Sample 3-33 illustrates the simplest type of curve (Fig. 1) in which the two main bands, one at 0.94 μm and the other at 2.0 μm, are contributed by Fe^{2+} in the clinopyroxene. The curve is generally similar to published polarized absorption spectra and diffuse reflectance spectra of orthopyroxenes (*2*). However, the band positions and the detailed shape of the curve are virtually identical with unpublished curves (*7*) of pigeonite concentrates from some basaltic achondrites, Pasamonte, for example.

The Pasamonte sample and other achondrites have the two absorption bands shifted to slightly longer wavelengths from the orthopyroxenes. Adams (*8*) found that there is a progressive shift to longer wavelengths of the two orthopyroxene bands as composition changes toward pigeonite and diopside. The lunar sample bands at 0.94 and 2.0 μm are here assigned to Fe^{2+} in sixfold coordination, with the splitting resulting from the distorted octahedral site, as occurs in orthopyroxenes. The shift of the bands to slightly longer wavelengths relative to orthopyroxene is ascribed to minor Ca^{2+} in the pyroxene lattice, which requires an adjustment of Fe-O bond distances on the octahedral sites.

The reflectivity curve for sample 3-33 also shows a faint (<5 percent) band at 0.5 μm and a flattening of the curve at 1.3 μm. These structures occur at frequencies that correspond to titanium in the pyroxene. The shallowness of the bands suggests that TiO_2 comprises less than 1 percent of the pyroxene. The curve shows no evidence for chromium.

Sample 3-33 is a useful reference for the interpretation of bands in other lunar materials because it contains essentially no olivine or glass. The other major minerals present, plagioclase and ilmenite, do not contribute any measurable band structure to the rock curve.

Sample 20-38 differs mineralogically from sample 3-33 in that the pyroxene appears to be richer in Ca and minor olivine is present. Olivine alone has a single strong absorption band at 1.03 μm that arises from Fe^{2+} in sixfold coordination. In the curve for sample 20-38 (Fig. 1) the band at 1.00 μm is attributed to a combination band representing the unresolved pyroxene and olivine bands. The band at 2.0 μm is assigned to the pyroxene, because olivine has no selective absorption in this wavelength region. The clinopyroxene alone may have a band between 0.94 and 1.00 μm due to the increased Ca; however, it has not been possible to verify this band frequency without access to a separate pyroxene concentrate. A more calcic pyroxene would also explain the weaker band at 2.0 μm in this sample. The curves for sample 20-38 also show an enhancement of the titanium bands at 0.5 and 1.3 μm, implying a higher TiO_2 content in the pyroxene.

The samples of fines and breccias also contain pyroxene and traces of olivine, but less olivine than sample 20-38. The fines and breccias show a weak (<5 percent) band at approximately 0.95 μm and only a faint suggestion of a dip near 2 μm (Fig. 1). The band at 0.95 μm is consistent with the observed proportions of pyroxene and olivine and does not require the presence of other minerals. The virtual disappearance of the 2.0-μm band in the fines and breccias is related to the overall darkness of these materials and to a smaller pyroxene content.

The fines and breccias are darker, have shallower absorption bands, and have steeper overall curves than the crystalline rocks. We attribute these features to the presence of the dark red-brown glass. Although the glass selectively absorbs shorter wavelengths, it does not contribute measurable band structure in the lunar materials. We suggest that the pigmentation in the glass is iron and titanium that was derived largely from ilmenite through melting, probably by impact, of the crystalline rock. The decrease in reflectivity from crystalline rock to glass-rich rock or fines can be explained by more efficient use of the available metal ions as light absorbers. In the oxide form, ferrous iron absorbs nearly all incident light in the outer 5 μm of a mineral grain; hence the interior portion of an ilmenite grain does not further attenuate light. If, however, the ilmenite grain is dissolved in a silicate melt, there is greater opportunity for light absorption by each iron or titanium ion.

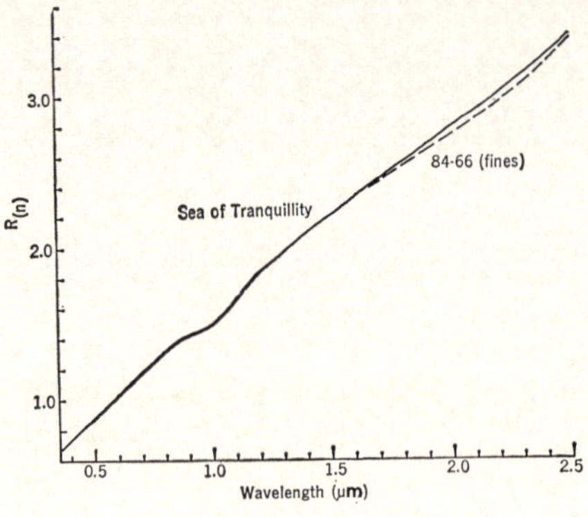

Fig. 2. Comparison of McCord's telescopic curve for an 18-km area that includes the Apollo-11 site with the curve for sample 84-66, surface fines. $R_{(n)}$, normalized reflectance.

Microbrecciation of mineral grains (presumably by impact shock) is observed on the surfaces of many of the Apollo-11 rocks. Microfractured silicate minerals reflect more light than undisturbed grains. Each fracture is an optical discontinuity with the potential of reflecting incident light. The effects of microbrecciation are closely analogous to the increase in reflectivity that results from grinding silicate minerals to smaller particle sizes (*9*). Comparison of reflectivity measurements of outside and of interior surfaces of the lunar samples shows that, although microbrecciated surfaces have up to a 30 percent higher reflectivity, there is no change in the frequency of any of the absorption bands. This observation is in agreement with theory, because the band frequencies are determined by Fe-O bond distances which in turn are little affected by mechanical deformation on a scale larger than the unit cell.

More extreme shock leads to vitrification of mineral grains, and it is expected that absorption bands will shift in frequency when new metal-oxygen bonds are formed. In the samples observed glass-lined pits covered too small an area and the glass was too dark to yield a definitive measurement of any new absorption bands. We observed that absorption-band frequencies are unaffected by the presence of glass-lined pits or microbrecciated zones on the surfaces of the Apollo-11 rocks.

We have found no evidence for metal coatings or other deposits (other than glass) on the surfaces of mineral grains. If such coatings are present, they are not thick enough to cause a measurable attenuation of light and can be disregarded for purposes of interpreting mineral absorption bands.

Reflectivity measurements also were made of surface fines in vacuum, in N_2, and in air to test for possible changes arising from removal of the samples from the lunar environment. No differences in reflectivity were observed between 87-13 in vacuum and 84-66 which had been exposed to dry nitrogen. In addition, 0.5 g of 84-66 was exposed to air for 3 hours at 25°C. No departures from the previous reflectivity measurements were noted.

We have compared the laboratory reflectivity curves of the Apollo-11 rocks with earth-based telescopic measurements of the landing site (*10*). The reflectivity curve for an 18-km area that includes the Apollo-11 site agrees very closely with the laboratory curve for the bulk surface fines (Fig. 2). We conclude that: (i) Sample 84-66 (fines) appears to be representative of the main lunar surface material for at least tens of kilometers around the Apollo-11 site. (ii) Crystalline rocks are not abundant enough at and around this site to impose the pyroxene double-band structures on the telescopic reflectivity curves. (iii) The single weak band at 0.95 μm in the telescopic curve is the clinopyroxene-olivine combination band. The telescopic curves, therefore, are recording mineralogical information. Earlier predictions (*3*) of mineralogy based on a reported band near 1 μm for the Sea of Tranquillity are generally substantiated. (iv) The low albedo of the Sea of Tranquillity, the steepness of the reflectivity curve, and the weakness of the absorption-band structure can now be accounted for by the presence of glass rich in iron and titanium. (v) Interpretation of telescopic curves for other parts of the moon can be expected to yield information on areal differences in mineralogy and in glass content.

JOHN B. ADAMS
Caribbean Research Institute, College of the Virgin Islands, St. Croix 00820

ROBERT L. JONES
Manned Spacecraft Center, Houston, Texas 77058

References and Notes

1. W. B. White and K. L. Keester, *Amer. Mineralogist* **51**, 774 (1966); R. G. Burns and W. S. Fyfe, in *Researches in Geochemistry*, P. H. Abelson, Ed. (Wiley, New York, 1967), vol. 2, p. 259.
2. G. M. Bancroft and R. G. Burns, *Amer. Mineralogist* **52**, 1278 (1967); W. B. White and K. L. Keester, *ibid.*, p. 1508.
3. J. B. Adams, *Science* **159**, 1453 (1968).
4. T. B. McCord and J. B. Adams, *ibid.* **163**, 1058 (1969); J. B. Adams and T. B. McCord, *J. Geophys. Res.* **74**, 4851 (1969); D. P. Cruikshank, *Science* **166**, 215 (1969).
5. Lunar Sample Preliminary Examination Team, *Science* **165**, 1211 (1969).
6. J. B. Adams and R. L. Jones, *Geochim. Cosmochim. Acta*, in press.
7. J. B. Adams and L. T. Silver, in preparation.
8. J. B. Adams, in preparation.
9. J. B. Adams and A. L. Filice, *J. Geophys. Res.* **72**, 5705 (1967).
10. See also T. B. McCord, T. V. Johnson, H. H. Kieffer, *ibid.* **74**, 4385 (1969).
11. Supported by NASA (NAS 9-9578). We thank J. Larsen for his assistance, and Dr. T. McCord for new telescopic data in advance of publication.

5 January 1970

Apollo 11 Soil Mechanics Investigation

Abstract. *The fine-grained surface material at the Apollo 11 landing site is a brownish, medium-gray, slightly cohesive granular soil, with bulky grains in the silt-to-fine-sand range, having a specific gravity of 3.1 and exhibiting adhesive characteristics. Within the upper few centimeters, the lunar soil has an average density of about 1.6 grams per cubic centimeter and is similar in appearance and behavior to the soils studied at the Surveyor equatorial landing sites. Although considerably different in composition and in range of particle shapes, it is similar in its mechanical behavior to terrestrial soils of the same grain size distribution.*

The upper few centimeters of surface material in the vicinity of Tranquillity Base consist of a brownish, medium-gray, slightly cohesive granular soil largely composed of bulky grains in the silt-to-fine sand size range. Angular-to-subrounded rock fragments ranging up to 1 m in size are distributed throughout the area. Some of these fragments are lying on the surface, some are partly buried, and others are barely exposed.

Fig. 1. Penetration of loosely placed lunar soil sample at the nitrogen cabinets of the LRL Biological Preparation Laboratory. [LRL photo S-69-47484]

The soil grains vary in shape from blocky angular with smooth plane surfaces to completely spherical; some of the larger particles are vesicular. No shards, needles, or filaments have been observed. A large portion of the soil grains consists primarily of glasses, with chemical composition covering a wide range. The color of these particles varies from clear, to dark reddish-brown, to dark gray.

The grain size distribution of fine-grained material collected with the documented sample, the core tubes, and the bulk sample was found, in all cases, to be that of a silty fine sand. However, aggregation of individual particles may have biased the analyses toward the large size range.

The following observations and measurements were made (*1, 2*) on lunar soil samples placed in the nitrogen cabinets of the Lunar Receiving Laboratory's Biological Preparation Laboratory at room temperature and at normal atmospheric pressure.

1) Visual examination of core-tube soil samples 10004 and 10005 revealed that the soil was remarkably uniform in color and texture; fine reflecting surfaces over about 10 percent of the area produced a sparkling appearance. There was no variation in structure with length along the samples, although there was a very slight color difference between the upper and lower halves of sample 10004; however, probing indicated no discernible differences in mechanical properties. Sample 10005 contained numerous small cracks and voids. Its average bulk density, not taking into account the voids and cracks, was about 1.66 g/cm^3, as compared with 1.54 g/cm^3 for sample 10004.

2) The specific gravity of the soil in the core tubes was found to be 3.1, as measured by a gas comparison pycnometer. This is considerably higher than the typical value of 2.7 for terrestrial soils and may be attributable to the fact that the lunar soil is composed mainly of the basic igneous minerals (for example, plagioclase, olivine, and pyroxene), as well as relatively large amounts of titanium and iron oxides.

3) On the basis of the specific gravity and bulk-density measurements, the void ratios of the core-tube samples were found to be 1.01 and 0.87. The respective porosities are 50.1 and 46.5 percent. It should be noted that the core bit of both core samplers was flared inward at 15°, the reverse of the direction for most terrestrial samplers. Thus, the soil was probably deformed considerably during sampling, and the measured bulk densities and porosities may not necessarily be indicative of the bulk density and porosity of the undisturbed lunar soil.

4) The bulk density of bulk sample material finer than 1 mm, placed as loosely as possible in a container, was determined to be 1.36 g/cm^3, corresponding to a void ratio of 1.28 and a porosity of 56 percent. Under the 1/6-*g* lunar gravity, the adhesive forces between the particles would probably

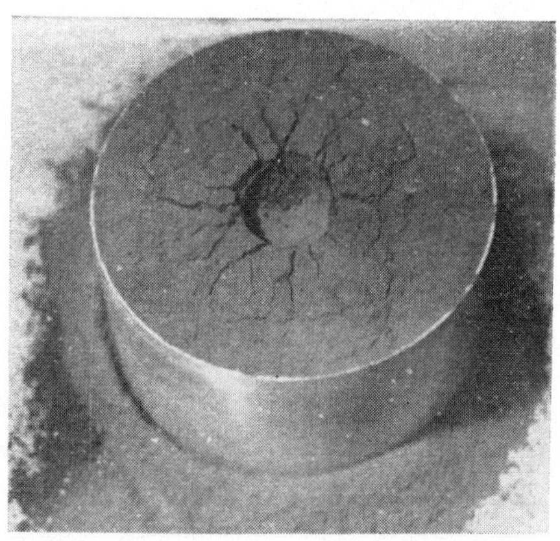

Fig. 2. Penetration of compacted lunar soil sample at the nitrogen cabinets of the LRL Biological Preparation Laboratory. [LRL photo S-69-47489]

Table 1. Results of lunar soil bulk sample penetration test.

Test	Density (g/cm³)	Force (lb)	Area (in.²)	Pressure (lb/in.²)	Penetration (in.)	Pressure/penetration ratio (lb in.⁻² in.⁻¹)
1	1.36	< .41*	0.049	< 8.3	0.25	
2	1.36	< .41*	.049	< 8.3	.77	
3	1.36	< .41*	.049	< 8.3	.77	
4	1.36	< .41*	.049	< 8.3	.77	
5	1.36	.69	.416	1.65	.79	2.1
6	1.77	.41	.049	8.3	.32	25.9
7	1.77	1.22	.049	24.8	.67	37.0
8	1.77	< .41*	.049	< 8.3	.25	
9	1.77	2.20	.049	44.7	1.00	44.7
10	1.77	1.30	.049	26.5	.83	31.9
11	1.77	8.70	.416	20.7	.67	30.9
12a	1.80	6.50	.416	15.6	.26†	60.9
12b	1.80	17.95	.416	43.1	.77†	84.5

* Penetrometer did not meet with sufficient resistance to compress the spring. Tabulated force is weight of penetrometer. † Penetrometer was removed after achieving pressure of 15.6 lb/in.² at a penetration of 0.26 inch; it was applied again at the same place and pushed into the soil until the applied unit load reached the value 43.1 lb/in.² at a penetration of 0.77 inch from the original surface of the sample.

permit a looser soil structure to develop. In a second test series, the soil was compacted in several layers by rodding, tamping, and compressing to a maximum bulk density of 1.80 g/cm³, corresponding to a void ratio of 0.72 and a porosity of 41.8 percent. In each compaction state the resistance to penetration of the soil sample was measured by means of a small, spring-loaded penetrometer. Results of these tests are shown in Table 1. In the loose state the soil failed in a punching mode, as shown in Fig. 1; in the dense state the failure mode, shown in Fig. 2, indicated classical incompressible shear failure.

The physical characteristics and mechanical behavior of the lunar soil as deduced from extravehicular activity data can be summarized as follows (*1*, *2*).

1) The loose, powdery, surficial fine-grained material ranging up to fine-sand size tended to adhere to any object with which it came into contact. Fine, powdery material adhered to lunar rock samples brought back to earth, and left a trace of fine dust coating the core tubes that were returned to the Lunar Receiving Laboratory. This adhesion, however, was not of sufficient magnitude to offer any resistance to pulling of staffs inserted into the lunar surface.

2) The lunar surface is relatively soft to depths ranging between 5 and 20 cm. It can be easily scooped, offers low resistance to penetration, and provides low lateral support for staffs, poles, or core tubes. Beneath this relatively soft surface, the penetration resistance of the material increases considerably.

Results of tests made at the LRL to study penetration resistance of the lunar soil (Table 1) indicate that, within the bulk-density range of 1.36 to 1.80 g/cm³, the resistance to penetration increases by a factor of 20. Therefore, if the lunar soil at the Apollo 11 landing site is compacted to a density somewhat greater than the minimum, it should offer sufficient static resistance per inch of penetration, estimated at 3 lb/in.² per inch (*2*, *3*), to account for the behavior observed by the astronauts when they pushed various tools into the surface, without postulation of the presence of an overconsolidated layer, a cemented layer, or bedrock. It does not follow, however, that no such layer existed at any or all of the locations where various penetrators were inserted into the lunar surface during extravehicular activity.

3) Most of the astronauts' footprints caused compression of the lunar surface soil, although in some instances, bulging and cracking of soil adjacent to a footprint occurred. The latter observation indicates shearing rather than

Fig. 3 (above). Typical astronaut footprints on relatively level lunar surface in the vicinity of the flagpole. [NASA AS 11-40-5874] Fig. 4 (right). Typical astronaut footprints at the top edge of a soft-rimmed crater. Combined weight of astronaut and Early Apollo Scientific Experiments Package produced a unit bearing pressure of 1.4 lb/in.² on one boot area in the lunar gravity field. [NASA AS 11-40-5946]

compressional deformation of the soil.

4) The soil possesses a small but finite amount of cohesion, evidenced by the following observations: (i) fine grains stick together, and sometimes soil clumps (about 10 percent of the blocky material protruding above the surface) could not be distinguished from rock fragments; (ii) natural clods of fine-grained soil crumbled under the astronauts' boots, indicating some cementation between the grains, although, in LRL tests, soil grains were found to cohere again to some extent after being separated; (iii) initially loose and fluffy material readily compacted under load and retained the detail of a deformed shape (Fig. 3), thus enhancing the astronauts' mobility on previously traversed surfaces; (iv) the material could stand unsupported on vertical slopes at least a few inches high; (v) during the bulk-sample collection, it was observed that, as the scoop cut through the lunar soil, the remaining material left a sharp solid edge, but the material that went into the scoop crumbled with no evidence of particle aggregation; (vi) the holes made by the core tubes appeared to remain intact upon the removal of the tubes; (vii) there was no tendency for the material collected in the core tubes to pour out, and the material was similar in appearance to a terrestrial moist soil; and (viii) during the terminal stages of the lunar module landing, the height at which soil erosion caused by exhaust from the spacecraft engine first became noticeable was approximately 100 feet, indicating that even the loose, powdery, top surficial layer of lunar soil possesses some cohesion, although possibly of lower magnitude than that exhibited by the underlying material, because cohesionless soil of the same grain-size range would be moved at a much higher elevation.

5) Confinement of the loose surface material leads to a significant increase in resistance to deformation, which is characteristic of soils deriving a large portion of their strength from interparticle friction. The relevant material properties can be assessed from the following observations. (i) Available information indicates that the lunar module landing was achieved under essentially static conditions as far as the landing gear was concerned and the relatively small penetrations (1 to 3 inches) of the lunar module footpad correspond to static bearing pressures exerted by the footpads on the lunar surface in the range of 2.1 to 0.8 lb/in^2. (ii) The relatively small depth of typical astronaut footprints, shown in Fig. 3, was about ½ inch when the static bearing pressure exerted by the astronaut weight on one boot was about 1 lb/in^2. (iii) Soft spots encountered during the extravehicular activity were generally located at the rims of small, fresh craters which consisted of loose, very-fine-grained material with essentially no large rock fragments. Close to the rim and especially on the upper edge of inside slopes, where the material was loose and relatively unconfined, the astronauts sank as much as 6 to 8 inches, as may be seen in Fig. 4.

In all of the above cases, calculations (*2*) based on the Terzaghi equation for ultimate bearing capacity (*4*), with appropriate adjustments to account for circular contact areas, as in the case of the lunar module footpads, and for sloping ground (*5*), as in the case of the soft-rimmed craters, indicate the following.

1) For a Surveyor soil model (*6*) with a density of 1.5 g/cm^3, a cohesion ranging between 0.05 and 0.1 $lb/in.^2$, and a friction angle of 35°, the assumed soil properties account reasonably well for the observed behavior.

2) Because of the small lunar gravity, the relatively small critical dimension of loaded areas, the shallow subsurface depth at which the loads were applied, and the fact that the lunar soil is predominantly frictional, it appears that the cohesion of the lunar soil, though slight, is a key characteristic contributing to the bility of the soil to support bearing loads.

Various observations and measurements at the LRL indicate that the cohesion of the lunar soil is not affected by short-term exposure to nitrogen atmosphere. (i) The core-tube samples retained their cylindrical shape upon removal of the top halves of the inner split-tube liners, and disturbance revealed that cohesion was small but sufficient to hold small clumps of fine particles together. (ii) Numerous cracks and voids, developed in the core samples as a result of disturbance, were retained. (iii) During sieving, very fine particles tended to form, break, and reform into lumps when shaken, as though the soil were slightly damp. (iv) When material finer than 1 mm was being scooped from the bulk sample container, the scoop cut clean, smooth vertical walls approximately 3 inches deep, in the soil mass, similar to the trenches dug in the lunar surface by the surface samplers of Surveyors III and VII (*6*). (v) Using bearing-capacity analysis and penetration-resistance test data from the test performed on the maximum-density soil (Table 1; Fig. 2), the cohesion of the lunar soil sample was estimated to vary between 0.20 and 0.05 $lb/in.^2$, values corresponding to angles of internal friction of the soil ranging between 35° and 45°. These strength parameters are in the range deduced from Surveyor data analyses (*6*) and are in reasonable agreement with results from shear strength tests performed on simulated lunar soils having the same grain-size distribution and compacted to the same void ratio as the lunar soil sample (*3, 7*).

The nature of the cohesion of the lunar soil in place requires more detailed investigation. Although the various observations and measurements suggest that it may not be affected by short-term exposure of the soil in nitrogen atmosphere, it has appeared to degrade after a few weeks of storage. Information on the actual pressure and composition of the lunar atmosphere, or on any outgassing from the interior of the moon occurring at the lunar surface, may shed new light on the mechanisms and processes governing this important mechanical property of lunar soil.

N. C. Costes
G. C. Marshall Space Flight Center, Huntsville, Alabama 35812
W. D. Carrier
Manned Spacecraft Center, Houston, Texas
J. K. Mitchell
University of California, Berkeley
R. F. Scott
California Institute of Technology, Pasadena

References and Notes

1. Lunar Sample Preliminary Examination Team, *Science* 165, 1211 (1969).
2. N. C. Costes, W. D. Carrier, J. K. Mitchell, R. F. Scott, *NASA (Nat. Aeronaut. Space Admin.) SP-214* (1969), pp. 85.
3. J. K. Mitchell, R. E. Goodman, W. N. Houston, P. A. Witherspoon, *Lunar Surface Engineering Properties Experiment Definition*, final report on NASA contract NAS8-21432 with G. C. Marshall Space Flight Center, Univ. of California, Berkeley (1969).
4. R. F. Scott, *Principles of Soil Mechanics* (Addison-Wesley, Reading, Mass., 1963).
5. G. G. Meyerhoff, *Geotechnique* 2, No. 2, 301 (1951).
6. R. F. Scott and F. I. Roberson, *Jet Propulsion Lab. Calif. Inst. Technol. Tech. Rep. TR 32-1264* (1968), pp. 135-185.
7. N. C. Costes *et al.*, "Lunar Soil Simulation Studies in Support of Apollo 11 Mission" (NASA technical report), in preparation; R. F. Scott and T. D. Lu, "Surveyor Surface Sampler Post Mission Simulation Studies" (California Institute of Technology technical report), in preparation.

7 January 1970

Interferometric Examination of Small Glassy Spherules and Related Objects in a 5-Gram Lunar Dust Sample

Abstract. *Over two hundred spherules and cylinders were extracted from the lunar dust sample. Sizes ranged from 0.75 to 0.03 millimeters, and most were shiny glassy objects, which were studied by interferometry. This study reveals very high specular reflection, frequent perfect sphericity, and clear evidence in some objects of microcracking and microchipping. Many spheres were once projectiles. Some have impacted in free flight with much smaller pieces of rocky material, which embedded in the surface. It is conjectured that the glassy spherules originated as a gas-blown shower from a pool of molten glass.*

This report is devoted to an examination of shapes and evidence for shock in glassy spherules in a presieved (1-mm sieve) lunar sample.

The sieved sample was sorted under the microscope. Over two hundred objects were found, some of which are shown in Fig. 1. They may be divided into the following groups: (i) specular, slightly transparent, red-brown glassy spherules (Fig. 1a), (ii) somewhat larger grayish spherules, which have a distinctly metallic luster but which are opaque, (iii) roughly spherical whitish nonglassy spherules, (iv) three faintly green and one deep blue, very clear, glassy sphere, (v) numerous specular, slightly transparent cylinders of red-brown glass (Fig. 1b).

The objects are so small that their total weight is little more than 0.01 percent of the whole sample, yet, if the sample is representative, this number, extracted from a mere 5 g, implies no less than 40,000 such objects per kilogram of moon dust!

The specular objects were studied by interferometry, using magnifications up to $\times$ 700 with a 4-mm microscope objective (λ 5460). All the numerous spherical shiny, glassy objects give good interference fringe patterns and, over most of their whole surfaces, show very regular "Newton's rings." The good fringe visibility and the circular perfection show that the surfaces are nearly spherical and are also highly specular, with refractive indices exceeding 1.6. A typical circular fringe pattern from a small spherule is shown in Fig. 1c. So nearly spherical are many of the objects that they must have been formed free from restraints, perhaps blown from a melt as fine droplets or perhaps as a spray of molten glass; thus they were able to solidify in free flight under influence of surface tension forces.

A number of the spherules, mainly of the gray metallic-luster type, have a tail-like attachment that is able to resist scraping action (Fig. 1d). The material in the tail is sintered gray dust particles. Its appearance strongly suggests that tails have been formed by impacting sintering action, as if the spherule had landed as a projectile. Except for the tail region, such particles are spherical. A few objects have several tails and give an impression of having rolled, while hot, in moon dust.

Many of the spheres have clearly been shocked. Some originally spherical objects have pieces broken off, and some show the existence of microcracks (Fig. 1e) or microchips (Fig. 1f). Several have on their surfaces collision impact craters. Tails are also evidence of violent impact.

A number of the larger spheres have been struck violently by much smaller rocky pieces of matter (Fig. 1g). In each case a small foreign object is embedded in the surface. Usually it has chipped away a circular fragment of the spherule. In several cases the incoming object has raised a circular crater. Most of the foreign embedded objects are reddish brown, but they look crystalline and give the impression of coming from the same parent material. The impact craters are found mainly on the grayish spherules with a metallic luster but are occasionally found on the red-brown glassy spherules.

The interferograms also show that

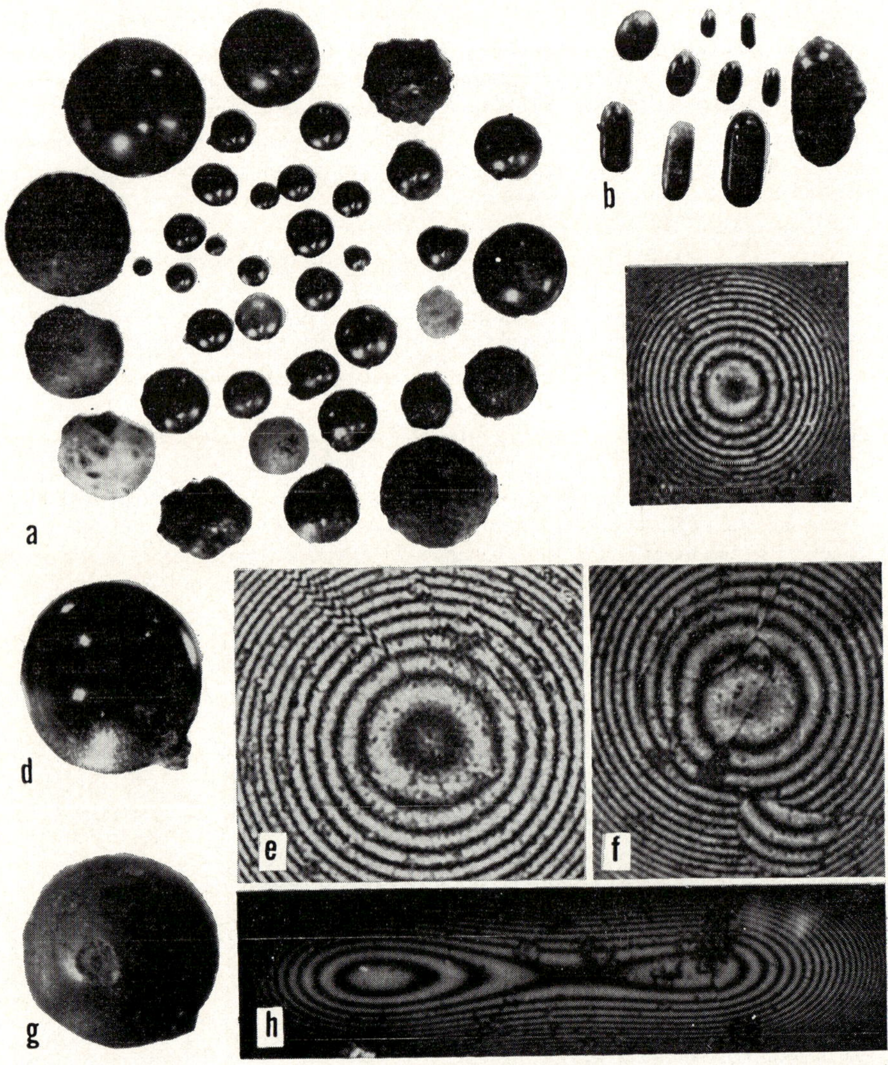

Fig. 1. Photomicrographs of objects found in a sieved sample of lunar dust. (a) Spherules; (b) cylinders; (c) typical fringe pattern in spherical shiny, glassy objects; (d) typical tail-like attachment on a spherule; (e) microcracks in a broken spherical object; (f) microchips in a broken spherical object; (g) a sphere with a small foreign object embedded in its surface; (h) fringe pattern in a typical cylinder.

there are numerous, and even smaller, depressions. It is evident that crater formation and tail production are due to two quite separate impact mechanisms. The crater appears due to impact in flight with a particle, the tail to impact from landing on the dust-covered lunar surface.

A number of red-brown specular cylinders were found, some more transparent than others but all with hemispherical ends. From fringe patterns (Fig. 1h), it is safe to conjecture that the cylindrical object in its initial molten state was part of a breakup of a thin jet and was subdividing into two droplets but solidified before it divided. One long cylinder gives a fringe pattern that indicates onset of breakup into three droplets.

A number of the grayish metallic-like spherules exhibit vacuole regions within their otherwise solid interiors, and in each case this region has created a small opening in the surface. The interiors of the vacuoles are highly specular and spherical in shape. It is likely that small gaseous or liquid inclusions have in each case caused a blowout. One sample has a 0.5-cm radius of curvature.

The spherules bear no resemblance to tektites. However, this study was restricted to a fine mesh sample, and a comparison is not valid.

Two possible sources of origin for the lunar spherules may be considered. (i) If large lunar craters are due to meteoric impact, such impact could partly remelt the rock struck and could scatter droplets over a wide area. Such a mechanism should produce many small spheres and, also, many larger spheres, perhaps of the size of tektites (australites). (ii) Let us postulate that the inner floors of some large craters have at one time experienced volcanic reheating to create in effect a large pool of molten material. Further, let us suppose that this volcanic reheating is followed by violent (either explosive or prolonged) gas blowout from below. Remembering the low gravity, such a hot gas blowout could create an enormous fountain of fine molten droplets. It would also produce filaments of fine jets or threads, which would break up into cylinders. Furthermore, this same explosive mechanism could simultaneously create considerable neighboring microshatter and throw up a dense cloud of microparticles from the surrounding solid regions through which the glassy spherules could pass. These conditions would favor the production of microimpacts. Of course, a massive meteoric impact could also produce minute droplets, as well as a cloud of solid particles. If extralunar micrometeorites created the microcraters on the spherules, then a high concentration will need to be postulated for collision to occur in free flight. It might be argued that the glassy spherules were at some time in orbit around the moon for periods long enough to create the probability of collision of a tiny object by other micrometeorites. It is not known why the impacted particles resemble in color and appearance the kind of material from which so many spherules appear to have originated. At present, both meteoric impact or volcanic blowout appear to be equally plausible as causes of origin.

S. TOLANSKY

Royal Holloway College, Egham, Surrey, England

4 January 1970

Surface Properties of Lunar Samples

Abstract. *Fine-grained samples disrupted after exposure to oxygen and oxygen with 3.5 percent water above 2 torr. Chemical etching revealed plastic deformation in some samples, adhesion due to impact melting in others, dislocations in crystalline phases and evidence that some glasses were partially devitrified. Specimens of rock that were fractured in ultrahigh vacuum exhibited a time-dependent adhesion and a network of localized electrostatically charged areas.*

We studied processes of agglomeration and disruption of lunar material in order to determine those processes primarily responsible for the present state of the lunar environment. We investigated the effects of long exposure of surfaces of lunar material to the lunar environment. The approach used was (i) to study the chemical effects of exposure of nearly pristine lunar material to gases, (ii) to etch mounted and polished specimens chemically to study the defect structure and microchemical composition distribution at particle interfaces, and (iii) to measure adhesion forces in ultrahigh vacuum (UHV) between fractured and cleaved rocks (or both), studying electrostatic phenomena which could give long-range or short-range adhesional effects.

Exposure of lunar material to gases should start with material kept as close to lunar conditions as possible. Unfortunately, the best specimen type, the UHV sample, was not returned on Apollo 11. An attempt was thus made to use the high vacuum sample, but this could not be done because there was a leak in the contingency shipping container vacuum seal (that from

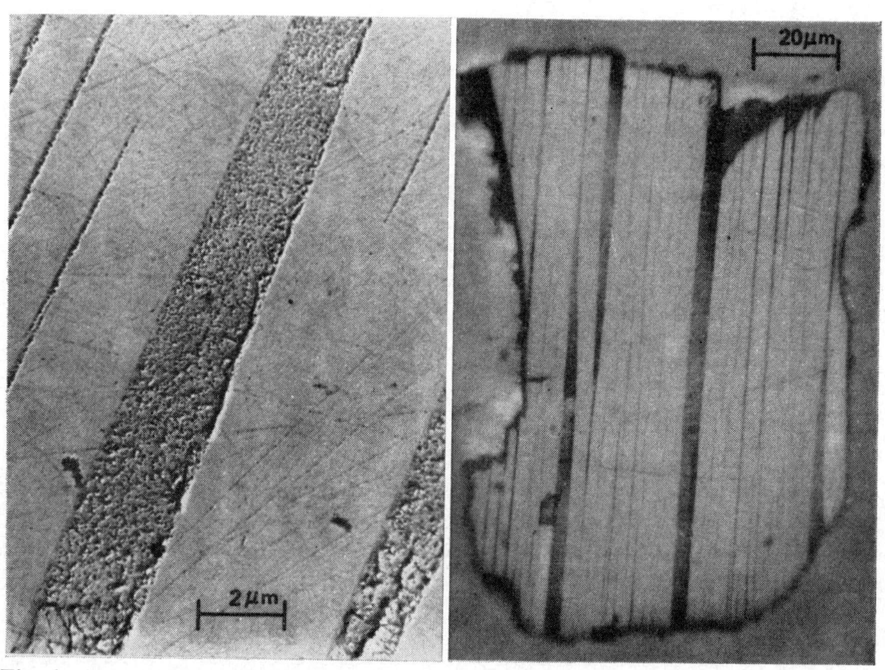

Fig. 1. A pyroxene grain from powder sample 10084,93. The particle composition is: SiO_2, 53.5 percent; MgO, 18.9 percent; FeO, 27.9 percent; CaO, 2.2 percent (hypersthene).

chamber F-201). Accordingly, a sample of 10084,93 fine powder maintained in N_2 was finally used. It was transferred to a UHV micromanipulator chamber and then treated as if it were a UHV sample. Small samples were transferred to optically flat walled reaction tubes, sealed, and finally attached to a specially designed vacuum manifold for preparation and metering of gases at controlled pressures from 10^{-10} torr to greater than 1000 torr, and at controlled temperatures from 20° to 400°C. To date, particles and bonded particle pairs have been exposed to O_2 and O_2(3.5 percent H_2O) at 2, 500, and 760 torr, and at 27°C and 200°C.

Disruption was definitely observed in three cases. On two occasions, translucent crystals attached to large opaque particles broke off, the first when exposed to O_2, and the second to O_2(3.5 percent H_2O). In the third case, a small section of a large particle broke off when exposed to the same O_2-H_2O mixture but exposed for about 7 hours. Furthermore, after an exposure of 6 days to O_2 and 9 days to the O_2-H_2O mixture, further disaggregation of the large particles was observed, appearing as a rounding of the corners. The disruption appears to have occurred at microbreccia interfaces. Microphysical and microchemical analyses of these fractured interfaces have not yet been made.

The coherence of the fine, lunar material is well known (*1*). In our etching experiments, where about 100 mg of powder sample 10084,93 was etched with a slow fluorosilic acid etch (H_2SiF_6 : HCl : citric acid; 1 : 1 : 1), most of the particles were dispersed after reaction. Only a few aggregates of black and white crystallites remained intact. Oriented etch figures, random etch figures, mottled surfaces, and unaffected areas were found. Larger particles which could be picked up with tweezers were found to be microbreccia. Another sample of the powder was etched using HBF_4 in place of H_2SiF_6. Except for a slower etching rate of glass-like material, the results were similar.

A portion of 10084,93 was mounted in plastic, lapped, polished and etched. Figure 1 shows a plastically deformed pyroxene in which the dislocation density varies between 10^{10} dislocations per square centimeter in the bands and a density equivalent to 6×10^{10} dislocations per cm² along slip planes. Also,

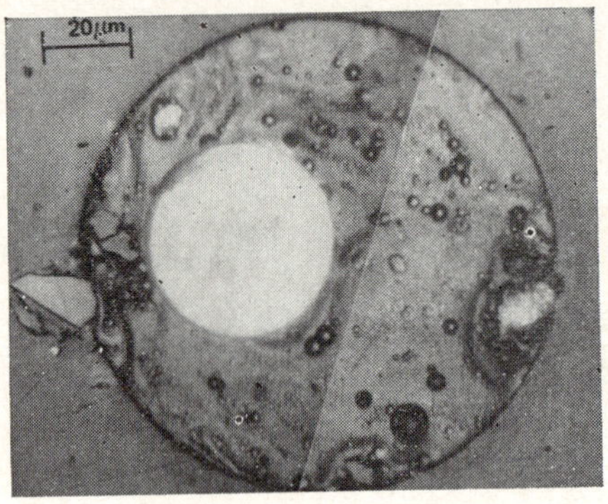

Fig. 2. Lapped and etched section of a glass sphere from powder sample 10084, 93. Impact adhesion of small grain on left side is seen with etched interface. The light circle is a spherical cavity.

microcracks are found along the band borders where the fracture stress was exceeded due to dislocation pileup (*2*). Figure 2 is a photomicrograph of a glass sphere section, to which a small fragment is attached (left side). Deformation of the sphere brought out by etching in the vicinity of the fragment indicates contact while still slightly fluid or more possibly local heating due to the impact. The white circle is a spherical void in the particle, and the dark to light dots are satellite voids and etch figures.

Some particles had shear deformed platelet inclusions and appeared to be fused to a glass-like microbreccia. One glass sphere, which also had been impact deformed exhibited a regular etch pattern suggesting partial recrystallization. Similar etch patterns were found on nonspherical particles.

A rock specimen of 10058,40 was mounted, lapped, polished, and etched with HF, HCl, and citric acid. Figure 3 shows bands in a crystal that was sheared in a direction normal to the surface. It is attached to a very heavily etched crystal that is tentatively identified as a plagioclase by its optical and etching characteristics. Other crystals from this rock specimen (probably pyroxenes) etched with the H_2SiF_6 etch exhibit band-like lines which form sharp angles enclosing three- and four-sided polygons. As yet, their origin is not known. However, a deformation feature, namely cross slip, has been observed in electron micrograph replicas of these etched surfaces.

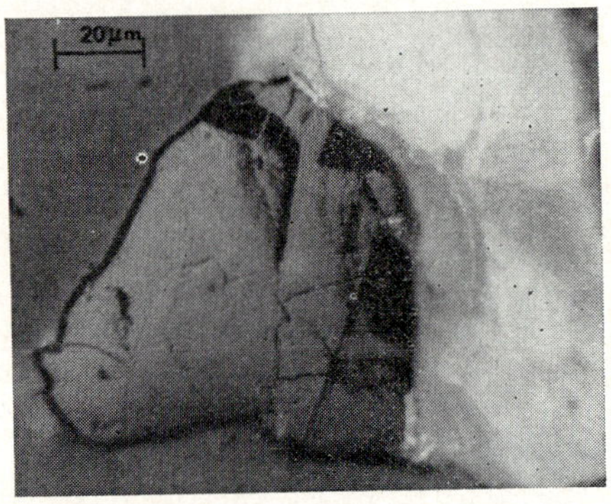

Fig. 3. Photomicrograph of chemically etched pyroxene deformation bands. The heavily etched crystal is a plagioclase.

The experimental apparatus for performing ultrahigh vacuum cleavage-fracture of silicates (*3*) has been used to fracture one lunar rock fragment (from 10065,33) to date. The fragment is a very friable, highly vesicular, igneous rock. Its dimensions prior to fracture were about 0.6 by 0.4 by 0.4 cm, and it was fractured roughly perpendicular to its long axis. Fracture was made at a system pressure of 7×10^{-10} torr. The halves were then brought together with contact made over only a very small area. The resultant adhesion force was about 8×10^2 dynes. This decreased to 2×10^2 dynes in 4 minutes and dropped below measurement capabilities (1×10^2 dynes) about 15 minutes later.

Microscopic examination of the fracture surface, after the system was brought to atmospheric pressure, revealed a large number of particles standing on end. These remained so even when moved laterally by a needle probe. Other particles, lying horizontally, would in some cases jump to the vertical position when moved. Movement of particles also, on occasion, caused others to jump to new positions and orientations. Attraction of particles to the needle was common. These observations indicate electrostatic charging of surfaces. The fractured specimens were then suspended in a covered container with the fractured faces downward. Water vapor was then introduced, and maintained, to determine whether the charge on the adhering particles and fracture faces could be neutralized. After 4 days, several particles had fallen from the bottom fracture face. All these observations are in accord with findings (*3, 4*) that cleavage of terrestrial silicates produces electrostatic charging that persists, though very much reduced when the system is brought to atmospheric pressure. The adhesional forces observed for the lunar sample are of the same order of magnitude as those found for terrestrial silicates (*3*).

The x-ray analyses of traverses across particle interfaces of lapped and polished crystalline specimens are typical of material formed from high-temperature phases. Low-temperature sintering (*5*) has not been found in the fine grained vesicular material from the powder 10084,93, in the coarse fines 10085,36, or in the rock specimen 10065,33. However, the Apollo 11 samples that were exposed to N_2 may have already disrupted at sintered interfaces.

JACK J. GROSSMAN, JACK A. RYAN
NALIN R. MUKHERJEE
MARY W. WEGNER
McDonnell Douglas Astronautics Company, Western Division, Santa Monica, California 90406

References and Notes

1. Lunar Sample Preliminary Examination Team, *Science* **165**, 1211 (1969); Surveyor Program, NASA SP-184 (1969).
2. C. Young III, *Amer. J. Sci.* **267**, 841 (1969); C. B. Raleigh, *J. Geophys. Res.* **73**, 5391 (1969); M. W. Wegner, J. J. Grossman, *Trans. Amer. Geophys. Union* **50**, 676 (1969).
3. J. A. Ryan, J. J. Grossman, W. M. Hansen, *J. Geophys. Res.* **73**, 6061 (1968).
4. P. J. Bryant, L. H. Taylor, P. L. Gutshall, Transactions of the 10th National Vacuum Symposium, (Macmillan, New York, 1963), p. 21; B. V. Deryagin, M. S. Metsik, *Soviet Phys., Solid State* **1**, 1393 (1960); J. J. Grossman, *J. Vac. Sci. Technol.* **6**, 233 (1969); J. A. Ryan, *J. Geophys. Res.* **71**, 4413 (1966); J. W. Salisbury, P. E. Glaser, B. A. Stein, B. Vonnegut, *ibid.* **69**, 235 (1964).
5. R. Smoluchowski, *Science* **150**, 1025 (1965).
6. We thank W. M. Hansen, A. D. Pinkul, W. B. Edwards, L. V. Marmaduke, G. Prather, Mrs. M. J. Dobson, A. R. Lopez, E. L. Miller, A. Phillips, Mrs. B. Tooper, and Dr. N. N. Greenman for their contributions to the program. Supported by NASA contract NAS9-8082 and NAS 7-307.

4 January 1970

Solar Radiation Effects in Lunar Samples

Abstract. *Optical properties of the pulverized crystalline rocks from the Apollo 11 samples are different from the optical properties of lunar soil. Changes in these properties were induced in the samples by ultraviolet and x-irradiation, standing, and heating. The albedo and spectrum of the soil differed significantly from expected values.*

The purpose of our work is to ascertain the possible role that various types of radiation may have played in determining the optical properties of the lunar surface. We investigated the effects of low-energy protons, ultraviolet and x-ray irradiation, heating, particle size, composition, and packing on the albedos and spectra of the samples.

The particle size distribution in fines sample 10084,81 was measured by sieving and sedimentation. The median particle size, weighted by mass, was about 40 μm, with over 35 percent of the mass contained in particles smaller than 20 μm. These results are in agreement with a previous size determination (*1*) and with estimates of "mean particle size of a few tens of microns or less" (*2*), based on telescopic optical data.

According to Wildey (*3*) the 0° phase angle albedo of the moon in the vicinity of the Apollo 11 site is approximately 0.099. Correcting this value for photometric function to a 6° phase angle gives 0.079. However, the subsurface soil appears to be darker than the surface material by 5 to 30 percent, depending on location (*4*). Since most of the sample presumably came from below the surface, a 6° albedo in the range of 0.060 to 0.076 is thus to be expected. The actual measured value relative to MgO is 0.090. The degree of compaction of the material cannot reduce this value sufficiently. We conclude that the returned sample of fines may not be representative of the general region of the moon around Tranquillity Base, at least as far as the factors that determine albedo are concerned.

The reflection spectrum of the fines, as measured by a Cary 14 spectrophotometer, is shown in Fig. 1 along with the general spectrum of the moon (*5*). The general lunar spectrum is considerably redder than the spectrum of the fines. We are not certain of the reason for this discrepancy; it may reflect errors in our measuring techniques, or it may indicate that this material is not typical of the surface of the moon as a whole, possibly because of alteration processes acting on the lunar surface or changes which occurred during or after collection. There is no sign in the fines of the 290-nm band reported in the lunar spectrum. We caution that the apparent peak at 250 nm in our data may be spurious. However, an absorption band at 900 to 1000 nm due to Fe^{2+} is clearly present in the spectrum of the fines.

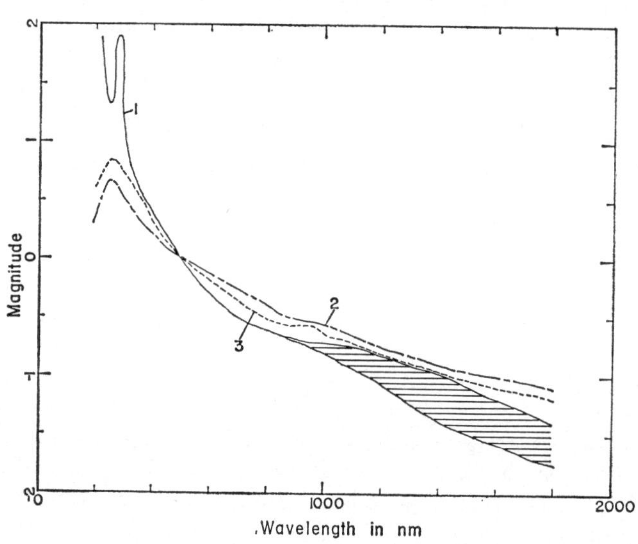

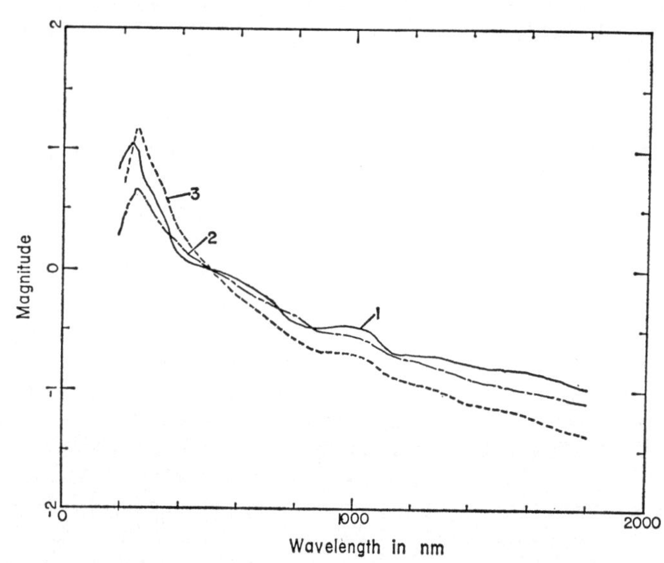

Fig. 1 (left). Relative reflection spectra, in magnitudes, of lunar materials, normalized at 500 nm; curve 1, average moon; curve 2, fines; curve 3, fines after ultraviolet irradiation. Fig. 2 (right). Relative reflection spectra, in magnitudes, of lunar materials, normalized at 500 nm; curve 1, pulverized crystalline rock 10022; curve 2, fines; curve 3, powdered rock after H-ion irradiation.

The lunar soil is commonly thought to be derived from larger fragments of lunar surface material by meteorite comminution. We attempted to ascertain whether the parent rocks of the soil could be identified by reflection spectra. Chips of all rocks received in our original allocation (numbers 10017 and 10022, crystalline, and 10048 and 10065, breccia) were ground to finer than 37 μm. The spectra and albedos of the powdered breccias are similar to those of the fines; however, these rocks are probably daughter products of the soil, formed by impact, rather than the parent rocks.

The crystalline rock powders are twice as bright as the fines, with albedos of 0.183; also the spectra of the two materials show significant differences (Fig. 2). However, the rocks and the fines are chemically quite similar (*1*). We emphasize that four rocks constitute too small a sample on which to base firm conclusions. However, if the crystalline rocks are indeed representative of the parents of the fines, then the optical properties of the crystalline rock powder must have been altered by some factor in the lunar environment.

It is suggestive that the albedo of the crystalline rock flour is similar to the albedo of the brighter rays on the moon. We also point out that several of the small differences between the spectra of various regions on the moon (*6*) can be accounted for if varying amounts of unaltered crystalline rocks or flour are exposed on the lunar surface.

It has been suggested that the solar wind is responsible for the low albedo and reddish color of the lunar soil (*7*). Therefore, a pulverized sample of crystalline rock 10022 was irradiated by approximately 20 coulomb/cm^2 of 2-kev hydrogen ions. This treatment reduced the albedo of the flour from 0.18 to 0.12, and the spectrum became more red, as shown in Fig. 2. This irradiation was conducted in a vacuum system constructed only of metal and ceramic components and evacuated by a sputter-ion pump with molecular-sieve roughing.

A sample of fines was subjected to intense ultraviolet and visible radiation from a 200-watt Oriel xenon-mercury lamp for 16 hours. This treatment (the dose has not yet been calibrated) increased the albedo from 0.090 to 0.096 and altered the spectrum of the soil (Fig. 1). Probably Fe^{2+} ions in the material were oxidized to Fe^{3+} by radiation, with a resulting decrease of absorption in the visible and near-infrared and increase in the ultraviolet.

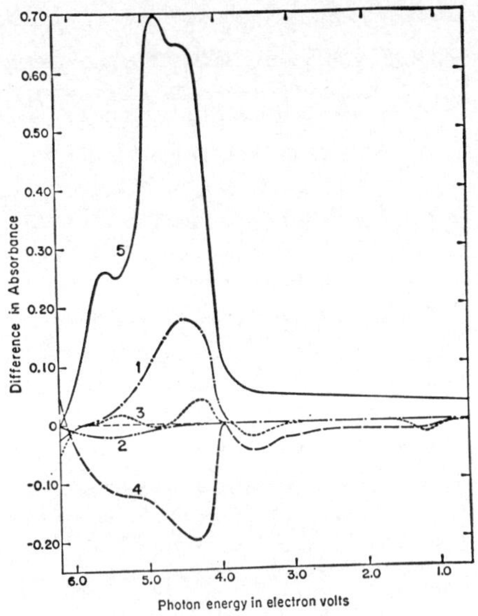

Fig. 3. Changes in absorption spectrum of lunar rock 10017 caused by radiation; curve 1, difference after ultraviolet treatment; curve 2, difference after x-irradiation from ultraviolet treatment; curve 3, difference after standing 330 hours after x-irradiation; curve 4, difference after heating at 200°C from standing; curve 5, difference after heating at 500°C from heating at 200°C.

The ultraviolet irradiation was carried out in air, and during this time the sample was heated to 160°C. To ascertain that the oxidation was not due to heating, another sample of soil was heated to 200°C in air; no change in albedo or in reflection spectrum occurred.

These radiation-induced changes were studied more quantitatively by measuring the changes in the absorption spectrum of a polished thin section of crystalline rock 10017, approximately 1 by 1 by 0.001 cm (*8*). The polished specimen of crystalline rock contained opaque ilmenite and three major transparent minerals: colorless plagioclase and two varieties of light brown pyroxenes. The spectra reported here are most probably spectra of the pyroxenes.

The spectrum of the sample before any treatment other than cutting and polishing showed the typical spectrum of ferrous ions in the near-infrared region and intense absorption due to ferric ions in the far-ultraviolet.

Because these spectra were similar to spectra of ferrous and ferric ions in synthetic quartz (*9*), we decided to use quartz as a lunar model and to perform experiments similar to earlier experiments.

Treatment of the specimen with unfiltered light of the xenon-mercury lamp for 23 hours resulted in the change shown in Fig. 3, curve 1. By subtracting the spectrum obtained before this ultraviolet treatment from the spectrum taken after treatment, we obtained the difference spectrum, which illustrates changes in the 1- to 6-ev region of the spectrum. Although no noticeable decrease had taken place in the ferrous region between 1 and 1.5 ev, there was a growth in the ferric region of the spectrum centered near 4.5 ev. The lunar rock wafer was next x-irradiated at 45-kv peak, 35 ma, with a Machlett AEG50T tube for 6 hours about 2 cm below the beryllium window. The difference induced by the x-ray treatment is shown in Fig. 3, curve 2. A decrease in the absorption spectrum took place in the ferric region centered near 5.5 ev. The specimen was allowed to stand at room temperature for 330 hours in subdued light. A decrease in the ferrous band occurred, accompanied by increase in two bands in the ferric region as shown in Fig. 3, curve 3. Heating the specimen for 2 hours at 200°C caused a large decrease in the region of the same two bands that increased previously (Fig. 3, curve 4). The specimen on standing showed decreased absorption on the deep ultraviolet and in the 3- to 4-ev region. This region decreased further on heating although the far ultraviolet showed an increase. As shown in Fig. 3, curve 5, heat treatment at 500°C in air for 2 hours caused an intense development of the α, β, and γ bands of the ferric ion (*9*). We interpret these results as follows.

If we take iron in quartz as a model, (i) ultraviolet treatment caused oxidation of ferrous ion with development of the β and γ ferric bands; (ii) x-ray treatment caused a reduction of ferric ion as shown by decrease in the α ferric band; (iii) standing caused slight oxidation of ferrous to ferric as shown by increase of α and γ ferric bands and decrease in the ferrous band; (iv) heating at 200°C caused a readjustment in the structure with possible oxidation of titanium, as indicated by increase at the ultraviolet cutoff, and decrease in the intensity of the ferric α, β, and γ bands; (v) heating at 500°C for 2 hours caused oxidation of probably most of the remaining ferrous ion with appearance of intense α, β, and γ bands of ferric ion. The trace ferric ion with high oscillator strength shows greater total change in absorbance than the abundant ferrous ion of low oscillator strength.

We conclude that these effects take

place primarily in the iron-containing pyroxenes, which are quite sensitive to radiation and temperature (in air) and should be a dynamic system when subjected to the direct solar wind even in vacuum.

Ever since the successful landing of Surveyor 1 it has been observed that gentle disturbance of the lunar soil, except by compression, results in a material which is as much as 30 percent darker than the undisturbed surface (*4*). We attempted to test several of the proposed explanations for this effect.

1) *Ultraviolet bleaching* (*10*). Ultraviolet irradiation of the soil caused albedo increases of approximately 6 percent; this value is of the same order as the disturbance effect observed by Surveyor 5.

2) *Packing.* The albedo of the soil is affected by the micromorphology of the surface in the sense that the looser the packing, the lower the albedo. The following albedos were obtained, depending on surface preparation: (i) sifted, 0.085; raked with a needle, 0.087; spooned onto a plate, 0.090; jarred by dropping the container from a height of about 5 cm, 0.094; tamped by pressing gently with the flat side of a spatula, 0.10. Since the lunar soil is continually stirred by micrometeorites it is doubtful that the natural condition of the surface is similar to that produced by tamping. Thus packing effects apparently are capable of altering the albedo by up to 10 percent.

3) *Particle size* (*11*). The albedos of fractions of fines separated as to particle size by dry sieving were measured with the following results: >149 μm, 0.089; 74 to 149 μm, 0.081; 37 to 74 μm, 0.080; < 37 μm, 0.079; unseparated, 0.090. Hence albedo differences of about 10 percent might be possible if the upper layer of particles are larger on the average than those below the surface.

4) *Lunar varnish* (*12*). Samples of the fines were examined with an optical microscope, a scanning-electron microscope, and an electron microprobe for coatings of dark material which might be scrubbed off by solar wind sputtering, but no definite evidence for such coatings was found. However, under high magnification the surfaces of many grains exhibited peculiar textures, which we were unable to interpret. These may indicate the presence of an amorphous coating on parts of some grains. In order to test the possibility that coatings of carbonaceous material might be present, a thermogravimetric analysis was carried out by heating a soil sample to 1200°C in air and measuring the change in weight (*13*). The weight of the sample increased continuously and monotonically with temperature because of oxidation, and no decrease, such as might be caused by ignition of a carbonaceous coating, was observed at any temperature.

5) *Conclusion.* Although several effects can alter the albedo of the lunar soil by up to 10 percent, none of the hypotheses yet advanced to explain the disturbance effect seem capable of completely accounting for all observations, but any or all of the hypotheses discussed here may contribute.

B. W. HAPKE, A. J. COHEN
W. A. CASSIDY, E. N. WELLS
Department of Earth and Planetary Sciences, University of Pittsburgh, Pittsburgh, Pennsylvania 15213

References and Notes

1. Lunar Sample Preliminary Examination Team, *Science* **165**, 1211 (1969).
2. B. Hapke, *Planet. Space Sci.* **16**, 101 (1968).
3. R. Wildey, private communication.
4. L. Jaffe *et al.*, *J. Geophys. Res.* **73**, 7165 (1968); E. Shoemaker *et al.*, *ibid.* **74**, 6081 (1969).
5. B. Hapke, in *Physics and Astronomy of the Moon*, Z. Kopal, Ed. (Academic Press, New York, ed. 2, 1970), vol. 1, p. 72.
6. T. McCord, *J. Geophys. Res.* **74**, 3131 (1969); ———, T. Johnson, H. Kiefer, *ibid.*, p. 4358; T. McCord and T. Johnson, *ibid.*, p. 4395.
7. B. Hapke and H. Van Horn, *J. Geophys. Res.* **68**, 4545 (1963).
8. Sample preparation by G. Moreland, U.S. National Museum.
9. A. J. Cohen and F. Hassan, *Science* **167**, 176 (1970).
10. A. J. Cohen and B. Hapke, *ibid.* **161**, 1237 (1968).
11. A. Felice, *J. Geophys. Res.* **72**, 5721 (1967).
12. E. Shoemaker *et al.*, *J. Geophys. Res.* **73**, 3989 (1968).
13. We thank G. Aubele and the Fisher Scientific Company for performing this analysis.
14. We thank B. Greenspan and F. Hassan for assistance. Supported by NASA contract NAS 9-9942.

4 January 1970; revised 6 January 1970

Determination of Manganese-53 by Neutron Activation and Other Miscellaneous Studies on Lunar Dust

Abstract. *A highly sensitive determination of spallogenic ^{53}Mn (T = 2 × 10^6 yr) was accomplished in 0.99 g of lunar soil. The chemical yield of Mn is determined with "carrier-free" ^{52}Mn tracer. During a 23-day reactor irradiation the ^{53}Mn is transformed into ^{54}Mn (T = 300 days). Appropriate chemical recycling was done by ion exchange and distillation. Interferences of the (n,p) and the (n,2n) nuclear reactions were carefully studied. A ^{53}Mn disintegration rate of 30.3 ± 5.5 dpm/kg results. This extremely economic method is proposed for further detailed lunar profile measurements. The Re content, which is of possible cosmochemical interest, was determined to be 11 ppb. Appropriate separation techniques were used. The rather weak and complex thermoluminescence properties made a more basic study advisable. Thermogravimetric analysis, mass spectroscopy, and Moessbauer spectroscopy were applied. The presence of ilmenite, metallic Fe etc., and of an unidentified Fe^{2+}-containing compound was deduced. Natural thermoluminescence could not be proved with certainty in our surface sample. However, the complexity of the artificial thermoluminescence demands better defined mineral fractions. The fission track method was used to measure U distribution in glass spherules etc.*

Manganese-53, favored by its high production cross section and long half-life of about 2 × 10^6 years, is one of the most interesting nuclides in space research. However, detection of the ^{53}Mn nuclide is rather difficult because of its soft x-rays, and only sophisticated "low-level" techniques succeed. This contrasts with ^{26}Al, which can be detected nondestructively by its characteristic γ-cascade with fairly high sensitivity. (Even with our 5-g sample of lunar dust, we were able to detect the stronger γ-emitters.)

Working with meteorites, a very sensitive and therefore economic method was found in which we convert ^{53}Mn by neutron capture into γ-emitting ^{54}Mn (*1–3*), as Millard (*4*) had proposed some years ago.

In 1957 Sheline and Hooper (*5*) predicted that, in meteorites, the long-lived ^{53}Mn isotope should be produced in relatively high yield. Honda *et al.* (*6*) thereafter confirmed this experimentally. Although the half-life of ^{53}Mn is not yet fully established, a half value of (1.9 ± 0.5) × 10^6 years was suggested by Kaye and Cressy (*7*). They compared the ^{53}Mn activity of an iron meteorite of high terrestrial age with the ^{53}Mn activity of a chondrite with short radiation age. However, the method involves a number of uncertainties, which must be allowed for. Millard showed that the product of the

^{53}Mn (n,γ) activation cross section, σ, and the ^{53}Mn half-life, T, is (350 $\pm$ 100) $\times$ 10^6 barns $\times$ years.

From this follows an activation cross section σ = 170 barns, on the basis of an estimated ^{53}Mn half-life of $T = 2 \times 10^6$ years. The latter value is accepted as the first approximation and has been used in the present study.

The spallation product ^{53}Mn is converted by an (n,γ)-reaction into the ^{54}Mn nuclide, and it has been proved that the latter can be identified conveniently by its characteristic 0.84 MeV γ-line. Because stable ^{55}Mn is present in the "lunar fines" with an abundance of about 0.2 percent, we preferred to work without any Mn carrier.

"Lunar fines" (0.990 g) were fused with NaOH in a Ni crucible. In order to establish the chemical yield of Mn a practically "carrier-free" ^{52}Mn ($T \simeq 5$ days) tracer was used. Dowex 1 $\times$ 8 ion-exchange resin served for purification. For better handling 50 mg of MgO and 5 mg of ZnO (the latter used as internal standard) were added. The chemical yield was 96.0 percent, as determined by means of a short reactor irradiation and nondestructive γ,γ-coincidence spectroscopy. Our dust sample (grain size $\leq$ 100 μ) was found to contain 1690 $\pm$ 80 ppm normal ^{55}Mn. The sample was then reactor irradiated for 23 days, together with 10 mg of MnO_2, 5 mg of ZnO, and 2 Fe-

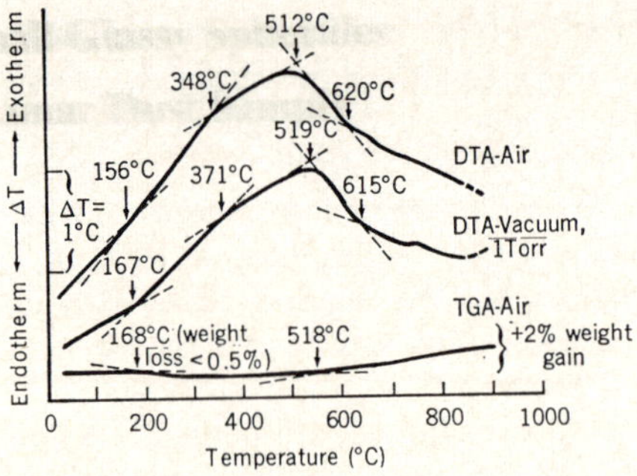

Fig. 1. Differential thermal and thermogravimetric analysis of lunar dust (sample 10084). Heating rate, 20°C/min; reference, Al_2O_3; sample weight, 5 to 7 mg.

standards in a thermal flux of 8.10^{12} n cm^{-2} sec^{-1} in the FRJ-2 reactor (Dido type) at Juelich. It is important to note that, in the appropriate position, the flux ratio ϕ(fast)/ϕ(therm) $\leq$ 0.001. Hence, neutron activation results in an activity ratio of

$$\frac{^{53}\text{Mn [dpm]}}{^{54}\text{Mn [dpm]}} = \frac{1}{6000}$$

At the end of the irradiation the relative flux deviations were controlled by counting the ^{65}Zn annihilation radiation in a coincidence circuit. Absolute flux calibrations were done with a standardized ^{22}Na sample.

The recycling of probe and standards consisted in taking up in 12N HCl, oxidizing with H_2O_2, adding 47.5 mg Fe (inactive carrier) and treating with a Dowex 1 $\times$ 8 ion exchanger. The eluate was brought to dryness, taken up in 2N HCl and loaded on a cation exchanger (Dowex 50). The eluate was digested with strong H_2SO_4, 1 g of $NaIO_4$ was added, and finally the manganese distilled at 40° to 50°C under reduced pressure. The distillation was repeated.

Influence of the (n,p) and the (n,2n) interfering reactions is best demonstrated in the Table 1. While the interference of the ^{54}Fe (n,p) ^{54}Mn process can be suppressed largely by radiochemistry and is only in the range of 2 percent, the other nuclear reaction ^{55}Mn (n,2n) is more serious, contributing up to about 40 percent. Clearly the latter interference is dependent mainly on the reactor irradiation conditions, and may be reduced by choosing the highest available ratio of ϕ(therm)/ϕ(fast).

Whereas the ^{53}Mn content of most of the investigated smaller meteorite is in the range of 300 to 600 dpm/kg, the ^{53}Mn content in lunar dust is found to be only 30.3 $\pm$ 5.5 dpm/kg. However, we have to relate this figure to the amount of 13 percent iron present, the main source of spallation nuclides. So, we get a value of ^{53}Mn = 230 dpm/kg Fe of lunar surface material. This still is only about 50 percent of what is found and expected in smaller iron meteorites.

On the other hand, from studies of production rates, one knows that ^{53}Mn is mainly formed by secondary reactions and lower energy primaries and that it shows a pronounced depth effect. An increase of ^{53}Mn up to a 25 cm distance beneath the surface of a spherical iron target is known. Evidently we shall have to account for growth of ^{53}Mn activity if we take samples beneath the moon's surface. Moreover, differing from meteorites, we have to consider a 2π instead of 4π "falling in" of cosmic rays.

Still another point of interest is the relatively long half-life of ^{53}Mn, compared to ^{36}Cl and ^{26}Al. From the high ^{26}Al values (up to 100 dpm/kg) one is inclined to assume that this nuclide is in radioactive equilibrium. If the accretion and mixing rate of lunar dust is found to be high, we should have some doubt that radioactive equilibrium is fully established in the same way for the longer lived ^{53}Mn. Further studies on selected mineral probes—for example, samples of bore-holes, should bring the necessary information about spallation profiles. Since the activation method described is so sensitive, manganese-53 at even greater depths (up

Table 1. Determination of spallogenic ^{53}Mn in lunar dust.

		Lunar dust fr. V (0.9901 g)	Mn standard	Fe standard
Weight		1.62 mg Mn	8.03 mg Mn	8.25 mg Fe
Chem. yield	Pre-irrad.	95.9%	—	—
	Recycl.	89.3%	69.0%	65.2%
Carrier after irrad.		47.5 mg Fe	—	10 mg Mn
$I_{spez_{54}}$ [cpm/mg Mn]	(n, γ)	7.64	—	—
	(n, 2n)	3.14	3.14*	0
	(n, p)	0.16†	0	(8.92* per mg Fe)
$A_0 spez_{54}$ [dpm/mg Mn] at end of irrad.	(n, γ)	110.87	—	—
A_{53}	[dpm/kg]	30.3 $\pm$ 5.5	—	—

* Neutron flux normalized. † Originating from Fe-traces in isolated lunar Mn.

Table 2. Iron-containing compounds of lunar dust.

Source	Fraction of total iron (%)		
	Normal	Magnetic fraction	Nonmagnetic fraction
Olivinic Fe	19.3	18.8	20.1
Pyroxenic Fe	34.0	31.7	43.2
Fe^{2+} in ilmenite	26.0	18.6	24.9
Fe^{3+} in ilmenite	<3.5	<4.6	<3.0
Unknown compound	10.2	16.8	8.9
Metallic Fe	>7.0	>9.3	—

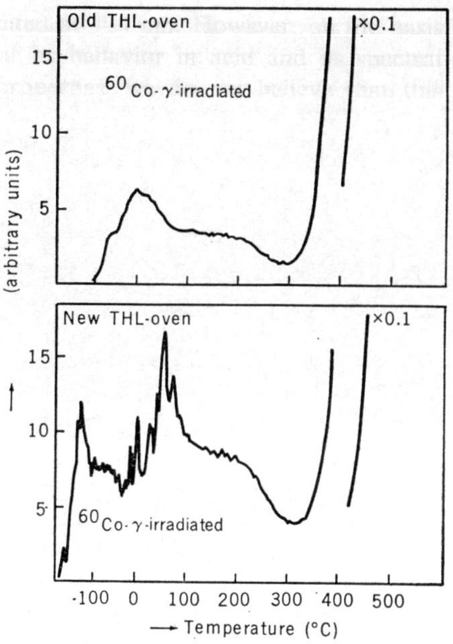

Fig. 2. Thermoluminescence glow curves of lunar dust (sample 10084); (upper curve) 40-mg sample; irradiated for 15 minutes; heating rate, 100°C/min; (lower curve) 47-mg sample; irradiated for 15 minutes; heating rate, 100°C/min.

several meters) should be measurable.

During the last decade the natural β^- activity of ^{187}Re was used by us successfully for geological age determinations (*8*). We showed that the ^{187}Os abundance varies considerably in the Re-containing minerals and in iron meteorites. It was pointed out, and D.D. Clayton showed this later in a detailed study, that the larger part of the terrestrial (stable) ^{187}Os should be regarded as having resulted from ^{187}Re decay and that most of that decay occurred before the formation of the solar system. These ideas are based on the now accepted s-process theory.

The β^- energy of ^{187}Re is extremely weak (~2 keV) and a problem arises about the mode of this decay. Quite recently Clayton (*9*) outlined the possibility that the radioactivity of ^{187}Re could be dependent on temperature and isotopic measurements of Re from lunar dust ("solar wind") could clarify the interesting problem of ^{187}Re cosmochronology. In view of these cosmological questions, we found it necessary to establish the presence of Re in lunar dust. As the abundance in Fe-meteorites is normally below 1 ppm, only neutron activation with chemistry could succeed. "Fines" (176 mg) were irradiated together with Re standards for 3 days, $\phi = 7 \times 10^{13}$ in the FRJ-2 reactor. Thereafter the mineral was fused by melting with NaOH + Na_2O_2 in the presence of 30 mg of Re-carrier. Several $Fe(OH)_3$ precipitations were followed by repeated dry distillation of Re_2O_7. The decay of ^{186}Re is observed. The Re content of soil (sample 10084, grains <100 μ) is calculated to be 11.2 ± 0.4 ppb if one assumes a normal terrestrial isotope abundance.

In the next step it is important to check the isotope ratio ^{187}Re/^{185}Re.

The thermodifferential analysis (TDA) and thermogravimetric analysis (TGA) curves (Fig. 1) show distinct temperature regions where gas-loss and recrystallization or annealing occurs. A critical temperature is seen at about 510°C. Oxidation at 800°C leads to a gain in weight of about 2 percent (see the lower TGA curve).

By mass spectroscopy, ratios of ^{3}He/^{4}He = 724 ± 118, ^{20}Ne/^{22}Ne = 2.34 ± 0.04, and ^{22}Ne/^{21}Ne = 22.6 ± 6 were measured in dust. Only the ^{20}Ne/^{22}Ne ratio is found to be very constant.

Nuclear γ-resonance spectroscopy (with ^{57}Fe) was done on magnetic and nonmagnetic fractions of lunar fines. No trace of Fe^{3+} was detected. Troilite was practically not observed in the soil but metallic Fe, ilmenite, pyroxene, and olivine were present. The spectra were compared with those of tektites and the existence of an unknown Fe^{2+}-containing compound in the lunar soil is proposed (see Table 2).

For the thermoluminescence study different soil fractions were used. However, after 8 weeks of "storing" time we did not detect any natural THL with certainty. Probably the skin layer of dust is already strongly annealed on the moon. Highest glow intensities occur in the temperature range of 100° to 150°C, if irradiated at 30°C. At −196°C γ-irradiation the intensity is considerably greater. A ^{60}Co-γ-saturation experiment shows that saturation for the soil is reached at a dose of about 10^6 rads (see Fig. 2).

These preliminary results demand better defined and selected material probes.

W. Herr, U. Herpers, B. Hess
B. Skerra

Institut für Kernchemie der Universität Köln, Cologne, Germany

R. Woelfle

Institut für Radiochemie der Kernforschungsanlage, Jülich, Germany

References and Notes

1. U. Herpers, W. Herr, R. Woelfle, "Radioactive Dating and Methods of Low-Level-Counting" (IAEA, Vienna, 1967), p. 199.
2. W. Herr, U. Herpers, R. Woelfle. *J. Radioanalyt. Chem.* **2**, 197 (1969).
3. U. Herpers, W. Herr, R. Woelfle, "Meteorite Research" (IAEA, Vienna, 1969), p. 387.
4. H. T. Millard, Jr., *Science* **147**, 503 (1965).
5. R. K. Sheline and J. E. Hooper, *Nature* **179**, 85 (1957).
6. M. Honda, J. P. Shedlovsky, J. R. Arnold, *Geochim. Cosmochim. Acta* **22**, 133 (1961).
7. J. H. Kaye and P. J. Cressy, *J. Inorg. Nucl. Chem.* **27**, 1889 (1965).
8. W. Herr, R. Woelfle, P. Eberhardt, E. Kopp, "Radioactive Dating and Methods of Low-Level Counting" (IAEA, Vienna, 1967), p. 499.
9. D. D. Clayton, *Nature* **224**, 56 (1969).

6 January 1970

Specific Heats of Lunar Surface Materials from 90 to 350 Degrees Kelvin

Abstract. *The specific heats of lunar samples 10057 and 10084 returned by the Apollo 11 mission have been measured between 90 and 350 degrees Kelvin by use of an adiabatic calorimeter. The samples are representative of type A vesicular basalt-like rocks and of finely divided lunar soil. The specific heat of these materials changes smoothly from about 0.06 calorie per gram per degree at 90 degrees Kelvin to about 0.2 calorie per gram per degree at 350 degrees Kelvin. The thermal parameter $\gamma = (k\rho C)^{-1/2}$ for the lunar surface will accordingly vary by a factor of about 2 between lunar noon and midnight.*

The surface temperature of the moon varies from approximately 90° to 385°K between lunar noon and midnight (*1, 2*).

Cooling of the lunar surface depends on the so-called thermal parameter, $\gamma = (k\rho C)^{-1/2}$, where k is the thermal conductivity, ρ is the density, and C is the specific heat. Most calculations of lunar surface temperature that use earth-based eclipse data (*1, 3*) have assumed a constant value of γ. However, specific heats of silicates vary by a factor of 3 or 4 in the temperature range 100° to 400°K. Accordingly, γ should show a nearly twofold variation over the range of lunar surface temperatures.

In order to verify the temperature dependence of γ, we have determined the specific heats at approximately 8°K intervals between 90° and 350°K for the samples 10057 and 10084 from Apollo 11.

The low-temperature adiabatic cal-

Table 1. Specific heats in cal g−1 deg−1 (smoothed values) for lunar samples 10057 (vesicular basalt) and 10084 (regolith). Values in parentheses are extrapolated.

Temperature (°K)	Sample 10057	Sample 10084
90	(0.0571)	(0.0615)
100	.0633	.0665
120	.0771	.0802
140	.0922	.0955
160	.1075	.1108
180	.1217	.1235
200	.1343	.1348
220	.1451	.1446
240	.1546	.1534
260	.1632	.1617
280	.1711	.1696
300	.1786	.1771
320	.1853	.1845
340	.1917	.1916
360	(.198$_3$)	(.197$_0$)

orimeter used in this investigation and details of the measurement technique and data reduction are discussed in detail by Robie and Hemingway (*4*).

Special sample holders were built for the lunar sample measurements from aluminum rod (T 6061 alloy). They are 5.1 cm deep with a 3.2-cm inside diameter, and they have a central reentrant well of 0.48-cm internal diameter with a length of 3.8 cm. The containing walls are 0.05 cm thick. The surface has a 10-microinch (mirror) finish. The sample holder has a mass, including the top, of 20.5 g and an internal volume of 42.2 cm^3.

A miniature platinum resistance thermometer (Minco 1059) with an ice point resistance of 100.09 ohms and sensitivity of 0.4 ohm deg^{-1} was used to determine the absolute temperatures (*5*). It is inserted within an aluminum bobbin externally wound with a 53-ohm "Evanohm" heater attached with Glyptal (G.E. 7031) and is brought into good thermal contact with the sample container using Apiezon T stopcock grease. The bobbin with the thermometer in place slip-fits within the reentrant well of the sample holder body and is brought into thermal contact using Apiezon T stopcock grease.

To preserve the samples in as nearly their pristine state as possible the samples were loaded into the aluminum containers at the Lunar Receiving Laboratory and sealed under dry nitrogen gas at 5 psi. The heat capacity of the empty sample containers was determined in a separate set of experiments. The raw data (electrical energy/ ΔT) were corrected for small deviations from true adiabatic conditions (that is, zero heat exchange between sample container and its surroundings) and for the heat capacities of the aluminum sample holder, N_2 exchange gas, and Teflon tape seal. The use of aluminum for the sample containers instead of copper, and of dry nitrogen instead of helium gas, reduces the precision of specific heat measurements by increasing the time constant of the calorimeter. The empty calorimeter contributed about 50 percent to the total observed heat capacity at 90°K and about 30 percent of the total at 350°.

The experimental data for sample 10057 are shown in Fig. 1. The results for the regolith sample, 10084, are similar. The temperature rise for each point was between 3.7° and 10.2°K. The accuracy of the data is ±0.4 percent. The corrected specific heats were smoothed using a form of least-squares orthogonal polynomials. The smoothed specific heat values are listed, at 20°K intervals, in Table 1.

If we neglect the small temperature variation of the thermal conductivity, we can calculate the temperature dependence of γ from the specific heats listed in Table 1, the measured densities (*6*), and reasonable estimates for the thermal conductivities. Scott (*7*) lists values for the thermal conductivity of evacuated, porous, silica-based materials of between 2.4 and 5.2 μcal cm^{-1} deg^{-1} sec^{-1}. These should be comparable to the conductivity of the lunar regolith. Using a bulk density of 1.6 g cm^{-3} as determined from the Apollo 11 core tube samples (*6*) and a mean conductivity of 4.0 μcal deg^{-1} cm^{-1} sec^{-1}, we obtain for the thermal parameter γ of the lunar soil (sample 10084), values of 1600 cm^2 sec$^{1/2}$ °K cal^{-1} at 90°K and 898 cm^2 sec$^{1/2}$ °K at 350°K.

In contrast, a large lunar outcrop of type A rock that has a conductivity of the order of 0.004 cal cm^{-1} deg^{-1}

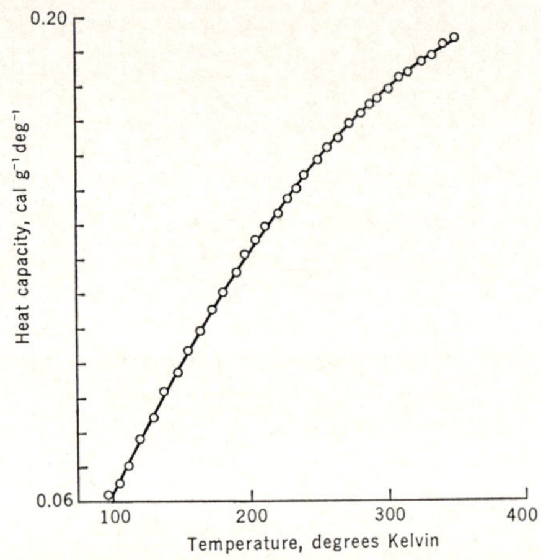

Fig. 1. Specific heat of Apollo 11 sample 10057. Circles indicate experimental observations. The full line is the least-squares fit to the data.

sec^{-1} would have a γ at 90°K of 35.9 cm^2 sec$^{1/2}$ °K cal^{-1}, which would decrease to 19.4 cm^2 sec$^{1/2}$ °K cal^{-1} at 350°K. In this calculation we have used Robertson's data (*8*) for the conductivity of vesicular basalt. The difference in the thermal parameter of the lunar soil and lunar outcrop values is due to the very large difference in the thermal conductivities between the porous soil (in a high vacuum) and the solid rock. Inasmuch as γ depends on the rate of cooling, the dust-covered portions of the lunar surface will cool much more rapidly than bare exposed outcrop, and thermal anomalies (*2*) will quickly develop over the exposed solid rock.

In Table 2, we list values for γ at several temperatures between 100° and 350°K for the lunar soil (10084) and for the vesicular basalt (10057).

Table 2. Thermal constant γ (in cm^2 sec$^{1/2}$ °K cal−1 for lunar samples 10057 and 10084. Values for sample 10057 were obtained by using $k = 0.004$ cal cm^{-1} deg^{-1} sec^{-1} and $\rho = 3.4$ g cm^{-3}. Values for sample 10084 were obtained by using $k = 0.000004$ cal cm^{-1} deg^{-1} sec^{-1} and $\rho = 1.6$ cm^{-3}.

Temperature (°K)	Sample 10057	Sample 10084
100	34.33	1543
150	27.12	1231
200	23.40	1078
250	21.51	1000
300	20.29	941
350	19.41	898

RICHARD A. ROBIE
BRUCE S. HEMINGWAY
WILLIAM H. WILSON

U.S. Geological Survey,
Silver Spring, Maryland 20910

References and Notes

1. J. W. Lucas, W. A. Hagemeyer, J. M. Saari, L. D. Stimpson, J. M. F. Vickers, "Surveyor Program Results" [*NASA SP-184* (1969), p. 181].
2. D. A. Allen and E. P. Ney, *Science* **164**, 419 (1969).
3. J. C. Jaeger, *Aust. J. Phys.* **6**, 10 (1953).
4. R. A. Robie and B. S. Hemingway, in preparation.
5. W. D. Johnston and G. W. Lindberg, *Rev. Sci. Instrum.* **39**, 1925 (1968).
6. Lunar Sample Preliminary Examination Team, *Science* **165**, 1211 (1969).
7. R. B. Scott, *Cryogenic Engineering* (Van Nostrand, Princeton, N.J., 1959), p. 206.
8. E. C. Robertson, unpublished data.

9. We thank our U.S.G.S. colleagues Raymond Truli and R. W. Werre for fabricating the sample holders, and Rudolph Raspet for many helpful suggestions. We are also grateful to Dr. D. R. Anderson for his help at L.R.L. and to NASA contract T-75405 for support of this work.

4 January 1970

Organic Chemistry

Lunar Organic Compounds: Search and Characterization

Abstract. *The carbon concentration in Apollo 11 lunar fine material is of the order of 200 ppm. By far the largest single amount of this carbon appears to be in carbon monoxide. Some of this seems to be in the form of gas bubbles in the glass spheres, but most of it could be in some complex form other than gas. This result would be consistent with the idea that most of the fines passed through a high temperature and that the carbon was oxidized by mineral oxides at that time.*

The search for carbon-containing material on the lunar surface is not only a component part of the study of the origin and history of the moon, but an important step in our understanding of the early stages of chemical evolution leading to the origin of life. It represents the first possibility of examining nonterrestrial material uncontaminated by the biological products found everywhere on earth.

We have therefore proceeded with three sequences of operations: (i) direct pyrolysis into the mass spectrometer at temperatures up to 1150°C; (ii) a preliminary extraction with a clean mixture of benzene and methanol, followed by direct introduction into the mass spectrometer; (iii) solution of the fines in 20 percent HF, purging with helium to trap any volatile materials, then solvent extraction and pyrolysis of the residue.

Two types of direct pyrolysis experiments on moon dust were performed. In the first, the sample was treated at 150°C, and any volatile material was removed (*1*). The sample was then heated directly to fusion at 1150°C for 1 hour. The evolved gases were introduced into a MS-902 high resolution mass spectrometer (*2*) and consisted of CO (156 ppm C), CO_2 (12 ppm C), traces of HCN, H_2S, SO_2, Ne, Ar, C_2H_2, N_2, and O_2 (*3*).

A second experiment involved a preliminary 150°C heating for 30 minutes, followed by direct heating to 500°C for 30 minutes. The gases evolved were CO (9 ppm C), CO_2 (<1 ppm C), and traces of the same series given above, but including some CH_4. A third experiment performed with the same temperature sequence as the second experiment, but involving passage of evolved gases through the GC-MS 12 combination (*4*), confirmed the presence of CH_4.

Because the sample was already so finely divided, we decided that any further reduction in size could only increase possible contamination. An optical examination of the material confirmed that it was a mixture of angular birefringent particles with a small amount of spherical or near-spherical nonbirefringent particles (Fig. 1), as well as some angular glassy particles. In addition, a proportion of the glass spheres appeared to contain vesicles and some optically dense inclusions (Fig. 1, b and c).

Table 1. Composition of the benzene-methanol extract of 19.8 g of lunar fines from Apollo 11 determined by high-resolution mass spectrometry (only ions representing possible molecular species are shown).

Nominal mass	Component	Average ion current (relative)	Observed mass (amu)
17	NH_3 *	260	17.02629
18	H_2O	85,000	
26	C_2H_2	650	
27	HCN *	300	27.01103
28	CO	18,000	27.99490
	N_2	210,000	28.00615
30	NO *	1,100	29.99769
32	CH_2O	1,220	
	O_2	40,300	
42	C_2H_2O	300	
	C_3H_6	420	
43	HCNO *	280	43.00797
	HN_3	40	43.01846
44	CO_2	6,100	
	CH_2NO *	270	44.01581
	C_2H_4O	1,250	
	C_2H_6N *	30	44.05208
46	CH_2O_2	710	
50	C_4H_2	250	
52	C_4H_4	100	
54	C_4H_6	200	
55	CHN_3—C_3H_3O	100	55.01682
56†	C_4H_8	650	
58	CH_4N_3—C_3H_6O	100	58.03816
59	C_2H_5NO *	550	59.03489
60	C_2H_6NO *	110	60.04380
149	$C_8H_5O_3$	3,400	

* Composition attributable to retro-rocket exhaust from lunar module (*5*). † Hydrocarbons were found to C_{10} but were not listed above C_4.

The dust (19.8 g) was extracted in 100 ml of a mixture of benzene and methanol (3:1), following the procedure of Han *et al.* (*4*). A part of this extract, after concentration to 3 μl, was examined by capillary gas chromatography (FID detector), and no organic compounds were detected. The remaining 1.7 μl was introduced into the MS-902 high resolution mass spectrometer (*2*), on-line to a computer; the results are given in Table 1. Materials resulting from retro-rocket exhaust are present (*5*), as are a number of contaminants which are derived from sample handling procedures (*6*). In order to complete the examination of the lunar samples for organic matter, the 19.8 g of cleaned dust were then dissolved in 20 percent HF prepared as described by Han *et al.* (*4*). The resulting residue (12 g) was mostly a white flocculate with some heavy black particles (less than 1 percent of the residue).

The benzene-methanol extract of this residue, after vacuum drying at 120°C for 12 hours, yielded no detectable material above solvent background either by gas chromatography or by direct introduction into the MS-902 ionization chamber. A 22-mg sample of this residue was subjected to sequential pyrolysis at 150°, 500°, and 1100°C. The principal product in each case was CO at 18 ppm C, 13 ppm C, and 88 ppm C, respectively, by weight of the original dust, adding up to a total of 119 ppm C.

The relatively low value of carbon, together with the presence of vesicles in the glass spheres, suggested that the dust might contain volatile material which would be liberated and lost upon solution in HF. We therefore designed an apparatus which would allow the dissolution of the sample with simultaneous trapping of any liberated gases. This involved a dissolver followed by a sequence of cold traps at the temperatures of Dry Ice–acetone, liquid nitrogen, and liquid helium, respectively. After dissolved and absorbed gases were eliminated, the apparatus was calibrated by dissolving approximately 20 g of silica gel, to which had been added 1 ml (3.5×10^{-5} mole), at standard temperature and pressure, of CO. Carbon monoxide (2×10^{-5} mole) was recovered from the liquid helium trap, indicating a 58 percent efficiency of recovery. When 17.8 g of moon dust as received from the Lunar Receiving Laboratory was dissolved in this equipment, a total of 66 ppm C was recovered as CO from the liquid helium trap. When added to the 119 ppm C as CO obtained by pyrolysis of residual silica after HF digestion, this totaled 185 ppm C, which can be compared with the value of 168 ppm C (CO value plus CO_2 value) found in the dust by direct pyrolysis at 1150°C. There remains the possibility that a small amount of volatile C has been lost during preliminary drying of the residual silica.

It is of interest to estimate the amount of gaseous CO which might be present in the vesicles of the glass spheres. A rough statistical analysis of their distribution would require that they contain 10^4 atmospheres partial pressure CO in order to be equivalent to the 66 ppm C liberated upon solution. This seems excessive, and we therefore feel a considerable amount of CO liberated by HF must be in other than gaseous form in the original moon dust, for example as carbonyls.

Calibration of the method with a Hawaiian basalt containing 120 to 150 ppm C by combustion (7) produces the corresponding amount of CO, whereas the addition of graphite to the fused fines, followed by pyrolysis, does not convert more than 3 percent of that graphite to CO. We therefore conclude that there will be only a small amount, if any, of graphite present in the fines.

The largest chemical component form of carbon in the lunar fines which we have so far analyzed is carbon monoxide. This could be the result of the oxidation of any reduced carbon by the heat of meteorite impact or other sources of energy available to the lunar surface, mineral oxides being the oxidizing agent. Most of the carbon monoxide so formed would have escaped the lunar gravitational field, thus accounting for the extremely low concentrations of total carbon in any of the samples yet analyzed (<0.02 percent). We might expect much more carbon in those lunar materials which have never been molten. These would be either the primordial aggregates from which the moon might have been formed, or the collection of infall of meteorites at low energy.

A. L. BURLINGAME
MELVIN CALVIN
J. HAN, W. HENDERSON
W. REED, B. R. SIMONEIT

Space Sciences Laboratory,
Departments of Chemistry and *Geology,*
University of California, Berkeley 94720

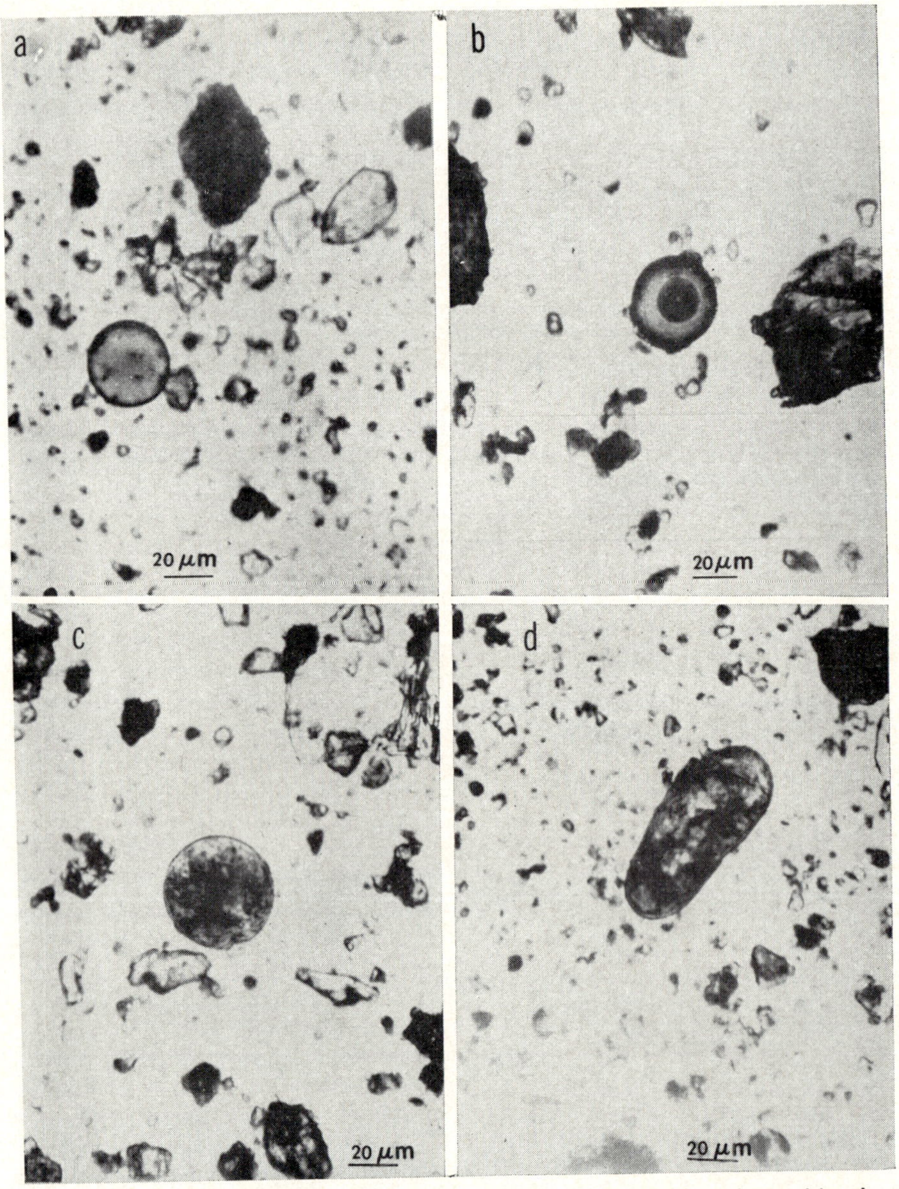

Fig. 1. Oil mounts (n = 1.569) of Apollo 11 dust. (a) Clear glass sphere with minor dark inclusions; (b) clear glass sphere containing vesicle; (c) cloudy glass sphere showing granular dark inclusions; (d) cloudy elongate glass "spherule" showing granular dark inclusions, and vesicle.

References and Notes

1. This consisted of air, water, CO, CO_2, and some methane.
2. A. L. Burlingame, D. H. Smith, T. O. Merren, R. W. Olsen in *Computers in Analytical Chemistry*, C. H. Orr and J. Norris, Eds. (Plenum Press. New York, 1969), p. 13; A. L. Burlingame, D. H. Smith, F. C. Walls, R. W. Olsen, *Proceedings of the 17th Annual Conference on Mass Spectrometry and Allied Topics* (Dallas, Texas, 18–23 May 1969), p. 28.
3. Nitrogen is 20 times more abundant than oxygen.
4. J. Han, B. R. Simoneit, A. L. Burlingame, M. Calvin, *Nature* **222**, 364 (1969).
5. B. R. Simoneit, A. L. Burlingame, D. A. Flory, I. D. Smith, *Science* **166**, 733 (1969).
6. Lunar Sample Preliminary Examination Team, *Science* **165**, 1211 (1969); D. A. Flory, B. R. Simoneit, D. H. Smith, *Apollo 11 Organic Contamination History*, NASA Manned Spacecraft Center Report, August 1969.
7. Samples obtained from I. Kaplan, University of California, Los Angeles.
8. We acknowledge the contributions of G. Thomas (metallurgy), T. Eberhart (electrical engineering), T. Hayes (medical physics) for electron microscopy; G. Somorjai (chemistry) for auger spectroscopy; D. H. Smith, M. Petrie, F. C. Walls, P. Harsanyi, C. Sprowls (Space Sciences Laboratory) for technical assistance in high resolution mass spectrometry and computation. Supported by NASA MSC contract NAS 9-7889.

4 January 1970

Search for Alkanes of 15 to 30 Carbon Atom Length

Abstract. *A 50-gram sample of lunar fines was subjected to stepwise extraction in a mixture of benzene and methanol while intact, after being pulverized, and after being digested in hydrofluoric acid. None of these three extracts contained detectable quantities of C_{15} to C_{30} alkanes. No C_{15} to C_{30} alkane was present in this lunar sample at a concentration exceeding 1 part per billion by weight.*

Alkanes are saturated organic compounds that are composed solely of carbon and hydrogen atoms. These hydrocarbons are ubiquitous but minor constituents of the waxes, fats, and oils of plants or animals (*1*). Because alkanes are less readily assimilated than other food substances and are chemically less reactive than most biological compounds, alkanes are preferentially preserved relative to other organic materials in sedimentary environments. Natural gas and petroleum deposits commonly appear to be concentrates of alkanes partially derived from preexistent organisms (*2*). Many alkanes from petroleum have molecular structures that are either identical or similar to alkanes and related compounds in plants, animals, soils, marine sediments, and sedimentary rocks (*3*). The marked structural and distributional resemblances between certain alkanes in rocks and compounds in organisms have led to the widespread use of alkanes as molecular or chemical fossils (*4*) and to the recommendation that C_{15} to C_{30} alkanes be employed in exobiological research (*5*). This report deals with a search for C_{15} to C_{30} alkanes in a lunar sample.

All of the lunar fines allocated for organic analyses came from the 15-kg bulk sample. This sample was collected near the side of the landing module that faced to the lunar northeast (*6*). A 50-g portion of lunar fines with grain sizes in the 20 μm to 1 mm range was used. Repeated tests established that all solvents, reagents, and apparatuses were free of detectable amounts of organic contaminants with vapor pressures equivalent to C_{15} to C_{30} alkanes. The gas-liquid chromatograph at the sensitivity setting used to monitor contamination gave 5 to 12 percent of full scale deflections with 10^{-7} g of pristane (2,6,10,14-tetramethylpentadecane), a C_{19} alkane. Sensitivity checks were run on standard solutions of pristane before and after each gas-liquid chromatographic analysis of a lunar extract sample.

The lunar fines were initially extracted with an azeotropic mixture of benzene and methanol in a specially designed combination Soxhlet extractor and ball mill (*7*). Throughout these extractions, a positive pressure of nitrogen was maintained in the system. Nitrogen flowed through a drying tower filled with silica gel into the extractor, and the nitrogen stream was exhausted through a bubbler tube with a 1 cm head of mercury. After approximately 500 ml of solvent per hour had been refluxed through the lunar fines for a period of 170 hours, a 200 ml sample of solvent was collected as it flowed from the extractor. The volume of this sample was reduced to several microliters by removal of solvent in a stream of filtered nitrogen at 40°C, and all of the residual sample that could be taken into a 10-μl syringe was injected into a Apiezon L coated capillary gas-liquid chromatographic column. No organic material except solvent was detected in this sample. The extractor was then disconnected from the reflux condenser and distillation flask, sealed with two glass stoppers, and rolled for 48 hours on a ball mill. All the extraction solvent was transferred from the distillation flask, and the volume of the extract was reduced by removal of solvent in a stream of filtered nitrogen at 40°C to approximately 100 μl. After a 2-μl portion of the remaining extract failed to yield detectable C_{15} to C_{30} alkane peaks in the gas chromatograph, the volume of the extract was further reduced to a few microliters, which were injected into the gas chromatograph. Only a solvent peak was observed in this chromatogram.

The extractor, containing the ball-milled lunar sample, was reconnected to the reflux condenser and distillation flask, and azeotropic benzene and menthanol were refluxed through the pulverized sample at a rate of approximately 500 ml per hour for 72 hours. At the end of this period, the solvent bumped, causing the mercury within the bubbler to be drawn into the extractor. The extraction was discontinued at the time of this mishap, and gas chromatographic analyses of the extract of the pulverized lunar sample indicated an absence of detectable quantities of organic materials with vapor pressures equivalent to C_{15} to C_{30} alkanes.

All contents of the extractor, except for the stainless steel grinding balls, were transferred to a 2-liter Teflon beaker. Mercury was removed from the pulverized lunar minerals with an acid cleaned gold wire, and 500 g of 48 percent hydrofluoric acid was poured slowly with stirring into the beaker. The hydrofluoric acid layer in the beaker was covered with a 5-cm layer of benzene, and the contents of the beaker were stirred periodically until reactions ceased. The beaker was then warmed for a period of approximately 2 hours to 80°C. After most of the benzene layer evaporated, the reaction mixture was transferred to a 2-liter separatory funnel containing 400 ml of distilled water. This mixture was shaken thoroughly with three successive 200-ml portions of benzene. These benzene extracts were composed and concentrated. These extracts did not contain detectable quantities of volatile organic materials.

The lunar fines analyzed contained no C_{15} to C_{30} alkane at a concentration exceeding 1 ppb by weight, whereas the concentrations of C_{15} to C_{30} alkanes in rocks from the surface of the earth commonly exceed 100 ppm by weight.

W. G. MEINSCHEIN
Department of Geology, Indiana University, Bloomington
EUGENE CORDES, V. J. SHINER, JR.
Department of Chemistry, Indiana University

References and Notes

1. H. W. Gerarde and D. F. Gerarde, *Ass. Food Drug Officials U.S. Quart. Bull.* **26**, 65 (1962); J. D. Mold, R. E. Means, R. K. Stevens, H. M. Ruth, *Biochemistry* **3**, 1293 (1964); G. Eglinton and R. J. Hamilton, *Science* **156**, 1322 (1967).
2. E. G. Baker, *Science* **129**, 871 (1959); W. G. Meinschein, *Bull. Amer. Ass. Petrol. Geol.* **43**, 925 (1959); ———, Y. M. Sternberg, R. W. Klusman, *Nature* **220**, 1185 (1968).
3. W. G. Meinschein, *Bull. Amer. Ass. Petrol. Geol.* **43**, 925 (1959); *Geochim. Cosmochim. Acta* **22**, 58 (1961); J. G. Bendoraitis, B. L. Brown, L. S. Hepner, *Anal. Chem.* **34**, 49 (1962); R. A. Dean and E. V. Whitehead, *Tetrahedron Lett.* No. 21, 768 (1961); A. L. Burlingame, P. Haug, T. Belsky, M. Calvin, *Proc. Nat. Acad. Sci. U.S.* **54**, 1406 (1965); I. R. Hills and E. V. Whitehead, *Nature* **209**, 977 (1967); ———, D. E. Anders, W. E. Robinson, *Chem. Commun.* **1966**, 752 (1966); R. B. Johns, T. Belsky, E. D. McCarty, A. L. Burlingame, P. Haug, H. K. Schnoes, W. Richter, M. Calvin, *Geochim. Cosmochim. Acta* **30**, 1191 (1966).
4. W. G. Meinschein, *Bull. Amer. Ass. Petrol. Geol.* **43**, 925 (1959); B. Nagy, W. G. Meinschein, D. J. Hennessy, *Ann. N.Y. Acad. Sci.* **93**, 25 (1961); W. G. Meinschein, B. Nagy, D. J. Hennessy, *ibid.* **108**, 533 (1963); W. G. Meinschein, *Nature* **197**, 833 (1963); ———, E. S. Barghoorn, J. W. Schopf, *Science* **145**, 262 (1964); G. Eglinton, P. M. Scott, T. Belsky, A. L. Burlingame, M. Calvin, *ibid.*, p. 263; E. S. Barghoorn, W. G. Meinschein, J. W. Schopf, *ibid.* **148**, 461 (1965); J. Oro, D. W. Nooner, A. Zlatkis, S. A. Wikstrom, E. S. Barghoorn, *ibid.*, p. 77; P. E. Cloud, Jr., J. W. Gruner, H. Hagen, *ibid.*, p. 1718; T. Belsky, R. B. Johns, E. D. McCarty, A. L. Burlingame, W. Richter, M. Calvin, *Nature* **206**, 446 (1965); G. Eglinton and M. Calvin, *Sci. Amer.* **216**, 32 (1967); J. Oro and D. W. Nooner, *Nature* **213**, 1082 (1967).
5. W. G. Meinschein, "Preliminary proposal for government contract research on development

of hydrocarbon analyses as a means of detecting life in space, Esso Research and Engineering Company," submitted to NASA, 31 January 1962.
6. E. A. King, Jr., "Lunar Sample Information Catalog Apollo 11, Lunar Receiving Laboratory Science and Applications Directorate," Manned Spacecraft Center, Houston, Texas (31 August 1969).
7. W. G. Meinschein, C. Frondel, P. Laur, K. Mislow, *Science* **154**, 377 (1966).
8. Sponsored by NASA contract NAS 9-9974. Development of laboratory for biogeochemical analyses supported by NSF grant GB-6583. We thank J. M. Mitchell, T. J. Jackson, R. P. Newlin, and S. M. Scherer for technical assistance.

4 January 1970

Fluorometric Examination of a Lunar Sample

Abstract. *We have been unable to detect porphyrins in 13 grams of the bulk fine lunar sample from the Sea of Tranquillity under conditions in which less than 10^{-13} mole per gram of lunar sample could have been detected. By appropriate extraction, however, the lunar sample yields a material which exhibits absorption maxima at 310 and 350 nanometers and a fluorescence maximum at 410 nanometers.*

Since porphyrins are essential components of all life, their presence in lunar samples would suggest that life once existed on the moon. Porphyrins have characteristic absorption spectra, which we have measured by energy-corrected fluorescence activation on nanogram quantities in organic solvents. We have examined several solvent extracts of a lunar surface sample for the presence of porphyrins by spectrophotometry and the highly sensitive linear-energy spectrophotofluorometry (LEF) (*1*). The LEF method in general permits measurement of absorption spectra of fluorescent compounds at a sensitivity of the order of 100 to 1000 times that of absorption spectrophotometry.

Our LEF (Perkin-Elmer) apparatus automatically compensates for variations of energy with wavelength of both the excitation light source and the detector sensitivity in such a way that the spectrum is free from distortion. The instrument uses a high-pressure xenon arc source, an EMI 9558 photomultiplier (trialkali photocathode tube), and a photon counter. We have used it to record the complete absorption spectrum of 10^{-12} mole of porphin in 0.2 ml of methylene chloride, and similarly spectra of the nickel mesoporphyrin IX dimethyl ester at 10^{-11} mole and vanadyl tetra-meso-beta-naphthyl porphin at 10^{-12} mole in the same volume.

The extraction scheme is based on the use of sequential Soxhlet extractions with methanol and benzonitrile (*2*) for 24 hours to remove loosely held material, followed by a 10-hour period of ball-milling with a melt of methane sulfonic acid–naphthalene (1:1, by weight) —at 50°C in order to remove any material bound in solid carbon phases (*3*). Each extract was worked up separately for fluorescent residues.

Fluorometric grade methanol (Harleco) and scintillation grade naphthalene (Packard) were used without treatment, but reagent grade benzonitrile and methane sulfonic acid were distilled freshly before use. Glassware was cleaned with hot aqua regia and rinsed with water and fluorometric grade methanol before use. Teflon sleeves were used on all ground joints. All operations on the sample were carried out in a dry box under an argon atmosphere.

Tests for possible contaminants in the extraction solvents were run as follows. Identical volumes of the solvent systems used on the sample were used in the same way to treat a 13-g sample of clean 200-mesh optical quartz. The test extracts were treated and examined exactly as those from the sample were.

Samples 10002,4 (5.2 g) and LRS-1002,4 (7.9 g) of lunar surface powder were pooled as received, and the bulk sample was layered in layers 5 mm thick separated by layers of borosilicate glass wool, about 5 mm thick, in a microporous fluorocarbon polymer Soxhlet thimble. The layering of the powdery sample was necessary in order to promote free access and percolation of the extraction solvent. The sample was placed in a double-jacketed Soxhlet apparatus so designed that the thimble and contents were bathed in the vapor of the refluxing solvent and thus extracted efficiently by solvent at its boiling point. The sample was extracted first with 200 ml of fluorometric grade methanol for 24 hours. The methanol extract was then filtered through an ultrafine-porosity glass filter to remove trace solids and concentrated to near dryness in a Buchler evaporator. The extract was dried under a stream of nitrogen, then taken up in 2 ml of methylene chloride for examination of the methanol extract. The sample was next extracted with 200 ml of benzonitrile under reflux at about 67°C and a pressure of 4 to 5 mm-Hg, filtered as before, and finally concentrated to dryness by vacuum distillation and taken up in 2 ml of methylene chloride for examination of the benzonitrile extract. The sample was freed of benzonitrile by brief Soxhlet extraction with methanol and transferred to a 500-ml glass-stoppered borosilicate reagent bottle. Alundum balls were added, together with methane sulfonic acid (28.9 g) and naphthalene (28.9 g), and the mixture was ball-milled as a melt at 54 rev/min for 10 hours at a temperature kept at 40° to 50°C by means of a heat lamp. The methane sulfonic acid–naphthalene sample slurry was transferred by water washes into a round-bottom flask, and excess naphthalene and water were removed on a Buchler evaporator at a pot temperature of 42°C and pressure of 0.5 mm-Hg. Water (200 ml) and methylene chloride (100 ml) were used, with scraping as needed to transfer the black mineral slurry to a separatory funnel equipped with a Teflon stopcock. The glass wool used in layering the sample in the thimble was removed into an ultrafine-porosity sintered glass filter apparatus and washed with water until the washes were approximately neutral; these washes were added to the main body of the acidic aqueous mineral slurry in the funnel. The aqueous phase was washed three times with 100 ml of methylene chloride, the yellow fluorescent organic phase was discarded, and the aqueous phase was centrifuged in 150-ml glass centrifuge bottles. The clear acidic aqueous supernatant was returned to the separatory funnel. The precipitate was washed twice with 50 ml of water, and these washes were added to the contents of the separatory funnel. The clear aqueous phase, with its presumed dicationic metal-free porphyrins, was brought to *p*H 8 by the addition of 1*N* KOH and extracted four times with 100 ml of methylene chloride. The methylene chloride phase was then concentrated to 2 ml under a stream of nitrogen for examination as the "methane sulfonic acid–naphthalene extract."

Spectral study of all extracts and solvent blanks before and after addition of methane sulfonic acid to demetallate any metallic porphyrins present (*4*) showed that the benzonitrile extract contained a material which exhibited a fluorescence peak at 690 nm when excited at 410 nm. However, on the basis of its behavior in acid and its spectral properties, we do not believe that this compound is a porphyrin. We therefore believe that porphyrins were not present in the lunar sample above the detectability limits of our method—that is, above 10^{-12} mole/g of sample (linear energy spectrofluorometer) or 10^{-13} mole/g of sample (conventional spectrofluorometer).

We have, however, found other fluorescent materials in the lunar sample but not in the solvent control blanks. Thus the benzonitrile and methane sulfonic acid–naphthalene extracts exhibit absorption maxima at 310 and 350 nm with a fluorescence maximum at 410 nm. This material is present in the lunar sample at concentrations of the order of 10 parts per billion (quinine sulfate equivalent). Because of the low concentration of the fluorescent material in the sample we have been unable to identify it by high resolution mass spectrometry. Extraction of an ancient terrestrial sample (Gunflint chert) by the methods outlined above yields a fluorescent material similar in its properties to those of the material present in the lunar sample.

JOON H. RHO, A. J. BAUMAN
Jet Propulsion Laboratory, California Institute of Technology, Pasadena 91103

T. F. YEN
University of Southern California, Los Angeles 90033

JAMES BONNER
California Institute of Technology

References and Notes

1. W. Slavin, R. W. Mooney, D. T. Palumbo, *J. Opt. Soc. Amer.* **51**(1), 93 (1961).
2. T. F. Yen, L. J. Boucher, J. P. Dickie, E. C. Tynan, G. B. Vaughan, *J. Inst. Petroleum* **55**, 87 (1969).
3. T. F. Yen, unpublished results.
4. J. G. Erdman, U.S. Patent No. 3-190-829 (June 1965).
5. We thank J. R. Thompson for technical assistance. This paper presents the results of one phase of research carried out at Jet Propulsion Lab under NASA contract NAS 7-100.

5 January 1970

Search for Organic Material in Lunar Fines by Mass Spectrometry

Abstract. *Three kinds of experiments were performed in an effort to detect and identify organic compounds present in the lunar material (fines 10,086): (i) vaporization of the volatilizable components directly into the ion source of a high-resolution mass spectrometer, and (ii and iii) extraction of the material with organic solvents before and after dissolving most of the inorganic substrate in hydrochloric and hydrofluoric acid. The extracts were investigated by a combination of gas chromatography and mass spectrometry. Although a number of organic compounds or compound types have been detected, none appears to be indigenous to the lunar surface.*

In an effort to characterize the organic compounds present in the surface covering the Apollo 11 landing site, a portion of fines (sample 10,086) has been subjected to mass spectrometric analysis. Three different types of experiments were undertaken: (i) vaporization of organic compounds present in lunar material; (ii) solvent extraction of lunar material; and (iii) dissolution of lunar material in acid followed by extraction of the aqueous phase.

The first experiment involved placing the lunar material (0.4 g) in a small glass bulb mounted directly on the ion source chamber of a high-resolution, double-focusing mass spectrometer (CEC 21-110B). The bulb was positioned in such a manner that any material evolving from the sample would vaporize directly into the ionizing electron beam (*1*). This approach involves the least risk of contamination and provides maximum sensitivity. The mass spectra were recorded in a number of consecutive exposures on a photographic plate (*2*); thus an integrated permanent record was obtained of all ions produced in the 20-minute period during which the sample was heated from ambient temperature to 400°C. Comparison of the resulting exposures with the spectra obtained from known amounts of *n*-decane introduced through the batch-inlet system of the spectrometer indicates that the 0.4 g of lunar material produced about 0.1 to 1.0 μg of organic material—that is, about 0.25 to 2.5 parts per million.

The composition of the ions produced is given in Table 1. Most of the species are hydrocarbon ions of various degrees of unsaturation. In addition to these, a few series of ions were found that contained one, two, or three oxygen atoms (mono-oxygenated ions predominating), and some ions containing one nitrogen atom. Furthermore, a few ions contained nitrogen and oxygen, and some rather abundant ions contained sulfur either alone or in combination with oxygen. The sulfur-containing species were either free of carbon (elemental sulfur, hydrogen sulfide, and sulfur dioxide) or had only one carbon atom (carbon disulfide and carbon oxysulfide).

A few of the ions may represent compositions with three nitrogen atoms, but these differed by only 1.3 millimass units from those having C_2H_2O instead of N_3, and the two species were therefore difficult to resolve. The absence of species with two nitrogens would argue against the presence of molecules containing more than one nitrogen atom.

Although we must emphasize that the species listed in Table 1 represent ions formed in electron impact and do not necessarily indicate molecules present as such, one can draw a number of conclusions from these data. First, there seems to be present a series of hydrocarbons having up to nine carbon atoms and ranging from almost completely saturated species (C_8H_{16} corresponds to a hydrocarbon with one double bond or ring) to more unsaturated ones. The aromatic ions (C_9H_7, C_8H_9, C_7H_8, and C_6H_6, which would correspond to an indenyl ion, styrene, toluene, and benzene) appeared to be present in much lower abundance, despite the high relative intensity of these ions in the mass spectra of the corresponding pure compounds.

Aside from the large quantities of CO_2, the most abundant oxygen-containing ions consisted of one or two carbons and one or two oxygens, as they may be produced upon ionization of small alcohols or glycols. Such ions might have come from either ethylene oxide or its hydrolysis product, ethylene glycol. (Ethylene oxide was used as a sterilizing agent at the Lunar Receiving Laboratory.) Similarly, the $C_2H_4O_2$ (the elemental composition of acetic acid) may be a reduction product of peracetic acid, also used as a sterilizing agent at LRL, whereas the ions containing three carbon atoms and one oxygen atom may have been derived from acetone which in turn may have been produced by pyrolysis of salts of acetic acid. The $C_8H_5O_3$ ions were probably a small trace of the common laboratory contaminants, dialkyl phthalates.

The nitrogen-containing ions all represent small molecules or fragments thereof, and some of them (indicated in Table 1 by an asterisk) had been previously found in the products of the lunar module retro-rocket exhaust (*3*). Last but not least, there appears to be a considerable amount of sulfur present, in the form of elemental sulfur, hydrogen sulfide, carbon disulfide, and sulfur dioxide.

Table 1. Ions observed upon heating lunar sample (0.4 g) to 400°C and volatilizing the material into the ion source of a high-resolution mass spectrometer. ×, Also present in background;*, found in rocket exhaust (*3*); +, relative intensity.

C,H	
++C_2H_4 ×, *	++C_5H_9 ×
++C_2H_5 ×	++C_5H_{10}
+C_2H_6	++C_5H_{11}
++C_3H	+C_6H_2
++C_3H_2 ×, *	+C_6H_3
+++C_3H_3 ×, *	+C_6H_4 *
++C_3H_4 *	+C_6H_5
++C_3H_5 ×	+C_6H_6 ×
++C_3H_6 ×	+C_6H_7
+++C_3H_7 ×	+C_6H_9
+C_3H_8	+C_6H_{10}
+C_4H	++C_6H_{11}
++C_4H_2 *	+C_6H_{12}
++C_4H_3	+C_6H_{13}
+C_4H_4 *	++C_7H_7
++C_4H_5	+C_7H_8
+C_4H_6	+C_7H_9
+++C_4H_7	+C_7H_{11}
+++C_4H_8 ×	+C_7H_{12}
+++C_4H_9	+C_7H_{13}
+C_5H_2	+C_8H_9
+C_5H_3	+C_8H_{13}
+C_5H_4	+C_8H_{15}
+C_5H_5	+C_8H_{16}
+C_5H_6	+C_9H_7
++C_5H_7	+C_9H_{11}
+C_5H_8	
C,H,O	
+++CO_2 *	+C_3H_5O
++CH_2O	++C_3H_6O
++CH_3O	+C_3H_7O
++CH_4O *	+C_4H_6O
+C_2O	+C_4H_8O
+C_2HO	+C_6H_6O
++C_2H_2O *	+C_7H_4O
+++C_2H_3O ×, *	+C_7H_5O
+++C_2H_4O *	+CH_2O_2
+++C_2H_5O	++$C_2H_4O_2$
+C_3HO	+$C_3H_5O_2$
+C_3H_2O	+$C_8H_3O_2$
++C_3H_3O ×	+$C_4H_2O_3$
+C_3H_4O	+$C_8H_5O_3$
C,H,N,O	
+CH_2N	+C_3H_3N
+CH_4N	+CNO *
+C_2N	+++HCNO *
++C_2HN *	+HNO
++C_2H_2N	+++NO_2 *
+C_2H_3N *	[++HN_3 *]
+C_2H_4N *	[+HCN_3 *]
+C_2H_6N	[+CH_3N_3 *]
C,H,S,O	
+++S	+COS
+HS	+++SO
++H_2S	+++SO_2
++CS	
+CS_2	

Thus it appears that the organic material which can be volatilized out of the sample represents small molecules which are probably of terrestrial origin, with the exception of CO_2, SO_2, H_2S, COS, CS_2, and elemental sulfur. These are probably indigenous to the sample, but it should be noted that a sulfur-base caulking compound was used to seal various partitions within the cabinet system at LRL. It should also be pointed out that most of these species began to appear in the spectra when the sample had reached temperatures of about 150° to 200°C and disappeared in later exposures taken at a sample temperature of 400°C. This makes it unlikely that they are pyrolysis products of otherwise nonvolatile material. Equally significant is the absence of a series of hetero-aromatic systems such as pyridines, furanes, and thiophenes, since they would be produced from the pyrolysis of more complex heteroatom-containing molecules which would be of greater interest to the organic chemist and biochemist. Particularly in view of the high relative intensity of the molecular ion of these types of compounds, it should have been possible to detect them in rather small amounts (a few nanograms).

The thermal release of organic material from an inorganic matrix is a simple, but drastic and sometimes inefficient, technique. Solvent extraction or removal of the inorganic material by dissolution in acid, followed by extraction of the liberated organic substances, is a commonly employed alternative to the direct volatilization described above. A 50-g portion of the fines (10,086) was extracted with 50 ml of benzene-methanol (1 : 1) in a sealed ampule by shaking in an ultrasonic bath for 48 hours at 60°C. After settling, the solvent was decanted and the residue was stirred with more solvent, then combined with the first batch and evaporated in a stream of purified helium to a fraction of a milliliter.

The insoluble residue was dissolved in (i) hydrochloric acid and (ii) hydrofluoric acid. The respective acid solutions were diluted, neutralized, and extracted with dichloromethane (*4*). These extracts were dried with magnesium sulfate and concentrated in a stream of helium to 0.1 to 0.2 ml each. Another aliquot (30 g) of the sample was treated with hydrochloric acid only, without prior extraction with benzene-methanol. The acidic solution was neutralized and extracted with dichloromethane. This extract was dried and concentrated as described above. It should be noted that, during the treatment of the sample with acids, hydrogen sulfide is evolved.

Since it was our major aim in these experiments to detect the presence of various types of organic compounds other than hydrocarbons, the resulting extracts were subjected to gas chromatography in which a general-purpose column (OV-17) was utilized rather than one that is specifically suited for a certain compound type. The gas chromatograph is operated in conjunction with a low-resolution mass spectrometer coupled to a computer, to enable one to permanently record all data in a form that then can be analyzed with the help of the computer (*5*). In none of the extracts discussed above, with the exception of elemental sulfur (from sulfides) were there any significant peaks in the gas chromatogram that could be attributed to material indigenous to the lunar sample. A number of compounds were identified, such as toluene, isopropylbenzene, diisopropyl disulfide, ethyl isopropyl disulfide, methyl isopropyl disulfide, phenol, diethyl phthalate, dibutyl phthalate, biphenyl, styrene, naphthalene, dihydronaphthalene, indane and anthracene. However, with the possible exception of the last six aromatic hydrocarbons, these compounds appear to be traces of contaminants introduced during the analytical procedure. The detection of the dialkyl disulfides is somewhat surprising, but these may be reaction products of the solvent (perhaps traces of higher alcohols in the methanol) with free sulfur or metal sulfides that were indigenous to the lunar sample. This possibility should be further investigated, but the relatively high volatility of these disulfides and their absence in the direct vaporization experiment (Table 1) seem to eliminate the possibility that they are indigenous to the moon. High-resolution mass spectra of the extracts discussed above have been obtained and are in process of being analyzed.

From our experiments we conclude that there are no volatile or extractable organic compounds exceeding the part-per-billion amounts region in the surface material returned from the Sea of Tranquillity by the Apollo 11 mission.

R. C. Murphy, G. Preti
M. M. Nafissi-V., K. Biemann
Department of Chemistry,
Massachusetts Institute of Technology,
Cambridge 02139

References and Notes

1. J. M. Hayes and K. Biemann, *Geochim. Cosmochim. Acta*, 32, 239 (1968).
2. Technical Operations, Inc., mass spectrometer plates (evaporated silver bromide) were used because of their relatively high sensitivity attributable to their lack of grain and fog.
3. B. R. Simoneit, A. L. Burlingame, D. A. Flory, I. D. Smith, *Science* 166, 733 (1969).
4. Organic solvents were Mallinckrodt "Nanograde." "Ultrex" hydrochloric and hydrofluoric acid were supplied through the courtesy of Dr. Morris Zief of J. T. Baker; water was doubly distilled from alkaline potassium permanganate.
5. R. A. Hites, thesis, Massachusetts Institute of Technology (1968); ——— and K. Biemann, in preparation.
6. This work is supported by NASA research contract NAS 9-8099.

5 January 1970

Organic Analysis of the Returned Lunar Sample

Abstract. *Lunar fines have been examined for organic compounds by crushing, programmed heating, hydrofluoric acid etching, and solvent extraction. Products were examined by mass spectroscopy. A variety of small organic molecules, including methane and other hydrocarbons, accompanied the release of the rare gases when the sample was heated in a stepwise fashion to 900°C under vacuum. Methane is more abundant (abundance on the order of 1 part per million) than argon in the matrix-entrapped gases liberated by hydrofluoric acid etching of lunar fines. Methane is also present in a dark portion of the gas-rich meteorite Kapoeta.*

We have undertaken an analytical procedure for detecting and identifying carbon compounds over a wide range of molecular weights. In particular we have attempted to locate any compounds of carbon formed as a result of bombardment of the surface of the moon by solar wind and cosmic rays (*1*). The solar-wind flux of hydrogen, carbon, nitrogen, oxygen, and other elements might be expected to result in the formation of organic compounds in the shallow surface layer of the exposed particles in the lunar fines and surface rocks. We have made approximate calculations of the ratios of the amounts of the elements that might be expected, but it seems impossible at this stage to judge the extent of the interaction between them on implantation in the mineral matrix. Methane is one possible product.

The sample container was opened and the aliquots of sample 10086.19 were transferred, in the flow of a clean air cabinet, to glass ampules which were then sealed under purified nitrogen. Standard clean room equipment, tools, and apparatus were used throughout. All the equipment had been previously cleaned through treatment with hot concentrated chromic acid, followed by washing with a solvent or by heating to 500°C and cooling under nitrogen. Exposure times were kept as brief as possible in order to minimize laboratory contamination.

We brought about partial dissolution of lunar fines as follows. A vacuum-degassed sample of 40 percent aqueous HF (about 1 ml) and a sample of lunar fines (about 200 mg), both under vacuum, were mixed by breaking a thin glass diaphragm. The reaction was allowed to proceed at room temperature for about 30 minutes, after which time the breakseal connecting the reaction vessel with the thoroughly evacuated and degassed fractionating system was broken. The fractionating system consisted of two traps in series, cooled in liquid nitrogen, followed by a 5-Å molecular-sieve trap and an open U-tube trap, sealed onto the inlet valve of the mass spectrometer (Varian MAT CH-7). The molecular sieve trap could be bypassed. The two traps cooled by liquid nitrogen retained the reactants and the reaction products, and all organics other than methane. Incorporation of the molecular-sieve trap into the system permitted a fractionation step, with argon, methane, oxygen, and nitrogen being retained at liquid-nitrogen temperatures, but not when the sieve was warmed to 20°C. Portions of a standard gas mixture (80.7 percent Ne; 7.5 percent Ar; 1.9 percent Kr; 7.8 percent CH_4, and 2.1 percent C_2H_6) were admitted in order to check sensitivity and to obtain a calibration point for approximate quantitation of sample gases. The minimum detectable quantity of any single gas was about 10^{-6} cm³ STP. Methane was conclusively identified by the fragmentation pattern and by the *m/e* value at the resolution available (1000, 10 percent valley).

The etching procedure released 3×10^{-3} cm³ of methane (1 ppm by weight) and about 9×10^{-4} cm³ of ^{36}Ar from 1 g of the lunar fines, while about 6×10^{-5} cm³ of methane and 7×10^{-6} cm³ of ^{36}Ar were released from 1 g of gas-rich chondrite, Kapoeta. Smaller amounts of methane were observed in the Pueblito de Allende C3 chondrite. No methane was observed in an apparatus and reagent blank. A sample of baked out alumina which had been handled in parallel with the lunar sample since its arrival in Bristol yielded no methane. Nonterrestrial rare gases were observed from the meteorites and from the lunar sample. Amounts and isotopic abundances of hydrogen, helium, neon, and argon observed in the lunar fines were of the same order as those already measured (*2*).

Our initial studies with the lunar fines were directed toward pulverizing them under high vacuum and examining liberated gases by mass spectrometry (*3*). Solar gases were indeed liberated (for example, ^{36}Ar, ^{38}Ar, ^{40}Ar, ^{20}Ne, ^{22}Ne) but we were unable to degas the stainless steel capsule sufficiently to achieve a good blank for organic gases.

The Burlingame-Biemann experiment, which was part of the preliminary examination conducted at the Lunar Receiving Laboratory, involved the stepwise heating of small amounts (~ 100 mg) of lunar samples, the products being evolved directly into the ion source of a sensitive, fast-scanning mass spectrometer (*2*). These investigators announced the low levels (less than 1 ppm) of organic compounds in the Apollo 11 samples and also emphasized the variation in results which would arise from particulate contamination of small samples.

We have carried out a similar pyrolysis but on a larger scale, and with the introduction of two traps (solid carbon dioxide and acetone, and liquid nitrogen) to effect preliminary fractionation in a fashion similar to that described for the etching experiment. The temperature of the sample (2.6 g), which was contained in a quartz vessel held at 10^{-6} torr, was raised in 100°C steps (of ~ 5 minutes each) and then held for about 15 minutes; the baked-out head-space gases, both untrapped and trapped in liquid nitrogen, were examined at each step by opening the system to the mass spectrometer. The total ion current was recorded continuously, and mass spectra were taken at frequent intervals. The system blank gave N_2, O_2, CO_2, and NO (mass-to-charge ratio, 30). Gas evolution was incomplete at each step, but trends are

Table 1. Gas content of pyrolized lunar sample; +++, ≥ 20 percent of total ion current; ++, ≥ 10 percent; +, 0 to 10 percent for both trapped and untrapped fractions.

Gas	Temperature							
	130°–200°C	200°–300°C	300°–400°C	400°–500°C	500°–600°C	600°–700°C	700°–800°C	800°–900°C
H_2	+	+	++	++	++	++	+	+
^{4}He	+	+	+	+	+	+	+	+
^{20}Ne				+	+	+	+	+
^{22}Ne				+	+	+	+	+
^{36}Ar				+	+	+	+	+
^{38}Ar				+	+	+	+	+
^{40}Ar				+	+	+	+	+
N_2	+	+	+	+	+	+	+	+
O_2	+	+						+
CO	+++	+++	+++	+++	+++	+++	+++	+++
CO_2	+++	+++	+++	+++	+++	+++	+++	+++
NH_3		+	++	++	+			
CH_4	+	+	++	++	++	+		
C_2H_4	+	+	+	++	+			
C_4H_8	++	+	++	+++	+			
C_6H_6	+	+	++	+++	+++	++	+	+
C_7H_8	+	+	+	++	++	+	+	+
C_8H_8				+	+	+		
C_8H_{10}				+	+	+	+	+
C_9H_8					+	+	+	+
C_9H_{10}					+	+	+	+
C_9H_{12}						+	+	+
C_4H_4S		+	+	+	+			

clearly evident from the data (Table 1). The structures of Table 1, assigned by low-resolution mass spectrometry, will be examined in greater detail by high-resolution mass spectrometry. Proper quantitative data are yet to be derived, but the most striking features are:

1) Carbon monoxide and carbon dioxide are continuously evolved, and in especially large amounts at the higher temperatures.

2) Methane appears in substantial quantity. It is accompanied by a variety of unsaturated and aromatic hydrocarbons which may be indigenous. Thermal alteration of indigenous material or contamination would also produce these hydrocarbons. Although evolution of methane virtually ceases at temperatures above 700°C, it is still liberated in quite large quantities by subsequent HF etching of the pyrolyzed sample together with further amounts of the rare gases (other organic compounds liberated would not be seen on account of the trapping system used).

3) The rare gases first appear at temperatures near 300°C, and they continue to be evolved thereafter; there is some variation in the observed isotope ratios, however, presumably due to selective baking out of sites and the presence of terrestrial gases at the lower temperatures.

We have constructed a special, interconnected, glass and Teflon closed system within which to carry out, in a helium atmosphere, the classical geochemical procedures of solvent extraction, separation, evaporation, derivatization, and chromatography, using small volumes of solvent, with brief exposure of samples, extracts, and fractions to the atmosphere and the laboratory environment. We have monitored organic compounds extractable from the lunar fines (14.8 g) with a mixture of benzene and methanol (3:1, 15 ml) during brief sonication (30 minutes). The evaporated extract, a brown oily gum, was dissolved in methanol (2 μl), and a portion (0.5 μl) of the resulting yellow solution was subjected to combined gas chromatography and mass spectrometry (Varian MAT CH7 system; column, 3.6 m × 1.5 mm, stainless steel; 3 percent OV-17 on Chromosorb W), programmed from 120° to 280°C at 4°C per minute, with a helium flow of 5 ml/minute. The minimum level for detection of n-C_{20} alkane was about 25 ng. Control runs with the same volume of solvent, first alone and then, when used for the extraction of a standard hydrocarbon (1-^{14}C-labeled n-octadecane), adsorbed on powdered alumina (grade 1, neutral) as support, demonstrated (i) the background level of contamination (equivalent to no peak > 20 ng) and (ii) the efficiency of recovery (> 70 percent at the 100-ng level) under operating conditions standardized for extraction and evaporation. We conclude that individual alkanes and other relatively nonpolar compounds, if present, do not occur in concentrations higher than 5 parts per billion in the molecular weight range ~ C_{12} to C_{32} which would be revealed by this study. The remaining portion of the extract was derivatized by heating for 45 minutes at 60°C with the reagent $CF_3CON(SiMe_3)_2$. The product, when analyzed by combined gas chromatography and mass spectroscopy, gave only two distinct peaks, which derived from the reagent alone.

We do not know to what extent the carbon compounds herein reported are truly indigenous to the lunar surface. However, we do believe that the methane is present as such in the inorganic matrix. The higher hydrocarbons, observed only in the pyrolysis experiment, might also be in the matrix, but until a refined HF etching experiment is performed we cannot be sure. Methane has been reported in carbonaceous chondrites by Anders *et al.* (*5*), but one analysis of Allende by us did not reveal significant quantities of this component, although the gas-rich chondrite, Kapoeta, which has a very low carbon content, contains appreciable quantities of methane. Thus, our limited data indicate that the quantity of methane is proportional to the solar wind rare gas content but not to the carbon content. There appears to be no previous report of the presence of methane in gas-rich or ordinary chondrites. Methane, in addition to carbon dioxide and carbon monoxide, has also been obtained from terrestrial igneous rocks when baked at 1000°C (*6*).

The preliminary results obtained for the Apollo 11 rocks and fines (*2*) point to an origin from high-temperature melts subsequently exposed on the surface for some millions of years. The presence of biogenic products is therefore extremely improbable, and the predominant sources of carbon are likely to be the solar wind and meteorite impact. However, two factors make it unlikely that meteorite impact contributes much of the carbon: (i) vaporization and consequent loss on impact, and (ii) low frequency of impact.

The rare gas values for lunar material apparently indicate saturation of the outer few hundred angstroms of the crystals and glasses (*2*). The losses by diffusion and shock events are not known but must have been considerable. The loss of carbon compared to that of the low molecular weight rare gases would, of course, be less, owing to retention by reaction with the matrix; bearing in mind the solar corona proportions, one arrives at a rough estimate of something of the order of 1 to 1000 ppm for carbon attributable to solar-wind implantation. On the basis of these calculations, one might even conjecture that all the carbon, as measured by Moore's experiments (*4*)—that is, of the order of 100 ppm—is provided by the solar wind.

Major organic contaminants expected in the sample from collection, handling in the Biopreparation Facility, and packaging are likely to be (*1*, *2*, *7*) butyl rubber, plasticizers, hydrocarbon oils, Teflon, cellulose fibers, contingency-bag contaminants, peracetic acid, and ethylene oxide and its products from sterilization. However, we have not yet detected these contaminants in our experiments. The bulk fines were collected over a fairly wide area within about 9 meters of the module engine, where they must certainly have received considerable effluent during the final few seconds of the descent of the spacecraft. Oxidizer (N_2O_4) was also vented on that side of the craft. No evidence of rocket exhaust products could be detected on baking a sample at 150°C under vacuum (10^{-6} torr) and examining the trapped gases by mass spectrometry.

Aromatic products similar to those reported here have been obtained from carbonaceous chondrites when heated (*8*). Despite the possibility of contamination both of the lunar samples and of the carbonaceous chondrites, the occurrence of aromatic compounds in the pyrolysis products of both materials may indicate that these compounds or their precursors are truly extraterrestrial.

The indegenous lunar methane could have one or more origins: (i) primordial, (ii) entrapped during crystallization of the rocks, (iii) a primary product of the solar wind, (iv) a secondary product of indigenous or solar wind carbon arising from "hot atom" chemistry (*9*) induced by high energy cosmic rays. We hope to distinguish between some of the above possibilities by more detailed work. The reaction between the solar wind and the inorganic matrix to give matrix-O-C products remains to be studied.

PAUL I. ABELL, G. H. DRAFFAN
GEOFFREY EGLINTON, JOHN M. HAYES
JAMES R. MAXWELL
COLIN T. PILLINGER

Organic Geochemistry Unit,
School of Chemistry,
The University, Bristol, England

References and Notes

1. G. H. Draffan, G. Eglinton, J. M. Hayes, J. R. Maxwell, C. T. Pillinger, *Chem. Britain* **5**, 296 (1969).
2. Lunar Sample Preliminary Examination Team, *Science* **165**, 1211 (1969).
3. R. Kranz, *Organic Geochemistry. Methods and Results*, G. Eglinton and M. T. J. Murphy, Eds. (Springer, Heidelberg, 1969).
4. C. B. Moore, *Science*, this issue.
5. M. H. Studier, R. Hayatsu, E. Anders, *ibid.* **149**, 1455 (1965).
6. C. G. Barker, *Nature* **201**, 1002 (1965).
7. D. A. Flory, B. R. Simoneit, D. H. Smith, personal communication.
8. J. M. Hayes, thesis, Massachusetts Institute of Technology (1966).
9. C. MacKay and R. Wolfgang, *Science* **148**, 899 (1965).
10. We thank the Science Research Council and the Natural Environment Research Council for financial assistance, and the National Aeronautics and Space Administration for a subcontract covering organic geochemical studies (subcontract NGL-05-003-003, made through the University of California, Berkeley). The award of fellowships from the following bodies is gratefully acknowledged: Science Research Council (to G.H.D. and C.T.P.); Shell International Petroleum Co., Ltd. (to J.R.M.); National Science Foundation (to J.M.H.); and the Petroleum Research Fund of the American Chemical Society (to P.I.A.). The work was conducted during leaves of absence from the universities of Indiana (J.M.H.) and Rhode Island (P.I.A.). We also thank Dr. R. Hutcheson of the British Museum for the sample of Kapoeta, and Mr. B. Simoneit of the Space Sciences Laboratory of the University of California, Berkeley, for the sample of Allende.

5 January 1970

Pyrolysis-Hydrogen Flame Ionization Detection of Organic Carbon in a Lunar Sample

Abstract. *The lunar samples were analyzed for total organic carbon content by pyrolysis at 850°C and subsequent detection of the resulting volatilized organic compounds in a hydrogen flame ionization detector. The organic content ranged from 10 to 126 parts per million, depending primarily upon the handling in the sample processing. No correlation of organic content with rock type has been shown. The small quantities of organic matter are primarily contamination, and the indigenous organic content is estimated to be less than 10 parts per million.*

During the preliminary examination of the Apollo 11 lunar samples at the Lunar Receiving Laboratory, 14 different specimens of lunar material were analyzed for total organic carbon content (*1*, *2*). One additional related sample was analyzed as part of the subsequent lunar sample distribution (*3*). The results of these analyses are given in Table 1.

The analytical method is based upon the response of the hydrogen flame ionization detector (*4*) to the volatile products arising from pyrolysis of the solid sample at 800° to 900°C. Inorganic compounds, including those with carbon as carbonate, graphite, and carbides (silicon and zirconium), give essentially no response under the conditions of the analysis. While the detector is sensitive and specific to organic compounds, the overall analysis is approximate, since both the degree of char formation and the exact re-

Table 1. Organic carbon content of Apollo 11 lunar samples.

Sample		Organic carbon (ppm)*	Comments
Sample number	Type of material		
		Contingency sample	
10,5	Fines	16 ± 7	Sampled immediately
10,26	Fines	48 ± 18	Sampled after 11 days
21,21	Aggregate	44 ± 31	Handled extensively
24,4	Crystalline	17 ± 6	Handled moderately
32,1	Crystalline	10 ± 9	Unhandled previously
		Documented box	
15,10	Fines	126 ± 64	From gas reaction cell
61,4	Aggregate	62 ± 19	Vacuum chamber
64,4	Aggregate	96 ± 30	Vacuum chamber
68,3	Aggregate	39 ± 18	Vacuum chamber (handled specially)
		Bulk box	
02,73	Fines	18 ± 8	Unhandled
86,A	Fines	40 ± 8	Distributed to investigators
46,3	Aggregate	80 ± 26	Handled routinely
48,3	Aggregate	88 ± 43	Handled routinely
49,7	Crystalline	54 ± 23	Handled routinely
50,3	Crystalline	28 ± 19	Handled routinely
56,12	Aggregate	10 ± 4	Handled specially

* The values are obtained from fitting to the log-log calibration plot. The errors quoted are derived from 90 percent confidence limits.

sponse factors for the pyrolysis products are variable with type of sample. The quantitation is based upon comparison with standard samples (*5*) and is confirmed by agreement with results of classical analyses of many terrestrial soils and rocks. Results showing low amounts are obtained only from very ancient sediments where most of the organic matter is present as intractable graphitic material.

From the first organic analysis it was apparent that the organic content of the lunar samples was extremely low and easily obscured by trace contamination. The samples from the documented box processed in the vacuum chamber were the most heavily contaminated. The remaining samples, which were processed under nitrogen with simpler procedures, were less contaminated. In general the organic content correlated directly with the degree and type of sample processing (*2*). The major sources of contamination appeared to be pump oil, fibers, Teflon, rubber-glove fragments, sterilizing agents, and unclean tools and surfaces.

Five of the samples listed in Table 1 had organic contents of less than 20 parts per million. All were processed under nitrogen with minimal handling. All other samples were treated less carefully during the processing scheme. On this basis we have placed our estimate of the indigenous organic matter at less than 10 ppm, realizing that even much of this may be due to trace contamination. It is not possible from these results to compare the organic content of the different rock types and the fines.

The pyrolysis-hydrogen flame ionization detector method is capable of determining the amount of organic carbon but not its source. The organic mass spectrometer used during the preliminary examination (*1*) can resolve individual organic fragments and permit their identification. The results reported from the mass spectra indicated less than 5 ppm of total organic matter including contamination, and less than 1 ppm of indigenous organics. These results are lower by at least 1 order of magnitude than those reported here. The samples were introduced into the mass spectrometer by gradual heating of the sample in a punctured nickel capsule at 500°C in the vacuum of the inlet system. The quantitation of that method was based upon comparison of the ionization current with that produced by known amounts of *n*-tetracosane (boiling point, 320°C) in clean sand. The use of a relatively low pyrolysis temperature, coupled with a quite volatile calibration standard, seemed inappropriate for accurate quantitation of organic compounds, particularly those that have withstood the thermal and vacuum environment of the moon.

This problem was investigated by heating soil samples low in organic content to $500° \pm 50°C$ at 7×10^{-6} torr for 30 minutes, simulating the inlet system of the mass spectrometer. Using our analytical method to compare the organic contents of these samples with portions of the untreated original sample, we were able to show that up to 90 percent of the organic matter could remain unvolatilized in the inlet of the mass spectrometer. Thus, when based upon a volatile standard, the mass spectrometer results could be low by 1 order of magnitude.

Although we have been unable to do more than place a 10 ppm upper limit on the possible indigenous organic matter, we feel that the values from the mass spectrometric analysis may be too low, and that the true levels of indigenous organic matter may lie between these two results.

RICHARD D. JOHNSON
CATHERINE C. DAVIS

NASA Ames Research Center, Moffett Field, California 94035

References and Notes

1. The Lunar Sample Preliminary Examination Team, *Science* **165**, 1211 (1969).
2. "Apollo 11 Preliminary Science Report," *NASA (Nat. Aeronaut. Space Admin.) SP-214* (1969), p. 136.
3. C. Ponnamperuma *et al.*, *Science*, this issue.
4. I. G. Mc William and R. A. Dewar, *Nature* **181**, 760 (1958).
5. The calibration standards consisted of *Serratia marcescens* (43.77 ± 0.67 percent carbon) freeze-dried into incinerated sand at concentrations from 10 to 1000 parts of carbon per million.

5 January 1970

Search for Organic Compounds in the Lunar Dust from the Sea of Tranquillity

Abstract. *A sample of lunar dust was examined for organic compounds. Carbon detected in concentrations of 157 micrograms per gram had a $\delta^{13}C$ per mil (PDB) value of +20. Treatment with hydrochloric acid yielded hydrocarbons of low molecular weight, suggesting the presence of carbides. The gas chromatogram of the acylated and esterified derivatives of the hydrolyzate was similar to that obtained for the Pueblito de Allende meteorite. There were no detectable amounts of extractable high-molecular-weight alkanes, aromatic hydrocarbons, isoprenoid hydrocarbons, normal alkanes, fatty acids, amino acids, sugars, or nucleic acid bases. Traces of porphyrins were found, perhaps arising from rocket exhaust materials.*

A useful approach to the study of chemical evolution is the examination of organic matter in ancient rocks and sediments. The oldest known microfossils are about 3×10^9 years old, indicating that life was already well established at this period (*1*). The possibility, however, of finding rocks on the earth older than 3½ billion years appears to be small. On the other hand, the samples from the moon may give us some clues to prebiotic organic synthesis in the solar system (*2*).

This report contains the results obtained by a group of investigators established as the NASA Ames Research Center Consortium to analyze the sample labeled "10086 bulk A fines." Since this sample consisted of a relatively fine powder, no further grinding was undertaken before analysis.

Our analysis included an examination of the lunar material for total carbon, organic carbon, isotope fractionation, microfossils, and mineralogy. Sequential treatment of the sample by a benzene-methanol mixture, water, and hydrochloric acid provided extracts for examination by chromatographic and spectrometric methods (Fig. 1).

To monitor every stage of the analysis, parallel experiments were conducted on an interior sample from a 6-kg piece of the Pueblito de Allende meteorite (*3*), and a sand blank was prepared by heating a sample of Ottawa sand for 48 hours at 1000°C. To minimize contamination, the analyses were carried out in a clean laboratory ventilated with filtered air, and the entire sequence of solvent extractions of the lunar dust was accomplished in a single glass vessel.

Total carbon was determined by measuring the volume of CO_2 evolved when a 1-g sample was outgassed at

150°C at a pressure of less than 1 μm-Hg and burned at 1050°C. The values ranged from 140 μg/g to 200 μg/g. The most consistent values were between 140 and 160 μg/g. The comparable figure for a sample of the Pueblito de Allende meteorite was 3000 μg/g. The sand blank showed no detectable carbon.

The amount of carbon that could be converted into volatile carbon-hydrogen compounds was determined by pyrolyzing about 30 mg of the dust at 800°C in an atmosphere of hydrogen and helium (*4*). The resulting volatile compounds were estimated by a hydrogen flame ionization detector. The average value obtained was 40 μg/g. For a sample of the Pueblito de Allende meteorite this amounted to approximately 14 μg/g, and for the sand blank, 1 μg/g.

Isotope measurements on the total sample resulted in a $\delta^{13}C$ value of +20 relative to the PDB standard and the $\delta^{34}S$ of +8.2 relative to the Canyon Diablo meteorite. These figures are considerably higher than those generally reported for intact meteorites ($\delta^{13}C$, −4 to −20; $\delta^{34}S$, −2 to +2). A detailed account of these findings and their significance are described by Kaplan and Smith (*5*).

A portion (20 mg) of the sample was pyrolyzed from 80° to 300°C in the ion source of a CEC 110 high-resolution mass spectrometer. The analysis was repeated at 700°C with the high-resolution mass spectrometer (MS 9) at the Jet Propulsion Laboratory in Pasadena. Ions indicating the possible presence of formic acid, acetic acid, and SO_2 were found above the background concentration. However, no conclusive identification was possible. Carbon dioxide was identified in the pyrolysis products at 250°C to 750°C. Observations of dust, surfaces of the microbreccias, and thin sections of microbreccias made by light and electron microscopy yielded no evidence of indigenous biological structures (*6*).

To detect extractable organic compounds, 54.6 g of the lunar dust were treated with a mixture of benzene and methanol (9 : 1). Twenty-three percent of the extract was examined for porphyrins (*7*). Fluorescence excitation and emission spectra of these extracts are suggestive of porphyrins. Since similar spectral responses were observed in exhaust products from tests of the lunar descent rocket engines, it is possible that these pigments were formed from

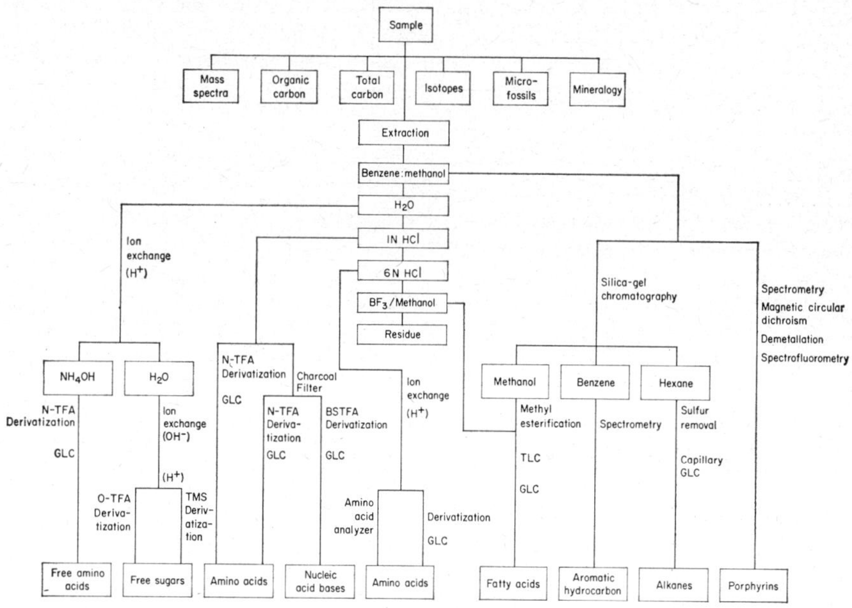

Fig. 1. Scheme of analysis.

components of the rocket exhaust (*8*). The amount of porphyrins detected was approximately 10^{-4} μg/g.

The remainder of the benzene-methanol extract was chromatographed on a silica-gel column and eluted with hexane, benzene, and methanol. Capillary gas-liquid chromatography of the hexane fraction showed that no single *n*-alkane from C_{12} to C_{32} was present at concentrations of 2×10^{-5} μg/g. The absorption spectrum of the benzene eluate showed bands at 224, 274, and 280 nm. Since these absorption bands were also present in the sand and solvent blanks, the presence of aromatics in the lunar sample cannot be inferred. The methanol eluate from the silica-gel chromatography was esterified and examined for fatty acids by gas-liquid chromatography. The C_{12} to C_{32} fatty acids were not detected at levels of 10^{-2} μg/g.

After treatment with benzene-methanol, the bulk fines were dried in a rotary evaporator and extracted with water. This extract was desalted and analyzed for free amino acids and carbohydrates by gas-liquid chromatography as *N*-trifluoroacetyl-*n*-butyl esters of amino acids, as trimethylsilyl derivatives of sugars, and as trifluoroacetyl derivatives of sugar alcohols. In the water extracts, the amino acids were not detected at a concentration of 10^{-5} μg per gram of sample. Sugars were not present at concentrations of 6×10^{-4} μg/g.

The residue, after the water extraction, was hydrolyzed with HCl. Hydrogen sulfide evolved during the hydrolysis was collected. The concentration was about 700 μg/g, with an average $\delta^{34}S$ value of +8.0. The 1*N* hydrolyzate was treated with Norite charcoal to absorb any bases that may have been present. The charcoal was extracted with formic acid, and the extract was treated with bis(trimethylsilyl)trifluoroacetamide and analyzed for the bases as trimethylsilyl derivatives. Purines and pyrimidine bases were not present at concentrations of 4×10^{-3} μg/g.

The hydrolyzate obtained after refluxing the sample with 6*N* HCl at 125°C for 19 hours was desalted and examined for amino acids by ion-exchange chromatography and gas chromatography of the *N*-trifluoroacetyl-*n*-butyl ester derivatives. Although as little as 2×10^{-3} μg/g could be detected by this method, none of the amino acids commonly found in protein appeared to be present.

However, in the HCl hydrolyzates several compounds were found which did not appear on the gas chromatogram unless the hydrolyzates were both esterified and acylated. These molecules were present in a similarly treated acid hydrolyzate of the Pueblito de Allende meteorite but not in the sand blank or the rocket exhaust material (Fig. 2). Organic compounds, as derivatives, appear to be present in both samples.

Since approximately 100 μg of carbon per gram of sample still remained to be accounted for, an attempt was made to determine whether some of the residual carbon was present as carbides. A fresh sample, outgassed in a vacuum at 150°C, was therefore hydrolyzed with 6*N* HCl in a sealed tube, and the resulting gases were distilled or extracted with *n*-hexane. Analysis by gas-liquid chromatography and mass spectrometry revealed the presence of C_1, C_2, C_3, and C_4 hydrocarbons. The total added up to almost 20 μg per gram of the lunar sample. A sample of Mighei meteorite and cohenite (Fe_3C) from the Canyon Diablo meteorite were similarly analyzed (Fig. 3), hydrocarbons being identified in each case.

Our examination of the Apollo 11 sample from the Sea of Tranquillity leads us to conclude that the concentration of carbon is about 157 μg per gram of sample. If we assume that the hydrocarbons generated from the acid hydrolysis come from carbides, the amount of carbon accounted for is approximately 20 μg/g. The isotopic compositions of carbon and sulfur are significantly different from the composition determined for other extraterrestrial samples. The $\delta^{13}C$ value of +20 relative to the PDB standard and the $\delta^{34}S$ value of +8.2 relative to the Canyon Diablo meteorite are unusual by comparison with both terrestrial and meteorite samples. Although the limits of detectability of our techniques were in the nanogram range, normal alkanes, isoprenoids, hydrocarbons, aromatic hydrocarbons, fatty acids, amino acids, sugars, and nucleic acid bases were not present at this level of concentration. There is a striking but yet unexplained similarity between the chromatogram of the compound that formed the derivative in the acid hydrolyzates of the lunar sample and of the Pueblito de Allende meteorite. These findings should be considered to be specific for a single surface sample from the Sea of Tranquillity. Samples from the highlands, or core samples, may be expected to yield different results.

CYRIL PONNAMPERUMA
KEITH KVENVOLDEN
SHERWOOD CHANG
RICHARD JOHNSON
GLENN POLLOCK
DELBERT PHILPOTT
NASA Ames Research Center, Moffett Field, California
ISAAC KAPLAN, JOHN SMITH
J. WILLIAM SCHOPF
University of California, Los Angeles
CHARLES GEHRKE
University of Missouri, Columbia
GORDON HODGSON
University of Calgary, Calgary, Alberta, Canada
IRVING A. BREGER
U.S. Geological Survey, Washington, D.C.
BERTHOLD HALPERN, ALAN DUFFIELD
KONRAD KRAUSKOPF
Stanford University, Stanford, California
ELSO BARGHOORN
Harvard University, Cambridge, Massachusetts
HEINRICH HOLLAND
Princeton University, Princeton, New Jersey
KLAUS KEIL
University of New Mexico, Albuquerque

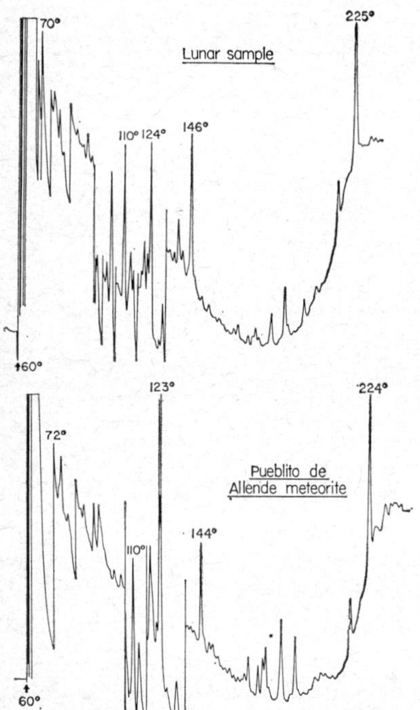

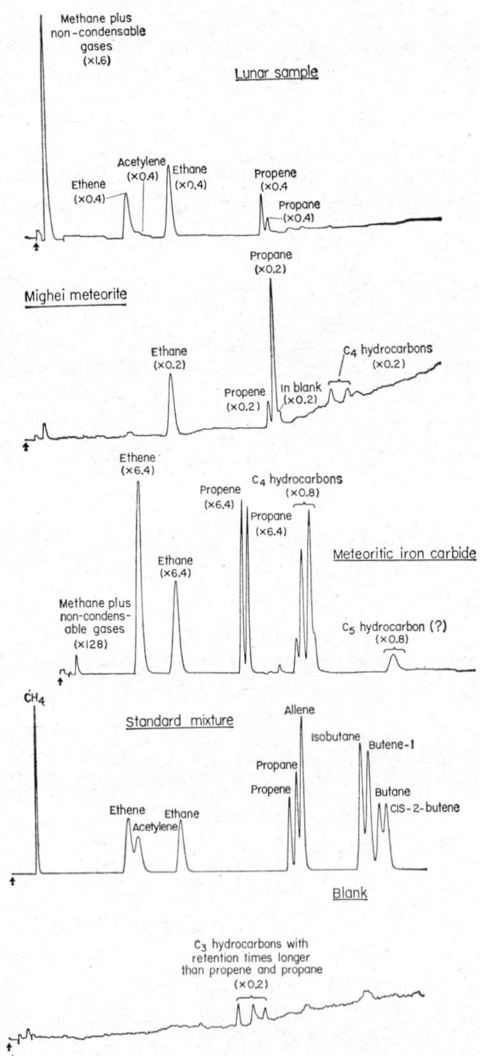

Fig. 2 (left). Gas chromatograms of *N*-trifluoroacetyl-*n*-butyl esters of 1*N* hydrolyzates of the Pueblito de Allende meteorite and the lunar sample, on a glass column (1 m by 6 mm), packed with 1.5 percent (by weight) OV-17, on heat-treated 80/100 mesh Chromosorb G, temperature programmed from 60°C to 250°C, at 4°C per minute. Vertical lines represent attenuation changes. Fig. 3 (right). Gas chromatograms of hydrocarbons from 6*N* HCl treatment of the lunar sample (0.1 g), Mighei meteorite (0.05 g), and Fe_3C (0.0014 g), on a stainless steel column (2 m by 3 m), packed with 150 mesh Poropack Q, programmed from ambient temperature to 150°C, at 10°C per minute. Standard mixture attenuated at × 160.

References and Notes

1. J. W. Schopf and E. S. Barghoorn, *Science* **156**, 508 (1967).
2. C. Sagan, *Nat. Acad. Sci.–Nat. Res. Council Publ. No. 49* (1961).
3. E. C. King, Jr., E Schonfeld, K A. Richardson, J. S. Eldridge, *Science* **163**, 928 (1969).
4. Preliminary Examination Team, *ibid.* **165**, 1211 (1969); R. C. Johnson and C. C. Davis, *ibid.*, this issue.
5. I. Kaplan and J. W. Smith, *ibid.*, this issue.
6. E. S. Barghoorn, D. E. Philpott, G. Turnhill, *ibid.*, this issue; J. W. Schopf, *ibid.*, this issue.
7. G. W. Hodgson *et al.*, *ibid.*, this issue.
8. B. R. Simoneit, A. L. Burlingame, D. A. Flory, I. D. Smith, *ibid.* **166**, 733 (1969).
9. We thank J. Flores, J. Lawless, J. Mazzurco, K. Pering, E. Peterson, D. Quinsland, D. Ramey, C. Saxinger, J. Williams, and F. Woeller of the Ames Research Center, W. Aue and R. Zumwalt of the University of Missouri, D. Stalling of the U.S. Bureau of Fisheries, and E. Ruth of the University of California, Los Angeles, for their invaluable assistance.

5 January 1970

Search for Porphyrins in Lunar Dust

Abstract. *Evidence for porphyrins was obtained in the Apollo 11 bulk sample of lunar dust by fluorescence spectrometry and analytical demetallation. The indicated porphyrins showed major fluorescence excitation at 390 nanometers. Abundance was about 10^{-10} gram of porphyrin per gram of dust. Similar pigments were found in exhaust products from tests of a lunar descent engine. The similarity of results suggests that most if not all of the indicated porphyrin aggregate of the lunar sample probably was synthesized from rocket fuel during the landing of the lunar module. These compounds may be the product of a novel high-temperature synthesis of cosmochemical interest.*

The present study was undertaken to determine whether porphyrins, tetrapyrrole organic pigments, were present in the Apollo 11 bulk sample and whether their presence in the sample was indicative of a biological origin (*1*). The results of the analyses show that pigments with spectral and chemical characteristics resembling porphyrins are present in the lunar bulk fines, but it appears that the compounds may have been introduced by the lunar descent rocket engine fueled with unsymmetrical dimethylhydrazine. Thus, while the analyses reported here evidently failed to reveal porphyrins which were unequivocally indigenous, the findings suggest that the porphyrins can perhaps be synthesized through a novel reaction involving simple compounds containing carbon, nitrogen, hydrogen, and oxygen. The pigments have also been detected in experiments simulating primitive (prebiotic) planetary conditions (*2*), and similar compounds evidently exist extraterrestrially, because porphyrins have been detected in a number of samples of carbonaceous chondrites (see *3*).

The analyses reported here are part of a comprehensive analytical program established by the Chemical Evolution Branch of the NASA Ames Research Center for the examination of all organic matter in the returned lunar samples (*4*). The group arranged to pool samples of lunar dust received by the principal investigators involved in the program. The particular sample discussed here was designated "10,086 bulk A fines" by the Lunar Receiving Laboratory.

Prior to receipt of the lunar samples, a simulation experiment was carried out on a sample of the Pueblito de Allende meteorite. Although analyses on a number of specimens of this meteorite showed porphyrins to be present in very small amounts (*5*), no porphyrins were detected in the particular sample used in the control run, and the results were substantially the same for a blank of Ottawa sand which had been fired at 1000°C for 48 hours.

The overall analytical procedure for the lunar sample is described by Ponnamperuma *et al.* (*4*). The lunar dust (54.6 g) was extracted directly with benzene-methanol (9:1), and 23 percent of the extract was made available for porphyrin analysis.

On concentration of the solution to a volume of 1 ml, weak emission bands at 600 and 630 nm for excitation at 400 and 425 nm (Turner 210 spectrofluorometer) appeared. Clear-cut excitation peaks for emission at 600 nm were observed at 390 and 455 nm, the latter being much more intense. Corresponding readings for the blank of Ottawa sand showed only a low background and no discernible peaks. Absorption spectrometry (Cary 14, 0 to 0.1 optical density) showed no distinct spectral features, only a gently rising background with decreasing wavelength. The solution was examined in an improved magnetic circular dichrometer with a superconducting magnet of 50,000 gauss; the data were evaluated by computer. Detection limits for the system were estimated to be 7 ng/ml for recognition of porphyrin compounds in metal-complexed form. Magnetic circular dichroism spectra of the concentrated extract solution, however, did not show the presence of metalloporphyrins above detection limits.

Free-base porphyrins fluoresce freely, but most metal complexes fluoresce only faintly. Thus, controlled demetallation is useful for the detection of traces of porphyrin metal complexes (*6*). The extract solution was divided into two portions, one four times the volume of the other. Each was reduced to dryness in a spectrofluorometer cell and taken up in 1 ml of diethyl ether. The smaller sample showed a faint excitation band at 410 nm for emission measured at 660 nm upon addition of 2 percent of the demetallating reagent, methanesulfonic acid (MSA). The intensity of the band increased at 7 percent MSA, and at 19 percent MSA the band was replaced by an excitation peak at 388 nm, which became more intense on further addition of MSA.

Recovery of the presumed free-base porphyrins was accomplished by neutralizing the solution with sodium acetate and extracting with ether. The ether layer showed faint excitation bands at 385 and 450 nm. Transfer to 6*N* HCl also showed a weak peak at 390 nm for emission measured at 600 nm, and this peak was interpreted tentatively as a Soret band of porphyrin pigments. To extend the identification of the apparent porphyrin, it was complexed with copper to quench its fluorescence. This was done by transferring the porphyrin to glacial acetic acid and adding cupric sulfate. The excitation band disappeared after a few minutes at room temperature, in accordance with the behavior of authentic porphyrins.

The same analysis was repeated on the other portion of the lunar extract, which corresponded to 10 g of lunar dust. Essentially the same results were

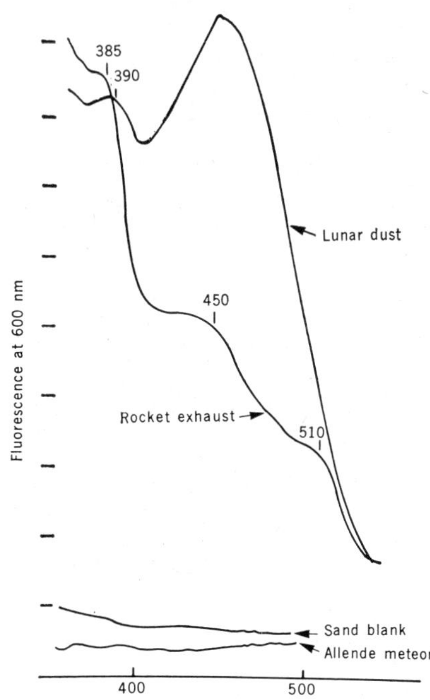

Fig. 1. Fluorescence excitation spectra for extracts of (i) lunar dust from the bulk sample, (ii) dunite exposed to the exhaust of a rocket engine fired at NASA White Sands Test Facility, (iii) crushed Pueblito de Allende carbonaceous chondrite, and (iv) Ottawa sand ignited at 1000°C for 48 hours. Instrument operating parameters were set for highest sensitivity in all cases except that of the rocket exhaust, in which fluorescence was suppressed by a factor of 20.

obtained, with the emergence of a band at 390 nm for emission at 600 nm on the addition of 20 percent MSA. After demetallation, the fluorescence of the pigments in ether was still weak, but definite emission could be detected at 630 nm for excitation features at 390 and 415 nm. Extraction with 6*N* HCl produced an aqueous layer with emission peaks at 600, 628, and 685 nm. A strong excitation peak was observed at 390 nm, and it appeared to be accompanied by a much weaker peak at 550 nm. The "porphyrins" were transferred back to ether and reextracted with HCl of increasing concentration. The acid strength necessary to transfer the pigments from the organic layer was 4*N* HCl; from this we tentatively concluded that few polar substituents were present to promote the solubility of the pigment in the aqueous phase. Formation of the copper complex resulted in the complete suppression of fluorescence. On the assumption that the pigment was porphyrin, the amount present was estimated to be 1 ng, for a concentration of 10^{-10} g of porphyrin per gram of lunar dust.

The sand blank was examined in the same manner, and it showed no indication of porphyrins above background. In addition, reagents, glassware, and handling procedures were checked for possible contaminants.

A similar demetallation analysis was carried out on the benzene eluate from silica-gel chromatography of the major portion of the benzene-methanol extract of the lunar dust. The hexane and methanol eluates were not available for porphyrin analysis. The benzene eluate left after the completion of other tests appeared to show the presence of a small amount of porphyrin pigments.

Samples of rocket exhaust were made available from the test firing reported by Simoneit *et al.* (7). These comprised samples from trap A, a trap for the exhaust gases cooled to liquid-nitrogen temperatures, and trap E, one containing dunite. Material from trap A showed an intense fluorescence background on which a broad excitation band (in methanol and ether) appeared at 450 nm for emission in the range 600 to 700 nm. In HCl, the exhaust products turned a bright yellow, and an excitation band at 390 nm for emission at 610 nm was clearly present. Absorption spectrometry showed only the yellow color. While the foregoing results suggested porphyrins in the free-base state, MSA demetallation of the total exhaust material—that is, without any chemical fractionation—showed evidence for metal complexes with a 412-nm excitation band appearing at 4 percent MSA. This spectral feature persisted to a concentration of 23 percent MSA; at higher concentrations of MSA a much stronger band emerged at 385 nm and shifted gradually to 392 nm. Attempts to recover free-base porphyrins from the foregoing aqueous acid layers were unsuccessful, however.

Material from trap E gave similar results. Extraction of the sample was carried out in the same manner as that for the lunar sample. A broad excitation band was evident at 450 nm, as in the case of trap A material, but on an intense background. Addition of MSA gave typical demetallation results for metalloporphyrins. At a concentration of 25 percent MSA a strong peak emerged; it increased in intensity with increasing acid content up to 50 percent MSA. Final location of the band was at 387 nm, and the corresponding emission peak was at 595 nm. Magnetic circular dichroism analysis of material extracted from trap E showed a large number of compounds, but none could be definitely identified as porphyrins.

To confirm and extend the previous results and to demonstrate adequate

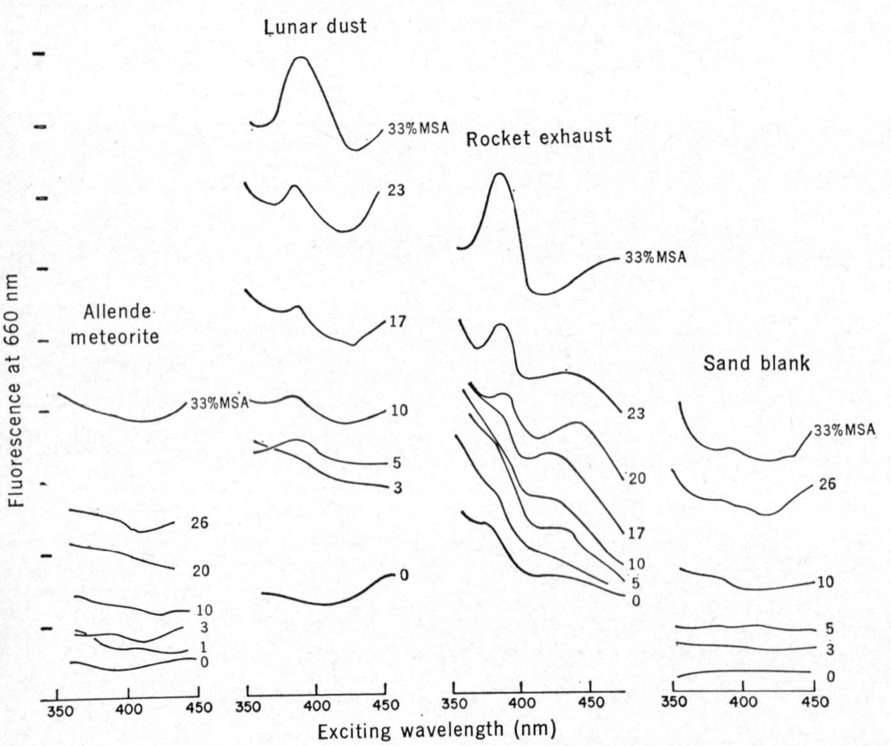

Fig. 2. Demetallation analysis of extracts showing emergence of excitation peaks in the region of the Soret band of porphyrins with increasing content of methanesulfonic acid (MSA). Operating parameters of the instrument were the same in all cases; portions taken for the analyses of lunar dust and rocket exhaust were, respectively, 75 and 11 percent of the total extracts.

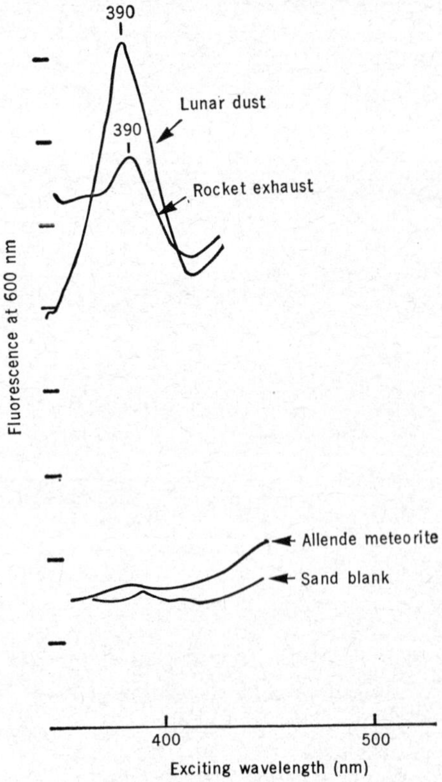

Fig. 3. Fluorescence excitation spectra in Soret region for pigments recovered from MSA demetallation analysis. Operating parameters for spectrofluorometer were the same in all cases. Solvent, 6*N* HCl.

control over laboratory procedures, an additional set of samples was analyzed. Approximately 10 g each of the same lunar dust, trap E dunite, Pueblito de Allende meteorite, and Ottawa sand were extracted with a solution of benzene-methanol (9:1). The extracts were identified by a code unknown to the analysts (G.W.H. and E.P.), and complete analyses were carried out. Figure 1 shows fluorescence excitation curves for the extracts. Demetallation caused by the addition of MSA is illustrated in Fig. 2, and Fig. 3 shows the results of procedures for the recovery of free-base porphyrins. The results from the lunar and exhaust samples confirmed the earlier analyses, and what appear to be free-base porphyrins were recovered from the products of MSA demetallation of both the lunar dust and the rocket exhaust. Results for extracts from both the sand and meteorite were negative.

Organic matter is undoubtedly synthesized extraterrestrially, as indicated for example by the carbonaceous matter of chondrites and the generally accepted graphitic interstellar dust (*8*). Simple compounds containing nitrogen, carbon, hydrogen, and oxygen are reported in interstellar dust clouds (*9*) and in stellar atmospheres (*10*). Recent theories (*11*) postulate that much of interstellar space is occupied by thermally stable gas phases at 10^4 °K, which suggests that high-temperature synthesis of organic matter including porphyrins throughout space is within reason.

GORDON W. HODGSON
University of Calgary, Calgary 44, Alberta, Canada
ETTA PETERSON
KEITH A. KVENVOLDEN
NASA Ames Research Center, Moffett Field, California 94035
EDWARD BUNNENBERG
B. HALPERN
Stanford University, Stanford, California 94305
C. PONNAMPERUMA
NASA Ames Research Center

References and Notes

1. J. E. Falk, *Porphyrins and Metalloporphyrins* (Elsevier, New York, 1964); C. Ponnamperuma and N. W. Gabel, *Space Life Sci.* **1**, 64 (1968); K. A. Kvenvolden and G. W. Hodgson, *Geochim. Cosmochim. Acta* **33**, 1195 (1969).
2. G. W. Hodgson and C. Ponnamperuma, *Proc. Nat. Acad. Sci. U.S.* **59**, 22 (1968).
3. G. W. Hodgson and B. L. Baker, *Geochim. Cosmochim. Acta* **33**, 943 (1969).
4. C. Ponnamperuma, K. Kvenvolden, S. Chang, R. Johnson, G. Pollock, D. Philpott, I. Kaplan, J. Smith, W. Schopf, C. Gehrke, G. Hodgson, I. Breger, B. Halpern, A. Duffield, K. Krauskopf, E. Barghoorn, H. Holland, K. Keil, *Science*, this issue.
5. G. W. Hodgson, E. Peterson, K. A. Kvenvolden, C. Ponnamperuma, in preparation.
6. G. W. Hodgson, E. Peterson, B. L. Baker, *Mikrochim. Acta* 805 (1969).
7. B. R. Simoneit, A. L. Burlingame, D. A. Flory, I. D. Smith, *Science* **166**, 733 (1969).
8. B. Donn, N. C. Wickramasinghe, J. P. Hudson, T. P. Stecher, *Astrophys. J.* **153**, 451 (1968).
9. L. E. Snyder, D. Buhl, B. Zuckerman, P. Palmer, *Phys. Rev. Lett.* **22**, 679 (1969); A. C. Cheung, D. M. Rank, C. H. Townes, D. D. Thomson, W. J. Welch, *ibid.* **21**, 1701 (1968).
10. M. S. Vardya, *Mon. Notic. Roy. Astron. Soc.* **134**, 347 (1966).
11. G. B. Field, D. W. Goldsmith, H. J. Habing, *Astrophys. J.* **155**, L149 (1969).
12. This work is supported by NASA grant NGR-05-020-004.

4 January 1970

Organogenic Elements and Compounds in Surface Samples from the Sea of Tranquillity

Abstract. *Organogenic elements, mainly carbon, nitrogen, phosphorus, and sulfur are present in the particulate material and in a breccia rock from Tranquillity Base in amounts ranging from 5 to 4200 parts per million. The major compounds of carbon released by heating are carbon monoxide and carbon dioxide; the former predominates. Small amounts of other compounds of carbon have also been observed. Sulfur can be released as hydrogen sulfide by treatment with acid. The carbon isotopic* $\delta^{13}C$ *values are definitely nonterrestrial (+ 13 to + 18.5 per mil).*

An igneous origin has been considered likely for the Sea of Tranquillity (*1*), and has since been confirmed by a preliminary examination of its surface rocks (*2*). This precludes the existence of large amounts of organic matter on this part of the lunar surface. Therefore, it was considered pertinent to analyze for those elements and compounds that are necessary for prebiological or biological synthesis of organic molecules (*3*). The compositions of the fine particulate material (type D) and a small specimen of breccia (type C) were studied with different methods of analysis. To avoid contamination and the introduction of artifacts, the samples were analyzed as directly as possible and with a minimum of handling. Thus the fines were neither sieved nor pulverized and the breccia was cleaved only once.

Spark-ionization mass spectrometry was used to determine the concentration of hydrogen, carbon, nitrogen, phosphorus, and sulfur (*4*). We used a mass spectrograph (CEC X21-110) that operated at a resolution of approximately 3000, and at a relative sensitivity that ranged from 0.0005 to 0.03 ppm. Each sample, of about 250 mg, was sparked against a pure gold electrode. Up to 20 mg of material from each sample was actually analyzed. If the sample is not homogenous, variations in the analytical values will be observed depending on the composition of the area sparked. All the samples were isolated from the laboratory environment by means of glove boxes up to and including insertion into the mass spectrometer vacuum system. Cryogenic rough pumping, ion pumping, and preliminary sparking assured a minimum of contamination present during actual measurements. The fine material was vacuum encapsulated, hydrostatically pressed to 4570 kg/cm² and then cleaved for analysis. The small breccia rock was cleaved and immediately analyzed at selected surfaces. Calibration analyses with six National Bureau of Standards specimens of Special Ingot

Table 1. Spark ionization mass spectrographic analysis for organogenic elements in type D (fines, sample 10086,6) and type C (breccia, sample 10002,54) lunar samples. Values expressed in ppm.

Element	m/e	Bulk fines			Breccia		
		Ia	Ib*	II	Fresh inner cut	Outer surface (dark area)	Outer surface (white spot)
H	1	1.2	1.2	0.41	0.13	0.011	0.083
C	12	9(5)†	24	41	27	6	5.5
N	14(7)	14	20	7	15	10	10
P	31	270	187	110	110	250	59
S	32	4200	590	1000	2800	2100	1000

* Same as Ia analyzed after 2 weeks of exposure to the terrestrial environment. † Second analysis performed the following day on the same sample without its removal from the instrument.

Table 2. Summary of data obtained by application of several techniques to the analysis of type D (fines) and type C (breccia) lunar samples.

Analysis	Description of method	Measurements
Biogenic elements	Spark ionization mass spectrography	H,C,N,P,S
Major volatile and pyrolytic compounds or both	Stepwise heating to 750°C and quadrupole mass spectrometry	CO (and N_2) CO_2 (SO_2 and/or S_2?)
	Gradual heating from 100° to 1200°C and high resolution mass spectrometry	$CO/CO_2 \simeq 10$ or higher noble gases
	Stepwise heating to 750°C and gas chromatography (Porapak column)–mass spectrometry	CO/CO_2 ($\simeq 10$ at 750°C) Traces of CH_4 above 400°C
Trace volatile and/or pyrolytic organic products	Stepwise heating to 600°C and flame ionization gas chromatography (SE-30 column)	$\simeq 0.6$ ppm (about 10 components)
Trace organic solvent–extractable organic compounds	Flame ionization–high resolution–gas chromatography (Polysev column)	<1 ppb
Volatile products generated by acid	Acid treatment and gas chromatography (Porapak column)–mass spectrometry	H_2S (H_2 not measured)
Carbon-13	Carbon isotopic ratio mass spectrometry	$\delta^{13}C = +13$ to $+18.5$ per mil PDB
and sulfur-34 isotopes	H_2S mass spectrometry	4.2 per cent S-34

Iron with carbon concentration ranging from 370 to 4000 ppm were previously carried out, and agreed to within 20 percent of the NBS certified values. The samples were degassed by preliminary sparking. Thus, the measured values corresponded to solid or chemically bound species of the elements analyzed.

Negligible amounts of hydrogen, small amounts of carbon and nitrogen, and larger amounts of phosphorus and sulfur were observed in the fines and the breccia (Table 1). Since the samples were neither pulverized nor homogenized, it was not surprising to see variations of less than one order of magnitude for C, N, P, and S. As could be expected, if only strongly bound species were measured, the values for C and N were found lower than those obtained for total C and N (*5*).

Several experiments were carried out by direct thermal volatilization–mass spectrometry of small amounts of fines. A specially designed T-type stainless steel oven, which could be heated quickly at preselected temperatures (up to 750°C) by means of a molten metal bath, was attached directly into a vacuum chamber–quadrupole mass spectrometric system, which included a monopole mass filter for total pressure measurements (*6*). Samples of about 100 to 450 mg were used. The pressures developed by the samples were at least two orders of magnitude above background. Evidence for the presence of parent ion masses 28 (CO and N_2) and 44 (CO_2), and traces of other compounds, was obtained with this instrument.

Programmed thermal volatilization of small samples of fines into a CEC 21-110B high resolution mass spectrometer provided gas release versus temperature curves for helium, neon, argon, CO, and CO_2. The noble gases first appeared at temperatures from 200° to 400°C and reached maxima from 500° to 800°C depending on the atomic mass of each gas and the heating rate. The appearance of CO, CO_2 and N_2 varied according to time and temperature. Carbon monoxide was the predominant gas at high temperatures, by approximately one order of magnitude over CO_2. The composition of the carbon oxides released may not necessarily reflect their state in the solid material. Differential thermal analysis carried out simultaneously with this mass analysis indicated melting of the bulk fines at about 1130°C.

Stepwise volatilization of the fines in a quartz, stopped-flow, pyrolyzer tube attached to an LKB 9000 gas chromatograph–mass spectrometer yielded similar data. The chromatographic column was a stainless steel tube, 120 cm long by 0.2 cm i.d., packed with 200-mesh Porapak (cross-linked polystyrene). It was operated at ambient temperature and 843 g/cm² of helium. Multiple mass scans were obtained of the effluent gases released by heating for 5 minutes at each preselected temperature up to 750°C. No significant amounts of gases were observed below 200°C. From 300° to 500°C the gas released was mainly CO_2 (and some CO), but at higher temperatures, 600° to 750°C, the predominant gas was CO (and some CO_2). This is in general agreement with the direct volatilization mass spectrometry data. Traces of methane were also seen at these higher temperatures.

Since the major compound of sulfur was presumed to be a sulfide (*2*) a small amount of fines (0.2 g) was treated with concentrated phosphoric acid in a special glass reactor attached to the LKB 9000 gas chromatograph–mass spectrometer. The gas chromatographic column and the chromatographic conditions were essentially the same as described above, except that the temperature of this reaction vessel was approximately 100°C. The main volatile product thus measured was identified as hydrogen sulfide.

The stable carbon isotope content of about 10-g samples of fines was also measured according to methods developed in this and other laboratories (*7*). Carbon dioxide was formed by heating at high temperatures in the presence of oxygen and cupric oxide. One sample was heated at 800°C for 1 hour and another sample at 900°C for 2 hours. The amounts of CO_2 obtained from these two samples corresponded to 42 ± 5 and 100 ± 5 ppm of carbon, respectively. These amounts are within the range of those reported for total carbon for the Apollo 11 lunar samples (*5*, *8*). The $\delta^{13}C$ values of these two samples were +13 and +18.5 per mil relative to the PDB isotope standard. The $\delta^{13}C$ data demonstrate that the carbon measured in these two samples is definitely nonterrestrial since such positive values have never been found on earth. Similar, or more positive, $\delta^{13}C$ figures have only been found in carbonates from carbonaceous chondrites (*7*).

With the information presently available it is not possible to assign an unequivocal origin to the carbon found in the lunar samples. This would include lunar igneous sources and extralunar sources, such as the solar wind, comets, and meteorites. The data on the solar abundances of carbon and on the presence of noble gases of solar composition in type C and D lunar samples (2) suggest that the solar wind is a possible contributor to the lunar carbon. This suggestion appears to be in line with the larger amounts of this element found in the finer particulate material (5). Whereas some solar carbon isotopic ratios (9) are consistent with such an interpretation, these are not yet known with sufficient accuracy. Determination of the abundances, isotopic ratio, and capturing characteristics of the carbon in the solar wind reaching the lunar surface may help to clarify this problem.

Nongraphitic compounds ionizable by a hydrogen flame were released from the fines (35 to 55 mg) by heating from ambient temperature to 600°C in a flow-through pyrolyzer. It consisted of a silver core heater and a quartz tube, 12 mm long by 0.3 mm o.d. The products were condensed in a capillary column (40 m by 0.5 mm i.d., coated with silicone oil, SE 30) that was cooled to approximately −60°C. The condensed products were eluted by programming the column temperature from −60° to about 130°C using helium as a carrier gas and analyzed by a sensitive flame ionization gas chromatograph. Simple gas chromatographic patterns of about ten components were obtained. However, quantitatively the totality of the organic compounds amounted to only 0.6 ppm. Measurement of the carbon isotope ratio of these compounds will be necessary before one can consider them as indigenous to the lunar samples.

Extraction of 1 g of fines with a benzene-methanol mixture, evaporation and analysis by flame ionization gas chromatography (10) yielded less than 0.001 ppm of organic compounds. A 1.4 g sample of fines was hydrolyzed. The hydrolyzate was deionized, and analyzed by gas chromatography for amino acids after their conversion to *N*-trifluoroacetyl isopropyl esters (11). No measurable amounts of amino acids were found under conditions which would have detected less than 0.1 ppm of any amino acid. The results of the above three different analyses together with the carbon isotope ratios demonstrate that at least some of the samples have not been grossly contaminated by terrestrial organic matter, and give confidence of the adequacy of procedures followed by NASA for the handling and distribution of the lunar samples at the LRL.

In summary, the evidence presented here (Table 2) shows that the elements that are necessary for the synthesis of organic compounds are present in type C and D lunar samples. Some of these elements can be released as gases in a partially reduced state. In line with the spark ionization data, others (12) have shown that carbon and phosphorus exist in the form of solid compounds, such as graphite and calcium phosphate, the latter of particular interest for prebiotic syntheses.

J. Oró, W. S. Updegrove
J. Gibert, J. McReynolds
E. Gil-Av*, J. Ibanez, A. Zlatkis
Departments of Biophysical Sciences and Chemistry, University of Houston, Houston, Texas 77004

D. A. Flory
NASA, Manned Spacecraft Center, Houston, Texas 77058

R. L. Levy
C. Wolf
McDonnell Research Laboratories, St. Louis, Missouri 63166

References and Notes

1. A. L. Turkevich, E. J. Franzgrote, J. H. Patterson, *Science* **165**, 277 (1969).
2. Lunar Sample Preliminary Examination Team, *ibid.*, p. 1211.
3. J. Oró, in *Current Aspects of Exobiology*, G. Mamikunian and M. H. Briggs, Eds. (Pergamon Press, New York, 1965), p. 131.
4. For comparative purposes a survey quantitative analysis of 35 other elements was also carried out.
5. Values ranging from 15 to 353 ppm for total C and 28 to 247 ppm for total N in the Apollo 11 lunar samples (all types) were obtained. Different fractions of the bulk fines gave values for carbon which are inversely proportional to the particle size of the fraction. Communications of C. B. Moore to the LRL.
6. W. S. Updegrove and J. Oró, in *Research in Physics and Chemistry*, F. J. Malina, Ed. (Pergamon, London, 1969), p. 53.
7. D. A. Flory, thesis, University of Houston, Houston, Texas (1969); R. N. Clayton, *Science* **140**, 192 (1963).
8. A 10 percent additional CO_2 was obtained by combusting the second sample 4 hours more at 900°C. The $\delta^{13}C$ of this CO_2 was not measured.
9. J. L. Greenstein, R. Richardson, M. Schwarzchild, *Publ. Astron. Soc. Pacific* **62**, 15 (1951); L. Herzberg, D. Delboville, G. Roland, *Astrophys. J.* **147**, 697 (1967).
10. J. Oró, D. W. Nooner, and R. J. Olson, in *Chromatographic Analysis of Lipids*, G. V. Marinetti, Ed. (Dekker, New York, 1968), vol. 2, p. 479.
11. E. Gil-Av, B. Feibush, R. Charles-Sigler, in *Gas Chromatography 1966*, A. B. Littlewood, Ed. (Institute of Petroleum, London, 1967), p. 227; E. Gelpi, W. A. Koenig, J. Gibert, J. Oró, *J. Chromatogr. Sci.* **7**, 604 (1969).
12. Private communications about graphite from G. Arrhenius and about apatite from R. A. Schmitt. Also, other work from this laboratory.
13. We thank J. Socha (Electronics Materials Div., Bell & Howell, Inc., Pasadena, Calif.); F. M. Wachi and W. K. Stuckey (Aerospace Corp., El Segundo, Calif.); R. Lewis (Analytical Instruments Div., Bell & Howell, Inc., Monrovia, Calif.); D. W. Nooner (Texaco, Houston); R. S. Becker, R. Segura, and F. Trifilio (Univ. of Houston); P. Parker (Univ. of Texas); and D. Baker, N. Feux, J. L. Margrave, S. J. Steck, and D. W. Muenow (Rice Univ.) for their cooperation in different phases of this work. Supported by NASA contract NAS-9-8012.
* Dr. Gil-Av is on leave of absence from the Weizmann Institute, Rehovoth, Israel.

5 January 1970

Bio-organic Compounds and Glassy Microparticles in Lunar Fines and Other Materials

Abstract. *Acid hydrolyzates of aqueous extracts indicated: glycine, alanine, glutamic acid, serine, aspartic acid, threonine, and α,β-diaminopropionic acid. The sources of these amino acids may be terrestrial contaminants, fuel exhaust products, or indigenous lunar material. The lunar particles found in the sample include: (i) spherules, rotational ellipsoids, dumbbells, tear-drops, rings, and crescents which have (ii) diameters of 0.1 to 500 microns; (iii) budlike features on the particles; and (iv) chemical inhomogeneity (electron probe). Some of the processes and results inferred for the moon have been simulated by the electron beam and low pressure of the electron microscope. Basalt (Kilauea-Iki) and chondrite (Orgueil) have been found to behave similarly.*

Samples of lunar fines collected during the Apollo 11 mission were examined in order to determine whether the material contained (i) amino acids or polymers thereof or (ii) microparticles with resemblance to terrestrial cells (*1*), proteinoid microspheres (*1*), microfossils (*1*), organic particles from hydrothermal veins (*2*), or microparticles from carbonaceous meteorites (*1, 2*). Other naturally occurring rounded particles of similar morphology and more certain inorganic nature are tektites and microtektites (*3*) and the products of terrestrial volcanism (*2*) and of meteoritic impact (*4*). Comparisons with these were also sought. The planned procedure, if substances and regular structures were found, involved identification of the chemical composition of the particles.

Other objectives that developed during the investigation included determining whether the oxidation of 1,1-dimethylhydrazine (rocket fuel) could produce amino acids and a quantitative assessment of terrestrial contaminants in amounts less than a nanomole.

For assay of amino acids, 10 to 15 g of lunar fines was extracted with 30 ml of especially purified water by gentle refluxing for 17 hours. The water had been purified by fresh distillation followed by filtration through washed Millipore filters (GSWPO 4700, pore size 0.22 μm). The cooled extract was filtered through a Millipore filter, and the filtrate was concentrated to dryness in a vacuum desiccator. Some of the residues were hydrolyzed for 24 hours in evacuated sealed tubes in an oil bath at 110°C, with 6*N* HCl purified by redistillation. Others were dissolved in 0.01*N* HCl and analyzed directly. All operations in the open were conducted in an Edcraft laminar flow hood in a clean room. Glassware was cleaned with dichromate-sulfuric acid, washed copiously with water, and baked at 500°C overnight. Control substances were run in the same manner with the lunar material omitted.

Suitable aliquots were first analyzed on a Phoenix K-5000 Amino Acid Analyzer but mainly on the ultrasensitive modified amino acid analyzer (*5*), which employs a 1.5-mm bore ion-exchange column. This ultrasensitive apparatus permits quantitation of amino acids in the range of 1 part per billion.

In the search for spray particles, separation was attempted by magnetism, flotation, and sieving. The last method proved most practical, particularly when combined with sedimentation in chloroform followed by further separation by rolling in a glass tube (2.5 cm bore).

Fields were examined in Permount resin (Fisher) under a Wild petrological microscope. Particles were mounted on Formvar-coated grids, shadowed with gold, and examined with the electron microscope (Philips 300). The experiments on fusion of particles were done in the same beam. Examination of glassy particles with the microspectrophotometer was carried out with the apparatus described by Strother and Wolken (*6*).

In order to determine whether amino acids could be formed by oxidation of the fuel, 0.05 ml of 1,1-dimethylhydrazine was treated under nitrogen with 0.04 ml of concentrated HNO_3. The yellow liquid changed to red and back to yellow during a 1-hour period. The mixture was then concentrated in a vacuum desiccator, dissolved in 0.01*N* HCl, and analyzed.

Analysis of hydrolyzates of lunar extracts on the Phoenix instrument revealed a peak with the retention time (RT) of ammonia, another in the region of basic amino acids, and a minimal third peak at about the RT of glycine.

Other samples were analyzed on the ultrasensitive apparatus. Seven peaks corresponding to amino acids were observed in two chromatograms. The one peak for which the identity was most uncertain corresponded to the RT of α,β-diaminopropine acid. Six others in one chromatogram had RT's in common with proteinogenous amino acids. The amino acids tentatively identified and their amounts in nanomoles are: aspartic acid, 0.03; threonine, 0.01; serine, 0.05; glutamic acid, 0.05; glycine,

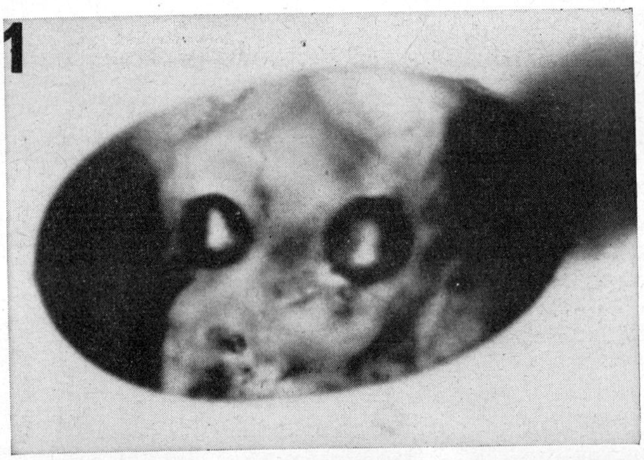

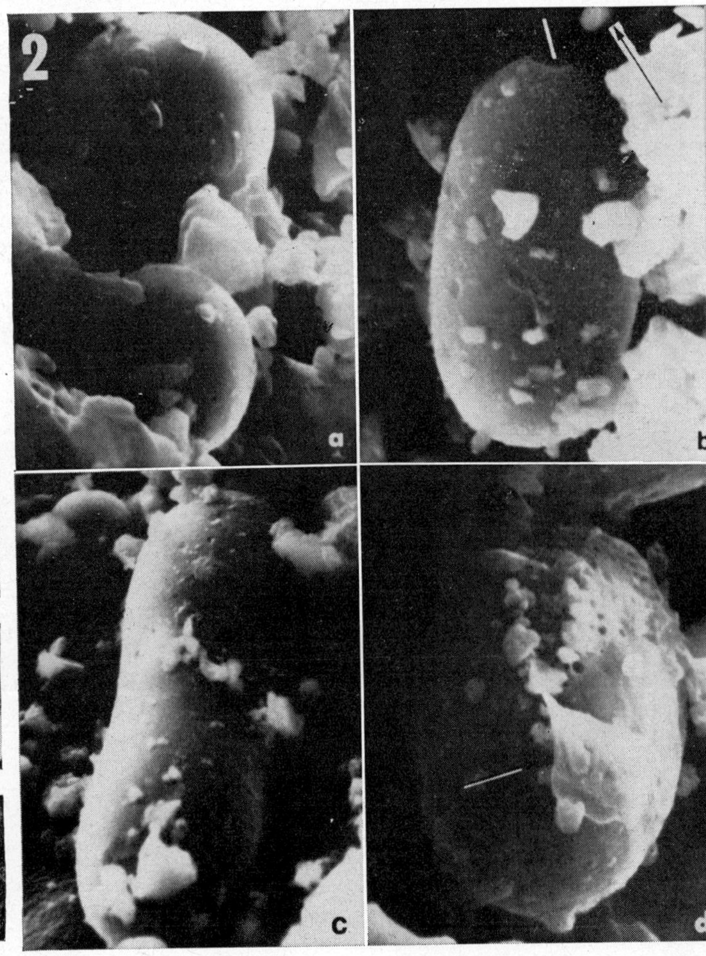

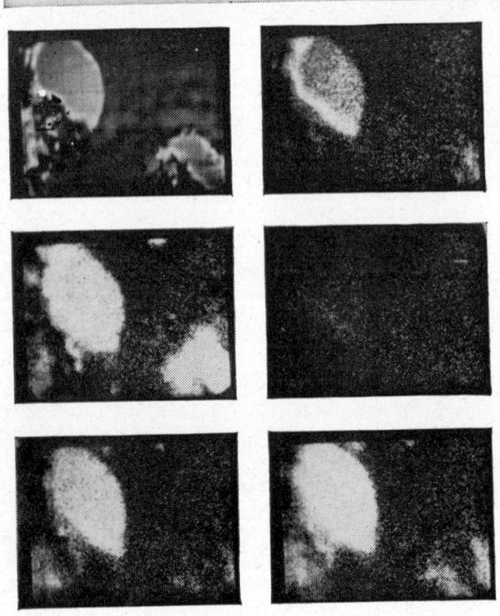

Fig. 1. Glass spherule having two bubbles in light cloudy central area. Extremities are formed of a red-brown glass (× 1500). Fig. 2. Typical forms of rounded glass particles. (a) Spherule (× 5500), (b) ellipsoid with hole on top (see line) from which a small sphere fell out (see arrow) (× 6500); (c) dumbbell (× 2100); (d) teardrop on ellipsoid (arrow) (× 4100). Small budlike protuberances frequently appear on surface of the particles. Fig. 3. Electron-probe analysis of particles shown in upper left frame. From left to right: (top row) distribution of Mg and Al; (second row) Si, K, Ca; (bottom row) Ti, Fe, Zr.

0.28; and alanine, 0.14, calculated to represent 1.0 g of lunar fines.

The controls, which were carried through all steps of the analytical sequence, contained significantly smaller amounts of amino acid (<0.01 nmole), and different patterns. Samples of the unhydrolyzed extracts showed relatively small amounts of amino acids. Analyses of other aliquots of these samples after 4 weeks in storage showed slightly increased amino acid contents in all cases. These latter results indicate the necessity of thoroughly evaluating the effect of storage on determinations of amounts less than a nanomole.

Analyses of fuel exhaust products supplied by Dr. A. L. Burlingame and by Dr. D. Flory, showed only traces of amino acids when hydrolyzed. Samples of 1,1-dimethylhydrazine oxidized by nitric acid showed a number of as yet unidentified ninhydrin-positive peaks.

Upon microscopic examination, the lunar fines were observed to be predominantly irregular fragments. The number of regular, or spray, particles was estimated as 5×10^6 per gram. Approximately 100 slides were prepared.

The spray particles in the lunar fines are present as symmetrical spheres, oblate and prolate spheroids, ellipsoids, dumbbells, teardrops, crescents, and rings. Most of the particles have characteristic minute budlike protrusions (Fig. 2, a–d). Sintered spray particles are relatively rare. The prolate spheroids and dumbbells often show central portions containing one or more gas bubbles (Fig. 1). In many of these, a convex or concave blob of a different glass with a darker and redder color than that of the central zone appears at one or both ends. The end of the spray particle may be tipped with a minute sphere of dark brown to opaque material (Fig. 1). Opaque, magnetic, hexagonal plates have been observed in fractions having a density greater than 3.4; these closely resemble the troilite plates of type 1 carbonaceous chondrites.

The diameters of the globular particles range between 0.1 and 500 μm; the dumbbells range between 5 and 500 μm. The smallest dumbbells have a twisted, propeller-like habit.

The bulk of the spray particles in the lunar fines contains no crystalline phases. The lunar tuffs contain sprays with a 2- to 20-μm surface zone of crystallites, and a spray particle with 1 to 3 μm diameter octahedra of a colorless, nonbirefringent and high refractive index mineral has been observed, at the Lunar Receiving Laboratory in Houston.

The spray particles may be colorless, opaque, yellow, green, blue, or brown. The microspectrophotometer indicated that the visible and near ultraviolet absorption spectra also show marked differences from particle to particle, in contrast to the relatively greater uniformity of volcanic, impact, and meteoritic glasses.

Electron-probe results also indicate that the elemental composition of the spray particles varies within a broad range; particles rich in Fe, Ca, or Al can be distinguished, and the concentration of Ti is highly variable (Fig. 3). The enrichment of Zr and other elements along a crinkled surface zone of a spray particle (Fig. 3) demonstrates compositional heterogeneity even within a single unit.

On focusing and intensifying the beam of the Philips 300 electron microscope, most of the lunar particles formed globules. These have often been seen to coalesce or divide, and they also reveal on their surface the minute budlike protrusions which are characteristic of lunar spray. Powders of basalts from Kilauea-Iki and of chondrites have been found to fuse to smooth globules, under similar conditions. Some of the lunar particles tend to decrepitate and sublime under the electron beam. The electron microscope may thus be useful in studying mineral morphogenesis under conditions resembling those on the moon (low pressure and electron irradiation) and, indeed, of formation of organic molecules.

The morphology of lunar spray resembles that of the tektites and microtektites (*3*), which tend to be, however, chemically more homogeneous than the former. These forms have so far not been found in volcanic (*2*) nor in meteoritic impact (*4*) sprays, nor amongst the chrondrules of the meteorites (*2*).

The amino acids found in the chemical analysis of bulk fines could be present as the result of terrestrial contamination, as the result of chemical conversion of fuel, or as the result of indigenous chemical processes. Stronger likelihood of contamination may be provided by tests for dominant L isomers. Stronger likelihood of indigenous synthesis on the moon would be suggested by concordant analyses of subsurface samples. Resolution of these alternative possibilities requires further samples and analyses.

The amino acids appear to be predominantly in a polymerized structure. Finding of amino acids in the quantities recorded required recognition of the possibility that the conditions on the moon could be conducive to formation and preservation of their polymers (*7*). This result is consistent also with terrestrial contamination.

The accumulated data on the spray particles strongly indicate that these particles have been subjected to rotational turbulence. In the aspects of their appearance which suggest such rotation, they differ from particles found in meteorites or volcanic zones. The marked chemical heterogeneity of the lunar spray also appears to preclude the processes of homogenization of any crystallized igneous rock through volcanic or meteorite impact melting. The indications are, therefore, that at least the bulk of the lunar spray was formed through the fusing of individual mineral grains on and above the lunar surface near the equator, probably in the course of exceptionally intensive solar flare activity. In this connection the elimination of hydrocarbons, Na, and K by solar wind and radiation from the zones close to the surface of the moon have been predicted (*8*) on a priori grounds. Gold (*9*) interpreted certain structures on the lunar surface, photographed during the Apollo 11 mission, as zones of surface fusion resulting from solar flares of unusual intensity.

The foregoing considerations may explain the relatively low ratio of organic carbon to the total carbon found in the surface material retrieved by the Apollo 11 mission, through a process of decomposition and distillation of organic substances. The possibility exists, therefore, that the abundance of organic substances on the moon would tend to increase toward the polar regions, or toward the deeper zones beneath the surface of the equatorial regions.

The morphology of the lunar spray particles resembles in several aspects those of the proteinoid microspheres, terrestrial cells, meteoritic "organized elements," and microfossils (*1, 2*). This includes shapes in various sizes and stages of fission. The processes are different in several ways. The chemical composition is, also, predominantly different, inasmuch as spray particles are largely siliceous while the microspheres and the polymers of cells are largely polyamino acids. The differences reemphasize the fact that living systems possess (i) a characteristic morphology and (ii) special kinds of organic chemistry. Each of these two is necessary for life, but neither alone is sufficient.

Our results may be relevant to the formation of judgments of morphological entities from data which are to be telemetered to the earth from the Viking missions in the absence of returned samples.

S. W. Fox
K. Harada
Institute of Molecular Evolution, University of Miami, Coral Gables, Florida 33134
P. E. Hare
Geophysical Laboratory, Carnegie Institution of Washington, D.C. 20008
G. Hinsch
G. Mueller
Institute of Molecular Evolution, University of Miami

References and Notes

1. S. W. Fox, *Nature* **205**, 328 (1965); *Naturwissenschaften* **56**, 1 (1969).
2. G. Mueller, *Bol. Soc. Biol. Concepcion* **40**, 161 (1968); *Astrophys. Space Sci.* **4**, 3 (1969).
3. B. Glass, *Nature* **214**, 372 (1967).
4. F. W. Wright and P. W. Hodge, Meteoritical Society, 32nd Annual Meeting, Houston, 29 to 31 October 1969.
5. P. E. Hare, *Fed. Proc.* **25**, 709 (1966).
6. G. K. Strother and J. J. Wolken, *Science* **130**, 1084 (1959)
7. S. W. Fox, in *Biology and the Exploration of Mars*, C. S. Pittendrigh, W. Vishniac, J. P. T. Pearman, Eds. (NAS-NRC, Washington, D.C., 1966), p. 223.
8. G. Mueller, *Nature* **215**, 1149 (1967).
9. T. Gold, *Science* **165**, 1345 (1969).
10. Supported by NASA grant NAS 9-8101. We thank Dr. W. J. Humphreys, University of Georgia, for use of his scanning electron microscope, Dr. J. J. Wolken of Carnegie-Mellon University for use of his microspectrophotometer, The Philips Electronics Application Laboratory, Mount Vernon, N.Y., for use of an electron probe and scanning electron microscope, C. R. Windsor for aid in amino acid assays, and Miss S. Robertson for aid in photography.

4 January 1970

Organic Compounds in Lunar Samples: Pyrolysis Products, Hydrocarbons, Amino Acids

Abstract. *Lunar fines and a chip from inside a rock pyrolyzed in helium at 700°C gave methane, other gases, and aromatic hydrocarbons. Benzene/methanol extracts of fines yielded traces of high molecular weight alkanes and sulfur. Traces of glycine, alanine, ethanolamine, and urea were found in aqueous extracts. Biological controls and a terrestrial rock, dunite, subjected to exhaust from the lunar module descent engine showed a different amino acid distribution. Interpretation of the origin of the carbon compounds requires extreme care, because of possible contamination acquired during initial sample processing.*

The amount of organic matter in lunar samples returned during the Apollo 11 mission is very small. Various sources of possible contamination (*1*) make a significant analysis difficult. Pyrolysis, solvent extraction, and microscopy were applied to lunar fines (10086) and chips (inside and outside) from a lunar breccia (10002,54) (*2*).

Lunar fines (148.6 mg, untreated) were pyrolyzed for 5 minutes at 510°C. Trace quantities of methane, CO_2, ethylene, propylene, benzene, and toluene were detected. Pyrolysis at 700°C, for 7½ minutes of (i) untreated lunar fines (623.4 mg), (ii) solvent-extracted fines (717.2 mg), from which remaining traces of solvent were removed by heating for 6 hours at 134°C and pressure of 35 mm-Hg), and (iii) a powdered rock chip (675.8 mg), from inside a rock, yielded these same compounds in higher concentrations, and additional products. The product distributions from these three pyrolyses at 700°C were almost identical and amounted to less than 100 parts per million. Again methane was the main component, followed by benzene, toluene, CO_2, and smaller amounts of saturated hydrocarbons of low molecular weight, alkyl benzenes, thiophene and methyl thiophenes, indene and methyl indene, naphthalene and methyl naphthalenes, diphenyl, styrene and methyl styrene (Fig. 1).

In these experiments a Hamilton pyrolysis unit consisting of a quartz tube (43 cm by 4 mm internal diameter), heated externally by an electric furnace and utilizing a helium flow of 3 to 5 ml/min, was connected by a heated transfer line to a Perkin-Elmer model 226 gas chromatograph coupled (*3*) with a Watson-Biemann molecular separator to a Hitachi RMU-6E single-focusing mass spectrometer. During pyrolysis, products were collected in a liquid N_2 trap, which was then heated rapidly to 250° to 300°C with an adjustable oven for instantaneous introduction of the trapped products onto the gas chromatographic column [50-foot by 0.02-inch internal diameter, capillary; polyphenyl ether). Individual components were identified by mass spectrometry.

The pyrolysis products may result from several sources, including degraded polymeric substances, the synthesis from low-molecular-weight compounds, or release by heat treatment from mineral enclosures. Scanning electron and light microscopy suggested possible sources of the pyrolysis products. Two 0.5 mg quantities of untreated lunar fines, and one HF- and HCl-treated sample were distributed on the surface of polished aluminum pedestals with a drop of absolute ethanol, which dispersed the samples and facilitated adhesion to the metal surface. An outside chip of lunar rock was fractured into two pieces, one of which was mounted on the sample pedestal by means of double adhesive tape. Sample mounting for scanning electron microscopy was performed in the clean room facilities of the Naval Weapons Center, China Lake, California. After mounting, the specimens were coated with gold-palladium alloy approximately 500Å thick, in vacuum. Coating artifacts, such as granular surfaces, were absent on the lunar samples. After coating, the samples were examined with a Stereoscan Mark II (Cambridge Instruments Company, Ltd.) electron microscope. Light microscopic studies were performed with a Leitz Ortholux microscope; untreated fines as well as samples pyrolyzed at 700°C were examined in Permount preparations under transmitted light.

The light and the scanning electron micrographs of Fig. 2 show enclosures in glass beads (Fig. 2B) and broken beads (Fig. 2, A and C). Electron micrographs of the interior of a small crater, taken at different magnifications (Fig. 2, D, E, and F), show hollow, bubble-like structures, which suggest that perhaps a positive vapor pressure was behind the once molten surface. Figure 2G shows a particle of similar character. The insoluble residue from lunar fines treated with hot 48 percent HF and hot 6*N* HCl showed a folded, sheet-like object, (Fig. 2H). This has been best seen on examination of stereopairs of photographs.

For the analysis of soluble organic substances, 22.81-, 22.61-, and 16.44-g samples of lunar fines were removed from the top, middle, and bottom of the sample container. It appeared that coarser-grained material, including a small rock, was distributed nearer the bottom of the container. The fines were extracted with benzene and methanol (4:1, vol/vol) for 1 hour with 28-kilocycle ultrasonic energy yielding 1.7, 0.3, and 0.2 mg of residue, respectively. The yellow extracts were chromatographed on prewashed thin-layer plates coated with Silica Gel G, with two solvent systems: (i) *n*-hexane and (ii) *n*-hexane, diethyl ether, acetic acid (95:5:2, vol/vol/vol). The plates, visualized with aqueous Rhodamine 6G (0.0005 percent) under ultraviolet light (*4*), revealed traces of saturated hydrocarbons (2.0, 0.5, and 0.3 parts per million, respectively), elemental sulfur, and a yellow fluorescent substance which remained at the origin even in the more polar solvent system. Procedural controls showed less than 0.05 parts of hydrocarbons per million.

Approximately half of the total extract described above was inserted into the heated, direct-inlet system of the mass spectrometer. At a probe temperature of 100°C it was possible to identify sulfur and hydrocarbons in all three extracts. As the temperature was raised to 110° to 115°C, the series of alkyl ions, C_nH_{2n+1}, appeared in the extract of the 22.81-g sample; at this and increasingly higher temperatures, C_{25} to C_{40} ions were noted. Prominent odd-mass ions suggesting branching of carbon skeletons were absent. The mass spectra of the second and third extracts showed substantially lower quantities of alkyl ions. A second analysis of each extract gave duplicate results, while sulfur and alkyl ions were absent in procedure controls. Optical rotation measurements made on the second extract showed no optical activity above the limits of detection of the polarimeter (0.3 × 10^{-3} degree of rotation).

Standard ion exchange chromatographic procedures of high sensitivity

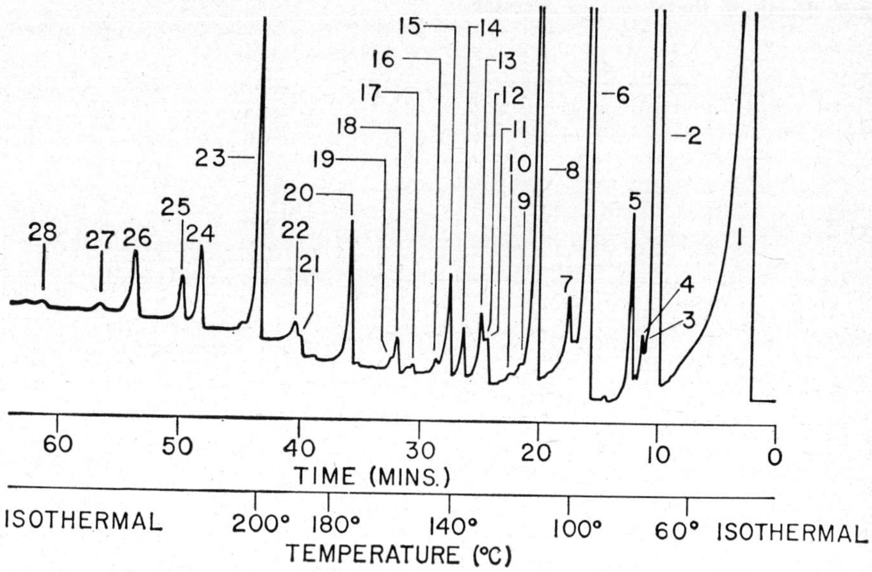

Fig. 1. Gas chromatogram of the pyrolysis products of a 675.8 mg powdered chip from the inside of a lunar breccia (No. 10002, 54), heated to 700°C in He. Peak 1, CH_4; 2, CO_2, C_2H_4, and C_2H_6; 3, 4, and 5, C_5H_6, C_5H_8, and C_5H_{10}; 6, benzene; 7, thiophene; 8, toluene; 9 and 10, methyl thiophenes; 11, ethyl benzene; 12, 13 and 14. *p*-, *m*-, and *o*-xylenes; 15 and 18, styrene and methyl styrene; 16, 17, and 19, C_3 alkyl benzenes; 20, indene; 21 and 22, methyl indenes; 23, naphthalene; 24 and 25, methyl naphthalenes; 26, diphenyl; 27 and 28, C_2 alkyl naphthalenes. Identifications made with a Hitachi RMU-6E mass spectrometer connected to a Perkin-Elmer 226 gas chromatograph, (polyphenyl ether capillary column, 50-foot by 0.02 inch, internal diameter).

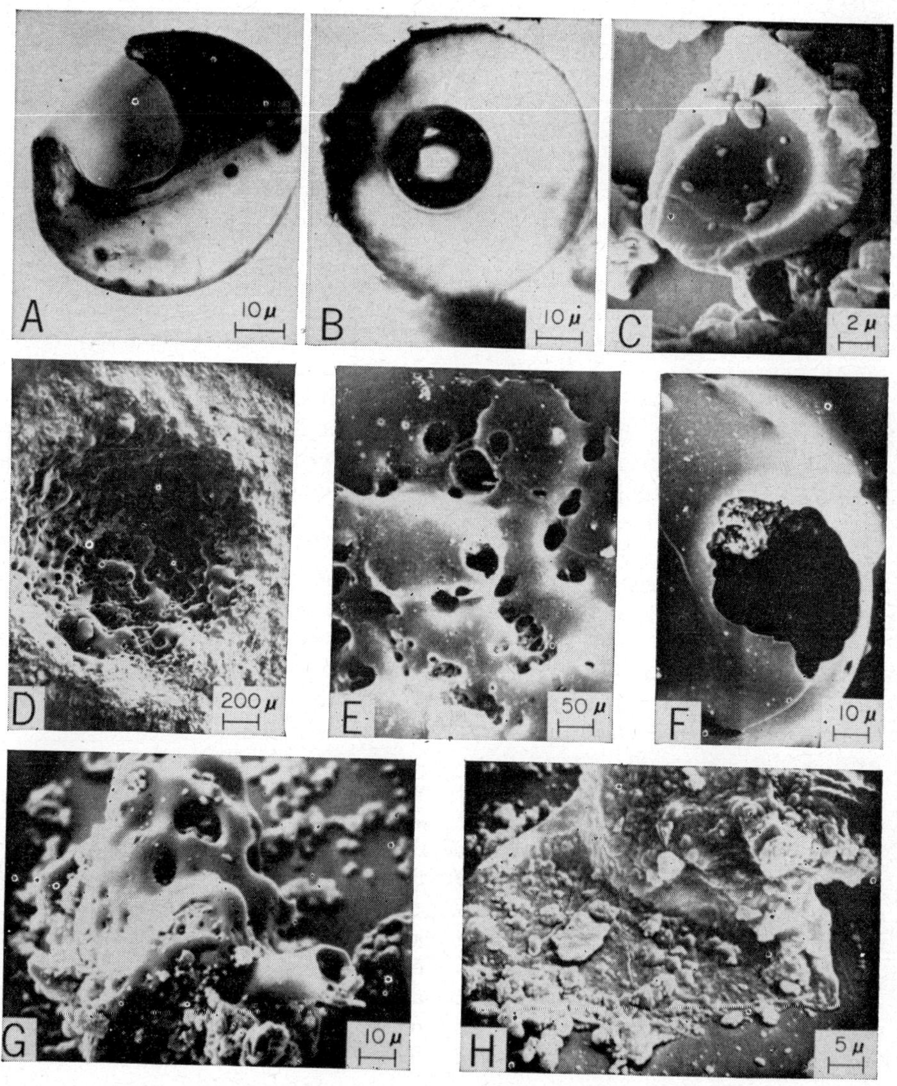

Fig. 2. Photomicrographs (A and B) and scanning electron micrographs (C–H) of lunar fines (10086) and the outer surface of a lunar breccia (10002, 54). (A and B) Particles that had been heated to 700°C for 7½ minutes; (A) broken hollow glass bead with the cavity exposed, (B) hollow glass bead with the interior intact. (C) Scanning electron micrograph of broken, hollow glass bead. (D, E, and F) Small crater on surface of lunar breccia showing hollow, thin-walled bubbles indicating an initial, positive vapor pressure behind the once molten surface. (G) Glass bead showing "bubble-like" structure. (H) Sheet-like folded material, with mineral enclosures, resembling possible polymeric substances, in the residue of fines treated with 48 percent hot HF and hot 6*N* HCl.

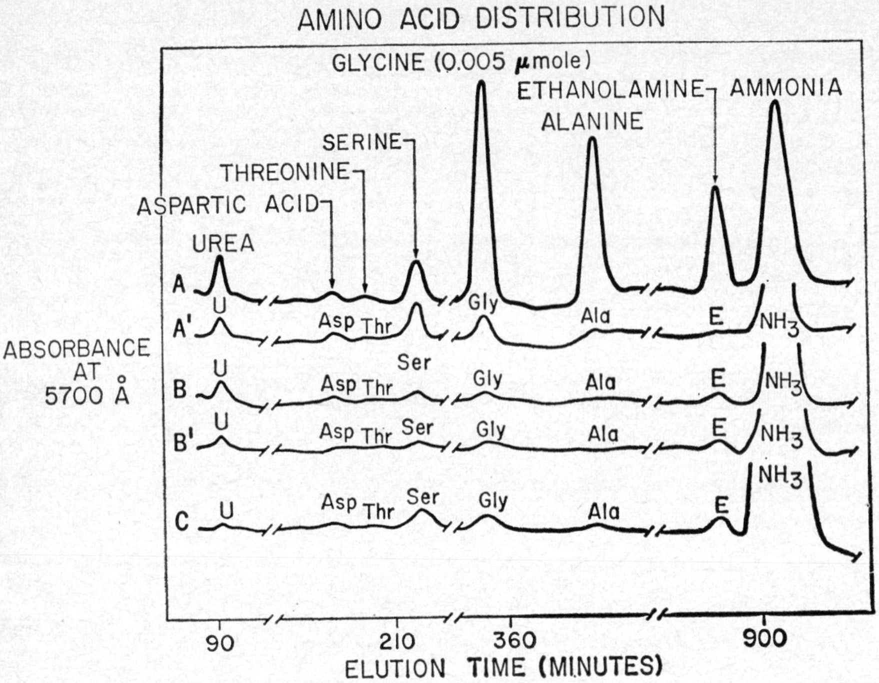

Fig. 3. High-sensitivity and high-resolution ion exchange chromatograms of (*A*) lunar fines (10086) extract; (*A'*) corresponding procedure blank; (*B*) extract of terrestrial dunite rock subjected to the exhaust from the lunar module descent engine in an initial vacuum; (*B'*) corresponding procedure blank; (*C*) hand contamination at 10^{-10} mole level.

and high resolution (*5*) were used for the analysis of free amino acids from 32.66 g of lunar fines extracted in a Soxhlet with 150 ml of freshly triple-distilled water. Precautions were taken to prepare and keep water and glassware free from hand and microbial contamination. After extraction, the contents of the flask were dried on a rotary evaporator. The supernatant in the thimble was transferred to a separate flask and evaporated similarly. In a procedure blank, the thimble contents were added to the extraction flask. The same lunar material was extracted a second time, and the corresponding procedure blank was obtained.

Immediately prior to chromatographic analysis, 2.0 ml of deionized water passed through a 0.22-μm pore filter was added to the flask, and 0.1 ml was removed for bacterial colony count by plating; all transfers of fluid were carried out aseptically. Three flasks showed no bacterial growth, and three flasks had very low colony counts. The source of this contamination was proven to be the added water. The remaining 1.9 ml of water was then evaporated. Subsequently 1.0 ml of 0.1*N* hydrochloric acid was added to the flasks to dissolve soluble residues, and 0.33 ml of the acid solution was placed on the ion exchange column. Prior to and between analyses of the lunar material, the analyzer was tested to show that amino acids were not originating from within the column, that recoveries of known amino acids were satisfactory, and that total operational contamination could be consistently kept at approximately 10^{-10} mole of each amino acid. Cysteic acid, taurine, aspartic acid, threonine, serine, glycine, alanine, and ornithine were readily detected at the 10^{-10} mole level, whereas other common amino acids were barely discernible or were not detected at all.

The only findings of interest appeared in the analysis of the first lunar Soxhlet extract. The following compounds (corrected for blank values) were detected: urea, 65 μg per kilogram of lunar material; glycine, 32 μg/kg; alanine, 36 μg/kg; ethanolamine, 22 μg/kg (Fig. 3). Apparently the first Soxhlet extraction of the lunar sample was essentially quantitative.

Although the position of a peak on the chromatogram is of great qualitative value, it is not an absolute criterion of identity. Neither is symmetrical overlap by a known compound co-chromatographed with a peak of interest an absolute criterion of identity, but it does provide confirmatory evidence. Accordingly, 0.065 ml, or one-fifth of the first extract, was co-chromatographed with a known amino acid mixture at the 10^{-9} mole level. Symmetrical peaks were obtained, and glycine, alanine, and ethanolamine completely overlapped the corresponding peaks in the lunar sample. Moreover, the net positive differences between the lunar sample plus test mixture and the test mixture were approximately one-fifth those determined in the original run. The above identifications are therefore reasonable. Identity must still remain tentative pending confirmation from other laboratories by different techniques.

The presence of these four substances (if the above identifications are correct) is inconsistent with either hand (*6*) or microbial contamination. Furthermore, it does not seem likely that they have arisen from presently existing materials of biological origin. In abiotic syntheses, simulating primitive earth conditions in the laboratory, glycine and alanine have usually been formed in greater abundance than other amino acids (*7*). One is led to consider abiotic synthesis of trace amounts of these substances from the exhaust from the lunar module descent engine, or abiotic synthesis on the lunar surface from indigenous organic materials under the influence of ultraviolet or ionizing radiation. Neither the hot aqueous extract of a 27.3-g sample of the terrestrial rock dunite, subjected by NASA (*8*) to exhaust from the lunar module descent engine in an initial vacuum of 10^{-4} torr, nor a portion (one-third) of the 150-ml extract, analyzed by the identical procedure, showed the same amino acid distribution as the lunar sample (Fig. 3). Furthermore, no exhaust gases from the lunar module descent engine were found in the lunar samples analyzed during the investigation at the Lunar Receiving Laboratory (*1*).

During the extraction of the lunar fines with hot water a strong odor, resembling H_2S, was noticed, and a positive $AgNO_3$ test was obtained. In one experiment, a spot approximating the position of sulfur, or of certain sulfur compounds, was observed on a thin-layer chromatogram when the plate was placed under the effluent gas.

Interpretation of the origin of the organic compounds in the lunar samples requires extreme caution because of the possibility that contamination occurred during acquisition and initial processing (*1*). It has been shown that the pyrolysis products found in lunar fines can be synthesized by the pyrolysis of methane in the laboratory (9). It is also possible that these compounds were synthesized from methane on the primitive moon. A definite conclusion is not warranted until the results are confirmed by subsequent analyses.

BARTHOLOMEW NAGY
University of Arizona, Tucson 85721

CHARLES M. DREW
Naval Weapons Center, China Lake, California 93555

PAUL B. HAMILTON
Alfred I. Du Pont Institute, Wilmington, Delaware 19899

VINCENT E. MODZELESKI
University of Arizona, Tucson

SISTER MARY E. MURPHY
St. Joseph College, West Hartford, Connecticut 06117

WARD M. SCOTT
University of Arizona, Tucson

HAROLD C. UREY
University of California at San Diego, La Jolla, California 92037

MARIA YOUNG
University of Arizona, Tucson

References and Notes

1. D. A. Flory, B. R. Simoneit, D. H. Smith, "Apollo 11 Organic Contamination History," *NASA Tech. Rep.* (1969); D. A. Flory and B. R. Simoneit, "A Summary of the Apollo 11 Organic Contamination History," *NASA Tech. Rep.* (1969); B. R. Simoneit, A. L. Burlingame, D. A. Flory, I. D. Smith, *Science* **166**, 733 (1969).
2. D. A. Flory, personal communication; Lunar Sample Preliminary Examination Team, *Science* **165**, 1211 (1969).
3. V. E. Modzeleski, W. D. MacLeod, Jr., B. Nagy, *Anal. Chem.* **40**, 987 (1968).
4. M. T. J. Murphy, B. Nagy, G. Rauser, G. Kritchevsky, *J. Amer. Oil Chem. Soc.* **42**, 475 (1965).
5. P. B. Hamilton, *Anal. Chem.* **35**, 2055 (1963).
6. ———, *Nature* **205**, 284 (1965).
7. S. L. Miller, *J. Amer. Chem. Soc.* **77**, 2351 (1955); ——— and H. C. Urey, *Science* **130**, 245 (1959); J. Oro, in *The Origins of Prebiological Systems*, S. W. Fox, Ed. (Academic Press, New York, 1965), p. 146.
8. D. A. Flory, personal communication.
9. J. Oro and J. Han, *J. Gas Chromatog.* **5**, 480 (1967).
10. We thank A. Lumpkins, J. Modzeleski, and L. A. Nagy of the University of Arizona, Tucson, R. DeFiore, J. C. Dickinson, J. Greenquist, and M. L. and T. T. Myoda of the Alfred I. duPont Institute, Wilmington, Delaware; and J. E. Thomas and R. Ward of the Naval Weapons Center, China Lake, California, for their capable assistance in the experiments. This investigation was supported by NASA at the University of Arizona, with additional contributions from the Naval Weapons Center, China Lake, NSF at St. Joseph College, West Hartford, and from the Alfred I. duPont Institute, Wilmington. This is contribution No. 197, department of geochronology, School of Earth Sciences, University of Arizona.

4 January 1970

A Search for Viable Organisms in a Lunar Sample

Abstract. *The hypothesis that the moon could harbor viable life forms was not verified on analysis of the first samples from the Apollo 11 mission. Biological examination of 50 grams of the bulk fines confirm the negative results obtained by the Manned Spacecraft Center quarantine team. No viable life forms, including terrestrial contaminants, were found when the sample was tested in 300 separate environments. Only colored inorganic artifacts, resembling microbial clones, appeared around some particles.*

This study to elicit the growth of organisms from lunar matter is the first in a series to determine the existence of extraterrestrial life. The discovery of extraterrestrial life and its characteristics will be important in constructing theories regarding the origin, evolution, distribution, and frequency of life in the universe, when these findings are related to cosmology and the physical and chemical nature of the particular body from which the samples are derived.

The moon is a testing ground for the concept of panspermia—that is, the theory that life can be disseminated through space without the intervention of man (*1*). The moon is hardly capable of the kind of chemical evolution, proposed for the earth, that might result in the synthesis of organic matter sufficient in amount and residence time to allow life to evolve; because of its mass, the moon cannot hold an atmosphere for long (*2*). At the time of its formation the moon must have been a sterile body, and it must subsequently have served as a repository—it appears from its impact history—for the collection of debris from space.

At the NASA Ames Research Center, we designed and supervised the construction of a biological barrier system to perform microbiological experiments on extraterrestrial samples. The system is maintained at a positive pressure (0.2 inch of water) to protect the sample from terrestrial contamination and is housed in a class-100 room [particle (> 0.5 μm) count less than 100 per cubic foot of air] to protect personnel from exposure to potential hazards of the sample.

The biological barrier system contains a preparation area and an incubation area. Twelve adjoining glove boxes serve as the preparation area. In this

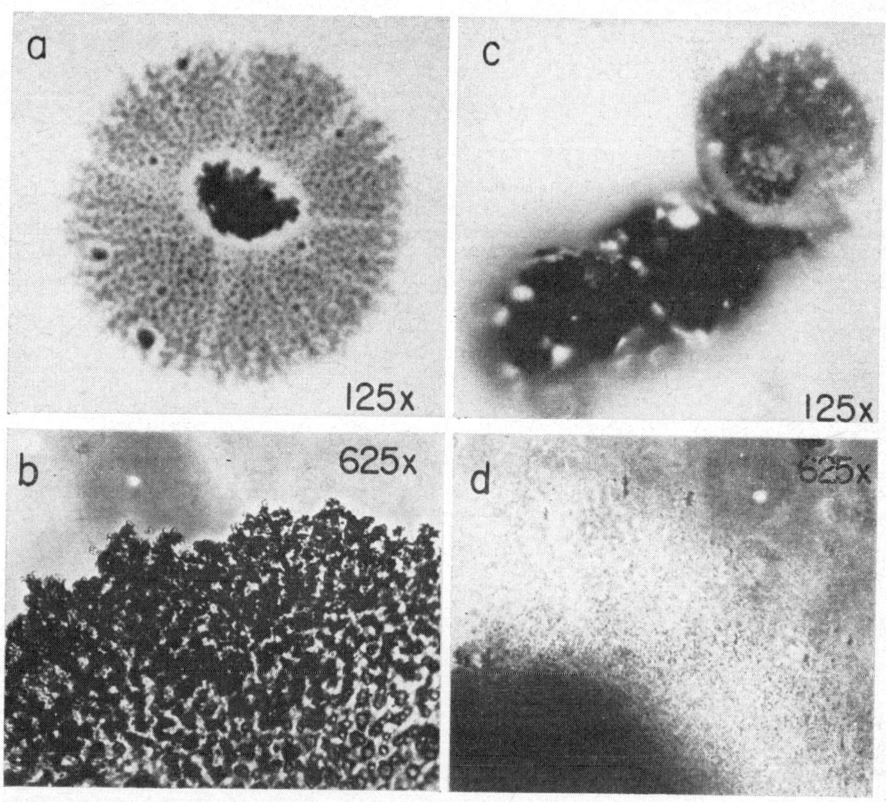

Fig. 1. Typical artifacts resembling microcolonies. (a) Type I (× 125); (b) type I (× 625); (c) type II (× 125); (d) type II (× 625).

Table 1. Atomic absorption analysis of extracts of lunar sample.

Extract*	Total in extract (mg/g of sample)						
	Li	Na	K	Fe	Mn	Ti	Cr
Water	<0.0001	1.76	0.062	<0.001	<0.0006	<0.002	<0.0003
1*N* HCl				8.75	.112	1.50	.256
6*N* HCl				8.00	.106	1.99	.120
1*M* sodium citrate				2.07	.0115	0.077	.0115

* Extractions of a 1-g portion of sample were made with 10 ml of water followed by 1*N* HCl, then by 6*N* HCl. A fresh sample was used with 1*M* sodium citrate. Extractions were performed with shaking for 30 minutes, at 80°C for water and sodium citrate, at room temperature for the other extracts. The citrate and 6*N* HCl extracts were yellow; the others were colorless.

area molten agar media are dispensed and hardened; the sample is weighed, sieved, and distributed onto the surface of the agar media. To this preparation area, aseptic means of removing material from, or introducing it to, the preparation area are provided by a pass-through sterilizer, equipped for steam or gaseous sterilization; a transfer box, equipped for gaseous or liquid sterilization; a dunk tank containing a solution of sterilant; and a sterile filter assembly, for passing thermolabile solutions into the barrier system. The preparation area is equipped with gas lines for introducing ethylene oxide for sterilizing, and dry-filtered nitrogen for purging and maintaining an oxygen-free working environment.

The incubation area consists of three wings of incubator glove boxes arranged in banks of four each. Each wing has an incubator set at 10°, 20°, 35°, and 55°C. Every incubator is divided vertically into three compartments for maintaining humidity. Each compartment has nine removable trays, each of which contains ten petri dishes, and a Teflon-lined bottom tray with a wick assembly for humidity control by means of water or saturated brine solutions. Fluorescent lamps, mounted behind each incubator, illuminate the petri dishes. The compartments are moved by an elevator; each tray can be removed or inserted through an opening into a glove-box working area. This working area has built-in objective lenses, to which a stereoscope body can be mounted. Magnifications up to × 32 are obtainable, and photographs can be taken. Each incubator has valved gas lines for the introduction of ethylene oxide, nitrogen, or mixed gases.

The gases provided pass through a pair of biological filters and through sterile tubing before entering the biological barrier. All gases leaving the barrier system, with the exception of ethylene oxide, are exhausted to the atmosphere through a furnace.

Prior to receipt of the sample, the barrier system was treated for 24 hours with a mixture of ethylene oxide and Freon, and purged with dry sterile nitrogen. These gases were monitored with a gas chromatograph during the purging cycles. The barrier system was then tested for sterility. All interior surfaces were swabbed, and the swabs were incubated in trypticase soy broth and fluid thioglycollate broth for 7 days. No viable organisms were found.

Nine media containing 2 percent agar were sterilized in the connecting autoclave, and 30 ml was dispensed into sterile glass petri dishes. The media selected were a synthetic complex medium with a low *p*H (*3*), a neutral synthetic complex medium (*4*), a synthetic complex medium with a high *p*H (*5*), a dilute synthetic complex medium (*6*), a basal medium containing only sea salts (*7*), a medium containing only inorganic forms of nitrogen, phosphorus, and sulfur (*8*), a medium containing elemental sulfur (*9*), a medium containing formose sugars (*10*), and a medium containing spark-discharge products (*11*). Three different concentrations of a synthetic sea salt formulation (*7*) were used in the preparation of all media; the concentrations selected were 0.1, 3.0, and 20.0 percent.

The 3000 petri dishes containing media were placed on trays and incubated at 10°, 20°, 35°, and 55°C under one of three different gas mixtures: 78 percent nitrogen, 20 percent oxygen, 2 percent carbon dioxide; 97 percent nitrogen, 1 percent oxygen, 2 percent carbon dioxide; and 98 percent nitrogen, 2 percent carbon dioxide. After a minimum 72-hour incubation, all the petri dishes were examined and found to be sterile. The barrier system was then purged with dry sterile nitrogen.

Upon delivery to Ames Research Center, the sample container was wiped with 5 percent peracetic acid solution and treated with ultraviolet light before insertion into the barrier system. The sample container was opened and the sample (10089,1) was weighed and sieved, and the bulk density was determined (*12*). During the week of 6 October 1969, 12 g of the sample were distributed uniformly on the 3000 petri dishes, by means of a small settling tower within the preparation area. The inoculated petri dishes were returned to the incubators, which were again flushed with one of the three gas mixtures. These gas mixtures were maintained for the duration of the incubation period. The plates were observed for evidence of colony formation until the week of 8 December 1969.

By the week of 1 December 1969 no microorganisms had been detected in the 12 g of lunar sample. An enrichment procedure was next used on the remainder of the sample. A disk of sterile filter paper was placed upon the surface of agar media [the media of (*4*), (*8*), (*10*), and (*11*), combined], and a 1-g sample was placed on the center of the filter paper. These plates were incubated at 20°, 35°, and 55°C. After 10 days the sample and filter paper were moved to a plate containing fresh medium. No microorganisms were detected, but incubation is continuing.

Although no microorganisms were detected in the 12 g of sample distributed on 3000 petri dishes, numerous artifacts resembling microcolonies were seen after 4 days of incubation. Two types were distinguished. Type I was circular, flat, granular, and yellow-to-brown, surrounding a central lunar particle (Fig. 1a). Type II was irregularly lobed, raised, moist, and orange-to-dark-brown, and often completely covered a lunar particle (Fig. 1c). Observations at higher magnifications revealed the type I artifacts to be composed of a more coarsely crystalline material than the type II artifacts (Fig. 1, b and d). After 6 weeks of incubation, transplants were made to identical media and environmental conditions, but they did not give rise to similar artifacts.

A series of spot tests (*13*) was performed on both types of artifacts for those elements present in the lunar material (*14*) which might give rise to colored inorganic precipitates. Type I artifacts gave negative results when tested for Ti, Mn, and Cr, but equivocal results for Fe. The artifacts were insoluble in 3*N* HCl, 6*N* HCl, 6*N* H_2SO_4, 0.05*N* NaOH, and concentrated NH_4OH. They dissolved partially in NaOBr, but complete dissolution was achieved when NaOBr treatment was followed by treatment with 6*N* H_2SO_4. Type II artifacts gave a negative result when tested for Ti but gave strong positive reactions for Fe when tested with

$K_4Fe(CN)_6$ or α,α'-dipyridyl in thioglycollate.

Extracts of the lunar sample were prepared and analyzed by atomic absorption spectrophotometry. The results (Table 1) showed that minerals containing Fe—the most abundant of the ions analyzed—were not solubilized by water alone, but required H^+ or Na^+. The results are consistent with cation exchange phenomena. The spot-test results obtained on the Type II artifacts support the cation exchange results. The reason for the equivocal response by the type I artifacts to the iron reagents is not known, but it may be that the brown color arising around the particle is an iron mineral at concentrations barely detectable by the two reagents. We believe that the higher concentrations of Fe in some lunar particles may well produce the lobular structures of the type II artifacts.

We conclude for this sample of the moon that there was no viable life present. Our conclusion agrees with that of the biological quarantine team at the Manned Spacecraft Center, Houston.

VANCE I. OYAMA, EDWARD L. MEREK, MELVIN P. SILVERMAN

NASA Ames Research Center, Moffett Field, California 94035

References and Notes

1. See A. I. Oparin in "The Origin and Initial Development of Life," *NASA Tech. Trans. NASA TTF-488* (1966), pp. 20–22.
2. C. Sagan, "Organic Matter and the Moon," *Nat. Acad. Sci.–Nat. Res. Counc. Publ. 757* (1961), pp. 13, 14.
3. The medium contained 1.0 g/liter of glycerol, sodium salts of lactate, pyruvate, and acetate, methanol, ethanol, K_2HPO_4, KNO_3, and $(NH_4)_2SO_4$; 0.2 g/liter of 18 DL-amino acids, glycine, 4-hydroxy-L-proline, β-alanine; 0.01 g/liter of seven purines and pyrimidines; and trace quantities of 16 vitamins; *p*H adjusted to 3.0 with HCl.
4. Same as described in (*3*), but with *p*H 7.0.
5. Same as described in (*3*), but with *p*H adjusted to 10–11 with NaOH.
6. Ingredients same as (*4*), but at 1/10 concentration.
7. Synthetic sea salt mixture (Aquarium Systems, Inc., Wickliffe, Ohio).
8. Same as (*7*), supplemented with 1.0 g/liter each of K_2HPO_4, KNO_3, and $(NH_4)_2SO_4$.
9. Same as (*8*), but with *p*H adjusted to 3.0 with H_2SO_4. After the agar solidified, 0.5 ml of a solution containing elemental sulfur (100 g) dissolved in Na_2S solution (180 g of $Na_2S\cdot 9H_2O$ per 330 ml of H_2O) was spread on the surface; elemental sulfur then precipitated.
10. Same as (*8*), supplemented with 10 g of formose sugars per liter. (Prepared by Dr. J. Shapira, NASA SP-134, 1967).
11. Contained 500 ml/liter of a solution prepared by the reaction of CH_4, NH_3, and water vapor in a spark discharge apparatus [J. Rabinowitz, F. Woeller, J. Flores, R. Krebsbach, *Nature* **224**, 796 (1969)] consuming 90 watt-hours. The solution contained (in grams per liter) organic carbon, 2.3; cyanide carbon, 1.3; NH_3–N, 7.3; other N, 1.9.
12. The sieve distributions of the 50-g sample (10089,1) were as follows: 16/35, 6.7%; 35/60, 6.3%; 60/115, 13.1%; 115/250, 15.8%; 250/325, 12.4%; and < 325, 45.7%. The sample color closely matched Munsell color index 5YR.
13. F. Feigl, *Spot Tests in Inorganic Analysis* (Elsevier, Amsterdam, 5th English ed., 1958); Fe (prussian blue reaction, p. 161; α,α'-dipyridyl reaction, p. 162), Cr (diphenylcarbazide reaction, p. 168). Mn (benzidine blue reaction, p. 175), Ti (chromotropic acid reaction, p. 197).
14. Lunar Sample Preliminary Examination Team, *Science* **165**, 1211 (1969).
15. We thank B. J. Berdahl, C. W. Boylen, G. C. Carle, J. O. Coleman, C. C. Davis, G. Hamilton, D. A. Jermany, P. J. Kirk, M. E. Lehwalt, A. K. Miyamoto, E. F. Munoz, G. E. Pollock, B. J. Tyson, and O. Whitfield for assistance; E. L. Craig and C. W. Boylen for photography, and G. E. Peterson and his staff for design and construction of the facility.

5 January 1970

Micropaleontological Study of Lunar Material

Abstract. *Samples of the lunar dust, rock chips, and thin sections of rocks from Tranquillity Base have been examined by use of white light and electron optics. In transmitted and in dark- and bright-field incident light and in the scanning electron beam the material reveals no indication of biological morphology. It is inferred that the lunar regolith has always been devoid of life.*

The hope of finding evidence of a preexistent biosphere, or even of stages in chemical evolution of organic systems, on the moon began to fade even as Astronauts Armstrong and Aldrin's early observations began to return to earth from Tranquillity Base. The assumption that no such evidence would be found has been fully confirmed, at least with respect to microorganic structures that might represent extinct or extant microorganisms on the lunar surface. The conclusion bears the limitation that the lunar maria, of which the Sea of Tranquillity is regarded as typical, are representative of the moon as a whole during its entire history. It can be inferred that the lunar regolith has never possessed life and is inimical to life.

The samples investigated in this study consisted of the following: (i) lunar fines (10086,8) from the bulk sample container, sieved to various size fractions; (ii) rock chips of microbrecciated structure, from an outside chip (10091,6) and an inside chip (10091,7), respectively; and (iii) thin sections of microbreccia (10059,32; 10065,25; 10046,56; 10021,29).

The lunar dust was examined as free powder, on the untreated surfaces of glass slides, by normal and polarized, transmitted, and indirectly reflected light (dark-field illumination), and by transmitted light on whole mounts employing various mounting media, such as microscope immersion oil, gum damar, and diaphane.

Rock chips were examined by scanning electron microscopy after deposition of a 300- to 500-Å layer of gold.

Thin sections were examined by polarized and nonpolarized transmitted light, by polarized and nonpolarized direct incident light, and by polarized indirect reflected light (dark-field illumination). (Thin sections bearing cover glasses can be examined by transmitted light only.)

The morphology and optical properties of discrete objects in the lunar fines and in the thin sections of the rocks, at all magnifications accessible to white-light microscopy, indicate total absence of structure that can be interpreted as biological in origin.

Studies of the rock chips, both of outside and inside surfaces, with the scanning electron microscope likewise demonstrate complete absence of biological morphology.

It is appropriate to note that certain of the ubiquitous glass beads in the Apollo 11 dust, especially those of <10-μm size range and of spheroidal and ellipsoidal shape, curiously resemble microorganisms. These "pseudofossils" occasionally appear to possess double "walls," owing to spherical refraction of light around their surfaces and false "organelles" resulting from minute inclusions within the glass and to minute particles of adhering dust. Their mineral origin can readily be determined by optical properties in different combinations of light response in the white-light microscope.

All samples examined in this study were virtually devoid of terrestrial contaminants, with the exception of occassional birefringent fibers occurring in the bulk dust sample and occasional chemically induced artifacts within the epoxy mounting medium of the thin section preparations.

E. S. BARGHOORN

Department of Biology, Harvard University, Cambridge, Massachusetts

DELBERT PHILLPOTT
CHARLES TURNBILL

NASA Ames Research Center, Moffett Field, California

References and Notes

1. We express our appreciation to Umesh Banerjee, Harvard University, for assistance in this investigation, and to Sheldon Moll, Advanced Metals Research, Burlington, Massachusetts, for assistance in the scanning electron microscopy studies. Supported by NASA contract NAS 9-8060.

4 January 1970

Micromorphology and Surface Characteristics of Lunar Dust and Breccia

Abstract. *Although nothing of direct biologic interest was observed in the sample studied, small shaped glass particles and glazed pits resemble objects which elsewhere have been described as fossils. These features, although nonbiological, do bear on processes of lunar weathering and outgassing. The glazed pits are impact features. Fusion of their surfaces released gases. Electron microscopy of the glasses, pits, and angular microfractured mineral grains indicates a prevalence of destructive weathering processes—thermal expansion and contraction, abrasion by by-passing particles, and, of course, impact.*

Our sample consisted of 7.4 g of fine black surface dust (sample 10086,15), 0.92 g of fragments from the interior of a breccia block (10091,4), a single 1.22-g fragment showing an external surface of the breccia (10091,5), four thin sections (10019,15; 10046,56; 10061,27; 10065,25), and three probe mounts (10059,32; 10059,37; 10067,28).

Both breccia and fines consist of a matrix of mostly very fine grained, mainly angular particles of mineral, rock, and glass, within which are scattered larger mineral and rock fragments as well as spheroids and fragments of glass. The main difference is the induration of the breccia, which shows a strong (shock?) coherence between grains and probably some welding between glass particles. Point counts show the order of abundance to be glass, rock fragments, plagioclase, pyroxene, ilmenite, olivine; but identification of minerals and estimates of their abundance and chemistry were not the goal of this study.

Our mission was to search for evidences of present or former life on the moon and, failing that, to consider aspects of the lunar sample that might bear indirectly on the question of life processes.

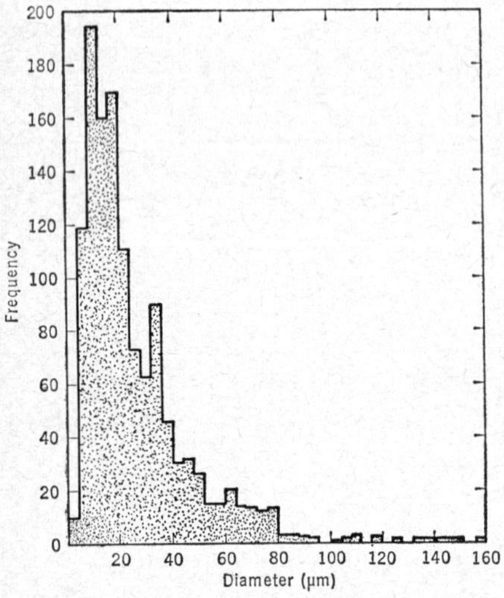

Fig. 1. Size distribution of glass spheroids from Apollo 11 thin sections 10059,37; 10061,15; 10061,27; and 10067,28.

Toward this end a variety of sample preparations was employed, resulting in end-products that were examined under varied modes of light microscopy, by scanning and transmission electron microscopy, and in the mass spectrometer. In addition to the study of optical thin sections prepared in Houston, our procedures involved the preparation of strewn slides of the dust, both before and after HCl-HF maceration; the preparation and vacuum-coating of standard plug mounts, grids, and replications of dust and rock fragments for electron microscopy; and the fusion (under vacuum) of dust at temperatures above 1200°C, followed by studies similar to the above and by mass spectrometry of volatiles obtained. Techniques employed are familiar or obvious ones.

As it became evident that our sample contained neither life nor fossils, interest focused on the abundant tiny glass spheroids and ovoids which, if carbonaceous, could be mistaken for microfossils. We also studied features indicative of outgassing and weathering processes (by techniques of reference (*1*) which might suggest the former presence of a lunar atmosphere in which biogenic or prebiogenic processes could have occurred.

Although glass appears to be the most abundant component of our sample, it is hard to estimate just how much there is. The minute size of many of the matrix grains, the degree of overlap in thin section, and the abundance of opaque minerals make it difficult to employ standard point-count procedures with confidence. Nevertheless, it seems that from 20 to 50 percent or more of the total dust and breccia is glass. This glass has an index of refraction mostly greater than 1.64, implying a density greater than 2.94 (*2*). Its color ranges from dark amber through yellow to clear, presumably varying with iron content. Small glass particles adhere to a weakly magnetized knife blade. Some are embayed, fractured, crazed, cracked, and locally devitrified, and they adhere to surrounding mineral particles as if hot at the time of deposition. One dumbbell-shaped fragment consists of two fused glass spheres, implying the possible significance of fusion of hot glass as a cementation mechanism.

What is of interest here, however, is that between 1 and 5 percent of the total rock consists of mostly tiny spheroids, ovoids, and other shaped particles of glass (Figs. 1–3). The abundant spheroids and ovoids are similar in shape to some algal and bacterial unicells, and the smaller ones are comparable in size. Indeed, were such glassy particles to be encased in carbon and then dissolved or altered (as might happen in a parent body such as that of the carbonaceous chondrites), they would make impressive pseudomicrofossils.

The size distribution of the glass spheroids in fact resembles the mortality curves of some microorganisms, ranging from diameters of well under 1 micron to 392 microns in populations counted and skewing so sharply toward smaller sizes (Fig. 1) as to suggest that, with better statistics in the smaller grain sizes, the curve would approach both x and y axes asymptotically. Indeed, examination of the lunar dust under the scanning and transmission electron microscopes shows a multitude of submicron sized spheroids going right down to a few tens of angstroms. Such a size distribution, however, is not like that usual for the morphologically simple procaryotic unicells which the glass spheres resemble in a crude way, but rather like the mortality curves of sexually reproducing microorganisms. This is *not* to propose that there are or were solid glass Protozoa on the moon but to add one more warning to the many that have already been given about a too-ready interpreta-

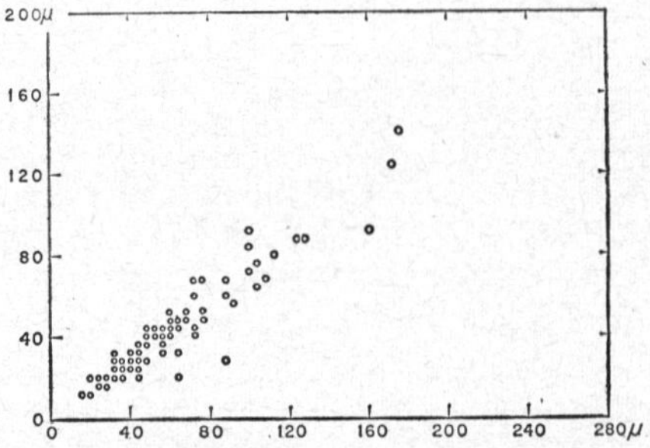

Fig. 2. Size distribution of glass ovoids from Apollo 11 thin sections 10059,37; 10061,15; 10061,27; and 10067,28.

tion of exotic objects as of vital origin on the basis of gross morphology alone.

The lunar glasses also contain many internal structures of interest—bubbles, "organelle"-like mineral inclusions, fine tubular structures, and microlites such as are well known in natural glasses on earth and some of which make rather spectacular pseudofossils (*3*). Some of the glasses involve complex composite structures.

The warning deserves emphasis. Elsewhere on the lunar or Martian surface may be lifelike artifacts that will be harder to discriminate from the real thing.

In addition to the glass observed as shaped particles and fragments, crusts of glass partly envelop rock or mineral particles, completely envelop dust aggregates of different shapes (including spheres), and glaze the surfaces of small rimmed pits or microcraters (Fig. 3, i–j). Such glazed surfaces, moreover, show bubble aggregates indicative of outgassing as well as local microfracture patterns consistent with origin of the microcraters by impact.

Although less than 1 square centimeter of external surface (10091,5) was available for study, examination of this and of smaller fragments from a dust sample (10086,15) revealed a number of roughly circular depressions with raised rims and glazed surfaces (Fig. 3, i–j). Pits observed range from 5 mm to 50 microns in diameter, average 0.5 mm, and occur with a frequency of about 10 per square centimeter. Curiously, no overlapping of such pits was observed. Examination by scanning electron microscopy shows these glazed surfaces to be pitted with innumerable smaller pits and bubble-like features, many of which contain still smaller "bubbles." The bubbles, both large and small, usually also have slightly raised margins, but they also commonly show their largest diameter beneath an overhang, are preferentially located on elevated surfaces, and are clearly burst bubbles and evidence of outgassing.

The glazed surfaces of the larger pits within which the bubbles are located are only a few microns thick, and the glazing extends the full width of the rim, sometimes bubbling over slightly onto the unglazed surrounding rock. Subconcentric microfractures may parallel the margins between glazed pit and unglazed rock, and microfractures within the glazed areas run between the bubbles with no apparent displacement. The sides of the pits are occasionally terraced (Fig. 3, j). Depth is roughly proportional to pit diameter. The bubbles within the pits tend to cluster and have rounded bottoms as well as (commonly) overhanging edges. They are comparable to the bubbles of scoriaceous terrestrial basalts.

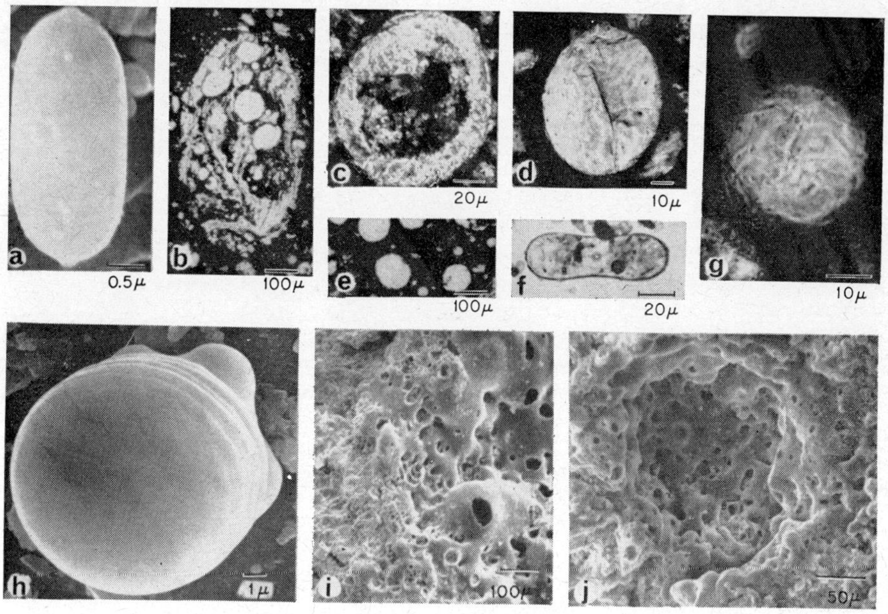

Fig. 3. Shaped glass particles and glazed microcraters from Apollo 11 dust and breccia. (Bar scales give magnification as indicated. Photographs b–g with optical microscope, a and h–j with scanning electron microscope). (a) Glass ellipsoid with terminal nipples from dust sample 10086,15. (b) Large composite glassy ovoid including matrix material, circular voids (where spheres ripped out during preparation) and glassy spheroids within a framework of wispy glass. Thin section 10019,15. (c) Section of a spheroid showing glassy rind around matrix and mineral core. Thin section 10019,15. (d) Slightly oblate and fractured glass spheroid with nipple on upper end. Thin section 10019,15. (e) Glass spheroids surrounded by fine grained matrix. Thin section 10019,15. (f) Dumbbell-shaped glass object with bubbles and mineral inclusions. From strewn slide of dust sample 10086,15b. (g) Glass spheroid with textured surface. Slide 10019,15. (h) Glass spheroid with protuberances and concentric flanges. Grain in dust sample 10086,15. (i) Margin between glazed pit (at right) and unglazed portion of microbreccia (at left) showing burst bubbles with overhanging rims. Surface of chip picked from dust sample 10086.15. (j) Crater-like glazed pit showing terracing and bubbles. External surface of breccia sample 10091,5.

We interpret the glazed pits as impact craterlets. The bubbles are just that—the product of outgassing resulting from high-velocity impact and local fusion of the lunar surface. The gas may have come from the projectile itself (for example, cometary particles) or, less likely, from fusion of the lunar breccia on impact. The subconcentric fractures at the margins, as well as the terraces within some of the glazed pits, were probably caused by the initial impact.

On the "biological" side these glazed pits are comparable to some impressions that have been described as pre-Paleozoic fossils. Of more interest, however, are their implications for a lunar atmosphere.

We wanted to know the source of the gas evaded during surface fusion and whether the glass spheroids and other shaped particles observed could be derived from fusion of local materials. With the help of Mark Stein, therefore, a small quantity of the lunar dust was fused under vacuum in a tantalum crucible after outgassing to 10^{-7} torr. Fusion took place at 1200° to 1400°C, producing glass spheroids having identical appearance and physical properties to those of the lunar dust and breccia. Vigorous bubbling during the fusion process indicates a sufficient quantity of indigenous volatilizable materials to account for the observed vesiculation of the glassy crusts, although very little of this remained gaseous at room temperature.

PRESTON CLOUD
STANLEY V. MARGOLIS
MARY MOORMAN, J. M. BARKER
G. R. LICARI

Department of Geology, University of California, Santa Barbara

DAVID KRINSLEY

Queens College, City University of New York, Flushing

V. E. BARNES

Bureau of Economic Geology, University of Texas, Austin

References and Notes

1. D. Krinsley and S. V. Margolis, *N.Y. Acad. Sci. Trans.* Ser. II, **31**, No. 5, 457 (1969).
2. V. E. Barnes, *North American Tektites* (Univ. of Texas Publ. 3945, 1940), p. 521.
3. C. S. Ross, *Amer. Mineral.* **47**, 723 (1962); M. N. Bramlette, *Science* **158**, 673 (1968).

4. Supported by NASA contract No. 9-7882 for major equipment and travel, NASA grant No. NGR 05-010-035 for operations, University of California for facilities and partial services, and NSF grant No. GB-7851 for equipment and general support of the Biogeology Clean Laboratory.

4 January 1970

Analysis of Lunar Material for Organic Compounds

Abstract. *A sample of lunar material from Apollo 11 was subjected to analysis by several techniques, which included mass spectrometry, gas chromatography, liquid chromatography, and nuclear magnetic resonance and their variations, in an effort to detect the presence of organic compounds. None were found. On the basis of the sensitivity ascribed to certain of the methods employed, it is assumed that if organic matter were present it would exist in concentrations less than 1 part per million.*

Three sample-handling procedures were utilized during efforts to determine the existence of organic compounds in the Apollo 11 lunar sample 10086.5 (class D). First, an aliquot of untreated sample was directly analyzed by gas chromatography and mass spectrometry. Second, the sample was subjected to extraction by certain organic solvents (*1*), the solvents were removed, and portions of the residue, if any, were analyzed by gas chromatography, mass spectrometry, and nuclear magnetic resonance. Third, an aliquot of the sample was acid-hydrolyzed, and the hydrolyzate was extracted and subjected to analysis by high-performance liquid chromatography on pellicular resins, by gas chromatography, and by mass spectrometry.

All solvents were of spectroscopic grade and were redistilled when necessary. A series of blanks was run prior to assay of the material in all instances.

For the nuclear-magnetic-resonance analysis a 23-g sample of lunar material was placed in an extraction thimble and extracted with a solution containing 80 ml of a 4:1 benzene-methanol (spectrograde, redistilled solvents) mixture in a Soxhlet extraction apparatus for 24 hours and then was stirred as a slurry for 24 hours. The contents were decanted into four test tubes and centrifuged for 2 hours. The supernatants were combined, and an aliquot was evaporated to dryness in a stream of pure nitrogen. The contents were triturated 18 hours with 0.8 ml of spectrograde carbon tetrachloride, and the carbon tetrachloride solution was added to a Wilmad Imperial 507PP sample tube (5 mm). A small amount of Matheson tetramethylsilane was added as a reference signal for the field-frequency lock. A blank was determined with identical materials and apparatus.

The nuclear-magnetic-resonance study was conducted with a Bruker HFX-3 spectrometer operating at 90 Mhz (H^1). The sensitivity was increased by use of a 4096 channel computer of average transients (CAT). The spectrum was determined over a sweep range from 30 to 1166 hz downfield with tetramethylsilane. The sweep speed, determined by the CAT, was 80 msec per channel, and the signal was filtered with an input integrator (time constant = 80 msec per channel).

After 27 hours, the spectrum showed peaks at 133 hz and a multiplet at 305 hz. The multiplet was also present in the blank and is a solvent impurity. No other peaks were present. The peak at 133 hz was probably due to the intense tetramethylsilane peak due to repeated scans by the signal-averaging device as evidenced by the spectrum of the blank.

The sensitivity of the Bruker HFX-3 spectrometer for a 1-percent (by volume) ethylbenzene solution, using a 5-mm sample tube, is given as a 50-to-1 signal-to-noise ratio. The CAT further increases the signal-to-noise ratio by a factor of 40. Therefore the sensitivity of a sample with the same molecular weight and number of hydrogen atoms as ethylbenzene will be approximately 5×10^{-4}. Any material present in a concentration of less than 1×10^{-4} will be difficult to detect.

An analysis of the sample extract was made with a high-performance liquid chromatography system (*2*) equipped with an ultraviolet detector. With this technique, minute quantities of substances, such as nucleic acid constituents, which have absorbtivity at 250 nm, can be separated and determined.

Ten grams of the finely ground lunar sample were digested with 50 ml of 2.5*N* hydrochloric acid at 50°C for 2 days. After centrifugation the extract was lyophilized. Then 1 ml of distilled water was added to the vessel in order to prepare the sample solution for the chromatographic investigation. Another portion of this sample solution was then processed for analysis by mass spectrometry and gas chromatography.

The LCS 1000 liquid chromatograph (Varian Aerograph) with 1-mm inside-diameter columns packed with pellicular cation exchange or anion exchange resins (particle diameter 50 μm) was used in a fashion described earlier (*2*). The length of both columns was 150 cm. Dilute potassium hydrogen phosphate solutions having different *p*H values were used as eluents. In some experiments gradient elution (concentrated KH_2PO_4 served as strong eluent) was employed. The flow rate was 10 ml/hour, and the column temperature was 60°C in all experiments. A 10- or 20-μl sample solution was injected onto the chromatographic columns with a Hamilton No. 701 microsyringe.

Chromatographic runs were made on both cation exchanger and anion exchanger columns with eluents of various *p*H values in the range 2 to 7. No peaks attributable to retarded solutes have been found. In gradient elution with both columns small peaks appeared regularly at a certain point of the chromatogram, but these peaks were also obtained when a solvent blank was analyzed. Thus it is concluded that under the experimental conditions described no nucleic acid constituents or ultraviolet-absorbing materials of similar chromatographic behavior have been extracted from the lunar sample in an appreciable quantity. Considering the sensitivity of the technique, therefore, the amount of individual nucleic acid constituents, if such were present, must be less than 10^{-10} mole per gram of sample.

For the gas-chromatography analysis a 25-mg sample of the lunar material without prior treatment was placed in a small quartz tube containing quartz wool and attached to a supported coated open tubular (SCOT) column, 15.1 m long with a 0.058-cm inside diameter, containing silicone fluid (SF96, 1000 cs) and Igepal CO-880 (nonylphenoxy-polyethyleneoxy-ethanol). The column was connected to a flame ionization detector (sensitivity 10^{-12} g/sec), and the entire system was flushed with carrier gas (nitrogen) for 30 minutes. The sample was heated to 150°C, then to 250°C, and finally to 500°C in the flash heater zone of the gas chromatograph. The temperature of the sample tubes was maintained at each temperature level while the column was temperature-programmed from 50° to 180°C at the rate of 3°C per minute. A control containing solvent-washed glass beads was substituted for the lunar sample and analyzed in a similar fashion. There was no evidence of organic compounds' being eluted from the sample. Similar results were obtained when these analyses were repeated on a 1.5-m by 0.058-cm SCOT column coated with Apiezon L and programmed to 210°C.

Aliquots from both the acid hydrolyzate and the organic solvent extracts of the sample and reagent blanks were concentrated and then subjected to the gas chromatographic procedures described above. Again, the findings were negative.

In all mass-spectrometry experiments, the spectrometer (M-S 9, Associated Electronics Industries, Manchester, England) was operated at 70 ev and at 100 or 300 μa trap current. The resolution was approximately 1:1000.

All samples, extracts, and reagent blanks were introduced via the direct insertion probe. Prior to sample introduction, the operating parameters of the instrument were optimized with perfluorotributylamine as a reference compound. After this material was pumped off, the ion source was baked at 250°C for 4 days.

A 10-mg portion of the lunar sample was loaded into a special quartz capillary tube attached to the direct-insertion probe. The probe was inserted into the mass spectrometer and spectra were recorded at various temperatures as the temperature of the ion source was increased from 50° to 250°C. The various spectra were dominated by peaks due to H_2O, N_2, O_2, and CO_2. The mass spectral patterns of the remaining peaks were indicative of very small quantities of low-molecular-weight hydrocarbons consistent with that noted as background. A repeat analysis gave similar results.

Studies of the extracts and reagent blanks by mass spectrometry did not reveal the presence of any components which could be considered to be indicative of endogenous material derived from the lunar sample.

In conclusion it is assumed that the principal components of the various spectra were contaminant gases and traces of low-molecular-weight hydrocarbons which had adsorbed onto the surface of the lunar sample during handling and exposure to the atmosphere.

S. R. Lipsky
R. J. Cushley
C. G. Horvath
W. J. McMurray

Section of Physical Sciences,
Yale University School of Medicine,
New Haven, Connecticut

References and Notes

1. W. G. Meinschein, *Science* **150**, 601 (1965).
2. C. Horvath, B. Preiss, S. R. Lipsky, *Anal. Chem.* **39**, 1422 (1967); C. Horvath and S. R. Lipsky, *ibid.* **41**, 1227 (1969); ——, *J. Chrom. Sci.* **7**, 109 (1969).
3. Supported by NASA grants NAS 9-8072 and NGL 07-004-008.

4 January 1970

Micropaleontological Studies of Lunar Samples

Abstract. *Optical and electron microscopic studies of rock chips and dust from the bulk sample box returned by Apollo 11 and of petrographic thin sections and acid-resistant residues of lunar material have yielded no evidence of indigenous biological activity.*

Although the present lunar environment is inimical to known biological systems, more favorable conditions may have existed in the geologic past. Urey (*1*) has suggested that the moon may have become "contaminated" with terrestrial organic matter early in the evolution of the earth-moon system. If this suggestion is correct, and if life became established, evidence of fossil organisms might be detectable in lunar rocks. It is even conceivable that such organisms might have been the progenitors of an extant biota, adapted to the harsh conditions of the lunar surface; such organisms probably could not survive in the terrestrial environment and therefore would not be recognized in studies designed to detect vital processes (such as metabolism, growth, and pathogenicity). The approach and techniques successfully used in Precambrian paleobiology (*2*), and the criteria developed to establish the indigenous and biogenic nature of Precambrian microfossils, seem well suited for the detection, characterization, and interpretation of any fossil or recently dead microorganisms that might occur in lunar materials (*3*).

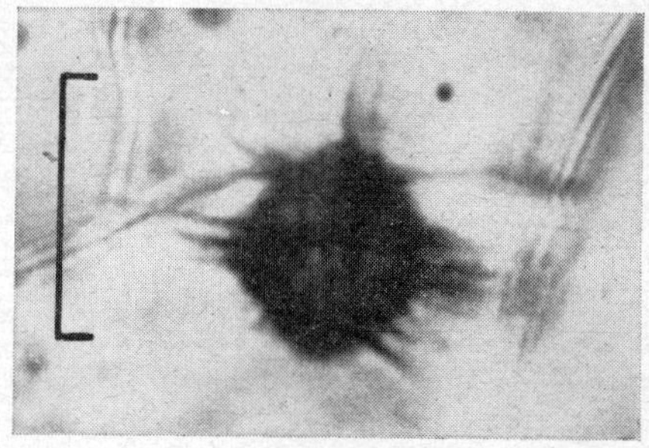

Fig. 1. Optical photomicrograph showing actinomorphic pseudofossil, apparently produced by partial devitrification of the surrounding glassy matrix, in a petrographic thin section of a microbreccia (sample 10046,56); the scale represents 10 μm.

In an effort to detect evidence of lunar organisms in the Apollo 11 samples, studies were made with a light microscope (L) at magnifications ranging from 4 to 1500, and, after the specimens had been coated with a thin gold-palladium film, with a scanning electron microscope (SEM) at magnifications ranging from 30 to 30,000. I examined samples as follows: (i) lunar dust (sample 10086,18 from the bulk sample box), divided into four size-fractions by sieving (>246 μm, 246 to 124 μm, 124 to 74 μm, and < 74 μm)—L and SEM; (ii) residue resulting from dissolution of lunar dust in hydrofluoric and hydrochloric acids—L; (iii) surfaces of rock chips from the exterior and interior of a microbreccia (sample 10002,54 from the bulk sample box), and fragments of these chips—L and SEM; (iv) petrographic thin sections of microbreccias (samples 10019,15, 10046,56, 10059,32, 10059,37, 10061,27, 10061,28, and 10065,25)—L; (v) as a member of the Ames Lunar Sample Consortium, I studied (L) samples being investigated by Ponnamperuma *et al.* (*4*) (sample 10086, bulk A fines); (vi) as a member of the Lunar Sample Preliminary Examination Team for the Apollo 11 mission, I studied (L) rocks, chips, dust, and bioquarantine samples (including portions of both cores) (*5*).

Several thin sections (among them 10046,56, 10059,32, and 10061,27) contain elongate, spheroidal, spinose, or actinomorphic structures (Fig. 1) that superficially resemble terrestrial microfossils; many of these mineralogic "pseudofossils" are the result of partial devitrification of glassy inclusions. During preliminary studies at the Lunar Receiving Laboratory I detected birefringent organic fibers, a few microns in diameter, in the lunar dust and bioquarantine samples; a few similar fibers were noted in samples i and ii listed above, and in the mounting medium (but *not* within mineral grains) of several petrographic thin sections. With the exception of these terrestrial contaminants, apparently derived from lens tissue or similar substances, no biogenic materials were detected in these investigations.

J. WILLIAM SCHOPF

Department of Geology, University of California, Los Angeles 90024

References and Notes

1. H. C. Urey, *Science* **151**, 157 (1966).
2. J. W. Schopf, *Biol. Rev.*, in press.
3. ———, paper presented at the Annual Meeting of the Geological Society of America, Atlantic City (1969).
4. C. Ponnamperuma *et al.*, *Science*, this issue.
5. Lunar Sample Preliminary Examination Team, *ibid.* **165**, 1211 (1969).
6. I thank Mrs. Carol Lewis for assistance and G. Oertel and J. Christie for suggestions. This work was supported by NASA contract NAS 9-9941.

4 January 1970

Numbering System for Moon Samples

Numbers are assigned to the lunar samples in the Lunar Receiving Laboratory (LRL) as soon as the samples have been photographed. The numbers have two parts: the generic, or first part, and the specific, or second part. The generic number is a 5-digit number assigned to each discrete piece as it is received; it remains with all portions of that piece. The generic numbers for Apollo 11 rocks are all in the 10-thousand series; those for Apollo 12 are in the 12-thousand series.

The specific number is a sequentially assigned integer used primarily for bookkeeping and is essentially the number used to designate a piece or fraction of the original sample. It is essentially a sample split number. Thus, if the piece labeled 10017,14 were cut into three pieces, the new pieces might be labeled 10017,72; 10017,73; and 10017,74 if these were the next unassigned numbers. Number 10017,14 would no longer be assigned to a piece, and the records would indicate that the piece was cut into smaller pieces.

The investigators who have received lunar material have used their own nomenclature for subsplits, producing a three-part number. The third part is the identification assigned by a particular investigator. The third part of the number is attached to the LRL number by any convenient punctuation. Some investigators have used dashes, others have used commas, slashes, or periods. In papers in this issue, some authors have omitted the first three digits of the generic number. Thus, an investigator working with a subsplit of 10017,72 may have designated it 10017,72-1 and may refer to it as 17,7201 or 17,72/1, or simply as 72-1. When the samples are returned to the Lunar Receiving Laboratory, new specific numbers will be assigned to the subsplits, yielding two-part numbers again.

In addition to specific numbers, samples are also identified by letters that indicate the type of material. Type A is fine-grained vesicular crystalline igneous rock; type B, medium-grained vuggy crystalline rock; type C, breccia; and type D, fines.

DANIEL H. ANDERSON
Curator, Manned Spacecraft Center, National Aeronautics and Space Administration, Houston, Texas

Acknowledgments

Publication of this issue of *Science* was accomplished with the help of many persons who are not members of the editorial staff. Wilmot Hess was instrumental in developing the broad outlines of the publication procedures that were ultimately adopted for the Apollo 11 Lunar Science Conference by NASA. After his departure from NASA, Gene Simmons and Anthony J. Calio helped complete the detailed arrangements.

The publication plan developed by the *Science* staff within this framework called for reviewing, editing, and revision of papers during the period of the conference. The results of the effort are shown in Table 1. Refereeing was accomplished by a group of 50 conference participants, who provided more than 250 reviews in less than 4 days. Authors whose papers required revision were called in to discuss the revisions with the referee; revisions were completed before the end of the conference. Reviewing was facilitated by the work of six reviewers who also served as topic chairmen: Stanley Hart and George R. Tilton, geochronology and geochemistry; Ian MacGregor and David Wones, mineralogy and petrology; David Strangway, physical properties; and Thomas Hoering, organic geochemistry. The chairmen enlisted additional reviewers, who were invited to look over the manuscripts grouped on a table by subject and take their pick. These additional reviewers were John B. Adams, Edward Anders, Orson Anderson, James A. Arnold, Gustaf Arrhenius, Elso Barghoorn, Peter M. Bell, Francis R. Boyd, Malcolm Campbell, Preston Cloud, Alvin J. Cohen, Herbert Diamond, Geoffrey Eglinton, Samuel Epstein, Larry W. Finger, Kurt Fredriksson, Clifford Frondel, G. F. J. Garlick, Paul Gast, J. E. Geake, Gordon Goles, Stefan Hafner, Stephen Haggerty, Larry A. Haskin, C. E. Helsley, H. Kanamori, Donald H. Lindsley, Warren Meinschein, Arnulf Muan, John A. O'Keefe, Robert O. Pepin, Dean Presnall, K. A. Richardson, James M. Schopf, S. Fred Singer, Joseph V. Smith, D. Tozer, Robert M. Walker, Louis S. Walter, G. J. Wasserburg, G. W. Wetherill, M. T. Yates, and Leonard P. Zill.

All papers submitted were screened for editorial problems during the conference, and editing was completed on a third of them. Authors were invited to examine the edited manuscripts and make changes if they wished to do so before the end of the conference.

Eleven members of the editorial staff worked in Houston. They were joined

Table 1. Dates of start and finish of each step in the publication of the Apollo 11 Lunar Science Conference issue of *Science*.

Step	Dates
Receipt of manuscripts from authors	4– 7 January
Reviewing of manuscripts	4– 8 January
Authors' responses to reviews	5– 8 January
Style editing and marking for printer	5–12 January
Redrafting and relettering illustrations	5–12 January
Authors' responses to style editing	5–23 January
Preparation of engravings	8–22 January
Typesetting	9–19 January
Proofreading of galley proofs	10–19 January
Pasteup of page dummies	13–20 January
Correction of galleys and makeup of pages	13–21 January
Proofreading of page proofs	17–24 January
Correction of pages	20–26 January
Proofreading of revised page proofs	21–27 January
Printing	23–28 January
Binding	28–30 January
Mailing	28–31 January

by Mary Jane Miles and Rowena Peoples for editing and by Kathleen Blake for manuscript tracking. Earl Rubenstein and John Harris of the Manned Spacecraft Center arranged for plentiful space for editing and refereeing, and with the assistance of Roy Magin and Stanley Jacobsen they quickly provided typists, copying machines, pencil sharpeners, dictionaries, draftsmen, photography, and other equipment and services whenever required.

The editing begun in Houston was finished in Washington with help for the staff from Murrie Burgan, Mary Eichhorn, Marcus Hairstone, Eleanor Johnson, Jerold Last, Helen Olney, Rowena Peoples, Stephen Petropulos, Horace Porter, Mary Porter, and Lucile Stryker. Proofreading help was provided by all these and by Helen Carter, Wanda Jenkins, Mary Ann Ormes, Barbara Porter, Iris Sexter, Diane Tremitiere, and Helen Wolfle.

Others who helped in various tasks from keeping track of manuscripts to mailing them at the post office and keeping the building open after regular hours were Shirley Bain, Kathleen Blake, Janet Bragg, Carol Brown, Mattie Fauntroy, Mattie Gardner, Elma Goss, Fannie Groom, Veronica Groom, Rose Lowery, Faye Lynch, Allan Sims, Isaac Smalls, Ethel Smith, James Stickley, James Walke, Albert Wright, and Marion Zeiger.

The front cover was designed by James White who based his work on a picture provided by Kurt Fredriksson and E. P. Henderson. The illustrations on page 451, left to right, were provided by E. C. T. Chao, Paul W. Gast, and Robert L. Fleischer.

The work of typesetting, page make up, printing, binding, and mailing was arranged for or done by:

H. Abdill, C. Adams, C. Albany, C. R. Altfather, R. E. Baeuerlein, W. B. Baker, B. Babik, J. J. Bankert, R. Beuchert, J. A. Cangalosi, R. B. Chapman, R. Colliere.

D. Cooper, J. Cooper, C. D. Cullum, B. Daisher, E. Dandridge, Wm. de Lorge, W. Doerk, A. Early, T. Evans, S. R. Gilbert, W. Giordano, J. E. Gundling.

J. D. Hare, W. H. Harmon, E. O. Harrison, Jr., A. K. Hutcherson, W. James, W. Jones, J. R. Keagy, M. Kendall, J. Kline, F. F. Kothman, J. Langsdale, J. Liverman.

W. McCawley, J. McDermott, H. McDonnell, N. C. Mueller, F. H. Nagel, J. Papadopulo, J. Patterson, C. H. Preston, O. Robertson, L. Ruegg, V. Sampson, L. Schiro.

E. Schulte, E. J. Schultz, J. E. Severn, L. C. Shomo, J. Simms, P. M. Smith, J. Stanley, P. R. Staub, N. Sutton, E. Sweeney, M. J. Trump, R. Via, B. Waterhouse.

JOHN E. RINGLE, ROBERT V. ORMES

List of Abbreviations

The explanations are those supplied by the authors of the papers in this issue.

-a combining form for ampere, as μa (microampere).
A artificial glow curve area
Å angstrom unit; 10^{-8} centimeter
a$_o$ unit-cell *a* parameter
A$_r$ *usu ital:* angle of rotation
AA atomic absorption
AA activation analysis
Ab albite ($NaAlSi_3O_8$)
a-c alternating current
AE aeon; 10^9 years
A.E.I. Associated Electrical Industries
AF alternating field
Ag silver
alk. alkalic
Allende Pueblito de Allende
Al aluminum
amp ampere
amu atomic mass unit
An anorthite ($CaAl_2Si_2O_8$)
An$_{100}$, anorthite molecule content of 100 percent
Ap apatite [$Ca_5(PO_4)_3(OH,F,Cl)$]
Ar argon
atm atmosphere
Au gold
A.U. astronomical unit
AVCC average carbonaceous chondrite composition
AWRE atomic weapon research establishment

b *usu ital:* mean diffusion lengths
b$_o$ *usu ital b:* unit-cell *b* parameter
Ba barium
Be beryllium
Bi bismuth
Br bromine
b.y. billion years; 10^9 years

c* *usu ital:* reciprocal *c* axis
°C degrees Celsius
c$_o$ *usu ital c:* unit-cell *c* parameter
C2/c *usu ital C, c:* space group
Ca calcium
ca about
cal calorie
cc cubic centimeter
ccSTP/g cubic centimeters at standard temperature and pressure per gram
cm^3 STP/g *ibid.*
cm^3/g(STP) *ibid.*
ccstp/g *ibid*
cstp *ibid.*
Cd cadmium
CEC Consolidated Electrodynamics Corporation
Chr chromite ($FeCr_2O_4$)
CIPW norm system, from initials of the originators, Cross, Iddings, Pirsson, and Washington
C.I.T. California Institute of Technology
Cl chlorine
class 100 room particle ($>$ 5 μm) count less than 100 per cubic foot of air
Cm curium
cm centimeter
cm^3 cubic centimeter
Co cobalt
count/min counts per minute
c$_p$ *usu ital c:* specific heat
cph counts per hour
cpm counts per minute
Cr chromium
Cr cristobalite (SiO_2)
Cs cesium
Cu copper

d day
d *usu ital:* particle diameter
d *usu ital:* density; for example, g/cm^3
D deuterium
d^4(t$_{2g}$)^3(e$_g$)^1 *usu ital d, e, g, t:* Group notation for high spin state Cr^{2+}, indicates an odd number of electrons in the e_g orbitals
d-c direct current
de *usu ital:* energy per unit area and unit time
Di diopside ($CaMgSi_2O_6$)
dph distintegrations per hour
dpm/kg disintegrations per minute per kilogram

E *usu ital:* activation energy
E emission spectrography
EASEP Early Apollo Scientific Experiments Package
emu electromagnetic units
emu/g electromagnetic units per gram
EMX electron microprobe x-ray analysis
En enstatite ($MgSiO_3$)
En$_{70}$ enstatite molecule content of 70 percent
EOB end of bombardment
EPR electron paramagnetic resonance
Eu europium
EVA extravehicular activity

F fluorine
Fa fayalite (Fe_2SiO_4)
Fe iron
f$_{H_2O}$ *usu ital f:* fugacity of water
FID flame ionization detector
f$_{O_2}$ *usu ital f:* fugacity of oxygen

Fo forsterite (Mg_2SiO_4)
Fo_{65} forsterite molecule content of 65 percent
Fs ferrosilite ($FeSiO_3$)
fs_2 *usu ital f:* fugacity of sulfur

g gram
g *usu ital:* gyromagnetic ratio
g/cc grams per cubic centimeter [more often g/cm^3]
Ga gallium
GC-MS combined analysis by gas chromatography and mass spectrometry
Gd gadolinium
Ge germanium
Gev 10^9 electron volts; gigaelectron volts
Gi mole percent $MgTiO_3$
G.M.T. Greenwich Mean Time

H *usu ital:* intensity of a magnetic field
H hydrogen
H^+ proton
HCl hydrochloric acid
Hd hedenbergite ($CaFeSi_2O_6$)
He helium
Hed hedenbergite ($CaFeSi_2O_6$)
Hf hafnium
HF hydrofluoric acid
Hg mercury
$Hg_{>450}$ concentration of mercury volatilized above 450°C
Hg_{total} total concentration of mercury
h0l *usu ital h, l:* reciprocal lattice net
HP Hewlett-Packard
hr hour
hz hertz

I iodine
I *usu ital:* intensity
I *usu ital:* specific intensity of magnetization
$I\bar{1}$, $C\bar{1}$ *usu ital I, C:* crystallographic space groups
IAA instrumental activation analysis
i.d. inner diameter
Il ilmenite
Ilm ilmenite ($FeTiO_3$)
in. inch
In indium
INAA instrumental neutron activation analysis
ir infrared
IRM isothermal remanent magnetization
Ir iridium

J *usu ital:* irradiation parameter
J_i *usu ital J:* induced magnetization per unit mass
J_r *usu ital J:* isothermal remanent magnetization per unit mass
J_s *usu ital J:* saturation magnetization per unit mass

k *usu ital:* absorption coefficient
k *usu ital:* Boltzmann's constant
k *usu ital:* thermal conductivity (for example, watt/m=°K)
k *usu ital:* thermal diffusivity
K *usu ital:* bulk modulus
K *usu ital:* thermal conductivity
K potassium
°K degrees Kelvin
kc kilocycle
kb kilobar
kev kiloelectron volt
kg kilogram
Kr krypton
kv kilovolt

L light microscope
La lanthanum
LEF linear-energy spectrophotofluorometry
Li lithium
LM lunar module
ln natural log
LRL Lunar Receiving Laboratory
LSAPT Lunar Sample Analysis Planning Team
LSPET Lunar Sample Preliminary Examination Team
Lu lutetium

m meter
m- milli-, 10^{-3}
M *usu ital:* molar
M sum of metal ions, Fe + Mn + Ca + Mg + Ti
M1, M2 *usu ital M:* lattice position (site)
Mb megabar
mb millibar
mbarn millibarn, 10^{-27} square centimeter
MCD magnetic circular dichroism
m/e *usu ital:* mass to charge ratio
Mev million electron volt
Mg magnesium
mg milligram
mgal milligal
Mhz megahertz
μamp microampere
μl microliter
μm micrometer
min minute
M.I.T. Massachusetts Institute of Technology
ml milliliter
mm millimeter
Mn manganese
modal, mode volume percent, meaured by statistical point count of a thin section
mole molecular
MSA methanesulfonic acid
MSC NASA Manned Spacecraft Center, Houston
MSID mass spectrometric isotope dilution
Mt magnetite [$(Fe^{2+})(Fe^{3+})_2O_4$]
mu mass units
m.y. million years; 10^6 years

N natural glow curve area
N *usu ital:* normal
N; N_2 nitrogen
n neutron
n *usu ital:* index of refraction
n_β *usu ital n:* intermediate index of refraction
n_{546} *usu ital n:* refractive index at 546 nm
n_D *usu ital n:* index of refraction at sodium D line
n_x, n_y, n_z *usu ital n:* index of refraction for the x, y, and z rays, respectively
N, N_D *usu ital N:* refractive index
n- nano-; 10^{-9}
Na sodium
NAA neutron activation analysis
NASA National Aeronautics and Space Administration
NBS National Bureau of Standards
Ne neon
ng nanogram
nicol polarizing prism
nlumen nanolumen
nm nanometer
NMR nuclear magnetic resonance
NRM natural remanent magnetization

O oxygen
O.D. optical density
o.d. outer diameter
oe oersteds (measure of magnetic field strength)
Ol olivine [$(Mg,Fe)_2SiO_4$]
Or orthoclase ($KAlSi_3O_8$)
ORB oceanic ridge basalts
Os osmium

P phosphorus
P *usu ital:* degree of polarization
P *usu ital:* pressure
P_{H_2O} *usu ital P:* water pressure
p- pico-; 10^{-12}
p proton
$P2_1/c$ *usual ital P, c:* space group
Pbnm *usu ital:* space group
Pb lead
$(^{206}Pb/^{204}Pb)_0$ atomic ratio of ^{206}Pb to ^{204}Pb in initial lead
$(^{206}Pb/^{204}Pb)_p$ atomic ratio of ^{206}Pb to ^{204}Pb at present
Pc plagioclase
Pd palladium
PDB Pee Dee belemnite standard; an isotopic standard for carbon
Permount microscopic preparation mounting medium
PET Preliminary Examination Team (Lunar Sample . . . .)
Plag plagioclase
P_{max} *usu ital P:* maximum of polarization curve
Po polonium
ppb parts per billion
ppm parts per million
psi pounds per square inch
Pu plutonium
Px pyroxene
Pyx pyroxene

Q *usu ital:* quality factor
Q quartz (SiO_2)
Qz quartz (SiO_2)

R *usu ital:* reflectivity, in air unless otherwise stated
R regolith
R_1, R_2 *usu ital R:* uniradial reflectivities
R_ω, R_ε *usu ital R:* reflectivities corresponding to the ω and ε directions of a uniaxial crystal
r>v *usu ital:* dispersion of axial angle with $2V$ for the red end of the spectrum larger than that for the violet end
r<v same as above, but $2V$ violet greater than $2V$ red
Rb rubidium
REE rare earth elements
rev/min revolutions per minute
RGAA radiochemical gamma activation analysis
Rh rhodonite ($MnSiO_3$)
R. I. refractive index
r. i. refractive index
RNAA radiochemical neutron activation analysis
rpm revolutions per minute
RT retention time
Ru ruthenium

s *usu ital:* frequency factor
Sc scandium

s soil

S *usu ital:* viscosity coefficient; magnetic moment per gram

S.D. standard deviation

S.E. standard error

sec second

SEM scanning electron microscope

S.E.M. standard error of the mean

Si silicon

Sm samarium

SMOW standard mean ocean water. An isotopic standard for ^{18}O and deuterium

SNOW standard normal ocean water

Sp. g. specific gravity

Sr strontium

sr steradian

SSM spark source mass spectrography

STP standard temperature and pressure

SWC solar wind composition experiment

T *usu ital:* temperature

T_p *usu ital T:* temperature of glow curve peak

T *usu ital:* age

T *usu ital:* mean life of ^{40}K

$t_{1/2}$ *usu ital t:* half-life

t *usu ital:* potassium-argon age of sample

t_s *usu ital t:* potassium-argon age of monitor

TB Tranquillity Base

TDA thermodifferential analysis

Te tellurium

TGA thermogravimetric analysis

Th thorium

THL thermoluminescence

thol. tholeiitic

Ti titanium

Tl thallium

TL thermoluminescence

torr Torricelli (measure of pressure)

Tr troilite

TRM thermoremanent magnetization

U uranium

UF ultrafine

UHV ultrahigh vacuum

USGS United States Geological Survey

U.T. Universal Time

UTC universal time coordinated

uv ultraviolet

V *usu ital*: phase angle

V_p *usu ital V:* compressional sound velocity

V_s *usu ital V:* shear sound velocity

2V *usu ital V:* optic axial angle

$2V_z$ *usu ital V, Z:* Optic axial angle around the slow (*Z*) vibration direction

VRM viscous remanent magnetization

Wo wollastonite ($CaSiO_3$)

wt weight

XRF x-ray fluorescence spectrometry

X,Y,Z *usu cap X, Y, Z:* Vibration directions of the principal refractive indices a, β, and γ

y year

Y yttrium

Yb ytterbium

Z *usu ital:* number of formula units in unit cell

Z$\wedge$c *usu ital Z, c:* angle between *c* (crystallographic axis) and the principal indicatrix axis

Zn zinc

Zr zirconium

α alpha particle

β angle of monoclinic cell

β volume compressibility

β intermediate refractive index

β^+/m-kg positrons per minute per kilogram

γ reciprocal of thermal inertia

δ isotopic abundance relative to some agreed standard. For example

$$\delta^{13}C = 1000\frac{[(^{13}C/^{12}C)\text{sample} - (^{13}C/^{12}C)\text{ std})]}{(^{13}C/^{12}C)\text{std}}$$

ε dielectric constant

θ angle of viewing

λ wavelength

λ radioactive decay constant

λ_ε decay constant for electron capture

λ_β decay constant for beta decay

μ- micro-; 10^{-6}

μ micron

$\rho(\psi)$ directional reflectance

$\rho_b(\psi, \theta)$ bidirectional reflectance

ρ_0 intrinsic density

ρ bulk density

ρ density

σ_m standard deviation of the mean

σ Poisson's ratio

Γ, B "ordering" parameters

χ_r reversible susceptibility per unit mass

ψ angle of illumination

ω frequency of periodic variation of moon surface temperature

$d\Omega$ solid angle

(001),(100) crystal planes, Miller indices

[001], [0112] crystallographic symbols; braces and brackets have special significance.

Errata

In "Argon-40/argon-39 dating of lunar rock samples" by G. Turner (p. 466), the J values (Table 1) should be 0.0341 ± 0.0007 for samples 10003,-43 to 10022,46 and 0.0319 ± 0.0007 for the remainder. The correct expression for J (p. 467, column 3, line 19) is

$$J = \exp(t_s/T) - 1/(^{40}\text{Ar}/^{39}\text{Ar})_s$$

(The correct expression for J has been used to calculate the ages presented in the report.)

Scale point 0.5 should appear on the right half of the abscissa of Fig. 1.

In "Tritium and argon radioactivities in lunar material" by E. L. Fireman, J. C. D'Amico, and J. C. DeFelice (p. 566), there should have been a reference 7 to indicate that "This work was supported in part by contract NAS 9-8105 from the National Aeronautics and Space Administration."

In "Petrology of a fine-grained igneous rock from the Sea of Tranquillity" by D. F. Weill *et al.* (p. 635), Fig. 1 is so poorly printed that much of the data is lost. See facing page for a better reproduction. The legend is at the bottom of column 1, this page.

In "Pattern of bombardment-produced radionuclides in rock 10017 and in lunar soil" by J. P. Shedlovsky *et al.* (p. 574), Table 1 contained errors in the data for ^{22}Na and ^{26}Al in rock 10017. In column 6 (T3SI, 60 mm) the counting results, in disintegrations per minute per kilogram for ^{22}Na and ^{26}Al were 31 ± 5 and 63 ± 12, respectively; in column 7 (Houston RCL) the counting results for the same

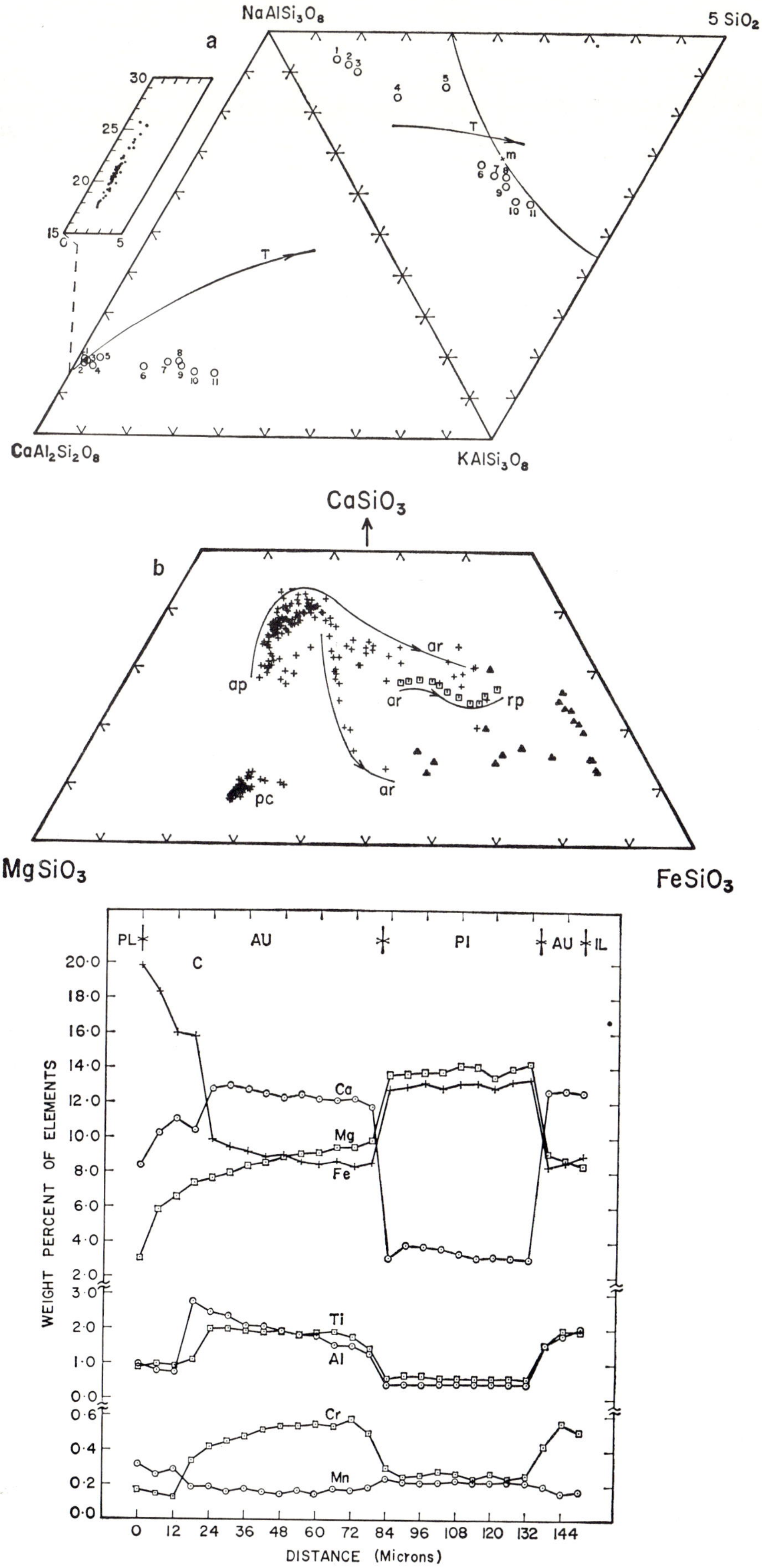

Fig. 1 (left). Plagioclase, pyroxene, and residual phase compositions. (a) Small filled circles (detail) represent the results of 48 plagioclase analyses; Nos. 1 through 11 represent a profile from a plagioclase grain through a residual material of quartzofeldspathic composition (anorthite + albite + orthoclase + quartz $\geqslant$ 88 percent by weight of total); T = Tasmanian dolerite salic trend (*4*); m = albite-orthoclase-quartz minimum at a water pressure of 500 kg/cm^2. (b) Pigeonite cores, *pc*; profiles from augite-pigeonite boundary to augite rim, *ap-ar*; profile from augite rim through residual "pyroxene," *ar-rp*; residual "pyroxene" analyses, triangles. (c) Profile across composite pyroxene; *PL*, plagioclase; *AU*, augite; *PI*, pigeonite; *IL*, ilmenite.

nuclides were 39 ± 7 and $73(\pm 13)$. No results were obtained for ^{36}Cl in these samples.

In "Crystallization of some lunar mafic magmas and generation of rhyolitic liquid" by I. Kushiro *et al.* (p. 610), Fig. 4 is incorrect. The published illustration shows two microphotographs of one of the mesostasis analyzed by the authors. The correct figure and legend are at the right.

In "Specific heats of lunar surface materials from 90 to 350 degrees Kelvin" by R. A. Robie, B. S. Hemingway, and W. H. Wilson (p. 749), the last sentence in paragraph 1, column 3, page 750, should read "Inasmuch as the rate of cooling depends on γ, the dust-covered portions of the lunar surface will cool much more rapidly than bare exposed outcrop, and thermal anomalies (*2*) will quickly develop over the exposed solid rock."

In "Organic analysis of the returned lunar sample" by P. I. Abell *et al.* (p. 757), the following paragraph was omitted because of an oversight during editing and preparing the manuscript for the printer. This paragraph should be inserted as the last one in the report.

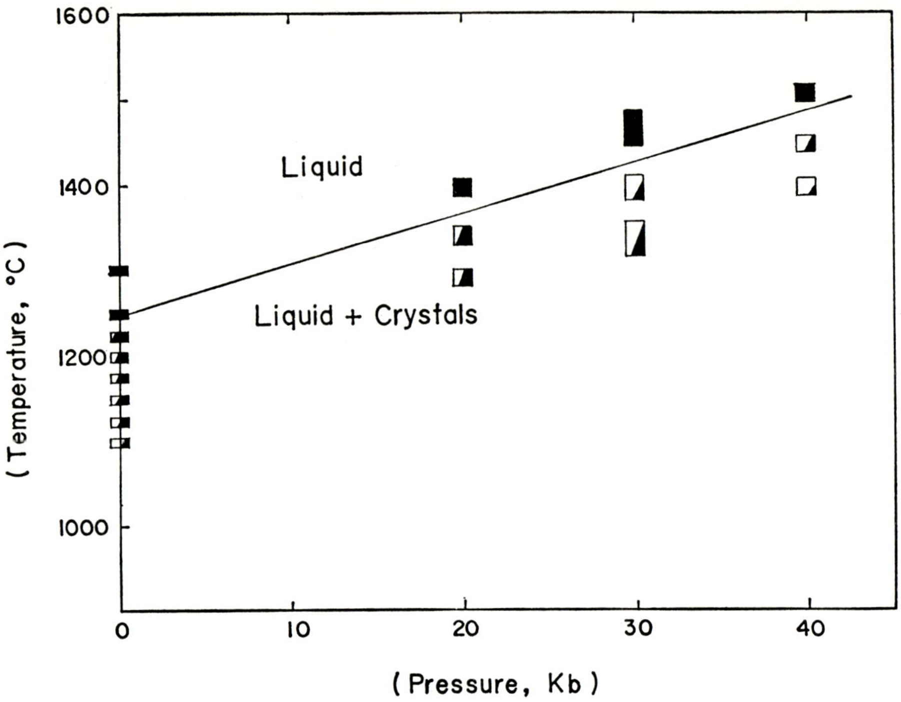

Fig. 4. Melting relations of a lunar crystalline rock (type A).

"We believe that significant information relevant to the study of solar wind implantation and cosmic ray irradiation of the moon and meteorites will become available by the kind of approach herein described. The compounds of carbon and other elements liberated from the samples simultaneously with the rare gases should be identified and quantified. In particular, it will be important to do this by careful etching studies designed to reveal precise locations and state of carbon compounds in the matrix."

Index

Index

Author Index

J

K

L

M

N

O

P

Q

R

S

T

U

V

W

Y, Z

Subject Index

A

B

C

D

E

F

G

H

O

P

Q

R

S

T

U

V

W

X–Z